T0137327

Lecture Notes on Data Engineering and Communications Technologies

Volume 22

Series editor

Fatos Xhafa, Technical University of Catalonia, Barcelona, Spain
e-mail: fatos@cs.upc.edu

The aim of the book series is to present cutting edge engineering approaches to data technologies and communications. It publishes latest advances on the engineering task of building and deploying distributed, scalable and reliable data infrastructures and communication systems.
The series will have a prominent applied focus on data technologies and communications with aim to promote the bridging from fundamental research on data science and networking to data engineering and communications that lead to industry products, business knowledge and standardisation.

More information about this series at http://www.springer.com/series/15362

Leonard Barolli · Natalia Kryvinska
Tomoya Enokido · Makoto Takizawa
Editors

Advances in Network-Based Information Systems

The 21st International Conference on Network-Based Information Systems (NBiS-2018)

Editors
Leonard Barolli
Department of Information and Communication Engineering, Faculty of Information Engineering
Fukuoka Institute of Technology
Fukuoka, Japan

Natalia Kryvinska
Faculty of Management
Comenius University in Bratislava
Bratislava, Slovakia

Tomoya Enokido
Faculty of Business Administration
Rissho University
Tokyo, Japan

Makoto Takizawa
Department of Advanced Sciences
Hosei University
Tokyo, Japan

ISSN 2367-4512 ISSN 2367-4520 (electronic)
Lecture Notes on Data Engineering and Communications Technologies
ISBN 978-3-319-98529-9 ISBN 978-3-319-98530-5 (eBook)
https://doi.org/10.1007/978-3-319-98530-5

Library of Congress Control Number: 2018950534

This Springer imprint is published by the registered company Springer Nature Switzerland AG
The registered company address is: Gewerbestrasse 11, 6330 Cham, Switzerland

Welcome Message from NBiS Steering Committee Co-chairs

Welcome to the 21st International Conference on Network-Based Information Systems (NBiS-2018), which will be held at Comenius University in Bratislava, Slovakia, from September 5 to 7, 2018.

The main objective of NBiS is to bring together scientists, engineers, and researchers from both network systems and information systems with the aim of encouraging the exchange of ideas, opinions, and experiences between these two communities.

NBiS started as a workshop and was held for 12 years together with DEXA International Conference as one of the oldest among DEXA workshops. The workshop was very successful, and in 2009 edition, the NBiS was held at IUPUI, USA, as an independent international conference supported by many international volunteers. In the following years, NBiSs was held at Takayama, Gifu, Japan (2010); Tirana, Albania (2011); Melbourne, Australia (2012); Gwangju, Korea (2013); Salerno, Italy (2014); Taipei, Taiwan (2015); Ostrava, Czech Republic (2016); and Toronto, Canada (2017).

In this edition of NBiS, many papers were submitted from all over the world. They were carefully reviewed, and only high-quality papers will be presented during the conference days. In conjunction with NBiS-2018 are held also nine international workshops where specific and hot topics are deeply discussed.

Many volunteer people have kindly helped us to prepare and organize NBiS-2018. First of all, we would like to thank General Co-chairs, Program Co-chairs, and Workshops Co-Chairs for their great efforts to make NBiS-2018 a very successful event. We would like to thank all NBiS-2018 organizing committee members, program committee members, and other volunteers for their great help and support. We have special thanks also for Finance Chair and Web Administrator Co-chairs. Finally, we thank the Local Organization Team at Comenius University in Bratislava, Slovakia, for their good arrangements.

We do hope that you will enjoy the conference and have a good time in Bratislava, Slovakia.

NBiS Steering Committee Co-chairs

Leonard Barolli Fukuoka Institute of Technology (FIT), Japan
Makoto Takizawa Hosei University, Japan

Welcome Message from NBiS-2018 General Co-chairs

We would like to welcome you to the 21st International Conference on Network-Based Information Systems (NBiS-2018), which will be held at Comenius University in Bratislava, Slovakia, from September 5 to 7, 2018.

It is our honor to chair this prestigious conference, as one of the important conferences in the field. Extensive international participation, coupled with rigorous peer reviews, has made this an exceptional technical conference. The technical program and workshops add important dimensions to this event. We hope that you will enjoy each and every component of this event and benefit from interactions with other attendees.

Since its inception, NBiS has attempted to bring together people interested in information and networking, in areas that range from the theoretical aspects to the practical design of new network systems, distributed systems, multimedia systems, Internet/Web technologies, mobile computing, intelligent computing, pervasive/ubiquitous networks, dependable systems, semantic services, grid, P2P, and scalable computing. For NBiS-2018, we have continued these efforts as novel networking concepts emerge and new applications flourish. NBiS-2018 consists of the main conference and nine workshops. We received 135 papers, and out of them, 38 were accepted (about 28% acceptance ratio), which will be presented during the three days of the conference.

The organization of an international conference requires the support and help of many people. A lot of people have helped and worked hard for a successful NBiS-2018 technical program and conference proceedings. First, we would like to thank all the authors for submitting their papers. We are indebted to Track Co-chairs, program committee members, and reviewers who carried out the most difficult work of carefully evaluating the submitted papers.

We would like to give our special thanks to Prof. Leonard Barolli and Prof. Makoto Takizawa, the Chairs of the Steering Committee, for giving us the opportunity to hold this conference and for their guidance on organizing the conference. We would like to thank Program Co-chairs and Workshops Co-chairs for their excellent work. We would like to express our great appreciation to our keynote speakers for accepting our invitation as keynote speakers of NBiS-2018.

We would like to thank the Local Arrangements Chairs and Local Organization Team at Comenius University in Bratislava, Slovakia, for making good local arrangements for the conference.

We hope that you have an enjoyable and productive time during this conference and have a great time in Bratislava, Slovakia.

NBiS-2018 General Co-chairs

Natalia Kryvinska Comenius University in Bratislava, Slovakia
Tomoya Enokido Rissho University, Japan
Markus Aleksy ABB AG Corporate Research Centre, Germany

Welcome Message from NBiS-2018 Program Committee Co-chairs

Welcome to the 21st International Conference on Network-Based Information Systems (NBiS-2018), which will be held at Comenius University in Bratislava, Slovakia, from September 5 to 7, 2018.

The purpose of NBiS conference is to bring together developers and researchers to share ideas and research work in the emerging areas of network and information systems.

The contributions included in the proceedings of NBiS-2018 cover all aspects of theory, design, and application of computer networks and information systems. There are many topics of information networking such as cloud computing, wireless sensor networks, ad hoc networks, peer-to-peer systems, grid computing, social networking, multimedia systems and applications, security, distributed and parallel systems, and mobile computing.

In this edition, 135 submissions were received from all over the world. Each submitted paper was peer-reviewed by program committee members and external reviewers who are experts in the subject area of the paper. Then, the program committee accepted 38 papers (about 28% acceptance ratio). In the conference program are also included the distinguished keynote addresses.

The organization of an international conference requires the support and help of many people. First, we would like to thank all authors for submitting their papers. We would like to thank all Track Chairs and program committee members, who carried out the most difficult work of carefully evaluating the submitted papers. We also would like to thank NBiS-2018 Workshops Co-chairs for organizing excellent workshops with the NBiS-2018 conference.

We would like to give special thanks to Prof. Leonard Barolli and Prof. Makoto Takizawa, the Chairs of the Steering Committee of NBiS, for their strong encouragement, guidance, insights, and for their effective coordination of conference organization. We would like to greatly thank General Co-chairs for their great support and invaluable suggestions to make the conference a very successful event.

We hope you will enjoy the conference and have a great time in Bratislava, Slovakia.

NBiS-2018 Program Committee Co-chairs

Michal Greguš	Comenius University in Bratislava, Slovakia
Fumiaki Sato	Toho University, Japan

Welcome Message from NBiS-2018 Workshop Co-chairs

Welcome to the NBiS-2018 workshops to be held in conjunction with the 21st International Conference on Network-Based Information Systems (NBiS-2018) at Comenius University in Bratislava, Slovakia, from September 5 to 7, 2018.

The goal of NBiS-2018 workshops is to provide a forum for international researchers and practitioners to exchange and share their new ideas, research results, and ongoing work on leading-edge topics in the different fields of network-based information systems and their applications. Some of the accepted workshops deal with topics that open up perspectives beyond the ordinary, thus enriching the topics usually addressed by the NBiS-2018 conference.

For this edition, the following workshops will be held with NBiS-2018.

1. The 13th International Workshop on Network-based Virtual Reality and Tele-existence (INVITE-2018)
2. The 12th International Workshop on Advanced Distributed and Parallel Network Applications (ADPNA-2018)
3. The 9th International Workshop on Heterogeneous Networking Environments and Technologies (HETNET-2018)
4. The 9th International Workshop on Intelligent Sensors and Smart Environments (ISSE-2018)
5. The 9th International Workshop on Trustworthy Computing and Security (TwCSec-2018)
6. The 8th International Workshop on Information Networking and Wireless Communications (INWC-2018)
7. The 7th International Workshop on Advances in Data Engineering and Mobile Computing (DEMoC-2018)
8. The 7th International Workshop on Web Services and Social Media (WSSM-2018)
9. The 6th International Workshop on Cloud and Distributed System Applications (CADSA-2018)

We would like to thank the research community for their great response to NBiS-2018 workshops. The excellent technical program of the workshops was the result of a professional work from Workshop Chairs, workshop program committees, reviewers, and authors.

We would like to give our special thanks to Prof. Leonard Barolli and Prof. Makoto Takizawa, the Steering Committee Chairs of NBiS conference, for their strong encouragement and guidance to organize the NBiS-2018 workshops. We would like to thank NBiS-2018 General Co-chairs and Program Co-chairs for their advice to make possible organization of NBiS-2018 workshops.

We would like to express special thanks to NBiS-2018 Web Administration Chairs for their timely unlimited support for managing the conference systems. We also would like to thank NBiS-2018 Local Arrangement Co-chairs and Local Organization Team at Comenius University in Bratislava, Slovakia, for making the local arrangement for the conference and workshops.

We wish all of you entertaining and rewarding experience in all the workshops and the NBiS-2018 conference.

NBiS-2018 Workshop Co-chairs

Peter Balco Comenius University in Bratislava, Slovakia
Yusuke Gotoh Okayama University, Japan
Isaac Woungang Ryerson University, Canada

NBiS-2018 Organizing Committee

General Co-chairs

Natalia Kryvinska	Comenius University in Bratislava, Slovakia
Tomoya Enokido	Rissho University, Japan
Markus Aleksy	ABB AG Corporate Research Centre, Germany

PC Co-chairs

Michal Greguš	Comenius University in Bratislava, Slovakia
Fumiaki Sato	Toho University, Japan

Workshop Co-chairs

Peter Balco	Comenius University in Bratislava, Slovakia
Yusuke Gotoh	Okayama University, Japan
Isaac Woungang	Ryerson University, Canada

Award Co-chairs

Minoru Uehara	Toyo University, Japan
David Taniar	Monash University, Australia
Hui-Huang Hsu	Tamkang University, Taiwan

Publicity Co-chairs

Kin Fun Li	University of Victoria, Canada
Akio Koyama	Yamagata University, Japan
Irfan Awan	University of Bradford, UK
Farookh Hussain	University Technology Sydney, Australia

International Liaison Co-chairs

Wenny Rahayu	La Trobe University, Australia
Hiroaki Kikuchi	Meiji University, Japan
Marek R. Ogiela	AGH University of Science and Technology, Poland
Arjan Durresi	Indiana University–Purdue University Indianapolis (IUPUI), USA

Local Arrangement Co-chairs

Peter Balco	Comenius University in Bratislava, Slovakia
Martina Drahošová	Comenius University in Bratislava, Slovakia

Finance Chair

Makoto Ikeda	Fukuoka Institute of Technology (FIT), Japan

Web Administrator Co-chairs

Kosuke Ozera	Fukuoka Institute of Technology (FIT), Japan
Yi Liu	Fukuoka Institute of Technology (FIT), Japan
Donald Elmazi	Fukuoka Institute of Technology (FIT), Japan
Miralda Cuka	Fukuoka Institute of Technology (FIT), Japan

Steering Committee

Leonard Barolli	Fukuoka Institute of Technology (FIT), Japan
Makoto Takizawa	Hosei University, Japan

Track Areas and PC Members

Track 1: Mobile and Wireless Networks

Track Co-chairs

Tetsuya Shigeyasu	Prefectural University of Hiroshima, Japan
Vamsi Krishna Paruchuri	University of Central Arkansas, USA
Evjola Spaho	Polytechnic University of Tirana, Albania

PC Members

Nobuyoshi Sato	Iwate Prefectural University, Japan
Kanunori Ueda	Kochi University of Technology, Japan
Masaaki Yamanaka	Japan Coast Guard Academy, Japan
Takuya Yoshihiro	Wakayama University, Japan
Tomoya Kawakami	Nara Institute of Science and Technology, Japan
Masaaki Noro	Fujitsu Laboratories, Japan
Admir Barolli	Aleksander Moisiu University of Durres, Albania
Makoto Ikeda	Fukuoka Institute of Technology (FIT), Japan
Keita Matsuo	Fukuoka Institute of Technology (FIT), Japan
Elis Kulla	Okayama University of Science, Japan
Noriki Uchida	Fukuoka Institute of Technology (FIT), Japan
Arjan Durresi	Indiana University–Purdue University Indianapolis (IUPUI), USA
Sriram Chellappan	University of South Florida (USF), USA

Track 2: Internet of Things and Big Data

Track Co-chairs

Nik Bessis	Edge Hill University, UK
Chun-Wei Tsai	National Ilan University, Taiwan
Patrick Hung	University of Ontario Institute of Technology, Canada

PC Members

Ella Perreira	Edge Hill University, UK
Sergio Toral	University of Seville, Spain
Stelios Sotiriadis	University of Toronto, Canada
Eleana Asimakopoulou	Hellenic National Defence College, Greece
Xiaolong Xu	University of Posts and Telecommunications, China

Kevin Curran	Ulster University, UK
Kamen Kanev	Shizuoka University, Japan
Shih-Chia Huang	National Taipei University of Technology, Taiwan
Jorge Roa	UTN, Santa Fe, Argentina
Alvaro Joffre Uribe	Universidad Militar Nueva Granada, Colombia
Marcelo Fantinato	University of Sao Paulo, Brazil
Marco Zennaro	Wireless and T/ICT4D Laboratory, Italy
Priyanka Rawat	University of Avignon, France
Francesco Piccialli	University of Naples "Federico II", Italy
Chi-Yuan Chen	National Ilan University, Taiwan

Track 3: Cloud, Grid and Service-Oriented Computing

Track Co-chairs

Ciprian Dobre	Polytechnic University of Bucharest, Romania
Olivier Terzo	Istituto Superiore Mario Boella (ISMB), Italy
Muhammad Younas	Oxford Brookes University, UK

PC Members

Zia Rehman	COMSATS Institute of Information Technology, Pakistan
Walayat Hussain	University of Technology, Sydney, Australia
Farookh Hussain	University of Technology, Sydney, Norway
Adil Hammadi	Sultan Qaboos University, Oman
Rui Pais	University of Stavanger, Norway
Raymond Hansen	Purdue University, USA
Antorweep Chakravorty	University of Stavanger, Norway
Rui Esteves	National Oilwell Varco, Norway
Constandinos X. Mavromoustakis	University of Nicosia, Cyprus
Ioan Salomie	Technical University of Cluj-Napoca, Romania
George Mastorakis	Technological Educational Institute of Crete, Greece
Sergio L. Toral Marín	University of Seville, Spain
Marc Frincu	West University of Timisoara, Romania
Alexandru Costan	IRISA/INSA Rennes, France
Xiaomin Zhu	National University of Defense Technology, China
Radu Tudoran	Huawei, Munich, Germany
Mauro Migliardi	University of Padua, Italy
Harold Castro	Universidad de Los Andes, Colombia
Andrea Tosatto	Open-Xchange, Germany
Rodrigo Calheiros	Western Sydney University, Australia

Track 4: Multimedia and Web Applications

Track Co-chairs

Takahiro Uchiya	Nagoya Institute of Technology, Japan
Tomoyuki Ishida	Fukuoka Institute of Technology (FIT), Japan
Nobuo Funabiki	Okayama University, Japan

PC Members

Shigeru Fujita	Chiba Institute of Technology, Japan
Yuka Kato	Tokyo Woman's Christian University, Japan
Yoshiaki Kasahara	Kyushu University, Japan
Rihito Yaegashi	Kagawa University, Japan
Kazunori Ueda	Kochi University of Technology, Japan
Ryota Nishimura	Keio University, Japan
Shohei Kato	Nagoya Institute of Technology, Japan
Shinsuke Kajioka	Nagoya Institute of Technology, Japan
Atsuko Muto	Nagoya Institute of Technology, Japan
Kaoru Sugita	Fukuoka Institute of Technology (FIT), Japan
Noriyasu Yamamoto	Fukuoka Institute of Technology (FIT), Japan
Hiroaki Nishino	Oita University, Japan

Track 5: Ubiquitous and Pervasive Computing

Track Co-chairs

Chi-Yi Lin	Tamkang University, Taiwan
Joan Arnedo Moreno	Open University of Catalonia, Spain

PC Members

Jichiang Tsai	National Chung Hsing University, Taiwan
Chang Hong Lin	National Taiwan University of Science and Technology, Taiwan
Meng-Shiuan Pan	Tamkang University, Taiwan
Chien-Fu Cheng	Tamkang University, Taiwan
Ang Chen	University of Pennsylvania, USA
Santi Caballe	Open University of Catalonia, Spain
Evjola Spaho	Polytechnic University of Tirana, Albania
Elis Kulla	Okayama University of Science, Japan
Makoto Ikeda	Fukuoka Institute of Technology (FIT), Japan

Track 6: Network Security and Privacy

Track Co-chairs

Takamichi Saito	Meiji University, Japan
Sriram Chellappan	University of South Florida (USF), USA
Feilong Tang	Shanghai Jiao Tong University, China

PC Members

Satomi Saito	Fujitsu Laboratories, Japan
Kazumasa Omote	University of Tsukuba, Japan
Koji Chida	NTT, Japan
Hiroki Hada	NTT Security (Japan) KK, Japan
Hirofumi Nakakouji	Hitachi Ltd., Japan
Na Ruan	Shanghai Jiao Tong University, China
Chunhua Su	Osaka University, China
Kazumasa Omote	University of Tsukuba, Japan
Toshihiro Yamauchi	Okayama University, Japan
Masakazu Soshi	Hiroshima City University, Japan
Bagus Santoso	University of Electro-Communications, Japan
Laiping Zhao	Tianjin University, China
Jingyu Hua	Nanjing University, China
Xiaobo Zhou	Tianjin University, China
Yuan Zhan	Nanjing University, China
Yizhi Ren	Hangzhou Dianzi University, China
Arjan Durresi	Indiana University–Purdue University Indianapolis (IUPUI), USA
Vamsi Krishna Paruchuri	University of Central Arkansas, USA

Track 7: Database, Data Mining and Semantic Computing

Track Co-chairs

Wendy K. Osborn	University of Lethbridge, Canada
Eric Pardade	La Trobe University, Australia
Akimitsu Kanzaki	Shimane University, Japan

PC Members

Asm Kayes	La Trobe University, Australia

Ronaldo dos Santos Mello	Universidade Federal de Santa Catarina, Brazil
Saqib Ali	Sultan Qaboos University, Oman
Hong Quang Nguyen	Ho Chi Minh City International University, Vietnam
Irena Holubova	Charles University, Prague, Czech Republic
Prakash Veeraraghavan	La Trobe University, Australia
Carson Leung	University of Manitoba, Canada
Marwan Hassani	Aachen University, Germany
Tomoki Yoshihisa	Osaka University, Japan
Tomoya Kawakami	NAIST, Japan
Atsushi Takeda	Tohoku Gakuin University, Japan
Yoshiaki Terashima	Soka University, Japan
Yuuichi Teranishi	NICT, Japan
Jackie Rice	University of Lethbridge, Canada
Yllias Chali	University of Lethbridge, Canada
John Zhang	University of Lethbridge, Canada

Track 8: Network Protocols and Applications

Track Co-chairs

Irfan Awan	University of Bradford, UK
Sanjay Kumar Dhurandher	NSIT, University of Delhi, India
Hsing-Chung Chen	Asia University, Taiwan

PC Members

Samia Loucif	ALHOSN University, UAE
Abdelhamid Mammeri	Ottawa University Ontario, Canada
Jun He	University of New Brunswick, Canada
Peyman Kabiri	University of Science and Technology, Iran
Chen Chen	University of Texas, USA
Ahmed Abdelgawad	Central Michigan University, USA
Wael Elmedany	University of Bahrain, Bahrain
Behrouz Maham	School of Electrical and Computer Engineering, Iran
Rubem Pereira	Liverpool John Moores University, UK
Carlos Juiz	University of the Balearic Islands, Spain
Faheem Ahmed	Thompson Rivers University, Canada
Paulo Gil	FCT-UNL, Portugal
Michael Mcguire	University of Victoria, Canada
Steven Guan	Xian Jiaotong-Liverpool University, China
Gregorio Romero	Universidad Politecnica de Madrid, Spain

Amita Malik	Deenbandhu Chhotu Ram University of Science and Technology, India
Mayank Dave	NIT Kurukshetra, India
Vinesh Kumar	University of Delhi, India
R. K. Pateriya	MANIT, Bhopal, India
Himanshu Aggarwal	Punjabi University, India
Neng-Yih Shih	Asia University, Taiwan
Yeong-Chin Chen	Asia University, Taiwan
Hsi-Chin Hsin	National United University, Taiwan
Ming-Shiang Huang	Asia University, Taiwan
Chia-Cheng Liu	Asia University, Taiwan
Chia-Hsin Cheng	National Formosa University, Yunlin County, Taiwan
Tzu-Liang Kung	Asia University, Taiwan
Gene Shen	Asia University, Taiwan
Jim-Min Lin	Feng Chia University, Taiwan
Chia-Cheng Liu	Asia University, Taiwan
Yen-Ching Chang	Chung Shan Medical University, Taiwan
Shu-Hong Lee	Chienkuo Technology University, Taiwan
Ho-Lung Hung	Chienkuo Technology University, Taiwan
Gwo-Ruey Lee	Lung-Yuan Research Park, Taiwan
Li-Shan Ma	Chienkuo Technology University, Taiwan
Chung-Wen Hung	National Yunlin University of Science and Technology, Taiwan
Yung-Chen Chou	Asia University, Taiwan
Chen-Hung Chuang	Asia University, Taiwan
Jing-Doo Wang	Asia University, Taiwan
Jui-Chi Chen	Asia University, Taiwan
Young-Long Chen	National Taichung University of Science and Technology, Taiwan

Track 9: Intelligent and Cognitive Computing

Track Co-chairs

Lidia Ogiela	AGH University of Science and Technology, Poland
Farookh Hussain	University of Technology Sydney, Australia
Hae-Duck Joshua Jeong	Korean Bible University, Korea

PC Members

Yiyu Yao	University of Regina, Canada
Daqi Dong	University of Memphis, USA
Jan Platoš	VŠB-Technical University of Ostrava, Czech Republic
Pavel Krömer	VŠB-Technical University of Ostrava, Czech Republic

Urszula Ogiela	AGH University of Science and Technology, Poland
Jana Nowaková	VŠB-Technical University of Ostrava, Czech Republic
Hoon Ko	Chosun University, Korea
Chang Choi	Chosun University, Korea
Gangman Yi	Gangneung-Wonju National University, Korea
Wooseok Hyun	Korean Bible University, Korea
Hsing-Chung Jack Chen	Asia University, Taiwan
JongSuk Ruth Lee	KISTI, Korea
Hyun Jung Lee	Yonsei University, Korea
Ji-Young Lim	Korean Bible University, Korea
Omar Hussain	UNSW Canberra, Australia
Saqib Ali	Sultan Qaboos University, Oman
Morteza Saberi	UNSW Canberra, Australia
Sazia Parvin	UNSW Canberra, Australia
Walayat Hussain	University of Technology Sydney, Australia

Track 10: Parallel and Distributed Computing

Track Co-chairs

Kin Fun Li	University of Victoria, Canada
Bhed Bista	Iwate Prefectural University, Japan

PC Members

Deepali Arora	University of Victoria, Canada
Kosuke Takano	Kanagawa Institute of Technology, Japan
Masahiro Ito	Toshiba Lab, Japan
Watheq ElKharashi	Ain Shams University, Egypt
Martine Wedlake	IBM, USA
Jiahong Wang	Iwate Prefectural University, Japan
Shigetomo Kimura	University of Tsukuba, Japan
Chotipat Pornavalai	King Mongkut's Institute of Technology Ladkrabang, Thailand
Danda B. Rawat	Howard University, USA
Gongjun Yan	University of Southern Indiana, USA
ShuShaw Wang	Dallas Baptist University, USA
Naonobu Okazaki	Miyazaki University, Japan
Yoshiaki Terashima	Soka University, Japan
Atsushi Takeda	Tohoku Gakuin University, Japan
Tomoki Yoshihisa	Osaka University, Japan
Akira Kanaoka	Toho University, Japan

NBiS-2018 Reviewers

Ali Khan Zahoor
Barolli Admir
Barolli Leonard
Bista Bhed
Caballé Santi
Chellappan Sriram
Chen Hsing-Chung
Chen Xiaofeng
Cui Baojiang
Di Martino Beniamino
Durresi Arjan
Enokido Tomoya
Ficco Massimo
Fun Li Kin
Gotoh Yusuke
Hussain Farookh
Hussain Omar
Javaid Nadeem
Jeong Joshua
Ikeda Makoto
Ishida Tomoyuki
Kikuchi Hiroaki
Kohana Masaki
Koyama Akio
Kulla Elis
Lee Kyungroul
Matsuo Keita
Nishino Hiroaki
Ogiela Lidia
Ogiela Marek
Palmieri Francesco
Paruchuri Vamsi Krishna
Rahayu Wenny
Rawat Danda
Shibata Yoshitaka
Sato Fumiaki
Spaho Evjola
Sugita Kaoru
Takizawa Makoto
Taniar David
Terzo Olivier
Uchida Noriki
Uehara Minoru
Venticinque Salvatore
Waluyo Agustinus Borgy
Woungang Isaac
Xhafa Fatos
Yim Kangbin
Younas Muhammad

Welcome Message from INVITE-2018 Workshop Organizers

Welcome to the 13th International Workshop on Network-based Virtual Reality and Tele-existence (INVITE-2018), which will be held in conjunction with the 21st International Conference on Network-Based Information Systems (NBiS-2018) from September 5 to 7, 2018, at Comenius University in Bratislava, Slovakia.

The INVITE-2018 International Workshop serves as a forum for the exchange of information and ideas regarding network-based virtual reality (VR) and tele-existence, 2D and 3D computer graphics, computer animation, multimedia, object-oriented approach, Web technology, and e-learning. All participants of this workshop can share ideas and research work in the emerging areas of network-based VR and tele-existence.

Many people have kindly helped us to prepare and organize the INVITE workshop. First, we would like to thank the NBiS-2018 organization committee for their support, guidance, and help for making the workshop. We would like to express our special thanks to all of the INVITE program committee members and reviewers for organizing the workshop and reviewing the submitted papers, respectively.

Finally, we would like to give our special thanks to all of the members of N3VR research committee in Japan for supporting and promoting INVITE-2018.

We hope you will enjoy the workshop and have a great time in Bratislava, Slovakia.

INVITE-2018 Workshop Organizers

Yoshitaka Shibata	Iwate Prefectural University, Japan
Tomoyuki Ishida	Fukuoka Institute of Technology (FIT), Japan

INVITE-2018 Organizing Committee

Workshop Co-chairs

Yoshitaka Shibata	Iwate Prefectural University, Japan
Tomoyuki Ishida	Fukuoka Institute of Technology (FIT), Japan

Program Committee Members

Koji Koyamada	Kyoto University, Japan
Yasuo Ebara	Kyoto University, Japan
Hideo Miyaji	Tokyo City University, Japan
Tetsuro Ogi	Keio University, Japan
Akihiro Miyakawa	Nanao City, Japan
Kaoru Sugita	Fukuoka Institute of Technology (FIT), Japan
Noriki Uchida	Fukuoka Institute of Technology (FIT), Japan

Welcome Message from ADPNA-2018 Workshop Co-chairs

Welcome to the 12th International Workshop on Advanced Distributed and Parallel Network Applications (ADPNA-2018), which will be held in conjunction with the 21st International Conference on Network-Based Information Systems (NBiS-2018) from September 5 to 7, 2018, at Comenius University in Bratislava, Slovakia.

The purpose of this workshop is to bring together developers and researchers to share ideas and research work in the emerging areas of distributed and parallel network applications.

The papers included in ADPNA-2018 cover aspects of theory, design, and application of distributed and parallel systems and applications. ADPNA-2018 contains high-quality research papers. Each submitted paper was peer-reviewed by program committee members who are experts in the subject area of the paper.

For organizing a workshop, the support and help of many people are needed. First, we would like to thank all of the authors for submitting their papers. We also appreciate the support from program committee members and reviewers who carried out the most difficult work of carefully evaluating the submitted papers.

We would like to thank NBiS-2018 Co-chairs, Program Co-chair, and Workshop Co-chair, for their support and strong encouragement and guidance in organizing the workshop. We would like to express special thanks to NBiS-2018 Web Administrator Chairs for their timely unlimited support for managing the conference system.

We also would like to thank NBiS-2018 local arrangement members for making the local arrangement for the conference and workshops.

We hope you will enjoy ADPNA-2018 and NBiS-2018 International Conference and have a great time in Bratislava, Slovakia.

ADPNA-2018 Workshop Organizers

Makoto Takizawa	Hosei University, Japan
Leonard Barolli	Fukuoka Institute of Technology (FIT), Japan

ADPNA-2018 Workshop PC Chair

Tomoya Enokido	Rissho University, Japan

ADPNA-2018 Organizing Committee

Workshop Organizers

Makoto Takizawa	Hosei University, Japan
Leonard Barolli	Fukuoka Institute of Technology (FIT), Japan

Workshop PC Chair

Tomoya Enokido	Rissho University, Japan

Program Committee

Makoto Ikeda	Fukuoka Institute of Technology (FIT), Japan
Elis Kulla	Okayama University of Science, Japan
Aniello Castiglione	University of Salerno, Italy
Akio Koyama	Yamagata University, Japan
Admir Barolli	Aleksander Moisiu University of Durres, Albania
Wenny Rahaya	La Trobe University, Australia
Evjola Spaho	Polytechnic University of Tirana, Albania
Fumiaki Sato	Toho University, Japan
Minoru Uehara	Toyo University, Japan
Fatos Xhafa	Technical University of Catalonia, Spain
David Taniar	Monash University, Australia
Keita Matsuo	Fukuoka Institute of Technology (FIT), Japan
Tomoyuki Ishida	Fukuoka Institute of Technology (FIT), Japan
Kaoru Sugita	Fukuoka Institute of Technology (FIT), Japan

Welcome Message from HETNET-2018 Workshop Organizers

Welcome to the 9th International Workshop on Heterogeneous Networking Environments and Technologies (HETNET-2018), which will be held in conjunction with the 21st International Conference on Network-Based Information Systems (NBiS-2018) from September 5 to 7, 2018, at Comenius University in Bratislava, Slovakia.

Due to advances in networking technologies, and especially in wireless and mobile communication technologies, and devices with different networking interfaces, the networking systems are moving toward the integration of wired and wireless resources leading to Heterogeneous Networking Environments and Technologies (HETNET). Such environments comprise and integrate wireless LANs, multiple MANETs, cellular networks, wireless sensor networks, wireless mesh networks, and so on. Besides, other types of networking systems such as grid and P2P systems are going wireless and mobile, adding yet another dimension to the HETNET.

The integration of various networking paradigms into one networking system is requiring new understandings, algorithms and insights, frameworks, middleware and architectures for the effective design, integration, and deployment aiming at achieving secure, programmable, robust, transparent, and ubiquitous HETNETs.

The aim of this workshop is to present innovative researches, methods, and techniques related to HETNETs and their applications.

Many people contributed to the success of HETNET-2018. First, we would like to thank the organizing committee of NBiS-2018 International Conference for giving us the opportunity to organize the workshop. We would like to thank all authors for submitting their research works and for their participation. We are looking forward to meet them again in the forthcoming editions of the workshop.

We hope you will enjoy the workshop and have a great time in Bratislava, Slovakia.

HETNET-2018 Workshop Chair

Leonard Barolli	Fukuoka Institute of Technology (FIT), Japan

HETNET-2018 Workshop PC Chair

Jugappong Nawichai	Chiang Mai University, Thailand

HETNET-2018 Organizing Committee

Workshop Chair

Leonard Barolli	Fukuoka Institute of Technology (FIT), Japan

Workshop PC Chair

Jugappong Nawichai	Chiang Mai University, Thailand

Program Committee Members

Markus Aleksy	ABB AG Corporate Research Centre, Germany
Xiaofeng Chen	Xidian University, China
Paul Chung	Loughborough University, UK
Ciprian Dobre	Polytechnic University of Bucharest, Romania
Antonio Gentile	University of Palermo, Italy
Hui-Huang Hsu	Tamkang University, Taiwan
Makoto Ikeda	Fukuoka Institute of Technology (FIT), Japan
Axel Korthaus	Victoria University, Melbourne, Australia
Kin Fun Li	University of Victoria, Canada
John Mashford	CSIRO, Melbourne, Australia
Hiroaki Nishino	Oita University, Japan
Wenny Rahayu	La Trobe University, Australia
Makoto Takizawa	Seikei University, Japan
David Taniar	Monash University, Australia

Runtong Zhang	Beijing Jiao Tong University, China
Pruet Boonma	Chiang Mai University, Thailand
Paskorn Champrasert	Chiang Mai University, Thailand
Yuthapong Somchit	Chiang Mai University, Thailand

Welcome Message from ISSE-2018 Workshop Co-chairs

To automatically provide human-centric services, it is desired to build smart environments for such purposes. The background system needs to first detect and collect needed information in the environment via various sensors like video cameras, RFIDs, infrared sensors, motion sensors and pressure sensors. The information is then processed for understanding the context and intention of the people in the environment. Finally, the services are provided in a timely manner. Sensing technology, embedded systems, wireless communications, computer vision and intelligent methods are the core to build integrated smart environments. Previous ISSE's were held at Fukuoka, Japan (2010); Barcelona, Spain (2011); Melbourne, Australia (2012); Gwangju, Korea (2013); Salerno, Italy (2014); Taipei, Taiwan (2015); Ostrava, Czech Republic (2016); and Toronto, Canada (2017).

The 9th International Workshop on Intelligent Sensors and Smart Environments (ISSE-2018) will provide a platform for researchers to meet and exchange their thoughts. ISSE-2018 will be held in conjunction with the 21st International Conference on Network-Based Information Systems (NBiS-2018) at Comenius University in Bratislava, Slovakia, from September 5 to 7, 2018.

Many people contributed to the CFP and paper review of ISSE-2018. We wish to thank the program committee members for their great effort. We also would like to express our gratitude to Steering Committee Co-chairs, Program Co-chairs and Workshops Co-Chairs of NBiS-2018. Last but not least, we would like to thank and congratulate all the contributing authors for their support and excellent work.

ISSE-2018 Workshop Co-chairs

Hui-Huang Hsu	Tamkang University, Taiwan
Leonard Barolli	Fukuoka Institute of Technology (FIT), Japan

ISSE-2018 Organizing Committee

Workshop Co-chairs

Hui-Huang Hsu	Tamkang University, Taiwan
Leonard Barolli	Fukuoka Institute of Technology (FIT), Japan

Program Committee Co-chairs

Takahiro Uchiya	Nagoya Institute of Technology, Japan
Chuan-Yu Chang	National Yunlin University of Science and Technology, Taiwan

Program Committee Members

Keita Matsuo	Fukuoka Institute of Technology (FIT), Japan
Evjola Spaho	Polytechnic University of Tirana, Albania
Markus Aleksy	ABB AG Corporate Research Centre, Germany
Fatos Xhafa	Technical University of Catalonia, Spain
Makoto Ikeda	Fukuoka Institute of Technology (FIT), Japan
Chien-Cheng Lee	Yuan Ze University, Taiwan
Jose Bravo	Castilla-La Mancha University, Spain
Ching-Lung Chang	National Yunlin University of Science and Technology, Taiwan
Chu-Song Chen	Academia Sinica, Taiwan
Hsu-Yung Cheng	National Central University, Taiwan
Ding-An Chiang	Tamkang University, Taiwan
Wu-Chih Hu	National Penghu University, Taiwan
Lei Jing	University of Aizu, Japan
Wei-Ru Lai	Yuan Ze University, Taiwan
Wen-Ping Lai	Yuan Ze University, Taiwan
Jiann-Shu Lee	National University of Tainan, Taiwan
Kin Fun Li	University of Victoria, Canada
Chi-Yi Lin	Tamkang University, Taiwan
Jiming Liu	Hong Kong Baptist University, Hong Kong
Wei Lu	Keene University, USA
Mihai Sima	University of Victoria, Canada
Wen-Fong Wang	National Yunlin University of Science and Technology, Taiwan
Zhiwen Yu	Northwestern Polytechnical University, China

Welcome Message from TwCSec-2018 Workshop Co-chairs

It is our great pleasure to welcome you to the 9th International Workshop on Trustworthy Computing and Security (TwCSec-2018), which will be held in conjunction with the 21st International Conference on Network-Based Information Systems (NBiS-2018) at Comenius University in Bratislava, Slovakia, from September 5 to 7, 2018.

This international workshop is a forum for sharing ideas and research work in the emerging areas of trustworthy computing, security, and privacy. Computers reside at the heart of systems on which people now rely, both in critical national infrastructures and in their homes, cars, and offices. Today, many of these systems are far too vulnerable to cyber attacks that can inhibit their operation, corrupt valuable data, or expose private information. Future systems will include sensors and computers everywhere, exacerbating the attainment of security and privacy. Current security practices largely address current and known threats, but there is a need for research to consider future threats.

We encouraged contributions describing innovative work on security of next-generation operating systems (OSs), secure and resilient network protocols, theoretical foundations and mechanisms for privacy, security, trust, human–computer interfaces for security functions, key distribution/management, intrusion detection and response, secure protocol configuration and deployment, and improved ability to certify and analyze system security properties, and integrating hardware and software for security.

Many people contributed to the success of TwCSec-2018. First, we would like to thank the organizing committee of NBiS-2018 International Conference for giving us the opportunity to organize the workshop. We would like to thank all the authors of the workshop for submitting their research works and for their participation. We are looking forward to meet them again in the forthcoming editions of the workshop.

Finally, we would like to thank the Local Arrangement Chairs for the local arrangement of the workshop.

We hope you will enjoy the workshop and have a great time in Bratislava, Slovakia.

TwCSec-2018 Workshop Co-chairs

Leonard Barolli	Fukuoka Institute of Technology (FIT), Japan
Arjan Durresi	Indiana University–Purdue University Indianapolis (IUPUI), USA
Hiroaki Kikuchi	Meiji University, Japan

TwCSec-2018 Organizing Committee

Workshop Co-chairs

Leonard Barolli	Fukuoka Institute of Technology (FIT), Japan
Arjan Durresi	Indiana University–Purdue University Indianapolis (IUPUI), USA
Hiroaki Kikuchi	Meiji University, Japan

Advisory Co-chairs

Makoto Takizawa	Hosei University, Japan
Raj Jain	Washington University in St. Louis, USA

Program Committee Members

Sriram Chellappan	University of South Florida (USF), USA
Koji Chida	NTT, Japan
Qijun Gu	Texas State University, San Marcos, USA
Tesuya Izu	Fujitsu Ltd., Japan
Youki Kadobayashi	Nara Institute of Science and Technology, Japan
Akio Koyama	Yamagata University, Japan
Michiharu Kudo	IBM Japan, Japan
Sanjay Kumar Madria	Missouri University of Science and Technology, USA
Masakatsu Morii	Kobe University, Japan
Masakatsu Nishigaki	Shizuoka University, Japan
Vamsi Paruchuri	University of Central Arkansas, USA
Hiroshi Shigeno	Keio University, Japan
Yuji Suga	Internet Initiative Japan Inc., Japan
Keisuke Takemori	KDDI Co., Japan
Ryuya Uda	Tokyo University of Technology, Japan

Xukai Zou	Indiana University–Purdue University Indianapolis (IUPUI), USA
Wenye Wang	North Carolina State University, USA
Hiroshi Yoshiura	University of Electro-Communications, Japan
Farookh Hussain	University of Technology Sydney, Australia

Welcome Message from INWC-2018 Workshop Organizers

It is our great pleasure to welcome you to the 8th International Workshop on Information Networking and Wireless Communications (INWC-2018), which will be held in conjunction with the 21st International Conference on Network-Based Information Systems (NBiS-2018) at Comenius University in Bratislava, Slovakia, from September 5 to 7, 2018.

This international workshop is a forum for sharing ideas and research work in the emerging areas of information networking and wireless communications. Nowadays, information networks are developing very rapidly and evolving to heterogeneous networks. Thus, we have an increasing number of applications and devices. To optimize the communication in these heterogeneous systems, we need to propose and evaluate more complex schemes, algorithms, and protocols for wired and wireless networks.

Wireless communications are characterized by high bit error rates and burst errors, which arise due to interference fading, shadowing, terminal mobility, and so on. Since the traditional design of the algorithms, methods, and protocols of the wired Internet did not take wireless networks into account, the performance over wireless networks is largely degraded. Especially, the multi-hop communication aggravates the problem of wireless communication even further. To solve these problems, there has been increased interest to propose and design new algorithms and methodologies for wireless communication.

The aim of this workshop is to present the innovative researches, methods, and numerical analysis for wireless communication and wireless networks.

Many people contributed to the success of INWC-2018. First, we would like to thank the organizing committee of NBiS-2018 International Conference for giving us the opportunity to organize the workshop. We would like to thank all the authors of the workshop for submitting their research works and for their participation. We are looking forward to meet them again in the forthcoming editions of the workshop.

Finally, we would like to thank the Local Arrangement Chairs for the local arrangement of the workshop.

We hope you will enjoy the workshop and have a great time in Bratislava, Slovakia.

INWC-2018 Workshop Co-chairs

Leonard Barolli	Fukuoka Institute of Technology (FIT), Japan
Hiroshi Maeda	Fukuoka Institute of Technology (FIT), Japan

INWC-2018 Workshop PC Chair

Makoto Ikeda	Fukuoka Institute of Technology (FIT), Japan

INWC-2018 Organizing Committee

Workshop Co-chairs

Leonard Barolli	Fukuoka Institute of Technology (FIT), Japan
Hiroshi Maeda	Fukuoka Institute of Technology (FIT), Japan

Workshop PC Co-chair

Makoto Ikeda	Fukuoka Institute of Technology (FIT), Japan

Program Committee Members

Arjan Durresi	Indiana University–Purdue University Indianapolis (IUPUI), USA
Shinichi Ichitsubo	Kyushu Institute of Technology, Japan
Elis Kulla	Okayama University of Science, Japan
Zhi Qi Meng	Fukuoka University, Japan
Irfan Awan	University of Bradford, UK
Tsuyoshi Matsuoka	Kyushu Sangyo University, Japan
Fatos Xhafa	Technical University of Catalonia, Spain
Kiyotaka Fujisaki	Fukuoka Institute of Technology (FIT), Japan
Noriki Uchida	Fukuoka Institute of Technology (FIT), Japan
Keita Matsuo	Fukuoka Institute of Technology (FIT), Japan

Welcome Message from DEMoC-2018 Workshop Chair

Welcome to the 7th International Workshop on Advances in Data Engineering and Mobile Computing (DEMoC-2018) which will be held in conjunction with the 21st International Conference on Network-Based Information Systems (NBiS-2018) at Comenius University in Bratislava, Slovakia, from September 5 to 7, 2018.

This international workshop is to bring together practitioners and researchers from both academia and industry in order to have a forum for discussion and technical presentations on the current researches and future research directions related to these hot research areas: data engineering and mobile computing. We encouraged contributions describing innovative work on the DEMoC-2018 workshop.

Many people contributed to the success of DEMoC-2018. First, we would like to thank the organizing committee of NBiS-2018 International Conference for giving us the opportunity to organize the workshop. We would like to thank all the authors of the workshop for submitting their research works and for their participation. We are looking forward to meet them again in the forthcoming editions of the workshop.

Finally, we would like to thank the Local Arrangement Chairs for the local arrangement of the workshop.

We hope you will enjoy the workshop and have a great time in Bratislava, Slovakia.

DEMoC-2018 Workshop Chair

Yusuke Gotoh Okayama University, Japan

DEMoC-2018 Organizing Committee

Workshop Organizer

Yusuke Gotoh	Okayama University, Japan

Program Committee

Toshiyuki Amagasa	University of Tsukuba, Japan
Akihito Hiromori	Osaka University, Japan
Akiyo Nadamoto	Konan University, Japan
Kenji Ohira	Nara Institute of Science and Technology, Japan
Shusuke Okamoto	Seikei University, Japan
Wenny Rahayu	La Trobe University, Australia
David Taniar	Monash University, Australia
Hideo Taniguchi	Okayama University, Japan
Tsutomu Terada	Kobe University, Japan
Tomoki Yoshihisa	Osaka University, Japan

Welcome Message from WSSM-2018 Workshop Co-chairs

Welcome to Bratislava and the 7th International Workshop on Web Services and Social Media (WSSM-2018), which is held in conjunction with the 21st International Conference on Network-Based Information Systems (NBiS-2018) at Comenius University in Bratislava, Slovakia, from September 5 to 7, 2018.

Web system with IoT technology produces big data, and it has become increasingly important in our society. Using some AI engine enables us to build a novel Web services for home and business use. We can also imagine the next generation of social network services where not only consumer-generated contents but also IoT-generated contents drive our activity on the Internet.

The International Workshop on Web Services and Social Media (WSSM) encompasses a wide range of topics in the research, design, and implementation of Web-related things, especially, social activity on the Internet. The 1st edition of the workshop (WSSM-2012) was held at Melbourne, Australia; the 2nd edition in Gwangju, Korea; the 3rd edition in Salerno, Italy; the 4th edition in Taipei, Taiwan; the 5th edition in Ostrava, Czech Republic; and the 6th edition in Toronto, Canada. This time, we are honored to hold the 7th edition of the workshop.

This workshop provides an international forum for researchers and participants to share and exchange their experiences, discuss challenges, and present original ideas in all aspect of Web Services and Social Media.

Many people contributed to the success of WSSM-2018. First, we would like to thank the organizing committee of NBiS-2018 International Conference for giving us the opportunity to organize the workshop. We would like to thank our program committee members and all authors for submitting their research works and for their participation.

Finally, we would like to thank the Local Arrangement Chairs of the NBiS-2018 conference. We hope you will enjoy WSSM-2018 workshop and NBiS-2018 International Conference and have a great time in Bratislava, Slovakia.

WSSM-2018 Workshop Co-chairs

Shusuke Okamoto	Seikei University, Japan
Masaki Kohana	Ibaraki University, Japan

WSSM-2018 Organizing Committee

Workshop Co-chairs

Shusuke Okamoto	Seikei University, Japan
Masaki Kohana	Ibaraki University, Japan

Program Chair

Masayuki Ihara	NTT Service Evolution Laboratories, Japan

Program Committee Members

Fatos Xhafa	Technical University of Catalonia, Spain
Hiroki Sakaji	University of Tokyo, Japan
Jun Iio	Chuo University, Japan
Kin Fun Li	University of Victoria, Canada
Leonard Barolli	Fukuoka Institute of Technology (FIT), Japan
Makoto Ikeda	Fukuoka Institute of Technology (FIT), Japan
Makoto Takizawa	Hosei University, Japan
Masaru Kamada	Ibaraki University, Japan
Masaru Miyazaki	NHK, Japan
Tatsuhiro Yonekura	Ibaraki University, Japan
Tomoya Enokido	Rissho University, Japan
Yoshihiro Kawano	Tokyo University of Information Sciences, Japan
Yusuke Gotoh	Okayama University, Japan

Welcome Message from CADSA-2018 Workshop Organizer

Welcome to the 6th International Workshop on Cloud and Distributed System Applications (CADSA-2018), which is held in conjunction with the 21st International Conference on Network-Based Information Systems (NBiS-2018) at Comenius University in Bratislava, Slovakia, from September 5 to 7, 2018.

This International Workshop on Cloud and Distributed System Applications brings together scientists, engineers, and students for sharing experiences, ideas, and research results about domain-specific applications relying on cloud computing or distributed systems.

This workshop provides an international forum for researchers and participants to share and exchange their experiences, discuss challenges, and present original ideas in all aspects related to the Cloud and Distributed System Applications' design and development.

We have encouraged innovative contributions about cloud and distributed computing, like

- Distributed Computing Applications
- Cloud Computing Applications
- Collaborative Platforms
- Topologies for Distributed Computing
- Semantic Technologies for Cloud
- Modeling and Simulation of Cloud Computing
- Modeling and Simulation of Distributed System
- Distributed Knowledge Management
- Distributed Computing for Smart Cities
- Distributed Computing for e-Health
- Quality Evaluation of Distributed Services

Many people contributed to the success of CADSA-2018. First, I would like to thank the organizing committee of NBiS-2018 International Conference for giving us the opportunity to organize the workshop. Second, I would like to thank our program committee Members and, of course, I would like to thank all the authors of the workshop for submitting their research works and for their participation.

Finally, I would like to thank the Local Arrangement Chairs of the NBiS-2018 conference.

I hope you will enjoy CADSA workshop and NBiS International Conference, find this a productive opportunity for sharing experiences, ideas, and research results with many researchers, and have a great time in Bratislava, Slovakia.

CADSA-2018 Workshop Chair

Flora Amato	University of Naples "Federico II", Italy

CADSA-2018 Program Committee

Workshop Chair

Flora Amato	University of Naples "Federico II", Italy

Program Committee Members

Antonino Mazzeo	University of Naples "Federico II", Italy
Nicola Mazzocca	University of Naples "Federico II", Italy
Carlo Sansone	University of Naples "Federico II", Italy
Beniamino di Martino	Second University of Naples, Italy
Antonio Picariello	University of Naples "Federico II", Italy
Valeria Vittorini	University of Naples "Federico II", Italy
Anna Rita Fasolino	University of Naples "Federico II", Italy
Umberto Villano	Università degli Studi del Sannio, Italy
Kami Makki	Lamar University, Beaumont (Texas), USA
Valentina Casola	University of Naples "Federico II", Italy
Stefano Marrone	Second University of Naples, Italy
Alessandro Cilardo	University of Naples "Federico II", Italy
Vincenzo Moscato	University of Naples "Federico II", Italy
Porfirio Tramontana	University of Naples "Federico II", Italy
Francesco Moscato	Second University of Naples, Italy
Salvatore Venticinque	Second University of Naples, Italy
Emanuela Marasco	West Virginia University, USA
Massimiliano Albanese	George Mason University, USA
Domenico Amalfitano	University of Naples "Federico II", Italy
Massimo Esposito	Institute for High Performance Computing and Networking (ICAR), Italy

Alessandra de Benedictis	University of Naples "Federico II", Italy
Roberto Nardone	University of Naples "Federico II", Italy
Mario Barbareschi	University of Naples "Federico II", Italy
Ermanno Battista	University of Naples "Federico II", Italy
Mario Sicuranza	Institute for High Performance Computing and Networking (ICAR), Italy
Natalia Kryvinska	Comenius University in Bratislava, Slovakia
Moujabbir Mohammed	Université Hassan II Mohammedia, Casablanca, Morocco

NBiS-2018 Keynote Talks

IoT and CPS—Two Worlds Growing Together: The Embedded Software Point of View

Peter Hellinckx

University of Antwerp, Antwerp, Belgium

Abstract. IoT and CPS are growing together. Autonomous vehicles are probably the best example. The car as a CPS will communicate or even be part of an IoT environment, enriching environment knowledge and optimizing the CPS control. However, new innovations come with new challenges. In this talk, I will focus on different challenges in the distributed embedded software context. More specifically, I will talk about WCET—and schedulability—analysis, (real-time) resource-oriented real-time code placement/movement, and testing of emergent behavior.

SAP Leonardo, Your Digital Innovation System

Juraj Laifr

SAP Education, Bratislava, Slovakia

Abstract. Never before have there been so many promising breakthrough technologies available—and so many businesses ready to capitalize on them. SAP Leonardo delivers new capabilities in future-forward technologies, which add tremendous value to the company's digital journey. SAP Leonardo is a holistic digital innovation system that seamlessly integrates future-facing technologies and capabilities into the SAP Cloud Platform, using our Design Thinking Services. This powerful portfolio enables people to rapidly innovate, scale new models, and continually redefine the business.

Centralized e-Health Nationwide Concept to Reduce the Complexity of Integration

Peter Linhardt

Partner FineSoft Ltd., Kosice, Slovakia

Abstract. Implementation of e-solution in the field of health care is the must in order to achieve the efficiency, quality of care, and implementation of new methods for care provisioning. All subsystems and points of care have to communicate with EHR silos, central, and local applications, and it underlines the importance of proper protocols, data format and integration issues, as well the importance of proper project management for testing and delivery in real-time conditions of cloud and infrastructure. All this effort is essential in order to ensure the wide portfolio of the benefit for all groups of healthcare stakeholders and for the delivery of the future application in e-commerce and m-commerce environment.

Contents

The 8th International Workshop on Information Networking and Wireless Communications (INWC-2018)

The 7th International Workshop on Advances in Data Engineering and Mobile Computing (DEMoC-2018)

The 7th International Workshop on Web Services and Social Media (WSSM-2018)

The 6th International Workshop on Cloud and Distributed System Applications (CADSA-2018)

The 21st International Conference on Network-Based Information Systems (NBiS-2018)

A Fuzzy-Based System for Actor Node Selection in WSANs for Improving Network Connectivity and Increasing Number of Covered Sensors

Donald Elmazi[1](✉), Miralda Cuka[1], Makoto Ikeda[2], and Leonard Barolli[2]

[1] Graduate School of Engineering, Fukuoka Institute of Technology (FIT), 3-30-1 Wajiro-Higashi, Higashi-Ku, Fukuoka 811-0295, Japan
donald.elmazi@gmail.com, mcuka91@gmail.com

[2] Department of Information and Communication Engineering, Fukuoka Institute of Technology (FIT), 3-30-1 Wajiro-Higashi, Higashi-Ku, Fukuoka 811-0295, Japan
makoto.ikd@acm.org, barolli@fit.ac.jp

Abstract. Wireless Sensor and Actor Network (WSAN) is formed by the collaboration of micro-sensor and actor nodes. The sensor nodes have responsibility to sense an event and send information towards an actor node. The actor node is responsible to take prompt decision and react accordingly. In order to provide effective sensing and acting, a distributed local coordination mechanism is necessary among sensors and actors. In this work, we consider the actor node selection problem and propose a fuzzy-based system that based on data provided by sensors and actors selects an appropriate actor node. We use 4 input parameters: Size of Giant Component (SGC), Distance to Event (DE), Remaining Energy (RE) and Number of Covered Sensors (NCS) as new parameter. The output parameter is Actor Selection Decision (ASD). The simulation results show that by increasing SGC to 0.5 and 0.9, the ASD is increased 12% and 68%, respectively.

1 Introduction

Recent technological advances have lead to the emergence of distributed Wireless Sensor and Actor Networks (WSANs) wich are capable of observing the physical world, processing the data, making decisions based on the observations and performing appropriate actions [1].

In WSANs, the devices deployed in the environment are sensors able to sense environmental data, actors able to react by affecting the environment or have both functions integrated. Actor nodes are equipped with two radio transmitters, a low data rate transmitter to communicate with the sensor and a high data rate interface for actor-actor communication. For example, in the case of a fire, sensors relay the exact origin and intensity of the fire to actors so that they

L. Barolli et al. (Eds.): NBiS 2018, LNDECT 22, pp. 3–15, 2019.
https://doi.org/10.1007/978-3-319-98530-5_1

can extinguish it before spreading in the whole building or in a more complex scenario, to save people who may be trapped by fire [2–4].

To provide effective operation of WSAN, it is very important that sensors and actors coordinate in what are called sensor-actor and actor-actor coordination. Coordination is not only important during task conduction, but also during network's self-improvement operations, i.e. connectivity restoration [5,6], reliable service [7], Quality of Service (QoS) [8,9] and so on.

Sensor-Actor (SA) coordination defines the way sensors communicate with actors, which actor is accessed by each sensor and which route should data packets follow to reach it. Among other challenges, when designing SA coordination, care must be taken in considering energy minimization because sensors, which have limited energy supplies, are the most active nodes in this process. On the other hand, Actor-Actor (AA) coordination helps actors to choose which actor will lead performing the task (actor selection), how many actors should perform and how they will perform. Actor selection is not a trivial task, because it needs to be solved in real time, considering different factors. It becomes more complicated when the actors are moving, due to dynamic topology of the network.

In this paper, different from our previous work [10], we propose and implement a simulation system which considers also the Number of Covered Sensors (NCS) parameter. The system is based on fuzzy logic and considers four input parameters for actor selection. We show the simulation results for different values of parameters.

The remainder of the paper is organized as follows. In Sect. 2, we describe the basics of WSANs including research challenges and architecture. In Sect. 3, we describe the system model and its implementation. Simulation results are shown in Sect. 4. Finally, conclusions and future work are given in Sect. 5.

2 WSAN

2.1 WSAN Challenges

Some of the key challenges in WSAN are related to the presence of actors and their functionalities.

- *Deployment and Positioning:* At the moment of node deployment, algorithms must consider to optimize the number of sensors and actors and their initial positions based on applications [11,12].
- *Architecture:* When important data has to be transmitted (an event occurred), sensors may transmit their data back to the sink, which will control the actors' tasks from distance or transmit their data to actors, which can perform actions independently from the sink node [13].
- *Real-Time:* There are a lot of applications that have strict real-time requirements. In order to fulfill them, real-time limitations must be clearly defined for each application and system [14].
- *Coordination:* In order to provide effective sensing and acting, a distributed local coordination mechanism is necessary among sensors and actors [13].

- *Power Management:* WSAN protocols should be designed with minimized energy consumption for both sensors and actors [15].
- *Mobility:* Protocols developed for WSANs should support the mobility of nodes [6,16], where dynamic topology changes, unstable routes and network isolations are present.
- *Scalability:* Smart Cities are emerging fast and WSAN, as a key technology will continue to grow together with cities. In order to keep the functionality of WSAN applicable, scalability should be considered when designing WSAN protocols and algorithms [12,16].

2.2 WSAN Architecture

A WSAN is shown in Fig. 1. The main functionality of WSANs is to make actors perform appropriate actions in the environment, based on the data sensed from sensors and actors. When important data has to be transmitted (an event occurred), sensors may transmit their data back to the sink, which will control the actors' tasks from distance, or transmit their data to actors, which can perform actions independently from the sink node. Here, the former scheme is called Semi-Automated Architecture and the latter one Fully-Automated Architecture (see Fig. 2). Obviously, both architectures can be used in different applications.

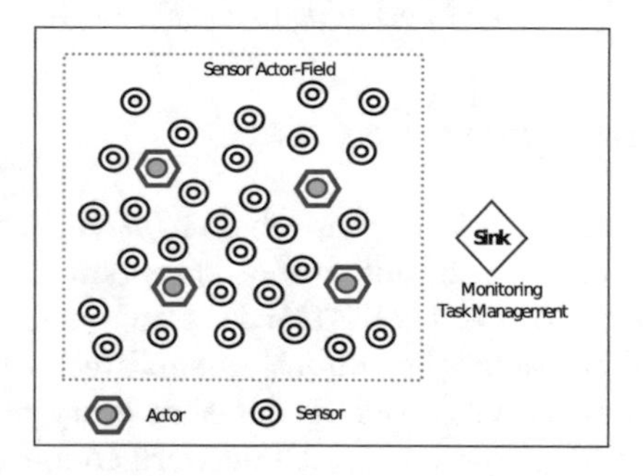

Fig. 1. Wireless Sensor Actor Network (WSAN).

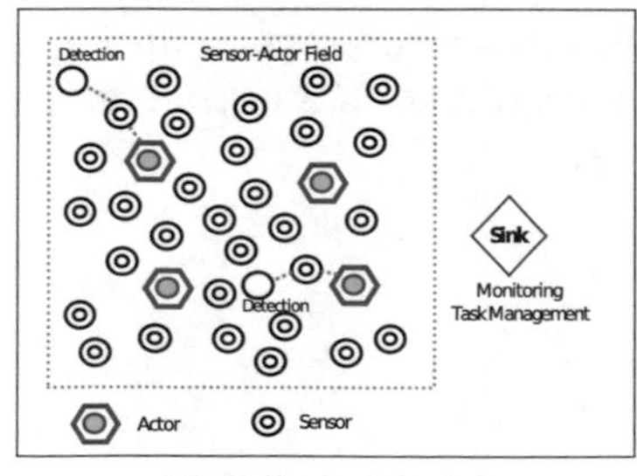

(a) Fully-Automated

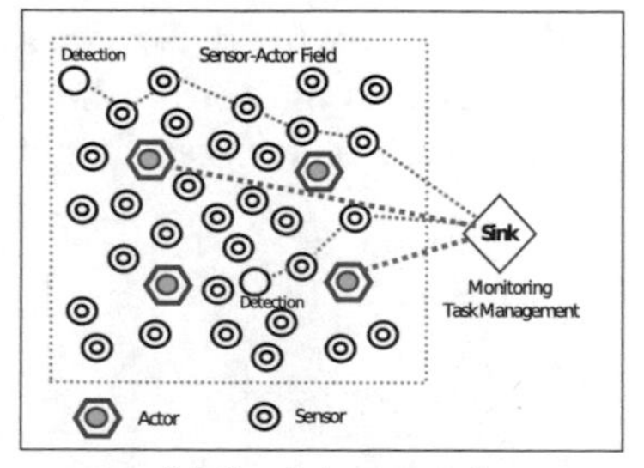

(b) Semi-Automated

Fig. 2. WSAN architectures.

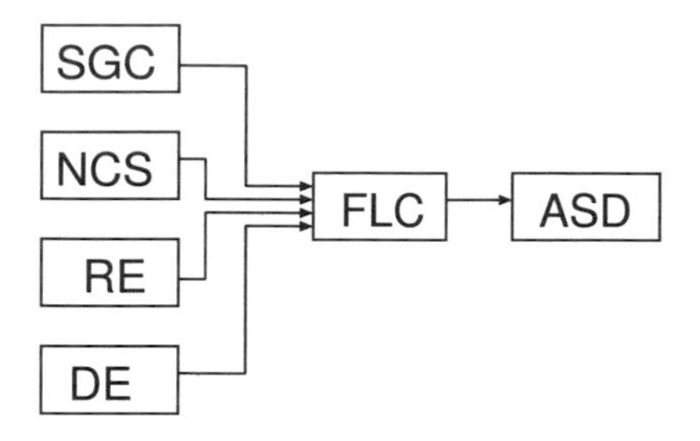

Fig. 3. Proposed system.

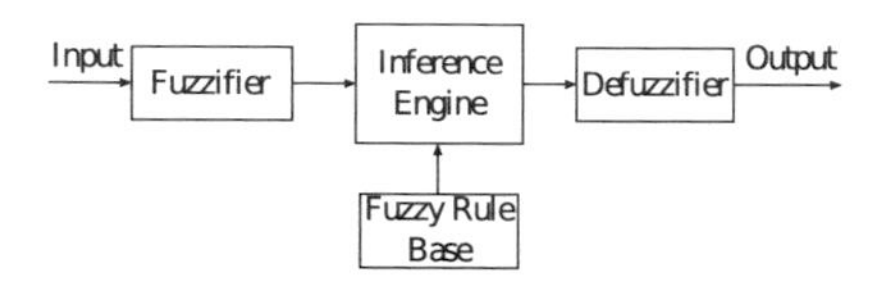

Fig. 4. FLC structure.

In the Fully-Automated Architecture are needed new sophisticated algorithms in order to provide appropriate coordination between nodes of WSAN. On the other hand, it has advantages, such as *low latency*, *low energy consumption*, *long network lifetime* [1], *higher local position accuracy*, *higher reliability* and so on.

3 Proposed Fuzzy-Based System

Based on WSAN characteristics and challenges, we consider the following parameters for implementation of our proposed system.

Size of Giant Component (SGC): A component is a group of nodes that are all connected to each other directly or indirectly. If the network has a maximal Giant Component (GC), this means that every node is reachable from every other nodes. The SGC is related with network connectivity. As can be seen in Fig. 6, there is a GC and other connected nodes, but they make islands and are not connected together.

Number of Covered Sensors (NCS): Sensor coverage is an important issue in WSANs. So in order to have a better coordination and communication, the sensors should be covered by actor nodes.

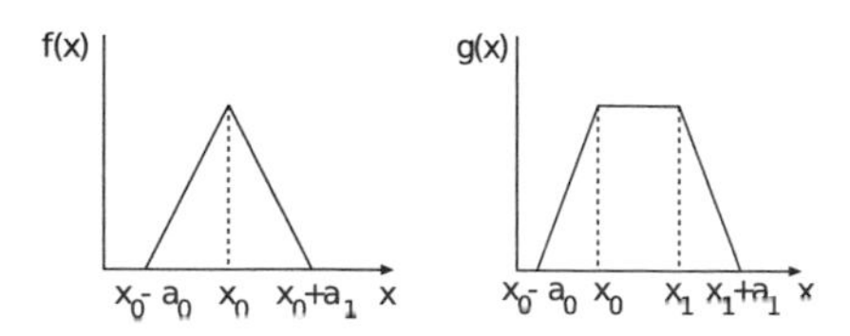

Fig. 5. Triangular and trapezoidal membership functions.

Fig. 6. An Example of SGC.

Remaining Energy (RE): As actors are active in the monitored field, they perform tasks and exchange data in different ways. Thus some actors may have a lot of remained power and other may have very little, when an event occurs. It is better that the actors which have more power are selected to carry out a task.

Distance to Event (DE): The number of actors in a WSAN is smaller than the number of sensors. Thus, when an actor is called for action near an event, the distance from the actor to the event is important because when the distance is longer, the actor will spend more energy. Thus, an actor which is close to an event, should be selected.

Actor Selection Decision (ASD): Our system is able to decide the willingness of an actor to be assigned a certain task at a certain time. The actors respond in five different levels, which can be interpreted as:

- Very Low Selection Possibility (VLSP) - It is not worth assigning the task to this actor.
- Low Selection Possibility (LSP) - There might be other actors which can do the job better.
- Middle Selection Possibility (MSP) - The Actor is ready to be assigned a task, but is not the "chosen" one.
- High Selection Possibility (HSP) - The actor takes responsibility of completing the task.
- Very High Selection Possibility (VHSP) - Actor has almost all required information and potential and takes full responsibility.

Fuzzy sets and fuzzy logic have been developed to manage vagueness and uncertainty in a reasoning process of an intelligent system such as a knowledge based system, an expert system or a logic control system [17–30].

The structure of the proposed system is shown in Fig. 3. It consists of one Fuzzy Logic Controller (FLC), which is the main part of our system and its basic elements are shown in Fig. 4. They are the fuzzifier, inference engine, Fuzzy Rule Base (FRB) and defuzzifier.

As shown in Fig. 5, we use triangular and trapezoidal membership functions for FLC, because they are suitable for real-time operation [31]. The x_0 in $f(x)$ is the center of triangular function, $x_0(x_1)$ in $g(x)$ is the left (right) edge of

Table 1. Parameters and their term sets for FLC.

Parameters	Term sets
Size of Giant Component (SGC)	Small (Sm), Medium (Md), Large (Lg)
Number of Covered Sensors (NCS)	Low (Lw), Medium (Me), High (Hg)
Remaining Energy (RE)	Low (Lo), Medium (Mdm), High (Hi)
Distance to Event (DE)	Near (Ne), Moderate (Mo), Far (Fa)
Actor Selection Decision (ASD)	Very Low Selection Possibility (VLSP), Low Selection Possibility (LSP), Middle Selection Possibility (MSP), High Selection Possibility (HSP), Very High Selection Possibility (VHSP)

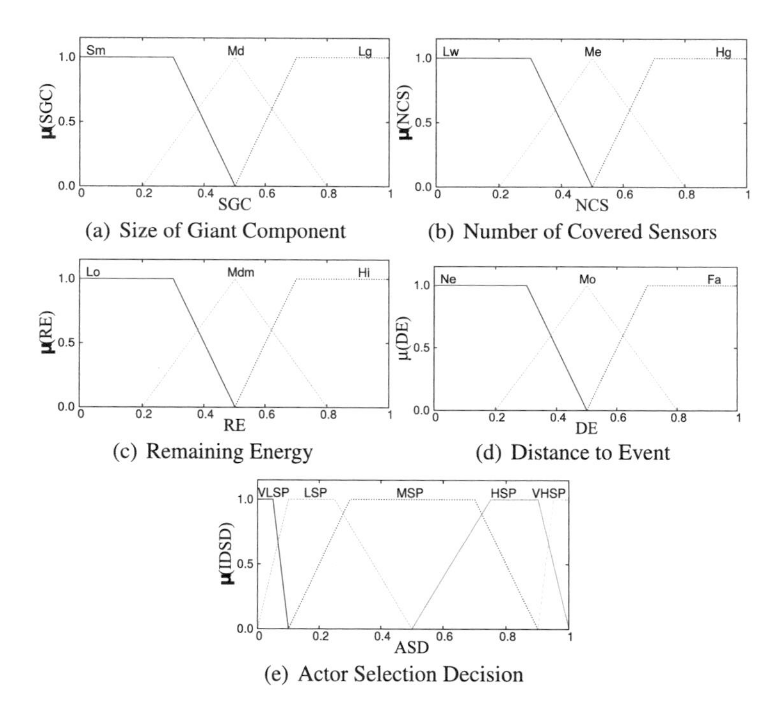

(a) Size of Giant Component (b) Number of Covered Sensors (c) Remaining Energy (d) Distance to Event (e) Actor Selection Decision

Fig. 7. Fuzzy membership functions.

trapezoidal function, and $a_0(a_1)$ is the left (right) width of the triangular or trapezoidal function. We explain in details the design of FLC in following.

We use four input parameters for FLC:

- Size of Giant Component (SGC);
- Number of Covered Sensors (NCS);
- Remaining Energy (RE);
- Distance to Event (DE).

The term sets for each input linguistic parameter are defined respectively as shown in Table 1.

$$T(SGC) = \{Small(Sm), Medium(Md), Large(La)\}$$
$$T(NCS) = \{Low(Lw), Medium(Me), High(Hg)\}$$
$$T(RE) = \{Low(Lo), Medium(Mdm), High(Hi)\}$$
$$T(DE) = \{Near(Ne), Moderate(Mo), Far(Fa)\}$$

The membership functions for input parameters of FLC are defined as:

$$\mu_{Sm}(SGC) = g(SGC; Sm_0, Sm_1, Sm_{w0}, Sm_{w1})$$
$$\mu_{Md}(SGC) = f(SGC; Md_0, Md_{w0}, Md_{w1})$$
$$\mu_{Lg}(SGC) = g(SGC; Lg_0, Lg_1, Lg_{w0}, Lg_{w1})$$
$$\mu_{Lw}(NCS) = g(NCS; Lw_0, Lw_1, Lw_{w0}, Lw_{w1})$$
$$\mu_{Me}(NCS) = f(NCS; Me_0, Me_{w0}, Me_{w1})$$
$$\mu_{Hg}(NCS) = g(NCS; Hg_0, Hg_1, Hg_{w0}, Hg_{w1})$$
$$\mu_{Lo}(RE) = g(RE; Lo_0, Lo_1, Lo_{w0}, Lo_{w1})$$
$$\mu_{Mdm}(RE) = f(RE; Mdm_0, Mdm_{w0}, Mdm_{w1})$$
$$\mu_{Hi}(RE) = g(RE; Hi_0, Hi_1, Hi_{w0}, Hi_{w1}).$$
$$\mu_{Ne}(DE) = g(DE; Ne_0, Ne_1, Ne_{w0}, Ne_{w1})$$
$$\mu_{Mo}(DE) = f(DE; Mo_0, Mo_{w0}, Mo_{w1})$$
$$\mu_{Fa}(DE) = g(DE; Fa_0, Fa_1, Fa_{w0}, Fa_{w1})$$

The small letters *w0* and *w1* mean left width and right width, respectively.

The output linguistic parameter is the Actor Selection Decision (ASD). We define the term set of ASD as:

$$\{Very\ Low\ Selection\ Possibility\ (VLSP),$$
$$Low\ Selection\ Possibility\ (LSP),$$
$$Middle\ Selection\ Possibility\ (MSP),$$
$$High\ Selection\ Possibility\ (HSP),$$
$$Very\ High\ Selection\ Possibility\ (VHSP)\}.$$

The membership functions for the output parameter *ASD* are defined as:

$$\mu_{VLSP}(ASD) = g(ASD; VLSP_0, VLSP_1, VLSP_{w0}, VLSP_{w1})$$
$$\mu_{LSP}(ASD) = g(ASD; LSP_0, LSP_1, LSP_{w0}, LSP_{w1})$$
$$\mu_{MSP}(ASD) = g(ASD; MSP_0, MSP_1, MSP_{w0}, MSP_{w1})$$
$$\mu_{HSP}(ASD) = g(ASD; HSP_0, HSP_1, HSP_{w0}, HSP_{w1})$$
$$\mu_{VHSP}(ASD) = g(ASD; VHSP_0, VHSP_1, VHSP_{w0}, VHSP_{w1}).$$

Table 2. FRB of proposed fuzzy-based system.

No	SGC	NCS	RE	DE	ASD	No	SGC	NCS	RE	DE	ASD
1	Sm	Lw	Lo	Ne	LSP	41	Md	Me	Mdm	Mo	LSP
2	Sm	Lw	Lo	Mo	VLSP	42	Md	Me	Mdm	Fa	VLSP
3	Sm	Lw	Lo	Fa	VLSP	43	Md	Me	Hi	Ne	LSP
4	Sm	Lw	Mdm	Ne	MSP	44	Md	Me	Hi	Md	HSP
5	Sm	Lw	Mdm	Mo	VLSP	45	Md	Me	Hi	Fa	HSP
6	Sm	Lw	Mdm	Fa	HSP	46	Md	Hg	Lo	Ne	LSP
7	Sm	Lw	Hi	Ne	HSP	47	Md	Hg	Lo	Mo	MSP
8	Sm	Lw	Hi	Mo	VHSP	48	Md	Hg	Lo	Fa	MSP
9	Sm	Lw	Hi	Fa	VHSP	49	Md	Hg	Mdm	Ne	LSP
10	Sm	Me	Lo	Ne	LSP	50	Md	Hg	Mdm	Mo	HSP
11	Sm	Me	Lo	Mo	HSP	51	Md	Hg	Mdm	Fa	HSP
12	Sm	Me	Lo	Fa	HSP	52	Md	Hg	Hi	Ne	HSP
13	Sm	Me	Mdm	Ne	MSP	53	Md	Hg	Hi	Mo	VHSP
14	Sm	Me	Mdm	Mo	HSP	54	Md	Hg	Hi	Fa	VHSP
15	Sm	Me	Mdm	Fa	VHSP	55	La	Lw	Lo	Ne	VLSP
16	Sm	Me	Hi	Ne	HSP	56	La	Lw	Lo	Mo	VLSP
17	Sm	Me	Hi	Mo	VHSP	57	La	Lw	Lo	Fa	VLSP
18	Sm	Me	Hi	Fa	VHSP	58	La	Lw	Mdm	Ne	VLSP
19	Sm	Hg	Lo	Ne	HSP	59	La	Lw	Mdm	Mo	VLSP
20	Sm	Hg	Lo	Mo	VHSP	60	La	Lw	Mdm	Fa	LSP
21	Sm	Hg	Lo	Fa	VHSP	61	La	Lw	Hi	Ne	MSP
22	Sm	Hg	Mdm	Ne	HSP	62	La	Lw	Hi	Mo	HSP
23	Sm	Hg	Mdm	Mo	VHSP	63	La	Lw	Hi	Fa	VHSP
24	Sm	Hg	Mdm	Fa	VHSP	64	La	Me	Lo	Ne	VLSP
25	Sm	Hg	Hi	Ne	VHSP	65	La	Me	Lo	Mo	VLSP
26	Sm	Hg	Hi	Mo	VHSP	66	La	Me	Lo	Fa	LSP
27	Sm	Hg	Hi	Fa	VHSP	67	La	Me	Mdm	Ne	VLSP
28	Md	Lw	Lo	Ne	VLSP	68	La	Me	Mdm	Mo	VLSP
29	Md	Lw	Lo	Mo	VLSP	69	La	Me	Mdm	Fa	MSP
30	Md	Lw	Lo	Fa	LSP	70	La	Me	Hi	Ne	LSP
31	Md	Lw	Mdm	Ne	VLSP	71	La	Me	Hi	Mo	MSP
32	Md	Lw	Mdm	Mo	LSP	72	La	Me	Hi	Fa	HSP
33	Md	Lw	Mdm	Fa	LSP	73	La	Hg	Lo	Ne	VLSP
34	Md	Lw	Hi	Ne	LSP	74	La	Hg	Lo	Mo	VLSP
35	Md	Lw	Hi	Mo	MSP	75	La	Hg	Lo	Fa	VLSP
36	Md	Lw	Hi	Fa	HSP	76	La	Hg	Mdm	Ne	LSP
37	Md	Me	Lo	Ne	VLSP	77	La	Hg	Mdm	Mo	VLSP
38	Md	Me	Lo	Mo	LSP	78	La	Hg	Mdm	Fa	MSP
39	Md	Me	Lo	Fa	LSP	79	La	Hg	Hi	Ne	VLSP
40	Md	Me	Mdm	Ne	VLSP	80	La	Hg	Hi	Mo	LSP
						81	La	Hg	Hi	Fa	MSP

The membership functions are shown in Fig. 7 and the Fuzzy Rule Base (FRB) is shown in Table 2. The FRB forms a fuzzy set of dimensions $|T(SGC)| \times |T(NCS)| \times |T(RE)| \times |T(DE)|$, where $|T(x)|$ is the number of terms on $T(x)$. The FRB has 81 rules. The control rules have the form: IF "conditions" THEN "control action".

4 Simulation Results

We present the simulation results in Figs. 8, 9 and 10. From simulation results, we found that as DE increases, the ASD decreases, because when actors move to a long distance they spend more energy. In Fig. 8 are shown the simulation results for DE=0.1, we can see that the ASD increases when the NCS parameter increases. Also, when RE is increased, the ASD is increased.

We can see that in Fig. 8(a) for DE = 0.6 when SGC = 0.1-NCS = 0.1 and RE = 0.9, the ASD is 0.3. Comparing Fig. 8(b) with Fig. 8(a), we can see that for DE = 0.6 the ASD is increased 15%. Also comparing Fig. 8(c) with Fig. 8(a), the ASD for DE = 0.6 is increased 32%.

In Fig. 9 are shown the simulation results for SGC = 0.5. Comparing Fig. 9(b) with Fig. 9(a), for DE = 0.7 when SGC = 0.5-NCS = 0.5 and RE = 0.9, the ASD is increased 11%. In Fig. 9(c) with Fig. 9(a), for DE = 0.7 when SGC = 0.5-NCS = 0.9 and RE = 0.9, the ASD is increased 28%. In Fig. 9, we can see that the performance

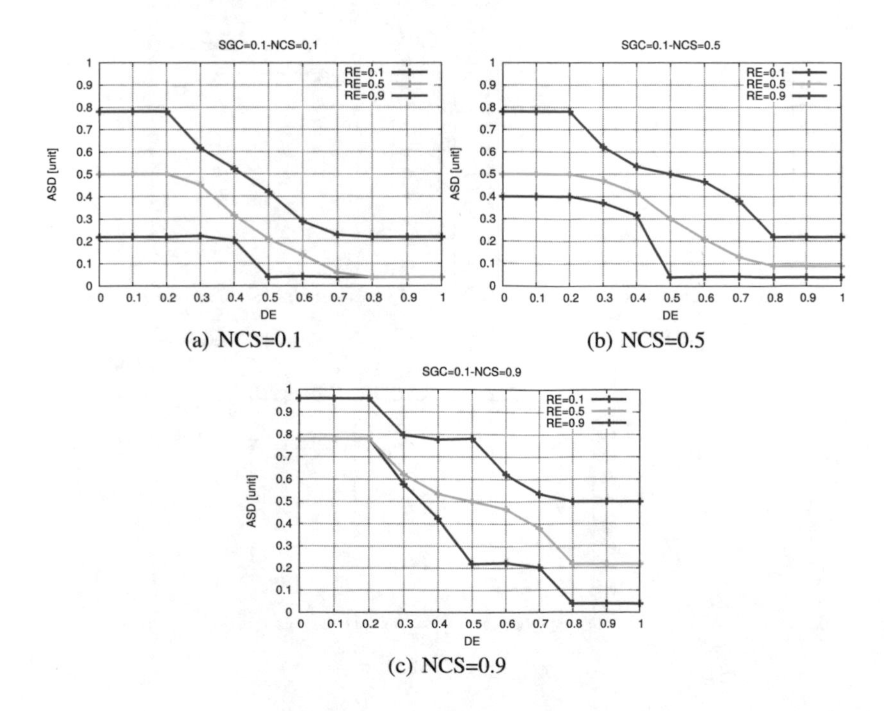

(a) NCS=0.1

(b) NCS=0.5

(c) NCS=0.9

Fig. 8. Results for $SGC = 0.1$.

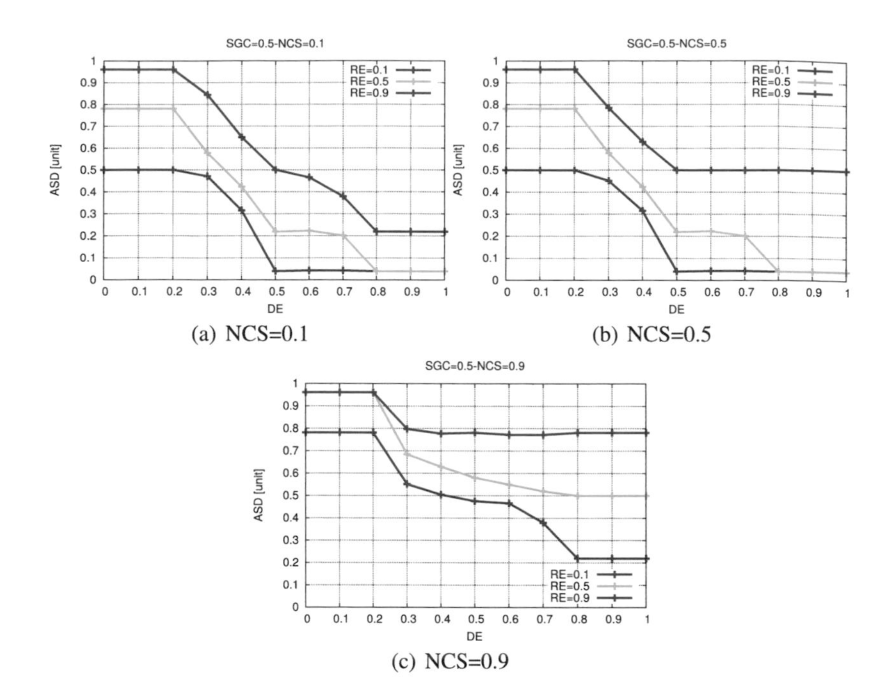

(a) NCS=0.1 (b) NCS=0.5

(c) NCS=0.9

Fig. 9. Results for $SGC = 0.5$.

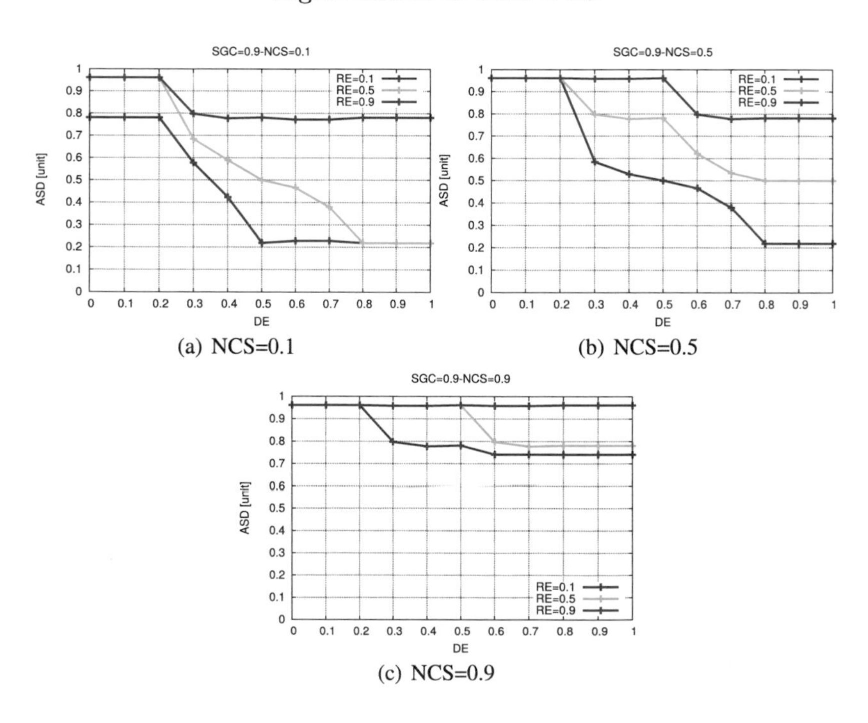

(a) NCS=0.1 (b) NCS=0.5

(c) NCS=0.9

Fig. 10. Results for $SGC = 0.9$.

is lower than in Fig. 8, because the DE is increased. The actors closest to the job place use less energy to reach the job position. Also in these graphs the performance continues to increase as NCS increases.

In Fig. 10, we can see the results for SGC = 0.9. Comparing Fig. 10(b) with Fig. 10(a), for DE = 0.5, when SGC = 0.9-NCS = 0.5 and RE = 0.5, the ASD is increased 29%. Also comparing Fig. 10(c) with Fig. 10(a), for DE = 0.5, when SGC = 0.9-NCS = 0.9 and RE = 0.5, the ASD is increased 46%.

Comparing Fig. 9(b) with Fig. 8(b) and Fig. 10(b) with Fig. 8(b), by increasing SGC to 0.5 and 0.9, the ASD is increased 12% and 68%, respectively.

5 Conclusions and Future Work

In this paper, we proposed and implemented a fuzzy-based simulation system for WSAN, which takes into account four input parameters (including NCS) and decides the actor selection for a required task in the network. From simulation results, we conclude as follows.

- When SGC, RE and NCS parameters are increased, the ASD parameter is increased, so the probability that the system selects an actor node for the job is high.
- When the DE parameter is increased, the ASD parameter is decreased, so the probability that an actor node is selected for the required task is low.

In the future work, we will consider also other parameters for actor selection and make extensive simulations to evaluate the proposed system.

References

1. Akyildiz, I.F., Kasimoglu, I.H.: Wireless sensor and actor networks: research challenges. Ad Hoc Netw. J. **2**(4), 351–367 (2004)
2. Akyildiz, I., Su, W., Sankarasubramaniam, Y., Cayirci, E.: Wireless sensor networks: a survey. Comput. Netw. **38**(4), 393–422 (2002)
3. Boyinbode, O., Le, H., Takizawa, M.: A survey on clustering algorithms for wireless sensor networks. Int. J. Space-Based Situated Comput. **1**(2/3), 130–136 (2011)
4. Bahrepour, M., Meratnia, N., Poel, M., Taghikhaki, Z., Havinga, P.J.: Use of wireless sensor networks for distributed event detection in disaster managment applications. Int. J. Space-Based Situated Comput. **2**(1), 58–69 (2012)
5. Haider, N., Imran, M., Saad, N., Zakariya, M.: Performance analysis of reactive connectivity restoration algorithms for wireless sensor and actor networks. In: IEEE Malaysia International Conference on Communications (MICC-2013), pp. 490–495, November 2013
6. Abbasi, A., Younis, M., Akkaya, K.: Movement-assisted connectivity restoration in wireless sensor and actor networks. IEEE Trans. Parallel Distrib. Syst. **20**(9), 1366–1379 (2009)
7. Li, X., Liang, X., Lu, R., He, S., Chen, J., Shen, X.: Toward reliable actor services in wireless sensor and actor networks. In: 2011 IEEE 8th International Conference on Mobile Ad Hoc and Sensor Systems (MASS), pp. 351–360, October 2011

8. Akkaya, K., Younis, M.: Cola: a coverage and latency aware actor placement for wireless sensor and actor networks. In: IEEE 64th Conference on Vehicular Technology (VTC-2006) Fall, pp. 1–5, September 2006
9. Kakarla, J., Majhi, B.: A new optimal delay and energy efficient coordination algorithm for WSAN. In: 2013 IEEE International Conference on Advanced Networks and Telecommuncations Systems (ANTS), pp. 1–6, December 2013
10. Elmazi, D., Cuka, M., Oda, T., Ikeda, M., Barolli, L.: Effect of node density on actor selection in WSANs: a comparison study for two fuzzy-based systems. In: 2017 IEEE 31st International Conference on Advanced Information Networking and Applications (AINA), pp. 865–871. IEEE (2017)
11. Akbas, M., Turgut, D.: Apawsan: actor positioning for aerial wireless sensor and actor networks. In: 2011 IEEE 36th Conference on Local Computer Networks (LCN), pp. 563–570, October 2011
12. Akbas, M., Brust, M., Turgut, D.: Local positioning for environmental monitoring in wireless sensor and actor networks. In: 2010 IEEE 35th Conference on Local Computer Networks (LCN), pp. 806–813, October 2010
13. Melodia, T., Pompili, D., Gungor, V., AkyildizZX, I.: Communication and coordination in wireless sensor and actor networks. IEEE Trans. Mobile Comput. **6**(10), 1126–1129 (2007)
14. Gungor, V., Akan, O., Akyildiz, I.: A real-time and reliable transport (rt2) protocol for wireless sensor and actor networks. IEEE/ACM Trans. Netw. **16**(2), 359–370 (2008)
15. Selvaradjou, K., Handigol, N., Franklin, A., Murthy, C.: Energy-efficient directional routing between partitioned actors in wireless sensor and actor networks. IET Commun. **4**(1), 102–115 (2010)
16. Nakayama, H., Fadlullah, Z., Ansari, N., Kato, N.: A novel scheme for wsan sink mobility based on clustering and set packing techniques. IEEE Trans. Autom. Control **56**(10), 2381–2389 (2011)
17. Inaba, T., Sakamoto, S., Kolici, V., Mino, G., Barolli, L.: A CAC scheme based on fuzzy logic for cellular networks considering security and priority parameters. In: The 9th International Conference on Broadband and Wireless Computing, Communication and Applications (BWCCA-2014), pp. 340–346 (2014)
18. Spaho, E., Sakamoto, S., Barolli, L., Xhafa, F., Barolli, V., Iwashige, J.: A fuzzy-based system for peer reliability in JXTA-overlay P2P considering number of interactions. In: The 16th International Conference on Network-Based Information Systems (NBiS-2013), pp. 156–161 (2013)
19. Matsuo, K., Elmazi, D., Liu, Y., Sakamoto, S., Mino, G., Barolli, L.: FACS-MP: a fuzzy admission control system with many priorities for wireless cellular networks and its performance evaluation. J. High Speed Netw. **21**(1), 1–14 (2015)
20. Grabisch, M.: The application of fuzzy integrals in multicriteria decision making. Eur. J. Oper. Res. **89**(3), 445–456 (1996)
21. Inaba, T., Elmazi, D., Liu, Y., Sakamoto, S., Barolli, L., Uchida, K.: Integrating wireless cellular and ad-hoc networks using fuzzy logic considering node mobility and security. In: The 29th IEEE International Conference on Advanced Information Networking and Applications Workshops (WAINA-2015), pp. 54–60 (2015)
22. Kulla, E., Mino, G., Sakamoto, S., Ikeda, M., Caballé, S., Barolli, L.: FBMIS: a fuzzy-based multi-interface system for cellular and ad hoc networks. In: International Conference on Advanced Information Networking and Applications (AINA-2014), pp. 180–185 (2014)

23. Elmazi, D., Kulla, E., Oda, T., Spaho, E., Sakamoto, S., Barolli, L.: A comparison study of two fuzzy-based systems for selection of actor node in wireless sensor actor networks. J. Ambient Intell. Humanized Comput., pp. 1–11 (2015)
24. Zadeh, L.: Fuzzy logic, neural networks, and soft computing. Commun. ACM, 77–84 (1994)
25. Spaho, E., Sakamoto, S., Barolli, L., Xhafa, F., Ikeda, M.: Trustworthiness in P2P: performance behaviour of two fuzzy-based systems for JXTA-overlay platform. Soft Comput. **18**(9), 1783–1793 (2014)
26. Inaba, T., Sakamoto, S., Kulla, E., Caballe, S., Ikeda, M., Barolli, L.: An integrated system for wireless cellular and ad-hoc networks using fuzzy logic. In: International Conference on Intelligent Networking and Collaborative Systems (INCoS-2014), pp. 157–162 (2014)
27. Matsuo, K., Elmazi, D., Liu, Y., Sakamoto, S., Barolli, L.: A multi-modal simulation system for wireless sensor networks: a comparison study considering stationary and mobile sink and event. J. Ambient Intell. Humanized Comput., 1–11 (2015)
28. Kolici, V., Inaba, T., Lala, A., Mino, G., Sakamoto, S., Barolli, L.: A fuzzy-based CAC scheme for cellular networks considering security. In: International Conference on Network-Based Information Systems (NBiS-2014), pp. 368–373 (2014)
29. Liu, Y., Sakamoto, S., Matsuo, K., Ikeda, M., Barolli, L., Xhafa, F.: A comparison study for two fuzzy-based systems: improving reliability and security of JXTA-overlay P2P platform. Soft Comput., 1–11 (2015)
30. Matsuo, K., Elmazi, D., Liu, Y., Sakamoto, S., Mino, G., Barolli, L.: FACS-MP: a fuzzy admission control system with many priorities for wireless cellular networks and its perforemance evaluation. J. High Speed Netw. **21**(1), 1–14 (2015)
31. Mendel, J.M.: Fuzzy logic systems for engineering: a tutorial. Proc. IEEE **83**(3), 345–377 (1995)

A Delay-Aware Fuzzy-Based System for Selection of IoT Devices in Opportunistic Networks

Miralda Cuka[1(✉)], Donald Elmazi[1], Keita Matsuo[2], Makoto Ikeda[2], and Leonard Barolli[2]

[1] Graduate School of Engineering, Fukuoka Institute of Technology (FIT), 3-30-1 Wajiro-Higashi, Higashi-Ku, Fukuoka 811-0295, Japan
mcuka91@gmail.com, donald.elmazi@gmail.com

[2] Department of Information and Communication Engineering, Fukuoka Institute of Technology (FIT), 3-30-1 Wajiro-Higashi, Higashi-Ku, Fukuoka 811-0295, Japan
kt-matsuo@fit.ac.j, makoto.ikd@acm.org, barolli@fit.ac.jp

Abstract. In opportunistic networks the communication opportunities (contacts) are intermittent and there is no need to establish an end-to-end link between the communication nodes. The enormous growth of devices having access to the Internet, along the vast evolution of the Internet and the connectivity of objects and devices, has evolved as Internet of Things (IoT). There are different issues for these networks. One of them is the selection of IoT devices in order to carry out a task in opportunistic networks. In this work, we implement a Fuzzy-Based System for IoT device selection in opportunistic networks. For our system, we use four input parameters: IoT Message Timeout Ratio (MTR), IoT Contact Duration (IDCD), IoT Device Storage (IDST) and IoT Device Remaining Energy (IDRE). The output parameter is IoT Device Selection Decision (IDSD). The simulation results show that the proposed system makes a proper selection decision of IoT devices in opportunistic networks. The IoT device selection is increased up to 18% and 28% by increasing IDST and IDRE, respectively.

1 Introduction

Future communication systems will be increasingly complex, involving thousands of heterogeneous devices with diverse capabilities and various networking technologies interconnected with the aim to provide users with ubiquitous access to information and advanced services at a high quality level, in a cost efficient manner, any time, any place, and in line with the always best connectivity principle. OppNet can provide an alternative way to support the diffusion of information in special locations within a city, particularly in crowded spaces where current wireless technologies can exhibit congestion issues. The efficiency of this diffusion relies mainly on user mobility. In fact, mobility creates the opportunities for contacts and, therefore, for data forwarding [1]. OppNets have

L. Barolli et al. (Eds.): NBiS 2018, LNDECT 22, pp. 16–29, 2019.
https://doi.org/10.1007/978-3-319-98530-5_2

appeared as an evolution of the MANETs. They are also a wireless based network and hence, they face various issues similar to MANETs such as frequent disconnections, highly variable links, limited bandwidth etc. In OppNets, nodes are always moving which makes the network easy to deploy and decreases the dependence on infrastructure for communication [2]. The concept of IoT is traffic going throught different networks. Hence, IoT can seamlessly connect the real world and cyberspace via physical objects that embed with various types of intelligent sensors. A large number of Internet-connected machines will generate and exchange an enormous amount of data that make daily life more convenient, help to make a tough decision and provide beneficial services. The IoT probably becomes one of the most popular networking concepts that has the potential to bring out many benefits [3,4].

OppNets are the variants of Delay Tolerant Networks (DTNs). It is a class of networks that has emerged as an active research subject in the recent times. Owing to the transient and un-connected nature of the nodes, routing becomes a challenging task in these networks. Sparse connectivity, no infrastructure and limited resources further complicate the situation [5,6]. Routing methods for such sparse mobile networks use a different paradigm for message delivery; these schemes utilize node mobility by having nodes carry messages, waiting for an opportunity to transfer messages to the destination or the next relay rather than transmitting them over a path [7]. Hence, the challenges for routing in OppNet are very different from the traditional wireless networks and their utility and potential for scalability makes them a huge success.

In mobile OppNet, connectivity varies significantly over time and is often disruptive. Examples of such networks include interplanetary communication networks, mobile sensor networks, vehicular adhoc networks (VANETs), terrestrial wireless networks, and under-water sensor networks. While the nodes in such networks are typically delay-tolerant, message delivery latency still remains a crucial metric, and reducing it is highly desirable [8].

However, most of the proposed routing schemes assume long contact durations such that all buffered messages can be transferred within a single contact. For example, when hand-held devices communicate via Bluetooth that has a typical wireless range of about 10 m, the contact duration tends to be as short as several seconds if the users are walking. For high speed vehicles that communicate via WiFi (802.11g), which has a longer range (up to 38 m indoors and 140 m outdoors), the contact duration is still short. In the presence of short contact durations, there are two key issues that must be addressed. First is the relay selection issue. We need to select relay nodes that will contact the message's destination long enough so that the entire message can be successfully transmitted. Second is the message scheduling issue. Since not all messages can be exchanged between nodes within a single contact, it is important to schedule the transmission of messages in such a way that will maximize the network delivery ratio [9].

In an OppNet, when nodes move away or turn off their power to conserve energy, links may be disrupted or shut down periodically. These events result in intermittent connectivity. When there is no path existing between the source and the destination, the network partition occurs. Therefore, nodes need to communicate with each other via opportunistic contacts through store-carry-forward operation. Since these types of networks require the IoT devices to store some information, storage is an important parameter in evaluation of their performance. However, the storage capacity of the device is limited which makes storage a requirement to be considered.

The Fuzzy Logic (FL) is unique approach that is able to simultaneously handle numerical data and linguistic knowledge. The fuzzy logic works on the levels of possibilities of input to achieve the definite output. Fuzzy set theory and FL establish the specifics of the nonlinear mapping.

In this paper, we propose and implement a delay-aware Fuzzy-based simulation system for selection of IoT devices in OppNet. For our system we consider four parameters for IoT device selection. We show the simulation results for different values of parameters.

The remainder of the paper is organized as follows. In the Sect. 2, we present a brief introduction of IoT. In Sect. 2.2, we describe the basics of OppNet including research challenges and architecture. In Sect. 3, we introduce the proposed system model and its implementation. Simulation results are shown in Sect. 4. Finally, conclusions and future work are given in Sect. 5.

2 IoT and OppNet

2.1 IoT

Internet of Things (IoT) allows to integrate physical and virtual objects. Virtual reality, which was recently available only on the monitor screens, now integrates with the real world, providing users with completely new opportunities: interact with objects on the other side of the world and receive the necessary services that became real due the wide interaction [10]. The IoT will support substantially higher number of end users and devices. In Fig. 1, we present an example of an IoT network architecture. The IoT network is a combination of IoT devices which are connected with different mediums using IoT Gateway to the Internet. The data transmitted through the gateway is stored, processed securely within cloud server. These new connected things will trigger increasing demands for new IoT applications that are not only for users. The current solutions for IoT application development generally rely on integrated service-oriented programming platforms. In particular, resources (e.g., sensory data, computing resource, and control information) are modeled as services and deployed in the cloud or at the edge. It is difficult to achieve rapid deployment and flexible resource management at network edges, in addition, an IoT system's scalability will be restricted by the capability of the edge devices [11].

2.2 Opportunistic Networks

In Fig. 2 we show an OppNet scenario. OppNets comprises a network where nodes can be anything from pedestrians, vehicles, fixed devices and so on.

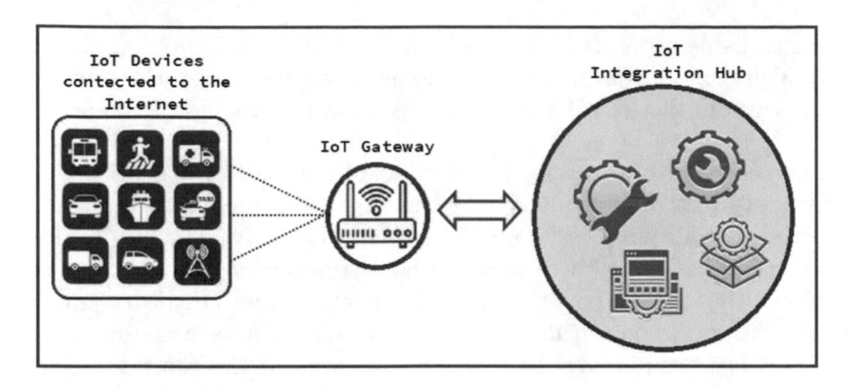

Fig. 1. An Iot network architecture

The data is sent from the sender to receiver by using communication opportunity that can be Wi-Fi, Bluetooth, cellular technologies or satellite links to transfer the message to the final destination. In such scenario, IoT devices might roam and opportunistically encounter several different statically deployed networks and perform either data collection or dissemination as well as relaying data between these networks, thus introducing further connectivity for disconnected networks. For example, as seen in the figure, a car could opportunistically encounter other IoT devices, collect information from them and relay it until it

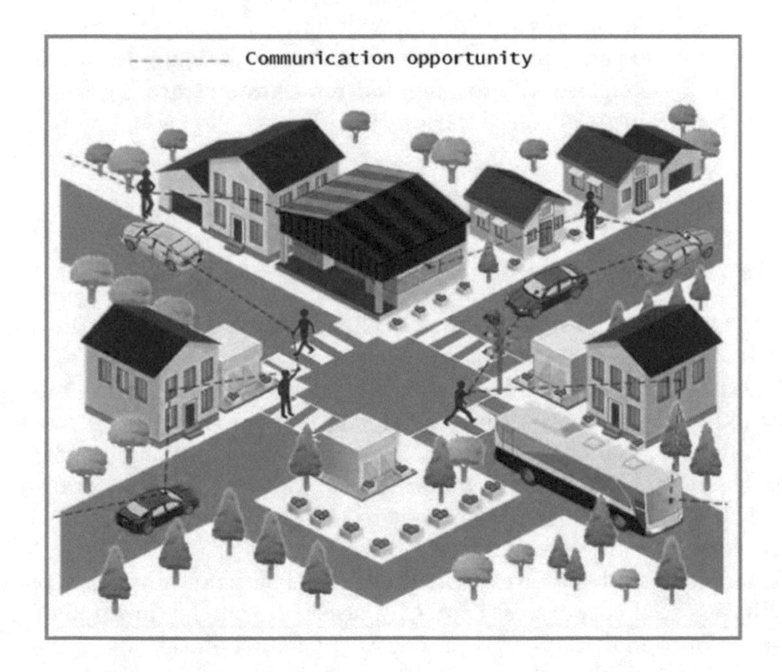

Fig. 2. Opportunistic Network Scenario

finds an available access point where it can upload the information. Similarly, a person might collect information from home-based weather stations and relay it through several other people, cars and buses until it reaches its intended destination [12].

Opportunistic Networks are not limited to only such applications, as they can introduce further conectivity and benefits to IoT scenarios. In an OppNet, due to node mobility network partitions occur. These events result in intermittent connectivity. When there is no path existing between the source and the destination, the network partition occurs. Therefore, nodes need to communicate with each other via opportunistic contacts through store-carry-forward operation. There are two specific challenges in an OppNet: the contact opportunity and the node storage.

- *Contact Opportunity:* Due to the node mobility or the dynamics of wireless channel, a node can make contact with other nodes at an unpredicted time. Since contacts between nodes are hardly predictable, they must be exploited opportunistically for exchanging messages between some nodes that can move between remote fragments of the network. Mobility increases the chances of communication between nodes. When nodes move randomly around the network, where jamming signals are disrupting the communication, they may pass through unjammed area and hence be able to communicate. In addition, the contact capacity needs to be considered [13, 14].
- *Node Storage:* As described above, to avoid dropping packets, the intermediate nodes are required to have enough storage to store all messages for an unpredictable period of time until next contact occurs. In other words, the required storage space increases as a function of the number of messages in the network. Therefore, the routing and replication strategies must take the storage constraint into consideration [15].

3 Proposed Fuzzy-Based System

3.1 System Parameters

Based on OppNets characteristics and challenges, we consider the following parameters for implementation of our proposed system.

IoT Message Timeout Ratio (MTR): Due to IoT device limited buffer size, the older messages are dropped. So it is very important that the message doesn't stay in the buffer for a long time. We use this parameter to show that if message has been staying in the buffer for too long, it should be transferred to another IoT device in order to deliver it to the destination. In the absence of a neighbor node, the message will be dropped.

IoT Device Contact Duration (IDCD): This is an important parameter in mobility-assisted networks as contact times represent the duration of message communication opportunity upon a contact. Contact durations is the time in which all buffered messages can be transferred within a single contact.

IoT Device Storage (IDST): In delay tolerant networks data is carried by the IoT device until a communication opportunity is available. Considering that different IoT devices have different storage capabilities, the selection decision should consider the storage capacity.

IoT Device Remaining Energy (IDRE): The IoT devices in OppNets are active and can perform tasks and exchange data in different ways from each other. Consequently, some IoT devices may have a lot of remaining power and other may have very little, when an event occurs.

IoT Device Selection Decision (IDSD): The proposed system considers the following levels for IoT device selection:

- Very Low Selection Possibility (VLSP) - The IoT device will have very low probability to be selected.
- Low Selection Possibility (LSP) - There might be other IoT devices which can do the job better.
- Middle Selection Possibility (MSP) - The IoT device is ready to be assigned a task, but is not the "chosen" one.
- High Selection Possibility (HSP) - The IoT device takes responsibility of completing the task.
- Very High Selection Possibility (VHSP) - The IoT device has almost all the required information and potential to be selected and then allocated in an appropriate position to carry out a job.

3.2 System Implementation

Fuzzy sets and fuzzy logic have been developed to manage vagueness and uncertainty in a reasoning process of an intelligent system such as a knowledge based system, an expert system or a logic control system [16–29]. In this work, we use fuzzy logic to implement the proposed system.

The structure of the proposed system is shown in Fig. 3. It consists of one Fuzzy Logic Controller (FLC), which is the main part of our system and its basic

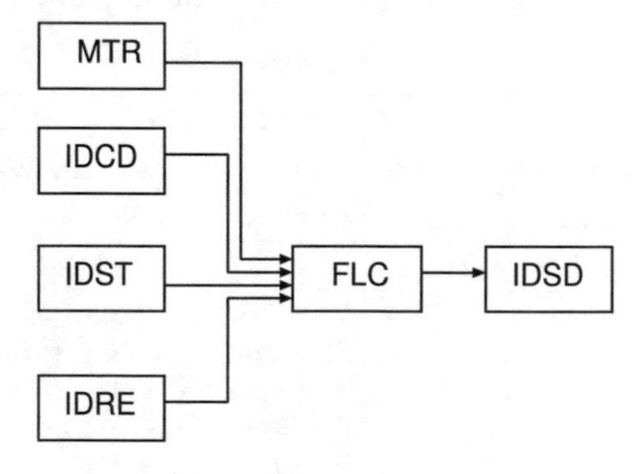

Fig. 3. Proposed system model.

elements are shown in Fig. 4. They are the fuzzifier, inference engine, Fuzzy Rule Base (FRB) and defuzzifier.

As shown in Fig. 5, we use triangular and trapezoidal membership functions for FLC, because they are suitable for real-time operation [30]. The x_0 in $f(x)$ is the center of triangular function, $x_0(x_1)$ in $g(x)$ is the left (right) edge of trapezoidal function, and $a_0(a_1)$ is the left (right) width of the triangular or trapezoidal function. We explain in details the design of FLC in following.

We use four input parameters for FLC: IoT Message Timeout Ratio (MTR), IoT Device Contact Duration (IDCD), IoT Device Storage (IDST), IoT Device Remaining Energy (IDRE).

The term sets for each input linguistic parameter are defined respectively as shown in Table 1.

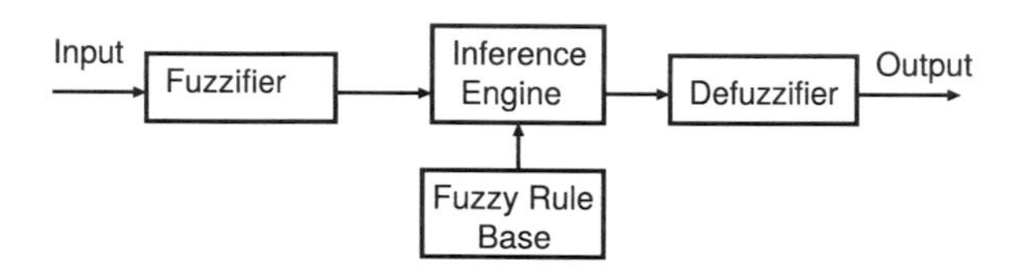

Fig. 4. FLC structure.

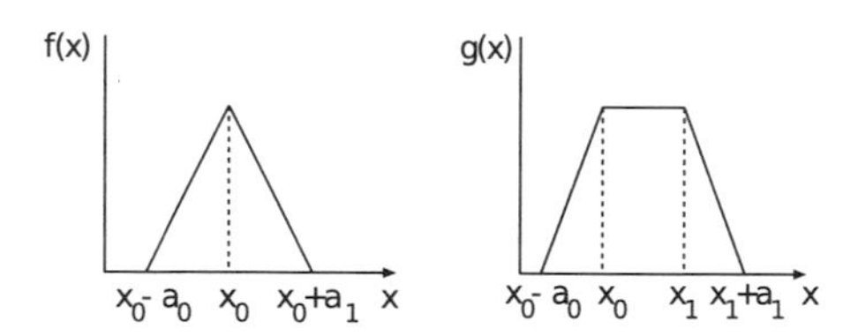

Fig. 5. Triangular and trapezoidal membership functions.

Table 1. Parameters and their term sets for FLC.

Parameters	Term Sets
IoT Message Timeout Ratio (MTR)	Low (Lw), Medium (Mi), High (Hg)
IoT Device Contact Duration (IDCD)	Short (Sho), Medium (Med), Long (Lg)
IoT Device Storage (IDST)	Small (Sm), Medium (Me), High (Hi)
IoT Device Remaining Energy (IDRE)	Low (Lo), Medium (Mdm), High (Hgh)
IoT Device Selection Decision (IDSD)	Very Low Selection Possibility (VLSP),Low Selection Possibility (LSP), Medium Selection Possibility (MSP), High Selection Possibility (HSP), Very High Selection Possibility (VHSP)

Table 2. FRB2.

No	MTR	IDCD	IDST	IDRE	IDSD	No	MTR	IDCD	IDST	IDRE	IDSD
1	Lw	Sho	Sm	Lo	VLSP	41	Mi	Med	Me	Mdm	VHSP
2	Lw	Sho	Sm	Mdm	LSP	42	Mi	Med	Me	Hgh	HSP
3	Lw	Sho	Sm	Hgh	MSP	43	Mi	Med	Hi	Lo	HSP
4	Lw	Sho	Me	Lo	LSP	44	Mi	Med	Hi	Mdm	VHSP
5	Lw	Sho	Me	Mdm	MSP	45	Mi	Med	Hi	Hgh	VHSP
6	Lw	Sho	Me	Hgh	HSP	46	Mi	Lg	Sm	Lo	VLSP
7	Lw	Sho	Hi	Lo	MSP	47	Mi	Lg	Sm	Mdm	LSP
8	Lw	Sho	Hi	Mdm	HSP	48	Mi	Lg	Sm	Hgh	MLSP
9	Lw	Sho	Hi	Hgh	VHSP	49	Mi	Lg	Me	Lo	LSP
10	Lw	Med	Sm	Lo	MSP	50	Mi	Lg	Me	Mdm	LSP
11	Lw	Med	Sm	Mdm	HSP	51	Mi	Lg	Me	Hgh	HSP
12	Lw	Med	Sm	Hgh	VHSP	52	Mi	Lg	Hi	Lo	MSP
13	Lw	Med	Me	Lo	HSP	53	Mi	Lg	Hi	Mdm	HSP
14	Lw	Med	Me	Mdm	VHSP	54	Mi	Lg	Hi	Hgh	VHSP
15	Lw	Med	Me	Hgh	VHSP	55	Hg	Sho	Sm	Lo	VLSP
16	Lw	Med	Hi	Lo	VHSP	56	Hg	Sho	Sm	Mdm	VLSP
17	Lw	Med	Hi	Mdm	VHSP	57	Hg	Sho	Sm	Hgh	VLSP
18	Lw	Med	Hi	Hgh	VHSP	58	Hg	Sho	Me	Lo	VLSP
19	Lw	Lg	Sm	Lo	LSP	59	Hg	Sho	Me	Mdm	VLSP
20	Lw	Lg	Sm	Mdm	MSP	60	Hg	Sho	Me	Hgh	VLSP
21	Lw	Lg	Sm	Hgh	HSP	61	Hg	Sho	Hi	Lo	LSP
22	Lw	Lg	Me	Lo	MSP	62	Hg	Sho	Hi	Mdm	MSP
23	Lw	Lg	Me	Mdm	HSP	63	Hg	Sho	Hi	Hgh	VLSP
24	Lw	Lg	Me	Hgh	VHSP	64	Hg	Med	Sm	Lo	LSP
25	Lw	Lg	Hi	Lo	HSP	65	Hg	Med	Sm	Mdm	LSP
26	Lw	Lg	Hi	Mdm	VHSP	66	Hg	Med	Sm	Hgh	MSP
27	Lw	Lg	Hi	Hgh	VHSP	67	Hg	Med	Me	Lo	LSP
28	Mi	Sho	Sm	Lo	VLSP	68	Hg	Med	Me	Mdm	LSP
29	Mi	Sho	Sm	Mdm	VLSP	69	Hg	Med	Me	Hgh	HSP
30	Mi	Sho	Sm	Hgh	LSP	70	Hg	Med	Hi	Lo	MSP
31	Mi	Sho	Me	Lo	VLSP	71	Hg	Med	Hi	Mdm	HSP
32	Mi	Sho	Me	Mdm	LSP	72	Hg	Med	Hi	Hgh	VHSP
33	Mi	Sho	Me	Hgh	MSP	73	Hg	Lg	Sm	Lo	VLSP
34	Mi	Sho	Hi	Lo	LSP	74	Hg	Lg	Sm	Mdm	VLSP
35	Mi	Sho	Hi	Mdm	MSP	75	Hg	Lg	Sm	Hgh	LSP
36	Mi	Sho	Hi	Hgh	HSP	76	Hg	Lg	Me	Lo	VLSP
37	Mi	Med	Sm	Lo	LSP	77	Hg	Lg	Me	Mdm	VLSP
38	Mi	Med	Sm	Mdm	MSP	78	Hg	Lg	Me	Hgh	LSP
39	Mi	Med	Sm	Hgh	HSP	79	Hg	Lg	Hi	Lo	LSP
40	Mi	Med	Me	Lo	MSP	80	Hg	Lg	Hi	Mdm	MSP
						81	Hg	Lg	Hi	Hgh	HSP

$$T(MTR) = \{Low(Lw), Medium(Mi), High(Hg)\}$$
$$T(IDCD) = \{Short(Sho), Medium(Med), Long(Lg)\}$$
$$T(IDST) = \{Small(Sm), Medium(Me), High(Hi)\}$$
$$T(IDRE) = \{Low(Lo), Medium(Mdm), High(Hgh)\}$$

The membership functions for input parameters of FLC are defined as:

$$\mu_{Lw}(MTR) = g(MTR; Lw_0, Lw_1, Lw_{w0}, Lw_{w1})$$
$$\mu_{Mi}(MTR) = f(MTR; Mi_0, Mi_{w0}, Mi_{w1})$$
$$\mu_{Hg}(MTR) = g(MTR; Hg_0, Hg_1, Hg_{w0}, Hg_{w1})$$
$$\mu_{Sho}(IDCD) = g(IDCD; Sho_0, Sho_1, Sho_{w0}, Sho_{w1})$$
$$\mu_{Mi}(IDCD) = f(IDCD; Med_0, Med_{w0}, Med_{w1})$$
$$\mu_{Lg}(IDCD) = g(IDCD; Lg_0, Lg_1, Lg_{w0}, Lg_{w1})$$
$$\mu_{Sm}(IDST) = g(IDST; Sm_0, Sm_1, Sm_{w0}, Sm_{w1})$$
$$\mu_{Me}(IDST) = f(IDST; Me_0, Me_{w0}, Me_{w1})$$
$$\mu_{Hi}(IDST) = g(IDST; Hi_0, Hi_1, Hi_{w0}, Hi_{w1})$$
$$\mu_{We}(IDRE) = g(IDRE; Lo_0, Lo_1, Lo_{w0}, Lo_{w1})$$
$$\mu_{Mo}(IDRE) = f(IDRE; Mdm_0, Mdm_{w0}, Mdm_{w1})$$
$$\mu_{St}(IDRE) = g(IDRE; Hgh_0, Hgh_1, Hgh_{w0}, Hgh_{w1})$$

The small letters *w0* and *w1* mean left width and right width, respectively.

The output linguistic parameter is the IoT device Selection Decision (IDSD). We define the term set of IDSD as:

$$\{Very\ Low\ Selection\ Possibility\ (VLSP),$$
$$Low\ Selection\ Possibility\ (LSP),$$
$$Middle\ Selection\ Possibility\ (MSP),$$
$$High\ Selection\ Possibility\ (HSP),$$
$$Very\ High\ Selection\ Possibility\ (VHSP)\}.$$

The membership functions for the output parameter *IDSD* are defined as:

$$\mu_{VLSP}(IDSD) = g(IDSD; VLSP_0, VLSP_1, VLSP_{w0}, VLSP_{w1})$$
$$\mu_{LSP}(IDSD) = g(IDSD; LSP_0, LSP_1, LSP_{w0}, LSP_{w1})$$
$$\mu_{MSP}(IDSD) = g(IDSD; MSP_0, MSP_1, MSP_{w0}, MSP_{w1})$$
$$\mu_{HSP}(IDSD) = g(IDSD; HSP_0, HSP_1, HSP_{w0}, HSP_{w1})$$
$$\mu_{VHSP}(IDSD) = g(IDSD; VHSP_0, VHSP_1, VHSP_{w0}, VHSP_{w1}).$$

The membership functions are shown in Fig. 6 and the Fuzzy Rule Base (FRB) for our system are shown in Table 2.

The FRB forms a fuzzy set of dimensions $|T(MTR)| \times |T(IDCD)| \times |T(IDST)| \times |T(IDRE)|$, where $|T(x)|$ is the number of terms on $T(x)$. We have four input parameters, so our system has 81 rules. The control rules have the form: IF "conditions" THEN "control action".

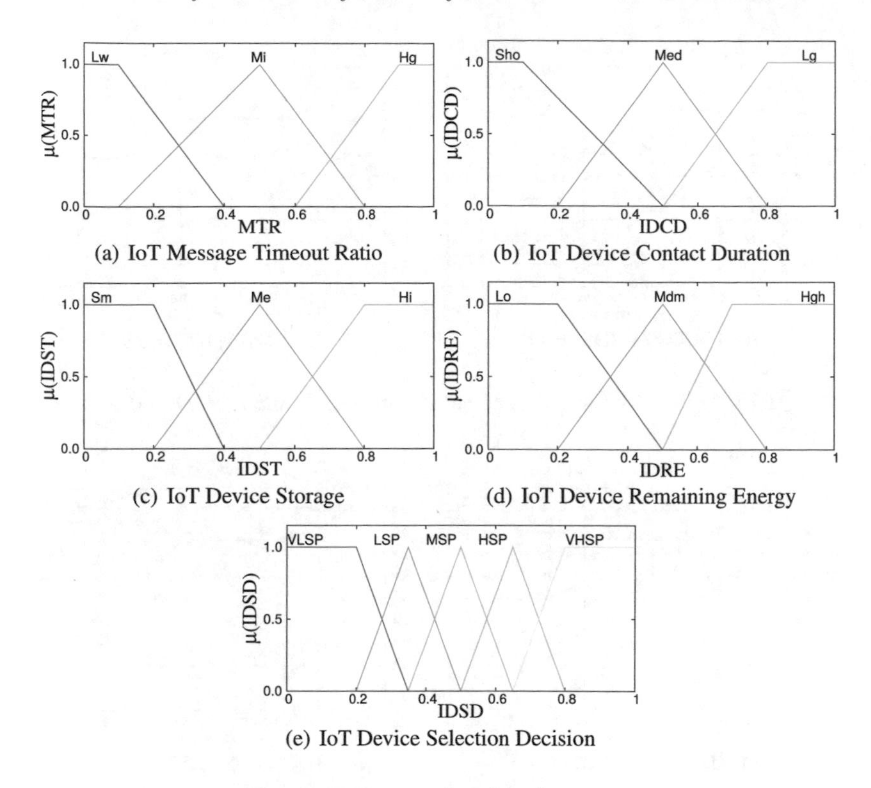

(a) IoT Message Timeout Ratio
(b) IoT Device Contact Duration
(c) IoT Device Storage
(d) IoT Device Remaining Energy
(e) IoT Device Selection Decision

Fig. 6. Fuzzy membership functions.

4 Simulation Results

We present the simulation results in Figs. 7, 8 and 9. In these figures, we show the relation between the probability of an IoT device to be selected (IDSD) to carry out a task, versus MTR, IDCD, IDST and IDRE. We consider IDCD and IDST constant and change the values of MTR and IDRE. We see that IoT devices with more remaining energy, have a higher possibility to be selected for carrying out a job. In Fig. 7(a), when IDRE is 0.1 and MTR is 0.3, the IDSD is 0.18. For IDRE 0.5, the IDSD is 0.27 and for IDRE 0.9, IDSD is 0.46, thus the IDSD is increased about 9% and 28%, for IDRE 0.5 and IDRE 0.9, respectively.

From the simulation results, we compare Figs. 7(a), 8(a) and 9(a) for MTR 0.1. For IDRE 0.9, comparing Fig. 7(a) with Fig. 8(a) and Fig. 8(a) with Fig. 9(a), we see that the IDSD is increased 32% and decreased 12% respectively. The duration of a contact is the total time that IoT devices are within reach of each other, and have thus the possibility to communicate. This parameter directly influences the capacity of OppNets because it limits the amount of data that can be transferred between nodes. A short time of contact means that two devices may not have enough time to establish a connection. While, when the contact time is long, the OppNet loses the mobility. Also, the neighbor IoT device remaining buffer capacity will be decreased.

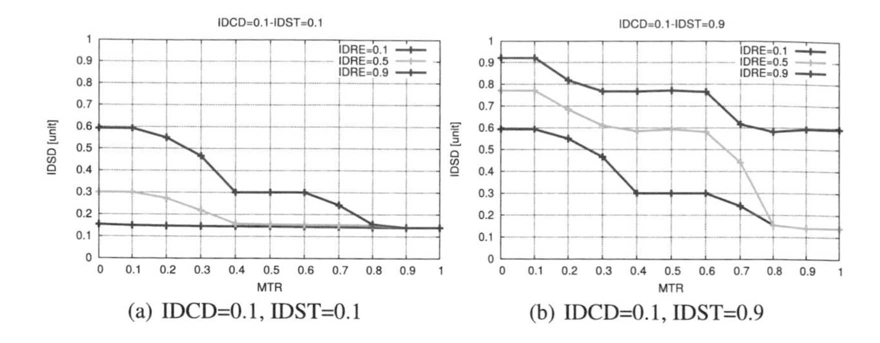

(a) IDCD=0.1, IDST=0.1 (b) IDCD=0.1, IDST=0.9

Fig. 7. Results for different values of $IDST$ and $IDCD = 0.1$.

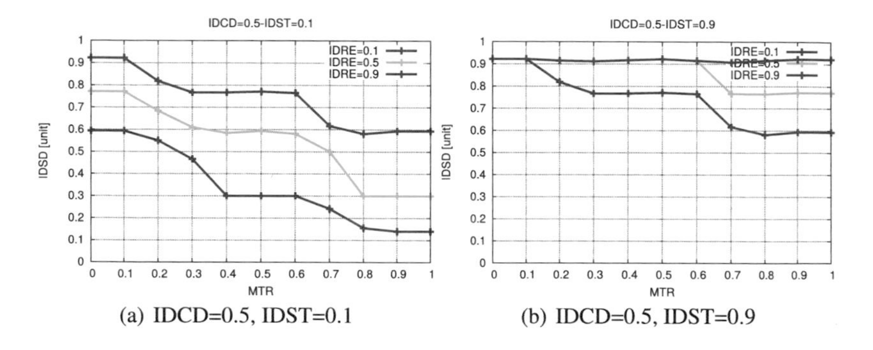

(a) IDCD=0.5, IDST=0.1 (b) IDCD=0.5, IDST=0.9

Fig. 8. Results for different values of $IDST$ and $IDCD = 0.5$.

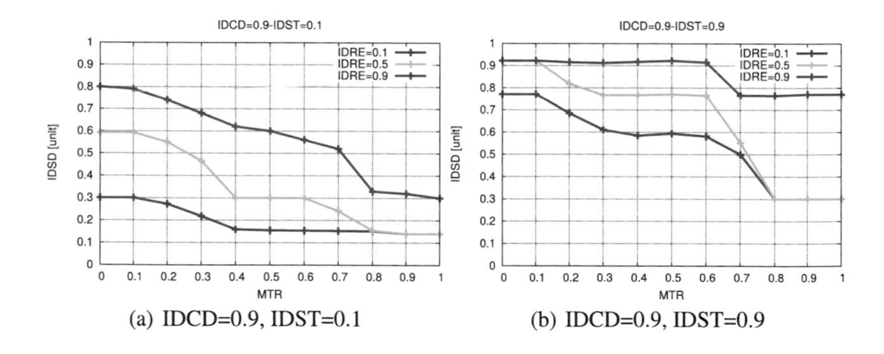

(a) IDCD=0.9, IDST=0.1 (b) IDCD=0.9, IDST=0.9

Fig. 9. Results for different values of $IDST$ and $IDCD = 0.9$.

In Fig. 7(a) and (b), we increase the IDST value to 0.1 and 0.9, respectively and keep IDCD constant. From the figures we can see that for MTR 0.3 and IDRE 0.9 the IDSD is increased 31%. By increasing the IDST value, the IDSD is also increased as shown in Fig. 7(b), because devices with more storage capacity are more likely to carry the message until there is a contact opportunity.

5 Conclusions and Future Work

In this paper, we proposed and implemented a fuzzy-based IoT device selection system for OppNets, which is used to select an IoT device for a required task.

We evaluated the proposed system by computer simulations. The simulation results show that the devices with high contact opportunity, are more likely to be selected for carrying out a job, so with the increase of IDCD the possibility of an IoT device to be selected increases. We can see that by increasing MTR, IDST and IDRE, the IDSD is also increased. But for the IDCD parameter, we need to find an optimal time that IoT devices have contact with each other.

In the future work, we will also consider other parameters for IoT device selection such as Node Computational Time, Interaction probability and make extensive simulations to evaluate the proposed system.

References

1. Mantas, N., Louta, M., Karapistoli, E., Karetsos, G.T., Kraounakis, S., Obaidat, M.S.: Towards an incentive-compatible, reputation-based framework for stimulating cooperation in opportunistic networks: a survey. IET Netw. **6**(6), 169–178 (2017)
2. Sharma, D.K., Sharma, A., Kumar, J., et al.: KNNR: K-nearest neighbour classification based routing protocol for opportunistic networks. In: 10th International Conference on Contemporary Computing (IC3), pp. 1–6. IEEE (2017)
3. Kraijak, S., Tuwanut, P.: A survey on internet of things architecture, protocols, possible applications, security, privacy, real-world implementation and future trends. In: 16th International Conference on Communication Technology (ICCT), pp. 26–31. IEEE (2015)
4. Arridha, R., Sukaridhoto, S., Pramadihanto, D., Funabiki, N.: Classification extension based on IoT-big data analytic for smart environment monitoring and analytic in real-time system. Int. J. Space-Based Situated Comput. **7**(2), 82–93 (2017)
5. Dhurandher, S.K., Sharma, D.K., Woungang, I., Bhati, S.: HBPR: history based prediction for routing in infrastructure-less opportunistic networks. In: 27th International Conference on Advanced Information Networking and Applications (AINA), pp. 931–936. IEEE (2013)
6. Spaho, E., Mino, G., Barolli, L., Xhafa, F.: Goodput and PDR analysis of AODV, OLSR and DYMO protocols for vehicular networks using cavenet. Int. J. Grid Utility Comput. **2**(2), 130–138 (2011)
7. Abdulla, M., Simon, R.: The impact of intercontact time within opportunistic networks: protocol implications and mobility models. TechRepublic White Paper (2009)
8. Patra, T.K., Sunny, A.: Forwarding in heterogeneous mobile opportunistic networks. IEEE Commun. Lett. **22**(3), 626–629 (2018)
9. Le, T., Gerla, M.: Contact duration-aware routing in delay tolerant networks. In: International Conference on Networking, Architecture and Storage (NAS), pp. 1–8. IEEE (2017)
10. Popereshnyak, S., Suprun, O., Suprun, O., Wieckowski, T.: IoT application testing features based on the modelling network. In: 2018 XIV-th International Conference on Perspective Technologies and Methods in MEMS Design (MEMSTECH), pp. 127–131. IEEE (2018)

11. Chen, N., Yang, Y., Li, J., Zhang, T.: A fog-based service enablement architecture for cross-domain IoT applications. In: 2017 IEEE Fog World Congress (FWC), pp. 1–6. IEEE (2017)
12. Pozza, R., Nati, M., Georgoulas, S., Moessner, K., Gluhak, A.: Neighbor discovery for opportunistic networking in internet of things scenarios: a survey. IEEE Access **3**, 1101–1131 (2015)
13. Akbas, M., Turgut, D.: Apawsan: actor positioning for aerial wireless sensor and actor networks. In: IEEE 36th Conference on Local Computer Networks (LCN-2011), pp. 563–570, October 2011
14. Akbas, M., Brust, M., Turgut, D.: Local positioning for environmental monitoring in wireless sensor and actor networks. In: IEEE 35th Conference on Local Computer Networks (LCN-2010), pp. 806–813, October 2010
15. Melodia, T., Pompili, D., Gungor, V., Akyildiz, I.: Communication and coordination in wireless sensor and actor networks. IEEE Trans. Mobile Comput. **6**(10), 1126–1129 (2007)
16. Inaba, T., Sakamoto, S., Kolici, V., Mino, G., Barolli, L.: A CAC scheme based on fuzzy logic for cellular networks considering security and priority parameters. In: The 9th International Conference on Broadband and Wireless Computing, Communication and Applications (BWCCA-2014), pp. 340–346 (2014)
17. Spaho, E., Sakamoto, S., Barolli, L., Xhafa, F., Barolli, V., Iwashige, J.: A fuzzy-based system for peer reliability in JXTA-overlay P2P considering number of interactions. In: The 16th International Conference on Network-Based Information Systems (NBiS-2013), pp. 156–161 (2013)
18. Matsuo, K., Elmazi, D., Liu, Y., Sakamoto, S., Mino, G., Barolli, L.: FACS-MP: a fuzzy admission control system with many priorities for wireless cellular networks and its performance evaluation. J. High Speed Netw. **21**(1), 1–14 (2015)
19. Grabisch, M.: The application of fuzzy integrals in multicriteria decision making. Eur. J. Oper. Res. **89**(3), 445–456 (1996)
20. Inaba, T., Elmazi, D., Liu, Y., Sakamoto, S., Barolli, L., Uchida, K.: Integrating wireless cellular and ad-hoc networks using fuzzy logic considering node mobility and security. In: The 29th IEEE International Conference on Advanced Information Networking and Applications Workshops (WAINA-2015), pp. 54–60 (2015)
21. Kulla, E., Mino, G., Sakamoto, S., Ikeda, M., Caballé, S., Barolli, L.: FBMIS: a fuzzy-based multi-interface system for cellular and ad hoc networks. In: International Conference on Advanced Information Networking and Applications (AINA-2014), pp. 180–185 (2014)
22. Elmazi, D., Kulla, E., Oda, T., Spaho, E., Sakamoto, S., Barolli, L.: A comparison study of two fuzzy-based systems for selection of actor node in wireless sensor actor networks. J. Ambient Intell. Humanized Comput. **6**(5), 635–645 (2015)
23. Zadeh, L.: Fuzzy logic, neural networks, and soft computing. ACM Commun., 77–84 (1994)
24. Spaho, E., Sakamoto, S., Barolli, L., Xhafa, F., Ikeda, M.: Trustworthiness in P2P: performance behaviour of two fuzzy-based systems for JXTA-overlay platform. Soft Comput. **18**(9), 1783–1793 (2014)
25. Inaba, T., Sakamoto, S., Kulla, E., Caballe, S., Ikeda, M., Barolli, L.: An integrated system for wireless cellular and ad-hoc networks using fuzzy logic. In: International Conference on Intelligent Networking and Collaborative Systems (INCoS-2014), pp. 157–162 (2014)

26. Matsuo, K., Elmazi, D., Liu, Y., Sakamoto, S., Barolli, L.: A multi-modal simulation system for wireless sensor networks: a comparison study considering stationary and mobile sink and event. J. Ambient Intell. Humanized Comput. **6**(4), 519–529 (2015)
27. Kolici, V., Inaba, T., Lala, A., Mino, G., Sakamoto, S., Barolli, L.: A fuzzy-based CAC scheme for cellular networks considering security. In: International Conference on Network-Based Information Systems (NBiS-2014), pp. 368–373 (2014)
28. Liu, Y., Sakamoto, S., Matsuo, K., Ikeda, M., Barolli, L., Xhafa, F.: A comparison study for two fuzzy-based systems: improving reliability and security of JXTA-overlay P2P platform. Soft Comput. **20**(7), 2677–2687 (2015)
29. Matsuo, K., Elmazi, D., Liu, Y., Sakamoto, S., Mino, G., Barolli, L.: FACS-MP: a fuzzy admission control system with many priorities for wireless cellular networks and its perforemance evaluation. J. High Speed Netw. **21**(1), 1–14 (2015)
30. Mendel, J.M.: Fuzzy logic systems for engineering: a tutorial. Proc. IEEE **83**(3), 345–377 (1995)

A Fuzzy-Based Approach for Improving Peer Awareness and Group Synchronization in MobilePeerDroid System

Yi Liu[1(✉)], Kosuke Ozera[1], Keita Matsuo[2], Makoto Ikeda[2], and Leonard Barolli[2]

[1] Graduate School of Engineering, Fukuoka Institute of Technology (FIT), 3-30-1 Wajiro-Higashi, Higashi-Ku, Fukuoka 811-0295, Japan
ryuui1010@gmail.com, kosuke.o.fit@gmail.com

[2] Department of Information and Communication Engineering, Fukuoka Institute of Technology (FIT), 3-30-1 Wajiro-Higashi, Higashi-Ku, Fukuoka 811-0295, Japan
{kt-matsuo,barolli}@fit.ac.jp, makoto.ikd@acm.org

Abstract. In this work, we present a distributed event-based awareness approach for P2P groupware systems. The awareness of collaboration will be achieved by using primitive operations and services that are integrated into the P2P middleware. We propose an abstract model for achieving these requirements and we discuss how this model can support awareness of collaboration in mobile teams. We present a fuzzy-based system for improving peer coordination quality according to four parameters. This model will be implemented in MobilePeerDroid system to give more realistic view of the collaborative activity and better decisions for the groupwork, while encouraging peers to increase their reliability in order to support awareness of collaboration in MobilePeerDroid Mobile System. We evaluated the performance of proposed system by computer simulations. From the simulations results, we conclude that when AA, SCT and GS values are increased, the peer coordination quality is increased. With increasing of NFT, the peer coordination quality is decreased.

1 Introduction

Peer to Peer technologies has been among most disruptive technologies after Internet. Indeed, the emergence of the P2P technologies changed drastically the concepts, paradigms and protocols of sharing and communication in large scale distributed systems. As pointed out since early 2000 years [1–5], the nature of the sharing and the direct communication among peers in the system, being these machines or people, makes possible to overcome the limitations of the flat communications through email, newsgroups and other forum-based communication forms.

L. Barolli et al. (Eds.): NBiS 2018, LNDECT 22, pp. 30–41, 2019.
https://doi.org/10.1007/978-3-319-98530-5_3

The usefulness of P2P technologies on one hand has been shown for the development of stand alone applications. On the other hand, P2P technologies, paradigms and protocols have penetrated other large scale distributed systems such as Mobile Ad hoc Networks (MANETs), Groupware systems, Mobile Systems to achieve efficient sharing, communication, coordination, replication, awareness and synchronization. In fact, for every new form of Internet-based distributed systems, we are seeing how P2P concepts and paradigms again play an important role to enhance the efficiency and effectiveness of such systems or to enhance information sharing and online collaborative activities of groups of people. We briefly introduce below some common application scenarios that can benefit from P2P communications.

Awareness is a key feature of groupware systems. In its simplest terms, awareness can be defined as the system's ability to notify the members of a group of changes occurring in the group's workspace. Awareness systems for online collaborative work have been proposed since in early stages of Web technology. Such proposals started by approaching workspace awareness, aiming to inform users about changes occurring in the shared workspace. More recently, research has focussed on using new paradigms, such as P2P systems, to achieve fully decentralized, ubiquitous groupware systems and awareness in such systems. In P2P groupware systems group processes may be more efficient because peers can be aware of the status of other peers in the group, and can interact directly and share resources with peers in order to provide additional scaffolding or social support. Moreover, P2P systems are pervasive and ubiquitous in nature, thus enabling contextualized awareness.

Fuzzy Logic (FL) is the logic underlying modes of reasoning which are approximate rather then exact. The importance of FL derives from the fact that most modes of human reasoning and especially common sense reasoning are approximate in nature [6]. FL uses linguistic variables to describe the control parameters. By using relatively simple linguistic expressions it is possible to describe and grasp very complex problems. A very important property of the linguistic variables is the capability of describing imprecise parameters.

The concept of a fuzzy set deals with the representation of classes whose boundaries are not determined. It uses a characteristic function, taking values usually in the interval [0, 1]. The fuzzy sets are used for representing linguistic labels. This can be viewed as expressing an uncertainty about the clear-cut meaning of the label. But important point is that the valuation set is supposed to be common to the various linguistic labels that are involved in the given problem.

The fuzzy set theory uses the membership function to encode a preference among the possible interpretations of the corresponding label. A fuzzy set can be defined by examplification, ranking elements according to their typicality with respect to the concept underlying the fuzzy set [7].

In this paper, we propose a fuzzy-based system for MobilePeerDroid system considering four parameters: Activity Awareness (AA), Sustained Communication Time (SCT), Group Synchronization (GS) and Number of Failed

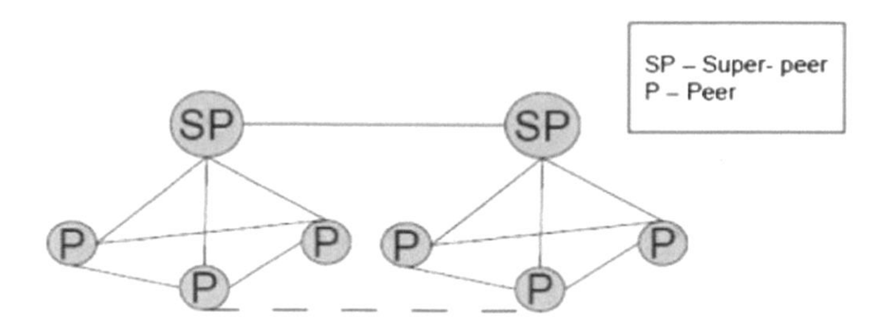

Fig. 1. Super-peer P2P group network.

Tasks (NFT) to decide the Peer Coordination Quality (PCQ). We evaluated the proposed system by simulations.

The structure of this paper is as follows. In Sect. 2, we introduce the group activity awareness model. In Sect. 3, we introduce FL used for control. In Sect. 4, we present the proposed fuzzy-based system. In Sect. 5, we discuss the simulation results. Finally, conclusions and future work are given in Sect. 6.

2 Group Activity Awareness Model

The awareness model considered here focuses on supporting group activities so to accomplish a common group project, although it can also be used in a broader scope of teamwork [8–14]. The main building blocks of our model (see also [15,16] in the context of web-based groupware) are described below.

Activity awareness: Activity awareness refers to awareness information about the project-related activities of group members. Project-based work is one of the most common methods of group working. Activity awareness aims to provide information about progress on the accomplishment of tasks by both individuals and the group as a whole. It comprises knowing about actions taken by members of the group according to the project schedule, and synchronization of activities with the project schedule. Activity awareness should therefore enable members to know about recent and past actions on the project's work by the group. As part of activity awareness, we also consider information on group artifacts such as documents and actions upon them (uploads, downloads, modifications, reading). Activity awareness is one of most important, and most complex, types of awareness. As well as the direct link to monitoring a group's progress on the work relating to a project, it also supports group communication and coordination processes.

Process awareness: In project-based work, a project typically requires the enactment of a workflow. In such a case, the objective of the awareness is to track the state of the workflow and to inform users accordingly. We term this process awareness. The workflow is defined through a set of tasks and precedence relationships relating to their order of completion. Process awareness targets the information flow of the project, providing individuals and the group with a partial view (what they are each doing individually) and a complete view (what they are doing as a group), thus enabling the identification of past, current and next states of the workflow in order to move the collaboration process forward.

Communication awareness: Another type of awareness considered in this work is that of communication awareness. We consider awareness information relating to message exchange, and synchronous and asynchronous discussion forums. The first is intended to support awareness of peer-to-peer communication (when some peer wants to establish a direct communication with another peer); the second is aimed at supporting awareness about chat room creation and lifetime (so that other peers can be aware of, and possibly eventually join, the chat room); the third refers to awareness of new messages posted at the discussion forum, replies, etc.

Availability awareness: Availability awareness is useful for provide individuals and the group with information on members' and resources' availability. The former is necessary for establishing synchronous collaboration either in peer-to-peer mode or (sub)group mode. The later is useful for supporting members' tasks requiring available resources (e.g. a machine for running a software program). Groupware applications usually monitor availability of group members by simply looking at group workspaces. However, availability awareness encompasses not only knowing who is in the workspace at any given moment but also who is available when, via members' profiles (which include also personal calendars) and information explicitly provided by members. In the case of resources, awareness is achieved via the schedules of resources. Thus, both explicit and implicit forms of gathering availability awareness information should be supported.

3 Application of Fuzzy Logic for Control

The ability of fuzzy sets and possibility theory to model gradual properties or soft constraints whose satisfaction is matter of degree, as well as information pervaded with imprecision and uncertainty, makes them useful in a great variety of applications [17–23].

The most popular area of application is Fuzzy Control (FC), since the appearance, especially in Japan, of industrial applications in domestic appliances, process control, and automotive systems, among many other fields.

In the FC systems, expert knowledge is encoded in the form of fuzzy rules, which describe recommended actions for different classes of situations represented by fuzzy sets.

In fact, any kind of control law can be modeled by the FC methodology, provided that this law is expressible in terms of "if ... then ..." rules, just like in the case of expert systems. However, FL diverges from the standard expert system approach by providing an interpolation mechanism from several rules. In the contents of complex processes, it may turn out to be more practical to get knowledge from an expert operator than to calculate an optimal control, due to modeling costs or because a model is out of reach.

A concept that plays a central role in the application of FL is that of a linguistic variable. The linguistic variables may be viewed as a form of data compression. One linguistic variable may represent many numerical variables. It is suggestive to refer to this form of data compression as granulation [17].

The same effect can be achieved by conventional quantization, but in the case of quantization, the values are intervals, whereas in the case of granulation the values are overlapping fuzzy sets. The advantages of granulation over quantization are as follows:

- it is more general;
- it mimics the way in which humans interpret linguistic values;
- the transition from one linguistic value to a contiguous linguistic value is gradual rather than abrupt, resulting in continuity and robustness.

FC describes the algorithm for process control as a fuzzy relation between information about the conditions of the process to be controlled, x and y, and the output for the process z. The control algorithm is given in "if ... then ..." expression, such as:

$$\text{If x is small and y is big, then z is medium;}$$
$$\text{If x is big and y is medium, then z is big.}$$

These rules are called *FC rules*. The "if" clause of the rules is called the antecedent and the "then" clause is called consequent. In general, variables x and y are called the input and z the output. The "small" and "big" are fuzzy values for x and y, and they are expressed by fuzzy sets.

Fuzzy controllers are constructed of groups of these FC rules, and when an actual input is given, the output is calculated by means of fuzzy inference.

4 Proposed Fuzzy-Based System

The P2P group-based model considered is that of a superpeer model (see Fig. 1). In this model, the P2P network is fragmented into several disjoint peergroups (see Fig. 2). The peers of each peergroup are connected to a single superpeer. There is frequent local communication between peers in a peergroup, and less frequent global communication between superpeers.

To complete a certain task in P2P mobile collaborative team work, peers often have to interact with unknown peers. Thus, it is important that group members must select reliable peers to interact.

In this work, we consider four parameters: Activity Awareness (AA), Sustained Communication Time (SCT), Group Synchronization (GS) and Number of Failed Tasks (NFT) to decide the Peer Coordination Quality (PCQ). The structure of this system called Fuzzy-based Coordination Quality System (FCQS) is shown in Fig. 3. These four parameters are fuzzified using fuzzy system, and based on the decision of fuzzy system the peer coordination quality is calculated. The membership functions for our system are shown in Fig. 4. In Table 1, we show the Fuzzy Rule Base (FRB) of our proposed system, which consists of 108 rules.

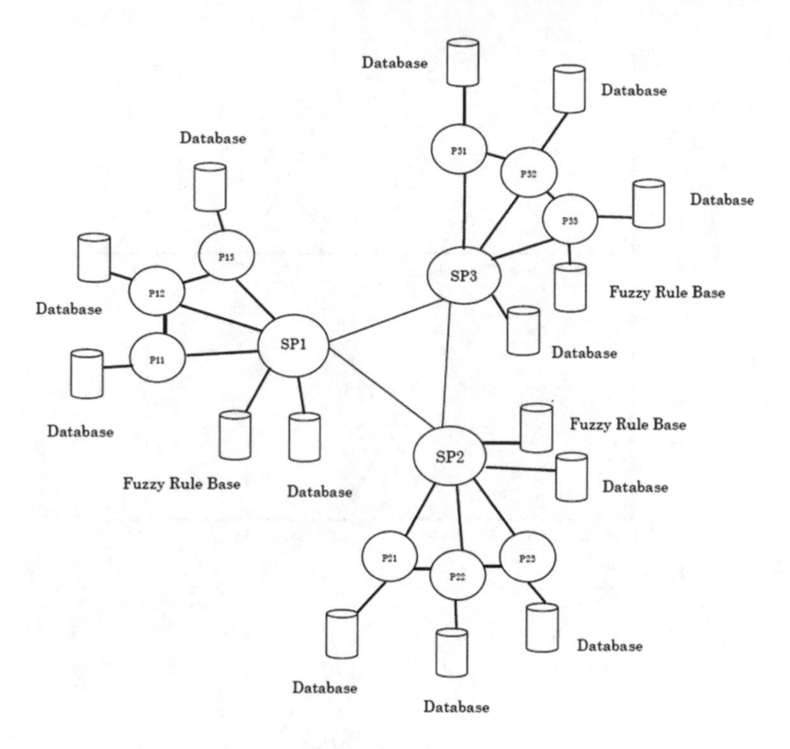

Fig. 2. P2P group-based model.

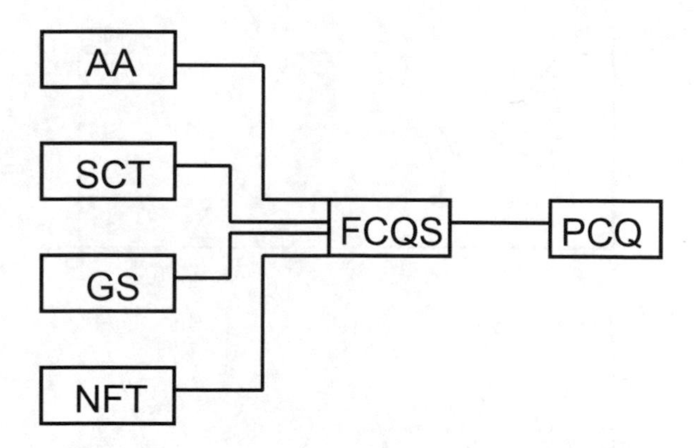

Fig. 3. Proposed system of structure.

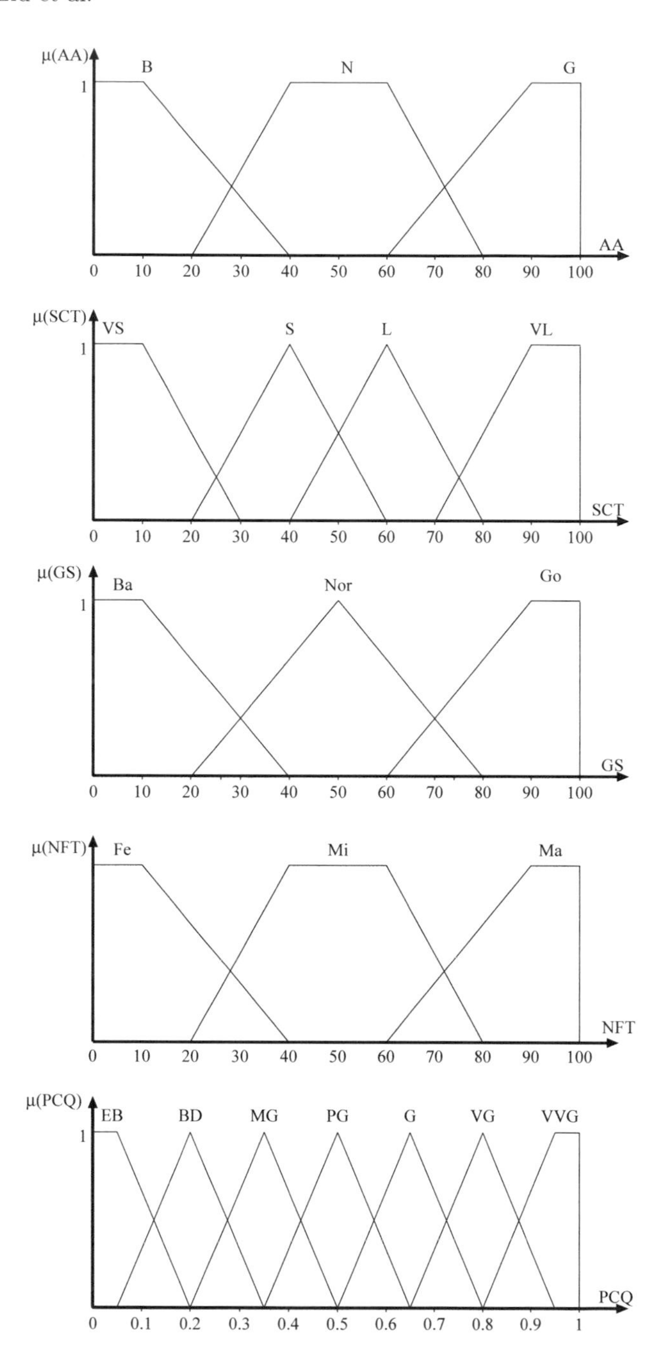

Fig. 4. Membership functions.

Table 1. FRB.

Rule	AA	SCT	GS	NFT	PCQ	Rule	AA	SCT	GS	NFT	PCQ	Rule	AA	SCT	GS	NFT	PCQ
1	B	VS	Ba	Sm	EB	37	N	VS	Ba	Sm	BD	73	G	VS	Ba	Sm	MG
2	B	VS	Ba	Mi	EB	38	N	VS	Ba	Mi	EB	74	G	VS	Ba	Mi	BD
3	B	VS	Ba	Ma	EB	39	N	VS	Ba	Ma	EB	75	G	VS	Ba	Ma	EB
4	B	VS	Nor	Sm	BD	40	N	VS	Nor	Sm	PG	76	G	VS	Nor	Sm	G
5	B	VS	Nor	Mi	EB	41	N	VS	Nor	Mi	BD	77	G	VS	Nor	Mi	PG
6	B	VS	Nor	Ma	EB	42	N	VS	Nor	Ma	EB	78	G	VS	Nor	Ma	BD
7	B	VS	Go	Sm	PG	43	N	VS	Go	Sm	G	79	G	VS	Go	Sm	VG
8	B	VS	Go	Mi	MG	44	N	VS	Go	Mi	PG	80	G	VS	Go	Mi	G
9	B	VS	Go	Ma	EB	45	N	VS	Go	Ma	MG	81	G	VS	Go	Ma	PG
10	B	S	Ba	Sm	BD	46	N	S	Ba	Sm	MG	82	G	S	Ba	Sm	PG
11	B	S	Ba	Mi	EB	47	N	S	Ba	Mi	BD	83	G	S	Ba	Mi	MG
12	B	S	Ba	Ma	EB	48	N	S	Ba	Ma	EB	84	G	S	Ba	Ma	BD
13	B	S	Nor	Sm	MG	49	N	S	Nor	Sm	G	85	G	S	Nor	Sm	VG
14	B	S	Nor	Mi	BD	50	N	S	Nor	Mi	MG	86	G	S	Nor	Mi	G
15	B	S	Nor	Ma	EB	51	N	S	Nor	Ma	BD	87	G	S	Nor	Ma	MG
16	B	S	Go	Sm	G	52	N	S	Go	Sm	VG	88	G	S	Go	Sm	VVG
17	B	S	Go	Mi	PG	53	N	S	Go	Mi	G	89	G	S	Go	Mi	VG
18	B	S	Go	Ma	BD	54	N	S	Go	Ma	PG	90	G	S	Go	Ma	G
19	B	L	Ba	Sm	MG	55	N	L	Ba	Sm	PG	91	G	L	Ba	Sm	G
20	B	L	Ba	Mi	EB	56	N	L	Ba	Mi	MG	92	G	L	Ba	Mi	PG
21	B	L	Ba	Ma	EB	57	N	L	Ba	Ma	EB	93	G	L	Ba	Ma	MG
22	B	L	Nor	Sm	PG	58	N	L	Nor	Sm	VG	94	G	L	Nor	Sm	VG
23	B	L	Nor	Mi	MG	59	N	L	Nor	Mi	PG	95	G	L	Nor	Mi	VG
24	B	L	Nor	Ma	BD	60	N	L	Nor	Ma	MG	96	G	L	Nor	Ma	PG
25	B	L	Go	Sm	VG	61	N	L	Go	Sm	VVG	97	G	L	Go	Sm	VVG
26	B	L	Go	Mi	G	62	N	L	Go	Mi	VG	98	G	L	Go	Mi	VVG
27	B	L	Go	Ma	MG	63	N	L	Go	Ma	G	99	G	L	Go	Ma	VG
28	B	VL	Ba	Sm	PG	64	N	VL	Ba	Sm	G	100	G	VL	Ba	Sm	VG
29	B	VL	Ba	Mi	BD	65	N	VL	Ba	Mi	PG	101	G	VL	Ba	Mi	G
30	B	VL	Ba	Ma	EB	66	N	VL	Ba	Ma	BD	102	G	VL	Ba	Ma	PG
31	B	VL	Nor	Sm	G	67	N	VL	Nor	Sm	VG	103	G	VL	Nor	Sm	VVG
32	B	VL	Nor	Mi	PG	68	N	VL	Nor	Mi	G	104	G	VL	Nor	Mi	VG
33	B	VL	Nor	Ma	MG	69	N	VL	Nor	Ma	PG	105	G	VL	Nor	Ma	G
34	B	VL	Go	Sm	VG	70	N	VL	Go	Sm	VVG	106	G	VL	Go	Sm	VVG
35	B	VL	Go	Mi	VG	71	N	VL	Go	Mi	VG	107	G	VL	Go	Mi	VVG
36	B	VL	Go	Ma	PG	72	N	VL	Go	Ma	VG	108	G	VL	Go	Ma	VG

The input parameters for FCQS are: AA, SCT, GS and NFT the output linguistic parameter is PCQ. The term sets of AA, SCT, GS and NFT are defined respectively as:

$$
\begin{aligned}
AA &= \{Bad,\ Normal,\ Good\} \\
&= \{B,\ N,\ G\}; \\
SCT &= \{Very\ Short,\ Short,\ Long,\ Very\ Long\} \\
&= \{VS,\ S,\ L,\ VL\};
\end{aligned}
$$

$$\begin{aligned} GS &= \{Bad,\ Normal,\ Good\} \\ &= \{Ba,\ Nor,\ Go\}; \\ NFT &= \{Few,\ Middle,\ Many\} \\ &= \{Fe,\ Mi,\ Ma\}. \end{aligned}$$

and the term set for the output PCQ is defined as:

$$PCQ = \begin{pmatrix} Extremely\ Bad \\ Bad \\ Minimally\ Good \\ Partially\ Good \\ Good \\ Very\ Good \\ Very\ Very\ Good \end{pmatrix} = \begin{pmatrix} EB \\ BD \\ MG \\ PG \\ G \\ VG \\ VVG \end{pmatrix}$$

5 Simulation Results

In this section, we present the simulation results for our FCQS system. In our system, we decided the number of term sets by carrying out many simulations.

From Figs. 5, 6 and 7, we show the relation between AA, GS, SCT and PCQ. In these simulations, we consider the GS and NFT as constant parameters. In Fig. 5, we consider the NFT value 10 units. We change the GS value from 10 to 90 units. When the GS increases, the PCQ is increased. Also, when the SCT and AA are high, the PCQ is increased. In Figs. 6 and 7, we increase the NFT values to 50 and 90 units, respectively. We see that, when the NFT increases, the PCQ is decreased.

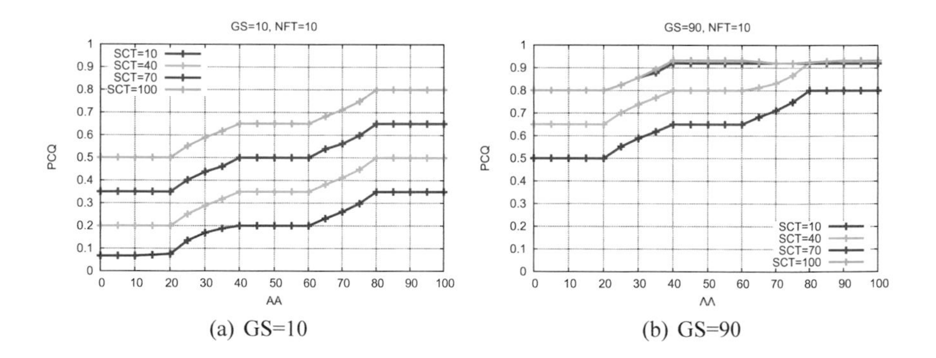

(a) GS=10 (b) GS=90

Fig. 5. Relation of PCQ with AA and SCT for different GS when NFT = 10.

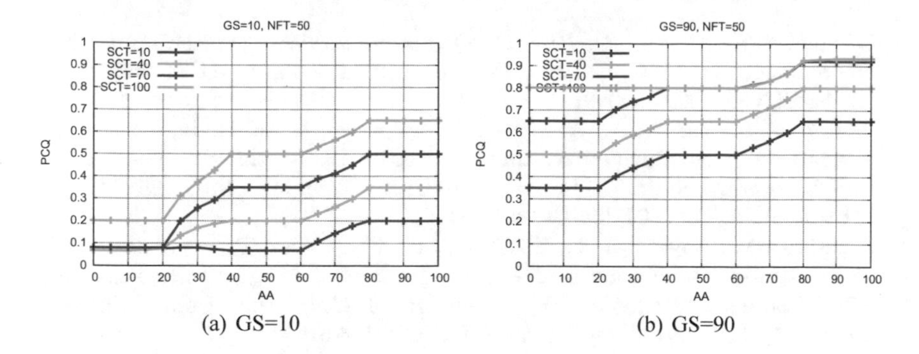

(a) GS=10 (b) GS=90

Fig. 6. Relation of PCQ with AA and SCT for different GS when NFT = 50.

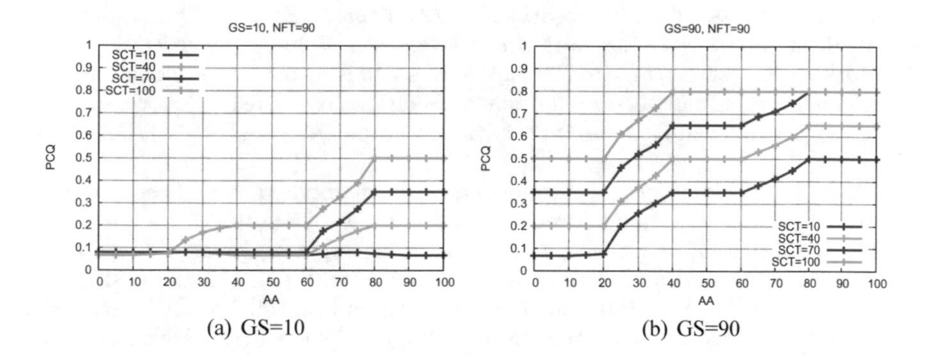

(a) GS=10 (b) GS=90

Fig. 7. Relation of PCQ with AA and SCT for different GS when NFT = 90.

6 Conclusions and Future Work

In this paper, we proposed a fuzzy-based system to decide the PCQ. We took into consideration four parameters: AA, SCT, GS and NFT. We evaluated the performance of proposed system by computer simulations. From the simulations results, we conclude that when AA, SCT and GS values are increased, the PCQ is increased. But, by increasing NFT, the PCQ is decreased.

In the future, we would like to make extensive simulations to evaluate the proposed systems and compare the performance with other systems.

References

1. Oram, A. (ed.): Peer-to-Peer: Harnessing the Power of Disruptive Technologies. O'Reilly and Associates, CA (2001)
2. Sula, A., Spaho, E., Matsuo, K., Barolli, L., Xhafa, F., Miho, R.: A new system for supporting children with autism spectrum disorder based on IoT and P2P technology. Int. J. Space-Based Situated Comput. **4**(1), 55–64 (2014). https://doi.org/10.1504/IJSSC.2014.060688

3. Di Stefano, A., Morana, G., Zito, D.: QoS-aware services composition in P2PGrid environments. Int. J. Grid Utility Comput. **2**(2), 139–147 (2011). https://doi.org/10.1504/IJGUC.2011.040601
4. Sawamura, S., Barolli, A., Aikebaier, A., Takizawa, M., Enokido, T.: Design and evaluation of algorithms for obtaining objective trustworthiness on acquaintances in P2P overlay networks. Int. J. Grid Utility Comput. **2**(3), 196–203 (2011). https://doi.org/10.1504/IJGUC.2011.042042
5. Higashino, M., Hayakawa, T., Takahashi, K., Kawamura, T., Sugahara, K.: Management of streaming multimedia content using mobile agent technology on pure P2P-based distributed e-learning system. Int. J. Grid Utility Comput. **5**(3), 198–204 (2014). https://doi.org/10.1504/IJGUC.2014.062928
6. Inaba, T., Obukata, R., Sakamoto, S., Oda, T., Ikeda, M., Barolli, L.: Performance evaluation of a QoS-aware fuzzy-based CAC for LAN access. Int. J. Space-Based Situated Comput. **1**(1) (2011). https://doi.org/10.1504/IJSSC.2016.082768
7. Terano, T., Asai, K., Sugeno, M.: Fuzzy Systems Theory and Its Applications. Academic Press INC., Harcourt Brace Jovanovich, Publishers (1992)
8. Mori, T., Nakashima, M., Ito, T.: SpACCE: a sophisticated ad hoc cloud computing environment built by server migration to facilitate distributed collaboration. Int. J. Space-Based Situated Comput. **1**(1) (2011). https://doi.org/10.1504/IJSSC.2012.050000
9. Xhafa, F., Poulovassilis, A.: Requirements for distributed event-based awareness in P2P groupware systems. Proc. AINA **2010**, 220–225 (2010)
10. Xhafa, F., Barolli, L., Caballé, S., Fernandez, R.: Supporting scenario-based online learning with P2P group-based systems. Proc. NBiS **2010**, 173–180 (2010)
11. Gupta, S., Kaiser, G.: P2P video synchronization in a collaborative virtual environment. In: Proceedings of the 4th International Conference on Advances in Web-Based Learning (ICWL 2005), pp. 86–98 (2005)
12. Martnez-Alemn, A.M., Wartman, K.L.: Online Social Networking on Campus Understanding What Matters in Student Culture. Taylor and Francis, Routledge (2008)
13. Puzar, M., Plagemann, T.: Data sharing in mobile ad-hoc networks a study of replication and performance in the MIDAS data space. Int. J. Space-Based Situated Comput. **1**(1) (2011). https://doi.org/10.1504/IJSSC.2011.040340
14. Spaho, E., Kulla, E., Xhafa, F., Barolli, L.: P2P solutions to efficient mobile peer collaboration in MANETs. In: Proceedings of 3PGCIC 2012, pp. 379–383, November 2012
15. Gutwin, C., Greenberg, S., Roseman, M.: Workspace awareness in realtime distributed groupware: framework, widgets, and evaluation. In: BCS HCI 1996, pp. 281–298 (1996)
16. You, Y., Pekkola, S.: Meeting others supporting situation awareness on the WWW. Decis. Support Syst. **32**(1), 71–82 (2001)
17. Kandel, A.: Fuzzy Expert Systems. CRC Press (1992)
18. Zimmermann, H.J.: Fuzzy Set Theory and Its Applications, Second Revised Edition. Kluwer Academic Publishers (1991)
19. McNeill, F.M., Thro, E.: Fuzzy Logic. A Practical Approach. Academic Press Inc. (1994)
20. Zadeh, L.A., Kacprzyk, J.: Fuzzy Logic For The Management of Uncertainty. Wiley (1992)
21. Procyk, T.J., Mamdani, E.H.: A linguistic self-organizing process controller. Automatica **15**(1), 15–30 (1979)

22. Klir, G.J., Folger, T.A.: Fuzzy Sets, Uncertainty, and Information. Prentice Hall, Englewood Cliffs (1988)
23. Munakata, T., Jani, Y.: Fuzzy systems: an overview. Commun. ACM **37**(3), 69–76 (1994)

A Hybrid Simulation System Based on Particle Swarm Optimization and Distributed Genetic Algorithm for WMNs: Performance Evaluation Considering Normal and Uniform Distribution of Mesh Clients

Admir Barolli[1(✉)], Shinji Sakamoto[2], Leonard Barolli[3], and Makoto Takizawa[4]

[1] Department of Information Technology, Aleksander Moisiu University of Durres, L.1, Rruga e Currilave, Durres, Albania
admir.barolli@gmail.com

[2] Department of Computer and Information Science, Seikei University, 3-3-1 Kichijoji-Kitamachi, Musashino-shi, Tokyo 180-8633, Japan
shinji.sakamoto@ieee.org

[3] Department of Information and Communication Engineering, Fukuoka Institute of Technology, 3-30-1 Wajiro-Higashi, Higashi-Ku, Fukuoka 811-0295, Japan
barolli@fit.ac.jp

[4] Department of Advanced Sciences, Faculty of Science and Engineering, Hosei University, Kajino-Machi, Koganei-Shi, Tokyo 184-8584, Japan
makoto.takizawa@computer.org

Abstract. The Wireless Mesh Networks (WMNs) are becoming an important networking infrastructure because they have many advantages such as low cost and increased high speed wireless Internet connectivity. In our previous work, we implemented a Particle Swarm Optimization (PSO) based simulation system, called WMN-PSO, and a simulation system based on Genetic Algorithm (GA), called WMN-GA, for solving node placement problem in WMNs. In this paper, we implement a hybrid simulation system based on PSO and distributed GA (DGA), called WMN-PSODGA. We analyze the performance of WMNs using WMN-PSODGA simulation system considering Normal and Uniform client distributions. Simulation results show that the WMN-PSODGA has good performance for Normal distribution compared with the case of Uniform distribution.

1 Introduction

The wireless networks and devises are becoming increasingly popular and they provide users access to information and communication anytime and anywhere [3, 8–11, 14, 20, 26, 27, 29, 33]. Wireless Mesh Networks (WMNs) are gaining

L. Barolli et al. (Eds.): NBiS 2018, LNDECT 22, pp. 42–55, 2019.
https://doi.org/10.1007/978-3-319-98530-5_4

a lot of attention because of their low cost nature that makes them attractive for providing wireless Internet connectivity. A WMN is dynamically self-organized and self-configured, with the nodes in the network automatically establishing and maintaining mesh connectivity among them-selves (creating, in effect, an ad hoc network). This feature brings many advantages to WMNs such as low up-front cost, easy network maintenance, robustness and reliable service coverage [1]. Moreover, such infrastructure can be used to deploy community networks, metropolitan area networks, municipal and corporative networks, and to support applications for urban areas, medical, transport and surveillance systems.

Mesh node placement in WMN can be seen as a family of problems, which are shown (through graph theoretic approaches or placement problems, e.g. [6,15]) to be computationally hard to solve for most of the formulations [37]. In fact, the node placement problem considered here is even more challenging due to two additional characteristics:

(a) locations of mesh router nodes are not pre-determined, in other wards, any available position in the considered area can be used for deploying the mesh routers.
(b) routers are assumed to have their own radio coverage area.

Here, we consider the version of the mesh router nodes placement problem in which we are given a grid area where to deploy a number of mesh router nodes and a number of mesh client nodes of fixed positions (of an arbitrary distribution) in the grid area. The objective is to find a location assignment for the mesh routers to the cells of the grid area that maximizes the network connectivity and client coverage. Node placement problems are known to be computationally hard to solve [12,13,38]. In some previous works, intelligent algorithms have been recently investigated [4,7,16,18,21–23,31,32].

In our previous work, we implemented a Particle Swarm Optimization (PSO) based simulation system, called WMN-PSO [24]. Also, we implemented another simulation system based on Genetic Algorithm (GA), called WMN-GA [19], for solving node placement problem in WMNs.

In this paper, we design and implement a hybrid simulation system based on PSO and distributed GA (DGA). We call this system WMN-PSODGA. We analyze the performance of WMNs using WMN-PSODGA simulation system considering Normal and Uniform client distributions.

The rest of the paper is organized as follows. The mesh router nodes placement problem is defined in Sect. 2. We present our designed and implemented hybrid simulation system in Sect. 3. The simulation results are given in Sect. 4. Finally, we give conclusions and future work in Sect. 5.

2 Node Placement Problem in WMNs

For this problem, we have a grid area arranged in cells we want to find where to distribute a number of mesh router nodes and a number of mesh client nodes of fixed positions (of an arbitrary distribution) in the considered area. The objective is to find a location assignment for the mesh routers to the area that maximizes the network connectivity and client coverage. Network connectivity is measured by Size of Giant Component (SGC) of the resulting WMN graph, while the user coverage is simply the number of mesh client nodes that fall within the radio coverage of at least one mesh router node and is measured by Number of Covered Mesh Clients (NCMC).

An instance of the problem consists as follows.

- N mesh router nodes, each having its own radio coverage, defining thus a vector of routers.
- An area $W \times H$ where to distribute N mesh routers. Positions of mesh routers are not pre-determined and are to be computed.
- M client mesh nodes located in arbitrary points of the considered area, defining a matrix of clients.

It should be noted that network connectivity and user coverage are among most important metrics in WMNs and directly affect the network performance.

In this work, we have considered a bi-objective optimization in which we first maximize the network connectivity of the WMN (through the maximization of the SGC) and then, the maximization of the NCMC.

In fact, we can formalize an instance of the problem by constructing an adjacency matrix of the WMN graph, whose nodes are router nodes and client nodes and whose edges are links between nodes in the mesh network. Each mesh node in the graph is a triple $\boldsymbol{v} =< x, y, r >$ representing the 2D location point and r is the radius of the transmission range. There is an arc between two nodes $\boldsymbol{u}$ and $\boldsymbol{v}$, if $\boldsymbol{v}$ is within the transmission circular area of $\boldsymbol{u}$.

3 Proposed and Implemented Simulation System

3.1 Particle Swarm Optimization

In PSO a number of simple entities (the particles) are placed in the search space of some problem or function and each evaluates the objective function at its current location. The objective function is often minimized and the exploration of the search space is not through evolution [17]. However, following a widespread practice of borrowing from the evolutionary computation field, in this work, we consider the bi-objective function and fitness function interchangeably. Each particle then determines its movement through the search space by combining some aspect of the history of its own current and best (best-fitness) locations with those of one or more members of the swarm, with some random perturbations. The next iteration takes place after all particles have been moved. Eventually the swarm as a whole, like a flock of birds collectively foraging for food, is likely to move close to an optimum of the fitness function.

Each individual in the particle swarm is composed of three $\mathcal{D}$-dimensional vectors, where $\mathcal{D}$ is the dimensionality of the search space. These are the current position $\vec{x}_i$, the previous best position $\vec{p}_i$ and the velocity $\vec{v}_i$.

Algorithm 1. Pseudo code of PSO.

```
/* Initialize all parameters for PSO */
Computation maxtime:= Tp_max, t := 0;
Number of particle-patterns:= m, 2 ≤ m ∈ N^1;
Particle-patterns initial solution:= P_i^0;
Particle-patterns initial position:= x_ij^0;
Particles initial velocity:= v_ij^0;
PSO parameter:= ω, 0 < ω ∈ R^1;
PSO parameter:= C_1, 0 < C_1 ∈ R^1;
PSO parameter:= C_2, 0 < C_2 ∈ R^1;
/* Start PSO */
Evaluate(G^0, P^0);
while t < Tp_max do
  /* Update velocities and positions */
  v_ij^{t+1} = ω · v_ij^t
        +C_1 · rand() · (best(P_ij^t) − x_ij^t)
        +C_2 · rand() · (best(G^t) − x_ij^t);
  x_ij^{t+1} = x_ij^t + v_ij^{t+1};
  /* if fitness value is increased, a new solution will be accepted. */
  Update_Solutions(G^t, P^t);
  t = t + 1;
end while
Update_Solutions(G^t, P^t);
return Best found pattern of particles as solution;
```

The particle swarm is more than just a collection of particles. A particle by itself has almost no power to solve any problem; progress occurs only when the particles interact. Problem solving is a population-wide phenomenon, emerging from the individual behaviors of the particles through their interactions. In any case, populations are organized according to some sort of communication structure or topology, often thought of as a social network. The topology typically consists of bidirectional edges connecting pairs of particles, so that if j is in i's neighborhood, i is also in j's. Each particle communicates with some other particles and is affected by the best point found by any member of its topological neighborhood. This is just the vector $\vec{p}_i$ for that best neighbor, which we will denote with $\vec{p}_g$. The potential kinds of population "social networks" are hugely varied, but in practice certain types have been used more frequently. We show the pseudo code of PSO in Algorithm 1.

In the PSO process, the velocity of each particle is iteratively adjusted so that the particle stochastically oscillates around $\vec{p}_i$ and $\vec{p}_g$ locations.

3.2 Distributed Genetic Algorithm

Distributed Genetic Algorithm (DGA) has been focused from various fields of science. DGA has shown their usefulness for the resolution of many

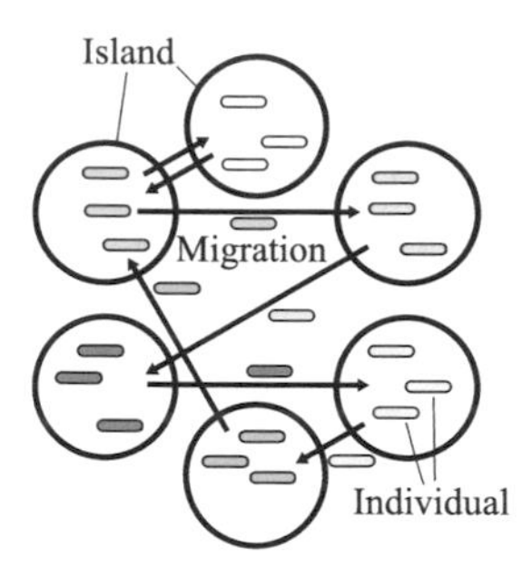

Fig. 1. Model of Migration in DGA.

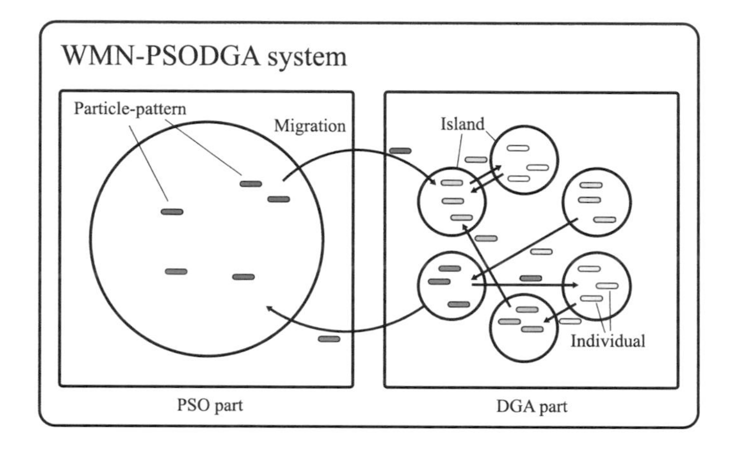

Fig. 2. Model of WMN-PSODGA migration.

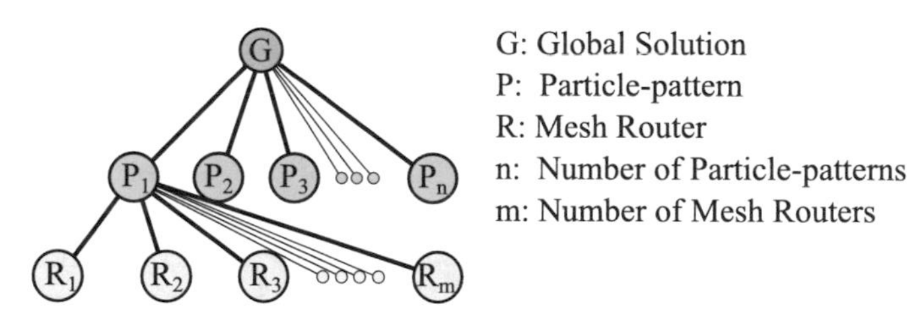

Fig. 3. Relationship among global solution, particle-patterns and mesh routers in PSO part.

computationally hard combinatorial optimization problems. We show the pseudo code of DGA in Algorithm 2.

Population of individuals: Unlike local search techniques that construct a path in the solution space jumping from one solution to another one through local perturbations, DGA use a population of individuals giving thus the search a larger scope and chances to find better solutions. This feature is also known

Algorithm 2. Pseudo code of DSA.

```
/* Initialize all parameters for DGA */
Computation maxtime:= Tg_max, t := 0;
Number of islands:= n, 1 ≤ n ∈ N^1;
initial solution:= P_i^0;
/* Start DGA */
Evaluate(G^0, P^0);
while t < Tg_max do
   for all islands do
      Selection();
      Crossover();
      Mutation();
   end for
   t = t + 1;
end while
Update_Solutions(G^t, P^t);
return Best found pattern of particles as solution;
```

as "exploration" process in difference to"exploitation" process of local search methods.

Fitness: The determination of an appropriate fitness function, together with the chromosome encoding are crucial to the performance of DGA. Ideally we would construct objective functions with "certain regularities", i.e. objective functions that verify that for any two individuals which are close in the search space, their respective values in the objective functions are similar.

Selection: The selection of individuals to be crossed is another important aspect in DGA as it impacts on the convergence of the algorithm. Several selection schemes have been proposed in the literature for selection operators trying to cope with premature convergence of DGA. There are many selection methods in GA. In our system, we implement 2 selection methods: Random method and Roulette wheel method.

Crossover operators: Use of crossover operators is one of the most important characteristics. Crossover operator is the means of DGA to transmit best genetic features of parents to offsprings during generations of the evolution process. Many methods for crossover operators have been proposed such as Blend Crossover (BLX-α), Unimodal Normal Distribution Crossover (UNDX), Simplex Crossover (SPX).

Mutation operators: These operators intend to improve the individuals of a population by small local perturbations. They aim to provide a component of randomness in the neighborhood of the individuals of the population. In our system, we implemented two mutation methods: uniformly random mutation and boundary mutation.

Escaping from local optima: GA itself has the ability to avoid falling prematurely into local optima and can eventually escape from them during the search process. DGA has one more mechanism to escape from local optima by considering some islands. Each island computes GA for optimizing and they migrate its gene to provide the ability to avoid from local optima (See Fig. 1).

Convergence: The convergence of the algorithm is the mechanism of DGA to reach to good solutions. A premature convergence of the algorithm would cause that all individuals of the population be similar in their genetic features and thus the search would result ineffective and the algorithm getting stuck into local optima. Maintaining the diversity of the population is therefore very important to this family of evolutionary algorithms.

3.3 WMN-PSODGA Hybrid Simulation System

In this subsection, we present the initialization, particle-pattern, fitness function. Also, our implemented simulation system uses Migration function (see Algorithm 3) as shown in Fig. 2. The Migration function swaps solutions among lands including PSO part.

Initialization

We decide the velocity of particles by a random process considering the area size. For instance, when the area size is $W \times H$, the velocity is decided randomly from $-\sqrt{W^2 + H^2}$ to $\sqrt{W^2 + H^2}$. In our system, many kinds of client distributions are generated. In this paper, we consider Normal and Uniform distributions for mesh clients.

Particle-pattern

A particle is a mesh router. A fitness value of a particle-pattern is computed by combination of mesh routers and mesh clients positions. In other words, each particle-pattern is a solution as shown is Fig. 3.

Gene coding

A gene describes a WMN. Each individual has its own combination of mesh nodes. In other words, each individual has a fitness value. Therefore, the combination of mesh nodes is a solution.

Fitness function

One of most important thing in PSO algorithm is to decide the determination of an appropriate objective function and its encoding. In our case, each particle-pattern has an own fitness value and compares it with other particle-pattern's fitness value in order to share information of global solution. The fitness function follows a hierarchical approach in which the main objective is to maximize the SGC in WMN. Thus, the fitness function of this scenario is defined as

$$\text{Fitness} = 0.7 \times \text{SGC}(\boldsymbol{x}_{ij}, \boldsymbol{y}_{ij}) + 0.3 \times \text{NCMC}(\boldsymbol{x}_{ij}, \boldsymbol{y}_{ij}).$$

Routers replacement method for PSO part

A mesh router has x, y positions and velocity. Mesh routers are moved based on velocities. There are many moving methods in PSO field, such as:

Constriction Method (CM)
: CM is a method which PSO parameters are set to a week stable region ($\omega = 0.729$, $C_1 = C2 = 1.4955$) based on analysis of PSO by M. Clerc et. al. [2,5,35].

Random Inertia Weight Method (RIWM)
: In RIWM, the ω parameter is changing randomly from 0.5 to 1.0. The C_1 and C_2 are kept 2.0. The ω can be estimated by the week stable region. The average of ω is 0.75 [28,35].

Linearly Decreasing Inertia Weight Method (LDIWM)
: In LDIWM, C_1 and C_2 are set to 2.0, constantly. On the other hand, the ω parameter is changed linearly from unstable region ($\omega = 0.9$) to stable region ($\omega = 0.4$) with increasing of iterations of computations [35,36].

Linearly Decreasing Vmax Method (LDVM)
: In LDVM, PSO parameters are set to unstable region ($\omega = 0.9$, $C_1 = C_2 = 2.0$). A value of V_{max} which is maximum velocity of particles is considered. With increasing of iteration of computations, the V_{max} is kept decreasing linearly [30,34].

Rational Decrement of Vmax Method (RDVM)
: In RDVM, PSO parameters are set to unstable region ($\omega = 0.9$, $C_1 = C_2 = 2.0$). The V_{max} is kept decreasing with the increasing of iterations as

$$V_{max}(x) = \sqrt{W^2 + H^2} \times \frac{T - x}{x}.$$

where, W and H are the width and the height of the considered area, respectively. Also, T and x are the total number of iterations and a current number of iteration, respectively [25].

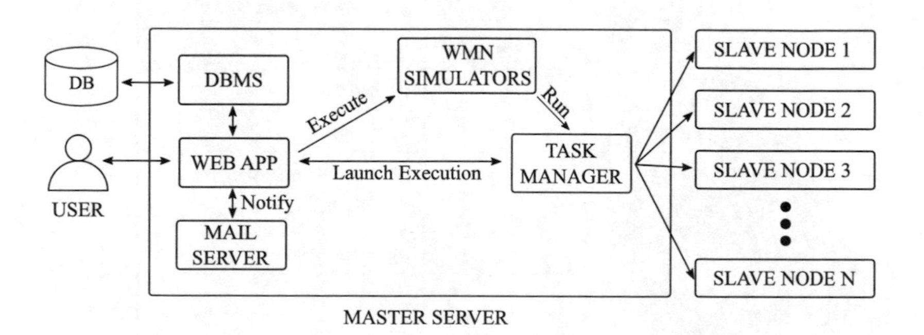

Fig. 4. System structure for web interface.

3.4 WMN-PSODGA Web GUI Tool and Pseudo Code

The Web application follows a standard Client-Server architecture and is implemented using LAMP (Linux + Apache + MySQL + PHP) technology (see Fig. 4). Remote users (clients) submit their requests by completing first the parameter setting. The parameter values to be provided by the user are classified into three groups, as follows.

- Parameters related to the problem instance: These include parameter values that determine a problem instance to be solved and consist of number of router nodes, number of mesh client nodes, client mesh distribution, radio coverage interval and size of the deployment area.
- Parameters of the resolution method: Each method has its own parameters.
- Execution parameters: These parameters are used for stopping condition of the resolution methods and include number of iterations and number of independent runs. The former is provided as a total number of iterations and depending on the method is also divided per phase (e.g., number of iterations in a exploration). The later is used to run the same configuration for the same problem instance and parameter configuration a certain number of times.

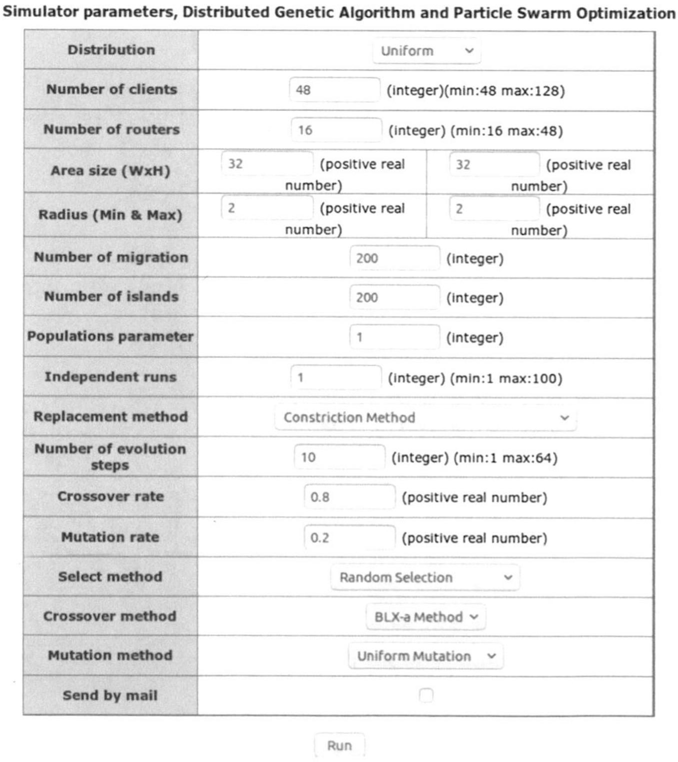

Simulator parameters, Distributed Genetic Algorithm and Particle Swarm Optimization

Distribution	Uniform	
Number of clients	48 (integer)(min:48 max:128)	
Number of routers	16 (integer) (min:16 max:48)	
Area size (WxH)	32 (positive real number)	32 (positive real number)
Radius (Min & Max)	2 (positive real number)	2 (positive real number)
Number of migration	200 (integer)	
Number of islands	200 (integer)	
Populations parameter	1 (integer)	
Independent runs	1 (integer) (min:1 max:100)	
Replacement method	Constriction Method	
Number of evolution steps	10 (integer) (min:1 max:64)	
Crossover rate	0.8 (positive real number)	
Mutation rate	0.2 (positive real number)	
Select method	Random Selection	
Crossover method	BLX-a Method	
Mutation method	Uniform Mutation	
Send by mail	☐	

Run

Fig. 5. WMN-PSODGA Web GUI Tool.

We show the WMN-PSODGA Web GUI tool in Fig. 5. The pseudo code of our implemented system is shown in Algorithm 3.

Algorithm 3. Pseudo code of WMN-PSODGA system.

Computation maxtime:= T_{max}, $t := 0$;
Initial solutions: $\boldsymbol{P}$.
Initial global solutions: $\boldsymbol{G}$.
/* Start PSODGA */
while $t < T_{max}$ **do**
  Subprocess(PSO);
  Subprocess(DGA);
  WaitSubprocesses();
  Evaluate($\boldsymbol{G}^t$, $\boldsymbol{P}^t$)
  /* Migration() swaps solutions (see Fig. 2). */
  Migration();
  $t = t + 1$;
end while
Update_Solutions($\boldsymbol{G}^t$, $\boldsymbol{P}^t$);
return Best found pattern of particles as solution;

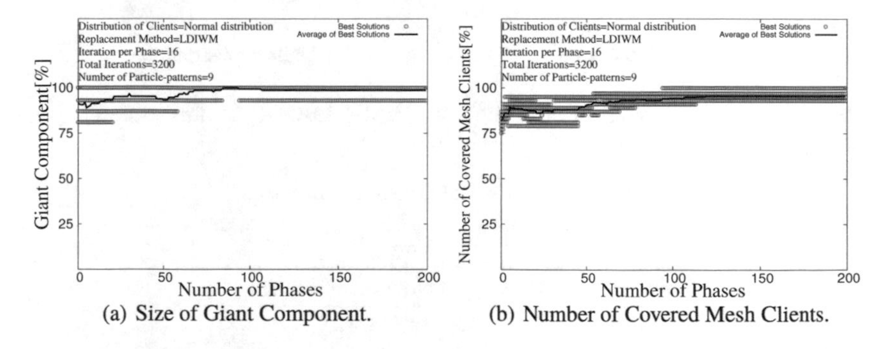

(a) Size of Giant Component. (b) Number of Covered Mesh Clients.

Fig. 6. Simulation results of WMN-PSODGA for Normal distribution of mesh clients.

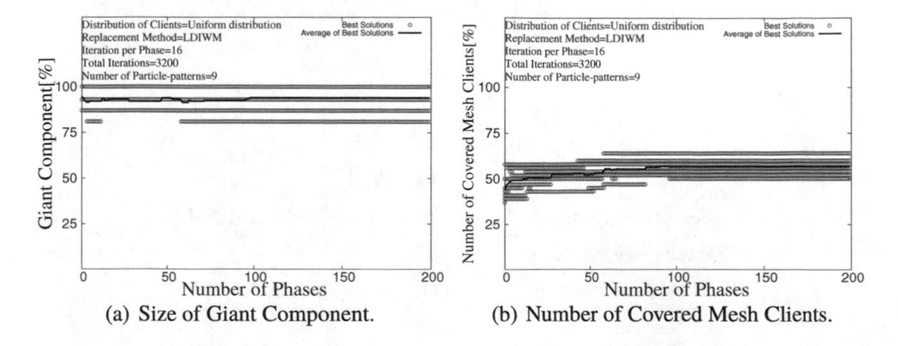

(a) Size of Giant Component. (b) Number of Covered Mesh Clients.

Fig. 7. Simulation results of WMN-PSODGA for Uniform distribution of mesh clients.

4 Simulation Results

In this section, we show simulation results using WMN-PSODGA system. In this work, we analyse the performance of WMN-PSODGA system considering Normal and Uniform distribution of mesh clients. The number of mesh routers is considered 16 and the number of mesh clients 48. We conducted simulations

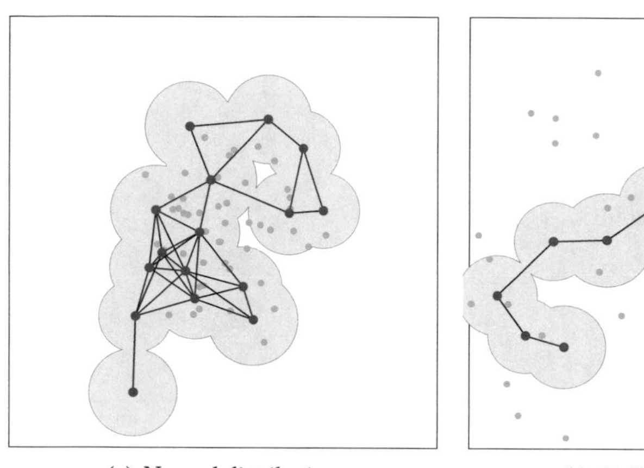

(a) Normal distribution. (b) Uniform distribution.

Fig. 8. Visualized simulation results of WMN-PSODGA for different client distributions.

Table 1. WMN-PSODGA Parameters.

Parameters	Values
Clients distribution	Normal, Uniform
Area size	32.0×32.0
Number of mesh routers	16
Number of mesh clients	48
Number of migrations	200
Evolution steps	9
Number of GA islands	16
Radius of a mesh router	2.5 - 3.5
Replacement method	LDIWM
Selection method	Roulette wheel method
Crossover method	SPX
Mutation method	Boundary mutation
Crossover rate	0.8
Mutation rate	0.2

10 times, in order to avoid the effect of randomness and create a general view of results. We show the parameter setting for WM-PSODGA in Table 1.

We show simulation results from Figs. 6, 7 and 8. In Fig. 6(a) and 7(a), we see that WMN-PSODGA performs better when the client distribution is Normal. For Normal distribution, WMN-PSODGA can find the maximum SGC for all cases. However, for Uniform distribution, the average of best solutions does not reach the maximum. Also, in Fig. 6(b) and 7(b), the WMN-PSODGA system perfoms better for Normal distribution. We show the visualized simulation results in Fig. 8. For Normal distribution, only one mesh client is not covered. However, many mesh clients are not covered for Uniform distribution.

5 Conclusions

In this work, we evaluated the performance of WMNs using a hybrid simulation system based on PSO and DGA (called WMN-PSODGA). Simulation results show that the WMN-PSODGA has good performance for Normal distribution compared with the case of Uniform distribution.

In our future work, we would like to evaluate the performance of the proposed system for different parameters and patterns.

References

1. Akyildiz, I.F., Wang, X., Wang, W.: Wireless mesh networks: a survey. Comput. Netw. **47**(4), 445–487 (2005)
2. Barolli, A., Sakamoto, S., Ozera, K., Ikeda, M., Barolli, L., Takizawa, M.: Performance evaluation of WMNs by WMN-PSOSA simulation system considering constriction and linearly decreasing Vmax methods. In: International Conference on P2P, Parallel, Grid, Cloud and Internet Computing, pp. 111–121. Springer (2017)
3. Barolli, A., Sakamoto, S., Barolli, L., Takizawa, M.: Performance analysis of simulation system based on particle swarm optimization and distributed genetic algorithm for WMNs considering different distributions of mesh clients. In: International Conference on Innovative Mobile and Internet Services in Ubiquitous Computing, pp. 32–45. Springer (2018)
4. Barolli, A., Sakamoto, S., Ozera, K., Barolli, L., Kulla, E., Takizawa, M.: Design and implementation of a hybrid intelligent system based on particle swarm optimization and distributed genetic algorithm. In: International Conference on Emerging Internetworking, Data & Web Technologies, pp. 79–93. Springer (2018)
5. Clerc, M., Kennedy, J.: The particle swarm-explosion, stability, and convergence in a multidimensional complex space. IEEE Trans. Evol. Comput. **6**(1), 58–73 (2002)
6. Franklin, A.A., Murthy, C.S.R.: Node placement algorithm for deployment of two-tier wireless mesh networks. In: Proceedings of Global Telecommunications Conference, pp. 4823–4827 (2007)
7. Girgis, M.R., Mahmoud, T.M., Abdullatif, B.A., Rabie, A.M.: Solving the wireless mesh network design problem using genetic algorithm and simulated annealing optimization methods. Int. J. Comput. Appl. **96**(11), 1–10 (2014)
8. Goto, K., Sasaki, Y., Hara, T., Nishio, S.: Data gathering using mobile agents for reducing traffic in dense mobile wireless sensor networks. Mobile Inf. Syst. **9**(4), 295–314 (2013)

9. Inaba, T., Elmazi, D., Sakamoto, S., Oda, T., Ikeda, M., Barolli, L.: A secure-aware call admission control scheme for wireless cellular networks using fuzzy logic and its performance evaluation. J. Mobile Multimed. **11**(3&4), 213–222 (2015)
10. Inaba, T., Obukata, R., Sakamoto, S., Oda, T., Ikeda, M., Barolli, L.: Performance evaluation of a QoS-aware fuzzy-based CAC for LAN access. Int. J. Space-Based Situated Comput. **6**(4), 228–238 (2016)
11. Inaba, T., Sakamoto, S., Oda, T., Ikeda, M., Barolli, L.: A testbed for admission control in WLAN: a fuzzy approach and its performance evaluation. International Conference on Broadband and Wireless Computing, Communication and Applications, pp. 559–571. Springer (2016)
12. Lim, A., Rodrigues, B., Wang, F., Xu, Z.: k-center problems with minimum coverage. Comput. Comb., 349–359 (2004)
13. Maolin, T.: Gateways placement in backbone wireless mesh networks. Int. J. Commun. Netw. Syst. Sci. **2**(1), 44 (2009)
14. Matsuo, K., Sakamoto, S., Oda, T., Barolli, A., Ikeda, M., Barolli, L.: Performance analysis of WMNs by WMN-GA simulation system for two WMN architectures and different TCP congestion-avoidance algorithms and client distributions. Int. J. Commun. Netw. Distrib. Syst. **20**(3), 335–351 (2018)
15. Muthaiah, S.N., Rosenberg, C.P.: Single gateway placement in wireless mesh networks. In: Proceedings of 8th International IEEE Symposium on Computer Networks, pp. 4754–4759 (2008)
16. Naka, S., Genji, T., Yura, T., Fukuyama, Y.: A hybrid particle swarm optimization for distribution state estimation. IEEE Trans. Power Syst. **18**(1), 60–68 (2003)
17. Poli, R., Kennedy, J., Blackwell, T.: Particle swarm optimization. Swarm Intell. **1**(1), 33–57 (2007)
18. Sakamoto, S., Kulla, E., Oda, T., Ikeda, M., Barolli, L., Xhafa, F.: A comparison study of simulated annealing and genetic algorithm for node placement problem in wireless mesh networks. J. Mobile Multimed. **9**(1–2), 101–110 (2013)
19. Sakamoto, S., Kulla, E., Oda, T., Ikeda, M., Barolli, L., Xhafa, F.: A comparison study of hill climbing, simulated annealing and genetic algorithm for node placement problem in WMNs. J. High Speed Netw. **20**(1), 55–66 (2014)
20. Sakamoto, S., Kulla, E., Oda, T., Ikeda, M., Barolli, L., Xhafa, F.: A simulation system for WMN based on SA: performance evaluation for different instances and starting temperature values. Int. J. Space-Based Situated Comput. **4**(3–4), 209–216 (2014)
21. Sakamoto, S., Kulla, E., Oda, T., Ikeda, M., Barolli, L., Xhafa, F.: Performance evaluation considering iterations per Phase and SA temperature in WMN-SA system. Mobile Inf. Syst. **10**(3), 321–330 (2014)
22. Sakamoto, S., Lala, A., Oda, T., Kolici, V., Barolli, L., Xhafa, F.: Application of WMN-SA simulation system for node placement in wireless mesh networks: a case study for a realistic scenario. Int. J. Mobile Comput. Multimed. Commun. (IJMCMC) **6**(2), 13–21 (2014)
23. Sakamoto, S., Oda, T., Ikeda, M., Barolli, L., Xhafa, F.: An integrated simulation system considering WMN-PSO simulation system and network simulator 3. International Conference on Broadband and Wireless Computing, Communication and Applications, pp. 187–198. Springer (2016)
24. Sakamoto, S., Oda, T., Ikeda, M., Barolli, L., Xhafa, F.: Implementation and evaluation of a simulation system based on particle swarm optimisation for node placement problem in wireless mesh networks. Int. J. Commun. Netw. Distrib. Syst. **17**(1), 1–13 (2016)

25. Sakamoto, S., Oda, T., Ikeda, M., Barolli, L., Xhafa, F.: Implementation of a new replacement method in WMN-PSO simulation system and its performance evaluation. In: The 30th IEEE International Conference on Advanced Information Networking and Applications (AINA-2016), pp. 206–211 (2016). https://doi.org/10.1109/AINA.2016.42
26. Sakamoto, S., Obukata, R., Oda, T., Barolli, L., Ikeda, M., Barolli, A.: Performance analysis of two wireless mesh network architectures by WMN-SA and WMN-TS simulation systems. J. High Speed Netw. **23**(4), 311–322 (2017)
27. Sakamoto, S., Ozera, K., Barolli, A., Ikeda, M., Barolli, L., Takizawa, M.: Implementation of an intelligent hybrid simulation systems for WMNs based on particle swarm optimization and simulated annealing: performance evaluation for different replacement methods. Soft Comput., pp. 1–7 (2017)
28. Sakamoto, S., Ozera, K., Barolli, A., Ikeda, M., Barolli, L., Takizawa, M.: Performance evaluation of WMNs by WMN-PSOSA simulation system considering random inertia weight method and linearly decreasing vmax method. International Conference on Broadband and Wireless Computing, Communication and Applications, pp. 114–124. Springer (2017)
29. Sakamoto, S., Ozera, K., Ikeda, M., Barolli, L.: Implementation of intelligent hybrid systems for node placement problem in WMNs considering particle swarm optimization, hill climbing and simulated annealing. Mobile Netw. Appl., pp. 1–7 (2017)
30. Sakamoto, S., Ozera, K., Ikeda, M., Barolli, L.: Performance evaluation of WMNs by WMN-PSOSA simulation system considering constriction and linearly decreasing inertia weight methods. In: International Conference on Network-Based Information Systems, pp. 3–13. Springer (2017)
31. Sakamoto, S., Ozera, K., Oda, T., Ikeda, M., Barolli, L.: Performance evaluation of intelligent hybrid systems for node placement in wireless mesh networks: a comparison study of WMN-PSOHC and WMN-PSOSA. In: International Conference on Innovative Mobile and Internet Services in Ubiquitous Computing, pp. 16–26. Springer (2017)
32. Sakamoto, S., Ozera, K., Oda, T., Ikeda, M., Barolli, L.: Performance evaluation of wmn-psohc and wmn-pso simulation systems for node placement in wireless mesh networks: a comparison study. International Conference on Emerging Internetworking, Data & Web Technologies, pp. 64–74. Springer (2017)
33. Sakamoto, S., Ozera, K., Barolli, A., Barolli, L., Kolici, V., Takizawa, M.: Performance evaluation of WMN-PSOSA considering four different replacement methods. International Conference on Emerging Internetworking, Data & Web Technologies, pp. 51–64. Springer (2018)
34. Schutte, J.F., Groenwold, A.A.: A study of global optimization using particle swarms. J. Global Optim. **31**(1), 93–108 (2005)
35. Shi, Y.: Particle swarm optimization. IEEE Connections **2**(1), 8–13 (2004)
36. Shi, Y., Eberhart, R.C.: Parameter selection in particle swarm optimization. Evol. Program. **VII**, 591–600 (1998)
37. Vanhatupa, T., Hannikainen, M., Hamalainen, T.: Genetic algorithm to optimize node placement and configuration for WLAN planning. In: Proceedings of the 4th IEEE International Symposium on Wireless Communication Systems, pp. 612–616 (2007)
38. Wang, J., Xie, B., Cai, K., Agrawal, D.P.: Efficient mesh router placement in wireless mesh networks. In: Proceedings of IEEE Internatonal Conference on Mobile Adhoc and Sensor Systems (MASS-2007), pp. 1–9 (2007)

Evaluation of Table Type Reader for 13.56 MHz RFID System Considering Distance Between Reader and Tag

Kiyotaka Fujisaki(✉)

Fukuoka Institute of Technology, 3-30-1 Wajiro-higashi, Higashi-ku, Fukuoka 811-0295, Japan
fujisaki@fit.ac.jp

Abstract. RFID system is one of the key technology to bring the efficiency to the automatic rental of goods or the automatic adjustment of shopping. In this paper, we evaluate the basic performance of a table type RFID reader which is a key device to offer these services. Furthermore, in order to increase the communication performance, we consider the use of a parasitic element on the table type RFID reader and show its usefulness.

1 Introduction

The wireless communication is a technique which realize the long-distance communications. The progress of the radio technology enabled various wireless services and changed our life dramatically. As a result, we can use the communication services anytime and anywhere. Recently, it is used in the near field wireless communication, because it easily enables the non-contact data exchange.

One of the near field wireless communication techniques is Radio Frequency Identification (RFID) technique [3]. This technique uses electromagnetic coupling for data exchange between the reader/writer and the tag. The RFID system using this technique enables us to manage the goods efficiently in the case that a large quantity of goods is managed. For example, in the library or the rental shop, by using RFID system, we expect the following services: (1) rental of goods and the return, (2) search of the goods, (3) collection inventory, (4) access control of users [4,14]. Furthermore, if RFID system is integrated with smartphones and wireless sensor networks, the library system can trace books and users in the library. Also, the system may send useful information to users. A lot of applications using RFID are proposed by many researchers because RFID system is a simple system [8,12,15,17]. Moreover, the development of RFID device and performance evaluation using RFID system are performed to realize a reliable RFID system [1,2,5–7,9–11,13,16].

When using the tag in the library, it is not easy to change the size and the shape of tag because the tag is stuck on a book. Therefore, the improvement of the tag is not easy. On the other hand, the improvement of the reader is possible

L. Barolli et al. (Eds.): NBiS 2018, LNDECT 22, pp. 56–64, 2019.
https://doi.org/10.1007/978-3-319-98530-5_5

because there is a small limitation to the shape. In [6], as a first step of the development of the reader which is most suitable for the use in library, we evaluated the performance of table type RFID reader using 13.56 MHz. Furthermore, when RFID reader is on the metallic plate which affected the performance of RFID, we investigated the relation between the reading area rate and the distance between metallic plate and RFID reader.

In this paper, at first, we evaluate the basic performance of table type RFID reader in details. Furthermore, in order to increase the communication performance we add a parasitic element to the table type RFID reader and show its usefulness.

The structure of this paper is as follows. In Sect. 2, the RFID system is introduced and the experimental method is shown. In Sect. 3, the basic performance of table type RFID reader is evaluated in details. In Sect. 4, the effect of a parasitic element to the performance of table type RFID reader is discussed. In Sect. 5, we conclude this paper.

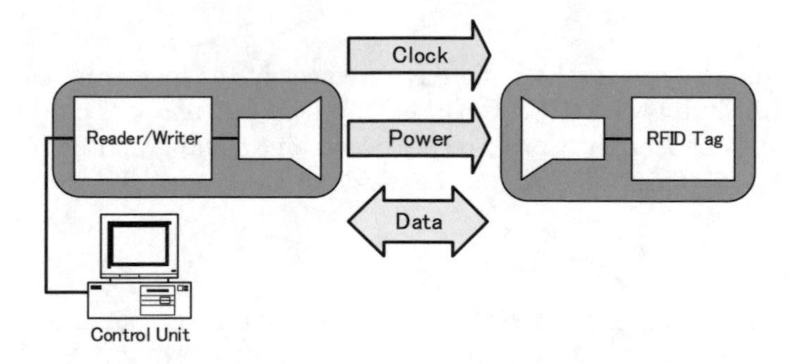

Fig. 1. Basic concept of RFID system

2 RFID System and Experimental Method

An RFID system is one of the technique used for the automatic identification in wireless communication. The automatic identification means to "automatically input bar-code, magnetic-card and RFID data with the use of hardware and software and not human intervention in order to recognise the content of the data". Also, the biometrics, OCR, the machine vision are included in this technique. Because RFID system uses the wireless communication is able to get the ID from the tag without touching the tag.

The RFID system is made up of two components as shown in Fig. 1 [3]. One is the RFID tag, which is located on the object to be identified and another is the reader/writer. The RFID tag normally does not have the power supply to work, so the reader/writer not only exchange the data, but also supply the power and clock signal to the RFID tag.

In our work, ST-RW01 produced by SOFEL was used as the table type RFID reader. The size of the housing of this reader is 25 cm × 35 cm. The probe and

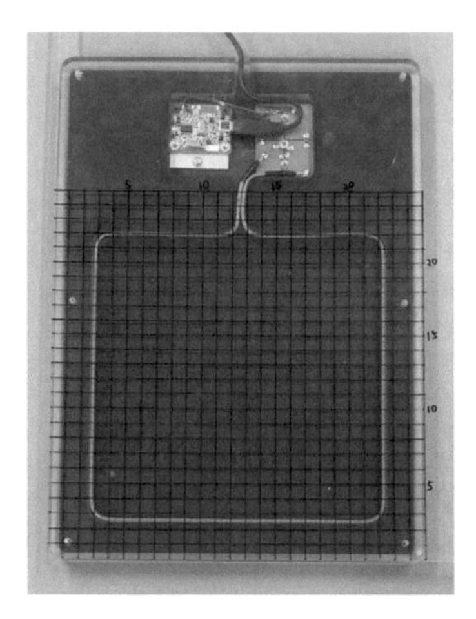

Fig. 2. Photo of table type RFID reader ST-RW01 and observation area

the spectrum analyzer used for the measurement of the magnetic field are EM-6993 of ELECTRO-METRICS Corp. having loop antenna of 1 cm in diameter and MS2601B of Anritsu Corp., respectively. The photograph of ST-RW01 is shown in Fig. 2. This photo shows the inside of the reader. This reader has an

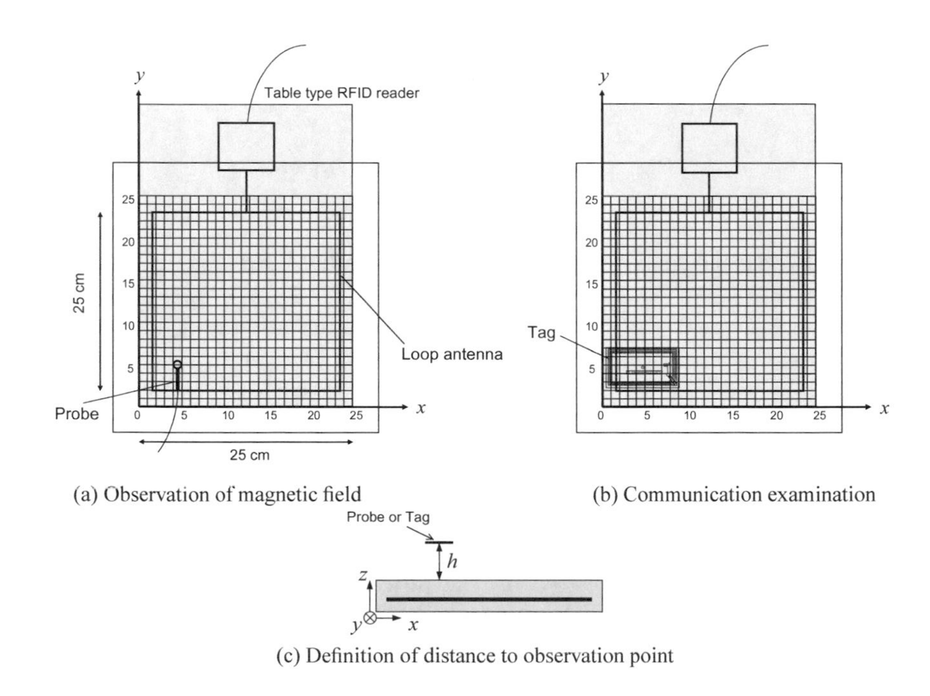

Fig. 3. Measurement setup and observation image

approximately 20 cm square loop antenna and this system communicates using electromagnetic induction. By using the reader, we carried out two experiments. One is the measurement of the power obtained from the intensity of magnetic field using a probe. Another one is the evaluation of the reading performance using a tag.

In [6], we have observed the magnetic field at each point on the reader and have shown the relation between the distance from the reader and the received power. In the experiment, in order to measure the distribution of the magnetic field, we used the loop antenna of 6 cm in diameter. On the other hand, in this paper, to evaluate the magnetic field in each point, a smaller loop antenna of 1 cm in diameter is used.

We define the axis and the distance of probe or tag from the reader as shown in Fig. 3 and the measurement is done at 1 cm intervals in 25 cm^2.

3 Basic Performance of Table Type RFID Reader

Figure 4 shows the observed power at each point and the distance between the probe and the reader. From these figures, we see that the observed power decreases as the probe height increases, but the attenuation of the power near the center of the antenna is lower than that of the other areas. To show this situation clearly, we evaluated the distribution of the observed power of the x-axis direction in the case of $y = 5$ cm and 10 cm. The results are shown in Fig. 5. When the distance between the probe and the reader is about 10 cm, compared with the results of $h = 0$ cm, the observed power near the center of antenna attenuate 10 dB–15 dB and the observed power near the edge of antenna attenuate about 20 dB. In this evaluation, we used a probe having a smaller loop antenna, but these results had the similar characteristics in [6]. As a result, because there is not a large difference between the characteristics of power distributions in these results, the magnetic field via the neighborhood of center of the probe becomes dominant.

Next, the distribution of the points where a reader was able to communicate with tag was evaluated in each communication distance between the reader and a tag. In this experiment, the coil of tag is put on parallel to the loop antenna of reader. The results are shown in Fig. 6. In these figures, the painting area means the point that the reader could communicate with a tag. As shown in these figures, communication area decreases and becomes near the center of antenna of reader as the distance between the tag and the reader increases. Table 1 shows the relation between the reading area rate and the communication distance. In this table, the reading area rate is the ratio of the number of the points which were able to communicate on the observation area to all number of observation points. In this experiment, when communication distance exceeded 14 cm, the point that could read a tag disappeared.

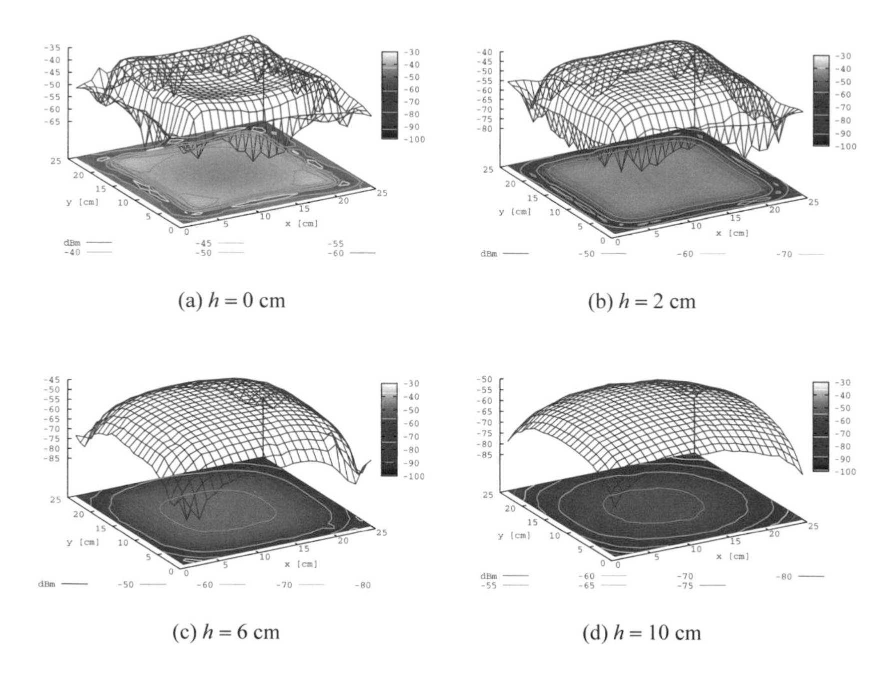

(a) $h = 0$ cm
(b) $h = 2$ cm
(c) $h = 6$ cm
(d) $h = 10$ cm

Fig. 4. Observed power at each point and distance to observation point

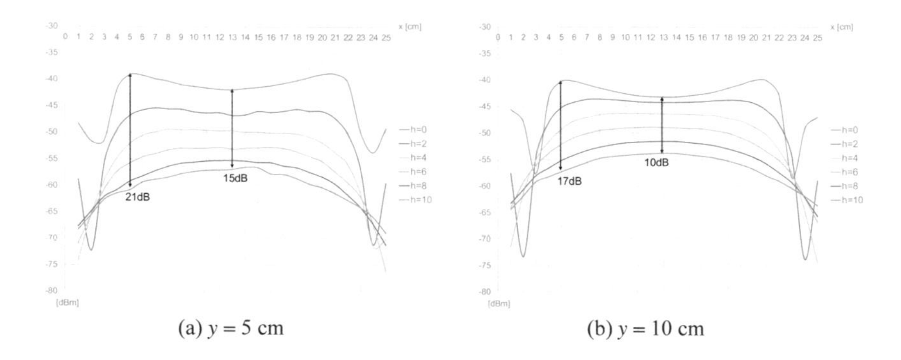

(a) $y = 5$ cm
(b) $y = 10$ cm

Fig. 5. Observed power at $y = 5\,\text{cm}$ and $10\,\text{cm}$, respectively

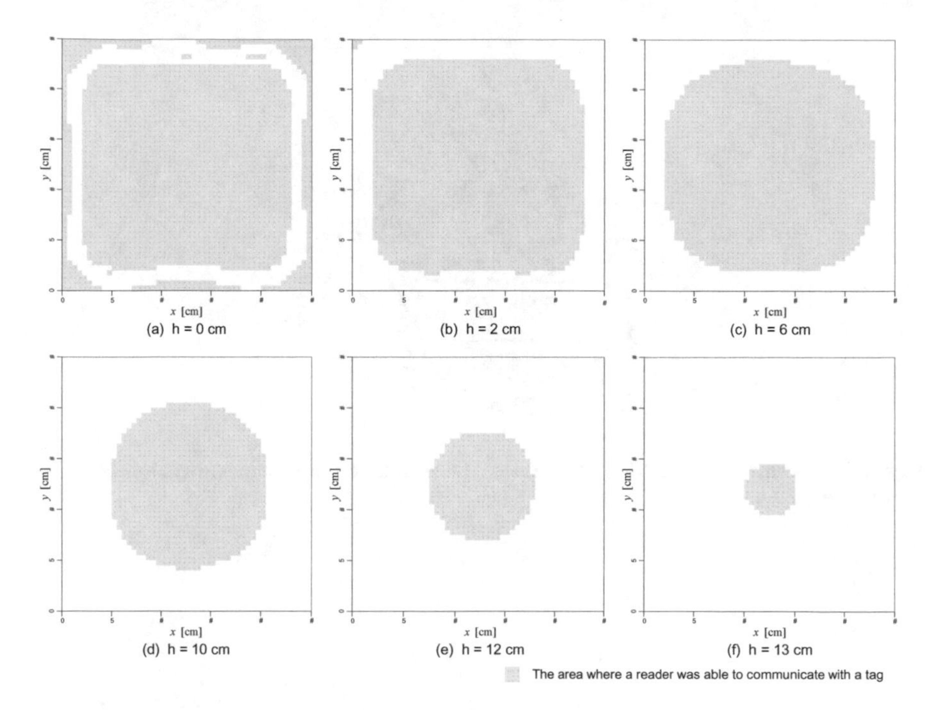

Fig. 6. Distribution of points where reader is able to communicate with tag

Table 1. Reading area rate vs communication distance between reader and tag

h [cm]	0	2	4	6	8	10	12	13	14
Reading area rate [%]	85.89	65.83	61.56	54.09	40.39	24.40	7.15	1.11	0.0

4 Effect of a Parasitic Element

In order to expand the communication distance between the table type RFID reader and the tag, we consider the use of a parasitic element on the table type RFID reader. Figure 7 shows the experimental image of the communication performance evaluation between the reader and the tag. As a parasitic element, a loop coil is placed on half position between the reader and the tag, and we evaluate the points where communication is enabled on the observation area at each communication distance. The structure of the parasitic element is 20 cm square loop coil. In this experiment, as a parasitic element, we used three kinds of coil: one turn of coil, two turns of coil, and three turns of coil.

The results are shown in Fig. 8. In these figures, the painting area means the points that the reader could communicate with a tag. The experiments are carried out three times using a certain sample tag. From Fig. 8(a)–(d), we see that the points where communication is enabled increase as the number of turns of coil increase when the communication distance is fixed.

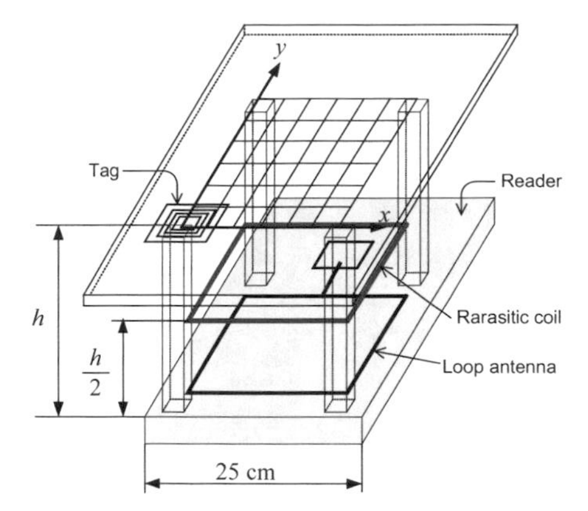

Fig. 7. Experimental image of evaluation of points where communication is enabled on observation area

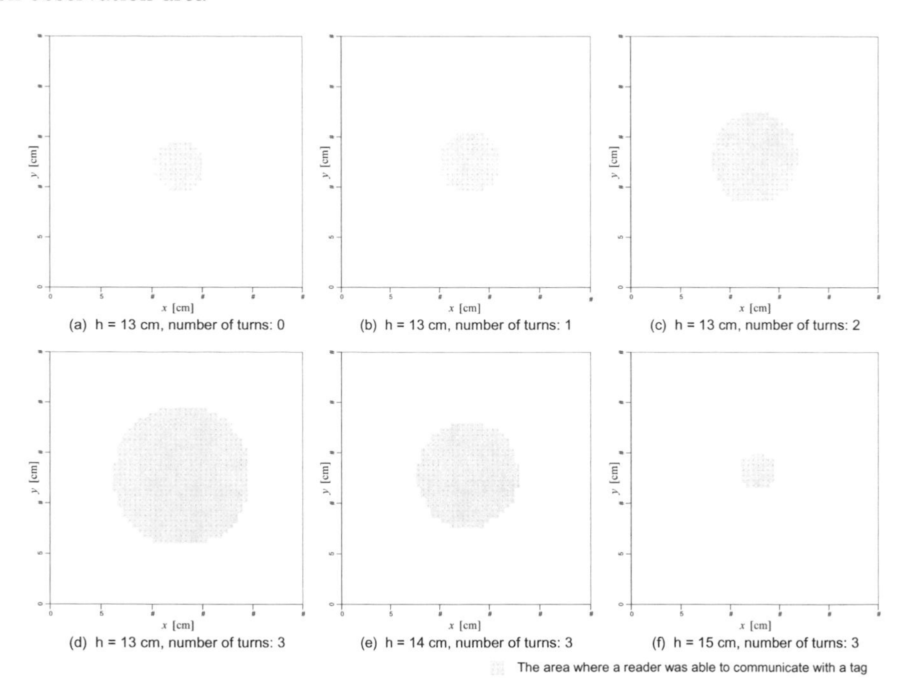

Fig. 8. Distribution of the points where reader is able to communicate with tag

Furthermore, Fig. 8(e) and (f) show the points where communication is enabled when the communication distance are 14 cm and 15 cm, respectively and the turns of coils is 3. In previous section, we reported that the point where communication is enabled with tag disappeared when communication distance exceeded 14 cm. However, when we used three turns of coil as a parasitic element, the communication length was expanded.

Finally, we show the ratio of points where the reader was able to communicate with the tag in Fig. 9. By using parasitic element, it is clearly shown that the reading area can be expanded and the communication length can be extended.

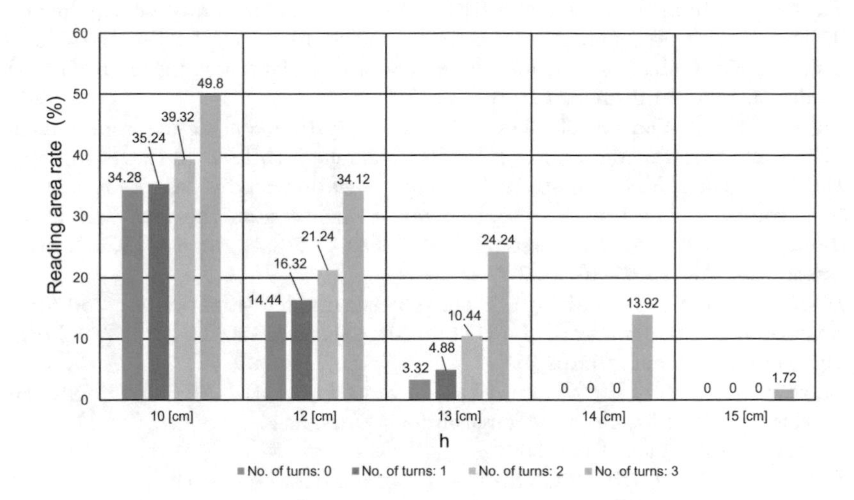

Fig. 9. Ratio of points where reader is able to communicate with tag

5 Conclusions

In this paper, we show the basic performance of table type RFID reader. In order to expand the communication performance, we consider the use of a parasitic element. By using a loop coil between the reader and the tag as a parasitic element, the points where communication is enabled with tag increased as the number of turns of coil increased. Furthermore, the parasitic element expanded the communication length.

In the future work, we want to evaluate in details the influence of parasitic element and to suggest a method to improve the performance.

Acknowledgements. I am indebted to Hiroto Iwasaki and Misaki Etoh for their assistance in experiments.

References

1. Bolomey, J.C., Capdevila, S., Jofre, L., Romeu, J.: Electromagnetic modeling of RFID-modulated scattering mechanism. In: Proceedings of IEEE, Application to Tag Performance Evaluation (2010). https://doi.org/10.1109/JPROC.2010.2053332
2. Cantatore, E., Geuns, T.C.T., Gelinck, G.H., et al. A 13.56-MHz RFID system based on organic transponders. IEEE J. Solid-State Circ. (2007). https://doi.org/10.1109/JSSC.2006.886556

3. Finkenzeller, K.: RFID Handbook. 3rd edn. Wiley (2010). https://doi.org/10.1002/9780470665121
4. Fujisaki, K.: The implementation of the RFID technology in the library, and electromagnetic compatibility. Monthly EMC **183**, 86–94 (2003). (in Japanese)
5. Fujisaki, K.: Implementation of a RFID-based system for library management. Int. J. Distrib. Syst. Technol. (2015). https://doi.org/10.4018/IJDST.2015070101
6. Fujisaki, K.: Evaluation and measurements of main features of a table type RFID reader. J. Mob. Multimed. **11**, 21–33 (2015)
7. Fujisaki, K.: Evaluation of 13.56 MHz RFID system considering communication distance between reader and tag. In: Proceedings of IMIS 2018 (2018, in press)
8. Ha, O.K., Song, Y.S., Chung, K.Y., et al.: Relation model describing the effects of introducing RFID in the supply chain: evidence from the food and beverage industry in South Korea. Personal Ubiquitous Comput. Archive (2014). https://doi.org/10.1007/s00779-013-0675-x
9. Kuoa, S.K., Hsub, J.Y., Hungb, Y.H.: A performance evaluation method for EMI sheet of metal mountable HF RFID tag. Measurement (2011). https://doi.org/10.1016/j.measurement.2011.02.018
10. Li, N., Gerber, B.B.: Performance-based evaluation of RFID-based indoor location sensing solutions for the built environment. Advanced Eng. Inf. (2011). https://doi.org/10.1016/j.aei.2011.02.004
11. Potyrailo, R.A., Morris, W.G., Sivavec, T., et al.: RFID sensors based on ubiquitous passive 13.56-MHz RFID tags and complex impedance detection, Wirel. Commun. Mob. Comput. (2009). https://doi.org/10.1002/wcm.711
12. Prasad, N.R.K., Rajesh, A.: RFID-based hospital real time patient management system. Int. J. Comput. Trends Technol. **3**, 1011–1016 (2012)
13. Basat, S.S., Kyutae, L., Laskar, J., Tentzeris, M.M.: Design and modeling of embedded 13.56 MHz RFID antennas. In: Proceedings of IEEE International Symposium on Antenna Propagation (2005). https://doi.org/10.1109/APS.2005.1552740
14. Sing, J., Brar, N., Fong, C.: The State of RFID applications in libraries. Inf. Technol. Libr. (2006). https://doi.org/10.6017/ital.v25i1.3326
15. Symonds, J., Seet, B.C., Xiong, J.: Activity inference for RFID-based assisted living applications. J. Mob. Multimed. **6**, 15–25 (2010)
16. Uysal, D.D., Gainesville, F., Emond, J., Engles, D.W.: Evaluation of RFID performance for a pharmaceutical distribution chain: HF vs. UHF. In: Proceedings of 2008 IEEE International Conference RFID (2008). https://doi.org/10.1109/RFID.2008.4519382
17. Zhonga, R.Y., Dai, Q.Y., Qu, T., et al.: RFID-enabled real-time manufacturing execution system for mass-customization production. Robot. Comput. Integr. Manuf. (2013). https://doi.org/10.1016/j.rcim.2012.08.001

A Position Detecting System Using Supersonic Sensors for Omnidirectional Wheelchair Tennis

Keita Matsuo(✉) and Leonard Barolli

Department of Information and Communication Engineering,
Fukuoka Institute of Technology (FIT),
3-30-1 Wajiro-Higashi, Higashi-Ku, Fukuoka 811-0295, Japan
{kt-matsuo,barolli}@fit.ac.jp

Abstract. The wheelchair with good performance for the aged and disabled is attracting attention from the society. Also, the wheelchair can provide the user with many benefits, such as maintaining mobility, continuing or broadening community social activities, conserving energy and enhancing quality of life. The wheelchair body must be compact enough and should be able to make different movements in order to have many applications. In our previous work, we presented the design and implementation of an omnidirectional wheelchair. In this paper, we propose a position detecting system using supersonic sensors. The proposed system can find correctly the wheelchair position for collision avoidance.

1 Introduction

Recently, the convenient facilities and equipments have been developed in order to satisfy the requirements of elderly people and disabled people. Among them, wheelchair is a common one which is used widely. A wheelchair can provide the user with many benefits, such as maintaining mobility, continuing or broadening community social activities, conserving energy and enhancing quality of life.

Because of aged tendency of population and rapid growth in the number of the disabled caused by diseases or injuries, the wheelchair with good performance for the aged and disabled is attracting attention from the society. There are many research works on wheelchairs including wheelchair for recovery, climbing stairs, playing sport and multifunction [11]. Therefore, it is necessary to design a wheelchair with the feature of easy-walking, convenient-use, and small-radius-swerving because the wheelchair is often used in a relatively narrow and small room [12,13].

The wheelchair tennis has been recognized in the world and players using wheelchair tennis are increasing in many countries. But, when playing tennis a wheelchair user is required to quickly and accurately control the wheelchair. Also, there are needed sophisticated techniques to move the wheelchair and strong physical strength for playing tennis. In particular, for the beginner who has not strong strength to move the wheelchair is very difficult to hit the ball. Thus, it is

L. Barolli et al. (Eds.): NBiS 2018, LNDECT 22, pp. 65–75, 2019.
https://doi.org/10.1007/978-3-319-98530-5_6

necessary to prepare an easy environment for tennis beginners. For this reason, we proposed an omnidirectional wheelchair for playing tennis. In this paper, we propose a position detecting system using supersonic sensors. The proposed system can find correctly the wheelchair position for collision avoidance.

The structure of this paper is as follows. In Sect. 2, we introduce the related work. In Sect. 3, we present our proposed omnidirectional wheelchair system for playing tennis. In Sect. 4, we discuss some implementation issues. In Sect. 5, we show the experimental environment and results. Finally, conclusions and future work are given in Sect. 6.

2 Related Work

Most of the work, for mobile robots has be done for improving the quality of life of disabled people. One of important research area is robotic wheelchairs. The persons having physical impairment often find it difficult to navigate the wheelchair themselves. The reduced physical function associated with the age or disability make independent living and playing sport more difficult.

Many research works have been undertaken to reduce the problem of navigation faced by the physically and mentally challenged people and also older age persons. One of the suggestive measures are the development of a Brain Control Interface (BCI), that assist an impaired person to control the wheelchair using his own brain signal. The research proposes a high-frequency SSVEP-based asynchronous BCI in order to control the navigation of a mobile object on the screen through a scenario and to reach its final destination [3]. This could help impaired people to navigate a robotic wheelchair. The BCIs are systems that allow to translate in real time the electrical activity of the brain in commands to control devices, provide communication and control for people with devastating neuromuscular disorders, such as the Amyotrophic Lateral Sclerosis (ALS), brainstem stroke, cerebral palsy, and spinal cord injury [8].

One of the key issue in designing wheelchairs is to reduce the caregiver load. Some of the research works deal with developing prototypes of robotic wheelchairs that helps the caregiver by lifting function or which can move with a caregiver side by side [10,14]. The lifting function equipment facilitates easy and safe transfer from/to a bed and a toilet stool by virtue of the opposite allocation of wheels from that for a usual wheelchair. The use of lifting function and the folding of frames makes it more useful in indoor environments. Robotic wheelchair based on observations of people using integrated sensors can move with a caregiver side by side. This is achieved by a visual-laser tracking technique, where a laser range sensor and an omnidirectional camera are integrated to observe the caregiver.

Another important issue for the design of wheelchair is the collision detection mechanism. The omnidirectional wheelchairs with collaborative controls ensures better safety against collisions. Such wheelchairs possess high level of ability when moving over a step, through a gap or over a slope [2,9]. To achieve omnidirectional motion, vehicles are generally equipped with an omniwheel consisting of a large number of free rollers or a spherical ball wheel. The development of such omniwheels attempts to replace the conventional wheel-type mechanism.

There are also other works which deal with vision design of robotic wheelchairs by equipping the wheelchair with camera for monitoring wheelchair movement and obstacle detection and pupil with gaze sensing [1,7]. Prototype for robotic wheelchairs have been suggested in various research works, which are exclusively controlled by eye and are used by different users, while proving robust against vibration, illumination change, and user movement [4,6].

To enable older person to communicate with other people the assisting devices have been developed. They can improve the quality of life for the elderly and disabled people by using robotic wheelchairs. The head gesture recognition is performed by means of real time face detection and tracking techniques. They developed a useful human-robot interface for RoboChair [16].

Also, the application of detecting an obstacle using supersonic sensors has been an active research area in robotics. One of the applications is the detecting system for approaching vehicles using supersonic sensors [5]. In [15] is presented a method of simultaneous and velocity measurement using supersonic sensors.

3 Proposed Omnidirectional Wheelchair for Playing Tennis

In this section, we describe the implementation of wheelchair for playing tennis. We show a conventional wheelchair in Fig. 1. The wheelchair should make 5 movements. This is only one example of using the wheelchair, but when the wheelchair is used for playing tennis is difficult to make movements.

In order to deal with these problems, we propose an omnidirectional wheelchair as shown in Fig. 2. The implemented omnidirectional wheelchair and

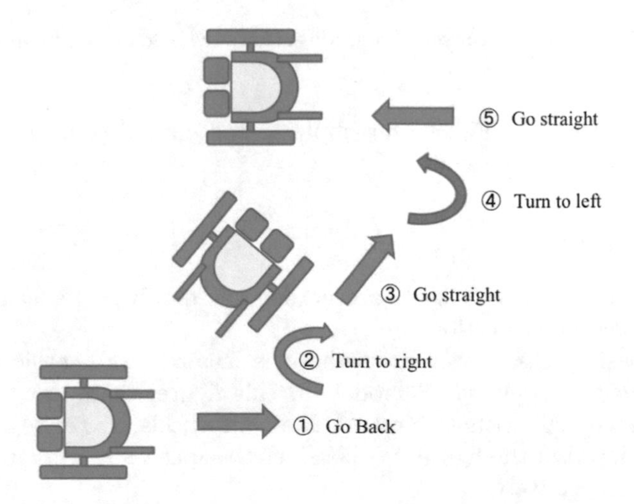

Fig. 1. Moving of conventional wheelchair.

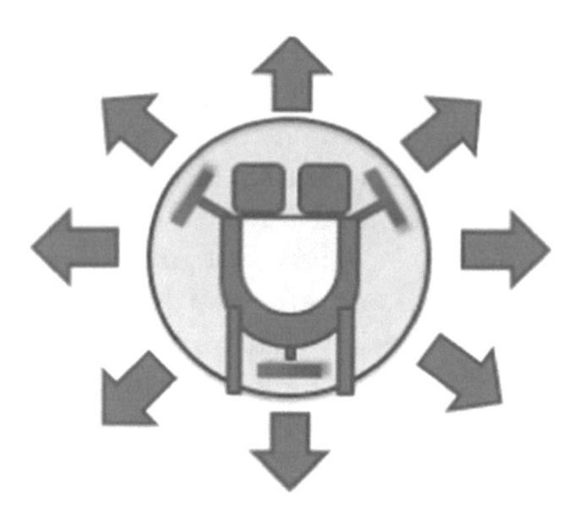

Fig. 2. Moving of our proposed wheelchair.

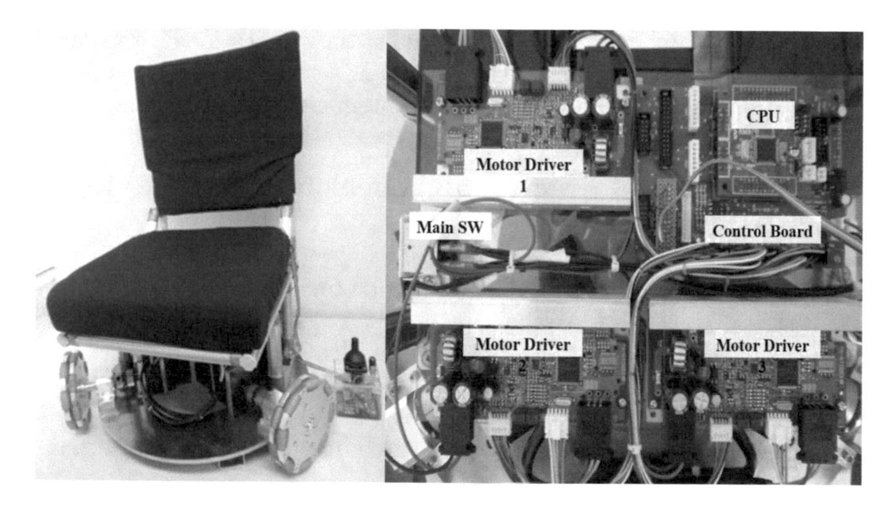

Fig. 3. Real implementation of omnidirectional wheelchair and control unit.

control unit is shown in Fig. 3. Our implemented omnidirectional wheelchair is very suitable for playing tennis.

3.1 Kinematics

For the control of the wheelchair are needed the omniwheel speed, wheelchair movement speed and direction.

Let us consider the movement of the wheelchair in 2 dimensional space. In Fig. 4, we show the omniwheel model. In this figure, there are 3 omniwheels which are placed 120° with each other. The omniwheels are moving in clockwise direction as shown in the figure. We consider the speed for each omniwheel M1, M2 and M3, respectively.

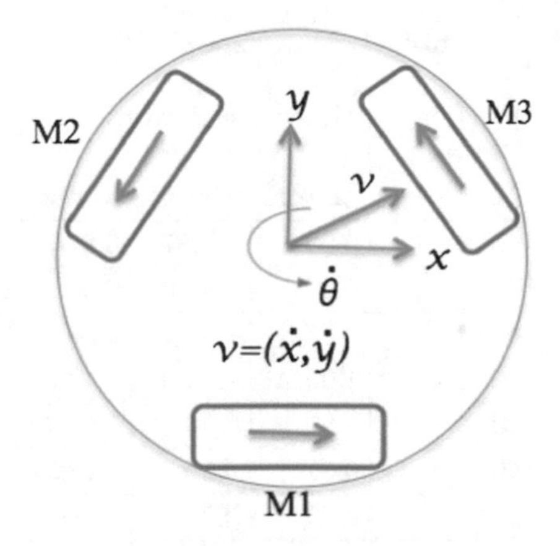

Fig. 4. Model of omniwheel.

As shown in Fig. 4, the axis of the wheelchair are x and y and the speed is $v = (\dot{x}, \dot{y})$ and the rotating speed is $\dot{\theta}$. In this case, the moving speed of the wheelchair can be expressed by Eq. (1).

$$V = (\dot{x}, \dot{y}, \dot{\theta}) \tag{1}$$

Based on the Eq. (1), the speed of each omniwheel can be decided. By considering the control value of the motor speed ratio of each omniwheel as linear and synthesising the vector speed of 3 omniwheels, we can get Eq. (2) by using Reverse Kinematics, where (d) is the distance between the center and the omniwheels. Then, from the rotating speed of each omniwheel based on Forward Kinematics, we get the wheelchair moving speed. If we calculate the inverse matrix of Eq. (2), we get Eq. (3). Thus, when the wheelchair moves in all directions (omnidirectional movement), the speed for each motor (theoretically) is calculated as shown in Table 1.

$$\begin{vmatrix} M_1 \\ \\ M_2 \\ \\ M_3 \end{vmatrix} = \begin{vmatrix} 1 & 0 & d \\ \\ -\frac{1}{2} & -\frac{\sqrt{3}}{2} & d \\ \\ -\frac{1}{2} & \frac{\sqrt{3}}{2} & d \end{vmatrix} \begin{vmatrix} \dot{x} \\ \\ \dot{y} \\ \\ \dot{\theta} \end{vmatrix} \tag{2}$$

$$\begin{vmatrix} \dot{x} \\ \\ \dot{y} \\ \\ \dot{\theta} \end{vmatrix} = \begin{vmatrix} \frac{2}{3} & -\frac{1}{3} & -\frac{1}{3} \\ \\ 0 & -\frac{1}{\sqrt{3}} & \frac{1}{\sqrt{3}} \\ \\ \frac{1}{3d} & \frac{1}{3d} & \frac{1}{3d} \end{vmatrix} \begin{vmatrix} M_1 \\ \\ M_2 \\ \\ M_3 \end{vmatrix} \tag{3}$$

Table 1. Motor speed ratio.

Direction (Degrees)	Motor speed ratio		
	Motor1	Motor2	Motor3
0	0.00	−0.87	0.87
30	0.50	−1.00	0.50
60	0.87	−0.87	0.00
90	1.00	−0.50	−0.50
120	0.87	0.00	−0.87
150	0.50	0.50	−1.00
180	0.00	0.87	-0.87
210	−0.50	−1.00	−0.50
240	−0.87	0.87	0.00
270	−1.00	0.50	0.50
300	−0.87	0.00	0.87
330	−0.50	−0.50	1.00
360	0.00	−0.87	0.87

3.2 Control System of the Proposed Omnidirectional Wheelchair

For the control of the proposed omnidirectional wheelchair, we considered R8C38 CPU board from Renesas Electronics Corporation. This CPU board has a small size and high speed processing time. The core of the CPU has a maximum frequency of 20 MHz. It is equipped with a flash memory, which is easy to rewrite. The R8C38 board has the following features:

- 8 bit multi functions timer: 2,
- 16 bit output competition timer: 5,
- Real time clock timer: 1,
- UART/clock synchronization type serial interface: 3 channels,
- 10 bit A/D converter: 20 channels,
- 8 bit D/A converter: 2 circuits,
- Voltage detected circuit,
- Number of output and input port: 75,
- External interrupt input: 9.

In Fig. 5 is shown the control system for the proposed omnidirectional wheelchair. The direction movement of the wheelchair is decided by the Joystick. The Analog-Digital Converter changes the analog value to a digital value needed for R8C38 board. The R8C38 board based on the Eq. (2) calculates the motors control value. Based on this value, the Pulse Width Modulation (PWM) generator generates an appropriate value for the control of each motor. The number of rotation of each motor is detected by Pulse Counter and is sent to the R8C38 board in order to make a correct feedback control.

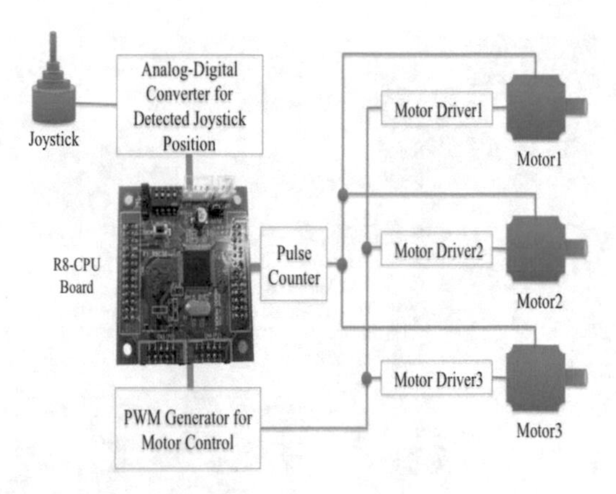

Fig. 5. Control system for omnidirectional wheelchair.

4 Implementation and Application Issues

In Fig. 6, we show a schematic illustration of position detecting system using supersonic sensors for omnidirectional wheelchair tennis. In order to realize the position detecting system, we implemented a testbed as shown in Fig. 7. In our proposed omnidirectional wheelchair tennis, we would like to avoid collision during doubles games. So, the proposed position detecting system should avoid collision between wheelchairs.

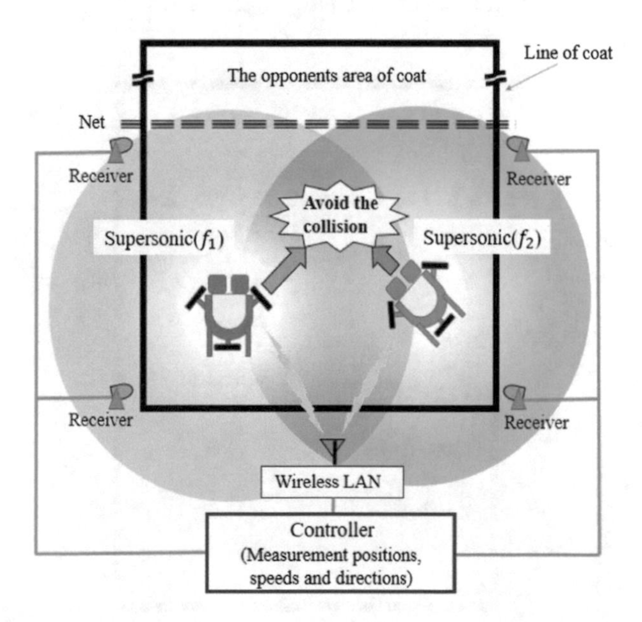

Fig. 6. Position detecting system using supersonic sensors for omnidirectional wheelchair tennis.

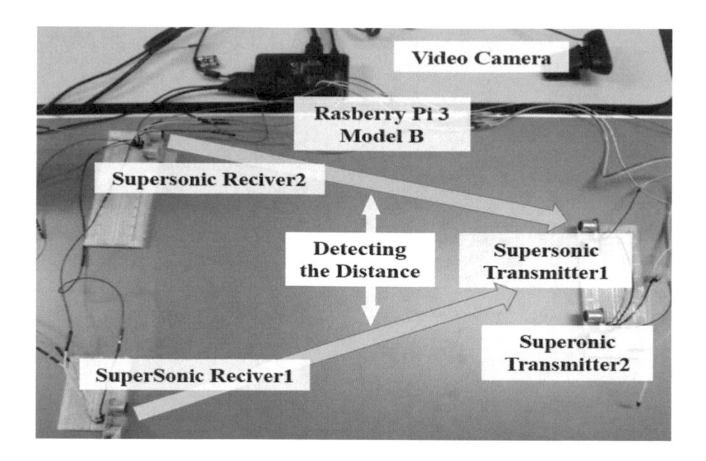

Fig. 7. Implemented testbed of position detecting system with supersonic sensors.

Firstly, we carried out position detecting in a 2D space by supersonic sensors. We used 2 supersonic receivers for detecting the distance from the supersonic receivers to wheelchair (supersonic transmitter). In order to detect 2 omnidirectional wheelchairs, we need to install the transmitter on the omnidirectional wheelchair. Each wheelchair (transmitter) is emitting the signal alternately. The controller can get the difference of receiving signal time from each transmitter and it can calculate the distance between transmitter and receiver. Thus, the position of omnidirectional wheelchair can be decided.

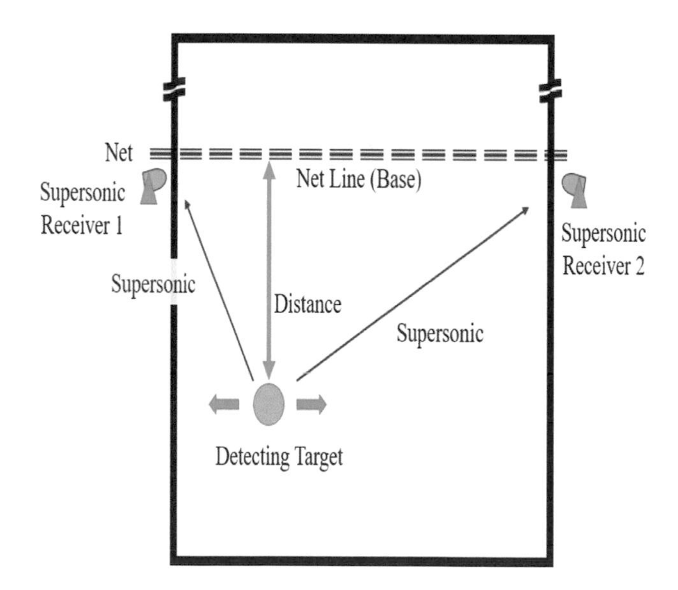

Fig. 8. Experimental environment using testbed.

5 Experimental Environment and Results

We have shown the experimental environment using testbed in Fig. 8. In Fig. 8, we consider as a base line the net line and the distance from base to detecting target is calculated. The Detecting Target sends the supersonic signal to Receiver1 and Receiver2 every 0.1 s. When each receiver get the supersonic signal, it immediately calculates the distance using Heron's formula. We carried out experiments using the implemented testbed when the distance was considered 100 cm and 120 cm. The experimental results are shown In Fig. 9. The results of experiments are almost correctly, considering these two distances. We would like to control the omnidirectional wheelchair to avoid collision by using the target position data.

Now, our implemented testbed can measure the distance with supersonic sensors for short length. This is because we are using cheap supersonic sensors

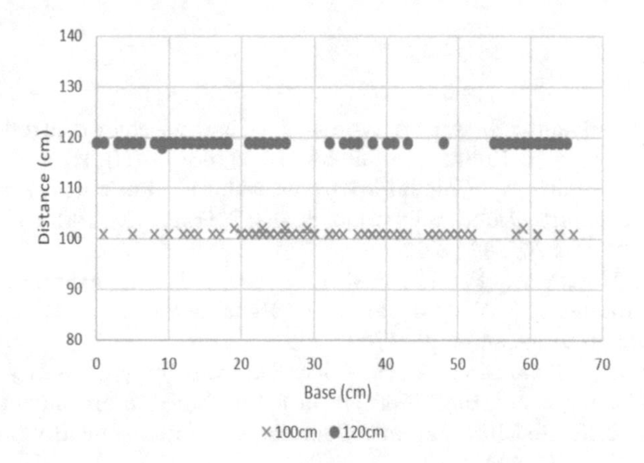

Fig. 9. Experiment results.

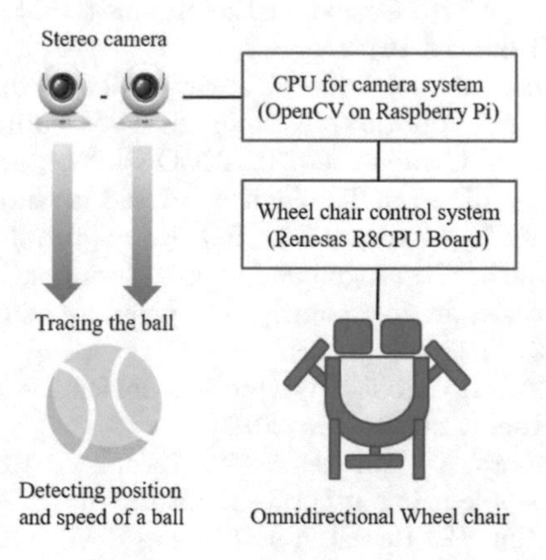

Fig. 10. Tennis ball and opponent detecting system.

for the testbed. However, the proposed system could correctly detect the target position.

6 Conclusions and Future Work

In this paper, we introduced our implemented omnidirectional wheelchair for tennis. We showed some of the previous works and discussed the related problems and issues. Then, we presented in details the kinematics and the control system for our implement omnidirectional wheelchair. We proposed a position detecting testbed using supersonic sensors. The testbed could correctly detect the target position for avoiding collision.

In the future work, we want to implement an automatic control (see Fig. 10) in order that the system not only avoid the collision, but the player can move to an accurately position for the shot.

References

1. Arai, K., Mardiyanto, R.: A prototype of electric wheelchair controlled by eye-only for paralyzed user. J. Robot. Mechatron. **23**(1), 66–74 (2011)
2. Carlson, T., Demiris, Y.: Collaborative control for a robotic wheelchair: evaluation of performance, attention, and workload. IEEE Trans. Syst. Man Cybern. Part B (Cybern.) **42**(3), 876–888 (2012)
3. Diez, P.F., Mut, V.A., Perona, E.M.A., Leber, E.L.: Asynchronous BCI control using high-frequency SSVEP. J. Neuroeng. Rehabil. **8**(1), 1–8 (2011). http://www.jneuroengrehab.com/content/8/1/39
4. Escobedo, A., Spalanzani, A., Laugier, C.: Multimodal control of a robotic wheelchair: using contextual information for usability improvement. In: Proceedings of the IEEE/RSJ International Conference on Intelligent Robots and Systems 2013, pp. 4262–4267 (2013)
5. Furuno, Y., Tanaka, Y., Nakatoh, Y.: Detecting system of approaching vehicles using an ultrasonic wave. In: Consumer Electronics (ICCE) on IEEE International Conference, pp. 393–394 (2016)
6. Gonzalez, J., Muaeoz, A., Galindo, C., Fernandez-Madrigal, J., Blanco, J.: A description of the SENA robotic wheelchair. In: Proceedings of the IEEE Mediterranean Electrotechnical Conference (MELECON-2006), pp. 437–440 (2006)
7. Gray, J., Jia, P., Hu, H.H., Lu, T., Yuan, K.: Head gesture recognition for hands-free control of an intelligent wheelchair. Ind. Robot Int. J. **34**(1), 60–68 (2007)
8. Grigorescu, S.M., Lüth, T., Fragkopoulos, C., Cyriacks, M., Gräser, A.: A BCI-controlled robotic assistant for quadriplegic people in domestic and professional life. Robotica **30**(03), 419–431 (2012)
9. Ishida, S., Miyamoto, H.: Collision-detecting device for omnidirectional electric wheelchair. ISRN Robot. **2013**, 1–8 (2012)
10. Kobayashi, Y., Kinpara, Y., Shibusawa, T., Kuno, Y.: Robotic wheelchair based on observations of people using integrated sensors. In: IROS, pp. 2013–2018 (2009)
11. Lu, T., Yuan, K., Zhu, H.: Research status and development trend of intelligent wheelchair. Appl. Technol. Robot **2**, 1–5 (2008)

12. Matsuo, K., Barolli, L.: Design and implementation of an omnidirectional wheelchair: control system and its applications. In: Proceedings of the 9th International Conference on Broadband and Wireless Computing, Communication and Applications (BWCCA-2014), pp. 532–535 (2014)
13. Matsuo, K., Liu, Y., Elmazi, D., Barolli, L., Uchida, K.: Implementation and evaluation of a small size omnidirectional wheelchair. In: Proceedings of the IEEE 29th International Conference on Advanced Information Networking and Applications Workshops (WAINA-2015), pp. 49–53 (2015)
14. Mori, Y., Sakai, N., Katsumura, K.: Development of a wheelchair with a lifting function. Adv. Mech. Eng. **2012**, 1–9 (2012)
15. Thong-un, N., Hirata, S., Orino, Y., Kurosawa, M.K.: A linearization-based method of simultaneous position and velocity measurement using ultrasonic waves. Sens. Actuators A Phys. **233**, 480–499 (2015)
16. Wang, H., Grindle, G.G., Candiotti, J., Chung, C., Shino, M., Houston, E., Cooper, R.A.: The personal mobility and manipulation appliance (PerMMA): a robotic wheelchair with advanced mobility and manipulation. In: 2012 Annual International Conference of the IEEE Engineering in Medicine and Biology Society, pp. 3324–3327 (2012)

Distributed Approach for Detecting Collusive Interest Flooding Attack on Named Data Networking

Tetsuya Shigeyasu(✉) and Ayaka Sonoda

Prefectural University of Hiroshima, Hiroshima, Japan
sigeyasu@pu-hiroshima.ac.jp

Abstract. Recently, network consumers use Internet for getting contents: videos, musics, photos, and other contents created by many producers. Those contents accelerate the increasing traffic volumes. For reducing the increasing traffic volume to keep stabilities of broadband network, realizing the concept of CCN (Contents Centric Networking) is strongly required. The NDN (Named Data Networking) which is the most popular network architecture have been proposed to realize the concept of CCN. However, it have been also reported that the NDN is vulnerable to CIFA (Collusive Interest Flooding Attack). In this paper, we propose a novel distributed algorithm for detecting CIFA for keep availabilities of NDN. The results of computer simulations confirm that our proposal can detect and mitigate the effects of CIFA, effectively.

1 Introduction

Recently, network consumers use Internet for getting contents: videos, musics, photos, and other contents created by many producers. Those contents accelerate the increasing traffic volumes. For reducing the increasing traffic volume to keep stabilities of broadband network, realizing the concept of CCN (Contents Centric Networking) [1] is strongly required. The NDN (Named Data Networking) [2–7] which is the most popular network architecture have been proposed to realize the concept of CCN. However, it have been also reported that the NDN is vulnerable to security attacks. On the traditional networks, attacks called Dos (Denial of Service) is well-known security attack which disables network services by sending large amount of invalid packets to the targeted servers. As similar as the DoS attack, it has been reported that the IFA (Interest Flooding Attack) [11] may invalidates the contents deliveries on NDN. IFA is an attack which sends large amount of Interest packet consists of unreal contents name, to NDN. Naturally, these Interests does not hit the cache/contents on CR (Contents Router)/ Original server, then, entries on PIT (Pending Interest Table) will over flows and valid contents requested by legitimate users will not be returned. The IFA attack, however, can be prevented by predicting the data returning ratio for each Interest. Data returning ratio for the Interests from malicious users, becomes

L. Barolli et al. (Eds.): NBiS 2018, LNDECT 22, pp. 76–86, 2019.
https://doi.org/10.1007/978-3-319-98530-5_7

0 cause the invalid Interest does not have a contents name of existing content in the NDN.

By the way, it has been also reported that another version of newly IFA called CIFA (Collusive Interest Flooding Attack) [10,11] may wreak enormous damage to NDN. As against IFA, in case of the CIFA, malicious users partner up with the malicious servers in the same network. The malicious users send invalid Interest with content name for invalidating the PIT entry on valid CRs. Although there is no content corresponding to the invalid Interest, malicious server respond with the invalid Interest with its content consisting of random bits. Hence, it is hard to detect simply, by calculating the data returning ratio of each users. By CIFA, overflows both of cache on CS (Contents Store) and PIT entries, significantly degrades performance of NDN.

In this paper, we propose a novel distributed algorithm for detecting CIFA for keep availabilities of NDN. Our proposal detects illegal attacks by CIFA by deriving count for cache reference on each CR, instead of data returning ratio for typical IFA. The results of computer simulations confirm that our proposal can detect and mitigate the effects of CIFA, effectively.

2 Named Data Networking

2.1 Packet Formats on NDN

Named Data networking is one of network architecture realizing the concept of CCN. NDN uses two types of packets named Interest and Data for its content acquisition phase. Packet formats of the Interest and Data are shown in Fig. 1.

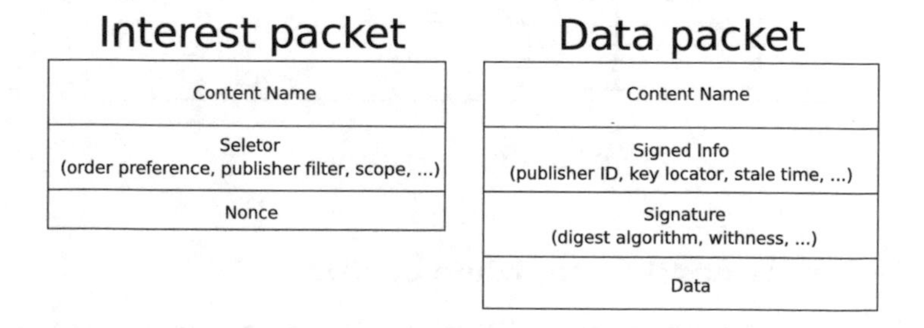

Fig. 1. Frame formats of Interest and Data in NDN.

2.2 Structure of CR

Figure 2 shows the structure of CR on NDN. Each CR on NDN consists of three tables for realizing NDN content acquisition.

- CS CS stores copy of contents traversing over the each CR. Stored copy is treated as cache and it will be returned instead of content from original server. By increasing ratio of cache utilization, response time for contents acquisition becomes low on NDN.

- PIT PIT stores the entry which indicates the detailed information of the content request. The PIT entry consists of interface number and contents name, those indicate the arrival number of interface and name of requested content, respectively.
- FIB (Forwarding Information Table) FIB consists of entry indicating next hop address for arrival Interest. The FIB entry consists of interface number and content name. When the new Interest arrived at a CR, the CR forwards the Interest to the specified number of interface according to the FIB entry corresponding to the content name of the Interest.

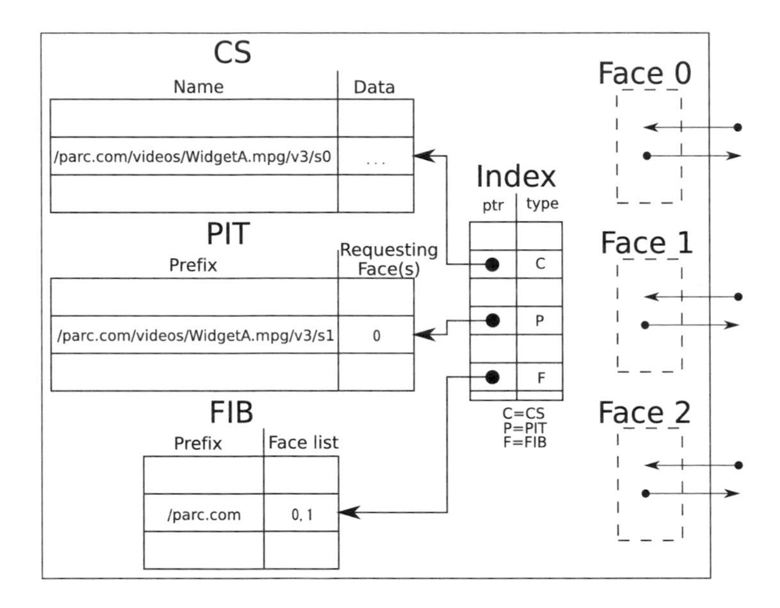

Fig. 2. Structure of CR on NDN

2.3 Content Acquisition Procedure of NDN

Figure 3 shows overview of NDN. When a CR received an Interest, the CR checks if the corresponding content is in its CS or not. If there is corresponding cache, the CR simply returns it to the interface where the Interest has arrived. Otherwise, the CR checks the corresponding entry in PIT. If the entry exists, the CR adds the interface number to the entry, and discards the Interest without forwarding. If the corresponding entry exists, the CR forwards the Interest according to the interface number of the entry. The Interest continues to be forwarded until it arrive at the CR/servers having corresponding cache/content. When the CR/server having the corresponding cache received the Interest, the CR/server returns the cache/content according to the PIT entry. The PIT entry removed after returning the cache/content. CRs relaying the returning content, caches the content in its CS.

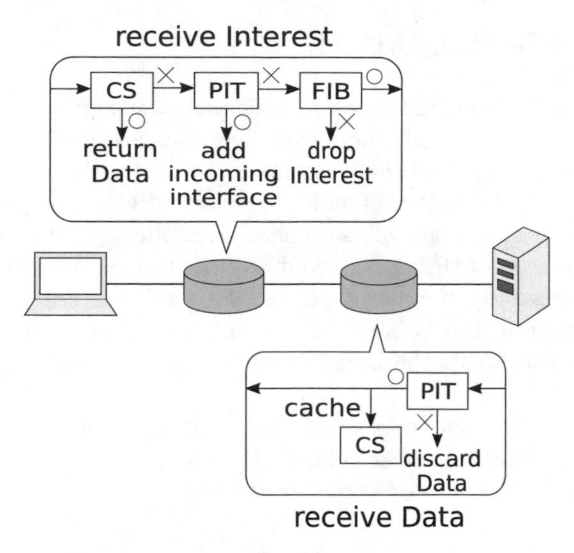

Fig. 3. Content acquisition procedure of NDN

The cached contents will be used for responding to the Interest from the other users request. This mechanism is called in-network caching, and the cache hit ratio for the in-network caching much influences the performance of NDN.

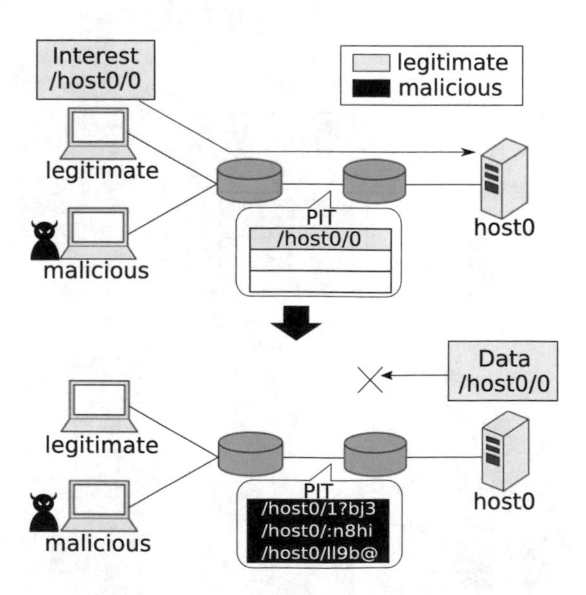

Fig. 4. Overview of IFA attack on NDN

3 Interest Flooding Attack

By sending a bunch of invalid Interests, malicious attackers based on IFA disables a content acquisition on NDN. On relay CRs, entries of PIT drop due to the arrivals of Interests from malicious attackers.

For the IFA attacks, Interests will be forwarded with the given invalid content name. Where, the name will be generated randomly without care of actual existence. The reason of that is, to avoid cache hit of the invalid Interest on the relay CRs for increasing number of PIT entries for fast overflows. If the Interest hit the cache of a CR, the Interest will not be forwarded to any more upstream CRs. Hence, in such case, the invalid Interest can not increase PIT entries of those upstream CRs.

Figure 4 shows the overview of overflows on PIT entries by IFA. As this figure shows, legitimate Interest colored as grey will be dropped by incoming large number of invalid Interests colored as black.

4 Collusive Interest Flooding Attack

As described as the above, IFA overflows the PIT entries on relay CRs, and this disables the contents request on NDN. The IFA, however, have been reported

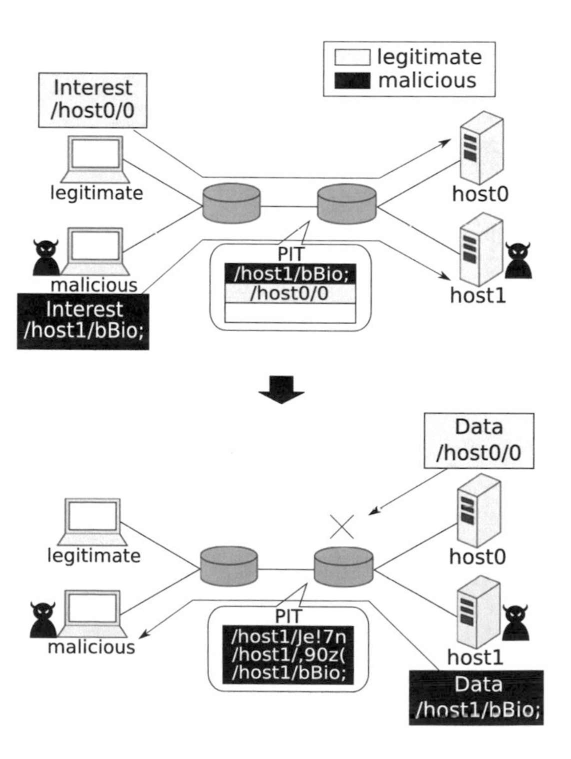

Fig. 5. Overview of CIFA attack on NDN

that this type of security attack is easy to deal. Literatures have proposed the methods destined for IFA employing the procedure calculating returning ratio of Interest for each prefix of content name.

However, recently, it has been reported that Collusive Interest Flooding Attack (CIFA) may degrades performance of content acquisition on NDN, as newly threat. In case of the CIFA, malicious attackers attack the NDN system together with the malicious servers returning fake contents with regarding to the invalid Interest from malicious attackers.

In the CIFA, malicious attackers send the Invalid Interest as same as IFA, however, when the Interest have arrived at the malicious servers, the server return the contents with fake Interest. Hence, the relay CR among malicious attackers and malicious servers can not detect the CIFA attack by calculating the returning ratio of each Interest.

Figure 5 shows the overview of overflows on PIT entries by IFA. As this figure shows, legitimate Interest colored as grey will be dropped by incoming large number of invalid Interests colored as black. However, in CIFA, all contents corresponding to the invalid Interests will be returned unlike the case of IFA.

5 Detection and Prevention Method for CIFA

5.1 Overview of Poposal

In this paper, we propose a new detection and prevention method for CIFA. Our proposal in this section, detects the security attacks based on CIFA by calculating the number of cache reference on relay CRs, and discards the incoming Interests having content name which detect to be used as the security attacks.

For more concretely, to detect the invalid Interests, our proposal employs the three stages detection: PIT overflow state, Incoming Interest rate and number of cache reference.

Figure 6 shows a procedures to detect security attacks based on CIFA. The procedure shown in the figure, is implemented in all relay CRs on our proposal. As shown in the figure, a relay CR firstly checks if the its PIT is in overflow state and entries of the PIT are replaced frequently. This inspection detects the actual damages by security attacks. The reason for that is if there are overflows of PIT entries ignores the content return of contents requested by legitimate users.

If the overflow state on PIT is detected by the previous inspection, a relay CR checks the incoming Interest rate for each prefix of content name, next. In the case of security attacks based on CIFA, malicious attackers use prefix of the content name for their sending Interest in order to cooperate with their malicious servers. The reason for this inspection is that without using the same prefix of the content name, invalid Interests can not reach at the malicious server. Hence, with those invalid Interest could not serve to overflow the PIT entries on intermediate relay CRs.

For checking the incoming Interest rate, our proposal uses α for the threshold. With regard to the value of the α will be given by operator in consideration of bandwidth, delay of each link.

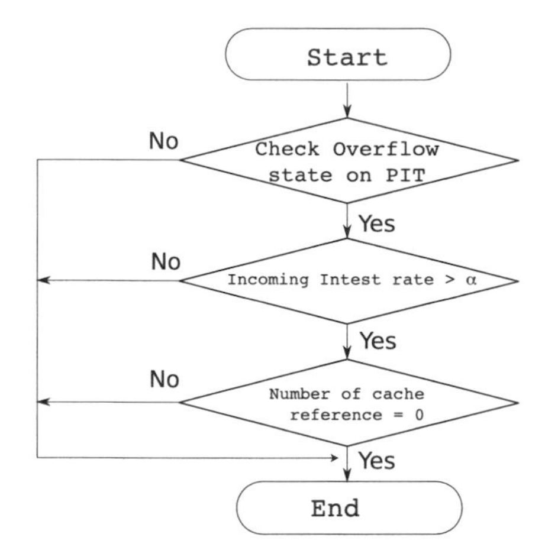

Fig. 6. A prediction procedure for Collusive IFA.

When the incoming rate of Interest is recognized to exceed the given threshold α, our proposal checks the number of cache reference for the prefix of content name. As the previous section described that, attackers based on the IFA and CIFA will never request content having same content name with its previous request. Therefore, number of cache reference becomes exceedingly small or zero. Then, if so, our proposal detects as the prefix of the content name is used as malicious attacks based on CIFA.

6 Performance Evaluations

6.1 Simulation Settings

This section, we report the results of the computer simulations for clarifying the availabilities of our proposal. In the simulations, we evaluate the value of ratio of contents acquisition.

Simulation topology and the parameters those used for the evaluations are shown in Fig. 7 and Table 1, respectively. The node marked as "0" in the network, is a legitimate Provider which has original $n+1$ contents: "$/host0/0$" $\sim$ "$/host0/n$". The node marked as "1" is a malicious server in cooperate with the attackers based on CIFA. The node 1 returns fake contents as a response to the received invalid content requests. The nodes marked as "8" and "10" are the legitimate users. These users request $n+1$ contents, named, "$/host0/0$" $\sim$"$/host0/n$" with random probabilities according to the Zip'f law [12]. The nodes marked as "7", "9", "11" are the security attackers (malicious user) based on the CIFA. These attackers transmit invalid Interest having content name consist of the prefix "$/host1/$" and randomly generated postfix.

Interest arrival rate of each legitimate user and malicious user are 10 [pkt/sec] and 320-3,020 [pkt/sec], respectively. Algorithm for entry replacement for CS and

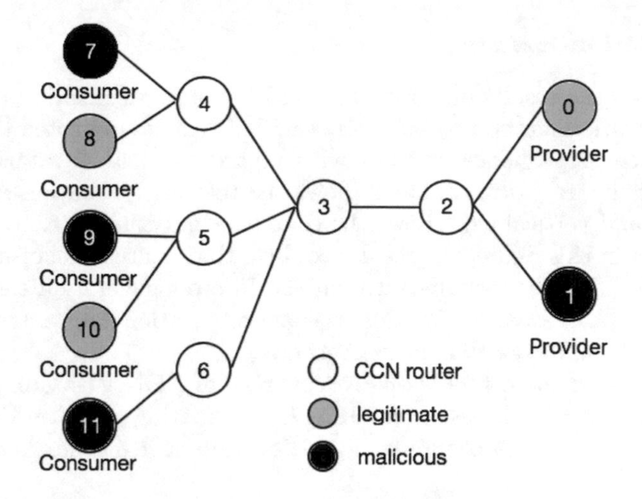

Fig. 7. Simulation topology.

Table 1. Simulation conditions.

Parameter	Value
Legitimate provider ID	0
Malicious provider ID	1
Legitimate consumer ID	8, 10
Malicious consumer ID	7, 9, 11
NDN router ID	2, 3, 4, 5, 6
Interest generation rate (legitimate)	10 [pkt/sec]
Interest generation rate (malicious)	320–3020 [pkt/sec]
Interest Packet	1024 [bytes]
Data Packet	1024 [bytes]
CS size	10
PIT size	50
link rate	1 [Mbps]
Link delay	1 [msec]
α	110
Simulation period	100 [sec]

PIT are FIFO (First-In-First-Out). In addition, prediction procedure for CIFA will be executed in a every one second. And threshold α is set as 110 [pkt] due to the link bandwidth[1]

[1] According to the packet size and link bandwidth, each communication link can carries up to 100 packets during one second.

6.2 Simulation Results

Figure 8 shows the results of characteristics of content acquisition rate of NDN without mitigation method by our proposal. This results indicated that the general resistance performance of NDN with regard to security attacks based on CIFA. In this figure, horizontal axis shows the Interest generation rate on malicious users, and vertical axis shows the content acquisition rate.

As shown in this figure, the results confirm that content acquisition rates of both users are degraded significantly on the Interest generation rate is around 1,020 [pkt/sec]. The reason for this is that information for content returning paths are lost due to the PIT entry overflows.

Figure 9 shows the contents acquisition rate of NDN with our proposal for detecting security attack based on CIFA. By comparing with the Figs. 8 and 9, it can be confirmed that contents acquisition rate of legitimate users on NDN

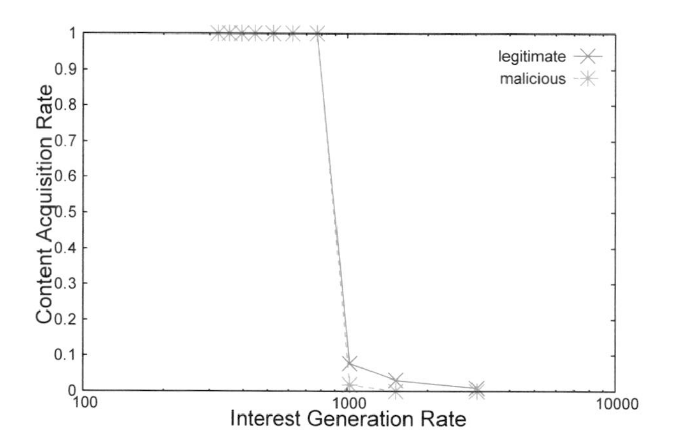

Fig. 8. Content acquisition rate without mitigation method.

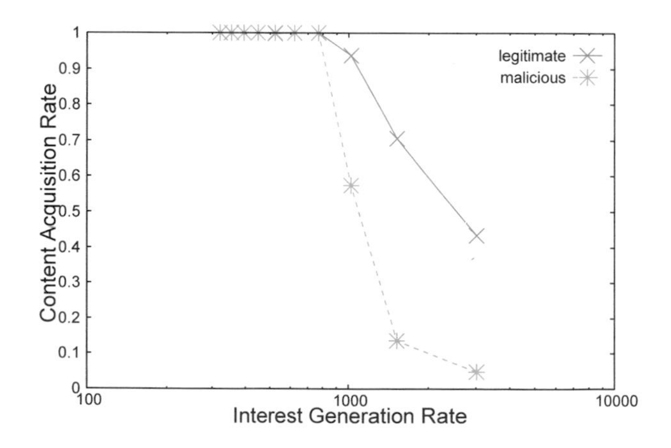

Fig. 9. Content acquisition rate.

with our proposal is improves significantly while keeping low rate of malicious users.

Hence, we can be thought that our proposal effectively works for the CIFA even though the security attacks based on CIFA is hard to detect compared with typical IFA on NDN. Our proposal, however, can not improve the content acquisition rate over 50%. The reason for that is, our proposal repeatedly checks to detect security attack and invalid Interest discarding for every one second, according to the statistical information of last one second. Then, invalid Interest from malicious attackers will be delivered during one second, after prevention period until next prevention period.

7 Conclusion

This paper discussed and proposed the security attack based on CIFA and prevention method for the attacks. The results of the computer evaluation confirmed that our proposal based on the calculation of number of cache reference, effectively works for improving the content acquisition rate even under CIFA.

Remaining tasks for our project is to develop to mitigate effects of CIFA more hardly by introducing method for keeping prevention state for malicious users.

Acknowledgements. This work was supported by JSPS KAKENHI Grant Number JP17K00132.

References

1. Dannewwitz, C., Imbrenda, C., Kutscher, D., Ohlman, B.: A survey of information-centric networking. IEEE Commun. Mag. **50**(7), 26–30 (2012)
2. Jacobson, V., Smetters, D., Thornton, J., Plass, M., Briggs, N., Braynard, R.: Networking named content. In: Proceedings of ACM CoNEXT 2009, pp. 1–12 (2009)
3. Named Data Networking (NDN) – A Future Internet Architecture. https://named-data.net. Accessed May 2017
4. Soniya, M., Kumar, K.: A survey on named data networking. In: Proceedings of 2015 2nd International Conference on Electronics and Communication Systems (ICECS 2015), Coimbatore, pp. 1515–1519 (2015)
5. Chen, Q., Xie, R., Yu, F., Liu, J., Huang, T., Liu, Y.: Transport control strategies in named data networking: a survey. IEEE Commun. Surv. Tutorials **18**(3), 2052–2083 (2016)
6. Amado, M., Campolo, C., Molinaro, A., Mitton, N.: Named data networking: a natural design for data collection in wireless sensor networks. In: Wireless Days (WD), 2013 IFIP, Valencia, pp. 1–6 (2013)
7. Hail, M., Amado, M., Molinaro, A., Fischer, S.: Caching in named data networking for the wireless internet of things. In: Proceedings of 2015 International Conference on Recent Advances in Internet of Things (RIoT), Singapore, pp. 1–6 (2015)

8. Xin, Y., Li, Y., Wang, W., Li, W., Chen, X.: A novel interest flooding attacks detection and countermeasure scheme in NDN. In: IEEE Global Communications Conference (GLOBECOM), Washington, DC 2016, pp. 1–7 (2016)
9. Shinohara, R., Kamimoto, T., Sato, K., Shigeno, H.: Cache control method mitigating packet concentration of router caused by interest flooding attack. In: IEEE Trustcom/BigDataSE/ISPA, Tianjin 2016, pp. 324–331 (2016)
10. Sarah, H., Strufe, T.: Evaluating and mitigating a collusive version of the interest flooding attack in NDN. In: Proceedings of 2016 IEEE Symposium on Computers and Communication (ISCC), pp. 938-945 (2016)
11. Xin, Y., Li, Y., Wang, W., Li, W., Chen, X.: Detection of collusive interest flooding attacks in named data networking using wavelet analysis. In: MILCOM 2017–2017 IEEE Military Communications Conference (MILCOM), Baltimore, MD, pp. 557–56 (2017)
12. Breslan, L., Cao, P., Fan, L., Phillips, G., Shenker, S.: Web caching and Zipf–like distributions: evidence and implications. In: Proceedings of 18th Annual Joint Conference of the IEEE Computer and Communications Societies (IEEE INFOCOM 1999), vol. 1, pp. 126–134 (1999)

An Energy-Efficient Dynamic Live Migration of Multiple Virtual Machines

Dilawaer Duolikun[1(✉)], Shigenari Nakamura[1], Tomoya Enokido[2], and Makoto Takizawa[1]

[1] Hosei University, Tokyo, Japan
dilewerdolkun@gmail.com, nakamura.shigenari@gmail.com, makoto.takizawa@computer.org
[2] Rissho University, Tokyo, Japan
eno@ris.ac.jp

Abstract. In this paper, we propose an algorithm to migrate virtual machines to reduce the total electric energy consumption of servers. Here, virtual machines are dynamically resumed and suspended so that the number of processes on each virtual machine can be kept fewer. In addition, multiple virtual machines migrate from a host server to a more energy-efficient guest server. In our previous studies, time to migrate virtual machines is assumed to be zero. The more often a virtual machine migrates, the longer time it takes to perform processes on the virtual machine. We propose a model to estimate the electric energy consumption of servers by considering the migration time of each virtual machine. By using the model, virtual machines to migrate and to perform processes are selected so that the total electric energy consumption can be reduced. In the evaluation, we show the total electric energy consumption of servers can be reduced compared with other algorithms.

1 Introduction

Electric energy consumed in scalable systems like cloud computing systems [19] and IoT (Internet of Things) [17] has to be reduced to realize eco society [8]. Power consumption and computation models are so far proposed [9–14] to estimate electric power consumption of each server to perform application processes and the estimated execution time of each application process. Based on the models, types of algorithms [10–12, 15] are proposed to select an energy-efficient server to perform an application process issued by a client. It may take a longer time to perform processes on a host server than estimated. Accordingly, the host server consumes more electric energy. A process migration approach [1, 2, 5–7] is also discussed where virtual machines with application processes migrate from host servers to energy-efficient guest servers in the live manner [20, 21].

In our previous studies, static and dynamic migration algorithms are proposed [2–4]. In the static migration EAMV (Energy-aware Migration of Virtual Machines) algorithm [2], a cluster supports the fixed number of virtual machines. The size of a virtual machine means the number of application processes

L. Barolli et al. (Eds.): NBiS 2018, LNDECT 22, pp. 87–98, 2019.
https://doi.org/10.1007/978-3-319-98530-5_8

performed on the virtual machine. The more number of application processes are issued, the larger each virtual machine is. Here, the size of a virtual machine gets too large to migrate to another server. The dynamic DMMV [3] and DMMV2 [4] algorithms are also proposed where virtual machines are dynamically suspended and resumed depending on the number of application processes. In the DMMV algorithm, one virtual machine migrates from a host server to a guest server. In the DMMV2 algorithm, multiple virtual machines migrate from a host server to a guest server.

In the previous algorithms, it is shown the electric energy consumption of servers and the average execution time of processes can be reduced by migrating virtual machines. However, time to migrate a virtual machine is assumed to be zero. It may take a longer time to perform processes since it takes time to migrate a virtual machine. In this paper, we newly propose a model to estimate the electric energy to be consumed by servers by taking into consideration the migration time of a virtual machine. By using the estimation model, we propose a dynamic migration (DM) algorithm where virtual machines to migrate and to perform application processes are selected so that the total electric energy to be consumed by all the servers can be reduced. Here, DM(v) shows a DM algorithm where at most v ($\geq$1) virtual machines migrate from a host server to a guest server. DM(1) stands for the DMMV2 algorithm [4]. DM(*) shows as many virtual machines migrate as possible. In the evaluation, we show the total electric energy consumption of servers and the average execution time of application processes in the DM(*) algorithm can be reduced.

In Sect. 2, we present a system model. In Sects. 3 and 4, we propose the estimation model and DM algorithm. In Sect. 5, we evaluate the DM algorithm.

2 System Model

A cluster S is composed of multiple servers $s_1, \ldots, s_m$ ($m \geq 1$). Each server s_t is equipped with np_t ($\geq$1) homogeneous CPUs. Each CPU is composed of cc_t ($\geq$1) homogeneous cores, each of which supports ct_t ($\geq$1) homogeneous threads. The total number nt_t (=$np_t \cdot cc_t \cdot ct_t$) of threads on nc_t (=$np_t \cdot cc_t$) cores are supported by a server s_t. A thread is a unit of computation resource to perform processes. Processes on different threads are performed independently of one another. A thread is *active* if and only if (iff) at least one process is performed, otherwise *idle*. In this paper, a *process* means an application process to be performed on a server, which uses CPU resource.

Suppose virtual machines $vm_1, \ldots, vm_v$ ($v \geq 0$) on the servers are supported. We assume every process is performed on some virtual machine and is not directly performed on a host server. $VCP_h(\tau)$ shows a set of processes on a virtual machine vm_h at time τ. The size $|vm_h|$ of a virtual machine vm_h shows the number $|VCP_h(\tau)|$ of processes. A virtual machine vm_h is *smaller* than a virtual machine vm_k ($vm_h < vm_k$) iff $|vm_h| < |vm_k|$. $CP_t(\tau)$ is a set of all the processes performed on a server s_t at time τ.

In this paper, we assume every server s_t holds a memory image of each virtual machine so that every virtual machine can migrate to any server.

A virtual machine vm_h is *dormant*, where any process can be neither performed nor issued. Every dormant virtual machine is in the virtual machine pool DVM_t of some server s_t. A virtual machine vm_h is *ready* iff a client can issue processes. VM_t is a set of ready virtual machines on a server s_t. A ready virtual machine vm_h is *active* at least one process is performed, otherwise *idle*. *Resuming* and *suspending* a virtual machine vm_h mean that vm_h transits from dormant to idle and from idle to dormant, respectively. A server s_t is *engaged* iff there resides at least one ready virtual machine, otherwise *free*. A server s_t is *active* iff at least one virtual machine is active.

We assume processes are fairly allocated with CPUs, cores, and threads in each server, e.g. in the round-robin algorithm. In the MLPCM model [15,16], the electric power consumption $NE_t(n)$ [W] of a server s_t to perform n processes is given as follows:

$$NE_t(n) = minE_t + nap_t(n) \cdot bE_t + nac_t(n) \cdot cE_t + nat_t(n) \cdot tE_t. \quad (1)$$

Here, $nap_t(n) = n$ if $n \leq np_t$, else np_t. $nac_t(n) = n$ if $n \leq nc_t$, else nc_t. $nat_t(n) = n$ if $n \leq nt_t$, else nt_t. The maximum electric power consumption $maxE_t$ [W] of a server s_t is $minE_t + np_t \cdot bE_t + nc_t \cdot cE_t + nt_t \cdot tE_t$ where every thread is active. The electric energy consumed by a server s_t from time st [tu (time unit)] to time et is defined to be $\sum_{\tau=st}^{et} NE_t(|CP_t(\tau)|)$ [Wtu] in this paper.

If only a process p_i is exclusively performed on a server s_t without any other process, the execution time T_{ti} [tu] of the process p_i is shortest, $T_{ti} = minT_{ti}$. In a cluster S, $minT_i$ is $minT_{fi}$ on the fastest thread which is on a server s_f, i.e. $minT_i \leq minT_{ti}$ for every server s_t. Here, the server s_f is *fastest*. One virtual computation step [vs] is assumed to be performed on a thread of the fastest server s_f for one time unit [tu] [12,14]. This means, the thread computation rate TCR_f of a fastest server s_f is one [vs/tu]. For another server s_t ($\neq s_f$), $TCR_t \leq TCR_f$ (=1). The total number VC_i [vs] of virtual computation steps of a process p_i is defined to be $minT_i$ [tu] $\cdot TCR_f$ [vs/tu] $= minT_i$ [vs] since $TCR_f = 1$ for a fastest server s_f. Thus, $minT_i$ shows the total number of virtual computation steps to be performed in a process p_i. If only one process p_i is performed without any other process on a thread of a server s_t, the maximum number of virtual computation steps of the process p_i are performed for one time unit. The maximum process computation rate $maxPCR_{ti}$ of a process p_i on a server s_t is $VC_i/minT_{ti} = minT_i/minT_{ti}$ [vs/tu] ($\leq$1). On a fastest server s_f, $maxPCR_{fi} = TCR_f = 1$. For every pair of processes p_i and p_j on a server s_t, $maxPCR_{ti} = maxPCR_{tj} = TCR_t$ ($\leq$1) [14]. The server computation rate $SCR_t(\tau)$ of a server s_t at time τ is $at_t(\tau) \cdot TCR_t$ where $at_t(\tau)$ ($\leq nt_t$) is the number of active threads at time τ. The maximum server computation rate $maxSCR_t$ of a server s_t is $nt_t \cdot TCR_t$. Here, $at_t(\tau)$ is $nat_t(n)$ where $n = |CP_t(\tau)|$ at time τ. The server computation rate $NSR_t(n)$ [vs/tu] of a server s_t to perform n processes in the MLCM [16] model is given as follows:

$$NSR_t(n) = \begin{cases} n \cdot TCR_t \text{ if } n \leq nt_t. \\ maxSCR_t (= nt_t \cdot TCR_t) \text{ if } n > nt_t. \end{cases} \quad (2)$$

Here, each process p_i is performed at rate $NPR_{ti}(n) = NSR_t(n)/n$ ($\leq TCR_t$) [vs/tu] on a server s_t where totally n ($=|CP_t(\tau)|$) processes are concurrently performed. Here, the process computation rate $PCR_{ti}(\tau)$ [vs/tu] of a process p_i on a server s_t at time τ is $SCR_t(\tau)/n = NSR_t(n)/n$ ($=NPR_{ti}(n)$). Suppose a process p_i on a server s_t starts at time st and ends at time et. Here, $VC_i = \sum_{\tau=st}^{et} NPR_{ti}(|CP_t(\tau)|) = minT_i$ [vs]. If a process p_i starts on a server s_t at time τ, p_i gets active and the variable plc_i is VC_i. If each active process p_i is performed at time τ, plc_i is decremented by $NPR_{ti}(|CP_t(\tau)|)$. Then, p_i terminates at time τ if $plc_i \leq 0$.

3 Estimation Models

In order to select a host server to perform a process issued by a client, we have to estimate the execution time of each current process on each server. The simple estimation model is proposed [6,7]. Here, each process p_i is assumed to be composed of the same number of virtual computation steps, $VC_i = VC = 1$. The half of the total number VC_i of virtual computation steps of every current process p_i is assumed to be performed, i.e. $VC/2 = 1/2$. Here, suppose n ($=|CP_t(\tau)|$) processes are performed on a server s_t at time τ. The total amount of computation to be performed by the n processes is $n/2$. If k processes newly start, the total computation VC of each new process is to be performed, i.e. totally k virtual computation steps. Here, it takes $(n/2 + k)/NSR_t(n + k)$ time units [tu] to perform $(n/2 + k)$ virtual computation steps are to be performed. Hence, the expected termination time $SET_t(n, k)$ [tu] and energy consumption $SEE_t(n, k)$ [W tu] of each server s_t to perform both n current and k new processes are $(n/2+k)/NSR_t(n+k)$ and $SET_t(n,k) \cdot NE_t(n+k) = (n/2+k) \cdot NE_t(n + k)/NSR_t(n + k)$, respectively.

Suppose n_t (≥ 0) processes are performed on a server s_t at time τ. A virtual machine vm_t with nv_h processes starts migrating to the server s_t at time τ. As discussed, every current process terminates by time $\tau + ET_t$ ($=SET_t(n_t, 0)$). If $mt \leq ET_t$, not only n_t current processes are performed but also nv_h processes on the virtual machine vm_h start after time $\tau + mt$ as shown in Fig. 1(1). Here, the n_t processes already finish the computation $(n_t/2)(mt/ET_t)$ and has to still perform $(n_t/2)$ (1- mt/ET_t). Here, it still takes $((1 - mt/ET_t)\ n_t/2 + nv_h/2)/NSR_t(n_t + nv_t)$ [tu] to perform all the processes. Hence, if the virtual machine vm_h migrates to a server s_t, it takes $NT_t = mt + ((1 - mt/ET_t)n_t/2 + nv_h/2)/NSR_t(n_t + nv_t)$ [tu]. Here, the server s_t consumes the electric energy $NE_t(n_t) \cdot mt + NE_t(n_t + nv_h)$ $(NT_t - mt)$.

Next, suppose $mt > ET_t$. Here, every current process terminates on the virtual machine vm_h of the server s_t as shown in Fig. 1(2). After time $\tau + mt$, only nv_h processes on vm_h are performed. Hence, $NT_t = m_t + (nv_h/2)/NSR_t(nv_h)$. Here, s_t consumes the electric energy $NE_t(n_t) \cdot ET_t + minE_t \cdot (mt - ET_t) + NE_t(nv_h) \cdot (NT_t - mt)$. Thus, the expected termination time $MET_t(n_t, nv_h)$ and electric energy $MEE_t(n_t, nv_h)$ of a server s_t where a virtual machine vm_h

with nv_h processes migrates from s_t to s_u are given where $ET_t = SET_t(n_t, 0)$ as follows:

$$MTE_t(n_t, nv_h) = \begin{cases} mt + ((1 - mt/ET_t)n_t/2 + nv_h/2)/NSR_t(n_t + nv_t) \ \textbf{if}\ ET_t > mt. \\ mt + (nv_h/2)/NSR_t(nv_h) \quad \textbf{otherwise}. \end{cases} \tag{3}$$

$$MEE_t(n_t, nv_h) = \begin{cases} mt \cdot (1 - mt/ET_t)NE_t(n_t) + (MTE_t(n_t, nv_h) - mt) \cdot NE_t(n_t + nv_h) \\ \textbf{if}\ ET_t > mt. \\ ET_t \cdot NE_t(n_t) + minE_t \cdot (mt - ET_t) \\ + (MTE_t(n_t, nv_h) - ET_t) \cdot NE_t(nv_h) \quad \textbf{otherwise}. \end{cases} \tag{4}$$

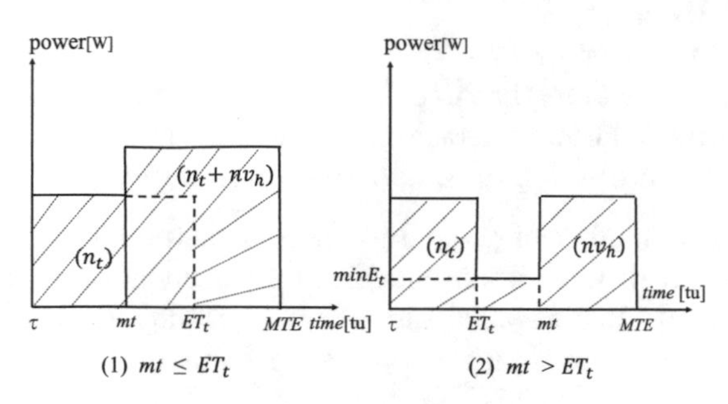

Fig. 1. Electric energy consumption MEE.

4 A Dynamic Algorithm to Find and Migrate Virtual Machines

4.1 Virtual Machine Selection

In the static migration algorithms [18,20,21], the total number v of virtual machines is invariant and every virtual machine is ready in a cluster. In the dynamic migration algorithms [4,6,7], virtual machines are dynamically resumed and suspended. Here, the more number of processes, the more number of ready virtual machines.

Suppose a cluster S is composed of servers $s_1, \ldots, s_m$ ($m \geq 1$). A variable VM is a set of ready virtual machines in the clusters. A variable VM_t ($\subseteq VM$) shows a set of ready virtual machines on each server s_t ($t = 1, \ldots, m$). Initially, $VM_t = \phi$ for every server s_t and $VM = \phi$. Let n_t and nv_h denote the numbers $|CP_t(\tau)|$ and $|VCP_h(\tau)|$ of processes on a server s_t and a virtual machine vm_h, respectively, at current time τ. DVM_t is a set of dormant virtual machines on a server s_t.

Suppose a process p_i is issued to a cluster S. One server s_t is first selected to perform the new process p_i where the expected electric energy consumption $SEE_t(n_t, 1)$ is minimum in the cluster S. The execution time of each process p_i

depends on the total number $n_t + 1$ of processes to be concurrently performed on a host sever s_t but is independent of the number $|VM_t|$ [20]. Hence, each virtual machine vm_k is kept smaller than some size $maxNVM_t$. If the size of every virtual machine is smaller than $maxNVM_t$, a smallest ready virtual machine vm_h is selected in VM_t. Otherwise, a dormant virtual machine vm_h in DVM_t is resumed on the server s_t. Then, the process p_i is performed on the virtual machine vm_h.

[**VM selection**] A process p_i is issued to a cluster S.
  select a host server s_t whose $SEE_t(n_t, 1)$ is minimum;
  if $VM_t = \phi$ /* the server s_t is not engaged */
    resume a dormant virtual machine vm_h in the pool DVM_t of s_t;
  else /* $VM_t \neq \phi$ */
  if $n_t/|VM_t| \leq maxNVM_t$,
    select a smallest ready virtual machine vm_h in the set VM_t;
  else resume a dormant virtual machine vm_h in the pool DVM_t;
  perform the process p_i on the virtual machine vm_h;

Thus, the more number of processes are issued, the more number of virtual machines are resumed. Idle virtual machines are suspended as follows.

[**VM resumption**] Each engaged server s_t is periodically checked as follows:
 if $|VM_t| > 1$ and there is an idle virtual machine on the engaged server s_t,
  if $n_t/(|VM_t| - 1) < maxNVM_t$, {
   select one idle virtual machine vm_h;
   suspend the virtual machine vm_h; };

Thus, virtual machines are dynamically suspended and resumed depending on the number of processes performed. The more number of processes are performed on a server, the more number of ready virtual machines reside on the server.

4.2 Virtual Machine Migration

One or more than one virtual machine migrates from a host server s_t to a guest server s_u in order to reduce the total electric energy consumption of servers. ET and EU are the electric energy to be consumed by the servers s_t and s_u to perform every process, respectively. Let MT be $minE_t$ MU be $minE_u$. TT and TU show the expected termination time of the servers s_t and s_u, respectively. If $TT \leq TU$, every process on s_t terminates before s_u. Here, s_t just consumes the minimum electric power $minE_t$ after time TT. The total electric energy to be consumed by the servers s_t and s_u to perform every process are give by the following function TE:

$$TE(ET, TT, MT, EU, TU, MU) = \begin{cases} ET + EU + (TT - TU)\, MU \textbf{ if } TT \geq TU. \\ ET + EU + (TU - TT)\, MT \textbf{ if } TT < TU. \end{cases} \quad (5)$$

Each engaged server s_t is periodically checked. Here, active virtual machines in the set $VM_t = \{vm_{t1}, \ldots, vm_{tv_t}\}$ ($v_t \geq 0$) on the server s_t are ordered in terms of size as $vm_{t1} < \ldots < vm_{tv_t}$. In the DM(*) and DM(1) algorithms, a virtual machine vm_h is selected to perform a new process in the VM selection algorithm. In the DM(1) and DM(*) algorithms, only a smallest virtual machine and multiple virtual machines migrate from a host server to a guest server, respectively. The DM(1) algorithm is the DMMV [3].

[VM migration (VMM) DM(*)]

$EE_t = SEE_t(n_t, 0)$; $ET_t = SET_t(n_t, 0)$; /* energy and time on s_t */
for each other server s_u ($\neq s_t$), { /* n_u = number of current processes of s_u */
    $EE_u = SEE_u(n_u, 0)$; $ET_u = SET_u(n_u, 0)$;
    $CEE_{tu} = TE(EE_t, ET_t, minE_t, EE_u, ET_u, minE_u)$; /* energy consumption */
}; /* **for** s_u end */
$nv = 0$; $NE = \infty$; $s =$ NULL;
for $i = 1, \ldots, v_t$, { /* for each virtual machine VM_{ti} on s_t */
    if $nv_{ti} > 0$, { /* vm_{ti} is active */
      $nv = nv + nv_{ti}$; /* total number of processes on $vm_{t1}, \ldots, vm_{ti}$ */
      /* nv processes migrate from s_t to s_u */
      $NE_t = SEE_t(n_t - nv, 0)$; $NT_t = SET_t(n_t - nv, 0)$;
      for each other server s_u ($\neq s_t$), { /* $vm_{t1}, \ldots, vm_{ti}$ to migrate to s_u */
        $NE_{tu} = MEE_u(n_u, nv)$; $NT_{tu} = MTE_u(n_u, nv_h)$;
        $NEE_{tu} = TEE(NE_t, NT_t, minE_t, NE_{tu}, NT_{tu}, minE_u)$;
        if $NEE_{tu} < CEE_{tu}$ and $NEE_{tu} < NE$,
         { /* energy can be reduced by taking s_u as a guest server */
           $NE = NEE_{tu}$; $s = s_u$; $v = i$;
         } **else break**; /* **if** end */
      } /* **for** s_u end */
    } **else break**; /* **if** end */
}; / ***for** i end */
if $s \neq$ NULL /*guest server is found */ **migrate** $vm_{t1}, \ldots, vm_{tv}$ **from** s_t **to** s;

First, the expected electric energy consumption $EE_u = SEE_u(n_u, 0)$ and termination time $ET_u = SET_u(n_u, 0)$ to perform all the n_u current processes on each server s_u ($u = 1, \ldots, n$) are obtained. The electric energy CEE_{tu} to be consumed by a pair of the servers s_t and s_u is calculated by using the function TE (EE_t, ET_t, $minE_t$, EE_u, ET_u, $minE_u$) (4) where no virtual machine on the server s_t migrates.

First, a smallest virtual machine vm_{t1} is selected on the host server s_t. Here, the number nv of processes migrate with the virtual machine vm_h from the host server s_t to another server, i.e. $nv = nv_{t1}$. The host server s_t is expected to consume the electric energy $NE_t = SEE_t(n_t - n_v, 0)$ by time $NT_t = SET_t(nt$ - n_v, 0). The server s_u consumes the electric energy $NE_{tu} = MEE_u(n_u, nv)$ by time $NT_{tu} = MTE_u(n_u, nv_h)$ where nv processes on the virtual machine vm_h are additionally performed. As discussed in the preceding section, the migration time mt is considered in MEE_u. The electric energy consumption NEE^1_{tu} of the servers s_t and s_u is calculated as $NEE^1_{tu} = TE(NE_t, NT_t, minE_t, NE_{tu}, NT_{tu}, minE_u)$. If $CEE_{tu} > NEE^1_{tu}$, the virtual machine vm_{t1} can migrate from s_t to

s_u because the total electric energy consumption of s_t and s_u can be reduced. Secondly, a next smallest virtual machine vm_{t2} is taken, i.e. $vm_{t1} < vm_{t2}$. A pair of vm_{t1} and vm_{t2} are tried to migrate from the host server s_t to another guest server. Here, totally nv ($=nv_{t1}+nv_{t2}$) processes migrate from s_t to another server. $NE_{tu} = MEE_u(n_u, nv_{t1} + nv_{t2})$ and $ET_{tu} = MTE_u(n_u, nv_{t1} + nv_{t2})$. NEE^2_{tu} is also calculated for each server s_u ($s_u \neq s_t$) by using the function TE. A guest server s_u to which the virtual machines vm_{t1} and vm_{t2} migrate from s_t is found where the electric energy consumption NEE^2_{tu} is minimum as discussed in the first virtual machine vm_{t1}. If $NEE^1_{tu} < NEE^2_{tu}$, the algorithm terminates and only one virtual machine vm_{t1} migrates. Otherwise, three virtual machines vm_{t1}, vm_{t2}, and vm_{t3} are tried to migrate. Thus, these steps are iterated until $NEE^v_{tu} < NEE^{v+1}_{tu}$. Here, $NEE^v_{tu} < NEE^i_{tu}$ for every $i < v$ and $i = v + 1$. Then, a collection of the v virtual machines $vm_{t1}, \ldots, vm_{tv}$ ($v \leq v_t$) migrate to a guest server s_u. Here, totally $nv = nv_{t1} + \ldots + nv_{tv}$ processes on the virtual machines $vm_{t1}, \ldots, vm_{tv}$ migrate from s_t to s_u.

5 Evaluation

The DM(1) and DM(*) algorithms are evaluated in terms of the total electric energy consumption TEE [Wtu] and total active time TAT [tu] of servers $s_1, \ldots, s_m$ and the average execution time AET [tu] of processes $p_1, \ldots, p_n$ compared with the non-migration type, random (RD), round robin (RR), and SGEA [14], and the static migration SM(1) and SM(*) algorithms. We consider four servers $s_1, \ldots, s_4$ ($m = 4$) in our laboratory. The power consumption and performance parameters of each server s_t are shown in Table 1. In the static algorithms, a set VM of virtual machines include sixteen virtual machines $vm_1, \ldots, vm_v$ ($v = 16$). Each server s_t hosts four virtual machines.

In the RD algorithm, one virtual machine vm_h is randomly selected to the process p_i in the set VM. In the RR algorithm, a virtual machine vm_h is selected after a virtual machine vm_{h-1} is selected. In the SGEA algorithm, a host server is selected so that the total electric energy of all the servers can be minimized. The SM(1), SM(*), DM(1), and DM(*) are migration algorithms. In the DM(1) and DM(*), there is initially no virtual machine on each server s_t. Idle virtual machines are resumed and ready virtual machines are suspended depending on number of processes on the virtual machines. In the SM(1) and DM(1), only one virtual machine migrates. In the SM(*) and DM(*), multiple virtual machines migrate. The SM(1) is the EAMV [2]. The SM(*) is the EAMV except that multiple virtual machines migrate in the same way as the DM(*). The DM(1) is the DMMV [3], where one virtual machine migrate. In the DM(*), multiple virtual machines migrate. The migration time mt is assumed to be 5 [tu]. Each engaged server is checked every σ time units and virtual machines migrate to a guest server if the electric energy consumption of the servers can be reduced by migrating the virtual machine. In the evaluation, $\sigma = 5$.

The number n ($\geq$1) of processes $p_1, \ldots, p_n$ are randomly issued to the cluster S. One time unit [tu] is assumed to be 100 [ms]. In each process configuration

PF_{ng}, the minimum execution time $minT_i$ of each process p_i is randomly taken from 15 to 25 [tu]. VC_i [vs] of each process p_i is $minT_i$. In the evaluation, the half $n/2$ of n processes randomly start from time $xtime/4 - xtime/40$ to $xtime/4 + xtime/40$. One quarter $n/4$ of the processes randomly start from time $xtime/4 - xtime/20$ to $xtime/4 - xtime/40$ and from $xtime/4 + xtime/40$ to $xtime/4 + xtime/20$. The other $n/4$ processes randomly start from time 0 to $xtime/4 - xtime/20$ and from $xtime/4 + xtime/20$ to $xtime$ - 1. The execution time ET_i of each process p_i is $etime_i - stime_i + 1$ where the process p_i terminates at time $etime_i$. The simulation time $xtime$ is 1,000 [tu] (=100 [s]). We randomly generate eight process configurations $PF_{n1}, \ldots, PF_{n8}$ for each number n of processes.

There are variables EE_t and AT_t for each server s_t. In the simulation, EE_t is incremented by the power consumption $NE_t(n) - minE_t$ for $n = |CP_t(\tau)|$ and AT_t is incremented by one if $n > 0$ at each time τ. When the simulation ends, EE_t shows the total electric energy s_t and AT_t stands for the total active time of s_t.

Figure 2 shows the total electric energy consumption $TEE = EE_1 + \ldots + EE_4$ [Wtu] of the servers $s_1, \ldots, s_4$ for number n of processes. $maxNVM_t = 10$ in the DM(1) and DM(*) algorithms. The total electric energy TEE of the RD algorithm is almost the same as the RR algorithm. The total electric energy consumption TEE of the DM(*) algorithm is smaller than the other algorithms.

Figure 3 shows the total active time $TAT = AT_1 + \ldots + AT_4$ [tu] of the servers $s_1, \ldots, s_4$ for the number n of processes. The total active time TAT of the RR and RD algorithms are almost the same. The total active time TAT of the DM(*) algorithm is about 50% shorter than the RD and RR algorithms and a little bit shorter than the DM(1) algorithm. The servers are more lightly loaded in the dynamic migration DM(*) algorithm than the non-migration algorithms.

Figure 4 shows the average execution time AET [tu] of the number n of processes. $AET = (ET_1 + \ldots + ET_n)/n$. The average execution time AET of the DM(*) algorithm is longer than the RD and RR algorithms for $n < 500$ but shorter for $n \geq 500$. By dynamically migrating virtual machines, the average execution time AET of the processes can be thus reduced if the servers are more loaded.

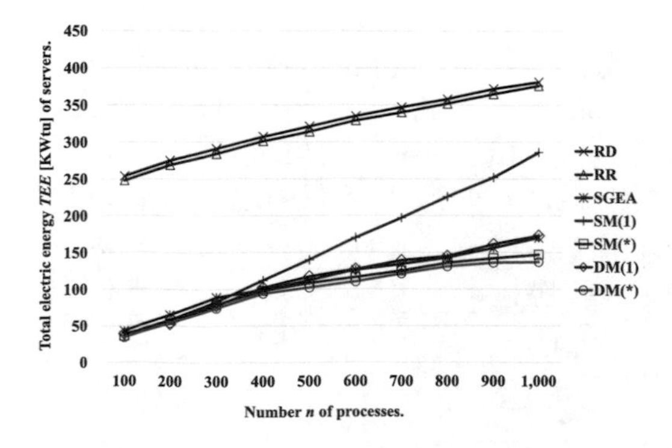

Fig. 2. Total electric energy consumption ($m = 4$, $\sigma = 5$, $maxNVM_t = 10$).

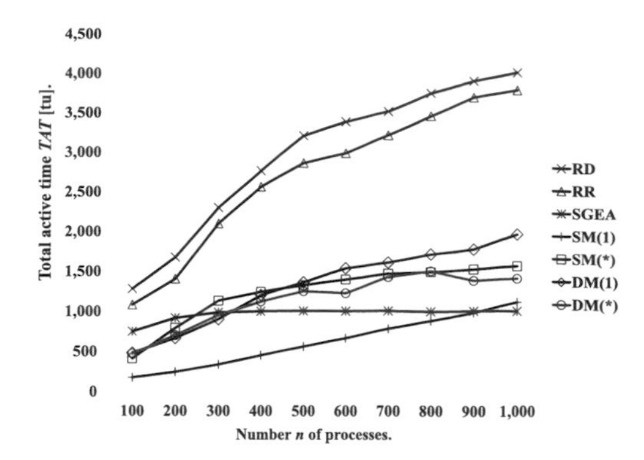

Fig. 3. Total active time TAT of servers ($m = 4$, $\sigma = 5$, $maxNVM_t = 10$).

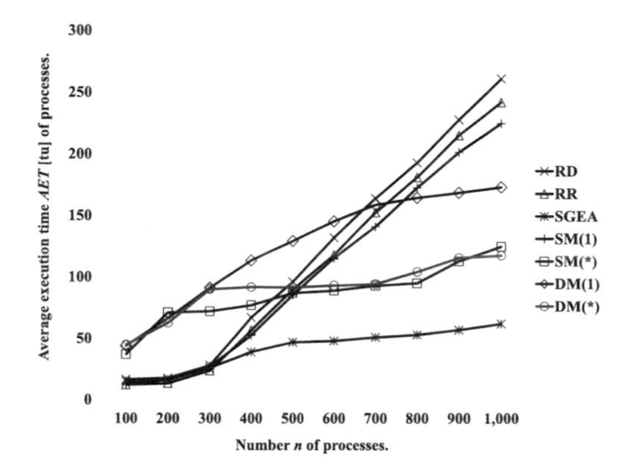

Fig. 4. Average execution time AET of processes ($m = 4$, $\sigma = 5$, $maxNVM_t = 10$).

Table 1. Parameters of servers.

Parameters	s_1	s_2	s_3	s_4
np_t	2	1	1	1
nc_t	8	8	6	4
nt_t	32	16	12	8
CRT_t [vs/tu]	1.0	1.0	0.5	0.7
$maxCR_t$ [vs/tu]	32	16	6	5.6
$minE_t$ [W]	126.1	126.1	87.2	41.3
$maxE_t$ [W]	301.3	207.3	136.2	89.6
bE_t [W]	30	30	16	15
cE_t [W]	5.6	5.6	3.6	4.7
tE_t [W]	0.8	0.8	0.9	1.1

6 Concluding Remarks

In this paper, we proposed the model to estimate the execution time of processes and electric energy consumption of servers by taking into consideration the migration time of each virtual machine. In this paper, the DM(*) algorithm is proposed by using the estimation model to reduce the total electric energy consumption of servers in a cluster. Here, virtual machines are dynamically resumed and suspended as the number of processes increases and decreases, respectively. In addition, one or more than one virtual machine migrates from a host server to a guest server so that the total electric energy consumed by the host and guest servers can be reduced. In the evaluation, we showed the total electric energy consumption of servers can be reduced in the DM(*) algorithm compared with other algorithms.

References

1. Dilawaer Duolikun; Tomoya Enokido; Makoto Takizawa: An energy-aware algorithm to migrate virtual machines in a server cluster. Int. J. Space Based Situat. Comput. **7**(1), 32–42 (2017)
2. Duolikun, D., Nakamura, S., Watanabe, R., Enokido, T., Takizawa, M.: Energy-aware migration of virtual machines in a cluster. In: Proceedings of the 11th International Conference on Broadband and Wireless Computing, Communication and Applications (BWCCA 2016), pp. 21–32 (2016)
3. Duolikun, D., Watanabe, R., Enokido, T., Takizawa, M.: An eco migration algorithm of virtual machines in a server cluster. In: Proceedings of IEEE the 32nd International Conference on Advanced Information Networking and Applications (AINA 2018)
4. Duolikun, D., Watanabe, R., Enokido, T., Takizawa, M.: Energy-efficient replication and migration of processes in a cluster. In: Proceedings of the 12th International Conference on Complex, Intelligent and Software Intensive Systems (CISIS 2018)
5. Duolikun, D., Watanabe, R., Enokido, T., Takizawa, M.: A model for migration of virtual machines to reduce electric energy consumption. In: Proceedings of the 19th International Conference on Network-based Information Systems (NBiS 2016), pp. 50–57 (2016)
6. Duolikun, D., Watanabe, R., Enokido, T., Takizawa, M.: An eco algorithm for dynamic migration of virtual machines in a server cluster. In: Proceedings of the 20th International Conference on Network-Based Information Systems (NBiS 2017), pp. 42–54 (2017)
7. Duolikun, D., Watanabe, R., Enokido, T., Takizawa, M.: Energy-aware dynamic migration of virtual machines in a server cluster. In: Proceedings of the 12th International Conference on Broadband and Wireless Computing, Communication and Applications (BWCCA 2017), pp. 161–172 (2017)
8. Elnozahy, E.N., Kistler, M., Rajamony, R.: Energy-efficient server clusters. Power Aware Comput. Syst. **2325**, 179–197 (2003)
9. Enokido, T., Aikebaier, A., Deen, M., Takizawa, M.: Power consumption-based server selection algorithms for communication-based systems. In: Proceedings of the 13th International Conference on Network-Based Information Systems (NBiS 2010), pp. 201–208 (2010)

10. Enokido, T., Aikebaier, A., Takizawa, M.: A model for reducing power consumption in peer-to-peer systems. IEEE Syst. J. **4**(2), 221–229 (2010)
11. Enokido, T., Aikebaier, A., Takizawa, M.: Process allocation algorithms for saving power consumption in peer-to-peer systems. IEEE Trans. Ind. Electron. **58**(6), 2097–2105 (2011)
12. Enokido, T., Aikebaier, A., Takizawa, M.: An extended simple power consumption model for selecting a server to perform computation type processes in digital ecosystems. IEEE Trans. Ind. Inform. **10**(2), 1627–1636 (2014)
13. Kataoka, H., Nakamura, S., Duolikun, D., Enokido, T., Takizawa, M.: Multi-level power consumption model and energy-aware server selection algorithm. Int. J. Grid Util. Comput. **8**(3), 201–210 (2017)
14. Kataoka, H., Duolikun, D., Enokido, T., Takizawa, M.: Multi-level computation and power consumption models. In: Proceedings of the 18th International Conference on Network-Based Information Systems (NBiS 2015), pp. 40–47 (2015)
15. Kataoka, H., Duolikun, D., Enokido, T., Takizawa, M.: Energy-aware server selection algorithm in a scalable cluster. In: Proceedings of IEEE the 30th International Conference on Advanced Information Networking and Applications (AINA 2016), pp. 565–572 (2016)
16. Kataoka, H., Sawada, A., Duolikun, D., Enokido, T., Takizawa, M.: Simple energy-efficient server selection algorithm in a scalable cluster. In: Proceedings of the 11th International Conference on Broadband and Wireless Computing, Communication and Applications (BWCCA 2016), pp. 45–46 (2016)
17. McEwen, A., Cassimally, H.: Designing the Internet of Things. Wiley, Hoboken (2013)
18. Watanabe, R., Duolikun, D., Enokido, E., Takizawa, M.: A simply energy-efficient migration algorithm of processes with virtual machines in server clusters. Int. J. Wirel. Mob. Netw. Ubiquit. Comput. Dependable Appl. **8**(2), 1–18 (2017)
19. Rafaels, R.J.: Cloud Computing: From Beginning to End, Create Space Independent Publishing Platform (2015)
20. Watanabe, R., Duolikun, D., Enokido, T., Takizawa, M.: An eco model of process migration with virtual machines in clusters. In: Proceedings of the 18th International Conference on Network-Based Information Systems (NBiS 2016), pp. 292–297 (2016)
21. Watanabe, R., Duolikun, D., Enokido, T., Takizawa, M.: Energy-aware virtual machine migration models in a scalable cluster of servers. In: Proceedings of IEEE the 31st International Conference on Advanced Information Networking and Applications (AINA 2017), pp. 85–92 (2017)

Evaluation of an Energy-Efficient Tree-Based Model of Fog Computing

Ryuji Oma[1(✉)], Shigenari Nakamura[1], Dilawaer Duolikun[1], Tomoya Enokido[2], and Makoto Takizawa[1]

[1] Hosei University, Tokyo, Japan
ryuji.oma.6r@stu.hosei.ac.jp, nakamura.shigenari@gmail.com, dilewerdolkun@gmail.com, makoto.takizawa@computer.org
[2] Rissho University, Tokyo, Japan
eno@ris.ac.jp

Abstract. A huge number of devices like sensors are interconnected in the IoT (Internet of Things). In the cloud computing model, processes and data are centralized in a cloud. Here, networks are congested and servers are overloaded due to heavy traffic from sensors. In order to reduce the delay time and increase the performance, data and processes to handle the data are distributed to not only servers but also fog nodes in fog computing models. On the other hand, the total electric energy consumed by fog nodes increases to process a sensor data. In this paper, we newly propose a tree-based fog computing model to distribute processes and data to servers and fog nodes so that the total electric energy consumption of nodes can be reduced in the IoT. In the evaluation, we show the total electric energy consumption of nodes in the tree-based model is smaller than the cloud computing model.

Keywords: Energy-efficient fog computing
IoT (Internet of Things) · Energy-efficient IoT
Tree-based fog computing model

1 Introduction

A huge number and various types of nodes including not only computers like servers but also devices like sensors and actuators are interconnected in the Internet of Things (IoT) [10]. Sensor data is transmitted to servers and processed to make a decision on actions to be done by actuators. Here, networks are congested and servers are overloaded due to heavy traffic. In order to realize the IoT, an intermediate layer named *fog* layer [10] is introduced between devices and clouds. The fog layer [10] is composed of fog nodes which are interconnected with other fog nodes, devices, and servers in networks. Fog nodes not only exchange data with sensors and other fog nodes, i.e. do the routing functions but also process the data received from other nodes. Fog nodes deliver processed data to servers in clouds via other fog nodes. In addition, a fog node

L. Barolli et al. (Eds.): NBiS 2018, LNDECT 22, pp. 99–109, 2019.
https://doi.org/10.1007/978-3-319-98530-5_9

makes a decision on actions to be done by actuators and sends the actions to actuators via other fog nodes. Thus, processes and data are distributed to not only servers but also fog nodes in the IoT while centralized to servers in cloud computing systems.

We have to reduce the electric energy consumed in information systems in order to realize green society [2]. Especially, the IoT is more scalable than the cloud computing systems [4] since a huge number and various types of nodes like sensors and actuators are included in addition to servers. While data and processes to handle the data are distributed to fog nodes in addition to servers in order to efficiently transmit and process sensor data, more electric energy is consumed by fog nodes. We have to reduce the electric energy consumed by not only servers but also fog nodes. Power consumption models of a computer are proposed to show how much electric power the computer consumes to perform application processes [8,9]. Computation models of a computer are also proposed, which give the expected execution time of each process on the computer [8,9].

The linear fog computing model is proposed to reduce the electric energy consumption of the nodes [11]. Here, fog nodes are linearly connected. In this paper, we propose a more general model, a tree-based fog computing model where processes and data are distributed to a tree structure of fog nodes whose root node shows servers in clouds and leaf nodes are edge nodes which communicate with devices [12]. We evaluate the tree-based fog computing model compared with the cloud computing model in terms of total electric energy consumption and total processing time of the nodes. We show the total electric energy consumption and total execution time can be reduced in the tree-based fog computing model compared with the cloud computing system.

In Sect. 2, we present the system model. In Sect. 3, we propose the tree-based fog computing model. In Sect. 4, we evaluate the tree-based model.

2 System Model

In addition to computers, a huge number and various types of devices like sensors and actuators are interconnected in the IoT (Internet of Things) [10,14]. In addition to requests issued by clients, a large volume of data including multimedia data generated by sensors are transmitted to servers in networks. In order to reduce the network traffic and satisfy the time constrains between sensors and actuators, the IoT is composed of three layers, cloud, fog, and device layers as shown in Fig. 1. Clouds are composed of servers [4]. Each server supports applications with computation and storage services [4].

The device layer is composed of sensor and actuator nodes. A sensor node collects data obtained by sensing events occurring in physical environment. Sensor data is forwarded to neighbor sensor nodes in wireless networks as discussed in wireless sensor networks (WSNs) [14]. Sensor data is finally delivered to edge nodes at the fog layer. Based on the sensor data, actions to be done by actuators are decided in the IoT. Actuator nodes receive actions from edge nodes and perform the actions on the physical environment.

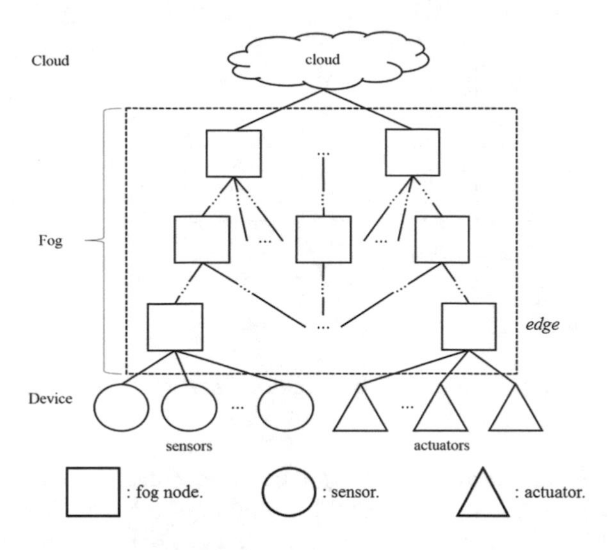

Fig. 1. IoT model.

Fog nodes are at a layer between the device and cloud layers [13]. Fog nodes are interconnected with other fog nodes in networks. A fog node supports the routing function where messages are routed to destination nodes, i.e. routing between servers and edge nodes like network routers [14]. Thus, fog nodes receive sensor data and forward the sensor data to servers via fog-to-fog communication. In addition, fog nodes do some computation on a collection of data sent by sensor nodes and other fog nodes. A fog node is also equipped with storages to buffer data. A fog node makes a decision on actions to be done by actuator nodes based on the sensor data. Then, the edge nodes issue the actions to actuator nodes.

3 Fog Computing Model in the IoT

3.1 Model of Fog Nodes

Fog nodes are interconnected with one another in networks. A fog node receives data from sensor nodes and other fog nodes and forwards data to the other fog nodes, i.e. does the routing functions. In addition, the data is processed and new data, i.e. processed data is generated by a fog node. For example, an average value calculated by summing of a collection of data obtained from sensor nodes is calculated. Data processed by a fog node is sent to neighbor fog nodes and servers finally receive data processed by fog nodes. In addition, a fog node makes a decision on what actions actuator nodes have to do based on sensor data. Thus, data and processes are distributed to not only servers but also fog nodes in fog computing systems while centralized to servers of clouds in cloud computing systems (Fig. 2).

In this paper, fog nodes are hierarchically structured. Let f_0 be a root node which shows a cloud of servers. The root node f_0 has child fog nodes

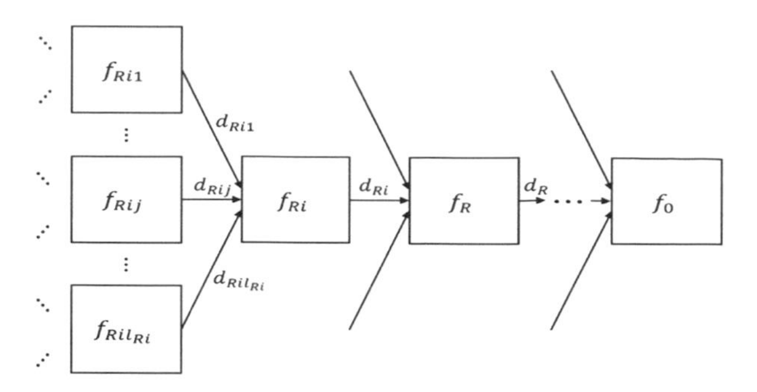

Fig. 2. Fog computing system.

$f_{01}, ..., f_{0l_0}$ ($l_0 \geq 1$). Each fog node f_{0i} has also child fog nodes $f_{0i1}, ..., f_{0il_{0i}}$ ($l_{0i} \geq 1$). Thus, f_{Ri} shows an ith child fog node of a fog node $f_R (i = 1, ..., l_R)$ where the suffix R is a sequence $0r_1...r_{m-1}$ of labels which denotes a path $f_0, f_{0r_1}, f_{0r_1r_2}, ..., f_{0r_1r_2...r_{m-1}}$ ($= f_R$) from a root f_0. Here, the length $|R|$ of the label R is m. A fog node f_R is at level $|R| - 1$ in the tree. Thus, there are child fog nodes $f_{R1}, ..., f_{Rl_R}$ ($l_R \geq 1$) of each fog node f_R where f_{Ri} is an ith child fog node of the fog node f_R. In turn, f_R is a parent fog node of f_{Ri}. There is at most one parent node of each node. Each fog node f_{Ri} has also child fog nodes $f_{Ri1}, ..., f_{Ril_{Ri}}$ ($l_{Ri} \geq 1$). An edge node f_{Ri} has no child node and a root node f_0 has no parent node.

A fog node f_{Ri} takes data d_{Rij} sent by each child fog node $f_{Rij} (j = 1, ..., l_{Ri})$. A process p_{Ri} of the fog node f_{Ri} does the computation on a collection $\{d_{Ri1}, ..., d_{Ril_{Ri}}\}$ of data obtained from the child fog nodes $f_{Ri1}, ..., f_{Ril_{Ri}}$ and generates output data d_{Ri}. Then, the fog node f_{Ri} sends the output data d_{Ri} to the parent fog node f_R.

Each process p_{Ri} of a fog node f_{Ri} is modeled to be composed of an input I_{Ri}, computation C_{Ri}, output O_{Ri}, and storage S_{Ri} modules [11] as shown in Fig. 3. The input module I_{Ri} receives data d_{Rij} from each child fog node $f_{Rij} (j = 1, ..., l_{Ri})$. Then, the computation module C_{Ri} processes a collection of the input data $d_{Ri1}, ..., d_{Ril_{Ri}}$ and generates the output data d_{Ri}. The output module O_{Ri} sends the data d_{Ri} to a parent fog node f_R in networks. The storage module S_{Ri} stores the data $d_{Ri1}, ..., d_{Ril_{Ri}}$ and d_{Ri} in the storage DB_{Ri}. That is, a collection of the output data d_{Ri} and input data $d_{Ri1}, ..., d_{Ril_{Ri}}$ are buffered in the storage DB_{Ri}. Since the volume of the storage DB_{Ri} is limited, some data, e.g. the most obsolete data is removed to make space to store new data if the storage DB_{Ri} is full. On the other hand, servers and devices are interconnected with networks in the cloud computing systems. Here, each fog node has just a routing function, i.e. input and output modules.

In each fog node f_{Ri}, input data is thus processed in addition to the routing function. A notation $|d|$ shows the size of data d. Thus, the size $|d_{Ri}|$ of the output data d_{Ri} is smaller than the input data $D_{Ri} = \{d_{Ri1}, ..., d_{Ril_{Ri}}\}, |d_{Ri}| < |D_{Ri}| (= |d_{Ri1}| + ... + |d_{Ril_{Ri}}|)$. The ratio $|d_{Ri}|/|D_{Ri}|$ is the *reduction ratio*

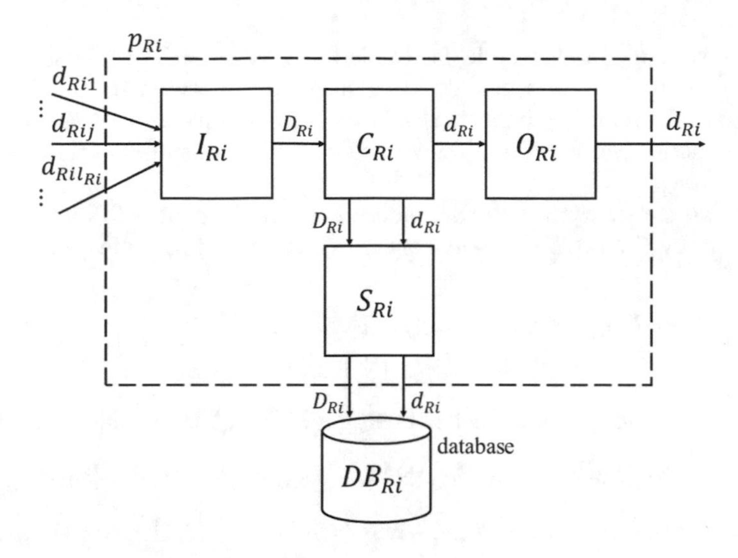

Fig. 3. Model of a process p_{Ri}.

ρ_{Ri} ($\leq$1) of a fog node f_{Ri}. For example, let D_{Ri} be a set $\{v_1, v_2, v_3, v_4\}$ of numbers showing temperature obtained by child fog nodes $f_{Ri1}, ..., f_{Ri4}$. If output data d_{Ri} is an average value v of the values $v_1, ..., v_4$, the reduction ratio ρ_{Ri} is $|d_{Ri}|/|D_{Ri}| = 1/4$ (Fig. 4).

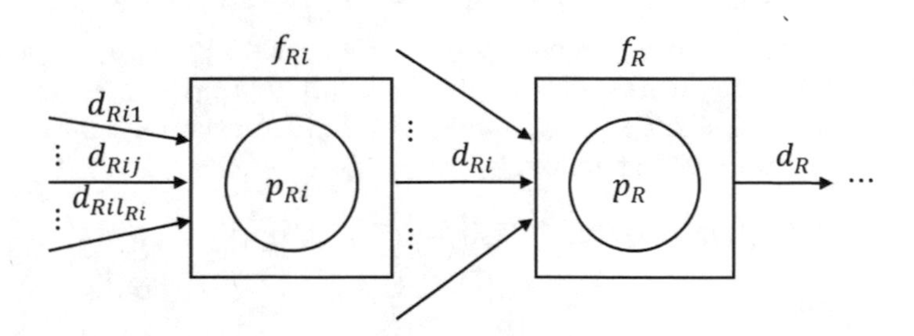

Fig. 4. Fog nodes.

3.2 Energy Consumption and Computation Models

We discuss the electric energy consumed by a fog node f_{Ri} to receive, process, store, and send data. In this paper, we assume modules I_{Ri}, C_{Ri}, S_{Ri}, and O_{Ri} of each fog node f_{Ri} are serially performed for each collection of input data. Let $EI_{Ri}(x)$, $EC_{Ri}(x)$, $EO_{Ri}(x)$, and $ES_{Ri}(x)$ show the electric energy [2] consumed by the input I_{Ri}, computation C_{Ri}, storage S_{Ri}, and output O_{Ri} modules of a fog node f_{Ri} to input, do the computation, output, and store data of size x, respectively. $TI_{Ri}(x)$, $TC_{Ri}(x)$, $TO_{Ri}(x)$, and $TS_{Ri}(x)$ denote time [sec] for a fog node f_{Ri} to input, do the computation, output, and store data of size x, respectively. Here, the fog node f_0 stands for a server s. The transmission time $TT_{R,Ri}(x)$ shows time to transmit data of size x [bit] between a child fog node f_R and a parent fog node f_{Ri} in networks. For the bandwidth $b_{R,Ri}$ [bps] between the parent fog node f_R and the child fog node f_{Ri}, the transmission time $TT_{R,Ri}(x)$ is $x/b_{R,Ri}$ [sec].

First, we consider the electric energy consumption of a fog node f_{Ri}. The electric energy $TE_{Ri}(|D_{Ri}|)$ and consumed by each fog node f_{Ri} is given as follows:

$$\begin{aligned} TE_{Ri}(|D_{Ri}|) = EI_{Ri}(|D_{Ri}|) + EC_{Ri}(|D_{Ri}|) \\ + ES_{Ri}(|D_{Ri}| + |d_{Ri}|) + EO_{Ri}(|d_{Ri}|). \end{aligned} \tag{1}$$

In the root node f_0, the electric energy $TE_0(|D_0|)$ is given as follows:

$$TE_0(|D_0|) = EI_0(|D_0|) + EC_0(|D_0|) + ES_0(|D_0| + |d_0|). \tag{2}$$

Here, $|D_{Ri}| = |d_{Ri1}| + ... + |d_{Ril_{Ri}}|$ and $|d_{Ri}| = \rho_{Ri}|D_{Ri}| = \rho_{Ri}(|d_{Ri1}| + ... + |d_{Ril_{Ri}}|)$.

In the cloud computing system, every sensor data is processed by the root node f_0, i.e. the servers. Each fog node just does the routing function, i.e. input and output modules are performed. The total electric energy $CE_{Ri}(|D_{Ri}|)$ and $CE_0(|D_0|)$ consumed by a fog node f_{Ri} and a root fog node f_0 is, respectively:

$$CE_{Ri}(|D_{Ri}|) = EI_{Ri}(|D_{Ri}|) + EO_{Ri}(|D_{Ri}|). \tag{3}$$

$$CE_0(|D_0|) = EI_0(|D_0|) + EC_0(|D_0|) + ES_0(|D_0| + |d_0|). \tag{4}$$

Here, the reduction ratio ρ_{Ri} is one ($\rho_{Ri} = 1$) for each fog node f_{Ri} since data is not processed by fog nodes. Hence, $|d_{Ri}| = |D_{Ri}| = |d_{Ril}| + ... + |d_{Ril_{Ri}}|$.

The execution time $ET_{Ri}(|D_{Ri}|)$ and $ET_0(|D_0|)$ of each fog node f_{Ri} and a root fog node f_0 are given as follows:

$$\begin{aligned} ET_{Ri}(|D_{Ri}|) = TI_{Ri}(|D_{Ri}|) + TC_{Ri}(|D_{Ri}|) \\ + TS_{Ri}(|D_{Ri}| + |d_{Ri}|) + TO_{Ri}(|d_{Ri}|). \end{aligned} \tag{5}$$

$$ET_0(|D_0|) = TI_0(|D_0|) + TC_0(|D_0|) + TS_0(|D_0| + |d_0|). \tag{6}$$

The execution time $TI_{Ri}(x)$, $TS_{Ri}(x)$, and $TO_{Ri}(x)$ [sec] of the input, storage, and output modules are linearly proportional to the size x of data, $i_{Ri} \cdot x$, $s_{Ri} \cdot x$, and $o_{Ri} \cdot x$, respectively, where i_{Ri}, s_{Ri} and o_{Ri} are constants. The execution time $TC_{Ri}(x)$ of the computation module C_{Ri} depends on an algorithm of the computation module C_{Ri}. In this paper, we consider two types of processes where computation complexities are $O(x)$ and $O(x^2)$. The execution time $TC_{Ri}(x)$ is $c_{Ri} \cdot x$ or $c_{Ri} \cdot x^2$, where c_{Ri} is a constant, depending on the computation complexity $O(x)$ and $O(x^2)$, respectively.

The power consumption model is proposed [5–8]. In this paper, we take the simple power consumption (SPC) model for simplicity [8]. Here, the power consumption PC_{Ri} [W] of a fog node f_{Ri} is maximum xE_{Ri} [W] if at least one process is performed. Otherwise, PC_{Ri} is minimum mE_{Ri} [W]. The electric energy consumption $EC_{Ri}(x)$ [J] of a fog node f_{Ri} to process data of size x is given as follows:

$$EC_{Ri}(x) = xE_{Ri} \cdot TC_{Ri}(x). \tag{7}$$

The electric power consumption PI_{Ri} and PO_{Ri} of the input I_{Ri} and output O_{Ri} modules are proportional to the receiving and transmission rates of a fog node f_{Ri}, respectively [9]. In this paper, we assume $TI_{Ri}(x) = TT_{Rij,Ri}(x)$ and $TO_{Ri}(x) = TT_{R,Ri}(x)$. Hence, the electric energy $EI_{Ri}(x)$ and $EO_{Ri}(x)$ to receive and send data of size x, respectively, are given as follows:

$$EI_{Ri}(x) = PI_{Ri} \cdot TI_{Ri}(x). \tag{8}$$

$$EO_{Ri}(x) = PO_{Ri} \cdot TO_{Ri}(x). \tag{9}$$

PS_{Ri} [W] shows the electric power of a fog node f_{Ri} to store data in a database DB_{Ri} which depends on the access rate a_{Ri} [bps]. Hence, the electric energy consumption $ES_{Ri}(x)$ of a fog node f_{Ri} to store data of size x is given as follows:

$$ES_{Ri}(x) = PS_{Ri} \cdot TS_{Ri}((1 + \rho_{Ri}) \cdot x). \tag{10}$$

4 Evaluation

We evaluate the tree-based fog computing model. In this paper, we consider a height-balanced k-ary tree of fog nodes. The reduction ratio ρ_{Ri} of each process p_{Ri} on a fog node f_{Ri} is assumed to be the same, i.e. $\rho_{Ri} = \rho$. We consider a server f_0 where the minimum electric power consumption mE_0 is 126.1 [W] and maximum electric power consumption xE_0 is 301.3 [W] with two Intel Xeon E5-2667 v2 CPUs [1]. Each fog node f_{Ri} is realized by a Raspberry Pi Model B [3]. Here, the minimum electric power mE_{Ri} is 2.1 [W] and the maximum electric power xE_{Ri} is 3.7 [W].

In order to make clear the computation rate, we perform a same C program p which uses only CPU on the server f_0 and the fog node f_{Ri}. It takes mT = 0.879 [sec] to perform the process p without any other process on f_0. The computation rate CR_0 of the server s is assumed to be one. If the same process p is performed without any other process on a fog node f_{Ri}, it takes 4.75 [sec]. Hence, the computation rate CR_{Ri} of each fog node f_{Ri} is 0.879/4.75 = 0.185 (Fig. 5).

We consider a balanced k-ary tree-based fog computing model. That is, each fog node f_R has k ($\geq$1) child nodes $f_{R1}, ..., f_{Rk}$ and every edge node is at the same level $h - 1$. Here, h shows the height of the tree. We assume a process p is realized as a sequence of subprocesses $p_0, p_1, ..., p_m (m \geq 1)$. Each process p_i receives data from a preceding process p_{i+1} and outputs data to a succeeding process p_{i-1}. The computation complexity of each process p_i is $O(x)$ or $O(x^2)$ for size x of input data. Each fog node f_{Ri} of level l performs a process p_{m-h+l}. A subsequence $p_0, ..., p_{m-h}$ of processes are performed on the node f_0. Each process $p_i (i = m - h + 1, ..., m)$ is performed on fog nodes at a level $i - m + h$. There are k^{h-1} edge nodes. Let x be the total amount of sensor data [B] collected by

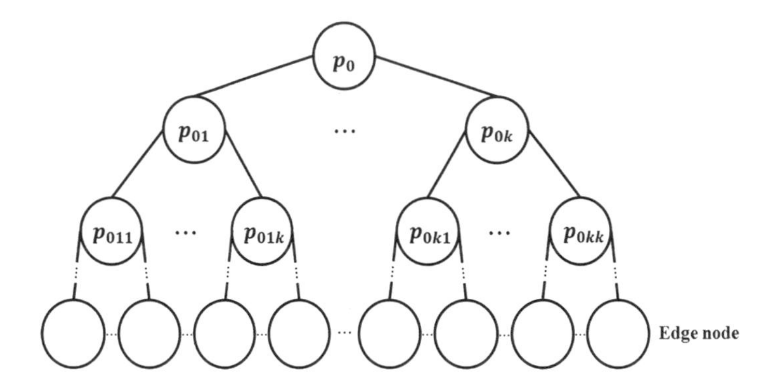

Fig. 5. k-ary tree of fog nodes.

sensor nodes. Each edge node receives sensor data of size x/k^{h-1}. In this evaluation, we assume $x = 1$[MB]. We also assume $k = 4$ or $k = 1$. "$k = 1$" means a linear model [11]. The network N supports every pair of nodes with the same bandwidth b. The bandwidth b is assumed to be 200 [Kbps].

In the cloud computing model, we consider a tree of fog nodes from sensor nodes to a server s. Each fog node f_{Ri} just forwards messages to a fog node f_R. Hence, each fog node f_{Ri} supports the input module I_{Ri} and output module O_{Ri}. The data d obtained from sensors is just forwarded from a fog node f_{Ri} to another fog node f_R.

Figure 6 shows the ratio of the total electric energy consumed by n nodes in the tree-based ($k = 4$) and linear ($k = 1$) fog computing models to the cloud computing model for the tree height h. Here, the computation complexity of

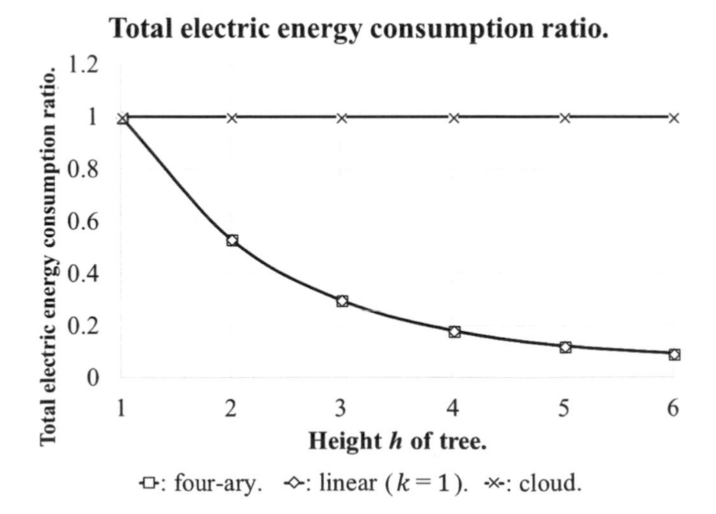

Fig. 6. Total electric energy consumption ratio for $O(x)$.

each process is $O(x)$. As shown in Fig. 6, the total electric energy consumption is the same and can be reduced in the tree-based model compared with the cloud computing model. The total electric energy consumption of the cloud model is independent of h. The total electric energy consumption of nodes monotonically decreases as the number n of fog nodes increases in the tree-based model for $k = 1$ and $k = 4$.

Figure 7 shows the ratio of the total electric energy consumption where the computation complexity of each process is $O(x^2)$. The total electric energy consumption ratio of a tree-based model with $k = 4$ is smaller than the linear model, i.e. $k = 1$.

Figures 8 and 9 show the ratio of the total execution time of nodes in the tree-based fog computing model to the cloud model for tree height h where the computation complexity of each process is $O(x)$ and $O(x^2)$, respectively. The total execution time can be reduced in the four-ary tree-based model compared with the cloud model for $h \geq 3$. For $h = 2$, it takes longer time to perform the process in the tree-based model than the cloud model. As shown in Fig. 8, the total execution time of the linear model ($k = 1$) monotonically increases as the tree height h increases for $O(x)$ processes. The total execution time of the cloud model is constant and shorter than the linear model. On the other hand, Fig. 9 shows the total execution time where the computation complexity of the process is $O(x^2)$. The total execution time monotonically increases in the linear model and is invariant in the cloud model. The total execution time of the four-ary tree-based model monotonically decreases and shortest for computation complexity $O(x^2)$ of the processes.

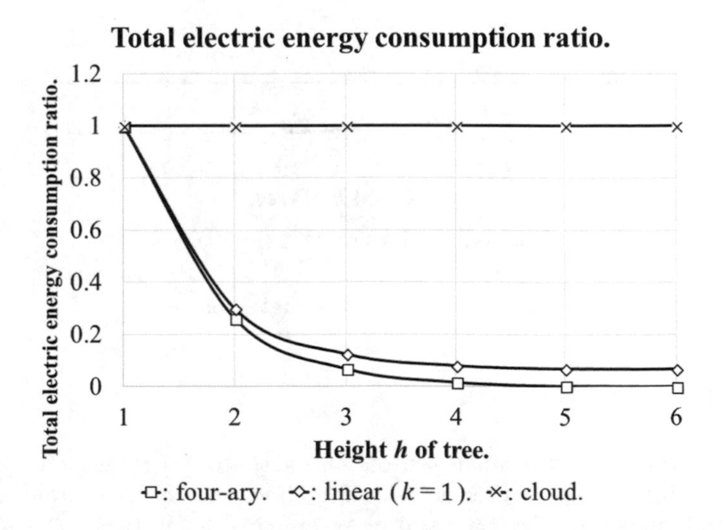

Fig. 7. Total electric energy consumption ratio for $O(x^2)$.

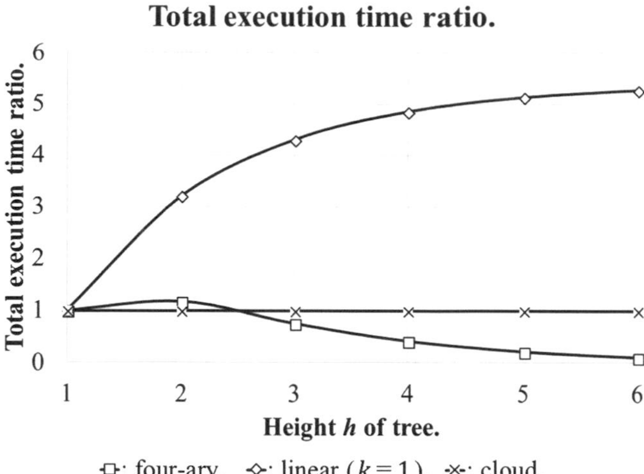

Fig. 8. Total execution time ratio for $O(x)$.

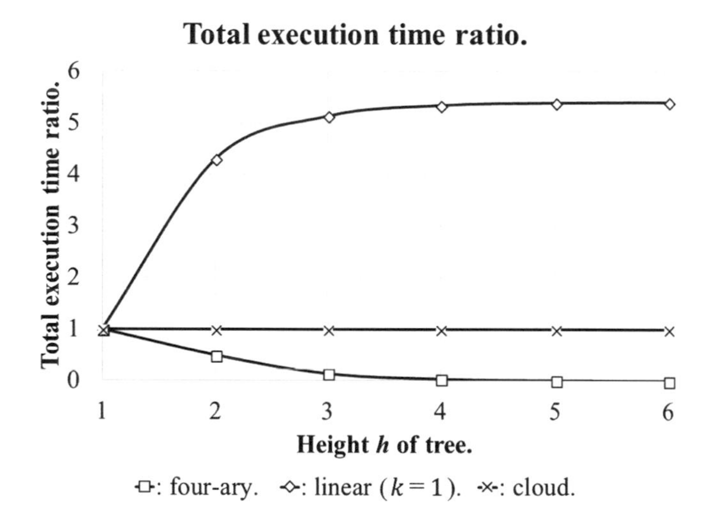

Fig. 9. Total execution time ratio for $O(x^2)$.

5 Concluding Remarks

The IoT is scalable and includes sensors and actuators. Processes and data are distributed to not only servers but also fog nodes in the fog computing model in order to reduce the delay time and processing overhead. Here, a huge amount of electric energy is consumed by a large number of nodes. Hence, it is critical to reduce the electric energy consumption of nodes in the IoT. In this paper, we proposed the tree-based fog computing model to reduce the total electric energy consumption and the total execution time of nodes in the IoT. We evaluated the tree-based fog computing model and consumed. We showed the total electric energy consumption of nodes and the execution time of nodes can be reduced in the tree-based model compared with the cloud computing model.

Acknowledgements. This work was supported by JSPS KAKENHI grant number 15H0295.

References

1. Dl360p gen8. www8.hp.com/h20195/v2/getpdf.aspx/c04128242.pdf?ver=2
2. Google green. http://www.google.com/green/2015
3. Raspberry pi 3 model b. https://www.raspberrypi.org/products/raspberry-pi-3-model-b/
4. Creeger, M.: Cloud computing: an overview. Queue **7**(5), 3–4 (2009)
5. Duolikun, D., Kataoka, H., Enokido, T., Takizawa, M.: Simple algorithms for selecting an energy-efficient server in a cluster of servers. Int. J. Commun. Netw. Distrib. Syst. (IJCNDS) (2017, accepted)
6. Enokido, T., Ailixier, A., Takizawa, M.: A model for reducing power consumption in peer-to-peer systems. IEEE Syst. J. **4**, 221–229 (2010)
7. Enokido, T., Ailixier, A., Takizawa, M.: Process allocation algorithms for saving power consumption in peer-to-peer systems. IEEE Trans. Ind. Electron. **58**(6), 2097–2105 (2011)
8. Enokido, T., Ailixier, A., Takizawa, M.: An extended simple power consumption model for selecting a server to perform computation type processes in digital ecosystems. IEEE Trans. Ind. Inform. **10**, 1627–1636 (2014)
9. Enokido, T., Ailixier, A., Takizawa, M.: An integrated power consumption model for communication and transaction based applications. In: Proceedings of IEEE the 25th International Conference on Advance Information Networking and Applications (AINA-2011), pp. 98–109 (2017)
10. Hanes, D., Salgueiro, G., Grossetete, P., Barton, R., Henry, J.: IoT Fundamentals: Networking Technologies, Protocols, and Use Cases for the Internet of Things. Cisco Press, Indianapolis (2018)
11. Oma, R., Nakamura, S., Enokido, T., Takizawa, M.: An energy-efficient model of fog and device nodes in IoT. In: Proceedings of IEEE the 32nd International Conference on Advanced Information Networking and Applications (AINA-2018), pp. 301–306 (2018)
12. Oma, R., Nakamura, S., Enokido, T., Takizawa, M.: A tree-based model of energy-efficient fog computing systems in IoT. In: Proceedings of the 12th International Conference on Complex, Intelligent, and Software Intensive Systems (CISIS-2018) (2018, accepted)
13. Yao, X., Wang, L.: Design and implementation of iot gateway based on embedded μtenux operating system. Int. J. Grid Util. Comput. **8**(1), 22–28 (2017)
14. Zhao, F., Guibas, L.: Wireless Sensor Networks: An Information Processing Approach. Morgan Kaufmann Publishers, San Francisco (2004)

Object-Based Information Flow Control Model in P2PPS Systems

Shigenari Nakamura[1(✉)], Tomoya Enokido[2], and Makoto Takizawa[1]

[1] Hosei University, Tokyo, Japan
nakamura.shigenari@gmail.com, makoto.takizawa@computer.org
[2] Rissho University, Tokyo, Japan
eno@ris.ac.jp

Abstract. In the P2PPS (P2P (peer-to-peer) type of topic-based PS (publish/subscribe)) model, each peer process (peer) publishes and subscribes event messages which are characterized by topics with no centralized coordinator. An illegal information flow occurs if an event message e_j published by a peer p_j carries information on some topics into the peer p_i, which the target peer p_i is not allowed to subscribe. In our previous studies, the SBS, TBS, and FS-H protocols are proposed to prevent illegal information flow among peers by banning event messages. In the protocols, the number of topics kept in every peer monotonically increases. Hence, most of the event messages are banned. In this paper, we newly consider the P2PPSO (P2PPS with object concept) model where the number of topics kept in every peer increases and decreases each time objects obtained by every peer are updated. In order to prevent illegal information flow from occurring in the P2PPSO system, we newly propose a TOBS (topics of objects-based synchronization) and TSOBS (topics and states of objects-based synchronization) protocols. In the TOBS protocol, it is simpler to detect illegal information flow than the TSOBS protocol. On the other hand, the fewer number of event messages are banned in the TSOBS protocol than the TOBS protocol.

1 Introduction

A distributed system is composed of peer processes (peers) which are cooperating with one another by manipulating objects and exchanging messages in networks. Here, messages carry objects in one peer to other peers. Even if a peer is not allowed to read data in an object o_i, the peer can read the data in another object o_j if the data is written into the object o_j [3]. Here, illegal information flow occurs. In order to keep information systems secure by preventing illegal information flow among objects, types of protocols [7–9] are proposed based on the role-based access control (RBAC) model [5]. On the other hand, content-based systems like PS (publish/subscribe) systems [1,2,4,15] are getting more important in various applications. Here, the PS model is an event-driven model of a distributed system and a process is modeled to be a sequence of publication and receipt events. Event messages carry objects in a peer to another peer.

L. Barolli et al. (Eds.): NBiS 2018, LNDECT 22, pp. 110–121, 2019.
https://doi.org/10.1007/978-3-319-98530-5_10

We consider a P2PPS (P2P (peer-to-peer) [16] of topic-based PS [14]) model [13] where each peer can play both publisher and subscriber roles with no centralized coordinator.

The TBAC (topic-based access control) model is proposed [12] in topic-based PS systems [1,2,4,15]. Here, only a peer granted publication and subscription rights is allowed to publish and subscribe topics, respectively. The topic sets $p_i.P$ and $p_i.S$ are publication and subscription topics of a peer p_i, respectively. An event message e_i published by a peer p_i is received by a target peer p_j if the subscription $p_j.S$ includes at least one common topic with the publication $e_i.P$. Here, topics in the subscription $p_i.S$ but not in the subscription $p_j.S$ are *hidden* in the event message e [12].

In our previous studies, the SBS (subscription-based synchronization) [12], TBS (topic-based synchronization) [10], and FS-H (flexible synchronization for hidden topics) [11] protocols are proposed based on the TBAC model to prevent illegal information flow. In the SBS protocol, it is checked whether or not an event message is illegal in terms of publication and subscription rights. Here, in addition to illegal event messages, even legal event messages may be banned. In the TBS protocol, only topics which each peer really manipulates are checked to decide whether or not an event message is illegal. Hence, all and only illegal event messages are banned. In the FS-H protocol [11], even if an event message carries hidden topics, if the hidden topics are strongly related with some subscription topics, a target peer receives the event message and adds the hidden topics to its subscription. The number of event messages banned is more reduced by using the learning mechanism.

In the protocols, every peer p_i keeps every topic carried into p_i. Here, the number of topics kept by every peer p_i monotonically increases. However, every topic stored in a peer is not necessarily permanently meaningful. For instance, an object in a peer is deleted or updated and topics of the object are changed. Thus, some topic may disappear in a peer. A temporarily meaningful topic may become meaningless after its term of validity. Such topics should be deleted from every peer because the more number of topics are kept in every peer, the more highly an illegal information flow causes.

In this paper, we newly consider a P2PPSO (P2PPS with object concept) model where the number of topics kept by every peer increases and decreases each time objects obtained by every peer are updated. In the P2PPSO system, information exchanged among peers is considered to be composed of objects. Here, each event message carries some objects. Each object has a set of topics which are related with the information in the object. Every peer keeps every topic of objects carried into p_i. In this paper, we assume that only the creator of an object can update data in the object. If a peer updates data in an object, every replica of the object obtained by the other peers is updated to keep to the mutual consistency among the replicas. For instance, if a peer deletes some information from an object, the topics of the information are also deleted from a topic set of both the object and replicas of the object. Thus, the number of topics kept by every peer can decrease in the P2PPSO model while only increasing in the previous studies.

In order to prevent illegal information flow from occurring in the P2PPSO system, we newly propose TOBS (topics of objects-based synchronization) and TSOBS (topics and states of objects-based synchronization) protocols. In the TOBS protocol, an event message which may cause an illegal information flow is banned at a target peer. Only comparison between the topics carried by an event message and the subscription of a target peer of the event message have to be done to check whether or not an event message causes an illegal information flow. This means, the mechanism to detect illegal information flow is simple. In the TSOBS protocol, if a target peer obtains the objects whose data is newer than the objects carried by an event message, the peer ignores the objects. Here, some event messages which are banned in the TOBS protocol are not banned in the TSOBS protocol. Hence, the number of event messages banned is more reduced than the TOBS protocol. However, in the TSOBS protocol, the state of every object carried by an event message is checked. Hence, the mechanism to detect illegal information flow is more complex than the TOBS protocol.

In Sect. 2, we discuss the object concept. In Sect. 3, we define the information flow relations based on the TBAC model and objects. In Sect. 4, we newly propose the TOBS and TSOBS protocols in the P2PPSO system.

2 System Model

In this paper, we consider the P2PPS (P2P (peer-to-peer) [16] of PS (publish/subscribe) [1,2,4,15]) model [13], which is composed of a set P of peer processes (peers) $p_1, \ldots, p_{pn} (pn \geq 1)$. Here, each peer p_i can play both publisher and subscriber roles. Let T be a set $\{t_1, \ldots, t_{tn}\} (tn \geq 1)$ of all topics in a system. In a topic-based PS system [14], a peer p_i publishes an event message e_i with publication $e_i.P$ ($\subseteq T$). A peer p_j specifies topics for its subscription $p_j.S$ ($\subseteq T$). An event message e_i is received by a peer p_j if $e_i.P \cap p_j.S \neq \phi$. Here, p_i and p_j are *source* and *target* peers of the event message e_i, respectively.

In the TBAC (*topic-based access control*) model [12], an access right $\langle t, op \rangle$ shows that a peer granted the access right $\langle t, op \rangle$ is allowed to issue an operation op ($\in$ {publish (pb), subscribe (sb)}) on a topic t ($\in T$). The publication $p_i.P$ ($\subseteq T$) and subscription $p_i.S$ ($\subseteq T$) of a peer p_i are subsets of topics which the peer p_i is allowed to publish and subscribe, respectively. A peer p_i can publish an event message e_i with publication $e_i.P$ only if the peer p_i is granted a publication right $\langle t, pb \rangle$ for every topic t in the publication $e_i.P$, i.e. $e_i.P \subseteq p_i.P$. An event message e_i published by a source peer p_i is received by a target peer p_j if $p_j.S \cap e_i.P \neq \phi$.

Example 1. Suppose there are three peers p_i, p_j, and p_k in a system. We also suppose $p_i.S = p_i.P = \{x, y\}$, $p_j.S = p_j.P = \{x, y, z\}$, and $p_k.S = p_k.P = \{y, z\}$. First, the peer p_i publishes an event message e_i with publication $e_i.P = \{x\} (\subseteq p_i.P)$. The event message e_i is received by the peer p_j since $e_i.P \cap p_j.S = \{x\} \neq \phi$. Here, the peer p_i may receive event messages about the topics x and y before publishing the event message e_i since the peer p_i is granted the access rights

$\langle x, sb \rangle$ and $\langle y, sb \rangle$. Here, the event message e_i may carry information on the topics x and y.

Then, suppose a peer p_j publishes an event message e_j with publication $e_j.P = \{z\}(\subseteq p_j.P)$. The event message e_j is received by the peer p_k since $e_j.P \cap p_k.S = \{z\} \neq \phi$. Since the event message e_j is published by the peer p_j after the event message e_i is delivered to the peer p_j, the event message e_i causally precedes the event message e_j according to the causality theory [6]. This means, the event message e_j may carry event information of the event message e_i, i.e. event information on not only the topic z in the subscription $p_j.S$ but also the topics x and y of the event message e_i. However, the peer p_k is not granted the subscription right $\langle x, sb \rangle$. This means, information on the topic x which the peer p_k is not allowed to subscribe can be delivered to the peer p_k via the peer p_j. Here, information illegally flows to the peer p_k from the peer p_j.

In our previous studies, the SBS (subscription-based synchronization) [12], TBS (topic-based synchronization) [10], and FS-H (flexible synchronization for hidden topics) [11] protocols are proposed in order to prevent every illegal information flow. In the protocols, every topic carried into a peer p_i is stored in the variable $p_i.T$, i.e. $p_i.T$ indicates topics of information obtained by the peer p_i. On receipt of an event message, the number of topics in the variable $p_i.T$ of each peer p_i monotonically increases. The variable $p_i.T$ is carried by the event message e_i as a variable $e_i.T$ and whether or not an illegal information flow causes is checked by using $e_i.T$. However, every topic stored in a peer is not necessarily permanently meaningful. For instance, a topic which is temporarily meaningful may become meaningless after its term of validity. Such topics should be deleted from the variable $p_i.T$ because the more number of topics are included in the variable $e_i.T$, the more often an illegal information flow causes. In the Example 1, if the topic x becomes invalid and is deleted from $p_i.T$ and $e_i.T$ between after receiving e_i and before publishing the event message e_j, an illegal information flow from the peer p_j to the peer p_k does not occur. In our previous studies, dynamically decreasing the number of topics in the variable $p_i.T$ and $e_i.T$ is not considered.

In this paper, we newly propose a P2PPSO (P2PPS with object concept) model to reduce the number of topics stored in each peer. Each event message e_i carries some objects in a set $p_i.D$ which is composed of objects obtained by the source peer p_i. Let $e.O$ be a set of objects carried by the event message e. Let o_i^j be an object whose creator is the peer p_i obtained by the peer p_j, i.e. $o_i^j \in p_j.D$. We assume only the creator p_i can update the data in the objects o_i created by the peer p_i. A set $o_i^j.T$ of each object o_i^j indicates topics of the information included in the object o_i^j. Here, a set $e.T$ shows topics of the information which the event message e carries, i.e. a set of topics in the objects carried by the event message e, i.e. $e.T = \{t | t \in o.T \text{ and } o \in e.O\}$.

Example 2. We consider the same system in the Example 1. First, a pair of peers p_i and p_j create objects o_i^i and o_j^j and then store them in their storage $p_i.D$ and $p_j.D$, respectively. The objects o_i^i and o_j^j include information on a pair of topics

x and y and a pair of y and z, i.e. $o_i^i.T = \{x, y\}$ and $o_j^j.T = \{y, z\}$, respectively. Next, the peer p_i publishes an event message e_i where $e_i.P = \{x\}, e_i.O = \{o_i^i\}$, and $e_i.T = o_i^i.T = \{x, y\}$. Since $e_i.P \cap p_j.S \neq \phi$, the peer p_j receives the event message e_i. The peer p_j stores the object o_i^i in its storage $p_j.D$. Here, $p_j.D = \{o_i^j, o_j^j\}$.

Then, suppose the information on the topic x becomes meaningless in the peer p_i. The peer p_i deletes the information on the topic x from the object o_i^i. Here, $o_i^i.T = \{y\}$. Since o_i^j obtained by the peer p_j is the same object as the object o_i^i, o_i^j is also updated as well as o_i^i. Hence, $o_i^j.T = o_i^i.T = \{y\}$.

Next, the peer p_j includes a pair of objects o_i^j and o_j^j into an event message e_j and publishes e_j where $e_j.P = \{z\}, e_j.O = \{o_i^j, o_j^j\}$, and $e_j.T = o_i^j.T(= \{y\}) \cup o_j^j.T(= \{y, z\}) = \{y, z\}$. Since $e_j.P \cap p_k.S \neq \phi$, the peer p_k receives the event message e_j. Here, it is noted that the event message e_j does not carry the information on the topic x which the peer p_k is not allowed to subscribe differently from the Example 1. Hence, the information does not illegally flow to the peer p_k from the peer p_j. The peer p_k stores the objects o_i^j and o_j^j in its storage $p_k.D$. Here, $p_k.D = \{o_i^k, o_j^k\}$.

3 Information Flow Relations

In the Example 2, each object includes some information and is carried to other peers by event messages. Each object o_i^j has a variable $o_i^j.T$ which indicates what information the object o_i^j includes as some topics. The variable $e_j.T$ of the event message e_j also shows what information every object o_i^j in the event message e_j includes as some topics. The set of topics $p_i.S$ of the peer p_i indicates the topics which the peer p_i is allowed to subscribe. Hence, information flow relations are defined in terms of the topics.

We define the information flow relations based on the TBAC model and objects. First, the information flow relation ($\rightarrow$) is defined as follows:

Definition 1. An event message e_i published by a peer p_i *flows* to a peer $p_j(e_i \rightarrow p_j)$ iff (if and only if) $e_i.O \neq \phi$ and $e_i.P \cap p_j.S \neq \phi$.

If an event message e_i flows to a peer $p_j(e_i \rightarrow p_j)$, an event message e_i published by the peer p_i can be received by the peer p_j. Here, some information obtained by the peer p_i flows into the peer p_j. Otherwise, no information from the peer p_i flows into the peer p_j because the event message e_i is not received by the peer p_j.

In the Example 2, the event message e_i flows to the peer $p_j(e_i \rightarrow p_j)$ since $e_i.O = \{o_i^i\} \neq \phi$ and $e_i.P \cap p_j.S = \{x\} \neq \phi$. Here, the event message e_i includes an object o_i^i which is related with the pair of topics x and y. Since the peer p_j is allowed to subscribe the topic x, the event message e_i is received by the peer p_j. Here, the object o_i^i including some information on the topics x and y flows into the peer p_j.

Next, the legal information flow relation ($\Rightarrow$) is defined as follows:

Definition 2. An event message e_i published by a peer p_i *legally flows* to a peer $p_j (e_i \Rightarrow p_j)$ iff $e_i \rightarrow p_j$ and $e_i.T \subseteq p_j.S$.

The condition $e_i.T \subseteq p_j.S$ shows that every topic in the variable $e_i.T$ is also in the subscription $p_j.S$. This means, the event message e_i carries no information on the topics which the target peer p_j is not allowed to subscribe. This means, no information illegally flows into the peer p_j.

In the Example 2, the event message e_i legally flows to the peer $p_j (e_i \Rightarrow p_j)$ since $e_i \rightarrow p_j$ and $e_i.T (= \{x, y\}) \subseteq p_j.S (= \{x, y, z\})$. The event message e_i carries an object o_i^i whose variable $o_i^i.T = \{x, y\}$. This means, the information in the object o_i^i is related with the topics x and y. Since the event message e_i carries the only object o_i^i, the variable $e_i.T = o_i^i.T = \{x, y\}$. Here, the information on the topics x and y flows into the peer p_j. The peer p_j is allowed to subscribe the topics x and y, i.e. there is no information on the topics which the peer p_j is not allowed to subscribe. Hence, the information legally flows into the peer p_j from the peer p_i.

Finally, the illegal information flow relation ($\mapsto$) is defined as follows:

Definition 3. An event message e_i published by a peer p_i *illegally flows* to a peer $p_j (e_i \mapsto p_j)$ iff $e_i \rightarrow p_j$ and $e_i.T \not\subseteq p_j.S$.

The condition $e_i.T \not\subseteq p_j.S$ means that the event message e_i carries the information on the topics into the target peer p_j, which the target peer p_j is not allowed to subscribe.

In the Example 2, if the topic x is not deleted from the object o_i^j and is carried to the peer p_k by the event message e_j, the peer p_k can get the information on the topic x although the peer p_k is not allowed to subscribe the topic x. Here, the information related with the topic x illegally flows into the peer p_k from the peer p_j.

4 Synchronization Protocols

In Sect. 3, we define the illegal information flow relation ($\mapsto$) based on the TBAC model and objects. An illegal information flow occurs if an event message carries some information into the target peer which a target peer is not allowed to get in terms of topics. In this section, we newly propose TOBS (topics of objects-based synchronization) and TSOBS (topics and states of objects-based synchronization) protocols to prevent illegal information flow in the P2PPSO system based on the information flow relations. In both the protocols, objects obtained by each peer are synchronized so that no illegal information flow occurs through exchanging event messages among peers.

4.1 TOBS (Topics of Objects-Based Synchronization) Protocol

First, we consider the TOBS (topics of objects-based synchronization) protocol. In the P2PPSO system, a peer p_i is granted topics in its subscription $p_i.S$ and publication $p_i.P$. A peer p_i is allowed to subscribe and publish a topic in $p_i.S$ and $p_i.P$, respectively. An event message e carries objects. A set of topics carried by the event message e is $e.T$ which is composed of topics of the objects included in e. In the TOBS protocol, on receipt of an event message e_i, the condition $e_i.T \subseteq p_j.S$ is checked in the target peer p_j to make clear whether or not the event message e_i causes an illegal information flow. If a peer p_i updates data in an object o_i^i, the peer p_i publishes an *update* event message ${}_ue_i$ to synchronize replicas of the object o_i^i obtained by other peers. In this paper, we assume only the creator p_i of the object o_i^i can update the data in the objects o_i. An object o_i^j which is created by a peer p_i but is obtained by a peer p_j is a replica of an object o_i^i iff $o_i^i \xrightarrow{r} o_i^j$.

[TOBS protocol]

1. A peer p_i publishes an event message e_i:
   a. $e_i.O$ = objects included in e_i from $p_i.D$, i.e. $e_i.O \subseteq p_i.D$;
      $e_i.T = \{t | t \in o^i.T \text{ and } o^i \in e_i.O\}$;
      $e_i.P$ = publication topics of e_i, i.e. $e_i.P \subseteq p_i.P$;
   b. p_i publishes the event message e_i;
2. A peer p_i receives an event message e_j from a peer p_j, i.e. $e_j \rightarrow p_i$ holds:
   a. If $e_j \Rightarrow p_i$, e_j is delivered to p_i and
      i. if e_j is an update event message, i.e. $e_j = {}_ue_j$, for every object o_j^i such that $o_j^j \xrightarrow{r} o_j^i$ and $o_j^j \in {}_ue_j.O$, $o_j^i = o_j^j$;
      ii. otherwise,
         A. for every object o_k^i such that $k \neq i$, $o_k^k \xrightarrow{r} o_k^i$, and $o_k^j \in e_j.O$, $o_k^i = o_k^j$;
         B. every object o_m^j such that $m \neq i$, $o_m^m \xrightarrow{r} o_m^i$, $o_m^j \in e_j.O$, and $o_m^i \notin p_i.D$, o_m^j is added to the storage $p_i.D$;
   b. Else if $e_j = {}_ue_j$ and
      i. if there is no o_j^i in $p_i.D$ such that $o_j^j \xrightarrow{r} o_j^i$ and $o_j^j \in {}_ue_j.O$, p_i ignores ${}_ue_j$;
      ii. otherwise, o_j^i is deleted from $p_i.D$;
   c. Otherwise, the event message e_j is banned at the peer p_i;
3. A peer p_i updates data in an object o_i^i in $p_i.D$:
   a. p_i makes an update event message ${}_ue_i$ where ${}_ue_i.O = \{o_i^i | o_i^i \text{ is being updated by } p_i\}$, ${}_ue_i.T = \{t | t \in o_i^i.T \text{ and } o_i^i \in {}_ue_i.O\}$, and ${}_ue_i.P = {}_ue_i.T$;
   b. Then, p_i updates data in an object o_i^i so that $o_i^i.T \subseteq p_i.S$;
   c. ${}_ue_i.O = \{o_i^i | o_i^i \text{ is an updated object}\}$;
      ${}_ue_i.T = \{t | t \in o_i^i.T \text{ and } o_i^i \in {}_ue_i.O\}$;
   d. Next, p_i publishes an update event message ${}_ue_i$;

In the P2PPSO system, if a peer p_i updates data in an object o_i^i, every replica object o_i^k obtained by every other peer p_k, i.e. $i \neq k$, is synchronized with o_i^i. In the TOBS protocol, such updates of objects are realized in the same way as the exchanges of event messages among peers.

Example 3. Suppose there are three peers p_i, p_j, and p_k as shown in Fig. 1. We also suppose $p_i.S = p_i.P = \{x, y\}, p_j.S = p_j.P = \{x, y, z\}$, and $p_k.S = p_k.P = \{y, z\}$. First, a pair of the peers p_i and p_j create objects o_i^i where $o_i^i.T = \{x\}$ and o_j^j where $o_j^j.T = \{y, z\}$ and then store them in their storages $p_i.D$ and $p_j.D$, respectively. Next, the peer p_i publishes an event message e_i where $e_i.O = \{o_i^i\}$, $e_i.T = o_i^i.T = \{x\}$, and $e_i.P = \{x\}$. Since $e_i \rightarrow p_j$, the peer p_j receives the event message e_i. The event message e_i is delivered to the peer p_j and the object o_i^i is stored in the storage $p_j.D$ of the peer p_j because $e_i.T \subseteq p_j.S$, i.e. $e_i \Rightarrow p_j$. Here, $p_j.D = \{o_i^j, o_j^j\}$.

Then, suppose the peer p_i updates data in the object o_i^i. The object o_i^i newly includes information on the topic y and $o_i^i.T$ is changed with topics $\{x, y\}$. The peer p_i publishes an update event message $_ue_i$ to make other peer p_m synchronize the object o_i^m with o_i^i. Here, the publication $e_i.P$ is same as the variable $o_i^i.T$ of the unupdated object o_i^i, i.e. $e_i.P = \{x\}$. Since $e_i \Rightarrow p_j$, the event message e_i for update of the object o_i^j is delivered to the peer p_j. Hence, data in the object o_i^j is updated and $o_i^j.T(= \{x\})$ is changed with $\{x, y\}$.

Next, the peer p_j publishes an event message e_j where $e_j.O = \{o_i^j, o_j^j\}$, $e_j.T = o_i^j.T \cup o_j^j.T = \{x, y, z\}$, and $e_j.P = \{z\}$. Since $e_j \rightarrow p_k$, the peer p_k receives the event message e_j. However, the topic x in the variable $e_j.T$ is not included in the subscription $p_k.S$ of the peer p_k, i.e. $e_j \mapsto p_k$ holds. Here, an illegal information flow from the peer p_j to the peer p_k occurs. Hence, the event message e_j is banned at the peer p_k.

4.2 TSOBS (Topics and States of Objects-Based Synchronization) Protocol

First, we consider the following example with TOBS protocol:

Example 4. We consider the same system in the example 3. The publish and subscribe operations of each peer are issued as shown in Fig. 2. First, a pair of the peers p_i and p_j create objects o_i^i where $o_i^i.T = \{x\}$ and o_j^j where $o_j^j.T = \{y, z\}$ and then store them in their storages $p_i.D$ and $p_j.D$, respectively. Next, the peer p_i publishes an event message e_i where $e_i.O = \{o_i^i\}$, $e_i.T = o_i^i.T = \{x\}$, and $e_i.P = \{x\}$. Since $e_i \Rightarrow p_j$, the event message e_i is delivered to the peer p_j and the object o_i^i is stored in the storage $p_j.D$ of the peer p_j. Hence, $p_j.D = \{o_i^j, o_j^j\}$.

Then, suppose the peer p_i updates data in the object o_i^i. The information on the topic x is changed with the information on the topic y. Here, $o_i^i.T$ is changed with $\{y\}$. The peer p_i publishes an update event message $_ue_i$ to make other peer p_m synchronize the object o_i^m with o_i^i. Here, the publication $e_i.P$ is same as the

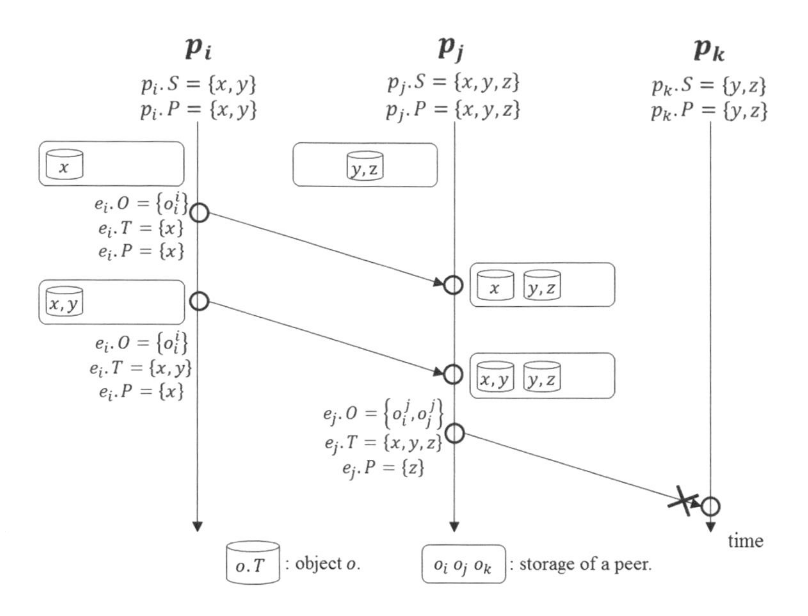

Fig. 1. TOBS protocol.

variable $o_i^i.T$ of the unupdated object o_i^i, i.e. $e_i.P = \{x\}$. Since $e_i \Rightarrow p_j$, the object o_i^j in the storage $p_j.D$ of the peer p_j is updated. Hence, $o_i^j.T(= \{x\})$ is changed with $\{y\}$.

Next, the peer p_i publishes an event message e_i where $e_i.O = \{o_i^i\}$, $e_i.T = o_i^i.T = \{y\}$, and $e_i.P = \{y\}$. Since $e_i \Rightarrow p_k$, the object o_i^i is added to the storage $p_k.D$. Hence, $p_k.D = \{o_i^k\}$.

Suppose the peer p_j publishes an event message e_j before receiving the event message e_i for updating the object o_i^j in $p_j.D$ and the event message e_j reaches the peer p_k after the peer p_k receives the updated object o_i^i. Here, since $e_j \mapsto p_k$, the event message e_j is banned at the peer p_k.

In the Example 4, the event message e_j carries the unupdated object o_i^j to the peer p_k and is banned. If the state of the object o_i^j in the event message e_j is up-to-date, i.e. $o_i^j = \{y\}$, the event message e_j is not banned. The peer p_k already knows the object o_i^j is not up-to-date when p_k receives e_j because p_k receives the updated object o_i^i from p_i before receiving e_j.

In this paper, we newly propose the TSOBS (topics and states of objects-based synchronization) protocol. Here, each peer p_i ignores the unupdated objects carried by an event message e_j when p_i receives e_j to reduce the number of event messages banned in the system. For this aim, we consider the time stamp $o_i^j.ts$ of each object o_i^j which indicates time when an object o_i^j is finally updated.

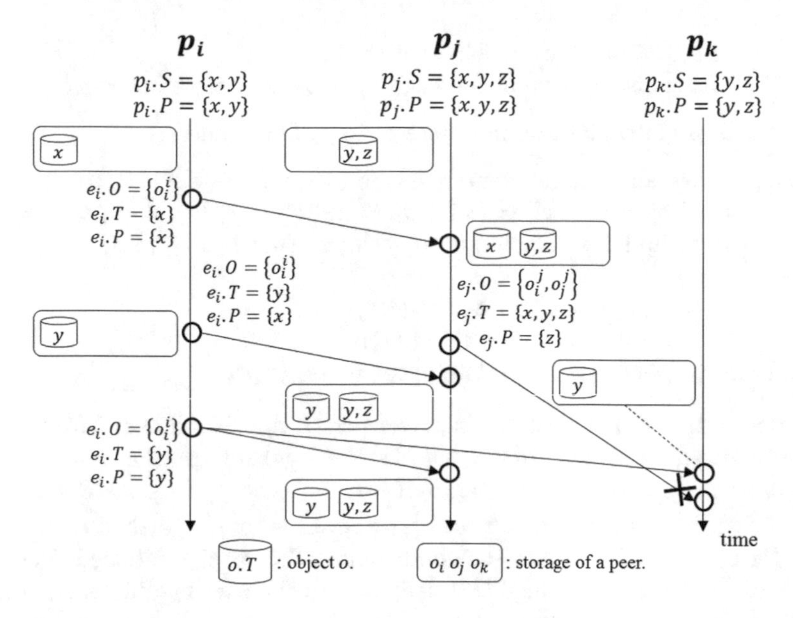

Fig. 2. Illegal information flow caused by unupdated object delivery in the TOBS protocol.

[TSOBS protocol]

1. A peer p_i publishes an event message e_i:
   a. $e_i.O =$ objects included in e_i from $p_i.D$, i.e. $e_i.O \subseteq p_i.D$;
      $e_i.T = \{t | t \in o^i.T \text{ and } o^i \in e_i.O\}$;
      $e_i.P =$ publication topics of e_i, i.e. $e_i.P \subseteq p_i.P$;
   b. p_i publishes the event message e_i;
2. A peer p_i receives an event message e_j from a peer p_j, i.e. $e_j \rightarrow p_i$ holds:
   a. If $e_j \Rightarrow p_i$, e_j is delivered to p_i and
      i. if e_j is an update event message, i.e. $e_j = {}_ue_j$, for every object o_j^i such that $o_j^j \xrightarrow{r} o_j^i$ and $o_j^j \in {}_ue_j.O, o_j^i = o_j^j$;
      ii. otherwise,
         A. for every object o_k^i such that $k \neq i$, $o_k^k \xrightarrow{r} o_k^i$, and $o_k^j \in e_j.O$, $o_k^i = o_k^j$;
         B. every object o_m^j such that $m \neq i$, $o_m^m \xrightarrow{r} o_m^i$, $o_m^j \in e_j.O$, and $o_m^i \notin p_i.D$, o_m^j is added to the storage $p_i.D$;
   b. Else if $e_j = {}_ue_j$ and
      i. if there is no o_j^i in $p_i.D$ such that $o_j^j \xrightarrow{r} o_j^i$ and $o_j^j \in {}_ue_j.O$, p_i ignores ${}_ue_j$;
      ii. otherwise, o_j^i is deleted from $p_i.D$;
   c. Otherwise, for every object o_k^j such that $o_k^j \in e_j.O$ and $o_k^i.ts > o_k^j.ts$, $e_j.O - o_k^j$;
      Then, $e_j.T$ is calculated based on $e_j.O$;

i. If $e_j.T \subseteq p_i.S$, e_j is delivered to p_i;
ii. Otherwise, the event message e_j is banned at the peer p_i;

3. A peer p_i updates data in an object o_i^i in $p_i.D$ at time τ:
   a. p_i makes an update event message ${}_u e_i$ where ${}_u e_i.O = \{o_i^i | o_i^i$ is being updated by $p_i\}$, ${}_u e_i.T = \{t | t \in o_i^i.T$ and $o_i^i \in {}_u e_i.O\}$, and ${}_u e_i.P = {}_u e_i.T$;
   b. Then, p_i updates data in an object o_i^i so that $o_i^i.T \subseteq p_i.S$;
   $o_i^i.ts = \tau$;
   c. ${}_u e_i.O = \{o_i^i | o_i^i$ is an updated object$\}$;
   ${}_u e_i.T = \{t | t \in o_i^i.T$ and $o_i^i \in {}_u e_i.O\}$;
   d. Next, p_i publishes an update event message ${}_u e_i$;

Suppose the TSOBS protocol is performed in the Example 4. After receiving the event message e_j, the peer p_k checks the legal information flow condition. Since $e_j \mapsto p_k$, p_k checks time stamp of every object o_m^j in $e_j.O$. Since $o_i^k.ts > o_i^j.ts$, the peer p_k ignores the object o_i^j, i.e. $e_j.O - \{o_i^j\}$, and then $e_j.T$ is calculated. Here, $e_j.T = \{y, z\}$ and the condition $e_j.T \subseteq p_k.S$ is satisfied. Hence, the event message e_j is not banned and delivered to the peer p_k differently from the TOBS protocol.

5 Concluding Remarks

In our previous studies, the SBS [12], TBS [10], and FS-H [11] protocols are proposed in order to prevent illegal information flow among peers by banning event messages which may cause illegal information flows. Here, the number of topics kept by every peer monotonically increases. Hence, almost all the event messages are banned. In this paper, we newly consider the P2PPSO (P2PPS with object concept) model where the number of topics kept by every peer increases and decreases each time objects obtained by every peer are updated. In order to prevent illegal information flow from occurring in the P2PPSO system, we newly propose the TOBS and TSOBS protocols. The mechanism of the TOBS protocol to detect illegal information flow is easier than the TSOBS protocol. On the other hand, the fewer number of event messages are banned in the TSOBS protocol than the TOBS protocol.

Acknowledgements. This work was supported by Japan Society for the Promotion of Scienc (JSPS) KAKENHI 15H0295 and Grant-in-Aid for JSPS Research Fellow grant 17J00106.

References

1. Google alert. http://www.google.com/alerts
2. Blanco, R., Alencar, P.: Event models in distributed event based systems. In: Principles and Applications of Distributed Event-Based Systems, pp. 19–42 (2010)
3. Denning, D.E.R.: Cryptography and Data Security. Addison Wesley (1982)

4. Eugster, P.T., Felber, P.A., Guerraoui, R., Kermarrec, A.M.: The many faces of publish/subscribe. ACM Comput. Surv. **35**(2), 114–131 (2003)
5. Ferraiolo, D.F., Kuhn, D.R., Chandramouli, R.: Role-based Access Controls, 2nd edn. Artech (2007)
6. Lamport, L.: Time, clocks, and the ordering of event in a distributed systems. Commun. ACM **21**(7), 558–565 (1978)
7. Nakamura, S., Duolikun, D., Enokido, T., Takizawa, M.: A flexible read-write abortion protocol to prevent illegal information flow among objects. J. Mobile Multimedia **11**(3&4), 263–280 (2015)
8. Nakamura, S., Duolikun, D., Enokido, T., Takizawa, M.: A write abortion-based protocol in role-based access control systems. Int. J. Adapt. Innovative Syst. **2**(2), 142–160 (2015)
9. Nakamura, S., Duolikun, D., Enokido, T., Takizawa, M.: A read-write abortion (RWA) protocol to prevent illegal information flow in role-based access control systems. Int. J. Space-Based Situated Comput. **6**(1), 43–53 (2016)
10. Nakamura, S., Enokido, T., Takizawa, M.: Topic-based synchronization (TBS) protocols to prevent illegal information flow in peer-to-peer publish/subscribe systems. In: Proceedings of the 11th International Conference on Broadband and Wireless Computing, Communication and Applications (BWCCA-2016), pp. 57–68 (2016)
11. Nakamura, S., Ogiela, L., Enokido, T., Takizawa, M.: A flexible synchronization protocol for hidden topics to prevent illegal information flow in P2PPS systems. In: Proceedings of the 12th International Conference on Broad-Band Wireless Computing, Communication and Applications (BWCCA-2017), pp. 138–148 (2017)
12. Nakamura, S., Ogiela, L., Enokido, T., Takizawa, M.: An information flow control model in a topic-based publish/subscribe system. J. High Speed Netw. (JHS) **24**(3), 243–257 (2018)
13. Nakayama, H., Duolikun, D., Enokido, T., Takizawa, M.: Reduction of unnecessarily ordered event messages in peer-to-peer model of topic-based publish/subscribe systems. In: Proceedings of IEEE the 30th International Conference on Advanced Information Networking and Applications (AINA-2016), pp. 1160–1167 (2016)
14. Setty, V., van Steen, M., Vitenberg, R., Voulgaris, S.: Poldercast: Fast, robust, and scalable architecture for P2P topic-based pub/sub. In: Proceedings of ACM/IFIP/USENIX 13th International Conference on Middleware (Middleware 2012), pp. 271–291 (2012)
15. Tarkoma, S.: Publish/Subscribe System: Design and Principles, 1st edn. Wiley (2012)
16. Waluyo, A.B., Taniar, D., Rahayu, W., Aikebaier, A., Takizawa, M., Srinivasan, B.: Trustworthy-based efficient data broadcast model for P2P interaction in resource-constrained wireless environments. J. Comput. Syst. Sci. (JCSS) **78**(6), 1716–1736 (2012)

Performance Evaluation of Energy Consumption for Different DTN Routing Protocols

Evjola Spaho[1(✉)], Klodian Dhoska[2], Kevin Bylykbashi[3], Leonard Barolli[4], Vladi Kolici[1], and Makoto Takizawa[5]

[1] Department of Electronics and Telecommunication, Faculty of Information Technology, Polytechnic University of Tirana, Mother Teresa Square, No. 4, Tirana, Albania
evjolaspaho@hotmail.com, vkolici@fti.edu.al

[2] Department of Production-Management, Faculty of Mechanical Engineering, Polytechnic University of Tirana, Mother Teresa Square, No. 4, Tirana, Albania
kdhoska@upt.al

[3] Faculty of Information Technology, Polytechnic University of Tirana, Mother Teresa Square, No. 4, Tirana, Albania
kevin.bylykbashi@fti.edu.al

[4] Department of Information and Communication Engineering, Fukuoka Institute of Technology (FIT), 3-30-1 Wajiro-Higashi, Higashi-Ku, Fukuoka 811-0295, Japan
barolli@fit.ac.jp

[5] Department of Advanced Sciences, Hosei University, 3-7-2, Kajino-cho, Koganei-shi, Tokyo 184-8584, Japan
makoto.takizawa@computer.org

Abstract. In this paper, we evaluate the energy consumption of different routing protocols in a Delay Tolerant Network (DTN). Seven groups with three stationary sensor nodes for each group sense the temperature, humidity and wind speed and send these data to a stationary destination node that collects them for statistical and data analysis purposes. The opportunistic contacts will exchange the sensed data to different relay nodes that are pedestrians and cyclist equipped with smart devices moving in Tirana city roads, until the destination node is reached. For simulations we use the Opportunistic Network Environment (ONE) simulator. Nodes in this DTN are energy constrained and play an important role in the success of delivering messages. When the energy of a node is low chance to deliver messages across the network. In this work, we evaluate and compare the performance of different routing protocols in order to find the energy-efficient routing protocol to be used for message transmission for our DTN application. We evaluate the nodes average remaining energy, the number of dead nodes, delivery probability and overhead ratio for different routing protocols.

L. Barolli et al. (Eds.): NBiS 2018, LNDECT 22, pp. 122–131, 2019.
https://doi.org/10.1007/978-3-319-98530-5_11

1 Introduction

Delay Tolerant Networks (DTNs) enable communication where connectivity issues like sparse connectivity, long delay, high error rates, asymmetric data rate, and no end-to-end connectivity exists. In order to handle disconnections and long delays, DTNs use store-carry-and-forward approach.

Smartphones equipped with different communication interfaces like Bluetooth and WiFi are the main computing and communication platform nowadays. Smartphones can be used to carry and forward messages in DTNs. However, most modern smartphones are powered by lithium-ion batteries and have a limited energy. The motion of nodes and lack of power connections is the reason for limited energy in DTNs. Networking requires energy for sending, receiving and storing messages and these deplete the battery and reduce the node's lifetime.

In this paper, we investigate and compare the performance of different routing protocols in a DTN considering energy consumption. For the simulations we use the Opportunistic Network Environment (ONE) [1] simulator. ONE is a simulation environment, capable of generating node movement using different movement models. ONE offers various DTN routing algorithms for routing messages between nodes. Its graphical user interface visualize both mobility and message passing in real time. ONE can import mobility data from real-world traces or other mobility generators. It can also produce a variety of reports from node movement to message passing and general statistics.

The simulation results show that best results in terms of average remaining energy and overhead ratio are achieved by Spray and Wait protocol.

The remainder of this paper is as follows. Section 2 introduces DTN. A short description of routing protocols is shown in Sect. 3. The simulation system design and simulation scenarios are presented in Sect. 4. In Sect. 5 are shown the simulation results. Finally, the conclusions and future work are presented in Sect. 6.

2 DTN Overview

DTNs are occasionally connected networks, characterized by the absence of a continuous path between the source and destination [2,3]. The data can be transmitted by storing them at nodes and forwarding them later when a link is established. This technique is called message switching. Eventually the data will be relayed to the destination. DTN is the "challenged computer network" approach that is originally designed from the Interplanetary Internet, and the data transmission is based upon the store-carry-and-forward protocol for the sake of carrying data packets under a poor network environment such as space [2]. Different copies of the same bundle can be routed independently to increase security and robustness, thus improving the delivery probability and reducing the delivery delay. However, such approach increases the contention for network resources (e.g., bandwidth and storage), potentially leading to poor overall network performance.

In [4], authors have studied this model and found that it can provide substantial capacity at little cost, and that the use of a DTN model often doubles that capacity compared with a traditional end-to-end model. The main assumption in the Internet that DTNs seek to relax is that an end-to-end path between a source and a destination exists for the entire duration of a communication session. When this is not the case, the normal Internet protocols fail. DTNs get around the lack of end-to-end connectivity with an architecture that is based on message switching. It is also intended to tolerate links with low reliability and large delays. The architecture is specified in RFC 4838 [5].

Bundle protocol has been designed as an implementation of the DTN architecture. A bundle is a basic data unit of the DTN bundle protocol. Each bundle comprises a sequence of two or more blocks of protocol data, which serve for various purposes. In poor conditions, bundle protocol works on the application layer of some number of constituent Internet, forming a store-and-forward overlay network to provide its services. The bundle protocol is specified in RFC 5050. It is responsible for accepting messages from the application and sending them as one or more bundles via store-carry-and-forward operations to the destination DTN node. The bundle protocol runs above the TCP/IP level.

3 Routing Protocols

In order to handle disconnections and long delays in sparse opportunistic network scenarios, DTN uses store-carry-and-forward approach. A network node stores a bundle and waits for a future opportunistic connection. When the connection is established, the bundle is forwarded to an intermediate node, according to a hop-by-hop forwarding/routing scheme. This process is repeated and the bundle will be relayed hop-by-hop until reaching the destination node. In [6–12, 16] authors deal with routing in DTNs.

In this work, we will use three widely applicable DTN routing protocols Epidemic [13], Spray and Wait [14] and Maxprop [15].

Epidemic Routing Protocol: Epidemic [13] is a protocol that is basically a flooding mechanism. Each message spreads like a disease in a population without priority and without limit. When two nodes encounter each other they exchange a list of message IDs and compare those IDs to decide which message is not already in storage in the other node. The next phase is a check of available buffer storage space, with the message being forwarded if the other node has space in its buffer storage. The main goals of this protocol are: maximize the delivery ratio, minimize the latency and minimize the total resources consumed in message delivery. It is especially useful when there is lack of information regarding network topology and nodes mobility patterns.

Spray and Wait Routing Protocol: Spray and Wait [14] is a routing protocol that attempts to gain the delivery ratio benefits of replication-based routing as well as the low resource utilization benefits of forwarding-based routing. The Spray and Wait protocol is composed of two phases: the spray phase and the wait phase. When a new message is created in the system, a number L is attached to that message indicating the maximum allowable copies of the message in the network. During the spray phase, the source of the message is responsible for "spraying", or delivery, one copy to L distinct "relays". When a relay receives the copy, it enters the wait phase, where the relay simply holds that particular message until the destination is encountered directly.

Maxprop Routing Protocol: Maxprop [15] is based on prioritizing both the schedule of packets transmitted to other peers and the schedule of packets to be dropped. These priorities are based on the path likelihoods to peers according to historical data and also on several complementary mechanisms, including acknowledgments, a head-start for new packets, and lists of previous intermediaries.

Fig. 1. Tirana city map imported from osm.

4 Simulation System and Scenarios

The network scenario is based on the map-based model of a part of Tirana city in Albania. The map was imported from Open Street Map [17] (see Fig. 1). Simulations are carried out using the ONE simulator. We create a DTN with 300 nodes for environment monitoring. In our simulations, the nodes are divided into stationary and mobile nodes. As stationary nodes there are 7 groups with 3 nodes for each group that send data to a single destination node. These 21 nodes can be sensors that gather information about temperature, humidity and wind speed and sends their data to a node that collects these data for statistical purposes. Source and destination nodes are stationary and have a 100 MB buffer. The mobile nodes are pedestrians and cyclist. We simulated an urban scenario where pedestrians move according to shortest-path map-based movement model and cyclist according to map based movement for 4 h. The cyclist and pedestrians are equipped with a smart device with 100 MB buffer. We considered the distance of the source nodes with destination node 800 m–2000 m. The initial position of all nodes are shown in Fig. 2.

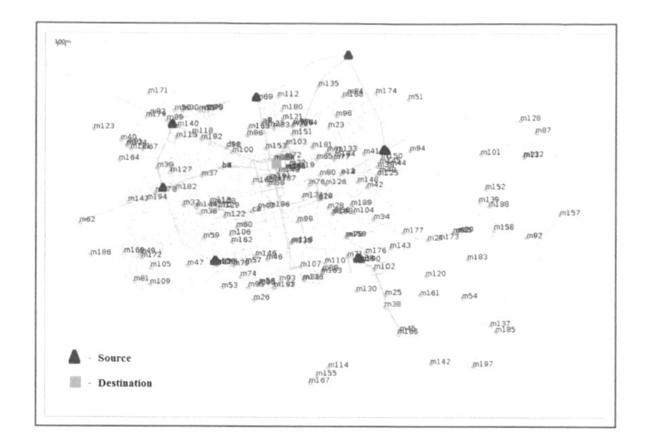

Fig. 2. Nodes initial positions.

All network nodes use a WiFi link connection with a transmission data rate of 250 KBps and the transmission range is considered 20 m. The event generator is responsible for generating bundles with sizes uniformly distributed in the ranges [10 kB, 50 kB]. A bundle is created every 30 s. The data bundles ttl is set 90 min. The simulation parameters are shown in Table 1.

In Table 2 are shown the energy parameters and their values. For the simulations, the stationary and mobile nodes have different initial energy. Scan energy represents the energy spent by the node to discover its neighbors. Scan response

Table 1. Simulation parameters and their values.

Parameters	Values
Number of stationary nodes	22
Number of mobile nodes	278
Simulation time	14400 s
Map size	5 km × 3.5 km
Buffer size	100 MB
Interface type	WiFi
Interface Transmission Speed	250 KBps
Interface Transmission Range	20 m
Message TTL	90 min
Pedestrians speed	5.4–9 km/h
Cyclists speed	5–22 km/h
Message size	10 k, 50 k
Warm up time	100 s
Events interval	30 s

Table 2. Energy parameters and their values.

Parameters	Values (units)
Stationary nodes initial energy	10000
Mobile nodes initial energy	5000
Scan energy	0.26
Scan response energy	0.26
Transmit energy	0.1
Base energy	0.01

energy represents the energy consumed while responding to neighbors on discovery phase. Transmit energy is energy used by nodes to send messages. Base energy is the energy consumed while the node is in idle mode.

We consider a network with 300 nodes and evaluate the performance of the system for 3 different routing protocols: Epidemic, Spray and Wait and Maxprop.

We use the following metrics to measure the performance of different routing protocols: delivery probability, average latency, nodes average remaining energy and number of dead nodes.

- **Delivery probability** is the ratio of number of delivered messages to that of created messages.
- **Overhead ratio** is the difference relayed and delivered messages upon the number of delivered messages.
- **Nodes average remaining energy** is the average of the nodes energy left after the completion of the simulation time.
- **Number of dead nodes** is the number of nodes whose energy reaches almost zero.

5 Simulation Results

In this section, we present the simulation results. We calculated the average remaining energy of nodes after 4 h of simulation to find which routing protocol performs better in terms of energy consumption. First, we set the initial energy of all nodes 5000 units. Since all stationary nodes died after 4 h, which is the end of our simulation time, we increased the initial energy from 5000 to 10000 units. Nodes with higher average remaining energy will live longer and can transfer messages for a longer time.

The simulation results of delivery probability and overhead ratio are shown in Table 3. In terms of delivery probability, Maxprop show the best performance. In terms of overhead ratio, Spray and Wait performs better because the number of copies of the message in the network is low. MaxProp is flooding-based in nature, so if a contact is discovered, all messages not held by the contact will attempt to be replicated and transferred.

Table 3. Simulation results I.

Routing protocol	Delivery probability	Overhead ratio
Epidemic	0.58	170.6
Spray and Wait	0.52	9.01
Maxprop	0.75	111.3

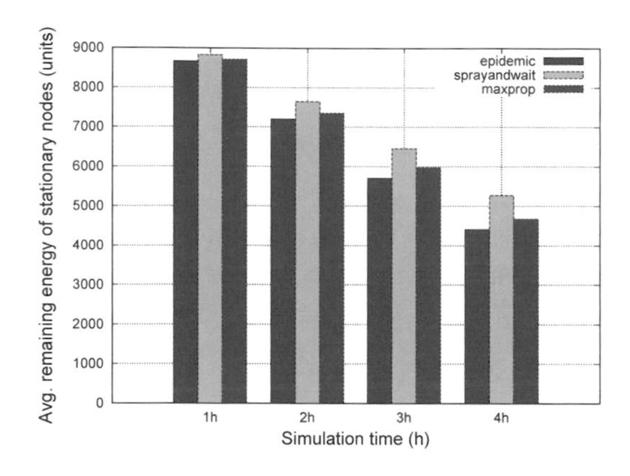

Fig. 3. Simulation results of average remaining energy for stationary nodes.

In Fig. 3 are shown the simulation results of the average remaining energy of stationary nodes every hour for 4 h of simulation time. With the increase of simulation time the average remaining energy for all routing protocols decreases. The average remaining energy of stationary nodes is highest for Spray and Wait protocol and lowest for Epidemic protocol. This is due to the fact that in Spray and Wait, other nodes will just have to wait and deliver the message after the source node has not found the destination. Thus, a small number of scans and transmissions with other nodes take place which results in low energy consumption while only giving energy strain to the source nodes and destination node. The maximum allowable copies of the message in the network for Spray and Wait is 6.

Epidemic is flooding-based in nature, as nodes continuously replicate and transmit messages to newly discovered contacts that do not already possess a copy of the message a lot of energy is consumed.

In Fig. 4 are shown the simulation results of the average remaining energy for mobile nodes. Also in this case the average remaining energy of mobile nodes decreases during the simulation time for all protocols. Spray and Wait performs better also in this case.

In Table 4 are shown the results of number of dead nodes for all routing protocols every hour of simulation. When Spray and Wait protocol is used, there is no dead node during all the simulation but for other protocols more than 180 nodes dies at the end of simulation.

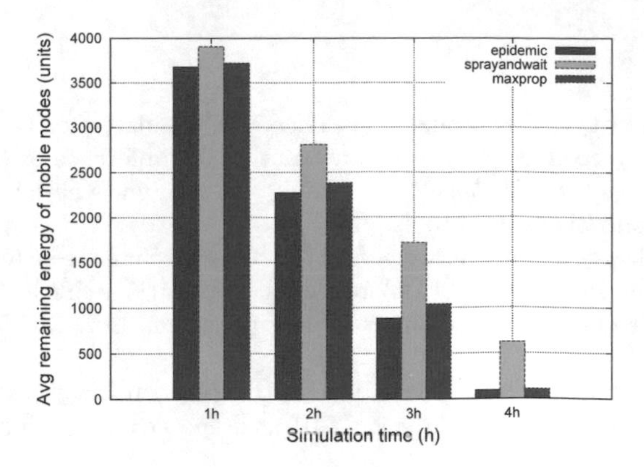

Fig. 4. Simulation results of overhead ratio for the second scenario.

Table 4. Simulation results of dead nodes.

Routing protocol	Dead nodes in 1 h	Dead nodes in 2 h	Dead nodes in 3 h	Dead nodes in 4 h
Epidemic	0	0	3	199
Spray and Wait	0	0	0	0
Maxprop	0	0	1	188

6 Conclusions

In this work, we evaluated and compare the performance of three routing protocols (Epidemic, Spray and Wait and Maxprop) in a many-to-one communication opportunistic network scenario. For evaluation we considered average remaining energy, number of dead nodes, delivery probability and overhead ratio. From the simulation results we conclude as follows.

- Best results in terms of average remaining energy, number of dead nodes and overhead ratio are achieved by Spray and Wait protocol.
- Epidemic performs poor for all metrics because it spreads a large number of messages in the network and consumes resources.
- Maxprop performs better than other protocols in terms of delivery probability, but the average energy consumption of nodes, overhead ratio and number of dead nodes is high.

In the future, we would like to create an energy-aware routing protocol for DTNs evaluate its performance and compare with different routing protocols considering different scenarios and parameters.

References

1. Keranen, A., Ott, J., Karkkainen, T.: The ONE simulator for DTN protocol evaluation. In: Proceedings of the 2-nd International Conference on Simulation Tools and Techniques (SIMUTools-2009) (2009). http://www.netlab.tkk.fi/tutkimus/dtn/theone/pub/theonesimutools.pdf
2. Fall, K.: A delay-tolerant network architecture for challenged Internets. In: Proceedings of the International Conference on Applications, Technologies, Architectures, and Protocols for Computer Communications, ser. SIGCOMM-2003, pp. 27–34 (2003)
3. Delay- and disruption-tolerant networks (DTNs) tutorial, NASA/JPL's Interplanetary Internet (IPN) Project (2012). http://www.warthman.com/images/DTNTutorialv2.0.pdf
4. Laoutaris, N., Smaragdakis, G., Rodriguez, P., Sundaram, R.: Delay tolerant bulk data transfers on the Internet. In: Proceedings of the 11-th International Joint Conference on Measurement and Modeling of Computer Systems (SIGMETRICS-2009), pp. 229–238 (2009)
5. Cerf, V., Burleigh, S., Hooke, A., Torgerson, L., Durst, R., Scott, K., Fall, K., Weiss, H.: Delay-tolerant networking architecture. IETF RFC 4838 (Informational), April 2007
6. Massri, K., Vernata, A., Vitaletti, A.: Routing protocols for delay tolerant networks: a quantitative evaluation. In: Proceedings of ACM Workshop PM2HW2N-2012, pp. 107–114 (2012)
7. Ishikawa, S., Honda, T., Ikeda, M., Barolli, L.: Performance analysis of vehicular DTN routing under urban environment. In: Proceedings of CISIS-2014, July 2014
8. Demmer, M., Fall, K.: DTLSR: delay tolerant routing for developing regions. In: Proceedings of the 2007 ACM Workshop on Networked Systems for Developing Regions, 6 pages (2007)
9. Ilham, A.A., Niswar, M., Agussalim: Evaluated and optimized of routing model on Delay Tolerant Network (DTN) for data transmission to remote area. In: Proceedings of FORTEI, Indonesia University Jakarta, pp. 24–28 (2012)
10. Jain, S., Fall, K., Patra, R.: Routing in a delay tolerant network. In: Proceedings of ACM SIGCOMM 2004 Conference on Applications, Technologies, Architectures, and Protocols for Computer Communication, Portland, Oregon, USA, 30 August–3 September 2004, pp. 145–158 (2004)
11. Zhang, Z.: Routing in intermittently connected mobile ad hoc networks and delay. IEEE Commun. Surv. Tutorials **8**(1), 24–37 (2006)
12. Soares, V.N.G.J., Rodrigues, J.J.P.C., Farahmand, F.: GeoSpray: a geographic routing protocol for vehicular delay-tolerant networks. Inf. Fusion **15**(1), 102–113 (2014)
13. Vahdat, A., Becker, D.: Epidemic routing for partially connected ad hoc networks. Technical report CS-200006, Duke University, April 2000
14. Spyropoulos, T., Psounis, K., Raghavendra, C.S.: Spray and Wait: an efficient routing scheme for intermittently connected mobile networks. In: Proceedings of ACM SIGCOMM 2005 – Workshop on Delay Tolerant Networking and Related Networks (WDTN-2005), Philadelphia, PA, USA, pp. 252–259 (2005)
15. Burgess, J., Gallagher, B., Jensen, D., Levine, B.N.: Maxprop: routing for vehicle-based disruption-tolerant networks. In: Proceedings of the IEEE Infocom, April 2006

16. Lindgren, A., Doria, A., Davies, E., Grasic, S.: Probabilistic routing protocol for intermittently connected networks. draft-irtf-dtnrg-prophet-09. (http://tools.ietf.org/html/draft-irtf-dtnrg-prophet-09)
17. Open street map. http://www.openstreetmap.org/

A Robot Gesture Framework for Watching and Alerting the Elderly

Akihito Yatsuda[1(✉)], Toshiyuki Haramaki[2], and Hiroaki Nishino[2]

[1] Graduate School of Engineering, Oita University, Oita, Japan
v17e3019@oita-u.ac.jp
[2] Faculty of Science and Technology, Oita University, Oita, Japan
{haramaki,hn}@oita-u.ac.jp

Abstract. Watching and medical care for the elderly is one of promising application fields of IoT. Humanoid robots are considered to be useful agents for not only relaxing the elderly but also watching and alerting them in a daily life. Detecting and preventing the risk of indoor heat stroke is an important issue especially for the elderly who live alone. A method for reliably conveying the possible danger of the indoor heat stroke to the elderly is a crucial factor to implement a practical system. In this paper, we describe the system for informing unusual conditions to the elderly by using a communication robot that normally gives users healing. We designed and implemented a set of normal and special motions for a desktop humanoid robot and evaluated whether the robot motions effectively make users aware of abnormal situations.

1 Introduction

Increasing the number of heat stroke patients has become a serious social problem coupled with weather abnormality due to global warming in recent years. About half of the total patients are aged 65 years or older, and half of these elder heat stroke patients are affected indoors. Because aging makes the insensitivity of temperature and the consciousness level for heat stroke of the elderly lower significantly [1]. The cause of the problem heavily depends on the physical characteristics of the elderly such as the difficulty of body temperature adjustment indoors and their blunt awareness to heat stroke. Developing a system for continuously monitoring and detecting any risky conditions of indoor heat stroke is a crucial issue. Additionally, such system needs to robustly alerting the detected conditions to the elderly to avoid risk of life.

In most existing elderly watching systems, when they detect a dangerous condition, they tend to automatically controlling an air conditioning system to adjust the temperature, and sending a notification to caregivers and family members. This approach, however, cannot be applied for supporting the elderly who live alone. If the systems automatically control temperature, the elderly easily gets accustomed to the automated function and their intensions to protect themselves from the dangerous situations are lost. Furthermore, we cannot expect for the elderly to adapt themselves in places without such assisting systems.

We, therefore, adopt a conversational robot for making the elderly awake on the risk of their own motive. The robot is proved to be a useful vehicle for not only

L. Barolli et al. (Eds.): NBiS 2018, LNDECT 22, pp. 132–143, 2019.
https://doi.org/10.1007/978-3-319-98530-5_12

assisting physical weakness of the elderly but also providing mental support through conversation. We use a desktop-style robot as an agent for composing smooth communications between the elderly and the risk detection system without enforcing psychological burdens [2]. The robot usually gives healing to the elderly with cute motions and nice conversations. However, when the system detects a dangerous condition, it immediately alerts the elderly with a series of strange behaviors that are very different from its normal motions being generated during daily conversations. We design and implement these robot motions to robustly notice the situation to the elderly, and verify how such robot motions can effectively convey the critical information to users through a preliminary experiment.

2 Related Work

There are several precedent research projects for using robots for assisting and supporting the elderly in daily life. Sabelli et al. explored a method for mentally supporting the elderly using a robot [3]. They installed a conversational robot at an elderly care center for 3.5 months and observed a process on how tenants gradually accept and live together with the robot in their daily life. According to their observations, the elderly tend to feel the robot becomes a closer and a mentally reliable partner through routine dialogue such as daily greetings and conversations with relatives. Kuwahara reported a trial for assisting elderly's recreational activities at nursing homes using a robot [4]. He set up a robot for interacting with dementia patients when they are involved in common recreational programs. He concluded even simple conversations with the robot were effective ways for the patients to be engaged in the activities. These are trials for using robots to mentally supporting the elderly through daily conversations. We found the potential ability of the robot to be a reliable partner for watching and assisting the daily life of the elderly from these activities.

There are some projects studying on the effectiveness of using robots for watching and assisting the elder drivers. Fujikake et al. examined a robot-based notification method for supporting aged drivers to prevent them from accidents through a subjective evaluation [5]. In this project, they conducted an experiment with 31 elderly people over 65 years old as subjects. The subjects were able to easily awakening to notifications with the robot better than other methods with images and sounds. Tanaka et al. analyzed the influence of different forms of an agent for elder driver's acceptance to assist their driving operations [6]. They prepared three assisting styles including a real robot agent, a video clip of a real robot, and a voice only assistant. They employed both of the elderly and non-elderly subject groups and asked them to compare these agents. As a result, the real robot agent acquired the highest score among the three methods in both groups. One of the reason for obtaining such a result was the presence of the real robot. The elderly subjects stated that the operating noise of the robot they heard during robot motions made them feel the presence and expect the following notification behaviors. On the other hand, an elder subject claimed he could not understand the message contents sufficiently in the case of voice only warning because he could not find no advance notice like the noise.

There is a precedent study for exploring ways to express emotions using robot motions. Takegoshi et al. developed an automatic motion generation system targeted for a humanoid robot [7]. They proposed a way to create a new robot motion by modifying preinstalled basic motion sets according to an emotional word specified by a user such as delight, anger, sorrow and pleasure. They conducted an experiment for generating robot motions with emotions and examined to break these motions into patterns as follows: a motion with raising one arm and lowering the other tends to express delight, a motion with stretching out both arms outward and strongly crouching tends to express anger, a motion with small body movement tends to express sorrow, and a motion with raising one arm and lowering the other and strongly crouching tends to express pleasure. The finding can be used as a framework for designing specific robot motions to robustly alerting the elderly.

Another project tries to explore the usefulness of the robot motions in a practical application system. Murakawa and Totoki examined the effectiveness of robot behaviors targeted for a service robot [8]. They conducted an experiment for guiding people in a public space and demonstrated the superiority of robot behaviors. They found that about 63% of visitors were attracted to the robot with behaviors while only 50% of visitors were attracted to the robot without behaviors. In the experiment, many visitors wanted the robot to make eye contact although the robot did not voluntarily try it. Therefore, making eye contact from the robot should be worth including in the robot motion design for supporting the elderly.

Our objective is to implement a system for watching the elderly and protecting them from affecting the heat stroke indoors. Because the system should enable the elderly to naturally live with it, we adopt a conversational robot as an agent for watching and comforting them in a daily life. The crucial function of the system is to reliably and safely make the elderly awaken to any risks when detected. We implemented a set of robot motions for that purpose. The motions are designed based on some findings acquired in the above-mentioned research activities and they are very different from normal behaviors to bring relief for the elderly. We develop a prototype system using a desktop humanoid robot and evaluate the effectiveness of the proposed method through a preliminary experiment.

3 System Implementation

3.1 System Organization

Figure 1 shows the overall organization of the proposed system. It consists of three modules: control, sensor, and alert modules. The sensor module is continuously measuring indoor environment parameters such as temperature and humidity. The control module is a system coordinator dealing with the main system procedure such as acquiring sensing data from the sensor node, detecting possible risk of heat stroke, and requesting the alert module to notify the risk when detected. Both of the control and sensor modules are running on a single Raspberry Pi board and the alert module is running on an embedded board inside the robot. These components can be placed in an

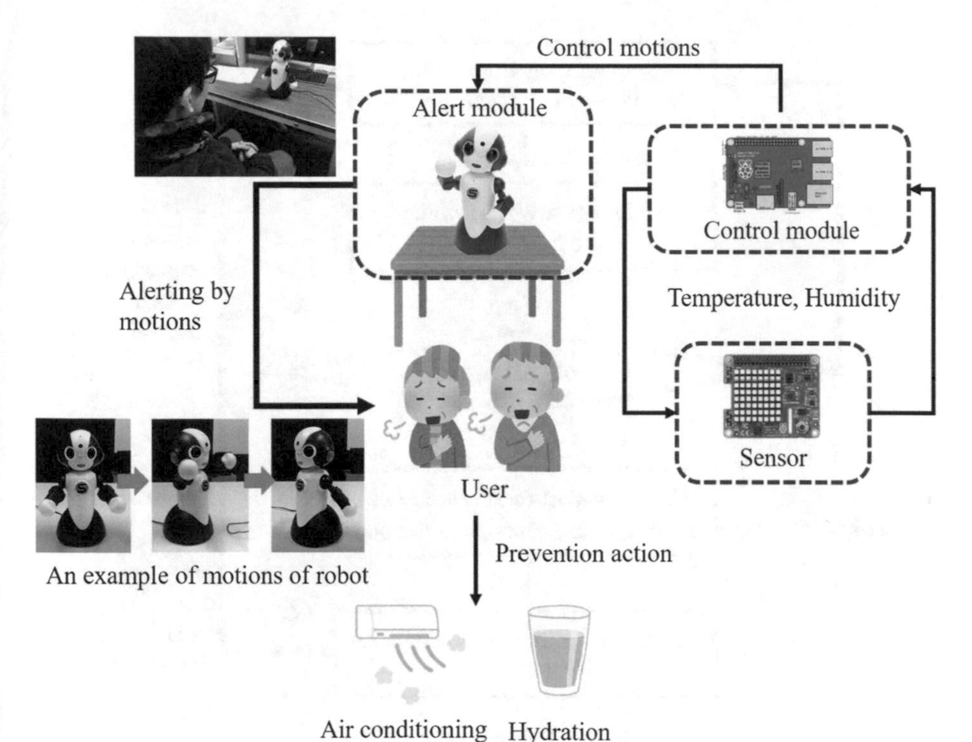

Fig. 1. Organization of the proposed system.

appropriate spot in the room for monitoring operation. Three modules exchange sensing data and request/response messages via a Wi-Fi connection.

The main system processing flow executed in the control module is shown in Fig. 2. After the user sets and starts the system, the control module periodically requests the environment data, the indoor temperature and humidity, to the sensor module. Then, the control module calculates a value called Wet Bulb Glove Temperature (WBGT) using the returned data from the sensor module. The WBGT value is an indoor heat index to be calculated by using a set of temperature and humidity as explained in Sect. 3.3. The system uses the derived WBGT value as a threshold for detecting the risk level of indoor heat stroke. As shown in Fig. 4, the control module judges the elder user is in a risky condition if the value exceeds 31°. After that, it requests the alert module for notifying the risk level to the user based on the robot motion and voice message. Then, the alert module activates the robot for notifying the risky condition by performing predefined gestures with a synthesized voice message and LED lights embedded in the robot. Because we found that aged users are uncomfortable to interact with unfamiliar devices in our previous study [9], we adopt a commercially available desktop humanoid robot called Sota [10] as a communication agent.

Our previous system adopted a different alert notification mechanism by directly deploying varied output devices as is such as an LED array, a set of speakers, and a vibrator. While it was an effective way for conveying an alert message in a multimodal

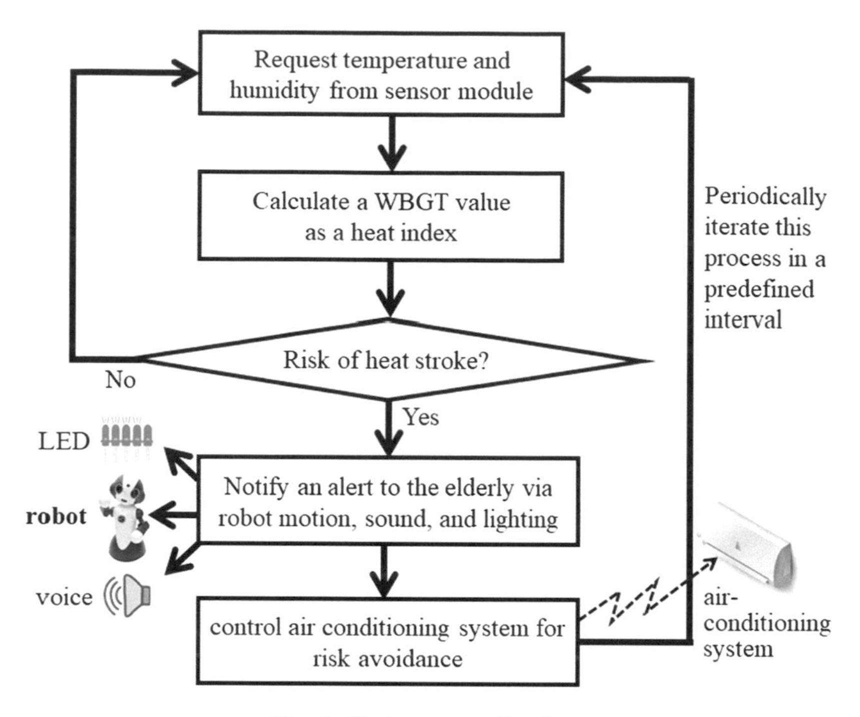

Fig. 2. System processing flow.

manner using light, sound, and vibration, it inflicts some levels of psychological burden on the elderly. They were reluctant to use because it looked a strange and unfamiliar gismo. Therefore, we replace these devices by a conversational robot for reducing the psychological burden and making the system to be a friendly entity. The system notifies the risk with gestures and glowing eyes of the robot to the elderly who cannot hear well, and with the synthesis speech sound from the robot's speaker to the elderly who have poor vision. Additionally, the system provides some extended functions for protecting the elderly from affecting the heat stroke when it detects the highest risk level. The elderly normally use the air conditioner in their residence and judge whether they should receive hydration. Regarding the air conditioner operation, if the elderly do not take any action to adjust the indoor environment, the system starts air conditioning by activating the infrared communication function of Raspberry Pi. This improvement makes the system more effective in watching the elderly.

3.2 Sensing Temperature and Humidity

Figure 3 shows the captured image of the sensor module hardware for monitoring the indoor temperature and humidity. This is an environmental observation device called Sense HAT that is equipped with an LED display and a sensing unit of temperature and humidity. We directly attach it on the control module (a Raspberry Pi board). It provides functions for not only monitoring the indoor temperature and humidity in real time to judge the risk level of indoor heat stroke, but also rendering the sensed

Fig. 3. Captured image of Sense HAT consisting of an LED display and a sensor.

temperature and humidity values as well as the derived WBGT value on the LED display. Additionally, the system can change the background color of the display according the risk level of heat stroke as shown in Fig. 4 and visually present the current risky condition in an easily understandable fashion.

Heat index value	Notes for the elderly
Danger (over31°C)	The elderly have **a high risk** of heat stroke even if they are rested.
Severe vigilance (28~31°C)	The elderly need to pay attention to room temperature rising.
Vigilance (25~28°C)	When the elderly exercise or work hardly, they need to take periodical rest.
Caution (under25°C)	Generally there is little risk of heat stroke.

Fig. 4. Guideline of risk levels and notes for the elderly.

3.3 WBGT Calculation and Risk Detection

The WBGT value is calculated as an approximate value using the following formula. In the formula, *Hu* and *Te* represent a temperature value and a humidity value, respectively. The unit of WBGT is degree.

$$\begin{aligned} WBGT &= (Hu - 20) \\ &\quad \times ((Te - 40) \times 2 \times (-0.00025) + 0.185) + \left(\frac{11.0}{15.0} \times (Te - 25)\right) + 17.8 \end{aligned}$$

Figure 4 shows a guideline of risk levels and notes for the elderly to judge the current condition based on the WBGT value. According to the figure, if the WBGT value exceeds 31°, the system judges the elderly have a possible danger of indoor heat stroke even if they are resting. Then, the system activates the alert procedure. The threshold is clearly defined for each stage of the risk level as shown in Fig. 4. We, therefore, devised a method for preventing any risks by changing the robot gestures and voice messages according to the risk level.

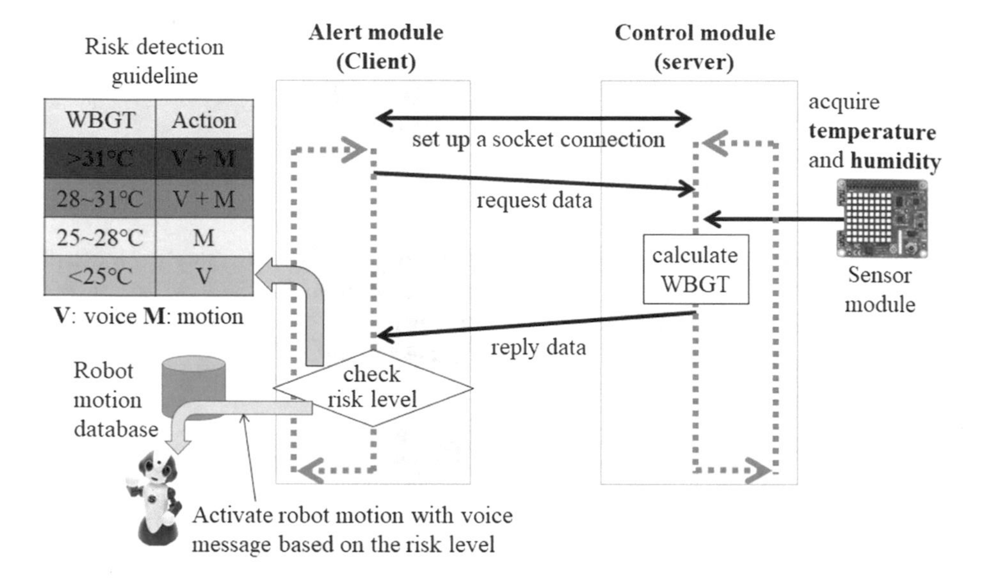

Fig. 5. Software implementation for alerting the risk using robot motion.

3.4 Alert Notification Using a Robot

The system controls the robot by a pair of client and server programs communicated via a Wi-Fi connection using socket interface. Figure 5 shows the processing flow. The server process is running on the control module (Raspberry Pi board) and the client process dealing with the robot is running on the Intel Edison processor embedded in the robot. The server program is written in Python and the client program is written in Java. After the processes initialize a connection using the socket interface, the client requests the indoor sensing data to the server. Next, the server acquires the indoor temperature and humidity from the sensor module and calculates the WBGT value. After that, the server replies these values to the client. Then, the client checks the WBGT value, selecting an appropriate set of a robot motion and an alert message to present based on the guideline as shown in Fig. 4, and actuating the robot for notifying the identified risk level with by brightening the robot eyes and uttering the selected alert messages.

As shown in Fig. 5, the alert motion and message can be changed based to the WBGT value. As a result, the system can adapt its alert contents based on user's demand. In this study, we designed and implemented three different robot motions usable for the alert notification as shown in Fig. 7. Because the robot has an embedded

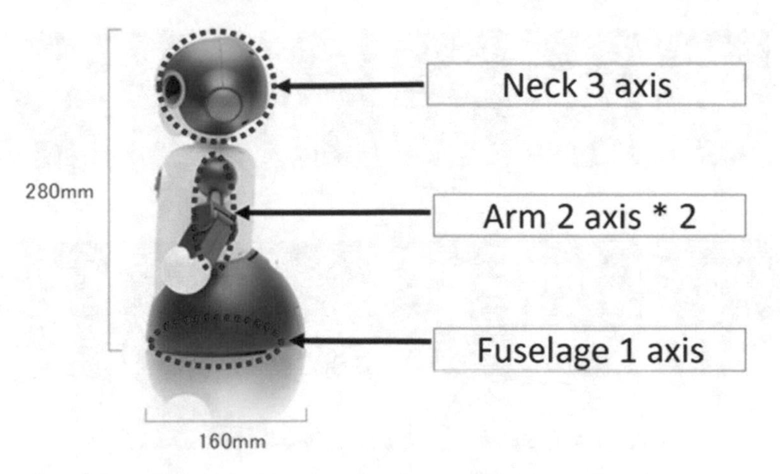

Fig. 6. Motion range of the robot "Sota".

camera, we are developing a function to follow and look toward the elderly when activating a motion and uttering a message.

A set of standard motions of the robot are provided for giving the elderly healing. Example motions are as follows: the robot speaks while moving the right and left hands alternately, and it combines multiple movements of its neck, hands, and uttering. Because our objective is to draw strong attention of the elderly using robot motions when risky situation is detected, we try to exaggerate the robot motions to be noticed by the elderly under varied conditions. Compared to the above-mentioned standard motions, we designed and implemented a set of greatly different motions as warning actions. Firstly, we investigated the motion range of each joint of the robot for making large movements to draw attention of the elderly. Figure 6 shows the robot's motion range. Since the robot can move its neck with 3 axes, the arm with 2 axes, and the torso with 1 axis, we can effectively utilize these parts. We applied the findings studied by Takegoshi et al. [7] on designing the robot motions and implemented the three motion sets as shown in Fig. 7. They found some operational features for expressing emotions by robot motions. Because reliably conveying the danger of indoor heat stroke to the elderly is the most crucial issue in the system, applying emotional behaviors such as healing or comfort are not appropriate for this purpose. We designed the motions for presenting emotions like anger and sorrow. It is easier to express a feeling of sadness by making body motions smaller ones, so we implemented a motion for finely moving both arms while swinging the neck twice as shown in motion A in Fig. 7. While extending and protruding both arms outwards as well as increasing crouching is easy to express a feeling of anger, we implemented the motion B largely pushing up the both arms upwards, and motion C which repeats protruding the both arms twice with moving the fuselage.

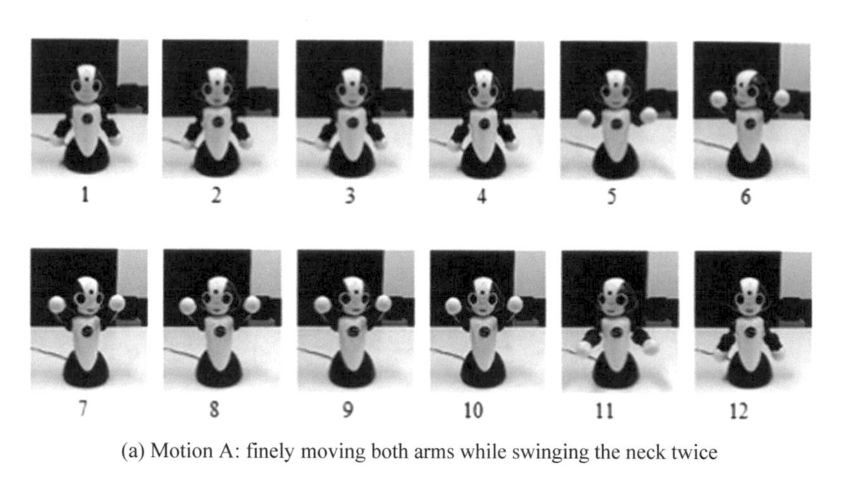

(a) Motion A: finely moving both arms while swinging the neck twice

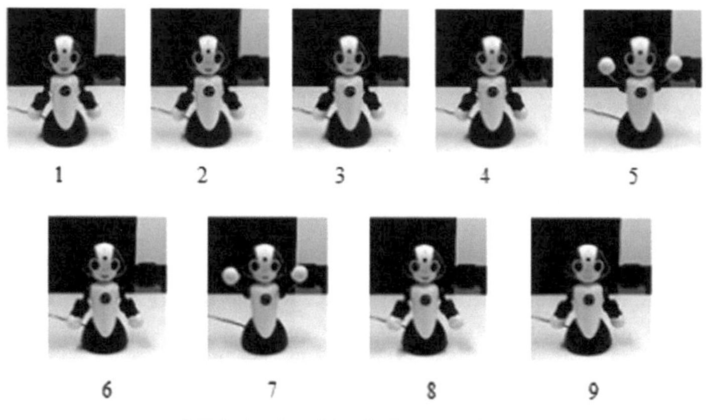

(b) Motion B: raising both arms twice

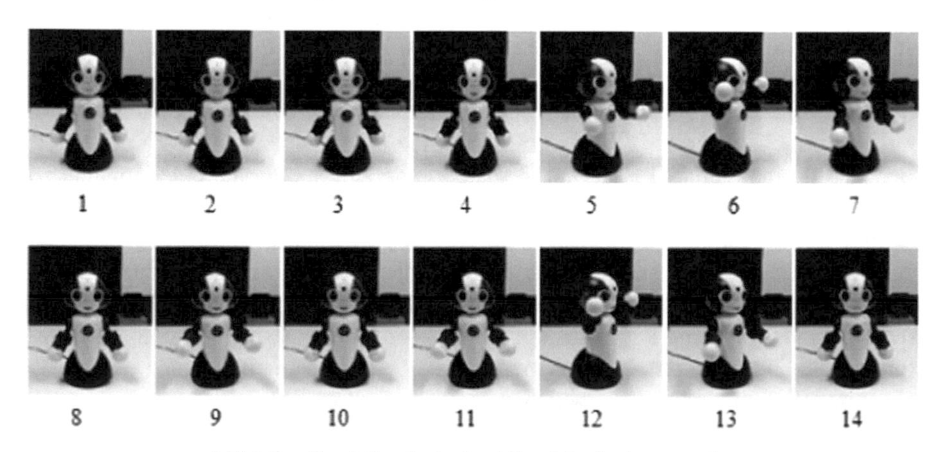

(c) Motion C: rotating the body while raising both arms twice

Fig. 7. Three types of robot motions used in alert notification.

4 Preliminary Experiment

4.1 Evaluation Procedure

We conducted a preliminary experiment to verify the effectiveness of three robot motions as explained in Sect. 3.4 for making users noticed. We employed ten university students as subjects who are in their early 20 s. We requested the subjects for observing three motions of the robot placed just in front of them as shown in Fig. 8. Firstly, the robot presents the motion A for 12 s, followed by motion B for 9 s, and finally motion C for 12 s. After they observed all the motions, we asked them to answer a questionnaire with the following three questions: "What is the most noticeable motion among the three motions?", "What is the most surprising one among the three motions?", and "What kind of impression did you have in each motion?" We also asked them to describe the reasons for why they gave the answer. Table 1 shows the results of the evaluation experiments.

Fig. 8. Experiment image using the robot.

Table 1. Results of the experiment.

Subject	*a*	*b*	*c*	*d*	*e*	*f*	*g*	*h*	*i*	*j*
Most aware	C	C	C	C	C	C	C	C	C	C
Most surprised	C	C	C	C	C	C	C	C	C	C
Good impression	A	B	C	A	C	B	B	B	A	A

a–j: subject IDs

4.2 Discussions

In the experiment, the most noticeable and surprised one was the motion C among the three motions. The subjects who indicated the motion C was noticeable as well as surprised one stated "it reminded me due to its big movements and strange behaviors". On the other hand, the motions A and B got the same number of supporters as an impressive behavior. Some subjects stated that "it seemed to be a human-like behavior". As a result, we found the big and exaggerated robot motions are effective for awaking users and conveying some important messages like risky conditions through robot behaviors.

While taking advantage of the results obtained in the experiment and the findings of previous studies, we would like to continue for examining more effective robot motions. Murakawa et al. found that about half of subjects using a robot were expecting eye contact with the robot in their experiment. They particularly confirmed that the subjects' attention can be drawn by the robot's utterance when they look into the robot's face [9]. Therefore, enabling the robot to make eye contact to users should be an effective way for drawing strong attention of human. We are currently developing an extended function for scanning its surrounding environment using an embedded camera in the robot, automatically detecting the direction of its user, and performing speech and motion toward that direction.

In designing the robot motion, it is a crucial issue for clearly distinguishing the two motions between the alert motions and the standard ones normally activated in the daily life. We would like to conduct more detailed experiments by employing elder users for investigating more effective way to watch them and provide useful support in their daily life.

5 Conclusions

We designed and implemented the system with a slightly different approach from general watching approach in order to prevent indoor heat stroke of the elderly. We adopted a conversational robot that is normally used for the purpose of healing to users. We made a set of new robot motions that are very different from its normal behaviors to robustly notifying the danger to the elderly. We elaborated the implementation method of the system and the motions for the alert notification. We conducted a preliminary experiment and observed the result was promising.

As a future work, we would like to continue the challenge for realizing more effective watching system for the elderly based on the findings acquired in this study.

References

1. Shibata, Y., Tobita, K., Matsubara, N., Kurazumi, Y.: Actual conditions of the recognition of heat disorders in the residential places and preventive measures for the elderly. Jpn. J. Biometeorol. **47**(2), 119–129 (2010). (in Japanese)
2. Yatsuda, A., Haramaki, T., Nishino, H.: A study on robot motions inducing awareness for elderly care. In: Proceedings of the 2018 IEEE International Conference on Consumer Electronics, Taiwan, pp. 391–392 (2018)
3. Sabelli, A.M., Kanda, T., Hagita N.: A conversational robot in an elderly care center: an ethnographic study. In: Proceedings of ACM HRI 2011, pp. 37–44 (2011)
4. Kuwahara, N.: Assessing the use of communication robots for recreational activities at nursing homes. In: Proceedings of the 9th EAI International Conference on Bio-inspired Information and Communications Technologies, pp. 61–64 (2015)

5. Fujitake, K., Taanaka, T., Yonezawa, T., Yamagishi, M., Inagami, M., Kinoshita, F., Aoki, H., Kanamori, H.: Subjective evaluation of elderly people on driver assistance of driver-agents. In: Proceedings of the 58th Conference of Japan Ergonomics Society, vol. 53, pp. 276–277 (2017). (in Japanese)
6. Tanaka, T., Fujikake, F., Noyekawa, T., Inagami, M., Aoki, H., Kanamori, H.: Analysis of relationship between form of driving support agent and gaze behavior—study on driver-agent for encouraging safety driving behavior of elderly drivers. In: Proceedings of the Human-Agent Interaction Symposium, P-2, pp. 1–9 (2017). (in Japanese)
7. Takegoshi, T., Hagiwara, M.: An automatic motion generating system for humanoid robots with emotions. Trans. Jpn. Soc. Kansei Eng. **17**(1), 21–29 (2018). (in Japanese)
8. Murakawa, Y., Totoki, S.: Evaluation of "behavior" of service robot "enon"—experimental operation of enon in a shopping center. IPSJ SIG technical report, ICS-146-131, pp. 31–36 (2006). (in Japanese)
9. Yatsuda, A., Haramaki, T., Nishino, H.: An unsolicited heat stroke alert system for the elderly. In: Proceedings of the 2017 IEEE International Conference on Consumer Electronics, Taiwan, pp. 347–348 (2017)
10. https://www.vstone.co.jp/english/index.html. Accessed Aug 2017

A Robot Assistant in an Edge-Computing-Based Safe Driving Support System

Toshiyuki Haramaki[1(✉)], Akihito Yatsuda[2], and Hiroaki Nishino[1]

[1] Faculty of Science and Technology, Oita University, Oita, Japan
{haramaki,hn}@oita-u.ac.jp
[2] Graduate School of Engineering, Oita University, Oita, Japan
v17e3019@oita-u.ac.jp

Abstract. In this paper, we describe a robot-based interface for presenting important information to assist safety driving. We have been developing a safe driving support system consisting of various devices for sensing the in-vehicle environment and driver's vital signals, a set of edge computing nodes for analyzing the sensed data, and actuators for presenting the analyzed results to the driver. Because visual and auditory messages are commonly used in an instrumental panel, an audio system, and a navigation system in the car, adding similar notification methods may hinder the driver's safety driving operations. We, therefore, propose to use robot motions with voice messages as a new way of delivering important information to the driver. We designed and implemented two sets of the driver assisting methods using a real robot placed in a vehicle and a visual robot aid moving on a monitor screen. We conducted a comparative experiment among the methods to verify their effectiveness and practicality.

1 Introduction

Research and development on ITS (Intelligent Transport System) represented by automatic driving technology have been progressed so rapidly in recent years. These are mainly performed under initiatives of major automobile manufacturers and IT companies. R & D on safe driving support technology aimed at restraining the occurrence of the traffic accident and alleviating accident damage has also been actively carried out. It monitors the surrounding traffic situation of a car, analyzes the existence of risk, and taking an action when it detects a possible danger. Such system called pre-crash safety system detects surrounding obstacles and cars in front by multiple sensors and cameras equipped outside, and then automatically braking or pursuing a front car with maintaining a safe distance.

Expectations on the safe driving technology in the society are very large. However, drivers tend to excessively depend on this technology and their awareness and skills of safe driving may significantly be reduced. For example, the drivers may completely rely on the operation of the pre-crash safety system, neglecting safety confirmation ahead, and operating their smartphone during driving. This is a scenario far from the assumed one to apply the pre-crash safety system.

Driver monitoring system (DMS) is one of the safe driving support technologies to solve the above-mentioned problems. It constantly monitors the state of a driver and

L. Barolli et al. (Eds.): NBiS 2018, LNDECT 22, pp. 144–155, 2019.
https://doi.org/10.1007/978-3-319-98530-5_13

presents appropriate information to support for safe driving and prevent accidents according to the driver's state. The DMS is a promising IoT application system expecting a great demand in the field of ITS. It installs multiple IoT devices equipped with a plurality of sensors for monitoring a vehicle and a driver by cooperating with each other, providing support functions useful for comfortable and safe driving.

We have been developing a system to provide safe driving support functions based on a distributed edge computer architecture consisting of multiple cooperating nodes. In this paper, we propose a mechanism to use a robot as an agent for providing safe driving information to drivers and passengers. Figure 1 shows the organization of the proposed system. It consists of three computing nodes including the sensor, control, and presenter nodes, dealing with different roles, and cooperatively performing various functions by asynchronously exchanging data and messages between the nodes. Among the three nodes, the presenter node conveys appropriate information to the driver via a conversational robot, so the driver continues safe driving operations by himself based on the notified information. Because modern cars have plenty of audio-visual interaction functions, the driver's cognitive ability should be overflowed if the system uses the same modality. Therefore, we implement the proposed functions using a set of robot motions which are proved to be an effective approach for supporting the elderly in safe driving. The proposed method can be used for watching in-vehicle environment and passengers, providing useful information for comfortable and safe driving even when the development of automatic driving technology completes.

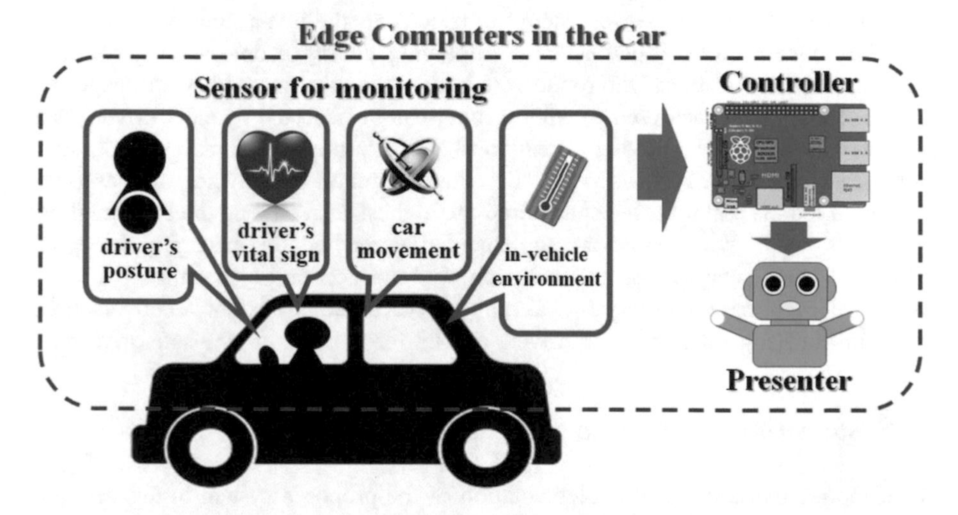

Fig. 1. Organization of the proposed system (real-time processing mode).

2 Related Work

There are some previous research projects for monitoring driver's activities, analyzing their state, and alerting for preventing accidents. There are mainly two types of research activities on the topic. One is to explore a method for sensing drivers and surrounding environments, and analyzing them for extracting useful information to assist drivers. The other is to devise a method for effectively notifying drivers of the analyzed results.

Li et al. proposed a method for predicting driver's condition and classifying it among several classes by monitoring their physiological data such as ECG and respiration wave signals [2]. Islinger et al. proposed a method for analyzing sensing data acquired by monitoring driver activities and modeling their state [3]. Kim et al. proposed a method to construct a machine learning classifier by collecting sensors and human annotated data including vehicle motion, traffic condition, physiological response and driver movement, to detect abnormalities of the driver [4]. Gupta et al. proposed a method of examining a feature from the input image of driving attitude and classifying whether it is a posture suitable for safe driving [5]. These studies focus on designing methods for predicting and judging the driver's conditions such as fatigue level and concentration degree based on multiple sensor data.

There are some researches that are also being conducted to recognize the traffic around the car using new sensing device. Koukoumidis et al. proposed a service that jointly detects and predicts a traffic signal schedule by utilizing the smartphone mounted on the windshield and its camera [6]. Sharing monitoring data between vehicles traveling in a limited area is a beneficial approach. Healey et al. proposed a method for sharing the driver's monitoring data and their predicted condition with neighboring vehicles and avoiding expected collision accidents [7].

There are some activities adopting robots as assistants for notifying important information to drivers. Fujikake et al. examined a robot-based notification method for supporting aged drivers to prevent them from accidents through a subjective evaluation [1]. They conducted an experiment by employing elderly people as subjects. They found that the notifications with robot were better than other methods with images and sounds. Tanaka et al. analyzed whether a robot can be accepted by elder drivers as an agent for assisting their driving operations [8]. They prepared three assisting styles including a real robot agent, a video clip of a real robot, and a voice only assistant. They reported the real robot agent acquired the highest score among the three methods.

In this paper, we propose an implementation method for managing the whole process of DMS such as sensing, analyzing, and notification based on a set of cooperating edge computers installed in a vehicle. We especially adopt a conversational robot for notifying information to drivers crucial for maintaining the safe driving.

3 System Implementation

We elaborate the design and implementation of the proposed system in this section.

3.1 Functional Overview

We have been working on some precedent studies related to the proposed system such as sensor data visualization [9] and knowledge model acquisition from sensing data [10]. We use the results and findings obtained from these efforts to design and develop the proposed system. It is a type of DMS consisting of a network of multiple edge computers, watching in-vehicle environment and driver state in real time. In designing the system, we define the two operation modes: real-time processing mode and safe driving knowledge acquisition mode.

The former is a mode in which all of the functions (sensing, state detection, and notification to a driver) are performed at high speed on an in-vehicle system as shown in Fig. 1. It can immediately notify the driver of information which is crucial for maintaining safe driving.

The latter is a mode in which all of the sensor data collected by the in-vehicle system are sent to a cloud server and acquiring an effective risk assessment procedure for supporting safe driving on the server by machine learning as shown in Fig. 2. The obtained procedure is periodically downloaded to the in-vehicle system and utilized in the real-time mode process. Sharing the server with multiple in-vehicle systems allows all participating vehicles to continue updating the model trained by various sensing data and deal with more complex situations.

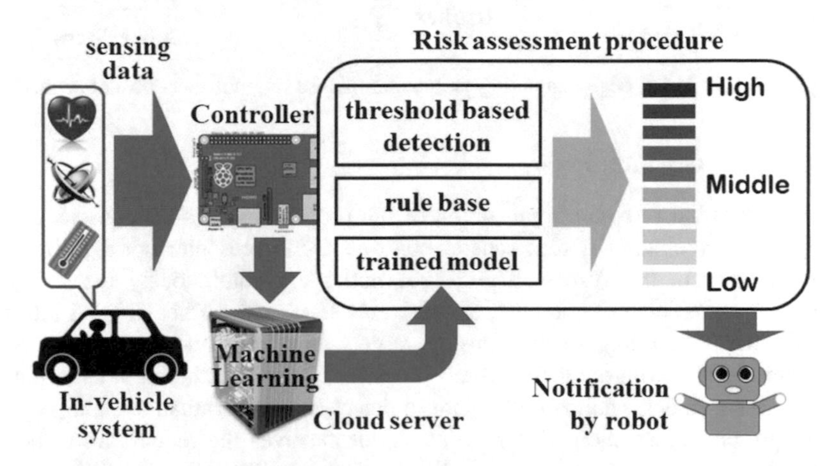

Fig. 2. Organization of the proposed system (safe driving knowledge acquisition mode).

3.2 Distributed Edge Computing Based on Pub-Sub Model

The system consists of multiple edge computing nodes dealing with their own assigned roles. Because the system fulfills its functions by exchanging a huge amount of sensing data and request/response messages between the nodes, the internode communication should be a lightweight one.

Therefore, we implement the system based on a communication architecture called publisher-subscriber (Pub-Sub) model as shown in Fig. 3 to satisfy the above-mentioned requirement. As shown in the figure, an arbitrary number of publishers and subscribers exchange messages via a broker. The publishers are nodes for sending messages and the subscribers are dedicated for receiving messages. They exchange each message in a specific class individually defined in advance by both sides and the broker manages the message exchange between publishing and subscribing nodes. Because publishers and subscribers are loosely coupled based on an asynchronous communication model via the broker, the Pub-Sub model enables the system to be efficient and flexible than popular synchronous protocols used in many IoT applications such as HTTP and FTP.

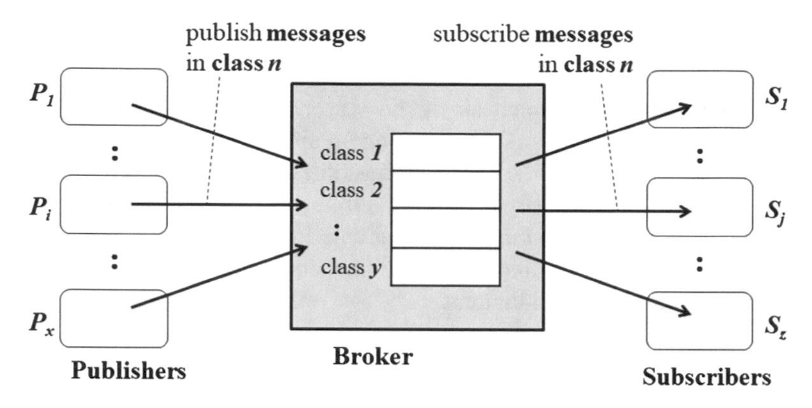

Fig. 3. A distributed edge computing architecture based on publisher-subscriber model.

3.3 Software Organization

Figure 4 shows the software organization of the proposed system. It consists of three kinds of edge computing nodes: sensor, controller, and presenter nodes cooperating with each other in the system. The sensor nodes are continuously measuring car movements, in-vehicle environment, and driver's state, and further divided into multiple sub-nodes according to the sensing targets. The controller node is a system coordinator dealing with the main system procedure such as gathering all sensing data from the sensor nodes, analyzing the data to detect any risky situations, and request a robot to present information. The presenter node receives the request from the controller and activates the robot for the notification to the driver.

The nodes can flexibly be deployed inside the car and exchanging information via wireless communication. We normally place the controller node near the driver, deploying the sensor nodes in accordance with the sensing targets, and arranging the presenter node near the robot, respectively. We adopt a light-weight network framework called MQTT (Message Queue Telemetry Transport) [11] which is a type of the Pub-Sub protocol as explained in Sect. 3.2. It is an application level protocol running on TCP/IP in the protocol stack. A publicly available MQTT broker module called mosquitto [12] is running on the controller node.

Figure 4 shows all the software modules running on the nodes. All the sensor nodes act as a publisher for sending sensing data to the controller. The subscriber module on the controller requests the broker for subscribing the sensing data and waits for them to be published. Each sensor node publishes its sensing data via a specific topic predefined for the sensing device. The controller also acts as a publisher for sending a notification request to the presenter node. Each sensor node has a pair of sensing and data processing modules for periodically driving its managing sensors and making messages to publish. The controller has the intelligent processing module for analyzing the sensing data and detecting any risky conditions. The controller has an additional

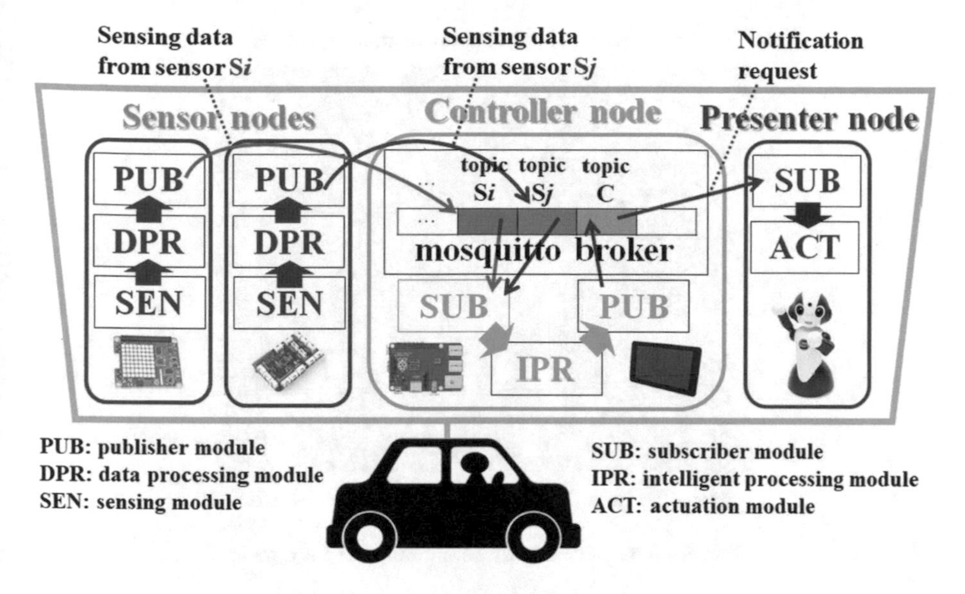

Fig. 4. Software organization of the proposed system.

function for communicating with the remote cloud server. The presenter node has the actuation module to activate specific robot behaviors according to requests from the controller.

3.4 Sensor Node Function

The system efficiently collects data by dividing the sensing targets into three parts including the car movement data, the in-vehicle environment data, and the driver's state data. Figure 5 shows the sensor node for monitoring the car movement data made by using a Raspberry Pi board. It uses an add-on type built-in sensor called Sense Hat directly connected to the Raspberry Pi board. The Sense Hat has a nine-axis IMU (inertial measurement unit) sensor consisting of gyro, acceleration, and magnetic sensors for detecting the car movements. It also has a set of sensors including temperature, humidity, atmospheric pressure sensors for monitoring the surrounding environment of the car. It also uses a GPS sensor to track the route of the car.

Figure 6 shows the sensor node for monitoring the in-vehicle environment data also made by using a Raspberry Pi board. It uses a connector for managing multiple sensors on the Raspberry Pi board including a temperature/humidity sensor, a light sensor, a sound sensor, and an ultrasonic sensor. In addition to the in-vehicle environment data, it also monitors the driver's state such as their upper-body movements by the ultrasonic sensor and the level of tiredness from their voice. We also use a wearable sensor for monitoring the driver's physiological data to detect more detailed conditions. The driver wears it like a wristwatch and the device constantly measure their vital signals such as heart rate.

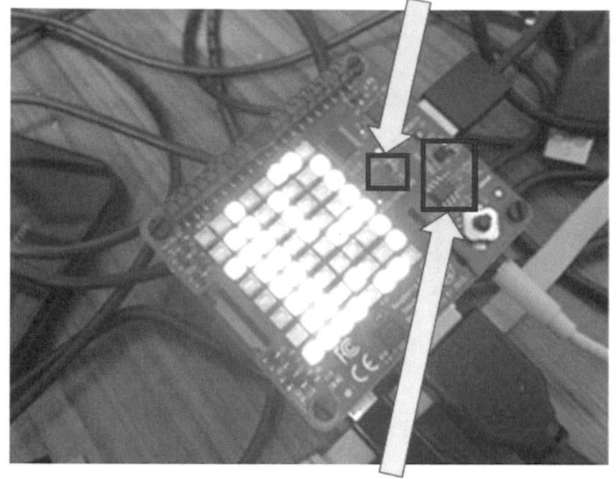

Fig. 5. A sensor node for monitoring car movement.

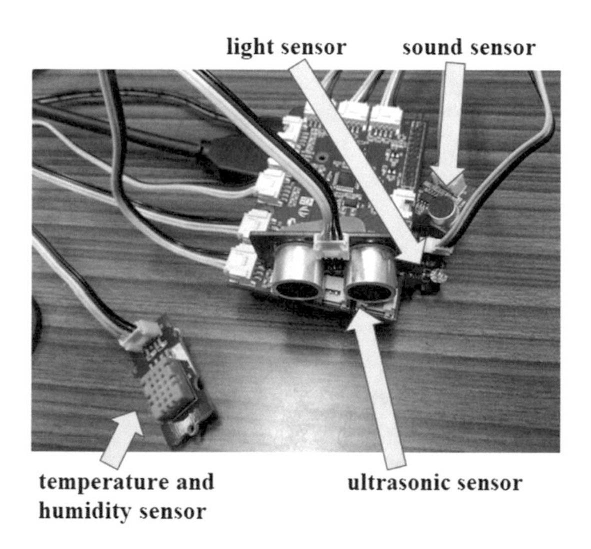

Fig. 6. A sensor node for monitoring in-vehicle environment.

3.5 Controller Node Function

The controller node processes the sensing data received from the sensor nodes and presents information crucial for assisting safe driving to the driver. The controller derives a risk level by estimating the possible danger of an accident based on the sensing value. There are three methods to derive the risk level. The first method is a simple one for determining it based on predefined threshold values. It judges the degree of influence on the driver's operation based on whether the in vehicle environment data such as temperature and humidity exceeds the normal range.

The second method estimates the risk level based on a rule base. It uses a knowledge base determined by the cloud server based on the sensing value. The knowledge base provides an estimation guideline of inhibiting factors for safe driving based on parameters such as the driver's physical condition classified by the driver's vital signals and the instability of car movements.

The third method uses a trained model generated by the cloud server with machine learning. It trains the model with collected sensing datasets as inputs from its participating vehicles. The trained model can be expected for deriving a more detailed risk level in a variety of situations. The system judges the risk level by selectively using a specific method or combining multiple methods. Then, the controller node requests the presenter node for notifying the detected risk through robot behaviors.

3.6 Alert Notification Using a Robot

We adopt a desktop robot as an agent for presenting the detected risk level to the driver. We use the humanoid type conversational robot called Sota developed and marketed by Vstone as shown in Fig. 7. It is 280 mm tall and has a camera and a microphone as inputs, and a speaker and LEDs in eyes and mouth as outputs. It can move its body, arm, and neck, and each part can be controlled by a program. Since it also has a voice synthesis function, the system generates voice messages in parallel with the gestures as the notification. The robot control program running on the presenter node always waits for the request from the controller as a subscriber and activates an appropriate set of gestures and voice messages upon the receipt of the request.

Fig. 7. A real robot agent to notify information to driver.

The system also uses a video robot agent from the viewpoint of easiness to place in the vehicle. This is a method for displaying the images and video clips of the robot Sota on the monitor of the presenter node as shown in Fig. 8. All images and video clips used in this function need to be edited and recorded in the system beforehand. It replays

Fig. 8. A video robot agent to notify information to driver.

appropriate images and video clips in the monitor according to the risk level detected by the controller.

We can investigate how the proposed function affects the driving operation by recording and analyzing the video robot information with some sensing data such as the car movements and the driver's state information. The presenter node synthesizes a voice message according to the risk level from the controller and emits it as a voice message.

4 Preliminary Experiment

4.1 Experimental Procedure

We conducted a preliminary experiment to verify the effect of the implemented system functions. Figure 9 shows a captured image of the experiment using a real vehicle. As shown in the figure, we installed all the computing nodes on the dashboard and verified the system is appropriately functioning. One sensor node monitors the car movement and observes the running speed of the vehicle using an acceleration sensor. We placed the other sensor node straight in front of a driver for measuring the distance to the driver using an ultrasonic sensor. The system gives advice on the driving attitude when the measured distance exceeds the predefined range. We examined two types of robot-based alert notification methods: one is to use a real robot agent sitting in the passenger seat and the other is a video robot agent displayed on the monitor of the presenter node on the dashboard. We fixed the real robot on the passenger seat so as not to move during traveling for safety.

We confirmed some basic functions while running in the experiment such as a greeting function when the robot recognizes the driver, a conversation function before and after the driving, an alert function in the case of excessive speed, a warning function about the driver's attitude, a function for the robot to praise the driver for their

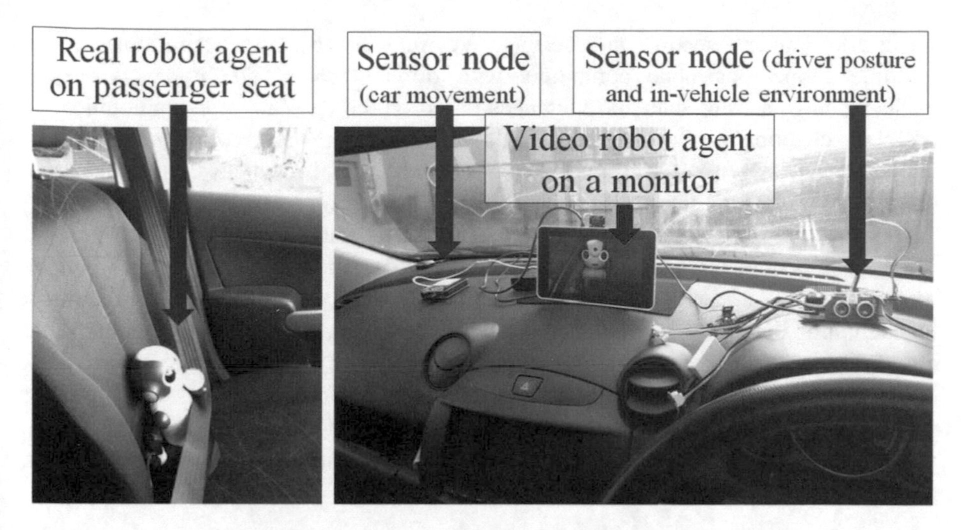

Fig. 9. Experiment image using a robot vehicle.

safe driving operation, and a conversation function on some topics unrelated to driving. We made a few tests run on the campus road.

4.2 Results and Discussions

We successfully confirmed both of the two types of robot agents were up and running in the experiment just as we intended. We, therefore, confirmed all the basic functions described in the previous subsection were functioning.

In the experiment, we got some findings and issues need to be considered towards its practical usage in the vehicle. The driver could feel the presence of the robot when driving with the real robot. While the driver could not look at the robot sitting in the front passenger seat at all during driving, he was able to sense its presence and motion precursor by hearing operating sounds. The driver, however, may doubt the normal operation of the system when he got no response at a certain time interval. Although nonresponse from the robot indicates a good sign for safe driving, giving periodical heartbeat information to the driver should be considered even when the system is normally functioning. Because inadvertently setting the robot may obstruct its motor operations, securing the motion range of the robot should be considered upon installation. Utilizing the system during actual driving is an urgent task to identify further issues and problems.

5 Conclusions

In this paper, we proposed a method for monitoring a driver and in-vehicle environment, detecting a risky condition leading to serious accidents, and alerting it to the driver through a robot agent as a safe driving support system. We designed a system based on a distributed edge computing architecture to enable real-time processing from sensing to danger detection and alerting. We implemented the system consisting of multiple cooperating node computers with different roles. The nodes wirelessly exchange huge sensing data and messages based on the Pub-Sub type communication protocol, enabling a high-speed yet lightweight massive data communication. We adopted a variety of sensors for monitoring car behaviors, in-vehicle environment, and driver's vital signs to detect a dangerous situation with high accuracy. These sensing values are quickly aggregated for analyzing the danger state and notifying to the driver. The system periodically uploads the sensing data to a remote cloud server and requests the server to update a risk detector by machine learning. We conducted an experiment by examining the prototype system using a real car and confirmed basic operations are effectively functioning.

Future tasks include considering a stable installation method for the in-vehicle robot agent, designing a method to use the real robot agent for other purposes such as vehicle theft prevention, and further considering the system customization based on gathered long-term vital data of a specific driver.

References

1. Fujikake, K., Tanaka, T., Yonekawa, T., Yamagishi, M., Inagami, M., Kinoshita, F., Aoki, H., Kanamori, H.: Subjective evaluation of elderly people on driver assistance of driver-agents. In: Proceedings of the 58th Conference of Japan Ergonomics Society, vol. 53, pp. 276–277 (2017). (in Japanese)
2. Li, N., Misu, T., Tawari, A., Miranda, A., Suga, C., Fujimura, K.: Driving maneuver prediction using car sensor and driver physiological signals. In: Proceedings of the 18th ACM International Conference on Multimodal Interaction, pp. 108–112 (2016)
3. Islinger, T., Kohler, T., Wolff, C.: Human modeling in a driver analyzing context: challenge and benefit. In: Proceedings of the 3rd International Conference on Automotive User Interfaces and Interactive Vehicular Applications, pp. 99–104 (2011)
4. Kim, S., Chun, J., Dey, A.: Sensors know when to interrupt you in the car: detecting driver interruptibility through monitoring of peripheral interactions. In: Proceedings of the 33rd Annual ACM Conference on Human Factors in Computing Systems, pp. 487–496 (2015)
5. Gupta, R., Mangalraj, P., Agrawal, A., Kumar, A.: Posture recognition for safe driving. In: Proceedings of the 3rd International Conference on Image Information Processing, pp. 141–146 (2015)
6. Koukoumidis, E., Martonosi, M., Peh, L.-S.: Leveraging smartphone cameras for collaborative road advisories. IEEE Trans. Mob. Comput. **11**(5), 707–723 (2012)
7. Healey, J., Wang, C-C., Dopfer, A., Yu, C-C.: M2M gossip: why might we want cars to talk about us? In: Proceedings of the 4th International Conference on Automotive User Interfaces and Interactive Vehicular Applications, pp. 265–268 (2012)
8. Tanaka, T., Fujikake, F., Yonekawa, T., Inagami, M., Aoki, H., Kanamori, H.: Analysis of relationship between form of driving support agent and gaze behavior-study on driver-agent for encouraging safety driving behavior of elderly drivers. In: Proceedings of the Human-Agent Interaction Symposium 2017, P-2, pp. 1–9 (2017). (in Japanese)
9. Haramaki, T., Shimizu, D., Nishino, H.: A wireless network visualizer based on signal strength observation. In: Proceedings of the IEEE International Conference on 2017 ICCE-TW, pp. 23–24 (2017)

10. Haramaki, T., Nishino, H.: An edge computer based driver monitoring system for assisting safety driving. In: Proceedings of the 6th International Conference on EIDWT-2018, pp. 639–650 (2018)
11. MQTT. http://mqtt.org/. Accessed June 2018
12. Mosquitto. https://mosquitto.org/. Accessed June 2018

The Improved Transmission Energy Consumption Laxity Based (ITECLB) Algorithm for Virtual Machine Environments

Tomoya Enokido[1(✉)], Dilawaer Duolikun[2], and Makoto Takizawa[2]

[1] Faculty of Business Administration, Rissho University, Tokyo, Japan
eno@ris.ac.jp

[2] Department of Advanced Sciences, Faculty of Science and Engineering, Hosei University, Tokyo, Japan
dilewerdolkun@gmail.com, makoto.takizawa@computer.org

Abstract. Various types of distributed applications are realized in server cluster systems equipped with virtual machines like cloud computing systems. On the other hand, a server cluster system consumes a large amount of electric energy since a server cluster system is composed of large number of servers and each server consumes the large electric energy to perform application processes on multiple virtual machines. In this paper, the improved transmission energy consumption laxity based (ITECLB) algorithm is proposed to allocate communication processes to virtual machines in a server cluster so that the total electric energy consumption of a server cluster and the average transmission time of each communication process can be reduced. We evaluate the ITECLB algorithm in terms of the total electric energy consumption of a server cluster and the average transmission time of each process compared with the transmission energy consumption laxity based (TECLB) algorithm.

Keywords: Energy-efficient information systems · Green computing
Virtual machine · Load balancing · Server cluster systems

1 Introduction

In current information systems, various types of scalable and high performance distributed applications like data centers [1,2] are realized. In order to provide these distributed application services, a huge volume of data is gathered from various types of system components like mobile devices, sensors, and home appliances as discussed in the Internet of Things (IoT) [3]. In addition, the gathered data is accessed by a huge number of clients. Hence, scalable and high performance computing system is required to provide these distributed application services. Server cluster system [4,5,7–10] equipped with virtual machine technologies [6] like cloud computing systems [1,2] are widely used to implement

L. Barolli et al. (Eds.): NBiS 2018, LNDECT 22, pp. 156–167, 2019.
https://doi.org/10.1007/978-3-319-98530-5_14

these distributed applications. Multiple virtual machines are performed on each server and application processes are performed on multiple virtual machines in a server cluster. Here, the computation and communication resources of each server in a server cluster can be more efficiently and easily utilized by using the virtual machine technologies. In order to satisfy application requirements like the data download time, processing load of virtual machines to perform application processes has to be balanced in a server cluster system. On the other hand, a large amount of electric energy is consumed by a server cluster since the server cluster system is composed of multiple servers and each server consumes electric energy to perform application processes on multiple virtual machines. Hence, not only the response time of each application process but also the total electric energy consumption of a server cluster to perform application processes have to be reduced to provide scalable and high performance distributed application services as discussed in Green computing [1,2].

In this paper, we consider the *communication type application processes* (*communication processes*) to be performed on virtual machines in a server cluster. A communication process transmits data to a requesting client, e.g. downloading a file. In our previous studies, the *transmission energy consumption laxity based* (*TECLB*) algorithm [14] is proposed to select a virtual machine for each request process so that not only the total electric energy consumption of a server cluster to perform communication processes on virtual machines but also the average transmission time of each communication process can be reduced. In the TECLB algorithm, the effective transmission rate of a server is uniformly allocated to every virtual machine performed in the server. If the total transmission rate allocated to a virtual machine is larger than the total receipt rates required by clients, a part of the transmission rate allocated to the virtual machine is not used. As a result, the more number of request processes are issued to a server cluster, the longer average transmission time it takes to perform each communication process in the TECLB algorithm.

In this paper, we newly proposed the *improved transmission energy consumption laxity based* (*ITECLB*) algorithm to select a virtual machine for each request process so that not only the total electric energy consumption of a server cluster to perform communication processes on virtual machines but also the average transmission time of each communication process can be reduced. In order to more efficiently utilize the effective transmission rate of each server, at the higher transmission rate a virtual machine would like to utilize, with the higher transmission rate the server allocates to the virtual machine in the ITECLB algorithm. Then, the effective transmission rate of each server in the server cluster can be more efficiently utilized. As a result, the average transmission time of each communication process can be reduced in the ITECLB algorithm than the TECLB algorithm. We evaluate the ITECLB algorithm in terms of the total electric energy consumption of a homogeneous server cluster and the average transmission time of each communication process compared with the TECLB algorithm. Evaluation results show the average transmission time of each communication process can be more reduced in the ITECLB algorithm than the

TECLB algorithm while the total electric energy consumption of a homogeneous server cluster to perform communication processes in the ITECLB algorithm is the same as the TECLB algorithm.

In Sect. 2, we discuss the transmission model and power consumption model of a server to perform communication processes on virtual machines. In Sect. 3, we discuss the ITECLB algorithm. In Sect. 4, we evaluate the ITECLB algorithm compared with the TECLB algorithm.

2 Transmission and Power Consumption Model

2.1 Transmission Model

Let S be a cluster of servers s_1, ..., s_n ($n \geq 1$). Let nc_t be the total number of cores in a server s_t and C_t be a set of cores c_{1t}, ..., c_{lt} ($l \geq 1$). We assume Hyper-Threading Technology [11] is enabled on a CPU in each server s_t. The TCP segmentation offload (TSO) [12] function is disabled in every server s_t. Let TH_t be a set of threads th_{1t}, ..., th_{qt} ($q \geq 1$) in a server s_t. Let ct_t be the number of threads on each core c_{ht} in a server s_t. Threads $th_{(h-1) \cdot ct_t+1}$, ..., $th_{h \cdot ct_t}$ ($1 \leq h \leq l$) are bounded to a core c_{ht} in a server s_t. Let nt_t be the total number of threads in a server s_t, i.e. $nt_t = nc_t \cdot ct_t$. Let V_t be a set of virtual machines VM_{1t}, ..., VM_{qt} ($q \geq 1$) in a server s_t. Each virtual machine VM_{kt} holds one virtual CPU and is bounded to a thread th_{kt} in a server s_t. Each virtual machine VM_{kt} holds a full replica of a file f. A virtual machine VM_{kt} is referred to as *active* if and only if (iff) at least one process is performed on the virtual machine VM_{kt}. A virtual machine VM_{kt} is *idle* iff the virtual machine VM_{kt} is initiated on a thread th_{kt} but no process is performed on the virtual machine VM_{kt}. A virtual machine is *stopped* iff the virtual machine is not initiated on any thread. Let $AVM_t(\tau)$ be a set of active virtual machines in a server s_t at time τ. Let $nv_t(\tau)$ be the number of active virtual machines in a server s_t at time τ, i.e. $nv_t(\tau) = |AVM_t(\tau)|$. A core c_{ht} is referred to as *active* iff at least one virtual machine VM_{kt} is active on a thread th_{kt} in the core c_{ht}. A core c_{ht} is *idle* if the core c_{ht} is not active. Let $ac_t(\tau)$ be the number of active cores in a server s_t at time τ.

In this paper, a process stands for a communication process which transmits a file f to a client. Suppose a client cl^i first issues a transmission request to obtain a file f to a load balancer K. A load balancer K selects one virtual machine VM_{kt} on a server s_t in a server cluster S for the transmission request and forwards the request to the virtual machine VM_{kt} as shown in Fig. 1. On receipt of the request, a process p^i_{kt} is created and performed on the virtual machine VM_{kt} to transmit a file f to the requesting client cl^i. Processes which are being performed and already terminate at time τ are *current* and *previous*, respectively. Let $CP_{kt}(\tau)$ be a set of current processes on a virtual machine VM_{kt} at time τ. Let $NC_{kt}(\tau)$ be $|CP_{kt}(\tau)|$. In this paper, we assume any virtual machine does not migrate to another server in a server cluster S.

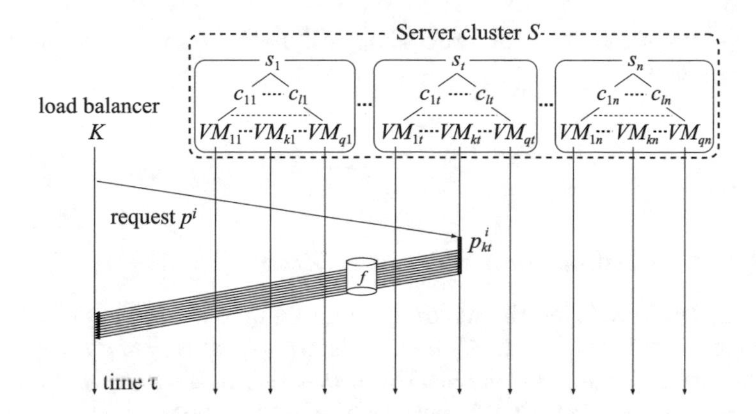

Fig. 1. Forwarding a request process.

Let b_t be the maximum network bandwidth [bps] of a server s_t to do the communication between virtual machines in the server s_t and clients. Let $maxtr_t(\tau)$ ($\leq b_t$) be the effective transmission rate [bps] of a server s_t at time τ. The effective transmission rate $maxtr_t(\tau)$ is calculated as $\gamma_t^{nv_t(\tau)} \cdot b_t$ [13]. Here, $\gamma_t^{nv_t(\tau)}$ is the *transmission degradation ratio* of a server s_t ($0 \leq \gamma_t \leq 1$). The more number of virtual machines are active, the lower effective transmission rate $maxtr_t(\tau)$ is allocated to a communication path between virtual machines and clients at time τ. Let $maxtr_{kt}(\tau)$ be the effective transmission rate [bps] of a virtual machine VM_{kt} at time τ. Here, $0 \leq maxtr_{kt}(\tau) \leq maxtr_t(\tau)$.

Let $tr^i_{kt}(\tau)$ be the transmission rate of a process p^i_{kt} performed on a virtual machine VM_{kt} at time τ. Let $maxrr^i$ be the maximum receipt rate of the client cl^i. At time τ, each client cl^i receives a file f at receipt rate $rr^i(\tau)$ ($\leq maxrr^i$). We assume each client cl^i receives a file f from at most one virtual machine VM_{kt} at rate $rr^i(\tau)$. The virtual machine VM_{kt} allocates the transmission rate $tr^i_{kt}(\tau)$ to a process p^i_{kt} to be $tr^i_{kt}(\tau) \leq maxrr^i$.

At time τ, the total transmission rate $tr_{kt}(\tau)$ of a virtual machine VM_{kt} and the total transmission rate $tr_t(\tau)$ of a server s_t are is given as follows [13]:

[Transmission rate of a virtual machine] If $\sum_{p^i_{kt} \in CP_{kt}(\tau)} tr^i_{kt}(\tau) \geq maxtr_{kt}(\tau)$, $tr_{kt}(\tau) = maxtr_{kt}(\tau)$. Otherwise, $tr_{kt}(\tau) = \sum_{p^i_{kt} \in CP_{kt}(\tau)} tr^i_{kt}(\tau)$.

Here, $0 \leq tr_{kt}(\tau) \leq maxtr_{kt}(\tau) \leq maxtr_t(\tau)$.

[Transmission rate of a server] If $\sum_{VM_{kt} \in AVM_t(\tau)} tr_{kt}(\tau) \geq maxtr_t(\tau)$, $tr_t(\tau) = maxtr_t(\tau) = \gamma_t^{nv_t(\tau)} \cdot b_t$. Otherwise, $tr_t(\tau) = \sum_{VM_{kt} \in AVM_t(\tau)} tr_{kt}(\tau)$.

Suppose a process p^i_{kt} starts and terminates transmitting a file f to the client cl^i at time st^i_{kt} and et^i_{kt}, respectively. The transmission time T^i_{kt} of the process p^i_{kt} is (et^i_{kt} - st^i_{kt}) and $\sum_{\tau=st^i_{kt}}^{et^i_{kt}} tr^i_{kt}(\tau) = |f|$. If the process p^i_{kt} is exclusively performed on the virtual machine VM_{kt} at time τ, $tr^i_{kt}(\tau) = min(maxtr_{kt}, maxrr^i)$ [bps]. The *transmission laxity* $lt^i_{kt}(\tau)$ of a process p^i_{kt} at time τ, i.e.

how many bits of the file f the process p^i_{kt} still has to transmit to the client cl^i, is given as follows:

$$lt^i_{kt}(\tau) = |f| - \sum_{x=st^i_{kt}}^{\tau} tr^i_{kt}(x). \tag{1}$$

2.2 Power Consumption Model of a Server

Let $maxE_t$ and $minE_t$ be the maximum and minimum electric power [W] of a server s_t, respectively [7]. Let cE_t be the electric power [W] of a server s_t if one core gets active. Let vE_t be the electric power [W] of a server s_t if one virtual machine gets active. Let $PC_t(tr_t(\tau))$ be the electric power [W] of a server s_t at time τ, which transmits files to clients at the total transmission rate $tr_t(\tau)$. The electric power $PC_t(tr_t(\tau))$ [W] of a server s_t to perform processes on virtual machines at time τ is given as follows [13]:

$$PC_t(tr_t(\tau)) = minE_t + \delta_t \cdot tr_t(\tau) + (ac_t(\tau) \cdot cE_t) + (nv_t(\tau) \cdot vE_t). \tag{2}$$

Here, δ_t [W/Mb] is the electric power of a server s_t to transmit one [Mbit]. The more number of cores are active in a server s_t, the larger electric power the server s_t consumes. The more number of virtual machines are active in a server s_t, the larger electric power the server s_t consumes.

The total electric energy consumption $TE_t(\tau_1, \tau_2)$ [J] of a server s_t from time τ_1 to τ_2 is $\sum_{\tau=\tau_1}^{\tau_2} PC_t(tr_t(\tau))$. The transmission power $TPC(tr_t(\tau))$ [W] of a server s_t at time τ is $PC_t(tr_t(\tau))$ - $minE_t$. The total transmission energy consumption $TTE_t(\tau_1, \tau_2)$ of a server s_t from time τ_1 to τ_2 is $\sum_{\tau=\tau_1}^{\tau_2} TPC_t(tr_t(\tau))$.

3 Virtual Machine Selection Algorithm

We discuss the *improved transmission energy consumption laxity based* (*ITECLB*) algorithm to select a virtual machine for each request process so that the total electric energy consumption of a server cluster and the average transmission time for each process can be reduced.

3.1 Transmission Rates

Each server s_t can transmit files at the effective transmission rate $maxtr_t(\tau)$ $(= \gamma_t^{nv_t(\tau)} \cdot b_t)$ at time τ. In the TECLB algorithm [14], the effective transmission rate $maxtr_t(\tau)$ of a server s_t is uniformly allocated to every active virtual machine VM_{kt} in the server s_t at time τ, i.e. $maxtr_{kt}(\tau) = maxtr_t(\tau)/nv_t(\tau)$. However, the total maximum receipt rate $TRR_{kt}(\tau)$ $(= \sum_{p^i_{kt} \in CP_{kt}(\tau)} maxrr^i_{kt})$ of every current process p^i_{kt} in a set CP_{kt} of a virtual machine VM_{kt} at time τ might be smaller than $maxtr_t(\tau)/nv_t(\tau)$. Here, the transmission rate $(maxtr_t(\tau)/nv_t(\tau)$ - $TRR_{kt}(\tau)$ is not used. In order to more efficiently utilize the effective transmission rate $maxtr_t(\tau)$ of a server s_t, at the higher rate a virtual machine VM_{kt} would like to transmit, with the higher transmission rate $maxtr_{kt}(\tau)$ a server s_t allocates to the virtual machine VM_{kt}. In the ITECLB algorithm, a server s_t allocates the effective transmission rate $maxtr_{kt}(\tau)$ at time τ to each virtual machine VM_{kt} by the **TRATE** procedure (Algorithm 1).

Algorithm 1. The TRATE procedure

```
Input: SS, TR, RR
  /* Calculation for maxtr_kt(τ): SS ← AVM_t(τ); TR ← maxtr_t(τ) / nv_t(τ); RR ← TRR_kt(τ);
     Calculation for tr^i_kt(τ): SS ← CP_kt(τ); TR ← maxtr_kt(τ) / NC_kt(τ); RR ← maxrr^i_kt; */
Output: transmission_rate /* maxtr_kt(τ) or tr^i_kt(τ) */
  procedure TRATE(SS, TR, RR)
     shortage ← 0; surplus ← 0;
     for each element in SS do
        if RR ≤ TR then
           transmission_rate ← RR;
           surplus ← surplus + (TR - RR);
        else
           shortage ← shortage + (RR - TR);
        end if
     end for
     for each element in SS do
        if RR > TR then
           transmission_rate ← TR + surplus · (RR - TR) / shortage;
        end if
     end for
     return (transmission_rate);
  end procedure
```

Suppose the maximum bandwidth b_t and transmission degradation factor γ_t of a server s_t are 500 [Mbps] and 0.8, respectively. Suppose a pair of virtual machines VM_{1t} and VM_{2t} are active in the server s_t at time τ, i.e. $nv_t(\tau) = 2$. Here, the effective transmission rate $maxtr_t(\tau)$ of the server s_t is $\gamma_t^{nv_t(\tau)} \cdot b_t = 0.8^2 \cdot 500 = 320$ [Mbps]. Suppose the total maximum receipt rates $TRR_{1t}(\tau)$ $(= \sum_{p^i_{1t} \in CP_t(\tau)} maxrr^i_{1t})$ and $TRR_{2t}(\tau)$ $(= \sum_{p^i_{2t} \in CP_t(\tau)} maxrr^i_{2t})$ are 120 and 200 [Mbps], respectively. In the TECLB algorithm [14], each virtual machine VM_{kt} $(k = 1, 2)$ is allocated with the same effective transmission rate $matr_{kt}(\tau)$ $= maxtr_t(\tau)/nv_t(\tau) = 320/2 = 160$ [Mbps] as shown in Fig. 2 (1). Here, the transmission rate 160–120 = 40 [Mbps] is not used for the virtual machine VM_{1t}. In addition, the virtual machine VM_{2t} cannot use the total maximum receipt rate $TRR_{2t}(\tau)$ (= 200 [Mbps]). On the other hand, in the ITECLB algorithm, the unused transmission rate of the virtual machine VM_{1t} (= 40 [Mbps]) can be used for the virtual machine VM_{2t} as shown in the Fig. 2 (2). This means that every current client cl^i performed on a pair of virtual machines VM_{1t} and VM_{2t} can receive a file f at the maximum receipt rate $maxrr^i$ at time τ.

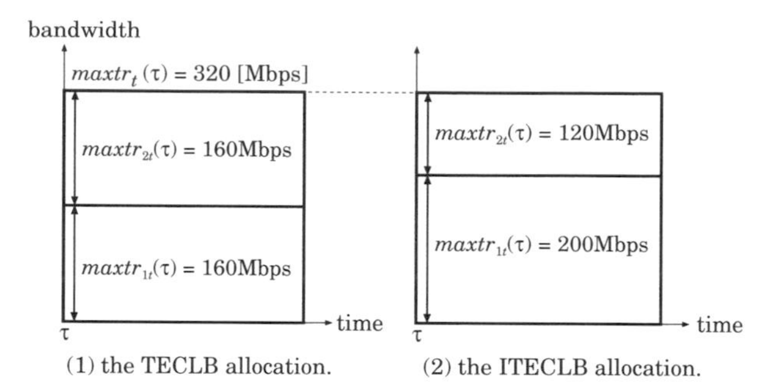

Fig. 2. Transmission rate of a virtual machine.

We discuss how to allocate the transmission rate $tr^i_{kt}(\tau)$ to each process p^i_{kt} performed on a virtual machine VM_{kt} at time τ. In the fair allocation algorithm, the transmission rate $tr^i_{kt}(\tau)$ for each process p^i_{kt} in the set $CP_{kt}(\tau)$ is the same, i.e. $tr^i_{kt}(\tau) = maxtr_{kt}(\tau)/NC_{kt}(\tau)$. If the maximum receipt rate $maxrr^i$ of a client cl^i is smaller than $maxtr_{kt}(\tau)/NC_{kt}(\tau)$, the transmission rate $(maxtr_{kt}(\tau)/NC_{kt}(\tau))$ - $maxrr^i$ is not used. In order to more efficiently utilize the effective transmission rate $maxtr_{kt}(\tau)$ of a virtual machine VM_{kt}, at the higher rate a client cl^i would like to receive, with the higher transmission rate $tr^i_{kt}(\tau)$ a virtual machine VM_{kt} allocates to a process p^i_{kt}. In the TECLB [14] and ITECLB algorithms, a virtual machine VM_{kt} allocates the transmission rate $tr^i_{kt}(\tau)$ at time τ to each process p^i_{kt} by the **TRATE** procedure (Algorithm 1) in a similar way to the transmission rate allocation of virtual machines.

3.2 Estimation of the Electric Energy Consumption Laxity

The total transmission energy consumption laxity $ttel_t(\tau)$ [J] shows how much electric energy a server s_t has to consume to perform every current process on every active virtual machine in the server s_t at time τ. We discuss how to estimate the total transmission energy consumption laxity $ttel_t(\tau)$ [J] of a server s_t where a new request process is allocated to a virtual machine VM_{kt} performed on the server s_t at time τ. Suppose a load balancer K receives a new request process p^{new} from a client cl^{new} and allocates the process p^{new} to a virtual machine VM_{kt} performed on a server s_t at time τ. Here, the new request process p^{new}_{kt} is added to the current process set $CP_{kt}(\tau)$ of a virtual machine VM_{kt}, i.e. $CP_{kt}(\tau) = CP_{kt}(\tau) \cup \{p^{new}_{kt}\}$. Let $\mathbf{CP}_t(\tau)$ be a family $\{CP_{1t}(\tau), ..., CP_{qt}(\tau)\}$ of current process sets of every virtual machine VM_{kt} in a server s_t at time τ. In our previous studies, the **TELAXITY** procedure [14] to estimate the total transmission energy consumption laxity $ttel_t(\tau)$ of a server at time τ is proposed. In the ITECLB algorithm, the load balancer K estimates the total transmission energy consumption laxity $ttel_t(\tau)$ of a server s_t at time τ by the **TELAXITY** procedure (Algorithm 2).

Algorithm 2. The TELAXITY procedure

```
Input: s_t, τ, ac_t(τ), nv_t(τ)
Output: ttel_t(τ)
  procedure TELAXITY(s_t, τ, ac_t(τ), nv_t(τ))
    if every CP_kt(τ) = φ in CP_t(τ) then
      return (0);
    end if
    for each VM_kt in a server s_t do
      maxtr_kt(τ) ← TRATE(AVM_t(τ), mactr_t(τ) / nv_t(τ), TRR_kt(τ));
      for each p^i_kt in CP_kt(τ) do
        tr^i_kt(τ) ← TRATE(CP_kt(τ), maxtr_kt(τ) / NC_kt(τ), maxrr^i_kt);
        tr_kt(τ) ← tr_kt(τ) + tr_k t^i(τ);
        lt^i_kt(τ + 1) ← lt^i_kt(τ) - tr^i_kt(τ);
        if lt^i_kt(τ + 1) ≤ 0 then
          CP_kt(τ) = CP_kt(τ) - {p^i_kt};
        end if
      end for
      tr_t(τ) ← tr_t(τ) + tr_kt(τ);
    end for
    ttel_t(τ) ← PC_t(tr_t(τ)) - minE_t; /* formula (2).*/
    return (ttel_t(τ) + TELAXITY(s_t, τ + 1, ac_t(τ + 1), nv_t(τ + 1)));
  end procedure
```

3.3 The Improved TECLB (ITECLB) Algorithm

Let $TEL_{kt}(\tau)$ [J] be the total transmission energy consumption laxity of a server cluster S where a new request process p^{new} is allocated to a virtual machine VM_{kt} at time τ. In the ITECLB algorithm, a virtual machine VM_{kt} where the total transmission energy consumption laxity TEL_{kt} of a server cluster S is the minimum is selected for a new request p^{new}. Suppose a load balancer K receives a new request process p^{new} at time τ. Then, the load balancer K selects a virtual machine VM_{kt} by the **ITECLB** algorithm (Algorithm 3):

Algorithm 3. The ITECLB algorithm

```
Input: p^new, τ
Output: vm : a virtual machine which p^new is allocated..
  procedure ITECLB(p^new, τ)
    for each VM_kt in a server cluster S do
      CP_kt(τ) ← CP_kt(τ) ∪ {p^new};
      TEL_kt(τ) ← Σ_{t=1}^{n} TELAXITY(s_t, τ, ac_t(τ), nv_t(τ));
    end for
    vm ← a virtual machine VM_kt where TEL_kt(τ) is the minimum;
    return (vm);
  end procedure
```

4 Evaluation

We evaluate the ITECLB algorithm in terms of the total transmission energy consumption of a homogeneous server cluster S and average transmission time of each process p^i compared with the TECLB [13] algorithm.

A homogeneous cluster S is composed of five servers s_1, ..., s_5 ($n = 5$). Every server s_t ($t = 1, ..., 5$) follows the same transmission model and the same power consumption model as shown in Table 1. The parameters of each server s_t are obtained from the experiment [13]. Every server s_t is equipped with a dual-core CPU ($nc_t = 2$). The Hyper-Threading Technology [11] is enabled on a CPU in every server s_t. Two threads are bounded for each core in a server s_t, i.e. $ct_t = 2$. The number nt_t of threads in each server s_t is four, i.e. $nt_t = nc_t \cdot ct_t = 2 \cdot 2 = 4$. $\gamma_t = 0.71$, $b_t = 500$ [Mbps], $\delta_t = 0.02$ [W/Mb], $minE_t = 15.1$ [W], $cE_t = 0.1$ [W], and $vE_t = 0.4$ [W]. Each virtual machine VM_{kt} is bounded to a thread th_{kt} in a server s_t ($k = 1, ..., 4$ and $t = 1, ..., 5$). Hence, there are twenty virtual machines in the server cluster S. Each virtual machine VM_{kt} holds a full replica of a data file f. The size of the data file f is 50 [Mbytes].

Table 1. Parameters of each server s_t.

Server	nc_t	ct_t	nt_t	γ_t	b_t	δ_t	$minE_t$	cE_t	vE_t
s_t ($t = 1, ..., 5$)	2	2	4	0.71	500 [Mbps]	0.02 [W/Mb]	15.1 [W]	0.1 [W]	0.4 [W]

The number m of clients randomly download replicas of the file f from one virtual machine VM_{kt} ($1 \leq m \leq 1{,}000$). The maximum receipt rate $maxrr^i$ of every client cl^i is between 80 and 90 [Mbps]. Each client cl^i issues one transfer request at time st^i in the simulation. The starting time of each client is randomly selected in a unit of one second between 1 and 300 [s]. In the evaluation, the TECLB and RR algorithms are performed on the same traffic pattern.

Let TEC_{tm} be the total transmission energy consumption [J] to perform the number m of processes ($1 \leq m \leq 1{,}000$) in the homogeneous server cluster S obtained in the tm-th simulation. The total transmission energy consumption TEC_{tm} is measured ten times for each number m of processes. Then, the average total transmission energy consumption $ATEC$ is calculated as $\sum_{tm=1}^{10} TEC_{tm}/10$ for each number m of processes. Figure 3 shows the average total transmission energy consumption $ATEC$ of the server cluster S to perform the number m of processes in the ITECLB and TECLB algorithms. In the ITECLB and TECLB algorithms, the average total energy consumption $ATEC$ of the server cluster S increases as the number m of processes increases. For $0 \leq m \leq 1{,}000$, the average total energy consumption $ATEC$ of the server cluster S in the ITECLB algorithm is the same as the TECLB algorithm.

The transmission time T^i_{kt} for each process p^i_{kt} ($1 \leq i \leq m$) performed on a virtual machine VM_{kt} in the homogeneous server cluster S is measured ten times for the total number m of processes ($1 \leq m \leq 1{,}000$). Let $AT^{i}_{kt,tm}$ be the average transmission time T^{i}_{kt} obtained in the tm-th simulation for total number m of processes. The average transmission time AT^{i} is $\sum_{tm=1}^{10}\sum_{i=1}^{m} AT^{i}_{kt,tm}/(m \cdot 10)$. Figure 4 shows the average transmission time AT^{i} in the homogeneous server cluster S for the total number m of processes in the ITECLB and TECLB algorithms. For $1 < m \leq 1{,}000$, the average transmission time AT^{i} increases as the number m of processes increases in the ITECLB and TECLB algorithms. The average transmission time AT^{i} in the ITECLB algorithm is shorter than the TECLB algorithm for $1 < m \leq 1{,}000$.

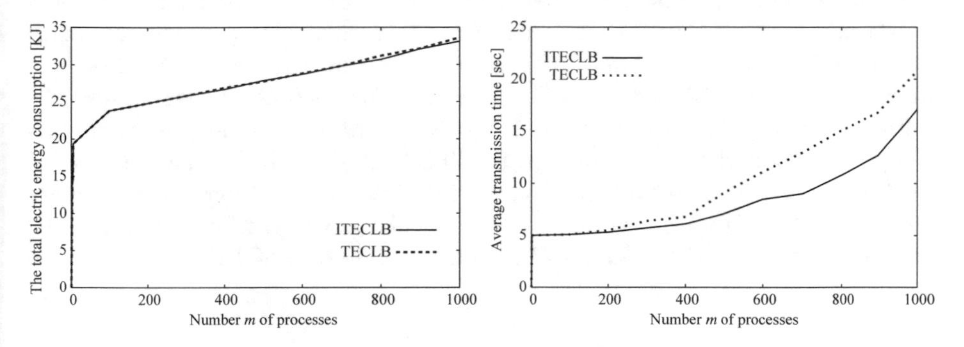

Fig. 3. The average total processing energy of the server cluster S.

Fig. 4. The average transmission time of each process.

The effective transmission rate $maxtr_t(\tau)$ of each server s_t is more efficiently utilized in the ITECLB algorithm than the TECLB algorithm as discussed in Subsect. 3.1. As a result, at the higher rate a client cl^{i} would like to receive, with the higher transmission rate $tr^{i}_{kt}(\tau)$ a virtual machine VM_{kt} can allocate to a process p^{i}_{kt} in the ITECLB algorithm than the TECLB algorithm. Hence, the average transmission time AT^{i} can be reduced in the ITECLB algorithm than the TECLB algorithm. This means that the active times of virtual machines and cores in each server can be more reduced in the ITECLB algorithm than the TECLB algorithm. The electric power $PC_t(tr_t(\tau))$ of a server s_t at time τ depends on the total transmission rate, the number of active cores, and the number of active virtual machines in the server s_t. The total transmission rate $tr_t(\tau)$ of each server s_t to perform processes on virtual machines for each time τ in the ITECLB algorithm is higher than the TECLB algorithm. However, the active times of cores and virtual machines can be reduced in the ITECLB algorithm than the TECLB algorithm. As a result, the average total electric energy consumption $ATEC$ of the server cluster S in the ITECLB algorithm is the same as the TECLB algorithm. Following the evaluation, we conclude the ITECLB algorithm is more useful than the TECLB algorithm.

5 Concluding Remarks

In this paper, we newly proposed the Improved TECLB (ITECLB) algorithm to select a virtual machine for each communication process so that not only the total electric energy consumption of a server cluster to perform communication processes on virtual machines but also the average transmission time of each process can be reduced. The effective transmission rate of each server in the server cluster is more efficiently utilized in the ITECLB algorithm than the TECLB algorithm. A virtual machine where the total transmission energy consumption laxity of a server cluster is the minimum is selected for each process in the ITECLB and TECLB algorithms. We evaluated the ITECLB algorithm in terms of the average total energy consumption of a server cluster to perform communication processes on virtual machines and the average transmission time of each process compared with the TECLB algorithm. The evaluation results show the average total energy consumption of a server cluster to perform communication processes on virtual machines in the ITECLB algorithm is the same as the TECLB algorithm. The average transmission time of each process can be more reduced in the ITECLB algorithm than the TECLB algorithm. Following the evaluation, we showed the ITECLB algorithm is more useful than the TECLB algorithm.

References

1. Natural Resources Defense Council (NRDS): Data center efficiency assessment - scaling up energy efficiency across the data center Industry: Evaluating key drivers and barriers (2014). http://www.nrdc.org/energy/files/data-center-efficiency-assessment-IP.pdf
2. Natural Resources Defense Council (NRDS): Is cloud computing always greener? finding the most energy and carbon efficient information tech nology solutions for small- and medium-sized organizations (2012). http://www.nrdc.org/energy/files/cloud-computing-efficiency-IB.pdf
3. Cuka, M., Elmazi, D., Bylykbashi, K., Spaho, E., Ikeda, M., and Barolli, L.: A fuzzy-based system for selection of IoT devices in opportunistic networks considering IoT device storage, waiting time and security parameters. In: Proceedings of the 6th International Conference on Emerging Internet, Data and Web Technologies (EIDWT-2018), pp. 94–105 (2018)
4. Kataoka, H., Nakamura, S., Duolikun, D., Enokido, T., Takizawa, M.: Multi-level power consumption model and energy-aware server selection algorithm. Int. J. Grid Util. Comput. (IJGUC) **8**(3), 201–210 (2017)
5. Duolikun, D., Enokido, T., Takizawa, M.: An energy-aware algorithm to migrate virtual machines in a server cluster. Int. J. Grid Util. Comput. (IJGUC) **7**(1), 32–42 (2017)
6. KVM: Main Page - KVM (Kernel Based Virtual Machine) (2015). http://www.linux-kvm.org/page/Mainx_Page
7. Enokido, T., Aikebaier, A., Takizawa, M.: Process allocation algorithms for saving power consumption in peer-to-peer systems. IEEE Trans. Ind. Electron. **58**(6), 2097–2105 (2011)

8. Enokido, T., Aikebaier, A., Takizawa, M.: A model for reducing power consumption in peer-to-peer systems. IEEE Syst. J. **4**(2), 221–229 (2010)
9. Enokido, T., Aikebaier, A., Takizawa, M.: An extended simple power consumption model for selecting a server to perform computation type processes in digital ecosystems. IEEE Trans. Ind. Inf. **10**(2), 1627–1636 (2014)
10. Enokido, T., Takizawa, M.: Integrated power consumption model for distributed systems. IEEE Trans. Ind. Electron. **60**(2), 824–836 (2013)
11. Intel: Intel xeon processor 5600 series: the next generation of intelligent server processors (2010). http://www.intel.com/content/www/us/en/processors/xeon/xeon-5600-brief.html
12. B. S. Ang: An evaluation of an attempt at offloading TCP/IP protocol processing onto an i960rn-based NIC, (technical report 2001-8) (2001). http://www.hpl.hp.com/techreports/2001/HPL-2001-8.html
13. Enokido, T., Takizawa, M.: Power consumption model of a server to perform communication type application processes on virtual machines. In: Proceedings of the 10th International Conference on Broadband and Wireless Computing, Communication and Applications (BWCCA-2015), pp. 275–282 (2015)
14. Enokido, T., Takizawa, M.: An energy-efficient load balancing algorithm for virtual machine environments to perform communication type application processes. In: Proceedings of the 30th IEEE International Conference on Advanced Information Networking and Applications (AINA-2016), pp. 392–399 (2016)

Continuous k-Nearest Neighbour Strategies Using the mqrtree

Wendy Osborn(✉)

Department of Mathematics and Computer Science,
University of Lethbridge, Lethbridge, AB T1K 3M4, Canada
wendy.osborn@uleth.ca

Abstract. In this paper, two strategies for processing a continuous k-nearest neighbor query for location-based services are proposed. Both use a spatial access method, the mqrtree, for locating a safe region. The mqrtree supports searching within the structure, so searches from the root are not required - a property which is exploited in the strategies. However, the proposed strategies will work with most spatial access methods. The strategies are evaluated and compared against a repeated nearest neighbor search. It is shown that both approaches achieve significant performance gains in reducing the number of times a new safe region must be identified, in both random and exponentially distributed points sets.

1 Introduction

A location-based service provides information to a user of a mobile device (e.g. smartphone or tablet) based on their location, interests and the type of query issued by the user [22]. One example of such a query is to find the nearest k restaurants to the user. This query is referred to as a k-nearest neighbour (k-NN) query. Another important aspect to processing queries for location-based services is the aspect that the user is moving around as their query is being processed. As the user moves, this affects the result for their query - in the above case, the k nearest restaurants - and therefore this results needs to be continually updated. Therefore, an efficient continuous k-nearest neighbour query processing strategy is important, especially when the query is initiated from a mobile device [22].

Many strategies have been proposed that process nearest neighbour queries for location-based services. Several utilize spatial access methods, including [1,2,4,5,10,20,21,23]. In addition, several strategies have been proposed for processing continuous k-nearest neighbour queries in static point sets, including [3,7,8,11–17,19]. These strategies have several limitations, including repeated searching of a spatial access method, caching more data on the device than is desired, requiring that safe regions - regions where a query remains valid when it moves around - need to be constructed from scratch whenever a new one is needed, and knowing the query trajectory in advance.

The mqrtree [18] is a spatial access method with desirable properties, including two-dimensional nodes and the maintenance of spatial relationships between

L. Barolli et al. (Eds.): NBiS 2018, LNDECT 22, pp. 168–181, 2019.
https://doi.org/10.1007/978-3-319-98530-5_15

objects. This structure lends itself to continuous spatial query processing that requires no repeated searching from the root, no requirement to know the trajectory in advance, and no requirement to construct safe regions from scratch, since potential ones exist in the index structure. The mqrtree has been applied to process k-nearest neighbour queries [20], and preliminary results are promising. However, when used to process a continuous k-nearest neighbour queries, many adjacent queries produce the same results, which is undesirable and unnecessary.

There, this paper proposes two strategies for continuous k-nearest neighbour processing that utilize an approximation-based spatial access method (e.g. mqrtree). The strategies obtain safe regions from the spatial access method. In addition, the strategies determine where to begin the search *within the structure* for an updated safe region, when a new one is needed. An experimental evaluation and comparison shows that the strategies significantly reduce the number of new safe regions needed for processing a user trajectory of queries, and also reduces the computation required for the employed k-nearest-neighbour strategy.

2 Background

This section presents some background relevant to the work to be proposed. An overview of the mqrtree is provided here, along with a summary of its k-nearest neighbour search strategy. More details on the mqrtree (including validity, insertion, construction, and basic region searching algorithms) can be found in [18], while details on the k-nearest neighbour search can be found in [20].

The mqrtree [18] is a approximation-based spatial access method that uses two-dimensional nodes to organize objects in two-dimensional space. This allows the existing spatial relationships to be maintained between objects and the regions of space that contain them. After an object or point is inserted, a validity test is performed to ensure that all spatial relationships are maintained, and any objects or regions that violate the spatial relationship rules are relocated. In addition to traditional region searching and point searches, the features of the mqrtree also allow it to support k-NN searching. A very nice feature of the mqrtree that will lend itself nicely to k-NN searching is that zero overlap of regions (on the same level of the tree) occurs when the mqrtree is used to solely index point data [18].

Figure 1 presents an mqrtree for the given dataset. A node has the quadrants NW, NE, SW and SE. Each node has a corresponding nodeMBR, which encompasses all objects, points and regions in the subtrees accessible from the node. All objects and regions containing other objects, are placed in the appropriate quadrant based on their relationship to the centroid of the nodeMBR. For example, in the leaf node containing m1, m2 and p9, we observe that m1 is NW of the centroid for the nodeMBR that contains it Similarly, m2 is SW and p9 is NE of the centroid, respectively. Therefore, these objects are placed in the NW, SW, and NE quadrants of the node, respectively. All nodes, including the root node, are organized in this manner.

The mqrtree k-nearest neighbour search strategy [20] requires two steps: (1) location of a candidate nodeMBR that leads to at least k points, which requires

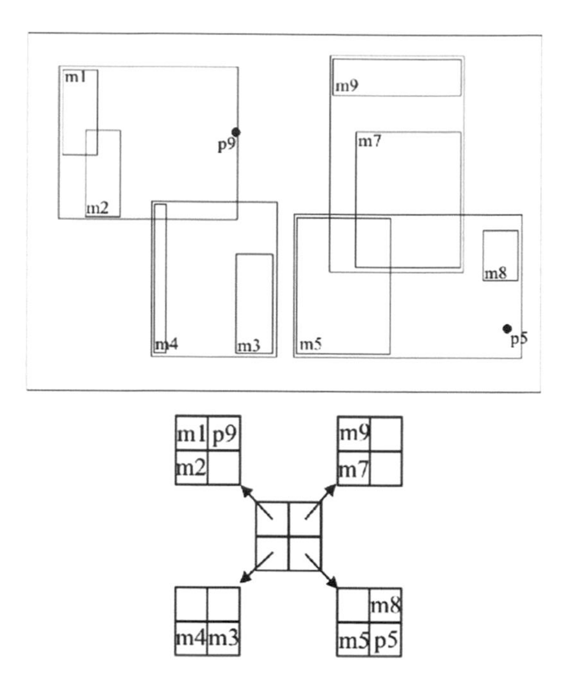

Fig. 1. mqrtree example [18]

a simple path search, and (2) a traversal of all nodes from that location in order to obtain the corresponding set of points that reside in the nodeMBR. If the obtained nodeMBR and point set contain a valid k-nearest neighbour result for the query, then the search is stopped. Otherwise, this process is repeated at the parent entry of the nodeMBR, until a suitable candidate is found. In the worst case, the nodeMBR at the Root node may be chosen. However, studies show that this case happens infrequently [20].

3 Continuous k-Nearest Neighbour Strategies

In this section, I propose two strategies for continuous k-nearest neighbour query processing. First, the concept of a safe region will be presented, along with how it can be obtained easily. Then, I propose two strategies - CKNN1 and CKNN2 - that deal with decisions on what to do when safe regions are no longer valid. I will first present the strategies respect to the mqrtree k-nearest neighbour strategy proposed in [20]. After, the application to any existing spatial access method will be discussed.

3.1 Safe Regions

Intuitively, a region of space is defined as a *safe region* if the query remains in the region, as this guarantees that the query answer is valid. With respect to a k-nearest neighbour query, if a query remains in the safe region, then the current result (i.e. k nearest neighbours) remains valid. This work also utilizes the concept of a *safe region*. However, whether or not a region of space is considered safe will be determined differently than in other safe-region approaches. The trade off is that existing minimum bounding rectangles within a spatial access method can serve as safe regions.

Given a region (lx, ly, hx, hy), a query point q and the kth nearest neighbour kq, a region of space is identified as a safe region if it meets the following criteria:

1. it contains at least k points, and
2. the distance between q and kq is less than or equal to the distance between q and each side of (lx, ly, hx, hy):

$$dist(q, kq) < min(dist(q, lx), dist(q, hx), dist(q, ly), dist(q, hy))$$

Figure 2 depicts a valid and invalid safe region respectively for a 1-nearest neighbour query. In Fig. 2a, the distance from the query point to the k point (where $k = 1$) is less than all distances to the four sides of the region. Therefore, this is a valid safe region for this query point. In Fig. 2b, the distance between the query point and the kth point is greater than the distance to the north side of the region. This means that there may be a closer nearest neighbour that resides outside of the north side. Therefore, this region is not a safe region since the query result is not guaranteed.

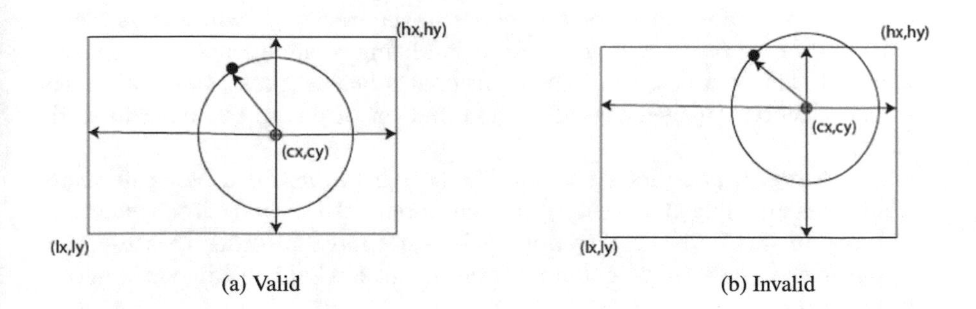

Fig. 2. Valid and invalid cases

To obtain a safe region, one approach is to construct a new one whenever that is required. This can become costly if being created from scratch. Another approach - and the one adopted for this work - is to utilize a data structure that already contains pre-defined regions, in order to locate a suitable safe region. A spatial access methods (i.e. spatial index) [6] is a hierarchical structure that indexes regions of objects, and regions of regions, using minimum bounding rectangles. Therefore, a spatial access method is very suitable for obtaining candidate safe regions that will support multiple queries along a user trajectory without having to be updated.

3.2 CKNN1 and CKNN2

I now present the two approaches for continuous k-nearest neighbour query processing that utilize minimum bound rectangles (MBRs) from a hierarchical approximation spatial access method. For presentation purposes here, I assume the use of the mqrtree [18] and its corresponding k-nearest neighbour strategy [20].

The first strategy, called CKNN1, works as follows. First, for the first query point p, an initial safe region and corresponding set of m points (where $m >= k$) that reside within the safe region are identified using the k-nearest neighbour strategy. For each subsequent query point along the user trajectory, it is handled based on one of the following situations:

1. The safe region is still valid and is not the minimum bounding rectangle from the root node of the mqrtree. There is nothing further that needs to be done here until the next query arrives.
2. The safe region is still valid, but was derived from the minimum bounding rectangle that corresponds to the root node. The search starts over again from the root node for a new safe region and corresponding point set that contains the k nearest neighbours. The reasons for handling this case in this manner are the following. First, a safe region that corresponds to the root node will have a corresponding set of $m >= k$ points that includes all points from within the mqrtree. This can be up to thousands or millions of points, which most mobile devices will not (likely) have the storage capacity to handle. Second, and related, a safe region corresponding to the root node will always remain valid and never result in a new safe region being sought, unless the situation is forced. This will result in some additional processing cost, but the tradeoff is that a more management safe region and set of points can be send to the client.
3. The safe region is no longer valid. The search begins for a new safe region (and corresponding point set) from the parent of the node from which the safe region was derived. This approach is proposed in order to attempt to optimize the k-nearest neighbour search by not having it begin at the root of the mqrtree.

This strategy continues while query points are sent from the client and remain within the general region of space covered by the mqrtree. Figure 3 depicts the pseudocode for the CKNN1 strategy. This pseudocode utilizes the following functions: (1) knn_query(starting node, query point, safe region, result set), to find and return a new safe region and points (i.e. result set) that reside in it, and (2) node(region) to return the node that represents the region passed in.

The second proposed strategy, called CKNN2, is a simplified version of CKNN1. It works as follows. An initial search for a safe region and corresponding set of points that reside in the safe region is still carried out. Then, for each subsequent query point along the user trajectory, it is handled in one of two ways:

```
knn = k (from client)
search_point = NULL (from client)
result_set = NULL (send to client when updated)
safe_region = NULL (sent to client when updated)

/*obtain first point on user trajectory*/
search_point = obtain_next_point();

While exists(search_point)
   /*is k-nn result still valid in the current node region?*/
   If exists(safe_region) and and node(safe_region) != Root
        and safe_region_valid(safe_region,search_point,knn)
        /*nothing to do - fetch the next point

   Else if not_exists(safe_region) or node(safe_region) = Root
        /*perform k-nn query from root node*/
        knn_search(Root, search_point, knn, safe_region,
                   result_set)

    Else     /*safe region not valid*/
        /*perform k-nn query from parent of node corresponding
           to safe_region*/
        knn_search(node(safe_region)->parent, search_point,
                   knn, safe_region, result_set)

    End If

    /*obtain next point on user trajectory*/
    search_point = obtain_next_point()
End While
```

Fig. 3. First continuous k-NN search strategy - CKNN1

- The safe region is still valid and is not the minimum bounding rectangle from the root node of the mqrtree. There is nothing further that needs to be done here until the next query arrives.
- For all other cases, the search for a new safe region and corresponding point set begin at the Root node.

The reason for proposing a simplified version of CKNN1 is to determine if starting new searches from within the mqrtree structure, as opposed to returning to the root, results in significant savings in computation.

3.3 Application to Other Spatial Access Methods

Although I assumed above that the mqrtree k-nearest neighbour strategy [20] is the ad-hoc strategy of choice, both strategies will also work with other spatial access methods that are hierarchical-based and utilize minimum bounding rectangles for approximations of regions of points and regions that contain other regions. The only requirement are: (1) a k-nearest neighbour strategy exist for a chosen spatial access method that can identify and return both a safe region and a corresponding superset of points that reside in the safe region, and (2) that pointers from all nodes (except the root) exist that point back to the parent nodes. If these requirement are met, then the above strategies CKNN1 and CKNN2 can be applied.

4 Evaluation

This section presents the empirical evaluation of the mqrtree-based k-NN strategy, including a comparison against repeated k-nearest neighbour searching. I first present the framework and evaluation methodology. Then, I will present the outcome of the evaluation and the resulting discussion of the outcome.

4.1 Methodology

The mqrtree-based k-NN strategy is implemented in C on a PC running the Centos 7 version of Linux. It was evaluated using several synthetic point sets for both the data and the queries. These were chosen so that certain characteristics, such as the number of points and distribution, could be controlled. Altogether, 18 points sets were used for the experiments. The first are 6 point sets of uniform distribution, each of which 500, 1000, 5000, 10000, 50000, and 100000 points respectively. Each set of n points is drawn from a two-dimensional region of space of dimensions $(\sqrt(n) * 10)x(\sqrt(n) * 10)$. For example, the points in the 1000-point set are drawn from a 310×310 region.

Next are 6 point sets of exponential (i.e., skewed) distribution. Each also contains the same number of points as the uniform datasets, and are drawn from the same sized regions of space as the uniformly distributed sets. Finally, there are 6 sets of query points. Each contain the square root of the number of points of the data set they are applied to. For example, the query point set corresponding to the 1000-point data sets contains 31 query points, while the one corresponding to the 100000-point data set contains 316 query points. The reason for this is because the query point set in each file represents a trajectory that proceeds diagonally through the region of space that contains the points.

The two continuous strategies proposed above are evaluated by using the mqr-based k-nearest neighbour strategy proposed in [20]. This strategy is also employed for comparison against the two strategies by performing repeated searching. This strategy was chosen for comparison due to promising performance improvements shown in its preliminary evaluation. The expectation is that using the continuous strategies with it will result in further performance improvements.

The following tests are performed by applying the point query sets to their respective point data sets, mentioned above:

- 1-NN, first continuous strategy. 12 tests, with 6 using the uniform point sets and 6 using the exponential point sets.

- 1-NN, second continuous strategy. 12 tests, with 6 using the uniform point sets and 6 using the exponential point sets.
- 1-NN, repeated individual search. 12 tests, with 6 using the uniform point sets and 6 using the exponential point sets.
- k-NN (k = 1 to 10), first continuous strategy. 20 tests, with 10 using the uniform 10000-point set and 10 using the exponential 10000 point set.
- k-NN (k = 1 to 10), second continuous. 20 tests, with 10 using the uniform 10000-point set and 10 using the exponential 10000 point set.
- k-NN (k = 1 to 10), repeated individual search. 20 tests, with 10 using the uniform 10000-point set and 10 using the exponential 10000 point set.

For all tests, the following performance factors are recorded:

- The CPU time is recorded for each query. Two average times are calculated and recorded: the average time over all queries, and the average time for the queries that have less than 2 µs CPU time. The reason for this will be explained later.
- The number of queries that have less than 2 µs CPU time.
- The number of pages hits. The number of page hits (i.e., nodes that are checked) for each query is recorded.
- The number of queries with invalid tests over the trajectory. Tt is possible that the initial set of candidate points that is fetched may not contain a valid set of k nearest neighbours, due to other members of the true result being outside of the corresponding superMBR. Therefore, the number of queries that produce at least one invalid result during the search is recorded.
- The number of queries that traverse from the root. In [20] it is possible that the chosen safe region corresponds to the root node, which means that the entire spatial access method must be traversed.

In addition, for the tests that utilize the continuous query processing approaches, the number of times a safe region remained valid when the next query point was evaluated is also recorded. This is not specifically recorded for repeated k-nearest neighbour searching, since all queries are executed from the beginning (i.e. Root).

4.2 Results

I first present the comparisons of most results across all of the performance factors above, followed by a comparison of how the retention of safe regions over many query points affect the number updates to the safe region that must be performed. In all charts presented here: strat is the strategy applied (cknn1, cknn2, knn is for repeated knn), k is the number of nearest neighbours #points is the number of points in the data set, #queries is the number of queries, max#phits is the number of page hits (i.e. node accesses) required by the worst performing query point, avg#phits is the average number of page hits over all queries, #valid is the number of times a safe region remained valid when one or more subsequent query points arrived, #invalid is the number of queries that had last least one invalid safe region generated during the search for a valid safe region (see [20] for more info), #root is the number of queries that ultimately chose a safe region that corresponded to the root of the mqrtree, avgtime is the average CPU time over all queries, #u2ms is the number of queries that required less than 2 μ of CPU time, and tu2ms is the average running time for the queries that ran in less than 2 μ.

Figures 1 and 2 present the results of the 1-nearest neighbour tests across all uniform and exponential point sets, respectively. Due to space limitations, we only present the results for 5000 to 100000 points. For the Uniform data sets, we observe a significantly high percentage of times that safe regions remained valid when subsequent queries were processed - from approximately 68% for queries in the 500 point set, up to approximately 95% for queries in the 100000 point set. In the Exponential data sets, we see similar trends, with the only exception being for the exponential 500-point set. This number also contributes to improvements in the number of queries that identified at least one invalid safe region during the search for one is significantly lower in both CKNN1 and CKNN2, over repeated knn searching.

We also see that for the Uniform data set, there exists no noticeable difference between the average number of page accesses between CKNN1 and CKNN2, although both have modest improvements over repeated knn searching. However, a surprise is with the difference in the average number of pages accesses for the Exponential data sets. The average for CKNN2 is actually lower than that for CKNN1.

Finally, we see the high maximum number of page accesses for both the Uniform and Exponential data sets. Although we observe a high number of times that a safe region remains valid, unfortunately in some cases performing a k-nearest neighbour query still incurs a high number of page hits. This does have more to do with the k-nearest neighbour strategy chosen [20] than the proposed approaches here.

Next, Figs. 3 and 4 present the results of the mqrtree-based k-NN tests on both the 10000-point Uniform and Exponential point sets, respectively. Due to space limitations, only the even nearest neighbour results (i.e. 2,4,6,8 and 10) are shown. However, this subset still represents the outcome across all 10 cases. Here, we can observe another surprising finding - that the number of nearest neighbours does not affect the number of times a safe region remains valid when subsequent queries are processed! In addition, although the number of queries that find at least one invalid safe region does increase with k, the increase is modest when compared to repeated k-nearest neighbour searching. Finally, the trends with the average number of page hits found in the 1-nearest neighbour tests (Tables 1 and 2) also exist here.

Finally, Figs. 4 and 5 depict the percentage of times that a new safe region is needed due to it being no longer valid. As the number of points increases, this percentage decreases - and is especially noticeable when compared to the 100% that is required for repeated 1-nearest neighbour searching. Also, the number of nearest neighbours does not affect the percentage of updates required to the

Table 1. 1-nn results - uniform distribution

#points	#queries	strat	max#phits	avg#phits	#valid	#invalid	#root	avgtime	#u2ms	tu2ms
5000	70	cknn1	3854	91.10	58	6	1	1.26	67	0.51
		cknn2	3854	91.41	58	6	1	1.25	67	0.51
		knn	3858	193.59	0	34	2	1.94	65	0.52
10000	100	cknn1	5759	157.23	87	4	2	3.26	96	0.52
		cknn2	5759	157.56	87	4	2	3.28	96	0.52
		knn	5759	199.63	0	38	2	3.30	96	0.56
50000	223	cknn1	28815	215.52	211	2	1	21.77	218	0.51
		cknn2	28815	215.64	211	2	1	21.71	218	0.51
		knn	28815	277.70	0	81	1	21.88	216	0.57
100000	316	cknn1	76980	403.92	301	9	1	63.10	309	0.50
		cknn2	76980	404.05	301	9	1	62.60	309	0.50
		knn	76980	531.31	0	136	1	63.58	299	0.53

Table 2. 1-nn results - exponential distribution

#points	#queries	strat	max#phits	avg#phits	#valid	#invalid	#root	avgtime	#u2ms	tu2ms
5000	70	cknn1	6626	630.50	51	14	8	5.08	59	0.50
		cknn2	6627	553.14	48	15	8	4.63	60	0.50
		knn	6627	993.74	0	39	13	7.79	53	0.51
10000	100	cknn1	13420	945.49	78	14	7	13.29	89	0.50
		cknn2	13420	788.22	76	13	7	11.70	90	0.50
		knn	13420	1061.04	0	50	8	14.90	87	0.51
50000	223	cknn1	69652	2597.97	198	17	6	165.65	212	0.50
		cknn2	69652	2006.54	194	15	6	140.46	214	0.50
		knn	69652	3114.56	0	115	7	209.86	210	0.54
100000	316	cknn1	137292	3173.13	296	14	5	410.00	305	0.50
		cknn2	137292	2526.88	292	15	5	349.52	306	0.50
		knn	137292	5761.48	0	171	7	788.15	286	0.56

safe region, and overall is significantly lower than required for repeated searching (Tables 3 and 4).

4.3 Discussion

Overall, we observe some significant improvements in performance when a continuous k-nearest neighbour query processing strategy is used in conjunction with a hierarchical approximation-based spatial access methods, and related k-nearest neighbour algorithm. Notably, that identifying appropriate safe regions from a spatial access method leads to significant savings in the number of times a new safe region needs to be found, as well as significant savings in the number of invalid attempts when trying to locate a safe region.

However, some surprising findings include a lower number of page hits in Exponential data when using what would appear to be a less efficient algorithm. CKNN2 restarts all searches for a new safe region from the root of the spatial

Table 3. knn results - uniform distribution

k	strat	avg#phits	max#phits	#valid	#invalid	#root	avgtime	#u2ms	tu2ms
2	cknn1	157.28	5759	87	9	2	3.27	96	0.52
	cknn2	157.61	5759	87	10	2	3.26	96	0.52
	knn	240.24	5759	0	88	2	3.45	95	0.59
4	cknn1	388.69	7713	84	26	5	8.56	93	0.52
	cknn2	388.92	7713	84	26	5	8.51	93	0.52
	knn	501.43	7713	0	120	5	8.80	91	0.60
6	cknn1	415.80	7713	88	36	5	8.66	93	0.52
	cknn2	396.86	7713	88	32	5	8.56	93	0.52
	knn	564.56	7713	0	144	5	9.12	89	0.65
8	cknn1	473.26	7713	88	34	6	10.32	92	0.52
	cknn2	473.33	7713	88	34	6	10.25	92	0.52
	knn	564.56	7713	0	144	5	9.12	89	0.65
10	cknn1	552.07	7713	84	40	7	12.87	91	0.52
	cknn2	552.22	7713	84	40	7	12.88	91	0.52
	knn	756.05	7713	0	142	7	13.47	86	0.67

Table 4. knn results - exponential distribution

k	strat	avg#phits	max#phits	#valid	#invalid	#root	avgtime	#u2ms	tu2ms
2	cknn1	1014.65	13420	78	15	8	14.35	88	0.50
	cknn2	857.36	13420	76	14	8	12.75	89	0.50
	knn	1061.35	13420	0	53	8	14.82	87	0.51
4	cknn1	1273.40	13420	81	16	10	17.72	85	0.50
	cknn2	1173.58	13420	81	15	10	16.77	86	0.50
	knn	1598.87	13420	0	73	11	21.88	82	0.52
6	cknn1	1339.14	13420	80	17	11	18.77	84	0.50
	cknn2	1239.32	13420	80	16	11	17.90	85	0.50
	knn	2341.27	13420	0	73	19	32.89	73	0.52
8	cknn1	1738.94	13416	77	21	15	29.41	80	0.50
	cknn2	1639.10	13416	77	20	15	23.45	81	0.50
	knn	2679.39	13416	0	74	21	37.69	70	0.52
10	cknn1	1946.78	13416	74	24	18	27.55	77	0.50
	cknn2	1846.94	13416	74	23	18	26.77	78	0.50
	knn	3076.22	13416	0	76	24	44.15	64	0.52

access method, where CKNN1 starts this same search from the parent of the node where the safe region came from. The problem lies in the fact that in the Exponential data set, points are clustered around the positive x- and y-axes, with very few elsewhere. When a search starts from the parent, a traversal to

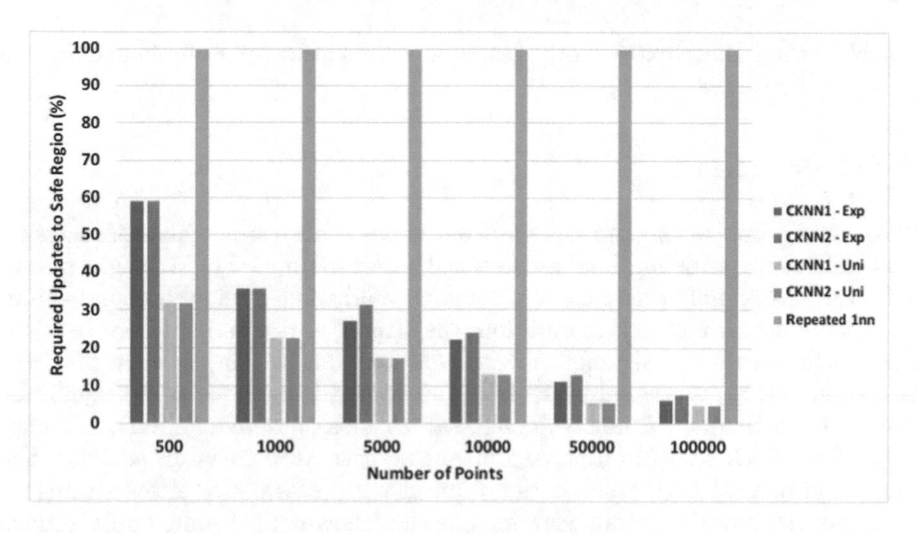

Fig. 4. Required update to safe region - 1NN

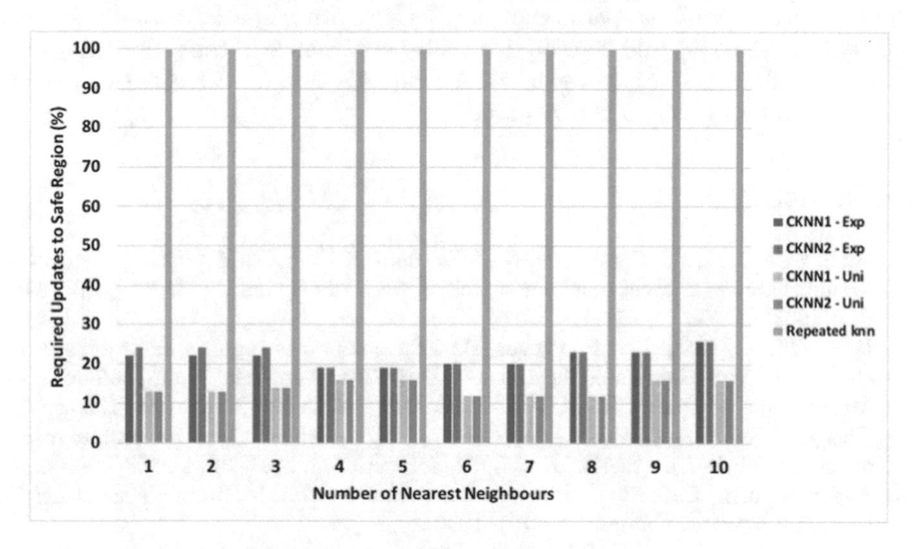

Fig. 5. Required update to safe region - kNN

obtain a point set may result in an invalid point set right from the start, and this search would proceed further up the tree in the same matter. Starting from the root will allow for drill-down to a lower node before the traversing needs to start. Therefore, this result is expected.

In addition, for the average CPU time it was noticeably lower for CKNN1 and CKNN2 than for repeated k-nearest neighbour searching. This is due to the elimination of multiple repetitions of really costly searches. If one search was costly but produced a safe region that was valid for several queries that followed, than these extra costly searches were successfully eliminated by the proposed strategies.

5 Conclusion

This paper proposes two strategies for continuous k-nearest neighbour processing that utilize an approximation-based spatial access method. The strategies obtain safe regions - regions where a query remains valid when it moves around - from the spatial access method. In addition, the strategies determine where to begin the search *within the structure* for an updated safe region, when a new one is needed. An experimental evaluation and comparison shows some significant results. In particular, the number of new safe regions that are required is not more than 20%, which result in other significant reductions in costs. In addition, the number of nearest neighbours does not affect the performance of the strategies.

Some directions of future work include the following: (1) some optimizations to both CKNN strategies and to the k-NN strategies in [20] have been identified and can further improve performance, (2) using other spatial access methods, such as the R-tree [9], and determining their performance in the proposed frameworks, and (3) a comparison versus other strategies that must create safe regions from scratch every time one is required.

References

1. Arya, S., Mount, D., Netanyahu, N., Silverman, R., Wu, A.: An optimal algorithm for approximate nearest neighbor searching fixed dimensions. J. ACM **45**(6), 891–923 (1998)
2. Brinkhoff, T., Kriegel, H.P., Seeger, B.: Efficient processing of spatial joins using R-trees. In: Proceedings of the 1993 ACM SIGMOD International Conference on Management of Data, SIGMOD 1993, pp. 237–246. ACM, New York (1993)
3. Cheng, R., Lam, K.Y., Prabhakar, S., Liang, B.: An efficient location update mechanism for continuous queries over moving objects. Inf. Syst. **32**(4), 593–620 (2007)
4. Friedman, J.H., Baskett, F., Shustek, L.J.: An algorithm for finding nearest neighbors. IEEE Trans. Comput. **24**(10), 1000–1006 (1975)
5. Fukunage, K., Narendra, P.M.: A branch and bound algorithm for computing k-nearest neighbors. IEEE Trans. Comput. **24**(7), 750–753 (1975)
6. Gaede, V., Günther, O.: Multidimensional access methods. ACM Comput. Surv. **30**, 170–231 (1998)
7. Gao, Y., Zheng, B.: Continuous obstructed nearest neighbor queries in spatial databases. In: Proceedings of the 2009 ACM SIGMOD International Conference on Management of Data, pp. 577–590. ACM (2009)
8. Gupta, M., Tu, M., Khan, L., Bastani, F., Yen, I.L.: A study of the model and algorithms for handling location-dependent continuous queries. Know. Inf. Syst. **8**(4), 414–437 (2005)
9. Guttman, A.: R-trees: a dynamic index structure for spatial searching. In: Proceedings of ACM SIGMOD International Conference on Management of Data, pp. 47–57 (1984)

10. Hjaltason, G.R., Samet, H.: Ranking in spatial databases. In: SSD 1995: Proceedings of the 4th International Symposium on Advances in Spatial Databases, pp. 83–95. Springer (1995)
11. Huang, Y.K., Chen, C.C., Lee, C.: Continuous k-nearest neighbor query for moving objects with uncertain velocity. GeoInformatica **13**(1), 1–25 (2009)
12. Huang, Y.K., Chen, Z.W., Lee, C.: Continuous k-nearest neighbor query over moving objects in road networks. In: Advances in Data and Web Management, pp. 27–38. Springer (2009)
13. Ilarri, S., Bobed, C., Mena, E.: An approach to process continuous location-dependent queries on moving objects with support for location granules. J. Syst. Softw. **84**(8), 1327–1350 (2011)
14. Iwerks, G.S., Samet, H., Smith, K.P.: Maintenance of k-nn and spatial join queries on continuously moving points. ACM Trans. Database Syst. (TODS) **31**(2), 485–536 (2006)
15. Ku, W.S., Zimmermann, R., Wang, H.: Location-based spatial query processing with data sharing in wireless broadcast environments. IEEE Trans. Mob. Comput. **7**(6), 778–791 (2008)
16. Lam, K.Y., Ulusoy, Ö.: Adaptive schemes for location update generation in execution location-dependent continuous queries. J. Syst. Softw. **79**(4), 441–453 (2006)
17. Liu, F., Hua, K.A.: Moving query monitoring in spatial network environments. Mob. Netw. Appl. **17**(2), 234–254 (2012)
18. Moreau, M., Osborn, W.: mqr-tree: a two-dimensional spatial access method. J. Comput. Sci. Eng. **15**, 1–12 (2012)
19. Mouratidis, K., Papadias, D.: Continuous nearest neighbor queries over sliding windows. IEEE Trans. Knowl. Data Eng. **19**(6), 789–803 (2007)
20. Osborn, W.: A k-nearest-neighbour query processing strategy using the mqr-tree. In: Proceedings of the 20th International Conference on Network-Based Information Systems (NBiS 2017), pp. 566–577 (2017)
21. Roussopoulos, N., Kelley, S., Vincent, F.: Nearest neighbor queries. SIGMOD Rec. **24**(2), 71–79 (1995)
22. Schiller, J.H., Voisard, A. (eds.): Location-Based Services. Morgan Kaufmann, San Francisco (2004)
23. Sproull, R.: Refinements to nearest-neighbor searching in k-dimensional trees. Algorithmica **6**(4), 579–589 (1991)

Developing a Low-Cost Thermal Camera for Industrial Predictive Maintenance Applications

Alda Xhafa[1(✉)], Pere Tuset-Peiró[2], and Xavier Vilajosana[3]

[1] Department of Electronics and Telecommunications, Polytechnic University of Tirana (UPT), Tirana, Albania
axhafa@fti.edu.al

[2] Computer Science, Multimedia and Telecommunications Department (EIMT), Universitat Oberta de Catalunya (UOC), Barcelona, Spain
peretuset@uoc.edu

[3] Internet Interdisciplinary Institute (IN3), Universitat Oberta de Catalunya (UOC), Castelldefels, Spain
xvilajosana@uoc.edu

Abstract. This paper presents the development and evaluation of a low-cost thermal camera based on off-the-shelf components that can be used to predict failures of industrial machines. On the sensing side the system is based on a LWIR thermal camera (FLiR Lepton), whereas for the data acquisition it uses an ARM Cortex-M4F micro-controller (Texas Instruments MSP432) running FreeRTOS. For the data communications the system uses a Wi-Fi transceiver with an embedded IPV4 stack (Texas Instruments CC3100), which provides seamless integration with the Cloud back-end (Amazon AWS) used to retrieve, store and process the thermal images. The paper also presents the calibration method used to obtain the relation between the camera raw output and the actual object temperature, as well the measurements that have been conducted to determine the overall energy consumption of the system.

Keywords: Thermal camera · Embedded system
Real-Time Operating System · Cloud computing
Predictive maintenance

1 Introduction

Traditionally, maintenance professionals of industrial equipment have applied corrective and preventive maintenance techniques [2] to avoid machine downtime. These techniques rely on understanding machine failure modes using both quantitative and qualitative data, and replacing sensible parts before they reach the end of their expected life and affect operation. However, corrective and preventive maintenance does not take into account real operation conditions, thus not allowing to predict machine failures in advance. Hence, over the last

L. Barolli et al. (Eds.): NBiS 2018, LNDECT 22, pp. 182–193, 2019.
https://doi.org/10.1007/978-3-319-98530-5_16

decades maintenance has evolved from applying corrective and preventive actions to becoming predictive. That is, focusing on obtaining operational data and creating mathematical models that allow to anticipate failures and allow to further reduce machine downtime.

There are many physical parameters that can be measured to determine the operating status of a machine and predict its failure using such mathematical models including (but not limited to): temperature, vibrations and energy consumption. Focusing on temperature, thermal cameras [3] have become one of the most valuable diagnostic tools for industrial applications over the past decade as they can provide accurate and non-obtrusive measurements. But despite its applicability, most thermal cameras that are available commercially today are not suitable to be applied to long-term deployments due to their design, cost and energy consumption. For instance, FLIR cameras in the test & measurements category (i.e., FLIR EX/EXX series) are engineered to be operated by hand, whereas devices in the automation & industrial safety category (i.e., FLIR AX8) have a high cost and fail to operate autonomously due to high energy consumption.

Given the current limitations of commercially available thermal cameras, the objective of this paper is to present the development and evaluation of a low-cost thermal camera that can operate autonomously for a long period of time, and that can seamlessly be integrated with a Cloud back-end to retrieve, store and process thermal images. The system will be applied to monitoring temperature in critical machinery (i.e., press machines), thus enabling to obtain long-term operational data that can be used to create predictive maintenance models [4] to further reduce machine downtime.

The remaining of the paper is organized as follows. Sections 2, 3 and 4 describe the overall system design including the hardware, the firmware and the Cloud back-end, respectively. Section 5 describes the system calibration to ensure accurate measurements, and Sect. 6 evaluates the system energy consumption to correctly size the battery capacity required for autonomous operation. Finally, Sect. 7 concludes the paper and outlines the future work.

2 Hardware

As presented earlier, the system is based on off-the-shelf components that allow to build a low-cost prototype that can acquire the thermal images and communicate directly to the cloud back-end for the data retrieval, storage and analysis. After analyzing various alternatives for each element, the system is composed of an infrared camera (FLIR Lepton) that is responsible of acquiring the thermal images, an ARM Cortex-M4F micro-controller (Texas Instruments MSP432) that processes the data taken from the camera and, finally, a Wi-Fi transceiver with an embedded IPv4 stack (Texas Instruments CC3100) that allows to communicate with the Cloud back-end. The overall system diagram, the general hardware architecture and the hardware prototype are depicted in Figs. 1 and 2 respectively, and the following subsections present the justification and technical characteristics of the sensing, processing and communication subsystems.

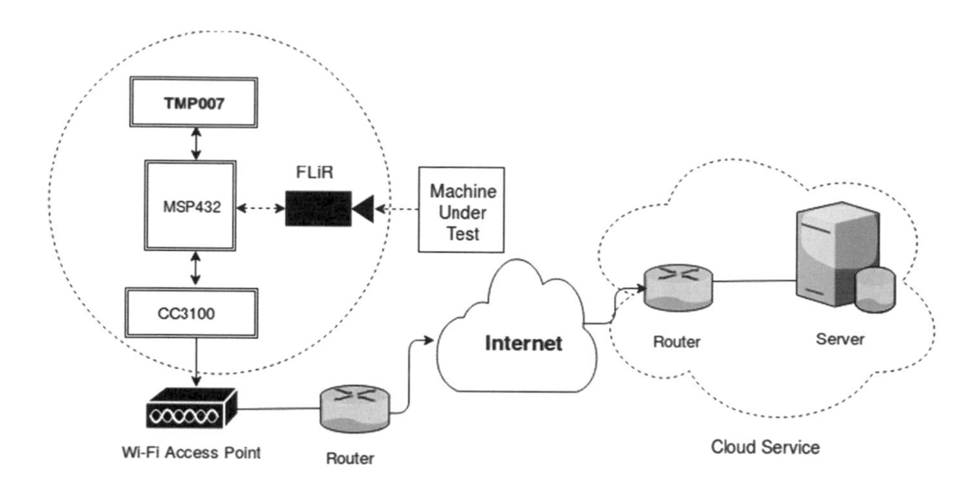

Fig. 1. System diagram

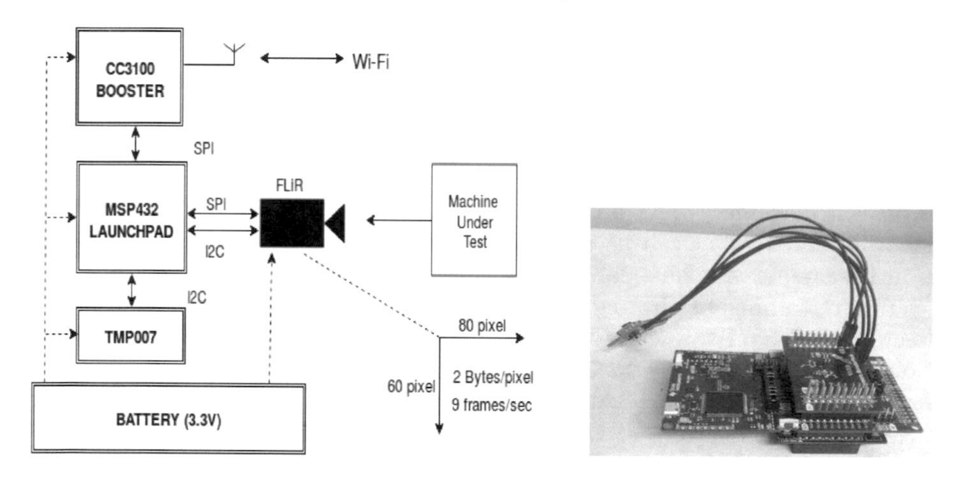

Fig. 2. System architecture (left) and hardware prototype (right).

2.1 Sensing Subsystem

For the sensing subsystem we have selected the FLIR Lepton [10] as it is the only low-cost LWIR (Long-Wave Infra-Red) camera with a sufficient resolution, i.e., 80×60 pixels, that was available at the beginning of the project. The camera integrates a fixed-focus lens assembly with a 51° horizontal FOV (Field Of View), an 80×60 pixels micro-bolometer sensor array, and signal-processing electronics. The sensor array is sensitive to wavelengths in the 8–14 μm range, has a thermal sensitivity of $\tilde{5}0$ mK, and provides a frame rate of 27 Hz. However, the true camera frame rate is limited to 9 Hz to be compliant with the United States export regulations. The camera can be interfaced [9]to a micro-controller using

I2C (Inter-Integrated Circuit) for the configuration and control, and SPI (Serial Peripheral Interface) for the data transfer.

There are two versions of the FLIR Lepton camera depending on radiometric mode availability. The version with radiometric mode allows to convert each pixel output directly into an absolute temperature with an accuracy of ±5 °C regardless of the ambient temperature, whereas the version without radiometric mode can only provide an output value that is near the middle range of the 14-bit ADC (Analog-to-Digital Converter) range when the scene temperature is equal to the temperature of the camera. Since our system required better accuracy than that provided by the radiometric-enabled FLIR, we decided to use the non-radiometric version and apply a calibration process that allows to obtain accurate temperature measurements.

Finally, the sensing subsystem also includes the Texas Instruments TMP007 [13] infrared thermopile, which is used during the system operation to determine the room temperature as required by the calibration process [1] (see Sect. 5).

2.2 Processing Subsystem

In order to acquire and process the thermal images from the FLIR camera we need a micro-controller that meets three main technical requirements. First, it has to provide enough RAM memory to acquire the images generated by the sensing subsystem, i.e., 9600 bytes/image, and buffer them before being transmitted. Second, it has to provide low-power operation modes to ensure that the system can operate autonomously for long periods of time. Third, it has to be available as a low-cost development platform that facilitates both development and integration with the sensing and the communication subsystems.

Considering the above technical requirements, we opted for the Texas Instruments MSP432P401R [11] micro-controller, which contains a 32-bit ARM Cortex-M4F processor (1.22 DMIPS/MHz) with an active current consumption of 80 μA/MHz (3.84 mA at 48 MHz system clock) and a low-power current consumption of 660 nA (full RAM retention and RTC operation). On the memory side the micro-controller includes 64 kybtes of SRAM, 256 kbytes of Flash, and 32 kbytes of ROM with the SimpleLink SDK, thus allowing for faster library code execution. Regarding peripherals, the micro-controller provides a 14-bit successive approximation ADC running at 1 Msps and with up to 24 input channels, and up to 8 serial communication peripherals supporting UART, SPI, I2C and IrDA serial protocols. Finally, the MSP432 micro-controller is available in the Texas Instruments MSP432 Launchpad, providing am integrated programming and debugging interface.

2.3 Communications Subsystem

For the communications subsystem we analyzed different low-power wireless communication technologies, i.e., Bluetooth (IEEE 802.15.1), Thread (IEEE 802.15.4) and Wi-Fi (IEEE 802.11). In the end we selected Wi-Fi

(IEEE 802.11) for several reasons. First, the average bandwidth required to send the thermal video stream is around 700 kbps (80×60 pixels, 2 bytes/pixels, 9 images/second), so IEEE 802.15.4 does not provide enough bandwidth (i.e., 250 kbps) to transmit the data. Second, the communication model of Bluetooth requires an additional device (i.e., a mobile phone) to act as bridge and provide access to the Internet, thus making the deployment more complicated. Contrarily, Wi-Fi is ubiquitous and provides seamless access to the Internet using IPv4, making it easy to deploy and integrate the system with the Cloud back-end. However, current consumption of Wi-Fi is higher compared to the other alternatives, which may place additional constraints on the system autonomy if low-power modes are not correctly handled.

Based on the selected technology, we opted for the Texas Instruments CC3100 [12] network processor to implement the communications subsystem. The CC3100 network processor embeds a Wi-Fi transceiver compatible with the IEEE 802.11g standard (see Table 1 for the transmit and receive characteristics) and an IPv4 stack with effective application throughputs of 13 Mbps and 16 Mbps for TCP (Transmission Control Protocol) and UDP (User Datagram Protocol), respectively. The CC3100 can be directly connected to a micro-controller using a serial port (i.e., SPI or UART), and the SimpleLink library provided by Texas Instruments allows for easy integration to support Wi-Fi connectivity and BSD sockets. The CC3100 network processor is available in the Texas Instruments CC3100 BoosterPack, which allows for direct interfacing with the Texas Instruments MSP432 Launchpad.

Table 1. CC3100 transmit and receive parameters.

Modulation technique	Transmit power (dBm)	Receiver sensitivity (dBm)	Current consumption (mA)
1 DSSS	18	−95.7	188–272
54 OFDM	14.5	−74	160–223

3 Firmware

This Section describes the firmware that runs on the MSP432 micro-controller and performs the data acquisition from the FLIR camera and the Texas Instruments TMP007 sensor, as well as the data transmission to the Cloud back-end infrastructure using the Texas Instruments CC3100 Wi-Fi transceiver.

The firmware is implemented in C and is composed of two different tasks that run concurrently and share the micro-controller resources, i.e., data acquisition and data transmission. To manage the execution and the communication of these tasks we have decided to use an RTOS (Real-Time Operating System) with preemption support instead of using a collaborative approach, i.e., super-loop, due to the stringent timing requirements. After evaluating different RTOS with support for the MSP432 micro-controller, i.e., TI-RTOS and FreeRTOS, we decided to use FreeRTOS because it is simple, robust and distributed under an open source license.

When FreeRTOS starts executing it creates the data acquisition and the data transmission tasks, as well as a semaphore to allow the data acquisition task to notify the data transmission task when an image is ready to be processed and send. In addition, it also creates a dual buffer structure, each having a capacity to store a thermal image (9600 bytes, 80×60 pixels, 2 bytes/pixel), to effectively decouple the data acquisition from the data transmission.

After that, the data acquisition task starts by synchronizing with the FLIR camera, whereas the data transmission task starts by establishing the Wi-Fi connection with the AP (Access Point), acquiring an IPv4 address using DHCP (Dynamic Host Configuration Protocol), and creating a UDP socket that points to the address and port tuple of the server located in the Cloud back-end.

Once the start-up process finishes, the data acquisition task starts reading images from the FLIR camera. As stated earlier, the camera can only provide a true frame rate of 9 Hz due to export regulations. However, the camera generates a frame rate of 27 Hz, meaning that each real image is transmitted three times before a new real image is generated. Moreover, since the camera does not provide an interrupt pin to indicate when the image is ready, the data acquisition task uses a hardware timer that generates a periodic interrupt (37 ms) to indicate when a new image has to be read to maintain synchronization. Hence, after each timer interrupt, the data acquisition task reads an image from the FLIR camera, reads the Texas Instruments TMP007 sensor to determine the ambient temperature at which the thermal image has been taken, and notifies the data transmission task using the semaphore.

Upon being notified by the semaphore, the data transmission task retrieves the thermal image from the image buffer, prepares the UDP packets and sends them to the Cloud back-end using the UDP socket. Given that each thermal image contains 9600 bytes of data and that the MTU (Maximum Transmission Unit) of a UDP packet has a typical value of 1480 bytes, we decided to fragment the image in 7 UDP packets before transmitting them. Thus, each UDP packet is 1440 bytes long and contains a 4 bytes header and 1440 bytes of payload, as depicted in Table 2. The UDP packet header is composed of the packet sequence number (1 byte), the camera identifier (1 byte) and the ambient temperature at which the thermal image has been taken (2 bytes). On its behalf, the UDP packet payload carries out a given number of lines of the thermal image: the first six UDP packets contain 9 lines each and the last UDP packet contains 6 lines.

Table 2. Structure of an UDP Frame.

Header			Payload
Packet ID	Camera ID	Ambient temperature	Thermal image (9/6 lines)
1 byte	1 byte	2 bytes	1440/960 bytes

4 Cloud Back-End

Data retrieval, storage and processing is implemented using a Cloud back-end to facilitate the application life cycle, from development to deployment. After analyzing and testing various Cloud service providers we selected Amazon AWS as it provides a scalable, reliable, and secure global computing infrastructure. In particular, for the computing we selected a t2.small instance that provides 1 CPU and 2 GBytes of RAM, and for the data storage we selected a S3 standard service with an initial capacity of 50 GBytes.

The t2.small instance runs an GNU/Linux distribution and executes a Python script that is responsible to retrieve and store the thermal images coming from the system. Specifically, the Python script creates a UDP socket and tries to receive all the UDP packets that conform a thermal image. Once all the packets that conform a thermal image are received the script consolidates the converts the raw values of the image to the true temperature using the calibration process described in Sect. 5, i.e., ambient temperature, calibration curves and object emissivity coefficient [5] of the target object. Finally, the thermal image is stored in a time-series database, i.e., InfluxDB, so that it can be later recovered and analyzed.

5 System Calibration

As described in Sect. 2, we use the non-radiometric version of the FLiR Lepton camera. Hence, in this Section we present the calibration method that we use to obtain the linear regression curves that allow to match the raw values extracted from the camera with the absolute temperature being measured by each pixel.

To calibrate the camera we use the method proposed in Bower et al. [6], which replaces the expensive commercial blackbodies with still water without sacrificing the calibration accuracy. Water can be considered as an ideal calibration source because: (1) in vapor state it does not absorb significant radiation, and, (2) in liquid state it has a well-known high emissivity coefficient ($\epsilon \approx 0.98$).

However, given the dependence of the camera sensor output with the room temperature [10], we need to perform the calibration process at five different room temperatures, i.e., 16 °C, 21 °C, 24 °C, 27 °C and 34 °C, to have sample of the typical operating conditions. Then, for each room temperature value we have to obtain 8 samples at different water reference temperatures, i.e., from 11 °C to 81 °C at 10 °C steps, to be able to generate each calibration curve.

To obtain the reference room and water temperatures we use the GM1312 thermometer [14], which uses a K-type thermocouple that can measure temperatures between −50–300 °C with an accuracy of ±1.5%. Since the GM1312 thermometer has two individual channels, we use the first channel to determine the room reference temperature and the second channel, with the probe placed 3 mm below the water surface, to determine the water reference temperature.

Once we have determined the room and water reference temperatures, we mount the thermal camera at 1 cm from the water surface to ensure that the lens aperture angle covers a sufficient area of the water. Since the calibration process is manual, a maximum of 10 s may elapse between measuring the room and the water reference temperatures and the actual data acquisition. However, since the thermal dissipation of the room and the water have a slow time response, we can assume that the error is below the thermometer sensitivity, thus becoming inappreciable.

Then, for each water reference temperature measured at a given room temperature, we perform both spatial and temporal averaging to obtain the true system output, i.e., the ADC value. That is, for each measurement we obtain the average value of the 20×20 pixels area located directly above the thermocouple probe, and we repeat the process 10 times, i.e., 10 pictures, to obtain the temporal average. The average values obtained during the calibration process for each room temperature (5) and each water reference temperature (8) are displayed in Table 3.

Table 3. Single gray levels for selected air room temperatures.

Temperature	**11 °C**	**21 °C**	**31 °C**	**41 °C**	**51 °C**	**61 °C**	**71 °C**	**81 °C**
Raw value (16 °C)	7971.8	8094.9	8494.1	8922.7	9412.9	9670.2	10155.1	10499.7
Temperature	**13 °C**	**23 °C**	**33 °C**	**43 °C**	**53 °C**	**63 °C**	**73 °C**	**83 °C**
Raw value (21 °C)	7900.9	8346.2	8633	9084.1	9406.5	9574.2	10197.1	10328.1
Temperature	16 °C	26 °C	36 °C	46 °C	56 °C	66 °C	76 °C	86 °C
Raw value (24 °C)	7937.9	8233.2	8545.2	9026.4	9371.1	9742.5	9818.4	9896.7
Temperature	**17 °C**	**27 °C**	**37 °C**	**47 °C**	**57 °C**	**67 °C**	**77 °C**	**84 °C**
Raw value (27 °C)	7843.6	8236.8	8530.2	8831.4	9247.6	9443.0	9848.1	9930.3
Temperature	**14 °C**	**24 °C**	**34 °C**	**44 °C**	**54 °C**	**64 °C**	**74 °C**	**84 °C**
Raw value (34 °C)	7815.4	8035.2	8379.2	8780.7	8978.3	9297.7	9692.1	10121.7

After completing the acquisition of the calibration data we use a MATLAB program to study the relationship between the sensor output value and the reference water temperature depending on the room ambient temperature. First, we check if a linear correlation exists between them. As expected, there is a linear relation between both variables for each room temperature, so we can calculate a regression line with the form $Y = aX - b$ that relates the sensor output value with the reference water temperature for each room temperature.

Table 4 summarizes the a and b coefficients for each room air temperature including its estimation error. As it can be observed, the maximum error for a and b is 6.6% and 7.5% respectively for a room air temperature of 24 °C. This error leads to an estimated accuracy of ±3 °C. However, the real temperature measurements made with the camera shows that the values are very close to the real ones and the worst case accuracy reading does not overcome ±1 °C.

Finally, the a and b coefficients for each room temperature can be stored in a computer program and be used to determine the absolute temperature value of

Table 4. Calibration curve a, b coefficients and its estimation error.

Room air temperature (°C)	a	a_{error}	b	b_{error}
16	0.026	±0.001	192.041	±9.18
21	0.028	±0.001	211.591	±1.10
24	0.030	±0.002	227.301	±17.04
27	0.030	±0.002	220.317	±7.97
34	0.030	±0.001	219.789	±7.58

each raw pixel obtained from the camera. The only requirement is that the room temperature at which the picture has been taken is known in advance. As an example, Fig. 3 displays two images acquired with the FLIR camera where the room temperature has been determined using the Texas Instruments TMP007 temperature sensor and the real temperature of each pixel has been obtained using the a and b coefficients.

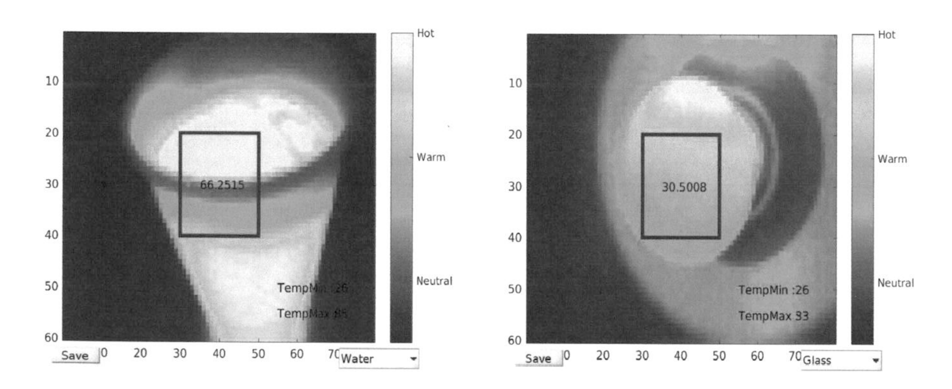

Fig. 3. Real images taken with the FLiR camera and processed using MATLAB.

6 Energy Consumption

Since the system is intended to operate autonomously [7,8], a critical parameter during the design and development phase is estimating and determining the overall energy consumption to correctly dimension the batteries required to operate for the intended period of time, i.e., one week. Hence, in this section we focus on characterizing the overall system energy consumption to determine the required battery capacity.

During the design phased we estimated the expected current consumption of the system by looking at the individual component datasheets. As it can be observed in Table 5, the maximum current consumption of the system during

Table 5. Estimated energy consumption

Hardware	Type	Energy consumption (mA, typical value)
FLiR Lepton	I_DDC+I_DD+I_DDIO	100
MSP432	LDO-Based, Active Mode	6.5
CC3100	TX Traffic (MCU Active)	272
CC3100	RX Traffic (MCU Active)	53
TMP007	Active Mode	0.27

transmission is around 330 mA, whereas the current consumption of the system during reception is expected to be around 160 mA.

As described earlier, the camera generates a total of 9600 bytes per image (80×60 pixels, 2 bytes/pixel) and a new image is generated every 111 ms (9 Hz). On the communication side the CC3100 Wi-Fi transceiver provides a theoretical maximum data rate of 16 Mbps for UDP packets. However, the effective data rate is limited by the SPI interface between the MSP432 micro-controller and the CC3100 transceiver to a maximum data rate of 6 Mbps.

Considering these values, the system will be in transmit state 12% of the time with a current consumption of 379 mA and in idle/receive state in 88% of the time with a current consumption of 160 mA. Hence, the average energy consumption of the system will be 187 mAh and the battery will need to have a minimum capacity of 32 Ah to allow for one week continuous operation.

Once the system prototype was build and the firmware development completed, we measured the real current consumption using a National Instruments cDAQ-9174 and the NI-9203/NI-9227 modules, which allow to measure both small and large currents accurately (i.e., 0–5 A, 2 μA, 50 ksamples/s, 16 bits).

As it can be observed in Fig. 4, the average system current consumption is around 273 mA, with a minimum of 116 mA and a maximum of 357 mA corresponding to the radio idle and transmit periods respectively. Hence, the mean energy consumption of the system is 273 mAh and the battery capacity required for the system to operate autonomously for a week would be 46 Ah.

The measurements conducted validate our initial estimation of the maximum and minimum current consumption of the overall system, although there is a deviation of 86 mA between the estimated and the real average current consumption. Such large difference in average current consumption needs to be further investigated to ensure proper battery dimensioning, but it is most probably caused by the state and the timing of the radio transceiver.

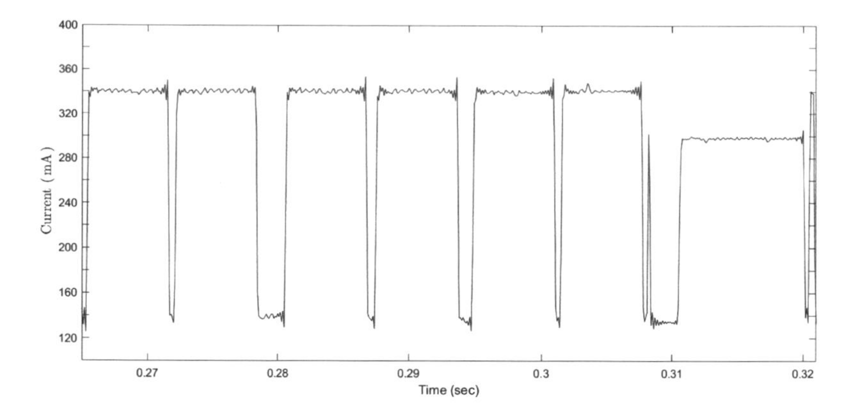

Fig. 4. Measured energy consumption of the system.

7 Conclusions

This paper has presented the development and evaluation of a low-cost infrared thermal camera for industrial predictive maintenance applications. The system is based on off-the-shelf components (i.e., FLIR Lepton, ARM Cortex-M4 and Wi-Fi transceiver) and communicates with a Cloud-based service (i.e., Amazon AWS) that retrieves, stores and processes the thermal images. Since the camera that we use operates in non-radiometric mode, we have also conducted a calibration process in different operating conditions to ensure that the temperatures that we obtain are accurate. Finally, we have measured the overall system energy consumption to be able to correctly dimension the batteries required for autonomous operation.

As for future work, we plan to investigate and optimize the overall system power consumption by introducing electronics that allow to activate/deactivate the different subsystems so that we can further improve system operation time. We are also planning to develop a back-end software base on machine learning techniques that can automatically predict machine failure based on temperature measurements. Finally, we are also considering to create a 3D-printed case that allows to deploy the system in real-world conditions and evaluate its operation.

References

1. Yu, Z., Lincheng, S., Dianle, Z., Daibing, Z., Chengping, Y.: Camera calibration of thermal-infrared stereo vision system. In: 4th International Conference on Intelligent Systems Design and Engineering Applications (ISDEA 2013), pp. 197–201, November 2013
2. Skog, I., Karagiannis, I., Bergsten, A.B., Gustafsson, L.: A smart sensor node for the internet-of-elevators-non-invasive condition and fault monitoring. IEEE Sens. J. **17**(16), 5198–5208 (2017)

3. Ferreira, U.M., Fortes, M.Z., Dias, B.H., Maciel, R.S.: Thermography as a tool in electric panels maintenance. IEEE Lat. Am. Trans. **13**(9), 3005–3009 (2015)
4. Ramirez-Nunez, J.A., Morales-Hernandez, L.A., Osornio-Rios, R.A., Antonino-Daviu, J.A., Romero-Troncoso, R.J.: Self-adjustment methodology of a thermal camera for detecting faults in industrial machinery. In: 42nd Annual Conference of the IEEE Industrial Electronics Society (IECON 2016), pp. 7119–7124, October 2016
5. Madding, R.P.: Emissivity measurement and temperature correction accuracy and considerations. In: Proceedings of SPIE Conference on Thermosense XXI (1999)
6. Bower, J.K.S.M., Saylor, J.R.: A method for the temperature calibration of an infrared camera using water as a radiative source. Rev. Sci. Instrum. **80**, 095107 (2009)
7. Schoenitz, N., Winterstein, W., Strzalkowski, B.: Optimizing power consumption of embedded systems using smart power supply solutions. In: International Exhibition and Conference for Power Electronics, Intelligent Motion, Renewable Energy and Energy Management (PCIM 2014), June 2014
8. Musaddiq, A., Zikria, Y.B., Hahm, O., Yu, H., Bashir, A.K., Kim, S.W.: A survey on resource management in IoT operating systems. IEEE Access **6**, 8459–8482 (2018)
9. Coster, E.J., Kerstens, W.C.M., Schroedel, O.: Implementation of an automatic FLIR-scheme in a 20 kV distribution grid. In: 12th IET International Conference on Developments in Power System Protection (DPSP 2014), Copenhagen, Denmark, 31 July 2014
10. FLIR Commercial Systems: FLiR Lepton. Datasheet, October 2014
11. Texas Instruments: MSP432P401R SimpleLink Mixed-Signal Micro-controllers. Datasheet, September 2017
12. Texas Instruments: CC3100 SimpleLink Wi-Fi Network Processor. Datasheet, January 2015
13. Texas Instruments: TMP007 Infrared Thermopile Sensor. Datasheet, May 2014
14. Labfacility: GM1312 General Purpose Digital Thermometer. Datasheet

Globally Optimization Energy Grid Management System

Abdul Rahman[1], Nadeem Javaid[1](✉), Muhammad Asad Zaheer[1], Maryam Bibi[2], Zoha Fatima[2], and Qurat Ul Ain[3]

[1] COMSATS University, Islamabad 44000, Pakistan
abdul0734@gmail.com, nadeemjavaidqau@gmail.com
[2] National University of Modern Languages, Islamabad 44000, Pakistan
maryambibi723@gmail.com, zohafatima689@gmail.com
[3] Department of Physics and Energy Harvest Storage Research Center (EHSRC), University of Ulsan, Ulsan 44610, Korea
aineee.87@gmail.com
http://www.njavaid.com/

Abstract. This paper focuses on the smart grid with cloud and fog computing to reduce the wastage of electricity. A traditional grid is converted into a smart grid to reduce the increase of temperature. A smart grid is the combination of traditional grid and information and communication technology (ICT). The Micro Grid (MG) is directly connected with fog and has small scale power. MG involves multiple sources of energy as a way of incorporating renewable power. The Macro Grid has a large amount of energy, and it provides electricity to the MG and to the end users. Clusters are the number of buildings having multiple homes. Some load balancing algorithms are used to distribute the load efficiently on the virtual machines and also helps the maximum utilization of the resources. However, a user is not allowed to communicate directly with MG, a smart meter is used with each cluster for the communication purpose. If the MG is unable to send as much energy as needed then fog will ask cloud to provide energy through macro grid. Optimized bubble sort algorithm is used and it is actually a sorting algorithm. The sorting, in this case, means that the virtual machine sorts on the basis of a load. The virtual machine which has the least load will serve the demand. In this way, the virtual machine works and this mechanism give least response time with high resources utilization. Cloud analyst is used for simulations.

Keywords: Cloud computing · Fog computing · Load balancing
Renewal sources · Smart grid · Utility

1 Introduction

Smart grid with cloud and fog computing is one of the encouraging solution to reduce the wastage of electricity. As the production of the electricity produces

L. Barolli et al. (Eds.): NBiS 2018, LNDECT 22, pp. 194–208, 2019.
https://doi.org/10.1007/978-3-319-98530-5_17

lot of Carbon Dioxyde which pollutes the environment and also increases the temperature of the environment as mention by author [1].

Smart grid is actually the combination of tradition grid and information and communication technology (ICT). ICT is actually a two way communication and it is the infrastructure and components that allows us for modern computing. As author [2] used ICT in mobile communication in backward areas.

Cloud is actually the system having very high processing speed and huge memory. Cloud computing is actually such type of computing that is computing resources.

The services which are provided by the cloud are:

- Software as a Service (SaaS):
  Saas is a software delivery service which provides the access to the program and also the functionality which is accessible remotely in the form of Web-Bases services.
- Platform as a Service (PaaS):
  PaaS is an external platform used by companies or data centers to purchase and manage their own software and hardware layer.
- Infrastructure as a Service (IaaS)
  Computer infrastructure, such as servers, warehouse and networks which is delivered in the form of a service. IaaS is very known in organizations that value the benefits of a cloud dealer manage who manages Information Technology (IT) infrastructure.

The visual representation of the services provided by Cloud are (Fig. 1):

1.1 Motivation

The major reason to implement smart grid with cloud and fog processing is to increase the user reliability and minimize the cost so that this concept is easy to install and the end user have to pay low. Byun et al. [4] discuss the cloud processing and energy management system. The author describe the cloud computing to balance energy efficiently and also tried to reduce the energy consumption through optimization algorithm. Writer also focus on the user friendly location and situation for the end users. The Author in Self-Sustained Smart Building in the Smart Cities [5] describe the best evolution of the Smart Buildings and the connectivity of it with smart devicesµ(IoT). The most challenging task he described was the exchange of information between various smart devices. Different smart devices have different processing power and processing of such large databases is the most challenging task. In papers [6–8] smart buildings are connected fog which is responsible for the processing of the requests send by the homes. The utilities are directly connected to the fog and the utilities provide the energy according to the demand of the homes. The author used round robin load balancing algorithm for the efficient use of the energy. The cost of the users are decrease and the efficient use of energy takes place.

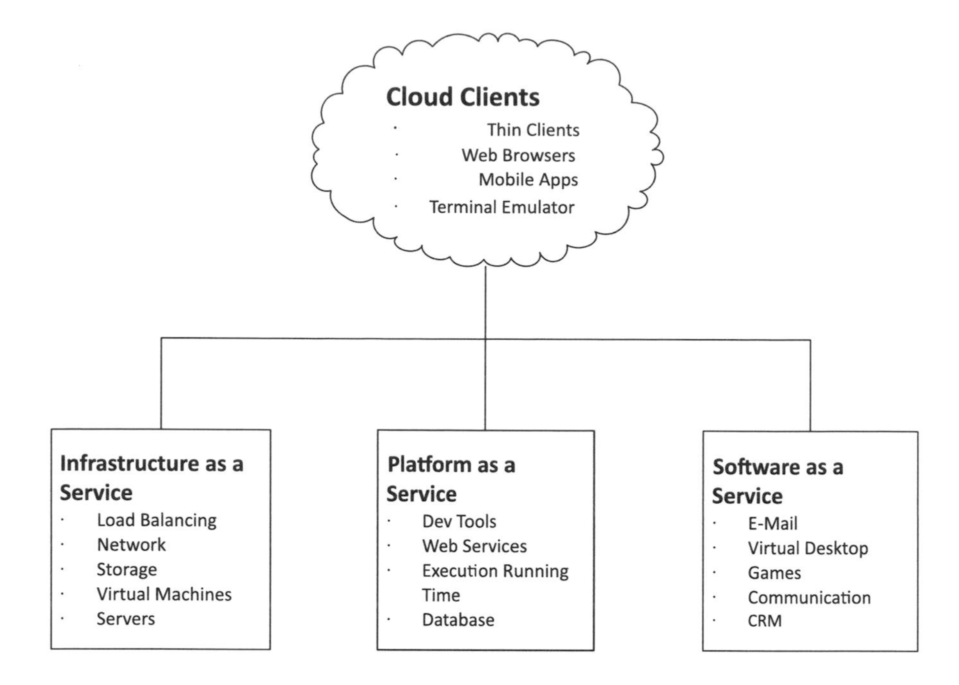

Fig. 1. Cloud services

1.2 Problem Formulation

The heterogeneous behavior of cloud computing is due to the connectivity of different types of devices connected with it. Due to heterogeneity load balancing becomes the complex task. Physical machines allocate cloud computing resources in an imbalanced way resulting in cloud services issues [22]. Therefore, allocation and scheduling of cloud computing resources for Virtual Machine (VM) load balancing in a cloud computing based system is significant task.

$$VirtualMachines = \{V_1, V_2, V_3, \ldots, V_n\} \tag{1}$$

Suppose there are VirtualMachines. Therefore, in Eq. 1 V denotes the of n number of VirtualMachines.

$$UserBase = \{UB_1, UB_2, UB_3, \ldots, UB_n\} \tag{2}$$

Equation 2 denotes the set of n number of User Base.

The relationship between request for resource generated by User Base and capacity of VirtualMachines. The resource request generated by User Base must be less than the capacity of the VirtualMachines.

$$\sum_{i=1}^{VirtualMachine} Z_{ik} \leq C_k \tag{3}$$

The Eq. 3 displays that the total resource request Z_{ik} cannot exceed the total capacity of C_k [20] When a datacenter controller requests for new VM allocation, the algorithm finds the efficient VM with minimum forecasted response time having least load for allocation. Then the algorithm returns id of that efficient VM to datacenter controller. This results in updation of allocation counter table for that VM. Datacenter controller receives the response following the VM processing completion. After that notication for VM deallocation is generated.

Equation 4 shows the formula for Bandwidth utilization. It dependancy lies on region, data transfer size and user request. N_{UR} is the number of user request which is to be transmitted in different regions and also encorporates the user requests from two regions.

$$Bandwidth_{util} = \frac{\text{Bandwidth}_{alloc}}{N_{UR}} \tag{4}$$

The Eq. 5 shows the model of datacenter processing time. T_{datpro} is calculated based on user request/hour in accordance to bandwidth allocation [21]

$$PT_{DC} = \frac{\text{UR}_{hour}}{BW_{alloc}} \tag{5}$$

Equation 6 depicts the execution of task v_i on VirtualMachines m_r, denoted by $EC^r{}_i$. h_r depicts the used building interval of VM m_r. C_r shows cost/interval unit of VM m_r. S_{ij} is the size of data which is to be transmitted from v_i to v_j. Tc_{ij} shows data transfer cost communication data from v_i to v_j[21]

$$ECS_i^r = \left(\frac{ET_i^r \times c_r}{h_r}\right) + \left(s_{ij} \times tc_{ij}\right) \tag{6}$$

2 Related Work

Cloud computing is a conceptualized environment. It provides resource sharing facilities to its clients. The resources are sharing in such a way that the use of physical machines minimizes whereas, the use of virtual machines increases so that it helps to reduce the energy consumption among the data centers. Most of the load balancing algorithms which have been used to reduce the load on the loud are Modified Bubble Sorting(MBS), Round Robin (RR), Throttled, Active Clustering, Biassed Random Sampling, Active Monitoring, and Honeybee Foraging. The authors in [9] proposed a Cloud Load Balancing Algorithm (CLB) which distributes the load of the cloud and automatically it compares the result with the mentioned algorithms.

The author in [10] discuss the resource allocation problem which is entirely based on resource allocation through Ant Colony System (ACS). The first objective was to minimize the energy consumption by using load balancing on cloud side. Author [11] proposed a new load balancing algorithm in which the proposed algorithm allocates the task efficiently to every virtual machine.

RR technique also provide very good and efficient results of load balancing in fog and cloud processing. The author [6] used RR in the paper. Load of the fog divided in to different Virtual Machines(VM). The request sent to the fog by the host or consumers are of electricity requirement.

The author in [9], proposed a cloud load balancing (CLB) algorithm for load balancing on cloud and compare the results with other algorithms discussed before. Authors present a new communication model using two models: the cloud based demand response model and distributed demand response model for computing the communication efficiency with reference to optimal resource allocation [12]. This scheme procures the high cost by incorporating the high demand response scenarios. Authors presented a cost oriented model for demand side management by optimally allocating the cloud computing resources [13]. This model gives consumers the flexibility in resource optimization which results in cost minimization.

The author in [14] proposed a certificate less provable data possession (CL-PDP) method. The method is proposed for cloud based SG applications. In addition, the authors in [15] used wavelet recurrent neural network (WRNN) predictors. The solutions are given for cloud distributed model and balance for energy management is also achieved for available devices by means of prediction.

Cloud computing facilitates the customers with: computing, storage and networking capacities. However, it suffers from latency, security and downtime issues. For improving the latency, flexibility, resiliency and reliability of the cloud computing services, a new scheme is proposed in and implemented using the profile of the devices for Web services.

Many scholars have deliberated how to use cloud-fog computing to support manage the SG. For demand response optimization in the SG, [16] examined the advantages of using cloud-fog platform and presented a example scenario. Also, a SG data management based on cloud computing model is presented in [17], which takes the benefit of dispersed data management for parallel processing, real-time data gathering, ubiquitous access for real-time information retrieval.

Moreover, for faster response time in large scale deployments a cloud based demand response architecture is presented in [12]. The studied demand responses are of two types: cloud based demand response and distributed demand response. The objective of this study is to minimize the convergence time and efficient bandwidth utilization. In [18], an efficient SG electric vehicle charging and discharging service at public supply stations based on cloud environment scheduling is proposed. It ensures the communication links between SG and cloud platforms.

3 System Model

Cloud with fog computing was introduced in 2016. The concept of cloud and fog is approximately same. The differences in cloud and fog are: size, distance from user, memory, and processing. The distance of cloud varies from ground level to thousands of kilometer whereas the fog must lie on ground level. The processing speed, memory of the cloud is much high than fog. The other components in this system model are: Macro grid, utility, micro grid, broker policy and cluster. The source of energy of macro grid is from fossil fuels, nuclear plants or wind and water turbine. Macro grid provides energy to the micro grid and its responsibility is to provide energy to the end users. Micro grid operate with or independently with main electric grid. The working of the broker policy is to route the traffic which comes from the end user to fog and after that from fog to cloud. Cluster are the number of building or buildings which contain many homes (Fig. 2).

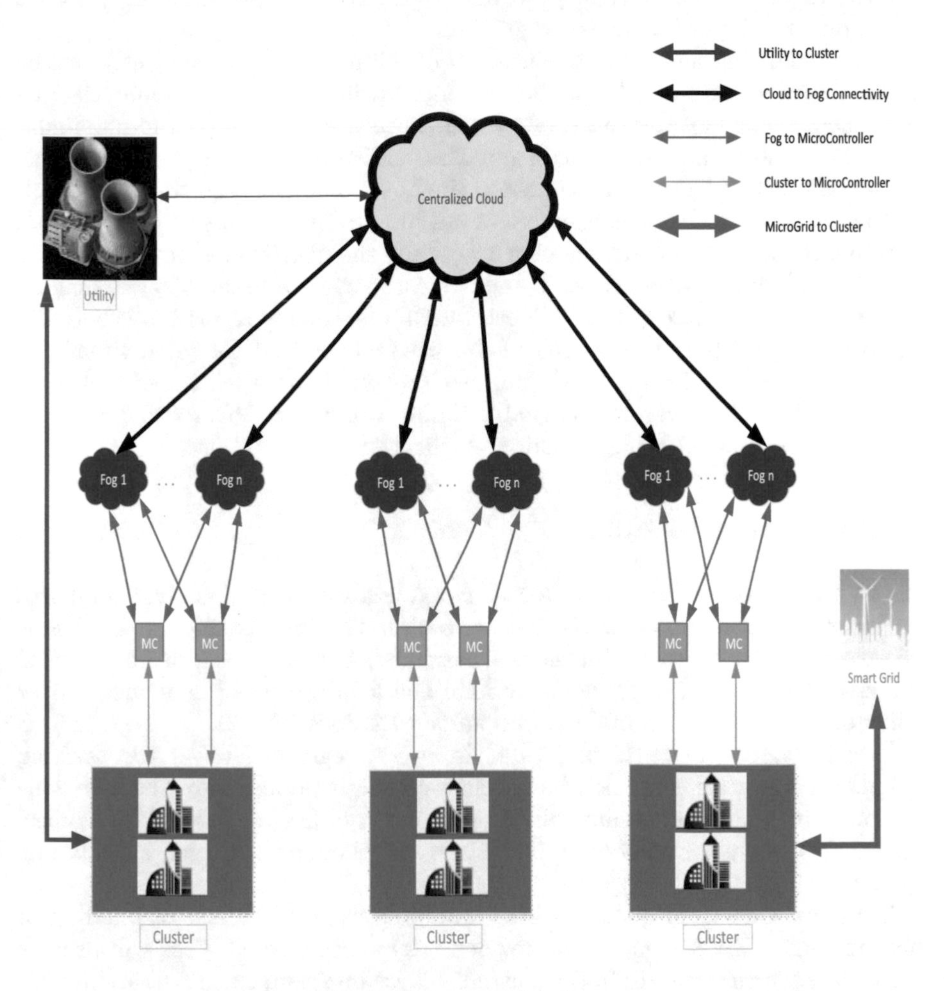

Fig. 2. System model diagram

The model has three layers. Third layer has clusters, Electric management controller (EMC) and micro grid. All the homes in the clusters are smart homes. Smart homes means that they have smart meter which are directly connected to the electric management controller. EMC controls all the requests which are sending from homes to fog. The fog fulfills all the requests. Now there are two more cases.

If the home demand for energy then fog first send the request to micro grid. If micro grid has enough energy then it will fulfill the requirement of the home. Another case is that, the home send to request to fog for more energy.

The fog will send the request to micro grid and if the micro grid is unable to fulfill the requirement then the fog will send the request to the cloud and cloud sent the request to the macro grid. Macro grid then supply the energy to the homes directly. Load Balancing Algorithm.

Load balancing algorithm are used to distribute the load efficiently on the virtual machines and also helps the maximum utilization of the resources. Virtual machines are located on servers and its major function is the maximum utilization of fog and cloud processing. The service broker policy is responsible for the allocation of virtual machines to the hosts. It depend on the service policy that whether it allocates one virtual machine to many hosts or many virtual machines to one host. Load balancing is actually the distribution of the workload to achieve minimum response time and avoid the overload request by managing the requests. Optimized Bubble Algorithm Bubble sorting algorithm is actually the sorting algorithm. The sorting in this case means that the virtual machine sorts on the basis of load. The virtual machine which have least load will serve the demand. In this way the virtual machine works and this mechanism give least response time with high resources utilization.

4 Simulation Results

The tool we used for the simulation is ‘cloud analyst’. Cloud analyst compares the broker policies and show the better results. The load balancing algorithms are also compared with all broker policies. Load balancing algorithm are used to distribute the load of all homes within the fogs whereas the service broker policy distribute the load of all homes among the fogs (Fig. 3).

In this paper the world is divided in to six regions however the working is done on four regions which have relatively high population. These regions contain a fog which manipulate all the processing of the requests of the homes. Each region of the world contain 3 clusters and they are managed by only one fog as shown in Table 1.

We have considered one fog which manipulates three cluster of building in three regions. All the fogs in each regions are interconnected. Fogs will process all the data which comes from the clusters. These interconnected fogs are further connected with the centralized cloud. The cloud have huge storage for data and also have the availability of macro grid. Now the services which is provided by the cloud are:

Development as a Service. Development as a service means the sharing of web development tools. Actually it is equivalent to the locally installed development tools. Data as a Service It is also a web based design which is responsible to the

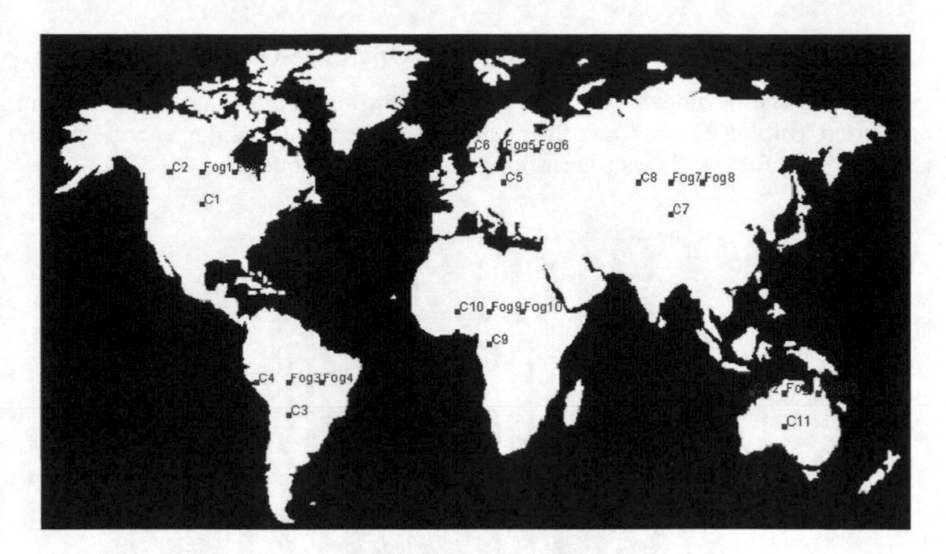

Fig. 3. Division of World.

Table 1. Division of world

Continents	ID
North America	0
South America	1
Europe	2
Asia	3

access of the defined API layer of the cloud data. It also act as a service offering of specialized subset of software.

Software as a Service. It provides the maintaining and installing of the software in the cloud and a user can access the software from the cloud anywhere over the internet. The cloud application runs on the server. Software as a service is scalable and the system administrators of the may load the application on several servers.

Platform as a Services. Platform as a service act as an equivalent to middle ware in the traditional delivery of the databases and application platforms.

Infrastructure as a Service. Infrastructure as a service actually act as a virtual machine. The companies pay fee to run virtual servers, networks and storage from the cloud. Fogs are based on the structure of the cloud. But it has less memory and processing speed as cloud.

4.1 Comparison of Round Robin and Optimized Bubble Algorithm

We apply same perimeters on both round robin load balancing algorithm and optimized Bubble Sorting algorithm having closest data center service policy (Fig. 4). The Table 2 shows the response time of the request.

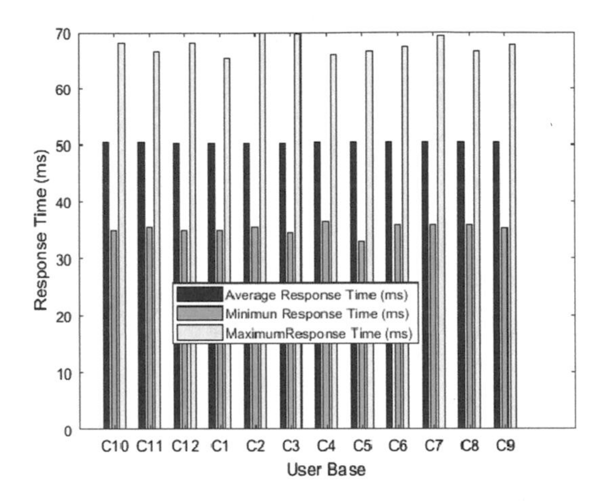

Fig. 4. Load on fog by clusters

Table 2. Response time by region of optimized Bubble algorithm

Region	Cluster of Building	Average (ms)	Minimum (ms)	Maximum (ms)
Region 0	C1	50.18	34.72	65.44
	C2	50.17	35.36	69.86
	C3	50.18	34.46	69.68
Region 1	C4	50.40	36.42	66.08
	C5	50.40	32.94	66.48
	C6	50.38	35.72	67.17
Region 2	C7	50.26	35.77	69.26
	C8	50.26	35.77	66.52
	C9	50.25	35.25	67.75
Region 3	C10	50.23	34.86	68.12
	C11	50.21	35.36	66.48
	C12	50.20	34.72	67.96

The Table 3 shows the response time of the request sent by the user to the fog and the response received against the request of Round Robin Optimization Algorithm.

Table 3. Response time by region Round Robin Optimization

Region	Cluster of Building	Average (ms)	Minimum (ms)	Maximum (ms)
Region 0	C1	50.18	34.72	65.44
	C2	50.18	35.36	69.86
	C3	50.18	34.46	69.68
Region 1	C4	50.41	35.10	66.08
	C5	50.40	32.94	66.48
	C6	50.39	35.72	67.17
Region 2	C7	50.27	35.77	69.26
	C8	50.26	35.77	66.52
	C9	50.25	35.25	67.75
Region 3	C10	50.23	34.86	68.12
	C11	50.22	35.43	66.48
	C12	50.20	34.72	67.96

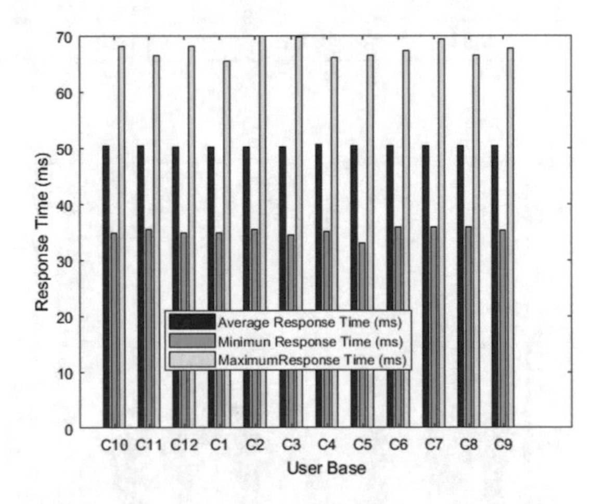

Fig. 5. Load on Fog by Clusters

This table shows the load on each fog send by the cluster.

These simulations shows that the average response time of optimized bubble algorithm is better than Round Robin load balancing algorithm.

4.2 Working

Optimized Bubble Algorithm is actually sorting the virtual machine which are present in the fogs. Fogs have their own process time and memory. First the bubble sorting algorithm sort the virtual machine which have low load and fog will send the request towards the low load virtual machine. After that the fog will respond back to the cluster and if the demand of the user is nt able to complete then the fog will send the request to the cloud. The cloud sends the request to the macro grid and then macro grid provide the demand energy.

4.3 Fog Request Service Time

The data center request servicing time of the fogs of optimized bubble algorithm are: This table shows the request service diagram of the each fog. Request service is the time in which a user sends the request to the fog and the time a fog needs to response (Table 4 and Fig. 6).

Table 4. Data center request servicing times optimized Bubble algorithm

Data center	Average (ms)	Minimum (ms)	Maximum (ms)
Fog1	0.50	0.01	1.01
Fog2	0.64	0.03	1.32
Fog3	0.59	0.02	1.21
Fog4	0.51	0.01	1.13

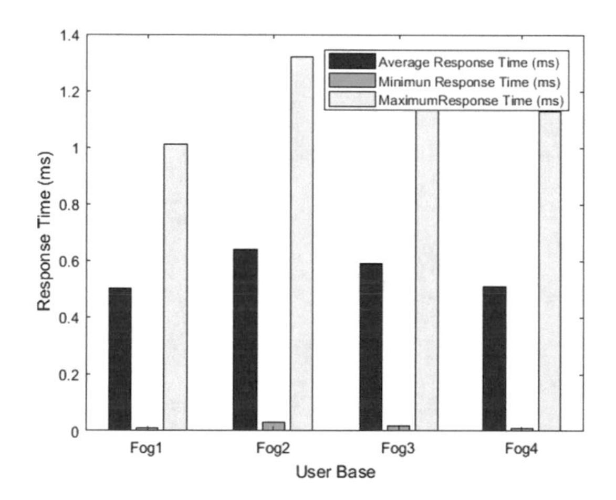

Fig. 6. Fog performance

The data center request servicing time of the fogs of Round Robin optimization algorithm are (Table 5) and Fig. 7:

This graph shows the values obtained from the paper [6].

The visual representation of request servicing time of the data center is shown in Fig. 5.

Table 5. Data center request servicing times Round Robin optimization algorithm

Data center	Average (ms)	Minimum (ms)	Maximum (ms)
Fog1	0.50	0.01	1.07
Fog2	0.65	0.03	2.17
Fog3	0.59	0.02	1.77
Fog4	0.51	0.01	1.32

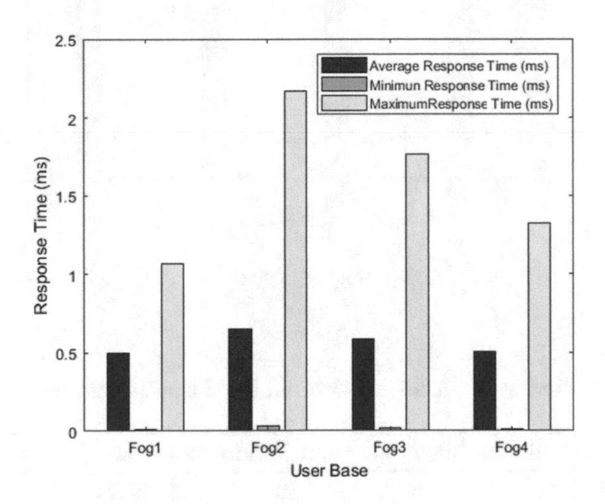

Fig. 7. Fog performance

4.4 Cost

The cost of virtual machine, micro grid through optimized bubble algorithm are (Table 6)

The total cost of Fog.1, Fog.2, Fog.3, Fog.4 are 1332.86, 1335.00, 1335.52, 1876.05 relatively (Fig. 8).

The cost of virtual machine, micro grid through Round Robin optimization algorithm are (Table 7)

The total cost of Fog.1, Fog.2, Fog.3, Fog.4 are 1332.86, 1335.00, 1335.52, 1876.05 relatively (Fig. 9).

Table 6. Total cost optimized bubble algorithm

Data Center	Virtual Machine cost	Micro Grid cost	Data Transfer Cost	Total
Fog1	19.20	1236.39	77.27	1332.86
Fog2	19.20	1238.40	77.40	1335.00
Fog3	19.20	1238.89	77.43	1335.52
Fog4	19.20	1754.40	109.65	1876.05

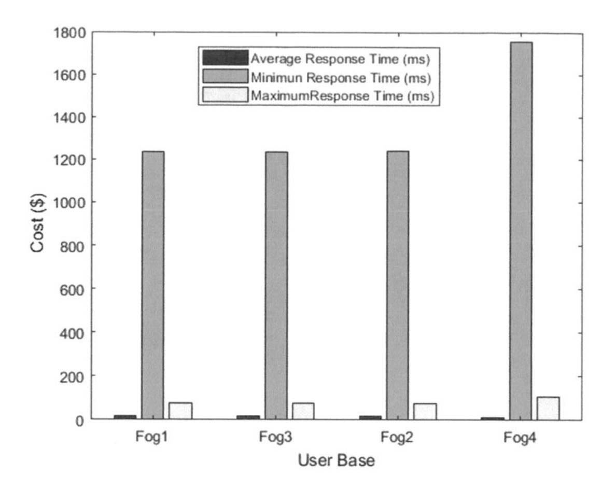

Fig. 8. Cost

Table 7. Total cost Round Robin optimization

Data center	Virtual Machine cost	Micro Grid cost	Data Transfer cost	Total
Fog1	19.20	1236.39	77.27	1332.86
Fog2	19.20	1238.40	77.40	1335.00
Fog3	19.20	1238.89	77.43	1335.52
Fog4	19.20	1754.40	109.65	1876.05

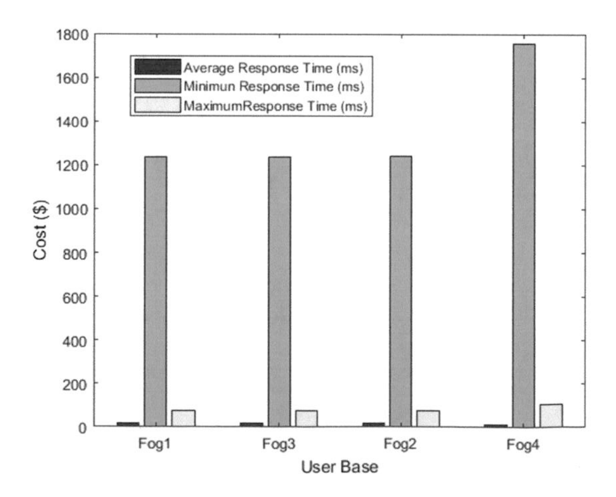

Fig. 9. Cost

5 Conclusion

In this paper, fog based environment is proposed which is integrated with SG. We assume multiple residential clusters having 150 to 250 homes which have Smart Meters(SM) and are directly connected with fog. Customers send the request of electricity to fog to complete their requirement. All the fogs are directly connected to the centralized Cloud. When the MG is unable to complete the demand of the customer then the fog send the request of the customer toward the cloud. Cloud is directly connected with the utility. The utility will then provide the electricity to that home.

Cloudsim emulator performs the simulation of the proposed system model. Simulations are using optimized response time service broker policy for fog selection. VM load balancing algorithms like Optimized bubble algorithm and RR load balancing algorithm considered to check the performance of the proposed system. However, one scenario is considered to check the behavior of the fog having different number of VMs installed on physical machines. It is analyzed from the results that the OBA algorithm has nearly same response time, processing time of the fog and cost. Furthermore, the proposed model may extended by increasing the users and the fog services. The major target is to reduce the RT and PT.

References

1. Abdallah, L., El-Shennawy, T.: Reducing carbon dioxide emissions from electricity sector using smart electric grid applications. J. Eng. **1–8**, 2013 (2013)
2. Meso, P., Musa, P., Mbarika, V.: In LDCs: the case of sub-Saharan Africa. pp. 119–146 (2005)
3. Yi, S., Li, C., Li, Q.: A survey of fog computing: concepts, applications and issues. In: Proceedings of the 2015 Workshop on Mobile Big Data - Mobidata 2015, pp. 37–42 (2015)
4. Byun, J., Kim, Y., Hwang, Z., Park, S.: An intelligent cloud-based energy management system using machine to machine communications in future energy environments. In: Digest of Technical Papers - IEEE International Conference on Consumer Electronics, pp. 664–665 (2012)
5. Kumar, N., Vasilakos, A.V., Rodrigues, J.J.P.C.: A multi-tenant cloud-based DC nano grid for self-sustained smart buildings in smart cities. IEEE Commun. Mag. (2017)
6. Fatima, I., Javaid, N.: An Efficient Utilization of Fog Computing for an Optimal Resource Allocation in IoT based Smart Grid Network
7. Coalition based Game Theory Energy Management System of a Building as-service-over Fog Computing
8. Saman, Z., Javaid, N.: A Cloud-fog based Smart Grid Model for Effective Information Management
9. Chen, Y.Y., Kuo, S.H., Shang Liang Chen: CLB: a novel load balancing architecture and algorithm for cloud services. Comput. Electr. Eng. **58**, 154–160 (2017)
10. Pham, N.M.N., Le, V.S.: Applying Ant Colony System algorithm in multi-objective resource allocation for virtual services. J. Inf. Telecommun. **1**(4), 319–333 (2017)

11. Razzaghzadeh, S., Navin, A.H., Rahmani, A.M.: Probabilistic modeling to achieve load balancing in Expert Clouds. Ad Hoc Networks **59**, 12–23 (2017)
12. Yaghmaee, M.H., Leon-garcia, A.: On the performance of distributed and cloud - based demand response in smart grid, 3053(c) (2017)
13. Cao, Z., Lin, J., Wan, C., Song, Y., Zhang, Y., Wang, X.: Optimal cloud computing resource allocation for demand side management, pp. 1–13 (2016)
14. He, D., Kumar, N., Zeadally, S., Wang, H.: Certificateless provable data possession scheme for cloud-based smart grid data management systems, **13**(9) (2017)
15. Capizzi, G., Lo Sciuto, G., Napoli, C., Tramontana, E.: Advanced and adaptive dispatch for smart grids by means of predictive models, **3053**(c), 1–8 (2017)
16. Okay, F.Y., Ozdemir, S.: A fog computing based smart grid model. In: 2016 International Symposium on Networks, Computers and Communications, ISNCC 2016, pp. 1–6 (2016)
17. Sofana Reka, S., Ramesh, V.: Demand side management scheme in smart grid with cloud computing approach using stochastic dynamic programming. Perspect. Sci. **8**, 169–171 (2016)
18. Chekired, D.A., Khoukhi, L.: Smart grid solution for charging and discharging services based on cloud computing scheduling. IEEE Trans. Industr. Inf. **13**(6), 3312–3321 (2017)
19. Chonglin, G., Fan, L., Wenbin, W., Huang, H., Jia, X.: Greening cloud data centers in an economical way by energy trading with power grid. Future Gener. Comput. Syst. **78**, 89–101 (2018)
20. Guo, M., Guan, Q., Ke, W.: Optimal scheduling of VMs in queueing cloud computing systems with a heterogeneous workload. IEEE Access **6**, 15178–15191 (2018)
21. Jena, S.R., Ahmad, Z.: Response time minimization of different load balancing algorithms in cloud computing environment. Int. J. Comput. Appl. **69**(17), 22–27 (2013)
22. Hemamalini, M., Srinath, M.V.: Response time minimization task scheduling algorithm. Int. J. Comput. Appl. **145**(1), 9–14 (2016)

Efficient Resource Distribution in Cloud and Fog Computing

Mubashar Mehmood[1], Nadeem Javaid[1](✉), Junaid Akram[2], Sadam Hussain Abbasi[1], Abdul Rahman[1], and Fahad Saeed[3]

[1] COMSATS University, Islamabad 44000, Pakistan
nadeemjavaidqau@gmail.com
[2] National University of Sciences and Technology (NUST), Islamabad 44000, Pakistan
[3] COMSATS University Islamabad, Wah Campus 47040, Pakistan
http://www.njavaid.com

Abstract. Smart Grid (SG) is a modern electrical grid with the combination of traditional grid and Information, Communication and Technology. SG includes various energy measures including smart meters and energy-efficient resources. With the increase in the number of Internet of Things (IoT) devices data storage and processing complexity of SG increases. To overcome these challenges cloud computing is used with SG to enhance the energy management services and provides low latency. To ensure privacy and security in cloud computing fog computing concept is introduced which increase the performance of cloud computing. The main features of fog are; location awareness, low latency and mobility. The fog computing decreases the load on the Cloud and provides same facilities as Cloud. In the proposed system, for load balancing we have used three different load balancing algorithms: Round Robin (RR), Throttled and Odds algorithm. To compare and examine the performance of the algorithms Cloud Analyst simulator is used.

Keywords: Fog computing · Cloud computing · Load balancing
Tasks scheduling · Response time · Smart Grid · Microgrid
Virtual machines

1 Introduction

Smart Grid (SG) as a communication system allows devices to interact with each other. SG is a modern electrical grid which includes various energy measures including smart meters, energy-efficient resources and renewable energy resources. SG uses two-way communication between service provider and customers enabled them to monitor their pricing and energy consumption. With the number of end users connected to SG increases, SG requires high processing and high storage. The Internet provides the solution for demand-side energy management which uses two-way flow of information which is considered as SG

L. Barolli et al. (Eds.): NBiS 2018, LNDECT 22, pp. 209–221, 2019.
https://doi.org/10.1007/978-3-319-98530-5_18

extension. Internet of Things is presented as extension of internet. As IoT services increases, IoT devices are used in various fields such as technology, industry and several domains.

The factors that affect the IoT growth rate are the cost of devices and reduction in size of IoT devices. IoT plays an important role in smart homes, business applications, industry, smart grid, smart city and e-health. IoT has potential to develop intelligent applications in almost every field. The integration of SG and IoT is known as the internet of energy (IoE). The primary goal of IoE is to distribute energy between various sources including industrial consumers, electric vehicles and distributed energy storage.

To control a large amount of data generated by numerous devices such as actuators, sensors and smart meter which need to be store process and managed by cloud computing. The emergence of cloud computing reduces the huge computing power [1]. SG through smart meter gets energy consumption as well as private data of customers which store on cloud data centre. The integration of IoT and cloud computing provide ubiquitous services and allow data to process intelligently. However, an increase in a number of IoE devices increases latency and response time in the cloud which lead to divergence of time demand for sensitive applications and devices. During data transmission to the cloud, SG must ensure privacy [2] which becomes harder when the number of consumers increases [3]. To overcome these problems fog concept has been proposed.

Fog extended to cloud computing provides an effective solution to get control of tremendously increasing smart devices by simply moving data to the cloud [4]. The fog is a middle layer between the cloud and the end user that allocate resources closer to the customers, which reduces response time. The most important features of fog are location awareness and low latency. Fog computing increases the reliability of SG, maximize security and reduce computation cost [5]. Buildings communicate with fog through the smart meter and send the request to fog. The fog finds the nearest Micro Grid (MG) and establishes a connection with MG. If MG has efficient power supply then respond back to fog otherwise fog interact with the cloud to find another nearest MG. Cloud and fog computing provides three services such as Software As A Service (SaaS), Platform As A Service (PaaS) and Infrastructure As A Service (IaaS).

Both Cloud and fog provide on-demand services on the internet. In this paper, Cloud and fog based SG environment are discussed. A system model is designed which covers two regions of the world based on continent of the world with a fog in each region. Load balancing algorithms are used for optimal scheduling of tasks. Therefore, these algorithms are tested for a scenario. The scenario presented is based on Virtual Machines (VMs) on every fog, to determine the number of VMs that are sufficient to balance load with better response time, cost and processing time.

1.1 Motivation

The rapid increase of technologies affects the computing power of SG so a centralized cloud platform is presented [6]. In the past decade, response time and storage in the cloud has been a crucial issue, to handle this fog computing concept is presented in [7]. The fog is an intermediate layer between the end user and the cloud [8] that distribute resources closer to the customers, which reduces response time. A large number of consumers are using cloud and fog which may cause load balancing issues [9]. To handle a large number of request load balancing techniques are used in this paper.

1.2 Contribution

Cloud and fog integration with SG provides benefits to end user. Optimal allocation of resources to consumers is tackle by load balancing techniques, following are the contribution of this paper:

- Smart devices are used to provide location awareness services.
- The residential area of the world is divided into six regions based on the six continents to manage the energy consumption in a balanced way.
- The fog computing concept is used to improve the latency (request time, response time, processing time) and to access internet services flexibly for IoT devices used by end users in the residential buildings across the world.
- Odds algorithm is presented to efficiently manage resources

2 Related Work

Cloud and fog computing provides shared resources which allocate VMs to reduce the load on the cloud. Different load balancing algorithms are applied on the cloud and fog to manage the tasks. The authors in [10], proposed dragonfly optimization algorithm. The efficiency of this algorithm is outperformed than others in the term of execution time, response time, tasks migration and load balance between machines is higher than other algorithms. However, the limitation of the proposed system is that there is no categorization of the cloud. In [11], the authors proposed self-similarity-based load balancing mechanism. The efficiency of SSLB is outperformed than other algorithms in the term of reducing cost and achieve the best allocation of resources. However, the limitation of the proposed system is that execution time increase. In [12], the authors presented cloud load balancing Algorithm. CLB performs better than other load balancing algorithms for reducing the processing time. However, the limitation of the proposed system is that cost does not balance.

In [13], authors proposed an Ant Colony Optimization. Ant Colony Optimization optimizes response time and balances the load. However, the limitation of the proposed system is that high execution time occurs. In [14], authors proposed Particle Swarm Optimization and Pattern Search Algorithm that achieve reliability and power utilization. The proposed algorithm performs better for reducing cost. However, the limitation is that the main function can be improved for better performance. In [15], authors proposed energy management system that

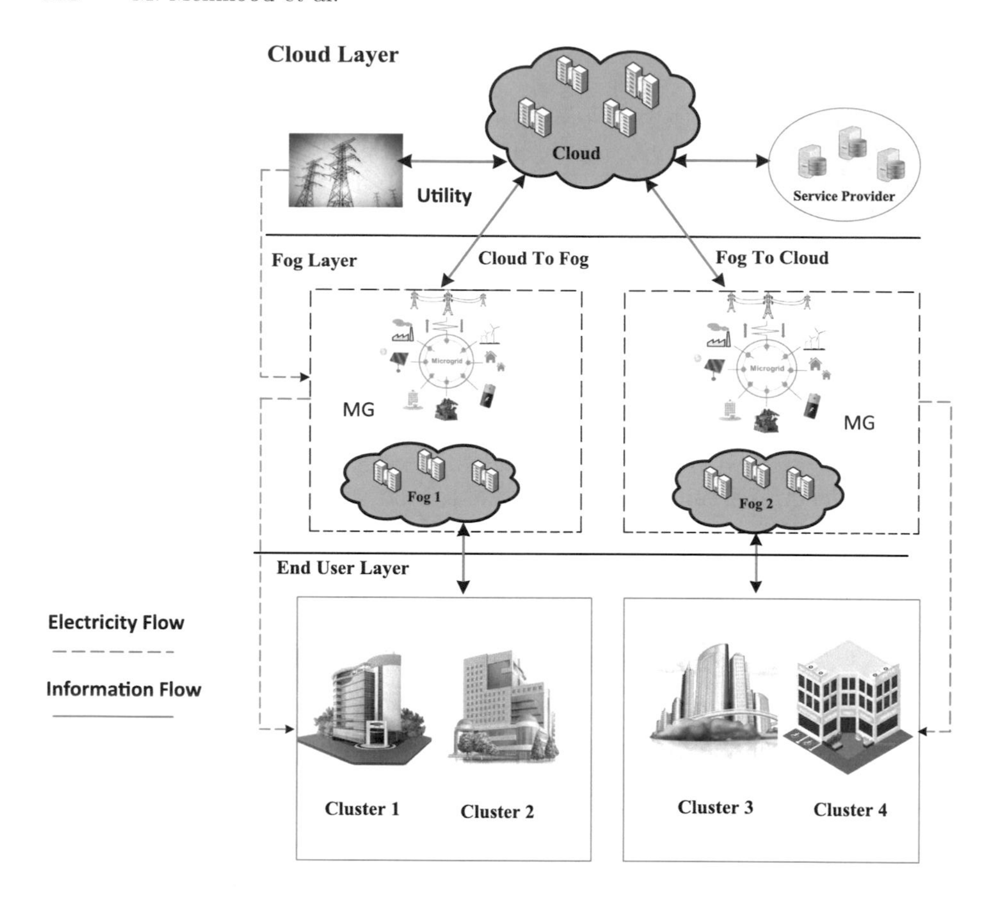

Fig. 1. Proposed system model.

perform well in energy management system. This algorithm is efficient as compared to others. However, limitation of the proposed system is that fuel emission cells can be replaced.

In [16], authors proposed a framework to minimize the wastage of resources through consolidation of resources which is achieved by assigning several requests on the same machine. To implement the semi-online workload consolidation bin packing is presented. The proposed algorithm is based on bin, each task assigned with a bin which is then it map to a machine. The proposed algorithm solve the problem of request consolidation in the real-time assignment of resources. The proposed algorithm get the information for a short time interval which helps to take more precise decisions.

In [17], authors presented MSLBL algorithm that places the budget limit of application to the task budget limit with the budget level which reduces the scheduling length of applications. The proposed algorithm reduces schedule length by preallocating the task with budget-level cost. The issue of reducing the scheduling length of applications on the heterogeneous system has two subproblems.

The first issue is the satisfying budget constraint and the second issue is to reduce the schedule length of applications. The small-time complexity is schedule to reduce schedule length. The experiment results demonstrate that the proposed algorithm produce low scheduling length in different condition and the different budget limit.

In [18], authors presented the tail matching sub-sequence mobility prediction based approach to predict the mobility pattern of users to put resources near to users. For optimal offloading decision improved genetic algorithm is proposed. To enhance the efficiency of genetic algorithm reproduction operations crossover and mutation are used. Integer encoding is developed to consider multiple cloudlet scenarios. In this scenario, several cloudlets for a user may be accessible. Experimental demonstration on prediction is performed by using mobility dataset of humans. Which is used for evaluation of cloudlets reliability and use as a feature in offloading scenario.

Table 1 shows the summary of related work.

Table 1. Summary of related work

Technique(s)	Objective (s)	Feature (s)	Achievement (s)	Limitation (s)
Dragonfly [10]	Minimize electricity cost	Schedule the load demand of home appliances	the cost is minimized and minimize response time	No categorization of appliances
SSLB [11]	Reduce PAR and electricity cost b	Allocation of resources	Cost and PAR is reduced according to user satisfaction	Increased Execution time
Cloud load balancing [12]	Minimization of electricity cost	Utilization of resources	Electricity cost is reduced	Cost is not balanced
Ant colony optimization [13]	Enhance fuzzy controller simulation	Use fuzzy logic to control solutions	Cost is reduced	High execution time
Particle swarm optimization-pattern search [14]	Reliability, power utilization and environmental adaptable	Load Balancing	Efficiency of energy management system	Main function could be modified
Energy management system [15]	Cost saving and load balancing	Confirm performance of Energy	Peak to average ratio minimization	Fuel emission cells can be replaced

Remaining paper is categories as; in Sect. 2 literature works has been discussed, Sect. 3 illustrates the proposed system model, Sect. 4 contains load balancing algorithm, Sect. 4 simulation results and discussion and Sect. 5 is based on conclusion.

3 System Model

In this paper, a cloud and fog based environment. Fog is an extension of the cloud. The cloud and fog combination provide enhanced services to end-users, helps to minimize latency and minimize load on cloud data centres. Fog computing is an environment, that provides data storage, computation and communication between devices and cloud.

Figure 1 shows a fog and the cloud-based platform that comprises three-layered architecture, the first layer contains cloud, the second layer contains fog and the third layer comprises of end users. The communication occurs between all layers. As end users communicate with fog intern the fog interact with the cloud. The cloud and fogs are connected with MG. The fogs transfer information with the cluster of buildings directly or through the smart meter.

At the end user layer two clusters of buildings are assumed and these clusters comprise of multiple homes. Buildings communicate with fog through the smart meter and send the request to fog. The fog finds the nearest MG and establishes a connection with MG. If MG has efficient power supply then respond back to fog otherwise fog interact with the cloud to find another nearest MG. Fogs contain multiple VMs to handle user requests to achieve minimum response time and enhance the computing power which is important for load balancing in MG.

The fog layer manages latency problems and manages resources in an effective way. Fog devices contain resources such as main memory, data storage and computation. To run applications on a single hardware platform, virtual machines contain processors for executing applications. In virtual machines, numerous applications on processors are running to execute services. Fog is a middle layer between cloud and end user.

Cloud layer contains data centres which provide on-demand storage, computation processing to end users and provide services with pay-as-you-go paradigm (pay only for necessary services and resources) in cloud architecture. The consumer only pays for the necessary services.

The world is divided into six regions, comprises of six continents as shown in Table 2. Fogs are placed in different regions of the world. In this paper region 3 and region 4 are assumed in each region two clusters are considered and one fog is placed per region. These two regions have the energy crisis and load balancing issues as a large number of end users send the request to MG. The two fogs are located in these two regions and respond to the cluster of buildings in the region where fog are located. Different load balancing algorithms are used to manage the user requests.

Table 2. Region distribution

Region	Region Id
North America	0
South America	1
Europe	2
Asia	3
Africa	4
Oceania	5

3.1 Round Robin (RR)

Round Robin use time sharing mechanism. Round Robin algorithm develope a table of users requests with respect to time arrival. Round Robin algorithm allocate virtual machines one by one to balance the load of multiple requests on virtual machines. Round robin algorithm assigns the request to the VMs on the base of time slice.

3.2 Throttled

Throttle algorithm contains the index table of all VMs which retains information of all VMs and assigns the requests to VM that can fulfill demand request. For allocating virtual machines requests sent to load balancer and identify which machine is able to perform tasks efficiently. If all VMs are busy then it waits for VMs in the queue.

3.3 Odds Algorithm

The odds algorithm works on optimal stopping strategy to get the best probability of VMs that can available to assign tasks. The odds algorithm increase the probability of identifying a VM. The odds-algorithm is the rule to observe the VMs one after the other and to stop on the first available VMs. The odds-algorithm count the VMs in reverse order until it reaches the last virtual machine.

3.4 Problem Formulation

In order to increase the stability of platform load of virtual resources on fog are balanced. Three main performance parameters are considered to enhance the stability of the system are processing time, cost and response time. These parameters are formulated by the linear programming [19] In this paper, cloud-fog based architecture with SG integration is considered. The fog contains n number of virtual machines $vm_1, vm_2, ...vm_n$. Using load balancing techniques virtual machine resources are efficiently allocated to optimize the workload on fog. In this model there are n numbers of buildings and h number of homes are considered. All virtual machines have the same capacity and are assigned to tasks base on users requests. The total numbers of virtual machines (m) and requests (R) are mathematically demonstrated by Eqs. 1 and 2

$$T_{VMs} = \sum_{i=1}^{m} (VM_i) \tag{1}$$

Let Total requests are represented in Eq. 2

$$T_R = \sum_{j=1}^{k} (R_i) \tag{2}$$

Performance parameters are affected by mapping requests to virtual machines. [20] User tasks are varying from 0 to n. tasks assign to different virtual machines. Assignment of VMs to URs is dependent of response time (RT) and processing time (PT) of VMs

3.4.1 Processing Time:

Processing time of VMs to process the user tasks and state of user tasks is defined in Eq. 3.

$$S = \begin{cases} 1, & \text{If VM is assigned to tasks} \\ 0, & \text{Otherwise} \end{cases} \tag{3}$$

Processing time can be represented by Eq. 4.

$$P_t = \frac{L_t}{C_{vm}} \tag{4}$$

where P_t represent processing time. L_t represent length of tasks C_vm represent capacity of virtual to perform tasks.

3.4.2 Response Time:

Response time (RT) is time taken by the fog to response the request is presented in Eq. 5.

$$RT = Delay_{Time} + Finish_{Time} - Arrival_{Time} \tag{5}$$

The total time taken by the VM as in Eq. 6.

$$Total_{time} = Finish_{Time} - Start_{Time} \tag{6}$$

Our objective function is to minimize response time (RT) and processing time P_t.

Algorithm 1. Odds Algorithm

```
 1: Input VMs list, I = VMs, Compute delay
 2: Output: VMs_id
 3: continue = true, stopping threshold = 0
 4:
 5: while count do
 6:     sum the VMs_id
 7:     if the sum of n VMs_id < 1 then
 8:         continue = false
 9:         else
10:         stopping threshold = stopping threshold + 1
11:         if stopping threshold= n then continue = false then
12:             VMs_id = stopping threshold
13:             return VMs_id
14:         end if
15:     end if
16: end while
```

4 Simulation and Discussion

In this paper, Cloud Analyst tool is used to perform the simulation. Cloud Analyst tool use to inspect the performance of load balancing algorithm in the same scenario. The comparison is done between RR, throttled and Odds algorithm, on the basis of fog's response time, processing time and total cost.

4.1 Discussions

Figure 2 shows comparison between response times of RR, Throttled and Odds Algorithm.

Figure 3 shows the Response time of each algorithm. Total Response time of throttled in the scenario is 56.94 ms, Round Robin Response time is 57.35 ms whereas the overall Response time of odds algorithm is 57.42 ms.

Figure 3 shows Response time of odds algorithm and throttled is much better as compared to Round Robin because, Round Robin allocate the tasks without looking the task allocation on VMs, which maximize the load on VMs and maximize the response time.

Figure 4 shows the average time a fog takes to fulfil a request when RR, Throttled and odd algorithm are used. RR has high processing time because of the same reason that is mentioned above.

Figure 5 shows the total cost of fogs processing RR, Priority-based and throttled algorithm, each fog manage multiple requests from clusters. The total cost includes VMs cost, the cost of the power station and data transfer cost.

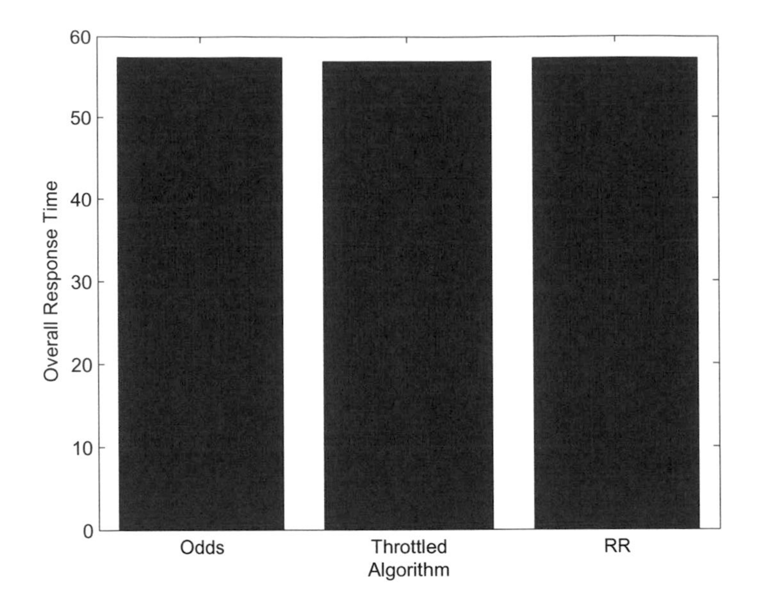

Fig. 2. Algorithms.

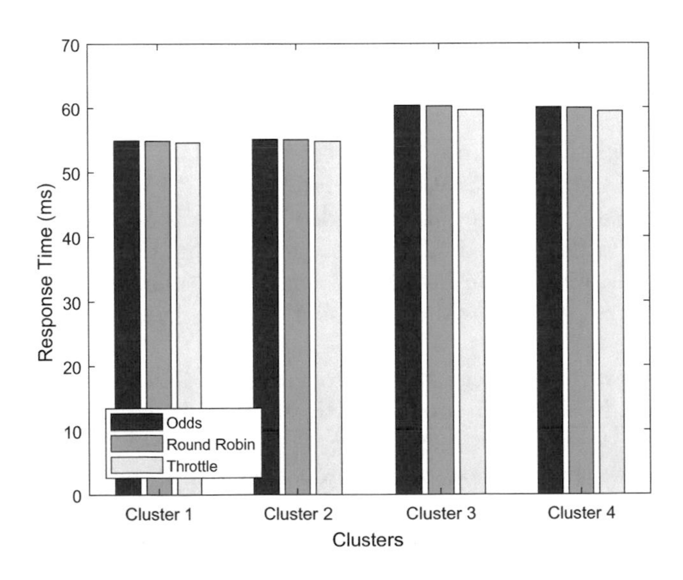

Fig. 3. Average response time.

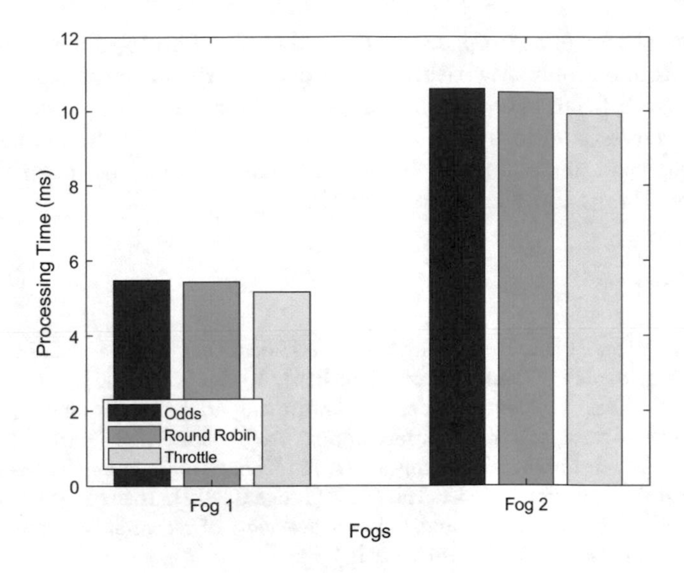

Fig. 4. Average processing time.

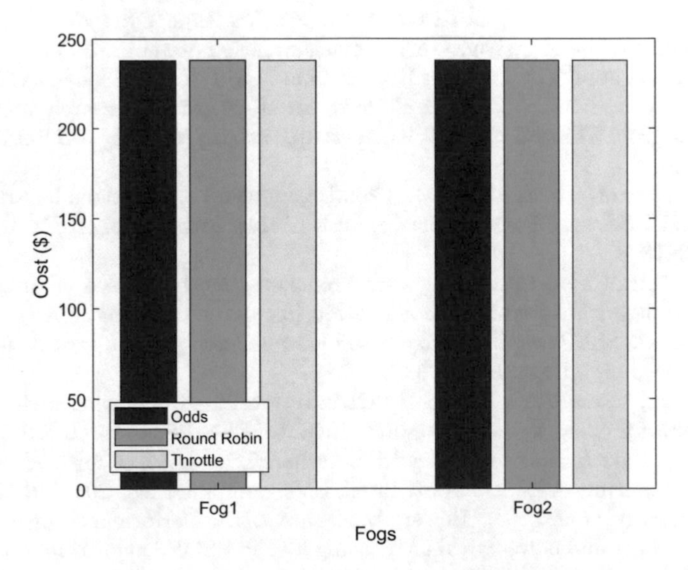

Fig. 5. Total cost.

5 Conclusion

In the given scenario, as proposed in the paper the Throttled algorithm outperforms the Round Robin Algorithm. The odds algorithm under some conditions dominates both Round Robin and Throttled algorithm. But as we increase the number of requests on fogs from different users residing in the different regions of the world the odds algorithm behave inefficiently and gets outperformed by both Round Robin and Throttled algorithm.

References

1. Cao, Z.: Optimal cloud computing resource allocation for demand side management in smart grid. IEEE Trans. Smart Grid **8**(4), 1943–1955 (2017)
2. Kim, J.Y., Kim, Y.: Benefits of cloud computing adoption for smart grid security from security perspective. J. Supercomput. **72**(9), 3522–3534 (2016)
3. Faruque, A., Abdullah, M., Vatanparvar, K.: Energy management-as-a-service over fog computing platform. IEEE Internet Things J. **3**(2), 161–169 (2016)
4. Chiang, M., Zhang, T.: Fog and IoT: an overview of research opportunities. IEEE Internet Things J. **3**(6), 854–864 (2016)
5. Hussain, Md., Alam, M.S., Beg, M.M.: Fog Computing in IoT Aided Smart Grid Transition-Requirements, Prospects, Status Quos and Challenges. arXiv preprint arXiv:1802.01818 (2018)
6. Ramadhan, G., Purboyo, T.W., Latuconsina, R.: Experimental model for load balancing in cloud computing using throttled algorithm. Int. J. Appl. Eng. Res. **13**(2), 1139–1143 (2018)
7. Luo, F., Zhao, J., Dong, Z.Y., Chen, Y., Xu, Y., Zhang, X., Wong, K.P.: Cloud-based information infrastructure for next-generation power grid: Conception, architecture, and applications. IEEE Trans. Smart Grid **7**(4), 1896–1912 (2016)
8. Chiang, M., Zhang, T.: Fog and IoT: an overview of research opportunities. IEEE Internet Things J. **3**(6), 854–864 (2016). https://doi.org/10.1109/JIOT.2016.2584538
9. Bera, S., Misra, S., Rodrigues, J.: Cloud computing applications for smart grid: a survey. IEEE Trans. Parallel Distrib. Syst. (2014). https://doi.org/10.1109/TPDS.2014.2321378
10. Branch, S.R., Rey, S.: Providing a load balancing method based on dragonfly optimization algorithm for resource allocation in cloud computing (2018)
11. Li, C., et al.: SSLB: self-similarity-based load balancing for large-scale fog computing. Arab. J. Sci. Eng., 1–12 (2018)
12. Chen, S.L., Chen, Y.Y., Kuo, S.H.: CLB: a novel load balancing architecture and algorithm for cloud services. Comput. Electric. Eng. **58**, 154–160 (2017)
13. Dam, S., et al.: An ant-colony-based meta-heuristic approach for load balancing in cloud computing. Appl. Comput. Intell. Soft Comput. Eng., 204–232 (2018)
14. Gabbar, H.A., Labbi, Y., Bower, L., Pandya, D.: Performance optimization of integrated gas and power within MG using hybrid PSOPS algorithm. Int. J. Energy Res. **40**(7), 971–982 (2016)
15. Varela Souto, A.: Optimization and Energy Management of a Microgrid Based on Frequency Communications (2016)

16. Armant, V., De Cauwer, M., Brown, K.N., O'Sullivan, B.: Semi-online task assignment policies for workload consolidation in cloud computing systems. Future Gener. Comput. Syst. (2018)
17. Chen, W., Xie, G., Li, R., Bai, Y., Fan, C., Li, K.: Efficient task scheduling for budget constrained parallel applications on heterogeneous cloud computing systems. Future Gener. Comput. Syst. **74**, 1–11 (2017)
18. Shi, Y., Chen, S., Xiang, X.: MAGA: a mobility-aware computation offloading decision for distributed mobile cloud computing. IEEE Internet Things J. **5**(1), 164–174 (2018)
19. Devi, D.C., Uthariaraj, V.R.: Load balancing in cloud computing environment using improved weighted round robin algorithm for nonpreemptive dependent tasks. Sci. World J. **2016**, 1–14 (2016)
20. Wickremasinghe, B., Buyya, R.: CloudAnalyst: a cloudsim-based tool for modelling and analysis of large scale cloud computing environments. MEDC Project Rep. **22**(6), 433–659 (2009)

Resource Allocation over Cloud-Fog Framework Using BA

Farkhnada Zafar, Nadeem Javaid(✉), Kanza Hassan, Shakeeb Murtaza, Saniah Rehman, and Sadia Rasheed

COMSATS University, Islamabad 44000, Pakistan
nadeemjavaidqau@gmail.com

Abstract. Edge computing or fog computing (FG) are introduced to minimize the load on cloud and for providing low latency. However, FG is specified to a comparatively small area and stores data temporarily. A cloud-fog based model is proposed for efficient allocation of resources from different buildings on fog. FG provides low latency hence, makes the system more efficient and reliable for consumer's to access available resources. This paper proposes an cloud and fog based environment for management of energy. Six fogs are considered for six different regions around the globe. Moreover, one fog is interconnected with two clusters and each cluster contains fifteen numbers of buildings. All the fogs are connected to a centralized cloud for the permanent storage of data. To manage the energy requirements of consumers, Microgrids (MGs) are available near the buildings and are accessible by the fogs. So, the load on fog should be balanced and hence, a bio-inspired Bat Algorithm (BA) is proposed which is used to manage the load using Virtual Machines (VMs). Service broker policy considered in this paper is closest data center. While considering the proposed technique, results are compared with Active VM Load Balancer (AVLB) and Particle Swarm Optimization (PSO). Results are simulated in the Cloud Analyst simulator and hence, the proposed technique gives better results than other two load balancing algorithms.

Keywords: Cloud computing · Fog computing
Requests processing time · Smart grid · Microgrid · Virtual machine
Resource allocation · Energy management

1 Introduction

With the popularity of internet of thing (IoT) over the last few years, the concept of smart buildings in smart cities has become popular [1]. SG consists of smart meters which are used to monitor the electricity consumption by different devices in a smart home [1]. Many authors have proposed a cloud-fog based framework with integrating SGs.

L. Barolli et al. (Eds.): NBiS 2018, LNDECT 22, pp. 222–233, 2019.
https://doi.org/10.1007/978-3-319-98530-5_19

Cloud computing is a geographically distributed network which works on shared computing resources instead of confined servers. Cloud computing lies on a very underlying principle of multipurpose usability of IT. Cloud computing, as compared to conventional 'grid computing' provides more widen perspectives across organizational boundaries [2]. Cloud provides delivery of hosted services on the internet. Cloud providers offer three services, Infrastructure as a Service (Iaas) Software as a Service (SaaS) and Platform as a Service (Paas). SaaS provides a complete application to the customer. Whereas, PaaS provides a platform to allow developers from different domains to build their applications and services. However, IaaS offers basic storage and computing facilities as systemize services.

FG concept is introduced at the edge of the network. FG is introduced to lower the latency delay of rapidly increasing applications and connected device. FG helps to minimize the load burden of cloud and directly communicate with users through the internet. For maintaining load and efficient working fog cannot store data permanently. For this purpose fog send data to the cloud for the permanent storage [3]. In the integration of cloud-fog framework with SG, it is important to consider demand from consumer's end. The demand may be of accessing any utility company or MG. The work in [4,5] has integrated cloud and FG with SG. This is possible only when a communication is established among consumers' end, fog and cloud.

1.1 Motivation

Information Technology (IT) plays a notable role in computing. Due to this, demand of computing and storage is increasing rapidly [6]. Consumers demand services and resources that can be provided at any time. The cloud computing incorporation with FG is a most likely solution towards this. FG broadens cloud computing by providing more efficient resources and locality-based services at the edge of the network to higher serve mobile traffics [7]. On the other hand, load balancing is one of the technique which is used for resource allocation and helps in utilization of resources and services. As authors in [8] proposed a fog-cloud based model for effective management of information by using load balancing techniques. They have used Artificial Bee Colony, Round Robin, PSO, Throttled and Ant Colony Optimization and proposed a hybrid approach of ACO and ABC (HABACO).

1.2 Contribution

Cloud-fog based framework is beneficial for SG to get numerous benefits like efficiency, reliability and low latency rate. A fog based model is designed, where we cover the six regions of the world with large number of consumers to send requests on fog to access the required resources The paper is contributed as follows:

1. Developed a cloud-fog based framework for the efficient management of the consumer's demands.

2. Fog provides low latency services as it is placed near the end user layer and can respond faster to user's requests.
3. MGs are placed near to each fog which fulfills the electricity requirements by maintaining the sustainability of buildings.
4. Response time and request servicing time is optimized using the BA. Results are compared with AVLB and PSO.

The remaining paper is assembled as follows; Related work is shown in Sect. 2. The proposed system model is presented in Sect. 3. However, simulation and discussions are described in Sect. 4 and Sect. 5 presents conclusion of the paper.

2 Related Work

For the optimization of the immense demand of the electricity, multiple methodologies have been proposed in the literature. Some methodologies are either cloud-based or fog based or both. Cloud and FG provides a virtual environment for efficient resource sharing to the connected consumers. These resources are assigned to increase the use of VMs instead of using physical machines for energy consumption reduction. The load on cloud and fog is balanced by applying multiple different algorithms on the cloud. These algorithms schedule the requests to manage the load. The load balancing algorithms which are used mostly are Round Robin (RR), Throttled, Active Load Balancing (ALB), PSO and Honey Bee Foraging etc.

The authors in [9], proposed a Cloud Load Balancing (CLB) algorithm for load balancing on cloud and compare the results with other algorithms. The proposed algorithm performs better than others and successfully balances the load efficiently when different users demand resources simultaneously. As authors in [10] proposed Ant Colony System (MORAACS) for the resource allocation. The authors compared processing time, energy consumption and standard deviation with RR algorithm. In this algorithm, minimization of energy consumption by balancing the load on the cloud is resolved. The authors used Cloudsim simulator for simulations. A new service broker policy is proposed for data center selection in [11]. The authors proposed Master Fuzzy Context (MFC) for provisional fuzzy logic. Fuzzy rules are designed for RT and data center priority.

In [12], a novel communication model is presented using two other models for optimizing the communication efficiently using optimized resource allocation. The proposed system attains high cost through the high demand scenarios. In addition, authors presented a cost computation model for cost reduction by optimized utilization of the cloud computing resources. Moreover, this model gives the opportunity to the consumers regarding optimized resource availability which helps in cost reduction. It decreases the total operation cost of the system. A novel cloud-based nano grid framework is proposed for energy efficiency in SG. Renewable Energy System (RES) is installed with building using nano grid. Cloud controllers are used to make decisions about energy automation. The proposed solution reduces the execution time of various jobs and performance of selected parameters [13]. In [14], authors proposed a scheme for efficient energy management of SG-based Cyber-Physical System (CPS). Physical plane consists of smart devices whereas controller is located on the cyber plane. Nash equilibrium approach is used to analyze the performance of proposed coalition. Proposed solution provides a cost-effective solution during peak hours for energy management to end user (Table 1).

Table 1. Related work survey

Reference(s)	Technique(s)	Feature(s)	Tool(s)
[9]	CLB	Load balancing on cloud	Microsoft Visual Studio 2010 C#, Microsoft SQL Server
[10]	MORA-ACS	Resource allocation for virtual services, Load balancing on virtual machines	Cloudsim simulation tool
[11]	Fuzzy Based Service Broker Policy	Increases the efficiency and minimizes the cost	Cloud Analyst and MATLAB
[15]	Stochastic dynamic programming	Demand side management (DSM) in MG-level energy management and Home Energy Management (HEM)	MATLAB
[16]	Job allocation mechanism (JAM)	Minimization of battery consumption at cyber-physical-social big data processing	JAVA SE8 Swing
[17]	Diffusion strategy	Scale-free, minimizing cost in SG	Not specified
[12]	Frequency control via randomized demand response	Minimize the frequency recovery time, Stabilize the system frequency during contingencies	IEEE 9-Bus test system, Ireland power system

The concept of nano grids is implemented in supportable smart buildings for multi-tenant cloud environment in [16]. In [17], authors proposed a cloud-based demand-side management (DSM) system which manages energy for the consumers in multiple different regions and micro-grids to reduce the utility's and consumers' cost. It also reduces the time and efforts by integrating the modularity feature in developing smart cities. They implemented Bi-level Optimization Algorithm using linear cost function. In [18], an efficient SG electric vehicle system is proposed for providing charging and discharging optimally in different stations. A stochastic model is used to schedule the approaches for shifting load in the cloud computing based framework for DSM in SG. The proposed method is used to create small energies hub for users. Cost is reduced by transferring load from hours where energy is highly consumed to less energy consumed hours by Monte Carlo method [12]. In [15], a model is presented using fog based framework for consumers to manage energy as a service. FG provides flexibility, data privacy, interoperability and real-time energy management. Using MG-level energy management and Home Energy Management (HEM) implementation cost and time to market are reduced significantly.

The work [19] focuses to examine economically power trading to main grid. Energy cost minimization, carbon emission minimization and to cope with hazardous emission renewable sources are goals of this work.

The above-mentioned techniques are discussed for the energy management of the specific set of buildings or appliances in cloud and fog based models. None of these techniques can tackle the energy management for optimization of resource allocation in the specific region of the world. This paper presents the bio-inspired algorithm for the efficient resource allocation of VMs in the residential buildings around the world using fog and cloud environment.

3 System Model

In this section, we present a cloud fog framework for efficient resource allocation among different clusters of buildings in different regions of the world. There are six regions which are scattered geographically as shown in Table 2. The system model is presented in Fig. 1 which depicts the interaction among buildings, MGs, fog and cloud.

Each region contains one fog which is connected to two clusters. Each cluster contains fifteen buildings and each of which contains the different multiple homes. Smart buildings communicate with the fog devices via a smart meter. The fog devices consist of network bandwidth, storage, main memory, processor and VMs. VM manager manages all the virtual machines. All the fogs are connected to a centralized cloud. All information about building's energy consumption, generation and schedule is stored in fogs. Fogs store data temporarily and if the data is to be stored permanently, it sends data to the cloud. Each fog is connected with MG to communicate for electricity demand and supply. The clusters of buildings communicate with fog via smart meter for its electricity requirements and in return fog communicates with nearby MG for the supply. The cluster cannot directly communicate with MG whereas electricity supply from MG to buildings is carried out directly. Table 3 shows clusters, fogs and regions distribution.

3.1 Problem Formulation

In this section, we provide a formal description of our problem for minimum response time and minimum processing time by formulating it.

We have considered the following components for resource allocation which are categorized as: set of regions which contains six regions from all over the world, six cluster of buildings with varying number of buildings in it, denoted by C, set of VMs denoted by VM, MGs, six fogs and a cloud data center. Requests from clusters are denoted by R. It is mathematically represented as:

$$C = \{c_1, c_2, ..., c_{12}\} \tag{1}$$

$$VM = \{vm_1, vm_2, ..., vm_n\} \tag{2}$$

$$R = \{r_1, r_2, ..., r_{12}\} \tag{3}$$

where, the set of total number of the clusters, VM and requests are varying from 1 to m, 1 to n and 1 to p respectively. In this environment, all VMs are working in parallel to each other and have the same capacity. Total requests denoted by $\mathrm{T_R}$, coming from clusters can be expressed as,

$$T_R = \sum_{i=1}^{N}(\sum_{j=1}^{M}(C_\mathrm{i} \times R_\mathrm{j})) \tag{4}$$

Equation 4 shows total requests coming from user's end. After calculating total requests comming from user's end, now we compute response time which is denoted by RT.

$$RT_{total} = F_\mathrm{T} - \mathrm{AT} + Delay \tag{5}$$

where $\mathrm{F_T}$ is task finishing time and $\mathrm{A_T}$ is arrival time (ms) of request from a cluster to the fog which includes network delay. The processing time P of allocating task i to VM j is $\mathrm{P}_{\mathrm{i,j}}$ and status of task is δ

$$\delta_{\mathrm{i,j}} = \begin{cases} \text{If task is asssigned,} & 1 \\ \text{else} & 0 \end{cases}$$

The objective function is

$$\mathrm{S_{minimize}} = \sum_{i=1}^{T_R}\sum_{j=1}^{N}(\delta_{\mathrm{i,j}} \times P_{\mathrm{i,j}}) \tag{6}$$

where

$$P_{\mathrm{i,j}} = \frac{\text{Length of task at ith place}}{\text{Capacity of VM at jth place}}$$

Table 2. Table to show regions distribution

Region ID	Region
0	North America
1	South America
2	Europe
3	Asia
4	Africa
5	Ocenia

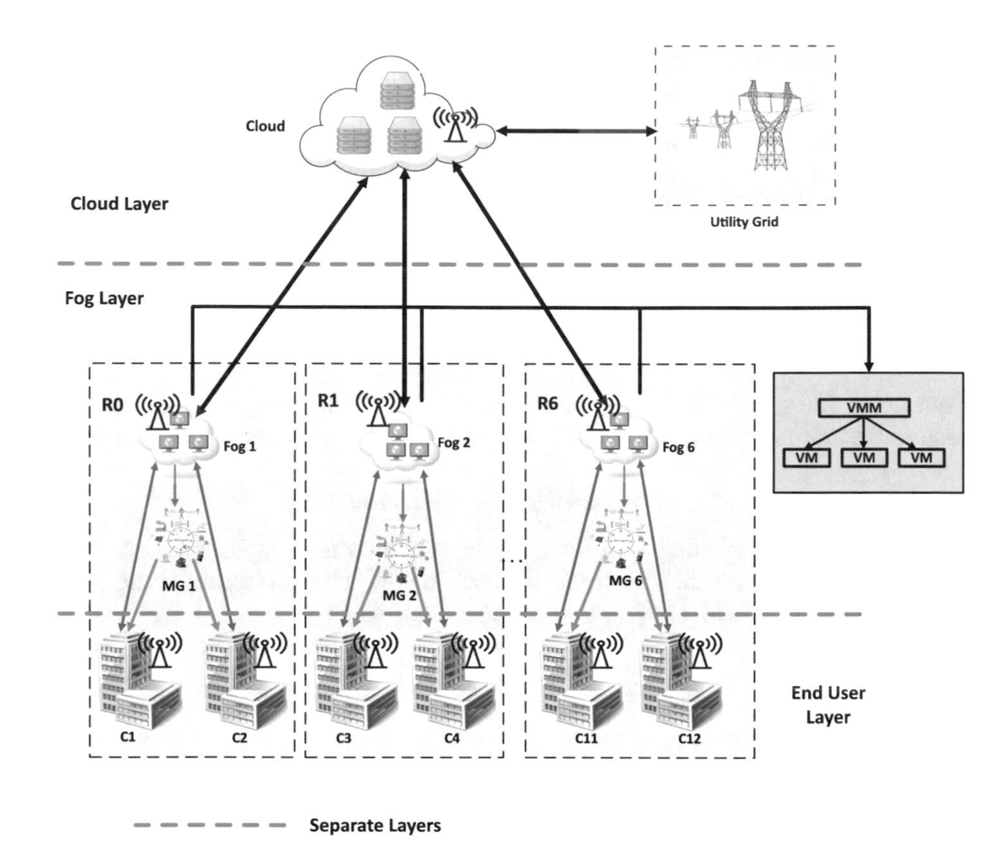

Fig. 1. Proposed system model

Table 3. Table to show clusters and fogs distribution

Clusters	Fogs	Regions
C 1	1	0
C 2	1	0
C 3	2	1
C 4	2	1
C 5	3	2
C 6	3	2
C 7	4	3
C 8	4	3
C 9	5	4
C 10	5	4
C 11	6	5
C 12	6	5

Algorithm 1. BA

```
1: Input: List of tasks, List of VMs
2: Initialize: bats, generations, velocity, echo=tasks
3: Calculate the load, capacity of VM
4: for i=1 to generations do
5:     echo[i].best=current position
6:     echo[i].bestfit=current fit
7: end for
8: Calculate Vbest and Pbest for each machine
9: Pbest= echo.best with lowest fit)
10: for j=1 to generations do
11:     for for t=1 to echo do
12:         frequency();
13:         Update velocity ();
14:         Update position();
15:         if current fitness < echo[t].bestfit then
16:             echo[t].best=current position
17:             echo[t].best=current fit
18:         end if
19:     end for
20:     Pbest= echo.best with lowest fit
21: end for
22: Return Pbest
```

3.2 BA

In this paper, for effective load balancing, bio-inspired BA is used. For load balancing of virtual machines BA is used as a novel intelligence technique in cloud computing. It is based on bat behaviour. It is used on the echo location behaviour of bats. Each bat flies randomly across an area with a velocity at any position x with different pulse rates. Search is strengthen by random walk. Selection for the best solution continues until defined stopping criteria is met.

3.3 Closest Data Center

Depending on the network delays, the shortest path is selected from consumer's end to the nearest cloud data center or fog data center for the quick response. This provides low latency rates and hence increases performance efficiency and user's satisfaction

4 Simulation and Discussion

FG makes communication easy and efficient as compared to cloud computing. It provides an easy way of communication to its consumers without interruption and with minimum delay. For this purpose, a fog-based environment is designed for six regions with two cluster of buildings considered for each region.

The regions are identified based on six continents in the world as shown in Table 2. In this paper, extensive simulations are employed in order to demonstrate the effectiveness of proposed system model, which relies on distributed fog framework and centralized cloud. Resource allocation policies used for the simulations is closest data center, optimized response time and proposed service broker policy. The load balancing algorithms are used are AVLB, PSO, and BA. Results of these policies are compared. These algorithms are used for optimal resource allocation and distribution to the consumers based on the requests.

4.1 Simulation Setup

Simulations are conducted for 24 h of the day and Cloud Analyst simulator is used for simulations to determine the dependency of performance parameters on the number of buildings, location-aware DCs and load balancing policies. In our setup, the world is categorized into 6 regions. Two load balancing policies are used for comparison such as AVLB and PSO with BA. For efficient allocation of VMs, Closest Data Center Proximity Policy is used.

4.2 Response Time

Figures 2, 3 and 4 present the response time of buildings with PSO, AVLB and BA, respectively. In Fig. 4, as we can see that the average, minimum and maximum response time with BA is less as compared to AVLB and PSO. Simulations are conducted for twelve clusters connected with six fogs. Figure 5 shows the overall collation of the response time of all of the three algorithms used in this paper i.e. BA, PSO, and AVLB.

4.3 Processing Time

Figures 6, 7 and 8 shows the hourly processing time of fogs with Active VM Load Balancer, PSO, and BA. In Fig. 7, as we can see that the average, minimum and

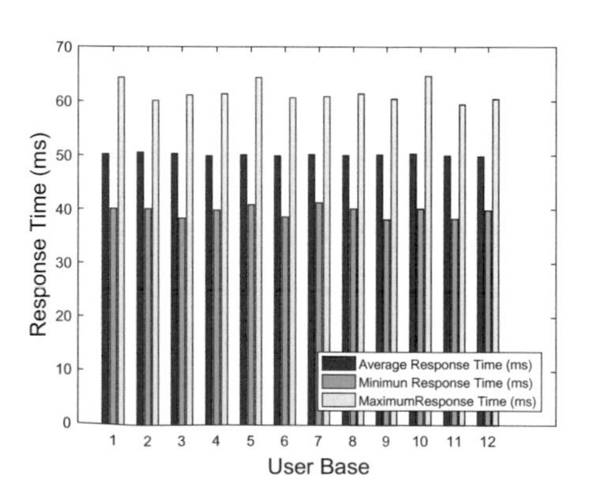

Fig. 2. Response time of active Vm load balancer

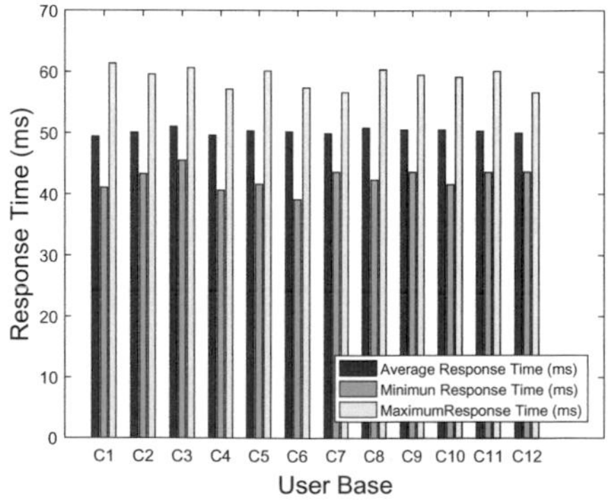

Fig. 3. Response time of PSO

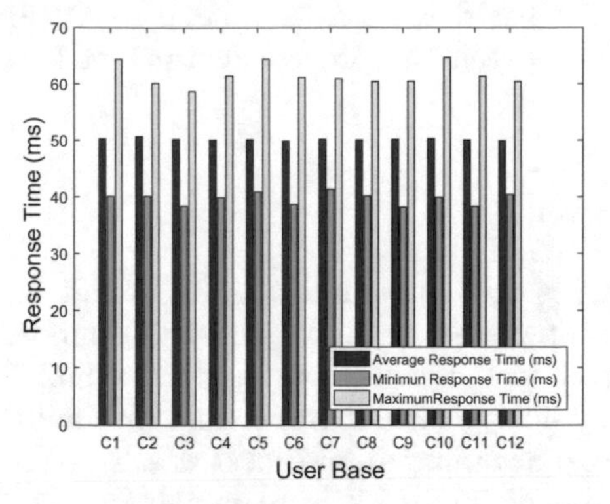

Fig. 4. Response time of BA

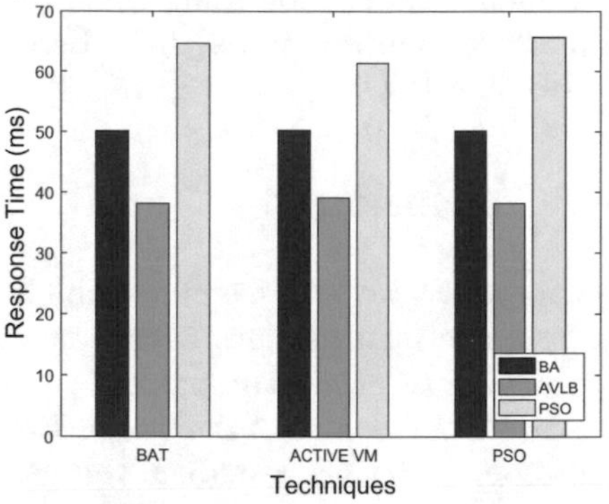

Fig. 5. Response time

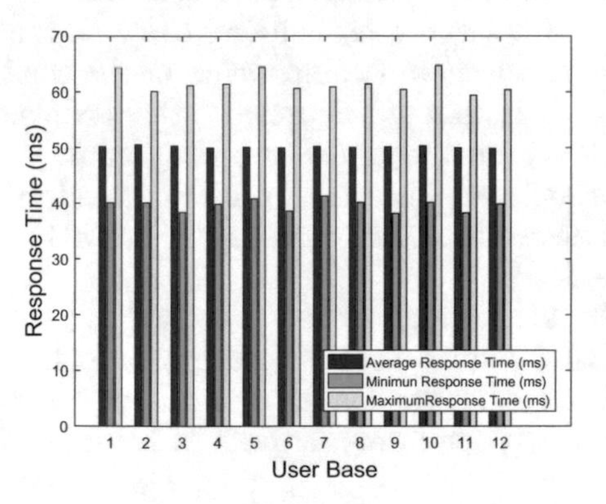

Fig. 6. Processing time of AVLB

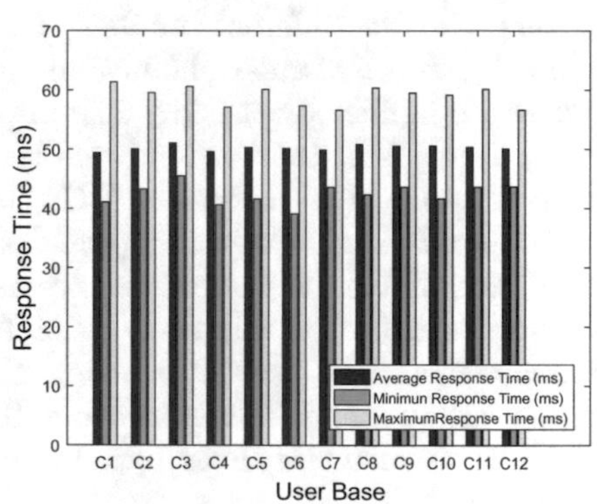

Fig. 7. Processing time of PSO

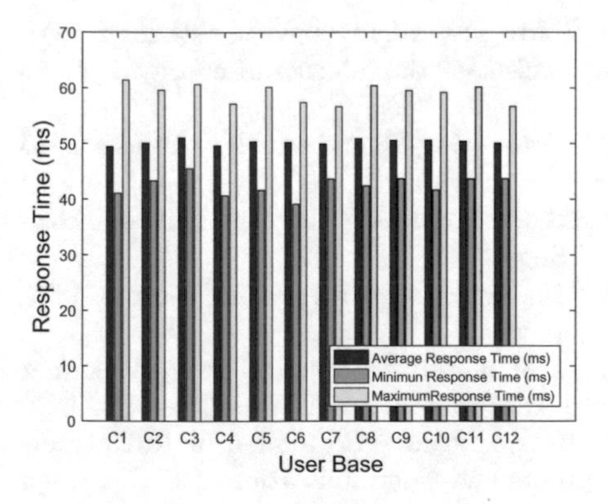

Fig. 8. Processing time of BA

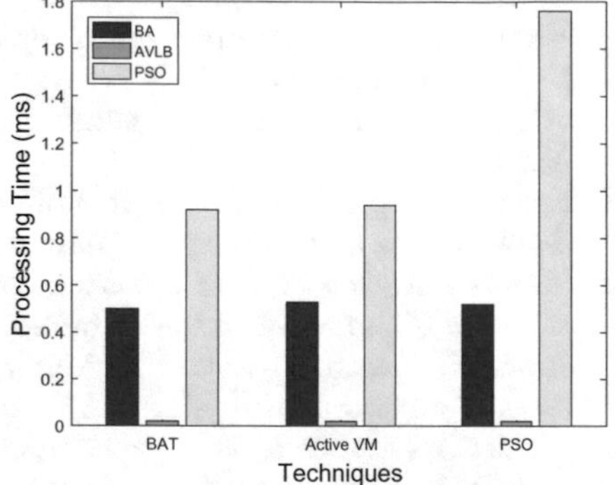

Fig. 9. Processing time

maximum request processing time with BA has reduced as compared to AVLB and PSO. Figure 9 shows the difference between the processing time of BA, AVLB, and PSO.

5 Conclusion

In this paper, we have examined the benefits and opportunities of cloud and fog for resource management. Moreover, an integrated fog and cloud-based model is proposed to efficiently allocate resources to the clusters of residential buildings in all the regions of the world. Energy management is important for homes, buildings, and MGs to reduce the wastage of resources and meet user's power requirements or resources so, the purpose of this work is to manage energy requirements of the smart homes. Hence, in this paper, a novel energy management bio-inspired technique, BA, is proposed for the efficient allocation of computational resources on the fog. In this model, each region contains two clusters with the different number of buildings and each building contains the different number of smart homes and smart devices. Requirements of the consumers are fulfilled by the fog. The implementation of FG provides the flexible, data privacy, fast response and real-time features required for energy management. Consumers get access to MGs through the fog. All fogs are managed by a centralized cloud which also provides access to utility company when MG is unable to fulfill the consumer requirement.

Cloud Analyst simulator is used for simulations of the proposed system model. Simulations are taken using closest data center service broker policy for fog selection and VMs allocation. The results are compared with PSO and AVLB. From results, it is analyzed that BA outperforms Active VM Load Balancer and PSO.

References

1. Blanco-Novoa, O., Fernandez-Carames, T.M., Fraga-Lamas, P., Castedo, L.: An electricity price-aware open-source smart socket for the internet of energy. Sensors **17**(3), 643 (2017)
2. http://www.thbs.com/downloads/Cloud-Computing-Overview.pdf. Accessed 15 Apr 2018
3. Fatima, I., Javaid, N.: Integration of Cloud and Fog Based Environment for Effective Resource Distribution in Smart Buildings
4. Yasmeen, A., Javaid, N.: Exploiting Load Balancing Algorithms for Resource Allocation in Cloud and Fog Based Infrastructures
5. Javaid, S., Javaid, N., Aslam, S., Munir, K., Alam, M.: A Cloud to Fog to Consumer Based Framework for Intelligent Resource Allocation in Smart Buildings, pp. 1–10
6. Okay, F.Y., Ozdemir, S.: A fog computing based smart grid model. In: 2016 International Symposium on Networks, Computers and Communications (ISNCC), pp. 1–6. IEEE, May 2016
7. Suryawanshi, R., Mandlik, G.: Focusing on mobile users at the edge of internet of things using fog computing. Int. J. Sci. Eng. Technol. Res. **4**(17), 3225–3231 (2015)

8. Zahoor, S., Javaid, N.: A Cloud-Fog Based Smart Grid Model for Effective Information Management (2018)
9. Chen, S.L., Chen, Y.Y., Kuo, S.H.: CLB: a novel load balancing architecture and algorithm for cloud services. Comput. Electr. Eng. **58**, 154–160 (2017)
10. Pham, N.M.N., Le, V.S.: Applying Ant Colony System algorithm in multi-objective resource allocation for virtual services. J. Inf. Telecommun. **1**(4), 319–333 (2017)
11. Islam, N., Waheed, S.: Fuzzy based efficient service broker policy for cloud. Int. J. Comput. Appl. **168**(4), 37–40 (2017)
12. Moghaddam, M.H.Y., Leon-Garcia, A., Moghaddassian, M.: On the performance of distributed and cloud-based demand response in smart grid. IEEE Trans. Smart Grid (2017, in press)
13. Fang, B., Yin, X., Tan, Y., Li, C., Gao, Y., Cao, Y., Li, J.: The contributions of cloud technologies to smart grid. Renew. Sustain. Energy Rev. **59**, 1326–1331 (2016)
14. Kumar, N., Zeadally, S., Misra, S.C.: Mobile cloud networking for efficient energy management in smart grid cyber-physical systems. IEEE Wirel. Commun. **23**(5), 100–108 (2016)
15. Reka, S.S., Ramesh, V.: Demand side management scheme in smart grid with cloud computing approach using stochastic dynamic programming. Perspect. Sci. **8**, 169–171 (2016)
16. Cao, Z., Lin, J., Wan, C., Song, Y., Zhang, Y., Wang, X.: Optimal cloud computing resource allocation for demand side management in smart grid. IEEE Trans. Smart Grid **8**(4), 1943–1955 (2017)
17. Yaghmaee, M.H., Moghaddassian, M., Leon-Garcia, A.: Autonomous two-tier cloud-based demand side management approach with microgrid. IEEE Trans. Ind. Inform. **13**(3), 11091120 (2017)
18. Chekired, D.A., Khoukhi, L.: Smart grid solution for charging and discharging services based on cloud computing scheduling. IEEE Trans. Ind. Inform. **13**(6), 3312–3321 (2017)
19. Gu, C., Huang, H., Fan, L., Wu, W., Jia, X.: Greening cloud data centers in an economical way by energy trading with power grid. Future Gener. Comput. Syst. **78**, 89–101 (2018)

Cloud-Fog Based Smart Grid Paradigm for Effective Resource Distribution

Muhammad Ismail, Nadeem Javaid(✉), Muhammad Zakria, Muhammad Zubair, Faizan Saeed, and Muhammad Asad Zaheer

COMSATS University, Islamabad 44000, Pakistan
nadeemjavaidqau@gmail.com
http://www.njavaid.com/

Abstract. Smart grid (SG) provides observable energy distribution where utility and consumers are enabled to control and monitor their production, consumption, and pricing in almost, real time. Due to increase in the number of smart devices complexity of SG increases. To overcome these problems, this paper proposes cloud-fog based SG paradigm. The proposed model comprises three layers: cloud layer, fog layer, and end user layer. The 1st layer consists of the cluster of buildings. The renewable energy source is installed in each building so that buildings become self-sustainable with respect to the generation and consumption. The second layer is fog layer which manages the user's requests, network resources and acts as a middle layer between end users and cloud. Fog creates virtual machines to process multiple users request simultaneously, which increases the overall performance of the communication system. MG is connected with the fogs to fulfill the energy requirement of users. The top layer is cloud layer. All the fogs are connected with a central cloud. Cloud provides services to end users by itself or through the fog. For efficient allocation of fog resources, artificial bee colony (ABC) load balancing algorithm is proposed. Finally, simulation is done to compare the performance of ABC with three other load balancing algorithms, particle swarm optimization (PSO), round robin (RR) and throttled. While considering the proposed scenario, results of these algorithms are compared and it is concluded that performance of ABC is better than RR, PSO and throttled.

Keywords: Smart grid · Fog computing · Cloud computing
Micro grid · Load balancing

1 Introduction

Merging of the traditional grid with the controlling devices and two-way communication technologies is called SG [1]. SG provides observable energy distribution where utility and consumers are enabled to control and monitor their production, consumption, and pricing, in almost real time by using smart meter interface (SMI) [2].

L. Barolli et al. (Eds.): NBiS 2018, LNDECT 22, pp. 234–247, 2019.
https://doi.org/10.1007/978-3-319-98530-5_20

As large number of smart devices such as SM, EVs, sensors, actuators and other smart devices are integrated with SG. The complexity of the SG increases as these devices generate the enormous amount of data, which is to be processed, stored and accessed.

To overcome these problems, powerful processing and storage resources are required which can be provided by cloud computing. Cloud computing is the collection of software and hardware resources which are available on the internet and third-party service providers managing it [3]. This shared infrastructure comprises of the large number of computers that are linked together. Cloud provides three types of services: Infrastructure as a Service (IaaS), Platform as a Service (PaaS) and Software as a Service (Saas). Users access cloud through a network and can obtain on-demand services.

As the number of IoT devices are increasing rapidly, which results in increased network traffic, service interruption, network latency and other issues [4], which is not desirable for delay sensitive devices. Also the degree of data privacy and security are also low as all data are stored at the central location. Therefore, in this era of big data, we need service computing model which emphasizes on the edge of computing devices.

To overcome these issues author in [5] proposed the concept of edge-computing which is also called fog-computing. Fog computing expands the network computing mode of cloud computing. Fog acts as a middle layer between cloud and end users. Data storage and processing concentrated at the edge of the local area network closest to the end users and this result in high scalability, low latency, and high efficiency. Also fog provides location aware, secure service to its users. However there are computation and storage constraints in fog-computing.

For the efficient utilization of edge-cloud resources like VM, efficient load balancing algorithm is needed to distribute the user requests among the VMs fairly, to minimize the service delay and to increase the performance of fog. To address these issues we proposed ABC optimization algorithm. On the arrival of user request our proposed algorithm [illegible] optimal [illegible] and assigned the user request to that VM. In this way the proposed algorithm distributed the request among VMs fairly that increase the computational time and minimize the information lose and ultimately reduces the response time.

To evaluate of performance of the ABC algorithm, experiment is performed using cloud analyst. The simulation result shows that ABC algorithm obtained high performance as compared to PSO, Throttled and RR.

1.1 Motivation

With the rapid increase of monitoring and controllable facilities in the demand side of the energy management system (EMS), efficient and centralized information and communication technology (ICT) based resources are required to support demand side management (DSM) [4].

The smart phones is the part of the cyber-physical system. Dynamic voltage scaling technique is emerged to minimize energy consumption by lowering voltage and frequency of processors [6]. The author proposed a novel energy-aware dynamic task scheduling algorithm in order to minimize the aggregated energy consumed by the smart phones.

As the number of IoT devices increasing rapidly day by day, this results in the increase of load on fog. Therefore an efficient load balancing algorithm has an important impact on the performance of fog. A good load balancing algorithm increases the performance of the fog. The Fog computing environment provides several on-demand services and resource sharing for clients [7]. The author used PSO, RR and throttled algorithms for efficient allocation of resources in fog. The author claimed that PSO outperformed RR and throttled in the proposed scenario.

Based on these studies, we have integrated the concept of fog and cloud computing with the smart grid for effective utilization of resources in residential buildings. We also consider user request processing time, cost and response time.

1.2 Contribution

In this paper, the cloud-fog based communication architecture is considered. In this scenario, two residential regions are considered on the basis of population and energy crises. The proposed scenario comprises of three layers; cloud layer, fog layer and cluster layer. Fog layer is the intermittent layer, which provides efficient service with low latency to the consumers, this result in maximization of throughput and minimization of response time. Micro Grid (MG) is connected with fog to meet the energy demand of consumers.

ABC load balancing algorithm is used for efficient allocation of fog resources. Finally, simulation is done to compare the performance of ABC with RR, throttled and PSO load balancing algorithms.

The rest of the paper is organized as follows: Sect. 2 consists of related work. System model is discussed in Sect. 3. Simulation and experimental results are discussed in Sect. 4. Finally, conclusions are drawn in Sect. 5.

2 Related Work

Novel cloud based communication architecture is proposed for charging and discharge of EVs at public supply stations (EVPSS) [8]. To minimize the waiting time of EVs at EVPSS, priority-based scheduling algorithms are proposed. The present study is based on public supply stations and authors ignored the home supply station.

Self-sustained nanogrid based smart homes proposed in [9], in which all buildings are capable of maintaining their own renewable energy sources. The authors considered DC-based energy sources. The extra energy can be stored in batteries and used during on-peak hours to maintain supply and demand. However, the authors ignored AC based renewable energy sources.

In [10], the authors proposed cloud to fog to customer (C2F2C) based communication model for demand side management (DSM). Four performance parameters response time, cost, processing time and request per hour are considered.

For efficient allocation of fog resources, shortest job first (SJF) task scheduling algorithm is proposed. SJF is compared with RR and equally spread current execution (ESCE). The author claimed that SJF outperformed RR and ESCE in terms of aforementioned parameters. However, in SJF longer requests lead to starvation if there are large numbers of shorter requests.

For effective energy management of buildings, the author proposed an integrated cloud and fog based environment in [11]. For the selection of fog, new dynamic service proximity, service broker policy is proposed. Two load balancing algorithms RR and throttled are used to validate the result of the proposed policy. Authors claimed that proposed policy gives the better result with throttled as compared to RR.

Authors proposed three-layered based clod-fog architecture in [12]. Three load balancing algorithms throttled, RR, and Particle Swarm Optimization with Simulated Annealing (PSOSA) are used for fog resource allocation. Authors claimed that PSOSA outperformed RR and throttled for the proposed scenario. In [13], the authors discussed the tradeoff between power consumption and transmission delay in the cloud-fog based computing environment and formulated the problem of workload allocation. The primary problem is divided into three sub-problems and then solved via existing optimization techniques. Finally, by simulation and numerical calculation, the author concluded that by reducing communication latency, fog computing can enhance the performance of cloud computing significantly.

To manage the load of the cloud, [14] proposed a novel load balancing paradigm and new algorithm. The proposed cloud load balancing algorithm (CLB) first compute the priority service value (PS) of VMs after every 0.1 s and stored in the database. When the user request is received by the cloud, the algorithm first obtained first half of the servers and then assigned user request to one of the servers on the basis of pooling.

The green scheduling of cloud data centers is discussed by [15]. The author focused on how to trade energy economically using the power grid. The objectives are to reduce energy costs and carbon emissions. As traditional power generation units emit greenhouse gases, which has an adverse effect on the climate.

Mobile cloud task offloading remains the main area of interest for researchers in the recent year. Most of the researchers proposed the idea of mobile cloud offloading to use the unlimited resources of the cloud. However, in [16], authors considered the single-site offloading problem.

In this paper, the author addressed these issues and proposed multi-site offloading to minimize the execution time of mobile devices.

According to the above mentioned works, we need efficient and reliable SG platform, based on cloud-fog architecture for efficient management of such a huge volume of information in order to provide cost oriented and efficient service to the consumers and utility of the SG.

3 Proposed Model

Authors in [17], considered total six regions for the whole world. In the proposed model, two regions are considered that are region number 3 and 4. As the energy crisis in these regions are critical, therefore an efficient EMS is required for effective utilization of electric energy. Cloud-fog based communication architecture is used to process such a large collection of data.

The proposed scenario assumes two regions, each region has two fogs and all fogs are connected to a central cloud as depicted in Fig. 1. The proposed system comprises of three layers.

The 1st layer consists of the cluster of buildings. Each building comprises of 80 to 100 homes. The renewable energy source is installed in each building so that buildings become self-sustainable with respect to the generation and consumption. The renewable energy sources comprise of photovoltaic (PV) panel and wind energy source. The electricity generated from renewable sources can be stored in batteries and can be used during peak hours to maintain the

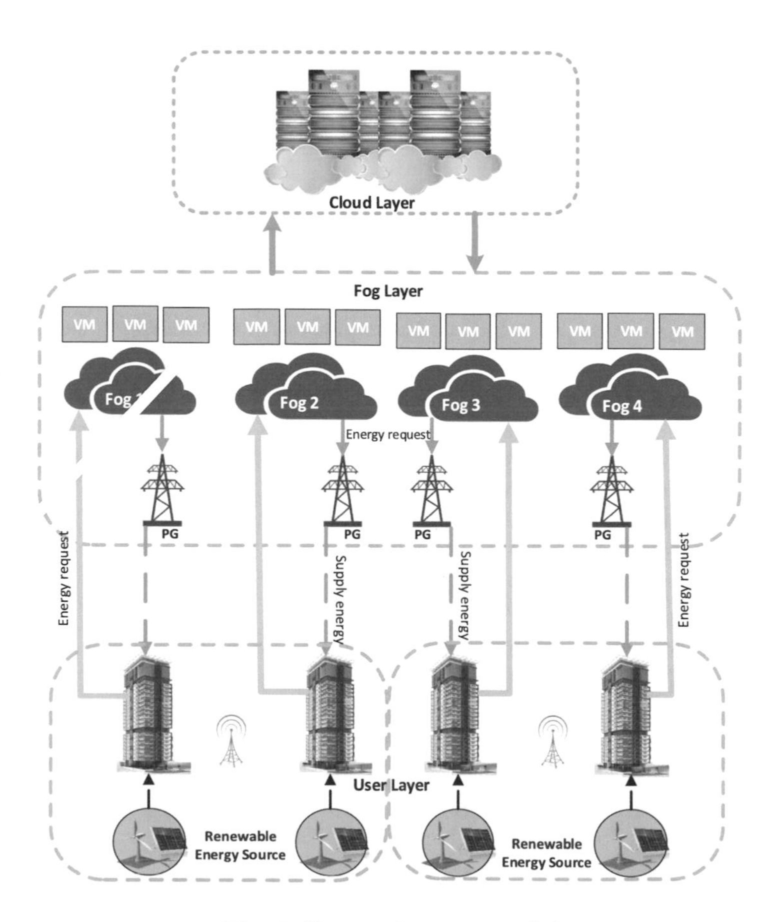

Fig. 1. Proposed system model

balance between demand and supply. Each building shares their deficit and surplus energy through fog via SM.

The second layer is fog layer which manages the user's requests, network resources and act as a middle layer between end users and cloud. Each building is connected with a single fog. Fog processes the consumer request and stores data temporarily. For permanent storage it sends data to the cloud. It provides all the services to its user closest to their location. By doing so, fog reduces the latency and improves quality of service. Fog creates VMs to process multiple user's request simultaneously, which increases the overall performance of the communication system. MG is connected with the fogs. When user's energy requirement exceeds than the energy generated by renewable energy sources then users send demand for extra energy to the MG through a fog, MG then provides the required energy to the consumers.

The top layer is cloud layer. All the fogs are connected with a central cloud. Fogs temporarily store the data and send to the cloud which is responsible to store data permanently. Cloud provides services to end users by itself or through fog.

3.1 Problem Formulation

Our proposed model comprises a set of fogs $\{\mathrm{f1, f2, ..., f_M}\}$ where $F = \{f_{\mathrm{i}}|1 <= \mathrm{i} <= \mathrm{M}\}$ is the collection of M fogs. Each fog contains a set of VMs $\{\mathrm{v1, v2, ..., v_N}\}$ where $V = \{v_{\mathrm{j}}|1 <= \mathrm{j} <= \mathrm{N}\}$ is the collection of N VMs and a set $\{\mathrm{r1, r2, ..., r_K}\}$ of user requests to be run on VMs such that $R = \{r_{\mathrm{k}}|1 <= \mathrm{k} <= \mathrm{K}\}$.

In this paper we proposed artificial colony (ABC) algorithm to distribute the user request among the VMs. Load balancing improves the performance and availability of cloud.

Execution Time: A matrix R is defined by [18] to ensure the assignment of a user request to VMs, such as:

$$r_{jk} = \begin{cases} 1 & \text{, if task k is executed on virtual machine j} \\ 0 & \text{, otherwise} \end{cases} \tag{1}$$

Execution time of user request depends upon the size of the request and processing speed of CPU of VMs. Execution time of a user request is calculated by:

$$E_t = \frac{w_r}{u} \tag{2}$$

Where $\mathrm{w_r}$ is the size of task and u is the frequency of VM CPU.

By [2], execution time $\mathrm{E_i}$ is the request running in VMs on the $\mathrm{i^{th}}$ fog can be defined as:

$$E_i^t = \sum_{j=1}^{n} \sum_{k=1}^{l} E_{ijk}^t * r_j k \tag{3}$$

Where E_{ijk} is the execution time of k requests running in j^{th} VMs on i^{th} fog. Hence the total execution time can be calculated as:

$$E^t_{total} = \sum_{i=1}^{M} E^t_i \tag{4}$$

Response Time. In this paper Response time is represented by R_t and can be calculated by:

$$R_t = TC_t - A_t + D_t \tag{5}$$

Where, arrival time of task is represented by A_t, task completion time by TC_t and transmission delay by D_t.

Delay time includes the difference of arrival and response time. It also depicts the delay time when user is located in different region than the fog.

$$D_t = \left(TC_t - TE_t\right) - \left(A_t - E_t\right) \tag{6}$$

Cost: As the total cost is equal to the sum of datacenter cost, transmission cost and cost of smart grid.

User request processing cost can be calculated by the following formula:

$$PC_r = (P_r * R_{arr} + p)T \tag{7}$$

Where PC $_r$ is the cost for processing user request. R_{arr} is request arrival rate. P_r is the energy attribution of request and T is the length of time slot and p is the electricity price at the fog in current time slot.

Transmission cost can be calculated as :

$$TC_r = T_r * R_{arr} * T * D_r \tag{8}$$

In Eg. 8 T_r is the unit cost. D_r is the distance between source and fog.

Daily energy cost of each user can be calculated as:

$$EC^u_{s,g} = \sum_{t=1}^{T} P_t l^t_{s,g} \tag{9}$$

In the above equation $l^t_{s,g}$ and P_t are the hourly power consumption of user s, in region g, and the energy price determine by the utility respectively. The total cost can be calculated by:

$$EC_{total} = PC_r + TC_r + EC^u_{s,g} \tag{10}$$

Our objective function is to:
minimize: EC_{total}
minimize: R_t
minimize: E^t_{total}

3.2 The ABC Model

ABC model consists of three types of bees: scouts, onlooker and employed bees. Scout bees perform the random search for new food sources. Employee bees collect food from already found source and onlooker bees observe the dance of the employed bees. Dance is the way of communication between bees, in this way the bees found the location of food. Scouts and onlooker are non-working bees where employee bees are working bees. Both working and non-working bees search foods near their hive. The employee bees whose solution cannot be improved after multiple attempts become a scout.

For the problem optimization, the possible solution is represented by the location of the food source and the quantity of nectars represents the quality of solutions.

When a task is created, ABC algorithm finds the suitable VM for the execution of that task while balancing the load of all VMs. Initially, VMs has no load. First ABC makes a list of all suitable VMs to which task can be assigned. A VM is suitable if it has enough capacity to execute a task. At each iteration, bees collect the load information of VMs and add to load history. While moving from VM to VM bee chooses the VM with minimum load. The employed bee of an abandoned food source becomes a scout and starts to search a new food source randomly. The VM has been abandoned by the bees will be replaced with the new virtual machine by the scouts. Then the scout finds a new VM to be replaced. Scout bees exit of the local optimum. So we can increase the search area and find new VMs having a better probability. Finally, the bee assigns the task to the appropriate VM and load information table is updated. This policy is repeated until all tasks are assigned to the VMs.

4 Results and Discussions

Based on these studies, we have integrated the concept off fog and cloud with smart grid for effective utilization of resources in residential buildings. We also consider user request processing time, cost and response time.

4.1 Simulation Parameters

For evaluation of performance of the load balancing algorithms extensive experiment is performed using cloud analyst, the simulation parameters are given in Table. 1. Dynamically reconfigure with load service broker policy is used for the selection of fog. This is the extension of service proximity policy. In this policy, fog is selected on the basis of current load on the available fogs.

4.2 Simulation Result

For experiment two regions with two fogs, two buildings and a cloud are considered. Response time, processing time and cost are taken as performance matrices. Performance of ABC is compared with PSO throttle and RR.

Algorithm 1. ABC

```
1: ABC parameters initialization: best solution, iteration, population size
2: Input: List of vm and task_k
3: Generate initial populaiton of vm_i
4: Evaluate the fitness (f_i) of the population
5: Maximum iteration, NoOfIteration
6: BestSol=Best vm_i
7: Iterate=0
8: while (Iterate <NoOfIteration) do
9:     for i=1:EmployeeBee do
10:        Select a random vm_i
11:        apply random Neighborhood structure
12:        Sort the solutions in ascending order Based on the penalty cost
13:        Determine the probability p_i for each vm_i
14:    end for
15:    for i=1:OnlookerBee do
16:        Solution* =Select the vm_i who has the higher probability
17:        Produce new solution vm_i
18:        Calculate the value of f_i
19:        if ( thenSolution**<BestSol )
20:            BestSol=Solution**
21:        end if
22:    end for
23:    Scoutbee find the food source
24:    Replace it with the new source
25:    Iterate++
26: end while
```

Table 1. Simulation parameters

Parameters	Values
Cloud	One cloud
Regions	2 regions are selected
Fogs	2 fogs each region
Buildings	2 building each fog

Response Time. Total time taken to respond a user request for service is called response time. Figure 2 shows the response time of buildings with ABC, PSO, throttle and RR. This figure shows that response time of Building1, Building2, Building3 and Building4 with ABC is 98.72, 99.87, 116.91 and 91.49. However, with PSO and throttled 100.63, 99.87, 116.96, 91.97 and 101.207, 101.38, 117.612, 92.029, respectively and with RR is 95.55, 131.99, 113.15 and 88.77.

Table 2 and Fig. 2 depicted average response time of this scenario, where performance of the ABC is better as compared to PSO, throttled and RR (Fig. 3).

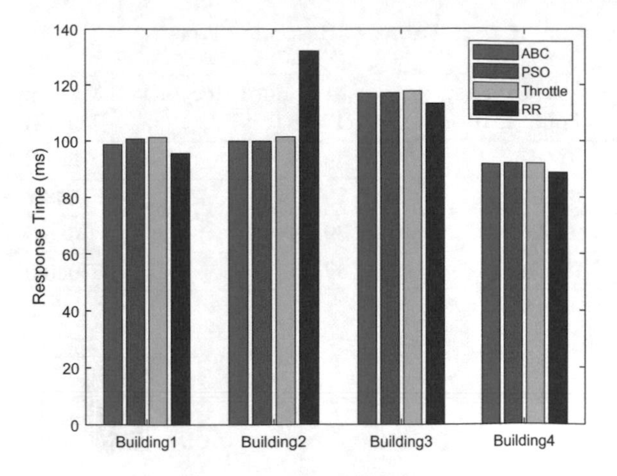

Fig. 2. Response Time

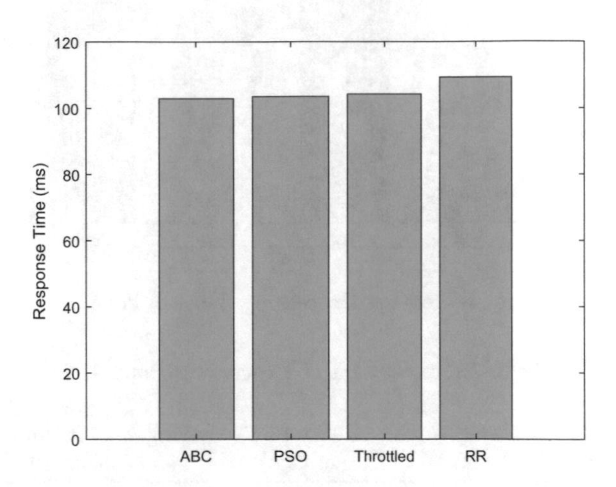

Fig. 3. Average Response Time

Processing Time. Figure 4 shows the average processing time of fogs with ABC, PSO, throttled and RR. In this figure, we can see that processing time of Fog1, Fog2, Fog3 and Fog4 with ABC 34.70, 65.26, 5.53 and 54.64. However, with PSO and throttled is 33.11, 68.95, 53.1, 45.39 and 31.71, 70, 71, 53.28, 57.2, respectively and with RR is 30.72, 97.35, and 47.01. Table. 3 shows average, minimum and maximum processing time of ABC, PSO, throttled and RR, whereas, Fig. 5 shows the average processing time of these algorithms. Facts show that performance of ABC is better as compared to PSO, throttled and RR.

Table 2. Response Time

Algorithm	Average Response Time (ms)	Minimum Response Time (ms)	Maximum Response Time (ms)
ABC	102.87	39.78	557.95
PSO	103.50	39.78	548.77
Throttled	104.26	39.78	685.87
RR	109.31	39.78	57050.01

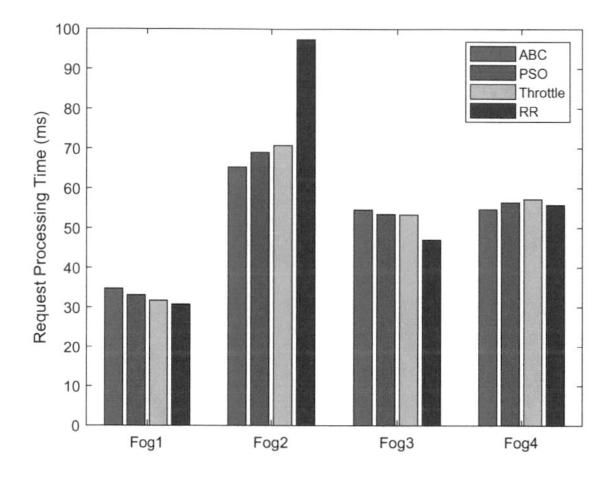

Fig. 4. Request Processing Time of Fogs

Table 3. Average Request Processing Time of Fogs

Algorithm	Average Processing Time (ms)	Minimum Processing Time (ms)	Maximum Processing Time (ms)
ABC	102.87	39.78	557.95
PSO	103.50	39.78	548.77
Throttled	104.26	39.78	685.87
RR	109.31	39.78	57050.01

Cost. For this scenario three types of costs are calculated that are, virtual machines (VMs) cost, data transfer cost and micro grid cost. Figure 6 shows that there is minor difference in cost of different algorithms. Detail cost in Table 4 shows that ABC allocates VMs more efficiently; this results in minimization of cost as compared to other load balancing algorithms.

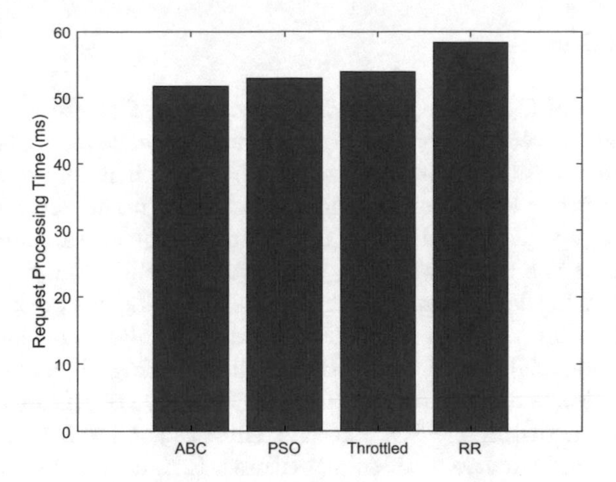

Fig. 5. Average Request Processing Time

Table 4. Average Cost

Cost($)	ABC	PSO	Throttled	RR
Total VMs:	944.82	945.58	945.72	945.94
Total Data Transfer:	14.30	14.30	14.30	14.30
MG	188.96	189.12	189.15	189.19
Grand Total:	1148.08	1149.00	1149.16	1149.42

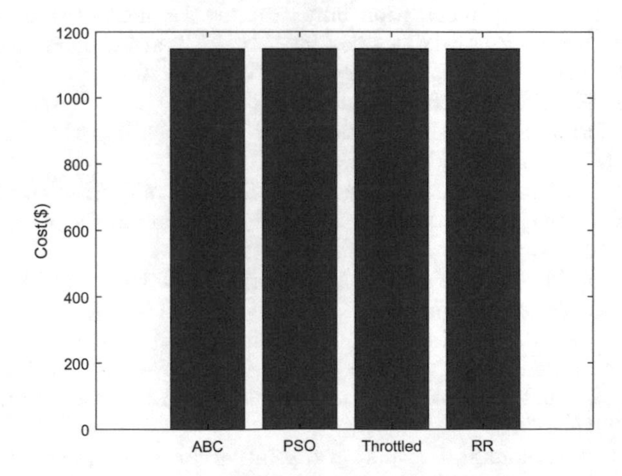

Fig. 6. Average cost

5 Conclusion

In this paper cloud-fog based SG model is proposed. The proposed model comprised three layers: cloud layer, fog layer, and end user layer. Residential buildings are considered at the 1st layer. Each building has their own renewable energy sources. When user's energy requirement exceeds than the energy generated by renewable energy sources, then users send demand for extra energy to the MG through a fog at the 2nd layer, MG then provides the required energy to the consumers. The fog layer manages the user's requests, network resources and acts as a middle layer between end users and cloud. For efficient allocation of fog resources, artificial bee colony (ABC) load balancing algorithm is proposed. Finally, simulation is done to compare the performance of ABC with three other load balancing algorithms, PSO, RR and throttled. Cloud analyst is used to evaluating the performance of these algorithms. Results show that ABC attains well load balance across virtual machines with minimum cost. It is also observed that ABC algorithm obtained high performance by maximizing the throughput and minimizing the response time as compared to other techniques.

References

1. Fang, X., Misra, S., Xue, G., Yang, D.: Smart grid the new and improved power grid: a survey. IEEE Commun. Surv. Tutorials **14**(4), 944–980 (2012). https://doi.org/10.1109/SURV.2011.101911.00087
2. Jing, J., Qian, Y.: Distributed communication architecture for smart grid applications. IEEE Commun. Mag. **54**(12), 60–67 (2016)
3. Luo, F.: Cloud-Based information infrastructure for next-generation power grid: conception, architecture, and applications. IEEE Trans. Smart Grid **7**(4), 1896–1912 (2016). https://doi.org/10.1109/TSG.2015.2452293
4. Al Faruque, M.A., Vatanparvar, K.: Energy management-as-a-service over fog computing platform. IEEE Internet Things J. **3**(2), 161–169 (2016). https://doi.org/10.1109/JIOT.2015.2471260
5. Bonomi, F., Milito, R., Zhu, J., Addepalli, S.: Fog computing and its role in the Internet of Things. In: Proceedings of the First Edition of the MCC Workshop on Mobile Cloud Computing, pp. 13-16. ACM (2012)
6. Li, Y., Chen, M., Dai, W., Qiu, M.: Energy optimization with dynamic task scheduling mobile cloud computing. IEEE Syst. J. **11**(1), 96–105 (2017)
7. Zahoor, S., Javaid, N., Khan, A., Muhammad, F.J., Zahid, M., Guizani, M.: A cloud-fog- based smart grid model for efficient resource utilization. In: 14th IEEE International Wireless Communications and Mobile Computing Conference (IWCMC-2018) (2018)
8. Chekired, D.A., Khoukhi, L.: Smat grid solution for charging and discharging services based on cloud computing scheduling. IEEE Trans. Industr. Inf. **13**(6), 3312–3321 (2017)
9. Kumar, N., Vasilakos, A.V., Rodrigues, J.J.P.C.: A multi-tenant cloud-based DC nano grid for self-sustained smart buildings in smart cities. IEEE Commun. Mag. **55**(3), 14–21 (2017). https://doi.org/10.1109/MCOM.2017.1600228CM

10. Javaid, S., Javaid, N., Tayyaba, S., Sattar, N. A., Ruqia, B., Zahid, M.: Resource allocation using Fog-2-Cloud based environment for smart buildings. In: IEEE International Wireless Communications and Mobile Computing Conference (IWCMC-2018) (2018)
11. Fatima, I., Javaid, N., Iqbal, M.N., Shafi, I., Anjum, A., Memon, U.: Integration of cloud and fog based environment for effective resource distribution in smart buildings. In: IEEE International Wireless Communications and Mobile Computing Conference
12. Yasmeen, A., Javaid, N., Iftkhar, H., Rehman, O., Malik, M.F.: Efficient resource provisioning for smart buildings utilizing fog and cloud based environment. In: IEEE International Wireless Communications and Mobile Computing Conference (IWCMC 2018) (2018)
13. Deng, R., Lu, R., Lai, C., Luan, T.H., Liang, H.: Optimal workload allocation in fog-cloud computing toward balanced delay and power consumption. IEEE Internet Things J. **3**(6), 1171–1181 (2016). https://doi.org/10.1109/JIOT.2016.2565516
14. Chen, S.L., Chen, Y.Y., Kuo, S.H.: CLB: a novel load balancing architecture and algorithm for cloud services. https://doi.org/10.1016/j.compeleceng.2016.01.029
15. Gu, C., Fan, L., Wu, W., Huang, H., Jia, X.: Greening cloud data centers in an economical way by energy trading with power grid. https://doi.org/10.1016/j.future.2016.12.029
16. Goudarzi, M., Zamani, M., Haghighat, A.T.: A fast hybrid multi-site computation offloading for mobile cloud computing. https://doi.org/10.1016/j.jnca.2016.12.031
17. Wickremasinghe, B., Buyya, R.: CloudAnalyst: a cloudsim-based tool for modeling and analysis of large scale cloud computing enviornments. MEDC project report 22, no. 6, pp. 433–659 (2009)
18. Idrissi, A., Zegrari, F.: A new approach for a better load balancing and a better distribution of resources in cloud computing. https://doi.org/10.14569/IJACSA.2015.061036

Effective Resource Allocation in Fog for Efficient Energy Distribution

Abdullah Sadam, Nadeem Javaid(✉), Muhammad Usman Sharif, Abdul Wasi Zia, Muhammad Yousaf, and Syed Muhammad Saleh Arfi

COMSATS University, Islamabad 44000, Pakistan
nadeemjavaidqau@gmail.com
http://www.njavaid.com

Abstract. Fog computing is used to distribute the workload from cloud, decrease Network Latency (NL) and Service Response Time (SRT). Cloud have the capability to respond to too many requests from consumer side, however, the physical distance between a consumer and cloud is far than the consumer and fog. Fog is limited to a specific location, moreover, fog is meant to deal requests locally and helps out in processing the Consumer's Requests (CRs) and provide efficient response. A fog holds the consumer's data temporarily, processes it and provides response then sends it to cloud for permanent storage. Apart from this, it also sends the consumer's data when Micro Grids (MGs) are not able to fulfill the consumer's energy demand. Cloud communicates with Macro Grid. Fog and cloud computing concepts are integrated to create an environment for effective energy management of a building in a residential cluster. Fog deals with requests at the consumer's end, because it is nearer to the consumer than a cloud. The theme of this paper is efficient allocation of Virtual Machines (VMs) in a fog, therefore, Insertion Sort Based Load Balancing Algorithm (ISBLBA) is used for this purpose. Simulations have been conducted, comparing ISBLBA to Round Robin (RR) technique and results regarding fog performance, cluster performance and cost are elucidated in the Sect. 5.

Keywords: Insertion Sort Based Load Balancing Algorithm (ISBLBA)
Cloud computing · Fog computing · Micro Grid (MG) · Macro Grid

1 Introduction

Efficient energy management is an intense need of today's world. The goal is to stop wastage of energy which is produced with resources and efforts. The consumption of energy is increased due to the massive demographic growth [1]. Most of the energy is wasted at the consumer's end and world needs such a system which conserves energy. Therefor, Home Energy Management System (HEMS) integrated with cloud, is introduced for effective energy management [2]. However To Increase SRT and decrease the NL, different kinds of fog con cepts are introduced in cloud environment [3–6]. Many IoT devices operate at

L. Barolli et al. (Eds.): NBiS 2018, LNDECT 22, pp. 248–259, 2019.
https://doi.org/10.1007/978-3-319-98530-5_21

the consumer's end and require energy. HEMS are introduced at the consumer's end to manage these devices. HEMS put restrictions on off-peak and on-peak hour on different devices. If a consumer wants to run an appliance, the request is generated via HEMS. A controller communicates with the fog, shares the consumer's data, Meter ID and the required load. A VM having its own operating system, memory and processing power is assigned to the request via ISBLBA. VM processes CR and forwards it to the MG which is a near by energy producing entity consists of Photovoltaic Cells (PVCs), batteries and generators. MG receives a request, and sends energy to the Meter ID provided by the fog. SG notifies the fog about its energy production status and amount of energy sent to the consumer, and fog decide whether the request needs to be sent further to the cloud.

Fog shares all the processed data to cloud for permanent storage so that it could be possible to keep track of CRs for services and provided services. If MG fails to fulfill the requirement of a consumer, then cloud requests to the Macro Grid to supply energy to the cluster from where the request came from. Macro Grid is a smart grid [3] which keep track of cloud's requests and produce energy as per requirements, so that resources wastage can be controlled. Macro Grid further sends the request and required energy to Utility. Utility is responsible for distribution and transmission of energy to geographical locations. Utility supplies energy to consumers and notifies the cloud. Cloud then keeps track of all the data.

2 Related Work

Cloud and fog computing environments are used to facilitate consumers in Energy Management (EM). Researchers are delivering their best to automate this system. It is the need of this era that we have an automated system. No extra machines and human involvement is there. For this purpose, allocation of resources is a problem which needs to be solved. Authors in [7], have proposed an effective solution to this problem.

To manage energy crises more efficiently, the idea of Nano Grid is introduced in [8]. Energy generation is empowered near to the consumer side. Nano Grids are energy generation grids at homes or buildings, which are meant to fulfill the energy demand of that building.

In [9], Smart Grid is used to monitor the electricity consumption at the consumer's end. It establishes many optimization approaches, where multiple numbers of energy consuming devices are scheduled in a organized manner. User comfort is considered in an on-peak and off-peak hours. The vital role of Smart Grid is elaborated.

In [10], authors have proposed an energy management system for smart devices. Energy Aware Dynamic Task Scheduling (EDTS) algorithm is used for reducing energy consumption. Their simulations and results show that the proposed technique delivers high energy efficiency.

In [11], it is described that to deal with too many consumers, greater computation power is required. The more devices are connected to the Utility Energy

Management System, more data is generated, which ultimately increases data and computational complexity.

In [12–16], authors explained that a platform is required where devices are made inter-operable, all operations are flexible and fast response is ensured. Such system is used for effective implementation of energy management system. It is also mentioned that the increasing number of consumers can cause the complexity of game theory based models with respect to space and time. To avoid such circumstances all the service providers are empowered to shift their services cloud [3].

In [17], authors have introduced a model which overcomes the overall service response time and network latency issues in the current cloud model. Their scenario is better for few limited devices when they request to cloud; however, it is really difficult to manage too many requests.

In [18], authors have created a system model for Mobile Cloud Network (MCN), to efficiently manage the computational and network resources of an MCN. It enables the globally optimized placement of static and Mobile applications. Resource management tasks are done efficiently.

In [19], authors have revealed the advantages of the combined Fog-to-Cloud architecture based on SRT, power consumption, service disruption probability and bandwidth occupancy parameters. The service disruption probabilities are addressed in terms of leaving and joining of a node.

In [20], authors have introduced a new concept of decentralized distributed computing of Mobile and edge devices, suitable for IoT computational requirement. Since IoT devices need a decentralized and peer-to-peer connectivity, this paper has addressed all those concepts which are related to IoT infrastructure.

2.1 Motivation

To acquiring high SRT and low NL, fog computing is the introduced solution to fulfill such need. Besides these features, fog computing provides scalability, reliability and virtualization. Fog is used to distribute the processing load from cloud, it responds to the CRs locally. As in [6], authors used a load balancing technique to distribute the work load of a cluster among different VMs of a fog to achieve high SRT.

In the near future we are entering to the IoT world which will be based upon Ubiquitous computing. All kind of IoT devices need energy consumption, an effective management of energy might save us from energy crises in the future. The predicted number of IoT devices are 26 billion according to the scientists, which will be consuming a very high amount of energy. Thus, there is a need to bring such solutions to the market that can help us in the coming future. The motivation is that for all these things we need a system where we can efficiently distribute energy so that we become able to conserve as much energy as we can.

3 System Model

In this paper addresses fog based environment, in three regions. Every region has three clusters, and each cluster contains a residential sector having five hundred buildings and one hundred houses per building. Multiple IoT devices operate in a

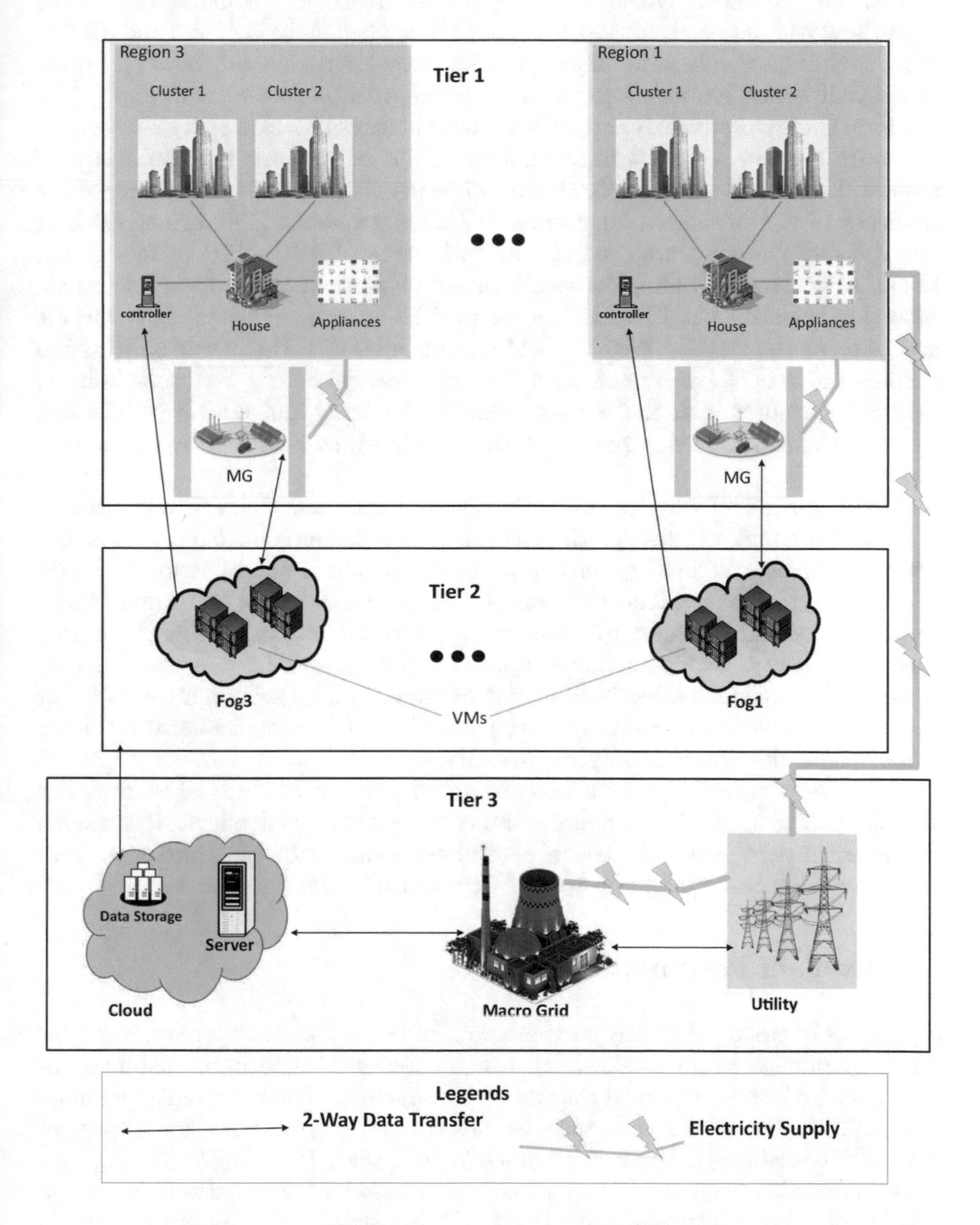

Fig. 1. Proposed system model

house, and these devices require energy. All IoT devices of a house are connected to a controller. Every controller takes demand from these devices, communicates with fog and requests for energy. Fog is used to distribute load from centralized cloud. Fog is nearer to the cluster which increases SRT and decreases NL.

The physical distance of a consumer to fog is less than the physical distance of consumer to cloud. Obviously, data packets reach fog in much lesser time than these packets reach to the cloud. So, NL is ultimately low, and the SRT of fog is high. Fog also has the functionalities of centralized cloud, however, these functionalities are limited. A fog communicates with MG for energy supply. MG is a local energy producer, consists of PVCs, batteries and electricity generators.

When fog gets a CR, it must allocate a VM or resource to processes that request. ISLBLA is used for effective resource distribution in fog. To increase the efficiency of fog, optimized allocation of VMs is necessary. Different algorithms like RR and Throttled are used to efficiently assign VMs to CR. In this paper, ISBLBA is introduced for effective allocation of VMs. In our proposed scenario, every fog has six VMs. ISBLBA considers it as an array of VMs, thus sorts it according to the state of VM. If a VM state is free, ISBLBA considers it as the smallest value in the array and puts it on the first index. If a VM state is busy, ISBLBA considers it as the biggest value in the array and puts it on the last index of the array. This is how all VMs are sorted, so it efficiently allocates a free VM to CR.

There is a minor physical distance between MG and clusters, and provide a limited amount of energy when requested. For exceeded energy demand, fog requests cloud. Cloud is considered as a centralized computational service and storage platform. Cloud requests the Macro Grid, which is a Smart Grid, responds to cloud's request by exchanging information and supplying required energy to Utility. Utility further supply energy to a specific cluster. Utility is considered as an electricity distributor through its transmission network all over the region. Utility supplies energy to the exact location via controllers and smart meters, thus, demand is managed efficiently.

All tiers communicate with each other in a series of steps. The first tier contains clusters, buildings, houses, smart meters and controllers. The second tier consists a fog network, and a centralized cloud is in the third tier. This whole model is based on 3 tier architecture which is illustrated in Fig. 1.

4 Problem Formulation

Allocation of resources to any service needs to be very efficient, sine the performance parameter require this task to be so. Services come from customers or consumers, which we assumed that a residential cluster contains multiple homes and requests are generated in these homes. Let assume that H represent a set of cluster containing n number of homes h, such that, $H = \{h_1, h_2, h_3, ..., h_n\}$.

A Service is a request from a consumer of house h for any particular job and it is divided into multiple of tasks tsk. Let S is a service consists of i number of tasks tsk came from house h, We can formulate this as,

$$S = \sum_{i=1}^{n}(tsk_i) \tag{1}$$

Every tsk has a status, and we denoted it as x. If tsk is allocated to any particular resource or VM, the status of that task is $tsk_x = 1$. If a tsk is not allocated to any resource the status is $tsk_x = 0$.

A fog consists of Virtual Machine (VMs) which are used to execute tasks. Let $fog = \{VM_1, VM_2, VM_3, VM_4, VM_5, VM_6\}$.

$$fog = \sum_{j=1}^{6}(VM_j) \tag{2}$$

The number of VMs must not exceed six, because we have provided only six VM to a fog. A Virtual Machine Manager (VMM) knows the state of each VM, which is denoted by y. A VM is considered available where $VM_y = 0$, and VM is considered busy when $VM_y - 1$.

Following are the two main problems that are identified and formulated consecutively.

4.1 Service Response Time

SRT is the time taken by the resources to respond to the consumer's request for service. A Service is a collection of tasks tsk as shown in Eq. 1. Every consumer request is responded, Let SRT be the Overall SRT, which is the sum of Processing Time (PT) and NL.

$$SRT = PT + NL \tag{3}$$

Furthermore, PT is the time taken by VM_j to execute a task tsk_i of Service S. We deduce that $P_{i,j}$ is the PT of a single tsk_i in a virtual Machine VM_j. Based on these we infer that

$$\sum_{i=1}^{x}\sum_{j=1}^{y}(P_{i,j})_{x,y} \tag{4}$$

is the total PT to complete a service S. The SRT can also be denoted as,

$$SRT = \sum_{i=1}^{x}\sum_{j=1}^{y}(P_{i,j})_{x,y} + NL. \tag{5}$$

4.2 Cost

Cost is a monetarily expenditure which is estimated by the usage of resources of a particular utility. In this particular scenario, $VMcost$ is calculated by the usage of VMs, $MGcost$ is the cost of MG, $DTcost$ is the cost of Data Transfer over the network. We can say that

$$Total_{cost} = VM_{cost} + MG_{cost} + DT_{cost} \tag{6}$$

5 Simulations and Results

In this paper, 'Cloud Analyst' platform is used for simulation purposes. Through this tool service broker policies are compared with load balancing algorithms in order to evaluate the results. For results purpose, ISBLBA is compared with the RR technique against Closest Data Center Policy (CDCP). CDCP is considered as the least proximity from consumers. Proximity means least NL. Three regions are taken, containing two clusters and one fog each as illustrated in Fig. 2. Region distribution is shown in the Table 1.

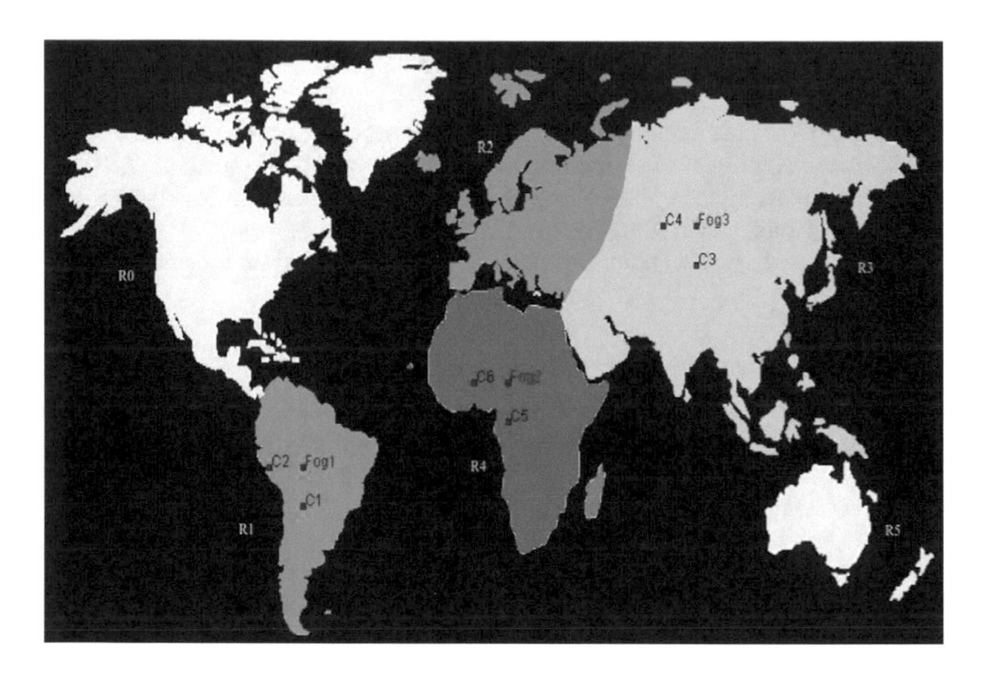

Fig. 2. Visual representation of clusters and fog in regions

Table 1. Regions

Regions	ID
North America	R0
South America	R1
Europe	R2
Asia	R3
Africa	R4
Australia	R5

5.1 Closest Data Center Policy (CDCP)

Two clusters in every region are considered, and single fog is offering services. The fog consists of VMs which are used for work load distribution. In our scenario, fog has six VMs having 8 GB of RAM, 10 TB HDD and 2.4 GHz processing power. To manage the coming requests from consumers, fog take some PT. We have considered a single service broker policy which is the CDCP.

5.2 Cluster Performance in Terms of SRT

Consumers have generated requests which are managed in a local fog. Every request gets a response in some time, by the fog, which is called SRT. ISBLBA and RR algorithms are simulated using our scenario and their comparative cluster SRT is shown in Table 2.

Table 2. Cluster performance in terms of SRT

Clusters	Avg (ms)		Min (ms)		Max (ms)	
	ISBLBA	RR	ISBLBA	RR	ISBLBA	RR
C1	49.77	49.82	38.32	38.36	63.79	63.79
C1	50.23	50.21	39.31	39.31	62.82	62.86
C1	50.06	50.04	38.52	38.52	66.27	66.29
C1	49.98	49.98	39.77	39.77	60.78	60.82
C1	50.19	50.26	41.81	41.08	61.80	61.83
C1	50.08	50.12	40.56	40.56	60.80	60.83

Figures 3 and 4 of ISBLBA and RR respectively illustrate cluster performance with respect to SRT from the fog.

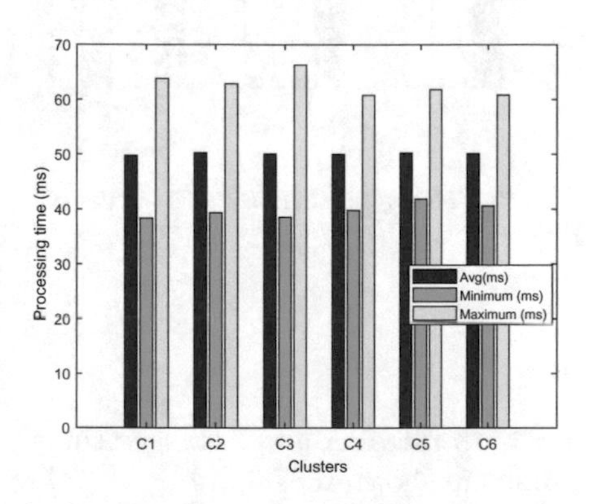

Fig. 3. ISBLBA cluster performance in terms of SRT

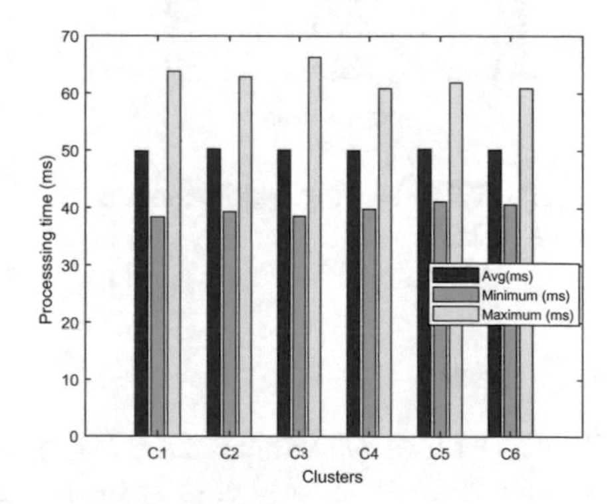

Fig. 4. RR cluster performance in terms of SRT

5.3 Fog Performance in Terms of PT

The data coming from a consumer is managed in the fog. Load balancing algorithms are defined to distribute the load. We have compared our proposed algorithm with RR technique and simulation shows that the proposed algorithm is working efficiently than RR technique. The comparative fog processing of both algorithms is shown in the Table 3.

Table 3. Fog performance in terms of PT

Data centers	Avg (ms)		Min (ms)		Max (ms)	
	ISBLBA	RR	ISBLBA	RR	ISBLBA	RR
Fog1	0.42	0.44	0.02	0.02	0.87	0.89
Fog2	0.42	0.43	0.02	0.02	0.83	0.83
Fog3	0.39	0.40	0.01	0.01	0.82	0.82

Figures 5 and 6 of ISBLBA and RR respectively show the fog performance with respect to SRT from the consumers.

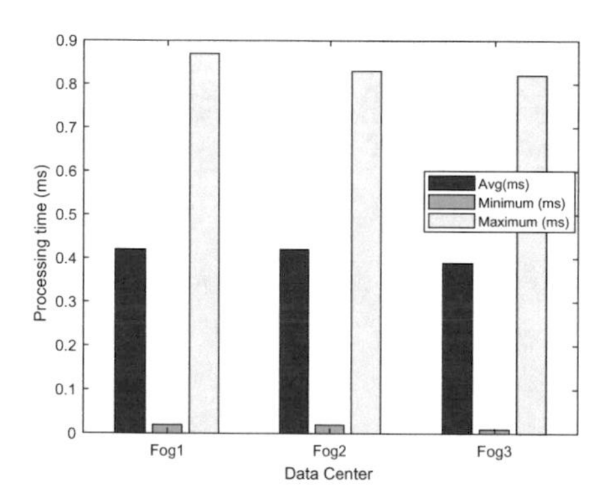

Fig. 5. ISBLBA fog performance in terms of PT

Fig. 6. RR fog performance in terms of PT

5.4 Cost

Cost of VM, MG, Data Transfer (DT) of using both ISBLBA and RR algorithms are the same. The total cost of both algorithms are shown in Table 4.

Figures 7 and 8 of ISBLBA and RR respectively shown. A fog provides services to multiple CRs hourly. Minimum requests are serviced in peak hours, which minimized the cost. A fog has VMs, data transfer and MG cost, and total cost is also shown in the Figs. 7 and 8.

Table 4. ISBLBA and RR comparative cost

Data centers	VM cost($)		MG Cost($)		DT Cost($)		Total cost($)	
	ISBLBA	RR	ISBLBA	RR	ISBLBA	RR		
Fog1	0.40	0.40	19.77	19.77	1.24	1.24	21.41	21.41
Fog2	0.20	0.20	19.83	19.83	1.24	1.24	21.26	21.26
Fog3	0.20	0.20	19.75	19.75	1.23	1.23	21.18	21.18

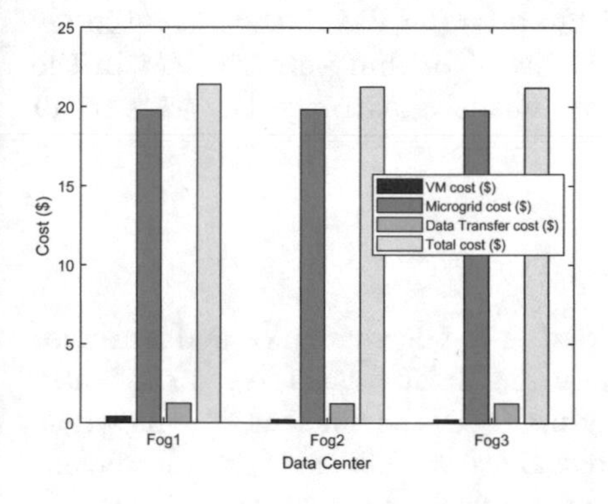

Fig. 7. ISBLBA cost

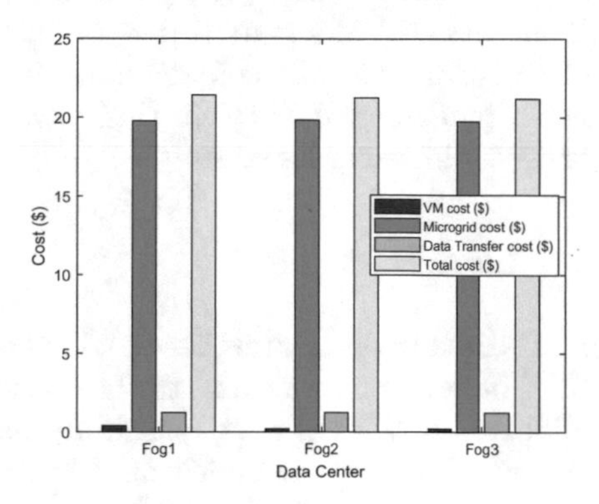

Fig. 8. RR cost

5.5 Discussions

In this paper, two load balancing algorithms are compared based on the distribution of load on a fog, which is coming from the consumer's side. Results related to processing time of fogs and clusters shows that our proposed algorithm performs well in the considered scenario. In ISBLBA, the Cost of VM, MG, and data transfer is exactly the same as RR.

When we talk about CDCP, we keep low latency in our mind, that how to lower the latency of the network. Closest proximity means the closest available services which eventually decrease the latency. Quick responses are assured and network congestion is avoided. Cloud gets a sigh of relief by not having too many requests from consumers. We have taken two clusters, however, when it comes to the real world then there will be enormous requests from consumers. There will be million of residential sectors which might request for services thus it would not be possible for a cloud to handle this much load at once. Fog computing is one of the best solutions to distribute load from cloud and in this paper, ISBLBA does the same. It reduces the request load from cloud by effectively managing the fog resources to get to every service request by any consumer. However, installing too many fog publicly require a huge monetarily disbursement from the utilities. In this paper, we have simulated that our proposed scheme shows a reduced cost, which can save a huge amount of investment to the utilities in the sector.

Another important discussion is about the performance of fog. As much as cloud needs to be relieved of the consumer's load, fog also needs to be relieved.

Cloud has a huge computing capacity and fog does not, so it is still difficult for the fog to get to every request efficiently unless we configure it this way. VMs are the essential power of any cloud, which ultimately responds to every request, thus, managing VMs efficiently means an increase in fog performance. For this purpose, our proposed algorithm for load balancing is effectively managing VMs inside the fog. This algorithm puts all VMs in an array, by considering its state. If VM's state is free, the algorithm puts the VM in the start index of the array, if VM's state is busy, the algorithm puts the VM in the nth index of the array. If the second VM's state is free, the algorithm puts the VM in the second index of the array, and if the VM's state is busy, the algorithm puts the VM in the nth-1 index of the array. In this way, it continuously rearrange the VMs in an array by knowing its state.

6 Conclusion

Fog is used to distribute the work load of cloud. A fog decreases NL and increases SRT because of its locality to the consumer's end. ISBLBA is used to efficiently allocate the VMs of a fog, which ultimately increases the efficiency of the whole system. CRs are responded quickly and resources are allocated in an efficient manner and services are provide to consumers which increases the consumer's satisfaction ultimately. Two load balancing algorithms are compared, and their results are discussed and elaborated in the respected sections. Though there is a very minor performance difference between ISBLBA and RR using the same scenario on CDCP, however, ISBLBA is performing slightly better than RR. RR algorithm is being used for load balancing purposes too many times, however, we have used Insertion Sort Algorithm for this purpose in this paper. It performs slightly better the RR. We can conclude that it is our contribution to use Insertion Sorting algorithm for load balancing purposes, to the best of our knowledge this is the first paper using this sorting algorithm for resource scheduling. So far this algorithm is performing well than RR algorithm already used for this purpose.

References

1. Melhem, F.Y., Moubayed, N., Grunder, O.: Residential energy management in smart grid considering renewable energy sources and vehicle-to-grid integration. In: IEEE Electrical Power and Energy Conference (EPEC), 12–14 October 2016, Ottawa, ON, Canada, pp. 1–6 (2016)
2. Yaghmaee, M.H., Moghaddassian, M., Garcia, A.L.: Power consumption scheduling for future connected smart homes using bi-level cost-wise optimization approach. In: Lecture Notes of the Institute for Computer Sciences, Social-Informatics and Telecommunications Engineering, LNICST (2016)
3. Markovic, D.S., Zivkovic, D., Branovic, I., Popovic, R., Cvetkovic, D.: Smart power grid and cloud computing. Renew. Sustain. Energy Rev. **24**, 566–577 (2013)
4. Chen, S., Zhang, T., Shi, W.: Fog computing. IEEE Internet Comput. **21**(2), 4–6 (2017)

5. Javaid, S., Javaid, N., Aslam, S., Munir, K., Alam, M.: A cloud to fog to consumer based framework for intelligent resource allocation in smart buildings, pp. 1–10
6. Fatima, I., Javaid, N.: Integration of cloud and fog based environment for effective resource distribution in smart buildings
7. Pham, N.M.N., Le, V.S.: Applying Ant Colony System algorithm in multi-objective resource allocation for virtual services. J. Inf. Telecommun. **1**(4), 319–333 (2017)
8. Cao, Z., Lin, J., Wan, C., Song, Y., Zhang, Y., Wang, X.: Optimal cloud computing resource allocation for demand side management. IEEE Trans. Smart Grid 1–13 (2016). https://doi.org/10.1109/TSG.2015.2512712
9. Faruque, M.A.A., Vatanparvar, K.: Energy management-as-a-service over fog computing platform. IEEE Internet Things J. **3**(2), 161–169 (2016)
10. Okay, F.Y., Ozdemir, S.: A fog computing based smart grid model. In: 2016 International Symposium on Networks, Computers and Communications (ISNCC) (2016)
11. Wang, W.Y.C., Rashid, A., Chuang, H.M.: Toward the trend of cloud computing. Electron. Commer. Res. **12**(4), 238–242 (2011)
12. Mondal, A., Misra S., Obaidat, M.: Storage in smart grid using game theory. IEEE Syst. J. 1–10 (2015)
13. Yu, M., Hong, S.H.: Supply - demand balancing for power management in smart grid: a Stackelberg game approach. Appl. Energy **164**, 702–710 (2016)
14. Hong, J.S., Kim, M.: Game-theory-based approach for energy routing in a smart grid network. J. Comput. Netw. Commun. **2016** (2016)
15. Ni, J., Ai, Q.: Economic power transaction using coalitional game strategy in micro-grids. IET Gener. Transm. Distrib. **10**(1), 10–18 (2016)
16. Mediwaththe, C.P., Stephens, E.R., Smith, D.B.: A dynamic game for electricity load management in neighborhood area networks. IEEE Trans. Smart Grid **7**(3), 1–8 (2016)
17. Alonso-Monsalve, S., García-Carballeira, F., Calderón, A.: A heterogeneous mobile cloud computing model for hybrid clouds. Future Gener. Comput. Syst. (2018). https://doi.org/10.1016/j.future.2018.04.005
18. Tärneberg, W., Mehta, A., Wadbro, E., Tordsson, J., Eker, J., Kihl, M., Elmroth, E.: Dynamic application placement in the Mobile Cloud Network. Future Gener. Comput. Syst. **70**, 163–177 (2017)
19. Ramirez, W., Masip-Bruin, X., Marin-Tordera, E., Souza, V., Jukan, A., Ren, G., Gonzalez de Dios, O.: Evaluating the benefits of combined and continuous Fog-to-Cloud architectures. Comput. Commun. **113**, 43–52 (2017)
20. Ciobanu, R.I., Negru, C., Pop, F., Dobre, C., Mavromoustakis, C.X., Mastorakis, G.: Drop computing: ad-hoc dynamic collaborative computing. Future Gener. Comput. Syst. (2017). https://doi.org/10.1016/j.future.2017.11.044

Optimized Load Balancing Using Cloud Computing

Wajahat Ali Gilani, Nadeem Javaid(✉), Muhammad KaleemUllah Khan, Hammad Maqbool, Sajid Ali, and Danish Majeed Qureshi

COMSATS University, Islamabad 44000, Pakistan
nadeemjavaidqau@gmail.com
http://www.njavaid.com

Abstract. The concept of fog computing is initiated to mitigate the load on cloud. Fog computing assists cloud computing services. It extends the services of cloud computing. The permanent storage of the data is come to pass in cloud. An environment based on fog and cloud is providd to manage the energy demand of the consumers. It deals with the data of buildings which are linked with clusters. To assist cloud, six fogs are deployed in three regions, which are found on three continents of the world. In addition, each fog is connected with clusters of buildings. There are eighty buildings in each cluster. These buildings are Smart Grid (SG) buildings. For the management of consumers energy demand, Micro Grids (MGs) are available near by buildings and reachable by fogs. The central object is to manage the energy requirements, so, fog assists consumers to attain their energy requirements by using MGs and cloud servers that are near to them. However, for balancing the load on cloud the implementation of an algorithm is needed. Virtual Machines (VMs) are also required. Pigeon hole algorithm is used for this purpose. Using proposed techniques results are compared with Round Robin (RR) which gives better results. The proposed technique in this paper is showing better results in terms of response time.

Keywords: Cloud computing · Fog computing · Response time
Smart grid · Micro-grid

1 Introduction

Whenever, the topic of load balancing or energy management is discussed, the term cloud computing involves somehow. Cloud computing concept is evolved very earlier. Cloud and fog computing are the approaches, which predominately use different researchers. Delivery of different services is the actual task of cloud. Services can not be provided easily without help of cloud computing. These services may be information, storage and other hardware devices. The concept of Smart Grid SG unification with fog and cloud computing is a new idea. SG is an electric grid that gather data by using information and communication

L. Barolli et al. (Eds.): NBiS 2018, LNDECT 22, pp. 260–272, 2019.
https://doi.org/10.1007/978-3-319-98530-5_22

technology. The monitoring of electricity that is used by consumers is monitored through smart meters. These smart meters are installed in homes. Multifarious reasons are behind the need of cloud computing. These needs compel various researches to use this concept. Some major reasons are needed to explain. One of the influential reason to use cloud computing in different organizations and business firms is the splendid effect of cost. Cloud computing is playing a key role in reducing the cost of different hardware and software and many other resources. Another, vital reason to use cloud computing is that it facilitates users to run applications and several other processes over Internet. Nobody has to use any other physical software or hardware. Cloud assists users to access data, no matter wherever and whenever you want to use it.

Cloud computing delivers various applications. There are innumerable cloud applications, few of them are SaaS, PaaS and IaaS that are most prominent. These are service providers, which provide different services. The acronym of the SaaS is "Software as a Service". SaaS exploits the internet to provide applications to its users. And it is managed by third party vendor, some glaring examples are Google Apps, Dropbox, Concur and CISCO WebEx. To do something, user get access to the specific software which is needed. User can only get access, he can not own it. To store different files Drop-box carries software. Microsoft 365 is a typical example of this, which gives access. The user have to pay for these services. PaaS stands for "Platform as a service". PaaS is responsible to provide run time environment. It is scalable and highly available. Heroku, Apache Stratos, OpenShift are examples of PaaS. They provide a platform to develop a software. By doing so, a user can develop its own software. This platform helps the user to grow software. The basic purpose of the PaaS is to convey application developer. PaaS provides different tools and access. It also deliver hardware resources on which application run. PaaS supports in building an application. OpenShift is the most conspicuous example of PaaS. Last but not least, IaaS is also service of cloud. IaaS abbreviation is "Infrastructure as a Service". IaaS is responsible to provide hardware resources to its users. IaaS is flexible model. Google Compute Engine (GCE), DigitalOcean and Linode are examples of IaaS [1]. The cardinal function of IaaS is to provide control over hardware resources. Even developers take help from IaaS [2]. Smart meters have replaced traditional meters. It also recognizes the consumption of electricity. Different authors in different papers used distinct scenario to combine SG with cloud and fog. With the amalgamation of MG, the SG facilitates consumers [3,4].

The dependency of the cloud computing is on shared resources, not on local servers and devices [5]. However, services like storage, computation and other application are obtained via the web and the cost of these services on internet is paid by cloud customers. For the enhancement of cloud computing, CISCO has introduced the concept of fog computing in January 2014. The applications and other devices are increasing swiftly, so, fog computing helps to carry cloud computing on the edge of networks to beat rapid increase. Fog computing is used to attenuate cloud burden the reason is that different devices and applications use cloud services and they consume great amount of data. However, the permanent storage of the consumers data is taken place in this way, fog stores the data temporarily and then fog contact with cloud that stores data permanently.

Whenever, discussions are made about the incorporation with cloud and fog based environment, demands from consumers side starts to come. The demand may be any type of like web page access or a MG. So, the communication among consumers can only be achieved, if there is connection establishment among them. This connection can be established between consumers side to fog and then from fog to cloud. The requests are made by consumers side to fog and fog seeks nearby obtainable resources to provide resources. These resources are able to attain requirements.

The services which are lodged over, helps to mitigate hardware and software cost. Cloud based services are popular because of cost casting strategy. There are some issues which deals cloud computing. Of late, issues like overloading, load balancing and task scheduling are such problem that can not be overlook. So, to resolve them researchers have to take the support of cloud computing the reason is that it handles these problems very properly [6,7].

In this paper, model is fractionated into three layers; first layer is comprised of group of buildings. This group of buildings is named as cluster and it is based on SG. Second layer consist of fog and MGs. Third layer contains cloud and macro-grid. Anyhow, clusters communicate with fog because they can not communicate with MG directly. VMs are used with each fog which include storage and applications. All clusters are linked with those fogs which are located near to them. Fog stores data temporarily while cloud stores it permanently. The users who want to communicate with fog first they need to make their profile. The profile may include information like their location and usage of electricity on daily base etc. It makes easy for cloud and fog to handle data.

Rest of the paper is organized as: Sect. 2 presents related work. Section 3 includes motivation. System model is illustrated in Sect. 4. Anyhow, simulation results and conclusion are in Sects. 5 and 6.

2 Related Work

The distribution and management of electricity are the issues, which face users globally. Moreover, the demand of energy is increasing in all domains. So, it is inevitable to adopt new mechanisms and techniques. Many researchers are doing work to cook up these issues. Different mechanisms and approaches have been implemented by taking help of cloud and fog computing. Cloud and fog computing is playing pivot role in this domain by offering services. The core object of cloud computing is to deliver facilities like resource sharing to its users. For the reduction of energy consumption, resources are provided by increasing VMs. Even so, the concept of fog computing expands cloud computing because it mitigates load on cloud.

Multiple algorithms can be used to balance the load on cloud and fog. The management of requests and scheduling of tasks is taken place through these algorithms. In [8], the author solves problem of resource allocation by using

Multi-Objective Resource Allocation-Ant Colony System (MORA-ACS). Allocation of resources on cloud is done over here. It is the prominent feature of this technique. With the help of RR algorithm, the author is doing comparison among processing time, standard deviation and energy consumption. However, the author is succeeded in achieving two objectives, which are energy consumption minimization and load on cloud.

Another author in [9] proposed a technique for load balancing. For the allocation of low power tasks a mathematical model is implemented. Therefore, the algorithm ameliorates the throughput by diminishing tardiness. HRs are used to achieve the targeted goal. In obtaining desired goals, human resources are like bones in a body because all the procedure is revolving around these sources. To escalate the performance, these human resources are serving. One HR is not enough, the reason is that one HR can not handle the whole situation. The workload needs to dispense on the basis of HRs. The collection of data from environment is not needed, the reason behind it is that different algorithms and policies are used for this purpose. So, data collection environment is not inevitable. To issue jobs, there is need of state of the system. They use current state of system for job issuance. In this way, the scheduling of the tasks occur. The allotment of the task was a major problem which is faced by different researcher. This paper intimates a new way for the allotment of task that leans on two main element: one of them is arrival rate and second one is service time. However, allocation of task via this method is a very important contribution. For load balancing, the authors presented a method that will not overlook cases like task priority and mapping while on the other hand, the other algorithm, which are used before this, they have not paid focus on such cases.

The method in [10] presents a Certificate Less Provable Data Possession (CL-PDP). This method is offered for SG applications which is based on cloud. The major purpose of the approach is to deliver adequate storage, security and processing capacity.

For the betterment of latency, security and reliability of cloud services, another technique is used in [11] with the help of Devices Profile for Web Services (DPWS). By using fog environment, it presents management of energy.

In [12] authors implement an approach by the name of Privacy Preserving Fog-enabled Aggregation (PPFA). Data collection of fog node from smart meters estimates aggregation of cost on the level of fog. However total cost is calculated by cloud.

Wavelet Recurrent Neural Network (WRNN) predictors are used in [13]. Cloud distributed model is stated in this paper. The purpose have achieved by using predictions.

Another technique is proposed by author in [14] by the name of Service Oriented Middle-ware (SOM) for the unification of cloud and usage of fog and Cloud of Things (CoT). Components and services that are used for accessed applications are encapsulated by smart ware city by using service oriented model.

In [15] authors implements an approach based on Game Theory (GT). In general, it deals with issues which arises in energy routing. The core purpose of the author is to reduce transmission cost. It efficiently uses surplus power.

Authors in [16,17] proposed a technique to reduce cost on supply and demand end. The delay comes in power supply, this is the limitation of these techniques. Therefore, they establish coordination between consumers and utility. Their aim is to increase profit and decrease electricity bill of the consumer.

As distribution of grids and automation is very obligatory to fix up energy problems. To deal with this matter there is a dire need of an architecture for smart metering. So, in [18–20] an infrastructure has been introduced that delivers a proper way for smart metering through which communication between smart meters and distribution grid occur. This infrastructure is hinge on solutions of cloud. Furthermore, this paper provides distinct layers. These layers are device integration, middle-ware and application layer. These are generally applications or services, which plays a major contribution in communication among different parts of the infrastructure to provide ease. These services help out to integrate devices and bidirectional communication, as well as it provides tools to design applications. Although this paper is providing a solution for present obstacles and complications but also for those issues, which are expected down the road with respect to automation and distribution grid. Interoperability arises as a problem when discussions are made about heterogeneous devices in regards of communication, integration and hardware. This paper entitles interoperability and deals with each communication model.

Monitoring in [21] is a pivot issue which needs to pay heed for user comfort, this paper assists in doing so. Independent System operators (ISO) and different transmission owners (TOs) are at the helm of monitoring this problem. Cloud computing is like a pith of the tree in the connection of running Grid-Cloud. The data that comes from different sources is stored through a data collector, which has own infrastructure. Grid operating companies provide data. State estimation is another key contribution of this paper. Different Phasor Measurement units (PMUs) are used in various points for data acquisition. These units take inputs and estimates state of the grid. After that these readings takes state estimator and give accurate output. Delivery visualization applications are even linked with it.

In [9], HRs are used to achieve the targeted goal. In obtaining desired goals, human resources are like bones in a body because all the procedure is revolving around these sources. One HR is not enough, the reason is that one HR can not handle the whole situation. The workload needs to dispense on the basis of HRs. The collection of data from environment is not needed, the reason behind it is that different algorithms and policies are used for this purpose. To issue jobs, there is need of state of the system. They use current state of system for job issuance. In this way, the scheduling of the tasks occur. The allotment of the task was a major problem which is faced by different researcher. This paper intimates a new way for the allotment of task that leans on two main element: one of them is arrival rate and second one is service time. However, allocation of task via this method is a very important contribution.

Authors in [22] are proposing a hybrid cloud computing paradigm. Development of new architecture is very essential. Dependency of development depends on newly formulated architecture. It enables users to flourish an architecture. This is a milestone in contribution. Another intelligent service to control workload is modeled in this paper. This service is factoring service. In a similar way, the pith of this service is to divide the workload into different slices. Two types of zones have been used in this paper, they namely as base zone and flash crowd zone. Both of these zones perform different tasks. Base zone and flash zone are quite different with respect to their work that facilitates in workload management. All the processes of applications run on base zone while flash crowd zone deals with transient periods. This end of the contribution also have great importance.

3 Motivation

In the present era, the demand of energy is increasing day by day. Researchers are trying to cook up this issue. Cloud and fog computing helps to improve latency, response and processing time. These techniques tackle the problems that are faced by users. The management and requirement of electricity is also a major issue which is handled by different authors. By using the infrastructure establishment of cloud and fog computing in the SG application system, the problem of load balancing is taken place. Authors use cluster system between two buildings to maintain and manage the energy requirement by distributing apartments into two categories: the surplus energy and deficit energy. Authors implement cloud and fog services to make it possible. In this paper, authors use coalition based game theory. For the improvement of latency, reliability, flexibility and processing time author presents concept of fog computing. Authors use MG to fulfill energy requirement. Author in [5] used a cluster based infrastructure scheme to mitigate energy load. Nano-grids are implemented to generate energy in buildings. After reviewing some papers, it is decided to work on cloud and fog computing to improve results specially, response and processing time.

4 Proposed System Model

In this paper, a cloud and fog based environment is introduced. First time, this concept was proposed by CISCO. The architecture is basically to extend cloud computing. The core function of fog is to reduce the load on cloud to obtain the targeted goal PHT is used. By using this PHT load balancing will be handle. The architecture is divided into three layers: first layer is comprised of controller which is attached with buildings, second layer contains fog network, third layer consists of centralized cloud as shown in Fig. 1 each layer within the framework communicates with each other, controller communicates with fogs and fogs simply contact with centralized cloud. The cloud confers services to consumers as it is co-ordinated with MG. Fog is connected with clusters of buildings. Each cluster is comprised of eighty buildings. For each cluster, there is a controller that will manage the electricity demand by a user from a smart

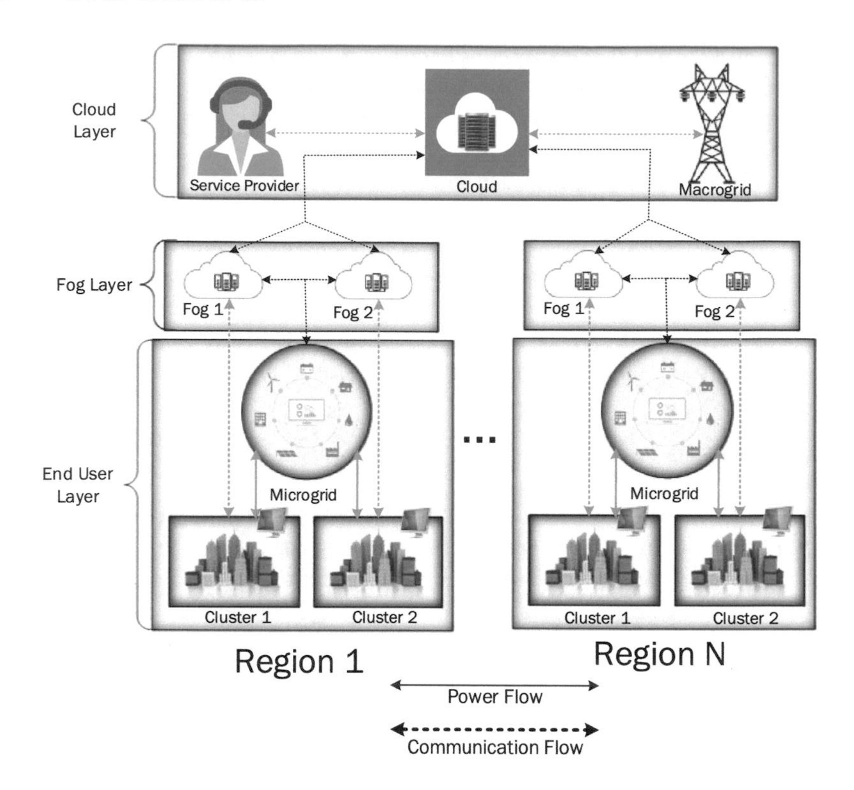

Fig. 1. System model

meter. Apartments send requests to fog for electricity through controller and fog communicates with MG. The MG responds to those apartments. If MG is not able to fulfill the demand then fog communicates with cloud to get electricity from MG. Here cloud is providing communication and storing the data coming from fog. Fogs are deployed in different regions of the world. There are six regions in the world that are comprised of six continents as shown in Table 1. Two fogs are deployed in each region. Fog forwards consumers requests and MG provides energy. If MGs are not able to fulfill requirements then through cloud energy is provided to consumers.

Table 1. Table 1 shows response time of fog

Data center	Avg (ms)	Min (ms)	Max (ms)
Fog1	4.06	0.52	48.54
Fog2	3.65	0.52	34.18
Fog3	3.77	0.53	23.65
Fog4	3.98	0.53	60.95
Fog5	3.53	0.52	44.47
Fog6	3.70	0.52	22.18

4.1 Problem Formulation

Each system, we proposed or design is comprised of some parameters. In this scenario, we have n numbers of fogs which are represented as

$$F = \{f_1, f_2, ...f_6\} \tag{1}$$

we have m numbers of requests

$$R = \{r_1, r_2, ..., r_m\} \tag{2}$$

similarly, we have l number of vm

$$VM = \{vm_1, vm_2, ..., vm_l\} \tag{3}$$

RT stands for Response Time which is obtained by calculating total subtracting delivery time and adding D time. Here D is universal constant for delay time.

$$RT = Total_{time} - RequestDelivey_{time} + D \tag{4}$$

Cost total is basically grand total which is obtained by taking sum of data transfer, MG and VM cost. The cost of MG and DT is calculated by taking their sum which is illustrated in Eq. 5. In this way, total cost is calculated.

$$Cost_{Total} = \sum_{i=1}^{n}(Cost_{DT}) + \sum_{j=1}^{n}(Cost_{MG}) + \sum_{k=1}^{n}(Cost_{VM}) \tag{5}$$

$$Cost_{VM} = \sum_{k=1}^{n}(VM_{Start} - VM_{Finish}) * VM_{Cost} \tag{6}$$

Cost Vm is total VM cost which is calculated by subtracting completion time from initial time and multiplying with per unit time cost. The sigma represents total number of VMs varying from 1 to n.
our objective function is to:
calculate response time: RT
Calculate total cost: Total cost

5 Simulations

In this paper, simulation is taken place through a tool cloud analyst. Here, six fogs are deployed in three regions, the reason behind is that there is high density in population. To manage load on cloud Pigeon Hole Technique PHT is applied. In Fig. 2 it is shown.

In each region, there are two fogs for two clusters of buildings. The fog is further connected with centralized cloud. Fogs are connected with VMs, storage and memory to manage consumers demand. We have done simulations and then compare results.

5.1 Average Response Time

First of all, both techniques RRT and PHT are applied to compare response time on proposed scheme scenario. After comparing results of both techniques (RRT and PHT), it would not be incorrect to say that PHT is showing better results than RRT. When RRT is applied on proposed scheme the average response time is 58.73 ms, minimum time 40.3 ms and maximum 124.50 ms. The simulation results are different from it, the average, minimum and maximum time is 53.43 ms, 42.09 ms and 111.48 ms. These results clearly depicts that PHT is demonstrating better results. Figure 2 shows Response Time of RRT. Table 1 demonstrates response time summary of applied technique (Fig. 3).

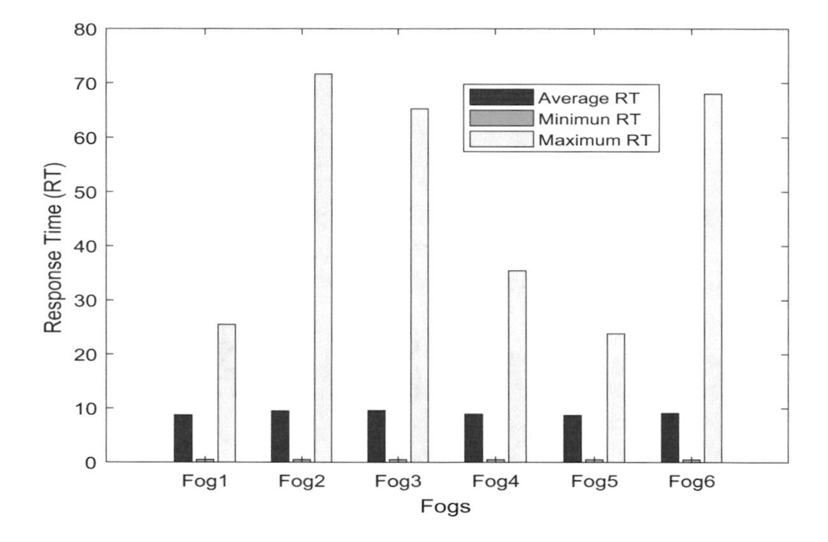

Fig. 2. Response time of RRT

5.2 Cost of Fog

Comparison between cost of fog of both techniques demonstrates minor difference. The cost of fog is near about equal to each other. The total cost of fog are included: the cost of VMs, (Micro-grids) and data transfer cost. The total cost of VMs is 2.40 dollar. Secondly, total cost of MGs is 38.44 dollar. Thirdly, the total cost of data transfer is 2.40 dollar. Lastly, the grand total is 43.24 dollar. However, total VMs cost of both techniques is 2.40 dollar. Similarly data transfer cost is same which is 2.40 dollar. The difference of the cost comes in the cost of MGs and grand total. Micro-grid cost is 38.48 and grand total is 43.28 dollar. So, difference of cost is very small which is bearable. Overall, it shows that results of PHT are better than RRT. Figures 4 and 5 which are illustrated below show comparison between them. The comparison of cost is taking place between RRT and PHT techniques, whose results are mentioned above (Table 2).

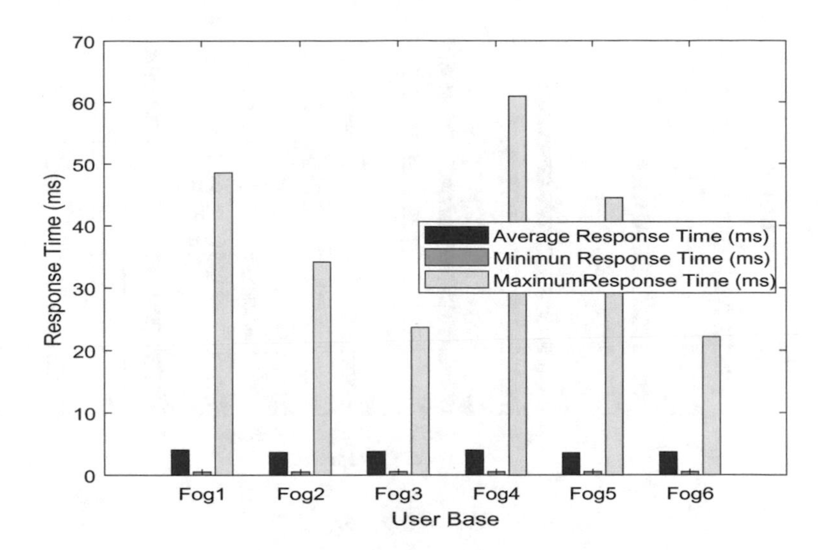

Fig. 3. Response time of PHT

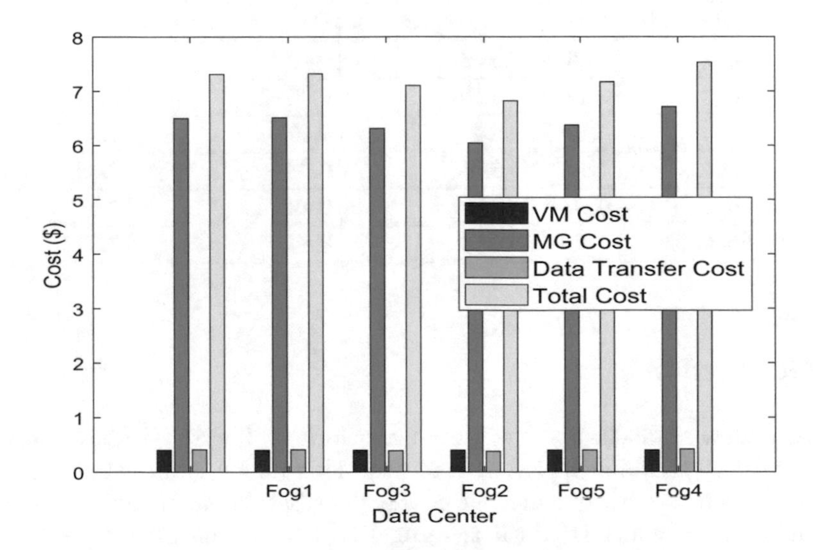

Fig. 4. Cost of fog RRT

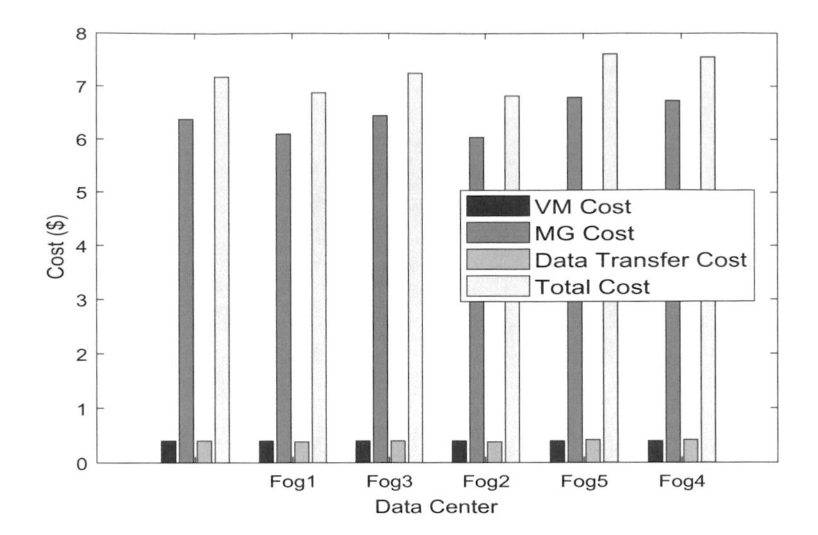

Fig. 5. Cost of fog using PHT

Table 2. Table 2 shows cost summary of fog

Data center	VM cost	Microgrid cost	Data transfer cost	Total
Fog1	0.40	6.38	0.40	7.18
Fog2	0.40	6.10	0.38	6.88
Fog3	0.40	6.45	0.40	7.25
Fog4	0.40	6.04	0.38	6.82
Fog5	0.40	6.79	0.42	7.61
Fog6	0.40	6.73	0.42	7.55

6 Conclusion

To put it in a nutshell, the system model which is illustrated above based on cloud and fog environment is implemented. This model is an extension of the model A which uses RRT to manage energy distribution among different clusters of buildings. In model B which is using PHT, load balancing is taken place by using cloud and fog. The primary goal of the approach is to manage load on cloud among three regions of the world. By using PHT, energy demand of consumers is fulfilled. Simulations are showing good results in respect of response time and processing time. This approach is providing features like flexibility and connectivity for energy management. Anyhow, results of PHT are better in terms of response time than RRT.

References

1. Zahoor, S., Javaid, N.: A Cloud-fog based Smart Grid Model for Effective Information Management
2. Alharbi, Y., Yang, K.: Optimizing jobs' completion time in cloud systems during virtual machine placement. In: 2016 3rd MEC International Conference on Big Data and Smart City (ICBDSC), pp. 1–6 (2016)
3. Kumar, S., Goudar, R.H.: Cloud computing - research issues, challenges, architecture, platforms and applications: a survey. Int. J. Future Comput. Commun. **1**(4), 356–360 (2012)
4. Chen, S., Zhang, T., Shi, W.: Fog computing. IEEE Internet Comput. **21**(2), 4–6 (2017)
5. Kumar, N., Vasilakos, A.V., Rodrigues, J.J.P.C.: A multi-tenant cloud-based DC nano grid for self-sustained smart buildings in smart cities. IEEE Commun. Mag. **55**, 14–21 (2017)
6. Nahir, A., Orda, A., Raz, D.: Replication-based load balancing. IEEE Trans. Parallel Distrib. Syst. **27**(2), 494–507 (2016)
7. Ananth, A.: Cooperative Game Theoretic Approach for Job Scheduling in Cloud Computing, pp. 147–156 (2015)
8. Pham, N.M.N., Le, V.S.: Applying ant colony system algorithm in multi-objective resource allocation for virtual services. J. Inf. Telecommun. **1**(4), 319–333 (2017)
9. Razzaghzadeh, S., Navin, A.H., Rahmani, A.M., Hosseinzadeh, M.: Probabilistic modeling to achieve load balancing in expert clouds. Ad Hoc Netw. **59**, 12–23 (2017)
10. He, D., Kumar, N., Zeadally, S., Wang, H.: Certificateless provable data possession scheme for cloud-based smart grid data management systems. IEEE Trans. Ind. Inf. **14**(3), 1232–1241 (2018)
11. Faruque, M.A.A., Vatanparvar, K.: Energy management-as-a-service over fog computing platform. IEEE Internet Things J. **3**(2), 161–169 (2016)
12. Lyu, L., Nandakumar, K., Rubinstein, B., Jin, J., Bedo, J., Palaniswami, M.: PPFA: privacy preserving fog-enabled aggregation in smart grid. IEEE Trans. Ind. Inf. **3203**(c), 1–11 (2018)
13. Capizzi, G., Sciuto, G.L., Napoli, C., Tramontana, E.: Advanced and adaptive dispatch for smart grids by means of predictive models. IEEE Trans. Smart Grid **3053**(c), 1–8 (2017)
14. Mohamed, N., Al-Jaroodi, J., Jawhar, I., Lazarova-Molnar, S., Mahmoud, S.: SmartCityWare: a service-oriented middleware for cloud and fog enabled smart city services. IEEE Access **5**(c), 17576–17588 (2017)
15. Hong, J.S., Kim, M.: Game-theory-based approach for energy routing in a smart grid network. J. Comput. Netw. Commun. **2016**, 8 (2016)
16. Mondal, A., Misra, S., Obaidat, M.S.: Storage in Smart Grid Using Game Theory, pp. 1–10 (2015)
17. Yu, M., Hong, S.H.: Supply - demand balancing for power management in smart grid: a Stackelberg game approach. Appl. Energy **164**, 702–710 (2016)
18. Pau, M., Patti, E., Barbierato, L., Estebsari, A., Pons, E., Ponci, F., Monti, A.: A cloud-based smart metering infrastructure for distribution grid services and automation. Sustain. Energ. Grids Netw. 1–12 (2017)
19. Jin, J., Gubbi, J.: An information framework for creating a smart city through internet of things. Internet Things J. **1**, 1–8 (2014)

20. Celli, G., Pegoraro, P.A., Pilo, F., Pisano, G., Sulis, S.: DMS cyber-physical simulation for assessing the impact of state estimation and communication media in smart grid operation. IEEE Trans. Power Syst. **29**(5), 2436–2446 (2014)
21. Anderson, D., Gkountouvas, T., Meng, M., Birman, K., Bose, A., Hauser, C., Litvinov, E., Luo, X., Zhang, F.: GridCloud: Infrastructure for Cloud-based Wide Area Monitoring of Bulk Electric Power Grids, vol. 3053(c) (2018)
22. Zhang, H., Jiang, G., Yoshihira, K., Chen, H.: Proactive workload management in hybrid cloud computing. IEEE Trans. Netw. Serv. Manage. **11**(1), 90–100 (2014)

A Cloud-Fog Based Smart Grid Model Using Max-Min Scheduling Algorithm for Efficient Resource Allocation

Sadia Rasheed, Nadeem Javaid(✉), Saniah Rehman, Kanza Hassan, Farkhanda Zafar, and Maria Naeem

COMSATS University, Islamabad 44000, Pakistan
nadeemjavaidqau@gmail.com
http://www.njavaid.com

Abstract. Cloud-fog infrastructure revolutionized the modern world, providing, low latency, high efficiency, better security, faster decision making, while lowering operational cost [1]. However, integration of Smart Grid (SGs) with cloud-fog platform provides high quality supply and secure generation, transmission and distribution of power; uninterrupted demand-supply chain management. In this paper, integration of SG uses cloud-fog based environment is proposed, for better resource distribution. Six fogs are considered in different geographical regions. Whereas, each fog is connected with clusters, each cluster consists of 500 smart homes. In order to fulfill energy demand of homes, fogs receive a number of requests, where different load balancing algorithms are used on Virtual Machines (VMs), in order to provide efficient Response Time (RT) and Processing Time (PT). However, in this paper, Max-Min algorithm is proposed, for load balancing with advanced service broker policy. Considering the proposed load balancing algorithm, results are compared with Round Robin (RR), from simulations, we conclude, proposed load balancing algorithms outperform than RR.

Keywords: Cloud computing · Fog computing · Smart Grid
Max-Min · Round Robin

1 Introduction

With the advancement of new technology, everything looks like operates on computing environment. For example, scientific experiments, checking weather conditions, storing and managing data which is far apart etc. The Internet has entirely changed the concept parallel computing to distributed communication environment [2]. The concept of cloud computing brings new opportunities in the worlds of distributed communication, providing storage space, efficient processing and on demand applications over the Internet with security. However, with increasing numbers of users in cloud computing platform, bring new challenges, latency and privacy. Whereas fog computing, overcome nearly all challenges of

L. Barolli et al. (Eds.): NBiS 2018, LNDECT 22, pp. 273–285, 2019.
https://doi.org/10.1007/978-3-319-98530-5_23

cloud computing, with its dynamic features. Fog operates as an intermediate layer between end users and cloud layer. However, with its dynamic environment, energy efficiency, better resource management and improved latency, fog computing meet nearly all the requirements of end users [3].

Recently, cloud-fog computing framework is the main focus of researchers. As, cloud-fog platform provides, efficiency, security, fast processing, easy to access, and huge database. Whereas, much research work has been done using this infrastructure. Using the cloud-fog platform, the researcher main focus is on energy management using cloud-fog infrastructure.

Although energy demand in residential areas has been increased from past few years [4]. According to International Energy Agency (IEA) report [5], energy used in the household sector increased by 28% between 1990 and 2007. With emerging of new technology, traditional grids are not coping up the needs of the customers. The concept of Smart Grids (SG) was proposed last few years, this requires communication technologies, in order to provide high quality supplies and secure generation, transmission and distribution of power [6]. Whereas SG helps user to meet their energy requirements.

In this paper, cloud-fog based integrated system with SG for load balancing is proposed. Whenever, energy demand from end users exceeds from providing limit, clusters sends request to SG to meet the demand of energy requirement. The SG sends request to fog to overcome the deficit energy. i.e. from end users. However, to meet the demand, fog checks the available solutions, how it fulfills the demand with minimum response. Although, each fog server comprises of a number of VMs, each VM receives a number of requests for resource allocation from end users. If fog finds any VM), with minimum response time and less load, it allocates end user requests to the VM on fog layer. However, in order to meet better Response Time (RT), efficient resource utilization and Processing Time (PT), different load balancing algorithm works on VMs In this paper, Max-Min algorithm for resource allocation is proposed, in order to achieve better resource utilization. However, if the fog is not able to find any VM on the fog layer, that can fulfill the demand, it request cloud layer, which is a service provider. Whereas, in the cloud layer, each user has its own profile, maintained by cloud, which stores data permanently. However, if cloud finds any VM, in underlying network, that is able to meet the demand, assigns that VM to the profile user. Whereas, if cloud is not able to find VM in underlying network, it sends request to neighboring clouds, to provide services for end users.

1.1 Motivation

Cloud computing has emerged as a new computing paradigm, aims to provide reliable, customized and QoS guaranteed dynamic computing environments for end-users [7]. However, with increasing numbers of customers, brings new challenges in term of latency, cost and safety. Whereas, fog computing reduces latency, improves security and provide reliability. Fog infrastructure enables users to access data quickly, more efficiently and securely.

Fog based infrastructures discussed in [8], where Electric Vehicles (EVs) uses fog infrastructure for charging and discharging. SG is next generation energy distribution system. SG offers fully observable energy distributed environment, where service providers and users, are enabled to monitor and control their billing price, production and consumption in almost real time [9]. Whereas, increasing number of users, resource management emerges as new problem. In cloud-fog based platform, many load balancing technique are discussed [10][11], where different load balancing algorithms are proposed, for efficient energy distribution among consumers. Motivated from these, max-min load balancing algorithm for SG consumers, using cloud-fog framework is proposed in this paper.

1.2 Contributions

In order to get benefit of cloud-fog platform for efficient resource allocation, SG integrated with cloud-fog framework is proposed. The contributions is paper is summed up as follows:

1. Proposed a model using Max-Min algorithm using cloud-fog framework for SG.
2. SG are integrated for efficient resource distributions.
3. Six fogs are placed under geographical regions near end users, for better response to the request of users and low latency.

The rest of the paper is organized as: Sect. 2 reviews Related Work. In Sect. 3, Proposed System Model, Load Balancing Algorithm and Service Broker Policy. Section 4 explains Experimental Results and Discussions. In the end Conclusion is drawn.

2 Related Work

During past few years, lots of work have been done on cloud-fog based infrastructure. Many authors proposed system models using cloud-fog based platform, for better performance, efficient resource utilization and minimum latency. To minimize the load of request on cloud and fog, different load balancing techniques are proposed.

The authors in [12] proposed a cloud-fog based infrastructure for resource allocation and distribution on demand of end users, with improved PT and minimizing cost. Authors in [11] proposed a model based on integration of SG with cloud-fog based platform for optimal resource allocation using First Come First Serve (FCFS) and Ant Colony Optimization (ACO) technique, improved over all response time. A new service broker policy is proposed for fog selection in [13], authors proposed a model for resource distribution on cloud-fog based infrastructure for effective energy management.

The authors compared service proximity service broker policy with an advanced service broker policy for fog section.

In [14], the author proposed a framework for data information management in SG using cloud-fog based model. The authors discuss the comparative analysis of throttled and particle swarm optimization (PSO) for information management. In [10], proposed an idea for information management in hierarchical structure of cloud-fog based framework. For this, the authors proposed a hybrid approach of ACO and ABC (HABACO), for data information management in SG. In another paper, the authors proposed a new cost-orient the model for demand side management (DSM) using cloud based Information and Communication Technology (ICT) infrastructure, helps to use resources in a cost effect way [15]. Cost oriented model, performed a comparative analysis with modified priority list algorithm and simulating annealing based algorithm.

In [16], Authors proposed Active-VM load balance algorithm, in order to distribute incoming requests among all available VMs in an efficient manner. The authors, defines task scheduling framework based on load balancing in cloud computing. Whereas, user's task scheduling needs resource utilization [17]. To increase the number of users, efficient use of computing resource is a challenge. The authors proposed a hybrid Ant colony, Honey bee with dynamic feedback (ACHBDF) for optimum resource utilization load balancing algorithm in cloud computing. Whereas, the feedback strategy is implemented to check system load after each phenomenon [18].

In [19], authors proposed scheduler based load balancing algorithm, using an artificial bee colony algorithm; however, proposed system distributes tasks on VMs. Whereas, goal is to design a well balanced load balancing algorithm for virtual machines in order to maximize throughput within time limit, with increases performance and resource efficiency. Authors in [20] wants to improve cloud computing utilization, proposing an Improved Ant Colony Optimization (IACO) algorithm. However, this model integrates an optimal value of the user, with proposed algorithm, in order to make searching optimal.

Altough Internet of Things (IoT) brings advantages in communication technology for SG; however, these implementations is not very common. Authors [21], proposed holistic framework, integrate with different components from IoT frameworks, in order to efficiently integrate smart home objects in a cloud-centric IoT based solution. In [22] Authors proposed two tier cloud based computing platform in order to optimize demand and response ratio for evaluating performance.

Authors in [23], distributes resource among all fogs, therefore, the aim was to propose a model for job scheduling and trade off between CPU execution time and memory required by fog computing, using fog infrastructure. To achieve this, the authors proposed a bio-inspired optimization Algorithm. Therefore, results show better performance and better CPU execution time. Cloud computing improves security of users' data, availability of data and performance of the system, using distributed environment. Therefore, in [24], new system is proposed in cloud computing. However, fog computing platform has several issues, such as QoS and load balancing. The authors proposed Load Distribution Algorithm using Fog infrastructure, to optimize the load balancing. However, the authors also present case study, that shows working of the proposed algorithm in comparison to Load Distribution.

Table 1. Related work survey

References	Technique	Features	Achievement	Limitations
[10]	HABAC, PSO, ABC, ACO	Information Management in SG using Cloud-Fog Platform	Better Performance with PSO	Cloud, Fog and VM Information is ignored
[11]	FCFS, ACO	Fog integration with SG for Resource Allocation	Increase Response and Processing Time	Lack of demand-supply Management in SG
[12]	PSOSA with New Service Broker Policy	Resource Allocation and distribution on Request	Better Response, Processing Time with Minimum Cost	Limited Number of Users
[13]	RR, Throttle	Cloud-Fog based Effective Energy Management	Consumer gets Fast Response with Minimum Delay	Load on VM and its performance is ignored
[14]	RR, Throttle, PSO	Data Information Management on Cloud-Fog platform	PSO Performance is Better	Limited numbers of Buildings
[15]	Modified Priority List Algorithm	Cost Effective Resource Allocation on Demand Side Management (DSM)	High Potential in Context of SG	Fixed Load

3 Proposed System Model

Cloud is a permanent data storage platform. However, it can communicate in two way directions. Cloud service providers generally have several Data Centers (DCs), scattered along geographical areas. However, the major issue of cloud infrastructure is latency. Whereas, fog platform provides best RT, works as an intermediate platform between cloud layer and end user layer. SG is digital technology, communicate with both end users and energy distributors, used to reduce energy consumption, reduces cost, and improves efficiency with the best supply chain management method. The proposed cloud-fog based system model have multiple DCs, both at cloud and fog sides respectively. The proposed cloud-fog based system model presented in Fig. 1, comprises of three layers: end user layer, fog layer and Core Cloud Layer (CCC).

End user layer comprises of 35 clusters, each cluster has 500 homes. Whereas, each cluster has its own Micro-grid (MG), helps to distribute energy among

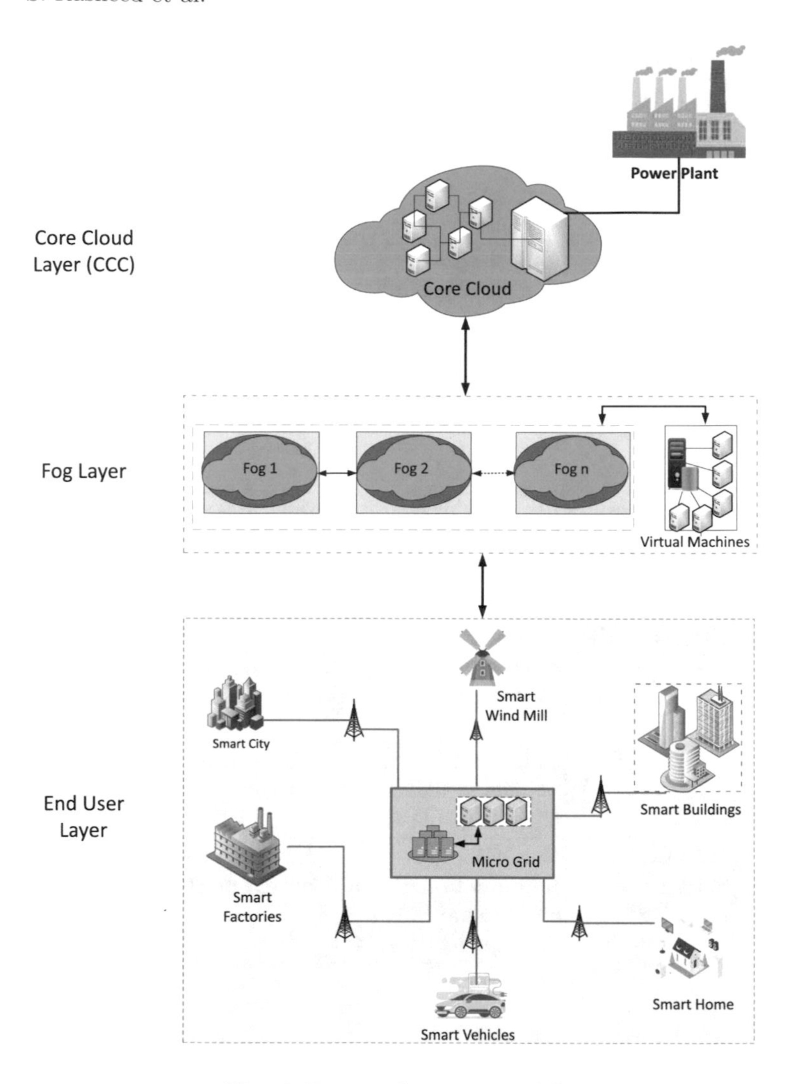

Fig. 1. Proposed system model

homes efficiently according to their requirements. Whereas each cluster can communicate with SG. Whenever, any home demand for electricity, it sends its request to MG, to overcome energy needs. MG, either respond to local energy distribution, for deficit energy, if full-fill the customer requirements, then, it supply the demand. Otherwise, it sends request to nearby fog, which may be one hop distance from the end user.

The fog layer comprises of 6 fogs, distributed among the geographical regions, one for each region. There are total 6 regions. Each fog works as an interface between end user layer of each cluster and CCC. Whereas fogs can communicate with each other to share resource information according to end user demand.

Fog comprise of a number of VMs. Whereas, VMs using a load balancing algorithm for resource allocation [25]. Whenever the fog receives a request from the end user layer, it starts locating VM that can serve the incoming request. All fog are interconnected to each other, if unable to find VM into its areas, it sends request to neighboring fog, to provide services to the incoming request.

CCC layer, consists of DCs, spread across scattered locations in geographical regions. Six clusters communicate with each other. Whenever the fog sends request for resource allocation, cloud first confirms in its geographical region, if found VM, having quick response time, it allocates the resource to VM. However, if the cloud cannot find VM, it requests nearby DC to allocate resources to the request.

3.1 Service Broker Policy

A service broker policy decides which DC provides service to the incoming request coming from end users. Thus, it defines the routing between end user layer and DCs. Whereas, in this paper Advanced Service Broker Policy [13] is used for routing.

3.2 Problem Formulation

Let we have a set of task T and virtual machines VM. Whereas VMs indicates the set of virtual machines and T are the set of task, so mathematically it is represented as in Eqs. 1 and 2:

$$T_i = \{T_1, T_2, T_3, ..., T_n\} \qquad \forall i = \{1, 2, 3, ..., n\} \tag{1}$$

$$VM_j = \{VM_1, VM_2, VM_3, ..., VM_m\} \qquad \forall j = \{1, 2, 3, ..., m\} \tag{2}$$

The set of tasks varying from 1 to n, where all assign to VMs using non-preemptive approach. The set of VMs is varying from 1 to m. Whereas all the VMs have the same capacity. The execution time represents with ET of any task is mathematically defined in Eq. 3. Where FT indicates finish time of task í ín VM j́ ánd ST represents start time of task í ín VM j́ ŕeceptively.

$$ET_{(ij)} = FT_{ij} - ST_{ij} \tag{3}$$

Sum of execution time of task T is represented in Eq. 4:

$$ET_{total} = \sum_{i=1}^{n} ET_{ij} \tag{4}$$

After calculating the total execution of tasks, Task which have maximum execution time is selected for resource allocation on VM. Equation 5 represents the VM task allocation status.

$$VMs_j = \begin{cases} 1, & \text{if VM is free.} \\ 0, & \text{otherwise.} \end{cases} \tag{5}$$

Equation 6 represents the total RT of the VM, which is subtracting task submission time from finish time of task. SUB represents task submission time to VM.

$$RT_T = FT_{ij} - SUB_{ij} \tag{6}$$

The objective of the proposed model is to minimize RT of the VMs, in order to efficiently use resources. Whereas, to increase RT of the VMs, the formula is given in Eq. 7.

$$RT_j = CT_{i,j} - AT_i + Delay \tag{7}$$

In the above equation, RT shows the response time of the VM, CT is the completion time of the task, AT is the arrival time of the assigned task and delay is the maximum delay of the VM, that occurs during task assignment and task execution. The RT is achieved by minimizing completion time and delay of the VM.

Total cost is calculated in Eq. 8.

$$TotalCost_{ij} = \sum_{i=1}^{n}\sum_{j=1}^{m}(ET_{total_{ij}} * VMcost_j) \tag{8}$$

The above equations shows that the total cost required for virtual machines are product of total time of tasks with cost of using VM resource.

3.3 Algorithm

In this paper load balancing algorithm is done with max-min load balancing algorithm. Tasks allocation is done on the basis to maximum execution time. This means that completion time of all tasks are calculated, then task having larger execution time is selected. Then task is assigned to selected VM, who's completion time is minimum among all available VMs.

Proposed Max-Min Algorithm is discussed in this section. It works as follows:

For all tasks T and Resource R, submit a task to a resource. Calculate estimate completion time of resources. Find the task having a maximum completion time from the list, and assign resources to the task. Remove the task entry from the list and look for the new resource.

Algorithm 1. Max-Min Algorithm

```
1: for all tasks 1:n do
2:     for all resources 1:m do
3:         CT = ET + ST
4:     end for
5: end for
6: Do, while the task set is not empty
7: Search for the task T maximum execution time
8: Allocate T to the resource R having least completion time
9: Delete T from the list
10: Upgrade tasks list
```

4 Experimental Results and Discussions

After performing simulation on Cloud Analyst, stimulations are performed in MATLAB. For experimental purposes 6 clusters having 6 fogs and 35 smart buildings are considered in a geographical area. Furthermore two load balancing policies Round Robin (RR) and Max-Min (MM) are considered in this paper. Results are taken on behalf of Cost, Processing Time and Response Time.

Simulations show that, proposed MM load balancing algorithms outperform than the RR algorithm. Processing time and response time of each cluster and fog are better than the RR algorithm. Simulation's results are shown below, comparing MM and RR load balancing algorithm.

Figures 2, 3 and 4 shows the hourly response time of clusters, that are taken in 24 h time slots with MM and RR load balancing algorithm.

Figure 2 shows minimum response time of each cluster respectively. Here we can see, with increase in processing time, response time of RR is more than MM on each cluster.

Figure 3 shows the average response time of each cluster respectively. Here we can see, as processing time increase, Average response time of the RR is more than MM on each cluster. As we can see it on cluster 4, it has maximum peak, reaches up to 250 ms on cluster 4.

Figure 4 shows the maximum responses time on each cluster respectively. Here we can observe, as processing time increase, maximum response time of RR increases as compared to MM on each cluster. Results show that on cluster 4, 5 and 6, it has maximum peak, crosses 900 ms peak on cluster5.

Figures 5, 6 and 7 shows the hourly processing time of each cluster, that are taken in 24 h time slots using MM and RR load balancing algorithm.

Figure 5 shows the hourly Processing time of fogs, taken in 24 h time slots, using MM and RR load balancing algorithm. The figure shows that the MM load balancing algorithm has minimum processing time as compared to RR. Figure 6 shows the hourly Processing time of fogs, taken in 24 h time slots, using MM and

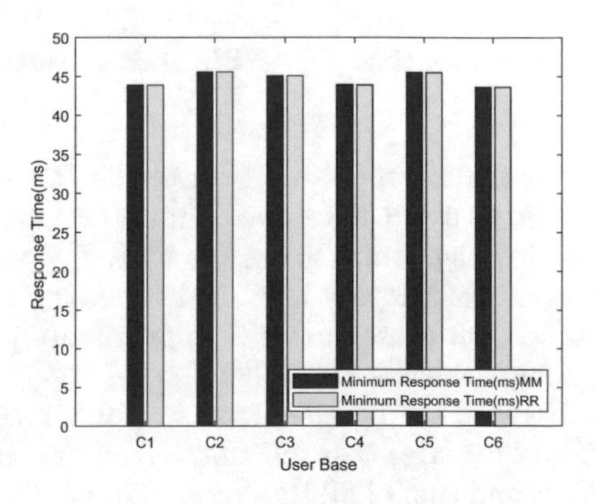

Fig. 2. Cluster average response time

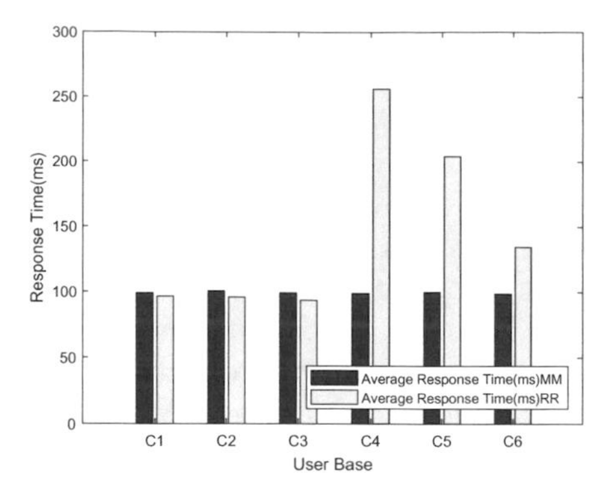

Fig. 3. Cluster minimum response time

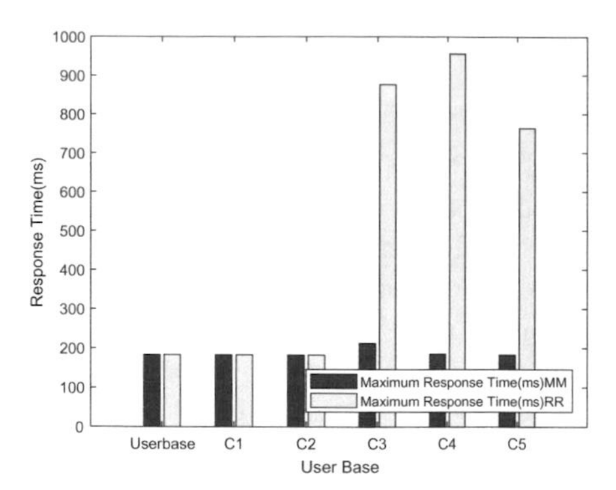

Fig. 4. Cluster maximum response time

RR load balancing algorithm. Here in figure we can see that average processing time of MM load balancing algorithm is less as compared to RR load balancing algorithm.

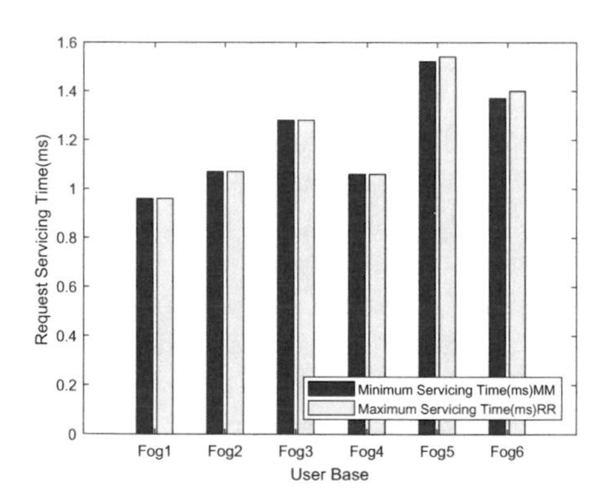

Fig. 5. Fog minimum processing time

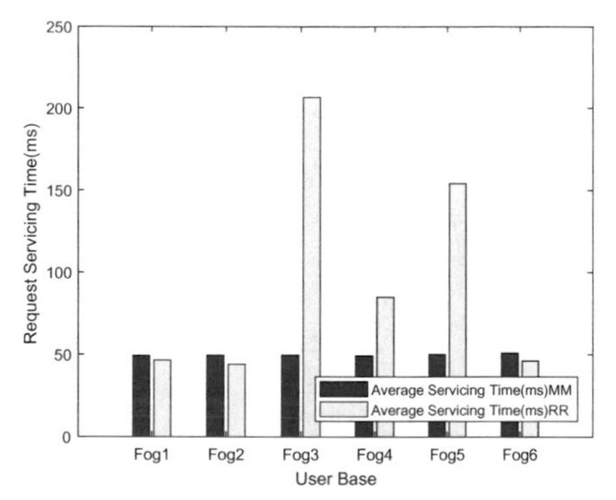

Fig. 6. Fog average processing time

Fig. 7 shows the maximum processing time of the RR and MM algorithm. Here in figure we can see that RR have maximum processing time as compared to the MM load balancing algorithm. It reaches up to 900 ms on fog 2, 3 and 5.

Figure 8 shows the cost of RR and MM load balancing algorithm. In figure, we can see the total cost on each cluster is approximately same on each fog. Maximum peak is on fog1 which is up to 250 $.

The cost of the RR and MM is depicted in Table 1 calculates from load balancing policies. The calculates cost includes Virtual Machine Cost, SG Cost and Data Transfer Cost and sum of all these cost (Grand Total). From the given simulation results, we conclude that the proposed load balancing policy performs better than other policy (Table 2).

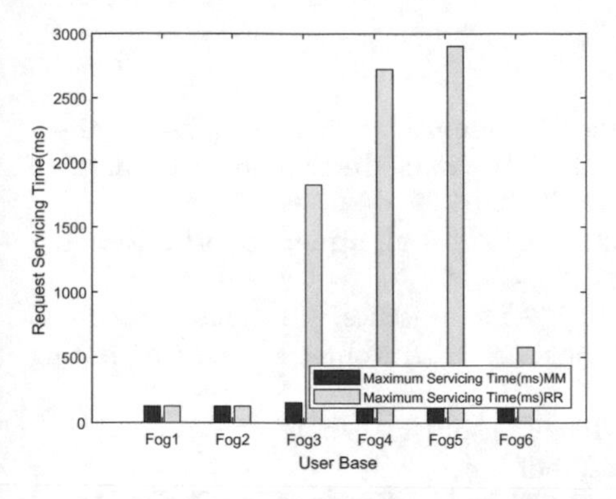

Fig. 7. Fog maximum processing time

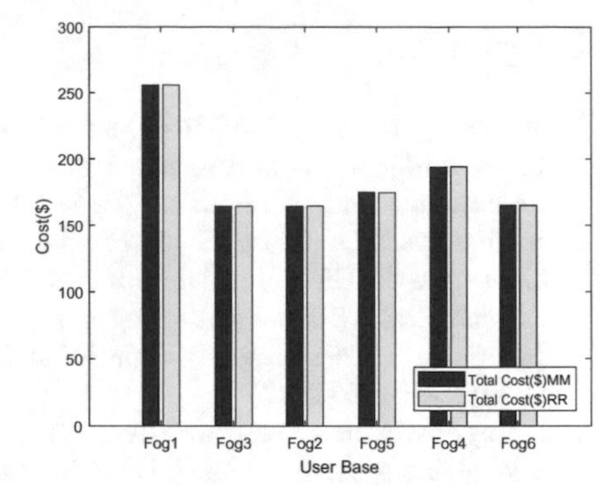

Fig. 8. Total cost

Table 2. Calculated cost

Fogs	VMCost	MicrogridCost	DataTransferCost	TotalCost
Fog1	186.62	65.12	4.07	255.82
Fog2	94.89	65.40	4.09	164.38
Fog3	94.77	65.51	4.09	164.38
Fog4	94.89	75.30	4.71	174.90
Fog5	94.77	93.34	5.83	193.94
Fog6	94.89	66.07	4.13	165.09

5 Conclusion

An integrated cloud-fog based model is proposed for efficient resource allocation of tasks. The purpose of this model is to manage energy distributions among end users, in all regions of the world. For this purpose, a novel resource allocation model is presented over cloud-fog platform. The proposed model comprises of three layers: end user layer, fog layer and core cloud layer. End user layer contains clusters of buildings, and each building has 500 apartments. And each home has a smart meter. Fog layers contain fog servers and VMs. Cloud layer contains cloud servers. The proposed algorithm in the scenario is max-min algorithm, for resource allocation on VMs, and the advanced service broker policy is used for routing. Simulations are performed on JAVA platform in eclipse. Results show that proposed Max-Min algorithm outperformed than RR, minimizing RT with minimum cost as well.

In future we plan to enhance the implementation of cloud-fog based SG, which include more features and services.

References

1. Varshney, P., Simmhan, Y.: Demystifying fog computing: characterizing architectures, applications and abstractions. In: 2017 IEEE 1st International Conference on Fog and Edge Computing (ICFEC). IEEE (2017)
2. Sadiku, M.N.O., Musa, S.M., Momoh, O.D.: Cloud computing: opportunities and challenges. IEEE Potentials **33**(1), 34–36 (2014)
3. Abbasi, B.Z., Shah, M.A.: Fog computing: security issues, solutions and robust practices. In: 2017 23rd International Conference on Automation and Computing (ICAC). IEEE (2017)
4. http://www.iiasa.ac.at/web/home/research/Flagship-Projects/GlobalEnergyAssessment/GEA_Chapter10.pdf
5. https://www.iea.org/publications/freepublications/publication/KeyWorld2017.pdf
6. Muyeen, S.M., Rahman, S.: Communication, Control and Security Challenges for the Smart Grid. Institution of Engineering and Technology (2017)
7. Chekired, D.A., Khoukhi, L.: Smart grid solution for charging and discharging services based on cloud computing scheduling. IEEE Trans. Ind. Inform. **13**(6), 3312–3321 (2017)
8. Wang, L., et al.: Cloud computing: a perspective study. New Gener. Comput. **28**(2), 137–146 (2010)
9. Fang, X.: The new and improved power grid: a survey. IEEE Commun. Surv. Tutor. Smart grid **14**(4), 944–980 (2012)
10. Zahoor, S., Javaid, N.: A Cloud-fog based Smart Grid Model for Effective Information Management
11. Fatima, I., Javaid, N.: An Efficient Utilization of Fog Computing for an Optimal Resource Allocation in IoT based Smart Grid Network
12. Yasmeen, A., Javaid, N.: Exploiting Load Balancing Algorithms for Resource Allocation in Cloud and Fog Based Infrastructures
13. Fatima, I., Javaid, N.: Integration of Cloud and Fog based Environment for Effective Resource Distribution in Smart Buildings
14. Zahoor, S., Javaid, N., Khan, A., Ruqia, B., Muhammad, F.J., Guizani, M.: A cloud-fog-Based Smart Grid Model for Efficient Resource Utilization
15. Cao, Z., et al.: Optimal cloud computing resource allocation for demand side management in smart grid. IEEE Trans. Smart Grid **8**(4), 1943–1955 (2017)
16. Domanal, S.G., Ram Mohana Reddy, G.: Optimal load balancing in cloud computing by efficient utilization of virtual machines. In: 2014 Sixth International Conference on Communication Systems and Networks (COMSNETS), pp. 1–4. IEEE (2014)
17. Fang, Y., Wang, F., Ge, J.: A task scheduling algorithm based on load balancing in cloud computing. In: International Conference on Web Information Systems and Mining. Springer, Heidelberg (2010)
18. Pawar, N., Lilhore, U.K., Agrawal, N.: A Hybrid ACHBDF Load Balancing Method for Optimum Resource Utilization in Cloud Computing (2017)
19. Rastkhadiv, F., Zamanifar, K.: Task scheduling based on load balancing using artificial bee colony in cloud computing environment. Int. J. Adv. Biotech. Res. (IJBR) **7**(5) (2016)
20. Nie, Q., Li, P.: An improved ant colony optimization algorithm for improving cloud resource utilization. In: 2016 International Conference on Cyber-Enabled Distributed Computing and Knowledge Discovery (CyberC). IEEE (2016)

21. Stojkoska, B.L.R., Trivodaliev, K.V.: A review of internet of things for smart home: challenges and solutions. J. Clean. Prod. **140**, 1454–1464 (2017)
22. Moghaddam, M.H.Y., Leon-Garcia, A., Moghaddassian, M.: On the performance of distributed and cloud-based demand response in smart grid. IEEE Trans. Smart Grid (2017)
23. Bitam, S., Zeadally, S., Mellouk, A.: Fog computing job scheduling optimization based on bees swarm. Enterp. Inf. Syst. **12**, 1–25 (2017)
24. Neto, E.C.P., Callou, G., Aires, F.: An algorithm to optimise the load distribution of fog environments. In: 2017 IEEE International Conference on Systems, Man, and Cybernetics (SMC). IEEE (2017)
25. Domanal, S.G., Ram Mohana Reddy, G.: An efficient cost optimized scheduling for spot instances in heterogeneous cloud environment. Future Gener. Comput. Syst. **84**, 11–21 (2018)
26. Wickremasinghe, B.: Cloud Analyst: A Cloud-Sim-Based Tool for Modeling And Analysis of Large Scale Cloud Computing Environments. MEDC Project (2010)

Efficient Energy Management Using Fog Computing

Muhammad KaleemUllah Khan, Nadeem Javaid(✉), Shakeeb Murtaza, Maheen Zahid, Wajahat Ali Gilani, and Muhammad Junaid Ali

COMSATS University, Islamabad 44000, Pakistan
nadeemjavaidqau@gmail.com
https://www.njavaid.com

Abstract. Smart Grid (SG) is a modern electricity network that promotes reliability, efficiency, sustainability and economic aspects of electricity services. Moreover, it plays an essential role in modern energy infrastructure. The main challenges for SG are, how can different types of front end smart devices, such as smart meters and power sources, be used efficiently and how a huge amount of data is processed from these devices. Furthermore, cloud and fog computing technology is a technology that provides computational resources on request. It is a good solution to overcome these obstacles, and it has many good features, such as cost savings, energy savings, scalability, flexibility and agility. In this paper, a cloud and fog based energy management system is proposed for the efficient energy management. This frame work provides the idea of cloud and fog computing with the SG to manage the consumers requests and energy in efficient manner. To balance load on fog and cloud a selection Base Scheduling Algorithm is used. Which assigns the tasks to VMs in efficient way.

Keywords: Energy Management Controller · Macro Grid
Micro Grid · Smart Grid · Internet of Things (IoT)
Fog computing · Cloud computing

1 Introduction

Cloud computing is a new concept in the area of information and communication technology. Cloud computing is a globally distributed network. It is the combination of memory, soft ware processing units, Physical Machines (PM) and some networks. There are many VMs in PM according to cloud infrastructure [1]. Cloud computing provides some kind of services to the consumers. Consumers get benefits from these services according to their demand. Tasks are coming from the consumer side. Cloud analyzes these task types and allocate the resources to the consumers, according to consumer demand.

Cloud computing concept firstly introduced in 2010 [2]. It is a geographical distributed network. It offers facilities that can be simply managed with less

L. Barolli et al. (Eds.): NBiS 2018, LNDECT 22, pp. 286–299, 2019.
https://doi.org/10.1007/978-3-319-98530-5_24

effort over the Internet. We can say, cloud computing provides delivery of hosted services on the Internet. It is merger of networks, storage and virtual machines to compute, process, store and sharing data. Each component of this system has its own importance. Authors in [3] work on cloud computing is based on three main services.

- Infrastructure as a Service (IaaS),
- Platform as a Service (PaaS),
- Software as a Service (SaaS).

IaaS helps to avoid the expense and complications of purchasing and managing his physical servers such as your data center infrastructure [4]. Each component has its own services. Customers can rent these services as needed. Basically, this service is based on infrastructure. In this service client has its own operating system, software and special applications.

PaaS is a complete cloud development and deployment environment [5]. Its resources allow, to provide everything from simple cloud applications to enterprise and cloud applications. Customers can purchase cloud resources from cloud providers as needed. Access these resources across the Internet. The PaaS environment, including IaaS. PaaS provides all IaaS services and some other services such as database management systems, deployment tools, business intelligence and middleware programs.

SaaS is a complete software environment that provides users with cloud-based applications. In this service system, all IaaS services are also included. Customers can purchase or rent these services through agreements or licenses within a certain period of time. The availability and security of customer data are the responsibility of the service provider. It provides different kind of services as Drop box, Slack and Docusign etc. [6].

All smart devices connected with the Internet, these devices are called Internet of Things (IoT) [7]. Nowadays, every person in this world is directly or indirectly connected to IoT. Authors in [8] say cloud users are increasing every day due to increasing no of IoT devices. These IoT devices generate large amounts of data. These numbers of users impact the efficiency of cloud. It is a big problem for cloud to compute, analyze and store this large amount of data, latency and delay also increase. It becomes a big task to handle this large amount of data.

To overcome the burden of cloud, fog computing introduced to extend cloud computing. Fog computing is a virtualized environment also called fogging or edge computing [9]. Cisco introduced fog computing. It becomes a middle layer between cloud and end user. Fog is not a centralized environment, with the merger of fog efficiency of cloud increases. Fog acts like cloud, fog provides data, storage, analyses and application services to the users. Fog computing is introduced to overcome the latency and delay. Some processes need very quick response like Smart Grid (SG), software defined network and smart light in vehicular networks. SG is an electric grid of modern era. There are many smart meters in a SG. Smart meters are used to identify the electricity usage and number of appliances in a home. The work in [10–12] has integrated cloud and fog computing with SG. In the real world scenario SG provides multiple facilities to users. When a Micro Grid (MG) is attached with this scenario, it becomes more helpful for users and utility.

We proposed cloud and fog based SG scenario. Requests are coming from consumers end to utility. When a user sends an electricity request, it makes a link between user, fog and cloud. Fog receives consumer request and fog explores the nearest MG and MG fulfill the demands of the user. In this paper users send a request to fog and fog refers this request to the MG of its region. MG analyzes the request of user, if MG is unable to fulfill the users need, this request is sent to a centralized cloud. Centralized cloud is attached with a Macro Grid. Cloud fulfill the demands of users with the help of Macro Grid. Large no of requests from the users end create a load on fog. Authors in [13] present cloud and fog based load balancing by use two techniques round robin and throttled algorithm for efficient load balancing. In this authors take six regions, each region has two fogs and two clusters of buildings.

In this paper six regions are considered, each region has two fogs and three clusters of buildings. Selection base scheduling algorithm is used for the reduction or balance the load on fog. Basic working of this algorithm is load balancing on fog side, by scheduling Virtual Machines (VM). This algorithm assigns task to VM by analyzing which VM is available or with minimum load.

1.1 Motivation

Now a day, energy becomes the basic need of the society and demand of energy increase with the passage of time. Renewable energy resources are used to overcome these demands [14]. Many authors work on cloud and fog based energy management in different scenarios. To achieve better response and processing time and efficient management. Cloud and fog computing is also a reliable and scalable environment. As [13] authors presents a cloud and fog based architecture for the distribution of resources in smart cities. Authors used one fog to manage one cluster of buildings. Author also use Round Robin Algorithm for efficient load management. Load is distributed to all virtual machines so that the RT of cloud and fog is decreased. In this paper two fogs manage three cluster of building. In this paper a selection base scheduling algorithm is used to achieve better RT and minimum cost. Rest of the paper based on: Sect. 2 is based on Literature Review, Sect. 3 is based on System Model, Sect. 4 is based on Simulations and Discussions and Sect. 5 portion of this paper is based Conclusion. References used in the paper are mentioned at the end of paper.

2 Related Work

Cloud computing is a virtualized environment including virtual machines, data centers and soft ware processing unit. Different types of processes are processed in cloud. Increasing number of users are main cause of delay and latency in processing and RT. To avoid delay and latency fog computing introduced, it shears the load of cloud. There are many algorithms used to overcome the load of cloud and fog. In this work [13] authors use Round Robin and Throttled algorithm for efficient management of load on cloud and fog. These algorithms schedule the tasks, requests and virtual machines for load balancing.

The authors of [15] wants to achieve the Quality of Service (QoS). First, for the heterogeneous cloud data-centers authors designed a QoS based energy consumption model to achieve the quality of the service. Second, authors purposed a consistent task scheduling technique to minimize the consumption of energy in a data center with the help of QoS aware Physical Machines (PM) selection method. Finally, analyze the performance of this proposed technique. Authors also work on some mathematical model to achieve the efficient response time and throughput. Response time is a time which is required to send a request to the cloud and get the response from the cloud.

In [16] authors present a technique Cloud Demand Response (CDR) to overcome the issues of Distributed Demand Response (DDR). CDR is a two layer cloud platform and DDR is a k cluster based environment. There are some issues in DDR, when a wireless communication network is used in communication channel, it becomes a cause of bit error rate. Due to this bit error rate they effect the overall demand response program. In clustered environment bit error rate is high. It requires high bandwidth. DDR depends on the iteration, and each time an update message is send among all users of the same group. After some iteration, the performance of this channel is not unique. It becomes the cause of message loss and delay. It also requires high bandwidth for the message delivery. Features of CDR is, it is an independent communication channel. It gives better cost analysis. It reduces the total communication cost, peak to average ratio, convergence time and use of bandwidth. If user demand high efficient response time it may cause high cost.

Author in [17] presents a distributed algorithm to respond to demands which are made from consumer end for real time. Each utility company and local users solve the sub-problems of the allocation process. For supply and demand balance, each competitive company evaluates clearing price.

In [18] authors want to achieve some goals. Author presents Home Energy Management (HEM). To implement power management system, a platform is needed, which offers interactive interoperability between devices and process elasticity. HEM is applied over a network platform to meet these requirements at lowest cost. However, the operating areas of the system are: scalability, heterogeneity and delay sensitive devices. Cost element not considered properly.

Authors in [19] present a fog base model for SG management. This system is divided in to three layers, first layer is based on smart grids, second layer is based on fog layer and the third layer is based on a centralized cloud layer. This system makes SG more reliable. It is a geographical distributed system. It provides locality and reliability to smart grid. This fog system increases the efficiency of cloud based SG system. This system increases privacy and reduce latency. This system is also beneficial in terms of control energy flow and balance energy load.

In this work [20] authors present a fog based energy management system. The term fog is introduced to divide the load of cloud. Peoples are directly or indirectly connected to the Internet through IoT. These IoT produce large amount of data. This large amount of data, create a problem for cloud. Author introduced fog in his system to overcome the delay, processing time, efficient sharing and management of resources and load balancing. Energy is managed on boat user and utility side. User side Energy management is called HEM and the utility side management is called Micro grid Level Management (MLM). In HEM energy is further managed in two different ways, HVAC controller and EV charging controller. Smart transfer controller manages MLM energy. Energy is managed on two sides. This two way energy management, reduces the setup cost, process computation time and power usage.

In this paper [21] authors proposed a bio inspired algorithm which is working on multi rumen. Tasks are allocated to related or suitable VM through anti grazing principle. In multi data center environment each data center is called one rumen. Rumens are the VMs, these VMs are used to complete the tasks. Authors design an algorithm for multi data centers load and task balancing in the cloud. The authors also analyze the some load balancing techniques through Task Completion Time (TCT) and Task Execution and Completion Time (TECT). Task completion time is a time required, when a virtual machine is assigned to a task and meet the task requirements of the task. TECT is a time required, when a task is coming from the consumer end including network delay and task completion time. The authors also analyze the randomly created datasets with this proposed algorithm. The authors also analyze the performance of this proposed technique with some existing techniques.

In above related work authors work on load balancing on cloud and fog through different techniques and algorithms. Some authors work on load balancing of cloud, some authors work on efficient management and allocation of VMs. Authors introduced some service broker polices for the efficient management of cloud resources. We are also working load balancing on cloud through efficient allocation of VMs to the consumers requests.

3 Proposed Model

This proposed system is based on three layer architecture, first layer is based on EMC (Energy Management Controller) on buildings and this EMC is connected with MG, second layer is based on fog and third layer is based on centralized cloud network. This centralized cloud network is connected with macro grid and service provider. Fog becomes the middle layer between cluster and cloud layer to operate the requests of cloud and reduce the burden on cloud. In other words EMC communicates with fog network and fog network communicates with cloud. Each cluster is comprised of 60 to 120 buildings and each building is comprised of 100 to 200 flats. Each cluster of the buildings has EMC to manage the requests of the consumers and respond to the requests. The demand of electricity of each flat request to EMC, then the EMC communicates with fog and fog refers this electricity request to MG. After this MG respond to the request and provides required electricity to the particular flat. When MG is not capable to meet the requirement of the customers demand, then fog communicates with cloud platform to deliver the closer Macro Grid facility to the particular consumer. Fogs are located in various regions of the world and the world is divided into six major regions and these regions consist of six continents as elaborated in Table 3 (Fig. 1).

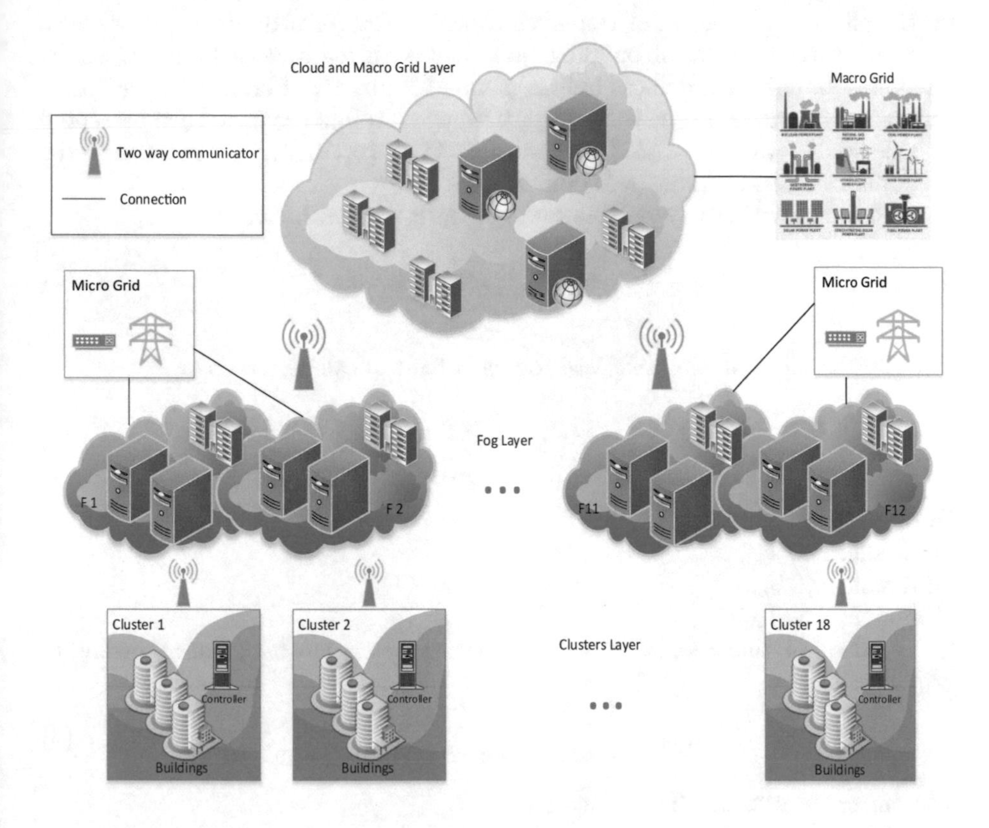

Fig. 1. Proposed system model

Two fogs are located in each region and these fogs are capable to entertain needs of the three clusters located within the same region. The MGs are attached withe the fogs, Consumers request to the fog to fulfill the energy need. Fogs are capable to forward the requirement of energy according the needs of the consumer. Consumers are not capable to deliver their request to MG directly to meet their energy requirement. The fog send consumer's requests to MG to fulfill their needs. The MG sends energy to meet the requirement of the consumer. When the consumer request to fog for the energy. The fog now choose the MG that requires power to meet the consumer demands or the fog will release a request to the neighboring MG. When a high demand sends to all MG they will not be able to meet the requirements, then the fog will request to the cloud, and cloud will fulfill the consumers demand with the help of Macro Grid.

3.1 Problem Formulation

This system is based on cloud and fog based Smart Grid (SG) model. Improving the efficiency of the system is the main task. Efficiency of the system is improved by the efficient allocation of resources to the tasks. According to this scenario there are three clusters of buildings attached with fog and each cluster has N number of buildings and these buildings have M number of homes. These homes generate requests and sent to the fog. Here the available requests are presented by this set $T = \{t_1, t_2, ..., t_n\}$ on each fog. $V = \{v_1, v_2, ..., v_m\}$ represents the available virtual Machines (VM) on each fog.

The total number of available tasks on each fog can be written as.

$$T\,total = \sum_{i=1}^{n} T_i \tag{1}$$

The total number of available VMs on each fog are calculated as.

$$V\,total = \sum_{j=1}^{m} V_j \tag{2}$$

Our objective function:
Minimum: PT_{total}
Minimum: RT_{total}
Minimum: $Total_c$.
If the value is 1 then assign the task otherwise task is not assigned according to Eq. 3.

$$T_{assing} = \begin{cases} 1; & \text{Task assing} \\ otherwise; & \text{Task not assign} \end{cases} \tag{3}$$

VM current status will be checked as.

$$V_{assign} = \begin{cases} 0, & \text{if the V is free} \\ 1, & \text{if the V is not free} \end{cases} \tag{4}$$

Total number of tasks divided by total available number of VM to get the processing time according to Eq. 5.

$$PT\ i, j = \frac{Ttotal}{Vtotal} \tag{5}$$

The total processing time formally represented as.

$$PT_{total} = \sum_{i=1}^{n} \sum_{j=1}^{m} (PT_{average} * V_{assign}) \tag{6}$$

Response time is obtain from Eq. 7 by subtracting the Arrival time of a task through the sum of task delay time and task finish time. In this equation time is taken in seconds s represents the seconds.

$$RT = Delay(s) + Finish(s) - Arrival(s) \tag{7}$$

Equation 8 gives the total cost of the proposed system model. Here c represents the cost. Sum of data-center cost, total VM cost and total MG cost represents total cost of the system.

$$Total_c = Datacenter_c + TotalV_c + TotalMG_c \tag{8}$$

4 Simulations

4.1 Setup and Parameters

Simulations of the scenario are completed by using a tool Cloud Analyst and the system specification is given below in Table 1. Six regions of the world are considered in this paper. Each region has two fogs and three building clusters. Fogs are directly connected to a central cloud system and clusters and fogs of a region are connected with each other. When a consumer of cluster building sends his request of electricity to fog. At that time, many other consumers also send their requests to fogs to fulfill the electricity demands. Huge number of requests are the problem for fog. As we know fog is based on virtual machines and virtual memory. All requests of the consumers are processed by the virtual machines. When the number of requests are very high. It is a difficult task for fog to decide which virtual machine is assigned to which request. To handle this situation we implement selection base scheduling to reduce the load of fog with the help of assigning virtual machines.

Table 1. System specification

Components	Power
Processor	i5-4200u CPU @ 1.60 GHz 2.30 Ghz
RAM	4.0 GB
Hard Drive 3	500 GB
Graphic Card	4.0 Gb

Table 2 shows the division of regions. In each region two fogs and three clusters are placed. Showing the regions one to six respectively and fogs 1 to 12 respectively, with the 1 to 18 clusters. According to this Table 2 fogs manage 3 clusters of buildings. Each cluster is connected with each fog with in a region. If one fog is busy with the other clusters requests, so, cluster will communicate with the other fog.

Table 2. Regions division

Regions	Fogs	Clusters
Region 1	F-1 F-2	C-1 C-2 C-3
Region 2	F-3 F-4	C-4 C-5 C-6
Region 3	F-5 F-6	C-7 C-8 C-9
Region 4	F-7 F-8	C-10 C-11 C-12
Region 5	F-9 F-10	C-13 C-14 C-15
Region 6	F-11 F-12	C-16 C-17 C-18

Table 3 shows the minimum RT, average RT and maximum RT of each fog. This table also show which fog requires maximum, minimum and average response time. According to this table fog 10 has maximum response time. Fog 1 has minimum average response time. This table shows how much time is required by fog to respond the consumer's request.

Table 3. Fogs performance

Fog	Average RT (ms)	Minimum RT (ms)	Maximum RT (ms)
Fog1	1.06	0.02	2.51
Fog2	1.58	0.02	4.06
Fog3	1.61	0.03	3.89
Fog4	1.63	0.03	3.99
Fog5	1.58	0.02	3.52
Fog6	1.58	0.02	4.05
Fog7	1.58	0.02	4.15
Fog8	1.56	0.02	4.08
Fog9	1.63	0.04	4.23
Fog10	1.60	0.04	4.27
Fog11	1.55	0.02	4.06
Fog12	1.58	0.02	4.07

The time required to connect is called RT. Figure 2 is a graphical representation of 12 fogs RT. That shows the minimum RT, average RT and maximum RT of each fog. This figure also show which fog requires maximum, minimum and average response time. According to this figure fog 10 has maximum response time. Fog 1 has minimum average response time.

Table 4 shows the minimum RT, average RT and maximum RT of each cluster. According to this table cluster 7 has maximum response time to the users

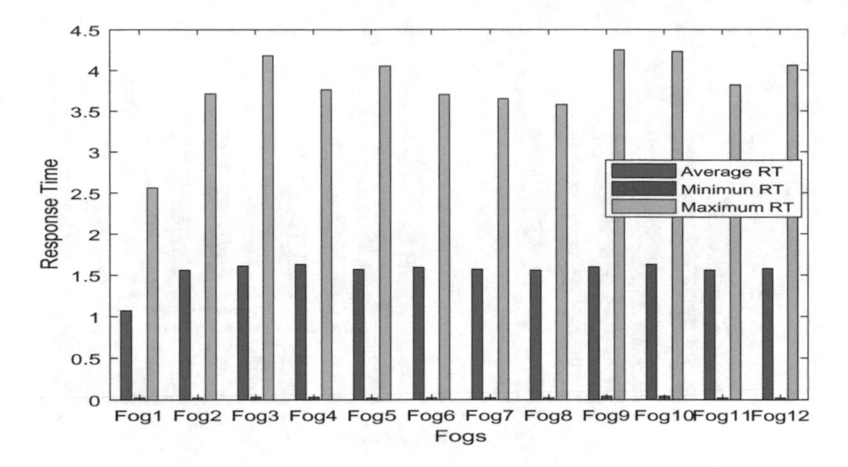

Fig. 2. RT of each fog

Table 4. Clusters performance

Cluster	Average RT (ms)	Minimum RT (ms)	Maximum RT (ms)
Cluster 1	50.82	38.95	61.32
Cluster 2	50.90	39.08	63.88
Cluster 3	51.06	39.03	63.36
Cluster 4	51.18	38.65	61.81
Cluster 5	51.34	38.57	63.76
Cluster 6	51.36	38.44	64.06
Cluster 7	51.27	37.57	67.85
Cluster 8	51.23	40.16	63.34
Cluster 9	51.37	39.13	64.83
Cluster 10	51.21	36.89	62.88
Cluster 11	51.11	41.12	65.04
Cluster 12	51.28	39.85	64.92
Cluster 13	51.31	37.72	66.89
Cluster 14	51.42	41.57	65.80
Cluster 15	51.52	40.48	64.68
Cluster 16	51.30	39.00	65.86
Cluster 17	51.22	39.37	61.62
Cluster 18	51.25	39.63	65.38

requests. According to this table cluster 1 has minimum average response time. This table shows how much time is required by cluster to respond the consumer's request.

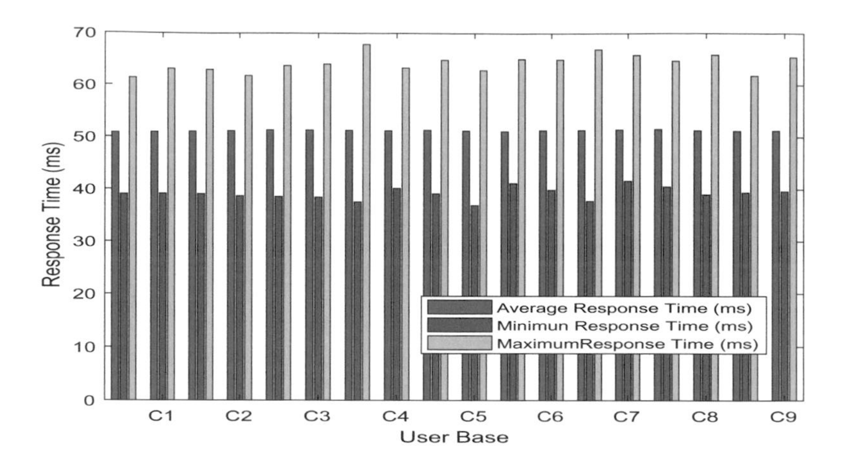

Fig. 3. RT of each cluster

Figure 3 is a graphical representation of 18 cluster of buildings RT of. The response time of each fog is also represented in this figure. That shows the minimum RT, average RT and maximum RT of each cluster. According to this figure cluster 7 has maximum response time to the users requests. This figure shows cluster 1 has minimum average response time. This figure shows how much time is required by cluster to respond the consumer's request. It also shows that which fog gives better RT to consumer's request.

Table 5. Cost of each fog

Data center	VM cost	Microgrid cost	Data transfer cost	Total cost
Fog1	16.80	196.74	12.30	225.83
Fog2	16.80	195.66	12.23	224.69
Fog3	16.80	270.14	16.88	303.82
Fog4	16.80	293.51	18.34	328.65
Fog5	16.80	231.09	14.44	262.33
Fog6	16.80	240.53	15.03	272.36
Fog7	16.80	258.27	16.14	291.21
Fog8	16.80	247.83	15.49	280.12
Fog9	16.80	277.25	17.33	311.37
Fog10	16.80	269.82	16.86	303.49
Fog11	16.80	222.70	13.92	253.42
Fog12	16.80	227.91	14.24	258.95

Table 5 shows the required cost of each data center which is presented on a fog. According to this table all fogs have same VMs cost. Fog 2 has minimum

MG, data transfer and total cost. Fog 4 has maximum MG, data transfer and total cost. In this table cost of VM, MG, data transfer cost and total cost which is required to create a data center on a fog.

Figure 4 is a graphical representation of cost of each fog, showing that which fog gives the minimum cost. This fig shows the required cost of each data center which is presented on a fog. According to this figure all fogs have same VMs cost. Fog 2 has minimum MG, data transfer and total cost. Figure 4 shows fog 4 has maximum MG, data transfer and total cost. In this table cost of VM, MG, data transfer cost and total cost which is required to create a data center on a fog.

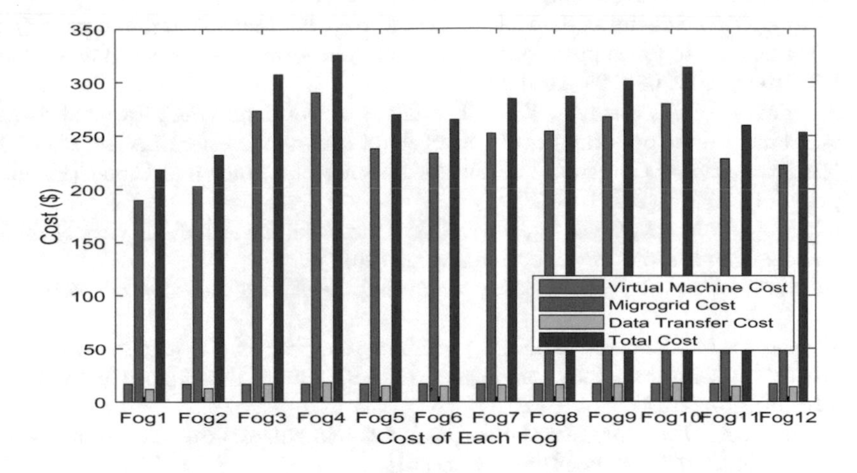

Fig. 4. Cost of each fog

5 Conclusion

Proposed system model is based on cloud and fog is presented to manage the huge number of requests from the consumers on fog. For this we used six regions of the world. The goal of the technique is to mitigate load on cloud. An EMC is used in the cluster of buildings for the efficient management of the energy. This EMC tells the demand of energy of each cluster to the fog. The Large Numbers of requests form the electricity consumers creates a load on the fog and cloud to over come this situation a selection based scheduling algorithm is used. This technique is used for efficient VMs allocations to the user's request and balance the load on fog and cloud. After applying this selection base scheduling algorithm on the proposed system model minimum RT, minimum request PT and minimum cost of the proposed system is achieved in short selection base scheduling algorithm is RT, PT and cost efficient. The implementation of proposed model also provides some features which are interoperability, connectivity and flexibility. Java platform is used to perform simulations in eclipse platform.

References

1. Hyser, C.D., McKee, B.A.: Starting up at least one virtual machine in a physical machine by a load balancer. US Patent 8,185,893, 22 May 2012
2. Armbrust, M., Fox, A., Griffith, R., Joseph, A.D., Katz, R., Konwinski, A., Lee, G., Patterson, D., Rabkin, A., Stoica, I., et al.: A view of cloud computing. Commun. ACM **53**(4), 50–58 (2010)
3. Kumar, S., Goudar, R.H.: Cloud computing - research issues, challenges, architecture, platforms and applications: a survey. Int. J. Future Comput. Commun. **1**(4), 356–360 (2012)
4. Sharma, P., Lee, S., Guo, T., Irwin, D., Shenoy, P.: Managing risk in a derivative IaaS cloud. IEEE Trans. Parallel Distrib. Syst. (2017)
5. Sharma, V.S., Sengupta, S., Mohamedrasheed, A.K.: Method and system for managing user state for applications deployed on platform as a service (PaaS) clouds. US Patent 9,635,088, 25 April 2017
6. Stavrinides, G.L., Karatza, H.D.: The effect of workload computational demand variability on the performance of a SaaS cloud with a multi-tier SLA. In: 2017 IEEE 5th International Conference on Future Internet of Things and Cloud (FiCloud), pp. 10–17. IEEE (2017)
7. Khan, A., Pohl, M., Bosse, S., Hart, S.W., Turowski, K.: A holistic view of the IoT process from sensors to the business value (2017)
8. Chen, S., Zhang, T., Shi, W.: Fog computing. IEEE Internet Comput. **21**(2), 4–6 (2017)
9. Vaquero, L.M., Rodero-Merino, L.: Finding your way in the fog: towards a comprehensive definition of fog computing. ACM SIGCOMM Comput. Commun. Rev. **44**(5), 27–32 (2014)
10. Yasmeen, A., Javaid, N.: Exploiting load balancing algorithms for resource allocation in cloud and fog based infrastructures
11. Javaid, S., Javaid, N., Aslam, S., Munir, K., Alam, M.: A cloud to fog to consumer based framework for intelligent resource allocation in smart buildings, pp. 1–10
12. Zahoor, S., Javaid, N.: A cloud-fog based smart grid model for effective information management
13. Fatima, I., Javaid, N.: Integration of cloud and fog based environment for effective resource distribution in smart buildings
14. Byun, J., Kim, Y., Hwang, Z., Park, S: An intelligent cloud-based energy management system using machine to machine communications in future energy environments. In: Digest of Technical Papers - IEEE International Conference on Consumer Electronics, pp. 664–665 (2012)
15. Xue, S., Zhang, Y., Xu, X., Xing, G., Xiang, H., Ji, S.: QET: a QoS-based energy-aware task scheduling method in cloud environment. Cluster Comput. **20**(4), 3199–3212 (2017)
16. Moghaddam, M.H.Y., Leon-Garcia, A., Moghaddassian, M: On the performance of distributed and cloud-based demand response in smart grid. IEEE Trans. Smart Grid **PP**(99) (2017)
17. Deng, R., Yang, Z., Hou, F., Chow, M.-Y., Chen, J.: Distributed real-time demand response in multiseller-multibuyer smart distribution grid. IEEE Trans. Power Syst. **30**(5), 2364–2374 (2015)
18. Vatanparvar, K., Chau, Q., Al Faruque, M.A.: Home energy management as a service over networking platforms. In: 2015 IEEE Power & Energy Society Innovative Smart Grid Technologies Conference (ISGT), pp. 1–5. IEEE, February 2015

19. Okay, F.Y., Ozdemir, S.: A fog computing based smart grid model. In: 2016 International Symposium on Networks, Computers and Communications (ISNCC) (2016)
20. Al Faruque, M.A., Vatanparvar, K.: Energy management-as-a-service over fog computing platform. IEEE Internet Things J. **3**(2), 161–169 (2016)
21. Sharma, S.C.M., Rath, A.K.: Multi-Rumen Anti-Grazing approach of load balancing in cloud network. Int. J. Inf. Technol. **9**(2), 129–138 (2017)

A New Generation Wide Area Road Surface State Information Platform Based on Crowd Sensing and V2X Technologies

Yoshitaka Shibata[1(✉)], Goshi Sato[2], and Noriki Uchida[3]

[1] Regional Corporate Research Center, Iwate Prefectural University, 152-52 Sugo, Takizawa, Iwate, Japan
shibata@iwate-pu.ac.jp
[2] National Institute of Information and Communications Technology, 2-1-3 Katahira, Aoba-ku, Sendai, Miyagi, Japan
sato_g@nict.go.jp
[3] Department of Information and Communication Engineering, Fukuoka Institute of Technology, 3-30-1 Wajiro-higashi, Higashi-ku, Fukuoka, Japan
n-uchida@fit.ac.jp

Abstract. In this paper, in order to keep safe and secure driving, a new generation wide area road surface state information platform based on crowd sensing and V2X Technologies is introduced. In crowd sensing, various environmental sensors including accelerator, gyro sensor, infrared temperature sensor, quasi electrical static sensor, camera and GPS are integrated to precisely detect the various road surface states and determine the dangerous locations on GIS. Those road information are transmitted the neighbor vehicles and road side server in realtime using V2X communication network. In V2X communication on the actual road, both the length of communication distance and the total size of data transmission must be maximized at the same time when vehicle are running on the road. The conventional single wireless communication such as Wi-Fi, IEEE802.11p, LPWA, cannot satisfy those conditions at the same time. In order to resolve such problems, N-wavelength cognitive wireless communication method is newly introduced in our research. Multiple next generation wireless LANS including IEEE802.11ac/ad/ah/ in addition to the current popular LANs with different wavelengths are integrated to organize a cognitive wireless communication. The best link of the cognitive wireless is determined by SDN. Driver can receive the road surface status information from the vehicle in opposite direction or road side server and eventually pay attentions to his/her driving before encountering the danger location. This technology can also apply for automatic driving car.

1 Introduction

Mobility is the most important means for economic activity being to carry persons, loads, feeds and other materials around world. Huge number of vehicles are being produced day by day and their qualities are developed year and year in the developed countries. Furthermore, self-driving cars have been produced and improved and

L. Barolli et al. (Eds.): NBiS 2018, LNDECT 22, pp. 300–308, 2019.
https://doi.org/10.1007/978-3-319-98530-5_25

running on public roads. However, on the other hand, the road infrastructures in any countries are not improved compared with vehicles. In particular, the road conditions in developing countries are so bad and dangerous due to luck of road maintenance, falling objects from other vehicles, overloaded trucks.

Second, in the cold or snow countries, most of the road surfaces are occupied with heavy snow and iced in winter. In both cases, traffic accidents are rapidly increased. Therefore, more safe and reliable road monitoring system for both drivers and self-driving cars. Furthermore, the information and communication environment along the roads in local areas are challenged and not well developed compared with urban areas. Once a traffic accidents or disaster occurred, information collection, transmission and sharing are delayed or even cannot be made. Eventually the resident's lives and reliabilities cannot be maintained. More robust and resilient information infrastructure and proper and quick information services with road environmental conditions are indispensable (Fig. 1).

Fig. 1. Road condition in various locations

On the other hand, as progress of sensor, image recognition and 3D GIS technologies, autonomous driving car technology have been developing in the world. According to the Japanese government press announcement, the autonomous driving cars will take part in many places at Tokyo Olympic game in 2020. However, since the current autonomous driving cars are developed based on ideal road conditions such as flat lined roads without obstacles on ideal weather condition, it is very difficult to run those autonomous cars on the most typical roads which are always exposed with rains, snowy, icy, rainy, or foggy weather conditions [1]. Even for well experienced human drivers, driving on the dangerous surface road conditions such as snow, icy, rainy, or foggy weather condition always encounter serious traffic accidents. In those cases, more and reliable information with road conditions have to be delivered to the driver to safely control the car.

In order to resolve those problem, so far we introduced in order to keep safe and secure driving, a new generation wide area road surface state information platform based on crowd sensing and V2X technologies is introduced. In crowd sensing, various environmental sensors including accelerator, gyro sensor, infrared temperature sensor, quasi electrical static sensor, camera and GPS are integrated to precisely detect the various road surface states and determine the dangerous locations on GIS. Those road information are transmitted the neighbor vehicles and road side server in realtime using V2X communication network. In V2X communication on the actual road, both the length of communication distance and the total size of data transmission must be maximized at the same time when vehicle are running on the road. The conventional single wireless communication such as Wi-Fi, IEEE802.11p, LPWA, cannot satisfy those conditions at the same time. In order to resolve such problems, N-wavelength cognitive wireless communication method is newly introduced in our research. Multiple next generation wireless LANS including IEEE802.11ac/ad/ah/ai/ay in addition to the current popular LANs with different wavelengths are integrated to organize a cognitive wireless communication. The best link of the cognitive wireless is determined by SDN.

In our system, road side wireless nodes and mobile smart nodes with various sensors and different wireless communication devices organize a large scale information infrastructure without conventional wired network such as Internet. The mobile smart node on the car collects various sensor data including acceleration, temperature, humidity and frozen sensor data as well as GPS data and carries and exchanges to other smart node as message ferry while moving from one end to another along the roads.

On other hand, the road side wireless node not only collects and stores sensor data from its own sensors in its database server but exchanges the sensor data from the mobile nodes when it passes through the road side wireless node by V2X protocol. Therefore, both sensor data at the road side wireless node and the mobile nodes are periodically uploaded to cloud system and synchronized. Thus, mobile node perform as mobile communication means even through the communication infrastructure is challenged environment or not prepared.

This network not only performs various road sensor data collection and transmission functions, but also performs Internet access network functions to transmit the various data, such as sightseeing information, disaster prevention information and shopping and so on as ordinal public wide area network for residents. Therefore, many applications and services can be realized.

2 System Configuration

The Fig. 2 shows a system configuration of our proposed a new generation wide area road surface state information platform which consists of multiple road side wireless nodes and mobile nodes. The road side wireless node is organized by different typed wireless network devices and various sensor devices such as semi-electrostatic field sensor, acceleration sensor, gyro sensor, temperature sensor, humidity sensor, infrared sensor and sensor server.

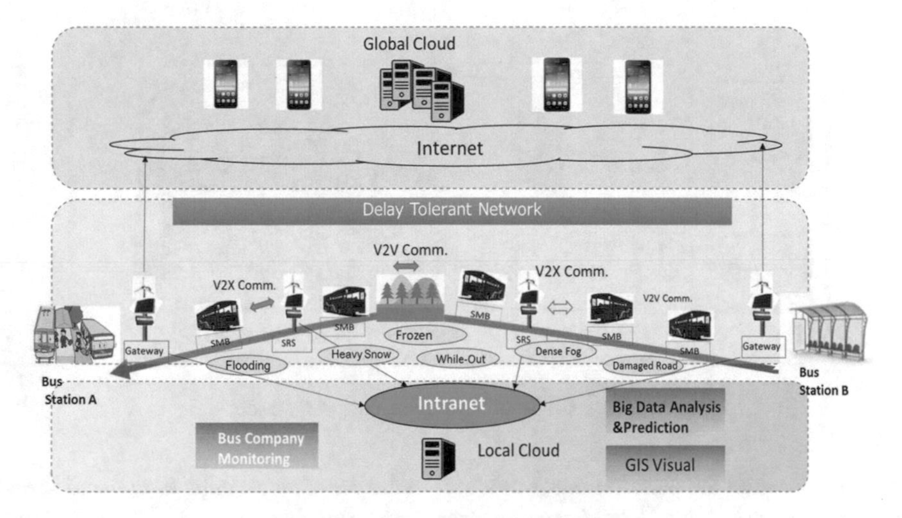

Fig. 2. System configuration

Figure 3 shows a mobile vehicle sensor wireless node which performs crowd sensing road condition functions on vehicle, communication function, storage function of the sensing data and exchange function of the sensor data from the road side node. The mobile vehicle sensor wireless node integrates multiple wireless network devices with different N-wavelength (different frequency bands) wireless networks such as IEEE802.11n (2.4 GHz), IEEE802.11ac (5.6 GHz), IEEE802.11ah (920 MHz) and organizes a cognitive wireless node. The network node selects the best link of cognitive wireless network depending on the observed network quality by Software Defined Network (SDN). If none of link connection is existed, those sensing data are locally and temporally stored until approaches to another mobile node or road side node, and starts to transmit sensor data by DTN Protocol. Thus, data communication can be attained even though the network infrastructure is not existed in challenged network environment such as mountain areas or just after large scale disaster areas.

3 Crowd Sensing

In order to detect the precise road surface conditions, such as dry, wet, dumpy, showy, frozen roads, various sensing devices including accelerator, gyro sensor, infrared temperature sensor, humidity sensor, quasi electrical static sensor, camera and GPS are integrated to precisely and quantitatively detect the various road surface states and determine the dangerous locations on GIS in sensor server. This decision process is shown in Fig. 4. The sensor data from various sensors are periodically sampled and filtered to reduce noise in Pre-filtering module, analyzed to estimate the mutual relation,

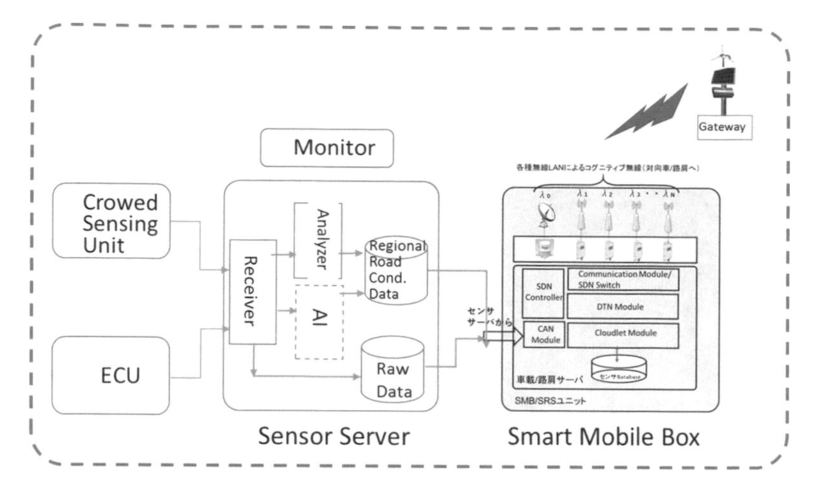

Fig. 3. Mobile vehicle sensor wireless node

then sent through the decision make module. In the decision module, the related sensor data are processed by Deep Learning processing function as shown in Fig. 5 and eventually, the correct road surface state can be decided.

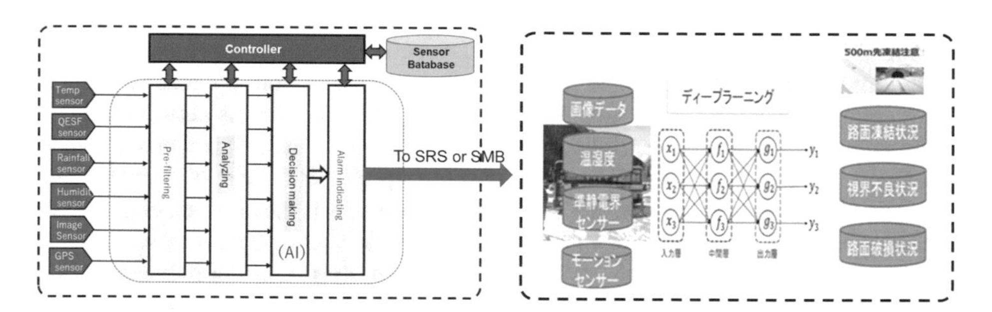

Fig. 4. Crowd sensing or system

Fig. 5. Deep learning for decision

4 V2X Communication

In Fig. 6 shows V2X communication method between the SMB of a vehicle and the SRS of road side server. One of the wireless network with the longest communication distance can first make connection link between SMB and SRS using SDN function. Through the this connection link, the communication control data of other wireless networks such as UUID, security key, password, authentication, IP address, TCP port number, socket No. are exchanged. As approaching each other, the second wireless network among the cognitive network can be connected in a short time and actual data transmission can be immediately started. This transmission process can be repeated during crossing each other as long as the longest communication link is connected.

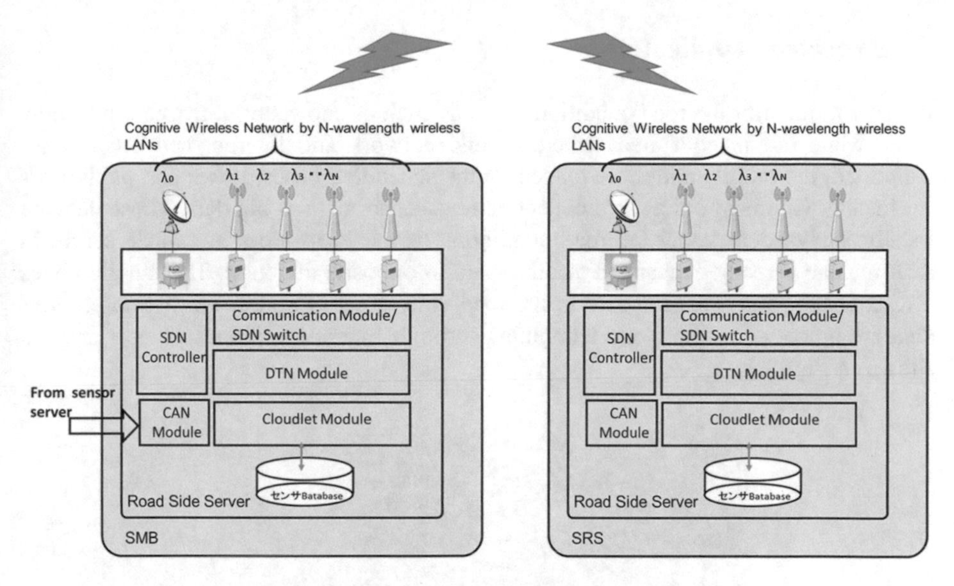

Fig. 6. V2X communication method between the SMB

This communication process between SMB and SRS is the same as the communication between SMB to other SMB except for using adhoc mode.

On the other hand, the V2x communication between vehicle and the global cloud server on Internet shows in Fig. 7. The sensor data from SMB is transmitted to the SRS in road side server. Then those data are sent to the gateway function unit and the address of those data from local to global address and sent to the global cloud server through Internet. Thus, using the proposed V2X communication protocol, not only Intranet communication among the vehicle network, but also Intranet and Internet communication can be realized.

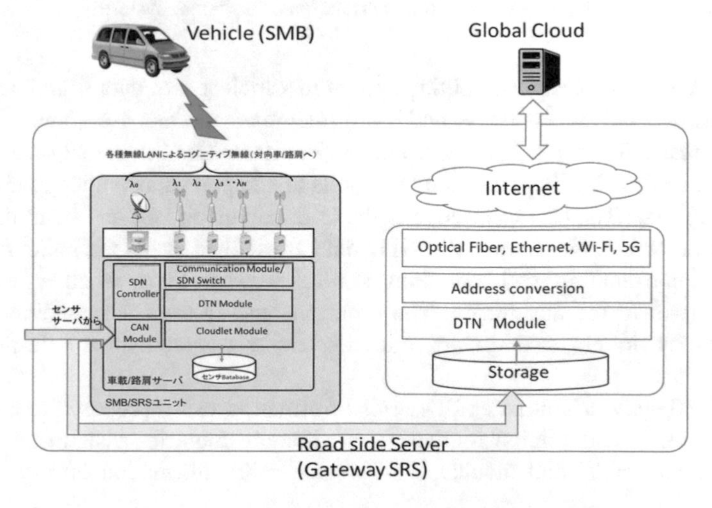

Fig. 7. V2X communication method between the SMB

5 Expected Applications

In order to identify the road conditions such as such as snow storm, frozen road, snow slide, white out in cold areas where mobile network and Internet network are not supported, the mobile nodes are loaded on the scheduled buses, taxies and public cars for local government offices and collect various sensor data as big data. Those data are locally analyzed to know the road conditions of the local areas at vehicle server in realtime and directly exchanged to other cars in opposite direction. The mobile nodes also exchange those sensor data to the road side wireless nodes. Furthermore, those data are uploaded to the cloud computing through Internet at the road side gateways shown in Fig. 8.

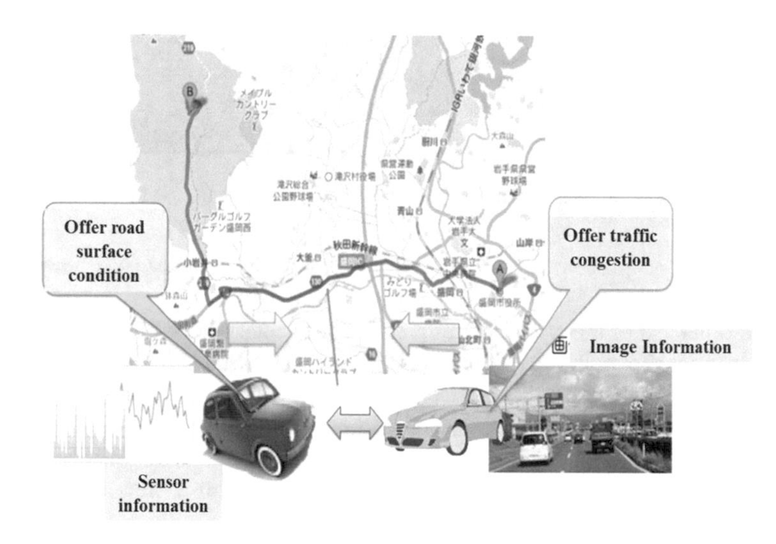

Fig. 8. Realtime road surface data exchange system

Figure 9 shows a cloud computing system in which sensor data from many mobile nodes and the road side wireless nodes are integrated to database as wide area road environmental information in cloud computing system. The integrated sensor data are filtered and analyzed as big data, and the results are displayed as unified road condition state on GIS map. The user can free refer the GIS map using PCs and smart phones and tablet terminals through Internet. Since big data are also analyzed to predict future road surface and surrounding road conditions so that users can know whether he can drive safely the road to the destination. Those current and future road condition data are useful for bus drivers, taxi drivers, track drivers to maintain their safe and proper operation plans.

Figure 10 show a Tourist and Disaster Information using proposed system. Many Japanese tourists and foreign tourists from different countries visit to the historical places, national parks and mountains where network information infrastructures are

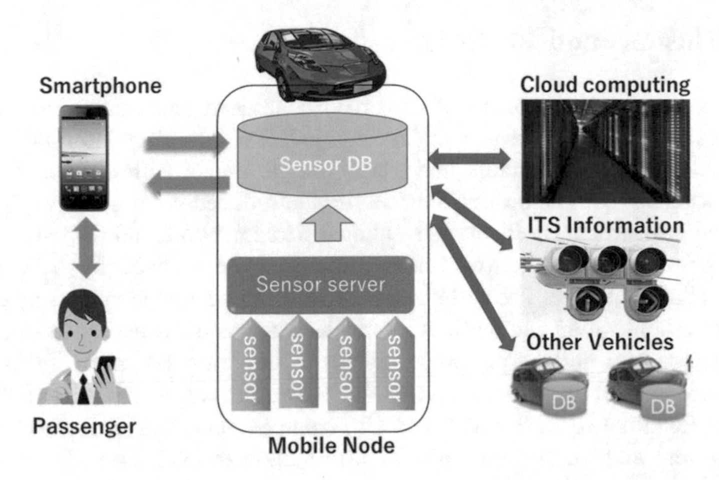

Fig. 9. Dig data road condition model

Fig. 10. Tourist information delivery on normal case

usually not well prepared in Japan. Even worse, it is difficult for foreigners to understand the information of Pints of Interest (POI) written in Japanese. However, by introducing the proposed mobility information infrastructure, and using smart phones and tablet terminals, the tourists can ask and speak by their own voices and obtain information POI by their own languages.

On the other hand, when disaster occurred, the push typed disaster information, evaluation information are automatically delivered to those tourists by their languages. Eventually, the tourist can safely evacuate to the proper shelter from the current location.

6 Conclusions and Remarks

In order to recover the problems of poor road conditions and challenged information communication infrastructure in aging society, IoT based mobility information network is proposed. Road side wireless nodes and mobile nodes with various sensors and different wireless communication devices are introduced to organize a large scale information infrastructure without conventional wired network and with very economic and low priced mobility network environment can be realized by V2V and V2X communication protocols. The system concept and architecture are precisely explained. The expected application and services for residents are also discussed. As future works, we will design and build a prototype system of our proposed mobility Information infrastructure in rural road between a city and a town in coast areas. We will design and implement Realtime Road Surface Data Exchange System, Dig Data Road Condition Sharing system and Tourist and Disaster Information to evaluate their functions and performance.

Acknowledgements. The research was supported by Strategic Information and Communications R&D Promotion Program Grant Number 181502003 by Ministry of Affairs and Communication.

References

1. A 2016 aging society white paper. http://www8.cao.go.jp/kourei/whitepaper/w-014/zenbun/s1_1_1.html
2. A 2016 decling birthrate white paper. http://www8.cao.go.jp/shoushi/shoushika/whitepaper/measures/english/w-2016/index.html
3. Ito, K., Hirakawa, G., Shibata, Y.: Experimentation of V2X communication in real environment for road alert information sharing system. In: IEEE AINA 2015, pp. 711–716 (2015)
4. Otomo, M., Sato, G., Shibata, Y.: In-vehicle cloudlet computing based delay tolerant network protocol for disaster information system. In: Advances on Broad-Band Wireless Computing, Communication and Application Applications. Series Lecture Notes on Data Engineering and Communications Technologies, vol. 2, pp. 255–266 (2016)
5. Hirakawa, G., Uchida, N., Arai, Y., Shibata, Y.: Application of DTN to the vehicle sensor platform CoMoSe. In: WAINA 2015, pp. 24–27 (2015)
6. Kitada, S., Sato, G., Shibata, Y.: A DTN based multi-hop network for disaster information transmission by smart devices. In: Advances on Broad-Band Wireless Computing, Communication and Application Applications. Series Lecture Notes on Data Engineering and Communications Technologies, vol. 2, pp. 601–611 (2016)
7. Goto, T., Sato, G., Hashimoto, K., Shibata, Y.: Disaster information sharing system considering communication status and elapsed time. In: IWDENS 2017 (2017)

Self-Outer-Recognition of OpenFlow Mesh in Cases of Multiple Faults

Minoru Uehara(✉)

Faculty of Information Sciences and Arts, Toyo University, Kawagoe 350-8585, Japan
uehara@toyo.jp

Abstract. Recently, renewable systems have received increasing attention. We propose a metabolic architecture that is suitable for the construction of renewable systems. A metabolic architecture-based system is one that can exchange all of its elements dynamically, similar to a multicellular organism. In this way, the system not only maintains homeostasis, but also adapts to environmental changes. We are developing OpenFlow Mesh as a system based on a metabolic architecture. OpenFlow Mesh is a 2 + 1D mesh of OpenFlow switches that recognizes the outer shape, i.e., physical allocations of elements, on the basis of the network structure. Previously, we proposed a propagation-based method for determining the outer shape in the case of single faults. In this paper, we describe a method that can determine the outer shape in the case of multiple faults.

1 Introduction

Recently, renewable systems have been recognized as important technological components. Renewable computing is defined as sustainable computing with disruptive innovation. We have previously proposed a metabolic architecture for renewable systems [1, 2]. A metabolic architecture is a computer architecture that maintains homeostasis by aggressively exchanging elements over a period of time. In addition, we have developed OpenFlow Mesh (OFM) as a renewable system based on this metabolic architecture [3, 4]. OFM is a 2D + 1D mesh network consisting of OpenFlow [5] switches. The 2 + 1D mesh is defined as a restricted 3D (XYZ) mesh in which 2D (XY) is added to 1D (Z). In a 2 + 1D mesh, the position of each node is described using the coordinates (x, y, z). However, unlike a complete 3D mesh, only the nodes at (x, y, 0), i.e., in the z = 0 plane, connect to the XY directions.

In previous research [4], we realized self-recognition in OFM. That is, whenever the topology changes, OFM is able to recognize the new topology correctly. Indeed, the OpenFlow Controller collects the network links of all switch nodes using the Link Layer Discovery Protocol defined in IEEE 802.1AB. This function enables the system to recover from failures.

However, to exchange a failed component with a new component, the system must identify the physical location of the failed component. The logical structure of the network, i.e., the topology, does not tell us the physical structure. Hence, in a previous study [6], we proposed a method of detecting the physical structure from the logical

L. Barolli et al. (Eds.): NBiS 2018, LNDECT 22, pp. 309–318, 2019.
https://doi.org/10.1007/978-3-319-98530-5_26

structure in OFM. Such an issue is regarded as self-outer-recognition. In [6], we discussed the case of a single fault. In this paper, we will discuss the case of multiple faults.

The remainder of this paper is organized as follows. Section 2 introduces previous research in this field in more detail. Sections 3–7 describe self-outer-recognition in various multiple-fault cases. Finally, Sects. 8 and 9 present a discussion of the results and our conclusions.

2 Self-Outer-Recognition in OFM

2.1 OFM

OFM is a 2 + 1D mesh network that consists of OpenFlow [5] switches. The 2 + 1D mesh is defined as a restricted 3D (XYZ) mesh in which 2D (XY) is added to 1D (Z). In a 2 + 1D mesh, the position of each node is given by the coordinates (x, y, z). However, unlike a complete 3D mesh, only the nodes at (x, y, 0), i.e. in the z = 0 plane, connect to the XY directions. Using such connections, it is easy to exchange failed components. For example, if there is a failed node at (x, y, z), OFM purges the nodes connected to (x, y, 0). After repairing node (x, y, z), OFM connects the new components to (x, y, 0).

When the number of nodes in the X, Y, and Z directions is n, the total number of nodes in OFM is n^3. In this paper, we represent such an OFM as OFM(n). Furthermore, to represent OFM easily, we introduce the 3-axis notation in which the first, second, and third axes are directed to the right, right and down, and down, respectively. For example, OFM(3) is shown in XYZ notation as follows, where 012 denotes a node at (0, 1, 2).

```
000 100 200
 010 110 210
  020 120 220
001 101 201
 011 111 211
  021 121 221
002 102 202
 012 112 212
  022 122 222
```

The following algorithm describes OFM(n) in XYZ notation.

```
Algorithm XYZ
  for z in range(n)
    for y in range(n)
      indent(y)
      for x in range(n)
        print_with_space(node(x, y, z))
    print_with_line()
```

In OFM, faults suffered by nodes in the z = 0 plane determine the connectivity. Therefore, in this paper, we consider only the z = 0 plane in XYZ notation.

2.2 Self-Outer-Recognition

A system based on a metabolic architecture aggressively exchanges components, even if they have not yet failed. This prevents the deterioration of elements, but requires a mechanism that simplifies the exchange of elements. In OFM, three-dimensional coordinates indicate the physical arrangement of the elements. In this way, OFM can easily identify failed nodes.

When OFM finds a failed node, the communication paths are reconfigured to avoid this node. In addition, once the failed node has been repaired, OFM reconfigures the network again to make the repaired node available. We call this reconfiguration of the network topology "self-inner-recognition."

In OFM, not all nodes are assigned coordinates. We call the process of coordinate assignment "self-outer-recognition."

The following CA0 algorithm assigns coordinates to all nodes in the case of no faults. CA0 consists of xy_line and z_line.

```
Algorithm CA0
  xy_line
  z_line
```

Here, xy_line determines the coordinates along the shortest path between nodes in front of edges after four corners have been detected in the z = 0 plane. The four corners can be detected from the number of links. Generally, nodes with 1, 2, 3, 4, and 5 links are located at the end of Z, the middle of Z, the corner of the z = 0 plane, the edge node of the z = 0 plane, and the inner node of the z = 0 plane, respectively. The z_line determines the coordinates along the Z axis, from the end to the z = 0 plane, after a node in the z = 0 plane has been assigned.

For example, assume that the following initial assignment is given. Here, "???" denotes an unassigned node. In this case, 000 and 400 are anchor nodes, which are initially given coordinates as hints.

```
000 ??? ??? ??? 400
 ??? ??? ??? ??? ???
  ??? ??? ??? ??? ???
   ??? ??? ??? ??? ???
    ??? ??? ??? ??? ???
```

The xy_line is determined in the following way. First, the corners are detected, with 040 and 440 being the corners closest to 000 and 400, respectively.

```
000 ??? ??? ??? 400
 ??? ??? ??? ??? ???
  ??? ??? ??? ??? ???
   ??? ??? ??? ??? ???
    040 ??? ??? ??? 440
```

Next, the corners are connected in the X direction.

```
000 100 200 300 400
 ??? ??? ??? ??? ???
  ??? ??? ??? ??? ???
   ??? ??? ??? ??? ???
    040 140 240 340 440
```

Next, the edges are connected in the Y direction.

```
000 100 200 300 400
 010 110 210 310 410
  020 120 220 320 420
   030 130 230 330 430
    040 140 240 340 440
```

Finally, the z_line assigns the coordinates in the Z direction.

2.3 Self-Outer-Recognition in the Case of Single Faults

The following CA1 algorithm assigns the coordinates in the case of a single fault. CA1 consists of an xyx_line, propagate instruction, and a z_line.

```
Algorithm CA1
  xyx_line
  propagate
  z_line
```

Here, xyx_line not only assigns the coordinates in XY order, but also in YX order. The xyx_line is equivalent to both an xy_line and a yx_line. Therefore, a yxy_line could be used instead. The propagate instruction collects the candidates from the neighbors of unassigned nodes, and then assigns the coordinates if the number of elements in the candidate set is 1.

The process of propagation is as follows. The initial state has already been applied in the xy_line. "N/A" denotes a failed node.

```
000 100 200 300 400
 010 ??? ??? ??? ???
  020 ??? ??? ??? ???
   030 ??? ??? ??? ???
    040 140 240 340 N/A
```

Node 100 sends a set of coordinates candidates {011, 101} to ???(011). Node 010 also sends the coordinate candidates {011, 011} to ???(011). In this case, the intersection has only one common element, 011. Therefore, the coordinate is determined as 011. In this way, the assignment propagates from the top left. Finally, the coordinates are assigned as follows.

```
000 100 200 300 400
 010 110 210 310 410
  020 120 220 320 420
   030 130 230 330 430
    040 140 240 340 N/A
```

CA1 can assign all nodes except the failed node xy0 and its Z nodes. Generally, CA1 can assign coordinates to $n^3 - n$ nodes for OFM(n) having a single fault in the z = 0 plane.

3 Self-Outer-Recognition in the Case of Two Faults

There is a pattern in which propagation does not work in the case of two faults. For example, if two diagonally opposite corners have failed, the xy_line does not work. Thus, the coordinates cannot be assigned. Furthermore, the process of propagation does not work. The following pattern is an example of two faults in XYZ notation.

```
000 ??? N/A
 ??? ??? ???
  N/A ??? 222
```

Here, N/A, ???, and underlined coordinates denote failed nodes, unassigned nodes, and anchors (pre-assigned nodes), respectively. In such a pattern, the xy_line cannot draw lines. Increasing the number of anchors can solve such an issue. An anchor is defined as a node for which the coordinates have been assigned manually. The fewer anchors there are, the better. However, in the above case, if the anchors are corner nodes, not all of the nodes can be assigned coordinates.

Next, we show a successful case. First, the anchors are selected as every other node along the four edges. The following is a case of two faults in OFM(5). In this case, the maximum number of unassigned nodes increases to 23 (= (5 × 5) − 2).

```
N/A ??? 200 300 400
 ??? ??? 210 310 410
  020 120 N/A 320 420
   030 130 230 330 430
    040 140 240 340 440
```

The neighbors of unassigned node 020 are {021, 010}. Therefore, propagation does not determine the coordinates of 010. Thus, propagation performs the z_line whenever a coordinate is assigned in the z = 0 plane. We call this new propagation process propagate_z. As a result, in the above case, all nodes except for the failed nodes can be assigned. This algorithm is shown as follows.

```
Algorithm CA2
  xyx_line
  propagate_z
```

However, even when using CA2, there are at most 18 (= (5 × 4) − 2) unassigned nodes in the following case.

```
000 ??? N/A 300 400
 ??? 110 210 310 410
  N/A 120 220 320 420
   030 130 230 330 430
    040 140 240 340 440
```

Next, we assume that the anchors are all of the edge nodes in the z = 0 plane. In this case, the maximum number of unassigned nodes decreases to 8 (= (5 × 2) − 2). This is the theoretical minimum value. As a result, it is better that every edge node is an anchor in the z = 0 plane. Additionally, in this case, CA1 is available.

4 Self-Outer-Recognition in the Case of Three Faults

In the case of three faults, even if all edge nodes are anchors, we cannot assign coordinates to all nodes. The following is a counter-example.

```
N/A 100 200 300 400
 010 ??? N/A 310 410
  020 N/A 220 320 420
   030 130 230 330 430
    040 140 240 340 440
```

Here, the result of CA1 is the same as that of CA2. The maximum number of unassigned nodes is 17 (= (5 × 4) − 3). In the propagation step, 110, 010, 120, and 210 provide hints for determining 110. However, 120 and 210 have failed, and so 010 and 100 cannot distinguish 110 from 000.

According to the above discussion, if no 000 node exists, node 110 may be determined. Thus, we discuss a proof of the non-existence of a node.

Non-existence means that there is no node at some particular coordinates. A non-existent node is regarded as a failed node. In other words, the proof of non-existence is the possibility of assigning coordinates to the failed node. An unassigned node is not always a non-existent node.

Here, we attempt to prove the non-existence of 210 using the above case. The neighbor set of 210 is {110, 200, 220, 310}. There are two shortest paths including 210, namely [110, 210, 310] and [200, 210, 220]. The path in which all nodes except 210 exist is [200, 210, 220]. If this path does not exist, node 210 also does not exist. If the path exists, the middle node of the shortest path from 200 to 220 is 210. In the above case, node 210 does not exist because there is no path. Node 120 also does not exist for the same reason. However, although there are two paths, namely [100, 000, 010] and [100, 110, 010], between 100 and 010, which is the neighbor set of 000, the coordinates of neither 000 nor 110 can be determined. Therefore, we cannot say that 000 does not exist and the coordinates of 110 cannot be assigned.

Now, we present algorithm CA3, including the proof of non-existence, as follows.

```
Algorithm CA3
  xyx_line
  non_existence
  propagate_z
```

In the case of three faults, CA3 is similar to CA1 in the sense of the maximum number of unassigned nodes. In this way, when there are three or more faults, the number of unassigned nodes may be greater than the number of failed nodes.

5 Self-Outer-Recognition in the Case of Four Faults

In the case of four faults, the proof of non-existence can reduce the maximum number of unassigned nodes. The following example is a result of CA1.

```
000 100 N/A 300 400
 N/A ??? ??? 310 410
  020 ??? ??? N/A 420
   030 N/A 230 330 430
    040 140 240 340 440
```

Similar to a rook in chess, a failed node propagates in four directions. The maximum unassigned area occurs when all failed nodes cover that area. In the above case, the non-existence of every failed node can be proved. Therefore, 100, 020, 230, and 310 determine the non-existence of 110, 120, 220, and 210, respectively. CA3 assigns the coordinates as follows.

```
000 100 N/A 300 400
 N/A 110 210 310 410
  020 120 220 N/A 420
   030 N/A 230 330 430
    040 140 240 340 440
```

In the case of using CA3, the pattern of the maximum number of unassigned nodes in the case of four faults is as follows. In this pattern, the maximum number of unassigned nodes is 26 (= $(6 \times 5) - 4$). This is essentially equivalent to the case of three faults.

```
000 N/A 200 300 400
 N/A ??? 210 310 410
  020 120 ??? N/A 420
   030 130 N/A 330 430
    040 140 240 340 440
```

6 Self-Outer-Recognition in the Case of Five Faults

Using CA3, the pattern of the maximum number of unassigned nodes in the case of five faults is as follows. In this pattern, the maximum number of unassigned nodes is 35 (= (8 × 5) − 5).

```
N/A 100 N/A 300 400
 010 ??? ??? N/A 410
  N/A ??? 220 320 420
   030 N/A 230 330 430
    040 140 240 340 440
```

7 Self-Outer-Recognition in the Case of Six Faults

Using CA3, the pattern of the maximum number of unassigned nodes in the case of six faults is as follows. In this pattern, the maximum number of unassigned nodes is 49 (= (11 × 5) − 6).

```
000 100 N/A 300 400
 010 N/A ??? 310 410
  N/A ??? ??? ??? N/A
   030 130 ??? N/A 430
    040 140 N/A 340 440
```

8 Discussion

Table 1 shows the results of the coordinate assignment in OFM(5). As indicated by this table, if the number of failed nodes increases, the number of assigned nodes decreases. This means that the number of unassigned nodes increases. When six nodes have failed in the z = 0 plane, almost half of the nodes are unavailable. An essential issue is #UxF in Table 1. The nodes in the area covered by the failed nodes and unassigned nodes become UxF.

Table 2 demonstrates the relationship between the number of faults and UxF in each algorithm. The UxF of CA2 is equal to that of CA1. However, as can be seen from Table 3, the average number of unassigned nodes is different. Here, the average number of unassigned nodes is defined as the ratio of unassigned nodes to the number of combinations of faults.

In the case where there is an initially failed node, the coordinates of the failed node are not assigned. However, if the neighbor nodes observe each other, the failed node may be found. At that point, the physical location of the failed node has not been detected, but candidate physical locations have been detected. It is permissible to exchange these candidates. In such a method, assume that a failed node is found. If it is a non-existent node, then the coordinates can be detected. Otherwise, each neighbor

Table 1. Coordinate assignment in OFM(5).

#F	#UxF	#U	#Ua	AR
1	0	1	4	96%
2	0	2	8	92%
3	1	4	17	84%
4	2	6	26	76%
5	3	8	35	68%
6	5	11	49	56%

#F: number of failed nodes in z = 0 plane;
#UxF: number of unassigned nodes, except failed nodes, in z = 0 plane;
#U: number of unassigned nodes in z = 0 plane (= #F + #UxF);
#Ua: total number of unassigned nodes (= #U × n − #F, where n = 5 in OFM(5));
AR: rate of available nodes (= $(n^3 - \#Ua - \#F)/n^3$).

Table 2. Relationship between the number of faults and UxF in each algorithm.

#F	CA1	CA2	CA3
1	0	0	0
2	0	0	0
3	1	1	1
4	4	4	2
5	8	8	3
6	9	9	5

Table 3. Average number of unassigned nodes in each algorithm.

#F	CA1	CA2	CA3
1	4.0	4.0	4.0
2	8.0	8.0	8.0
3	12.2	12.1	12.0
4	16.8	16.3	16.1
5	22.3	21.0	20.2
6	29.0	26.3	24.6

node notifies the OpenFlow Controller (OFC) that there is a failed node nearby. OFC then exchanges all candidates of failed nodes.

Even if a node fails in the z = 0 plane, it is possible to fix the problem by simply removing that node. Here, the set of z-axis elementary nodes stacked at *xy*0 is called the *xyz* stack. For instance, if node 120 fails in OFM(5), the set of nodes from 120 to 124 is the 12z stack. The problem is fixed as follows. All elements (124…120) are popped from 12z in this order. The failed node 120 is then taken to be repaired. Nodes 121 to

124 are then pushed to the 12z stack in this order. The old 121 is recognized as the new 120. Later, the repaired 120 returns to the 12z stack. This new node is recognized as 124. After self-outer-recognition is performed, if no coordinates are assigned to the new node, another failed node may cause the same problem. Thus, the other failed node should be fixed. In this way, the problem is fixed using a first-in–first-out queue in which non-existent nodes have higher priority.

9 Conclusions

In this paper, we have discussed self-outer-recognition in OFM in the case of multiple faults. Compared with a single fault, many more anchors are required to assign the coordinates. In addition, we proposed a new method based on the proof of non-existence of failed nodes. Using this method, we can enlarge the assigned area and improve recognition. However, if six failed nodes exist in the $z = 0$ plane, 56% of nodes cannot be assigned coordinates.

In future work, we will examine how to use unassigned nodes effectively. Even if a node is unassigned, it may not have failed. In this case, the node could be used as a computing resource.

Acknowledgments. We thank Stuart Jenkinson, Ph.D., from Edanz Group (www.edanzediting.com/ac) for editing a draft of this manuscript.

References

1. Uehara, M.: Proposal of an evolutional architecture for metabolic computing. In: Proceedings of the 3rd IEEE International Conference on Intelligent Networking and Collaborative Systems (INCoS 2011), Fukuoka, Japan, 30 November 2011 to 2 December 2011, pp. 287–292 (2011)
2. Uehara, M.: Metabolic computing: toward truly renewable systems. IGI Glob. Int. J. Distrib. Syst. Technol. IJDST **3**(3), 27–39 (2012)
3. Uehara, M.: OpenFlow mesh for metabolic computing. In: Barolli, L., et al. (eds.) Advances on Broad-Band Wireless Computing, Communication and Applications, Proceedings of the 18-th International Symposium on Multimedia Network Systems and Applications (MNSA-2016) in Conjunction with the 11th International Conference on Broad-Band Wireless Computing, Communication and Applications (BWCCA-2016), Soonchunhyang University, Asan, Korea, 5–7 November 2016, pp. 613–620
4. Yasui, S., Uehara, M. Self recognition and fault awareness in OpenFlow mesh. Computational Science/Intelligence and Applied Informatics, pp. 214–229 Springer, Cham (2018). https://doi.org/10.1007/978-3-319-63618-4_16
5. McKeown, N., et al.: OpenFlow: enabling innovation in campus networks. ACM SIGCOMM Comput. Commun. Rev. **38**(2), 69–74 (2008)
6. Uehara, M.: Coordinate assignment: self-outer-recognition in OpenFlow mesh. In: Proceedings of the 19th International Symposium on Multimedia Network Systems and Applications (MNSA-2017) in Conjunction with the 12th International Conference on Broadband, Wireless Computing, Communication and Applications (BWCCA-2017), Palau Macaya, Barcelona, Spain, 8–10 November 2017, pp. 608–615

Deterrence System for Texting-While-Walking Focused on User Situation

Megumi Fujiwara and Fumiaki Sato(✉)

Faculty of Science, Toho University, Funabashi, Chiba 274-8510, Japan
5514089f@nc.toho-u.ac.jp, fsato@is.sci.toho-u.ac.jp

Abstract. In recent years, users of smartphones are increasing explosively, and accidents and troubles due to texting-while-walking (TWW) are rapidly increasing along with this. Previous research on the TWW deterrence system was to display a warning on the screen, notify by voice or vibration, or make the screen dark and invisible when detecting TWW from the acceleration sensor. Although it is possible to detect TWW, but if the user feels the application troublesome, the application may be erased. This is because the TWW is forcibly suppressed even if user judges that the TWW is not dangerous at present or the short and important operation is required. In this research, we develop a suppression system focusing on user's situation, especially walking speed. During the TWW, the user is notified that the walking speed is low, and the user is urged to increase the walking speed within a safe range. In order to evaluate the effectiveness of this system, we prepared a conventional TWW deterrence system and our proposed system, and asked 17 peoples to do TWW. In the questionnaire after the experiment, 11 people answered that they would like to continue using the system that focused on walking speed.

1 Introduction

Accidents that are caused by "texting while walking" (TWW) are increasing. TWW increases the risk of accidents caused by a decline of viewable angles, decreased attentiveness, and so on. According to MMD research [1], 20% of the people who had done TWW admitted getting hurt during it, and 97.7% described TWW as risky. On the other hand, 32.3% continue to engage in TWW, suggesting that quite a few people cling to TWW's usefulness and necessity, even though they admit that this behavior is risky. Eradicating TWW will be difficult.

Walking while on the phone is a problem not only in Japan but also in all of the world. A regulation of "Distracted Walking Law" is passed in Honolulu, Hawaii, USA, in 25th October, 2016. According to the regulation, people are imposed a fine when they go across a road while watching a screen of electronic mobile devices such as smart phones or digital cameras. Moreover, many countries follow Hawaii.

In Japan, the mass media such as TV and newspapers enlighten people to stop TWW. Furthermore, posters for stopping TWW are put on the walls of railway stations, and tissue papers with the phrase are also distributed by railway companies. Moreover, railway companies are developing applications which prevent TWW. However, these existing applications force users to stop using phones when users are TWW. The force

L. Barolli et al. (Eds.): NBiS 2018, LNDECT 22, pp. 319–332, 2019.
https://doi.org/10.1007/978-3-319-98530-5_27

is not accepted for users and the applications never become popular. Therefore, in this paper, we propose a deterrence system that changes deterrence level according to user's situation.

By using the location information of the user, it is possible to estimate whether it is a good place to use a smartphone, such as a station or a pedestrian crossing. Further, by using the acceleration sensor, it is possible to detect whether or not a stairway is used. Also, the surrounding congestion degree can be estimated from the number of BLE terminals that can be acquired. Also, it is highly possible that the decrease in the walking speed of the user is aware of the surrounding risk. By using the situation of these users, it is judged whether or not it is a situation to suppress TWW. In this paper, we develop a suppression system focusing on walking speed. During the TWW, the user is notified that the walking speed is low, and the user is urged to increase the walking speed within a safe range. Evaluation experiments showed that the proposed method is more user-friendly than the conventional deterrent system.

2 Related Works

2.1 NTT DOCOMO TWW Deterrence Function

NTT DOCOMO, a mobile phone communication carrier company, provides an application which warns users not to use their smart phones by displaying a warning message when it detects they are walking while on the phone [2]. Users have to tap a button on the warning screen or press and hold the power button to erase the screen. The warning screen appears against any actions of smart phones such as using maps, emails and browsing while walking. Thus, the warning screen is not acceptable for users since activities of users are strongly restricted. As a result, users never use such kind of applications.

2.2 Collision Avoidance System of TWW by Using Range and Image Sensors

There is a research by Kodama et al. in which users are prevented from colliding with other users by range and image sensor when they are walking while on the phone [3]. Kodama et al. researched how dangerous walking while on the phone is by using range and image sensor. Kodama et al. acquired distance from objects around the phone and their angle by the sensor in order to know the relative speed from the objects and to know whether the objects are following the phone or not. The moving objects are often humans who are walking so that the phone can warn its user to watch out other people. Kodama et al. did experiments whether people collided or not with and without their warning system. Without their system, the collision rate was 13.3%. On the other hand, with their system, the rate was less than 10%. However, users cannot avoid the collision when something comes from the out of the range of the sensor. It is dangerous that the safety depends on only sensors.

2.3 Notification System of TWW

Another system is proposed by Tsuji et al. [4]. They used a tail lamp to warn people around a smart phone. In their system, not a smart phone user but people around the user are warned. Light module on the waist of a user glows when his smart phone detects the use of the phone and his walking. The system never prevents the use of the smart phone. On the other hand, other people around the phone can know there is a user who is TWW. Tsuji et al. mentioned that the system can decrease the rate of the collision of people. However, it is difficult to think that the user actually operates the smartphone wearing a tail lamp.

Kato and Uda [5] have proposed a system which gives a warning to the user by vibration of the smartphone. By giving vibration, we are not only warning the use of the user's walking smartphone, but also attracting the attention of the people around us by the vibration sound, trying to deter the use of walking smartphone. However, as with NTT's deterrent system, this method does not consider the user's situation, so there is a possibility that the function may be deleted without user support.

3 Proposed Texting While Walking Deterrence System

3.1 System Features

In this research, we propose a TWW deterrent system that changes deterrence level according to user's situation. As the first step of the research, we propose a system to stop TWW only when the walking speed drops below normal and the user makes surrounds danger or inconvenience.

Features of the proposed method are roughly divided into two. One method is to detect as TWW only the situation when troubling the surroundings which was not considered in the conventional method. This is to prevent the system from excessively deterring operation of the smartphone, so that users do not hate the TWW deterrence system. The second method is to prompt the user of the TWW to increase the walking speed by text display when the surroundings are safe before displaying the gradual dark layer to the user.

User's TWW detection and walking speed are performed using the acceleration sensor and the walking sensor mounted on the Android terminal. The system operation is the activation of the system for displaying the warning text and the warning layer, the display of the warning sentence by walking detection, and the stepwise display of the warning layer. The flow of the operation is shown in Fig. 1.

When the system detects the user's TWW, a warning message is displayed. When the TWW is detected continuously, the transparency of the background of the warning layer is gradually displayed dark according to the elapsed time. This layer has six levels of transparency, and these stages are defined as detection stages. When walking is not detected, for example when the user stops, the transparency of the layer is 100%.

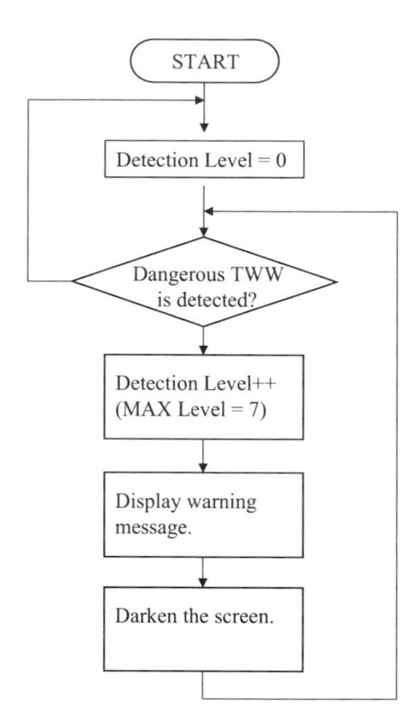

Fig. 1. Operation flow of the proposed TWW deterrence system.

3.2 Texting While Walking Detection

Detection of the TWW was carried out by combining an angle holding a smartphone and walking detection.

Detection of the holding angle of the smartphone was detected by the distribution of force applied to the terminal of the gravitational acceleration on the xyz axis. The way gravitational acceleration is applied varies depending on whether the smartphone is held vertically or horizontally, but it is not difficult to identify the holding state because a large force is applied in the vertical direction, and in the horizontal direction it hardly takes.

After identifying the holding state, by checking the magnitude of the acceleration in the z direction perpendicular to the screen and the acceleration in the y direction the angle can be determined. That is the case of holding in the vertical direction. The acceleration in the x direction is used in the case of holding in the horizontal direction.

In addition, in the case of walking, the vertical acceleration due to the user's movement other than gravitational acceleration is added. Therefore, in some cases, an acceleration opposite to the acceleration in the stationary state may occur. In this case, because it represents a situation where intense exercise is involved, it is excluded from calculation of holding angle.

Figure 2 shows the relationship between gravitational acceleration and the angle of the smartphone. The angle θ of the smartphone was calculated by *arctan (gy/gz)*, and it is judged that the user is operating the smartphone when $0 < \theta < 90°$.

We also detected whether the user is walking using Android's walk detection API. Also, in order to calculate the normal walking speed, the relationship between the

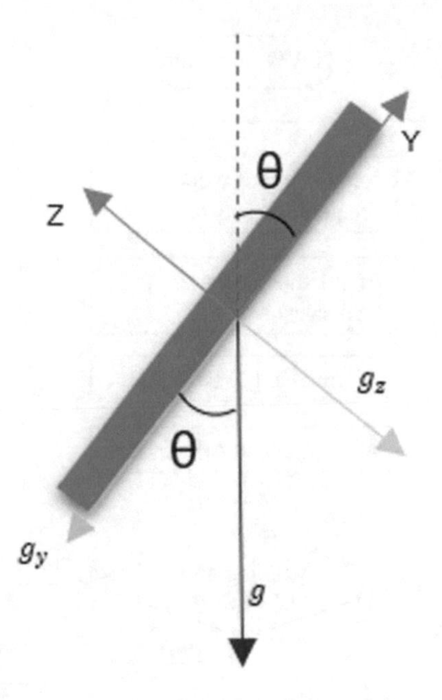

Fig. 2. Relationship between gravitational acceleration and the angle of the smartphone.

number of steps and the walking time in the state where the TWW is not detected is accumulated. Figure 3 shows the TWW detection algorithm.

3.3 Implemented System

We incorporated the proposed TWW detection function and suppression function into the original application that the user performs input operation. Figure 4 shows the main screen of the application. This is the first screen displayed when this system is activated.

The experiment is started by pushing the "START" button, "on the experiment" is displayed on the left, and elapsed time is displayed on the right.

The text for input is displayed at the top, and the user types the text correctly in the middle text box. When the user ends to enter, presses the "Input end" button under the text box. If the displayed text and the input text match, the upper text changes to another text and the character in the middle text box disappears. If they do not match, the text does not change and the text box text disappears.

By pressing the "START" button, the experiment is started and the values of the acceleration sensor and the walking sensor are stored. Click "END" button to finish the experiment. When entering the title name of the data in the left text box and pressing the "SAVE" button, the values of the acceleration sensor and the walk sensor are saved as text files.

As a feedback method of the proposed method, the feedback method of Negishi et al. [6] is adopted. The method of Negishi et al. is a method of superimposing layers having six levels of transparency on the screen, and an implementation example is

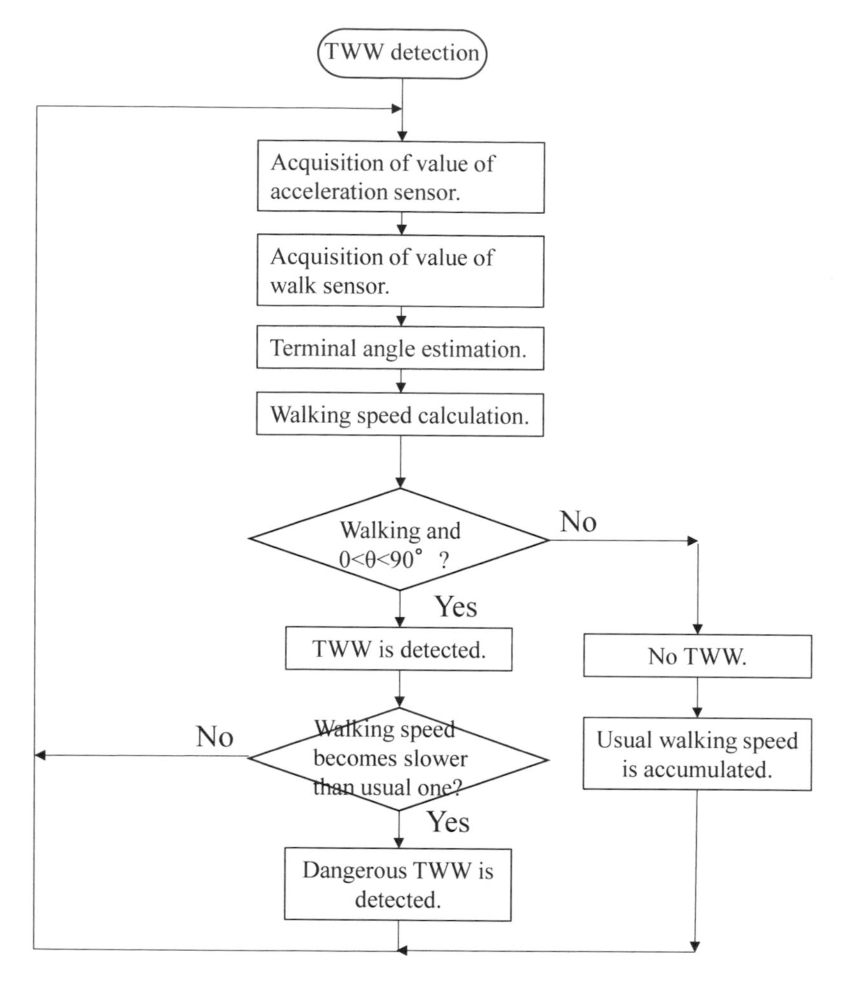

Fig. 3. TWW detection algorithm.

shown in Fig. 5. In the method proposed by us, we set a screen with a warning sentence whose walking speed has decreased, step by step. We change the display according to the detection stage. Figure 6 shows the warning layers of the stages, respectively. In the Negishi's method, when the TWW is detected, the screen becomes dark regardless of the walking speed, but in the proposed method the screen does not become dark if the walking speed does not decrease from the normal speed.

4 Evaluation Experiments

4.1 Preliminary Experiment

In the basic experiment, normal walking and TWW were repeated for 5 min in order to check whether TWW detection was correctly judged. As a result of the experiment,

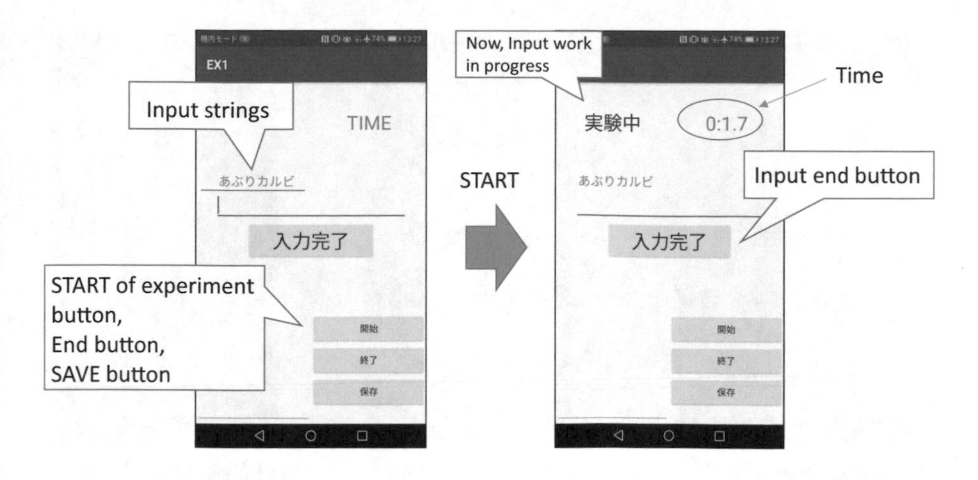

(a) Before the start. (b) During the experiment.

Fig. 4. Main screen of the implemented system.

3238 out of 3248 data matched the actual walking state and detection results to 99.6% (Fig. 7). The data that did not match was because walking detection in the change from normal walk to TWW was delayed by 0.2 s and the change from TWW to normal walk was delayed by 0.4 s. It can be judged that the accuracy of the walking detection is sufficient.

4.2 Evaluation Experiments

In this research, in order to compare the conventional method and the proposed method, 13 men and 4 women were asked to perform normal walking and TWW with the conventional method and the proposed method. A questionnaire survey was conducted after the experiment. The experiment was conducted at the main street of Toho University, Narashino campus.

(1) Change in time per step
The walking time during the experiment is compared with the conventional method and the proposed method. Table 1 shows the maximum time per step and difference between average time per step of TWW and normal walking. It is shown that the maximum time per step of the proposed method is 0.0699 [s] shorter than the conventional method. The average time per step of the proposed method has not increased much.

In the conventional method, when a walking smartphone is detected irrespective of the walking speed, a warning layer appears. As the time to do the TWW, it becomes difficult to see the screen gradually, and it is understood that the walking speed is lowered accordingly.

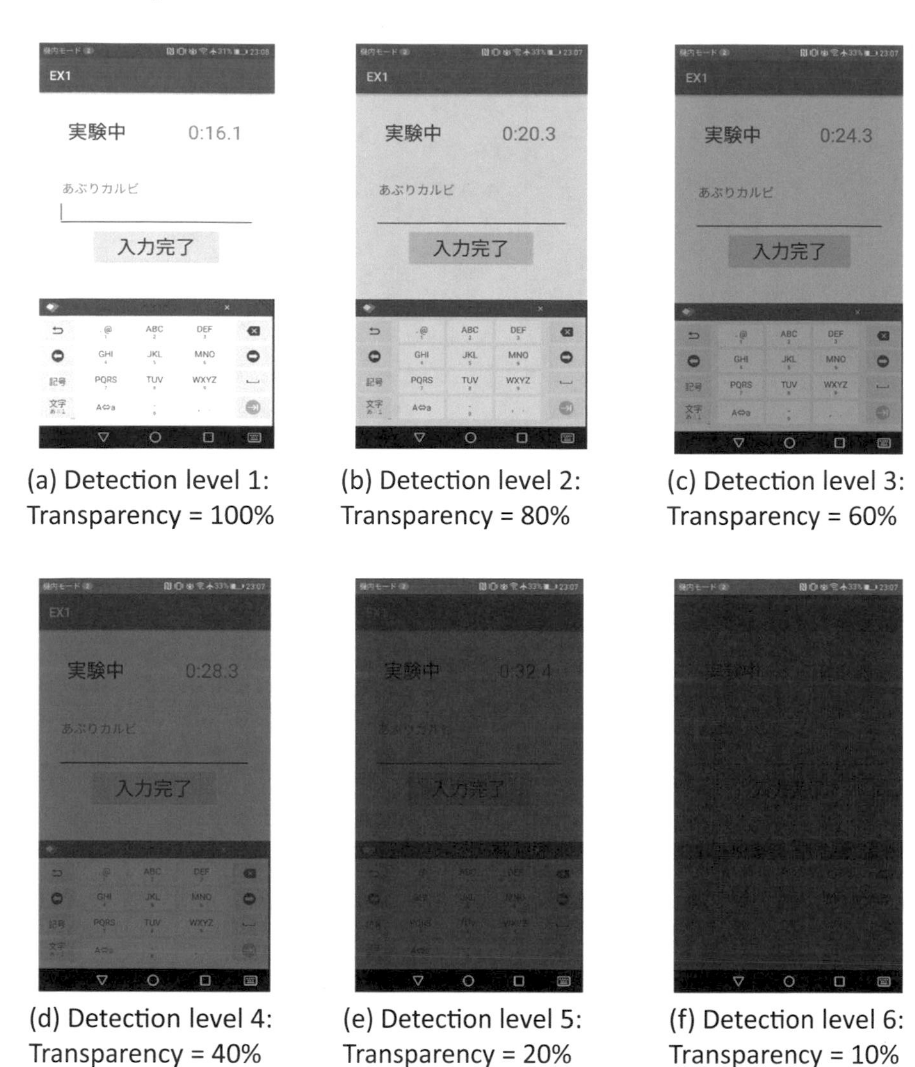

(a) Detection level 1: Transparency = 100%
(b) Detection level 2: Transparency = 80%
(c) Detection level 3: Transparency = 60%
(d) Detection level 4: Transparency = 40%
(e) Detection level 5: Transparency = 20%
(f) Detection level 6: Transparency = 10%

Fig. 5. Feedback method of Negishi et al.

On the other hand, the walking speed of the proposed method decreases as the warning message warning layer appears and it becomes difficult to see it, but the walking speed slows down accordingly, but in order to raise the walking speed to try to erase the warning layer, the walking speed in the average TWW is higher than the conventional method.

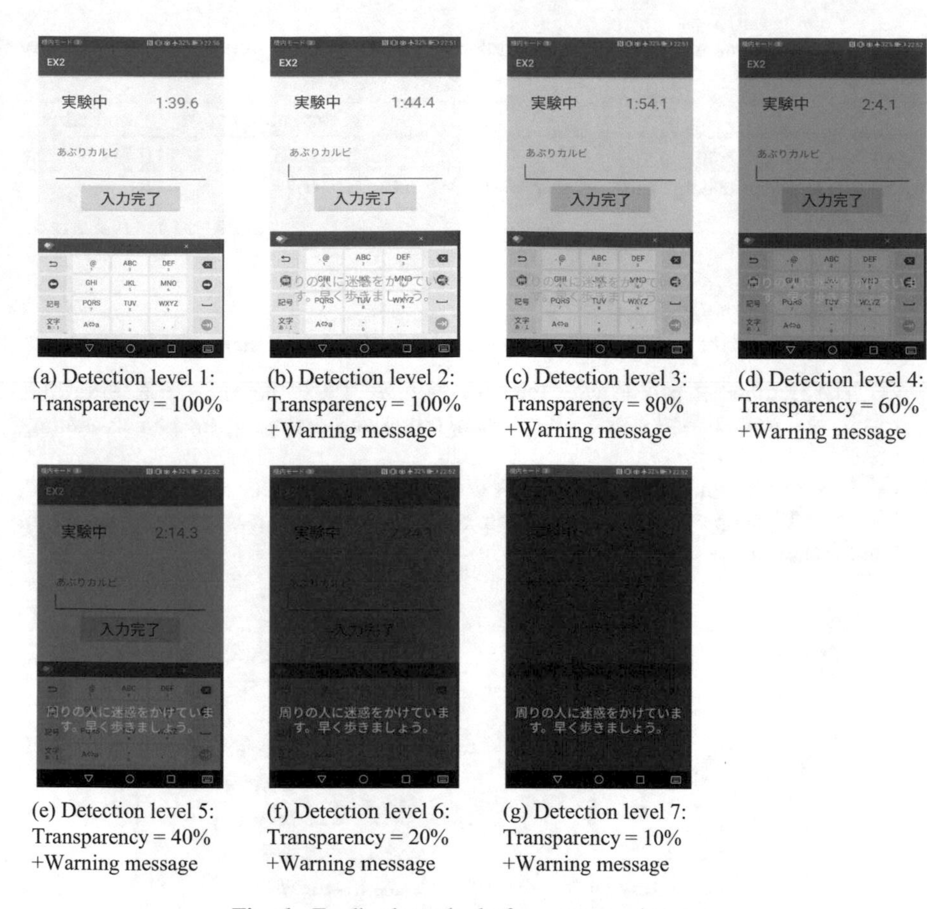

(a) Detection level 1: Transparency = 100%

(b) Detection level 2: Transparency = 100% +Warning message

(c) Detection level 3: Transparency = 80% +Warning message

(d) Detection level 4: Transparency = 60% +Warning message

(e) Detection level 5: Transparency = 40% +Warning message

(f) Detection level 6: Transparency = 20% +Warning message

(g) Detection level 7: Transparency = 10% +Warning message

Fig. 6. Feedback method of our proposed system.

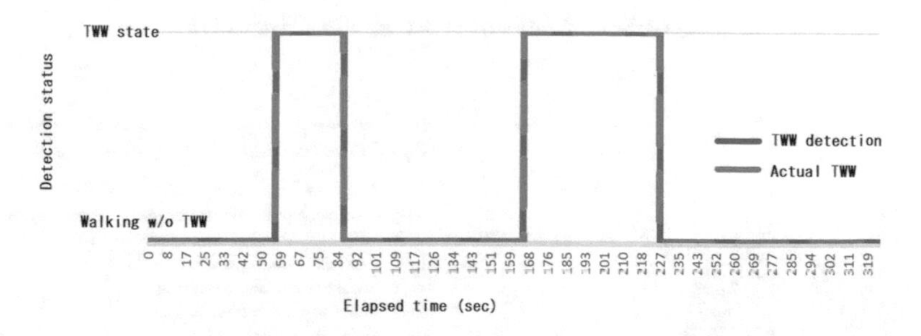

Fig. 7. Preliminary experiment result.

Table 1. The maximum time per step and difference between average time per step of TWW and normal walking.

	Conventional (s)	Proposed (s)
Maximum time per step	0.83333	0.71429
Difference between average time per step of TWW and normal walking	0.04674	0.0209

(2) Questionnaire evaluation

The results of the questionnaire conducted after the experiment are shown below.

Q1 The number of people who felt annoying by TWW became 76%. From this result, it is understood that TWW is dangerous and annoying to the surrounding area.

Q1-1 The most troublesome act by TWW is "A person walking my front was slow". 13 out of 17 people chose it. In addition, there was another answer that "A person in front has come to bump."

Fig. 8. Questionnaire survey Results of Q1 and Q1-1.

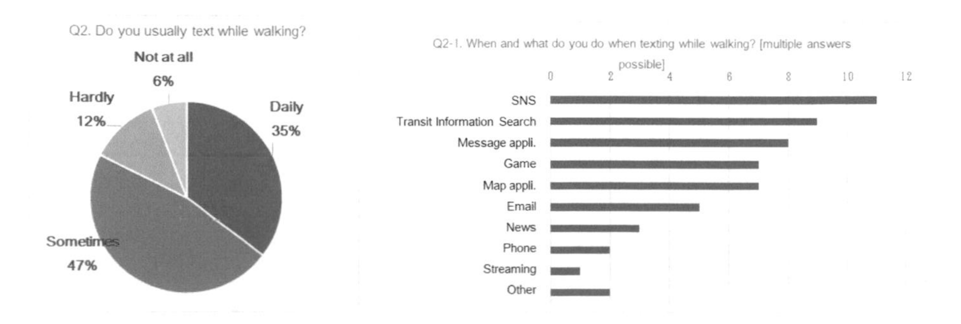

Fig. 9. Questionnaire survey Results of Q2 and Q2-1.

Q2 82% of the respondents said that they do TWW daily or sometimes (Figs. 8 and 9).
Q2-1 Most of the respondents said that they are using SNS or transit guide (Fig. 10).

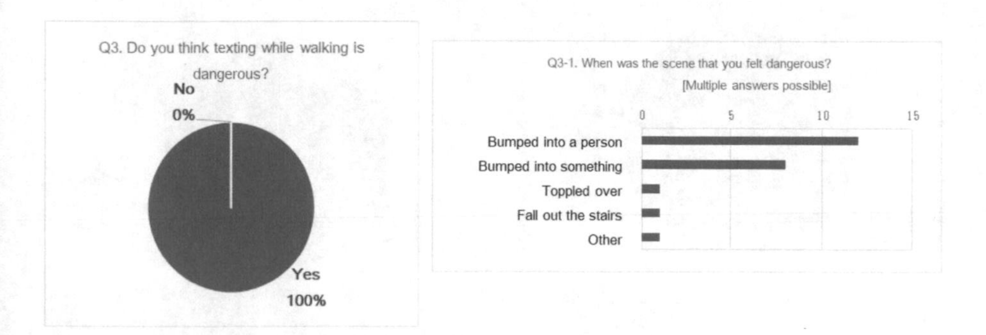

Fig. 10. Questionnaire survey Results of Q3 and Q3-1.

Q3 All 17 people answered TWW as dangerous. The people feel TWW is dangerous but 82% of the people do TWW from Q2 (Fig. 11).
Q3-1 Eleven from the people who had been inconvenienced by another person's TWW at Q1-1 said "I bumped a person" by his/her TWW.

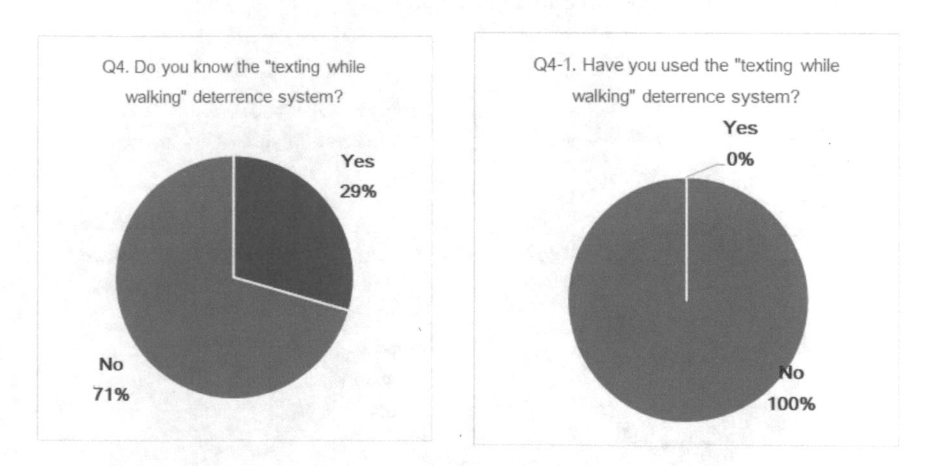

Fig. 11. Questionnaire survey Results of Q4 and Q4-1.

Q4 Despite the mobile companies providing TWW deterrent function, fewer than 30% knew of the existence of the TWW deterrence system.
Q4-1 No one actually used the TWW deterrence system among those who responded that they knew the TWW deterrence system (Fig. 12).

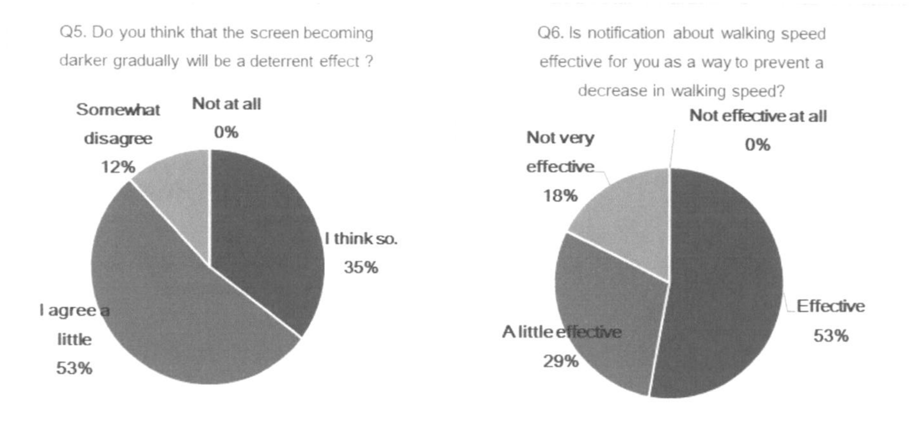

Fig. 12. Questionnaire survey Results of Q5 and Q6.

Q5 88% of respondents said feedback method that the screen becomes dark stepwise when detecting TWW used in conventional method and proposed method is effective for deterring walking smartphone.
Q6 82% of respondents said feedback methods for displaying walking speed drops in warning texts were effective before gradually darkening the screen after detection of TWW used in the proposed method (Fig. 13).

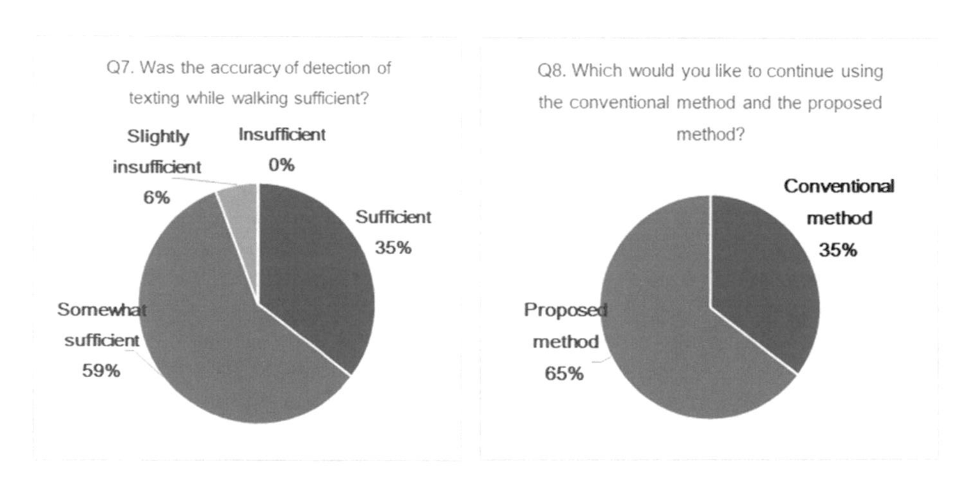

Fig. 13. Questionnaire survey Results of Q7 and Q8

Q7 94% answered that TWW detection accuracy was sufficient.
Q8 65% of people responded that they want to continue using the TWW deterrence system of the proposed scheme.

(3) Discussion

In this research, we proposed a system that suppresses only TWW where the walking speed decreases. Subject experiments showed that the proposed TWW deterrence system is a system easier for users to use than the conventional deterrence system. In the conventional method, when the TWW is detected and the warning layer is displayed, the warning layer does not disappear unless the walking is stopped or the using the smartphone is stopped. If a user are walking while watching a map, it is not convenient if the screen becomes invisible, and the user may delete the TWW suppression function. In our proposed system, the warning sentence and the warning layer disappear when the walking speed is returned to the normal walking speed according to the warning sentence. Since the user can continue the TWW without causing annoyance to the surroundings, the proposed method was chosen.

5 Conclusions

In this research, we proposed and implemented a system that suppresses texting-while-walking (TWW) focusing on user's situation. This system detected when the walking speed of TWW is lower than that during normal walking and inconvenience is caused around the surroundings. The feedback method in the case of detecting TWW urged the user to inform the user that the walking speed has declined with the text of the warning layer and increase the walking speed. We used a technique of gradually lowering the transparency of the background of the warning layer when TWW was detected even after warning. In the evaluation experiment, a basic experiment was carried out to confirm whether TWW detection is correctly judged. As a result of the experiment, the detection accuracy was as high as 99.6%. Evaluation experiment was conducted for 17 people and questionnaire survey was conducted after the experiment. 65% of subjects responded that they would like to continue using this system rather than the conventional system, and 82% of the subjects said that the warning text informing of the decrease in walking speed would be a trigger to increase walking speed. From this, it was found that the feedback method of detecting a decrease in walking speed and informing the user is effective as a suppression of the TWW when danger or troubles surrounding the user.

The future task is to make it more practical system. We plan to add a new method for detecting the situation of the user into a practical system.

References

1. Ministry of Internal Affairs and Communications: The Popularization of Internet. White papers of information and communication 2016 (Referred on 14 Jul. 2017) (in Japanese). http://www.soumu.go.jp/johotsusintokei/whitepaper/ja/h28/html/nc252110.html
2. NTT DoCoMo: The new approach to the prevention of walking while on the phone. Press Release (Referred on 14 Jul. 2017) (in Japanese). https://www.nttdocomo.co.jp/info/news_release/2013/12/03_00.html

3. Kodama, S., Enokibori, Y., Mase, K.: Examination of safe-walking support system for "texting while walking" using a time-of-flight range image sensor. SIG Technical Reports 2016-UBI-50(3), pp. 1–6 (2016) (in Japanese)
4. Tsuji, A., Ushida, K.: Development of a system alerting unsafe behaviors of texting while walking using a tail lamp. SIG Technical Reports 2017-CDS-19(22), pp. 1–6, 2017 (in Japanese)
5. Kato, H., Uda, R.: Texting while walking deterrence system by vibration of smartphone. In: Proceedings of the 12th International Conference on Ubiquitous Information Management and Communication (IMCOM 2018), Langkawi, Malaysia, 5–7 January 2018, 5 p
6. Negishi, T., Tanaka, J.: Development of "Nagara-Smartphone" prevention system. In: Proceedings of the 76th National Convention of IPSJ, 4W-5 (2014). (in Japanese)

Joint Deployment of Virtual Routing Function and Virtual Firewall Function in NFV-Based Network with Minimum Network Cost

Kenichiro Hida and Shin-ichi Kuribayashi(✉)

Seikei University, 3-3-1 Kichijoji-kitamachi, Musashino-shi, Tokyo 180-8633, Japan
kuribayashi@st.seikei.ac.jp

Abstract. It is essential for economical NFV-based network design to determine the place where each network function should be located in the network and what its capacity should be. The authors proposed an algorithm of virtual routing function allocation in the NFV-based network for minimizing the network cost, and provided effective allocation guidelines for virtual routing functions. This paper proposes the joint deployment algorithm of virtual routing function and virtual firewall function for minimizing the network cost. Our evaluation results have revealed the following: (1) Installing a packet filtering function, which is a part of the firewall function, in the sending-side area additionally can reduce wasteful transit bandwidth and routing processing and thereby reduce the network cost. (2) The greater the number of packets filtered by packet filtering function in the sending-side area, the more the reduction of network cost is increased. (3) The proposed algorithm would be approaching about 95% of the deployment with the optimal solution.

1 Introduction

Virtualization is a technique to make a single physical entity appear multiple logical entities or, conversely, to make multiple physical entities look like a single logical entity. Network functions virtualization (NFV) [1–4] represents an application of virtualization technology to network functions. It enables a piece of software that can conventionally operate only on a specific piece of hardware to operate on a general-purpose server, making it possible to build a network quickly and to operate it in a flexible manner. It has already been commercially introduced in public mobile networks [5].

In an NFV-based network, a variety of network functions is implemented in software on general-purpose servers. This makes it possible to select the capacity and location of each function without any physical constraints. It is essential to optimize the location and capacity of each network function for economical NFV-based network design. A new deployment method should be required as the existing deployment method that has limitations on capacity and deployment location cannot be used as it is. As a deployment method for NFV-based network, there is a method of efficiently allocating already installed physical resources to the virtual network function as

L. Barolli et al. (Eds.): NBiS 2018, LNDECT 22, pp. 333–345, 2019.
https://doi.org/10.1007/978-3-319-98530-5_28

short-term perspective. This has been studied as VNE (Virtual Network Assignment) problems [6]. In addition, the research from a medium-term to long-term perspective is required to decide the optimal physical resource capacity and deployment location, when expanding the network or adding new network functions. Many researches have been carried out on the former, but less research has not been done on the latter.

From a medium-term to long-term perspective, the authors previously focused on the routing function, which is an important network function, and proposed the algorithm for allocating virtual routing functions in a way that minimizes the network cost or the network power consumption [7]. The authors evaluated the proposed algorithm on a ladder model that simulates the shape of Japan, and developed multiple effective functional allocation guidelines. The guidelines show the trend for the capacity required for each network function and the optimal location of each function, which depend on the routing function cost relative to the bandwidth cost and the inter-area traffic distribution. The authors also clarified the influence of quality conditions such as the maximum allowable network delay on its deployment guidelines [8]. These guidelines provide network operators with critical information needed in designing and building an NFV-based network. For example, whether the total network cost can be reduced by 2% or by 20% can significantly affect the business decision. In an example presented in reference [8], it was shown that if the number of areas that packets are allowed to pass through is up to three, the network cost rises by about 40%, and the service fee would go up accordingly. If the network operator desires to limit a rise in the service fee to 20% or less, it is necessary to allow packets to pass through up to four areas in the above example. In this way, a quantitative evaluation can provide network design guidelines.

Assuming the deployment of firewall function in addition to routing function, this paper proposes the joint deployment algorithm of virtual routing function and virtual firewall function to minimize the NFV-based network cost, and derives effective allocation guidelines for two virtual network functions.

The rest of this paper is organized as follows. Section 2 explains related works. Section 3 proposes the joint deployment algorithm of virtual routing function and virtual firewall function to minimize the NFV-based network cost. In particular, it is proposed to place an additional packet filtering function, which is a part of the firewall function, in the sending-side area. Section 4 evaluates the proposed joint deployment algorithm and proposes the allocation guidelines for virtual routing function and the virtual firewall function. Finally, Sect. 5 provides the conclusions.

2 Related Work

One of the related studies is the VNE (virtual network resource allocation) problem, which is a problem of allocating virtual nodes and links to a physical network efficiently [6, 9, 10]. Reference [11] evaluates the allocation of multiple VNFs. Reference [12] proposes a functional allocation assuming a hybrid arrangement in which dedicated hardware and virtual functions are used. References [13, 14] proposes dynamic resource allocation and scheduling. Reference [15] evaluates the virtualization of S-GW (serving GW) and P-GW (packet data network GW) functions in a mobile network in terms of network load and network delay. Based on an evaluation made on a network that simulates the network structure of the U.S.A., it shows that four data centers that completely virtualize these GWs are required. Reference [16] proposes a method for solving the virtual DPI allocation problem with minimum cost.

The authors proposed an resource allocation method in a cloud environment in which computing power and access bandwidths are allocated simultaneously [17–20]. The key point of this method is how to pack as many requirements (simultaneously requiring two different types of resource of computing power and access bandwidths with service quality being not uniform) as possible into multiple sets of computing power and bandwidths.

Most of the above evaluations deal with how to pack (allocate) virtual network functions to finite physical network resources efficiently. In contrast, reference [8] and this paper try to develop allocation guidelines that indicate how much capacity is required for each virtual network function and where each function should be allocated in the network, both of which depend on the relative costs of the routing function and the bandwidth and the inter-area traffic distribution. These guidelines should be required for network carriers when they design and build an NFV-based network.

3 Allocation Algorithm of Routing and Firewall Functions with Minimum Network Cost

3.1 Conditions for Firewall Function Allocation and Proposed Allocation Algorithm

Network functions can be classified into three categories [21]: those that output a smaller volume of traffic than the volume of their input traffic, those that output a volume traffic equal to the volume of their input traffic, and those that output a larger volume of traffic than the volume of their input traffic. The firewall function is one of important network functions and belongs to the first category. It is necessary to deal with unauthorized intrusion prevention in accordance with the situation and condition. Therefore, firewall functions are generally placed on the receiving-side, not on the sending-side. Unlike the routing function, it is difficult to assume that the firewall function of one area will be integrated into that of another area. Thus, it is proposed that the firewall function is left deployed on the receiving-side area and only the packet filtering function, which is a part of its function, is placed in the sending-side area additionally, in order to reduce the network cost. As the packet filtering function filters packets based on only each packet's IP address and port number, and is irrespective of the particular traffic situation, it is easy to place it to another area.

Figure 1 illustrates an example of placing an additional packet filtering function in the sending-side area, assuming that a routing function is allocated in every area. This packet filtering function is the same as that in the firewall function which is placed in the receiving-side area. If each packet filtering function removes 30% of the traffic (100pps in Fig. 1) generated in the sending-side area, it reduces the volume of traffic that is sent to the receiving-side area by 30%. As a result, the bandwidth cost for carrying 60*L bps (L: packet length in bits), the routing function cost for handling 90pps, and the firewall function cost for handling 30pps are respectively saved compared to the case where no packet filtering function is placed in the sending-side area. That is, the additional packet filtering function will be placed only when the total cost reduction is greater than the cost of the additional packet filtering function for handling 100pps.

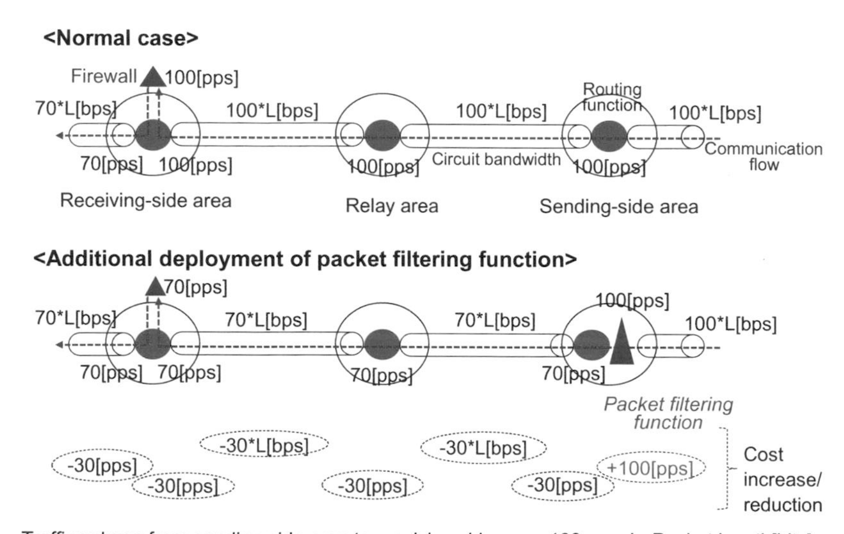

Fig. 1. Example of the effect of additional packet filtering function deployment

3.2 Additional Deployment Judgment on the Packet Filtering Function

Although the additional deployment (not moving) judgment on the packet filtering function could be performed on each sending-side area, it is assumed in this study that all target sending-side areas are collectively carried out. Here, the target sending-side area is an area where the traffic is generated for the receiving-side area.

The following judgment is performed independently for each receiving-side area from an area with a small area number. If the network cost can be reduced when the additional deployment is performed only for the packet filtering function in all target sending-side areas, additional deployment is made to all target sending-side areas. The capacity of packet filtering function additionally deployed is determined as the amount of traffic towards the receiving-side area from each sending-side area. If the network cost cannot be reduced, it is not additionally deployed. Figure 2 shows the additional deployment judgement flow of packet filtering function.

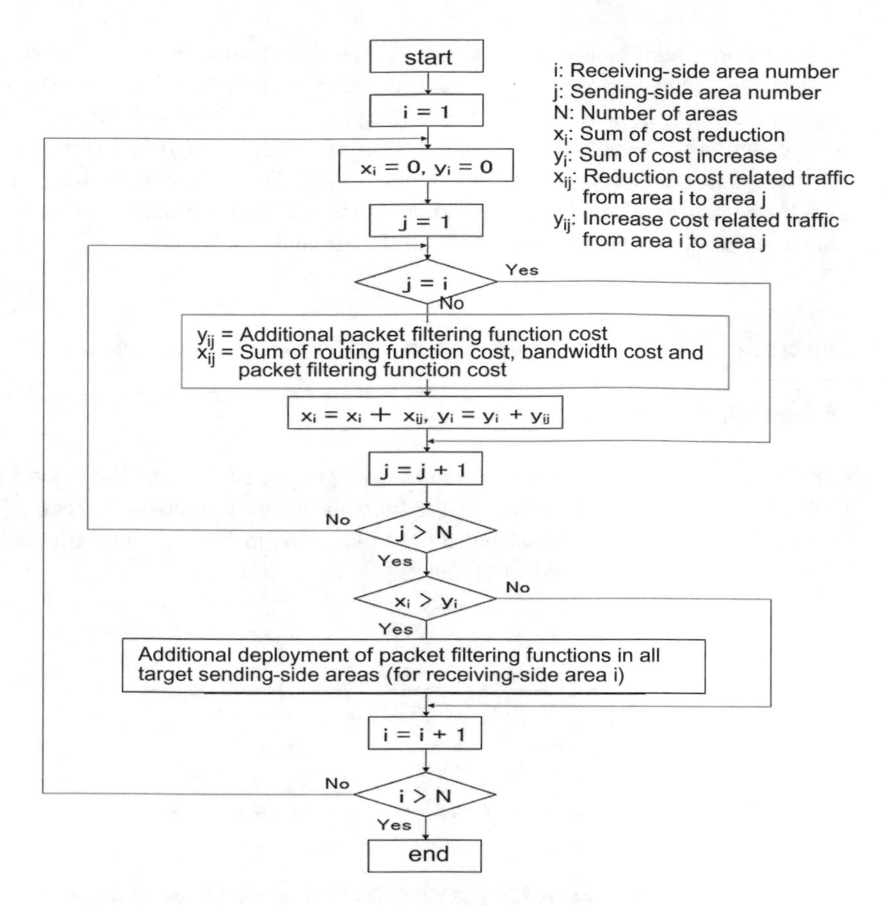

Fig. 2. Additional deployment judgement flow of packet filtering function

3.3 Guidelines for Routing Function Allocation

Guidelines for allocating routing functions need to be studied for each different traffic type.

(1) Input traffic to and output traffic from each area
As discussed in Sect. 3.1, a firewall function is placed in each receiving-side area. Therefore, a virtual routing function to handle the input traffic to and the output traffic from each area is placed in each area. This is different from the previous study [8] where only virtual routing functions were considered.

(2) Relay traffic between areas
As in the case where the allocation of only virtual routing functions was considered [8], virtual routing functions are placed in specific areas in such a way that the network cost is minimized. The statistical multiplexing effects, which can be gained by binding multiple traffic flows, are taken into consideration this time. The following algorithm for routing function allocation is proposed. First, a routing function is placed in every relay area based on the traffic flows between areas.

In other words, relay bandwidths are allocated on the assumption that the statistical multiplexing effect is large enough. Next, the relay area farthest from the network center is selected. If the removal of the routing function in the selected area results in a reduction in the network cost, that routing function is removed. This means that the statistical multiplexing effect is not sought. Then, the next farthest relay area from the network center is selected, and the same decision-making process as above is executed. This is repeated for all the remaining relay areas.

4 Evaluation of the Proposed Function Allocation Algorithm

4.1 Evaluation Conditions

1. We have developed a simulation program in C language that executes the algorithm for allocating the routing function and the firewall function proposed in Sect. 3.
2. Network structure: The ladder-shaped model illustrated in Fig. 3, which simulates the shape of Japan, is used as in Reference [8].

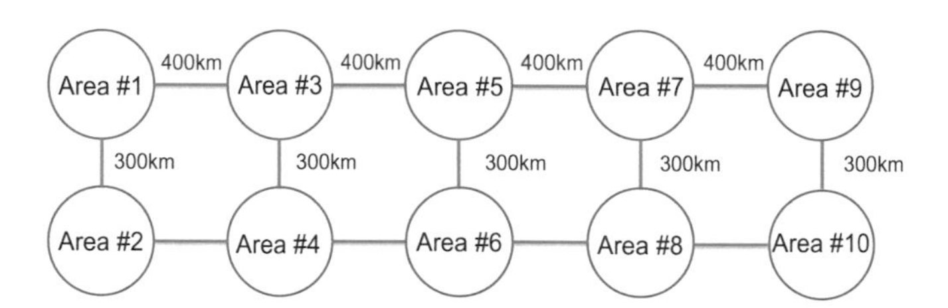

Fig. 3. Ladder-shaped network topology which simulates the shape of Japan

3. Traffic flow: The traffic flow is the same as that used in Reference [8]. It is also supposed that the amount of traffic between Area #5 and each of the remaining area is M times that of traffic between other areas [8].
4. Routing policy: The same policy as in reference [8] is adopted.
5. Network cost calculation: The network cost is greatly affected by the ratio (Z_c) of the routing function cost to the bandwidth cost, and the ratio (W_c) of the firewall function cost to the bandwidth cost. These ratios are defined as follows:

$$Z_C = \alpha_C / \beta_C \quad (1)$$

$$W_C = \gamma_C / \beta_C \quad (2)$$

where α_C is the routing function cost per packet per second, β_C is the bandwidth cost per Mbps per 10 km, and γ_C is the firewall function cost per packet per second. The fixed cost, which is the necessary cost even if traffic is 0, is not taken into consideration here, as in reference [8].

6. Filtering coefficient: The probability at which input traffic passes through a packet filtering function is defined as Pr. For example, Pr = 0.9 means that the 90% of entire traffic passes through while Pr = 0 means that all packets in the traffic are discarded. It is assumed in this study that the value of Pr is the same for all areas.
7. Filtering cost coefficient: The ratio of the packet filtering function cost to the entire firewall function cost is defined as **f**. For example, f = 0.3 means that 30% of the firewall function cost is a packet filtering function cost.
8. Statistical multiplexing coefficient: The ratio of the bandwidth resulting from multiplexing communication flows statistically to the bandwidth required before this multiplexing is defined as **g**. For example, g = 0.7 means that the bandwidth required after multiplexing is 70% of the bandwidth required before multiplexing.

4.2 Evaluation Results and Discussions

The evaluation results are shown in Figs. 4, 6, 7, 8, 10 and 11. Figure 4 shows the effect of Pr, Fig. 6 the effect of W_C, Fig. 7 the effect of f, Fig. 8 the effect of g, and Fig. 10 the effect of Z_C. The vertical axis of each figure shows a normalized total network cost. Figures 4, 6, 7, 8 and 10 also include the data for the solution with the minimum network cost and the details of network cost data. Figure 5 shows the final allocation of additional packet filtering functions. Figures 9 and 11 show the final allocations of routing functions for inter-area relay traffic as affected by g and Z_C respectively. This paper compares the proposed algorithm with the optimal solution which has the lowest cost ('solution with minimum network cost', in each Figure) and can be obtained by checking all possible cases.

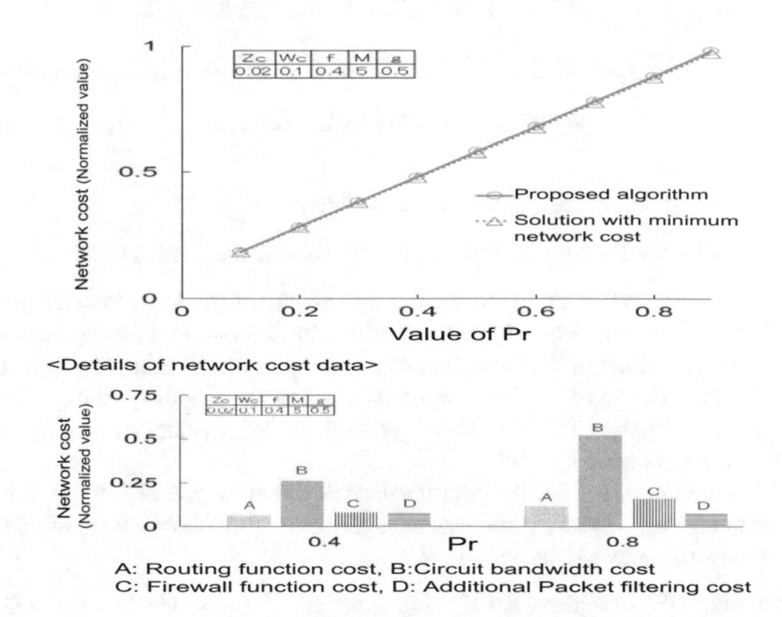

Fig. 4. Impact of Pr on network cost

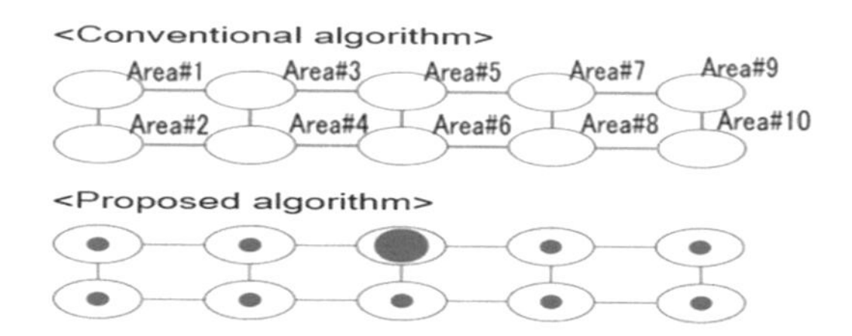

Fig. 5. Final topology of additional packet filtering function allocation to each area

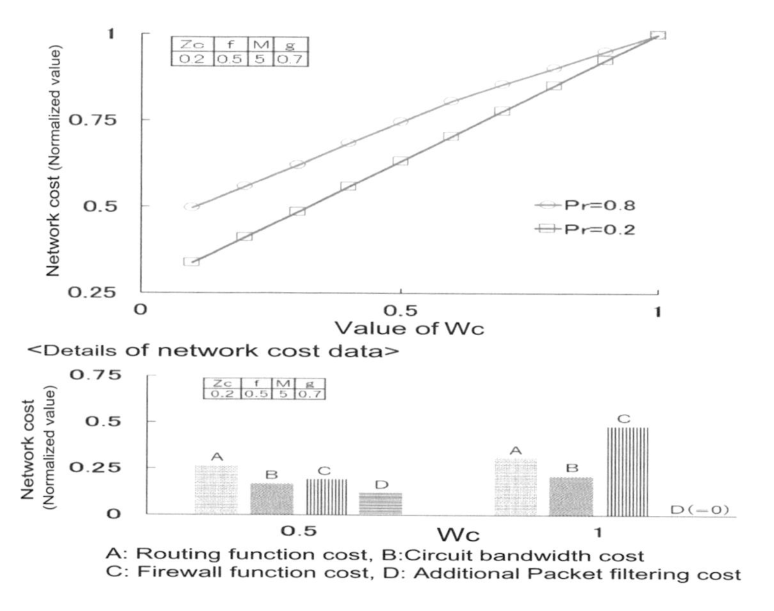

Fig. 6. Impact of Wc on network cost

The following points are clear from these Figures:

(a) The smaller the filtering coefficient, Pr, the smaller the network cost.

<Reason> The smaller Pr is, the smaller the volume of traffic that is passed by the packet filtering function. This reduces the routing function cost and the bandwidth cost. In Fig. 4, the total of the firewall function cost and the packet filtering function cost for the cases of Pr = 0.4 and Pr = 0.8 is more or less the same as that for the case of no additional packet filtering function. However, both the bandwidth cost and the routing function cost are reduced by half.

In the condition of Fig. 4, the additional deployment of packet filtering functions can be advantageous when Pr is not less than 0.7. One of main objectives of this paper is to provide such network design guidelines.

(b) Assuming NFV-based packet filtering function, it is possible to add small additional capacity which could not be realized by the conventional non-NFV based

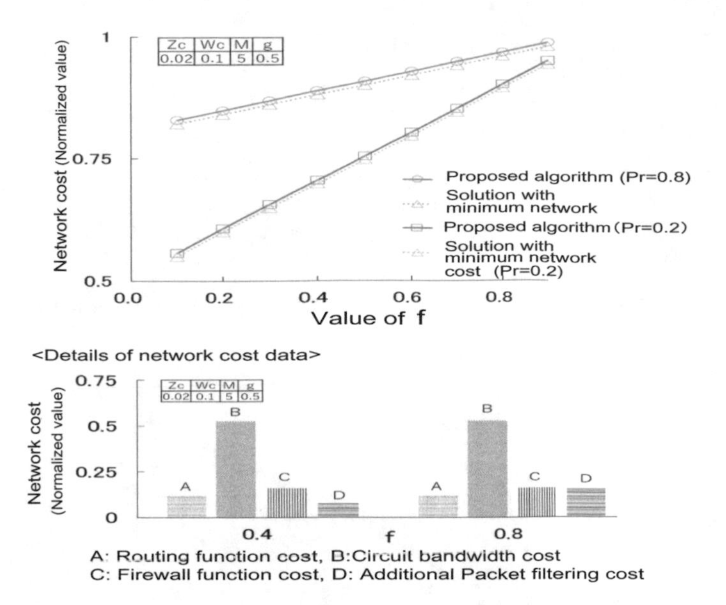

Fig. 7. Impact of **f** on network cost

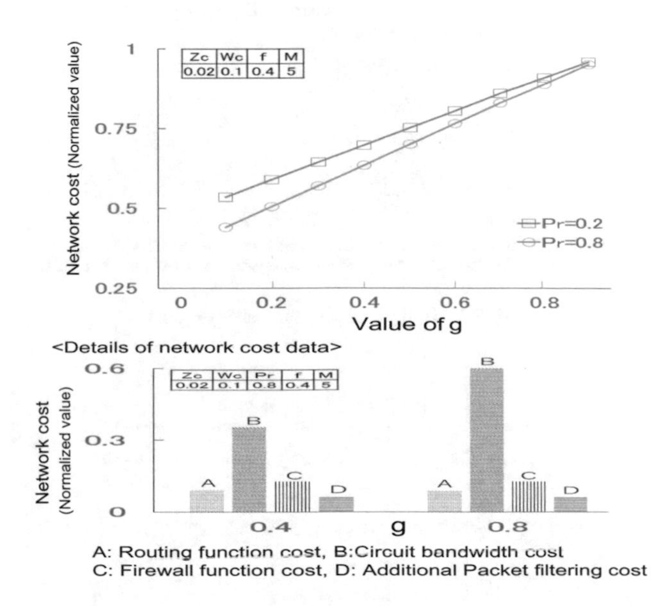

Fig. 8. Impact of **g** on network cost

method. In the example of Pr = 0.7 in Fig. 4, the network cost can be reduced by about 20% compared with the conventional network equipment, by the additional packet filtering deployment based on NFV. The additional packet filtering functions are not placed in all areas as it is not economical in the conventional network

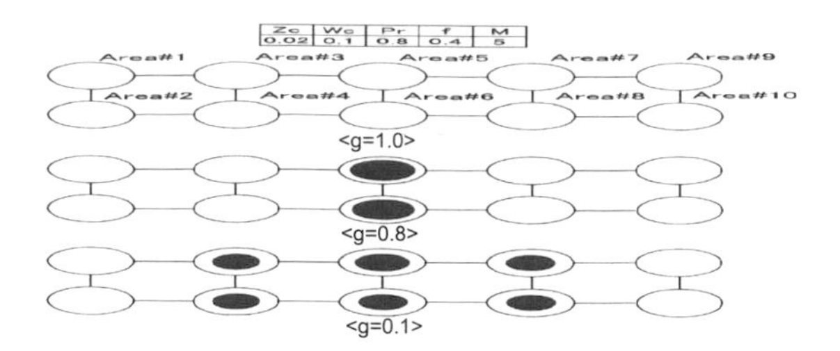

Fig. 9. Final topology of routing function allocation for inter-area relay traffic

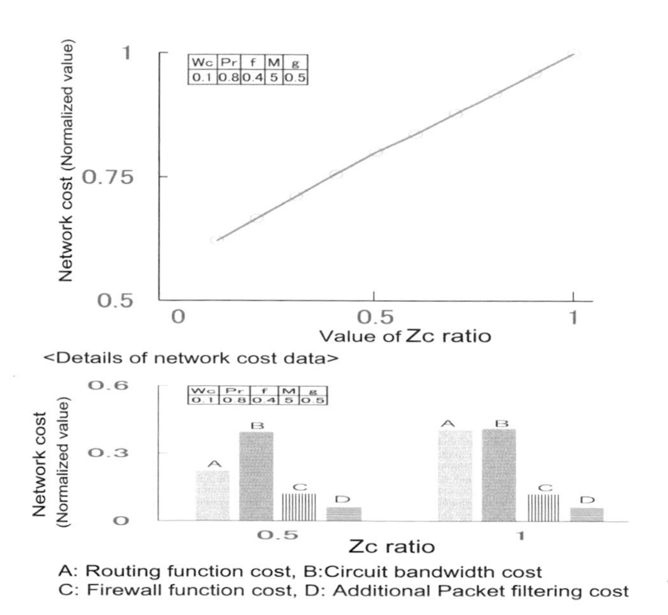

Fig. 10. Impact of Z_C on network cost

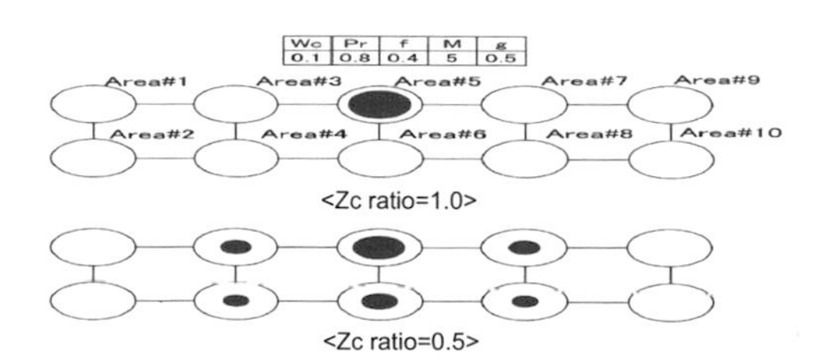

Fig. 11. Final topology of routing function allocation for inter-area relay traffic

design, as illustrated in Fig. 5. It should be an example of the effect unique to NFV.

(c) The smaller Wc is and the smaller f is, the smaller the network cost.

<Reason> As Wc becomes smaller, the costs of the firewall function and the additional packet filtering function become smaller, as shown in Fig. 6. Meanwhile, the routing function cost and the bandwidth cost remain unchanged. Thus, the total network cost is reduced. Moreover, in the example shown in Fig. 7, halving the value of **f** does not change the firewall function cost, the routing function cost or the bandwidth cost, but reduces the cost of the additional packet filtering function. Thus, the network cost can be reduced.

(d) The smaller the statistical multiplexing coefficient, **g**, and the smaller the cost ratio, Zc, the smaller the network cost. Also, the smaller g is and the smaller Zc is, the more distributed the final allocation of routing functions for relay traffic become.

<Reason> The smaller g is, the greater the effect of statistical multiplexing. This results in a smaller demand for bandwidth. As shown in Fig. 8, all types of cost other than the bandwidth cost for the case of g = 0.4 are more or less the same as those for g = 0.8. However, the bandwidth cost is almost halved. Moreover, as g becomes smaller, it becomes more advantageous to allocate routing functions for inter-area relay traffic. As a result, the final allocation of the routing functions for this type of traffic becomes more distributed, as shown in Fig. 9. As shown in Fig. 10, the smaller Zc is, the smaller the network cost. Moreover, as Zc becomes smaller, the amount of reduction in the routing function cost becomes larger, making it more advantageous to allocate routing functions for inter-area relay traffic, as shown in Fig. 11.

(e) The proposed algorithm would be approaching about 95% of the deployment with minimum network cost, as in Figs. 4 and 7.

The above evaluations are based on a ladder-shaped network model with 10 areas (Fig. 3). Even if the number of areas is increased to 100, for example, or a star-shaped or a loop-shaped model is used, the proposed function allocation algorithm can be applied basically. Therefore, the main trends discussed above and the proposed network design guidelines can be also applicable to such cases. However, the additional evaluations are necessary to decide where and how many virtual network functions should be allocated.

5 Conclusions

This paper has proposed the joint deployment algorithm of virtual routing function and virtual firewall function for minimizing the network cost. Our evaluation results have revealed the following: (1) Installing a packet filtering function, which is a part of the firewall function, in the sending-side area additionally can reduce wasteful transit bandwidth and routing processing and thereby reduce the network cost. (2) The greater the number of packets filtered by packet filtering function in the sending-side area, and the higher the bandwidth cost compared with the routing function cost, the more the reduction of network cost is increased. (3) There can be cases where distributing low-capacity routing functions or packet filtering functions in the network, which has been difficult to implement in conventional networks but is made possible by NFV, can dramatically reduce the network cost. (4) This paper has compared the proposed algorithm with the optimal solution which has the lowest cost and can be obtained by checking all possible cases. The proposed algorithm would be approaching about 95% of the deployment with the optimal solution. This paper has also provided the possible network design guideline based on the quantitative evaluation of how much total network cost actually requires.

It will be necessary to evaluate the proposed algorithm taking the fixed costs for routing function, firewall function and bandwidth into consideration. It will be also necessary to study the optimal function allocation for cases where virtual DPI function or virtual cache function is required.

References

1. Chiosi, M., et al.: Network Functions Virtualization—An Introduction, Benefits, Enablers, Challenges and Call for Action. ETSI NFV, October 2012
2. Network Functions Virtualisation (NFV); Architectural Framework. ETSI GS NFV 002 v1.2.1, December 2014
3. file:///C:/Users/Kuribayashi/Downloads/gs_NFV002v010201p.pdf
4. Mijumbi, R., Serrat, J., Gorricho, J., Bouten, N., Trurck, F.D., Boutaba, R.: Network function virtualization: state-of-the-art and research challenges. IEEE Commun. Surv. Tutor. **18**(1), 236–262 (2016)
5. NTTdocomo press release. https://www.nttdocomo.co.jp/english/info/media_center/pr/2016/0219_00.html
6. Herrera, J.G., Botero, J.F.: Resource allocation in NFV: a comprehensive survey. IEEE Trans. Netw. Serv. Manag. **13**(3), 518–532 (2016)
7. Hida, K., Kuribayashi, S.: Virtual routing function allocation method for minimizing total network power consumption. In: 18th International Conference on Information and Communication Systems (ICICS Venice 2016), August 2016
8. Hida, K., Kuribayashi, S.: NFV-based virtual routing function deployment under network delay. In: 19th International Conference on Computer Communications and Applications (ICCCA Kyoto 2017), November 2017
9. Fischer, A., Botero, J.F., Beck, M.T., Meer, H., Hesselbach, X.: Virtual network embedding: a survey. IEEE Commun. Surv. Tutor. **15**(4), 1888–1906 (2013)
10. Wei, X., Hu, S., Li, H., Yang, F., Jin, Y.: A survey on virtual network embedding in cloud computing centers. Open Autom. Control Syst. J. **6**, 414–425 (2014)
11. Mehraghdam, S., Keller, M., Karl, H.: Specifying and placing chains of virtual network functions. In: 2014 IEEE 3rd International Conference on Cloud Networking (CloudNet), pp. 7–13, October 2014
12. Moens, H., Turck, F.D.: VNF-P: a model for efficient placement of virtualized network functions. In: 2014 10th International Conference on Network and Service Management (CNSM), pp. 418–423, November 2014

13. Mijumbi, R., Serrat, J., Gorricho, J., Boutenz, N., Turckz, F., Davy, S.: Design and evaluation of algorithms for mapping and scheduling of virtual network functions. In: 1st IEEE Conference on Network Softwarization (NetSoft), pp. 1–9 (2015)
14. Clayman, S., Mainiy, E., Galis, A., Manzaliniz, A., Mazzocca, N.: The dynamic placement of virtual network functions. In: IEEE Network Operations and Management Symposium (NOMS), pp. 1–9 (2014)
15. Basta, A., Kellerer, W., Hoffmann, M., Morper, H.J., Hoffmann, K.: Applying NFV and SDN to LTE mobile core gateways; the functions placement problem. In: 4th Workshop on All Things Cellular: Operations, Applications and Challenges 2014, pp. 33–38, August 2014
16. Bouet, M., Leguay, J., Conan, V.: Cost-based placement of vDPI functions in NFV infrastructures. Int. J. Netw. Manag. **25**, 490–506 (2015)
17. Tsumura, S., Kuribayashi, S.: Simultaneous allocation of multiple resources for computer communications networks. In: APCC 2006, 2F-4, August 2006
18. Kuribayashi, S.: Optimal joint multiple resource allocation method for cloud computing environments. Int. J. Res. Rev. Comput. Sci. (IJRRCS) **2**(1), 1–8 (2011)
19. Kuribayashi, S.: Resource allocation method for cloud computing environments with different service quality to users at multiple access. Int. J. Comput. Netw. Commun. (IJCNC) **7**(6), 33–51 (2015)
20. Kuribayashi, S.: Evaluation of congestion control methods for joint multiple resource allocation in cloud computing environments. Int. J. Comput. Netw. Commun. (IJCNC) **9**(3), 41–54 (2017)
21. Mehraghdam, S., Keller, M., Karl, H.: Specifying and placing chains of virtual network functions. In: 2014 IEEE 3rd International Conference on Cloud Networking (CloudNet), pp. 7–13, October 2014

Consideration of Policy Information Decision Processes in the Cloud Type Virtual Policy Based Network Management Scheme for the Specific Domain

Kazuya Odagiri[1(✉)], Shogo Shimizu[2], Naohiro Ishii[3], and Makoto Takizawa[4]

[1] Sugiyama Jogakuen University, 17-3 Hosigaoka Motomachi Chiksa-ku, Nagoya, Aichi 464-8662, Japan
kodagiri@sugiyama-u.ac.jp
[2] Gakushuin Women's College, Tokyo, Japan
shogo.shimizu@gakushuin.ac.jp
[3] Aichi Institute of Technology, Toyota, Aichi, Japan
ishii@aitech.ac.jp
[4] Housei University, Tokyo, Japan
makoto.takizawa@computer.org

Abstract. In the current Internet system, there are many problems using anonymity of the network communication such as personal information leaks and crimes using the Internet system. This is why TCP/IP protocol used in Internet system does not have the user identification information on the communication data, and it is difficult to supervise the user performing the above acts immediately. As a study for solving the above problem, there is the study of Policy Based Network Management (PBNM). This is the scheme for managing a whole Local Area Network (LAN) through communication control for every user. In this PBNM, two types of schemes exist. As one scheme, we have studied theoretically about the Destination Addressing Control System (DACS) Scheme with affinity with existing internet. By applying this DACS Scheme to Internet system management, we will realize the policy-based Internet system management. In this paper, to realize management of the specific domain with some network groups with plural organizations, the policy information decision processes applied for this scheme are considered and described.

Keywords: Policy-based network management · DACS scheme
NAPT

1 Introduction

In the current Internet system, there are many problems using anonymity of the network communication such as personal information leaks and crimes using the Internet system. The news of the information leak in the big company is sometimes reported through the mass media. Because TCP/IP protocol used in Internet system does not have the user identification information on the communication data, it is difficult to

L. Barolli et al. (Eds.): NBiS 2018, LNDECT 22, pp. 346–360, 2019.
https://doi.org/10.1007/978-3-319-98530-5_29

supervise the user performing the above acts immediately. As studies and technologies for managing Internet system realized on TCP/IP protocol, those such as Domain Name System (DNS), Routing protocol, Fire Wall (F/W) and Network address port translation (NAPT)/network address translation (NAT) are listed. Except these studies, various studies are performed elsewhere. However, they are the studies for managing the specific part of the Internet system, and have no purpose of solving the above problems.

As a study for solving the problems, Policy Based Network Management (PBNM) [2] exists. The PBNM is a scheme for managing a whole Local Area Network (LAN) through communication control every user, and cannot be applied to the Internet system. This PBNM is often used in a scene of campus network management. In a campus network, network management is quite complicated. Because a computer management section manages only a small portion of the wide needs of the campus network, there are some user support problems. For example, when mail boxes on one server are divided and relocated to some different server machines, it is necessary for some users to update a client machine's setups. Most of computer network users in a campus are students. Because students do not check frequently their e-mail, it is hard work to make them aware of the settings update. This administrative operation is executed by means of web pages and/or posters. For the system administrator, individual technical support is a stiff part of the network management. Because the PBNM manages a whole LAN, it is easy to solve this kind of problem. In addition, for the problem such as personal information leak, the PBNM can manage a whole LAN by making anonymous communication non-anonymous. As the result, it becomes possible to identify the user who steals personal information and commits a crime swiftly and easily. Therefore, by applying the PBNM, we will study about the policy-based Internet system management.

In the existing PBNM, there are two types of schemes. The first is the scheme of managing the whole LAN by locating the communication control mechanisms on the path between network servers and clients. The second is the scheme of managing the whole LAN by locating the communication control mechanisms on clients. It is difficult to apply the first scheme to Internet system management practically, because the communication control mechanism needs to be located on the path between network servers and clients without exception. Because the second scheme locates the communication control mechanisms as the software on each client, it becomes possible to apply the second scheme to Internet system management by devising the installing mechanism so that users can install the software to the client easily.

As the second scheme, we have studied theoretically about the Destination Addressing Control System (DACS) Scheme. As the works on the DACS Scheme, we showed the basic principle of the DACS Scheme, and security function. After that, we implemented a DACS System to realize a concept of the DACS Scheme. By applying this DACS Scheme to Internet system, we will realize the policy-based Internet system management. Then, the Wide Area DACS system (wDACS system) [14] to use it in one organization was showed as the second phase for the last goal. As the first step of the second phase, we showed the concept of the cloud type virtual PBNM, which could be used by plural organizations [15]. As a step of the third phase, we showed the concept of the cloud type virtual PBNM, which could be used by plural organizations [16]. In this paper, as the progression phase, the policy information decision processes for the proposed scheme.is considered. In Sect. 2, motivation and related research for this study are described. In Sect. 3, the existing DACS Scheme and wDACS Scheme is described. In Sect. 4, after the user authentication processes for the scheme are explained, the policy information decision processes are described.

2 Motivation and Related Research

In the current Internet system, problems using anonymity of the network communication such as personal information leak and crimes using the Internet system occur. Because TCP/IP protocol used in Internet system does not have the user identification information on the communication data, it is difficult to supervise the user performing the above acts immediately.

As studies and technologies for Internet system management to be comprises of TCP/IP [1], many technologies are studied. For examples, Domain name system (DNS), Routing protocol such as Interior gateway protocol (IGP) such as Routing information protocol (RIP) and Open shortest path first (OSPF), Fire Wall (F/W), Network address translation (NAT)/Network address port translation (NAPT), Load balancing, Virtual private network (VPN), Public key infrastructure (PKI), Server virtualization. Except these studies, various studies are performed elsewhere. However, they are for managing the specific part of the Internet system, and have no purpose of solving the above problems.

As a study for solving the above problem, the study area about PBNM exists. This is a scheme of managing a whole LAN through communication control every user. Because this PBNM manages a whole LAN by making anonymous communication non-anonymous, it becomes possible to identify the user who steals personal information and commits a crime swiftly and easily. Therefore, by applying this policy-based thinking, we study about the policy-based Internet system management.

In policy-based network management, there are two types of schemes. The first scheme is the scheme described in Fig. 1. The standardization of this scheme is performed in various organizations. In IETF, a framework of PBNM [2] was established. Standards about each element constituting this framework are as follows. As a model of control information stored in the server called Policy Repository, Policy Core Information model (PCIM) [3] was established. After it, PCMIe [4] was established by extending the PCIM. To describe them in the form of Lightweight Directory Access Protocol (LDAP), Policy Core LDAP Schema (PCLS) [5] was established. As a protocol to distribute the control information stored in Policy Repository or decision result from the PDP to thc PEP, Common Open Policy Service (COPS) [6] was established. Based on the difference in distribution method, COPS usage for RSVP (COPS-RSVP) [7] and COPS usage for Provisioning (COPS-PR) [8] were established. RSVP is an abbreviation for Resource Reservation Protocol. The COPS-RSVP is the method as follows. After the PEP having detected the communication from a user or a client application, the PDP makes a judgmental decision for it. The decision is sent and applied to the PEP, and the PEP adds the control to it. The COPS-PR is the method of distributing the control information or decision result to the PEP before accepting the communication.

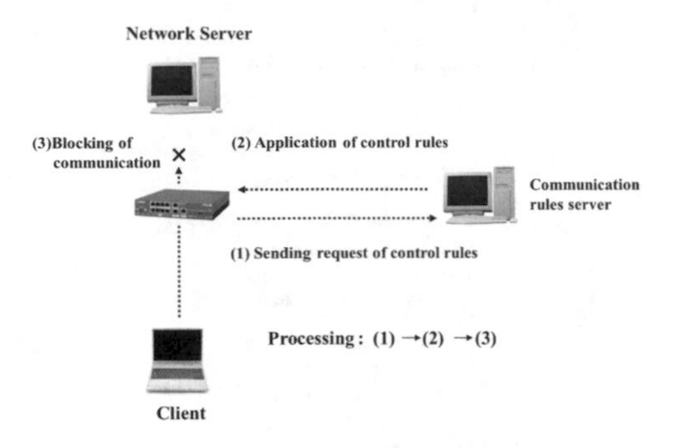

Fig. 1. Principle in first scheme

Next, in DMTF, a framework of PBNM called Directory-enabled Network (DEN) was established. Like the IETF framework, control information is stored in the server storing control information called Policy Server, which is built by using the directory service such as LDAP [9], and is distributed to network servers and networking equipment such as switch and router. As the result, the whole LAN is managed. The model of control information used in DEN is called Common Information Model (CIM), the schema of the CIM (CIM Schema Version 2.30.0) [11] was opened. The CIM was extended to support the DEN [10], and was incorporated in the framework of DEN. In addition, Resource and Admission Control Subsystem (RACS) [12] was established in Telecoms and Internet converged Services and protocols for Advanced Network (TISPAN) of European Telecommunications Standards Institute (ETSI), and Resource and Admission Control Functions (RACF) was established in International Telecommunication Union Telecommunication Standardization Sector (ITU-T) [13].

However, all the frameworks explained above are based on the principle shown in Fig. 1. As problems of these frameworks, two points are presented as follows. Essential principle is described in Fig. 2. To be concrete, in the point called PDP (Policy Decision Point), judgment such as permission or non-permission for communication pass is performed based on policy information. The judgment is notified and transmitted to the point called the PEP, which is the mechanism such as VPN mechanism, router and Fire Wall located on the network path among hosts such as servers and clients. Based on that judgment, the control is added for the communication that is going to pass by.

The principle of the second scheme is described in Fig. 3. By locating the communication control mechanisms on the clients, the whole LAN is managed. Because this scheme controls the network communications on each client, the processing load is low. However, because the communication control mechanisms needs to be located on each client, the work load becomes heavy.

When it is thought that Internet system is managed by using these two schemes, it is difficult to apply the first scheme to Internet system management practically. This is why the communication control mechanism needs to be located on the path between network servers and clients without exception. On the other hand, the second scheme locates the communication controls mechanisms on each client. That is, the software

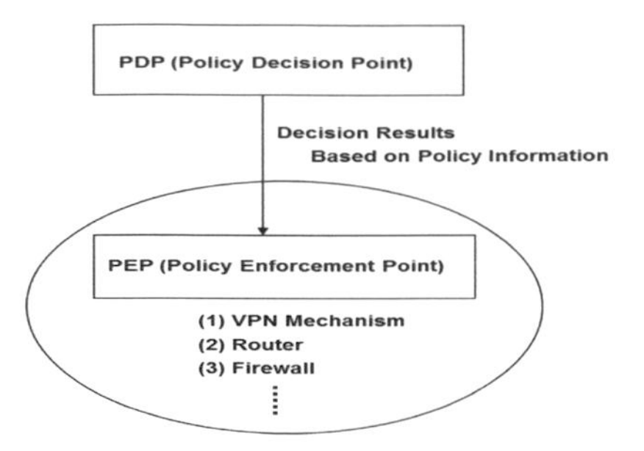

Fig. 2. Essential principle

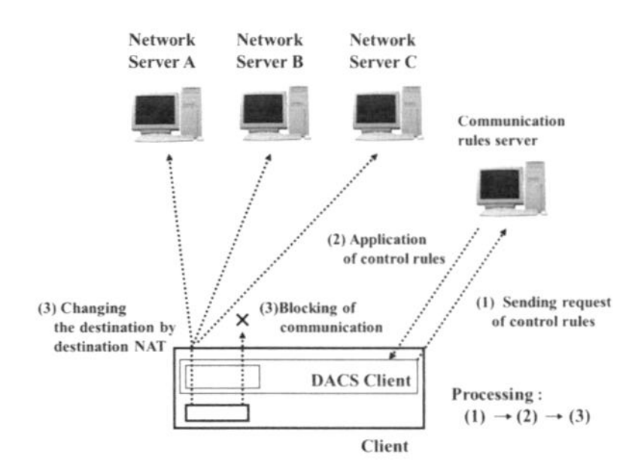

Fig. 3. Principle in second scheme

for communication control is installed on each client. So, by devising the installing mechanism letting users install software to the client easily, it becomes possible to apply the second scheme to Internet system management. As a first step for the last goal, we showed the Wide Area DACS system (wDACS) system [15]. This system manages a wide area network, which one organization manages. Therefore, it is impossible for plural organizations to use this system. In order to improve it, we showed the cloud type virtual PBNM, which could be used by plural organizations. After it, to expand its application area, the scheme to manage the specific domain and the user authentication processes for that scheme are examined. In this paper, the policy information decision processes which are performed after the user authentication are examined.

3 Existing DACS SCHEME and wDACS System

3.1 Basic Principle of the DACS Scheme

Figure 4 shows the basic principle of the network services by the DACS Scheme. At the timing of the (a) or (b) as shown in the following, the DACS rules (rules defined by the user unit) are distributed from the DACS Server to the DACS Client.

(a) At the time of a user logging in the client.

(b) At the time of a delivery indication from the system administrator.

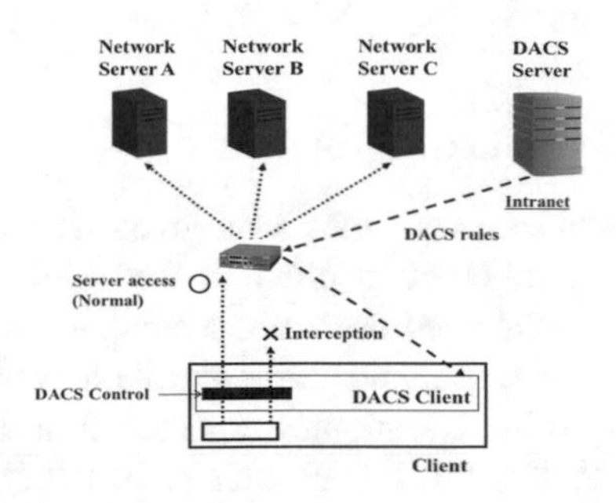

Fig. 4. Basic principle of the DACS Scheme

According to the distributed DACS rules, the DACS Client performs (1) or (2) operation are shown in the following. Then, communication control of the client is performed for every login user.

1. Destination information on IP Packet, which is sent from application program, is changed.
2. IP Packet from the client, which is sent from the application program to the outside of the client, is blocked.

An example of the case (1) is shown in Fig. 4. In Fig. 4, the system administrator can distribute a communication of the login user to the specified server among servers A, B or C. Moreover, the case (2) is described. For example, when the system administrator wants to forbid a user to use MUA (Mail User Agent), it will be performed by blocking IP Packet with the specific destination information.

In order to realize the DACS Scheme, the operation is done by a DACS Protocol as shown in Fig. 5. As shown by (1) in Fig. 5, the distribution of the DACS rules is performed on communication between the DACS Server and the DACS Client, which is arranged at the application layer. The application of the DACS rules to the DACS Control is shown by (2) in Fig. 5.

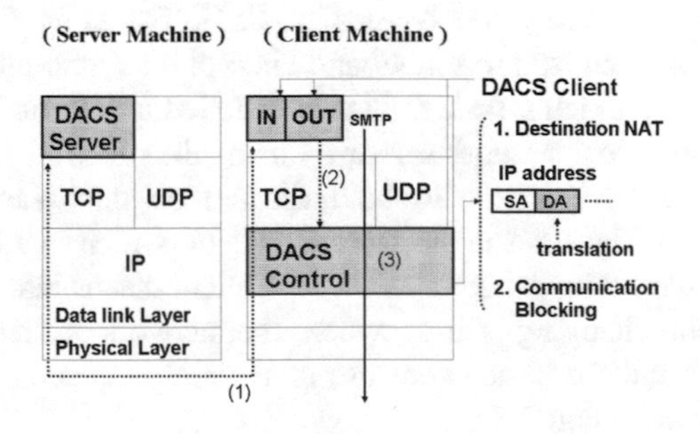

Fig. 5. Layer setting of the DACS Scheme

The steady communication control, such as a modification of the destination information or the communication blocking is performed at the network layer as shown by (3) in Fig. 5.

3.2 Communication Control on Client

The communication control on every user was given. However, it may be better to perform communication control on every client instead of every user. For example, it is the case where many and unspecified users use a computer room, which is controlled. In this section, the method of communication control on every client is described, and the coexistence method with the communication control on every user is considered.

When a user logs in to a client, the IP address of the client is transmitted to the DACS Server from the DACS Client. Then, if the DACS rules corresponding to IP address, is registered into the DACS Server side, it is transmitted to the DACS Client. Then, communication control for every client can be realized by applying to the DACS Control. In this case, it is a premise that a client uses a fixed IP address. However, when using DHCP service, it is possible to carry out the same control to all the clients linked to the whole network or its subnetwork for example.

When using communication control on every user and every client, communication control may conflict. In that case, a priority needs to be given. The judgment is performed in the DACS Server side as shown in Fig. 6. Although not necessarily stipulated, the network policy or security policy exists in the organization such as a university (1). The priority is decided according to the policy (2). In (a), priority is given for the user's rule to control communication by the user unit. In (b), priority is given for the client's rule to control communication by the client unit. In (c), the user's rule is the same as the client's rule. As the result of comparing the conflict rules, one rule is determined respectively. Those rules and other rules not overlapping are gathered, and the DACS rules are created (3). The DACS rules are transmitted to the DACS Client. In the DACS Client side, the DACS rules are applied to the DACS Control. The difference between the user's rule and the client's rule is not distinguished.

3.3 Security Mechanism of the DACS Scheme

In this section, the security function of the DACS Scheme is described. The communication is tunneled and encrypted by use of SSH. By using the function of port forwarding of SSH, it is realized to tunnel and encrypt the communication between the network server and the, which DACS Client is installed in. Normally, to communicate from a client application to a network server by using the function of port forwarding of SSH, local host (127.0.0.1) needs to be indicated on that client application as a communicating server. The transparent use of a client, which is a characteristic of the DACS Scheme, is failed. The transparent use of a client means that a client can be used continuously without changing setups when the network system is updated. The function that doesn't fail the transparent use of a client is needed. The mechanism of that function is shown in Fig. 7.

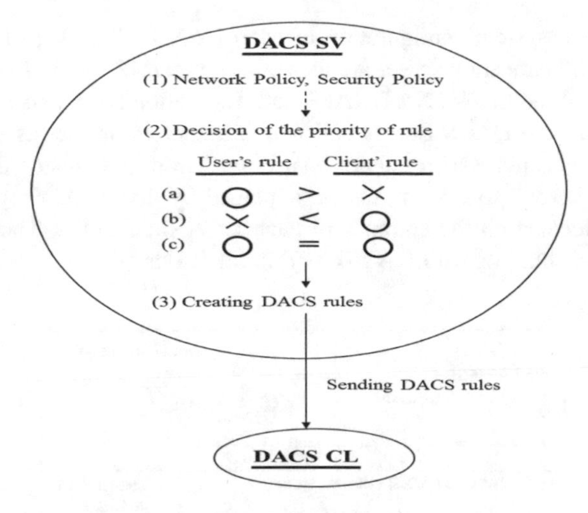

Fig. 6. Creating the DACS rules on the DACS Server

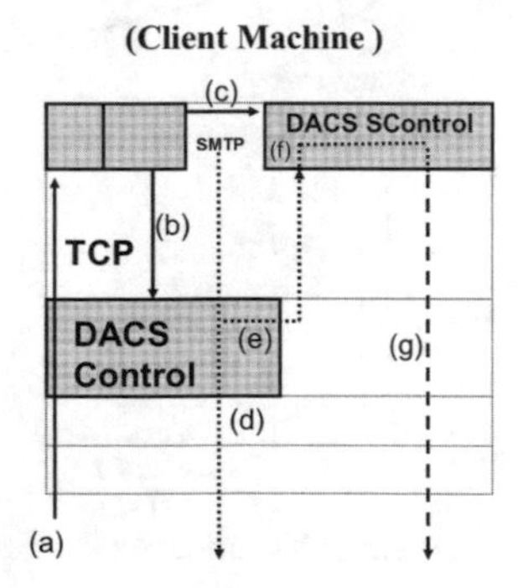

Fig. 7. Extend security function

3.4 Application to Cloud Environment

In this section, the contents of wDACS system are explained in Fig. 8. First, as preconditions, because private IP addresses are assigned to all servers and clients existing in from LAN1 to LAN n, mechanisms of NAT/NAPT are necessary for the communication from each LAN to the outside. In this case, NAT/NAPT is located on the entrance of the LAN such as (1), and the private IP address is converted to the global IP address towards the direction of the arrow. Next, because the private IP addresses are set on the servers and clients in the LAN, other communications except those converted by Destination NAT cannot enter into the LAN. But, responses for the communications sent from the inside of the LAN can enter into the inside of the LAN because of the reverse conversion process by the NAT/NAPT. In addition, communications from the outside of the LAN1 to the inside are performed through the conversion of the destination IP address by Destination NAT. To be concrete, the global IP address at the same of the outside interface of the router is changed to the private IP address of each server. From here, system configuration of each LAN is described. First, the DACS Server and the authentication server are located on the DMZ on the LAN1 such as (4). On the entrance of the LAN1, NAT/NAPT and destination NAT exists such as (1) and (2). Because only the DACS Server and network servers are set as the target destination, the authentication server cannot be accessed from the outside of the LAN1. In the LANs form LAN 2 to LAN n, clients managed by the wDACS system exist, and NAT/NAPT is located on the entrance of each LAN such as (1). Then, F/W such as (3) or (5) exists behind or with NAT/NAPT in all LANs.

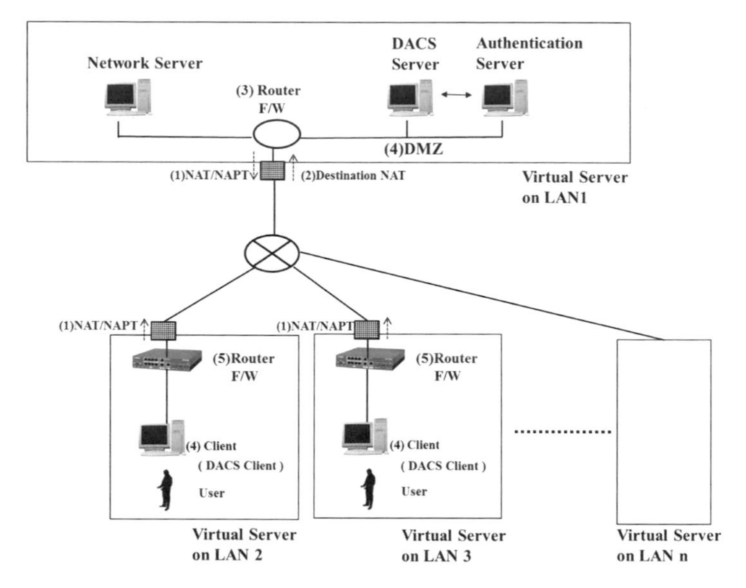

Fig. 8. Basic system configuration of wDACS system

3.5 The Cloud Type Virtual PBNM for the Common Use Between Plural Organizations

In this section, after the concept and implementation of the proposed scheme were described, functional evaluation results are described.

In Fig. 9 which is described in [16], the proposed concept is shown. Because the existing wDACS Scheme realized the PBNM control with the software called the DACS Server and the DACS client, other mechanism was not needed. By this point, application to the cloud environment was easy.

The proposed scheme in this paper realizes the common usage by plural organizations by adding the following elements to realize the common usage by plural organizations: user identification of the plural organizations, management of the policy information of the plural organizations, application of the PKI for code communication in the Internet, Redundant configuration of the DACS Server (policy information server), load balancing configuration of the DACS Server, installation function of DACS Client by way of the Internet.

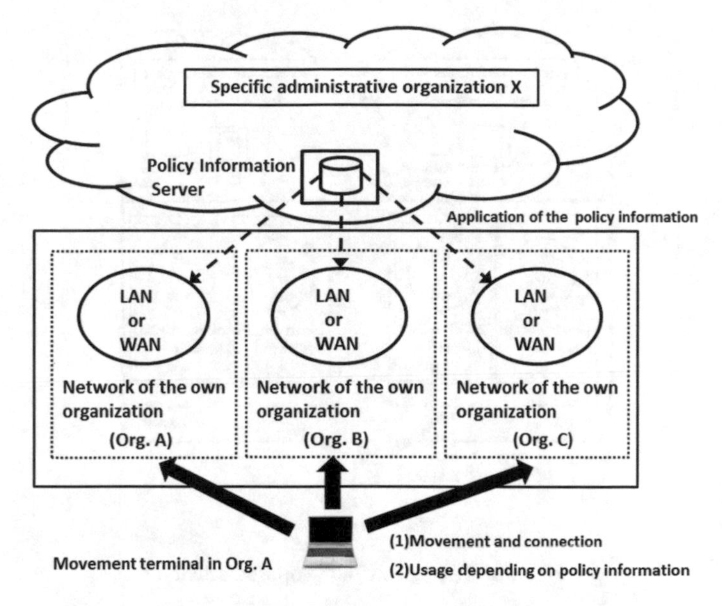

Fig. 9. Cloud type virtual PBNM for the common use between plural organizations

In the past study [14], the DACS Client was operated on the windows operation system (Windows OS). It was because there were many cases that the Windows OS was used for as the OS of the client. However, the Linux operating system (Linux OS) had enough functions to be used as the client recently, too. In addition, it was thought that the case used in the clients in the future came out recently. Therefore, to prove the possibility of the DACS Scheme on the Linux OS, the basic function of the DACS Client was implemented in this study. The basic functions of the DACS Server and DACS Client were implemented by JAVA language.

4 Policy Information Decision Processes for the Scheme to Manage the Specific Domain

In this section, after the user authentication processes applied for the scheme to manage the specific domain was shown in Fig. 10, the policy information decision processes are examined and described.

4.1 Management Scheme for the Specific Domain

This is a scheme to manage the plural networks group. In Fig. 10, the concept is explained. Specifically, as a logical range to manage organization A and organization B, network group 1 exists. Similarly, as a logical range to manage organization C and organization D, network group 2 exists. These individual network groups are existing methods listed in Fig. 9. When plural network groups managed by this existing scheme

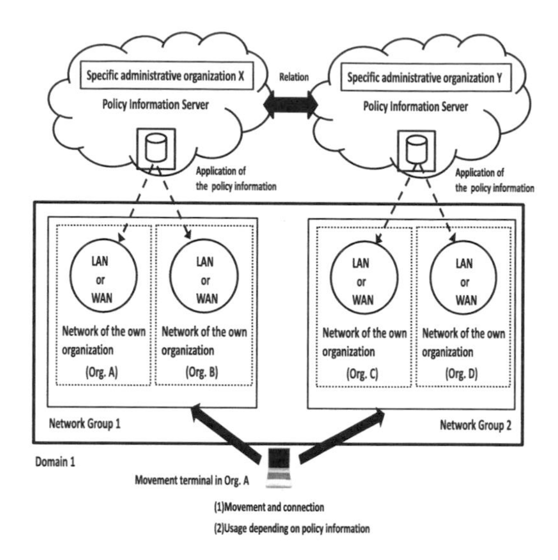

Fig. 10. Concept of the proposed scheme

exist, those plural network groups are targeted for management by this proposed method.

For example, when user A belonging to org. A in network group1 uses the network which org. C belonging to network group2 which is a different network group holds, administrative organization Y for network group2 refers for policy information of user A for administrative organization X of network group1 and acquires it. After it, in the form that policy information registered with Network Group2 beforehand is collated with the policy information, the final policy information is decided. As a result, the policy information is applied to the client that user A uses in network group2, and the communication control on the client is performed. When a user moves plural network groups as well as the specific network group, it is thought that the PBNM scheme to keep a certain constant management state is realized.

To realize this scheme, it is necessary to consider the following three factors (Fig. 11).

(Factor1) Method of user authentication
(Factor2) Determination method of the policy information
(Factor3) Distribution method of the policy information

Here, it is explained about the authentication processes when the client is connected to the network group2 by the user belonging to the network group1. First, between two administrative organization (organization X and Y), trust relationship is establised. As the result, the IP address of each other's DACS SV as policy information server are exchanged. This is done in advance before the authentication process. After it, authentication processes are performed as follows.

(1) Input of authentication information by the user

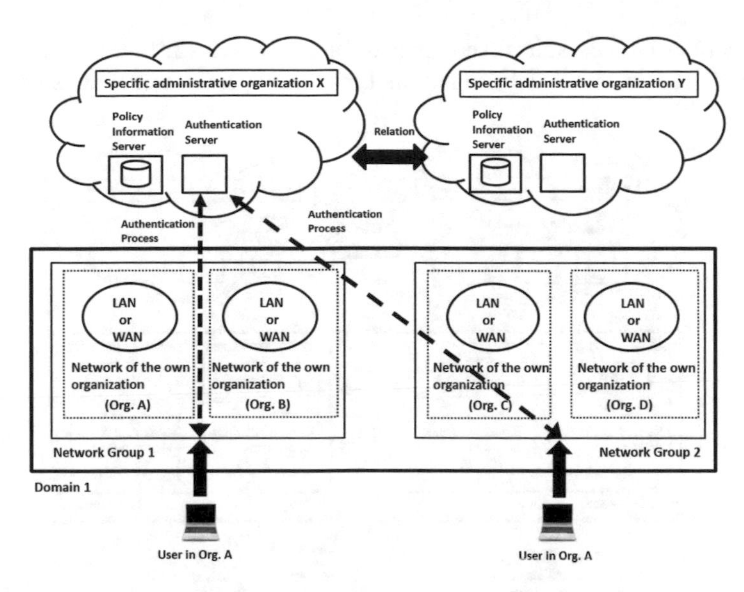

Fig. 11. The proposed user authentication method

When the client is connected to the network group2, and the power of it is turned on. After an operation system was started up, the DACS CL is started up. In the middle process, the input pox for entering authentication information is displayed on the screen of the client. The user inputs user name and password, name or IP address of the authentication server in network group1.

(2) Authentication request from the client
By use of the authentication information, authentication processes are performed in the form of encrypted communication by SSL. The authentication server is specified by the name of IP address the user inputs the input box.

(3) Authentication process at the authentication server side
Based on the user name and password, the user authentication is performed. In this scheme, as the authentication server, the ldap server which is constituted by the openldap is used. The user name and password are compared with the user account information. As a result of the comparison, when the authentication information matches, the IP address of the DACS SV in the network group 2 is notified to the client. When the authentication information does not match, the client is placed in a state in which no communication is possible based on the control by the DACS CL.

4.2 Policy Information Decision Processes for This Scheme

At first, concept of policy information decision method is described. In Fig. 12, as an example, the user belonging to the organization A connects the notebook client computer to another organization in Network Group 2. In this case, the DACS rules as policy information are extracted from the DACS SV as policy information server which are located in both network group (Network Group 1 and Network Group 2). After it, the DACS rules are determined and applied for the DACS CL in the client computer. When the DACS rules are determined, multiple points are considered as follows.

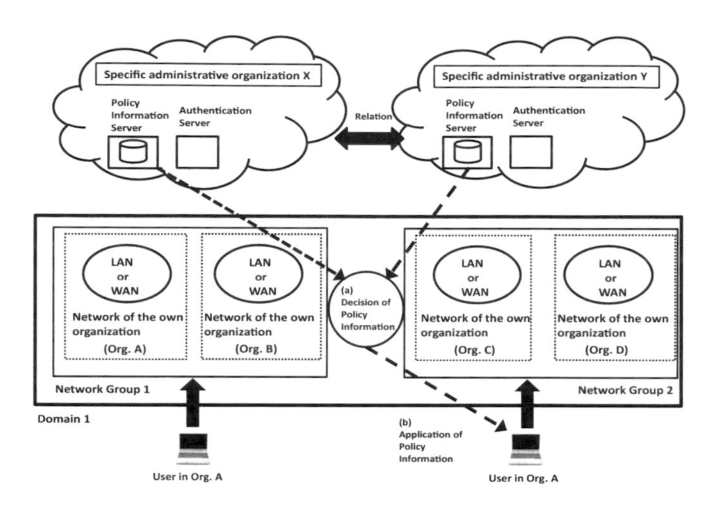

Fig. 12. Concept of policy information decision method

(Point1) Use of the information service of the organization to which the user belongs
(Point2) Use of the information service of the organization to which the user connects
(Point3) Use of the Packet Filtering Service

As the first example, with the user in organization A connected the own client to the network of network group 2, the user accesses the mail system in own organization. At that time, to enable the client to access the mail service, the communication must be permitted by packet filtering setting. In the setting of the mail client in the client, the server name of the mail system of its own organization is normally specified. As the result, it becomes possible to access the mail system from the client. In this case, as the packet filtering setting, the policy information transmitted from the policy information server of the management organization X is reflected on the client. As the second example, with the user in organization A connected the own client to the network of network group 2, the user accesses the mail system of the connected organization, like the first example, to enable the client to access the mail service, the communication must be permitted by packet filtering setting. In addition, because the server name of the mail system of its own organization is normally specified in the setting of the mail client in the client, it is necessary to change the mail system to be accessed by communication control by NAT. As the policy information for these two communication controls, the one transmitted from the policy information server of the connected organization is reflected to the client. After it, it becomes possible to access the mail system of the connected organization from the client. However, in the case of example 2, the policy information used in example 1 also exists, the policy information used in the above two examples is set in both management organizations. In this case, the fact that the policy information is set in the connected organization means that the service use of the connected organization is permitted. Therefore, the policy information sent from the connected organization should be given priority.

5 Conclusion

In this paper, the policy information decision processes for the proposed scheme are examined. Considering affinity with the Internet system, these processes were examined. As the future study, we will implement the prototype system including the user authentication and the policy information decision processes for this scheme.

Acknowledgments. This work was supported by the research grant of JSPS KAKENHI Grant Number 17K00113, and the research grant by Support Center for Advanced Telecommunications Technology Research, Foundation (SCAT). We express our gratitude.

References

1. Cerf, V., Kahn, E.: A protocol for packet network interconnection. IEEE Trans. Commun. **COM-22**, 637–648 (1974)
2. Yavatkar, R., Pendarakis, D., Guerin, R.: A Framework for Policy-based Admission Control. IETF RFC 2753 (2000)
3. Moore at el., B.: Policy Core Information Model—Version 1 Specification. IETF RFC 3060 (2001)
4. Moore, B.: Policy Core Information Model (PCIM) Extensions. IETF 3460 (2003)
5. Strassner, J., Moore, B., Moats, R., Ellesson, E.: Policy Core Lightweight Directory Access Protocol (LDAP) Schema. IETF RFC 3703 (2004)
6. Durham, D., et al.: The COPS (Common Open Policy Service) Protocol. IETF RFC 2748 (2000)
7. Herzog, S., et al.: COPS Usage for RSVP. IETF RFC 2749 (2000)
8. Chan, K., et al.: COPS Usage for Policy Provisioning (COPS-PR). IETF RFC 3084 (2001)
9. CIM Core Model V2.5 LDAP Mapping Specification (2002)
10. Wahl, M., Howes, T., Kille, S.: Lightweight Directory Access Protocol (v3). IETF RFC 2251 (1997)
11. CIM Schema: Version 2.30.0 (2011)
12. ETSI ES 282 003: Telecoms and Internet converged Services and protocols for Advanced Network (TISPAN); Resource and Admission Control Subsystem (RACS); Functional Architecture, June 2006
13. ETSI ETSI ES 283 026: Telecommunications and Internet Converged Services and Protocols for Advanced Networking (TISPAN); Resource and Admission Control; Protocol for QoS reservation information exchange between the Service Policy Decision Function (SPDF) and the Access-Resource and Admission Control Function (A-RACF) in the Resource and Protocol specification, April 2006
14. Odagiri, K., Shimizu, S., Takizawa, M., Ishii, N.: Theoretical suggestion of policy-based wide area network management system (wDACS system part-I). Int. J. Netw. Distrib. Comput. (IJNDC) **1**(4), 260–269 (2013)

15. Odagiri, K., Shimizu, S., Ishii, N., Takizawa, M.: Suggestion of the cloud type virtual policy based network management scheme for the common use between plural organizations. In: Proceedings of the International Conference on Network-Based Information Systems (NBiS-2015), pp. 180–186, September 2015
16. Odagiri, K., Shimizu, S., Ishii, N., Takizawa, M.: Suggestion of the cloud type virtual policy based network management scheme for the common use between plural organizations. In: Proceedings of the International Conference on International Conference on Network-Based Information Systems (NBiS-2015), pp. 180–186, September 2015

Mobile Interactions and Computation Offloading in Drop Computing

Radu-Ioan Ciobanu[1(✉)] and Ciprian Dobre[2]

[1] Faculty of Automatic Control and Computers, University Politehnica of Bucharest, Bucharest, Romania
radu.ciobanu@cs.pub.ro

[2] National Institute for Research and Development in Informatics, Bucharest, Romania
ciprian.dobre@ici.ro

Abstract. In recent years, the amount of data consumed by mobile devices has grown exponentially, especially with the advent of the Internet of Things and all its connected devices. For this reason, researchers are looking for methods of alleviating the congestion and strain on the network, generally through various means of offloading, or by bringing the data and computations closer to the devices themselves through edge and fog computing. Thus, in this paper we propose an extension to the Drop Computing paradigm, which introduces the concept of decentralized computing over multilayered networks. We present a novel offloading technique to be employed by Drop Computing nodes for increasing processing speed, reducing deployment costs and lowering mobile device battery consumption, by using the crowd of mobile nodes belonging to humans and the edge devices as opportunities for offloading data and computations. We compare our method with the initial Drop Computing implementation and with the default scenario for mobile applications and show that it is able to improve the overall network performance. We also perform an analysis of human interactions with two monitoring nodes located in an academic environment, to obtain realistic data and to extract behavior patterns regarding human habits and interactions, that aid us in developing an efficient offloading solution.

1 Introduction

Because the number of mobile devices connected to the Internet has been growing lately, new paradigms such as edge and fog computing have garnered plenty of attention. They alleviate the strain placed on the network by the large amounts of data exchanges by bringing information closer to where it is needed, at the edge of the network. Through offloading, edge devices are able to aid the mobile nodes (smart devices, Internet of Things sensors, etc.) by fulfilling requests faster.

In this paper, we aim to improve the performance of Drop Computing [1], a paradigm which offers decentralized computing for mobile nodes in networks that combine cloud and wireless technologies over a social crowd formed between

L. Barolli et al. (Eds.): NBiS 2018, LNDECT 22, pp. 361–373, 2019.
https://doi.org/10.1007/978-3-319-98530-5_30

mobile and edge devices. In the initial Drop Computing proposal, mobile nodes that had some computation tasks to solve would either offload them directly to a cloud, or would request help from other nodes in the network in an opportunistic device-to-device fashion. In this paper, we propose enhancing Drop Computing by adding edge devices to the paradigm, which can act as more powerful nodes that can perform computations, or as relays for the mobile nodes, helping them offload their computations to the cloud without needing to use up their data plan by employing mobile broadband communication. For this purpose, we present an experiment where we used two mobile nodes for collecting Wi-Fi interaction data. We analyze this data to observe offloading opportunities using edge devices and, based on the conclusions obtained, we propose an extended Drop Computing variant.

The remainder of this paper is structured as follows. In Sect. 2, we present the original Drop Computing paradigm and other similar solutions that perform data and computation offloading in mobile networks for various purposes, highlighting the benefits that our framework can bring. In Sect. 3, we perform an analysis on Wi-Fi data collected using two local static nodes, showing the behavior of humans in an academic environment and the possibilities for offloading that come up in such a scenario. Section 4 presents the edge-based offloading mechanism that we propose for Drop Computing, while the results we obtained are analyzed in Sect. 5. Finally, we extract some conclusions and discuss future work in Sect. 6.

2 Related Work

The solution we propose here is an extension of the Drop Computing paradigm [1], which advances the concept of decentralized computing over multi-layered networks, combining cloud and wireless technologies over a social crowd formed between mobile and edge devices. The need for Drop Computing arises from the insufficiency of the classic cloud model, which is not suitable anymore for the Internet of Things. When a large number of small devices communicate, they all need to send requests to the cloud, receive them back, and then process them. In Drop Computing, mobile devices and people interconnect to form ad-hoc dynamic collaborations to support the equivalent of a crowd-based edge multilayered cloud of clouds, where the capabilities of any mobile device are extended beyond the local technology barriers, to accommodate external resources available in the crowd of other devices. Thus, instead of every data or computation request going directly to the cloud, Drop Computing offloads requests towards the mobile crowd formed of edge nodes and devices in close proximity for quicker and more efficient access. Devices in the mobile crowd are leveraged for requesting already downloaded data or performing computations, and the cloud acts as the second (or even third) option.

There are two main approaches to traffic offloading in mobile networks [2,3], namely through Wi-Fi access points or in a terminal-to-terminal (T2T) fashion, where nearby nodes are used as a cache (if they already have the data stored) or as a relay. Several existing solutions employ a hybrid approach [4,5], where the two types of offloading are used concurrently. This is also what Drop Computing does, through the crowd of nodes and edge devices.

Several solutions that are somewhat similar to the idea of Drop Computing have been proposed in the literature, but they all have some drawbacks that we aim to address. One such solution is a framework for creating virtual mobile cloud computing providers [6], where users with common goals collaborate by computing parts of a task and merging their results. However, this solution does not account for user mobility and for heterogeneous devices. Another example is proposed in [7], where mobile devices owned by different people help each other perform computational tasks, but it has the drawback that nodes are only able to communicate with each other one hop away, which limits the collaboration opportunities considerably. Furthermore, in case offloading to another device cannot be performed, there is no cloud infrastructure or edge device that can take over, as is the case with Drop Computing. Further solutions include mCloud [8] (which runs resource-intensive applications on cooperating mobile devices while using the cloud as a backup, but only allows nodes to communicate in a single-hop fashion), cloudlets [9] (where trusted nodes are located in the vicinity of a mobile user and can help with computations, but the solution has some limitations with regard to keeping track of all the cloudlets and how to access them, while communication between nodes is only performed through Wi-Fi and 4G), and a code offloading platform that performs offloading through various wireless channels (Wi-Fi, 3G, Bluetooth, Wi-Fi Direct), employing a multi-criteria offloading decision-making algorithm, which takes into account energy consumption, execution time reduction, resource availability, network conditions, and user preferences. The main limitation of this last solution is that it does not allow nodes to communicate further than one hop. The issues presented here are addressed by the Drop Computing paradigm, as shown in Sects. 4 and 5.

3 Offloading Opportunities Analysis

In this section, we present a data collection experiment that we ran at our faculty in order to analyze the behavior of mobile nodes in terms of contacts with static access points that can be employed as edge devices in our proposed Drop Computing framework. We deployed two static nodes in two different locations in the campus that had their Wi-Fi interfaces set to monitor mode. The nodes were part of the MONROE platform [10,11], which is an open access hardware-based framework for large-scale experiments in mobile broadband and wireless scenarios. It offers a large set of static and mobile nodes that have one Wi-Fi and three MBB interfaces each, located in several countries. The platform allows users to schedule Linux-based containerized experiments through a Web interface. A MONROE node has a 1 GHz 64-bit quad-core AMD Geode APU, 4 GiB of RAM and a 16 GiB SSD.

We chose two locations for the MONROE nodes that would be different from each other in terms of mobile device interactions, in order to have a better view of the offloading opportunities, and also to be able to compare various situations. The first node was placed in a large amphitheater (of approximately 180 seats), where courses are being held daily. Furthermore, the hallway next to the amphitheater connects two large parts of the faculty building, so there is a constant flux of students passing nearby. The second MONROE node was placed in the Pervasive and Mobile Services Laboratory, where generally 4–5 people work from the morning until the evening. However, the laboratory also has a constant flux of traffic, because students and other faculty members come to discuss their work, research, etc.

We performed data collection on both nodes from the 11th of May (on a Friday) at 8 AM until the 21st of May (on a Monday) at 4 PM, so there is more than a full week that can show us how the contacts fluctuate based on the day of the week. Furthermore, we have collected data from two full Saturdays, Sundays and Mondays, in order to assess the predictability of human behavior on a weekly basis. It should also be noted that the collection was performed during the semester, when classes were being held and students were attending them.

Figure 1 shows the distribution of total and unique contacts for each four-hour interval of the data collection period. At first look, it can be observed that many of the contacts that the two MONROE nodes observe per four-hour intervals are unique, which means that other devices do not pass many times near the nodes. They either arrive in their range and stay for a longer period (for example, students attending classes for the amphitheater node, or the researchers working in the laboratory for the other scenario), or they simply pass through. The contacts that are not unique most likely belong to professors that have multiple courses in the amphitheater or to researchers moving about the laboratory. Repeating contacts are extremely useful for data offloading through device-to-device (D2D) communication, and is something we plan on investigating in more depth in the future. In Fig. 1, we computed the uniqueness of contacts per four-hour interval,

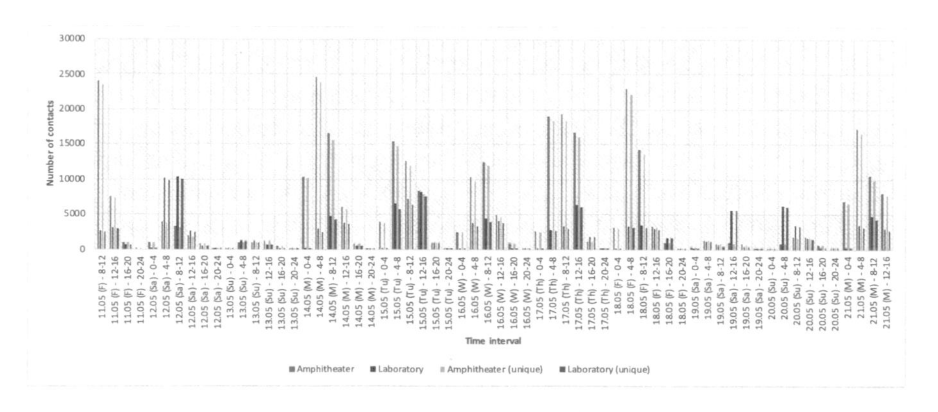

Fig. 1. Contacts of the two MONROE nodes with other devices.

but devices might repeat with larger inter-contact times, which is something we also wish to analyze in the future.

Another interesting conclusion that can be drawn from Fig. 1 is that, as expected when talking about a university scenario, there are far fewer contacts on the weekend. Furthermore, the time interval with the most contacts for the amphitheater node is between 4 AM and 8 AM. Most likely, this happens because the most courses in our faculty are in the morning, and that is the period when students arrive. On the other hand, the interval between 12 PM and 4 PM is the most popular one for the laboratory node, because the researchers working there arrive later. For data and computation offloading, it is important to know these intervals and the way they fluctuate. As shown in Fig. 1, this can be achieved by analyzing historical data, because there is a certain predictability in terms of number of contacts per day of the week (i.e., the number and distribution of contacts on 11 and 18 May are very close, and this is also true for 14 and 21 May). It has been previously shown [12] that opportunistic contacts in several types of environments can be modeled as Poisson processes, and in the future we wish to analyze the data collected here to see if this conclusion is true for this scenario. The difference in this case is that the MONROE nodes are static, as opposed to fully mobile nodes in opportunistic networks.

4 Mobile Data Offloading

The offloading mechanism that we propose in this paper is based on a network configuration as shown in Fig. 2. There are many mobile devices with mobile broadband connectivity that roam around the network area (or even exit it for periods of time). They can communicate with each other through close-range protocols such as Bluetooth or Wi-Fi Direct when they are in proximity, and they can also access the external cloud through 4G. Furthermore, there are several edge devices in the network that the nodes can connect to through Wi-Fi.

Fig. 2. Extended Drop Computing paradigm. The left arrow represents opportunistic communication between mobile nodes (where one node helps the other with computations), the middle arrow shows a mobile node communicating directly with the cloud, while the two arrows on the right show an edge device relaying a mobile node's requests to the cloud.

The mobile nodes run applications that require resources (such as CPU or battery) that they might not always posses. For this reason, in classic mobile applications, the main processing is performed in the cloud. The application is basically only a thin client for complex computations that are performed in the cloud and brought back to the user. However, such a scenario can prove somewhat costly, because the application developers have to maintain (or pay for) a cloud infrastructure that needs to serve all customers. In [1], we have shown that the Drop Computing paradigm is able to improve this situation by allowing other mobile nodes in range to help with the computations. Thus, large computations can be split into smaller tasks that can be passed on to encountered nodes, which can either process the tasks themselves or spread them even further in an opportunistic fashion [13]. This way, the cloud usage decreases because the number of requests is reduced. At the same time, the battery lifetime of the mobile devices increases, since they do not need to connect to the cloud through the mobile broadband interface, which is a notorious battery consumer [14] (especially when compared to the close-range protocols that Drop Computing employs for device-to-device communication).

In this paper, as shown in Fig. 2, we take the Drop Computing paradigm one step further, through the addition of edge devices that can either relay the computation requests to the cloud (and bring the results back), or that can even process some requests themselves. For the first case, the benefit would be brought by fast processing (since the transfer speed of Wi-Fi is higher than that of mobile broadband) and lower battery consumption. In addition, if the edge devices actually perform computations themselves, then the transfer latency between the edge device and the cloud would be gone, further reducing the total processing speed. Thus, a node that wants to compute some tasks has four possibilities: perform the computations itself, relay them to a device in range, send them to an edge node if there is one close by, or upload the tasks directly to the cloud. Naturally, the best option would be to use an edge node, as discussed above, but this can be done only when the node arrives in the edge device's Wi-Fi range.

The extended Drop Computing offloading mechanism is presented in detail in Algorithm 1. At certain intervals, the mobile nodes in the network generate tasks of various sizes and store them in a list. Until a mobile device encounters another node (either a mobile one or an edge device), it computes the tasks itself, starting from the beginning of the list (lines 44–50). When an edge node is encountered, the mobile device picks the first task from its list and sends it to the edge device, waiting for the answer (lines 14–16). When the task is computed, the edge device sends it back to the node if it is still in range. If that is the case, then the node picks the next task from the list, and repeats this operation until all tasks are successfully completed or the mobile node moves out of the edge device's range. When that happens, the node resumes computing itself the first task from its list. It should also be noted that, prior to sending tasks to the edge device, a mobile node first exchanges information about completed tasks with it. Thus, it first looks if the edge device has any completed tasks belonging to the current node, and then the mobile device also sends to the edge node information about the tasks it has completed itself for other nodes or that it has received from any encountered peers along the way (lines 8–11).

Algorithm 1. Extended Drop Computing offloading mechanism.

```
1: A - current node, L_A - list of tasks of node A (its own or belonging to others), F_A - completed
   tasks of other nodes
2:
3: for all time ticks do
4:     if A should generate tasks then
5:         generate tasks and add them to L_A
6:     end if
7:
8:     if edge device E in range then
9:         receive from E completed tasks belonging to A
10:        send F_A to E
11:    end if
12:
13:    while edge device E in range and L_A contains tasks do
14:        send first task T from L_A to E
15:        wait for completion of T
16:        mark T as completed and remove from L_A
17:
18:        if A is not T 's owner then
19:            add T to F_A
20:        end if
21:    end while
22:
23:    if mobile node B in range of A then
24:        receive from B completed tasks belonging to A
25:        send to B completed tasks from F_A belonging to B
26:
27:        if A and B are familiar then
28:            send F_A to B
29:            receive F_B from B and add to F_A
30:        end if
31:
32:        if L_A and L_B are not balanced then
33:            balance L_A and L_B
34:        end if
35:    end if
36:
37:    for all expired tasks T in L_A that belong to A do
38:        send T to cloud
39:        wait for completion of T
40:        receive T from cloud
41:        mark T as completed and remove from L_A
42:    end for
43:
44:    get first task T from L_A
45:    if T is completed at this tick then
46:        mark T as completed and remove from L_A
47:        if A is not T 's owner then
48:            add T to F_A
49:        end if
50:    end if
51: end for
```

Whenever another mobile node is encountered, the first step is to exchange information about completed tasks. Thus, the current node will receive the results for the tasks it had generated in the past that have been computed by other nodes (lines 24–25). Then, if the two nodes are familiar (i.e., are socially connected or they have encountered each other for at least a given number of times), they also exchange information about the completed tasks of other nodes (lines 27–30). In this way, they can help to opportunistically deliver the computation results to other nodes they may encounter. As shown by lines 32–34, for the next step the two nodes check if the tasks each has to execute are balanced (i.e., if the total estimated durations of their computations are roughly equal). If the nodes are not balanced, an attempt to balance them is made, taking into account the processing power of each node and the owner of each task. If the two nodes are familiar and they need to balance their tasks, they simply exchange tasks (i.e., a node that offloads a task will no longer have that task in its computation list). On the other hand, if the two nodes are not familiar but they need to balance their tasks, replicas of the tasks are made (so each node keeps its previous tasks and potentially adds some more).

There is also the possibility that a node has too many tasks to process and does not encounter any edge or mobile device for a long period of time. In this case, the only solution is to employ the cloud, as done in the default case (lines 44–50). In Sect. 5, we will show the benefits that our extended Drop Computing mechanism brings over simply offloading everything to the cloud.

5 Results

For testing the solution proposed in this paper, we have implemented it in MobEmu[1], which is an opportunistic network simulator that can run a user-created routing or dissemination algorithm on a real-life mobility trace or on a synthetic mobility model [15]. We have updated MobEmu with extra functionality that allows us to simulate the scenario presented previously in Fig. 2. For our tests, we considered four situations: 1) nodes exclusively use the cloud for performing computations for a mobile application (referred to as "cloud only"); 2) they compute the tasks themselves or opportunistically with the help of other nodes, with no access to an Internet infrastructure ("opportunistic only"); 3) they employ Drop Computing as presented in [1] and Sect. 2 ("Drop Computing"); 4) finally, we also considered the extended Drop Computing paradigm presented in Sect. 4 ("Ext. Drop Computing").

We have tested our Drop Computing implementation on the two scenarios presented in Table 1. The first scenario uses the HCMM mobility model [16] to simulate node behavior and interactions. We simulated a scenario with 40 nodes grouped in four communities, with 10 travelers (i.e., nodes that move between communities). The nodes move in area of 400×400 m with speeds between 1.25 and 1.5 m/s. We set the transmission radius of the devices to 10 m, since we assume that data exchanges between nodes are done on Bluetooth. For the second testing scenario, we used a real-life mobility trace collected in an office environment (entitled "UPB 2015") in Bucharest using the HYCCUPS Tracer

[1] https://github.com/raduciobanu/mobemu.

app[2], which was installed and run by the employees of an IT company. We selected a sample from this trace which contains contacts collected from 6 nodes during a 7-h window in a workday.

Table 1. Testing parameters.

Configuration	Scenario 1	Scenario 2
Mobility model/trace	HCMM mobility model	UPB 2015 mobility trace (office)
Duration (hours)	24	7
Number of nodes	40	6
Contacts	134034	324
Bluetooth transfer speed	3 MB/s	
Wi-Fi transfer speed	10 MB/s	
4G transfer speed	3 MB/s	
Mobile nodes	One core, 2.3 GHz	
Edge nodes	One core, 2.3 GHz	
Cloud VMs	One core, 3.3 GHz	
Types of tasks	Small (1 Mcycle), medium (1000 Mcycles), large (10000 Mcycles)	
Task size	5 MB	
Task expiration limit	1 ms for small tasks, 1 s for medium tasks, 10 s for large tasks	
Task generation	1–10 small, 0–10 medium, 0–10 large	

Both the HCMM mobility model and the UPB 2015 trace only contain contacts between mobile devices. However, our scenario also assumes the existence of static edge nodes that mobile devices can connect to through Wi-Fi and use as data relays and for task computation. For this reason, we have used the Wi-Fi data presented in Sect. 3 to simulate two edge nodes for each scenario. We have selected the nodes with the most contacts with the edge devices and mapped them onto the nodes from the simulation scenario (i.e., 40 nodes for HCMM and 6 for UPB 2015).

Figure 3 shows the results for the UPB 2015 scenario. It can be observed in Fig. 3(a) that Drop Computing manages to reduce the total computation time when compared to the case when devices only use the cloud or each other. For the cloud-only case, the total duration is 1.06 h, whereas the opportunistic-only case naturally has a much higher duration (about 1.92 h), since the devices have slower CPUs than the cloud virtual machines, while also having a limited battery life. By carefully selecting between using the cloud and collaborating with nearby devices, Drop Computing manages to reduce the computation time to about 0.92 h, for a reduction of 13.5% when compared to the cloud-only case, and 52% when compared to the opportunistic-only case. Furthermore, it can be observed that the extended version of Drop Computing presented here, that takes advantage of two edge devices to further offload computations, manages to

[2] http://www.smartrdi.net/2017/11/08/getting-started/.

reduce the total computation duration ever more, by decreasing it with a further 11% compared to the initial Drop Computing solution.

However, the benefits of employing Drop Computing do not stop here. Figure 3(b) shows the cloud usage time for the three cloud-based scenarios analyzed above (the opportunistic-only scenario assumes that nodes do not have access to the cloud). Firstly, it can be observed that both Drop Computing versions manage to reduce the total cloud computation time. This means that not only does Drop Computing perform computations faster, but it also reduces the costs of the application developers by decreasing the cloud usage. When compared with the cloud-only case, Drop Computing decreases the cloud usage with 38%, while the extended version presented in Sect. 4 obtains an even better reduction of 44.5%. This shows that, by carefully placing the edge devices in the network, the Drop Computing paradigm can bring clear benefits regarding several metrics.

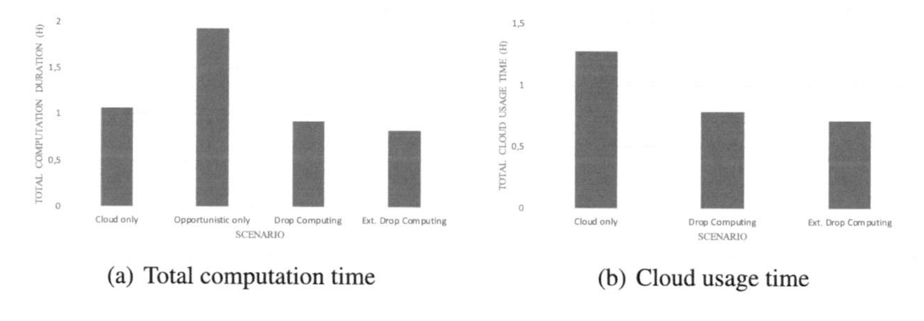

(a) Total computation time (b) Cloud usage time

Fig. 3. Results for the UPB 2015 scenario.

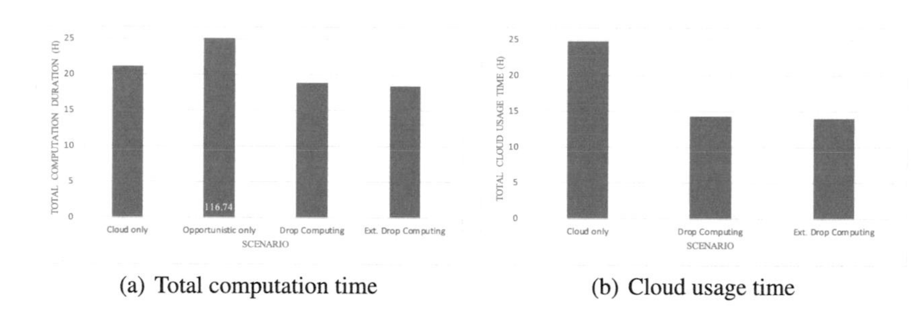

(a) Total computation time (b) Cloud usage time

Fig. 4. Results for the HCMM scenario.

A similar situation can be observed in Fig. 4 for the HCMM scenario. The first observation that can be made is that the opportunistic-only scenario performs much worse here than for the UPB 2015 trace. This happens because there are more nodes (40, as opposed to 6 at UPB 2015), which generate more tasks, making it more difficult for a node to compute both its own tasks and those of other peers. For the same reason, the total computation time for HCMM is much worse than for UPB 2015, with values as high as 116 for the opportunistic-only case. Again, Drop Computing manages to decrease both the total computation duration (as shown in Fig. 4(a)) and the cloud usage (seen in Fig. 4(b)). This means that computations are performed faster for the users and cheaper for the developers. When compared to the cloud-only case, the two Drop Computing versions improve the computation time by 11% and 12.8% respectively, while at the same time decreasing the cloud usage time by 41.5% and 43% respectively).

The reason that the improvement brought by extended Drop Computing when compared to the original version are not as high as for the UPB 2015 scenario is that there are 40 nodes instead of 6, and only two edges devices (corresponding to the two MONROE nodes that we used for collecting data, as presented in Sect. 3). We believe that, by strategically placing additional edge devices in the network, we can further improve the behavior of the extended version of Drop Computing presented in this paper, and we plan on verifying this assumption in future work.

6 Conclusions and Future Work

In this paper, we addressed the topic of computation offloading in mobile networks. More specifically, we discussed the Drop Computing paradigm, which introduces the concept of decentralized computing over multilayered networks. We presented a data collection experiment and analyzed it in terms of offloading opportunities from mobile to edge devices, and then we proposed an extension to Drop Computing that takes advantage of edge devices and of other mobile nodes to offload computation. Through experimental analysis on mobility and contact traces (including the one we presented in this paper), we showed that the solution proposed here is able to reduce the processing time of mobile node tasks, while at the same time reducing the total cloud usage. This leads to lower costs for app developers and a higher satisfaction for users.

In the future, we wish to perform a more in-depth analysis on the data collected using the two MONROE nodes acting as edge devices. It would be interesting to see not only the number of contacts per time interval, but also their recurrence, as well as the duration of the encounters between devices. Furthermore, an analysis on the type of devices observed based on the MAC would also yield interesting results. Finally, we wish to continue improving the Drop Computing solution presenting here, by adding more edge devices, as well as making more informed decision when offloading data. Using the MONROE nodes, we also want to add mobile broadband traces to our simulator, in order to have more realistic experiments with mobile nodes equipped with 4G capabilities.

Acknowledgements. This work is funded by the European Union's Horizon 2020 research and innovation programme under grant agreement No. 644399 (MONROE) through the open call project "Traffic and Data Offloading in Mobile Networks: TTOff". The views expressed are solely those of the authors. This research is also supported by University Politehnica of Bucharest, through the "Excellence

Research Grants" program, UPB - GEX 2017, identifier UPB- GEX2017, ctr. no. AU 11.17.02/2017.

References

1. Ciobanu, R.-I., Negru, C., Pop, F., Dobre, C., Mavromoustakis, C.X., Mastorakis, G.: Drop computing: Ad-hoc dynamic collaborative computing. Future Gener. Comput. Syst. (2017). http://www.sciencedirect.com/science/article/pii/S0167739X17305678
2. Rebecchi, F., de Amorim, M.D., Conan, V., Passarella, A., Bruno, R., Conti, M.: Data offloading techniques in cellular networks: a survey. IEEE Commun. Surv. Tutor. **17**(2), 580–603 (2015)
3. Aijaz, A., Aghvami, H., Amani, M.: A survey on mobile data offloading: technical and business perspectives. IEEE Wireless Commun. **20**(2), 104–112 (2013)
4. Dimatteo, S., Hui, P., Han, B., Li, V.O.: Cellular traffic offloading through wifi networks. In: 2011 IEEE Eighth International Conference on Mobile Ad-Hoc and Sensor Systems, pp. 192–201, October 2011
5. Pitkanen, M., Karkkainen, T., Ott, J.: Opportunistic web access via WLAN hotspots. In: 2010 IEEE International Conference on Pervasive Computing and Communications (PerCom), pp. 20–30, March 2010
6. Huerta-Canepa, G., Lee, D.: A virtual cloud computing provider for mobile devices. In: Proceedings of the 1st ACM Workshop on Mobile Cloud Computing & Services: Social Networks and Beyond, MCS 2010, pp. 6:1–6:5. ACM, New York (2010). https://doi.org/10.1145/1810931.1810937
7. Fernando, N., Loke, S.W., Rahayu, W.: Dynamic mobile cloud computing: ad hoc and opportunistic job sharing. In: Proceedings of the 2011 Fourth IEEE International Conference on Utility and Cloud Computing, UCC 2011, pp. 281–286. IEEE Computer Society, Washington, DC (2011). https://doi.org/10.1109/UCC.2011.45
8. Miluzzo, E., Cáceres, R., Chen, Y.-F.: Vision: Mclouds - computing on clouds of mobile devices. In: Proceedings of the Third ACM Workshop on Mobile Cloud Computing and Services, MCS 2012, pp. 9–14. ACM, New York (2012). http://doi.acm.org/10.1145/2307849.2307854
9. Verbelen, T., Simoens, P., De Turck, F., Dhoedt, B.: Cloudlets: bringing the cloud to the mobile user. In: Proceedings of the Third ACM Workshop on Mobile Cloud Computing and Services, MCS 2012, pp. 29–36. ACM, New York (2012). https://doi.org/10.1145/2307849.2307858
10. Alay, Ö., Lutu, A., García, R., Peón-Quirós, M., Mancuso, V., Hirsch, T., Dely, T., Werme, J., Evensen, K., Hansen, A., Alfredsson, S., Karlsson, J., Brunstrom, A., Khatouni, A.S., Mellia, M., Marsan, M.A., Monno, R., Lonsethagen, H.: Measuring and assessing mobile broadband networks with monroe. In: IEEE 17th International Symposium on A World of Wireless, Mobile and Multimedia Networks (WoWMoM), pp. 1–3. IEEE (2016)
11. Alay, Ö., Lutu, A., Peón-Quirós, M., Mancuso, V., Hirsch, T., Evensen, K., Hansen, A., Alfredsson, S., Karlsson, J., Brunstrom, A., Safari Khatouni, A., Mellia, M., Ajmone Marsan, M.: Experience: an open platform for experimentation with commercial mobile broadband networks. In: Proceedings of the 23rd Annual International Conference on Mobile Computing and Networking, pp. 70–78. ACM (2017)

12. Ciobanu, R.I., Dobre, C.: Predicting encounters in opportunistic networks. In: Proceedings of the 1st ACM Workshop on High Performance Mobile Opportunistic Systems, HP-MOSys 2012, pp. 9–14. ACM, New York (2012). https://doi.org/10.1145/2386980.2386983
13. Marin, R.-C., Ciobanu, R.-I., Dobre, C.: Improving opportunistic networks by leveraging device-to-device communication. IEEE Commun. Mag. **55**(11), 86–91 (2017)
14. Huang, J., Qian, F., Gerber, A., Mao, Z.M., Sen, S., Spatscheck, O.: A close examination of performance and power characteristics of 4g lte networks. In: Proceedings of the 10th International Conference on Mobile Systems, Applications, and Services, MobiSys 2012, pp. 225–238. ACM, New York (2012). https://doi.org/10.1145/2307636.2307658
15. Ciobanu, R.-I., Marin, R.-C., Dobre, C.: Mobemu: a framework to support decentralized ad-hoc networking. In: Modeling and Simulation in HPC and Cloud Systems, pp. 87–119. Springer (2018)
16. Boldrini, C., Passarella, A.: HCMM: modelling spatial and temporal properties of human mobility driven by users' social relationships. Comput. Commun. **33**(9), 1056–1074 (2010)

Lessons Learned in Tokyo Public Transportation Open Data APIs

Toshihiko Yamakami(✉)

CTO Team, ACCESS, 3 Kanda-Neribei-cho, Chiyoda-ku, Tokyo 101-0022, Japan
Toshihiko.Yamakami@access-company.com

Abstract. Open data is a vital part of new digital economy. It facilitates value creation from combining data from multiple sources. It also important to utilize a massive data flow from emerging IoT (Internet of Things) devices. Future smart cities will consist of a large aggregation of open data APIs. The size and variety of open data APIs provide challenges for usability, consistency, and integrity. The author analyzes issues in open data APIs in the Tokyo public transportation. The author discusses multiple aspects of issues. Then, the author presents a framework with three view models to deal with open data APIs: outcome, cause, and fixes. Finally, the author discusses lessons learned in the Tokyo public transportation open data APIs.

1 Introduction

There are two drivers in open data in the Internet. One is the reduced cost of storage and processing power. This encourages utilization of massive data with advanced data analytics and machine learning. The other is the emerging IoT (Internet of Things), which provides potentially massive incoming flow of data from devices.

These two trends combined, there is a large need to deal with open data today. Open data needs open data APIs. Engineering open data APIs are still in the early stage of evolution. There are many things to be learned in this domain.

Open data is fundamentally a part of open innovation and distributed collaboration among multiple stakeholders. It includes technical, social, organizational, and real-world issues.

Open data is a driving factor to make use of third party service providers with publicly usable data. One of the benefits of open data is that combined data from multiple sources create new values.

Tokyo has a complexity of intertwined transportation networks. Open data can be one of the solution to eliminate unnecessary complexity and to provide full potential of complicated transportation networks. Open data challenges in Tokyo provide an opportunity to consider the real-world challenges of open data. The author examines the difficulties of using open data in tokyo public transportation. And, the author provides a dimensional framework to parse the difficulties.

L. Barolli et al. (Eds.): NBiS 2018, LNDECT 22, pp. 374–384, 2019.
https://doi.org/10.1007/978-3-319-98530-5_31

2 Background

2.1 The Purpose of Research

The aim of this research is to provide a comprehensive framework to analyze complexity of usage of open data for public transportation in a large metropolitan area.

2.2 Related Work

Research on public transportation open data consists of three areas: (a) open data policies and implementations, (b) applications of open data, and (c) use of open data.

First, in regard to open data policies,Raggett discussed W3C plans of open data in IoT [10]. Not constraint to public transportation, W3C publishes general open data best practices [15]. Buranarach et al. discussed scalability of open government data catalog [5]. Telikicherla et al. discussed security requirement of open data APIs [12]. Bertot et al. discussed big data and e-government issues [2]. Osagie et al. discussed usability issues in an open government data platform [9].

Second, in regard to application of open data, Bouillet et al. discussed a real-time public transport awareness system in Dublin [3]. Maeda et al. discussed spatial human flows using smart card data [7]. Vandewiele et al. discussed prediction of train crowdedness [14].

Third, in regard to use of open data, Colpaert et al. discussed transportation flow analysis using open data [6]. Bourgois et al. discussed use of open data in air transport research [4].

The originality of this paper lies in its development of a framework to deal with open data APIs linked to the real world.

3 Research Method

The author performs as follows:

- analyzing issues of public transportation open data APIs
- identifying view models of open data issues to deal with these issues
- analyzing improvement measures of transportation open data APIs

4 Real-World Open Data

There was an open data challenge in Tokyo public transportation during December 2017 to March 2018 [1]. 26 public transportation companies supported this challenge, including train, bus, subway, and airport companies. The APIs provided access to a) static data, including train/station timetables, and b) dynamic data, such as train location information and status information.

This was a public event, therefore, there were multiple pieces of public feedback available [8,11,13]. The feedback provides a valuable source of insight for understanding challenges in wide use of open data APIs.

It should be noted that all feedbacks were at the time of publication. Therefore, issues may have been fixed after the identification of issues.

When open data is provided to third parties. There emerge visible issues of multiple aspects. The issues are:

- usability issues
- discrepancy with the real-world
- discrepancy with the standards
- user interface terminologies
- typos
- arbitrary extensions
- parameter conventions
- different interpretations among organizations
- non sequential IDs
- missing IDs
- differences of unit among organizations
- same key has different data among organizations
- difference of date formats
- multiple interpretations of non-existence values
- limitations of API
- obsolete data
- missing data in some semantic contexts
- differences in non-normal cases

The issues cover a wide variety of areas. And, they have different views from the cause viewpoint or the fix viewpoint.

The causes are:

- IT infrastructure issues
- Standards or API specifications issues
- non-conformant APIs issues
- organizational issues
- inter-organizational discrepancy issues
- maintenance issues

The fixes are:

- better API infrastructure
- improved standards
- verification of APIs
- API implementation guidelines
- education to API providers
- periodical check
- better exception rules

The variety of inconsistency and flaws in these instances of open data APIs indicates that the open data APIs are the reflection of the real-world. Even seemed-simple APIs include a significant complexity when the target real-world is complicated. Public transportation in Tokyo is one of such examples.

When the real world is complicated, we need to develop a design framework to deal with this complexity. From a list of ad hoc issues, it is important to set up a framework to capture the complexity.

5 A Framework to Deal with Open Data APIs

There are multiple aspects of issues of open data in Tokyo transportation open data. First, they are classified by location of issues. A classification is depicted in Table 1.

Table 1. Classification of open data API issues.

Infrastructure	Instability (Infrastructure was not reliable. The API server and the spec server were down simultaneously, which damages development work)
API spec protocol part	Typos (the part of spec was wrong, but the implementation was right.)
API encoding	Missing part (discrepancy between missing element and putting `null` to the element)
API content	Typos (wrong content, character error, obsolete content, error in time unit). Discrepancy with the real world (ID without real-world publication)
API usability	Inconsistent use of APIs among API providers) Missing API examples.
Exception handling	Exceptional operation days (operation days without specific day of week patterns) Missing trouble details (Operation irregularity without any explanation)

A classification from the cause viewpoint is depicted in Table 2. It should be noted the real causes may require further investigation. This table shows a variety of possible causes. A classification from the preventive measure viewpoint is depicted in Table 3.

From these findings, the author presents a three-part framework to capture open data APIs in public transportation of Tokyo as depicted in Fig. 1. It is not Tokyo-specific. This framework can be applied to any open data APIs with link to the complicated real-world.

The layer view helps increase the awareness of wide variety of inconsistency and pitfalls. The relationship view helps increase dynamic aspect of use of open data APIs. The support view leverages smooth knowledge transfer for improvement of open data APIs.

Table 2. Classification from the cause viewpoint.

Infrastructure design	Scalability, fault-tolerance. Incompatibility with use cases (Only 1000 query results can be returned where there are more than 1000 stations in one railway company. And there is no query continuation method.)
Standards	Lack of clarity, missing exception handling, missing ID rules.
API spec	Lack of certification, lack of consistency check (One bus operator returns time as ISO_DATE format at dct:valid (data guarantee period) where all time should be returned by ISO_OFFSET_DATE_TIME), lack of UI consideration (line direction names for UI display are not clear)
API implementation	Discrepancy with the real world. Lack of typo check (station names, part of parameter names such as `Train` and `Railway`). Lack of obsolescence check. Incompatibility with spec intention (A subway company returns `odpt:trainInformationStatus` only when trains operate as normal. Therefore, a required item `odpt:timeOfOrigin` is always missing.). Format violation (Some railway companies do not conform to the format of `owl:sameAs` at `odpt.RailwayFare:`). Inconsistency in APIs (Specified `odpt:RailDirection:` is replaced with `odpt:RailwayDirection:` in one API)
Local extension	Local rules may violate consistent design of general rules (Bus has only bus stop timetables. It is almost impossible to convert it to bus-specific route time table. `odpt:aircraftModel` does not exist in the spec and all values in implementation are `null`. One bus line uses `odpt.Calendar:Specific.OdakyuBus.Kichijouji.Weekday01` which possibly means weekdays.)
ID, numbering, and naming rules	Starting 0 or 1. Non sequential numbering of stations due to a historical reason of branching lines. Line names has three patterns (a) including line postfix in line names, (b) including line in English postfix in line names, (c) no line postfix. ID has a special one which does not appear in the real-world display
Inconsistency among providers	Lack of guidance, lack of mutual information sharing, lack of interaction at marginal cases. (In the spec `odpt:estimatedTime` cannot be obtained after departure. After departure, this element is missing or filled with `null`)
Complexity of real-world	The one-flight-one-object rule has discrepancy of the reality of code-share flights
Organizational conflicts	Lack of contacts between operation offices and IT offices. Discrepancy of update cycles in the operation side and in the API side. Lack of funding on API provisioning and maintenance

Table 3. Classification from the preventive measure viewpoint.

Preventive measure	Description
Strict standardization	Clarification of required items. Clarification of unit (time). Clarification of alias name use
Clear exception handling	Consistent guidelines of non-existing items and exceptional status description
Comprehensive local rule handling	Clarification of allowance of local extension
Training	Meaning and intention of standardized specs should be trained among API providers. Sharing conventions
Increased awareness of maintenance	Increasing awareness of maintenance (increased consistency with real-world changes, periodical verification of data)
Sharing lessons	Lessons learned in the past open data API experience should be shared among API providers

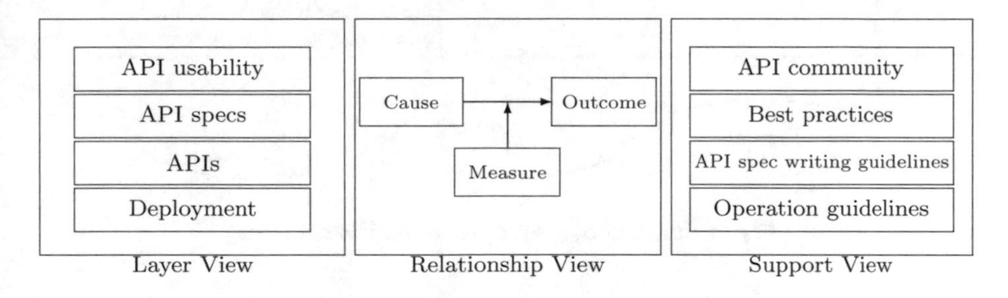

Fig. 1. A framework to deal with open data linked to real-world complexity

Large-scale open data APIs need to accomodate (a) complexity, (b) dynamic nature, and (c) organizational conflicts, (d) operation and maintenance domains of the real world business. In order to improve open data APIs, it is important to prepare challenges that arise from these aspects of open data APIs.

Technical people tend to think that tight guidelines can lead to clean and bullet-proof standards. From observation, it is not the case. Even Company-A IT team does not have full knowledge about what type of discrepancy may emerge from the real-world operation of their company's business. Community efforts with developers and other company's IT team are included in the support layer from this finding.

The use of open data is shifting from the static aspect to the dynamic aspect through increased deployment of IoT-based APIs. This leads to increased linkage of complexity of the real world. The dynamic aspect presents increased inconsistency in seni-normal states. One of such examples is operation irregularity of trains. Example of inconsistency is depicted in Table 4.

This inconsistency influences the whole process of implementing a service, as depicted in Fig. 2. When there is a trivial issue in parsing parameters, it is

Table 4. Example of inconsistency in operation irregularity of trains

Operator	Behavior per Line	Processing
Company-A	If there was irregularity today, a result is returned	When there is no data returned, operation is normal. When there is a result, it is necessary to check the `odpt:timeOfOrigin` to determine whether there is new irregularity
Company-B	It always returns a result and it contains `odpt:trainInformationStatus`	If there is `odpt:trainInformationStatus`, there is irregularity. If not, trains run normally
Company-C	If it is normal, it returns a result without `odpt:trainInformationStatus`. If there is irregularity, it does not return a result	If there is a result, it runs normally. If it does not receive a result, there is irregularity, but there is no method to know details. (Required `odpt:timeOfOrigin` is not received, so it needs careful processing at parsing)

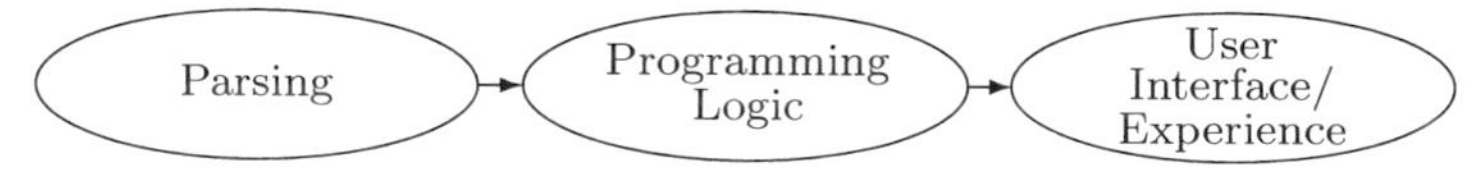

Fig. 2. Impacts of a sequence of implementation

possible to impact the later stage of user experience when the parsing cannot determine the existence of an event. The API specification requires consideration of whole user experience design process for dealing with missing parameters.

It even impacts marketing efforts of a service derived from open data APIs because it cannot claim that it always show the details of irregularity of operation to users. It should be noted that the operation irregularity may link to inter-organizational coordination between the operation and the IT division, or even with public relation division. When linked to the real world reality, open data APIs expose complexity of real-world operation and maintenance. Therefore, it is beyond the scope of static spec writing. An appropriate framework is necessary to deal with this.

6 Discussion

6.1 The Advantages of the Proposed Approach

The benefits of open data are increasing with increased connectivity, IoT and data analytics. The open data has extending in the direction of:

- large-scale
- dynamic nature

These two factors contribute to the increased challenges from the IT development viewpoint.

From lessons learned in Tokyo public transportation open data APIs, the author presents a framework to deal with these challenges. The framework with 3 view models provide a design framework for large-scale open data APIs, such as public transportation APIs.

The three-view framework can be applied to the software development in a heterogeneous environment.

Each view model facilitates the standardization process, API design, and API deployment.

Observation shows the real world operation brings naive assumption as well as complexity and dynamics. The semantics of the real world impact the API designers such as selection of second/minute or minute/day. It is inevitable that the real world reality influence of design process because its seemingly-apparent characteristics.

Also, the framework is based on the assumption that some of the issues are deep rooted in organizational and business aspects as depicted in Table 5.

Table 5. Examples of issues of deep-rooted organizational and business.

Item	Description
Tokyo's complexity	Tokyo greater area is served by many private companies. This presents conflicts with government-lead open data initiatives. Some issues arise when there are increasing instances of shared operation among multiple private railway companies
Business consideration	In another case, complexity of troubleshooting and reluctance of publicizing troubles may lead to not providing any details of operation at a operation trouble occasion
Power balance	Usually, IT division has less power than operation division in railway companies. This lead to frictions when there are open data provision in dynamic real-world operation data
Lack of awareness of technical people	IT teams have less awareness about long-term transition, obsolescence, and changes of real world operation. There are many irregular operations which occurs in rare cases, which usually does not catch much attention
Less communication	IT division can easily assign IDs, but in some cases, such IDs are not publicly available, therefore, using them may lead to confusion of users
Lack of awareness of operation division	Operation division does not care about impact from operation changes to open data APIs. This is a potential risk of long-term operation of open-data APIs

Some of problems can be avoided by building a design rule book. For example, the ID starting number can be fixed when it is officially declared. However, many problems have deep roots of business and operation reality. It is important to have awareness that the today's open data APIs directly deal with complexity of the real world in many aspects.

The dynamism of the real world introduced by IoT impacts standardization of open data APIs. It is difficult to anticipate all irregularity in advance in spec writing. The proposed framework accommodates evolutional and social development of API specs in a technical world tightly bound to the real-world complexity.

6.2 Limitations

The research is qualitative and exploratory. Verification of each dimension of the proposed view models remains for further studies.

The research is based on observation. A methodology to systematically collect and analyze transportation open data APIs is missing. Quantitative analysis is missing in this paper.

The analysis is empirical and in-depth exploration of causes of each issue is not performed in this paper.

The discussion is based on Tokyo public transportation open data APIs. Stage-specific, location-specific, infrastructure-specific and standard-specific factors are not analyzed. Cross-regional comparison remains future studies.

The fixing measures are not exhaustive and do not provide the concrete proposal.

Observation is based on an application contest. Consideration of deployment and operation of open data APIs with greater Tokyo 30-million population is out of scope of this paper.

7 Conclusion

The Web has become an integral part of our daily lives. With the penetration of the Internet, emergence of IoT, and the reduced costs of storage and processing, open data is a vital part of data-driven economy and smart society.

There is positive expectation of outcome of open data in smart society. However, open data APIs have exposed many technical, organizational, social, and real-world challenges. In order to make use of potentials of open data, it is important to provide a systematic methodological framework to analyze and deal with issues of open data APIs. The author analyzes issues of Tokyo public transportation open data APIs. Then, the author provides a framework with three view models to deal with open data APIs. This provides a building block for future improvement of open data APIs to empower open data-driven digital economy and smart societies.

Lessons indicate that the increased linkage to the real world using IoT leads to complexity and irregularity in development, operation and maintenance of open data APIs. It is necessary to accommodate those challenges.

The proposed framework is a stepping stone towards large-scale open data APIs of smart cities in order to cope with complexity of real-world daily operation and maintenance.

Acknowledgments. The research results have been achieved by "EUJ-02-2016: IoT/Cloud/Big Data platforms in social application contexts," the Commissioned Research of National Institute of Information and Communications Technology (NICT), JAPAN.

References

1. Association for Open Data of Public Transportation: Open data challenge for public transportation in tokyo, December 2017. https://tokyochallenge.odpt.org/en/index.html
2. Bertot, J.C., Choi, H.: Big data and e-government: issues, policies, and recommendations. In: Proceedings of the 14th Annual International Conference on Digital Government Research, dgo 2013, pp. 1–10. ACM, New York (2013)
3. Bouillet, E., Gasparini, L., Verscheure, O.: Towards a real time public transport awareness system: case study in Dublin. In: Proceedings of the 19th ACM International Conference on Multimedia, MM 2011, pp. 797–798. ACM, New York (2011)
4. Bourgois, M., Sfyroeras, M.: Open data for air transport research: dream or reality? In: Proceedings of The International Symposium on Open Collaboration, OpenSym 2014, pp. 17:1–17:7. ACM, New York (2014)
5. Buranarach, M., Krataithong, P., Hinsheranan, S., Ruengittinun, S., Supnithi, T.: A scalable framework for creating open government data services from open government data catalog. In: Proceedings of the 9th International Conference on Management of Digital EcoSystems, MEDES 2017, pp. 1–5. ACM, New York (2017). https://doi.org/10.1145/3167020.3167021
6. Colpaert, P., Chua, A., Verborgh, R., Mannens, E., Van de Walle, R., Vande Moere, A.: What public transit API logs tell us about travel flows. In: Proceedings of the 25th International Conference Companion on World Wide Web, WWW 2016 Companion, International World Wide Web Conferences Steering Committee, Republic and Canton of Geneva, Switzerland, pp. 873–878 (2016). https://doi.org/10.1145/2872518.2891069
7. Maeda, T.N., Mori, J., Toriumi, F., Ohashi, H.: Analysis of smart card data for understanding spatial changes in consumption-oriented human flows. In: Proceedings of the 2nd ACM SIGSPATIAL Workshop on Smart Cities and Urban Analytics, UrbanGIS 2016, pp. 2:1–2:4. ACM, New York (2016)
8. @ninten320: I touched the Tokyo public transportation open data (in Japanese), December 2017. https://qiita.com/ninten320/items/5e5ad4ca653177fff8fd
9. Osagie, E., Waqar, M., Adebayo, S., Stasiewicz, A., Porwol, L., Ojo, A.: Usability evaluation of an open data platform. In: Proceedings of the 18th Annual International Conference on Digital Government Research, dgo 2017, pp. 495–504. ACM, New York (2017)
10. Raggett, D.: W3C plans for developing standards for open markets of services for the IoT: the Internet of Things (ubiquity symposium). Ubiquity **2015**(October), 3:1–3:8 (2015)
11. Takeru-chan: Tokyo public transportation open data challenge (in Japanese), December 2017. https://www.junk-works.science/tokyo-public-transpotation-open-data-challenge/

12. Telikicherla, K.C., Choppella, V.: Enabling the development of safer mashups for open data. In: Proceedings of the 1st International Workshop on Inclusive Web Programming - Programming on the Web with Open Data for Societal Applications, IWP 2014, pp. 8–15. ACM, New York (2014). https://doi.org/10.1145/2593761.2593764
13. @teracy: This is tough, Tokyo public transport open data challenge API (in Japanese), January 2018. https://qiita.com/teracy/items/962d9feb3349824090e2
14. Vandewiele, G., Colpaert, P., Janssens, O., Van Herwegen, J., Verborgh, R., Mannens, E., Ongenae, F., De Turck, F.: Predicting train occupancies based on query logs and external data sources. In: Proceedings of the 26th International Conference on World Wide Web Companion, WWW 2017 Companion, International World Wide Web Conferences Steering Committee, Republic and Canton of Geneva, Switzerland, pp. 1469–1474 (2017)
15. W3C: Data on the web best practices, W3C recommendation, 31 January 2017. https://www.w3.org/TR/2017/REC-dwbp-20170131/

Slovak Broadcast News Speech Recognition and Transcription System

Martin Lojka(✉), Peter Viszlay, Ján Staš, Daniel Hládek, and Jozef Juhár

Department of Electronics and Multimedia Communications, Faculty Electrical Engineering and Informatics, Technical University of Košice, Boženy Němcovej 32, 042 00 Košice, Slovakia

{martin.lojka,peter.viszlay,jan.stas,daniel.hladek,jozef.juhar}@tuke.sk

Abstract. We have developed a working prototype of automatic subtitling system for transcription, archiving, and indexing of Slovak audio-visual recordings, such as lectures, talks, discussions or broadcast news. To go further in the development and research, we had to incorporate more and more modern speech technologies and embrace nowadays deep learning techniques. This paper describes transition and changes made to our working prototype regarding speech recognition core replacement, architecture changes and new web-based user interface. We have used the state-of-the art speech toolkit KALDI and distributed architecture to achieve better responsivity of the interface and faster processing of the audiovisual recordings. Using acoustic models based on time delay deep neural networks we have been able to lower the system's average word error rate from previously reported 24% to 15%, absolutely.

1 Introduction

Human speech is a natural way of communication between people and there has been always a desire to use the speech in human-machine interaction as well and this motivation has never weakened. The result is very strong community of researchers and research initiatives with one goal, to bring the speech to the machine. Nowadays we are surrounded by speech technologies that are used to perform many tasks, such as dictation, dialogue-based information retrieval systems or command and control. We can control many devices like smart phones, smart TVs by simply using a voice command. Those applications are apparent, but speech technologies found their way to more sophisticated usage. We live in time where we are overwhelmed by constant stream of information from which large amount is a simple speech. We are always trying to classify those big data to smaller chunks representing the same topic or find a way to search efficiently for information that we need. With increasing accuracy of current speech technologies, we can replace or make more efficient yet still manual transcription [2] of the TV shows (Broadcast news (BN) [3], discussion, talks or lectures [1]) to produce hidden subtitles for hearing impaired, to create written record of the meetings, to index large number of various archived audiovisual recordings.

L. Barolli et al. (Eds.): NBiS 2018, LNDECT 22, pp. 385–394, 2019.
https://doi.org/10.1007/978-3-319-98530-5_32

This urge is not strong only in the research area, but large number of companies are developing their own or trying to acquire speech technology to make use of them in their business model, make their products more attractive or simply to increase the efficiency of the firm production.

As mentioned above, using the speech technologies hidden subtitles for hearing impaired can be produced, which makes the technology highly attractive to television broadcasters that need to fulfill EU government quota. In Slovakia, in time of writing this paper, this amount of subtitled TV shows is 10% of broadcast for commercial companies and 50% for public broadcasters. Those quotas will be increasing with time demanding more human resources from the companies and broadcasters. The subtitling of the TV shows can be made efficient with use of speech technologies. This creates a new market for other companies focused primarily on speech technologies that are already offering their own solutions. The FP7 project SAVAS1 provide such applications for major EU languages, such as English, French, German, Italian, Spanish, or Portuguese. Unfortunately, software implementation is not sufficient for usable subtitling system. The heart of those systems are acoustic and language models that can be acquired by training process on large annotated data that are easier to collect for major languages than for minor ones as Slovak.

To cover the two main demands in Slovakia, the subtitling generation and the automatic meeting speech transcriptions, we have developed a working prototype that was introduced in our previous publications [12,15]. The system allows processing of multichannel audiovisual recording with automatic segmentation based on principal component analysis (PCA), gender and possibly speaker detection and diarization, parallel speech recognition for each channel with specialized acoustic models to match the input recording characteristics. The system makes use of several server-based speech recognizers and hypothesis combination to further increase the resulting accuracy.

In the development of the prototype and in the process of achieving better accuracy, more and more components and modules were added to the system. The system began to be more complex with little or no improvement. Other way to improve such a system is to improve the present modules itself and possibly remove those that are no longer necessary or bringing little improvement compared to their complexity and architecture requirements. With this settled, new prototype with replaced speech recognition core, new user interface, and distributed architecture was designed. However, the new design and speech technologies used made some of the previous modules unable to integrate into new architecture and thus they were removed. Despite of that the overall accuracy of the system was in the end improved. Which modules were removed, will be briefly discussed later in this paper.

In the following sections we will describe the architecture and main components of the system, resources used for training the models and we will introduce new user interface. The paper concludes with discussion about future work and research challenges.

2 Architecture

Our goal was to design a new architecture to further support and even improve the modularity keeping in mind fast deployment of the modules, distribution of the modules across multiple computing servers and clear separation of their concerns. The architecture meeting the requirements is depicted in the Fig. 1. The main three parts are the user interface, the task server, and finally the workers.

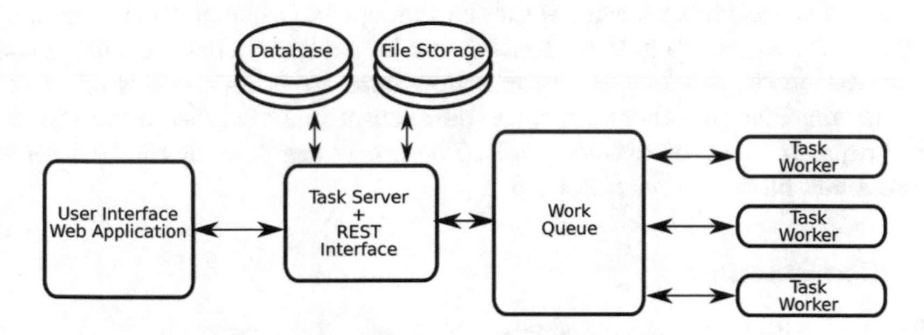

Fig. 1. Architecture of the proposed system

2.1 Task Server

The main function of the server is to provide interaction between user and task processing using private database and file storage. In other words, the system allows the user to form new task, upload files, query status of the tasks and results of the processing in case of successful competition. The server contains boxing system for organizing task into groups, while allows sharing those boxes between users with access control rules. For administrators the server contains user managing system. The server does not need to know any detail about the task just to provide the data to the workers for processing and collect output results that are provided to user.

In direction to workers, the server uses Advanced Message Queuing Protocol (AMQP) and a work queue. This way new workers can be created without changing or restarting the main task server. The tasks are then automatically split between workers based on their capabilities. This change helps to clearly separate concerns to the processing part and more general logistic part of the server.

In direction to the user, the server implements Representational State Transfer (REST) interface, allowing any user application complaining the interface to use server's functions. In this way, web, mobile and desktop application can be created later, or the system can be integrated to another larger one. With the new architecture design, also a new user web application was created, described later in this section.

2.2 Workers

The workers are independent copies on one or more computation servers listening for new tasks. The core of each worker is rather simple as their only one function is to run an external script processing the task and monitor its process. The advantage is that any script in any language can be run containing the chain of processing that needs to be done to complete the incoming task. Moreover, deployment of new modules is as easy as running them from command line without any need to implement any custom, nor special interface. Using work queue has also advantage in that each worker can independently control the amount of work that is assigned to it. This decision can be based on what computing capabilities the hosting server has and whether there are other workers running on the same machine. In other words, the parallelization is handled on the workers level by queuing method automatically. The processing done by each worker will be described further in next sections.

2.3 User Interface

Using new REST interface, a single web application was created. The application is independent from the server and rest of the processing and can be independently maintained. It makes full usage of the REST functions exposed by the server including administrator view, user management, task submission, processing status reporting and sharing the tasks between registered users. Two screenshots are depicted in the Figs. 2 and 3. The first one contains task list and processing screen. The later contains subtitles editor and viewer, where besides the video a list of segments, speakers and their speaking times are shown.

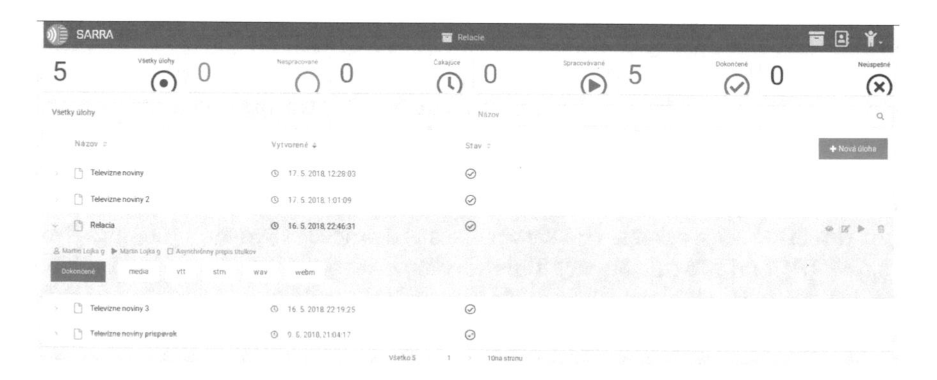

Fig. 2. Task manager screen

Fig. 3. Subtitles editor

3 Speech Segmentation

In general, the recognition of whole sentences is possible [3]. Unfortunately, during the recognition process of a sentence multiple hypotheses are maintained until the very end. The longer the sentence the larger the number of possible hypotheses that need to be explored and recomputed. It is more efficient to divide the audio recording into chunks that are holding enough relevant information for each model to describe it accurately and are short enough for recognizer to keep the number of hypotheses in manageable limits. For this purpose, we have adapted from the previous system voice activity detection and speaker diarization modules. Before the segmentation phase, the raw audio needs to be extracted and resampled to match acoustic model's training data. In this system each of the channels are extracted into separate file to support either single channel or multichannel processing at the same time. We will not describe this process in detail here.

3.1 Voice Activity Detection

The silence discrimination is performed by our unique VAD [14]. To determine the VAD labels, the waveform is processed in the time domain by overlapping blocks extracted by rectangular window with standard length of 25 ms and 10 ms frame step. After re-arranging the samples into sample matrix, the time domain principal component analysis (PCA) [4] is applied to each block. After that, N eigenvalues are computed for each block, where N is the dimension of PCA space. The eigenvalues are used to determine the nature of the i-th segment (voice/silence). Finally, the VAD coefficients are smoothed by sliding average window. For detailed description of the PCA-based VAD, see [14].

The VAD described was implemented in C++ for fast processing. The VAD does not split the original file into smaller files, but rather produces output file with relevant speech boundaries.

3.2 Diarization

To take advantage of the replaced speech recognition core the speaker diarization is necessary for better online adaptation process. The more data from one speaker are available on the beginning the better the adaptation is.

The speaker diarization is an automated annotation of speech recordings with labels that represent speakers. This task is performed without any prior information, neither the number of speakers, nor their identities, nor samples of their voices are available. The speaker diarization can be also employed for helping speech recognition, facilitating the searching and indexing of audio archives and increasing the richness of automatic transcriptions [9]. In our framework, the LIUM [9] was successfully applied for the segmentation and speaker clustering. The toolkit makes initially use of segmentation based on Bayes information criterion and later the Viterbi and GMM speaker clustering to further refine the boundaries. In the process it also detects the quality (spectrum width of the speech) and gender of the speaker. As we will describe in the acoustic modeling section, the speaker gender will be no longer used. Multichannel processing is achieved by processing each channel separately and in case of possibility that one speaker can be present in more than one channel also the cross-channel speaker clustering is supported.

4 Speech Recognition

We have used the KALDI toolkit for training and performing the speech recognition as complete solution [7]. The representation of the statistical resources is by Weighted Finite State Transducers (WFST). In previous prototype we have used multiple recognition processes whose outputs were merged by hypotheses combination technique. In this case we have omitted this function for now as will be described in the acoustic model section.

4.1 Acoustic Modeling

The presented automatic subtitling system makes use of triphone context-dependent AMs based on Deep Neural Networks (DNNs). As stated in the introduction, we had to embrace the new technologies and move forward to the deep learning techniques and modeling. The state-of-the art in speech recognition and acoustic modeling is nowadays represented by the KALDI toolkit that supports multiple types of DNNs [7]. One of the types is the Time Delay Neural Network (TDNN) supporting adaptations based on i-vectors and high resolution MFCCs.

The TDNNs acoustic model was obtained by sequence training on a single GPU based on a state-level variant of Minimum Phone Error (MPE) criterion, called sMBR, according to [5]. Also, the input feature vectors of 40 MFCCs without cepstral mean normalization (CMN) were used along with 100-dimensional i-vector for representation of the speakers and environmental characteristics. The advantage is that the adaptation of the later speech recognition can be done in one pass, instead of multiple passes like in case of Maximum Likelihood Linear Regression (MLLR). This also means that in speech recognition the more data from one speaker are available on the beginning the better the adaptation. From this point of view, clustering of the speakers is needed to achieve better accuracy.

The acoustic model was trained on gender-balanced acoustic databases that include: 100h of annotated speech recordings of judicial readings; 150h of read phonetically rich sentences, newspaper articles, and spelled items, recorded in conference rooms; 250h of annotated speech recordings acquired from BN TV shows; 90h of 90% male and 10% female annotated speech recordings realized in the main hall of the Slovak Parliament; and 80 h of annotated speech recordings from TV shows "Court Room" [6,10]. It is important to say that the speech recordings were downsampled to 16kHz/16bit PCM mono audio format for AM training. The training set also involves model of silence, short pause and additional noise events. In total, this makes about 600 hours of training data. It is logical to assume that the more data are available for training, the better the accuracy of the speech recognition is. To obtain the best from the available data, we used them all for training of one general acoustic model and skip training of gender dependent models as in previous prototype. This also includes removal of multiple settings for one speech recognition and combining their outputs. We will review this decision as more data will become available to us. As we will see in the experimental section, the achieved results justify the decision.

4.2 Language Modeling

The background language model was created using the SRILM toolkit [13]. It was restricted to the vocabulary size of about 500 thousand unique words and smoothed by the Witten-Bell algorithm. The trigram model was trained on the web-based corpus of Slovak written texts. The corpus size of about 1.89 billion tokens and 110.75 million of sentences was then segmented into 5.93 million paragraphs with approximately 315 words on average for better representation in the vector space.

After that, semantic indexing and vector space modeling were implemented to retrieve a subset of text documents from the background corpus relevant to the topic and speaking style of a speaker. Also, the authors proposed document retrieval approach based on using Paragraph Vectors [8] for topic-specific modeling to improve speech recognition accuracy for individual speakers [11].

In this approach, we select a subset of text documents semantically similar to the output hypotheses from recognized speech segments in the first decoding stage. A small topic-specific language model was then created from the relevant documents, interpolated with the background model, adapted to the current topic and speaking style of a speaker, and applied during the second decoding stage(s).

5 Subtitles

To make the system usable for subtitling of the BNs or any other audiovisual recordings, the subtitles module had to be improved and implemented as well. The main task is to divide the subtitles to meet mainly two requirements. Firstly, the amount of subtitles on a screen is limited and secondly, the subtitles need to be displayed as long as possible. Moreover, the subtitles must be divided by the speaker change and this is where the speaker change information from diarization also becomes helpful.

We created an algorithm that divides the subtitles iteratively by largest possible pause between segments while maintaining consistency of numbers. The output from the recognition is a sequence of time-aligned words. First step is to concatenate words that are representing numbers and converting them, but keeping the multipliers like millions and billions in textual form for better reading. For example, one million will be converted into "1 million". In the iteration process a stop condition is required in form of time that represents the minimal allowed time of the subtitles on a screen.

6 Experimental Results

We have conducted simple experiments to evaluate each of the processing steps and state its value to the whole system. The main flaw of the previous system was the recognition accuracy of BN interviewees, so in this section we focused on the reporters and interviewees performance (Table 1).

Based on the achieved results, we have been able to considerably increase the overall system accuracy. We have also conducted a test when the segmentation process was done by annotators and thus more reliable than automatically. As it is apparent speaker clustering is more than crucial to the system accuracy and speaker diarization module needs to be improved.

Table 1. Experimental results of a new prototype

	Previous system	VAD only	Speaker diarization	Language model adaptation	Segmentation by annotators
WER [%]	24.30	16.41	14.88	14.60	12.91

7 Conclusion and Future Work

We have introduced a new design of the working prototype that is usable for transcription of any kind of audio (or audiovisual) content, in general. Especially, the system is suitable for transcription of meeting speech and for broadcast news transcription/subtitling, because the transcription task in both cases is very similar. The new architecture is allowing us more flexibility and provides us with more production ready solution where backup system can be employed, and non-stop accessibility can be achieved by doubling the workers while the system is scalable by adding new computing servers and supporting easy deployment of new processing modules. The experimental results are showing us that we are on the right path, but there are bottlenecks that we need to revisit. One of them is the diarization module making difference in word error rate (WER) about 2% compared to manual segmentation.

Acknowledgments. The research in this paper was supported by the Ministry of Education, Science, Research and Sport of the Slovak Republic under the project VEGA 1/0511/17 and the Slovak Research and Development Agency under the project APVV-15-0517.

References

1. Akita, Y., Watanabe, M., Kawahara, T.: Automatic transcription of lecture speech using language model based on speaking-style transformation of proceeding texts. In: Proceedings of INTERSPEECH 2012, pp. 2326–2329, Portland, OR, USA (2012)
2. Álvarez, A., Mendes, C., Raffaelli, M., Luís, T., Paulo, S., Piccinini, N., Arzelus, H., Neto, J., Aliprandi, C., Del Pozo, A.: Automating live and batch subtitling of multimedia contents for several European languages. Multimedia Tools Appl. **75**(18), 10823–10853 (2016)
3. Gauvain, J.-L., Lamel, L., Adda, G.: The LIMSI broadcast news transcription system. Speech Commun. **37**(1–2), 89–108 (2002)
4. Jolliffe, I.T.: Principal Component Analysis. Springer, New York (1986)
5. Peddinti, V., Povey, D., Khudanpur, S.: A time delay neural network architecture for efficient modeling of long temporal contexts. In: Proceedings of INTERSPEECH 2015, pp. 3214–3218, Dresden, Germany (2015)
6. Pleva, M., Juhár, J.: TUKE-BNews-SK: Slovak broadcast news corpus construction and evaluation. In: Proceedings of LREC 2014, pp. 1709–1713, Reykjavik, Island (2014)
7. Povey, D., Ghoshal, A., Boulianne, G., Burget, L., Glembek, O., Goel, N., Hannemann, M., Motlicek, P., Qian, Y., Schwarz, P., Silovsky, J., Stemmer, G., Vesely, K.: The Kaldi speech recognition toolkit. In: Proceedings of ASRU 2011, Waikoloa, Hawaii, USA (2011)
8. Quoc, V.L., Mikolov, T.: Distributed representations of sentences and documents. In: Proceedings of the 31st International Conference on Machine Learning, pp. 1188–1196, Beijing, China (2014)
9. Rouvier, M., Dupuy, G., Gay, P., Khoury, E., Merlin, T., Meignier, S.: An open-source state-of-the-art toolbox for broadcast news diarization. In: Proceedings of INTERSPEECH 2013, pp. 1477–1481, Lyon, France (2013)

10. Rusko, M., Juhár, J., Trnka, M., Staš, J., Darjaa, S., Hládek, D., Sabo, R., Pleva, M., Ritomskỳ, M., Ondáš, S.: Recent advances in the Slovak dictation system for judicial domain. In: Proceedings of LTC 2013, pp. 555–560, Poznań, Poland (2013)
11. Staš, J., Hládek, D., Juhár, J.: Semantic indexing and document retrieval for personalized language modeling. In: Proceedings of ELMAR 2017, pp. 157–161, Zadar, Croatia (2017)
12. Staš, J., Viszlay, P., Lojka, M., Koctúr, T., Hládek, D., Kiktová, E., Pleva, M., Juhár, J.: Automatic subtitling system for transcription archiving and indexing of Slovak audiovisual recordings. In: Proceedings of LTC 2015, pp. 186–191, Poznań, Poland (2015)
13. Stolcke, A.: SRILM - an extensible language modeling toolkit. In: Proceedings of ICSLP 2002, Denver, CO, USA (2002)
14. Vavrek, J., Viszlay, P., Kiktová, E., Lojka, M., Juhár, J., Čižmár, A.: Query-by-example retrieval via fast sequential dynamic time warping algorithm. In: Proceedings of TSP 2015, pp. 453–457, Berlin, Germany (2015)
15. Viszlay, P., Stas, J., Koctúr, T., Lojka, M., Juhár, J.: An extension of the Slovak broadcast news corpus based on semi-automatic annotation. In: Proceedings of LREC 2016, pp. 4684–4687, Portorož, Slovenia (2016)

Web Based Interactive 3D Educational Material Development Framework Supporting 360VR Images/Videos and Its Examples

Yoshihiro Okada(✉), Akira Haga, and Wei Shi

Innovation Center for Educational Resources (ICER),
Kyushu University, Fukuoka, Japan
okada@inf.kyushu-u.ac.jp

Abstract. This paper treats one of the activities of ICER (Innovation Center for Educational Resources) in Kyushu University Library of Kyushu University, Japan. It is the development of educational materials using recent ICT for enhancing the educational efficiency in the university. Especially, this activity focuses on the development of attractive and interactive educational materials using 3D CG. So, the authors have already proposed a framework dedicated for the development of web-based interactive 3D educational materials and introduced a couple of practical educational materials actually developed using the proposed framework. For developing more and more attractive educational materials using Virtual Reality (VR)/Augmented Reality (AR), the authors have added newly functionalities to the framework that allows the development of web-based VR/AR applications. Recently, 360VR images/videos have become popular because 360VR recorders are released from several companies. Therefore, the authors also introduced new functionalities that support 360VR images/videos into the framework. This paper describes the details of the introduced functionalities and also shows a couple of example materials developed using the framework.

Keywords: e-Learning · 3D graphics · Educational materials · 360VR

1 Introduction

This paper proposes a web-based 3D material development framework supporting 360VR images/videos for attractive and interactive educational materials. This research is one of the activities of our center called ICER (Innovation Center for Educational Resources) [1] in Kyushu University Library of Kyushu University, Japan.

Efficient education may be achieved with attractive educational materials that would be realized by the most recent ICT like 3D CG because current students sometimes called video game generation are used to operate such contents [2, 3]. However, there is the problem that the development of such contents requires technological knowledge and programming skills but usual teachers do not have such knowledge and skills. Teachers need any tools to make it easier to develop attractive educational materials that use the most recent ICT. So, we have already proposed a framework dedicated for the development of web-based interactive 3D educational

L. Barolli et al. (Eds.): NBiS 2018, LNDECT 22, pp. 395–406, 2019.
https://doi.org/10.1007/978-3-319-98530-5_33

materials [4, 5]. In general, attractive educational materials should include various types of media data like audio, videos and 3D geometry data with their animation data besides texts. The proposed framework supports such media data. In some cases, there are story-based materials to follow an instructional design. The framework also supports such story information. As case studies, we have developed two educational materials with the proposed framework. One is for teaching certain ancient manners of the ceremony called 'Kanso' taken in the ancient imperial court called 'Gosyo' in Japanese history study, and the other one is as tourism information system about a certain ancient building called 'Korokan'. These are web-based contents so that they work on various platforms such as iPhone, iPad, and Android tablets besides standard desktop PCs. This is significant for BYOD (Bring Your Own Device) classes and Mobile Learning. However, for this, touch interfaces should be supported. Therefore, we have already introduced functionalities of touch interfaces into the framework.

Furthermore, to enable the development of more and more attractive contents using Virtual Reality (VR)/Augmented Reality (AR), we also introduced functionalities for that to the framework [6, 7]. Among of them are a stereo view, a device orientation and geolocation interfaces, and a camera interface for tablet devices and smart phones. By supporting these functionalities, it is possible to develop web-based VR/AR applications with the framework.

Recently, 360VR images/videos have become popular because 360VR recorders are released from several companies. 360VR images/videos are convenient for the development of immersive environments that have higher educational efficiency. Therefore, we also introduced new functionalities that support 360VR images/videos into the proposed framework. This paper describes details of the introduced functionalities and also shows a couple of example materials developed using the framework.

The remainder of this paper is organized as follows: next Sect. 2 describes related work. We briefly explain the proposed framework, its design and functional components in Sect. 3, and introduce the two actual contents we developed so far. In Sect. 4, we explain briefly the functionalities for supporting VR/AR, and newly introduced functionalities for supporting 360VR images/videos with simple example contents. Finally, we conclude the paper and discuss about future work in Sect. 5.

2 Related Work

There are many development systems and tools for 3D contents. Some of them are commercial products like 3D Studio Max, Maya, and so on. Although these products can be used only for creating 3D CG images or 3D CG animation movies, usually, these cannot be used for creating interactive contents. As a development system for 3D interactive contents, there is *IntelligentBox*, a constructive visual software development system for 3D graphics applications [8]. This system seems useful because there have been many applications actually developed using it so far. However, it cannot be used for creating web-based contents. Although there is the web-version of *IntelligentBox* [9], it cannot be used for creating story-based contents. With Webble world [10, 11], it is possible to create web-based interactive contents through simple operations for authoring and of course, possible to render 3D graphics assets. However, it does not have functionalities of *IntelligentBox*.

There are some electronic publication formats like ePub, EduPub, iBooks and their authoring tools. Of course, these contents are used as e-learning materials. However, basically, these are not available on the web and do not support 3D Graphics except iBooks. iBooks supports rendering functionality of a 3D scene and control interfaces of its viewpoint. However, story-based contents cannot be created using it.

From the above situation, for creating web-based interactive 3D educational contents, we have to use any dedicated toolkit systems. The most popular one is Unity, one of the game engines [12], that supports creating web contents. The use of Unity requires any programming knowledge and skills of the operations for it. Therefore, it is impossible to use Unity for standard end-users like teachers. As a result, in this paper, we proposed the framework [4–7] and extended it for supporting 360VR images/videos explained in the following sections.

3 Web Based Interactive 3D Educational Material Development Framework

Our first targets of the framework were contents of Japanese History study. In general, a history consists of several stories. So, the framework should support a story. Each story is realized with several 3D scenes consisting of several architecture objects like buildings and houses, and several moving characters like humans who have their own shape model and animation data. Firstly, such 3D assets data should be prepared. Next, contents creators have to define a story for the content as one JavaScript file called 'Story Definition File'. In our proposed framework, the requirements for creating a content are 3D assets data and 'Story Definition File'.

Figure 1 shows functional components of the proposed framework consisting of main components (Main.html) and sub components (AnimationCharacter.js). The main components include functions related to architecture objects and functions related to AnimationCharacter objects represented as moving characters. The sub components include the constructor of new AnimationCharacter class and its member functions. Besides main and sub components, our framework uses Three.js [13], one of the WebGL based 3D Graphics libraries as subsidiary components. When developing a story-based interactive 3D educational content with the proposed framework, a teacher has to prepare one story definition file and 3D assets for it. See the papers [4, 5] for more details.

As case studies, we have been developing two types of contents with the proposed framework. One is for learning certain events and ceremonies that were taken in the Imperial court called 'Gosyo', including ancient manners of the Emperor called 'Tennou' and Cabinet members called 'Daijin' and 'Daiben' and so on as Japanese history study. We have already finished the story called 'Kanso' and have been developing the story called 'Jimoku'. The other one is for teaching a certain ancient building called 'Korokan' to tourists visiting to our 'Fukuoka' city as one of the sightseeing spots. In fact, the building does not exist and only its remains exist.

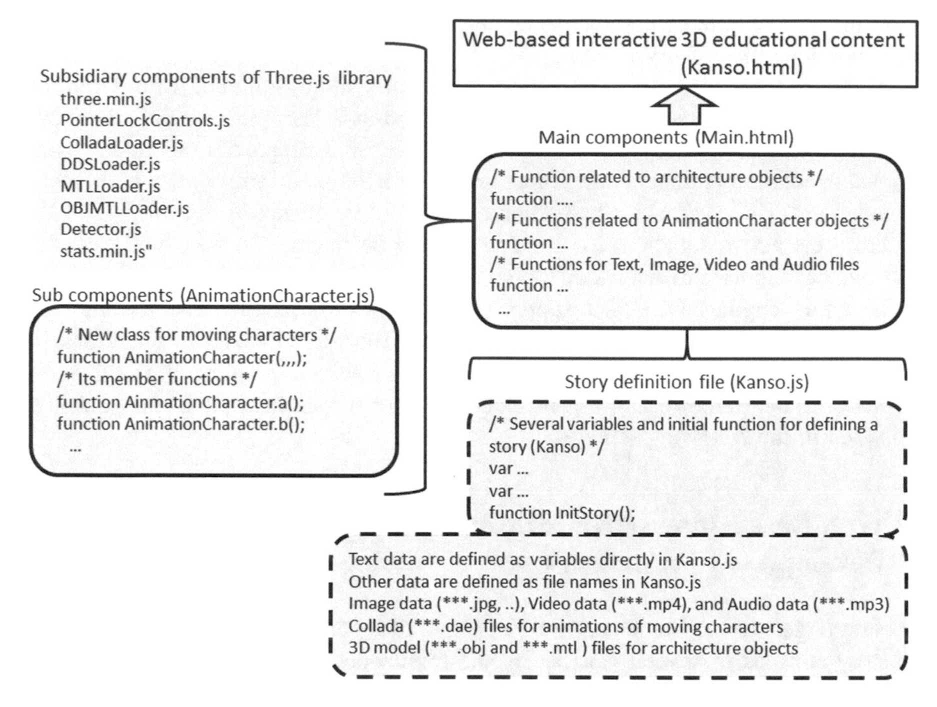

Fig. 1. Functional (Main and Sub) components of Framework, ex) Kanso.html.

'Kanso' of Japanese History Study
'Kanso' is one of the ceremonies that were taken in 'Gosyo'. 'Kanso' includes various types of manners of 'Tennou' and 'Daijin, etc. In Japanese history study, students have to learn such manners by reading old documents about 'Kanso'. However, it is very difficult to understand the manners from the old documents. Using this interactive 3D educational material of 'Kanso' shown in Fig. 2, it becomes easy to understand.

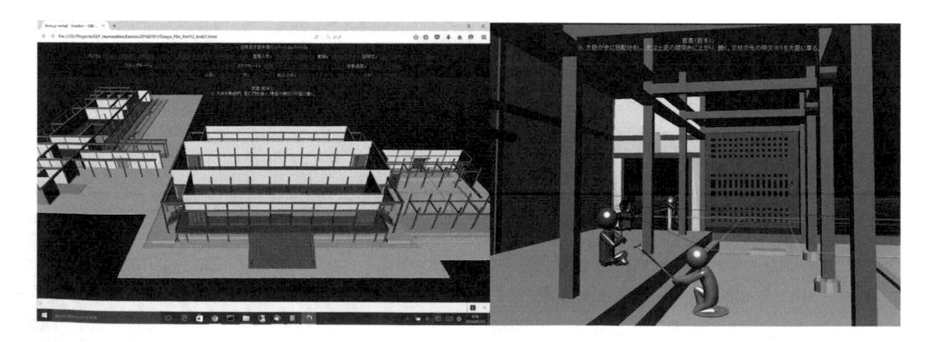

Fig. 2. Screen shots of 'Kanso' content, its architecture objects and some characters.

As shown in the left part of Fig. 2, there are three houses as architecture objects. The center one is ‘Shishinden’, the left one is ‘Seiryoden’, and the right one is ‘Jinnoza’. These architecture objects are prepared as three ‘obj’ format files with their material files. The right part of Fig. 2 shows a scene at ‘Jinnoza’. As moving characters of ‘Kanso’, there are seven characters, i.e., ‘Tennou’, ‘Daijin’, ‘Shi’, etc. Each moving characters are implemented as instances of AnimationCharacter class in JavaScript. AnimationCharacter has its several animation data prepared as ‘dae’ format files. For this content, we prepared around 100 dae format files. To reduce the cost to prepare these data files, we employ 3D human pictograms for the shape data of moving characters. Since their colours are different, it is possible to recognize each characters. For example, the colour of ‘Tennou’ is yellow, ‘Daijin’ is red and so on. There are seven poly-lines in a different colour shown in the right part of Fig. 2. Each is the moving path of the corresponding moving-character. Its colour is the same as the pictogram. At the upper and center position of the scene, texts data are displayed in white colour. These texts explain the behaviours of the moving characters. By reading the texts, the students can understand the manners in ‘Gosyo’ more deeply with looking at the manners as 3D CG animations.

‘Korokan’ Content for Teaching Tourists as Their Sight-Seeing Spot
Figure 3 shows screen shots of ‘Korokan’ content that is for teaching a certain ancient building called ‘Korokan’ to tourists visiting to our ‘Fukuoka’ city as one of the sight-seeing spots. Indeed, its building does not exist and only its remains exist. We have developed this content as one of the collaboration works with our ‘Fukuoka’ city government. The left part of Fig. 3 shows a certain scene of the building and the right part of Fig. 3 shows a screen shot of displaying one of the image files that explains some activities taken at that time in the building. Tourists can walk through the scene virtually using this content on their own PC and maybe become attracted to ‘Korokan’.

Fig. 3. Screen shots of ‘Korokan’ content, its architecture objects and an image.

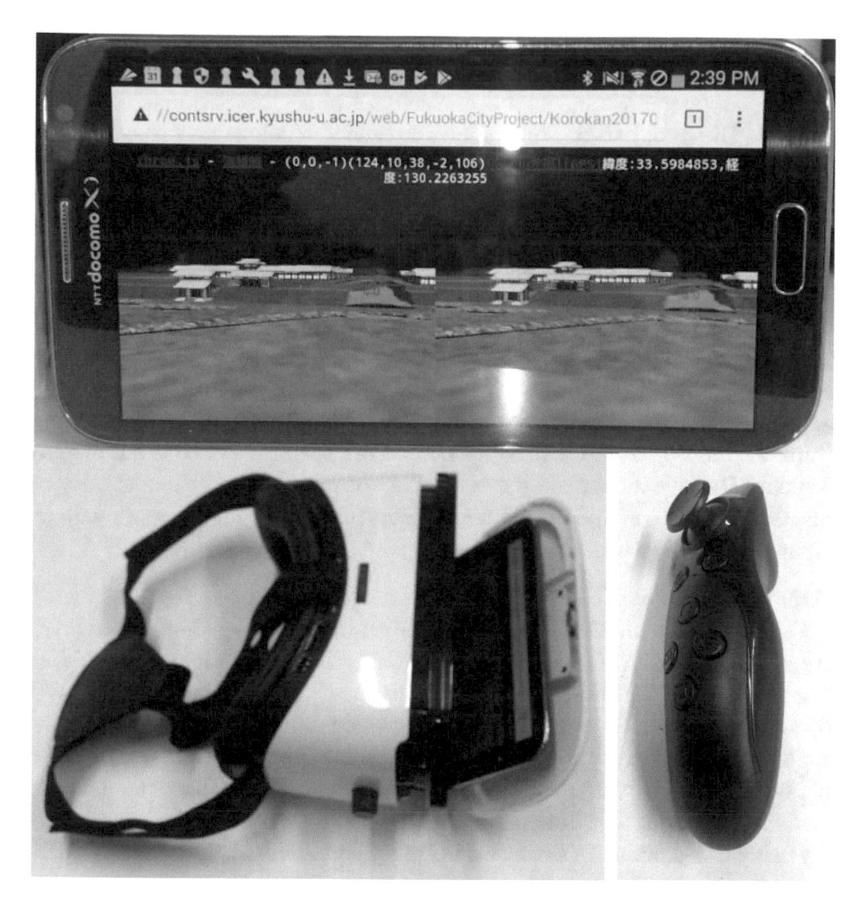

Fig. 4. Snap shot of 'Korokan' with a stereo view on a smart phone (upper), VOX + 3DVR goggle (lower left) and Poskey blue-tooth gamepad (lower right).

4 Web Based Interactive 3D Educational Material Development Framework Supporting VR/AR and 360VR Images/Videos

4.1 Web Based Interactive 3D Educational Material Development Framework Supporting VR/AR

The followings are functionalities to support VR/AR that are a stereo view, touch interfaces, device orientation/motion interfaces, geolocation interface and a camera interface for tablet devices and smart phones (Fig. 4). See the paper [6] for more detail.

1. **Stereo view support**
   JavaScript program for the stereo view is very simple within Three.js because Three.js provides StereoEffect.js and you can read this file in your HTML file and

can call effect.render(scene, camera) for the stereo view instead of calling renderer. render(scene, camera) for the standard view.

2. **Touch interfaces**
   JavaScript program for touch interfaces is also simple because HTML5 supports 'touchstart', 'touchend', 'touchmove' events and you can access x and y positions of your touch fingers by event.touches[*].pageX and event.touches[*].pageY, here, * means the indices of your fingers.
3. **Device orientation/motion interfaces**
   The device orientation/motion interfaces are simple but useful in a JavaScript program of HTML5 because there are 'deviceorientation' and 'devicemotion' events, and you can access the device orientation/motion of your smart phone by event.alpha, event.beta, event.gamma, event.acceleration.x, event.acceleration.y and event.acceleration.z, respectively.
4. **Geolocation interface**
   JavaScript program for geolocation interface is also simple within HTML5 because HTML5 supports navigator.geolocation variable and navigator.geolocation. getCurrentPosition() function, and you can access the device geolocation of your smart phone by position.coords.latitude and position.coords.longitude.
5. **Camera interface**
   For developing web-based AR applications, we have to manage video camera images in real-time on a web browser. There is WebRTC project that provides browsers and mobile applications with Real-Time Communications (RTC) capabilities via simple APIs. Using WebRTC, it is possible to manage video camera images on a web browser and to develop web-based AR applications.

4.2 Web Based Interactive 3D Educational Material Development Framework Supporting 360VR Images/Videos

The followings are newly introduced functionalities to support 360VR images/videos based on WebGL. Originally, there are example JavaScript programs of Three.js for displaying 360VR images/videos so that we extended those programs to make them more convenient. For displaying 360VR images/videos, we have to prepare Equirectangular projection images/videos as shown in Fig. 5. Using a special camera, e.g., RICHO Theta, it is possible to generate such projection images/videos from its captured images/videos shown in Fig. 6 by converting using the dedicated software provided by RICHO. Then, by the texture mapping of such projection image/video frame onto the surface of a sphere located in a virtual 3D space of WebGL. Figure 7 shows one frame of 360VR video.

In 360VR video of Fig. 7, one of the authors walked from one location to another location with bringing a RICHO Theta. In any video frame, it is possible to see 360-degree image. Therefore, if it is possible to adequately display any video frame anytime, we can walk through freely between the position of a first video frame and the position of a last video frame as shown in Fig. 8. Unfortunately, for playing a video on a Web browser using <video> tag of HTML5, it does not support the playback and we cannot walk through. For enabling the interactive display of any video frame on a web browser, once we made separated video frames as a certain image format from the

Fig. 5. Equirectangular projection image/video frame.

Fig. 6. 360VR camera, e.g., RICHO Theta and its captured image.

original video and added simple interfaces to 'go forward' and 'go backward' to display the next video frame and the previous video frame, respectively. Then, the walk through has become possible as shown in Fig. 9.

Although the proposed WebGL based 360VR video viewer is separated from the framework of Sect. 4.1, this viewer is also web-based and easy to use with the framework. For that, the information of link buttons was added into 'Story Definition File'. All one have to do is to write URL of the viewer for a certain 360VR video in 'Story Definition File'. Figure 10 shows a screen image of 'Korokan' content with some link buttons, one of them is URL of 360VR video shown in Fig. 9. 'Korokan' was one of the ancient buildings in 'Heiran' era and it does not exist at present. The 3D model of 'Korokan' shown in Fig. 10 is the imagination created from the information of ancient documents of 'Heian' era. The building of a 360VR video shown in Fig. 9 is 'Itsukushima' shrine and was built in 'Heian' era so that its floor, ceiling, pillars and those colors are used as references for the 3D model of 'Korokan'. Therefore, 360VR video of 'Itsukushima' shrine can be used as an educational material for teaching 'Korokan'.

Fig. 7. One frame of a certain 360VR video on WebGL based viewer.

Fig. 8. Several frames in backward (left) and forward (right) direction of 360VR video.

Fig. 9. Backward (left) and forward (right) direction images of a certain frame of 360VR video with 'go forward' and 'go backward' buttons for walk-through.

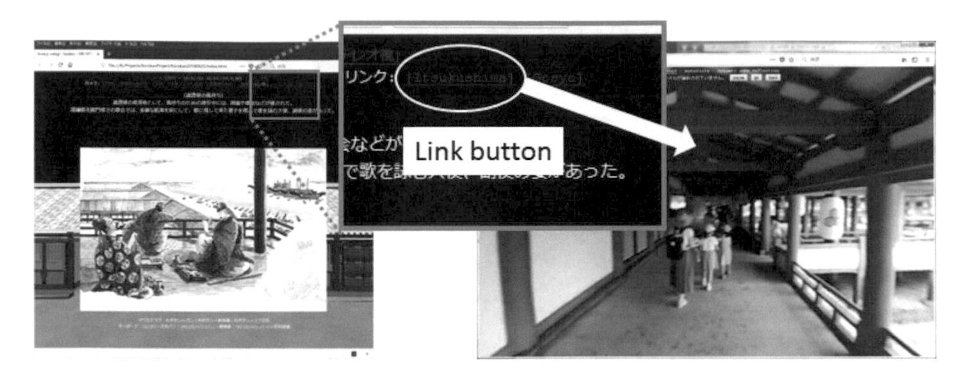

Fig. 10. Link button of 'Korokan' content for going to WebGL based viewer of 'Itsukushima shrine' 360VR video.

Figure 11 shows another example of the 360VR videos. In this example, there are four 360VR videos, each of which corresponds to one of the four directions at a crossroads. At the crossroads, the user can choose one of the four directions by taking one of the four operations, i.e., 'go forward', 'go backward', 'turn left & go forward' and 'turn right & go forward'. After that, the user can move along to the chosen direction by pushing 'go forward' button. Currently, we have been developing such interfaces that should appear at a crossroads.

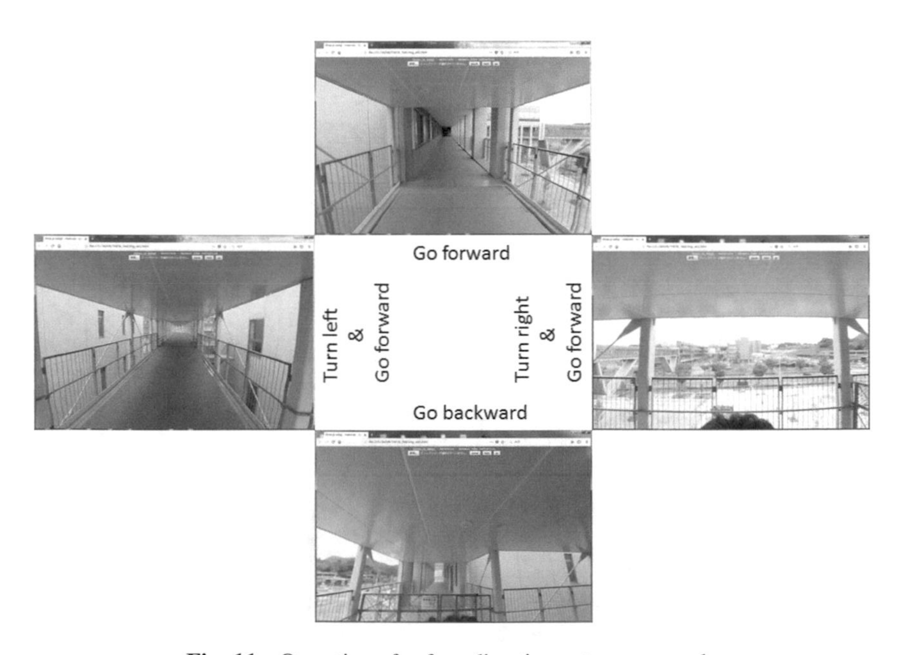

Fig. 11. Operations for four directions at a crossroads.

5 Conclusions

In this paper, we proposed the framework for the development of web-based interactive 3D educational materials. The framework supports the most recent Web technologies, i.e., HTML5 and various JavaScript libraries including WebGL. This paper explained the framework, its design and functional components. Furthermore, we explained briefly functionalities supporting VR/AR. Actually, the main contribution of this paper is that we added the WebGL based 360VR image/video viewer into the framework. Using 360VR images/videos, we can provide more and more attractive educational materials to students.

As future work, we will try to ask teachers to use the proposed framework for actually creating web-based interactive 3D educational materials of VR/AR and 360VR images/videos to clarify the usefulness of the framework. Furthermore, we will create several educational materials with the framework and ask students to learn using the materials. After that, we will consult the students for evaluating educational efficiency of the materials.

Acknowledgements. This research was partially supported by JSPS KAKENHI Grant Number JP16H02923 and JP17H00773.

References

1. ICER, 08 June 2018. http://www.icer.kyushu-u.ac.jp/en
2. Sugimura, R., et al.: Mobile game for learning bacteriology. In: Proceedings of IADIS 10th International Conference on Mobile Learning 2014, pp. 285–289 (2014)
3. Sugimura, R., et al.: Serious games for education and their effectiveness for higher education medical students and for junior high school students. In: Proceedings of the 4th International Conference on Advanced in Information System, E-Education and Development (ICAI-SEED 2015), pp. 36–45 (2015)
4. Okada, Y., Nakazono, S., Kaneko, K.: Framework for development of web-based interactive 3D educational contents. In: 10th International Technology, Education and Development Conference, pp. 2656–2663 (2016)
5. Okada, Y., Kaneko, K., Tanizawa, A.: Interactive educational contents development framework based on linked open data technology. In: 9th Annual International Conference of Education, Research and Innovation, pp. 5066–5075 (2016)
6. Okada, Y., Kaneko, K., Tanizawa, A.: Interactive educational contents development framework and its extension for web-based VR/AR applications. In: Proceedings of the GameOn 2017, Eurosis, pp. 75–79, 6–8 September 2017. ISBN: 978-90-77381-99-1
7. Ma, C., Srishti, K., Shi, W., Okada, Y., Bose, R.: E-learning material development framework supporting VR/AR based on linked data for IoT security education. In: Proceedings of 6th International Conference on Emerging Internet, Data and Web Technologies (EIDWT 2018), Tirana/Albania, pp. 479–491, 15–17 March 2018. ISBN: 978-3-319-75928-9
8. Okada, Y., Tanaka, Y.: IntelligentBox: a constructive visual software development system for interactive 3D graphic applications. In: Proceedings of the Computer Animation 1995, pp. 114–125. IEEE CS Press (1995)

9. Okada, Y.: Web Version of IntelligentBox (WebIB) and its integration with Webble World. Communications in Computer and Information Science, vol. 372, pp. 11–20. Springer, New York (2013)
10. Tanaka, Y.: Meme Media and Meme Market Architectures: Knowledge Media for Editing, Distributing, and Managing Intellectual Resources. Wiley, New York (2003)
11. Webble World 08 June 2018. https://github.com/truemrwalker/wblwrld3
12. Unity 08 June 2018. https://unity3d.com/jp
13. Three.js 08 June 2018. https://threejs.org/

On Estimating Platforms of Web User with JavaScript Math Object

Takamichi Saito(✉), Takafumi Noda, Ryohei Hosoya, Kazuhisa Tanabe, and Yuta Saito

Meiji University, Higashimita 1-1-1, Kawasaki, Japan
saito@cs.meiji.ac.jp, {ce185026,ce175024,ce175019,ce185016}@meiji.ac.jp

Abstract. Browser fingerprinting is a technique to identify a device using a combination of information that a server can gather from a browser. In general, the user-agent is known to be one of the most useful features for identification via browser fingerprinting. However, users can easily change the user-agent. This may lead to a decrease in the device identification accuracy. In this paper, we conducted two experiments. First, we proposed a method to estimate the platforms without using the user-agent. In particular, we used the fact that the computational result of a JavaScript math object varies depending on the platform. Using this method, we could classify 14 platforms into nine groups. Five of these uniquely identify the OS and the browser, two uniquely identify the OS and one uniquely identifies the browser. Second, we compared the accuracy of the browser fingerprint with user-agent (FP-A) and the browser fingerprint with our proposed method (FP-B). As a result, the identification accuracy rate with FP-B was only 0.4% lower than that with FP-A.

1 Introduction

Browser fingerprinting is a technique to identify a device using a combination of information that a server can gather from a browser. This technique is primarily used by advertising companies to track user behavior or for risk-based authentication. According to a survey by Englehardt et al. [3], 14,371 (1.6%) of Alexa's [4] TOP 1 million sites as of January 2016 use fingerprinting technology. Tanabe et al. [7] demonstrated the most useful features for device identification. According to their paper, the user-agent is as one of the most useful features for identification with browser fingerprinting. However, users can easily change the value of user-agent using browser add-ons, configurations, and tools. This may lead to decrease of device identification accuracy. In this paper, we attempt to estimate the platform without using the user-agent as an identification feature. In particular, we use the fact that the computational results of JavaScript math object vary depending on the platform. Using this method, we conducted the following two experiments.

L. Barolli et al. (Eds.): NBiS 2018, LNDECT 22, pp. 407–418, 2019.
https://doi.org/10.1007/978-3-319-98530-5_34

Experiment 1. *We examined the extent to which the platform can be correctly estimated.*

Experiment 2. *We compared the accuracy of the browser fingerprint with the user-agent (FP-A) to the browser fingerprint with our proposed method (FP-B).*

As a result of Experiment 1, we were able to classify 14 platforms into nine groups. Five of these uniquely identify the OSes and the browser, two uniquely identify the OSes, and one uniquely identify the browsers. Experiment 2 demonstrated that identification accuracy rate with FP-B was only 0.4% lower than that with FP-A.

2 Background Information

2.1 JavaScript Math Object

The math object [9] is a built-in JavaScript object that contains the properties and methods of mathematical constants and functions. According to the ECMAScript draft [5], which was defined on November 27, 2017, the implementation of IEEE 754-2008, which has floating-point standartan, atanh, atan2, cbrt, cos, cosh, exp, expm1, hypot, log, log1p, and ld is recommended. However, for acos, acosh, asin, asinh, atan, atanh, atan2, cbrt, cos, cosh, exp, expm1, hypot, log, log1p, log2, log10, pow, random, sin, sinh, sqrt, tan, and tanh, the computational result depends on the platform because they are implemented using the math library provided by the platform.

The Tor project trac.torproject.org [6], a site that manages tickets for Tor issues, reported that the results of Math.tan(−1e300) were different for each OS. In this paper, we conducted an exhaustive survey in reference to this result. We investigated the differences in the computational results for each platform using various methods involving math objects and arguments.

2.2 Related Work

Eckersley et al. [1] built a site to collect fingerprints and then analyzed those fingerprints. They found that 83.6% of the collected fingerprints were unique and that 94.2% were unique when limited to devices on which Flash or Java can run.

Mowery et al. [2] suggested that the result of depicting complicated characters and figures on a browser with the Canvas API varies depending on the combination of GPU, browser version, and OS. Further, 116 of the 294 samples collected in their survey were unique, and showing that this method could be applied to fingerprinting by preserving the hashed value of their result. In that case, the entropy was 5.73 bits and the browser could be correctly identified with a probability of 98.2%. They also showed that the GPU and model name could be collected using the WebGL API.

3 Experiments

3.1 Fingerprint Collection Site

In this paper, we publish a fingerprint collection site and collected fingerprint from the browsers on a device. The fingerprints are stored in the site database along with the HTTP cookie (UID). The UID is used to verify whether each access is from a browser on the same device. The site demonstrates that the information in the device is collected on the top page and that the participants can only cooperate with the experiment if they agree to the contents.

3.2 Collected Samples

We asked participants to access the site and used them as the sample. The collection period was from 11/27/2017 to 12/11/2017, and 417 samples were collected. Approximately 300 samples were collected by Lancers [8], a cloud sourcing service, and we requested the collection of samples not only from stakeholders but also from participants with diverse backgrounds.

3.3 Outline of Experiment 1

Using the fact that the computational result of a JavaScript math object varies depending on the platform, we verified the extent to which the platform could be correctly estimated. In addition, we extracted the version of the platform from the user-agent and verified the differences by version.

3.3.1 Samples of Experiment 1

We conducted experiments on 14 different platforms consisting of major OSes and browsers. Table 1 shows the number of samples for each platform.

Table 1. Number of samples for each platform

	Chrome	Firefox	Safari	Edge	IE
Android	16	6			
iOS	4	3	17		
Linux	1	5			
OS X	8	1	7		
Windows	100	29		24	52

Upon validation, it is necessary to correlate the computational results with the math object and the platforms. In this case, we used the user-agent to estimate the platform. However, as mentioned above, it is possible for users to disguise the user-agent. Therefore, when collecting samples on the site, we requested that the participants declare their platform. By comparing the user-agent with their declaration, we checked whether the disguise of the user-agent was disguised. Moreover, samples with no participant declaration and samples that it did not match user-agent were excluded. As a result, the number of samples in Experiment 1 was 273.

3.3.2 Evaluation Method of Experiment 1

In this paper, we used math object methods with a large number of digits in the computational result. In particular, we used Math.sin, Math.cos, Math.tan, Math.log, Math.exp, and Math.sqrt for the verification experiments. In addition, we conducted an exhaustive investigation using two types of method arguments.
Argument 1 10^0,10^1,...,10^{299},10^{300}
Argument 2 -10^0,-10^1,...,-10^{299}, -10^{300}

3.4 Outline of Experiment 2

Tanabe et al. [7] demonstrated the most useful combination of features for device identification. This combination has 18 features and includes the user-agent as one of the most useful features for identification with browser fingerprinting.

We compared the difference between the a case using the 18 features including the user-agent and a case using the 18 features with the user-agent replaced with a math object.

3.4.1 Samples of Experiment 2

The number of samples was 417. In these samples, the number of devices (UID) was 328, and the number of devices accessed multiple times was 55.

3.4.2 Evaluation Method of Experiment 2

In this paper, we defined the area under the curve (AUC) of the receiver operating characteristic (ROC) curve as the identification accuracy. The AUC is the area under the ROC curve. In this case, high AUC values indicate higher identification accuracies. By adopting the AUC value of the ROC curve as the identification accuracy, it is possible to confirm the TPR (true positive rate) and the FPR (false positive rate) at the same time.

4 Results

4.1 Result of Experiment 1

4.1.1 Estimating Number of Platforms

Comparing the number of identifiable platforms when using Argument 1 and Argument 2 in each math object method, Math.sin, Math.cos, and Math.tan were not changed between Argument 1 and Argument 2. Conversely, Math.exp, Math.log, and Math.sqrt resulted in more identifiable platforms when using

Argument 1. Hereinafter, we only describe the results of using Argument 1 as the argument. Of the math object methods that we used, Math.cos could identify the platforms in the most detail. We were able to classify the platforms into 13 groups. Table 2 shows the number of classifications per math object method.

In Table 4, we show a detailed classification result. However, when different platform versions are combined into one such as group numbers 1-1, and 1-2 in the Table 4, they can be classified into nine groups. In addition, in this experiment, trigonometric functions such as Math.tan, Math.sin, and Math.cos tended to produce more differences in the computational result depending on the platform.

Table 2. Number of classifications per math object method

Math object	Number of classifications
Math.cos	13
Math.sin	12
Math.tan	11
Math.exp	4
Math.log	2
Math.sqrt	2

4.1.2 Result of Estimating the Platforms

In this paper, we describe the estimation result for the platform based on the result of Math.cos, which has the most identifiable numbers. Table 3 shows the result of platform estimation with Math.cos.

Table 3. Result of Experiment 1

	Chrome	Firefox	Safari	Edge	IE
Android	△	○	NaN	NaN	NaN
iOS	•	•	•	NaN	NaN
Linux	△	○	NaN	NaN	NaN
OS X	△	•	•	NaN	NaN
Windows	△	○	NaN	○	○

○: OS and browser identifiable
•: Only OS identifiable
△: Only browser identifiable

However, there were some platforms that could identify neither the OS nor the browser depending on their versions. We described them in following fourth group. Based on these, we see that estimation results can be classified into the following four groups.

1. Both the OS and browser are identifiable.
   - Windows, IE
   - Windows, Edge
   - Windows, Firefox
   - Android, Firefox
   - Linux, Firefox

   We could identify the combination of OS and browser in the above five examples. Three of these are based on Windows or Firefox, accordingly, identification is relatively easy when using Windows or Firefox.
2. Only the OS is identifiable.
   - OS X
   - iOS

   When the above two OSes are used, the OS is identified but the browser cannot be identified. The computational results of (OS X, Safari) and (OS X, Firefox) were the same in all the arguments used in this experiment. Therefore, when OS X is used, it is not possible to distinguish between Safari and Firefox. Similarly, when iOS is used, (iOS, Safari), (iOS, Firefox), and (iOS, Chrome) cannot be identified, because their computational results of them are the same.
3. Only the browser is identifiable.
   - Chrome

   When using Chrome, the OS cannot be identified because the computational results are the same in the all arguments used in this experiment even if the OS is changed. The computational results of (Android, Chrome), (Linux, Chrome), (OS X, Chrome), and (Windows, Chrome) were the same, therefore, it is not possible to identify these OSes.

4. Neither the OS nor the browser are identifiable.
   - Windows, Chrome
   - Windows, Firefox
   - Android, Firefox

   Because the computational results are the same for the above three platforms in all the arguments used in this experiment, neither the OS nor the browser can be identified.

As a result, we could classify the 14 platforms into nine groups. Five of these uniquely identify the OS and the browser, two uniquely identify the OS and one uniquely identifies the browser. In addition, from the results of 1 and 2 above, we concluded that the OS could be estimated except when Chrome is used.

Table 4. Estimating platforms with Math.cos

Group number	Identifiable platform	Breakdown
1-1	Android, Firefox(1)	Android, Firefox/38.0 Android 4.2.2, Firefox/57.0
1-2	Android, Firefox(2)	Android 7.0, Firefox/57.0 Android 6.0.1, Firefox/54.0
2-1	iOS, Safari	iOS 9_3_5, Safari/601.1
2-2	iOS, (Safari, Firefox, Chrome)	iOS 11_1_2, Safari/604.1 iOS 11_1_2, Firefox/10.2b7796 iOS 11_1_2, Chrome/62.0 iOS 11_1_2, Firefox/10.3b7934 iOS 11_1_1, Safari/604.1 iOS 11_0_3, Safari/604.1 iOS 11_0_2, Safari/604.1 iOS 11_0_2, Chrome/62.0 iOS 10_3_3, Safari/602.1 iOS 10_3_3, Chrome/62.0 iOS 10_3_3, Firefox/10.3b7934 iOS 10_3_2, Safari/602.1 iOS 10_3_1, Safari/602.1 iOS 10_0_2, Safari/602.1
3	Linux, Firefox	Linux, Firefox/52.0 Linux, Firefox/28.0
4	OS X, (Safari, Firefox)	OS X 10_9_5, Safari/537.86.7 OS X 10_13_1, Safari/604.3.5 OS X 10_12_6, Safari/604.3.5 OS X 10_12_3, Safari/602.4.8 OS X 10.12 , Firefox/56.0
5	Windows, Edge	Windows NT 10.0, Edge/15.15063 Windows NT 10.0, Edge/16.16299 Windows NT 10.0, Edge/14.14393
6-1	Windows, Firefox(1)	Windows NT 6.3, Firefox/57.0 Windows NT 10.0, Firefox/57.0 Windows NT 10.0, Firefox/56.0 Windows NT 10.0, Firefox/52.0
6-2	Windows, Firefox(2)	Windows NT 6.3, Firefox/48.0 Windows NT 6.1, Firefox/35.0 Windows NT 5.1, Firefox/47.0 Windows NT 10.0, Firefox/35.0 Windows NT 10.0, Firefox/54.0
6-3	Windows, Firefox(3)	Windows NT 6.3 , Firefox/57.0 Windows NT 6.1, Firefox/55.0 Windows NT 6.1, Firefox/57.0 Windows NT 10.0, Firefox/57.0
7	Windows, IE	Windows NT 6.3, IE Windows NT 10.0, IE Windows NT 6.1 , IE

(*continued*)

Table 4. (*continued*)

Group number	Identifiable platform	Breakdown
8	(Windows, OS X, Linux, Android), Chrome	Windows NT 6.3, Chrome/62.0
		Windows NT 6.1, Chrome/62.0
		Windows NT 10.0, Chrome/64.0
		Windows NT 10.0, Chrome/62.0
		Windows NT 10.0, Chrome/61.0
		OS X 10_13_1, Chrome/62.0
		OS X 10_13_0, Chrome/62.0
		OS X 10_12_6, Chrome/62.0
		OS X 10_11_5, Chrome/61.0
		OS X 10_11_2, Chrome/62.0
		Linux, Chrome/61.0
		Android 7.1.1, Chrome/62.0
		Android 7.0, Chrome/62.0
		Android 6.0.1, Chrome/58.0
		Android 5.1.1, Chrome/62.0
		Android 5.1, Chrome/62.0
		Android 5.1, Chrome/56.0
		Android 5.0.2, Chrome/56.0
		Android 4.4.4, Chrome/59.0
		Android 4.2.2, Chrome/62.0
9	(Windows, Android), (Chrome, Firefox)	Windows NT 6.1, Firefox/33.0
		Windows NT 6.0, Chrome/49.0
		Windows NT 5.1, Chrome/49.0
		Windows NT 5.1, Chrome/40.0
		Windows NT 10.0, Chrome/59.0
		Android 4.4.2, Chrome/34.0
		Android 4.0.3, Chrome/42.0

4.1.3 Platforms Where Computation Errors Occur Depending on the Version

In this experiment, we found that we could also identify the platform version for some platforms. These platforms are shown below. We were not able to collect all platform versions; therefore, we compared the differences only for those that could be collected.

- Windows, Firefox
  This set is classified into three groups depending on the versions of Windows or Firefox. From this result, it is difficult to say whether the difference depends on Windows or Firefox. In addition, there were samples that had different computational results, despite the platform version being the same. We describe these further in the section of later Sect. 5.
- iOS, Safari
  It was assumed that this combination was classified into two groups depending on the version of iOS. iOS version 9 and iOS versions 10, and 11 could be identified.

- Android, Firefox
  It was assumed that this combination was classified into two groups depending on the version of Android. Android version 4 and 6, versions 6 and 7 could be identified as a result.

4.1.4 Identification Method

The results of this study were obtained by combining the arguments in Table 5. The gray part entries in Table 5 show the unique values of the argument. Because we can identify the platform when it has a unique value, it is not necessary to check the subsequent numbers. Therefore, we omitted subsequent numbers in that case. In addition, because of the length of the paper, we describe only the numbers at the beginning and the end of the computational result.

Table 5. Method to Identify platforms with Math.cos. (Gray areas correspond to unique values.)

	1e251	1e140	1e12	1e130	1e272	1e0	1e284	1e75
1-1	-0.374...155	-0.785...291	0.791...902	-0.767...913	-0.741...536	0.540...398	0.708...247	-0.748...322
1-2	-0.374...155	-0.785...291	0.791...902	-0.767...913	-0.741...536	0.540...398	0.708...247	-0.748...321
2-1	-0.374...155	-0.785...291	0.791...902	-0.767...131	-	-	-	-
2-2	-0.374...155	-0.785...292	-	-	-	-	-	-
3	-0.374...155	-0.785...291	0.791...902	-0.767...913	-0.741...535	-	-	-
4	-0.374...415	-	-	-	-	-	-	-
5	-0.374...154	-	-	-	-	-	-	-
6-1	-0.374...155	-0.785...291	0.791...902	-0.767...913	-0.741...536	0.540...398	0.708...246	-
6-2	0.782...917	-0.686...344	-	-	-	-	-	-
6-3	-0.374...155	-0.785...291	0.791...902	-0.767...913	-0.741...536	0.540...397	-	-
7	0.782...917	-0.686...343	-	-	-	-	-	-
8	-0.374...155	-0.785...291	0.791...903	-	-	-	-	-
9	-0.374...416	-	-	-	-	-	-	-

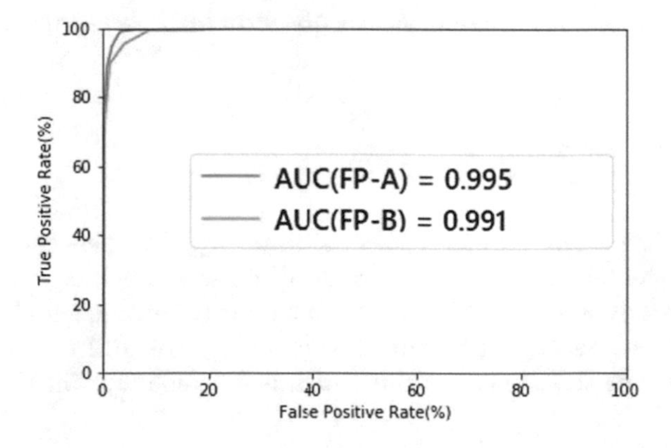

Fig. 1. AUC values of user-agent and Math.cos.

4.2 Result of Experiment 2

The results of Experiment 2 are shown in Fig. 1. We compared the accuracies of browser fingerprint with user-agent (FP-A) and browser fingerprint with our proposed method (FP-B). The rate of identification accuracy with FP-B was only 0.4% lower than that with FP-A.

5 Discussion

5.1 Discussion of Experiment 1

5.1.1 Suitable Math Objects for Platform Estimation

Math.sin and Math.cos are appropriate when to estimate the platform for the following reasons.

1. Experiment 1 indicates that many computational errors were seen occur on the platforms when using trigonometric functions.
2. These math object methods return a real number between −1 and 1 regardless of the argument. Therefore, it is possible to compare computation errors using any argument.

When using other methods, a value such as *infinity* may be returned depending on the argument. When the value becomes *infinity*, there is no computational error for each the platforms. Therefore, Math.sin, and Math.cos are suitable for platform estimation because it is possible to use them to examine all arguments.

5.1.2 Different Classification Results for the Same Version

In the computational result of Math.cos, samples classified into different groups existed even though the version of the platform was the same. Examples are shown below.

- Windows NT 6.3, Firefox/57.0
  - Windows, Firefox(1)
  - Windows, Firefox(3)

It is possible that a Tor browser could be disguising the user-agent in this case. Regardless of the OS, a Tor browser would disguise the user-agent as *Windows-Firefox*. Therefore, it is possible that we incorrectly recognized the OS and the browser. However, because this cannot be definitely concluded as resulting from the Tor browser at this time, the cause of this discrepancy will be pursued in a future study.

5.2 Discussion of Experiment 2

The accuracy with Math.cos likely decreased because we could make less estimates of the version of the platform with Math.cos than with the user-agent. In addition, the amount of information is small because the computational error of Math.cos is only 13 patterns at maximum. In future work, improvements in the accuracy can be expected if we identify the platform version in more detail. Our proposed method is valuable because it has the following positive points.

1. The difference between FP-A and FP-B is only 0.4%.
2. It is difficult for a user to disguise a math object.

Therefore, we conclude that Math.cos is valuable for browser fingerprinting.

6 Conclusion

In this paper, we attempted to estimate the platform, e.g., OS, browser, without using the user-agent, which is known as the most useful identification features available. Instead, we used the fact that the computational results of JavaScript math objects vary depending on the platform. We found two things in this experiment.

1. We could classify 14 platforms into nine groups. Five of these uniquely identify the OS and the browser, two uniquely identify the OS and one uniquely identifies the browser.
2. The rate of identification accuracy with the browser fingerprint with our proposed method (FP-B) was only 0.4% lower than that with the user-agent agent (FP-A).

As future work, we will attempt to shorten the measurement time and estimate the platform in more detail.

Acknowledgment. This work was supported by JSPS KAKENHI Grant Number 18K11305.

References

1. Eckersley, P.: How unique is your web browser? In: Proceedings of the 10th International Conference on Privacy Enhancing Technologies (PETS 2010) (2010)
2. Mowery, K., Shacham, H.: Pixel perfect: fingerprinting Canvas in HTML5. In: Proceedings of Web 2.0 Security and Privacy (W2SP) (2012)
3. Englehardt, S., Narayanan, A.: Online tracking: a 1-million-site measurement and analysis. In: Proceedings of ACM Conference on Computer and Communications Security (CCS 2016) (2016)
4. Alexa top sites, 18 June 2018. https://www.alexa.com/topsites/
5. Draft ECMA-262, 18 June 2018. https://tc39.github.io/ecma262/#sec-math-object
6. Tor issue ticket, 18 June 2018. https://trac.torproject.org/projects/tor/ticket/6119

7. Tanabe, K., Takahashi, K., Yasuda, K., Taneoka, M., Hosoya, R., Koshiba, R., Saito, Y., Saito, T.: Study on combination of features in browser fingerprinting. In: Computer Security Symposium 2017 Proceedings CD-ROM, pp. 1090–1097 (2017)
8. Lancers, 18 June 2018. https://www.lancers.jp
9. Math - JavaScript - MDN -Mozilla, 18 June 2018. https://developer.mozilla.org/ja/docs/Web/JavaScript/Reference/Global_Objects/Math

A Study on Human Reflex-Based Biometric Authentication Using Eye-Head Coordination

Yosuke Takahashi, Masashi Endo, Hiroaki Matsuno,
Hiroaki Muramatsu, Tetsushi Ohki, and Masakatsu Nishigaki(✉)

Shizuoka University, 3-5-1 Johoku, Naka,
Hamamatsu, Shizuoka 432-8011, Japan
nisigaki@inf.shizuoka.ac.jp

Abstract. Biometric information can be easily leaked and/or copied. Therefore, the biometric information used for biometric authentication should be kept secure. To cope with this issue, we have proposed a user authentication system using a human reflex response. It is assumed that even if people know someone's reflex characteristics, it is difficult to impersonate that individual, as anyone cannot control his/her reflexes. In this study, we discuss a biometric authentication system using eye-head coordination as a particular instance of reflex-based authentication. The availability of the proposed authentication system is evaluated through fundamental experiments.

1 Introduction

The application of biometric authentication is popular because there is no possibility of forgetting or losing the information used for authentication. However, biometric information, such as fingerprint or iris pattern data, can be easily leaked, leading to the threat of impersonation by unauthorized people [1, 2], which is a large drawback considering security. To overcome such problems, biometric authentication techniques that use the following types of biometric information have been proposed:

(1) Biometric information that is difficult to leak
(2) Behavioral biometric information

Authentication systems that use biometric information (1) have already been put into practice, for example, authentication using the vein of a finger or the palm [3, 4]. In addition, information such as handwritten signatures are used in authentication systems that utilize biometric information (2) [5].

However, there are threats, such as hidden sensors or phishing websites, that can steal the biometric information of a user under unexpected circumstances. Thus, it cannot be denied that biometric information can be extracted illicitly, even if we use biometric information (1), which is considered to be difficult to leak under normal conditions. Similarly, if we use behavioral biometric information (2), which is based on volitional behaviors such as a handwritten signature, we can consider various threats, such as an impostor practicing the handwriting of an authorized user to master it.

L. Barolli et al. (Eds.): NBiS 2018, LNDECT 22, pp. 419–429, 2019.
https://doi.org/10.1007/978-3-319-98530-5_35

There might be a demand for biometric authentication that is impossible to impersonate someone even when biometric information has been leaked, and cannot be mastered by practice or other means, particularly for situations such as authentication when accessing highly confidential information. A biometric reflex type of authentication has been proposed as an authentication method that could meet these requirements [6, 7]. Biometric reflexes provide involuntary biometric information that human beings find difficult to control; thus, if an impostor knows about them, we anticipate it would be difficult for him to impersonate a legitimate user.

However, there are not sufficient differences in certain reflexes (saccade response and pupillary light reflex) between different individuals, and hence it needs to be combined with physiological biometric information (blind spot) that differs for individual person, to enable its use in authentication [6, 7]. This study discusses a reflex with sufficient difference between individuals to realize a biometric authentication using human reflex itself. We use eye-head coordination as a particular instance for reflex-based authentication.

2 Authentication Based on Human Reflexes

Reflexes are involuntary responses that occur in muscles and other parts of the body when sensory organs are stimulated by external actions; they are always of a predetermined form and are expressed automatically, mechanically, and momentarily [8]. In this study, we define reflexes as bodily responses that human beings cannot control consciously.

The existence of various different reflexes has been confirmed in human beings and it is considered that such responses can be utilized for authentication, provided there is some degree of difference in response between different individuals. For example, it is known that when the intensity of light entering an eyeball increases suddenly, the pupil contracts as a reaction to the light [9].

Human reflexes are responses that are considered difficult for people to control consciously. Thus, it is predicted that even if authentication information, such as 'the pupil of user *P* contracts by *Q*% in response to a certain light stimulus *R*,' is leaked, a user other than user *P* would find it difficult to imitate (or master by practice) the human reflexes inherent to user *P* (the *Q*% contraction of the pupil with respect to that stimulus *R*). In other words, in user authentication based on differences in human reflexes between individuals, the biometric information should be something that is difficult for another person to imitate, even when the biometric information has been compromised.

In existing authentication methods, pupillary light reflexes are used only for liveness detection [10], which simply detects if there is a response; thus, there have been alarming reports that a colored contact lens with printed false iris patterns could pass through an iris authentication device [11]. Moreover, pupillary light reflex cannot be stably obtained because it varies with psychological factors, such as stress [12]. Note that the proposed biometric authentication uses the differences in human reflexes between individuals, and we make a clear distinction between that and the concept of liveness detection that simply detects presence of a response.

In this study, we propose a user authentication method that is based on the differences in human reflex itself. It is expected that spoofing will become further difficult, even when the biometric information has been leaked.

3 Authentication Method Using Eye-Head Coordination

3.1 Concept

As previously mentioned, in the existing human reflex-based authentication [6, 7], there are not sufficient differences in certain reflexes and therefore physiological biometric information is combinatorially used. In this study, we discuss a biometric authentication system using eye-head coordination as a particular instance of reflex-based authentication and evaluate the availability of the user authentication using human reflex response itself.

The eye-head coordination is a reflexive combination of eye and head movements when a human eye tracks a visual stimulus. It is known that "the ratio of sight line angle to head movement angle" is different for each subject. In the existing studies [13, 14], they use multiple sensors to calculate this feature. However, the use of such a rich sensor environment for authentication limits its application for the authentication method and is not realistic.

In this study, we use only the video of the eye taken by a corneal imaging camera shown in Fig. 1 and calculate the feature by the position of the eyeball when the user gazes at a stimulus. Consequently, the cost of authentication is reduced and the convenience of the user can be improved. It is expected that the application of this authentication method will also expand.

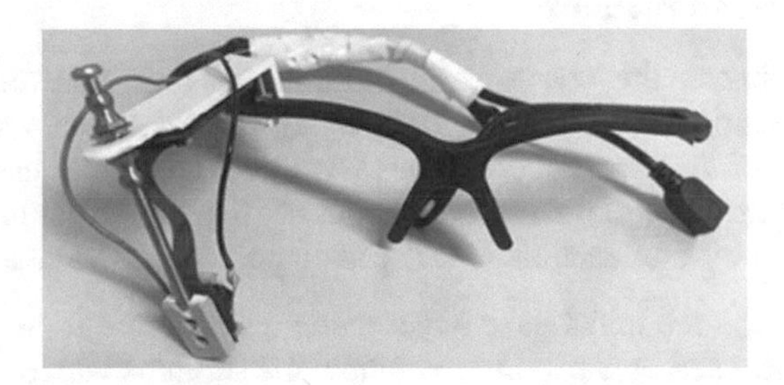

Fig. 1. Corneal imaging camera.

3.2 Feature Extraction

In this study, we apply "the ratio of sight line angle to head movement angle," which is called "the sharing rate," to the biometric authentication system. When a person gazes at a stimulus, the line-of-sight is represented by the vector sum of the eye and head movements. Therefore, when a user gazes at a stimulus, which is displayed to the right or left with the angle of θ degrees from the front, the angle of θ degree is the sum of the user's face and eye movement angles. Thus, when the user turns his/her face largely towards the stimulus, the viewing angle of his/her eyes is small and vice versa (Fig. 2). Therefore, by measuring the position of the cornea when gazing at the stimulus, it is possible to calculate the characteristic corresponding to "the sharing rate." Thus, we briefly calculate "the sharing rate" from only the video of the user's right eye taken by the corneal imaging camera.

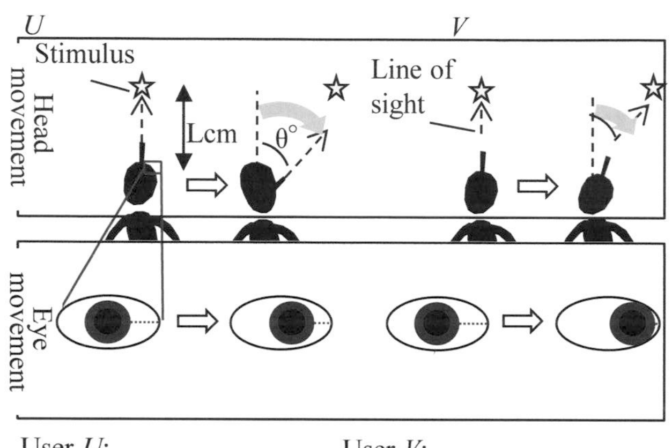

Fig. 2. Eye head coordination.

3.3 Authentication Procedure

During the experiment, the user wears a corneal imaging camera (Fig. 1) for the detection of the position of the cornea. The display was set so that the visual distance from the right eye of the user to the stimulus was about L cm, and the visual angle for the stimulus from the front of the user was about θ° (Fig. 2). The stimuli are displayed at the height of the eyes of each user. Our prototype system works as follows:

1. A system displays the initial gaze target (stimulus S) at the front of a user.
2. The user is instructed by the system to follow the stimulus with his eyes during the authentication.
3. Further, the system displays the stimulus at right (stimulus S'_R) or left (stimulus S'_L) randomly. The dynamic movement of the right eye of the user is captured in the video with the corneal imaging camera.
4. For each frame of the video obtained in step 3, the time series data calculated by measuring the position of the cornea is regarded as the feature.
5. If the features extracted in the authentication phase are close enough to the features extracted in the enrollment phase, the user is authenticated as the legitimate. The "closeness" used in this study is described in the following section.

3.4 Verification

We define the similarity between two data by distance, which is determined by the dynamic time warping (DTW) algorithm [15], and calculate the similarity between legitimate users as well as that between a legitimate user and others. The DTW algorithm measures similarity between two temporal sequences. The DTW algorithm produces an intuitive similarity measure, allowing similar shapes to match even if they are out of phase in the time axis.

In this system, the DTW score is obtained by comparing two sequences $X = (x_1, x_2, \ldots, x_\alpha)$ and $Y = (y_1, y_2, \ldots, y_\beta)$. A warping path is a sequence of grid points $F = f_1, f_2, \ldots, f_K$ on the $\alpha \times \beta$ plane. Let the distance between the two values, x_{ik} $(1 \le i \le \alpha)$ and y_{jk} $(1 \le j \le \beta)$, be $d(f_{(k)}) = |x_{ik} - y_{jk}|$. Further, we calculated the matching score by using DTW(X, Y), where X is the time series data of the enrollment phase and Y is the time series data of the authentication phase, which is calculated by the following equation:

$$\mathrm{DTW(X,Y)} = \frac{\min(\sum_{k=1}^{k} d(f(k)))}{k}. \tag{1}$$

The closer the score is to zero, the higher the similarity is. DTW$(X, Y) < t$, where t is a threshold for authenticating legitimate users.

Considering that the eye-head coordination has large variance even within individuals, each subject is asked to conduct several trials at the enrollment phase. The DTW algorithm is also used to select the optimal template from among the template data of N trials for each subject. Specifically, the optimal template data is compared with DTW$(T_{ij}, T_{ij'})$ $(j \neq j')$, where the trial number j $(1 \le j \le N)$ of subject i is denoted as $T_{i,j}$, and the one with the best score is selected.

3.5 Tolerance Against a Masquerade Attack

An active attacker would attempt to practice and imitate the eye and head movements of a legitimate user. However, the eye-head coordination is a remarkable involuntary movement for a human; head movements are generally dealt with successfully by counter-rotation of the eyes induced by the combined actions of the vestibulo-ocular reflex (VOR) and the optokinetic reflex [16]. Such complicated coordinated movements are biometric information that is considered to be very difficult for people themselves to control, and it makes a clear distinction between that and the behavioral biometrics which simply usc unconscious response. Thus, even if an impostor matches his/her eye and head position to "the position of the eye and head of the legitimate user when a legitimate user gazes at a stimulus," it is considered impossible for the attacker to imitate the feature at all times when gazing at a moving stimulus.

4 Basic Experiment

4.1 Experiment Purpose

To explore the feasibility of the proposed biometric authentication system based on human reflexes, we conducted a fundamental experiment on the differences between individual people in eye-head coordination. The test subjects were ten students from a university (subjects *A*–*J*). The experimental equipment has the camera only to the right side; thus, the experiment is conducted using the right-eye video acquired by the corneal imaging camera.

It is necessary to verify that it is impossible to imitate the eye-head coordination movements illicitly, even when the authentication information of a legitimate user is leaked. However, we have not conducted experiments on spoofing because our study is the verification stage that the availability of the proposed authentication system is evaluated by eye-head coordination.

4.2 Experiment Overview

Overviews of the system arrangement and the experiment environment are shown in Figs. 3 and 4. The details about each device are given below:

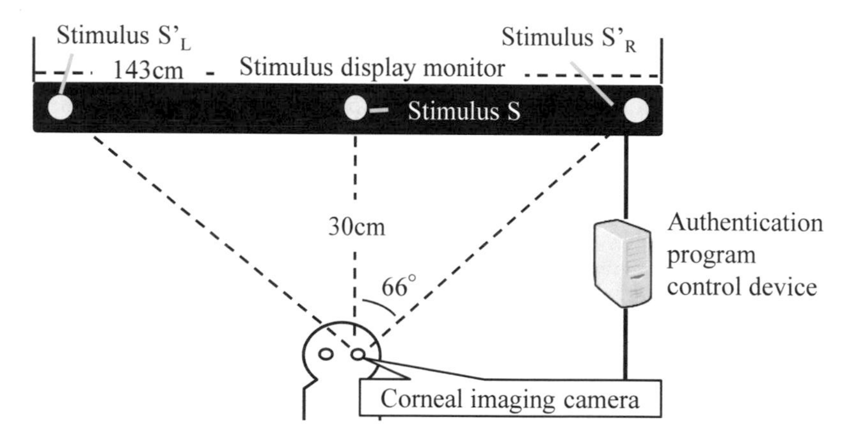

Fig. 3. System arrangement.

- Stimulus display monitor: This device displays the visual target. Using a 65 in. display TY-TP65P10S (developed by Panasonic), which has a resolution of 32,000 × 18,000 pixels.
- Corneal imaging camera: This device takes the video of the right eye of the subject. The subjects wear the corneal imaging camera (Fig. 1) during the experiment. We use NCM13-J (developed by Nippon Chemicon, 640 × 480 pixels, 15 fps).
- Authentication program control device: This device measures the corneal position. It contains a program written in JAVA, to implement each of the enrollment and

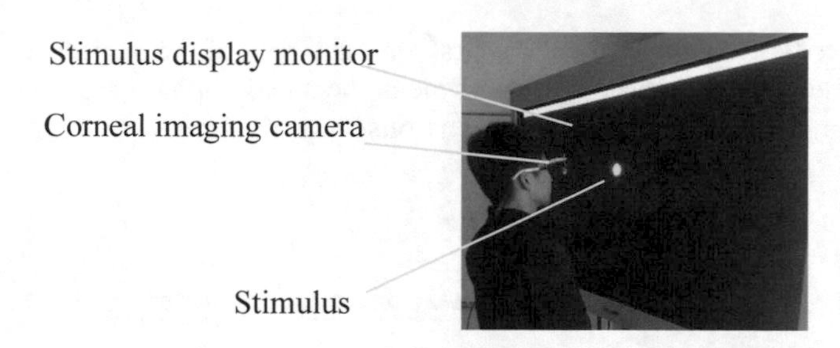

Fig. 4. Experiment environment.

authentication phases. This program runs on a PC (CPU: Intel Core i5 2.6 GHz, Memory: 8 GB, OS: Windows 10 HOME).

4.3 Experiment Method

This experiment was conducted with L = 30 and θ = 66 using the method described in Sect. 3.3, as shown in Fig. 3.

As described in Sect. 3.4, the eye-head coordination has a large variance even within individuals. To mitigate it, each subject was asked to undergo a total of five trials in the enrollment phase. The optimal template from among the five trials for each subject was selected and used in the authentication phase. This means we obtained 100 sample videos (ten subjects × five trials × two directions (S'_R or S'_L)) in the enrollment phase.

The experiment was conducted for two months. For the authentication phase, five trials were conducted at about two minutes after the enrollment phase, one month after, and two months after. Thus, we obtained 300 sample videos (ten subjects × five trials × two directions (S'_R or S'_L) × three days) in the authentication phase.

As described in Sect. 3.2, in this study, we calculate the position of the eyeball as a feature corresponding to "the sharing rate." Specifically, we measured the distance between the inner canthus and the cornea boundary (Fig. 5(a)) in each frame of the video and considered it as "the sharing rate." To eliminate the difference in eye size between subjects, this distance is normalized by dividing it by the distance between the inner canthus and center of the cornea (Fig. 5(b)).

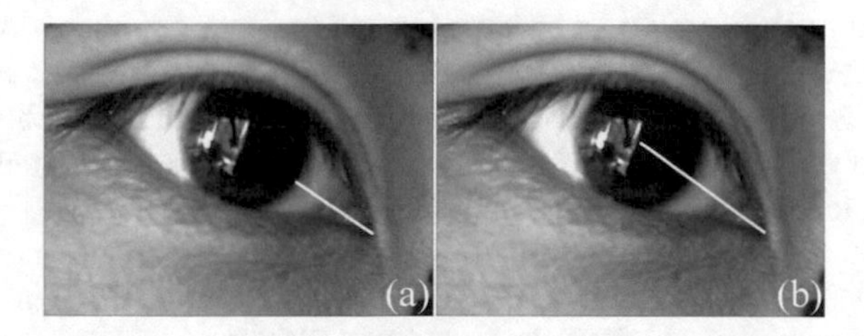

Fig. 5. Features. (a) Distance between the inner canthus and cornea boundary, and (b) Distance between the inner canthus and center of the cornea.

As a specific example of data, the first template data of the subject *A*'s enrollment phase and the first authentication data of the authentication phase are shown in Fig. 6. The time count starts when the stimulus is onset on the display.

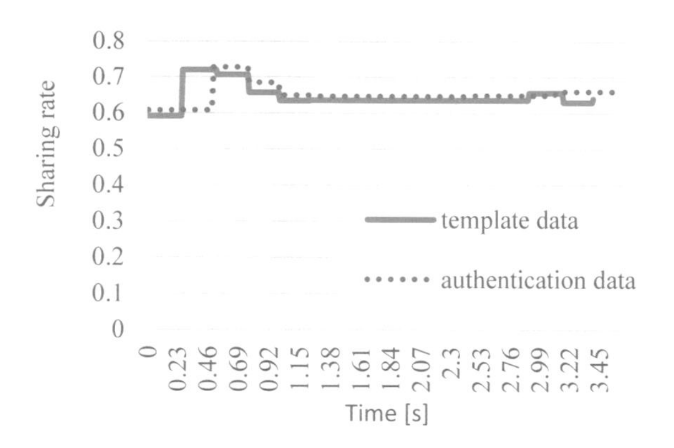

Fig. 6. Change of sharing rate for subject.

4.4 Evaluation

Using the optimum template data, the DTW score described in Sect. 3.4 along with each authentication data are calculated. Considering the experiment setting that the corneal imaging camera is put on the right eye of the subject, the authentication data of each subject is classified into two types: the right side stimulus S'_R and the left side stimulus S'_L. The authentication accuracy was evaluated in the following four decision modes, where the authentication data when the stimulus is presented on the right side of the subject is denoted as "authentication data S'_R" and the data when the stimulus is presented on the left side as "authentication data S'_L":

1. Use only authentication data S'_R
2. Use only authentication data S'_L
3. Use authentication data S'_R or S'_L
4. Use authentication data S'_R and S'_L

Authentication Accuracy. To investigate the authentication accuracy of the proposed method, the matching scores between each subject's template data and all subjects' authentication data acquired within two months (at two minutes after the template enrollment, and one month after, and two months after) were calculated. We evaluate our attempt in terms of false acceptance rate (FAR), false rejection rate (FRR), and equal error rate (ERR). The results with the four decision modes are illustrated in Table 1. Figure 7 shows the trade-off between FRR and FAR at different cut-off points; this is called as receiver operating characteristics (ROC).

The result is not a sufficient level of accuracy, however, Fig. 7 and Table 1 show the possibility that the eye-head coordination can be a feature for user identification.

Table 1. Authentication accuracy.

	(1)	(2)	(3)	(4)
FRR	24.0%	30.0%	28.7%	22.0%
FAR	23.9%	30.7%	28.3%	19.4%
EER	23.9%	30.3%	28.5%	21.8%

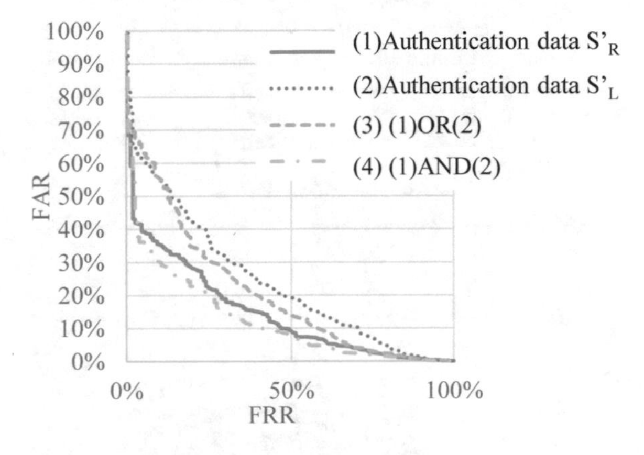

Fig. 7. FAR and FRR.

Moreover, Table 1 shows that the authentication data S'_R when presenting the stimulus on the right side has a higher authentication accuracy than that of the authentication data S'_L when presenting the stimulus on the left side. Generally, humans use the right eye for the stimulus on the right side and left eye for the stimulus on the left side. It seems probably because the corneal imaging camera is placed on the right-eye side, it can be confirmed that the identity is significantly found in the movement of the right eye that is mainly associated with gazing at a right side stimulus S'_R. Furthermore, it can be confirmed from Table 1 that the authentication accuracy is improved by using both the authentication data S'_R and S'_L. From the fact, it is considered that the identity is also included to a certain degree in the movement of the right eye when gazing at a left side stimulus S'_L.

Influence Against Change Over Time. To investigate the long-term stability of the proposed method, the matching scores between each subject's template data and all subjects' authentication data acquired at two minutes/one month/two months after the template enrollment were respectively calculated. The change of the average score over time for each subject is depicted in Fig. 8.

Thus, it was confirmed that the matching score of some subjects change over time. Therefore, at the current stage of this research, the application of our method will be limited to short-term authentication systems. A possible strategy is to update the template each time authentication is performed. It is necessary to continue discussing methods that can be verified over a longer term.

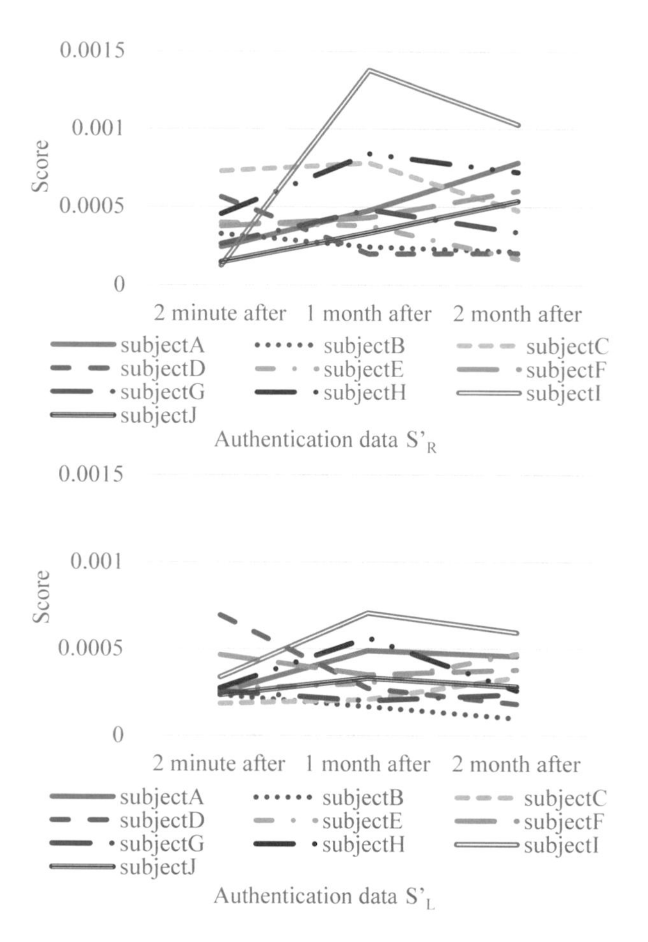

Fig. 8. Long term experimental result

5 Conclusion

In this study, we calculate the feature from the dynamic behavior of a user, when he/she follows a moving stimulus with his eyes, for the biometric reflex authentication using the individual difference of the eye-head coordination and verified it through a basic experiment. As described in Sect. 3.5, the eye-head coordination is a remarkable and complicated involuntary movement for humans; thus, it is considered impossible for an attacker to imitate it consistently with the movements of a legitimate user at all times while following a moving stimulus with his/her eyes.

It can be confirmed that the combining feature is improved by using an AND of authentication data when presenting the stimulus on the right side and that on the right side, and the best authentication accuracy is obtained with an EER of 21.8.

However, the difference between the user and others did not appear greatly in the biometric information of reflex; thus, further improvement is necessary. In addition, it is confirmed that the matching score of a user changes with time. These issues need to be addressed in future work.

Acknowledgments. We, the authors of this study, are grateful to Prof. Atsushi Nakazawa of Kyoto University, Japan for his advice on the authentication method and provision of experimental equipment in progressing this research.

References

1. Matsumoto, T.: Gummy and conductive silicone rubber fingers importance of vulnerability analysis. In: Zheng, Y. (ed.) Advances in Cryptology–ASIACRYPT 2002, pp. 574–575. Springer, Berlin (2002)
2. Daugman, J.: Recognizing persons by their iris patterns. In: Jain, A.K., Bolle, R., Pankanti, S. (eds.) Advances in Biometric Person Authentication, pp. 5–25. Springer, Boston (2004)
3. Watanabe, M.: Palm vein authentication. In: Ratha, N.K., Govindaraju, V. (eds.) Advances in Biometrics, pp. 75–88. Springer, London (2008)
4. Mulyono, D., Jinn, H.S.: A study of finger vein biometric for personal identification. In: Proceedings of the International Symposium on Biometrics and Security Technologies, ISBAST 2008. Islamabad, pp. 1–8 (2008)
5. Jain, A.K., et al.: On-line signature verification. Pattern Recogn. **35**(12), 2963–2972 (2002)
6. Nishigaki, M., Arai, D.: A user authentication based on human reflexes using blind spot and saccade response. Int. J. Biom. **1**(2), 173–190 (2008)
7. Nishigaki, M., Ozawa, Y.: A user authentication using blind and papillary light reflex. Inf. Process. Soc. Jpn. J. **48**(9), 3039–3050 (2007). (in Japanese)
8. Leis, A.A., et al.: Behavior of the H-reflex in humans following mechanical perturbation or injury to rostral spinal cord. J. Muscle Nerve **19**(11), 1373–1382 (1996)
9. Young, R.S., et al.: Transient and sustained components of the pupillary responses evoked by luminance and color. J. Vis. Res. **33**, 437–446 (1993)
10. Daugman, J.: Demodulation by complex-valued wavelets for stochastic pattern recognition. Int. J. Wavelets Multiresolut. Inf. Process. **1**(1), 1–17 (2003)
11. Sabarigiri, B., Suganyadevi, D.: Counter measures against iris direct attacks using fake images and liveness detection based on electroencephalogram (EEG). World Appl. Sci. J. **29**, 93–98 (2014)
12. Jomier, J. et al.: Automatic quantification of pupil dilation under stress. In: Proceedings of the IEEE International Symposium on Biomedical Imaging: Macro to Nano, pp. 249–252 (2004)
13. Gresty, M.A.: Coordination of head and eye movements to fixate continuous and intermittent targets. Vis. Res. **14**(6), 395–403 (1974)
14. Barns, G.R.: Vestibulo-ocular function during co-ordinated head and eye movements to acquire visual targets. J. Physiol. **287**(1), 121–147 (1979)
15. Sakoe, H., Chiba, S.: Dynamic programing algorithm optimization for spoken word recognition. IEEE Trans. Acoust. Speech Signal Process. **26**(1), 43–49 (1978)
16. Vercher, J.L., et al.: Eye-head movement coordination: vestibulo-ocular reflex suppression with head-fixed target fixation. J. Vestib. Res. **1**, 161–170 (1991)

Intrusion Detection Method Using Enhanced Whitelist Based on Cooperation of System Development, System Deployment, and Device Development Domains in CPS

Nobuhiro Kobayashi[1,2(✉)], Koichi Shimizu[2], Tsunato Nakai[2], Teruyoshi Yamaguchi[2], and Masakatsu Nishigaki[1]

[1] Shizuoka University, Shizuoka, Japan
kobayashi.nobuhiro.15@shizuoka.ac.jp, nisigaki@inf.shizuoka.ac.jp

[2] Mitsubishi Electric Corporation Information Technology R&D Center, Kanagawa, Japan
Kobayashi.Nobuhiro@ab.MitsubishiElectric.co.jp, Shimizu.Koichi@ea.MitsubishiElectric.co.jp, Nakai.Tsunato@dy.MitsubishiElectric.co.jp, Yamaguchi.Teruyoshi@cs.MitsubishiElectric.co.jp

Abstract. Cyber-physical systems (CPS), as fusions of virtual and real worlds, have attracted attention in recent years. CPS realize rationalization and optimization in various domains, by collecting real-world data in the virtual world, analyzing the information, then reacting to the real world. Moreover, attacks on CPS can cause direct damage in the real world. Therefore, with the NIST CPS framework in mind, this paper discusses an approach to designing countermeasures to risks, considering the interactions with a system. Applying the approach of intrusion detection for control systems, this paper also proposes an intrusion detection method using an enhanced whitelist.

1 Introduction

Cyber-physical systems (CPS) [1] have attracted attention in recent years as a method of fusing virtual space (cyberspace) with the real world (physical space). In The United States, the National Institute of Standards and Technology (NIST) has established the CPS Public Working Group (CPS PWG), which gathered a broad range of CPS experts from academia and industry, and created the Framework for Cyber-Physical Systems (CPS Framework) [2]. In the nine aspects of the above mentioned CPS Framework, the security of CPS was considered in detail from the viewpoint of an aspect of "trustworthiness". So far five security characteristics (properties) were addressed individually: safety, trustworthiness, privacy, cyber security, and resilience. The framework emphasizes the importance of the interactions of these properties and suggests to take an integrative, crosscutting perspective of these properties, promoting design approaches that can enhance an awareness of CPS risks and stem the flow of risks from the virtual world into the real world.

L. Barolli et al. (Eds.): NBiS 2018, LNDECT 22, pp. 430–444, 2019.
https://doi.org/10.1007/978-3-319-98530-5_36

Nevertheless, CPS Framework presents no concrete method for realizing this concept. In addition, the conventional CPS configuration is in a contradictory situation with this concept. In the configuration of an actual control system for a CPS, responsibility is taken for each of three domains: the "system development domain" for designing systems based on the CPS requirements; the "device development domain" for manufacturing sensors, actuators, and other devices based on the requirements for conceived normal operations; and a "system deployment domain" for allocating actual parameter (properties) to the device. This vertical structure of each domain makes the interactions of properties difficult. Therefore, how to ensure smooth linkages for security strategies within the actual control system architecture must be an important study. This is to ensure the best adaptation, so that the three domains involved in control system architecture are optimally adjusted with the five security characteristics (properties) of a CPS framework.

As one security strategy within a CPS control system, our attention has focused on a whitelist-type intrusion detection method. With this method, ordinary status is predefined in a whitelist, and abnormalities are detected when deviation from that ordinary status occurs. Within a control system where the usages and elements are limited and fixed, it is possible to formulate all the requirements in whitelist formats. Furthermore, whitelist-type intrusion detection method has an affinity with model-based development that is recently increasingly employed in control system design; whitelists can be generated automatically by translating from the models. For these reasons, it is considered that a whitelist-type intrusion detection method is particularly suitable for CPS control system designs.

Based on the above, this paper proposes the use of a whitelist as a means of enhancing linkages between the three domains of control system architecture in order to achieve interactions of the five security characteristics (properties) of a CPS framework. We define an enhanced whitelist to integrate, in supplemental fashion, respective knowledge of the requirements, conditions, and properties of the three domains in the CPS control system. This enables a whitelist-type detection method that can monitor cyberattacks on both virtual space and the real world.

2 Control System Cyberattacks and Response Strategies

The 2010 attack on nuclear facilities in Iran (via Stuxnet) attracted attention, as it was a serious attack on a control system. In this case, Stuxnet rewrote the ladder program of the programmable logic controller that operates the centrifuge, then destroyed the centrifuge by transmitting abnormal values to its frequency converter.

It only attacked the centrifuge, fulfilling its specific conditions, meaning that Stuxnet was created by a person (or persons) who must have had knowledge of each aspect of the system development, system deployment, and device development. To combat this type of attack, it is necessary to supplement the respective knowledge of the requirements, conditions, and properties of each of the three domains pertaining to the control system architecture, so as to successfully integrate these domains.

2.1 Security Strategy Design Within CPS

In the NIST CPS Framework, three activities are defined as key elements in system engineering processes in fields where CPS are deployed, and nine aspects are conceptually classified as items of concern. CPS employment fields are listed in accordance with differences in respective applications, and include production, transportation, and energy. The three activities of system engineering processes are: conceptualization (CPS model formulation), realization (including CPS design, manufacture, deployment, and operation), and assurance (guarantee of CPS reliability). The nine conceptual aspects are defined as: function, business, humans, trustworthiness, timing, data, interaction borders, configuration (architecture), and life cycle. Summarizing these relationships, for CPS deployment fields, the three activities of the system engineering process are to be pursued, while also considering the nine conceptual aspects that are considered items of concern.

Thus, when designing a security strategy for a CPS, it becomes essential that these nine conceptual aspects are seriously considered. The framework describes, in detail, the aspect of "trustworthiness" for CPS-dedicated security. It should be emphasized that it is necessary to have an awareness of mutual interactions, and a crossover perspective that takes into account all security characteristics that have conventionally been considered separately: safety, reliability ("trustworthiness"), privacy, cyber security, and resilience.

Since CPS failures can result in damaging consequences in the real world, it is especially important that CPS operations perform as expected. To secure CPS trustworthiness, an integrated risk approach is required that takes into account each of the five security properties. It was learned from the Stuxnet attack that the conventional approach, where each of these properties are considered individually in the investigation of system requirements, and in terms of system design, mounting, and deployment, leaves an exposed "gap" that is wide enough to be attacked and exploited.

Therefore, risk approaches that surpass property borders are essential. An integrative, cross-cutting perspective that is inclusive of safety, trustworthiness, privacy, cyber security, and resilience is indispensable in ensuring that design approaches for risk-related policies cover the entire span from the CPS virtual world into the real world. It thus becomes necessary for us to demonstrate an approach that enables the realization of concrete measures for a CPS security strategy.

2.2 Realization of Security Via Linkages Between CPS Domains

In the configuration of a CPS for an actual control system, multiple organizations must collaborate; each performing their work according to their respective duties. In this paper, division of the work duties, in accordance with the respective responsibilities of each unit of the organization, is divided into three distinct domains.

The first is the "system development domain," which is responsible for the design and development of device and systems based on the CPS requirements. This domain analyzes the requirements for the CPS, then designs and manufactures device, systems, and the network configuration accordingly. This domain also assigns conditions to devices (sensors, actuators, etc.) that will serve as connection points with the real world.

The second is the "device development domain," which designs and manufactures devices (sensors, actuators, etc.) based on the conditions for either conceived or preset normal operation, in accordance with the requirements as assigned by the system development domain.

The third is the "system deployment domain," which performs allocation of the actual setting parameters (property values) for operations of the CPS to the device, systems, and devices provided by both the system development and device development domains. Figure 1 shows the response-related relationships when the respective definitions of the three domains are applied to a control system.

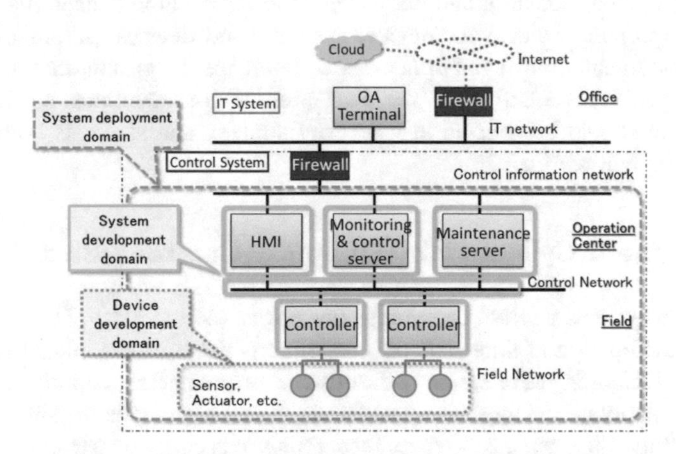

Fig. 1. One domain definition of control system.

Conventional security strategies for control systems are performed with the system development domain as the main factor. For interactions between the system development domain and the system employment domain, we present information on properties that must be set for: device and system operations, system deployment domain allocations to devices, and the actual system settings parameters (property values).

Meanwhile, for the sensors, actuators, etc. of a device that is a connection point with the real world, requirements are shown from the system development domain to the device development domain. Based on these requirements, the device development domain designs and manufactures the device after applying conditions that will guarantee normal operation of the device as "conceived" or "preset".

As above, in the design, manufacturing, and employment of conventional control systems, linkages between domains are limited to a one-way flow of information and demands, from the "system development domain" to the "device development domain," and from the "system development domain" to the "system deployment domain." There are no interactions in either both, or multiple directions.

Thus, when there is only individual consideration of the five security characteristics (properties) of a CPS (one by one), the fear is that "gaps" will occur in CPS security strategies, and these gaps could be exploited in a cyberattack. What is desired then, is an approach to CPS security strategies that will ensure mutual interactions between each domain, and includes multiple crossover communications. This method should both integrate and concentrate the security concerns of each of the three domains, enabling them to share a common understanding.

Thus, we propose a scheme that enables this shared understanding between each of the CPS domains which includes: knowledge and information concerning system requirements (as possessed by the system development domain); knowledge and information concerning the conditions for normal operations (as possessed by the device development domain); and knowledge and information concerning the settings parameters (property values) for device, systems, and devices (as possessed by the system deployment domain). Further, as a reference to ensure that CPS-directed operations perform as expected, we also present considerations on how mutual understanding should be realized in a security strategy that straddles both the virtual world and the real world.

3 Detection of Cyberattack Within the Control System

Some control systems perform continuous operations (24 h per day, 365 days per year) for an extended period of time (10 years or longer). Yet there are related risks in such operations, including termination of software maintenance support, inability to immediately apply necessary security and antivirus software updates. Since availability is a key priority, there are cases where responding to security updates, or deterioration of the antivirus software capabilities, or the ability to adopt real-time approaches, are unacceptable.

Our attention was therefore drawn to a system characterized by having fixed configuration elements and usages within a control system. Within such a control system, communication patterns with the network are fixed and stable, normal communications can be pre-defined, and deviant communications can be interpreted as attacks. To accomplish these goals, we believe that the best option is the introduction and deployment of a whitelist-type intrusion detection system.

3.1 Whitelist-Type Intrusion Detection System

One anomaly detection type of intrusion detection system is where normal operations are defined using whitelists. (i.e., a "whitelist-type intrusion detection method).

Statistics-Based. In this method, there is a computation of statistical quantities from the collection of large amounts of communications data amassed during normal operating times.

Machine Learning-Based. Here, there is learning of normal communications data through the use of self-organizing maps, neural networks, etc.

Knowledge-Based. Here, prior knowledge of protocols for system configuration and uses is applied. Approaches also exist where models for normal operation are configured using communication and system specifications.

Table 1. Comparison of whitelist generation methods

	Statistics-based	Machine Learning-based	Knowledge-based
Acquisition of actual data	✗	✗	✓
	Necessary	Necessary	Not necessary
Communication (alert, etc.) at time of emergency	✗	✗	✓
	Not support	Not support	Support
System and device specifications	✓	✓	✗
	Not necessary	Not necessary	Necessary
Detection error	✗	✗	✓
	Long-term small changes	Reasons for results	No errors

These comparison results are believed to demonstrate the appropriateness of a knowledge-based whitelist generation method for control systems, where the emphasis is for continuous, safe operation (Table 1). However, one problem still remains, regarding the important point of system and device specifications for whitelist generation. Currently, in actual control systems, where there are on-site arrangements for device-placement tasks (i.e., the system deployment domain), the whitelist remains incomplete in the system development domain with regards to the determination of final settings parameters (property values).

Therefore, the acquisition of settings parameters within a knowledge-based method is a difficult problem for the system development domain to resolve independently, as this domain lacks actual data. Therefore there is a fear that the whitelist will remain incomplete in such a situation, meaning that there can be no linkage with actual implementation. This is the problem of acquisition of settings parameters (property values).

Further complications occur with this issue for the special case of CPS. Earlier in the paper we described the generation of whitelists for defining "normal states" within existing whitelist-type intrusion detection methods. However, this is not enough, as demonstrated in the Stuxnet attack, where this method was unable to prevent an attack from the virtual world into the real world. To realize a security policy for attacks that are directly linked with negative effects in the real world, knowledge about sensors, actuators, and other devices, which the attacker seeks to use maliciously, must also be acquired and incorporated into the policy. This issue is not limited to the knowledge-based whitelist generation method, but is also a novel problem shared with the whitelist-type intrusion detection method. This is the problem of utilization of knowledge concerning devices. Solving these two problems is essential for the realization of an intrusion detection method using knowledge-based whitelists, which we consider to be an optimal security policy within a CPS.

4 Enhanced Whitelist-Type Intrusion Detection Method

Here, we explain solution strategies for two of the problems described in Sect. 3.

The first problem relates to the acquisition of settings parameters (property values). Through linkages between the "system development domain" for design and manufacture of device and system based on the requirements and the "system deployment domain," which performs, on site, allocations of the actual settings parameters (property values), a whitelist is realized from the existing knowledge-based via the sharing of information on each "requirement" and "property value."

The second problem concerns the active use of knowledge related to devices (sensors, actuators, etc.). We have made new provisions for collaborations between the "device development domain" that possesses knowledge, and the information regarding "conditions" for normal device operations, to enhance the existing whitelist knowledge base.

Thus, by linking the three domains an enhanced whitelist is realized that covers all aspects of a control system. (This linkage means that the "requirements" possessed by the "system development domain" are added to the "property values" possessed by the "system deployment domain," and the "conditions" possessed by the "device development domain").

In the conventional definition of "normal", based on individual "requirements" of the device and system such that "these are the normal operations for this apparatus," there is insufficient security for a CPS in regard to its connections with the real-world. Based on this awareness, our definition of "normal" includes "property values" secured when setting up the device and system, such that "this apparatus, with these settings, will operate normally under these conditions," in addition to "conditions for normal operations of devices that are connection points with the real world." This is the key point of our concept.

And yet, all this is not enough to ensure security in a CPS. Consideration must also be given to "property values" being set within device and systems, as well as to "conditions" for normal operation. It is also necessary to constantly check for deviations from these "properties" and "conditions". Without such considerations, it will be impossible to prevent attacks within "gaps" in the security policies for individual domains, as happened in the attack by Stuxnet.

Systems in critical infrastructure, etc., are characterized by fixed usage purposes and configuration elements. That is, the whitelist for "requirements" possessed by the "system development domain," the whitelist for "properties" possessed by the "system deployment domain," and the whitelist for "conditions" possessed by the "device development domain" are all "fixed" (preset). Further, no changes can be added to the whitelist of one domain that can then have an effect on the whitelists of other domains. It can thus be said that an extremely effective approach would be to integrate the whitelists possessed by each domain into a single whitelist that covers the entire control system. This system whitelist would specify that "apparatus with these settings will perform normal operations under these conditions."

In this paper, a whitelist that integrates the individual whitelists of the three domains, generated so that it covers the entire control system, is called an "enhanced whitelist." A key point is that linkages between the three domains can be organically accomplished via the mediation of this enhanced whitelist.

Below is an explanation that follows the framework for generating the enhanced whitelist in Fig. 2. Defined by the whitelist are: the portion of the "requirements" whitelist that states "normal operations are performed thus," the portion of the "properties" whitelist that identifies "the apparatus with these settings," and the portion of the "conditions" whitelist that describes "under so-and-so conditions."

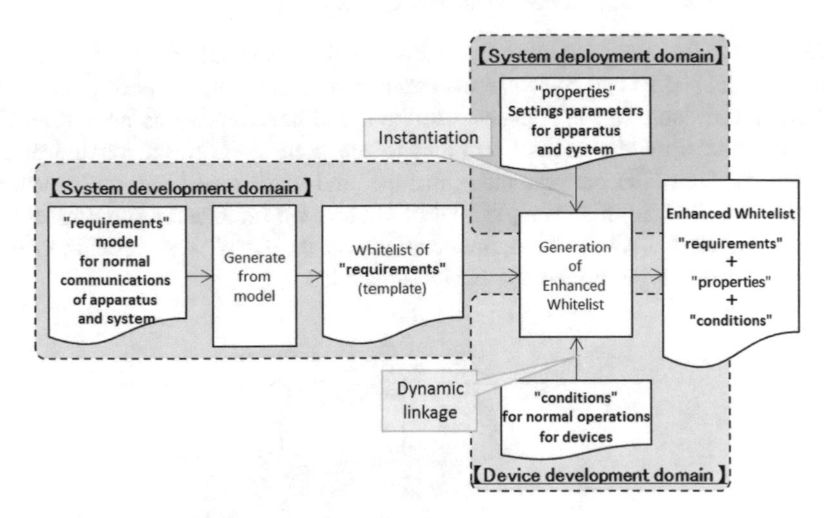

Fig. 2. Framework of enhanced whitelist generation.

Firstly, when the "system development domain" is to generate the "requirements" whitelist, normal communications are to be defined by an unambiguous model, such that the "conditions" whitelist can be automatically generated from this model. Nevertheless, this model does not include actual settings parameters (property values for parameters such as IP address, port number, etc.) that are to be allocated at the time of system deployment by the "system deployment domain." Further, this model does not include "conditions" that guaranteed normal device operations or "conditions" that are possessed by the "device development domain."

Therefore, so that the "system deployment domain" can subsequently add the settings parameters (property values) to be allocated to device and the system, the "requirements" whitelist is forced to automatically generate a template format with "blanks to be filled in" with these "properties." Into these "blank" areas of the template, the "system deployment domain" sets the actual "property values" for the device and system.

In the same way, so that the "device development domain" can subsequently add the "conditions" for normal device operations, there is an automatic generation of a "requirements" whitelist, which takes the form of a template with "blanks to be filled in" with said "conditions." Into these "blank" areas of the template, the "device development domain" inserts the actual "conditions" for normal device operations.

Here, with reference to object-oriented mechanisms, this "filling in" of the "property value" settings we call "instantiation." Further, with reference to the dynamic link library (DLL) used to perform dynamic linkages, when an application program is implemented, we call this "filling in" (insertion) of the "conditions" "DLL-ization." (as adjective, "DLL-ized"; below, said terms appear without italics).

4.1 Procedures for Enhanced Whitelist Generation

Figure 3 shows the control system that serves as the case example in this paper. The control terminal (HMI) and controller are connected via a LAN, and each is assigned an IP address. The controller is connected to the centrifuge (motor) as its device. Commands are sent from the control terminal to the controller. Based on the received command, the controller controls the centrifuge, and sends centrifuge sensor values to the control terminal. In this example control system, we imagined a dummy attack on the controller via a LAN, and communications on the LAN were monitored by the enhanced whitelist-type intrusion detection apparatus.

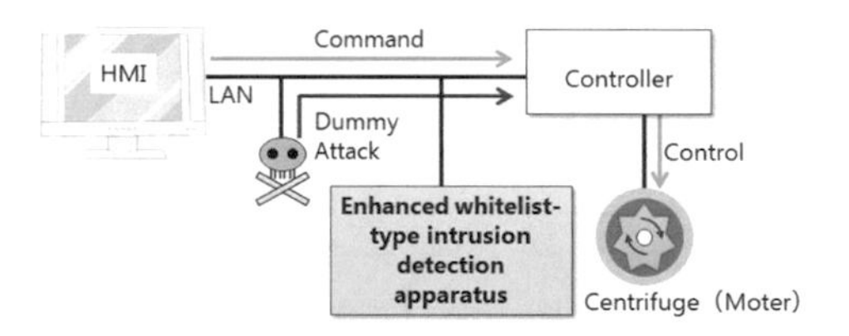

Fig. 3. Example of control system.

Generation of Whitelist Requirements. The generation of whitelist "requirements" is achieved by the "system development domain" defining "normal" in the "normal operations are performed thus" portion. Normal communications between target equipment and systems are defined with a non-ambiguous model.

Firstly, devices and network architecture in the control system of Fig. 3 are modeled with a modeling tool. In this example, we used Simulink, a modeling tool of MathWorks, Inc. In Simulink, devices are shown with subsystem blocks, and a line is used to show connections between the devices. Settings parameters (properties) are set to "fill-in-the-blanks," which are retrospectively allocated from the "system deployment domain." To facilitate this, the IP address (as the character string of the annotation block) and the properties for port numbers are both described. Then, to ensure that the "conditions" of normal operation of the devices (that are to be subsequently provided by the "device development domain") are inserted so as to "fill-in-the-blanks," the payload conditions are included.

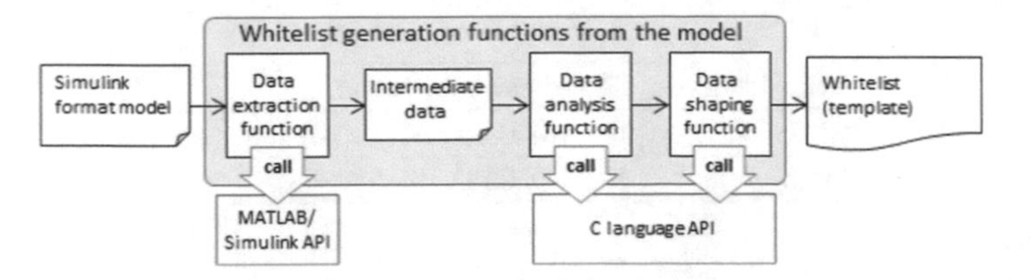

Fig. 4. Whitelist generation functions from the model.

Next, we describe the whitelist generation functions, from the Simulink format model. Transform functions from the whitelist model comprise of the three functions shown in Fig. 4.

Whitelist generation functions, from the model, take the Simulink format model as input, then process the data necessary for the whitelist through the data extraction function, the data analysis function, and the data shaping (formatting) function. Finally, the output of the whitelist template is generated. The data extraction function is realized using MATLAB/Simulink API, and intermediate data output is generated. Since MATLAB/Simulink is not adapted to process character strings, C language API is used to process the intermediate data, using the data analysis and data shaping functions of an external program. Table 2 shows the whitelist template format and a sample of data.

Table 2. Format of whitelist template and sample data

Protocol	Transmission source information	Destination information	Data length byte	Command	Payload condition	Communication cycle (seconds)
Control	<HMI>:XX	<PLC>:XX	1024	Read2	–	0.5

This template includes model-defined protocol, transmission-source information, transmission-destination information, data length, command, payload condition, and communication cycle. The “fill-in” portions (“Transmission source information” and “Destination information”) are placed in the template as device identifier (e.g., “HMI”) and as a temporary port number (“XX”), as shown in model subsystem blocks. Further, the “requirement” for normal operations that is to be inserted is also included in the template, so as to enable the subsequent insertion of a “payload condition.”

Generation of Whitelist Properties. Generation of whitelist properties is such that the “system deployment domain” defines “normal” in regard to “the settings of this apparatus” portion. The “system deployment domain” receives the whitelist template with its “filled-in blanks” from the “system development domain,” then inserts into the “blanks” the actual settings parameters (properties) for allocation, at the time of system deployment, then returns the instantiated whitelist to the “system development domain.”

With the inputting of the linked information, (the transmission device identifier included in the whitelist template of Table 2, the receiver device identifier, the

Table 3. Sample of properties

Device identifier	IP address	Port no.
<HMI>	192.168.0.10	ANY (not specified)
<PLC>	192.168.0.20	49153

Table 4. Whitelist with Properties

Protocol	Transmission source information	Destination information	Data length byte	Command	Payload condition	Communication cycle (seconds)
Control	192.168.0.10: ANY	192.168.0.20: 49153	1024	Read2	–	0.5

IP address and port numbers, etc.), the "blank" portion is filled in with the actual settings parameters (properties), and the instantiated whitelist is generated. Table 3 shows a sample of settings parameters (properties) data input.

Table 4 shows the instantiated whitelist with template "blanks" filled in with properties.

This instantiated whitelist is then sent back to the "system development domain" from the "system deployment domain."

Generation of Whitelist Conditions. The generation of whitelist "conditions" is such that the "device development domain" defines "normal" in the "perform under these conditions" portion. The "device development domain" receives the whitelist template with its "filled-in blanks" from the "system development domain," then inserts into the "blanks" the "conditions" for normal operation of the device, and returns the whitelist which has been subjected to DLL-ization back to the "system development domain."

However, to realize this DLL-ization, it is first necessary to have a shared awareness of how the device will be used, and also what conditions are required to ensure normal operation within this usage-purpose. To ensure that there is communication between the different domains, which is required for this shared awareness, it is first necessary to create a specific type of shared language.

Section 5 describes in detail the realization of this shared understanding for domain linkages, and the realization of the whitelist conditions.

5 Conditions for Normal Device Operations, and Domain Linkages

In order to realize a security system for attacks that are directly connected with negative impacts on the real world, it is essential that knowledge concerning devices (sensors, actuators, etc.), which the attackers seek to use maliciously, are also incorporated and utilized for defending against the attack. Described below is the acquisition of a shared understanding of knowledge and information concerning "conditions" for normal device operation, so as to enable linkages between the "system development domain" and the "device development domain" which possesses this information and knowledge.

While the "system development domain" and the "system deployment domain" are different domains, they are united in the goal of achieving a unified system architecture. Therefore, they already share a common awareness of the purposes of the system. Meanwhile, the "system development domain" and the "device development domain" share no such common awareness, in that the former deals with the system, and the latter deals with devices. In order to enhance a whitelist that forges linkages between these two domains, there must be a shared understanding between them regarding usage purposes, and it must be ensured that they acquire a common understanding regarding normal operating conditions. To achieve this, it is first necessary to stipulate a common language.

In the field of IT, the system usage purpose is expressed as a scenario, and it is recognized that the use of a Unified Modeling Language (UML) model is an effective tool for fostering a shared understanding between concerned parties having different standpoints. Here, a use case diagram is used as one type of UML model, to model the purpose of the centrifuge. Using a "UML diagram system" as the centrifuge, the controller is expressed as the "UML diagram actor," which performs exchanges of information with the centrifuge. The use case diagram describes changes in the operation of the centrifuge, in accordance with information from the controller. Figure 5 shows a use case diagram of a centrifuge which is being operated according to information from the controller.

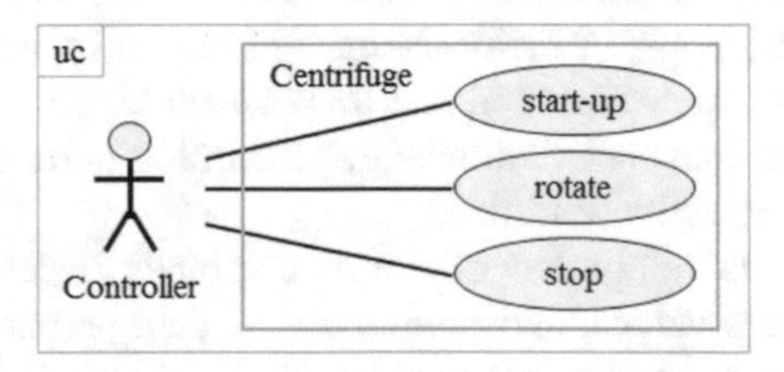

Fig. 5. Centrifuge use case diagram.

This use case diagram describes the "system development domain," and clarifies the purposes of the centrifuge. The "device development domain" interprets these purposes, and understands that the functions the device should be equipped with are "start-up," "rotate," and "stop" operations. Based on these functions, the "device development domain" changes internal operations occurring within the device in accordance with the usage purpose. Table 5 shows "start-up" is "power source ON," "stop" is "power source OFF," and "rotate" is "rotation" of the device.

Table 5. Purposes and Functions.

No.	Purpose	Function
1	Start-up	Power source ON
2	Stop	Power source OFF
3	Rotate	Rotation

The "device development domain" contains knowledge regarding the conditions that ensure normal operation of each of the device functions. Therefore, this can be defined as the whitelist. This is designed so that there is a sharing of knowledge and information regarding both the purposes of the "system development domain," and the conditions for normal operation of the "device development domain."

5.1 Conditions Whitelist

Next, normal operating conditions for each device function are defined. It is thought that device functions that impact the real world can be expressed through a combination of various feature values, related to each function.

Here, as an example case, we use the "motor" from a centrifuge. The "conditions" for normal operation are explained. Here computation is performed on such things as motor life, rotation speed, load, type of load, etc.

$$\text{life} = \text{rated life} \bullet \text{rotation speed coefficient} \div (\text{load factor}^3) \bullet \text{load type coefficient}. \quad (1)$$

In the motor life calculation, in the case where load type corresponds to frequent stopping and starting of the motor, the load type coefficient is added, which is linked to a shortening of the life of the motor. Additionally, when physical device or apparatus exceed their life expectancy, they enter a wear-out failure rate period according to their failure rate curve. In this period there is a remarkable rise in failure rate. Thus, where there are frequent starts, stops, or major changes in rotation speed accompanying these starts and stops, then it can be assumed that this constitutes an attack on the real world. Therefore, it is essential that such "knowledge" regarding normal states is incorporated within the whitelist, and actively utilized.

Thus, the "device development domain" combines necessary information that defines functions, feature values, and normal states, then prepares this as a conditions template (i.e., a "template of physical properties") as shown in Table 6.

Table 6. Template of physical phenomenon properties ("conditions")

No.	Purpose	Function	Feature value	Normal operating condition
1	Start-up	Power source ON	Voltage	• V
			Change frequency	• times/min
2	Stop	Power source OFF	Voltage	• V
			Change frequency	• times/min
3	Rotate	Rotation	No. of rotations	Max: •rpm
				Min: •rpm
			Displacement amount	• rpm/min

Portions marked with "•" within the "normal operating conditions" in Table 6 are portions that change according to the performance and durability of each physical device within a centrifuge, and with system usage purpose for example. These express those portions which, for each specific device selected by the "system development domain," are inserted as "normal operating conditions" according to the knowledge possessed by the "device development domain."

In regard to "Purpose" and "Function" within Table 7, these are formulated on the basis of Table 5 in a cooperative fashion between the "system development domain" and the "device development domain." As for "feature values" for "functions," as in the case of the "motor" explained above, these are formulated by the "device development domain" which has knowledge concerning coefficients that have an effect on life expectancy for normal operation. Thus, "feature value" items, in accordance with device types, are formulated by the "device development domain." Specific procedures therefore, are described below.

Table 7. Sample of conditions whitelist

No.	Purpose	Function	Feature value	Normal operating condition
1	Start-up	Power source ON	Voltage	100 V
			Change frequency	1 times/min
2	Stop	Power source OFF	Voltage	0 V
			Change frequency	1 times/min
3	Rotate	Rotate	No. of rotations	Max: 10,000 rpm
				Min: 0 rpm
			Displacement amount	500 rpm/min

The "device development domain" receives the whitelist with "blanks" (Table 4) from the "system development domain." Then, a "conditions template" (Table 6) is inserted for these blank portions ("whitelist payload conditions"). Next, portions marked with a "•" within the conditions template (i.e., normal operating conditions) are determined according to the specifications of the individual device selected by the "system development domain," and the conditions whitelist (Table 7) is completed. In this way, the Payload condition portion of Table 4 are "filled in" so as to create the enhanced whitelist with the inserted payload conditions of Table 7.

6 Summary and Future Prospects

This paper proposed the use of a whitelist as a means of enhancing both the linkages between three domains of control system architecture, and the five security characteristics (properties) of the CPS Framework introduced by NIST. Knowledge of the respective "requirements," "conditions," and "properties" possessed by the three domains are, in supplemental fashion, integrated so as to define the control system whitelist. We described how this enables the realization of an enhanced whitelist-type intrusion detection method that can monitor cyberattacks on both virtual space (cyberspace) and the real world (physical space).

Our future aim is to apply our method within actual systems. We aim to further strengthen security within a CPS, so as to contribute to the realization of a safe and secure society.

References

1. National Science Foundation: Cyber-Physical Systems (CPS), NSF 10-515 (2010)
2. The Cyber Physical Systems Public Working Group: Framework for Cyber-Physical Systems Release 1.0, Cyber-Physical Systems (CPS) Framework 1.0, CPS PWG (2016)

Innovative Protocols for Data Sharing and Cyber Systems Security

Urszula Ogiela[1], Makoto Takizawa[2], and Lidia Ogiela[3](✉)

[1] Cryptography and Cognitive Informatics Research Group, AGH University of Science and Technology, 30 Mickiewicza Avenue, 30-059 Kraków, Poland
ogiela@agh.edu.pl

[2] Department of Advanced Sciences, Hosei University, 3-7-2, Kajino-cho, Koganei-shi, Tokyo 184-8584, Japan
makoto.takizawa@computer.org

[3] Pedagogical University of Cracow, Institute of Computer Science, Podchorążych 2 Street, 30-084 Kraków, Poland
lidia.ogiela@gmail.com

Abstract. In this paper will be presented new classes of cryptographic secret sharing procedures dedicated for secure information division and transmission. In particular will be presented two classes of protocols, which allow to share information with application of grammar solutions, as well as personal or behavioral parameters. Some possible application of such technologies will also be presented especially with relation to secure data or services management in distributed structures or Cloud environment.

Keywords: Cryptographic protocols · Data security · Cloud computing Secure information management

1 Introduction

In modern computer systems very important role plays data security protocols. These types of cryptographic protocols are dedicated to guarantee data security, also by using splitting procedures. Data splitting approaches ensures protection by sharing information between different participants of the protocol. Data sharing protocols provide information protection by division procedures. Each information is divided between participants of the protocol, using different protocols, for example Tang's or Shamir's algorithms [1–4].

The novel solutions of data splitting and sharing algorithms include application of mathematical linguistic to split and share important/strategic data in various areas [4, 5]. The main idea of proposed methods is to use linguistic formalisms for data sharing procedures. The linguistic formalisms were taken from theory of formal languages [1, 4]. This theory was dedicated for semantical analysis, description and interpretation for data understanding. Application of formal languages allows to extract semantic information from the analyzed data sets.

L. Barolli et al. (Eds.): NBiS 2018, LNDECT 22, pp. 445–452, 2019.
https://doi.org/10.1007/978-3-319-98530-5_37

Cryptographic techniques dedicated to provide data security by using sharing protocols called threshold schemes [6–8]. New classes of these schemes are the linguistic threshold schemes [4, 9].

This paper presents a new algorithm dedicated to data security which use linguistic approach. Linguistic threshold schemes use traditional information sharing techniques for generation an additional information shadow. This additional linguistic shadow contains the linguistic information necessary to reconstruct the original secret information. The main idea of using formal linguistic (formal languages) to define a traditional threshold scheme consist [4]:

- selection one of the secret splitting schemes (e.g. Blakley's, Shamir's or Tang's algorithm) for data security processes,
- use of selected schemes for data division processes,
- conversion the shared data into a bit sequence form,
- definition of formal grammar, which describe bit positions for the shared data,
- the bit sequence is parsed with a parser defined for the introduced formal grammar,
- the parsing generates a sequence of production numbers (grammar rules) which allow the bit representation of the shared secret to be produced,
- the secret represented by a sequence of production numbers is divided using the selected threshold scheme (selected in step 1),
- distribution the secret shadows between all participants of this protocol.

This procedure consist the main steps of using linguistic threshold schemes for information security processes. This algorithm presents the basic steps necessary to secure information by shared it. Also, it's possible to use two different stages in linguistic threshold schemes, depends on protection of grammar rules [4]:

- if the grammar rules are known only to the trusted arbitrator, then we have defined a so-called mediatory protocol in which an arbitrator (or a trusted computer system) must always be involved to reconstruct the secret,
- if the grammar rules are public, then we have classical threshold scheme which can be executed by additional shadow containing the grammar rules.

In this paper will be present linguistic threshold schemes for multi-level information sharing dedicated to data security and services management in hierarchical structures and Cloud environment.

2 Linguistic Threshold Schemes for Information Sharing

This paper presents one of possible algorithm for information sharing procedure. This protocol can create a data representation in the form of n bit blocks. It's also possible create the less complicated bit blocks, as well as one bit blocks or two bit blocks. But in this paper Authors presents the most useful solution included n-bit blocks.

The main steps of linguistic threshold schemes for information sharing are the following [4]:

- selection of threshold scheme for secret sharing procedure,
- defining a binary representation and group it into n bit blocks,

- defining a grammar generating values of a pre-set n bit length,
- using a parser to convert n bit blocks into sequences of grammar rules' numbers,
- data sharing using the sequences of grammar rules, with selected threshold scheme,
- distribution all secret parts between all participants of the sharing protocol.

A formal definition converting n bit blocks to a new data/information representation can be defined by the following formal grammar [4]:

$$G_{l-shadow} = (NTERM,\ TERM,\ PS,\ STS)$$

where:

NTERM = {INFORMATION, BITBLOCK, B, 1B, 2B, 3B, 4B,…, NB} – is a set of non-terminal symbols,
TERM = {0, 1, nb, λ} – is a set of terminal symbols, which defines n-bit blocks,
$\{\lambda\}$ – is an empty symbol,
STS = INFORMATION – is a grammar start symbol,
PS – is a production set included the following grammar rules:

1. INFORMATION → BITBLOCK BITBLOCK
2. BITBLOCK → 1B | 2B | 3B | 4B, … | NB
   {BIT BLOCKS WITH VARIOUS LENGTH}
3. BITBLOCK → λ
4. 1B → {0, 1}
5. 2B → {00, 01, 10, 11}
6. 3B → {000, 001, 010, 011, 100, 101, 110, 111}
7. 4B → (0000,…,)
8. ………
9. NB → nB
10. B → {0, 1}

A presented formal grammar can decode the secret information, which will be shared between all protocol participants. A benefit of grouping bits into n bit (larger) blocks is that during the following steps of the secret sharing protocol we get shorter representations for the secret information that is split and then reconstructed. This is particularly visible when executing procedures that use excessive bit representations, i.e. when single-bit or several-bit values are saved and interpreted using codes in 8 or 16-bit representations [4].

Examples of application of linguistic threshold schemes for secret sharing presents Fig. 1.

Figure 1 presents an example of application of linguistic threshold scheme for secret sharing procedures, and generating shadow of secret information. This secret is "*top secret data*" sentence, which can be divided for 5 shadows. Each 4 of them can reconstruct the original secret. So, it's an example of (4, 5)-threshold scheme. In this example the length of bit blocks is equal 4 bits.

Fig. 1. Application of linguistic threshold scheme for secret sharing.

Application of linguistic threshold procedure for secret sharing ensure:

- data protection through their division,
- higher level of security through the use of linguistic grammar elements.

Linguistic threshold schemes distinguish the level of security. These levels are independent of the length of bit blocks converted with the defined formal grammar rules.

Presented methods of linguistic secret sharing can use bit blocks of various lengths. The easiest protocol can use one-bit blocks, the more advanced protocol can use two-bit blocks, and the most advanced protocols can use n-bit blocks. This solution show how information sharing protocols can be significantly enhanced by adding elements of linguistic and grammatical information description and analysis. The length of bit blocks has a major impact on the speed and the conciseness of the stage at which the input information representation is coded and secure. Also, at which stage information to be coded as a secret is prepared.

An algorithm of selecting the bit block length depends on the length of secret information which will be coded. Also, it's depends on the hardware and time available for all cryptographic processes as well as coding, encryption, sharing, reconstruction and decoding the original secret information.

3 Biometric Threshold Schemes for Information Sharing

The main application of linguistic threshold schemes is data security obtained by application of linguistic information. Beside linguistic threshold schemes it was also defined the biometric threshold schemes. In this class of schemes is possible to use personal information for data security, especially for mark secret shadows [10, 11]. The main idea of this protocols is creation of additional shadows included personal data.

Selection of types of personal data depends on the scheme being created. It's possible to use all personal information as individual data to mark secret information and shadows.

The main steps of biometric threshold schemes for information sharing are following [4, 11]:

- selection of threshold scheme for secret sharing procedure,
- defining the personal data, which can be used to mark secret shadows,
- selection of personal features to identify personal shadows,
- defining a binary representation and group it into n bit blocks,
- defining a grammar generating values of a pre-set n bit length,
- using a parser which convert n bit blocks into sequences of grammar rule numbers,
- sharing data using sequences of grammar rules with selected threshold scheme,
- distribution all secret parts between all participants of the sharing protocol.

This kind of threshold schemes provide unambiguous assign a shadow to the secret holder, by used his personal data [11, 12]. Also, this kinds of threshold schemes provide restore information by using the right personal data included in personal shadow.

Figure 2 presents the scheme of biometric threshold scheme for secret sharing procedures.

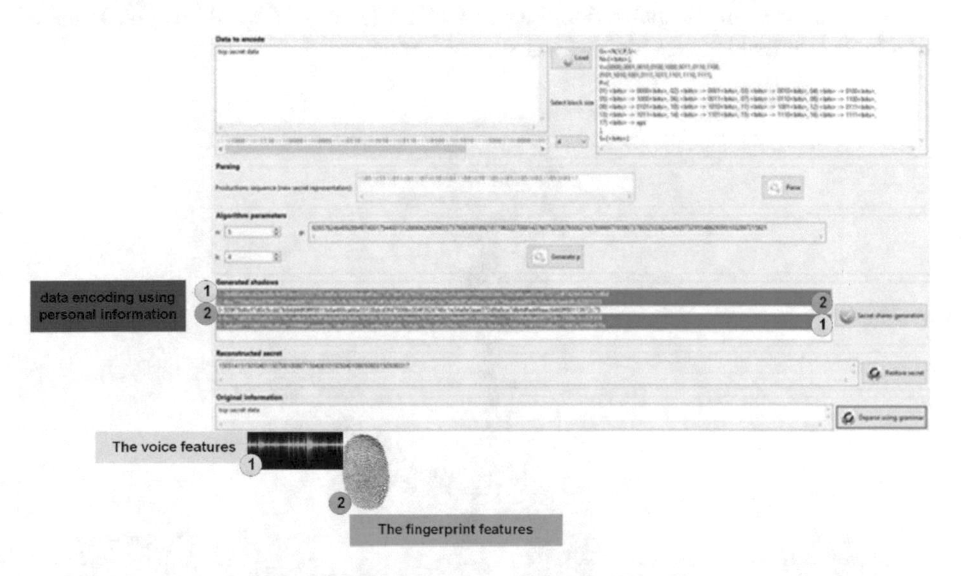

Fig. 2. Scheme of biometric threshold scheme.

Figure 2 presents the main idea of biometric threshold schemes using individual personal data for securing secret information. Figure 2 present two possible groups of biometric data, as well as voice parameters and fingerprint features. It's possible use different personal data for mark shadow at threshold schemes. Each of them consist an individual, unique features characteristic for a given person. Personal data can be used in the following form:

- as a mark of selected or all shadow, needed to mark shadows at restore original information procedure,
- as an additional secret shadow included all personal data, needed to restore original secret information.

Selection of right form of using biometric threshold schemes depends on the type of data protection. Used of biometrical threshold schemes for information security, ensures right distribution of secret parts and unambiguous original secret reproduction.

4 Examples of Applications

One of examples of applications of linguistic and biometric threshold schemes are hierarchical data division [1, 13]. Such division occurs most often in hierarchical structures. The most important feature of this division is the ability to data manage from different levels of hierarchy. Hierarchical structures occur in a different areas. For example, the most typical hierarchical structure is a hierarchical organization (company, enterprise – Fig. 3), but also different types of management levels create a hierarchical structure (organization, fog and cloud levels – Fig. 4). In this kinds of structures is possible used hierarchical secret sharing.

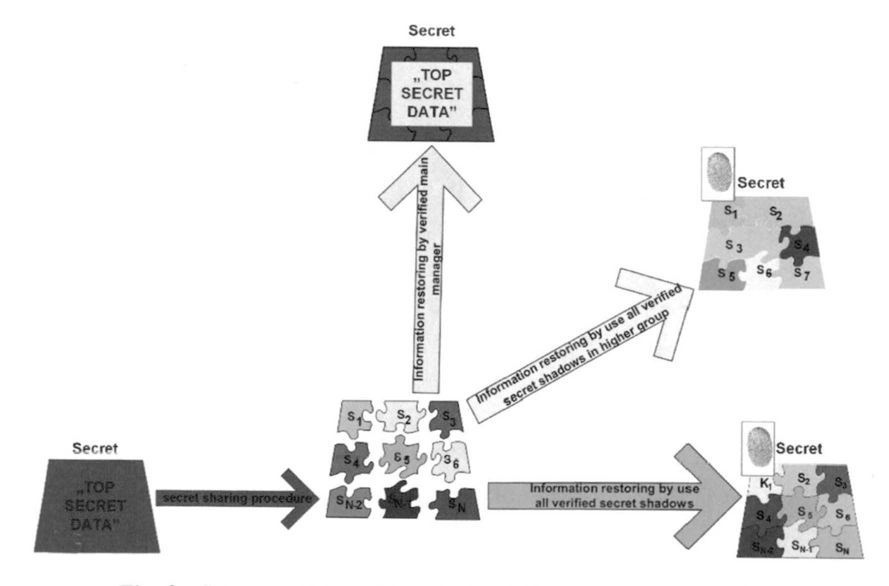

Fig. 3. Scheme of hierarchical sharing information in organization.

In hierarchical organization it's possible to manage secret information by split them. The sharing processes consist division of secret data between all participants of the protocols. But reconstruction of secret data is possible at different hierarchy levels.

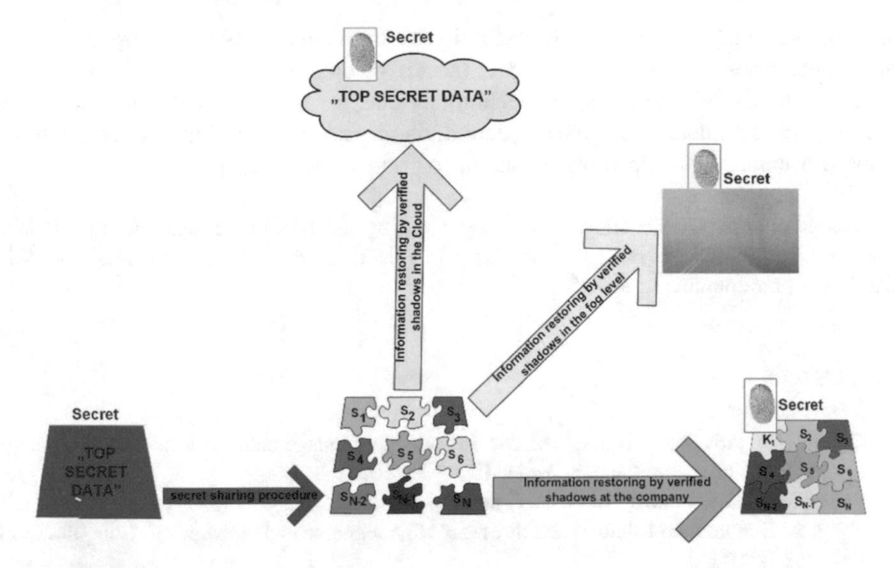

Fig. 4. Scheme of hierarchical sharing information at different management levels – at organization level, fog and Cloud levels.

For example at the lowest level of hierarchy all participants can restore original secret information after personal mark secret shadows. At the higher level another group of participants can restore secret information. But at the top of the hierarchy level is possible to restore secret information by main manager after right personal mark secret shadow.

Hierarchical structure means also different types of management levels, which create a hierarchical scheme, like organization and located on it fog and cloud levels. In this structure it's possible to manage secret information by split them, into organization level. The sharing processes consist division of secret information between all participants of the protocols included at organization. Reconstruction of original secret information is possible at different hierarchy levels. At the lowest level – organization level – all participants can restore original secret information after personal mark secret shadows. At the higher level – fog level – another group of participants can restore secret information. But at the top of the hierarchy level – Cloud level – is possible restore secret information by another group of secret holders.

The hierarchical structure may occur at different areas. This type of structure refers to structure possessing different hierarchies. This types of structures makes that is possible to use various secret sharing schemes at different levels of the hierarchy – privileges or not privileges [14].

5 Conclusions

This paper present new classes of cryptographic threshold schemes. The main classes are linguistic threshold schemes, dedicated to improving the division processes and accelerating its implementation. Another class of cryptographic schemes are biometric threshold schemes dedicated to data security by individual, personal information used for unambiguous marking of secret data (secret shadows).

These kinds of cryptographic techniques can be used for data protection and security, strategic data management, data division and transmission between different levels (for example between organization, fog and cloud levels).

Acknowledgments. This work has been supported by the National Science Centre, Poland, under project number DEC-2016/23/B/HS4/00616. This work was supported by JSPS KAKENHI grant number 15H0295.

References

1. Ogiela, L.: Advanced techniques for knowledge management and access to strategic information. Int. J. Inf. Manag. **35**(2), 154–159 (2015)
2. Shamir, A.: How to share a secret. Commun. ACM **22**, 612–613 (1979)
3. Tang, S.: Simple secret sharing and threshold RSA signature schemes. J. Inf. Comput. Sci. **1**, 259–262 (2004)
4. Ogiela, M.R., Ogiela, U.: Secure Information Management Using Linguistic Threshold Approach. Springer, Berlin (2014)
5. Ogiela, U., Takizawa, M., Ogiela, L.: Classification of Cognitive Service Management Systems in Cloud Computing. Lecture Notes on Data Engineering and Communications Technologies, vol. 12, pp. 309–313. Springer, Berlin (2018)
6. Beimel, A., Farras, O., Mintz, Y.: Secret-sharing schemes for very dense graphs. J. Cryptol. **29**(2), 336–362 (2016)
7. van Menezes, A., Oorschot, P., Vanstone, S.: Handbook of Applied Cryptography. CRC Press, Waterloo (2001)
8. Ogiela, L., Ogiela, M.R.: Insider threats and cryptographic techniques in secure information management. IEEE Syst. J. **11**, 405–414 (2017)
9. Ogiela, L.: Intelligent techniques for secure financial management in cloud computing. Electron. Commer. Res. Appl. **14**(6), 456–464 (2015)
10. Ogiela, L.: Cryptographic techniques of strategic data splitting and secure information management. Pervasive Mobile Comput. **29**, 130–141 (2016)
11. Ogiela, L., Ogiela, M.R.: Bio-inspired cryptographic techniques in information management applications. In: IEEE 30th International Conference on Advanced Information Networking and Applications (IEEE AINA), Switzerland, 23–25 March 2016, pp. 1059–1063 (2016)
12. Grossberg, S.: Adaptive resonance theory: how a brain learns to consciously attend, learn, and recognize a changing world. Neural Netw. **37**, 1–47 (2012)
13. Ogiela, M.R., Ogiela, U.: Secure information management in hierarchical structures. In: Kim, T.H., Adeli, H., Robles, R.J., et al. (eds.) 3rd International Conference on Advanced Science and Technology (AST 2011), Jeju Island, South Korea, 15–17 June 2011, Advanced Computer Science and Information Technology. Communications in Computer and Information Science, vol. 195, pp. 31–35 (2011)
14. Schneier, B.: Applied Cryptography: Protocols, Algorithms, and Source Code in C. Wiley, Chichester (1996)

Implementation of Mass Transfer Model on Parallel Computational System

Miron Pavluš[1], Tomáš Bačinský[1(✉)], and Michal Greguš[2]

[1] University of Prešov, Konštantínova 16, 080 01 Prešov, Slovakia
miron.pavlus@gmail.com , tomas.bacinsky@unipo.sk
[2] Faculty of Management, Comenius University in Bratislava, Odbojárov 10, 820 05 Bratislava, Slovakia
michal.gregus@fm.uniba.sk

Abstract. At present, parallel computational systems are being used more often for implementation of solutions of many phenomena. One of such phenomena is the mass transfer from one location to another that occurs in many porous building materials and the knowledge about this transfer is significant at least in civil engineering practice. In this paper, we consider a complex mass transfer diffusion model that involves beside water and water vapor, the air presence, and moreover, the phase transition of water to vapor or vice versa. The model was developed recently and its exact solution was found. In this paper, our intention is to implement this exact solution of the model on the parallel computational system HybriLIT. We suggest the sequential and parallel CUDA algorithms for the implementation of the solution. Thus, the speed-up of the parallel implementation is obtained and compared to the sequential implementation as a ratio of 44 s for sequential and 0.004 s for parallel implementation. GPU capacity is compared with total capacity used for problem solution on the sequential level. Input parameters of the problem are found so that the GPU capacity limits are reached.

1 Introduction

One of the thirteen large enough computer centers in the world (Tier-1) that store and process the Large Hadron Collider (LHC) data from source institution CERN in Switzerland (Tier-0) in the area of nuclear physics is the Laboratory of Information Technology in the Joint Institute for Nuclear Research in Dubna, Russian Federation. As a part of the worldwide LHC computing grid, the laboratory put into operation in 2014 the computational cluster HybriLIT [1] with a heterogeneous architecture. Information and software environment are created and implemented for the efficient use of the cluster that includes convenient services, debugging, and profiling of parallel applications. Project NICA (Nuclotron-based Ion Collider fAcility) [2] is another example which uses the cluster's options.

L. Barolli et al. (Eds.): NBiS 2018, LNDECT 22, pp. 453–463, 2019.
https://doi.org/10.1007/978-3-319-98530-5_38

In this paper, we use the cluster as a computational system for programming the exact solution of a complex mass transfer diffusion model which solution was derived in [3]. The model relates to transfer of moisture (liquid water and water vapor), and also air. Previous studies [4–8] were focused on the moisture transfer and the air presence in the material pores was usually neglected.

It is really a rare case when exact solution is found and thus no numerical analysis or numerical methods are needed. The model consists of three diffusion partial differential equations and one algebraic equation, and describes the law of conservation of masses (air, liquid water and water vapor). The masses pass through a porous material, and the transitions of liquid water to water vapor and vice versa are taken into account in the model, as well. The model involves the initial and boundary conditions that determine unique solution, and is formulated to describe the experiment conditions in the neutron transmission analysis [9] for detection of moisture transfer in a porous building material as accurate as possible. We consider the mass transfer in the sample of a porous wet material with dimensions $3 \times 9 \times 12$ cm^3 as in [9,10]. The sample is sealed on by a self-adhesive aluminum tape from all sides except right side 3×12 cm^2 which is open. As a consequence, the moisture and air diffusion occurs only in one direction x – along the width of the sample $0 \leq x \leq l$, where $l = 9$ cm.

The studied problem relates to the implementation of exact solutions of partial differential equations. Also other authors dealt with implementation that were related to solution of partial differential equations, for example, [11–13]. However, these works are concerned with implementations of various numerical methods for solution of the partial differential equations. Our study is focused on the implementation of exact solution which is expressed by means of the infinite series and this requires an approach, which is based on the quick addition of terms of partial sums. Algorithms for fast addition of sums were developed in the past [14] and use reduction method and shared memory in the framework of the threads.

In the beginning of this paper we show the basic hardware and software features of the cluster HybriLIT. Next, we formulate the mass transfer diffusion model including initial and boundary conditions and also the exact solution of the problem according to the work [3]. After, we present the sequential code in programming language C for implementation of the computational part of the exact solution. The main part of the paper involves Compute Unified Device Architecture (CUDA), Simple Linux Utility for Resource Management (SLURM), and Message Passing Interface (MPI) for implementation of the exact solution in C. Finally, in conclusions, we compare time realization of individual implementations for different parallel approaches.

2 Cluster HybriLIT

The heterogeneous cluster HybriLIT is used both to perform massively parallel computations and to learn how to use applied software packages and parallel programming technologies. The computing component of the cluster contains four nodes with NVIDIA Tesla K80 graphical processors and four nodes with NVIDIA Tesla K40 accelerators, a node with Intel Xeon Phi 7120P co-processors, and a node with two types of computing accelerators NVIDIA Tesla K20x and Intel Xeon Phi 5110P. All the nodes have two multicore processors Intel Xeon. Overall, the cluster contains 252 CPU cores, 77184 GPU cores, 182 PHI cores; it has 2.4 TB RAM and 57.6 TB HDD, and its total capacity is 142 TFLOPS for operations with single precision and 50 TFLOPS for double precision (Fig. 1).

All nodes of the heterogeneous computing cluster HybriLIT use Scientific Linux 6.7 as the operation system. Operating software of the cluster includes a set of CUDA Toolkit [18] for parallel programming and computations with the use of graphical accelerators (GPU). The documentation is available at NVIDIA site [19]. Intel Cluster Studio XE 2013 package [20] includes a set of tools for the parallel programming and computations with the use of processors (CPU) and coprocessors (PHI). The documentation is available at Intel site [21].

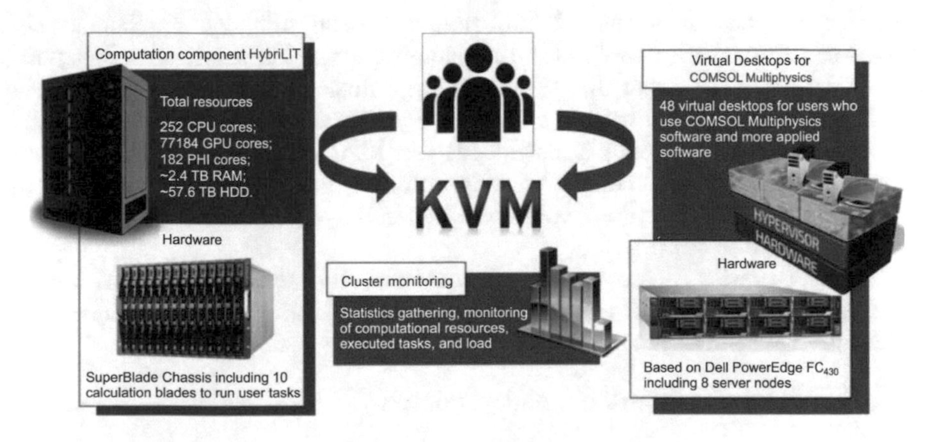

Fig. 1. Structure of heterogeneous computational cluster HybriLIT and recently included component [16].

3 Mass Transfer Diffusion Model

Let us consider a moist sample consisting of a solid phase while in the pores of the material some air, water, and vapor are present. Let us introduce the function of concentration of air $w_a(x,t)$, the function of concentration of liquid water $w_l(x,t)$, the function of concentration of vapor $w_v(x,t)$, and the source function $S(x,t)$ characterizing the rate of a phase transition which takes positive values if water is evaporating to vapor and negative values, if vapor condenses into water, while x is independent spatial variable and t is independent time variable. Let us denote by Π the porosity of the sample material, ρ_i and D_i the density and the diffusion coefficient for air $(i = a)$, for liquid water $(i = l)$, and for vapor ($i = v$). We shall assume that all these parameters are positive constants and that $\rho_l > \rho_a > \rho_v$.

Then we can describe the model of drying of a moist sample by the following system of four equations (all used quantities are dimensionless)

$$\frac{\partial w_a}{\partial t} = \frac{\partial}{\partial x}\left(D_a \frac{\partial w_a}{\partial x}\right), \quad 0 < x < 1, \quad t > 0, \tag{1}$$

$$\frac{\partial w_l}{\partial t} = \frac{\partial}{\partial x}\left(D_l \frac{\partial w_l}{\partial x}\right) - S, \quad 0 < x < 1, \quad t > 0, \tag{2}$$

$$\frac{\partial w_v}{\partial t} = \frac{\partial}{\partial x}\left(D_v \frac{\partial w_v}{\partial x}\right) + S, \quad 0 < x < 1, \quad t > 0, \tag{3}$$

$$\Pi = \frac{w_a}{\rho_a} + \frac{w_l}{\rho_l} + \frac{w_v}{\rho_v}, \quad 0 \le x \le 1, \quad t \ge 0. \tag{4}$$

Equation (1) describes the air diffusion with the diffusion coefficient D_a. Equation (2) describes the water diffusion with the diffusion coefficient D_l, Eq. (3) describes the vapor diffusion with the diffusion coefficient D_v, and these two Eqs. (2)–(3) are tied with the function S, ($S > 0$ is a source of vapor, resp. $S < 0$ is a source of water). Equation (4) can be understood so that the pores volume consists of three complementary volumes, namely, the air volume, the water volume, and the vapor volume. The similar equation was used also by other authors, e.g. see [17].

Finally, in order to ensure the uniqueness of the solution of (1)–(4) and to simulate the conditions of experiment presented in the work [9], we formulate the initial conditions

$$w_a(x, 0) = 0, \qquad w_l(x, 0) = \rho_l \Pi, \qquad 0 \le x \le 1, \tag{5}$$

which express that at the beginning of the process the pores are completely filled with water and neither vapor nor air are present in the pores of the sample. We also formulate the boundary conditions

$$\frac{\partial w_i}{\partial x}(0, t) = 0, \quad i = a, l, \quad t > 0, \tag{6}$$

$$w_a(1, t) = \Pi \rho_a (1 - e^{-\alpha_a t}), \qquad w_l(1, t) = \Pi \rho_l e^{-\alpha_l t}, \tag{7}$$

for $t > 0$, where $\alpha_l \ge \alpha_a > 0$, $\alpha_l \gg 1$. Conditions (6) mean that the left end of the sample is isolated and conditions (7) mean that after a relative short time the volume of water in the right end of the sample is decreasing to the minimal zero value.

4 Solution of the Model

As it was demonstrated in [3], the problem (1)–(7) has the following solution

$$w_a(x,t) = \Pi \rho_a \left[1 - e^{-\alpha_a t} + 2\alpha_a \sum_{k=1}^{\infty} \frac{(-1)^{k-1}(e^{-\alpha_a t} - e^{\lambda_k D_a t})}{\sqrt{|\lambda_k|}(\alpha_a + \lambda_k D_a)} X_k(x) \right] \quad (8)$$

$$\lambda_k = -\left[(2k-1)\frac{\pi}{2} \right]^2, \quad X_k(x) = \cos(\sqrt{|\lambda_k|}x), \quad 0 \le x \le 1, \quad t \ge 0,$$

$$w_l(x,t) = \Pi \rho_l e^{-\alpha_l t} + \sum_{k=1}^{\infty} T_k(t) X_k(x), \qquad 0 \le x \le 1, t \ge 0, \quad (9)$$

where

$$T_k(t) = \left[\frac{\varphi_k}{\alpha_l + \lambda_k D_0} - \psi_k \left(\frac{1}{\lambda_k (D_a - D_0)} + \frac{1}{\alpha_a + \lambda_k D_0} \right) \right] e^{\lambda_k D_0 t}$$

$$- \frac{\varphi_k}{\alpha_l + \lambda_k D_0} e^{-\alpha_l t} + \psi_k \left(\frac{e^{\lambda_k D_a t}}{\lambda_k (D_a - D_0)} + \frac{e^{-\alpha_a t}}{\alpha_a + \lambda_k D_0} \right), \quad t \ge 0,$$

and
$\varphi_k = 2\Pi \alpha_l \rho_l (-1)^{k-1}/\sqrt{|\lambda_k|}$, $\psi_k = 2\Pi \alpha_a \rho_a D_1 (-1)^{k-1} \sqrt{|\lambda_k|}/(\alpha_a + \lambda_k D_a)$,
$D_0 = (\rho_l D_l - \rho_v D_v)/(\rho_l - \rho_v)$, $D_1 = \rho_l \rho_v (D_a - D_v)/(\rho_l - \rho_v)/\rho_a$.

Finally, the last two unknowns are determined according to formulas

$$w_v(x,t) = \rho_v \left(\Pi - \frac{w_a}{\rho_a} - \frac{w_l}{\rho_l} \right), \quad 0 \le x \le 1, \quad t \ge 0, \quad (10)$$

$$S(x,t) = -\frac{\rho_l \rho_v}{\rho_l - \rho_v} \left[\frac{D_a - D_v}{\rho_a} \frac{\partial^2 w_a}{\partial x^2} + \frac{D_l - D_v}{\rho_l} \frac{\partial^2 w_l}{\partial x^2} \right], \; 0 < x < 1, \; t > 0. \quad (11)$$

5 Solution Accuracy and Convergence of Partial Sums

The solution (8)–(11) of the problem (1)–(7) is based on the infinite series which are expressed in formulas (8), (9) and (11). It can be shown by Weierstrass M-test [22] that these infinite series are uniformly convergent for $t \ge 0$ and $x \in (0,1)$. Next, for computational purposes, the infinite sums must be replaced by some finite partial sums. So, sufficiently large integer number N must be selected instead of infinity. The selection of N is sometime based on vanishing of the error estimation of the remaining terms of the series. Such selection guarantees a sufficient level of accuracy of the exact solution or a sufficient level of approximation of the exact solution. We note that the higher the value of N the more computational work is needed. Thus, N must be selected in a rational way.

Moreover, the following aspect must be taken to the account, as well. Choosing index k sufficiently large, all next terms in series (8), (9) and (11) can be in absolute value less than the machine zero. In this case there is no sense to add such terms to the series sum because the processor unit discerns such terms as zero values and operation of the adding does not change the previous value of the sum. Certainly, the question, which terms are discerned as a machine zero value and which of them are not, depends also on the floating point arithmetics incorporated in the considered processor unit for real number representation. For example, NVIDIA processors with CUDA architecture use the fused multiply-add operator, which was added to the IEEE 754 standard [15] in 2008. Thus, analysing and estimating, for example, the second derivative of the series (8), one can get the following upper bound

$$\frac{2\sqrt{|\lambda_k|}}{|\alpha_a + \lambda_k D_a|} \leq 10^{-p} \tag{12}$$

where p represents a positive machine zero exponent (for example, for double precission floating point numbers $p = 308$). Hence, we receive that if index

$$k \geq \frac{1}{2} + \frac{1}{\pi}\Big(2\frac{10^p}{D_a} - \sqrt{\frac{\alpha_a}{D_a}}\Big)$$

then the term, corresponding to this k, is recognized by the processor unit as a zero value.

In this paper, however, we select N according to a suitable approach of the unit function which was applied during the derivation of the solution in [3]. It can be easily derived integrating by per partes that

$$1 = \sum_{k=1}^{\infty} \frac{4(-1)^{k-1}}{(2k-1)\pi} \cos\frac{(2k-1)\pi x}{2} \qquad \text{for all x} \in (0,1).$$

Let us change the infinity in the last sum to a sufficiently large N. Then the norm

$$||1 - \sum_{k=1}^{N} \frac{4(-1)^{k-1}}{(2k-1)\pi} \cos\frac{(2k-1)\pi x}{2}|| = \max_{x\in(0,1)} \left|1 - \sum_{k=1}^{N} \frac{4(-1)^{k-1}}{(2k-1)\pi} \cos\frac{(2k-1)\pi x}{2}\right|$$

can be determined for individual N and results are shown in Table 1. We can see that the increasing N leads to decreasing norm value, which practically means the convergence of the corresponding partial sums to exact solution (8)–(11). We put $N = 10^4$ for further research purposes. Higher values of N are also possible but they can produce terms in series (8) and (9) which values are recognized by processor unit as a zero.

Table 1. Deviation of the partial sum from the unit function depending on N.

N	10^1	10^2	10^3	10^4	10^5	...
$\|\| \ \|\|$	0.8011	0.1790	0.0202	0.0020	0.0002	...

6 Sequential Implementation on One CPU And/or One GPU

At the beginning, we start to implement the solution (8)–(11) of the problem (1)–(7) by sequential way on one CPU and/or one GPU. For this purpose we classify all constants and all variables used including all accessory constants and accessory variables. Analysing the formulas (8)–(11) one can see that the constants are Π, π, ρ_i, D_i, for $i = a, l, v$, and α_i, for $i = a, l$, and the accessory constants are D_0, D_1, λ_k, ϕ_k, and ψ_k.

Figure 2 shows the tree of the accessory constants formating that starts from eigenvalue λ_k, (lak) in the bottom up to the solution variables w_a, w_l, S, and w_v in the top of the tree.

Algorithm 1. Sequential algorithm for one CPU.

Input: Π, α_a, α_l, ρ_l, ρ_v, ρ_a, D_l, D_v, D_a
Output: values of w_a, w_l, w_v and S for seven timesteps

```
1  calculate constants D0, D1
2  foreach timestep ∈ {0; 0.02; 0.05; 0.1; 0.3; 0.65; 1} do   days 0, ..., 20
3  foreach i ∈ {1; 2; ... 100} do                             steps of spatial var. x
4    Sa = Sa2 = Sl = Sl2 = 0;                                 initialization of sums
5    foreach k ∈ {1, 2, ..., 10^4} do
6      calculate lak, la_k, sc, arl, ara, ka
7      Sa += ka;
8      Sa2 += la_k × ka;
9      kl = sc × (arl − ara);
10     Sl += kl;
11     Sl2 += la_k × kl;
12   end % k
13   calculate wa[i], wad2[i], wl[i], wld2[i], wv[i], S[i];
14 end % i
15 savetofile
16 end % timestep
```

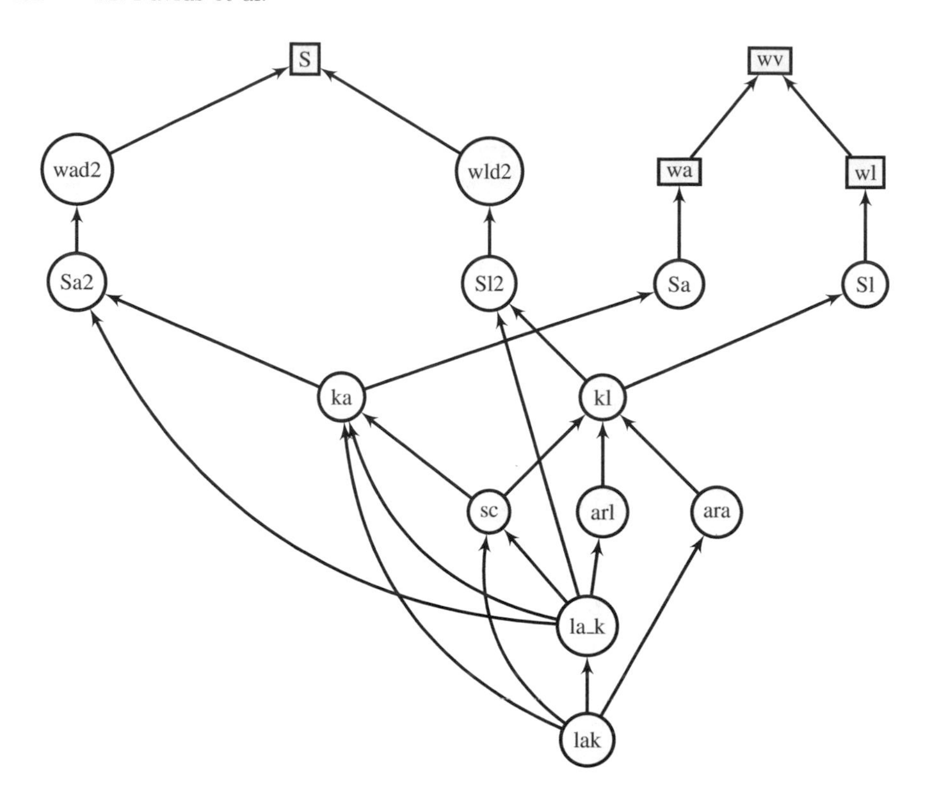

Fig. 2. Tree of constants and variables.

7 Parallel Implementation

Table 2 contains results for time realization of parallel algorithm with 606 blocks and different threads. We note that running time for one block and one thread (sequential algorithm on GPU) is 44 s.

Table 2. Running time in milliseconds for different numbers of running (r) and numbers of threads (t).

r/t	1	2	4	8	16	32	64	128	256	512
1	214, 17	108, 82	54, 41	27, 24	13, 34	6, 78	4, 74	4, 31	4, 42	6, 51
2	214, 16	108, 79	54, 36	27, 22	13, 45	6, 71	4, 76	4, 31	4, 43	6, 52
3	213, 77	108, 79	54, 37	27, 22	13, 45	6, 72	4, 76	4, 31	4, 44	6, 51
4	214, 17	108, 77	54, 39	27, 20	13, 34	6, 71	4, 85	4, 33	4, 42	6, 51
5	214, 11	108, 73	54, 36	27, 18	13, 33	6, 72	4, 77	4, 31	4, 43	6, 51
min	213, 77	108, 73	54, 36	27, 18	13, 33	6, 71	4, 74	4, 31	4, 42	6, 51

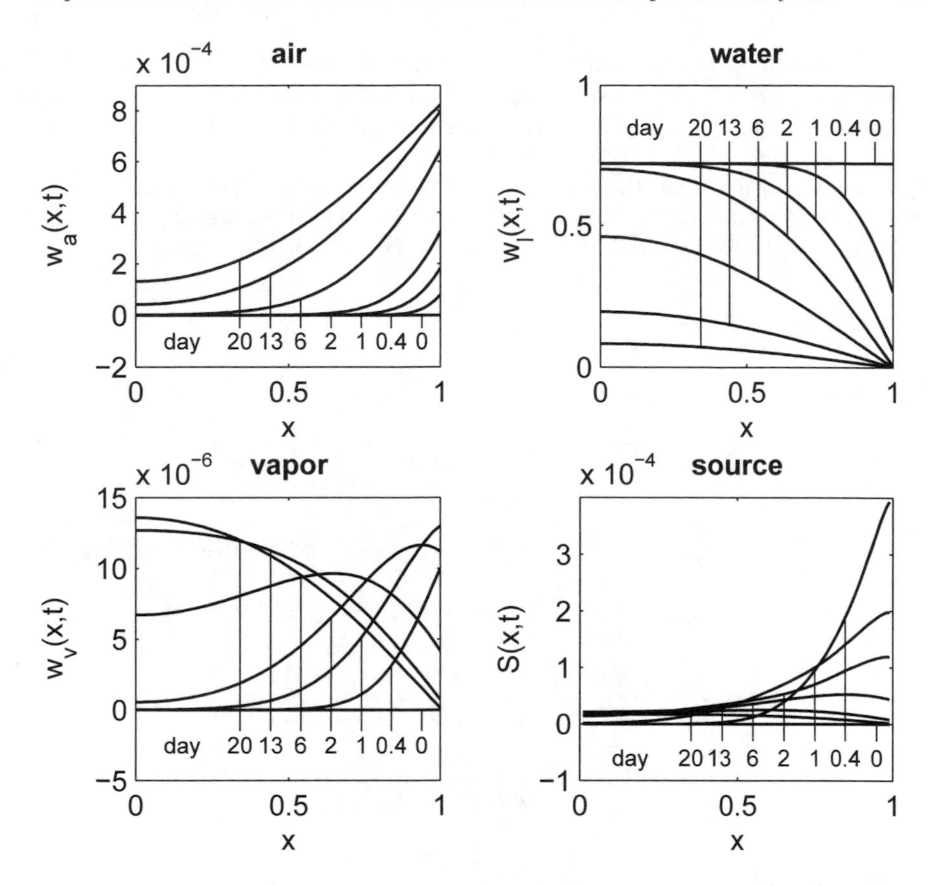

Fig. 3. Solutions received for air, water, and vapor concentrations w_a, w_l, w_v, and for source function S by means of the CUDA parallel algorithm.

8 Conclusions

The following results have been obtained by the CUDA parallel algorithm, see Fig. 3, which validate previous results received by sequential algorithm in [3].

In conclusion, an exact solution of the model was implemented on the parallel computational system. A sequential algorithm was suggested for the implementation of solution, particularly to calculate curves for seven time steps from the beginning of the process of drying of a moist sample. The algorithm was implemented on a parallel computational system, the running time for different thread counts was measured and notable speed-up was obtained.

Even if the problem considered in the paper is spatially one dimensional, the implementations and programming experiences received are intended to obtain the parallel realization for higher dimension problems in future.

It is worth mentioning, that to shorten computation time, we cannot increase the number of threads indefinitely. When taking more, than 128 threads, running time is becoming longer (Table 2). However, it is possible to further optimize running time by having steps (i, Algorithm 1.) of spatial variable and/or upper boundary of sum (N, Algorithm 1.) in form of power of two (2^n).

Acknowledgements. Authors thank Mr. Maxim Zuev from the Joint Institute of Nuclear Research (JINR) for consultations on CUDA. This work was supported by the JINR project No. 05-6-1118-2014/2019, protocol No. 4596-6-17/19.

Appendix

Used formulas

$$D_0 = \frac{D_l \rho_l - D_v \rho_v}{\rho_l - \rho_v}, \qquad D_1 = \frac{\rho_l \rho_v (D_a - D_v)}{\rho_a (\rho_l - \rho_v)},$$

$$lak = (2k-1)\frac{\pi}{2}, \qquad la\_k = -[(2k-1)\frac{\pi}{2}]^2, \qquad sc = \sin(lak)\cos(lak),$$

$$arl = \frac{\alpha_l \rho_l}{lak(\alpha_l + \lambda_k D_0)}(e^{\lambda_k D_0 t} - e^{-\alpha_l t}),$$

$$ara = \frac{\alpha_a \rho_a D_1 lak}{\alpha_a + \lambda_k D_a}\left[\frac{e^{\lambda_k D_0 t} - e^{-\alpha_a t}}{\alpha_a + \lambda_k D_0} + \frac{e^{\lambda_k D_0 t} - e^{\lambda_k D_a t}}{\lambda_k (D_a - D_0)}\right],$$

$$ka = \frac{sc \times (e^{-\alpha_a t} - e^{\lambda_k D_a t})}{lak \times (\alpha_a + \lambda_k D_a)}.$$

References

1. http://hybrilit.jinr.ru
2. Kekelidze, V.: Project NICA at JINR. Nucl. Phys. A **904–905**(2013), 945c–948c (2013)
3. Litavcová, E., Pavluš, M., Seman, J., et al.: Exact solution of system of mass transfer which includes air, water and vapor. Eur. Phys. J. (2015, to appear)
4. Litavcová, E., Korjenic, A., Korjenic, S.: Diffusion of moisture into building materials: a model for moisture transport. Energy Buildings **68**(PARTA), 558–561 (2014)
5. Antonyová, A., Korjenic, A., Antony, P.: Hygrothermal properties of building envelopes: reliability of the effectiveness of energy savings. Energy Buildings **57**, 187–192 (2013)
6. Amirkhanov, I.V., Puzynina, T.P., Puzynin, I.V., et al.: Numerical simulation of heat and moisture transfer subjected to the phase transition. Lecture Notes in Computer Science, vol. 7125, pp. 195–200 (2012)
7. Amirkhanov, I.V., Pavlušová, E., Pavluš, M., et al.: Materials and Structures, vol. 41, pp. 335-344 (2008). ISSN 1359-5997
8. Krus, M.: Moisture transport and storage coefficients of porous mineral building materials. Fraunhofer IRB Verlag, Stuttgart (1996). ISBN 3-8167-4535-0
9. Pleinert, H., Sadouki, H., Wittmann, F.H.: Determination of moisture distribution in porous building materials by neutron transmission analysis. Mater. Struct. **31**, 218 (1998)

10. Amirkhanov, I.V., Pavlušová, E., Pavluš, M., et al.: Preprint of the Joint Institute for Nuclear Research, Dubna, P11-2009-124, p. 11 (2009)
11. Cotronis, Y.: A comparison of CPU and GPU implementations for solving the convective diffusion equation using the local SOR method. Parallel Comput. **40**, 173–185 (2014)
12. Sánchez-Curto, J., Chamorro-Posada, P., McDonald, G.S.: Efficient parallel implementation of the nonparaxial beam propagation method. Parallel Comput. **40**, 394–407 (2014)
13. Sandroos, A.: Multi-GPU simulations of Vlasov's equation using Vlasiator. Parallel Comput. **38**(8), 306–318 (2013)
14. Harris, M.: Optimizing Parallel Reduction in CUDA, NVIDIA Developer Technology. http://developer.download.nvidia.com/compute/cuda/1.1-Beta/x86_website/projects/reduction/doc/reduction.pdf
15. IEEE 754-2008. Standart for Floating-Point Arithemetics, August 2008
16. Joint Institute for Nuclear Research 2016. Annual report, p. 111 (2017). ISBN 978-5-9530-0470-1
17. Reshetin, O.L., Orlov, S.Y.: Theory of heat and moisture transfer in a capillary porous body. Tech. Phys. **43**(2), 263 (1998)
18. https://developer.nvidia.com/cuda-toolkit
19. https://docs.nvidia.com/cuda/cuda-gdb/index.html?axzz373CGTIsQ
20. https://software.intel.com/en-us/intel-cluster-studio-xe
21. https://software.intel.com/sites/products/documentation/hpc/ics/icsxe2013sp1/lin/icsxe_gsg_files/
22. Rudin, W.: Principles of Mathematical Analysis. 3rd edn. McGraw-Hill Inc., New York (1976). ISBN 0-07-054235-X

The 13th International Workshop on Network-Based Virtual Reality and Tele-Existence (INVITE-2018)

Server System of Converting from Surface Model to Point Model

Hideo Miyachi[1(✉)] and Isamu Kuroki[2]

[1] Department of Information Systems, Faculty of Informatics, Tokyo City University, 3-1-1 Ushikubo-nishi, Tsuzuki-ku, Yokohama, Japan
miyachi@tcu.ac.jp

[2] Visualization Department, Cybernet Systems Co., Ltd., Kanda Neribei-cho, Chiyoda-ku, Tokyo, Japan
i-kuroki@cybernet.co.jp

Abstract. Handling the 3D CG model as a point clouds uniformly, it will be possible to easily perform data conversion, 3D shape search, and data downsizing. We have developed a system to convert a surface model to a point model and a system to downsize the point model. Then in order to publish these tools, we are currently proceeding to release these software as a service. In this paper, we introduce the service system.

1 Introduction

The 3D geometric model handled by computer graphics has been complicated and the data size has been larger and larger. A high-performance supercomputer outputs an enormous amount of simulation results. Therefore, the 3D model as the visualization result has also been bigger and bigger. Digitalization of 3D objects by laser measurement and photograph measurement also generates a large-scale 3D CG model.

Although such precise and detailed data is useful, but data reduction is needed when sharing quickly and efficiently the approximate appearance among a group. To meet such needs, we have been proposing a new data reduction method which converts surface data to point data [1–3]. When the overview of an object is displayed with a viewpoint at a certain distance away from the object, the object can be replaced with a point group generated with a density corresponding to the resolution of the window on which the object is displayed. By using this method, an arbitrary 3D surface model can be reduced to a data size corresponding to the display resolution without depending on the original data size. And also we have also demonstrated that the point data can be downsized efficiently by using the same principle [4].

By using these techniques, all 3D geometric models can be handled uniformly as point data. Since point data can be represented simply, we are considering that it has advantages such as easy data conversion and fast rendering, and if there is a possibility to facilitate 3D shape retrieval. Currently, we are beginning to offer two technologies as services so that many people can support the concept of uniformly handling all 3D models as points. In this paper, we would like introduce the service system. Below, in Sect. 2, we describe the data reduction method for converting surface data to points,

L. Barolli et al. (Eds.): NBiS 2018, LNDECT 22, pp. 467–475, 2019.
https://doi.org/10.1007/978-3-319-98530-5_39

and Sect. 3 describes the method to downsize point data. In Sect. 4, we introduce a system that provides those functions as a service.

2 Data Reduction Method for Converting Surface Data to Points

2.1 Data Reduction

Many methods of polygon reduction have been proposed in past [5]. The principle of these methods is to integrate multiple polygons into one at the area in which the polygons make almost flat plane. However, the limit of the reduction in this idea depends on the complexity of the geometry. Therefore, it has a possibility that it is not possible to achieve the data reduction to a specified size.

As a data reduction method that does not depend on geometric complexity before reduction, there is a method of mapping the geometry to voxel grid of a specified resolution [6]. This has the same goal as our proposal, which is to specify the target reduction size without depending on geometry complexity. However, for mapping the geometry to a voxel grid, it is necessary to completely interpret the 3D geometric information. In contrast, our proposal does not require its interpretation. The reason is because our proposal uses only the images and the depth data that remains on graphics memory when the geometry is rendered. The method to capture those data from an external process was implemented at Chromium project or others [7, 8]. Of course, our proposal has some limitations. The first is that only surface that can be seen from the outside can be reconstructed. The second is that it is difficult to handle the object with transparency attribute. The details of the proposal will be introduced in the next section.

2.2 Conversion from 3D Surface to 3D Point

Here, we introduce the surface to point conversion method, and a representative quality test result on our previous study [1–3]. Rendering with Z buffer algorithm, the rendering results are stored in graphics memory. At the time, the color and depth information are stored in frame buffer and Z buffer respectively like shown in Fig. 1. Combining those two kinds of information, we can obtain 2.5 dimensional point data.

Next, to reconstruct 3D point data to represent the entire object, 2.5 dimensional point data generated by rendering from various eye positions are obtained and those data are integrated to a single 3D model as shown in Fig. 2.

In the process of the data conversion, there are two parameters we can determine arbitrarily. One is the resolution of the window in which the object is rendered. Another is the number of the directions of the rendering. The vectors indicating the directions are also arbitrary parameters. But as we have not understood the impact on output by the parameters yet, those parameters are not considered in the work at present. Regarding with the resolution, it has been learned that a gap can be formed unless the resolution of the window when capturing an object is set not to double or more resolution in x and y direction respectively to the window in browsing. Regarding with

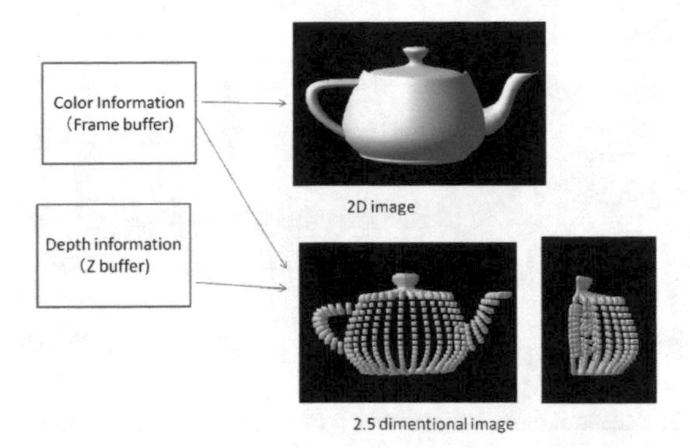

Fig. 1. Color and depth information in graphics memory after Z buffer rendering.

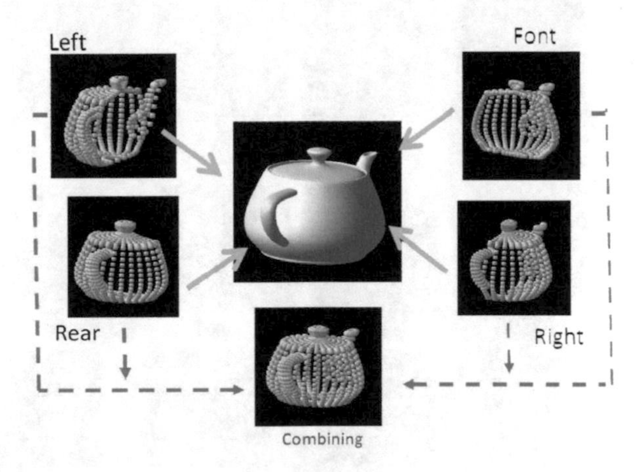

Fig. 2. Reconstructing 3D point data by combining multiple 2.5 dimensional point data.

the number of the directions of the rendering, we used 14 directions because it has been learned that 14 directions shown in Fig. 3 seems reasonable as an empirical know-how.

One of the results of the quality evaluation in existing work is shown in Fig. 4. The number of directions was set to 26 directions at this test. The 4 images in Fig. 4(1) and (2) are the rendering images of a sphere consist of surfaces and points respectively. The images in Fig. 4(3) present the differential images of (1) and (2). All of PSNR valucs between (1) and (2) indicated over 65 Db. Although 4 images are displayed in Fig. 4 as the samples, the evaluation was calculated at 16 directions, and all of them also indicated over 65 Db. It means that human cannot recognize the difference between two images. From these facts, it was found that the surface data which is too detailed with respect to the display resolution can be replaced by converting it into point by the proposed method, and the alternative point data has almost the same quality as the surface data. The data size can be smaller than the size before the conversion.

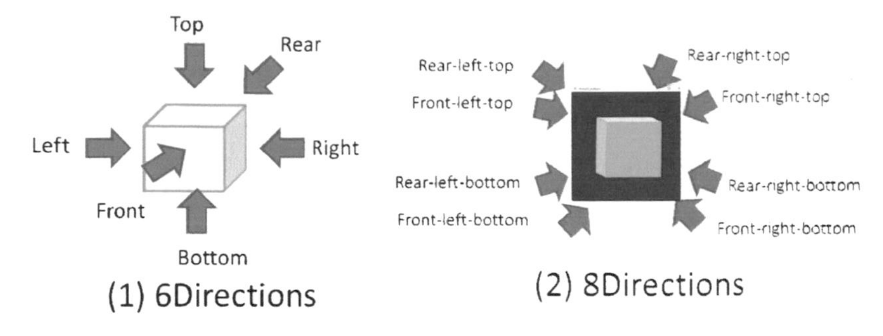

Fig. 3. Reasonable 14 directions ((1) + (2)) obtained as know-how.

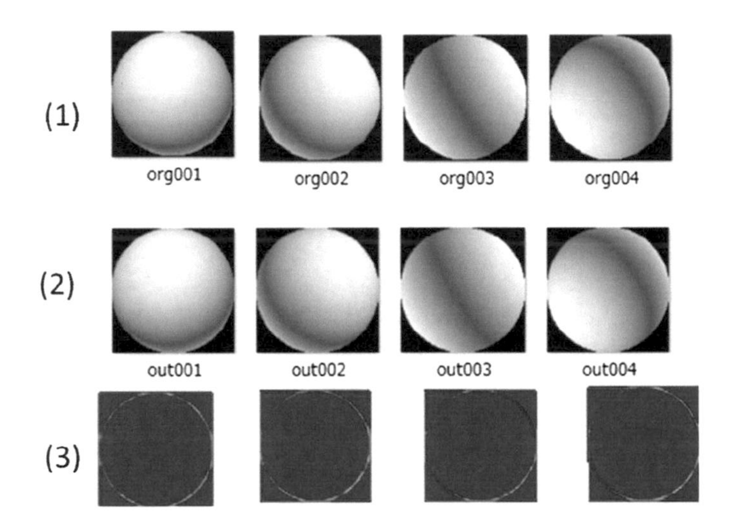

Fig. 4. An evaluation result about the image quality of 3D point data.

3 Downsize Point Data

In this section, we briefly introduce our previous study [4] of downsizing point data. Using the method described in Sect. 2, point data can be reduced as well as surface data. In addition, it has a risk to increase the data size in the case of conversion from surface to point, but it will be an extraction operation in the case of conversion from point to point, so the data size is always reduced. Since the process of point rendering is simple, we did not use the OpenGL graphics library, but we implemented the rendering program in C language, which can work at server computer without any graphics functions. The program supports point ID buffer that can identify each point. The ID buffer is used for recording the point ID of the foremost point. The ID buffer works as same as Z buffer in rendering process, but the stored information is not the depth, but point ID. After finishing the 1 frame rendering, the point ID of the most frontal point is remained at each pixel. The points with the same point ID which are captured from multiple directions can not be counted in duplicate.

We introduce an example that this method applied to point data. The test data provided from Prof. Tanaka, Ritsumeikan university is shown in Fig. 5. This is the data measured by a laser device which has about 1 G Byte data size on disk storage in PLY ascii format and includes about 26 million points. This is a festival car used in traditional GION festival at Kyoto City in Japan.

Fig. 5. Point data for a test of downsizing point data.

The quality performance of the proposed method to this data is shown in Fig. 6. The horizontal axis indicates the PSNR values between images before downsizing and after downsizing. It means a quality of the downsized point data. A higher value means higher quality. The vertical axis indicates the data size after downsizing as a percentage to the data size before downsizing. The figure contains five kinds of graph. As the legend shows, the each graph indicates the case that is the number of directions as 2, 6, 8, 14 and 26 respectively. Also the each graph contains eight plots. Each plot indicates the case that is the resolution of window in capturing as 300, 400, 600, 800, 1000, 1200, 1400 and 1600 pixels square from left to right. Using this graph, it is possible to specify the best parameters for downsizing to the target data size and quality.

For example, if you have to reduce the data size to 8% of the original, the leftmost graph(high quality size) crossing at the 8% line on the horizontal axis is the graph with 25 directions, and the nearest plot is fifth plot from the left. Therefore, we can understand that 26 directions and 1000 x 1000 resolution is the best parameter set.

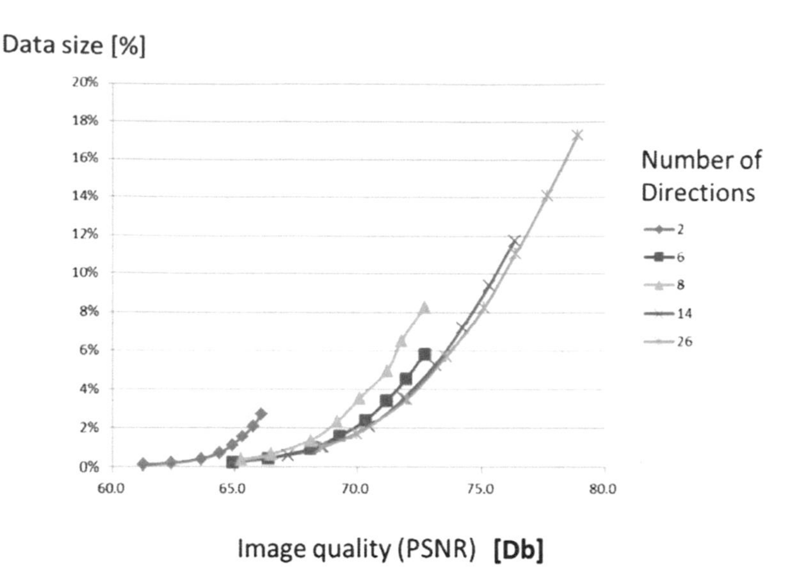

Fig. 6. 3D point data quality on the window resolution and the number of directions.

4 Service System on a Web Bases

We are aiming at uniformly managing all 3D models as point clouds. For that purpose, we are planning to publish the tools introduced in Sects. 2 and 3 in free as a Web service. The service system on developing plans to provide three functions as shown in Fig. 7. The first is the data conversion from STL surface data to PLY point data. The second is the downsizing of PLY point data. The third is to generate LOD (Level Of Detail) [9] point data set with 8 branch structure shown in Fig. 8(a). We will publish the LOD viewer written with OpenGL, which is implemented LOD control shown in Fig. 8(b). The viewer loads the point data at the scale level corresponding to the zoom ratio. Then, when the object is zoomed in, it sends the blocks of the point data only in the view frustum to the graphics pipeline.

The implemented Web user interface is shown in Fig. 9. The top area (Fig. 9①) is used for new data uploading to the system. The uploaded files are listed the file list area in the middle. Then, after clicking the check box the left side of the target file, the execution operation is selected from the three operations in Fig. 9②. By this operation, the target file is operated along to the diagram indicated in Fig. 7. After the conversion, downsizing or generation, the new files generated are also listed into the file list area.

To view the ply file listed, clicking the file name in the file list area invokes WebGL viewer embedded in local Web browser and the content in the ply file is displayed in the view. As all point data is downloaded in the WebGL viewer at client side, the point data can be rotated, translated or scaled in 3D space. The WebGL browser is not supported LOD viewing. Therefore, LOD data set should be downloaded to local storage. The LOD viewer, a point viewer with LOD function is separately distributed in

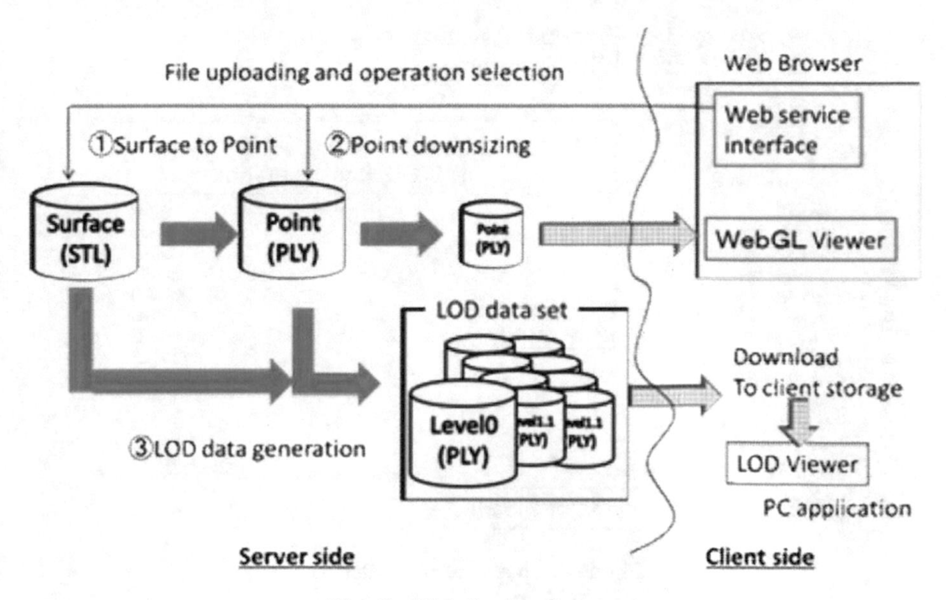

Fig. 7. Web Service diagram

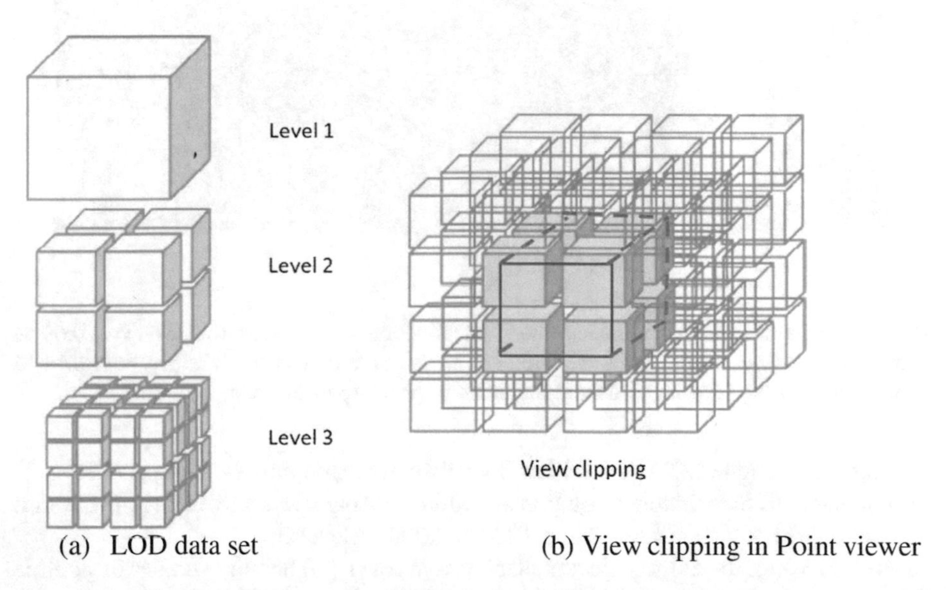

Fig. 8. LOD data set and LOD implementation to the point viewer.

free by our project. The LOD viewer installed in local in advance can display the LOD data set. Several snapshots while the LOD point viewer is working are shown in Fig. 10.

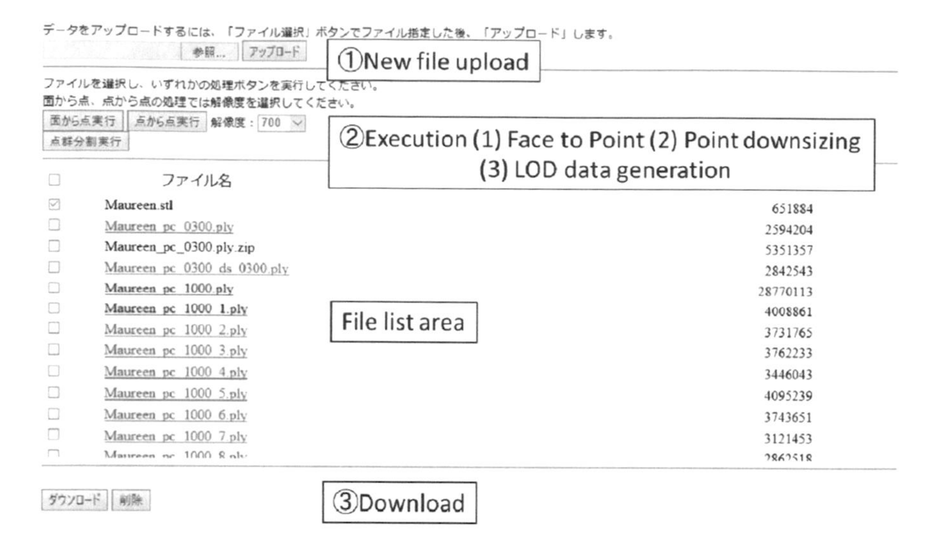

Fig. 9. Web service user interface

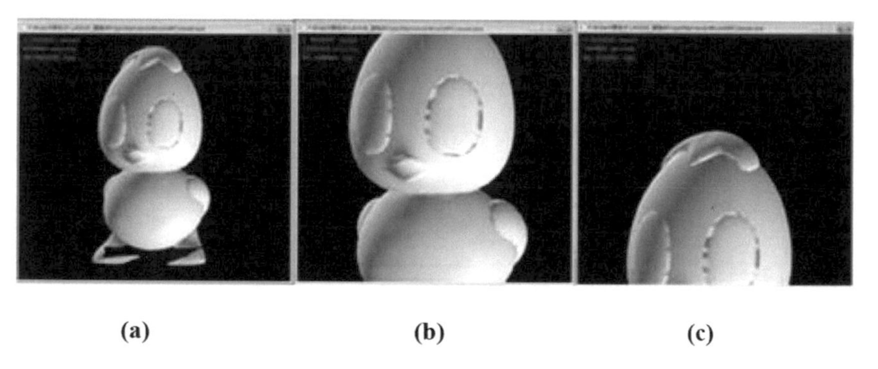

Fig. 10. Point viewer that was implemented LOD function. (a) Full view with level = 1, 0.68 M points (b) Zoom view with level = 2, 1.52 M points (c) Partial view with level = 2, 0.74 M points. The data was downloaded from the home page of AppliCraft Co., Ltd. [10]

Here, the sample LOD data set has 3 levels octree structure, in the case of level 1, the parameter of the window resolution at point capturing was set to 500 × 500. And it was set to 750 × 750 at level 2 and 1000 × 1000 at level 3.

In Fig. 10(a), the entire object is displayed at level 1. Then the number of point is about 0.68 million points. In Fig. 10(b), as the object is zoomed in, the level is set to 2 and the fine point data is loaded and displayed. The number of point is increased to 1.52 million points. However, in Fig. 10(c), most of the objects are out of view. The LOD viewer excludes points outside the view frustum, so the number of points to be displayed is reduced to about 0.74 million points. In this way, the LOD viewer displays the point data without gap even the object is zoomed in, and has a role of suppressing consumption of graphics memory.

5 Conclusion

Aiming to handle 3D geometric models as point data uniformly, we have been developing the tools which convert surface data to point data and downsize point data. In order to demonstrate the practicality of the concept, we are developing a web-based system that provides those tools as a service. In addition to the two functions: the surface-to-point conversion and the downsizing point data, the service includes LOD data set generation and the LOD viewer that effectively display the LOD data set. We are expecting that these tools will be utilized for sharing large-scale 3-D geometric data.

Acknowledgments. This research was supported by KAKENHI 17K00162.

References

1. Miyachi, H.: Data reduction of point data by using rendering images from multi directions. Trans. JSCES **2018**(1), 20181001–20181007 (2018). (in Japanese)
2. Miyachi, H.: Data reduction by points extraction in consideration of viewpoint positions of observer. Trans. VSJ **36**(8), 40–45 (2016). (in Japanese)
3. Miyachi, H: Quality evaluation of 3D image represented by points. In: The 21st International Symposium on Artificial Life and Robotics, pp. 686–689 (2016)
4. Miyachi, H: Quality evaluation of the data reduction method by point rendering. In: Proceedings of the Third International Symposium on Bio Complexity 2018 (ISBC 3rd 2018), pp. 986–990 (2018)
5. Heckbert, P.S., Garland, M.: Survey of Polygonal Surface Simplification Algorithms, Technical Report CMU-CS, Carnegie Mellon University (1997)
6. Szeliski, R.: Rapid octree construction from image sequences. Graph. Models Image Underst. **58**(1), 23–32 (1993)
7. Humphreys, G., Houston, M., Ng, R., Frank, R., Ahern, S., Kirchner, P.D., Klosowski, J.T.: Chromium: a stream processing framework for interactive graphics on clusters. ACM Trans. Graph. (SIGGRAPH 2002) **21**(3), 693–703 (2002)
8. Miyachi, H., Oshima, M., Ohyoshi, Y., Matsuo, T., Tanimae, T., Oshima, N.: Visualization PSE for multi-physics analysis by using OpenGL API fusion technique. In: Proceedings of IEEE 1st International e-Science and Grid Computing, pp. 530–535 (2005)
9. Clark, J.H.: Hierarchical geometric models for visible surface algorithms. Commun. ACM **19**(10), 547–554 (1976). https://doi.org/10.1145/360349.360354
10. AppliCraft Co. Ltd. Home page. https://www.applicraft.com/d3c/

Proposal of a Virtual Traditional Crafting System Using Head Mounted Display

Tomoyuki Ishida[1(✉)], Yangzhicheng Lu[2], Akihiro Miyakwa[3], Kaoru Sugita[1], and Yoshitaka Shibata[4]

[1] Fukuoka Institute of Technology, Fukuoka, Fukuoka 811-0295, Japan
{t-ishida, sugita}@fit.ac.jp
[2] Ibaraki University, Hitachi, Ibaraki 316-8511, Japan
18nm742y@vc.ibaraki.ac.jp
[3] Nanao-City, Nanao, Ishikawa 926-8611, Japan
a-miyakawa@city.nanao.lg.jp
[4] Iwate Prefectural University, Takizawa, Iwate 020-0693, Japan
shibata@iwate-pu.ac.jp

Abstract. Due to the spread of computers and Internet technologies in recent years, the traditional craft industry is presenting information using personal computer terminals for the purpose of market development. However, consumers must rely on imagination of the tastefulness, texture and scale of traditional crafts because information is presented on a flat display. Therefore, in this research, we have constructed the high presence immersive virtual traditional crafting presentation system. This system provides the high presence immersive virtual traditional crafting presentation experience by fusing "Japanese" and "Western". Furthermore, this system provides collaborative work functions by realizing remote sharing of space using network technology.

1 Introduction

In previous research, we have realized a high presence immersive VR presentation system by constructing the CAVE version and the tiled display version virtual traditional crafting presentation system [1–5]. However, these systems require large-scale facilities, and installation costs and running costs are very expensive. On the other hand, due to rapid broadbandization and development of communication technology in recent years, cellular phones have spread rapidly. Currently, mobile terminals such as smartphones and tablets with high processing capability are widely spread. Therefore, we have constructed a mobile version virtual traditional crafting presentation system in previous research [6, 7]. In this mobile version, we realized an AR presentation system that can be easily used on mobile terminals. However, users cannot experience a high presence presentation space using this mobile version.

L. Barolli et al. (Eds.): NBiS 2018, LNDECT 22, pp. 476–484, 2019.
https://doi.org/10.1007/978-3-319-98530-5_40

2 Research Objective

In this research, we construct the high presence immersive virtual traditional crafting presentation system based on an inexpensive head mounted display (HMD) as a platform. This system provides the high presence immersive virtual traditional crafting presentation experience by constructing an interior space that fusion of Japanese and Western using virtual reality technology. Furthermore, this system provides a space sharing function enabling remote users to cooperatively work in virtual space.

3 System Architecture

The architecture of this system is shown in Fig. 1. Traditional Crafting Space Control Function consists of User Activity Control Manager, Left Controller Manager, Right Controller Manager, Object Exchange Control Manager, Object Movement Control Manager, User Teleport Control Manager, Camera Rig, and Avatar Control Manager. Multi-user Space Sharing Function consists of Network Interface, Network Manager, and Network Identity Manager. Traditional Crafting Basic Space consists of 3D Presentation Scene, Mutable Objects, and Immutable Objects.

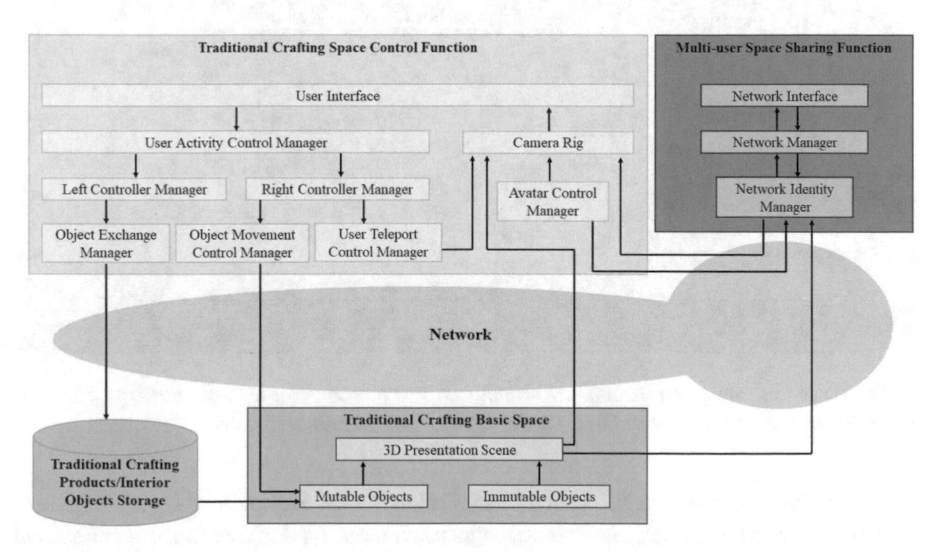

Fig. 1. The system architecture in our proposed virtual traditional crafting system. The system architecture of this system consists of Traditional Crafting Space Control Function, Multi-user Space Sharing Function, Traditional Crafting Basic Space, and Traditional Crafting Products/ Interior Objects Storage.

4 Prototype System

In this research, we use the head mounted display to provide virtual presentation of traditional crafts to users. The constructed virtual traditional crafting presentation system consists of a system start screen, a network connection screen, and a VR space presentation screen.

Two buttons "Traditional Craft Exhibition" and "EXIT" are placed in the system start screen. When the user selects the "Traditional Craft Exhibition" button, the screen transitions from the system start screen to the network connection screen. Also, When the user selects the "EXIT" button, the present system is terminated. The system start screen is shown in Fig. 2.

Fig. 2. After launching the virtual traditional crafting presentation system, the Unity logo is displayed for the first time, and after 4 s the system start screen is displayed.

The initial screen of the VR space presentation screen is shown in Fig. 3. The user can perform operations such as rotation and movement of the viewpoint, change and movement of the object.

This system provides the user with a traditional craft basic space composed of Japanese style houses and Japanese gardens. The Japanese-style house is a two-story structure, and the user can experience the whole inside of the first floor and the second floor. We also arranged space in accordance with modern lifestyles such as living room, kitchen, washroom, toilet and bathroom. The structure of the first floor of Japanese-style house is shown in Fig. 4.

The structure of the second floor of Japanese-style house is shown in Fig. 5.

Fig. 3. The VR space presentation screen provides the user with the traditional craft basic space and the avatar of the user through the head mounted display.

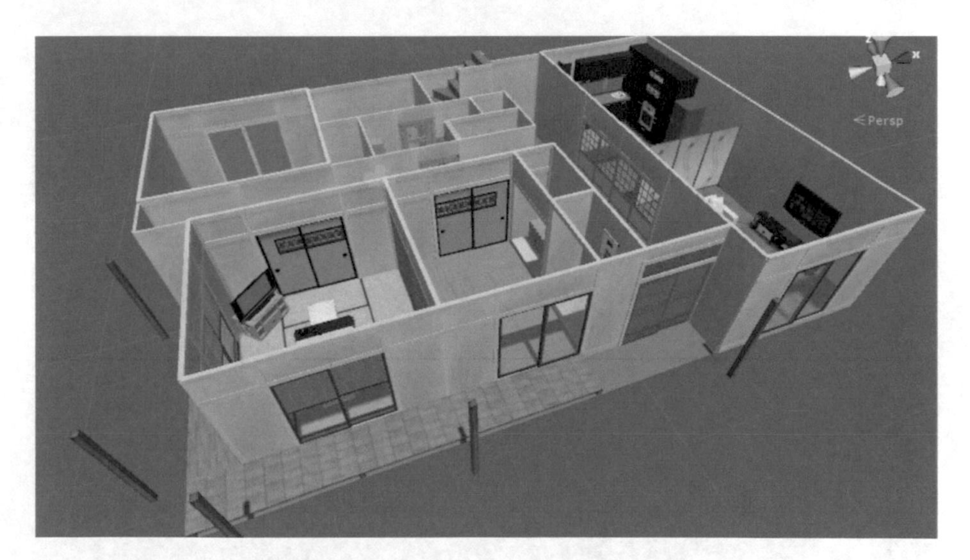

Fig. 4. The first floor of Japanese-style house is a Japanese traditional structure such as the Zashiki (Japanese-style drawing room), Chashitsu (tea room), Engawa (veranda), and entrance.

Users can experience the garden space that fusions Japanese traditional culture with modern lifestyle. The whole space of the Japanese garden is shown in Fig. 6.

In order to add the interior object, the user presses the trackpad of the left controller and displays the "Space Movement/Object Addition" panel. Thereafter, the user selects the "Object" button and displays the interior object panel. By the user selecting an arbitrary interior object, it is possible to add the interior object to the space. The addition of the interior object is shown in Fig. 7.

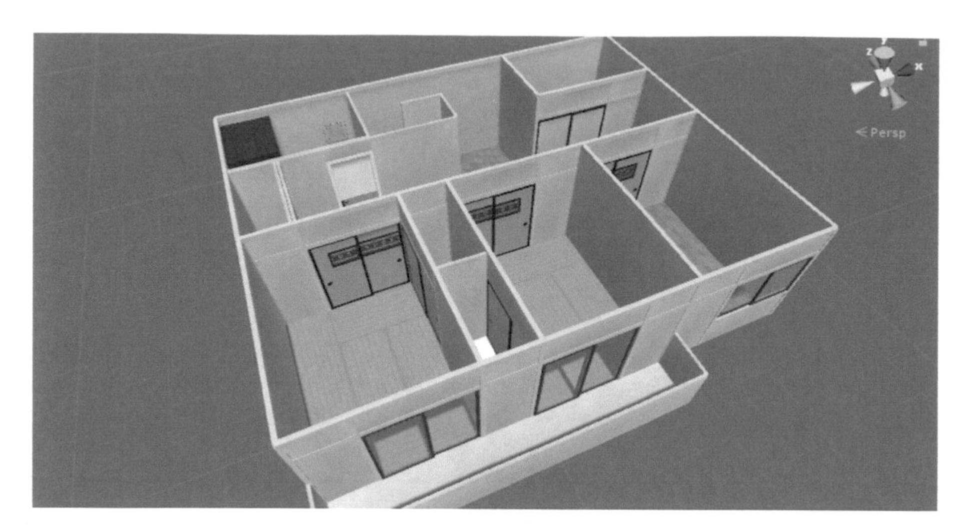

Fig. 5. The second floor of a Japanese style house consists of Western style room, Japanese style room, closet, veranda, toilet and shower room.

Fig. 6. The Japanese garden consists of walls, gates, warehouses, ponds, garden stones and plants to express the traditional Japanese garden space. In addition, we arranged parking lots and car objects tailored to modern lifestyles.

In order for the user to open and close the door object, the following operation is performed.

- STEP1. Point the door object with the controller from either the left or right.
- STEP2. While holding down "Trigger" of the controller, move the controller back and forth.

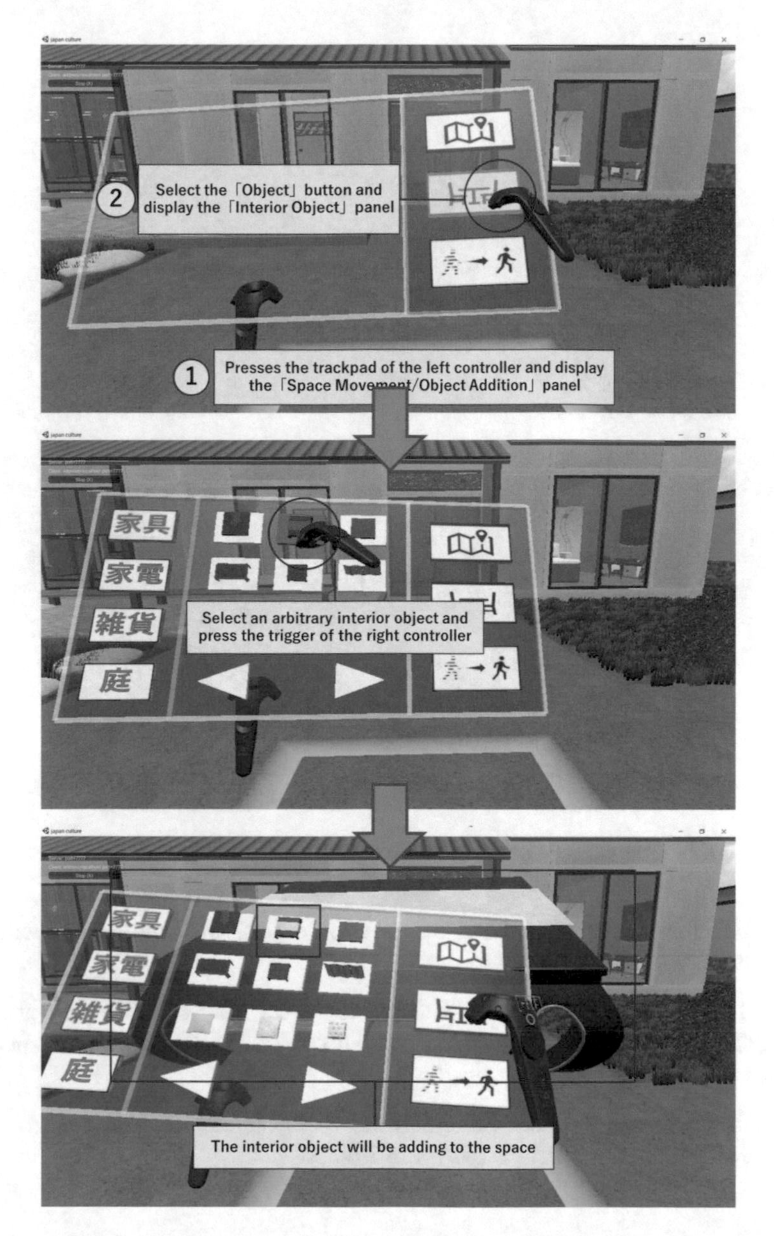

Fig. 7. The user can add an interior object constituting the space.

By this operation, the user can freely open and close the door object. The opening and closing operation of the door object is shown in Fig. 8.

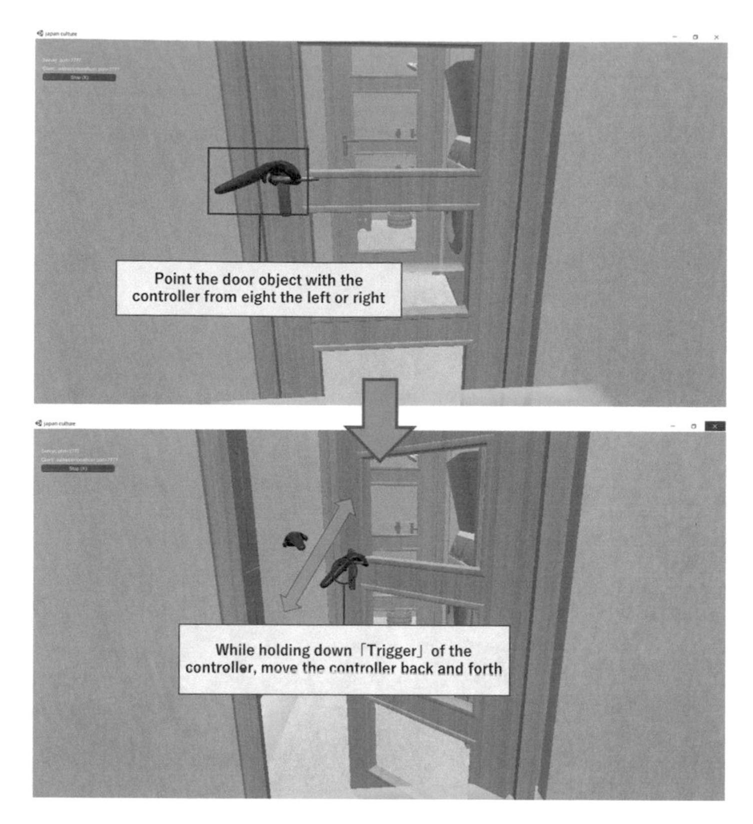

Fig. 8. The user can open and close a door object arranged in the space.

In order for the user to move the interior object, the following operation is performed.

- STEP1. Point the interior object with the controller from either the left or right.
- STEP2. While holding down "Grip" of the controller, move the controller in an arbitrary direction.

By this operation, the user can freely move the interior object. The movement operation of the interior object is shown in Fig. 9.

Fig. 9. The user can move the interior object arranged in the space.

5 Conclusion

In this research, we constructed the high presence immersive virtual traditional crafting presentation system using the head mounted display. This system consists of the system start screen, the network connection screen, and the VR space presentation screen. The user can perform operations such as rotation and movement of the viewpoint, change and movement of the object. The VR space consists of Japanese style houses and Japanese gardens, and users can experience various traditional craft interiors.

Acknowledgments. This work was supported by the research grand of The Telecommunications Advancement Foundation (TAF) of Japan.

References

1. Ishida, T., Miyakawa, A., Shibata, Y.: Virtual traditional Japanese crafting system using JGN II. In: Collaborative Work Environment. Proceedings of the IEEE 19th International Conference on Advanced Information Networking and Applications, pp. 229–234 (2005)
2. Ishida, T., Miyakawa, A., Ohhashi, Y., Shibata, Y.: Method of sharing virtual traditional crafting system based on high-speed network. In: Proceedings of the IEEE 20th International Conference on Advanced Information Networking and Applications, pp. 665–672 (2006)
3. Ishida, T., Yatsu, K., Shibata, Y.: Tele-immersion environment for video Avatar based CVE. In: Proceedings of the 12th International Conference on Network-Based Information Systems, Indianapolis, pp. 608–611 (2009)
4. Ishida, T., Shibata, Y.: Proposal of tele-immersion system by the fusion of virtual space and real space. In: Proceedings of the 13th International Conference on Network-Based Information Systems, pp. 408–413 (2010)
5. Ishida, T., Sakuraba, A., Shibata, Y.: Proposal of high realistic sensation system using the large scale tiled display environment. In: Proceedings of the 14th International Conference on Network-Based Information Systems, pp. 444–449 (2011)
6. Iyobe, M., Ishida, T., Miyakawa, A., Sugita, K., Uchida, N., Shibata, Y.: Proposal of a virtual traditional Japanese crafting presentation system mobile edition. In: Proceedings of the 10th International Conference on Innovative Mobile and Internet Services in Ubiquitous Computing, pp. 120–125 (2016)
7. Iyobe, M., Ishida, T., Miyakawa, A., Sugita, K., Uchida, N., Shibata, Y.: Development of a mobile virtual traditional crafting presentation system using augmented reality technology. Int. J. Space Based Situat. Comput. (IJSSC) **6**(4), 239–251 (2017)

Development of AR Information System Based on Deep Learning and Gamification

Tetsuro Ogi[1(✉)], Yusuke Takesue[1], and Stephan Lukosch[2]

[1] Graduate School of System Design and Management, Keio University, 4-1-1 Hiyoshi, Kouhoku-ku, Yokohama, Japan
ogi@sdm.keio.ac.jp, y.takesue@keio.jp

[2] Faculty of Technology, Policy and Management, Delft University of Technology, Jaffalaan 5, 2628 BX Delft, The Netherlands
s.g.lukosch@tudelft.nl

Abstract. Recently, several AR systems have been developed and used in various fields. However, in most AR systems, there are some restrictions caused by the usage of AR marker or location information. In this research, in order to solve these problems, AR information system that can recognize object itself based on deep learning was developed. In particular, this system was constructed using client-server model so that the machine learning can be updated while operating the system. In addition, the method of gamification was introduced to gather the learning data automatically from the users when they use the system. The prototype was applied to the AR zoo information system and the effectiveness of the proposed system was validated in the evaluation experiment.

1 Introduction

In recent years, several information systems using AR technology have been developed and used in various fields such as museum, entertainment, and tourism [1–3]. In these systems, AR markers, registered images, or location information are used as the method of recognizing target objects in the real world. In the method using AR marker, it may not be permitted to attach the marker to the target according to the places or objects. When the registered image is used as a marker, the user must look for the marker image and it cannot be used if the shape or the posture of the target is changed. In addition when the location information measured by GPS or beacon is used to specify the target, there is limitation in accuracy and it cannot be used for movable objects. In order to solve these problems, it is necessary to introduce a new method that identifies the target object itself.

On the other hand, AI technology such as deep learning is often applied to recognize objects as practical uses [4, 5]. This method can recognize object even if the target object moves or changes posture, because it can learn the concept of the object by performing machine learning based on huge data sets.

Therefore, in this research, AR information system that can recognize the target object itself was developed by using the deep learning method. In the following chapters, concept of the deep learning based AR information system, system

L. Barolli et al. (Eds.): NBiS 2018, LNDECT 22, pp. 485–493, 2019.
https://doi.org/10.1007/978-3-319-98530-5_41

configuration, application to zoo information system, operation using gamification, and evaluation experiment are described.

2 Deep Learning Based AR System and Related Works

In this research, the deep learning was applied to the AR information system as a method of recognizing target object itself without using an AR marker. The deep learning is one of the methods of machine learning, and it can identify target object by performing learning for multilayered neural network based on huge data set. In particular, CNN (convolutional neural network) is known to be effective neural network for the image recognition [6, 7]. In addition, though the deep learning requires a large computation load in training process, the computation load in testing process using the learned network model is small. Then it can be executed even on mobile devices. For example, the application of Aipoly Vision by Poly Ai Inc. and the software framework of Core ML by Apple Inc. provide interactive image recognition function based on the deep learning on the smartphone [8, 9].

However, in the case of the smartphone application, the neural network model of the deep learning cannot be updated while it is used, because the learned model of the neural network is installed beforehand. Therefore, this study aims at developing a deep learning based information system using a client-server model in which the smartphone executes application program while communicating with the server. As a platform of the deep learning function using the client-server model, Bluemix developed by IBM corporation is often used, but this system cannot be performed interactively because it needs the processes of transmitting captured image and receiving result between client and server [10].

In order to develop an AR information system based on the deep learning, the function of real-time interaction using the client-server architecture is very important. In particular, in the AR information system, it is necessary not only to identify the target object, but also to display the explanation information for the target by retrieving it from the database. Therefore, as a system configuration, it is required that the entire system executes in real time, including the data transmission among smartphone, deep learning server, and information database. Figure 1 shows the concept of the proposed AR information system based on the deep learning. The feature of this system is that the AR information system is constructed as an interactive system as well as using the client-server model based on the deep learning.

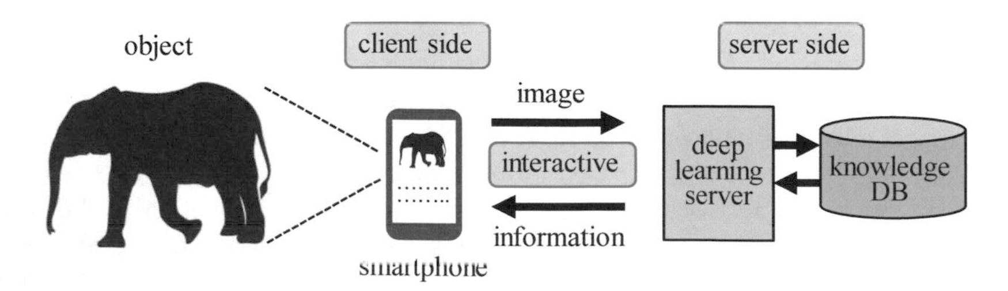

Fig. 1. Concept of deep learning based AR information system

3 System Construction

In this research, AR information system that runs on the user's smartphone while communicating with the deep learning server through the Internet was constructed. As a server for the deep learning, HPCT W114gs workstation (CPU: Intel Xeon E5-1620, GPU: TITAN X ×2, memory: 128 GB) of HPC TECH corporation was used, and as a software tool for the deep learning, Caffe was used. Caffe is an open source software developed at Berkeley Vision and Learning Center of the University of California, Berkeley, and especially it has been used successfully in the applications to image recognition using the CNN model [11].

In this system, a three-layered CNN structure shown in Fig. 2 was used as the neural network for image recognition. Each layer consists of a convolution layer, a pooling layer, and a ReLU layer, and the images of 64 × 64 pixels were inputted. The number of filters for feature extraction used in each convolution layer was 32, 32, and 64, respectively. The number of nodes in the output layer was 10, so that the network model can identify ten kinds of objects. The execution program of the CNN using Caffe was written in Python.

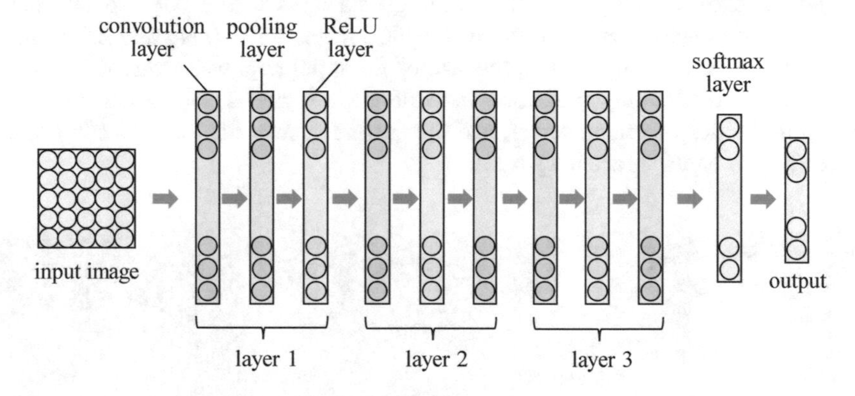

Fig. 2. Three-layered CNN structure used in this research

As a smartphone used on the client side, Apple's iPhone was used and the application software on the smartphone was developed using Swift 4.0. Images captured by the camera of the smartphone are sequentially transmitted to the deep learning server using HTTP protocol and they are overwritten and saved as an image file. In the deep learning server, when a new image is saved, it is inputted to the identification program of Caffe, and the result is outputted as a text file. On the client side, after transmitting the captured image to the deep learning server, it accesses to the server again to require AR information. In the deep learning server, when the client accesses to it, the php program reads the output result of Caffe and related information is retrieved from the database according to the identification result. This information is sent back to the smartphone and it is displayed under the camera image as AR information.

In this system, it is necessary to perform all the processes in real-time that includes capture of image by the smartphone, image recognition in the deep learning server, data retrieval from the database, and display of AR information on the smartphone. Therefore, the image was captured as small size of 240 × 320 pixels and it was transmitted to the server every one second so that the entire process is not delayed.

4 Application to AR Zoo Information System

As an application of the above mentioned deep learning based AR information system, the prototype of the AR zoo information system was developed. In developing the prototype system, 10 kinds of animals (Indian elephant, Lion, Red kangaroo, Red panda, Golden takin, Polar bear, South African fur seal, Humboldt penguin, Capercaillie, White eared-pheasant) were selected, and 120 photographs were taken for each animal. In order to increase the learning data, several image processing methods such as enlargement (1.2 times, 1.5 times), clockwise and counterclockwise rotation of 30°, flip, left and right movement, up and down movement, and blurring were used, and in total, 12,000 images were prepared as learning data set.

Figure 3 shows examples of the learning data used in the AR zoo information system. These images were divided into 10,000 images for training and 2,000 images for testing. In the training process, the size of the input data was reduced to 64 × 64 pixels, and 5,000 training processes were iterated. Figure 4 shows the progress of learning in the deep learning server, and the correct answer rate in the testing process was converged to 96% (accuracy = 0.96).

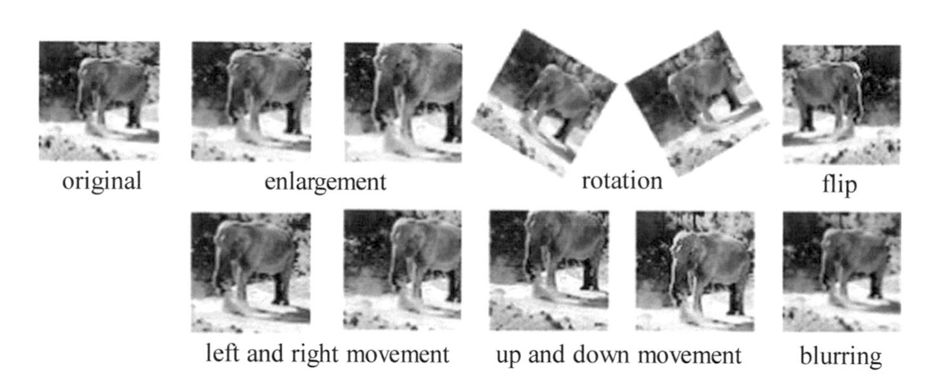

Fig. 3. Increased learning data set by various image processing methods

Figure 5 shows an example of the screen images of the application running on the smartphone. When the user starts the application (Fig. 5(a)) and taps the displayed image, the application is changed from off line mode to AR mode, and the captured image is sent to the deep learning server at one second intervals. Next, when the animal is identified on the deep learning server, the corresponding explanation information is

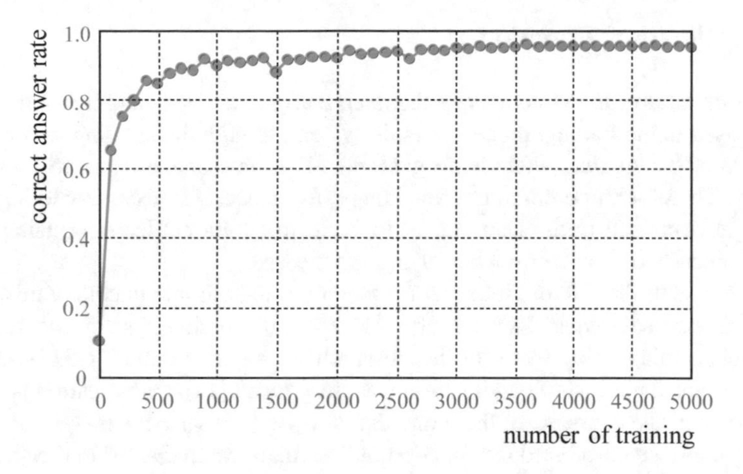

Fig. 4. Progress of learning in training process of deep learning

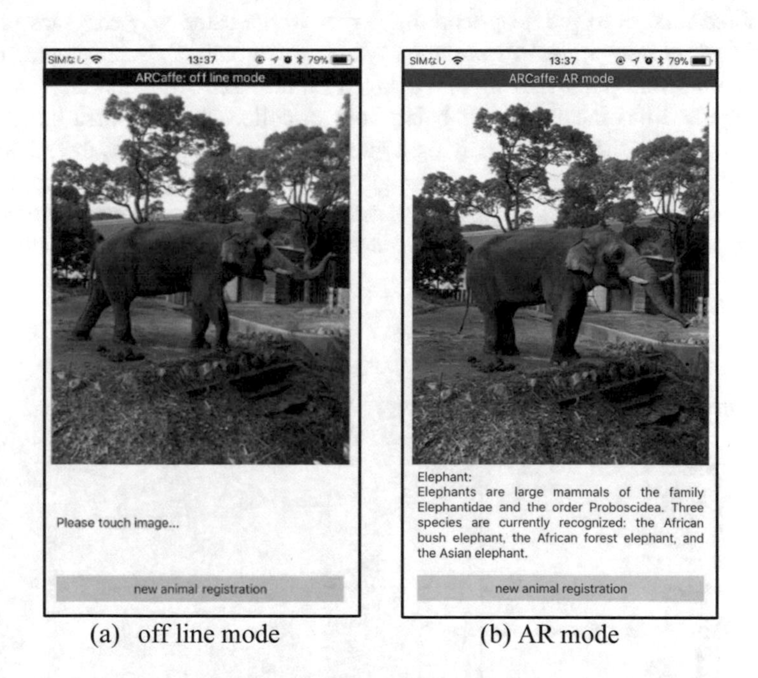

Fig. 5. Screen images of the AR zoo information system

retrieved from the database and it is displayed under the image (Fig. 5(b)). Thus, the user in the zoo can easily refer to the explanation information of the animal existing before his or her eyes through the smartphone.

5 Gamification Function

The performance of the system using the deep learning depends largely on the learning data set used in the training process. In this system, though the learning was performed using 12,000 image data, there is no guarantee that new image data can be identified correctly. Therefore, in order to improve the performance, it is necessary to increase the number of data used in the learning, or to perform additional learning using the data that was classified incorrectly while using the system.

In the case of the AR zoo information system, it is desirable to collect misclassified data from the users when they use the AR zoo information system and to perform additional learning using them. In this research, in order to collect misclassified data from the users, a method of gamification was introduced [12]. Gamification is a method that introduces the elements of the game that attracts the user's interest, to other fields. In this system, we introduced the function of "Animal Search Game" to the smartphone applications. This is a function of providing points to the user when the user finds a new animal that the system cannot identify. By using this function, users can send misclassified images to the deep learning server in the same way they are playing a game of finding new animal in the zoo. When a new animal is registered, the user receives points, and point ranking is displayed on the user's smartphone.

This method has the feature of being able to collect learning data automatically from the users with the sense of playing a game, and especially it has the advantage to intensively collect misclassified data. Figure 6 shows the system configuration to which the function of the gamification is added. Figure 7 shows an example of the screen images for the animal registration function in the smartphone application. When

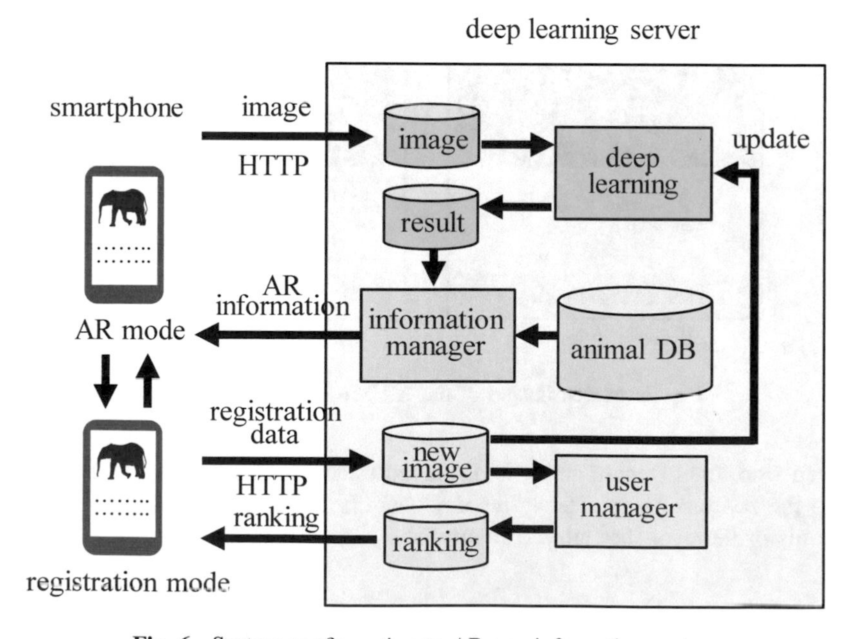

Fig. 6. System configuration to AR zoo information system

(a) animal registration

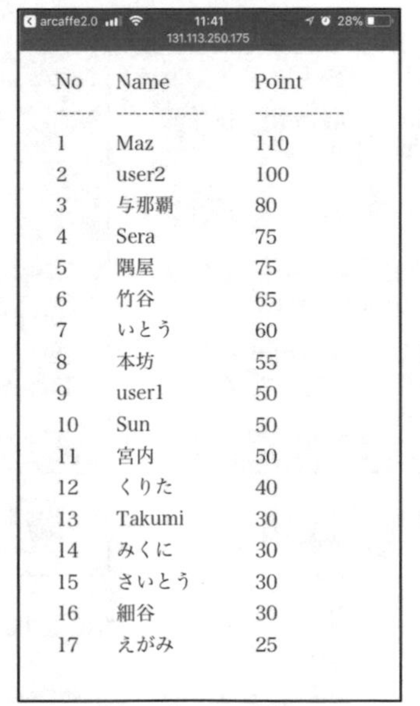

No	Name	Point
1	Maz	110
2	user2	100
3	与那覇	80
4	Sera	75
5	隅屋	75
6	竹谷	65
7	いとう	60
8	本坊	55
9	user1	50
10	Sun	50
11	宮内	50
12	くりた	40
13	Takumi	30
14	みくに	30
15	さいとう	30
16	細谷	30
17	えがみ	25

(b) point ranking

Fig. 7. Screen images of animal registration function

the image registration button is pressed on the smartphone, the screen is transited to the animal registration screen (Fig. 7(a)), and when the name of the animal is inputted and the registration is completed, a new point ranking list is displayed (Fig. 7(b)).

6 Evaluation Experiment

In this research, in order to evaluate the proposed AR information system based on deep learning, we conducted an evaluation experiment in the zoo of Zoorasia in Yokohama as shown in Fig. 8. The number of subjects in the experiment was 17, and

Fig. 8. Using the AR zoo information system in the experiment

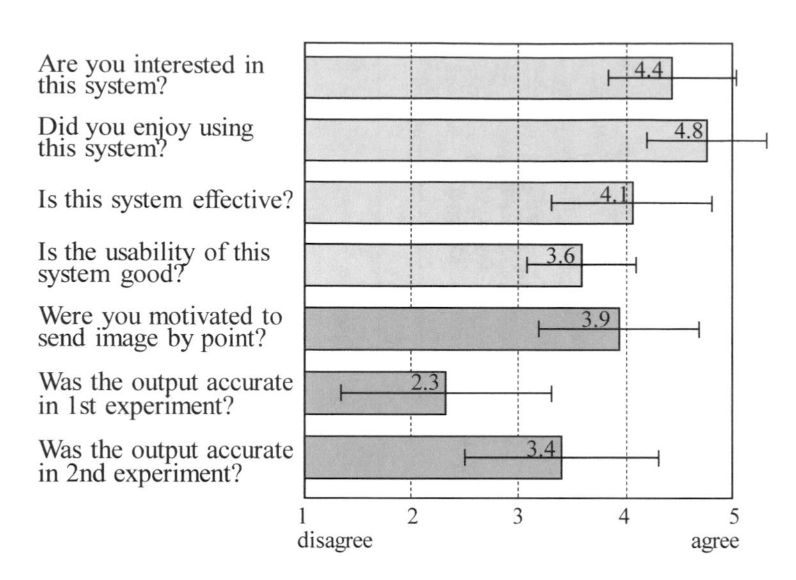

Fig. 9. Result of questionnaire in evaluation experiment

they were asked to freely walk around the zoo using the AR zoo information system for 30 min. The experiment was conducted using 2 days. On the first day, the experiment was conducted with 12 subjects, and after updating the deep learning network using the image data registered from the subjects, the experiment on the 2nd day was conducted with 5 subjects.

As the evaluation, the subjects were asked to answer the questionnaire after using the system. Figure 9 shows the results of the questionnaire answered using 5-point Likert scale. From the result, we can see that though the average of the answers for the question about "usability" was 3.6 and it should be improved, the answers for "interest", "enjoy" and "effectiveness" were 4.4, 4.8, 4.1, respectively and they are high. In addition, since the answer for "accuracy" in the animal identification was improved from 2.3 to 3.4 after the update of learning and the answer for "motivation" of image registration was 3.9, it is considered that the effect of the gamification was shown. In fact, during the usage in 30 min, the subjects registered 10.8 images on the average. Since more than one million people visit this zoo in one year, if 1% of visitors used this system, 100,000 learning data could be collected in one year. This result shows that it is possible to continuously improve the performance of the system by introducing the method of the gamification.

7 Conclusions

In this research, in order to realize the marker-less AR system, deep learning based AR system that can recognize target object itself was developed. In particular, in order to update the learning automatically while operating the system, client-server architecture was used and the method of gamification was introduced. A prototype of AR zoo information system was developed and the result of the evaluation experiment showed the effectiveness of the proposed system. The future work will include increasing the number of animals that can be identified and extending the displayed information to realize the practical use of the proposed system.

References

1. Grafe, M., Wortmann, R., Westphal, H.: AR-based interactive exploration of a museum exhibit. In: The First IEEE International Workshop Augmented Reality Toolkit (2002)
2. Rahman, H.R., Herumurti, D., Kuswardayan, I., Yuniarti, A., Khotimah, W.N., Fauzan, N.B.: Location based augmented reality game using Kudan SDK. In: 11th International Conference on Information and Communication Technology and System (ICTS), pp. 307–310 (2017)
3. Lim, Y., Park, Y., Heo, J., Yang, J., Kang, M., Byun, Y.-C.: A smart phone application based on AR for Jeju tourism. In: First ACIS/JNU International Conference on Computers, Networks, Systems and Industrial Engineering, pp. 271–272 (2011)
4. Luckow, A., Cook, M., Ashcraft, N., Weill, E., Djerekarov, E., Vorster, B: Deep learning in the automotive industry: applications and tools. In: International Conference on Big Data (Big Data), pp. 3759–3768 (2016)
5. Khan, S., Yong, S.-P.: A deep learning architecture for classifying medical images of anatomy object. In: Asia-Pacific Signal and Information Processing Association Annual Summit and Conference (APSIPA ASC), pp. 1661–1668 (2017)
6. Saatci, E., Tavsanoglu, V: Multiscale handwritten character recognition using CNN image filters. In: International Joint Conference on Neural Networks (IJCNN 2002), pp. 2044–2048 (2002)
7. Han, C., Tao, X., Duan, Y., Liu, X., Lu, J: A CNN based framework for stable image feature selection. In: Global Conference on Signal and Information Processing (GlobalSIP), pp. 1402–1406 (2017)
8. Aipoly Vision, Poly Ai Inc. https://www.aipoly.com/
9. Core ML, Apple Inc. https://developer.apple.com/documentation/coreml
10. Chanayot, S., Ogi, T.: Development of augmented reality system based on machine learning. In: 21st Annual Conference of the Virtual Reality Society of Japan, pp. 12E-03 (2016)
11. Cengil, E., Çınar, A., Özbay, E: Image classification with Caffe deep learning framework. In: International Conference on Computer Science and Engineering (UBMK 2017), pp. 440–444 (2017)
12. Thogersen, R.: Data quality in an output-agreement game: a comparison between game-generated tags and professional descriptors. In: International Conference on Collaboration and Technology (CRIWG 2013). Lecture Notes in Computer Science, vol. 8224, pp. 126–142. Springer (2013)

A Study on Practical Training with Multiple Large-Scale Displays in Elementary and Secondary Education

Yasuo Ebara[1(✉)] and Hiroshi Hazama[2]

[1] Academic Center for Computing and Media Studies, Kyoto University, Kyoto, Japan
eba@viz.media.kyoto-u.ac.jp
[2] Seisho High School, Osaka, Japan
hiroshi.ha@gmail.com

Abstract. To promote effective ICT utilization in the practical training to learn the basics of the industry subjects in elementary and secondary education, we believe that it is important to study the effective utilization of large-scale display systems. In this paper, we studied on the effective practical training by utilization of multiple large-scale display systems in conjunction with the interface for screen operation developed in our previous study. As the approach, we tried the practical training with g multiple large-scale displays using the interface for screen operation in combination with the SAGE2. From this trial, we showed that there is a high possibility that the effective practical training realize to conduct by combining the interface for screen operation and multiple displays environment.

1 Introduction

In the field of elementary and secondary education, the skill training to make a choice from a large amount of information and to utilize Information Communication Technology (ICT) such as computer and network as the effective method for information expression and communication is required. To promote the ICT utilization in the classroom of elementary and secondary education, the effective applications of audiovisual teaching material and educational display systems is important [1,2]. However, there are few cases where ICT is utilized in the practical training to learn the basics of the industry subjects (hereinafter referred to as "practical training") in elementary and secondary education. Therefore, we are trying various means to promote effective ICT utilization in the practical training.

In the practical training, it is fundamental to always use both hands in training using physical teaching materials such as parts, tools and equipment, and in explaining the outline of training contents by a teacher [3]. However, when switching from the outline explanation of training by electronic materials with a computer to the explanation of training using the physical thing, a teacher need

L. Barolli et al. (Eds.): NBiS 2018, LNDECT 22, pp. 494–502, 2019.
https://doi.org/10.1007/978-3-319-98530-5_42

to put the interface device for computer operation held in one hand at another place. By its action, when the teacher instructs by alternately presenting students with explanatory screen of the training and the physical teaching materials while maintaining the concentration of students, he/she miss the timing of important explanations.

To solve this problem, we have proposed and implemented the interface for screen operation which is possible to operate electronic teaching materials on the screen while walking around freely within the classroom and working with both hands by wearing the interface device in one hand [4]. In addition, we conducted practical training using ICT by using the interface for screen operation and evaluated the effectiveness of the proposed interface. From the evaluation results, we have shown the effectiveness of the proposed interface for screen operation [5].

On the other hands, in recent years, the installation of large-scale display systems such as projectors and electric boards in elementary and secondary education is progressing, there are the study cases of utilization in class by displaying the electric teaching materials on them. We believe that the effective utilization of large-scale display systems lead to the spreading of ICT utilization in the practical training. In this paper, we study on the effective practical training by utilization of multiple large-scale display systems in conjunction with the interface for screen operation developed in our previous study.

2 Related Works

As effective application of the large-scale display system, various technique of information utilization by simultaneous display of enormous multimedia contents by utilizing the ultra-wide display area have been studied. We have focused on the tiled display technology which enable to display enormous multimedia contents by configuring an ultra-wide display with two or more LCD panels to construct effective wide-area screen system [6,7].

In our previous study, we have constructed the collaborative work environment with tiled display wall [8], and have executed the cooperative class as the application in the field of education [9]. However, the tiled display wall exists the border area by bezel between each LCD from specific reasons of the system configuration. Therefore, as a part of audiovisual teaching material is displayed hiding by the border area, it is pointed out the problem which interfere with the browsing of contents and each work by viewers.

Recently, the low prices of high-definition LCD panel have progressing, 4K/8K resolution display system in standard home has been spreading, and the behavior for further spread has also been activated. In addition, the scale up on the screen size and the display resolution of educational display systems such as projectors and electric blackboards has been spreading, many products are on the market in recent years. For example, the electronic blackboard capable of touch panel operation with screen size exceeding 80 in. are sold as products [10,11]. Moreover, the product of the electric blackboard have various functions with ultra-short focus projector that can display in ultra-wide size of more than 100 in. are developed and sold [12,13].

However, from survey results in 2018 by MEXT (the Ministry of Education, Culture, Sports, Science and Technology) [14] the installation rate of electirc blackboard is about 25% of the number of classrooms in elementary and secondary education, there is no significant diffusion situation of electronic blackboards in Japan. Therefore, we consider that the latest high-spec electric blackboards and projectors is difficult to spread at an early date in elementary and secondary education. We study on the method of effective practical training by the utilization of low-spec projector and large-scale LCD display environments located in current classrooms.

3 Trial on Practical Training with Multiple Large-Scale Displays

3.1 Interface for Screen Operation in Practical Training Using ICT

As user interface to operate multimedia contents on the large-scale screens smoothly, the use of mobile terminal such as a smart phone and a tablet PC are a widespread approach to interacting with a large-scale display system [15–17]. However, in the practical training, a teacher is fundamental to always use both hands in training using physical teaching materials and in explaining the outline of training contents. Therefore, it is difficult to operate while viewing various contents on a large-scale display system with a mobile terminal. We consider to be important the development of the interface device that can be operated interactively by users while always viewing the screen of large-scale display system. We have proposed and implemented the interface for the screen operation to solve the requirements in class using large-scale displays in elementary and secondary education [5].

Figure 1 shows the configuration of the interface for screen operation, as designed in our study [18]. Figure 2 shows the body of the interface with the state worn in left hand. Table 1 shows the specifications of the interface for screen operation. A 16-bit one-chip microcomputer (PIC24FJ64GB002) is used in the main body of the interface for screen operation as shown in Fig. 2.

In the interface for screen operation, as shown in Fig. 2, the bracelet is worn on the base of the thumb, and between the base and the wrist of the index finger. The finger button is worn on the index finger and the middle finger. The bracelet has a function of pointing the position on the computer screen with the acceleration sensor (ADXL 345). The acceleration sensor is positioned at the tip of the interface. The finger button has a micro switch on the inside of the index finger and the middle finger, the screen operation by binary signal on/off of the micro switch is performed by pressing with the index finger and thumb or middle finger and thumb.

On the other hands, the explanation time is often prolonged in the usual practical training. Therefore, we have found out the problem which accompanied by a feeling of physical and mental tiredness by continuing the explanation while operating the interface in the actual practical training [5]. To reduce the feeling

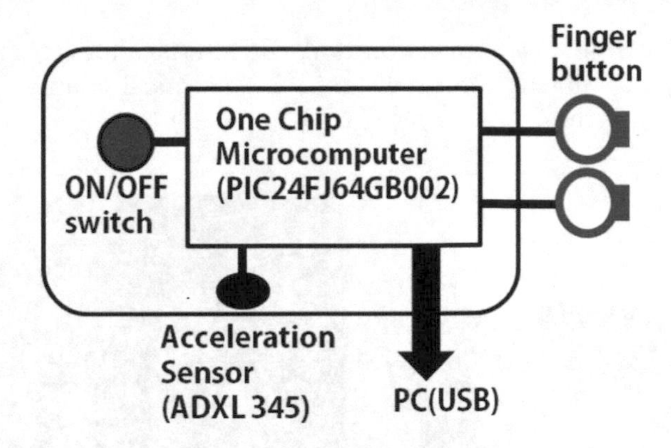

Fig. 1. Configuration of the interface for screen operation

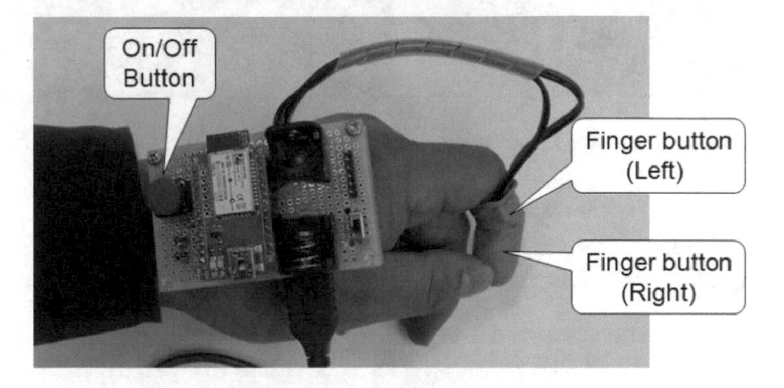

Fig. 2. Body of the interface for screen operation

Table 1. Specification of components to construct the interface for screen operation

Supported OS	Windows 10/8.1/7
PC connection	USB connection
Number of buttons	3
Movement range in room	20 m
Body size	65 [mm] × 50 [mm] × 33 [mm]
Body mass	64 [g]

of tiredness by the pointing operation, this interface for screen operation was designed so that the pointing operation can be performed at the wrist position in a handshake as shown in Fig. 3(a) to reduce that feeling of tiredness. Since the state of this wrist is in the same natural state as when the hand goes down during power down state as shown in Fig. 3(b). We are trying the reduction of tiredness to enable pointing smoothly even with few hand movements by operating in a part ahead of a wrist. We have shown that the interface for screen operation is the possibility of reducing the tiredness in the practical training compared to other interface devices.

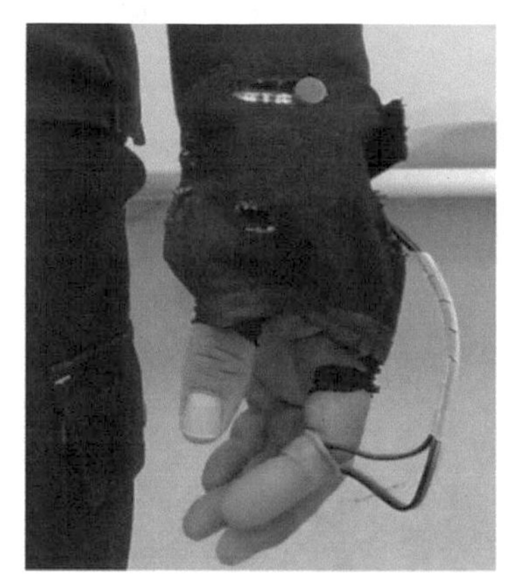

Fig. 3. Wearing situation of the interface for screen operation in natural state of wrist

3.2 Middleware for an Ultra-Wide Display Environment with Multiple Displays

Recently, despite the spread of the products of high spec electric blackboards and projectors, the installation of these systems has not progressed sufficiently in current elementary and secondary education. Therefore, it is considered that most classes are used by combining multiple low spec display systems such as general projector and LCD display. However, when multiple display systems are used in combination, the teacher become more difficult to continue the class smoothly by confusing to operate controller for each system is attached. We believe that it is desirable to use as an ultra-wide desktop environment of a single PC by virtually merging the display areas of these systems.

Many middleware to configure a large-scale screen by merging multiple displays have been proposed. In this study, we apply the SAGE2 (Scalable Amplified Group Environment) [19] was developed by the Electronic Visualization Laboratory. The SAGE2 is cloud-based and web-browser technologies in order to enhance data intensive co-located and remote collaboration by using a SRSD (Scalable Resolution Shared Displays). SAGE2's users can utilize the supported data contents and applications on the display environment to access the SAGE2 UI page with web browser. In addition, the interactive operation for data contents and applications with different types of input devices such as game controllers and touch overlays is supported. Figure 4 shows (a) the SAGE2's architecture and communication scheme, and (b) the example of use as the collaboration work with ultra-wide displays environment by SAGE2 [19].

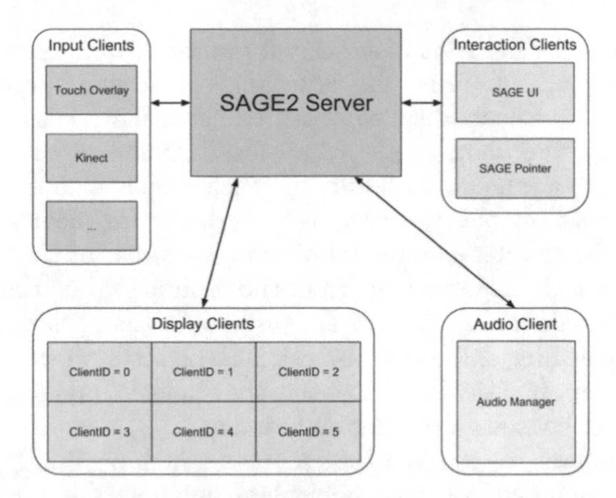

(a)SAGE2's architecture and communication scheme

(b)Example of use as the collaboration work

Fig. 4. Overview of SAGE2 [19]

3.3 Practical Training with Large-Scale Multiple Displays

Next, we examine the approach of the practical training with g multiple large-scale displays using the interface for screen operation in combination with the SAGE2 described in the previous section. In this study, as large-scale displays, we use a LCD display (60 in.) and a general projector located at practical room in Seisho High School (the author belongs). These displays are connected to the PC, an ultra-wide display environment is constructed with the SAGE2. In this display environment, multiple electrical teaching materials to explain the contents of the practical training are simultaneously displayed. Moreover, this display environment enables to perform interactive operation for electronic teaching materials using the interface for screen operation implemented in our previous study.

Figure 5 shows the situation of simulated practical training by a teacher with this display environment. From the situation, a teacher is possible to explain mutually by simultaneously displaying the teaching materials of the content of practical training, the photographs of physical thing and the design drawing on this display environment. In addition, teacher enable to operate freely the display size and position of these teaching materials using the interface for screen operation worn by teacher's one hand at every position of the practical room. We consider that it is easy to switch from the outline explanation the outline of training by electronic materials with a computer to the explanation of training using the physical thing. Moreover, by using the interface for screen operation, teachers can explain practical training near the students, and assume to increase in communication between a teacher and students.

From these points, we consider that there is a high possibility that the effective practical training enables to conduct by combining the interface for screen operation and multiple displays environment. However, the setting and the maintenance to use the SAGE 2 is not easy, and it seems to become a hard work for teachers in high school who do not use ICT. We consider that it is necessary to construct the mechanisms to support these works and can be easily set up as future work.

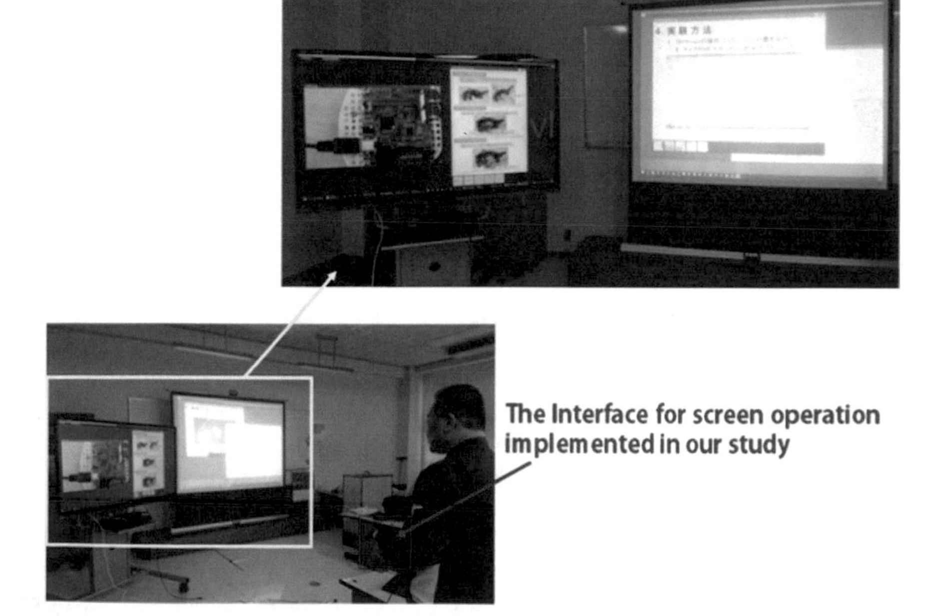

Fig. 5. Body of the interface for screen operation

4 Conclusion

In this paper, we studied on the effective practical training by utilization of multiple large-scale display systems in conjunction with the interface for screen operation developed in our previous study. As the approach, we tried the practical training in multiple large-scale displays environment using the interface for screen operation in combination with the SAGE2. From this trial, we showed that there is a high possibility that the effective practical training enables to conduct by combining the interface for screen operation and multiple displays environment.

As future work, we will study on the effectiveness and the optimal education method by conducting the actual practical training in elementary and secondary education by using this environment.

Acknowledgements. This work was supported by JSPS KAKENHI, Grant-in-Aid for Scientific Research(C) Grant Number JP16K01065.

References

1. MEXT, School course guidelines for high school in Japan - General rule Edition, December 2010. http://www.mext.go.jp/component/a_menu/education/micro_detail/__icsFiles/afieldfile/2010/12/28/1282000_01.pdf. Accessed 31 May 2018
2. MEXT, School course guidelines for high school in Japan - Information Edition, January 2012. http://www.mext.go.jp/component/a_menu/education/micro_detail/__icsFiles/afieldfile/2012/01/26/1282000_11.pdf. Accessed 31 May 2018
3. MEXT, School course guidelines for high school in Japan - Industrial Edition, January 2012. http://www.mext.go.jp/component/a_menu/education/micro_detail/__icsFiles/afieldfile/2010/06/01/1282000_13.pdf. Accessed 31 May 2018
4. Ebara, Y., Hazama, H.: Development and evaluation of operation interface for lesson using large-scale screen in elementary and secondary education. In: Proceedings of the 20th International Conference on Network-Based Information Systems (NBiS 2017), pp. 674–680 (2017)
5. Hazama, H., Ebara, Y., Ogasawara, T.: Development and evaluation of interface for screen operation for smooth practical training using ICT in elementary and secondary education. In: Proceedings of 4th IEEE International Symposium on Educational Technology (ISET 2018), pp. 104–107 (2018)
6. Ni, T., Schmidt, G.S., Staadt, O.G., Livingston, M.A., Ball, R., May, A.: A survey of large high-resolution display technologies, techniques, and applications. In: Proceedings of IEEE Virtual Reality Conference (VR 2006), pp. 223–236 (2006)
7. Jeong, B., Leigh, J., Johnson, A., Renambot, L., Brown, M.D., Jagodic, R., Nam, S., Hur, H.: Ultrascale collaborative visualization using a display-rich global cyberinfrastructure. IEEE Comput. Graph. Appl. **30**(3), 71–83 (2010)
8. Ebara, Y., Kukimoto, N., Koyamada, K.: Tele-immersive environment with tiled displays wall system for remote collaborative work, In: Proceedings of ASIAGRAPH 2008 in Tokyo, vol. 2, no. 2, pp. 130–133 (2008)
9. Ebara, Y.: Tele-immersive environment with tiled display wall for intuitive remote collaborative work. Int. J. Artif. Life Robot. **17**(3–4), 483–487 (2013)

10. Panasonic, Touch screen LCD TH-80BF1J. https://panasonic.biz/cns/prodisplays/products/bf1series/. Accessed 31 May 2018
11. iBoard Japan, StarBoard SN series. http://partner.iboardbiz.com/products/sn/. Accessed 31 May 2018
12. Hitachi, Ultra short projection model with electronic blackboard function CP-TW3506J. http://www.hitachi.co.jp/Prod/vims/proj/lineup/cp-tw3005j/. Accessed 31 May 2018
13. Sakawa, Ultra wide projector SP-UW360iR. http://www.sakawa.net/wiiide/features. Accessed 31 May 2018
14. MEXT, Survey results on actual situation of educational informatization in school (Summary), February 2018. http://www.mext.go.jp/component/a$_$menu/education/micro$_$detail/$_$$_$icsFiles/afieldfile/2018/03/07/1399330$_$01.pdf. Accessed 31 May 2018
15. Bauer, J., Thelen, S., Ebert, A.: Using smart phones for large-display interaction. In: 2011 International Conference on User Science and Engineering (i-USEr), pp. 42–47 (2011)
16. Cho, Y., Kim, M., Park, K.S.: LOTUS: composing a multi-user interactive tiled display virtual environment. Int. J. Comput. Graph. Vis. Comput. **28**(1), 99–109 (2012)
17. Kukimoto, N., Onoue, Y., Aoki, K., Fujita, K., Koyamada, K.: HyperInfo: interactive large display for informal visual communication. In Proceedings of 17th IEEE International Conference on Network-Based Information System (NBiS 2014), pp. 399–404 (2014)
18. Hazama, H., Ebara, Y., Ogasawara, T.: A study on interface for screen operation to reduce tiredness at manufacturing training using ICT in elementary and secondary education. In: Proceedings of 23rd International Symposium on Artificial Life and Robotics (AROB 23rd 2018), pp. 572–575 (2018)
19. Marrinan, T., et al.: SAGE2: a new approach for data intensive collaboration using Scalable Resolution Shared Displays. In: Proceedings of 10th IEEE International Conference on Collaborative Computing: Networking, Applications and Worksharing (2014)

Implementation and Evaluation of Unified Disaster Information System on Tiled Display Environment

Akira Sakuraba[1(✉)], Tomoyuki Ishida[2], and Yoshitaka Shibata[1]

[1] Regional Corporate Research Center, Iwate Prefectural University, Takizawa-City, Iwate-Prefecture, Japan
{a_saku, shibata}@iwate-pu.ac.jp

[2] Department of Information and Communication Engineering, Fukuoka Institute of Technology, Higashi-Ward, Fukuoka-City, Fukuoka-Prefecture, Japan
t-ishida@fit.ac.jp

Abstract. In the event of large disaster, local government will be charged of responsible activity over counter disaster operation. However, conventional procedure for understanding and sharing of disaster information at the headquarters is performed by paper-based documents and indication on paper-based map, it could be difficult to represent of understanding of disaster statuses for multiple headquarters personnel. This paper proposes design and implementation of disaster information system based on ultra high definition display environment using GIS. We have designed the system consists of large and ultra definition display environment as unified shared display at headquarters. Our system considers to implement directly status reporting function, detachment of displaying location and media content and user-system interaction method using smart devices as content controller. We had hands-on based experiment in order to assess usability of our system. The result shows relatively positive feedback for usability of the system by participant consist of non-specialist for counter disaster activity.

1 Introduction

In the event of large scale disaster in Japan, local government will establish Counter Disaster Headquarters (CDH) in effort to manage disaster fighting [1]. Their tasks include several activities related what happened in their jurisdictional areas, for example, understanding of damage states, status of rescue supplies, management of shelters or activity of cooperated organization. Particularly, understanding of disaster states is the basic knowledge to make decision what local government should measure to fight against disaster. In the early stage of disaster, damage related information is the most important to handle counter disaster operation in effort to realize rapid restoration. CDH in actual disaster uses paper based map for sharing of damage states at CDH. However, paper based disaster states indication system has difficulty for sharing among multiple CDHs, it is also difficult to record the difference of temporal sequence.

L. Barolli et al. (Eds.): NBiS 2018, LNDECT 22, pp. 503–513, 2019.
https://doi.org/10.1007/978-3-319-98530-5_43

GIS (Geographic Information System) based system turns into a powerful and commodity tool in disaster operation as well as management for infrastructure in normal time. GIS based disaster management system allows to connect spatial information and damage states and provides function to visualize what and where event happened. Generally, digital map based GIS is capable to manipulate freely at user's will. On the other hand, CDH operation requires entire view of disaster area to make common understanding all over CDH. Global view of entire disaster area requires less zoom level, which is equivalent to the scale in paper based map. But it brings lack of detail when system renders entire area. There is a trade-off in rendering of map between entire view and details. Especially this problem carries out when map is rendered on generic resolution display device. This type of display device has another challenge which is caused by physical size of display in effort to represent multiple users at once.

Tiled display wall (TDW) based ultra high definition display environment which consists of matrix-arranged display wall to realize unified large size and ultra high resolution display device. In the past decade, a lot of manufacturers launch LCD display wall with thin bezel. It realizes a seamless large sized display wall without awkwardness.

This paper introduces a unified disaster damage state management system for local government. The system is in charge of collection and representation of damage status related information. For scouting team who works at disaster area, our method provides disaster state reporting device which allows to report damage status by picture, voice annotation and additional information from disaster area in real-time. This system also uses TDW based large sized display wall placed in the headquarters room for decision makers. On TDW, this system can represent geospatial information by high resolution GIS map which describes entire states over disaster area, and digital whiteboard which renders images of disaster area in order to provide visual states of there.

We had experiment to measure how our method was designed and implemented easily to use by user hands-on in control functions. The experiment conducts real-time scenario which is similar to the method using in actual disaster drill. After hands-on, we requested subjects to answer questionnaire in order to measure how she/he thinks our method was easy to use or not. We had the result of experiment, in many functions which the subjects felt the system is well designed to use easily, but in some functions which have challenge for usability.

2 Related Works

GIS based system turns into necessary tool in case of disaster scene as well as normal time. Shizuoka Prefecture, located middle of Honshu Island in Japan, operates ASSIST-2 [2] which is integrated disaster information system across 24 tasks for example, disaster information, state of announce on evacuation advisory, operation states of relative organizations, etc. Many researchers have discussed and proposed GIS based counter disaster system like Tsai et al. system [3]. Humanitarian OpenStreetMap Team (HOT) [4] is an open project for humanitarian mission in large scale disaster using digital mapped GIS. The core objective of their mission statements includes connecting point between humanitarian actors and open mapping communities, providing remote mapping and several other factors, it had utilized for search and rescue operation during 2010 Haiti earthquake.

We can find several ultra high definition (UHD) display environments configured with multiple display arrangement. It is based on the distributed system which is illustrated by example as SAGE [5], Tiled++ [6], DisplayClusters [7], etc. in the past a decade. These methods have capabilities to display contents over 4 K or higher resolution without downscaling of it.

As the background of high bandwidth video interface to transmit from render device to display device such as DisplayPort, we can build unified multi display UHD environment on a generic PC or workstation without special equipment or intricate system. This breakthrough brings TDW environment into collaborative tool not only as large scale digital signage, but can cite as for SAGE2 [8] for example.

3 System Design

We have designed and implemented the gathering and displaying information system on ultra high definition environment called LIVEWall (*Large-scale disaster Interactive Visualization Environment on display Wall*) in order to share and organize current states of disaster area. This paper describes mainly method for operation for disaster information on ultra high definition wall. We introduce the detail of design concept illustrated in Fig. 1.

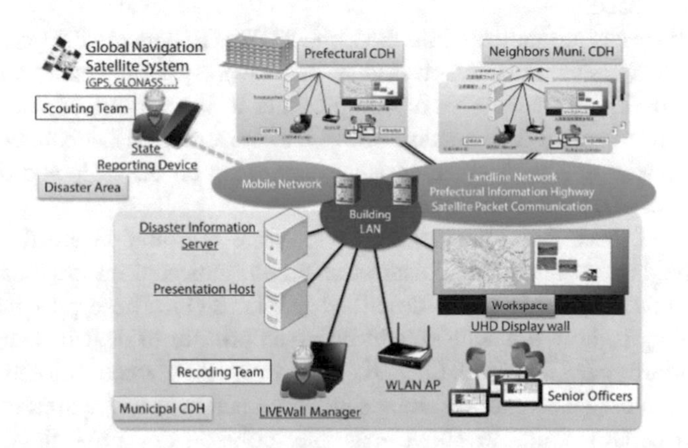

Fig. 1. System configuration of LIVEWall

3.1 System Overview

We divided functions of LIVEWall into, *reporting*, *sharing* and *posting*. In reporting phase, municipal CDH dispatches scout team into disaster area with State Reporting Device which conducts to obtain and report damage state information by photo, voice annotation and additional data called metadata. These information will be sent to CDH immediately, in the CDH, the information stored into database on subsystem called LIVEWall Server.

In the second sharing phase, the system provides function to represent disaster states on TDW based large scale and ultra high definition display located on the command room at CDH. Rendering TDW function is performed by Presentation Host to separate disaster states into single GIS map for displaying geological information, and single digital canvas like whiteboard for displaying image to describe disaster site's status. This system calls these information to render space as *workspace*. Sharing functions also required to control displayed content on TDW. Our system is designed to operate displayed items on TDW via smart device called LIVEWall Controller. The controller is designed to be used by executive members of CDH who have responsibility for decision-making.

Finally, damage state information will be distributed to other municipalities and prefecture in the sharing phase. Each CDH operates LIVEWall to observe their controlling area without paper based format. This allows to share every states among CDHs instantly and more precisely.

Additionally, there is subsystem to manage and modify metadata using by designated user at CDH called LIVEWall Manager which runs on generic desktop based host.

3.2 Workspaces

LIVEWall renders disaster states on TDW and displays states on different part of display area and format. For those functions, we designed Shared Workspace and Contents Workspace.

Both workspaces are indicated on the same TDW. Officers at CDH can understand current state of disaster area by observing these workspaces even many CDH personnels are working at there. This design provides to list geo-information and media information on a unified information wall. The current CDH separately deals and manages geospatial information using paper based map on the table and digital image which is projected on the other place.

Shared Workspace deals geolocation based state information are described what and where the damage happened. Disaster states are indicated on large and ultra high resolution digital map as drop-pin described in Fig. 2-(1). These pins have different colors according to how the state should be given priority to deal it. Our system has three-level priority, *HIGH/MIDDLE/LOW* with red/yellow/green colored pin respectively. Shared Workspace can also represent states across a wider area not only state associated to certain point. In above example, polygon describes flooding area by tsunami and filled with blue color. Shared Workspace on ultra high definition environment realizes functions to manage both wide area and detailed displaying although it is difficult to do it in generic display device due to limited resolution.

Contents Workspace conducts function to show multiple images of disaster area to have understanding of decision makers using digital whiteboard shown on Fig. 2-(2). This workspace also rendered on ultra high resolution display environment in effort to show image without downscaling or can compare multiple images with apposing them on it. Arrangement of each images area can be managed on the server as objects.

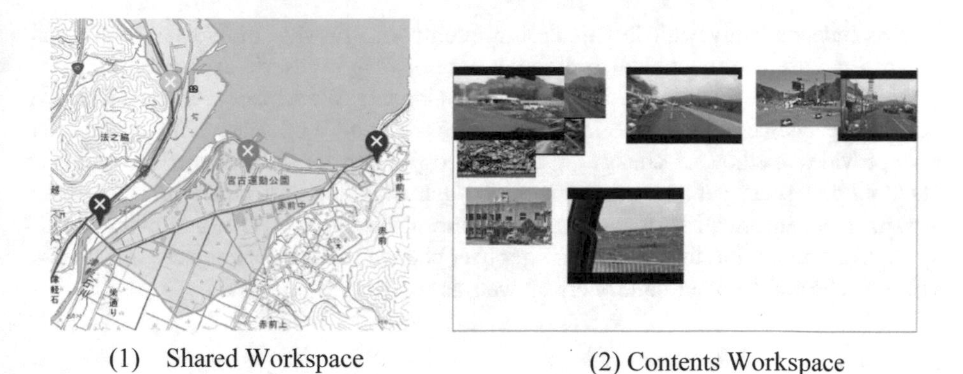

(1) Shared Workspace (2) Contents Workspace

Fig. 2. Design of Shared and Contents Workspace

3.3 Control Method to Workspace

LIVEWall implements controlling method for workspaces called Workspace Controller shown in Fig. 3.

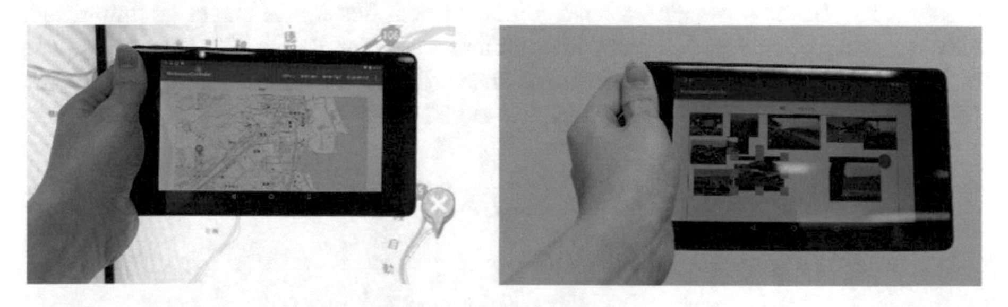

Fig. 3. Control screen of Shared Workspace (left) and control screen of Contents Workspace (right) on implementation of Workspace Controller

For Shared Workspace, Workspace Controller conducts operation for viewing map on TDW as one of functions. It allows to change zoom level and the center coordinate of map in order to change map view.

Workspace Controller also provides filtering function to reduce the number of states or highlight it on the map. We designed the latest, spatial, temporal, and priority for filtering. Filtering for the latest information allows to show only each states fitted metadata as the latest information on the database. Spatial filter puts a restriction of states overlaid on the map by geographically indicating where states includes in the circle. This type of filter highlights damages has been occurred into particular area. Temporal filter let only to indicate the time period from origin time to end time by user's designation to highlight the damages which has been broke out in period of time.

Workspace Controller has personal view of map. Overlaid states of Shared Workspace is common among TDW and any Workspace Controllers although viewing map is independently varied. This design attempts to provide different viewing map where each user desires to watch different area of map by user's expertise.

In Workspace Controller, operation for Content Workspace behaves just like controlling device. The device indicates entire Content Workspace with downscaled and provides method of change position of object on the workspace, the order of stacking and scale with intuitive operation such dragging and pinch-in/out for user. Arrangement information is completely synchronized among TDW and every Workspace Controllers. For this reason, if other user changed scale of an objects, this change will be reflected to other controllers as well as workspace on TDW.

3.4 System Architecture

Architecture of LIVEWall is based on typical client-server model. The detailed system configuration and its module configuration is shown in Fig. 4. Disaster Information Server consists of three modules including Workspace Composer, Workspace Manger and Session Manager. Workspace Composer provides geospatial location of disaster state stored on database for the Presentation Host and Workspace Controller. Workspace Manager performs to establish connection to Workspace Database and Content Store. Session Manager performs as interface between the other clients and server. Presentation Host takes charge of displaying the shared workspace corresponding to the disaster state. In the presentation host, primary Local Session Manager communicates with the other clients and servers to establish a session. Then Workspace Updater fetches the related contents from database and builds a workspace. Finally Workspace Renderer performs to render the workspace on TDW.

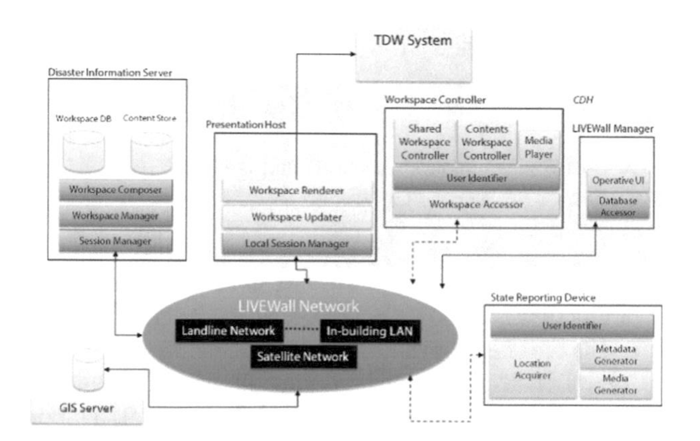

Fig. 4. System architecture of LIVEWall

Workspace Controller performs interface functions to manage the contents on both shared and contents workspaces. Media Player perform playout function of voice annotation.

State Report Device performs direct reporting function from disaster area. Location Acquirer obtains the location data using global navigation satellite system (GNSS).

Metadata Generator performs an interface function to determine the three levels of importance of the disaster information. Media Generator creates an image by taking by camera or revokes the saved image in the internal storage of smart device. In addition, this device also generates voice annotation. State Report Device includes the reporter's profile which includes the reporter name and affiliation.

LIVEWall Manager performs an interface to fix or create new state on the Disaster Information Server, and manages disaster session of LIVEWall, and consists of two modules including Database Accessor and Operative UI. Database Accessor deals with connections and modifications for database. Operative UI provides user interface function to Database Accessor so that operators can easily handle the disaster state information.

4 Prototype Configuration

In order to evaluate the functions and performance of our proposed system, we developed a prototype system and implemented each module of system components as shown in Fig. 5.

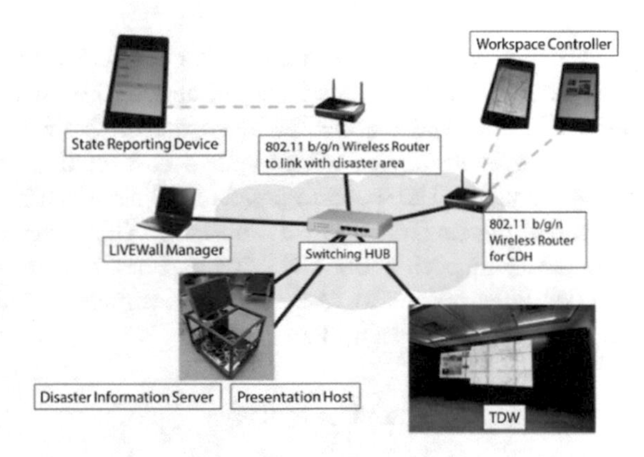

Fig. 5. Configuration of the prototype

As TDW, network based ultra definition display wall system Samsung UD with 27 tiles display arrays are used. The network connection among each component is set up by Gigabit Ethernet and wireless LANs for Workspace Controllers. As Presentation Host, a Windows PC with dual display compositions of both WUXGA and WQHD resolution is used. Disaster Information Server was built on OpenSUSE and runs with Apache and MySQL. The server also has GIS service and map tile images based on OpenLayers and CyberJapan Digital Map as the background map in this prototype. As State Report Device and Workspace Controller, Asus Nexus 7 (2013) tablet device is used to run Android application. Content workspace is web based product by Fabric.js to render media content and communication function is realized by Node.js library among workspace on this Workspace Controllers and Presentation Host.

5 Hands-On Experiment

5.1 Objective and Overview

This hands-on experiment intends to measure how our design and implementation has capability for easy-to-use, using actual developed environment. We discuss Workspace Controller for measuring objects in this paper. The experiment is based on disaster simulated scenario which progresses on assignment given by experimenter in real time. We have performed each one scenario for Workspace Controller, also requested to answer questionnaire after their experiment trials. Finally, we analyzed answers and feedbacked from subjects for assessment and suggestion for improvements of prototype. Questionnaire composes 5-levels of Likert scaled items which correspond to each functions of LIVEWall. User gives 5 points for the most positive score, conversely, 1 point brings the worst impression. We asked it for each functions.

The building of scenario is drawn on Disaster Imagination Game (DIG) [9] which is designed for counter disaster exercise in CDH established by local government. Following in DIG, *situation setting* stands for given assignments for subjects.

5.2 Scenario

We have set up a timeline to conduct hands-on, it consists of several controlling operations for content on TDW. The scenario begin about 10 h after earthquake and tsunami struck on the city located on the Pacific coast of Iwate Prefecture. These scenarios goes on the assumption that subject plays as a staff of headquarters in workspace controlling scenarios. The scenario targets on Shared Workspace consists of 4 filtering operations, two information (POINT and POLYGON, one each) adding to workspace, and manipulate map view on TDW twice. For Contents Workspace, the scenario required to add three images to workspace as instructed arrangement. Entire scenario is designed to complete 13 min in each trial.

5.3 Environment of Experiment and Subjects

Figure 6 describes environment of this experiment. We placed the subject's desk in front of TDW to simulate command room. Situation settings are assigned through a

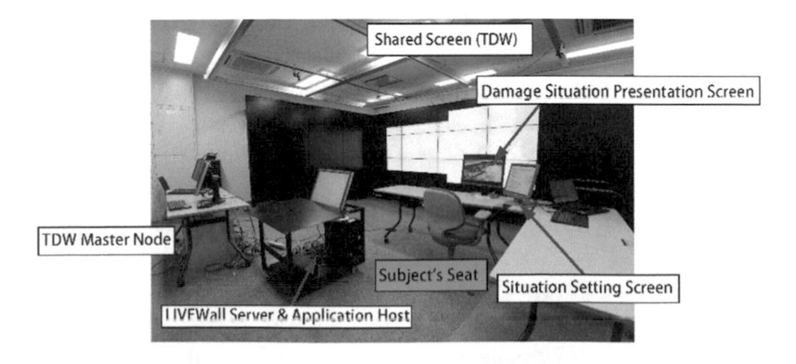

Fig. 6. Environment of hands-on experiment using TDW

card displayed on Situation Setting Screen where located at the right-side of subject's seat. At the left-side of subject's seat, we placed Damage Situation Presentation Screen in state reporting scenario. We also prepared instructions form on there.

Subjects were 6 persons who are composed of students or stuffs for administrative affairs in our university, aged from 10 s to 50 s, a female and 5 males. They were all non-professional regarding with counter disaster operation yet. Their in-use experience of smart devices was at least over 4 years and mode section was 4 years to 6 years.

5.4 Results

Average scores for operation on Shared Workspace are described in red items Table 1. The result shows that 7 of 8 main functions marked over three points, center of score, and above. Changing map view scored over 3.5 pts averagely, the result of invoking for filtering function may vary. We compared invoking spatial and temporal filters. The score difference between those two filters was 1.0 point averagely. Addition of a new POLYGON state makes user unsatisfied for easy-to-use design as compared with creation POINT state. In interaction on Shared Workspace, rendering time delay which is calculated the time from user's operations to displaying on TDW was reasonable for the users.

Table 1. Functions vs. average score and S.D. of user's impression

User's impression	Average score [pts]	S.D.
Easiness to change area on map view on TDW	4.0	0.6
Easiness to change zoom Lv. on map view on TDW	3.8	0.9
Easiness to invoke temporal filter	3.0	1.2
Easiness to invoke the latest filter	3.7	1.2
Easiness to invoke spatial filter	4.0	0.6
Shared workspace's rendering delay for operation to render on TDW	3.7	0.7
Easiness for addition of new POINT state	3.7	0.5
Easiness for Addition of New POLYGON State	2.8	0.7
Easiness for addition of new object to contents WS	3.5	1.3
Easiness for operation on objects	3.6	2.0
Rendering delay for operation to render on TDW	2.8	1.8
Managing both detailed view & individual view	4.4	2.1
Information relationship b/w shared & content WS	3.2	2.1
Overall: workspace controller	3.6	2.0

Result on Contents Workspace is described blue background row in Table 1. The system achieves 3.6 to 3.7 pts in adding and changing positions for object on workspace, Standard Deviations (SD) were respectively 1.3 to 2.0. Thus it could be different impression to do. Rendering delay on Contents Workspace is the largest challenging interaction for subjects. The average score was just 2.8 pts (SD = 1.8).

We developed separated design of viewing map with TDW and individual view on Workspace Controller, it was better than the center score. The scenario requested to the users to understanding the relationship between location information and media data on TDW into trials was scored 3.2 pts but SD spreads out in 2.1. For overall functions of Workspace Controller, subjects assessed 3.6 pts in average.

5.5 Discussion

As we analyzed the result, it is suggested that users felt the usability of our system was average. However Workspace Controller could be a complicated device for users due to a lot of functions. Especially in temporal filter configuration, they feel the difficulties to set the exact period of time to filter. Drawing polygon on the map has also the same problem, it is suggested that system should implement more simple UI design. On Contents Workspace, some users felt frustration with delay to render result by their operations on Workspace Controller. This delay time would be reduced if the proposed system is operated on modern TDW.

On the whole, their impression for our designed system was not bad especially in independent viewing map between TDW and controller, but the relationship between geo-location information and media contents are still challenged.

6 Conclusion

This paper introduces a disaster management GIS on ultra high definition display environment and input device for it. Our system has separated information workspaces which contains map based geo-location and media contents on unified display wall. This aims to show these disaster information with detailed view at once for multiple headquarters officers. We also developed damage reporting device from disaster area to the headquarters by image, voice and additional information, and controlling method for these workspaces using smart device. It considered intuitive human computer interaction.

We had hands-on based real time experiment subjected by non-professional for disaster management operation to assess usability of the design. We had observed that users did not feel bad impression for the proposed system relatively but they required more user easier and friendly interface. We have assessed based on these result and given feedback to only several users. As our future works, we will continue the functional evaluation with more numbers and age group of users to analyze functional the system evaluation from various viewpoints. Certainly, we are planning to examine the system experts of counter disaster operation in local government. We also would simultaneously redesign challenging functions using other method applies such above discussion.

Acknowledgments. The research was supported by Grant-in-Aid for Scientific Research (B) Number JSPS15H02693 and Grant-in-Aid for Scientific Research (C) Number JSPS 16K00119.

References

1. Iwate Regional Disaster Prevention Planning, Iwate Prefecture: http://www2.pref.iwate.jp/~bousai/shiryo/gakusyuu/bousaikeikaku/iwateken_chiikibousaikeikaku.pdf. (in Japanese). Accessed 01 May 2018
2. Iwata, T.: Information concerning the assessment of an earthquake disaster in Shizuoka Prefecture. J. Jpn. Assoc. Earthq. Eng. **4**(3), 138–142 (2004)
3. Tsai, H., Mmiller, L., Hua, M., Bojja, M.V.: Expanding the disaster management knowledge space through spatial mediation. In: Proceedings of 2012 45th Hawaii International Conference on System Sciences, pp. 3699–3708 (2012)
4. Humanitarian OpenStreetMap Team: https://www.hotosm.org/. Accessed 01 May 2018
5. Leigh, J., et al.: Scalable adaptive graphics middleware for visualization streaming and collaboration in ultra resolution display environments. In: Proceedings of Workshop on Ultrascale Visualization 2008, pp. 47–54 (2008)
6. Ebert, A., Thelen, S., Olech, P., Meyer, J., Hagen, H.: Tiled++: an enhanced tiled hi-res display wall. IEEE Trans. Vis. Comput. Graph. **16**(1), 120–132 (2010)
7. Johnson, G.P., et al.: DisplayCluster: an interactive visualization environment for tiled displays. In: Proceedings of 2012 IEEE International Conference on Cluster Computing, pp. 239–247 (2012)
8. Marrinan, T., et al.: SAGE2: a new approach for data intensive collaboration using scalable resolution shared displays. In: Proceedings of 10th International Conference on Collaborative Computing: Networking, Applications and Worksharing, pp. 177–186 (2014)
9. Komura, T., Hirano, A.: DIG (disaster imagination game). In: Proceedings of the Annual Conference of the Institute of Social Safety Science, pp. 135–139 (1997). (in Japanese)

3D Measurement and Modeling for Gigantic Rocks at the Sea

Zhiyi Gao[1(✉)], Akio Doi[1], Kenji Sakakibara[2], Tomonori Hosokawa[2], and Masahiro Harada[3]

[1] Iwate Prefectural University, Takizawa, Japan
g231o202@s.iwate-pu.ac.jp, doia@iwate-pu.ac.jp
[2] TOKU PCM Ltd., Morioka, Japan
k-sakakibara@tokupcm.com, t-hosokawa@toku-gp.com
[3] TAC Engineering Ltd., Tokyo, Japan
mas.harata@tac-e.co.jp

Abstract. In this research, we digitally archived large rocks at the sea, which are called "Sanouiwa" in Miyako city. We conducted two types of three-dimensional (3D) measurement techniques. The first is to take pictures by using drone. The second is to use Global Navigation Satellite System (GNSS). The point cloud data was generated from the high resolution camera images by using 3D shape reconstruction software. Finally, we integrated all point cloud data, and we constructed 3D triangular model by using these point cloud data.

1 Introduction

In Japan, there are many large rocks which are natural shapes called "strangely shaped rocks" or "gigantic rocks". Especially, in the large rocks at sea, the contrast between the sea and the rock is impressive, and it is a work of art created by nature.

Figure 1 shows an example of a huge rock at sea in Japan. It is difficult to install a laser scanning devices, because it is on the sea.

In order to archive digitally such large rocks at the sea, we utilized the Unmanned Aerial Vehicle (UAV), which is called "drone", and the Global Navigation Satellite System (GNSS) or Global Positioning Satellite System (GPSS). The drone is equipped with a high-resolution camera and capable of remote control. After placing several anti-aircraft signs, we fly drone around the rocks above the sea and shoot high resolution images. The point cloud data was generated from the high resolution images by using three-dimensional (3D) shape reconstruction software. The software finds common feature points from these photographic images and generates point group data of 3D coordinates. This technique is called Structure from Motion (SfM). Finally, we integrated all point cloud data, and we constructed 3D triangular model by using these point cloud data. We also modeled 3D printer model in order to print them by several 3D printers.

L. Barolli et al. (Eds.): NBiS 2018, LNDECT 22, pp. 514–520, 2019.
https://doi.org/10.1007/978-3-319-98530-5_44

Fig. 1. Three king rocks (“Sannouiwa”, Miyako-shi, Iwate-ken) and cat rock (“Nekoiwa”, Rebu-cho, Hokkaido)

2 Previous Works

We measured a cultural property garden, which is called “Nanshoso” in Morioka city. We conducted two types of 3D measurement techniques. The first is to take pictures by using UAV with a high resolution camera from the sky. The second is to use 3D laser measurement device from the ground. The point cloud data was generated from the high resolution camera images by using 3D shape reconstruction software with Structure of Motion technology. Next, we integrated both the point cloud data and a laser measurement data. Finally, we constructed 3D CAD model by using the integrated point cloud data [1, 2].

“Sannouiwa” in the “Taro” district of Miyako city is the most dominant place in “Sanriku” reconstruction national park, and “Sannouiwa” is the meaning of the three king rocks. It is a beautiful natural art piece formed. The three king rocks consist of three rocks, male rocks 50 m high in the center, female rock on the left side and drum rock on the right side. The direction of the drum rock shows the different striped pattern, because the drum rock was rolling down from the coast near the rock and was in the present position.

3 3D Measurement and Modeling for Gigantic Rocks at the Sea

3.1 3D Measurement Devices and the Tools

We used a small drone with high resolution camera and GNSS receiver device (Fig. 2). The drone is “MarvicPro” of DJI Ltd. The MarvicPro has 4K video and high resolution camera, and it can be remotely controlled from Apple iPad (Fig. 3). We use both Trimble R5 and Topcon GPT 9005A for measurement/verification of GNSS points. Measurement of the orientation points and the verification points is carried out by the “network type RTK method”. The anti-aircraft marking shall be 20 cm (5 pixels or more) on one side and adopt the shape (color) shown below (Fig. 4).

Fig. 2. Drone and GNSS measurement device

Fig. 3. Steering drone by iPad

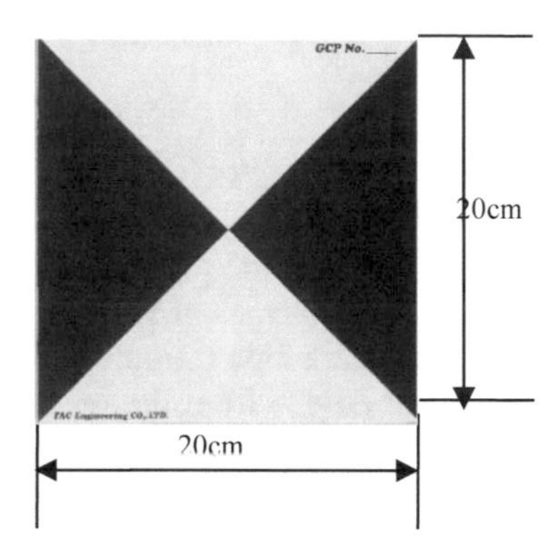

Fig. 4. Anti-aircraft sign

In the measurement of a gigantic rock at the sea, it was difficult to use 3D lazer devices from the ground. Therefore, we photographed the 1015 high-resolution pictures from both the sky and sides of the three king rocks. Before taking pictures, the five anti-aircraft sign were placed on the ground. Next, 3D point cloud data were generated from the photograph image using SfM technology. The software used for point group generation is "Pix 4 DMapper" of Pix 4 D Company. For 3D polygonal model generation, we used "ContextCapture" of Bentley Systems, which provide 3D modeling functions from point cloud data. Finally, we generated 3D closed polygonal model for 3D printing, since it is necessary to create a closed polygonal model. We used "POLYGONALmeister" of UEL Corporation for the polygonal closing step.

3.2 Coordinate Correction Using GNSS

GNSS (Global Navigation Satellite System/Global Positioning Satellite System) is a generic name for satellite positioning systems such as GPS, GLONASS, Galileo, Quasi-Zenith Satellite (QZSS). GPS (Global Positioning System) is a satellite positioning system operated by the United States (a system for measuring the current position on the earth). The principle of GPS positioning is that the light velocity c is constant in the local inertial system.

$$c = 2.99792458 \times 10^8 \text{ m/s} \tag{1}$$

If both the GPS satellite and the receiver have a clock which can be regarded as accurate, the distance can be found by multiplying the difference between the transmission time (measured value) T and the reception time t by c. Assuming that the position of the GPS satellite i is coordinates (X_i, Y_i, Z_i) and the position of the receiver is (x, y, z)

$$c^2(T_i - t)^2 = (X_i - x)^2 + (Y_i - y)^2 + (Z_i - z)^2 \tag{2}$$

In order to obtain the position of the GPS satellite, the navigation message signal superimposed on the received data is demodulated and combined with the transmission time. The reception time t is the value of the clock of the GPS receiver, and if it is accurate, there are at least three simultaneous equations to obtain three variables (unknowns) x, y, z of the receiver position I hope. However, the clock of the GPS receiver is not so accurate, and the reception time t also needs to be unknown. Therefore, by receiving from four satellites, these four unknowns can be obtained.

3.3 Structure from Motion Techniques

Structure from motion (SfM) is an imaging technique for estimating three-dimensional structures from two-dimensional image sequences[3]. It is studied in the fields of computer vision. SfM find object point of three-dimensional structures, and the position of the object point can be calculated by using the camera image positions.

4 Experimental Results

After taking a photograph of the three king rocks with the drone, we used 1015 pictures for generating the point cloud data by utilizing Structure from Motion (SfM) technology. Image analysis was carried out collectively without separating the photos of the sky (vertical) and the side. Although errors in the sky and the side naturally occur, the three king rocks itself photographed from every angle so as to reduce the step shift and the hole of the 3D data itself. Also, by setting an anti-aircraft sign on the land, real coordinates and real distance was generated.

In order to generate the point group from the obtained photographic image, it took about 2 days to generate the point group using the “Pix 4 DMapper”. In this software, feature points can be extracted from continuous photographs, and point cloud data model can be generated. It takes about 2.5 days to generate the point cloud data of the three king rocks. Since point cloud data has missing parts and noise data, it is difficult to fill polygons from point cloud data automatically. In order to create polygonal models from point cloud data, we used “ContextCapture”, and we utilized the automatic triangulation function from the point data set, and we removed the unnecessary faces. We also filled polygons for the triangle model interactively. It took about a week to generate the triangle model.

The workstation we used were HP Z 840, main memory 256 GB, CPU clock Xeon 2.10 GHz, and the GPU is 12 GB NVIDIA Quadro M 6000. Figure 6 is the textured triangle model of the three king rocks. The viewer is Acute 3D of Bentley Systems. Texture mapping is performed, and almost real time display is possible.

Figure 5 shows both point cloud data and the triangle model. Figure 6 shows textured triangulate model and the model with wireframe overlay. Figure 7 shows the side portions of the textured triangulate model. It was confirmed that the cave was modeled accurately. In Fig. 7, a wire frame display is over-layered on the textured model in order to show the size of a triangle. Figure 8 shows the 3D printer model, which consists of closed triangulate polygonal model. In order to create the 3D printer model, we used "POLYGONALmeister" of UEL Corporation, which has powerful modify/edit functions for polygonal models.

Fig. 5. Point cloud data and triangle model

Fig. 6. Textured triangle model and the model with wireframe overlay

Fig. 7. Side view of both textured model and the model with wireframe overlay

Fig. 8. 3D printer models (overview, clipped front, and clipped back)

Fig. 9. Overview of the three king rocks

5 Conclusion

By photographing the camera with a drone from the sea, we took a series of high resolution images of the three king rocks. From the obtained photographic images, the precise point cloud data were generated by using SfM techniques.

In order to create polygonal models from point cloud data, we utilized the automatic triangulation function from the point data set, and edited the triangular model interactively. Figure 9 shows the enlarged overview of the three king rocks. It was shown that modeling can be done with sufficient precision by photographing from the sky and sides.

Acknowledgements. For this research, we received research grants from Iwate Prefectural University Academic Research Fund (Regional Policy Research Center). We are deeply grateful to Mr. S. Daibou of specified nonprofit corporation "Tsunami Taro" for many advices.

References

1. Gao, Z., Satoh, K., Doi, A., Sakakibara, K., Hosokawa, T., Harada, M., Konno, T.: 3D modeling of cultural property gardens and utilization for acceleration of disaster reconstruction. In: 23rd International Symposium of Artificial Life and Robotics (2018)
2. Gao, Z., Doi, A., Sakakibara, K., Harada, M., Hosokawa, T., Konno, K.: 3D modeling of cultural property gardens and utilization for acceleration of disaster reconstruction. In: Virtual Reality Society of Japan, The 32nd Tele-Immergion Technology Workshop (2017)
3. Shapiro, L.G., Stockman, G.C.: Computer Vision. Prentice Hall (2001). ISBN: 0-13-030796-3

Adaptive Array Antenna Controls with Machine Learning Based Image Recognition for Vehicle to Vehicle Networks

Noriki Uchida[1](✉), Ryo Hashimoto[1], Goshi Sato[2], and Yoshitaka Shibata[3]

[1] Fukuoka Institute of Technology,
3-30-1 Wajirohigashi, Fukuoka, Higashi-ku, Fukuoka 811-0214, Japan
14b2048@bene.fit.ac.jp, n-uchida@fit.ac.jp

[2] Resilient ICT Research Center, National Institute of Information and Communications Technology,
2-1-3 Katahira, Aoba-ku, Sendai 980-0812, Japan
Sato-g@nict.go.jp

[3] Iwate Prefectural University, 152-52 Sugo, Takizawa, Iwate 020-0693, Japan
Shibata@iwate-pu.ac.jp

Abstract. With the developments of ITS technology, it is considered that the V2V communication is necessary for the new kinds of applications in the future. However, there are actually some subjects of wireless networks between vehicles caused by the fast movements or the radio noise of the moving vehicles. Thus, this paper proposes the Delay Tolerant Network System with the Adaptive Array Antenna controlled by the image recognition for the V2V Networks. In the proposed system, the target vehicle is recognized by the Machine Learning based image recognition system, the Kalman Filter algorithm to modify the influence of the vehicle's speed or the obstacles in the way of the road controls the direction of the Adaptive Array Antenna. The paper especially deals with the implemented image recognition system and the antenna direction controls from the experimental results of the prototype system, and the results indicate the effectiveness of the proposed system for the V2V networks.

1 Introduction

With the developments of ITS (Intelligent Transport System) technology, it is considered that the V2V (vehicle-to-vehicle) communication is necessary for the new kinds of applications in the future. For example, the paper [1] introduced the V2X (vehicle-to-everything) alert communication system by using the map-matching approaches with the GPS sensors on the smartphones and the vehicles. It proposed the map-matching algorithm notifies the status of the upcoming vehicles with the GPS values through the Internet from the vehicles, pedestrians, and traffic light poles, and the results showed the effectiveness for the driving safety system for the vehicles as well as the pedestrians.

However, there are actually some subjects of wireless networks between vehicles caused by the fast movements or the radio noise of the moving vehicles. First,

L. Barolli et al. (Eds.): NBiS 2018, LNDECT 22, pp. 521–530, 2019.
https://doi.org/10.1007/978-3-319-98530-5_45

currently, some researches are planning to apply the WiFi such as IEEE802.11a/b/g/n or IEEE802.11p [2], it is considered that the transmission range of these frequencies with non-directional antenna are not enough for the V2V communication. Especially, the transmission range of the 2.4 GHz or 5 GHz radio is about less than 100 m, and the transmission period is not enough if the vehicles run across in the opposite directions. Besides, there are many obstacles such as trees or buildings on the roads, and they might cause the radio noise such as the multipath or scattering. Secondly, if it is considered to apply the cellular networks such as W-CDMA [3] or LTE for V2V networks, there are the unavailable areas of these services in the mountain areas. This research especially focus on the northern mountain areas for the surveillance system of the winter frozen roads, and it still have some subjects to use the new LTE D2D (Device-to-device) technology for the V2V networks right now. At last, the LPWA (Low Power, Wide Area) [4] consisted of 920 MHz radio frequency has the longer transmission range, but the throughput is not enough if the movie or picture data exist in the content.

Thus, this paper proposes the Delay Tolerant Network System with the Adaptive Array Antenna controlled by the image recognition in order to realize the V2V Network. In the proposed system, the target vehicle is recognized by the Machine Learning based image recognition system, the Kalman Filter algorithm to modify the influence of the vehicle's speed or the obstacles in the way of the road controls the direction of the AAA (Adaptive Array Antenna). The paper especially deals with the implemented image recognition system and the antenna direction controls from the experimental results of the prototype system, and the results indicate the effectiveness of the proposed system for the V2V networks.

In the followings, the previous AAA studies and the proposed AAA controls with the Machine Learning based image recognition are discussed in Sects. 2, and 3 explains the Machine Learning based image recognition methods for the angle for the radio directional controls and the radio directional modification by the Kalman Filter algorithm. The implementation of the prototype system is introduced in Sect. 4. Finally, Sect. 5 introduces the experiments, and the conclusion and future studies are discussed in Sect. 6.

2 Adaptive Array Antenna

This research proposed the AAA controls for the longer transmission range with the broadband by the high frequency radio system such as 2.4 GHz or 5 GHz. The AAA [5, 6] is the smart beam-forming system with multiple antenna elements for the reduction of the fading problems and the increase of the antenna gain, and the radio direction is controlled by the adaptive weight of the radio shift or amplitude changes of each non-directional antenna element.

In general, the radio direction is decided by the DOA (Direction of Arrival), and the MMSE (Minimum Mean Square Error) or CMA (Constant Modulus Algorithm) is mainly used for the optimization of the radio arrival [7]. Then, with the results of the optimization of the angle of the radio arrival, the optimization algorithm such as the LMS (Least Mean Square) or the RLS (Recursive Least Square) controls the weight of each antenna element. The LMS is the algorithm based on the Gradient descent, steepest descent method, but the speed of the convergence is considered problems for the actual usages of the AAA. On the other hand, the RLS is the algorithm that based on the recursive mean square calculations by the previous sampling data, but the complexity of the calculations are supposed to be subjects although the convergence is faster than the LMS [8].

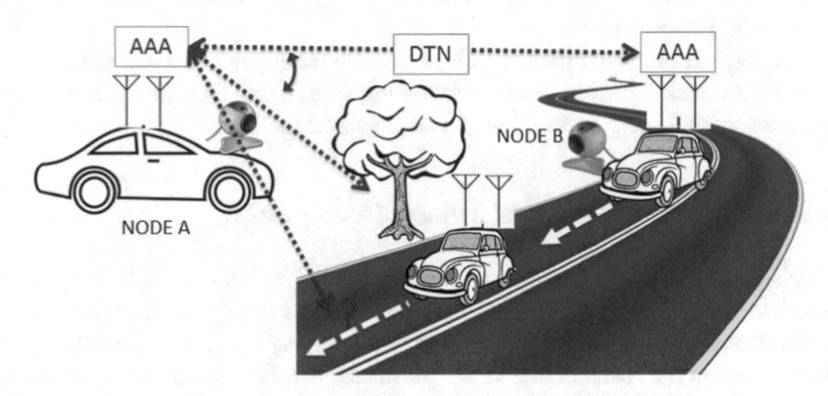

Fig. 1. The proposed networks in the proposed methods. The vehicles equips the broadband wireless networks system such IEEE802.11a/b/g/p, the AAA that consisted of the multiple antenna elements, and the camera for the detection of the vehicles.

Figure 1 presents the proposed AAA methods with the DTN (Delay Tolerant Networking) on the vehicles.

In the figure, the node A firstly captured the target node B with the camera, and the direction of the beamforming occurs for the proper connection. The transmission of data is confirmed by the DTN routing, and so the data of the node A is stored in the cache of node A if there is no transmittable nodes. Moreover, the modification by the forecast location is also supported by the Kalman Filter algorithm [9, 10] for the avoidance of the radio noise from the obstacles or the curve on the roads, If the vehicles run along the road as shown in the Fig. 1, the road might be curved or the tree might be disrupt the radio communication. If so, the radio direction of AAA is modified for the forecast location by the calculation of the proposed methods.

The DTN [11] is the robust network routing that is originally aim to realize the planetary network system. The data transmission is based on the stored-and-carried typed routing protocol, and the transmit data is stored in the cache if there is no transmittable nodes within the transmission range. Although it is currently known as one of the robust networking protocols for the disaster network system, the many researches proposed the enhanced routing methods such as the Spray and Wait [12] or the MaxProp [13] against the high latency and the low delivery rate about the DTN. These related researches are mainly related to deals with the hardware resources, the rearrangement of cached data, or node movements. This paper also introduced the Data Triage method [14] for the DTN routing because the road surveillance system in the winter mountain area is assumed for this project. The Data Triage Method is one of the queue ordering typed DTN protocols, and the messages in the stored storages are rearranged by the time and the user policy. Originally, this protocol assumed for the usages of the disaster network system, and the media types [14] or the conditions of the evacuators with the duration of the static body conditions [15] decides the priority values for the ordering queue. In this research, the dangerous levels of the frozen road conditions are planning to use the value of the priority values in the Data Triage Method in the future studies.

Moreover, the proposed methods also have the modification of the radio direction by the Kalman filter algorithm, and the following section explains the calculation of the target direction and the fore location in the proposed methods.

3 Machine Learning Based Image Recognition

The decision of the radio direction is decided by the captured vehicle images in the proposed methods. Although it is widely used for the safety driving system by the binocular camera or monocular camera with the land mark points such as the width of the white lines on the roads in recent, the proposed methods introduced the calculation of the pixel size of the captured vehicle image and the distance from the center point because the direction of the AAA do not need so much accuracy but need the quick responses for the V2V networks. Figure 2 presents the parameters of the proposed calculation for the distance the angle for the target vehicle [16]. Here, in the left figure of Fig. 2, total area of the actual vehicle is S_1, and total pixels in the captured image is S_2. The distance of the actual field stands for d_1, and the distance of the captured image is d_2. Then, there is the relations between the actual field and the captured image as shown in Formula (1).

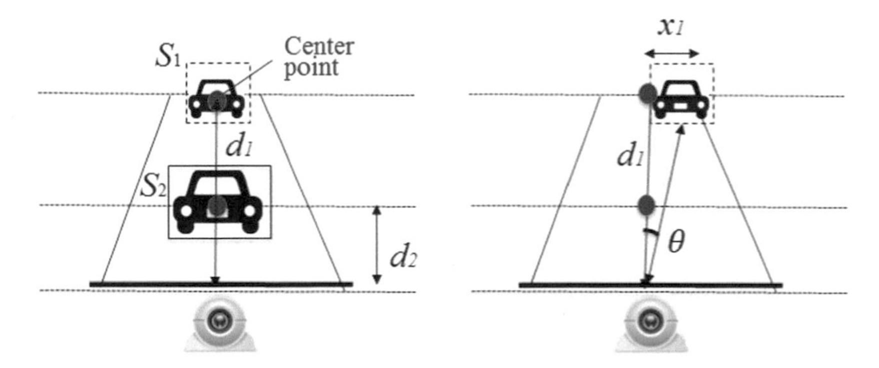

Fig. 2. The estimation of the distance and the angle for the target vehicle by the proposed monocular image recognitions. The left figure shows the parameters for the distance, and the right figure is the angle.

$$d_1 = d_2\sqrt{\frac{S_2}{S_1}} \tag{1}$$

Thus, if the camera previously knows the d2 and the ratio of S1and S2 in the proposed system, it is supposed that the distance d_1 can be calculated by the measurement of the total pixels S_2.

Also, in the right figure of Fig. 2, x_1 stands for the distance of the actual field between the center point and the moved point, and x_2 is for the distance of the captured image as the same way. If the angle from the center point is set as θ, there is a relationship between these parameters as shown in Formula (2).

$$\theta = \tan^{-1}\frac{x_1}{d_1} = \tan^{-1}\frac{x_2}{d_2} \tag{2}$$

Therefore, the angle of the proposed AAA is approximately calculated with the measurements of the captured images if the ratio of $S_1$and S_2 are previously prepared. Especially, in this paper, the authors additionally introduced the Machine Learning based image recognition for the vehicles in order to improve the accuracy of the recognitions. The Machine Learning is one of the artificial intelligent researches, and it analyze the some extends of the sample pictures since the learning of effective rules, regulations, knowledge, and decision standards. In this study, the Haar-Like classifier API in the OpenCV [16] is used for the implementation of the Machine Learning.

Next, the Kalman Filter confirms the modification of the AAA controls. As shown in Fig. 3, it is considered that the future predictions of the location are needed for the rapid movements of vehicles, and the next target point at the t + 1 becomes important if there is the obstacles such as trees or houses between the vehicles.

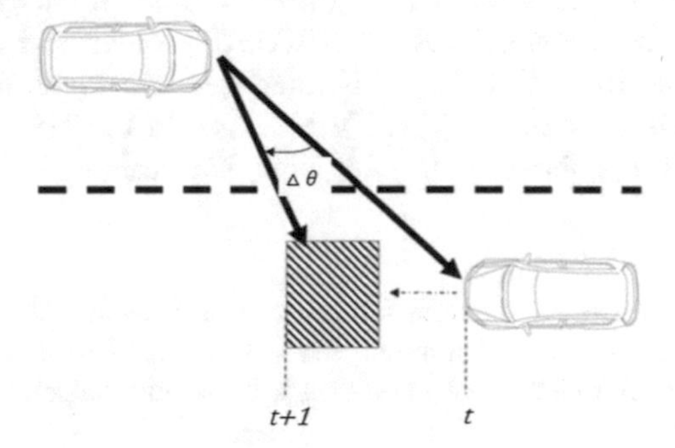

Fig. 3. The modification of the AAA angle between the vehicles. Although the current target vehicle locate in the point at t seconds, it will moves to the point at t + 1 s after the calculation of the image recognition. Besides, if there is the obstacles to the point at t seconds, the AAA should target to the next point at t + 1 s.

The Kalman filter is the estimation algorithm, and the process consists of two phases: One is the Time Update phase that is the update the current time and the prediction the values, and another process is the Measurement Update that is the update of the observed values and the adjustment [9, 17].

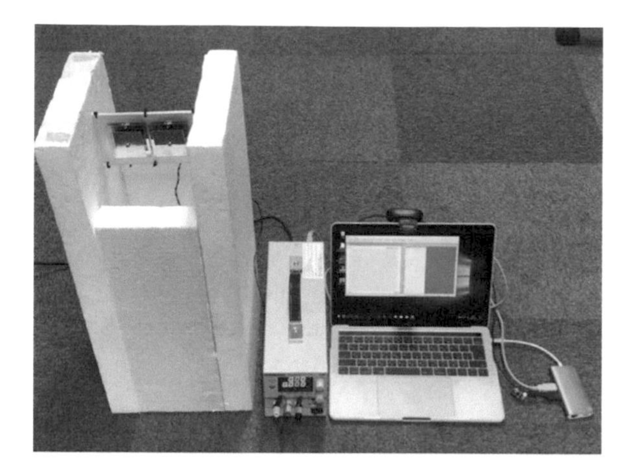

Fig. 4. The implemented prototype system. Each antenna element in the left of the picture has the radio phase shifter, and both phase shifters connect to the the voltage meter in the middle of the picture. PC recognize the vehicle by the web cam, and it controls the voltage meter through the LAN cable.

Then, the decisions of the predict angle of the AAA, the data transmission by the proposed DTN with the Data Triage Method is confirmed for the target vehicle.

4 The Prototype System

Figure 4 shows the prototype system of the proposed methods [18], and the Machine Learning based image recognition is newly implemented on the system.

The voltage meter is PMX18-5A by KIKUSUI Corporation and it supports the adjustments from the RS232C, USB, LAN interfaces. Moreover, the implementations use Windows 10, OpenCV2.4 [16], and Visual C++ in the MS Studio 2013 for the prototype system. The webcam is the Logicool Web camera c270 in the prototype system (Table 1).

Table 1. The voltage weights of the right and left antenna elements. According to the previous study [17], the radio direction of the proposed AAA system can be controlled by the following values. The right direction from the center is shown as the positive values, and the left direction is the negative values.

Radio direction (degree)	Voltage (right antenna element)	Voltage (left antenna element)
−45	0 V	+15 V
−30	0 V	+10 V
−15	0 V	+5 V
0	0 V	0 V
+15	+5 V	0 V
+30	+10 V	0 V
+45	+15 V	0 V

For the control of the AAA directions, the following table is previously introduced to the implementations because of the previous paper [17]. Then, the field experiment of the vehicle detection by the implemented Machine Learning recognition was held, and the results are discussed in the following section.

5 Experiments

The field experiments are held for the evaluations of the proposed methods and the future studies. Especially in this paper, the accuracy of the Machine Learning based image recognition was evaluated in this experiments. First of all, Fig. 5 is the captured window images of the prototype system.

Fig. 5. The captured windows of the prototype system. The prototype system recognized the running vehicles, and the rectangle marks are the areas of the recognized vehicles. Then, the pixel size of the area is used for the calculation of the distance from the prototype, and the pixel length from the center point as shown in the white line in the left picture use for the calculation for the angle of the radio direction.

The field experiments were held at the National Road No. 3 near the Fukuoka Institute of Technology in Japan, and the traffic along the road is usually heavy as shown in the picture. The number of the sample picture to identify the vehicle in the Harr-Like classifier was 50 pictures in the positive images and 50 pictures in the negative images. Table 2 presents the results of the recognized vehicles and the percentage of the success recognitions.

Table 2. The number of recognized vehicles and the percentage by the prototype system.

	Total vehicles	Recognized vehicles	Percentage
Front vehicle image	38	18	47.37
Back vehicle image	45	39	86.67

According to the results from Table 2, it is considered that the front captured images of the vehicles shows the worse recognitions in comparison with the back side images. It is supposed that the sample pictures for the Machine Learning were not enough in the prototype system, and also the shapes of the front images are more complex than the backside. Therefore, we are planning to increase more sample pictures for the future works.

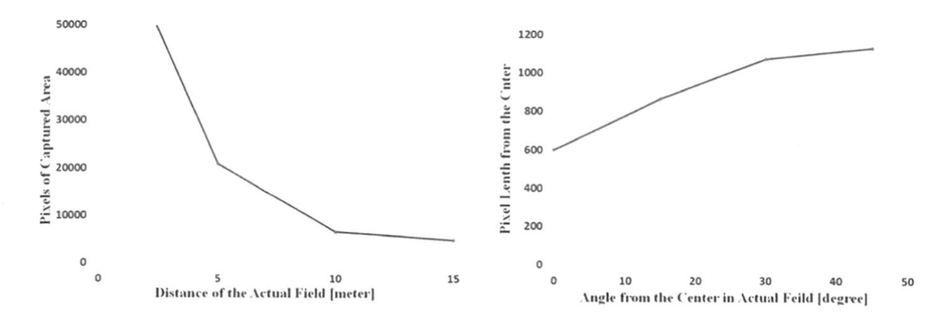

Fig. 6. The results of the calculated distance and angles from the captured images of the running vehicles in the experiments. The left figure is the results of the field distances and the captured pixel size, and the right is that of the field angles and the captured pixel length from the center.

Secondly, Fig. 6 shows the results of the calculated distances and the angles from the captured vehicle images.

The results indicate that there is a significant correlation between the captured images and the field differentials as the proposed methods, and the authors are now working for the additional implementations for the prototype system in the future works

6 The Conclusion and Future Study

There are actually some subjects of wireless networks between vehicles caused by the fast movements or the radio noise of the moving vehicles. Thus, this paper proposes the Delay Tolerant Network System with the Adaptive Array Antenna controlled by the image recognition for the V2V Networks. In the proposed system, the target vehicle is recognized by the Machine Learning based image recognition system, the Kalman Filter algorithm to modify the influence of the vehicle's speed or the obstacles in the way of the road controls the direction of the Adaptive Array Antenna. The paper especially deals with the implemented image recognition system and the antenna direction controls from the experimental results of the prototype system.

The prototype system is introduced the Machine Learning based image recognition function in this paper, and the experiments are curried out for the effectiveness of the proposed methods. The paper reports the results of the vehicle recognitions and the calculated results for the assumptions of the field distances and the angles, and the results shows the effectiveness of the proposed methods.

For the future studies, we are now working for the additional implementations of the prototype system including the predictions by the Kalman Filter algorithm in order to decrease the radio noise from the rapid movements of the vehicles and the obstacles in the roads. Also, the field experiments are planning for the future works.

Acknowledgement. This work was supported by SCOPE (Strategic Information and Communications R&D Promotion Programme) Grant Number 181502003 by Ministry of Internal Affairs and Communications in Japan.

References

1. Sakamoto, I., Hohuzu, H., Kozjima, R., Matumura, H., Takeuchi, T., Hasegawa, T., Tanaka, Y.: Fundamental study for inter-vehicle communication using smartphone. Natl. Traffic Saf. Environ. Lab. Forum. **2013**, 127–128 (2013)
2. Kato, S., Minobe, N., Tsugawa, S.: Experimental report of inter-vehicle communications systems: a comparative and consideration of 5.8 GHz DSRC, wireless LAN, cellular phone, IEICE Technical Report (2012), ITS 103(242), pp. 23–28
3. Aizawa, T., Shigeno, H., Yashiro, T., Matushita, Y.: A vehicle grouping method using W–CDMA for road to vehicle and vehicle to vehicle communication. IEICE Technical Report, ITS 2000(83(2000-ITS-002)), 67–72 (2000)
4. Ochoa, M.N., Guizar, A., Maman, M., Duda, A.: Evaluating LoRa energy efficiency for adaptive networks: from star to mesh topologies. In: 2017 IEEE 13th International Conference on Wireless and Mobile Computing, Networking and Communications (WiMob), pp. 1–8 (2017)
5. Takano, S., Fujii, T., Umebayashi, K., Kamiya, Y., Suzuki, Y.: Ad-hoc wireless system cooperated with MAC protocol and adaptive array antenna. IEICE Trans. Commun. **J91-B** (5), 550–557 (2008)
6. Chiba, I., Yonezawa, R., Kihira, K..: Adaptive array antenna for mobile communication. In: IEEE International Conference on Phased Array Systems and Technology, pp. 109–112 (2000)
7. Karasawa, Y., Sekiguchi, T., Inoue, T.: The software antenna: a new concept of Kaleidoscopic antenna in multimedia radio and mobile computing era. IEICE Trans. Commun. **E80-B**, 1214–1217 (1997)
8. Yokoi, T., Iki, Y., Horikoshi, J., et al.: Software receiver technology and its applications. In: IEICE Trans. Commun. **E83-B**(6), 1200–1209 (2000)
9. Welch, G., Bishop, G.: The Kalman Filter. http://www.cs.unc.edu/~welch/kalman/
10. Welch, G., Bishop, G.: An Introduction to the Kalman Filter. University of North Carolina at Chapel Hill Department of Computer Science (2001)
11. Fall, K., Hooke, A., Torgerson, L., Cerf, V., Durst, B., Scott, K.: Delay-tolerant networking: an approach to interplanetary internet. Commun. Mag. IEEE **41**(6), 128–136 (2003)
12. Spyropoulos, T., Psounis, K., Raghavendra, C.S.: Spray and wait: an efficient routing scheme for intermittently connected mobile networks. In: Proceedings of the 2005 ACM SIGCOMM Workshop on Delay-tolerant Networking, WDTN 2005, pp. 252–259 (2005)
13. Burgess, J., Gallagher, B., Jensen, D., Levine, B.: MaxProp: routing for vehicle-based disruption-tolerant networks. In: Proceedings of the 25th IEEE International Conference on Computer Communications, INFOCOM 2006, pp. 1–11 (2006)

14. Uchida, N., Kawamura, N., Shibata, Y., Shiratori, N.: Proposal of data triage methods for disaster information network system based on delay tolerant networking. In: The 7th International Conference on Broadband and Wireless Computing, Communication and Applications (BWCCA 2013), pp. 15–21 (2013)
15. Uchida, N., Shingai, T., Shigetome, T., Ishida, T., Shibata, Y.: Proposal of static body object detection methods with the DTN routing for life safety information systems. In: The 32nd International Conference on Advanced Information Networking and Applications Workshops (WAINA 2018), pp. 112–117 (2018)
16. Opencv: http://opencv.jp/cookbook/
17. Uchida N., Takahata, K., Shibata, Y., Shiratori, N.: Proposal of vehicle-to-vehicle based delay tolerant networks with adaptive array antenna control systems. In: The 8th International Workshop on Disaster and Emergency Information Network Systems (IWDENS 2016), pp. 649–654 (2016)
18. Uchida, N., Ichimaru, R., Ito, K. Ishida, T., Shibata, Y.: Implementation of adaptive array antenna controls with image recognitions for DTN based vehicle-to-vehicle networks. In: The 9th International Workshop on Disaster and Emergency Information Network Systems (IWDENS 2017), pp. 633–638 (2017)

Performance Evaluation of a Smartphone-Based Active Learning System for Improving Learning Motivation During Study of a Difficult Subject

Noriyasu Yamamoto(✉) and Noriki Uchida

Department of Information and Communication Engineering,
Fukuoka Institute of Technology, 3-30-1 Wajiro-Higashi, Higashi-Ku,
Fukuoka 811-0295, Japan
{nori,n-uchida}@fit.ac.jp

Abstract. In our previous work, we presented an interactive learning process in order to increase the students learning motivation and the self-learning time. We proposed an Active Learning System (ALS) for student's self-learning. We evaluated the proposed system for each level (low, middle and high level) class and checked the student concentration using our proposed ALS. We found that for a difficult subject, there are only few high level students. The other students are middle or low level students. However, it is important that when using ALS all students should keep their learning motivation. In this paper, we described the results of the performance evaluations of the proposed ALS for a difficult subject. The evaluation results show that the ALS increased the learning efficiently of the students for the difficult subject.

1 Introduction

There is a total of 777 universities today in Japan. The "middle-level" universities (which are ranked at around the middle in terms of academic level) form the largest group in Japan and their undergraduates students are the main workforces. Because the group represents the "middle-level", academic capacities of students often vary significantly. For this reason, the teaching speed should be controlled when deciding the understanding level. When a lesson offers advanced contents or is fast-paced, good students find it satisfying, while middle level or low level students fail to catch up. If a lesson is designed too simple or slow, the middle level and low level students find it satisfying, while good students get frustrated. To solve this problem, there are many e-learning systems [1–9]. Also, today's classes increasingly utilize information terminals, such as notebook PCs and projectors. These information tools have enabled lecturers to offer more information to students. But, their effectiveness has no positive impact on students who lack concentration and motivation. The students just look at the projector screen and get the information materials. Their satisfaction level for the lecture is increased, but their scores do not improve.

Usage of the desktop computers and notebook PCs for lecture may be inconvenient and they occupy a lot of space. Therefore, it will be better that students use small and

L. Barolli et al. (Eds.): NBiS 2018, LNDECT 22, pp. 531–539, 2019.
https://doi.org/10.1007/978-3-319-98530-5_46

lightweight terminals like Personal Digital Assistant (PDA) devices. Also, because of wireless networks are spread over university campuses, it is easy to connect mobile terminals to the Internet and now the students can use mobile terminals in many lecture rooms, without needing a large-scale network facility. Our idea has been to use various information terminals and tools to boost students' concentration and motivation.

We considered a method of acquiring/utilizing the study record using smartphone in order to improve the students learning motivation [10]. During the lecture the students use the smartphone for learning. The results showed that the proposed study record system has a good effect for improving students' motivation for learning.

For the professors of the university, it is difficult to offer all necessary information to the students. In addition, they cannot provide the information to satisfy all students because the quantity of knowledge of each student attending a lecture is different. Therefore, for the lectures of a higher level than intermediate level, the students should study the learning materials by themselves.

In our previous work, it was presented an interactive learning process in order to increase the students learning motivation and the self-learning time [11]. However, the progress speed of a lecture was not good. To solve this problem, we proposed an Active Learning System (ALS) for student's self-learning [12, 13]. Also, to improve student's self-learning procedure, we proposed some functions for the ALS [14]. We performed experimental evaluation of our ALS and showed that the proposed ALS can improve student self-learning and concentration [15]. Although, the self-learning time and the examination score were increased, the number of student that passed examination didn't increase significantly. To solve this problem, we proposed the group discussion procedure to perform discussion efficiently [16].

The learning system proposed so far only gives a feedback of students' learning record to lecturer. We propose a mechanism to enhance the learning effects by using a record of students' learnings in the whole process of learning system [17].

The previous interfaces for interactive learning had limited adequacy in maintaining participants' concentration when their skill levels vary. Thus, we proposed the interface of the mobile devices (Smartphone/Pad) on our ALS to improve the learning concentration [18]. We presented the performance evaluation of ALS when using the improved human interface on the mobile devices [19]. When the lecture uses ALS, the average dropping out was half compared with the conventional lecture.

In addition, we showed the flow of the group discussion and the performance evaluation of ALS for high level class [20]. The evaluation results showed that the interactive lecture by using ALS increases the students' concentration, and the average dropping out for the lecture was half compared with the conventional ALS. Also, the method of group discussion increased the students' concentration for high level class. However, in the group discussion, some students who understood the lecture could teach other students who did not understand "study points". But, many students complained that they did not feel like a lecture style.

To solve this problem, we proposed a new method for dynamic group formation for the group discussion [21]. After the system decides the level of lecture understanding, the students who did not understand the lecture make the presentation and show the points that they did not understand. Also, the students who understood the lecture explain by presentation the questions or the points that other students did not understand. So, in the group discussion, each group member presents the points that they understood or did not understand. Thus, most of students can study considering their understanding level using above group discussion and they can keep their learning motivation.

However, during study of a difficult subject, there are few high level students who understand the lecture and many other students are intermediate or low level students who do not understand the lecture. In this paper, we evaluate the performance of our ALS system during study of a difficult subject.

The paper structure is as follows. In Sect. 2, we introduce a smartphone-based ALS with Interactive Lecture and Group Discussion. Then, in Sect. 3, we present the performance evolution of a smartphone-based ALS on the difficult subject. Finally, in Sect. 4, we give some conclusions and future work.

2 A Smartphone-Based Active Learning System with Interactive Lecture and Group Discussion

We developed an ALS and show ALC (Active Learning Cycle) of the proposed ALS in Fig. 1. The system facilitates a learning cycle consisting of a lecture, students' self-learning at home and group discussions. The student and the lecturer confirm the movement by setting each cycle and information by their smartphones.

2.1 Interactive Lecture During Active Learning Cycle of ALS

At the beginning of ALC, the lecturer performs the interactive lecture by confirming the understanding degree of the student using their smartphone in real time. Prior to the lecture, the lecturer prepares "study points".

"Small examination" refers to a mini quiz prepared for each study point on students' smartphone. A mini quiz consists of simple multiple-choice questions. "Understanding level" is set by the result of the "Small examination". "Lecture speed" suggests whether students find the lecture progress too fast or too slow. By "Understanding level" and "Lecture speed", students' understanding can be judged. These two functions are used as feedbacks through the application on students' smartphone (students' application). During the lecture, these data are recorded in the database to be reflected in the application for the lecturer.

Based on the record accumulated through button operations, the system creates logs and updates the database on real-time. Furthermore, the log automatically updates the study points in the database to be reflected on the lecturer's screen.

2.2 Self-learning During Active Learning Cycle of ALS

A lecturer can transmit the knowledge to the student effectively. After the lecture is finished, the student can read the lecture log by their smartphone on their self-learning stage. The student can review the lecture content using the lecture log anywhere and anytime. If students' self-learning times were long, then, it means that the students could keep their learning motivations.

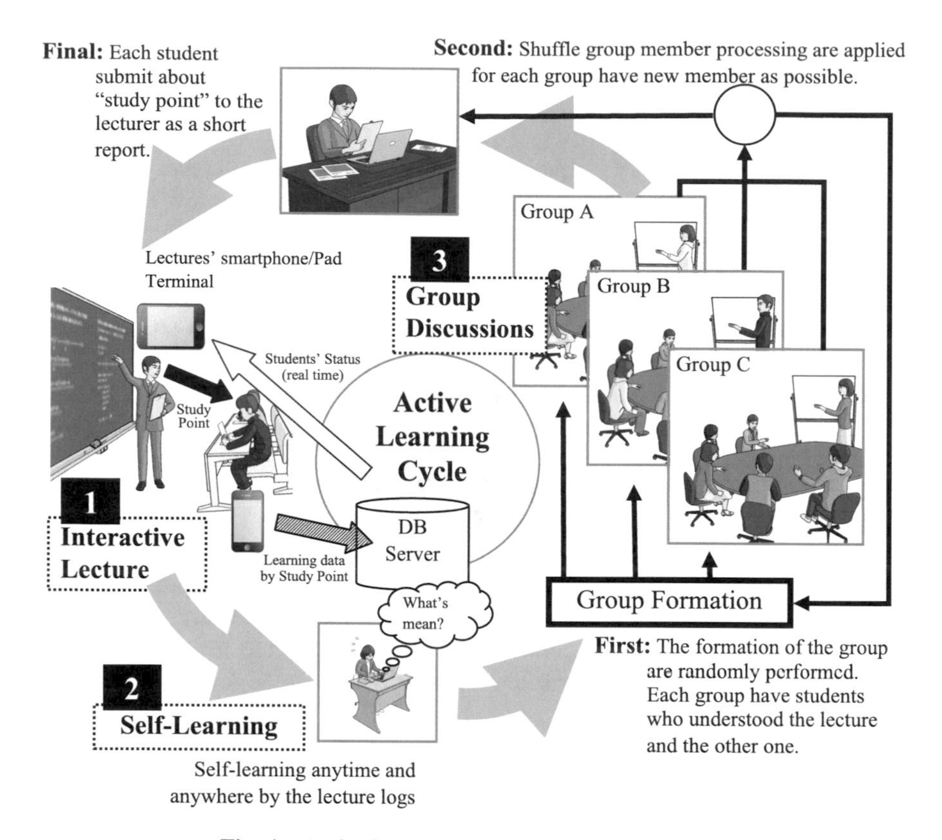

Fig. 1. ALC of proposed ALS using a smartphone.

2.3 Group Discussion During Active Learning Cycle of ALS

Then, after the self-learning stage, the student discusses the lecture content in small groups. There are the dynamic group formation after students' self-learning on ALS. It is a two steps discussion and the students send a short report to the lecturer.

Firstly, the formation of the group are randomly performed by the ALS server and the students are informed of the group members by their smartphone application. Each group is constructed by some students who understood the lecture and other students who did not understand the lecture. In the group discussion, each group member presents the points that understood or did not understand. Then, they will hold a short meeting for their presentations. After that, they submit the "study points" to the lecturer as a short report.

Secondly, shuffle group member processing is applied to each group in order to have new members by the ALS server. Then, the second group discussion is applied the same as first one.

Finally, each student submits the "study points" to the lecturer as a short report. Most of students can study by considering their understanding level using above group discussion and they can keep their learning motivation.

At the beginning of the next lecture, the student groups and the lecturer carry out open discussions based on the submitted reports and solve the problems that students may have. After the open discussion, the lecturer performs the next interactive lecture.

2.4 Performance of ALS

In this ALC, by adding the group learning and the open discussion, the understanding is increased and the lecturer can keep a fixed progress speed of the lecture. For each level (low, middle and high level) class, we showed that the students could keep high concentration using our proposed ALS [18–20]. However, for a difficult subject, there are few high level students. The other students are middle or low level students. However, it is important that all students keep their learning concentration and motivation. For this reason, we evaluate the performance of the proposed ALS for a difficult subject.

3 Performance Evaluation of Proposed ALS for a Difficult Subject

We performed experimental evaluation using the proposed ALS for a difficult subject. For the evaluation, we carried out experiments with a difficult programing subject using the conventional lecture and the proposed ALS. The number of students of these lectures were 20 and 22, for the conventional lecture and the proposed ALS, respectively. One of the lectures was performed for 90 min. We performed the lecture 13 times. The 14-th lecture is for the examination. For each class, 4 or 5 students were high level students (about 20% students), and the other students were middle or low level students (about 80% students) who could not understand the lecture for themselves.

We compared the students' concentration considering their self-learning time when using proposed ALS and not use it (the conventional lecture). Figure 2 shows the self-learning time comparing the high level students with middle or low level students during the conventional lecture. For the conventional lecture, the self-learning time is low and the middle or low level students could not keep their learning concentration.

In Fig. 3 is shown the self-learning time using the conventional ALS. The self-learning time for middle or low level students is almost the same with high level students.

In Fig. 4 is showing the self-learning time for the conventional lecture and the proposed ALS. The learning concentration using ALS is higher compared with the conventional lecture.

Table 1 shows the increasing rate of the self-learning time of proposed ALS compared with the conventional lecture. The average self-learning time for the high level students is 2 and for the middle or low level students is 3.9. Thus, the middle or low level student concentration is increased using ALS. These results indicate that the ALS increased the learning efficiently of the students for the difficult subject.

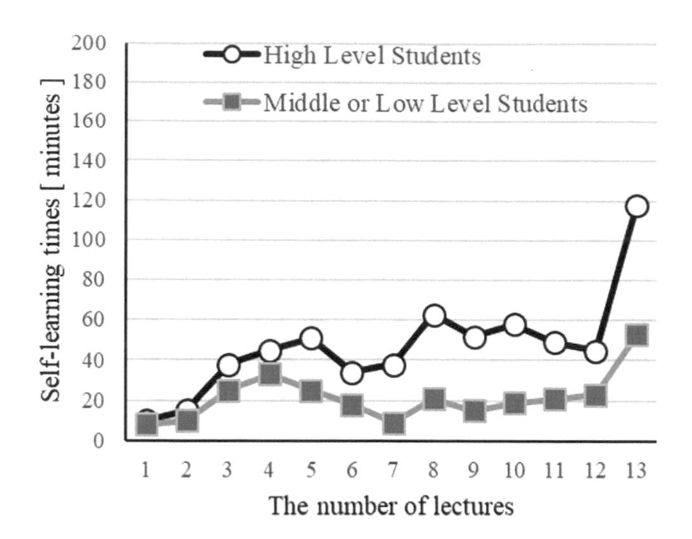

Fig. 2. Self-learning time for conventional lecture.

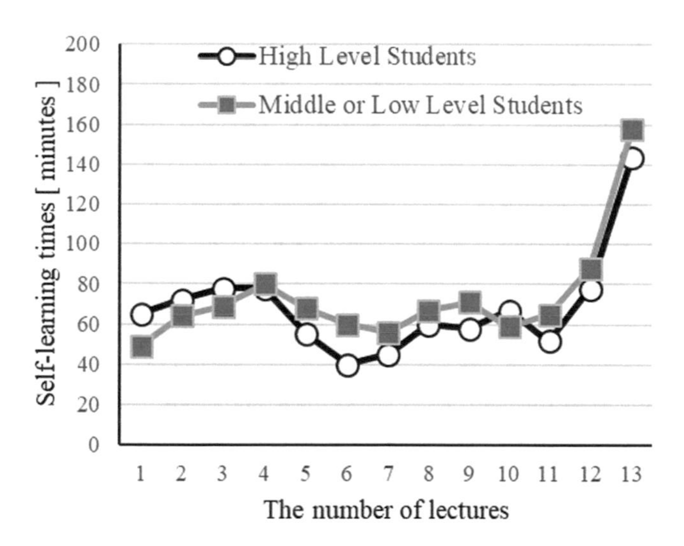

Fig. 3. Self-learning time using proposed ALS.

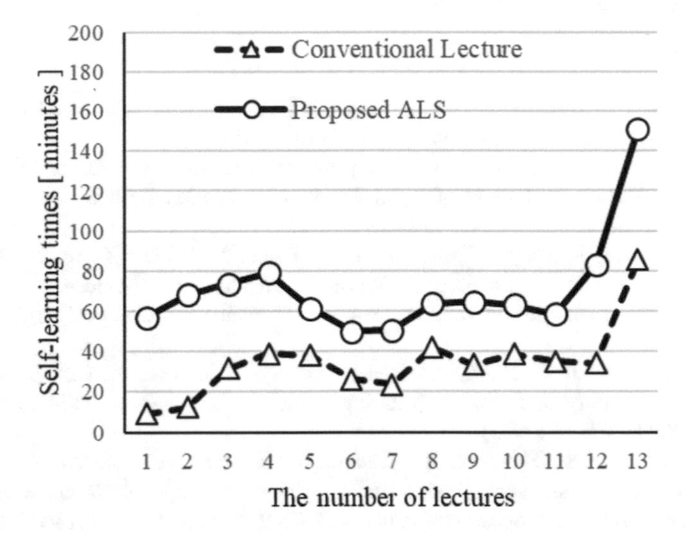

Fig. 4. The average of self-learning time for conventional lecture and proposed ALS.

Table 1. Increasing rate of the self-learning time on the proposed ALS compared with the conventional lecture.

	Number of Lectures													Average
	1	2	3	4	5	6	7	8	9	10	11	12	13	
High Level Students	6.5	4.8	2.1	1.7	1.1	1.2	1.2	1.0	1.1	1.2	1.1	1.7	1.2	**2.0**
Middle or Low Level Students	6.1	6.4	2.8	2.4	2.7	3.3	6.2	3.2	4.7	3.1	3.1	3.8	3.0	**3.9**
All Students	6.3	5.4	2.3	2.0	1.6	1.9	2.1	1.5	1.9	1.6	1.7	2.4	1.8	**2.5**

4 Conclusions

In our previous work, we presented an interactive learning process in order to increase the students learning motivation and the self-learning time. We proposed an ALS for student's self-learning. Also, we introduced the interface of the mobile devices on our ALS to improve the learning concentration. We evaluated the proposed system for each level (low, middle and high level) class and checked the student concentration using our proposed ALS. We found that for a difficult subject, there are only few high level students. The other students are middle or low level students. However, it is important that when using ALS all students should keep their learning motivation.

In this paper, we described the results of the performance evaluations of the proposed ALS for the difficult subject. The middle or low level student concentration is increased using ALS. The evaluation results indicate that the ALS increased the learning efficiently of the students for the difficult subject.

In the future, we plan to perform extensive evaluation of ALS. Also, we will develop an automatic algorithm to process students' reports.

References

1. Underwood, J., Szabo, A.: Academic offences and e-learning: individual propensities in cheating. Br. J. Educ. Technol. **34**(4), 467–477 (2003)
2. Harashima, H.: Creating a blended learning environment using Moodle. In: The Proceedings of the 20th Annual Conference of Japan Society of Educational Technology, pp. 241–242, September 2004
3. Brandl, K.: Are you ready to "Moodle"? Lang. Learn. Technol. **9**(2), 16–23 (2005)
4. Dagger, D., Connor, A., Lawless, S., Walsh, E., Wade, V.P.: Service-oriented e-learning platforms: from monolithic systems to flexible services. Internet Comput. IEEE **11**(3), 28–35 (2007)
5. Patcharee, B., Achmad, B., Achmad, H.T., Okawa, K., Murai, J.: Collaborating remote computer laboratory and distance learning approach for hands-on IT education. J. Inf. Process. **22**(1), 67–74 (2013)
6. Emi, K., Okuda, S., Kawachi, Y.: Building of an e-learning system with interactive whiteboard and with smartphones and/or tablets through electronic textbooks. In: Information Processing Society of Japan (IPSJ), IPSJ SIG Notes 2013-CE-118(3), 1–4, 2013-02-01 (2013)
7. Yamaguchi, S., Ohnichi, Y., Nichino, K.: An Efficient High Resolution Video Distribution System for the Lecture Using Blackboard Description. Technical Report of IEICE, 112 (190), pp. 115–119 (2013)
8. Hirayama, Y., Hirayama, S.: An analysis of the two-factor model of learning motivation in university students. Bulletin of Tokyo Kasei University, 1, Cultural and Social Science 41, pp. 101–105 (2001)
9. Ichihara, M., Arai, K.: Moderator effects of meta-cognition: a test in math of a motivational model. Jpn. J. Educ. Psychol. **54**(2), 199–210 (2006)
10. Yamamoto, N., Wakahara, T.: An interactive learning system using smartphone for improving students learning motivation. In: Information Technology Convergence Lecture Notes in Electrical Engineering, vol. 253, pp. 305–310 (2013)
11. Yamamoto, N.: An interactive learning system using Smartphone: improving students' learning motivation and self-learning. In: Proceedings of the 9th International Conference on Broadband and Wireless Computing, Communication and Applications (BWCCA-2014), pp. 428–431, November 2014
12. Yamamoto, N.: An active learning system using smartphone for improving students learning concentration. In: Proceeding of International Conference on Advanced Information Networking and Applications Workshops (WAINA-2015), pp. 199–203, March 2015
13. Yamamoto, N.: An interactive e-learning system for improving students motivation and self-learning by using smartphones. J. Mob. Multimedia (JMM) **11**(1&2), 67–75 (2015)
14. Yamamoto, N.: New functions for an active learning system to improve students self-learning and concentration. In: Proceedings of the 18th International Conference on Network-Based Information Systems (NBIS-2015), pp. 573–576, September 2015
15. Yamamoto, N.: Performance evaluation of an active learning system to improve students self-learning and concentration. In: Proceedings of the 10th International Conference on Broadband and Wireless Computing, Communication and Applications (BWCCA-2015), pp. 497–500, November 2015
16. Yamamoto, N.: Improvement of group discussion system for active learning using Smartphone. In: Proceedings of the 10th International Conference on Innovative Mobile and Internet Services in Ubiquitous Computing (IMIS-2016), pp. 143–148, July 2016

17. Yamamoto, N.: Improvement of study logging system for active learning using smartphone. In: Proceedings of the 11th International Conference on P2P, Parallel, Grid, Cloud and Internet Computing (3PGCIC–2016), pp. 845–851, November 2016
18. Yamamoto, N., Uchida, N.: Improvement of the interface of smartphone for an active learning with high learning concentration. In: Proceedings of the 31st International Conference on Advanced Information Networking and Applications Workshops (AINA-2017), pp. 531–534, March 2017
19. Yamamoto, N., Uchida, N.: Performance evaluation of a learning logger system for active learning using Smartphone. In: Proceedings of the 20th International Conference on Network-Based Information Systems (NBiS-2017), pp. 443–452, August 2017
20. Yamamoto, N., Uchida, N.: Performance evaluation of an active learning system using smartphone: a case study for high level class. In: Proceedings of the 6th International Conference on Emerging Internet, Data & Web Technologies (EIDWT-2018), pp. 152–160, March 2018
21. Yamamoto, N., Uchida, N.: Dynamic group formation for an active learning system using Smartphone to improve learning motivation. In: Proceedings of the 12th International Conference on Innovative Mobile and Internet Services in Ubiquitous Computing (IMIS-2018), pp. 183–189, March 2018

The 12th International Workshop on Advanced Distributed and Parallel Network Applications (ADPNA-2018)

A Causally Precedent Relation Among Messages in Topic-Based Publish/Subscribe Systems

Takumi Saito[1(✉)], Shigenari Nakamura[1], Tomoya Enokido[2], and Makoto Takizawa[1]

[1] Hosei University, Tokyo, Japan
takumi.saito.3j@stu.hosei.ac.jp, nakamura.shigenari@gmail.com, makoto.takizawa@computer.org
[2] Rissho University, Tokyo, Japan
eno@ris.ac.jp

Abstract. Event-driven publish/subscribe (PS) systems are widely used in various types of applications. In this paper, we consider a peer-to-peer (P2P) model of a topic-based PS system which is composed of peer processes (peers) with no centralized coordinator. Here, each peer publishes a message with publication topics while receiving messages whose publication topics are in the subscription topics of the peer. Each peer has to deliver every pair of messages related with respect to topics in the causal order of publication events. In this paper, a message is considered to carry objects whose meanings are denoted by topics. Based on the meanings of objects, we define an object-based-causally (OBC) precedent relation among messages. Based on the OBC precedent relation, we newly propose a protocol to topic-based-causally (TBC) deliver messages to peers. Here, each peer causally delivers event messages which are related with respect to topics. If a message m_1 OBC-precedes a message m_2, the message m_1 TBC-precedes m_2.

Keywords: Topic-based PS system · P2P model of PS system
OBC-precedent relation · TBC-precedent relation

1 Introduction

A publish/subscribe (PS) system [2,4,14,15] is a new event-driven model [3,5] of a distributed system which is composed of processes cooperating with one another in networks. In this paper, we consider a peer-to-peer (P2P) model of a topic-based P2P system (P2PPS system) [8–10,13] where multiple peer processes (peers) are cooperating with one another by exchanging messages in overlay networks. Here, each peer p_i subscribes a collection $p_i.S$ of interesting topics $p_i.ST$. This means, a peer p_i would like to receive messages on the subscription topics $p_i.ST$. The peer p_i receives a message m on some topics in the subscription

L. Barolli et al. (Eds.): NBiS 2018, LNDECT 22, pp. 543–553, 2019.
https://doi.org/10.1007/978-3-319-98530-5_47

$p_i.ST$. Here, the peer p_i is a target peer of message m. According to the distributed coordination, messages published by different peers may be received by different peers not in the same order while messages published by a peer are received by every peer in the publishing order.

In this paper, we consider a message carries objects and the meanings of objects are denoted by topics. A peer p_i includes objects stored in the peer p_i or objects created by the peer p_i to messages and then publishes the messages. In this paper, the topics, i.e. publication topics of a message are considered to be a collection of topics of objects carried by the message. Suppose a peer p_i receives a message m_1 and then publishes a message m_2. Here, the message m_1 causally precedes the message m_2 according to the traditional causality theory [7,12]. However, if a pair of the messages m_1 and m_2 are not related with each other, e.g. the messages m_1 and m_2 carry objects which are not related, the causally precedent relation among the messages m_1 and m_2 is meaningless. If objects of the message m_2 are related with objects of the message m_1, the message m_1 is considered to causally precede the message m_2. For example, if the message m_2 carries an object o_1 which is carried by the message m_1, the message m_2 is related with the message m_1. Here, the message m_1 causally precedes the message m_2 in terms of objects and every target peer is regard to receive the messages m_1 and m_2 in the causal order. We define an $object - based - causally$ (OBC) precedent relation among messages. Each target peer is required to receive a pair of messages in the OBC precedent order. Each peer p_i receives a message m whose publication topics interest the peer p_i, i.e. some publication topic of the message m is in the subscription topics of the peer p_i.

In order to order messages in the OBC-precedent order, each peer p_i has to keep in record every object of messages which the peer p_i receives. We need a simpler protocol to receive messages in the OBC-precedent order. Based on the OBC-precedent relation, we define a topic-based-causally (TBC) precedent relation among messages in terms of topics. A message m_1 TBC-precedes a message m_2 if the message m_1 OBC-precedes the message m_2. However, even if the message m_1 TBC-precedes the message m_2, the message m_1 may not OBC-precede the message m_2.

In Sect. 2, we present a system model of the P2PPS system. In Sect. 3, we newly discuss the OBC-precedent relation among messages. In Sect. 4, we discuss the TBC-precedent relation among messages. In Sect. 5, we propose a TBC protocol to causally deliver messages in P2PPS systems.

2 System Model

A peer-to-peer publish/subscribe (P2PPS) system [1,9–11,13] is composed of peer processes (peers) p_1, ..., p_n ($n \geq 1$) which are cooperating with one another by exchanging messages in overlay networks (Fig. 1). Here, there is no centralized coordinator among the peers. Each peer p_i can both publish messages and subscribe messages which the peer p_i would like to receive. We consider a topic-based P2PPS system. The publication and subscription of each peer are specified in terms of topics.

A peer publishes a message m and the message m is received only by a peer which is interested in the content of the message m. The content [6] of a message is considered to be a collection of objects in this paper. A peer who is interested in the content of a message m is a *target* peer of the message m. Messages published by a peer are received by a every target peer in the publication order. However, if a pair of messages m_1 and m_2 published by a pair of different peers p_i and p_k ($p_i \neq p_k$), respectively, the messages m_1 and m_2 may not be received by common target peers in different order since there is no centralized coordinator. Each peer has to receives every pair of related messages in the causally precedent order.

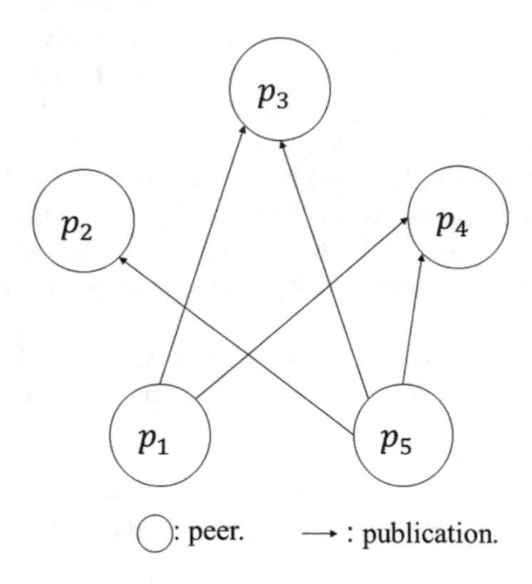

Fig. 1. P2PPS model.

3 An Object-Based-Causally (OBC) Precedent Relation

There are ways to describe the meanings of message contents. In one way, contents are denoted by topics [16]. In topic-based publish/subscribe (PS) systems [13,16], interesting topics are subscribed by a peer. This means, a peer p_i would like to receive messages whose topics interest the peer p_i, i.e. the peer p_i is interested in the topics. A peer p_i publishes a message m with topics which denote the meanings of the message m. A message is sent to target peers which are interested in the publication topics of the message. Let T be a set of topics in a P2PPS system.

Another way, the meanings of a message m are described by objects which are carried by the message m. Let O be a set of objects in the system S. Each peer p_i is equipped with a database $p_i.DB$ ($\subseteq O$) where objects are stored. Each object o_i is characterized in terms of topics. Here, $o_i.T$ ($\subseteq T$) is a subset of topics which denote the meanings of an object o_i. In the database $p_i.DB$, each peer p_i stores objects which are carried by messages which the peer p_i receives. In addition, objects which the peer p_i creates are stored in the database $p_i.DB$.

A peer p_i creates a message m by including objects in the database $p_i.DB$ and then publishes the message m. A message m thus carries a subset $m.O$ ($\subseteq p_i.DB$) of objects in the database $p_i.DB$. The meanings of a message m are characterized in terms of topics of objects in $m.O$ carried by the message m. $m.T$ is a subset of topics which denote the meanings of objects in $m.O$, i.e. $m.T = \cup_{o_i \in m.O}\ o_i.T$. The topic set $m.T$ is referred to as *publication* of a message m.

Each peer p_i subscribes a subset $p_i.ST$ ($\subseteq T$) of interesting topics. That is, the peer p_i would like to receive messages whose publication topics are topics subscribed by the peer p_i. The subset $p_i.ST$ of topics is referred to a *subscription* of the peer p_i. If a message m is published, a peer p_i receives only the message m whose publication topics interest the peer p_i, i.e. $m.T \cap p_i.ST \neq \emptyset$. Otherwise, the message m is not received by the peer p_i. Thus, a peer p_i where $p_i.ST \cap m.T \neq \emptyset$ is a target peer of a message m. On receipt of a message m, a target peer p_i stores objects in $m.O$ carried by the message m in the database $p_i.DB$.

Each peer p_i manipulates objects and creates a new object in the database $p_i.DB$. For example, a peer p_i updates values of an object o_i in the database $p_i.DB$. A new object o_i is created by manipulating objects in the database $p_i.DB$.

Suppose a peer p_i publishes a message m which includes objects in the database $p_i.DB$. A peer p_i receives the message m in the following two types of ways:

1. A peer p_i *explicitly* receives a message m, iff (if and only if) $m.T \subseteq p_i.ST$.
2. A peer p_i *implicitly* receives a message m, iff $m.T \cap p_i.ST \neq \emptyset$ and $m.T$ - $p_i.ST \neq \emptyset$

In the first explicit receipt, every publication topic t ($\in m.T$) of a message m is included in the subscription topics $p_i.ST$ of a peer p_i, i.e. $m.T \subseteq p_i.ST$. If some topic t in the publication $m.T$ of the message m is not in the subscription $p_i.T$ of a peer p_i, the message m is not received by the peer p_i, i.e. the peer p_i is not a target peer of the message m. If a peer p_i is interested in everything of the message m, the peer p_i explicitly receives the message m.

In the second implicit receipt, if some publication topic t ($\in m.T$) of a message m interests a peer p_i, i.e. $m.T \cap p_i.ST \neq \emptyset$, the peer p_i implicitly receives the message m. That is, a peer p_i receives a message m if the peer p_i is interested in something of the message m.

Each peer p_i publishes messages while receiving messages in the P2PPS system. Here, any pair of messages m_1 and m_2 published by each peer p_i are received by a target peer p_j of messages m_1 and m_2 in the publication order. However, some pair of messages published by different peers may be received by a pair of different target peers p_i and p_j not in the same order. For example, a peer p_i receives the message m_1 then the message m_2 while the other peer p_j receives the message m_2 then the message m_1. If a peer p_i publishes a message m_i after receiving another message m_j published by a peer p_j, the message m_i causally

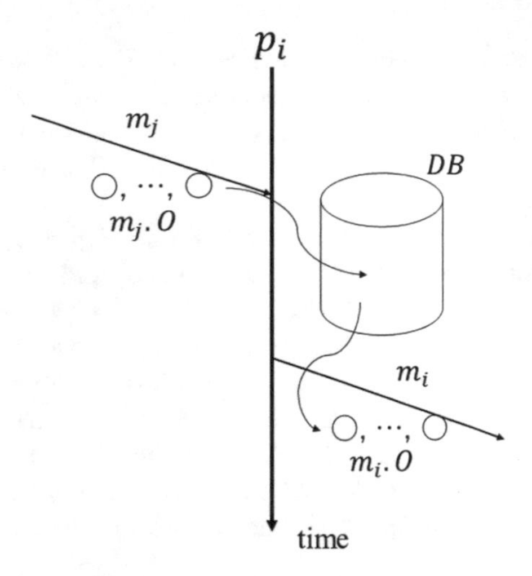

Fig. 2. Receipt and publication of messages.

precedes the message m_j ($m_i \rightarrow m_j$) as discussed in the causality theory [7]. Here, the message m_j carries a collection $m_j.O$ of objects to the peer p_i and the objects in $m_j.O$ are stored in the database $p_i.DB$ as shown in Fig. 2. After receiving the message m_j, the target peer p_i manipulates objects in the database $p_i.DB$. Here, the peer p_i may create new objects by manipulating objects in the database $p_i.DB$. Then, the peer p_i publishes a message m_i including objects in the database $p_i.DB$. Suppose an object o_j ($\in m_j.O$) carried by the message m_j is related with some object o_i ($\in m_i.O$) of the message m_i. Here, the message m_i is defined to be related with the message m_j. An object o_i of the message m is created by manipulating an object o_j in $m_j.O$. Here, the object o_i depends on the object o_j carried by the message m_j. The message m_i is also related with the message m_j, i.e. the message m_i can not exist without the message m_j. If no object of the message m_i is related with any object o_j of the message m_j, a pair of the messages m_i and m_j are not related. In this paper, a message m_i *object-based-causally* (*OBC*) precedes a message m_j iff not only the message m_i causally precedes the message m_j ($m_i \rightarrow m_j$) but also the message m_i is related with the message m_j in terms of objects.

We discuss how a pair of messages are related with each other in terms of objects. Let o_i be an object which a message m_i carries, i.e. $o_i \in m_i.O$. Let O_i be a subset of objects o_{i1}, ..., o_{il_i} ($l_i \geq 1$) in the database $p_i.DB$ of a peer p_i. We consider a function f_i^h: $O^h \rightarrow O$ in which an object o_i in the set O is related with objects $\langle$ o_{i1}, ..., o_{il_i} $\rangle \in O^h$ in the database $p_i.DB$, i.e. $o_i = f_i^h(o_{i1}$, ..., $o_{ili})$. There are the following types of a function f_i^h:

1. $o_i = f_i^1$ (o_j), where $o_i = o_j$.
2. $o_i = f_i^{li}$ $(o_{i1}$, ..., $o_{il_i})$. The object o_i does not exist unless the objects o_{i1}, ..., o_{il_i} exist.

In type 1, suppose a message m_j carries an object o_i to a target peer p_i, $o_i \in m_j.O$ and then the peer p_i publishes a message m_i which carries the same object o_i ($\in m_i.O$). Here, $o_i = o_j = f_i^1$ (o_j). That is, if the message m_i carries an object o_i in the message m_j, i.e. $m_i.O \cap m_j.O \neq \emptyset$, the message m_i is related with the message m_j.

In type 2, suppose a message m_j carries some object o_j which stands for a question. A peer p_i publishes a question message m_j and a target peer p_i receives the question message m_j. Then, a peer p_i makes an object o_i which denotes an answer of the question object o_j and publishes an answer message m_i with the answer object o_i. Here, $o_i = f_i^1$ (o_j), i.e. the answer object o_i does not exist without the question object o_j. Thus, an object o_i in a message m_i is created by manipulating objects o_{i1}, ..., o_{il_i}. The object o_i cannot exist without the objects o_{i1}, ..., o_{il_i}. Here, if a message m_j carries an object o_{ih} to a peer p_i and then peer p_i publishes a message m_i with the object o_i, the message m_i is related with the message m_j.

We define a types of causally precedent relation of messages which is meaningful in terms objects carried by the messages:

[Definition] A message m_i *object - based - causally* (*OBC*) *precedes* a message m_j ($m_i \Rightarrow m_j$) iff (if and only if) the following conditions are satisfied:

1. The message m_i causally precedes the message m_j ($m_i \rightarrow m_j$).
2. For some object o_j in $m_j.O$ and a subset of objects o_{i1}, ..., o_{il_i} in $m_i.O$, there is a function f_j such that $o_j = f_j^{l_i}$ $(o_{i1}, ..., o_{il_i})$.

It is not easy, almost impossible for each peer p_i to keep in record every object carried by messages received by the peer p_i. It takes time to check if objects of each message and objects in the database are related. We consider a topic-based PS system where each peer p_i receives the messages only based on publication topics of the message and the subscription topics of the peer p_i. It is simpler to decide which message causally precedes another message by using publication topics carried by messages and subscription topics of peers. We consider relations on objects and topics.

[Properties] Suppose $o_i = f_i^{l_i}$ $(o_{i1}, ..., o_{il_i})$. There are the following cases:

1. $o_i.T = o_{i1}.T \cup ... \cup o_{il_i}.T$.
2. $o_i.T \subseteq o_{i1}.T \cup ... \cup o_{il_i}.T$.
3. $o_i.T \supseteq o_{i1}.T \cup ... \cup o_{il_i}.T$.

In this paper, we assume $o_i.T = o_{i1}.T \cup ... \cup o_{il_i}.T$. This means, an object o_i is assumed to be composed of all the objects o_{i1}, ..., o_{il_i}.

4 A Topic-Based-Causally (TBC) Precedent Relation

In P2PPS systems, peers autonomously publish and receive messages. In the topic-based P2PPS system, a message m with publication topics $m.T$ is received

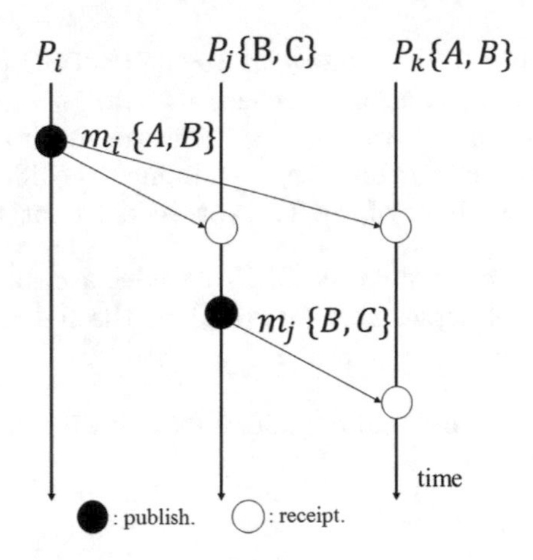

Fig. 3. Causally precedent relation on messages.

by a peer p_i if $m.T \cap p_i.ST \neq \phi$. Here, the peer p_i is a target peer of the message m. Suppose a pair of peers p_i and p_j publish messages m_i and m_j, respectively. Each of target peers p_h and p_k receives both the messages m_i and m_j since the peers p_h and p_k are the common target peers of the messages m_i and m_j. Here, the peer p_h receives the message m_i and then the message m_j. On the hand, the other peer p_k receives the message m_j and then the message m_i depending on network delay time because there is no centralized coordinator. Thus, messages published by different peers may be received by different target peers not in the same order. If the message m_i OBC-precedes the message m_j ($m_i \Rightarrow m_j$), each target peer p_k has to receive the message m_i before the message m_j.

Suppose there are three peers p_i p_j, and p_k where $p_j.ST = \{B, C\}$ and $p_k.ST = \{A, B\}$ as shown in Fig. 3. Suppose a peer p_i publishes a message m_i with publication $m_i.T = \{A, B\}$. A pair of peers p_j and p_k are target peers of the message m_i since $m.T \cap p_i.ST = \{A\}$ and $m_i.T \cap p_j.ST = \{A\}$. After receiving the message m_i, the peer p_j publishes a message m_j with publication $m_j.T = \{B, C\}$. Here, the message m_i causally precedes the message m_j ($m_i \rightarrow m_j$) according to the causality theory. Since $m_i.T \cap m_j.T = \{B\} \neq \emptyset$, a pair of the messages m_i and m_j are related with each other with respect to the topic B. That is, the message m_j is considered to carry objects on the topic B which are related with objects of the message m_j with respect to the topic B. Here, the message m_i OBC-precedes the message m_j ($m_i \Rightarrow m_j$). The peer p_k is required to receives the message m_i before the message m_j since $p_k.ST \cap m_j.T = \{B\} \neq \emptyset$.

[Definition] A message m_1 $topic-based-causally$ (TBC) precedes a message m_2 with respect to a subset A ($\subseteq T$) of topics ($m_1 \mapsto_A m_2$) iff one of the following conditions holds:

1. The message m_1 causally precedes the message m_2 ($m_1 \rightarrow m_2$).
2. $A \subseteq m_1.T \cap m_2.T$.

In the example of Fig. 3, the message m_i explicitly TBC-precedes the message m_j ($m_i \mapsto_B m_j$) since m_i causally precedes m_j ($m_i \rightarrow m_j$) and $m_i.T \cap m_j.T$ $= \{B\}$. Thus, since $m_i \Rightarrow m_j$, $m_i \mapsto_B m_j$. However, both the messages m_i and m_j carry object of the topic R but the objects may be different.

We discuss the transitivity of the TBC-precedent relation $\mapsto$.

[Definition] A message m_1 directly TBC-precedes a message m_2 with respect to a subset A ($\subseteq T$) of topics ($m_1 \hookrightarrow_A m_2$) iff the following conditions hold (Fig. 4):

1. A peer p_i receives the message m_1 and then publishes the message m_2.
2. $A \subseteq m_1.T \cap m_2.T$.

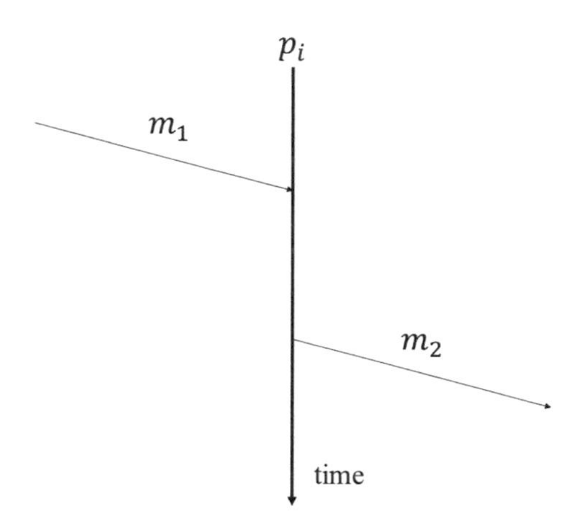

Fig. 4. Direct TBC precedence.

By using the direct TBC-precedent relation $\mapsto$, we define an explicitly TBC-precedent relation $\Rightarrow$ among messages.

[Definition] A message m_1 *explicitly-TBC (ETBC) precedes* a message m_2 with respect to a subset A of topics ($m_1 \Rightarrow_A m_2$) iff the message m_1 causally precedes the message m_2 ($m_1 \rightarrow m_2$), $A \subseteq m_1.T \cap m_2.T$, and one of the following conditions holds:

1. The message m_1 directly TBC-precedes the message m_2 with respect to the subset A ($m_1 \rightarrow_A m_2$).
2. There is some message m_3 such that $m_1 \Rightarrow_A m_3$ and m_3 directly TBC-precedes m_2 with respect to the subset A. ($m_3 \rightarrow_A m_2$).

Suppose there are four peers p_i, p_j, p_h and p_k as shown Fig. 5. The peer p_i first publishes a message m_i. Then, the peer p_j publishes a message m_j after receiving the message m_i. Lastly, the peer p_h publishes a message m_h after

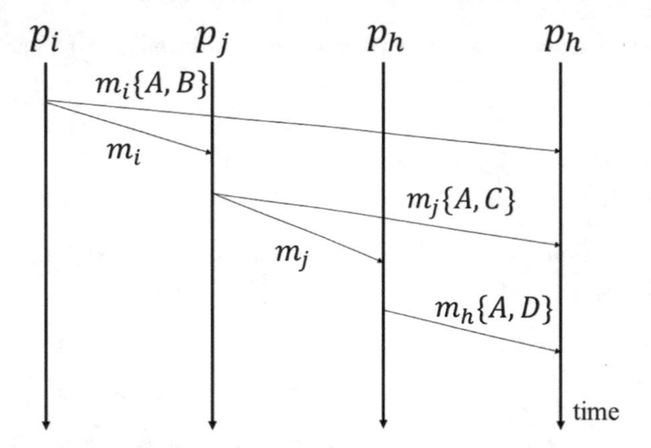

Fig. 5. Transitivity of TBC-precedent relation.

receiving the message m_j. Suppose the peers p_i, p_j and p_k subscribes topics, $m_i.T = \{A, B\}$, $m_j.T = \{A, C\}$, and $m_k.T = \{A, D\}$, respectively. Here, the message m_i directly TBC-precedes the message m_j ($m_i \hookrightarrow_{\{A\}} m_j$) since $m_i \rightarrow m_j$ and $\{A\} \subseteq m_i.ST \cap m_j.ST$. The message m_j directly TBC-precedes the message m_h ($m_j \hookrightarrow_{\{A\}} m_h$) since $m_j \rightarrow m_k$ and $\{A\} \subseteq m_j.T \cap m_k.T$. Hence, the message m_i explicitly TBC-precedes the messages m_j and m_k with respect to a topic A, i.e. $m_i \Rightarrow_A m_j$ and $m_i \Rightarrow_{\{A\}} m_h$. It is noted, every message m_i, m_j, and m_k includes a topic A and $m_i \hookrightarrow_{\{A\}} m_j$ and $m_j \hookrightarrow_{\{A\}} m_k$.

5 A TBC Protocol

We newly propose a TBC protocol to ETBC-deliver messages in a P2PPS system. In papers [8–11], the topic vector is proposed to ETBC-deliver messages. There are topics t_1, ..., t_m ($m \geq 1$), i.e. T $= \{t_1, ..., t_m\}$. Each peer p_i manipulates a topic vector $TV = \langle tv_1, ..., tv_m \rangle$ for a set t of the topics t_1, ..., t_m. Initially, each element TV_k is 0 for every topic t_k ($k = 1, ..., m$). We assume every message is broadcast to every peer p_i in a system. In this paper, messages are causally ordered in the linear clock [7]. Each message m carries the linear time $m.L$ which shows when a peer sends the message m. Suppose a peer p_i receives a pair of messages m_1 and m_2. If $m_1.L < m_2.L$, m_1 precedes m_2 in the peer p_i.

Suppose a peer p_i publishes a message m with publication $m.T$. The topic vector TV is manipulated and the message m is published by the peer p_i as follows:

[**Publication**]
1. $tv_k = tv_k + 1$ for every topic t_k in the publish $m.T$;
2. $m.TV = TV$;
3. **publish** m;

On arrived of a message m, a peer p_i manipulated the topic vector TV as follows.

[**Receipt**]
1. **receive** m if $m.T \cap p_i.ST \neq \phi$;
2. $TV_k = \max\ (tv_k,\ m.tv_k)$ for every topic t_k;

For a pair of topic vectors $A = \langle a_1, ..., a_m \rangle$ and $B = \langle b_1, ..., b_m \rangle$ of a topic set $T = \langle t_1, ..., t_m \rangle$ and a subset V ($\subseteq T$) of the topics;

1. $A >_V B$ iff $a_k > b_k$ for every topic t_k in V.
2. $A \equiv_V B$ iff $a_k = b_k$ for every topic t_k in V.
3. $A \geq_V B$ iff $A >_V B$ or $A \equiv_V B$.
4. $A \mid_v B$ (A and B are an comparable) iff neither $A \geq_V B$ nor $A \leq_V B$.

A pair of messages m_1 and m_2 received by a peer p_i are ordered in the TBC protocol as follows.

1. A message m_1 ETBC-precedes a message m_2 with respect to a subset V ($\subseteq T$) of the topics ($m_1 \Rightarrow_V m_2$) if $m_1.TV \leq_V m_2.TV$ and $m_1.L < m_2.L$.

For a pair of messages m_1 and m_2, if m_1 OBC-precedes m_2 ($m_1 \rightarrow m_2$), m_1 explicitly TBC-precedes m_2 with respect to some topic t in $m_1.ST$ and $m_2.ST$ ($m_1 \Rightarrow_{\{t\}} m_2$). However, even if $m_1 \Rightarrow_{\{t\}} m_2$, $m_1 \rightarrow m_2$ may not hold.

6 Concluding Remarks

In this paper, we discuss a peer-to-peer (P2P) model of a topic-base publish/subscribe (PS) system. First, we defined a causally precedent relation on messages in terms of objects carried by messages. A message m_1 object-based causally (OBC) precedes a message m_2 iff m_1 causally precedes m_2 and objects in the message m_2 are related with objects in the message m_2. Then, we define the topic-based causally (TBC) precedent relations among messages.

Acknowledgements. This work was supported by JSPS KAKENHI grant number 15H0295.

References

1. Google alert. http://www.google.com/alerts
2. Google cloud pub/sub. https://cloud.google.com/pubsub/docs/overview?hl=ja
3. Blanco, R., Alencar, P.: Event models in distributed event based systems. In: Principles and Applications of Distributed Event-Based Systems, pp. 19–42 (2010)
4. Eugster, P., Felber, P., Guerraoui, R., Kermarrec, A.: The many faces of publish/subscribe. ACM Comput. Surv. **35**(2), 114–131 (2003)
5. Hinze, A., Buchmann, A.: Principles and Applications of Distributed Event-Based Systems. IGI Global, Hershey (2010)

6. Jacobson, V., Smetters, D., Thornton, J., Plass, M., Briggs, N., Braynard, R.: Networking named content. In: Proceedings of the 5th International Conference on Emerging Networking Experiments and Technologies (CoNEXT 2009), pp. 1–12 (2009)
7. Lamport, L.: Time, clocks, and the ordering of event in a distributed systems. Commun. ACM **21**(7), 558–565 (1978)
8. Nakamura, S., Enokido, T., Takizawa, M., Ogiela, L.: An information flow control models in a topic-based publish/subscribe systems. J. High Speed Netw. **24**(3), 243–257 (2018)
9. Nakayama, H., Duolikun, D., Enokido, T., Takizawa, M.: Selective delivery of event messages in peer-to-peer topic-based publish/subscribe systems. In: Proceedings of the 18th International Conference on Network-Based Information Systems (NBiS-2015), pp. 379–386 (2015)
10. Nakayama, H., Duolikun, D., Enokido, T., Takizawa, M.: Reduction of unnecessarily ordered event messages in peer-to-peer model of topic-based publish/subscribe systems. In: Proceedings of IEEE the 30th International Conference on Advanced Information Networking and Applications (AINA-2016), pp. 1160–1167 (2016)
11. Nakayama, H., Ogawa, E., Nakamura, S., Enokido, T., Takizawa, M.: Topic-based selective delivery of event messages in peer-to-peer model of publish/subscribe systems in heterogeneous networks. In: Proceedings of the 18th International Conference on Network-Based Information Systems (WAINA-2017), pp. 1162–1168 (2017)
12. Raynal, M.: Distributed Algorithms for Message-Passing Systems. Springer, Heidelberg (2013)
13. Setty, V., van Steen, M., Vintenberg, R., Voulgais, S.: PolderCast: fast, robust, and scalable architecture for P2P topic-based pub/sub. In: Proceedings of ACM/IFIP/USENIX 13th International Conference on Middleware (Middleware-2012), pp. 271–291 (2012)
14. Tarkoma, S.: Publish/Subscribe System: Design and Principles, 1st edn. Wiley, New York (2012)
15. Tarkoma, S., Rin, M., Visala, K.: The publish/subscribe internet routing paradigm (PSIRP): designing the future internet architecture. In: Future Internet Assembly, pp. 102–111 (2009)
16. Yamamoto, Y., Hayashibara, N.: Merging topic groups of a publish/subscribe system in causal order. In: Proceedings of the 31st International Conference on Advanced Information Networking and Applications Workshops (WAINA-2017), pp. 172–177 (2017)

Fog-Cloud Based Platform for Utilization of Resources Using Load Balancing Technique

Nouman Ahmad[1], Nadeem Javaid[1](✉), Mubashar Mehmood[1], Mansoor Hayat[2], Atta Ullah[2], and Haseeb Ahmad Khan[2]

[1] COMSATS University, Islamabad 44000, Pakistan
nadeemjavaidqau@gmail.com
[2] Institute of Southern Punjab, Multan 66000, Pakistan
http://www.njavaid.com/

Abstract. Fog based computing concept is used in smart grid (SG) to reduce the load on the cloud. However, fog covers the small geographical area by storing data temporarily and send furnished data to the cloud for long-term storage. In this paper, a fog and cloud base platform integrated is proposed for the effective management of energy in the smart buildings. A request generated from a cluster of building at demand side end is to be managed by Fog. For this purpose six fogs are considered for three different regions including Europe, Africa an North America. Moreover, each cluster is connected to fog, comprises of the multiple number of buildings. Each cluster contains thirty buildings and these buildings consisted 10 homes with multiple smart appliances. To fulfill the energy demand of consumer, Microgrids (MGs) are used through fog. These MGs are placed nearby the buildings. For effective energy utilization in smart buildings, the load on fog and cloud is managed by load balancing techniques using Virtual Machines (VMs). Different algorithms are used, such as Throttled, Round Robin (RR) and First Fit (FF) for load balancing techniques. These techniques are compared for closest data center service broker policy. This service broker policy is used for best fog selection. Although using the proposed policy, three load balancing algorithms are used to compare the result among them. The results showed that proposed policy outperforms cost wise.

Keywords: Cloud computing · Fog computing · Virtual machine
Load balancing · Micro grids and Smart grid

1 Introduction

The integration of fog-cloud based computing concept is introduced in SG for effective utilization of energy [1]. SG is an upgradation of traditional grid with the integration of Information and Communication Technology (ICT). SG is a two way communication medium and uses smart meters at the user end.

L. Barolli et al. (Eds.): NBiS 2018, LNDECT 22, pp. 554–567, 2019.
https://doi.org/10.1007/978-3-319-98530-5_48

It is integrated with multiple energy resources and avoid energy wastage. SM is installed at various commercial, residential and industrial areas. It incorporates Internet of Things (IoTs) for the efficient communication. Hence, SM is the two way communication medium it monitors the energy consumption as well as energy generation.

Cloud computing is a platform which provides shared resources to all the users. Cloud provides three types of services namely; Infrastructure-as-a-Service (IaaS), Platform-as-a-Service (Paas) and Software-as-a-Service (SaaS) [1]. The Fig. 1 depicts the three services of cloud. The main idea of cloud computing is reusability of shared resources. SaaS provides model of cloud computing which allows usage of software application for end user. PaaS provides runtime environment for application deployment and deployment tools. IaaS provides access to fundamental resources; physical machine and virtual machine [3].

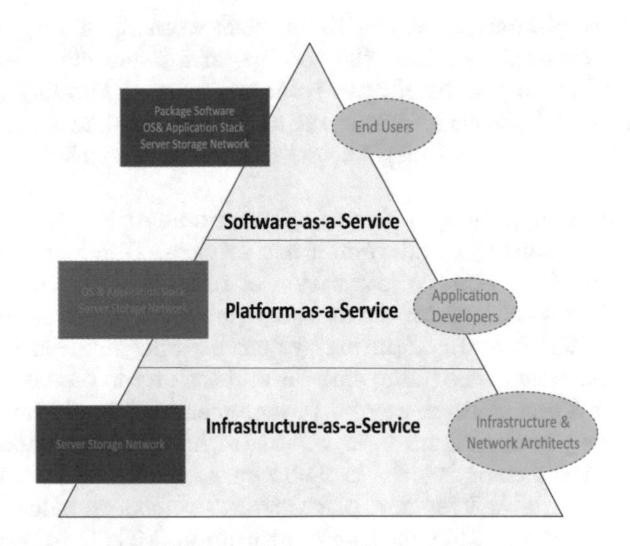

Fig. 1. Cloud computing services

The integration of fog-cloud computing makes SG more enhanced. Cloud computing, has shared resources with high latency rate, lack of security, However, fog provides the local resources rather than shared. [4]. As the number of users increase in the cloud [5] the security issues occurs more frequently. With the emergence of fog computing, the security issues reduce significantly. The integrated fog-cloud based computing platform makes SG make more efficient and reliable. The fog computing is more secure and less latency over network compared to cloud computing. It is capable to store data for short span of time. After n number of time, it transmit the furnished data to the cloud.

The paper is summarized, A load balancing optimization techniques is an essential factor for effective utilization of resources on fog computing. Load balancing is used for allocation VMs on basis of storage RAM configuration and requirement. These techniques used to achieve minimum response time. In this paper following techniques are discuss:

1. RR
2. Throttled
3. FF

Related work of this study is appended in Sects. 2 and 3 Comprised of system model of present study, Sect. 4 contains the proposed solution, Simulation and results of the study are given in Sect. 5, and finally, the conclusion of study given in Sect. 6.

1.1 Motivation

A fog-cloud based platform using the SG for effective energy management present in [1,2]. In [1], authors described the concept of fog and cloud base platform integrated together for the distributed resource in smart buildings. The author used different load balancing techniques for energy load management. These techniques are round robin (RR), throttled and compare these with multiple Broker polices.

In [2], the authors propose a cloud-fog base framework for data information management in SG using cloud computing services. Different load balancing techniques are used for effective utilization of resources. These load balancing techniques are throttles, round robin and particle swarm optimization. In [4], the author proposed the Cloud computing system for upcoming power generation grid, on the other hand, scheduling electric vehicles on the base of charging and discharging in a decentralized way to procurement the load shuffling facility. However, in this paper authors used FF algorithm for load balancing. FF is used to allocate the memory blocks to each request generated form user end. FF schedules the requests in VMs and places them in memory blocks. Afterwards it sends these requests to MGs for energy utilization. MGs then responds to the request and provide the energy to that cluster. If an MG is unable to fulfill the request then it will take that request to the cloud. Cloud then connects to the utility for energy. The utility provides the energy to the specific MG.

1.2 Contribution

In this paper, a fog-based framework integrated with SG is presented. The FF algorithm is used for load balancing. It selects the VMs on the basis of partitioning the memory blocks. A signing each block to the process job generated from the user end, if the number of jobs requests has maximum length, than it has placed in waiting queue. Each region consists of large consumers, which send requests on fog. The contributions of those are enlisted below: Three regions are considered for load balancing using fog as a middle layer. VMs are installed on the fogs and FF algorithm is used for selecting VMs. Consumer's data is stored temporarily on fog, after a specific span of time the data is sent to cloud for permanent storage. Each apartment includes renewable energy resources. FF is compared with RR and throttled algorithm. The proposed technique outperformed other two techniques on cost. However, RR and throttled outperformed proposed technique in processing time, response time.

2 Related Work

In [6], presented cloud to fog to consumer (C2F2C) framework used for resource management at residential buildings. The framework comprises of three layers. Fog layers is used as middle layer to assign the resources. It has low latency and high reliability as compared to cloud computing. Three optimization techniques used; FF, RR and equally spread current execution. Tradeoff for processing time occurred due to consumer's increasing demands. Authors in [5], presented SG model based on fog platform. Fog computing is of distributed nature, which is useful for separately collecting public, private data. SGs are used for integration of energy distribution system with green power resources, control power usage and load balancing of energy. Cloud computing is used due to the distribution and scalability. Authors in [7], presented a Hadoop based model for storage and processing of massive data for SG. In [8], authors applied load balancing techniques on data centers of SGs. Three level SGs network proposed in [9], with multi-agent cloud computing which performed efficiently for data processing.

Authors in [10], present that the need for enhancing SGs computing resources and data storage is rapidly growing. Authors in [11], proposed cloud computing based SG for charging and discharging of electric vehicles (EVs) at EV public supply stations (EVPSSs). Proposed model makes efficient management of SG operations communication between cloud and SG. Priority assigned algorithms used for minimizing the waiting to a plug-in time of EV at EVPSS.

Authors in [12], used fog computing for reduced high privacy and latency. In [13], author used three-tier architecture; cloud tier, fog tier, and microgrid layer. The fog based microgrid is proposed for effective energy utilization. Fog works as a middle layer between cloud and MG for instant data retrieval and storage, eventually making smart meters efficient. This architecture minimizes the communication complexity by taking swift decisions which consumes less time.

The core contributions of authors in [14], are as follows: Cloud provides different kind of services to user on demand. For this different services run on cloud required huge amount of energy. To minimize the consumption of energy for data center of cloud computing, authors develop a model called (ILP) integer linear Programming for task scheduling. Dynamic task scheduling concept use for energy minimization in data center of cloud computing author proposed a new Genetic technique called Adaptive Genetic Algorithm (GA). Results of cloud computing Environment infrastructures are proposed by analyzing and testing the performance and quality of algorithms. They compared this algorithm with First Come First Serve algorithm for finding the optimal solution.

Authors in [15], proposed Chaotic Social Spider Algorithm (CSSA) for task scheduling, load balance in cloud computing infrastructure. For task scheduling and load balancing the algorithm used by the authors was inspired by behavior of social spider species. This algorithm works to find the best optimized solution to find suitable VM for the user tasks. Various parameters of proposed algorithm simulated using Cloudsim tools. Improving throughput of cloud structure the main objective of this paper was minimizing the make-span of scheduling process in cloud. For this purpose authors measured various parameters of cloud such as: response time, make-span, average, response time, computational cost and degree of imbalance across the various types of VM. The author prove that the proposed algorithms has produce minimum make-span to 14.8 percent compare to the different algorithms such as GA, Particle swarm optimization, Artificial bees colony and HFKCS range for tasks between hundred to thousand.

Authors in [16], used the feasible Virtual Machine (VM) configuration for presenting the physical resource requirements of delay optimal scheduling of VMs formulated as a deciding process. Shortest Job First policy; an online low-complexity scheme is used for buffering the arriving jobs and using Min Min Best Fit algorithm for optimizing. Shortest Job First buffering is used with Reinforcement Learning (RL) to avoid starvation.

3 Problem Formulation

The performance of the cloud computing system depends upon the efficient resource utilization. In order to achieve these performance measure, load balancing must be applied on the available resources of fog-cloud in effective way. In the proposed system load balancing is applied on VMs.

In this paper study of load balancing algorithm FFA is done in cloud computing environment.

VM a set of n virtual machine is formally written as:

$$VM = \{vm_1, vm_2, vm_3, vm_4, \ldots, vm_n\} \tag{1}$$

Equation 2 denotes the set of n number of user base.

$$U_{base} = \{u_b1, u_b2, u_b3, \ldots, u_bn\} \tag{2}$$

Equation 3 shows the Transmission delay time on each VM which is equal to the latency and transaction time.

$$T_{delay} = T_{latency} + T_{transation} \tag{3}$$

The response time of VM is calculated in Eq. 4 which is equal to the network delay and total response time.

$$Response_{time} = N_{delay} + Total_{RT} \tag{4}$$

The total response time is mathematically written as 5.

$$Total_{RT} = RT_{finish} - RT_{arrival} \tag{5}$$

whereas network delay is calculated in 6.

$$Ntw_{delay} = \frac{\text{Bandwidth}}{\text{Size of data unit}} \times Trans_{rate} \tag{6}$$

$$R_{vk} \leq C_k \quad k \in \mathbb{K} \tag{7}$$

where K is the set of resources, R is the requested resources and C is the capacity of VM.

Equation 7 shows the relationship between request for resource generated by userbase and capacity of VMs. The resource request generated by userbase must be less than the capacity of the VM.

The Eq. 8 describes that the total resource request RR_{vk} cannot exceed the total capacity of Cap_k. When a datacenter controller requests for new VM allocation, the algorithm finds the efficient VM with minimum forecasted response time having least load for allocation. Then the algorithm returns id of that efficient VM to datacenter controller. The datacenter controller receives the response after the completion of task.

$$\sum_{i=1}^{v} RR_{ik} \leq Cap_k \tag{8}$$

Equation 9 shows the Total cost of the system. It is calculated as the sum of data transfer cost, VM cost, MG cost

$$Total_{Cost} = Data_{TC} + VM_{Cost} + MG_{Cost} \tag{9}$$

The data transfer cost is calculated as.

$$Data_{TC} = Data_{GBs} * Data_{\text{Cost per GBs}} \tag{10}$$

where GBs represent the gigabytes.

Equation 11 shows the formula for Bandwidth utilization. It dependancy lies on region, data transfer size and user request. N_{ur} is the number of user request which is to be transmitted in different regions and also encorporates the user requests from two regions.

$$Bandwidth_{util} = \frac{\text{Bandwidth}_{alloc}}{N_{ur}} \tag{11}$$

Equation 12 shows the model of datacenter processing time. T_{datpro} is calculated based on user request/hour in accordance to bandwidth allocation [17].

$$PT_{DC} = \frac{\text{UR}_{hour}}{BW_{alloc}} \tag{12}$$

The objective function of this paper is following as: reduce the cost 9, minimize the response time 4.

4 Proposed System Model

In this paper, cloud and fog environment presented. The proposed system model has three layered architecture. System model comprised of different components such as the cluster of buildings, home, microgrids (MGs), utility, fog and cloud server. It has the capability to share the information with a two-way communication system and utilize the energy efficiently. The proposed system model is presented in Fig. 2.

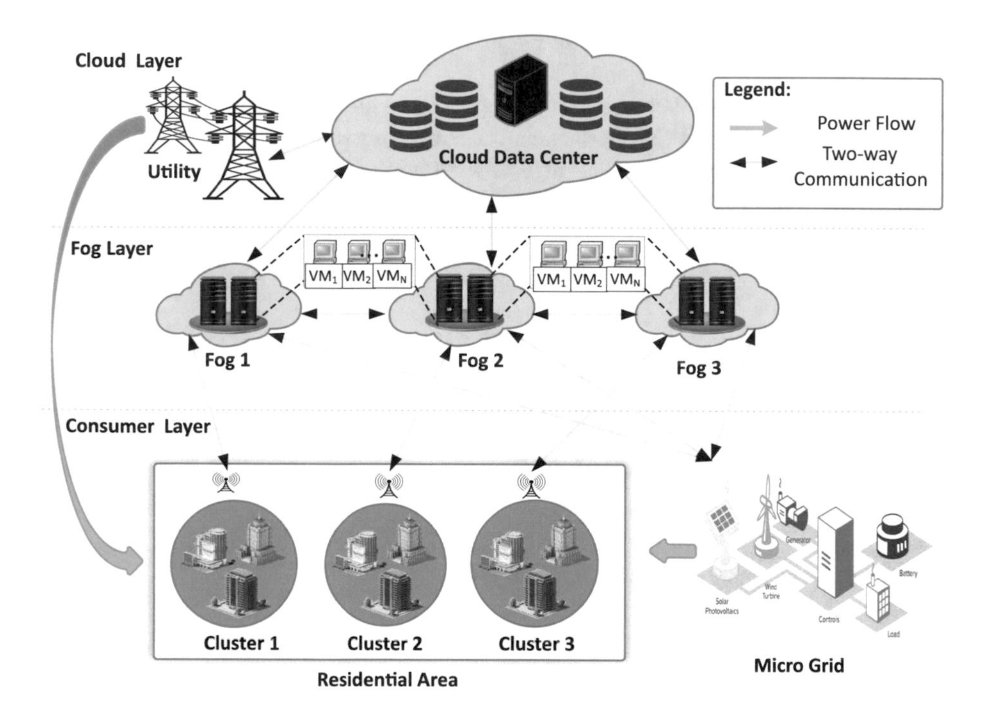

Fig. 2. System model of fog-cloud based platform

The first layer (consumer layer) contains a cluster of buildings. Each cluster consists of thirty buildings. Every building contains ten number of homes with multiple smart appliances. Each cluster is connected with one fog. MGs are placed nearby first layer, it provides the energy demand at the consumer side. The second layer contains fog network. Fogs are placed at near to the clusters and it works as a middle layer. With each fog multiple numbers of virtual machines are installed for handling request of consumers in an effective manner. VM works on fog as hardware and software called emulation of the system. It provide services for the execution OS. The energy demand request of user processed by fog, further fog communicate to the MG to fulfill the energy demand requirement.

However, when MGs are unable to meet the user demand then fog forwards that request to cloud that request contains user's energy requirement information. Thereafter cloud takes that request to utility generation and user request is served. The third layer is cloud layer which is mainly consists of Data Center. It connects with utility for fulfillment of deficient energy at user layer.

4.1 First Fit Algorithm

FFA is used to allocate the memory blocks to each request generated by end user. When request is generated FF handle that request to make load balancing more efficient. FF algorithm work on VM. It creates the memory blocks and fit the request in it. When the request is arrive it will check that if the size of request is smaller than memory block, It will processed the request by allocating memory blocks.

Algorithm 1 FFA

```
1: FFA parameters initialization: memory blocks size, process size, iteration
2: for users = 1:n do
3:     Initially the block of memory are unassigned
4:     Save block id of process
5:     initialized memory blocks set free
6:     pick each process and find suitable blocks
7:     for i=0:n (Iterations) do
8:         for j=0:n do
9:             if Blocksize>Processsize then
10:                allocate block j to p[i] process
11:                Reduce available memory in this block.
12:            end if
13:        end for
14:    end for
15:    for i=0:n (Iterations) do
16:        if Allocation ≠ 1 then
17:            Memory Allocation to jobs.
18:        else if
19:            then Memory Not Allocated.
20:        end if
21:    end for
22: end for
```

5 Simulation Results and Discussion

In this section, discussed the simulation results of the given scenario. For experimental purpose six fogs connected with six clusters, each cluster has thirty buildings which consists of ten home are considered. On the other hand a broker policy and cloud data center are also considered. Three load balancing techniques FF, RR, and Throttled are used to compare the simulation results. The number of the request generated at consumer end is sent to the service broker policy.

The closest data center service broker policy decides the best fog for consumer request, and the proposed techniques are run on VM of fog for task scheduling. According to the paper closest data center broker policy is used to handle the request of consumers. The performance metrics show that the results of these techniques are measured in; cost, processing time and response time.

Figures 3, 4 shows the response time of the buildings in each cluster and fogs. The graphs show the Average RT (ms), Minimum RT (ms) and Maximum

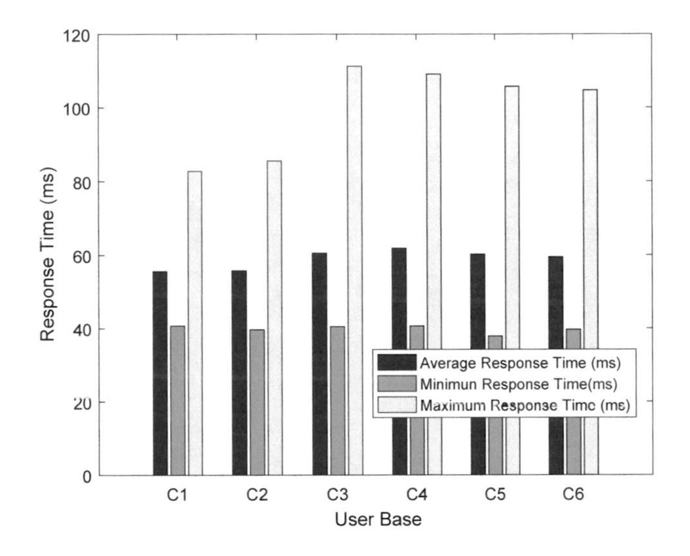

Fig. 3. Overall Response Time

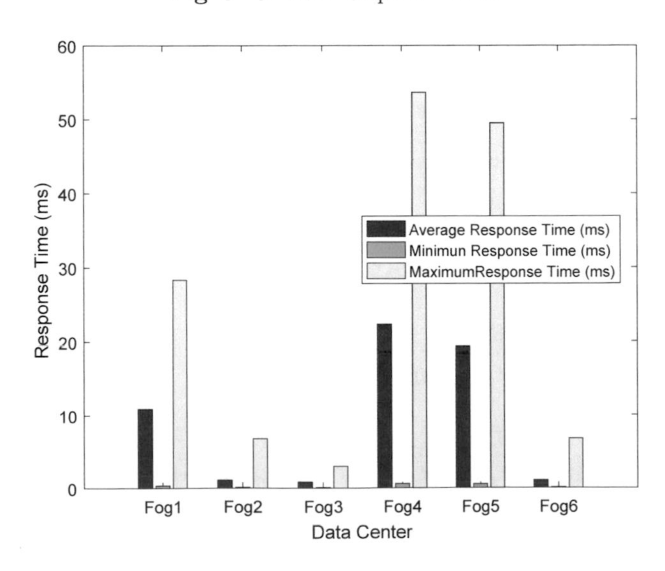

Fig. 4. Overall Response Time on each fog

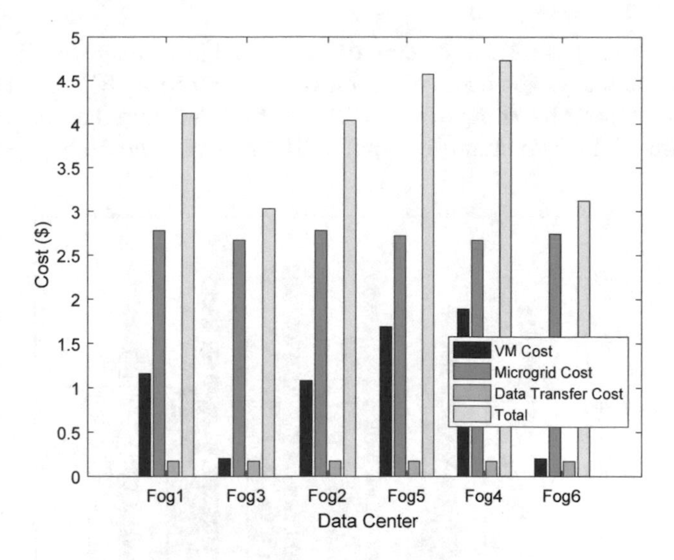

Fig. 5. Cost on each fog

RT (ms). The Fig. 3 has maximum values are 60.2, 40.73 and 111.28, respectively. Whereas, the Average RT (ms), Minimum RT (ms), and Maximum RT (ms) of Fig. 4 has maximum values 22.34, 0.59 and 53.64, respectively.

Figure 5 shows the cost on each fog. It is calculated by load balancing technique. The graph of Fig. 5 shows the VM cost, MG cost, Data Transfer cost and total cost of FF algorithms. The maximum value are 1.89, 2.78, 0.17 and 4.73, respectively.

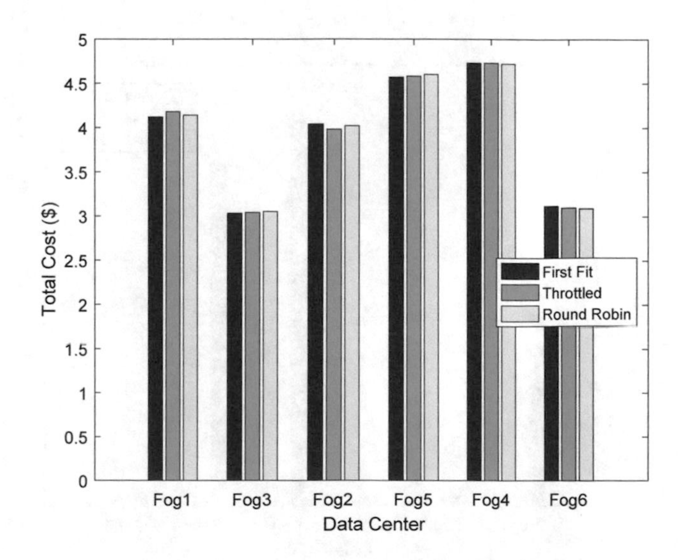

Fig. 6. Total Cost comparison

Figure 6 shows the total cost comparison of load balancing algorithms on each fog. The graph shows the comparison on total cost of FF, RR and Throttled.

Figure 7 shows the comparison of FF, Throttled and RR on the base of response time. The Maximum RT (ms) is 61.92, 50.67 and 50.67, respectively.

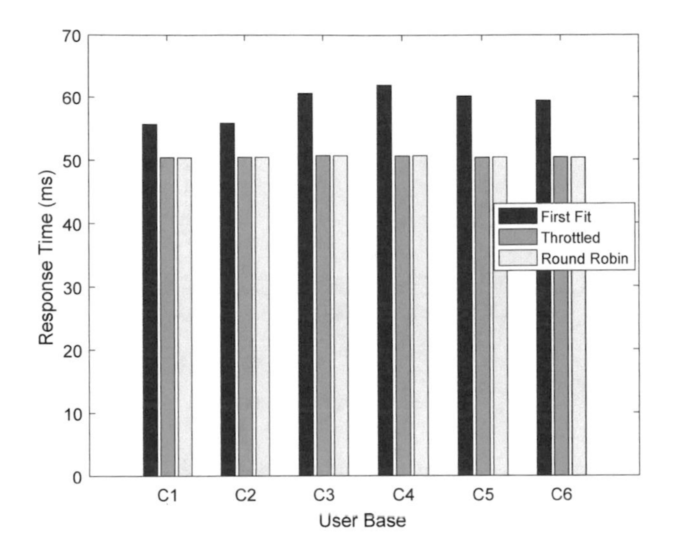

Fig. 7. Response Time comparison on each cluster

Figure 8 show the response time comparison of load balancing techniques such as FF, RR and Throttled on the base of Average RT (ms) 22.34, 1.09 and 1.09, respectively.

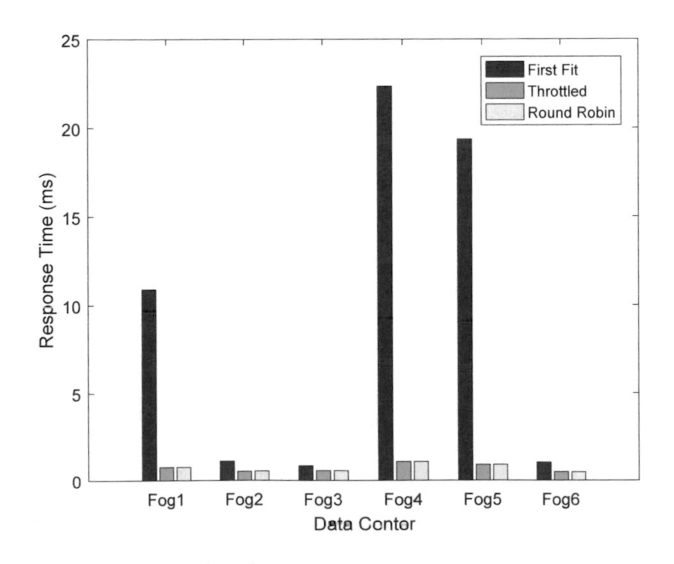

Fig. 8. Response Time comparison on each fog

Figure 9 show the processing time of FF on the base of Average RT (ms), Minimum RT (ms), Maximum RT (ms), respectively.

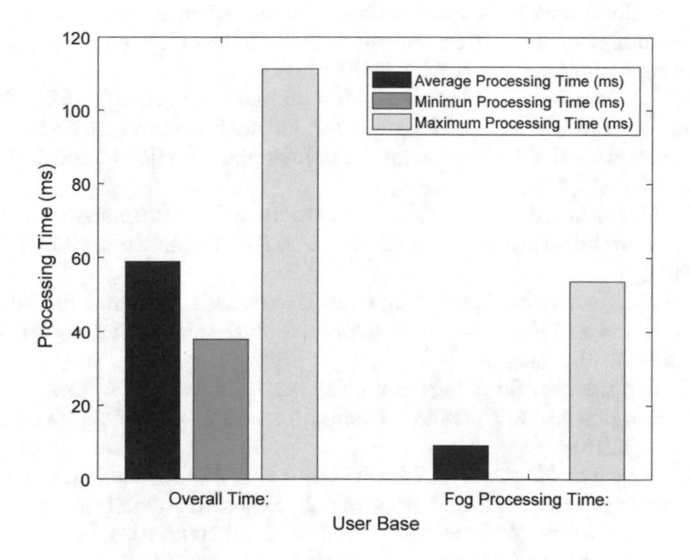

Fig. 9. Processing Time on each fog

6 Conclusion

In this paper fog-cloud based computing model is integrated for effective resource utilization management in SG. In fog-cloud based computing model, smart building with the multiple numbers of apartments consisting IoT devices which are considered. The fog-cloud based environment works between these apartments, when a consumer sends the request to fog for energy demand. The fog fulfills their electricity demand by further sending a request to the nearest MG. MG provides sufficient energy required by the particular consumer. The fog is further connected to the centralized cloud. When all the MGs run out of energy demand request of the consumer on fog then fog further sends a request to the cloud which provides access to the utility.

Simulations performed using Cloudsim simulator in eclipse; a java based platform. The closest data center service broker policy used to select the fog which is near to the clusters. It provides fast service response and minimum latency delay. RR, Throttled and FF used to depict results. The results show that the total cost of FF is better than RR and Throttled. However, response time and the processing time of other techniques are better than FF on various fogs.

References

1. Mohamed, N., Al-Jaroodi, J., Jawhar, I., Lazarova-Molnar, S., Mahmoud, S.: Integration of cloud and fog based environment for effective resource distribution in smart buildings. In: 14th IEEE International Wireless Communications and Mobile Computing Conference (IWCMC-2018) (2018)
2. Zahoor, S., Javaid, N., Khan, A., Muhammad, F.J., Zahid, M., Guizani, M.: A cloud-fog-based smart grid model for efficient resource utilization. In: 14th IEEE International Wireless Communications and Mobile Computing Conference (IWCMC-2018) (2018)
3. Luo, F.: Cloud-based information infrastructure for next-generation power grid: conception architecture and applications. IEEE Trans. Smart Grid **7**(4), 1896–1912 (2016)
4. Yasmeen, A., Javaid, N.: Exploiting Load Balancing Algorithms for Resource Allocation in Cloud and Fog Based Infrastructures. Institute of Information Technology, Islamabad 44000, Pakistan
5. Okay, F.Y., Ozdemir, S.: A fog computing based smart grid model. In: 2016 International Symposium on Networks, Computers and Communications (ISNCC), pp. 1–6. IEEE (2016)
6. Javaid, S., Javaid, N., Asla, S., Munir, K., Alam, M.: A cloud to fog to consumer based framework for intelligent resource allocation in smart buildings
7. Bai, H., Ma, Z., Zhu, Y.: The application of cloud computing in smart grid status monitoring. Internet of Things, pp. 460–465. Springer, Heidelberg (2012)
8. Mohsenian-Rad, A.H., Leon-Garcia, A.: Coordination of cloud computing and smart power grids. In: 2010 First IEEE International Conference on Smart Grid Communications (Smart- GridComm), pp. 368–372 (2010)
9. Jin, X., He, Z., Liu, Z.: Multi-agent-based cloud architecture of smart grid. Energy Procedia **12**, 60–66 (2011)
10. Bitzer, B., Gebretsadik, E.S.: Cloud computing framework for smart grid applications. In: 2013 48th International Universities Power Engineering Conference (UPEC), pp. 1–5 (2013)
11. Chekired, D.A., Khoukhi, L., Mouftah, H.T.: Decentralized cloud-SDN architecture in smart grid: a dynamic pricing model. IEEE Trans. Ind. Inf. **14**, 1220–1231 (2018). ISSN 1551-3203
12. Stojmenovic, I., Wen, S.: The fog computing paradigm: scenarios and security issues. In: Proceedings of the 2014 Federated Conference on Computer Science and Information Systems, pp. 1–8 (2014)
13. Barik, R.K.: Leveraging Fog Computing for Enhanced Smart Grid Network
14. Ibrahim, H., Aburukba, R.O., El-Fakih, K.: An Integer Linear Programming model and Adaptive Genetic Algorithm approach to minimize energy consumption of Cloud computing data centers (2018)
15. Xavier, V. A., Annadurai, S.: Chaotic social spider algorithm for load balance aware task scheduling in cloud computing. Cluster Computing, pp. 1-11 (2018)
16. Guo, M., Guan, Q., Ke, W.: Optimal scheduling of VMs in queueing cloud computing systems with a heterogeneous workload. IEEE Access **6**, 15178–15191 (2018)
17. Hussain, H.M., Javaid, N., Iqbal, S., Hasan, Q.U., Aurangzeb, K., Alhussein, M.: An efficient demand side management system with a new optimized home energy management controller in smart grid. Energies **11**(1), 190 (2018). ISSN 1996-1073
18. Javaid, N., Naseem, M., Rasheed, M.B., Mahmood, D., Khan, S.A., Alrajeh, N., Iqbal, Z.: A new heuristically optimized Home Energy Management controller for smart grid. Sustain. Cities Soc. **34**, 211–227 (2017)

19. Javaid, N., Ahmed, F., Ullah, I., Abid, S., Abdul, W., Alamri, A., Almogren, A.: Towards cost and comfort based hybrid optimization for residential load scheduling in smart grid. Energies **10**(10), 1546 (2017)
20. Javaid, N., Javaid, S., Abdul, W., Ahmed, I., Almogren, A., Alamri, A., Niaz, I.A.: A hybrid genetic wind driven heuristic optimization algorithm for demand side management in smart grid. Energies **10**(3), 319 (2017)

A Practical Indoor Localization System with Distributed Data Acquisition for Large Exhibition Venues

Hao Li(✉), Joseph K. Ng, and Shuwei Qiu

Department of Computer Science, Hong Kong Baptist University, Kowloon Tong, Hong Kong
{cshaoli,jng,csswqiu}@comp.hkbu.edu.hk

Abstract. In this paper, we focus on the Wi-Fi based indoor localization in large exhibition venues. We identify and describe the real-world problems in this scenario and present our system. We adopt a passive way to detect mobile devices with the consideration of users' preference and iOS devices' privacy issue, and collect signal strength data in a distributed manner which meets the practical demand in exhibition venues and save the power consumption of mobile devices. Since exhibition venues have many restrictions on traditional localization approaches, we propose our approach and solution to fit these special conditions. We propose the clustering and Gaussian process regression (GPR) method to improve localization accuracy. Series of experiments in Hong Kong Convention and Exhibition Centre (HKCEC) show our system's feasibility and effectiveness. Our approach has significant improvement in the localization accuracy when compared with traditional trilateration, fingerprinting and the state-of-the-art approaches.

Keywords: Wi-Fi localization · Exhibition venues
Distributed data acquisition · Gaussian process regression · Clustering

1 Introduction

Wi-Fi based indoor localization has attracted much attention in recent years. An accurate indoor location can provide better location based services in different application fields. In this paper, we focus on the Wi-Fi indoor localization system in exhibition venues which include a large number of visitors and exhibitors. Based on the estimated location in this scenario, many services can be provided, such as visitors' movement trajectories analysis, personalized recommendations for visitors, better booth rearrangement for organizers and so on. Because of these added values, it is necessary to study the localization techniques and approaches in this scenario. However, the exhibition scenario is a very special case. At first, we state some basic conditions in the exhibition scenario:

L. Barolli et al. (Eds.): NBiS 2018, LNDECT 22, pp. 568–577, 2019.
https://doi.org/10.1007/978-3-319-98530-5_49

1. Visitors are not willing to install additional applications on the smart phone.
2. The exhibition venues are always very dynamic in terms of floor plans because different types of fairs have their own arrangements for exhibitors.
3. The time gap between two different fairs is really short, normally one day.
4. The exhibition venues are huge normally with over 100000 m^2.

These basic facts will have some constraints on the traditional indoor localization system. In traditional Wi-Fi indoor localization system, mobile devices will scan access points (APs), capture the signal strength and then estimate the location. The fingerprinting and trilateration approaches are two common methods utilized in location estimation. We list the influences below:

1. We will have trouble in data collection part because no active participation of visitors, besides, only Android devices can capture the signal strength while iOS devices don't permit this because of protection of users' privacy.
2. Frequent scanning process will increase the power consumption of mobile devices which is not user friendly.
3. The traditional fingerprinting approach cannot meet our demands in exhibition venues since the dynamic floor plans. The old fingerprint is not suitable for new indoor environment after reconstruction of a new fair and recollection of the fingerprint is very time-consuming.
4. The trilateration approach is not suitable for our scenario because the indoor environment of exhibition venues is very complex and it is very difficult to formulate a signal strength to distance relationship.

In order to provide services for more visitors and reduce the error in the exhibition venues, we propose our customized system, which is easy and efficient to operate and highly increases the estimation accuracy. In our system, we adopt a passive way to estimate the location. We switch our APs from master mode into monitor mode which can sniff signals in the air and parse the signal data to obtain signal strength. We can call this special AP the *sniffer* according to its function. We propose a clustering and Gaussian process regression (GPR) method to estimate the location. With limited time and a huge venue, we can only collect a sparse fingerprint in an acceptable time, which means the positions need to be collected are sparse. After finishing the signal collection, we do clustering on the sparse fingerprint to make similar signal observations into clusters. Then we apply GPR to interpolate the fingerprint for each cluster. In online phase, when the server receives new signal vectors from different sniffers, it will map signal vectors into a cluster by comparing with the centroids of all clusters. Then it will use the corresponding trained GPR model to obtain the estimation location of the mobile device. We have conducted real-world experiments in Hong Kong Convention and Exhibition Centre (HKCEC). These experimental results show that our system has good performance with the sparse fingerprint.

The rest of this paper is organized as follows: In Sect. 2, the background and related works of indoor localization are described. Section 3 presents the system architecture. Our proposed approach for exhibition venues is given in Sect. 4. Section 5 compares our proposed approach with some other localization methods from the experiments conducted in HKCEC. And finally, conclusions are given in Sect. 6.

2 Background and Related Works

We briefly discuss the background and related works in this section. There are mainly two types of approaches based on signal strength of Wi-Fi, including the fingerprinting approach, and the trilateration approach.

2.1 The Fingerprinting Approach

The fingerprinting approach is the most popular method in indoor localization. Figure 1 illustrates this approach which includes two phases, the *offline* phase and the *online* phase. In the offline phase, different positions' signal vectors need to be collected together with their corresponding coordinates. The specific position where we collect signals is called the reference point. Our sniffers will collect all the signals sent from the mobile devices in different reference points. We use $R_i = \{RSSI_{i_1}, RSSI_{i_2}, RSSI_{i_3}, ..., RSSI_{i_n}\}$ to denote the signal strength captured from n sniffers in i-th reference point, and use $C_i = \{x_i, y_i\}$ to denote the coordinates of this reference point. Normally, the reference points are distributed uniformly and the higher density of reference points, the more accurate in the location estimation. Let's denote $S = \{R, C\}$ as the sparse fingerprint, where $R = \{R_1, R_2, R_3, ..., R_m\}$ are the signal vectors, $C = \{C_1, C_2, C_3, ..., C_m\}$ are coordinates of different reference points and m is the number of reference points. In the online phase, our distributed sniffers will collect signals around and then transfer to the central server. The server will summary all the packets and obtain the signal vectors of the different mobile devices. Then the server will compare the newly collected signal vector with the fingerprint in the offline phase, k-nearest neighbors (KNN), weighted k-nearest neighbors (WKNN) are the most common approaches utilized in this process to estimate the location of the mobile device.

Some other methods are also proposed considering the localization problem as a regression problem. Different regression methods are applied to this problem [1–3]. Among them, Gaussian process regression (GPR) is mostly employed to build the relationship between signal vectors and coordinates. GPR has been showed to be suitable to predict the nonlinear, spatial relation of signal strength in [4,5]. The main purpose of GPR in indoor localization is to interpolate the sparse fingerprint resulting in a continuous fingerprint to make further location estimation more accurate. Usually, the default GPR, with zero mean and the squared exponential kernel is chosen to make up the whole sparse fingerprint [6,7]. However, in a large-scale and complex exhibition venue, it is not suitable to interpolate the whole sparse fingerprint directly. Internal booths and pillars make the relationship between signal vectors and coordinates to be uncorrelated. For example, two reference points are very close in coordinates but separated by a pillar which will make the signal vectors to be not similar. If we directly apply

GPR into these coordinates and signal vectors, it will increase the localization error. Some methods have been proposed to solve this problem. Matic et al. [8] firstly use KNN to find most K similar signal vectors to the newly collected signal vector in the fingerprint and then use GPR to train the model online with K signal vectors and corresponding coordinates. This approach can be a little bit time consuming since GPR need to be trained for each online signal vector.

2.2 The Trilateration Approach

In trilateration approach, the mobile device's signal vector need to be transformed to distance vector based on the signal propagation model [9] in current indoor environment. If the dimension of the distance vector is over 3, the location of the mobile device can be estimated with the knowledge of sniffers' positions. Usually, the signal to distance relation can convert signal strength to distance precisely. Then the estimation can be determined by minimizing the summation of errors with all transformed distance. However, it is difficult to find such a good signal to distance relationship in various complex indoor environment and the signal strength itself also has fluctuation.

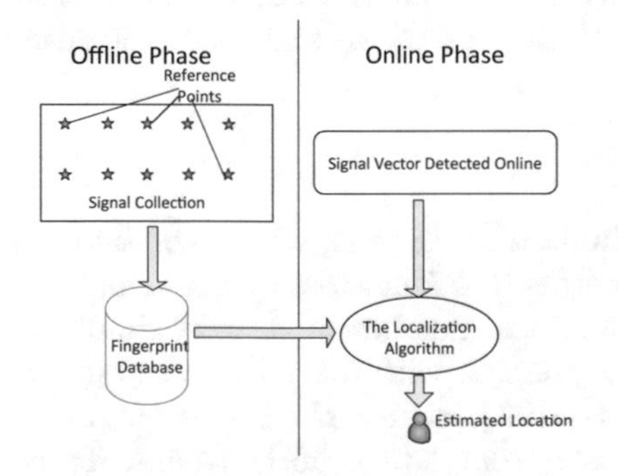

Fig. 1. The fingerprinting approach

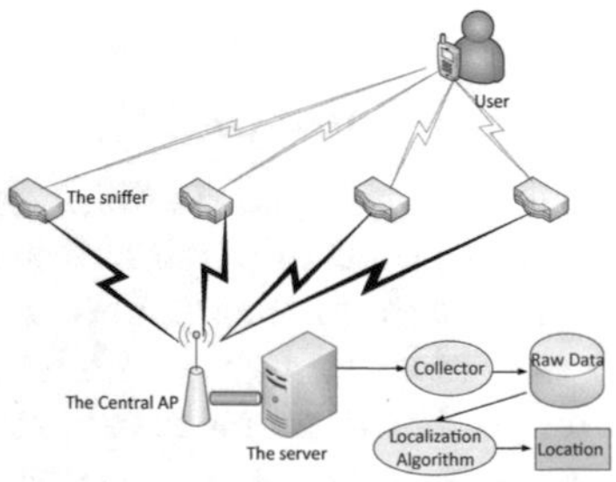

Fig. 2. Practical system architecture

3 Practical System Architecture

In this section, we show the system architecture. Traditional indoor localization system utilizes Android devices to collect signal strength from surrounding APs, which cannot meet the requirement of exhibition scenario. In our system, we adopt an opposite way. Figure 2 shows the overall system architecture. Each mobile device contains one unique media access control (MAC) address. When it emits packets, MAC address will be exposed. A program implemented by us in the sniffer can extract this MAC address and perceive the signal strength at the time when the signal is captured. Besides, our program can distinguish whether the signal is sent from the AP or the mobile device by the IEEE 802.11

frame format. We will only focus on the packets from mobile devices and abandon the packets from APs to save the computation resources of sniffers. Our customed-tailor programme in sniffers will generate new packets which contain the target mobile device's MAC address, signal strength and the timestamp. Our distributed sniffers will send all these newly generated packets to the server for processing and location estimation by wireless communication way. A central AP is set up to collect all packets sent by distributed sniffers and then deliver them to the server. The server will process the raw data and then do location estimation. In this mode, more Android and iOS devices can be detected and tracked without installing any additional application on the mobile devices. Besides, the power consumption of mobile devices can be reduced because of no scanning process.

4 Proposed Approach

In this section, we focus on our proposed approach. With limited setup time in exhibition venues because of the short gap between two exhibitions, we can only collect sparse fingerprint due to the huge coverage area of exhibition halls. The intervals between two reference points are usually more than 10 m. We propose our method to deal with the sparse fingerprint. It mainly contains three parts, including signal clustering, Gaussian process regression, and online location estimation.

4.1 Signal Clustering

After collecting the sparse fingerprint in limited setup time, we use the k-means clustering method to partition reference points into k clusters by the signal vector. Since our reference points are sparse so the number of clusters should be small, otherwise it may have impacts on regression with few reference points in one cluster. For the k-means algorithm, its initialization method is a randomized step because random signal vectors are chosen from data as initial means. Hence, the clustering results also have random property and can possibly obtain bad clustering results. This will also have impacts on regression and location estimation. Figure 3 shows the different clusterings' results with random initial means using the HKCEC sparse fingerprint. We can see that the clustering results are different with different choices of initial means.

With these considerations, we propose to do multiple times' clusterings to increase the amount of clusters without decreasing the size of clusters. We use the Euclidean metric to evaluate the similarity between signal vectors p and q. Then we aim to partition the reference points into k clusters $G = G_1, G_2, ..., G_k$ in one clustering step so as to minimize the within-cluster gap:

$$\arg\min_{G} \sum_{i=1}^{k} \sum_{x \in G_i} ||x_i - \alpha_i||^2, \tag{1}$$

where α_i is the mean of cluster G_i. In our approach, c times of clustering will be done, so totally $c \cdot k$ clusters will be generated.

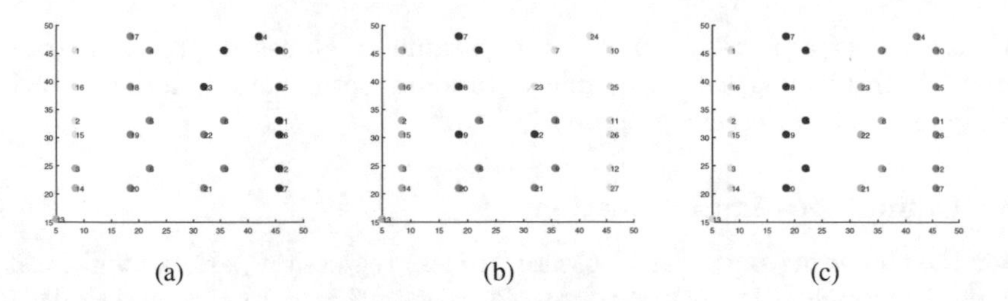

Fig. 3. Different clusterings' results

4.2 Gaussian Process Regression

We use GPR to model the relationship between signal vectors and coordinates of each cluster. We have the following observation model,

$$C_i = f(R_i) + \varepsilon_i, \tag{2}$$

where $R_i \in \mathbb{R}^n$, $C_i \in \mathbb{R}^2$, $\varepsilon_i \sim N(0, \sigma_n^2)$ represents the independent, identical, distributed (IID) noise, and f is the function to be modeled from the sparse fingerprint S. The key of GPR is that the covariance between any two output values, C_i and C_j, depends on the input samples R_i and R_j. This dependency can be described by covariance function or kernel function. There are many covariance functions. In our case, we adopt the most commonly used one, the squared exponential kernel,

$$kernel(R_i, R_j) = \sigma_f^2 \exp\left(-\frac{(R_i - R_j)^2}{2\ell^2}\right), \tag{3}$$

where ℓ is the length scale and σ_f^2 is the signal variance. Our observations are assumed to be with additive white Gaussian noise, so we have the covariance function,

$$cov(C_i, C_j) = kernel(R_i, R_j) + \sigma_n^2 \delta_{pq}, \tag{4}$$

where σ_n^2 is Gaussian noise, δ_{pq} is Kronecker delta, which means that if p equals to q, the output is 1, otherwise 0. Hence we have,

$$C \sim N(0, K + \sigma_n^2 I), \tag{5}$$

where K is $n \times n$ covariance matrix of R and I is the identity matrix.

Given a newly collected signal vector R_*, we want to make a prediction of position C_* using the GPR model. By assuming that the joint distribution of C and C_* is also a zero-mean Gaussian, we can condition on C using standard formula,

$$C_* | C, R, R_* \sim N(\mu_*, \sigma_*^2), \tag{6}$$

where

$$\mu_* = K(R_*, R)(K(R, R) + \sigma_n^2 I)^{-1} C, \tag{7}$$

$$\sigma_*^2 = K(R_*, R_*) - K(R_*, R)(K(R, R) + \sigma_n^2 I)^{-1} K(R, R). \tag{8}$$

Let's use $\theta = \{\ell, \sigma_n^2, \sigma_f^2\}$ to denote hyperparameter. Gaussian process regression learns θ from training data S by many standard optimization methods, such as conjugate gradient descent proposed in [10].

4.3 Online Location Estimation

After the clustering part and Gaussian process regression part, we will locate a mobile device when the signal vector R_* is transferred to the server. We first need to classify the new signal vector into one cluster. Since each cluster has a centroid α_i as the representative, we map R_* to one cluster by finding the nearest centroid as follows,

$$\arg\min_{\alpha_i} Eud(R_*, \alpha_i), \tag{9}$$

where $Eud()$ is Euclidean metric. Then, we use the corresponding Gaussian process regression model to do location estimation. In this way, we can focus on the most suitable regression model for the newly collected signal vector.

5 Experiments

To study the performance of our proposed systems, we have done a localization test for a Watch & Clock Fair in HKCEC in Sept. 2017. Figure 4 shows the layout of the HKCEC (effective size: 55 m × 40 m). We use twelve Complex MMZ558HV wireless APs as sniffers (marked with cyan APs) and their heights are about 3 m, at the positions indicated in the figure. The sniffers are enabled to detect Wi-Fi packets in 2.4 GHz frequency band and forward them to a central AP (marked with the red AP) by channels of 5 GHz frequency band and then to our server. 27 sparse reference points (marked with blue points) are acquired in offline phase. We finish the devices setting and the sparse fingerprint collection in half day which is acceptable for the practical demands. To test the effectiveness of our system quantitatively, we collect some points of signal vectors with known positions (marked with red points). We have applied different approaches to do localization with this data set and compared the results.

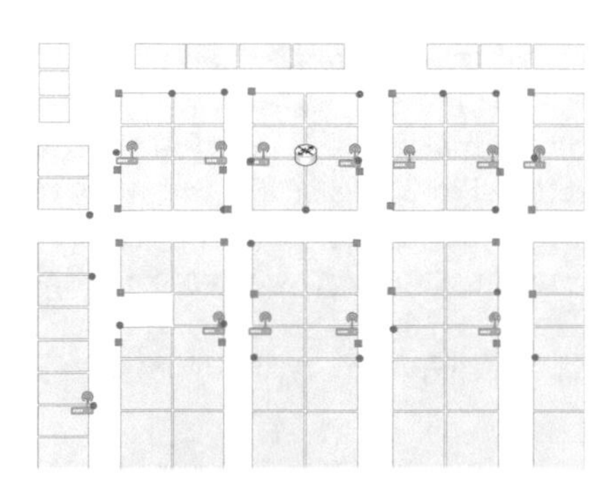

Fig. 4. Hong Kong convention and exhibition centre testbed

Table 1. Mean localization errors comparison with traditional approaches

	Mean localization error (meter)		
	Our approach	Fingerprinting	Trilateration
Mean error (m)	**5.09**	6.56	7.59

5.1 Experiments Results

5.1.1 Comparisons with Traditional Approaches

Table 1 shows the mean errors of the target device with our approach, the fingerprinting approach and the trilateration approach. For our proposed approach, clustering times is 15 and k is 5 in our k-means algorithm. For the fingerprinting, WKNN is implemented to do location estimation. K equals 4 and similarity metric used is Euclidean distance. For the trilateration method, we utilize sparse reference points and their signal vectors to generate a log distance path loss model for each sniffers. Results shows that our proposed approach has more accurate location results compared with the fingerprinting and trilateration approach. Figure 5 illustrates the cumulative distribution function (CDF) of the localization errors for these three approaches.

5.1.2 Impacts of Clustering Times

In order to study the localization performance of our approach with different clustering times, we test different times of clustering and obtain the corresponding location error. Figure 6 shows the mean errors of target devices with different clustering times. We can see that location error decreases when more clustering are completed and becomes stable when clustering times is large because later clusters are very similar to previous ones. This shows multiple clusterings can efficiently improve the performance of our system.

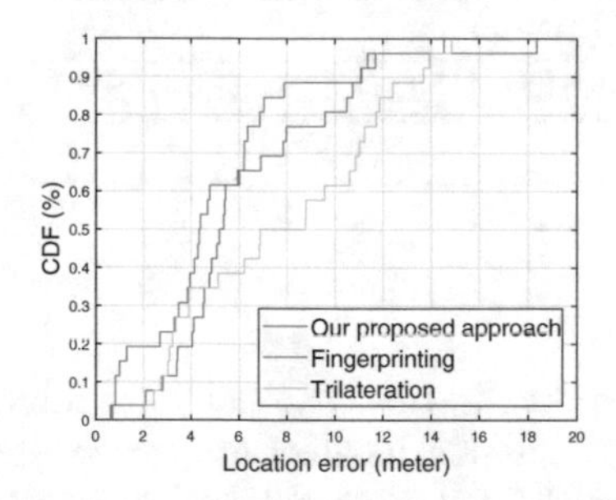

Fig. 5. Localization results comparison with traditional approaches

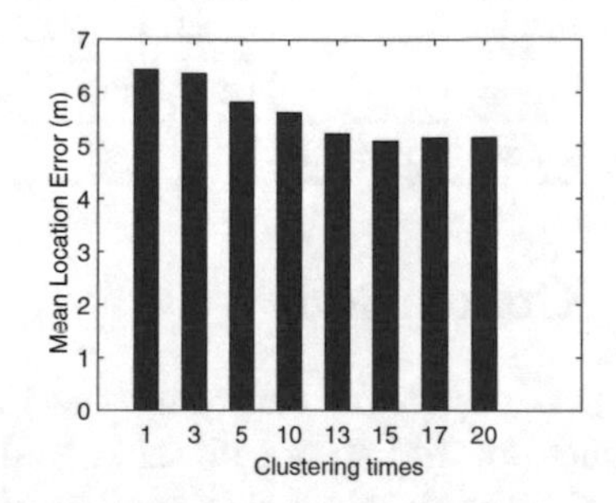

Fig. 6. Mean location error with different clustering times

5.1.3 Effects of Different Preprocessing Methods

As mentioned in Sect. 2.1, Matic et al. [8] use KNN to find most K similar signal vectors to the newly collected signal vector in the fingerprint, while we adopt clustering instead. Now we test both approaches together with no preprocessing approach which means GPR is directly used in the sparse fingerprint. Table 2 shows the mean localization errors of three preprocessing methods. Our approach has more accurate location estimation as compared to others in this dataset. It shows that the preprocessing step is needed due to the complex indoor environment and clustering is more suitable for the sparse fingerprint. More importantly, our GPR training is conducted in offline phase which can improve the computation efficiency.

Table 2. Mean localization error of different preprocessing methods

	Mean localization errors (meter)		
	Clustering + GPR	KNN + GPR	Only GPR
Mean error (m)	**5.09**	6.33	7.20

5.1.4 Comparisons of Different Regression Methods

In this set of experiments, we evaluate the effects of applying different regression models [1–3]. We keep the clustering step and just replace Gaussian process regression with linear regression (LR) and support vector regression (SVR). Both two regression models are trained with the sparse fingerprint S and do the further location estimation. Table 3 shows the mean localization errors of different regression methods. Gaussian process regression has more accurate location estimation compared to linear regression and support vector regression in this dataset. This set of experiments shows that Gaussian process regression is more suitable for describing the relationship between coordinates and signal strength.

Table 3. Mean localization error of different regression methods

	Mean localization errors (meter)		
	Clustering + GPR	Clustering + LR	Clustering + SVR
Mean error (m)	**5.09**	5.47	5.87

6 Conclusions

In this paper, we elaborate the practical problems that existed in exhibition venues for indoor localization and present a passive localization system with distributed data acquisition. The clustering and GPR approach we proposed for this scenario only needs sparse fingerprint for training which fits the practical demands. We have implemented our system in HKCEC and conducted experiments in one of its fairs. Experimental results show that the accuracy of the proposed approach is better when compared with traditional and the state-of-the-art approaches in terms of accuracy.

References

1. Shi, K., Ma, Z., Zhang, R., Hu, W., Chen, H.: Support vector regression based indoor location in IEEE 802.11 environments. In: Mobile Information Systems (2015)
2. Pasricha, S., Ugave, V., Anderson, C.W., Han, Q.: LearnLoc: a framework for smart indoor localization with embedded mobile devices. In: 2015 International Conference on Hardware/Software Codesign and System Synthesis (CODES+ISSS), pp. 37–44. IEEE (2015)
3. Zhuang, Y., Yang, J., Li, Y., Qi, L., El-Sheimy, N.: Smartphone-based indoor localization with bluetooth low energy beacons. Sensors **16**(5), 596 (2016)
4. Richter, P., Toledano-Ayala, M.: Revisiting Gaussian process regression modeling for localization in wireless sensor networks. Sensors **15**(9), 22587–22615 (2015)
5. Kumar, S., Hegde, R.M., Trigoni, N.: Gaussian process regression for fingerprinting based localization. Ad Hoc Netw. **51**, 1–10 (2016)
6. Atia, M.M., Noureldin, A., Korenberg, M.J.: Dynamic online-calibrated radio maps for indoor positioning in wireless local area networks. IEEE Trans. Mobile Comput. **12**(9), 1774–1787 (2013)
7. Yoo, J., Kim, H.J.: Target tracking and classification from labeled and unlabeled data in wireless sensor networks. Sensors **14**(12), 23871–23884 (2014)
8. Matic, A., Papliatseyeu, A., Osmani, V., Mayora-Ibarra, O.: Tuning to your position: FM radio based indoor localization with spontaneous recalibration. In: 2010 IEEE International Conference on Pervasive Computing and Communications (PerCom), pp. 153–161. IEEE (2010)
9. Wang, B., Zhou, S., Liu, W., Mo, Y.: Indoor localization based on curve fitting and location search using received signal strength. IEEE Trans. Ind. Electron. **62**(1), 572–582 (2015)
10. Rasmussen, C.E.: Gaussian processes in machine learning. In: Advanced Lectures on Machine Learning, pp. 63–71. Springer, Heidelberg (2004)

Secure Visible Light Communication Business Architecture Based on Federation of ID Management

Shigeaki Tanimoto[1(✉)], Chise Nakamura[1], Motoi Iwashita[1], Shinsuke Matsui[1], Takashi Hatashima[2], Hitoshi Fuji[2], Kazuhiko Ohkubo[2], Junichi Egawa[3], and Yohsuke Kinouchi[4]

[1] Faculty of Social Systems Science, Chiba Institute of Technology, Chiba, Japan
{shigeaki.tanimoto, iwashita.motoi, matsui.shinsuke}@it-chiba.ac.jp, s1542086ss@s.chibakoudai.jp

[2] NTT Secure Platform Laboratories, NTT, Tokyo, Japan
{hatashima.takashi, fuji.hitoshi, ohkubo.kazuhiko}@lab.ntt.co.jp

[3] EXGEN NETWORKS Co., Ltd., Tokyo, Japan
egawa@exgen.co.jp

[4] Tokushima University, Tokushima, Japan
kinouchi@tokushima-u.ac.jp

Abstract. With the progress of visible light communication (VLC) technology, opportunities for new mobile communication infrastructures and business creation are increasing. Specifically, there are many proposals for spot type broadcasting services linked to positions such as on-the-spot explanations at art museums and information delivery linked with digital signage. However, due to their limited capabilities, investigations into their security measures has been insufficient. On the other hand, in promoting VLC for business in the future, security measures are an important issue. We have previously proposed a secure business architecture through cooperation with VLC, public key encryption technology, and power line communication technology. This architecture provides an "position authentication ID" from a light source, such as an LED, and it is characterized by strict position authentication that encrypts these data using a public key infrastructure (PKI). In this paper, we propose a new business architecture to enable a more secure cooperation through cooperation with the ID management infrastructure, and contribute to the creation of new business by using VLC.

1 Introduction

The global trend towards environmental protection aiming at a low carbon society is also advancing the progress in the field of optical lighting. The systematic replacement of old incandescent lamps with LED electric lights has enabled substantial power savings along with lower CO_2 emissions. Their utilization in visible light communication (VLC)

L. Barolli et al. (Eds.): NBiS 2018, LNDECT 22, pp. 578–589, 2019.
https://doi.org/10.1007/978-3-319-98530-5_50

has also attracted attention, and these LEDs have enabled new telecom infrastructure development with power line communication (PLC) [1–4]. Mobile communication traffic, including that of cellular phones and smartphones, is increasing, so VLC has generated high interest. The amount of mobile communication traffic in 2016 was 18 times that in 2011 and 2.6 times that in 2014, and the demand for mobile communication is only expected to increase [5, 6]. Because of these trends, optical space communication using visible light has also attracted attention for implementation in mobile communication. Moreover, the communication path can be seen as light, unlike radio waves, providing a merit in which the secure side of the information disclosure range can be limited to VLC.

The Visible Light Communication Consortium (VLCC) in Japan is now focusing on standardization activities. A standard method for VLC was enacted by the Japan Electronics and Information Technology Industries Association (JEITA) in 2006 [7].

A business architecture utilizing this method has been proposed, and a communication infrastructure as a new source of VLC has been developed. Many broadcast service models linked to places (positions) such as on-the-spot explanations at art museums and information distribution linked with digital signage have been developed using this type of architecture and infrastructure [8–10]. However, these models do not have satisfactory security.

In previous research, we proposed a secure business architecture through cooperation with VLC, public key encryption technology, and PLC technology [11, 12]. Specifically, it features strict position authentication based on public key encryption of data called "position authentication ID" provided from a light source, such as an LED. In this paper, to cooperate with the ID management infrastructure, we propose a new business architecture to facilitate collaboration between the position authentication management infrastructure and service providers such as digital signage. These results will contribute to the creation of new business by VLC.

2 Issues with Current VLC System and Business Architecture

2.1 VLC System

As shown in Fig. 1(a1), the VLC system performs data communication by rapidly blinking LEDs at a speed high enough so that it cannot be perceived by the human eye [1, 2]. One of the features of VLC is the transmission and reception of the characteristic data (ID) of a position. As shown in Fig. 1(a2), a digital modulation signal receiver is placed on the spot where the area fixed by a light source is irradiated. Accordingly, position IDs can be transmitted from a light source [8, 13].

2.2 Business Architecture Issues

The business architecture using VLC shown in Fig. 1(b) is a broadcast service linked to a position such as on-the-spot explanations at a museum and information distribution

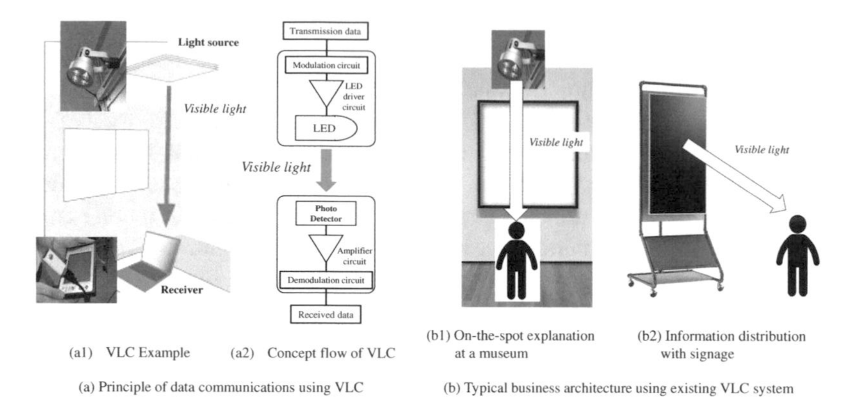

Fig. 1. Principle of data communications and typical business model using VLC

linked with digital signage [8–10]. Security measures are largely unnecessary because they are of limited use.

However, the use of position information, such as with car navigation systems and GPS pedestrian navigation using mobile phones, is rapidly spreading, and position information is becoming increasingly important [14]. That is, even in VLC, examining security measures is an important subject in promoting business applications in the future.

Incidentally, an international standard (ISO 15408, Common Criteria) is now used for computer security as a set of security countermeasures in software development [15, 16]. Moreover, the concept of security-by-design has been proposed, and it considers the importance of investigating how security is obtained from the upper process of a plan, a design phase, etc. [17]. Thus, proposals have been made with aims to reduce the cost of total security countermeasures and efficiently develop good software by investigating security countermeasures during the early stages of development. This paper proposes a secure business architecture for new VLC and presents an investigation on security countermeasures for it.

3 Proposal of Secure VLC Business Architecture

3.1 Requirement Specifications

The specifications are clarified from the viewpoint of promoting the diffusion of the VLC system into business. Specifically, authentication, encryption, and network security are essential for overall security. Existing functions are utilized to reduce development costs when introducing these functions. Table 1 shows the required specifications for this VLC system.

The following briefly describes the functions that satisfy the required specifications listed in Table 1.

Table 1. Requirement specifications and existing functions for VLC

Requirement specifications	Corresponding existing function
Authentication function	• ID management infrastructure, Public key infrastructure
Encryption function	• Public key cryptosystem
Network security	• Public key infrastructure • PLC

(1) Public key infrastructure (PKI)
PKI refers to the security infrastructure for a public key cryptographic system. This technology enables various security countermeasures, such as encryption, digital signatures, and certification. As shown in Fig. 2, a certificate authority that issues an electronic file called a certificate (public key certification) to users. This certificate is equivalent to a certificate of seal impression and an identification card in the real world [18, 19].

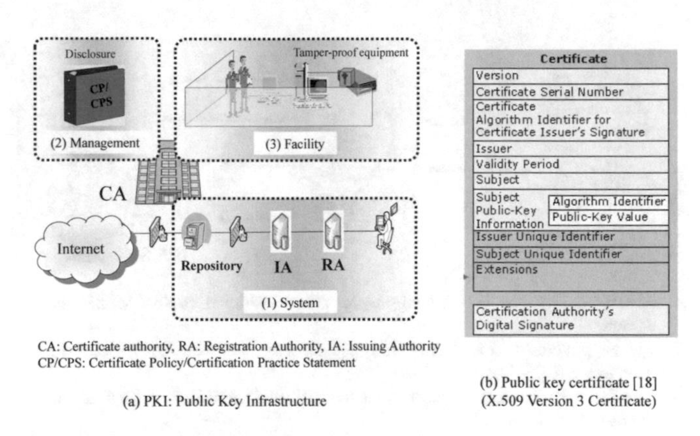

(a) PKI: Public Key Infrastructure

(b) Public key certificate [18] (X.509 Version 3 Certificate)

Fig. 2. System configuration of PKI

(2) Power line communication (PLC)
This type of communication refers to a technique of using a power line as a conduit for communication. Low-speed PLC uses frequencies of 450 kHz or less, and high-speed PLC uses frequencies of 2 to 30 MHz. In Japan, the Ministry of Internal Affairs and Communications revised a ministerial ordinance in October 2006 by adding an item that restricts indoor usage, accepting PLC using frequencies of 2 to 30 MHz. Currently, products corresponding to high-speed PLC have been distributed since December 2006 in response to this ministerial ordinance revision [20, 21]. Figure 3 demonstrates how PLC operates when connected to a normal outlet [21].

(3) ID management infrastructure
ID management guarantees the validity of the ID and provides a mechanism of access control (only people with the appropriate authority can gain access to need-to-know information). Combining the ID management infrastructure with the position authentication ID can guarantee the validity of the position authentication

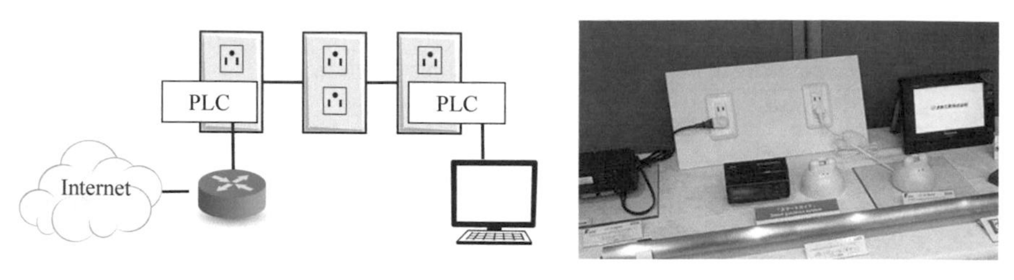

(1) Layout example of PLC (2) Application example of HD-PLC system [21]

Fig. 3. Usage pattern of PLC

ID and thorough control access by the administrator. As shown in Fig. 4, an ID management infrastructure using Shibboleth enables the authentication of a user ID of a system and enables unified management of authority [22].

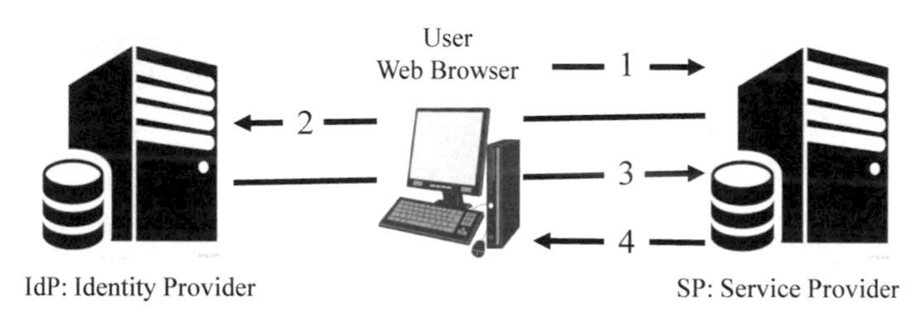

1. The SP detects the user attempting to access restricted content within the resource.
2. The SP generates an authentication request, then sends the request, and the user, to the user's IdP.
3. The IdP authenticates the user, then sends the authentication response, and the user, back to the SP.
4. The SP verifies the IdP's response and sends the request through to the resource which returns the originally requested content.

Fig. 4. Example of ID management by Shibboleth [22]

In addition, using an ID management base among multiple organizations requires creating a policy for the operation management of position authentication IDs. Furthermore, the structure of the operation, known as a trust framework, also has to be taken into account, including the audits to enable this policy to be observed.

3.2 Basic Architecture of Secure Position Authentication Infrastructure

As shown in our previous study [12], we present a secure business architecture plan with functions that satisfy the requirement specifications shown in Table 1 for a VLC system.

Figure 5 shows that the security of the system is guaranteed by cooperating with the new functions that utilize PKI, PLC, and ID management infrastructure. Specific procedures will be described for the service provision side and user side.

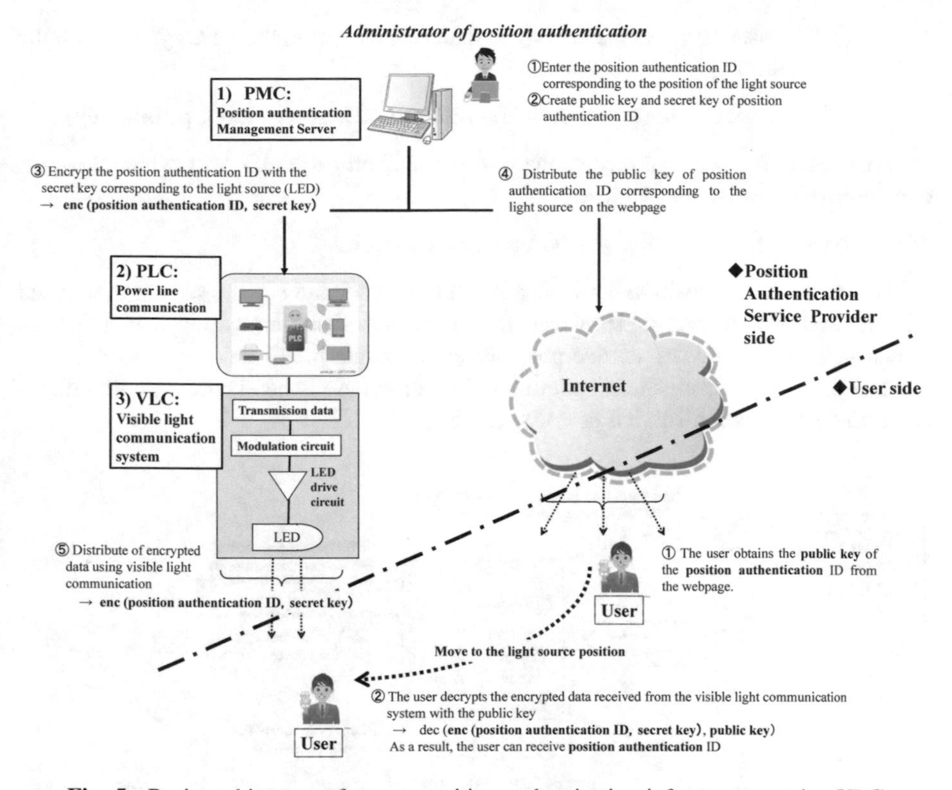

Fig. 5. Basic architecture of secure position authentication infrastructure using VLC

(1) From position authentication distribution of service provider to user reception

(Position authentication service provider side)

① The administrator inputs the position authentication ID to the server.

② Public and secret keys of the position authentication ID are created by the administrator.

③ The position authentication ID is encrypted with the secret key corresponding to the light source (i.e., LEDs).

-> **enc (position authentication ID, secret key)**

④ The public key corresponding to the light source (i.e., LEDs) is distributed on the webpage.

⑤ Encrypted data are distributed using VLC.

-> **enc (position authentication ID, secret key)**

(User side)

① The user obtains the public key of the position authentication ID from the webpage.

② The user decrypts the encrypted data received from the VLC system with the public key.

-> **dec (enc (position authentication ID, secret key), public key)**

As a result, the user can receive the position authentication ID, with which they can use the service associated with it.

(2) Service utilization using position authentication ID

By receiving the position authentication ID, the user can use the service associated with it as shown in Fig. 6. However, it is necessary for the position authentication infrastructure side and the service provider to cooperate in advance.

The position authentication system configuration base is hereinafter simplified as a position authentication service provider (PASP).

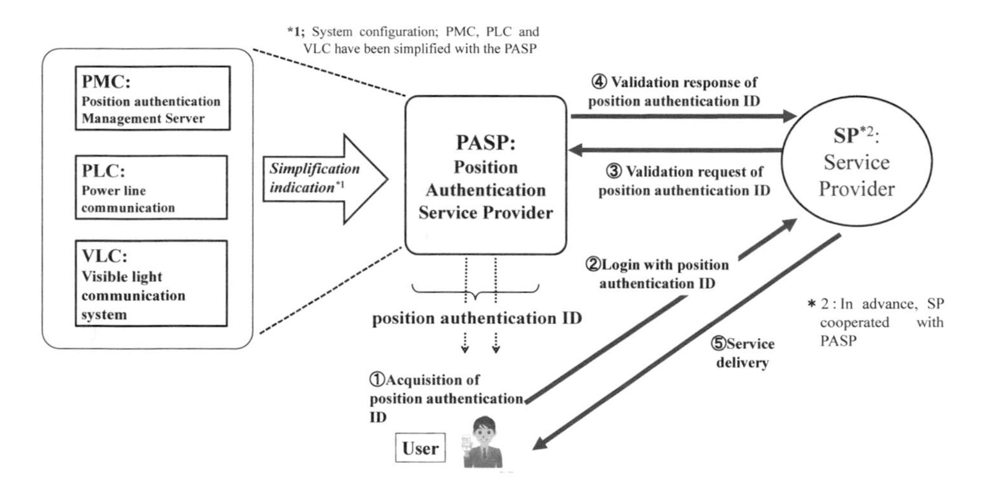

Fig. 6. Example of service utilization using position authentication ID

3.3 Expansion of Position Authentication Infrastructure

In a previous study, we clarified the basic architecture of a position authentication infrastructure using the position authentication ID for a light source. Here, we propose a cooperation with ID management from the viewpoint of scalability of a position authentication infrastructure.

Generally, position authentication needs to be installed in many positions from the viewpoint of service extensibility. In this case, multiple PASPs in the basic architecture outlined in Sect. 3.2 are also generated. On the other hand, considering the extensibility with various services using a position authentication ID, management becomes easier if intermediaries are used between the PASP and various service providers.

In general, as a plan of this intermediary, two methods of PKI utilization and ID management utilization are assumed.

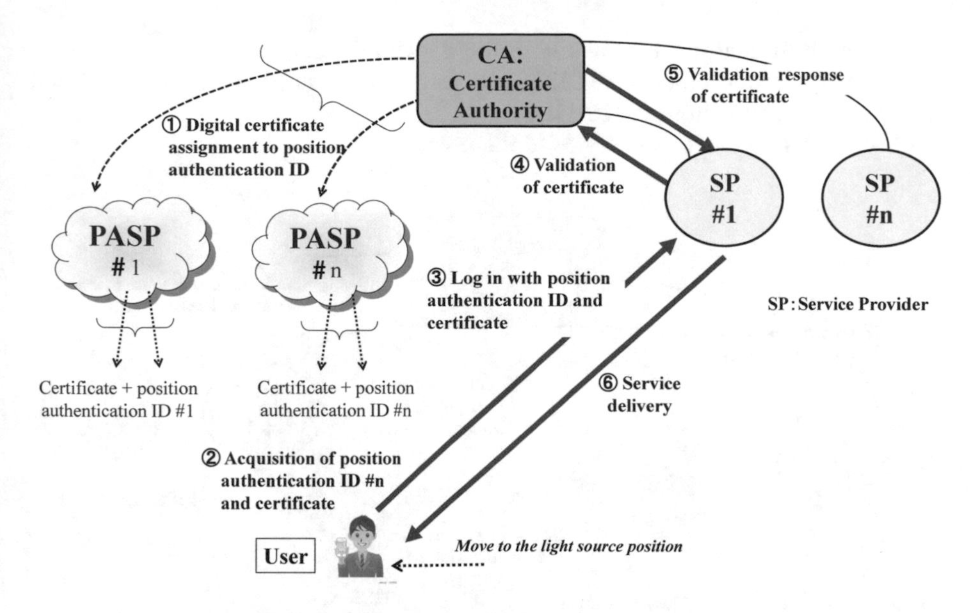

Fig. 7. Expansion of position authentication infrastructure service using PKI

3.3.1 PKI Utilization Method

As shown in Fig. 7, the multiplication scheme of the position authentication infrastructure by using the PKI is as follows.

① Issue a certificate to the PASP from the certificate authority
② The user decrypts the secret key of position authentication ID #n and obtains the ID's certificate.
③ The user logs into the SP (in this case, SP #1) to receive the service using the position authentication ID obtained in ② and the certificate.
④ SP #1 confirms the validity of the certificate with the certificate authority.
⑤ The verification result (in this case OK) of the certificate is received from the certificate authority.
⑥ SP #1 provides its service to the user

3.3.2 ID Management Utilization Method

In this section, we propose a multiplication scheme of position authentication infrastructure utilizing the ID management outlined in Sect. 3.1(3). As shown in Fig. 8, a new position authentication ID management provider (PA_IDP) is provided as the ID management of multiple position authentication infrastructures, in which its process is as follows with reference to Fig. 5.

① Each PASP registers the position authentication ID in the PA_IDP
② The user decrypts the secret key of position authentication ID #n.
③ The user logs in to the SP (SP #1 in this case) to receive the service using the position authentication ID obtained in ②.

④ SP #1 confirms the validity of the ID with the PA_IDP.
⑤ The verification result (OK in this case) of the position authentication ID # n is received from the PA_IDP.
⑥ SP #1 provides its service to the user.

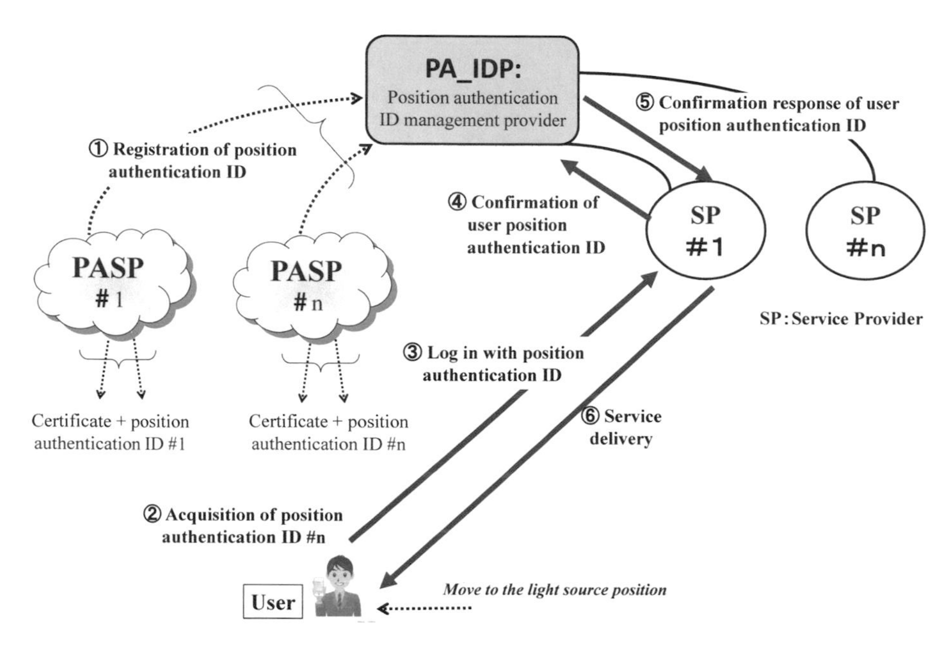

Fig. 8. Expansion of position authentication infrastructure service using ID management

3.3.3 Comparison of Extension Proposals

For the extension proposal outlined in Sect. 3.3.2, a qualitative comparative evaluation was conducted from the viewpoint of future business characteristics, namely, strictness of position authentication, convenience of users, and economics. The results are shown in Table 2.

Table 2. Features of the business architecture

Expansion proposal	Strictness of position authentication	Convenience of users	Economics	Evaluation
Case 1) PKI utilization	High	Medium	Low	Medium
Case 2) ID management Utilization	Medium	High	High	High

In the case of PKI utilization, the granting of a digital certificate by a certificate authority was excellent from the strictness of position authentication. However, there were difficulties in economics because users are required to send certificates to the service provider, and the position authentication service provider side is required to pay to issue certificates.

On the other hand, in the case of ID management utilization, despite the strictness of its position authentication ID management infrastructure being inferior to that of a certificate authority, the scheme was superior in all other characteristics. Therefore, as a proposal for expanding the position authentication ID, ID management utilization is preferable.

4 Consideration

In this section, we consider the business architecture proposed in Sect. 3.3 from the viewpoint of applying it to actual business.

As shown in Table 3, it is possible to provide a strict position authentication infrastructure on the basis of a new position authentication ID corresponding to a light source (LED) compared with the conventional VLC system.

Table 3. Features of the proposed business architecture

Viewpoint of modeling	Technical features	Application
Security guarantee	Strict identification of users by introducing position authentication ID management	PKI [17], ID management [22]
	Limited information distribution of VLC recipients	VLC technology [1]
	Information circulation limited to PLC network	PLC technology [19]
Ease of conducting business	ID distribution over the Internet	ID management [22]

As shown in Fig. 6, since the proposed position authentication service is assigned to a light source, it serves as a service for a fixed position. Therefore, it is easy to specify the position, and the value can be easily given to the position.

For example, by installing a light source in a sightseeing spot, it is possible to strictly prove that the user has visited the sightseeing spot. For this consideration, it is possible to easily add discount tickets (electronic value) to the facilities in tourist spots.

Furthermore, because the business architecture proposed in this paper is possible by combining existing functions, development of new functions and protocols is unnecessary and the development cost to secure a VLC system can be reduced.

5 Conclusion

In this paper, in view of the spread VLC development in the future, we proposed a secure business architecture using our previously researched concept, "position authentication ID" information, by combining an ID management function with a position authentication infrastructure on the basis of multiple VLC systems. A qualitative comparative evaluation demonstrated that it was possible to construct a secure business model.

In future work, we plan to evaluate the infrastructure in real world situations using an actual system configuration. An accounting function will also be added for increased functionality.

References

1. Komine, T., et al.: Fundamental analysis for visible-light communication system using LED lights. IEEE Trans. Consum. Electron. **50**(1), 100–107 (2004)
2. Grobe, L., et al.: High-speed visible light communication systems. IEEE Commun. Mag. **51**, 60–66 (2013)
3. Komine, T., Nakagawa, M.: Integrated system of white LED visible-light communication and power-line communication. IEEE Trans. Consum. Electron. **49**(1), 71–79 (2003)
4. Shimada, S., Takeda, Y.: Trends in visible light communication and application to ITS. Toshiba Rev. **64**(4), 27–30 (2009). (in Japanese)
5. Saini, H.: LI-FI (Light Fidelity)-the future technology in wireless communication. J. Comput. Appl. **7**(1), 13–15 (2016)
6. Psychic thought: The next "Li-Fi" coming after Wi-Fi - Relationship between Li-Fi and 5G in IoT era (2017). http://blog.livedoor.jp/utakknn/archives/2017-01-15.html. (in Japanese)
7. Visible Light Communication Consortium: About a standard (2014). http://vlca.jp/standard/. (in Japanese)
8. Panasonic: Started business of information linkage service using "optical ID" technology (2015). https://news.panasonic.com/jp/press/data/2015/12/jn151209-3/jn151209-3.html. (in Japanese)
9. Nakagawa Laboratories, Inc.: Products Information (2018). http://www.naka-lab.jp/product/index_e.html
10. Suzuki, K.: Visible light communication system for application to ITS. Toshiba Rev. **61**(8), 20–23 (2006). (in Japanese)
11. Tanimoto, S., et al.: Concept model of electronic value trading platform with visible light. In: Proceedings of LED Synthesis Forum 2018 in Tokushima, P-25, pp. 157–158 (2018)
12. Tanimoto, S., et al.: Proposal of secure business architecture by visible light communication system. In: Proceedings of 7th International Congress on Advanced Applied Informatics, Yonago, July 2018. (to appear)
13. Ueno, H., et al.: Visible light ID system. Toshiba Rev. **62**(5), 44–47 (2007). (in Japanese)
14. Takahashi, Y., et al.: The study for the position authentication and the security of location information. CSEC-38, pp. 1–6, IPSJ (2007). (in Japanese)
15. ISO: ISO/IEC 15408-1:2009. https://www.iso.org/standard/50341.html
16. Tsujii, S.: Paradigm of information security as interdisciplinary comprehensive science. In: Proceedings of the 2004 International Conference on Cyberworlds, pp. 1–12 (2004)

17. Kaneko, T.: Introduction to security by design (2016). https://www.ipa.go.jp/files/000055823.pdf. (in Japanese)
18. IPA: PKI related technical information (2012). https://www.ipa.go.jp/security/pki/. (in Japanese)
19. Microsoft TechNet: Digital Certificates. https://technet.microsoft.com/en-us/library/cc962029.aspx
20. Kobelco Systems: Internet connected to an outlet?! -What is PLC (Power Line Communications) (2007). http://www.kobelcosys.co.jp/column/itwords/42/. (in Japanese)
21. CNET Japan: Currently the "HD-PLC" communication standard using power lines (2016). https://japan.cnet.com/article/35089788/. (in Japanese)
22. Shibboleth: Component Interface (2017). https://wiki.shibboleth.net/confluence/display/CONCEPT/Home

Peer-to-Peer Data Distribution System with Browser Cache Sharing

Kazunori Ueda(✉) and Yusei Irifuku

Kochi University of Technology, Kochi, Japan
ueda.kazunori@kochi-tech.ac.jp

Abstract. Many services that provide video contents such have been available. Traffic caused by watching videos has been increasing and the traffic will consume network resources on the Internet or networks of contents delivery provides. In this paper, the authors proposed peer-to-peer data distribution system with browser cache sharing to decrease network resource consumption. Nodes that are watching the same video exchange fragments of video data by using WebRTC protocol.

1 Introduction

Recent years, web services for watching or sharing videos such as YouTube have been widely used. Forecast by the Cisco systems indicates the amount of such traffic caused by the web services will be increasing [2]. Traffic caused by these web services consumes network resources on the Internet or other wide area networks. Furthermore, services of watching or sharing videos require computing and networking resources for contents servers.

For these issues, peer-to-peer networking models are available as a method of constructing logical network for network applications. Although peers exchange and transfer messages to construct logical network or offer information about peers and contents, peers do not need to send the messages to a centralized server. This is one of advantages of peer-to-peer networking model in terms of load balancing or load distribution.

Peer-to-peer web proxy is a system that adopt peer-to-peer networking model. This system is available to avoid consumptions of computing or networking resources of a server. This is because peers can provide caches for other peers that require them. In the peer-to-peer web proxy system, peers have to store caches even if the caches are not necessary. To use the caches, the peers need to be in the network as active peers, which mean that the peers can provide services as servers. Furthermore, storages of peers are consumed if the amount of caches is large.

We focused the issues that peers have to store the caches continuously and computing and networking resources of contents servers are consumed. Moreover, we proposed a new cache sharing system based on peer-to-peer networking model. This proposed system enables peers that require the same contents to share caches of the contents. Contents servers manage a list that includes

L. Barolli et al. (Eds.): NBiS 2018, LNDECT 22, pp. 590–597, 2019.
https://doi.org/10.1007/978-3-319-98530-5_51

information about peers are downloading the contents which are managed by the contents servers. Peers included in the list are removed if the peers finish downloading the contents. This prevents peers from storing caches which are not needed by themselves. In addition, we showed effectiveness of the proposed method by comparing resource consumption of contents servers with the conventional model.

2 Contents Cache Sharing

2.1 Web Cache

Networking node stores data as caches and the caches are used without duplicate downloading when the data are required again. Since the node does not need to download the contents through the Internet or other networks, networking resources are saved and the contents can be used immediately. Existing peer-to-peer web proxy system enables peers to share caches that are downloaded by themselves.

2.2 Web Proxy

Web proxy is a function as a gateway server that requests web contents in charge of web clients and stores web contents [4]. Since web client receives contents from a web proxy server instead of a web server that has contents required by the web client, the web clients can receive the contents in shorter time and computing resources of servers and networking resources on a paths from the web proxy server to contents servers.

2.3 Peer-to-Peer Web Proxy

Peer-to-peer web proxy system is constructed by peers in order to share web contents cache on logical networks [5].

2.4 Parallel Download

Parallel download is technique to used to fetch contents data with multiple connections. Parallel download is achieved by HTTP Range Request [3]. Since maximum speed of data transfer is determined by settings of TCP, total speed of downloading can be improved if several parts of data blocks are fetched by individual connections. However, multiple connections consume computing and networking resource of web servers. Parallel download affects throughput of web server or other web clients. Our proposal system is based on ideas of parallel download technique.

3 Proposed System

In our proposed system [6,7], peers perform download by cooperating with other peers. A peer that is downloading certain data can receive caches of the same data from other peers that are downloading the same data at the same time. By cooperating each other, effective data transmission will be achieved. Actually, the proposal system has advantages about transmission time and consumption of network resources. This means that computing resource of contents servers and network resources in server networks will be saved. Following subsections describes basic concept of the proposed system briefly.

3.1 Framework of Proposal System

Figure 1 shows basic concept of the proposed system. When the peer at the bottom of the figure starts to download data from a server, the peers at the left or right side of the figure help the downloading peer, that is, the peers transmit cache of the data to the downloading peer. In this situation, peers at the left or right side of the figure are called cooperative peer(s). Since the cooperative peers started to download the same content before the downloading peer, the cooperative peers can offer caches of the same content. Since computers on which peers exist are in internal networks basically, the computers cannot communicate with other peers directly. We adopt WebRTC [1] and UDP hole punching technique as a way to let the computers communicate with other peers. The server, which provides contents, has functions as a signaling server and manages peer lists by using information about signaling.

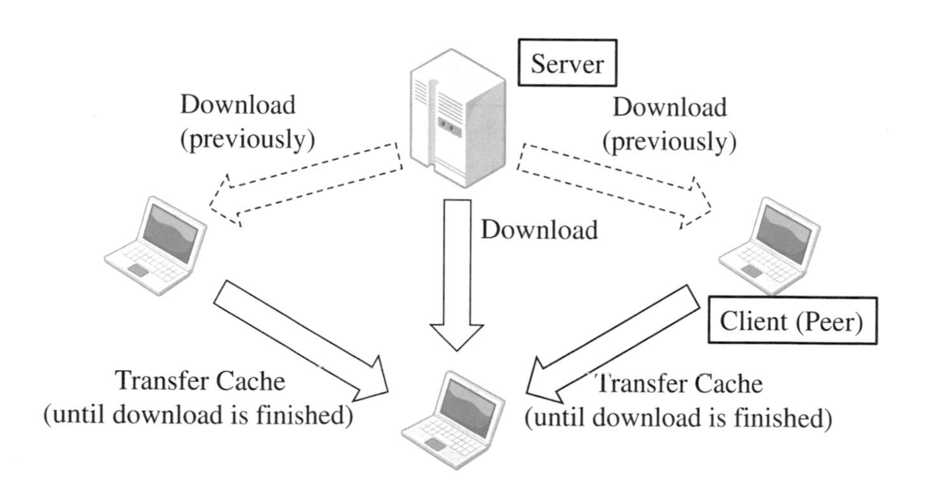

Fig. 1. Cooperative download

3.2 Management of Peer List

The server offers a peer list about a content which are required by the downloading peer to know information about peers that performed downloading the same contents in advance. The downloading peer is also added into the peer list by the server. The peer list is managed for each content by the server. Peers finished downloading will be removed from the peer list and the peer stop helping other peers. This is because the peer has to stay in the network and wait for requests if the peer offers many caches to other peers. These actions consume several resources.

4 Evaluation

In this paper, we simulate the proposed method and evaluate its performance. We verify the load on the server under each condition and compare it with the performance of client/server type communication which is the conventional delivery method. The number of blocks of content to be downloaded is set to 10, 100, 500, 1000, and verification is carried out when the peer's content acquisition start time is set at equal intervals and at random.

4.1 Conditions

Verify with the following conditions. The server load reduction rate of client/server type communication to be compared is assumed to be 100%.

- Condition 1
  - Number of blocks: 10, 100, 500, 1000, 10000
  - Number of peers: 2, 3, 5, 10, 25, 50, 100, 200, 500, 1000
  - Start time of each peer: Set at even intervals for each (content data amount/number of clients) with respect to the first start terminal
  - Verify the server's load reduction rate under the above conditions
- Condition 2
  - Number of blocks: 10, 100, 500, 1000, 10000
  - Number of peers: 2, 3, 5, 10, 25, 50, 100, 200, 500, 1000
  - Start time of each peer: randomly set from 1 to the number of setting blocks with respect to the first start terminal

4.2 Verification

The average reduction rate, the maximum reduction rate, and the minimum reduction rate when 1000 cycles were verified under the above conditions.

In condition 1 where peer start times are set at equal intervals, the load reduction rate increases as the number of peers to be downloaded increases, and the number of blocks is reduced to 1000 and the number of peers is reduced to

49.5%. In addition, almost the same effect was obtained regardless of the number of blocks except for the case where the number of blocks was 10. Figure 3 shows these results.

In addition, the minimum reduction rate when 1000 cycles were performed increased even in the case of the number of peers in which the increase in the average load reduction rate was almost disappeared. Figure 5 shows these results.

The simulation results showed that the load on the server can be expected in the proposed method compared to the client/server type communication which is the conventional video distribution method. Furthermore, it showed that if the number of blocks and the number of peers were more than a certain size, it was possible to obtain a larger effect and be stably reduced. Also, content is acquired in parallel from the Web server and collaborating peers, so content acquisition speed is also improved in proportion to the load of the server.

4.3 Discussions

Under Condition 1 where the starting time of each peer was set at equal intervals, it was possible to confirm the increase in the reduction rate as the number of peers increased up to about 50 peers. Regarding the number of blocks of contents, when the number of blocks is 10 as shown, the reduction rate has decreased by about 5% to 9% as compared with the case where the block size is large. When dealing with small files, it is thought that it is preferable to use it in the case of handling only large contents only when considering that the overhead becomes large and it is not possible to receive the benefits of parallel downloading.

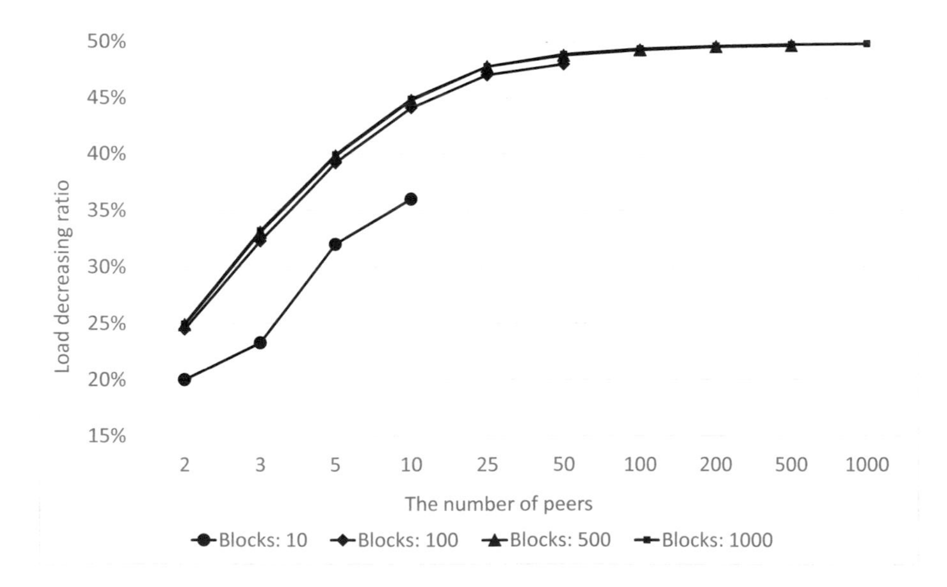

Fig. 2. Results of condition 1.

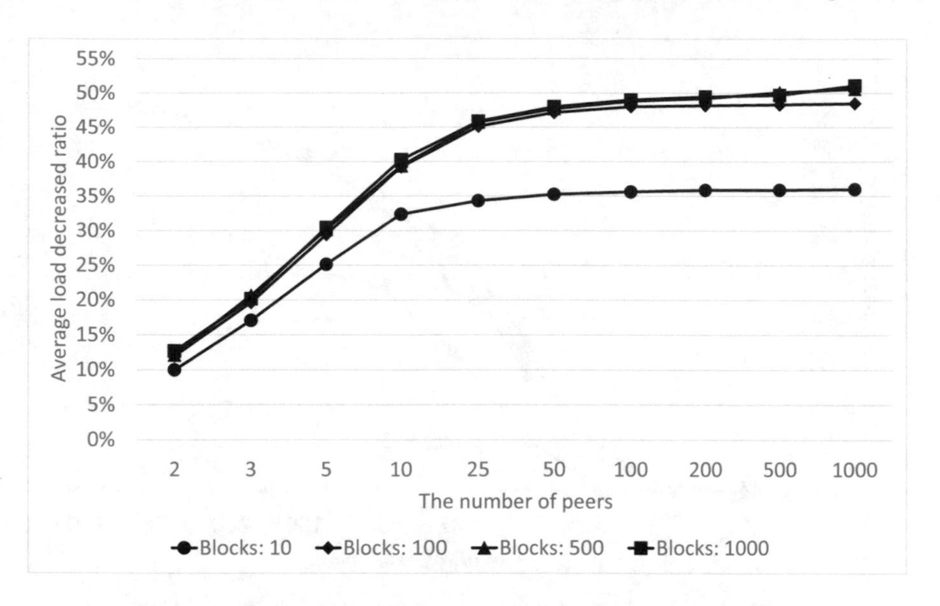

Fig. 3. Results of average load decreased ratio on condition 2.

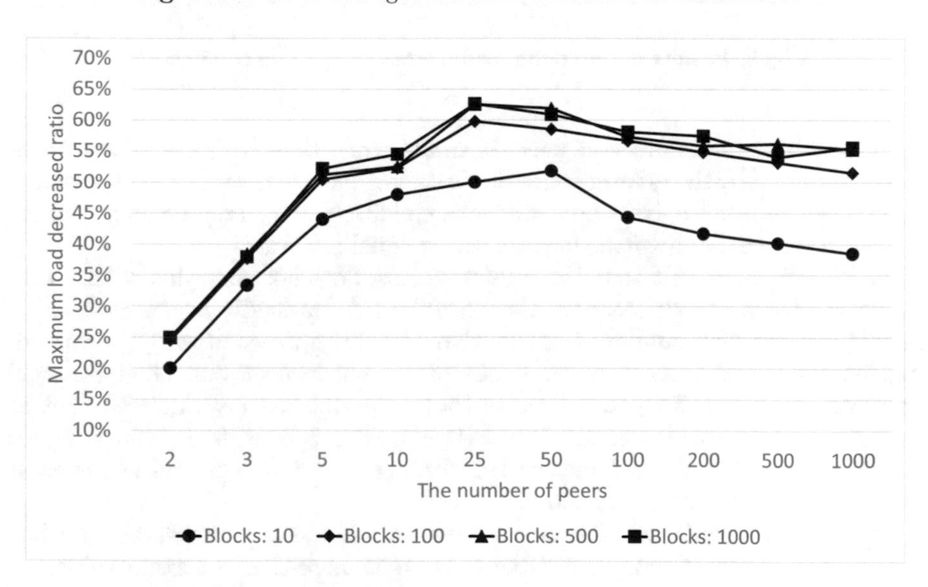

Fig. 4. Results of maximum load decreased ratio on condition 2.

In Condition 2 where the starting time of each peer was randomly set, the reduction rate improved similarly to Condition 1 as shown, and almost the same result was obtained (Figs. 2 and 4) with respect to the reduction rate when the number of peers increased became. Also, as in condition 1, when the number of blocks is 10, the reduction rate is lower than in the case of other blocks. In

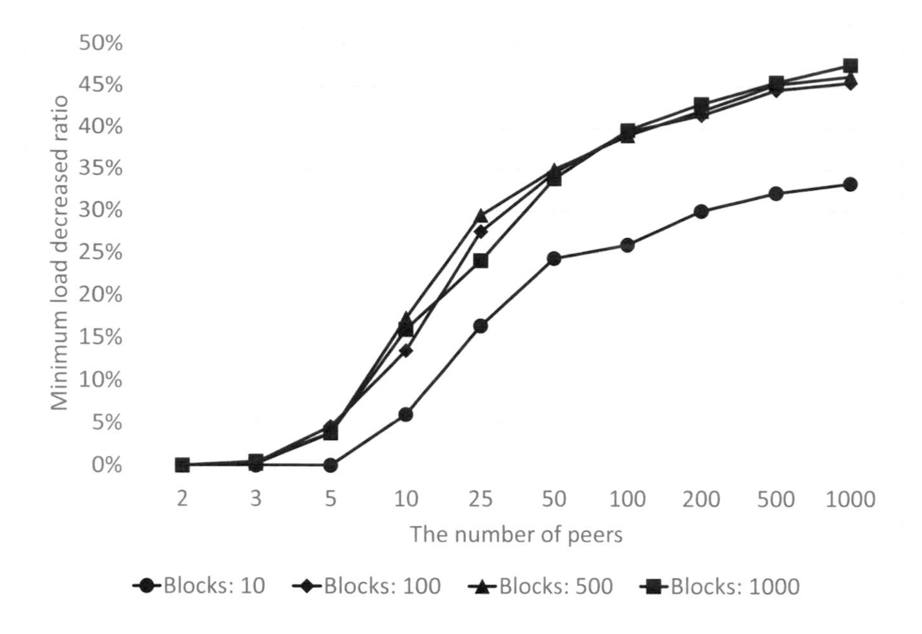

Fig. 5. Results of minimum load decreased ratio on condition 2.

the case where the number of peers is small, when the peer's start time is set at equal intervals, the greater reduction rate has been achieved. This is because it is considered that it is an advantageous condition that many peers are surely able to receive sharing when they are set at equal intervals.

Moreover, the result that the reduction rate does not rise when either condition reaches a certain peer number is obtained. Basically, as the number of peers increases, the number of blocks that the entire peer currently downloading increases, and the download speed of the subsequent download terminal improves. However, if there are peers that have finished acquiring content at high speed, The peer leaves the P2P network, so it is considered that the peer acquired thereafter can not receive the sharing from the peer and the content acquisition speed does not improve.

From these results, we showed that we can hit the load considerably when the number of blocks and the number of peers is more than a certain value.

When the number of blocks becomes equal to or larger than a certain size, even if the block size increases further, there is almost no influence on load reduction. However, in actual use, the larger the content, the larger the data size acquired from other than the server, so it is considered to be more effective in proportion to the content size.

If the number of peers is equal to or larger than a certain value like the influence due to the change in block size, even if the number of peers increases further, there is almost no influence on the reduction rate. However, even if the number of peers does not rise, the minimum mitigation rate continues to rise as shown. This is because the performance of the proposed method is stabilized as the number of peers increases.

Currently, verification is performed on the premise that all peers have the same performance, so it is a future task to verify in a real environment or in a real environment more realistically. Moreover, it is also necessary to investigate not only the performance of the client, but also the influence due to the popularity of the acquired content and the fluctuation of the communication bandwidth.

5 Conclusion

Many services that provide video contents such have been available. Traffic caused by watching videos has been increasing and the traffic will consume network resources on the Internet or networks of contents delivery provides. In this paper, the authors proposed peer-to-peer data distribution system with browser cache sharing to decrease network resource consumption. Nodes that are watching the same video exchange fragments of video data by using WebRTC protocol.

References

1. Bergkvist, A., Burnett, D.C., Jennings, C., Narayanan, A.: WebRTC 1.0: real-time communication between browsers (2015). http://www.w3.org/TR/webrtc/
2. Cisco System: Cisco visual networking index: Forecast and methodology, 2010–2015 (2011). http://www.cisco.com/en/US/solutions/collateral/ns341/ns525/ns537/ns705/ns827/white_paper_c11-481360.pdf
3. Fielding, R., Lafon, Y., Reschke, J.: Hypertext transfer protocol (HTTP/1.1): range requests (2014). http://www.ietf.org/rfc/rfc7233.txt
4. Fielding, R., Nottingham, M., Reschke, J.: Hypertext transfer protocol (HTTP/1.1): caching (2014). http://www.ietf.org/rfc/rfc7234.txt
5. Wang, J.Z., Pal, A., Srimani, P.K.: New efficient replacement strategies for P2P cooperative proxy cache systems. In: Proceedings of the 2004 Advanced Simulation Technologies Conference, pp. 97–103 (2004)
6. Matsushita, K., Nishimine, M., Ueda, K.: Cooperation P2P web proxy to reduce resource usage. In: The 8th International Workshop on Advanced Distributed and Parallel Network Applications (ADPNA 2014), pp. 420–425 (2014)
7. Nishimine, M., Ueda, K.: Design and implementation of a cache-less P2P web proxy. In: The 15th International Symposium on Multimedia Network Systems and Applications (MNSA 2013), pp. 500–505 (2013)

The 9th International Workshop on Heterogeneous Networking Environments and Technologies (HETNET-2018)

A Dynamism View Model of Convergence and Divergence of IoT Standardization

Toshihiko Yamakami(✉)

ACCESS, 3 Kanda-Neribei-cho, Chiyoda-ku, Tokyo 101-0022, Japan
Toshihiko.Yamakami@access-company.com

Abstract. As IoT (Internet of Things) continues to penetrate everyday life, we witness the increase in the number of IoT standardization activities. This is a kind of business and political conflicts. It can be also viewed as an emergence of new types of standardization. In a world where complicated cyber-physical systems come to exist, legacy view models such as layered view models are not adequate to the new landscape of IoT standardization. As a departure of static structural view of standardization, the author proposes a dynamism model for divergence and convergence.

1 Introduction

IoT (Internet of Things) has been increasing to become realities in everyday life. A CPS (Cyber-Physical System) is becoming a new reality of this planet. In order to cope with this new reality, it is important to ensure interoperability in a wide range of CPSs.

A total digitalization of economy leveraged by many advances technologies like mobile, cloud, analytics is both of the reasons and results of this wide adoption of CPSs.

Interoperability of CPSs is critically important, however, the increasing spawning of IoT standardization activities lead to a unclear landscape of trends of interoperability.

Telecom industry, Internet industry, software industry, electronic industry, machinery industry, appliance industry, energy industry, healthcare industry and others attempt to bring interoperability of IoT in place. IoT standardization includes oneM2M, ITU, ISO/IEC, OMA, W3C, IEEE, IETF, ETSI, and many other standardization bodies and industrial consortia.

Due to the characteristics of CPSs, it is often difficult to distinguish different IoT standardization activities by standardization bodies and industrial consortia.

The author discusses the limitations of static view models in standardization. Then, the author proposes a dynamism-oriented view model of convergence and divergence in IoT standardization.

L. Barolli et al. (Eds.): NBiS 2018, LNDECT 22, pp. 601–610, 2019.
https://doi.org/10.1007/978-3-319-98530-5_52

2 Background

2.1 Purpose of Research

The aim of this research is to develop a framework to understand the dynamism in convergence and divergence of IoT standardization.

2.2 Related Work

Research on public transportation with IoT standardization consists of three areas: (a) IoT standardization activities, (b) landscape of IoT standardization bodies, and (c) real-world examples of standard-based IoT deployment.

First, in regard to standardization activities, Yun et al. presented interworking in oneM2M [10]. da Silva et al. presented the OpenTOSCA architecture from OASIS's TOSCA standard [9]. Raggett presented a W3C's plan of IoT [8]. Abou-Zahra et al. discussed web standards for IoT [1]. Chen et al. presented the IoT architecture of Chinese standardization [4]. Cavalcante et al. presented an IoT architecture from WSO2, an OSS technology provider [3]. Zahariadis et al. presented the FIWARE initiative in EU for cloud infrastructure [11].

Second, in regard to landscape of IoT standardization bodies, Anthopoulos et al. discussed matching smart city projects and smart city-related standardization [2]. Costa-Perez presented trends of the latest telecommunication standardization including oneM2M [5].

Third, in regard to real-world examples of standard-based IoT deployment, Kubler et al. presented a smart city example based on standards from EU Horizon 2020 IoT project [7].

The originality of this paper lies in its identification of view models to compare IoT standardization.

3 Method

The author performs the following steps:

- analyze the diverging and converging patterns in IoT standardization.
- propose a dynamism view model of IoT standardization.

4 Legacy View Models

In 1980s, ISO's 7-layered view model was one of early development of standardization framework. It is a first layered framework to orchestrate different levels of standardization, as depicted in Fig. 1.

In 1980s, there was not sufficient understanding of function allocation among different communication entities for ensuring interoperability. This lack of common understanding lead to confusion in standardization in communication systems. The layered view is an outcome to develop common understanding to separate different functions performed in different kind of standardization.

Application layer
Presentation layer
Session layer
Transport layer
Network layer
Data Link layer
Physical layer

Fig. 1. ISO's OSI 7-layered view model

In the last 3 decades, this layered model provides a good foundation to arbitrage confusions among different communication layers. At the same time, this layered view is not appropriate as the technology advances. For example, SDN (Software-Defined Network) emerges and it does not fit to this 7-layered view model.

5 View Models

5.1 Perspectives of Segmentation

The author proposes a three-dimensional model of segmentation of standardization, as depicted in Fig. 2.

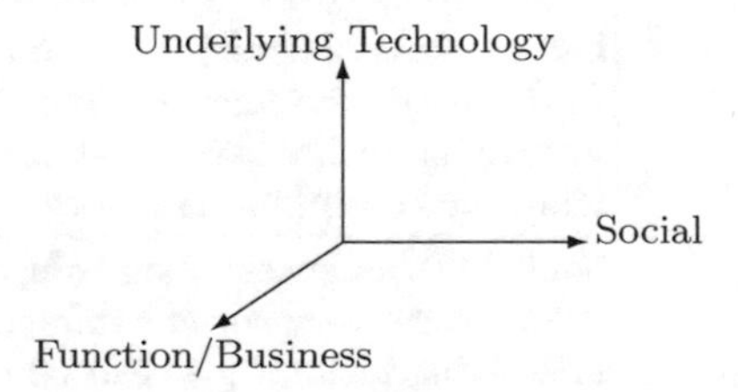

Fig. 2. Dimensional view of segmentation in standardization

This illustrates how people recognize segmentation of standardization.

Characteristics of each dimension are depicted in Table 1.

Scalability is one of technological issues that drive divergence. Scalability issues derived from use cases are depicted in Table 2.

When these scalability issues are built in an IoT standardization, another IoT standardization emerges in order to solve the built-in scalability issues.

5.2 Convergence Forces

The author presents the converging factors. First and foremost, converged standards will provide more consistent, reliable and well-defined interoperability. In addition to that, there are multiple organizational factors that drive convergence. Conditions to merge or enable coexistence of IoT standardization in a

Table 1. Characteristics of each dimension

Dimension	Description
Function/Business	Different use cases and provided functions will separate scopes and activities of standardization such as smart home, facility maintenance, energy delivery, et al.
Social	End users and service providers will separate activities of standardization. The channels to outreach users and industries will differentiate the activities. When service providers are under regulation, regulatory frameworks will separate the activities
Underlying technology	Underlying enabling technologies and problem solving paradigms will separate activities of standardization

Table 2. Scalability issues derived from target use cases.

Target use case	Implication to scalability
Local geographical location	Some of IoT instances need local limited geolocation coverage. It will limit the number to be supported. Such use cases easily lead to multicast and broadcast solutions, which is not scalable
No real-time control	Some IoT instances have use cases that only partial real-time factors are deployed. Such as an automotive system that require real-time monitoring but the data is used just for long-term maintenance. In this case, asymmetric nature requires specific underlying technologies that separate it from real-time controlling use cases
Single-device	Some IoT instances require single-device controlled. In this case, very specific infrastructure and underlying technologies can be mandate and it is isolated from other standards

technical perspective are depicted in Table 3. IoT standardization is immature, however, we are witnessing some convergence. The AllSeen Alliance agreed with the Open Connectivity Foundation to merge under the name of the Open Connectivity Foundation, in 2016. The Open Connectivity Foundation, established in February 2016, is a renamed version of the Open Interconnect Consortium to include key members of AllSeen Alliance. The Open Connectivity Foundation also inherited the UPnP technology invented by the UPnP Forum in 2016.

Converging and diverging factors will collide in many IoT standardization.

The author describes the technological perspective of convergence as above. The other two dimensions are depicted in Table 4.

Different dimensions of convergence play roles of competing forces against divergence factors.

Table 3. Conditions to merge or enable coexistence of IoT standardization

Direction	Pattern	Description
Merge	Common enemy	When there is a shared threat from a common enemy, they can merge
	Long-term trust	When there is trust among two organizations, they can merge
	Sharing membership	When there is shared membership of key players, they can merge
Co-existence	Existing umbrella framework	When there is an existing framework to correlate multiple standardization bodies, they coexist. Examples include 3GPP and OneM2M
	Technical encapsulation	When there is a good technical encapsulation such as layering, they can co-exist
	Complementarity	When multiple standards fill the vacancy of each other's features, they do not merge, but co-exist
	Working ecosystems	If there is a working ecosystem which is built on multiple standards, multiple standards do not merge, but co-exist

Table 4. Convergence in other dimensions.

Perspective	Converging factors
Player	Many influential enterprises occupy leading positions of multiple IoT organizations. These enterprises can trigger merge when they can use no more utilization of case-by-case use of each organization
Business	Shrinking organization or organizations without strong action items tend to be merged into more active organizations

5.3 Emergence of Cyber-Physical Systems

Emergence of cyber-physical systems lead to the penetration of API (Application Programming Interface) into a wide range of devices in the physical world.

In CPSs, there is nothing to block APIs, which means device-based segmentation of standardization is losing its border characteristics.

There is diversity of application areas of IoT. Ministry of Economy, Trade and Industry, Japan categorized major application areas, as depicted in Table 5 [6].

Table 5. Diversity of application areas of IoT.

Area	Application domains
Production process	Application domains to improve the manufacturing site with IoT. Factory automation, equipment maintenance efficiency etc.
Mobility	Application domains to control cars with IoT. Automatic operation, maintenance efficiency etc.
Distribution and Retail (including tourism)	Application domains to upgrade distribution, retail, tourism with IoT. Transport optimization, demand Advancement of forecasting inventory management, improvement of hospitality in tourism, etc.
Smart House	Application domains to manage home equipment and infrastructure with IoT. Remote operation of household appliances, energy consumption management etc.
Medical and health-care	Application domains that contribute to medical and health at IoT. Health management at wearable terminals, optimization of medical institutions, etc.
Infrastructure, energy and industrial operation	Application domains to maintain public utility infrastructure and industries by IoT or to improve energy development and production by IoT. Efficiency improvement of plant management of water supply, refinery, chemical factory, etc.
Public sectors	Application domains to improve public sectors by IoT. Increase efficiency of administrative affairs, publicize statistics/expedite data disclosure, etc.
Agriculture	Application domains to make agriculture more efficient by IoT. Plant factory, environmental sensing, traceability of agricultural crops, etc.

This indicates a wide range of different industry applications that require interoperability. It leads to a wild landscape of IoT standardization.

In the old days, there was a one-to-one mapping of industry and standardization. It was natural that different industrial players produce different standardization activities. Use cases, underlying networking and operational infrastructure, devices and business models are industry-specific.

Considering these converging and diverging factors, the author has come to recognize a three-dimensional model of IoT convergence of standardization, as depicted in Fig. 2. These three interrelated factors exist in any standardization. It is typical in any techno-business landscape.

5.4 Dynamism in Divergence and Convergence

The author proposes a two dimensional model of divergence and convergence, as depicted in Fig. 3.

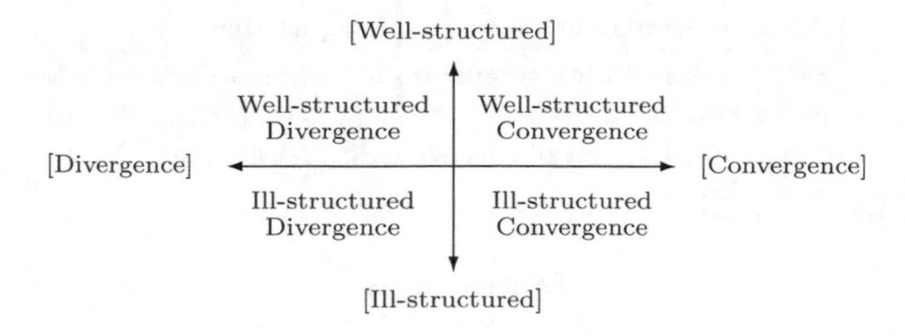

Fig. 3. A Two-dimensional model of divergence and convergence

Definitions of well-structured/ill-structured in this paper are defined in Table 6.

Table 6. Definition of well-structured/ill-structured in convergence and divergence in the IoT standardization landscape

Term	Definition
Well-structured	Different standardization activities are well defined in their scopes. It is desirable that there are well defined interfaces, or clearly defined as external entities as out of scope
Ill-structured	Different standardization activities are overlapped without well-defined scopes and interfaces. Operation among entities in different standardization activities are done in an ad hoc, or implicit, or case-by-case manner

Definitions of well-structured/ill-structured convergence/divergence are depicted in Table 7.

There are two vertical transformations as depicted in Fig. 4 as (a) and two horizontal transformations as (b).

It provides a guiding measure to analyze convergence of multiple standardization bodies and divergence from an existing standardization body.

6 Discussion

6.1 Advantages of the Proposed Method

Cross-standardization dynamism is an explored research field. The dimensional view model is useful to analyze the multiple aspects of spawning IoT standardization.

Table 7. Definitions of well-structured/ill-structured convergence/divergence.

Domain	Well-structured	Ill-structured
Convergence	IoT standardization is harmonized in a consistent and smooth manner	IoT standardization is aggregated to form a one with flaws and irregularities
Divergence	IoT standardization is split into a consistent and complementary set of diverse activities	IoT standardization is split into a conflicting, irregular, ad hoc, and illogical manner

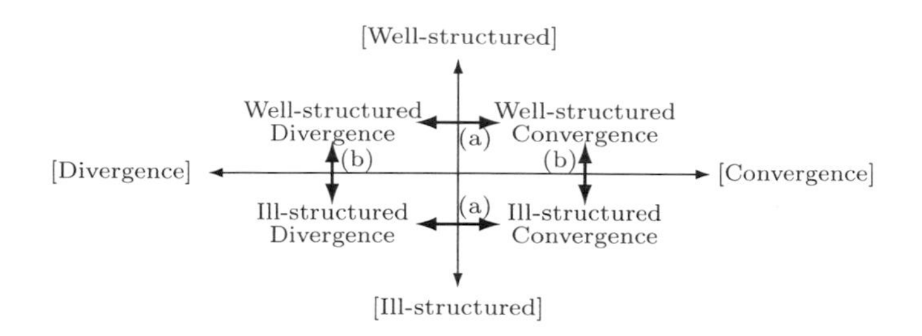

Fig. 4. Vertical and horizontal transformations

The author develops a taxonomy of well-structured/ill-structured convergence/divergence to analyze the interwined IoT activities. The close-bound-to-real-world nature of IoT has brought complexity of the real world into technical standardization. In order to clarify the parallel coexistence of multiple technology and difference use cases, it is necessary to develop a technology-agnostic and business-agnostic methodology to capture a wide range of transitions.

The author proposes a dynamism view model to understand the convergence and divergence of intertwined IoT standardization activities. Well-structured convergence indicates the maturity of IoT standardization in the long run. This is a stepping stone to capture the dynamism of IoT standardization which reflects the complexity and dynamism of the real world.

The proposed framework provides a method to analyze transition dynamism of multiple IoT standardization activities in an increasing interwined IoT world.

6.2 Limitations

This research is early and exploratory.

This paper is qualitative. Analytical measures of IoT standardization are not presented. Detailed analytical results are for future studies.

The real-world case studies of IoT standardization are not provided in this paper. In-depth analysis of reasons of convergence and divergence are not studied in this paper. A methodology to capture convergence and divergence for specific IoT standardization is not presented.

The comparison of multiple IoT standardization activities are not presented. Quantitative metrics to capture divergence and convergence are not defined. Measures to capture specific characteristics of convergence of divergence of standardization are not explored.

A concrete methodology to drive convergence of IoT standardization is not presented in this paper.

A methodology to promote well-structured convergence and divergence remains future studies.

7 Conclusion

CPSs (Cyber-Physical Systems) provides a unique opportunity to think about the technical framework surrounding physical realities. For securing interoperability of IoT, IoT standardization encounters new challenges of divergence and convergence.

In the past research, the building block to explore dynamism of convergence and divergence of IoT standardization is missing.

The proposed framework provides a stepping stone to capture the dynamism of divergence and convergence of IoT standardization. It will facilitate the planned convergence of IoT standardization activities to provide harmonized coordination of multiple aspects of IoT standardization.

Acknowledgments. The research results have been achieved by "EUJ-02-2016: IoT/Cloud/Big Data platforms in social application contexts," the Commissioned Research of National Institute of Information and Communications Technology (NICT), JAPAN.

References

1. Abou-Zahra, S., Brewer, J., Cooper, M.: Web standards to enable an accessible and inclusive Internet of Things (IoT). In: Proceedings of the 14th Web for All Conference on The Future of Accessible Work, W4A 2017, pp. 9:1–9:4. ACM, New York (2017)
2. Anthopoulos, L., Janssen, M., Weerakkody, V.: Smart service portfolios: do the cities follow standards? In: Proceedings of the 25th International Conference Companion on World Wide Web, WWW 2016 Companion, International World Wide Web Conferences Steering Committee, Republic and Canton of Geneva, Switzerland, pp. 357–362 (2016)
3. Cavalcante, E., Alves, M.P., Batista, T., Delicato, F.C., Pires, P.F.: An analysis of reference architectures for the Internet of Things. In: Proceedings of the 1st International Workshop on Exploring Component-based Techniques for Constructing Reference Architectures, CobRA 2015, pp. 13–16. ACM, New York (2015)
4. Chen, S., Xu, H., Liu, D., Hu, B., Wang, H.: A vision of IoT: applications, challenges, and opportunities with China perspective. IEEE Internet Things J. **1**(4), 349–359 (2014)

5. Costa-Pérez, X., Festag, A., Kolbe, H.J., Quittek, J., Schmid, S., Stiemerling, M., Swetina, J., van der Veen, H.: Latest trends in telecommunication standards. SIGCOMM Comput. Commun. Rev. **43**(2), 64–71 (2013)
6. Hitachi Consulting: Report on IoT-related international trend survey of standardization and de facto standards standards (in Japanese), February 2017. http://www.meti.go.jp/meti_lib/report/2016fy/000607.pdf
7. Kubler, S., Robert, J., Hefnawy, A., Cherifi, C., Bouras, A., Främling, K.: IoT-based smart parking system for sporting event management. In: Proceedings of the 13th International Conference on Mobile and Ubiquitous Systems: Computing, Networking and Services, MOBIQUITOUS 2016, pp. 104–114. ACM, New York (2016)
8. Raggett, D.: W3C plans for developing standards for open markets of services for the IoT: the Internet of Things (ubiquity symposium). Ubiquity **2015**, 3:1–3:8 (2015)
9. da Silva, A.C.F., Breitenbücher, U., Képes, K., Kopp, O., Leymann, F.: OpenTOSCA for IoT: automating the deployment of IoT applications based on the mosquitto message broker. In: Proceedings of the 6th International Conference on the Internet of Things, IoT 2016, pp. 181–182. ACM, New York (2016)
10. Yun, J., Choi, S.C., Sung, N.M., Kim, J.: Demo: towards global interworking of IoT systems – oneM2M interworking proxy entities. In: Proceedings of the 13th ACM Conference on Embedded Networked Sensor Systems, SenSys 2015, pp. 473–474. ACM, New York (2015)
11. Zahariadis, T., Papadakis, A., Alvarez, F., Gonzalez, J., Lopez, F., Facca, F., Al-Hazmi, Y.: FIWARE Lab: managing resources and services in a cloud federation supporting future internet applications. In: Proceedings of the 2014 IEEE/ACM 7th International Conference on Utility and Cloud Computing, UCC 2014, pp. 792–799. IEEE Computer Society, Washington, DC (2014)

Optimized Resource Allocation in Fog-Cloud Environment Using Insert Select

Muhammad Usman Sharif, Nadeem Javaid(✉), Muhammad Junaid Ali, Wajahat Ali Gilani, Abdullah Sadam, and Muhammad Hassaan Ashraf

COMSATS University, Islamabad 44000, Pakistan
nadeemjavaidqau@gmail.com
http://www.njavaid.com/

Abstract. Energy management in modern way is done using cloud computing services to fulfill the energy demands of the users. These amenities are used in smart buildings to manage the energy demands. Entertaining maximum requests in minimum time is the main goal of our proposed system. To achieve this goal, in this paper, a scheme for resource distribution is proposed for cloud-fog based system. When the request is made by the user, the allocation of Virtual Machines (VMs) to the Data Centers (DCs) is required to be done timely for DSM. This model helps the DCs in managing the VMs in such a way that the request entertainment take minimum Response Time (RT). The proposed Insert Select Technique (IST) tackle this problem very effectively. Simulation results depicts the cost effectiveness and effective response time (RT) achievement.

Keywords: Cloud Computing · Micro grid · Fog Computing
Macro Grid · Smart Grid

1 Introduction

Cloud Computing (CC) is a large scale user(s) facilitating paradigm which delivers services on the Internet as mentioned in [1]. It provides tremendous chances to Information Technology (IT) industry. There are three Cloud Computing Services (CCS) which play important role. in Cloud Computing Environment (CCE). These CCS are playing an important role in energy consumption techniques on large scales. These are Infrastructure as a Service (IaaS), Platform as a Service (PaaS), and Software as a Service (SaaS) as mentioned in [2,3]. CC is used in many industries Vast and most of the use of CC is in IT industry. CC and its architecture for resource allocation is well defined in [4].

CC has emerged as fast grown fragment of IT industry according to [5]. The data on cloud demands for the safety of data on cloud. Major issue of CC is it's security. [5] explains security issues and provides understanding of risks associated with cloud.

L. Barolli et al. (Eds.): NBiS 2018, LNDECT 22, pp. 611–623, 2019.
https://doi.org/10.1007/978-3-319-98530-5_53

Fog Computing also have security issues but it is preferred over cloud as it is secure as compared to CC. Data is permanently stored at cloud but the processing data is available at Fog. Authors in [6] explains the issues of Fog computing. The main purpose of placement of fog in the CCE is to minimize the load for DSM. There are number of consumers on DS that requests for different amount of energy. The time during which highest amount of energy is consumed by the users is known as Peak Time (PT). The major issue of the energy requirement and it0s management is faced by cloud during PT. In this paper, continents and the peak time considered for continent regions is shown in Table 1.

Table 1. Peak time considered for each continent

Continent	Peak start time (GMT)	Peak end time (GMT)
North America	01:00pm	05:00pm
South America	02:00pm	06:00pm
Europe	03:00pm	07:00pm
Asia	03:00pm	07:00pm
Africa	02:00pm	06:00pm
Oceania	04:00pm	08:00pm

Table 1 shows the continents with our peak start and end time (GMT) which we have assumed. In our assumption, we considered 4 h out of 24 as the peak hours for every continent.

CloudSim tool is used for simulation purpose. Service Broker Polices (SBPs) in the CloudSim manages the routing between DCs and clusters. SBPs described in [8] are: Closest Data Center Routing (CDC), Optimize RT (ORT) or Perfor-

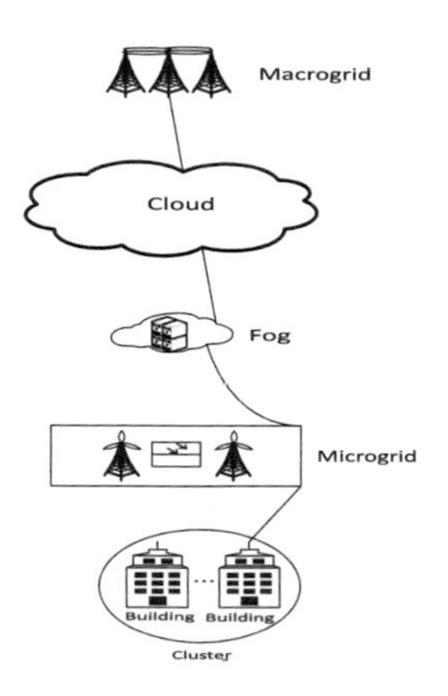

Fig. 1. Proposed system model

mance Optimized Routing, Reconfigure Dynamic Routing (RCD), and Advance SBP (ASBP). Cloud-Fog computing model is shown in Fig. 1

2 Motivation

Cloud and Fog services are used in such a way that all the requests are managed efficiently to allocate the resources to deal with the requests made by users on the system. Earlier, authors use different cloud and fog techniques to optimize the workload in large scale cloud environment by intelligent resource distribution techniques.

Fatima et al. in [2] used RR-algorithm with ASBP to minimize the load in fog environment. Authors in [3] use Particle Swarm Optimization (PSO) Load Balancing Algorithm to balance the load, and helps in maximize the services on cloud and fog. However, we also balanced the load on fog using a proposed IST technique. In our proposed model, we have considered a scenario in which there are 6 regions in a cloud-fog environment. Each region is actually a continent. Users submit tasks and to deal the tasks, VMs are required. IST allocates the task to VMs. The main focus of proposed model is on improving the PT and RT of DCs.

3 Problem Statement

As we are aware that there are several problems that may occur in CCE. Dealing with these problems is a challenging issue which is discussed in this section. The major problems that occurs in CCE are: First problem in cloud environment that occurs is to handle the coming requests from the user to the DCs in such manner so that the response time does not exceed too much so that there is a large delay in fulfilling the user request. User demands are the most important things in any system. Second problem that occurs in CCE is to check the status of resources i.e. whether available or busy. If any resource is busy then job to be done by VM is shifted to some other VM which is available and capable to perform the task. IST model works in such a way that it deals with the problems mentioned above very conveniently.

4 Related Work

For resource distribution, A cloud-fog based model is presented in [2] where authors use RR-algorithm and the SBP used here is Dynamic Service Proximity Service Broker Policy.

Valle et al. in [8] describes the use of PSO in power systems for solving non-linear optimizing problems. They use PSO and its variants for optimization of power system.

Wang et al. in [9] presented three layered model for traffic management. These techniques are implemented in CCE in such a way that they provide flexibility and reliability to the users as well as the system. Dehghanpour et al. in [10] uses a hybrid of Cuckoo Optimization algorithm and linear programming algorithm for dealing the overlays of Micro-grids (MGs).

In [11], Bogaraj et al. proposed ANFIS system for managing micro-grids. It is used for HRES. For implementation, MATLAB is used. There are many other techniques that are used for optimization in different prospectives.

Zachar and Daoutidis in [12] explains the scheduling of MGs. The focus is on minimizing the operational cost and surety of exchange of commitments.

Huld, Moner and Kriston in [13] presents a GIS-based model for estimating the mini grid system performance. Geo-spatial analysis and output energy mapping is the main methodology for GIS-based model. In [13], authors analyze small scale model which is based on photo-voltaic mini grid system.

Authors in [14] uses Enhanced Bee Colony Algorithm for minimizing the load on MGs. The analysis for such purposes is done in large as well as small scale system models.

Tumuluru, Huang and Tsang in [15] presents demand amount estimation model. The main purpose is to fulfill DS power request.

In [16], authors presents a framework for cloud data centers (DCs). This framework is named as StarCube. It is first system that handles fat-tree networks by allocating and de-allocating the requested services.

Earlier, the focus was on energy generation, but now the main focus is on minimizing the energy wastage in existing energy generation system. Author in [18] uses Genetic Algorithm for balancing the load in the CCE.

Nguyen, Thoai and Nam in [17] uses genetic algorithm for allocating VMs in private cloud and authors in [18] also uses GA for load balancing in cloud environment.

In [19], authors describes the importance of balancing the load in cloud environment and describes the survey made by the author. Authors also explain the challenges in cloud environment and techniques to overcome the load. Authors in [20] uses Dragon Fly Algorithm (DFA) for load management. The main purpose for DFA is to reduce the un-necessary and vague behavior in binary search problem.

Smart Grid (SG) concept uses the same technique of load balancing in Energy Management System (EMS) mentioned in [21].

Author in [22] also explains about the tools that are used for the analysis of performance.

Sahni et al. in [23] proposed heuristic scheduling Technique for handling the workflows that occurs in cloud environment. Monitor-control-loop manage the workflow in cloud environment in such a way that cost and time requirements are monitored continuously. They proposed a JIT-C algorithm which itself has 3 different algorithms for different tasks.

Wang, Zhang, et al. proposed a system FDALB which is described in [24]. Centralized and distributed load balancing techniques are used together in the proposed technique. Long and short both flows are scheduled differently in FDALB. Distributed switches handles the short flows.

Wang and Gelenbe in [25] proposed Task Allocation Platform (TAP) technique for tasks dispatching in cloud. TAP supports both dynamic and distinct static schemes. It provides solution to analytical models. Islam et al. in [26] proposed a model to conserve water for DCs. For such purpose they used GLB and Power proportionality techniques.

5 System Model

In this section we will discuss about our proposed system and its working. In this cloud-fog computing model, region 1, 2 and 6 are North America, South America and Oceana respectively. It comprises 3 clusters. Region 3, 4, and 5 are Europe, Asia and Africa respectively and they have 1 cluster in them. All of the 6 regions contains 1 fog in them hence there are 6 fogs i.e. F1,F2, ... , F6. There

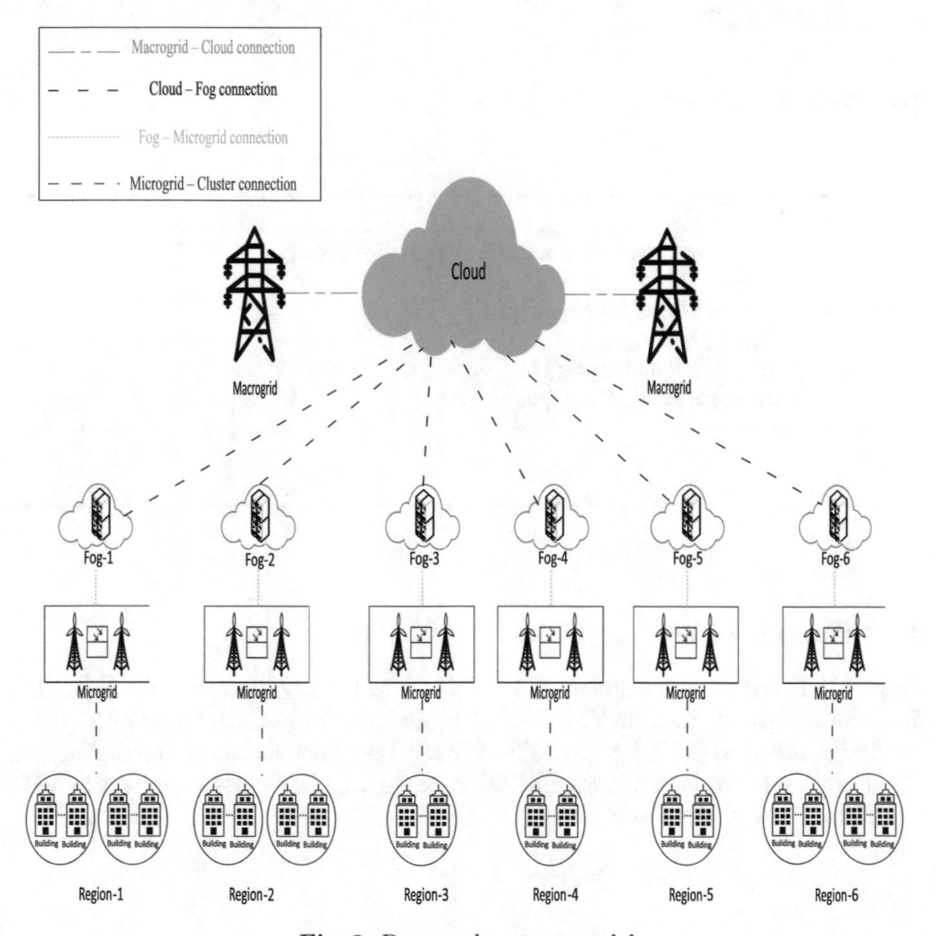

Fig. 2. Proposed system model

are total 9 clusters. They are C1, C2, ... , C9. Each cluster contains 50 buildings. Micro-grid provides electricity to the DS. VM allocation is done using IST.

The proposed scheme uses a Fog-Cloud based system which deals the users requests. Conferring to the system, there are 50 buildings considered in each cluster. Each building has 10 number of homes and 6 regions are divided into nine clusters. In this model, clusters are consist of group of buildings. Requests from consumers side are received at fog end, after this fog communicates with MGs. MGs deals with the requests and fulfill them, if not, fog contact with cloud which responds these requests. In this way, the proposed model operates. The simulation of presented technique is done in CloudSim (Fig. 2).

6 Algorithm and Working of Algorithm

This section describes the IST algorithm and its working. Working or IST is shown in the Fig. 3

6.1 IST Algorithm

Algorithm 1 IST Algo

$vmStatesList \leftarrow dataCenterController.getvmStateList()$
$getNextAvailableVM()$
$finalArray \leftarrow vmStateList.keySet()$
$doInsertionSort(finalArray)$
$whileN \leq finalArray : Length()do$
$ifvmStatesList.get \neq vmState.Busythen$
$returnfinalArray$
$endif$
$endwhile$

6.2 Working of IST Algorithm

Firstly, IST evaluates the status of VMs. If a VM is busy then the IST algorithm does not assign job to that VM. Tasks are assigned only to the idle VMs which are firstly added to the list of available VMs. A task queue is maintained. Keeping the queue under consideration, allocation of feasible resources is done and VMs handles the respective tasks.

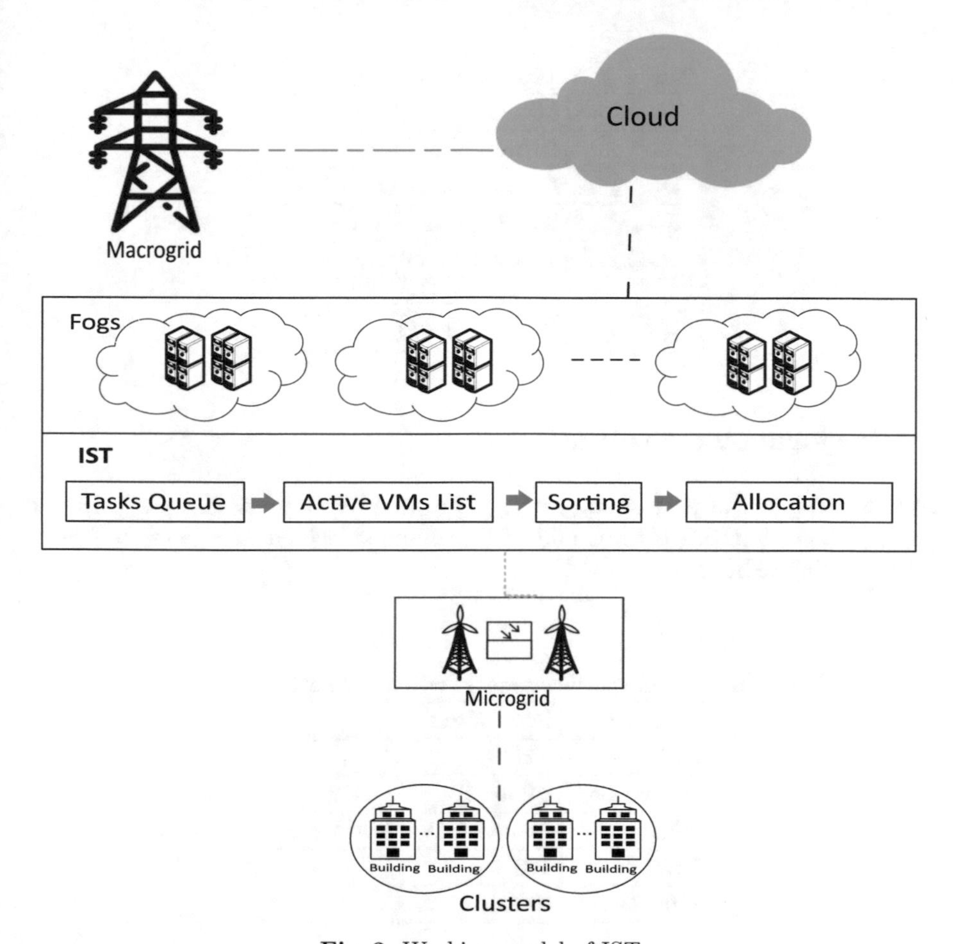

Fig. 3. Working model of IST

7 Problem Formulation

In this section, the problem formulation is discussed. As in this model, the main focus is on minimizing RT. For doing so, the whole mechanism for computing it, is it's formulation (Table 2).

The problem is given by:

$$P : \min\{RT\}$$
$$VM_{status} = \begin{cases} 1 & \text{if } VM = Available \\ 0 & \text{if } VM \neq Available \end{cases} \quad (1)$$

$$RT = \omega_{ftime} - \omega_{ftime} + \omega_{dtime} \quad (2)$$

Table 2. Notations and their meanings

Notations	Meanings
RT	DC response time
ω_{ftime}	Finishing time of the task
ω_{stime}	Starting time of the task
ω_{dtime}	Transmission delay time
Nt	Number of tasks
Nv	Number of VMs

8 Performance Evaluation

In this section, the performance of the proposed system is discussed. In this paper, plots of RR-algorithm and IST are shown which are generated using CDC routing (SBP).

Table 3. System specifications on which simulations performed

Processor	Intel Core i3
CPU	1.7GHz
System type	64bit-OS
Windows	Microsoft Windows 8.1 Pro
RAM	4GB
Hard Drive	500GB

Table 4 shows the DC RT using RR-algorithm.

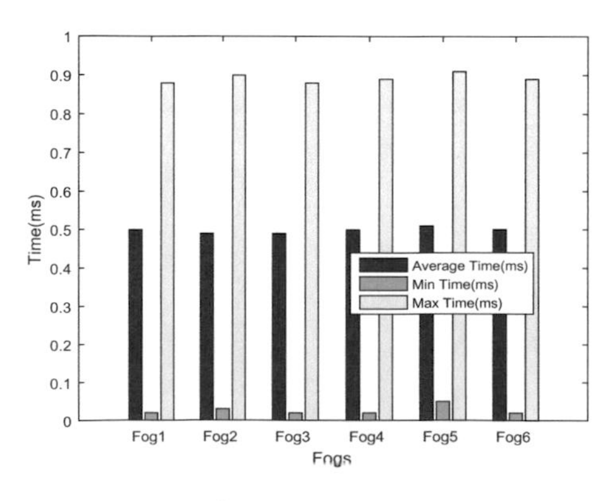

Fig. 4. DC RT using RR

Table 4. DC RT using RR

DC	Avg. (ms)	Min. (ms)	Max. (ms)
Fog1	0.5	0.02	0.88
Fog2	0.49	0.03	0.9
Fog3	0.49	0.02	0.88
Fog4	0.5	0.02	0.89
Fog5	0.51	0.05	0.91
Fog6	0.5	0.02	0.89

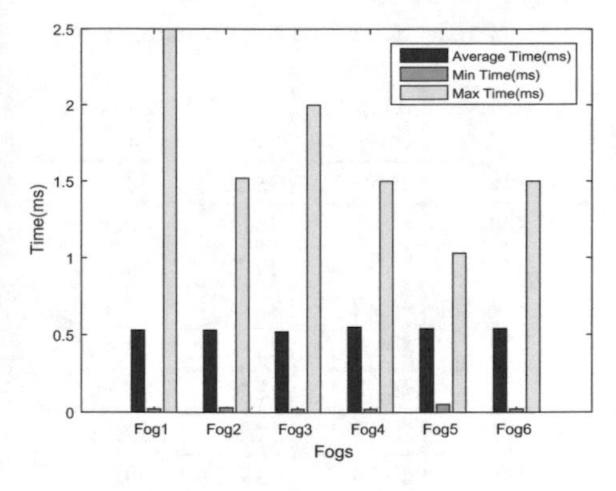

Fig. 5. DC RT using IST

Table 5. DC RT using IST

Fog	Avg. RT (ms)	Min. RT (ms)	Max. RT (ms)
Fog1	0.53	0.02	2.5
Fog2	0.53	0.03	1.52
Fog3	0.52	0.02	2
Fog4	0.55	0.02	1.5
Fog5	0.54	0.05	1.03
Fog6	0.54	0.02	1.5

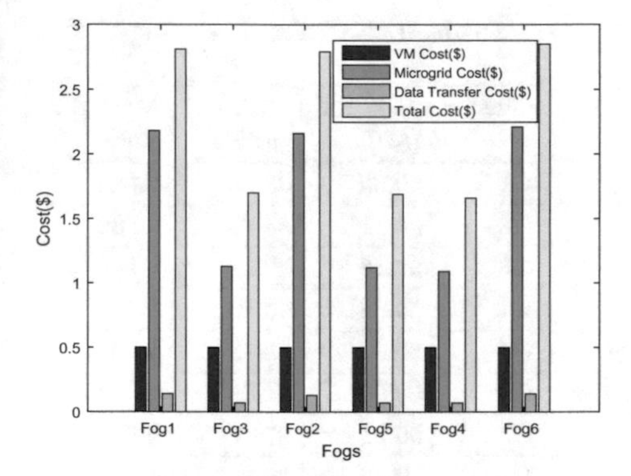

Fig. 6. Fog cost using RR

Table 6. Total fog cost using RR

DC	VM cost ($)	MG cost ($)	DT cost ($)	Total cost ($)
Fog1	0.5	2.18	0.14	2.81
Fog3	0.5	1.13	0.07	1.7
Fog2	0.5	2.16	0.13	2.79
Fog5	0.5	1.12	0.07	1.69
Fog4	0.5	1.09	0.07	1.66
Fog6	0.5	2.21	0.14	2.85

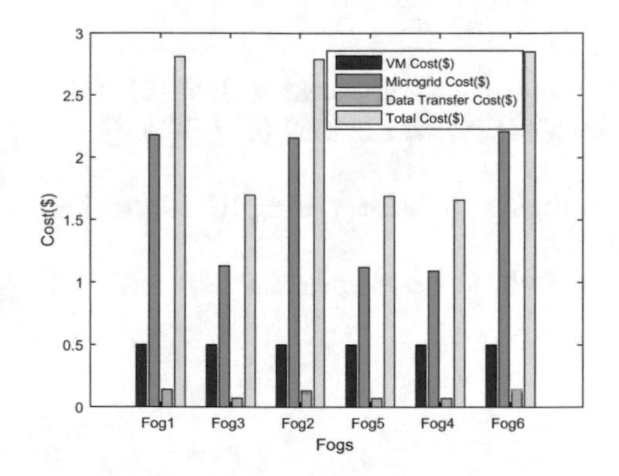

Fig. 7. Fog cost using IST

Table 7. Total cost using IST

DC	VM cost ($)	MG cost ($)	DT cost ($)	Total cost ($)
Fog1	0.5	2.18	0.14	2.81
Fog3	0.5	1.13	0.07	1.7
Fog2	0.5	2.16	0.13	2.79
Fog5	0.5	1.12	0.07	1.69
Fog4	0.5	1.09	0.07	1.66
Fog6	0.5	2.21	0.14	2.85

Table 5 shows the results of DC RT using IST. The average DC RT is 0.53ms Table 6 shows Total Fog Cost using RR-algorithm.
Table 7 shows the Total Cost of Fog using IST.

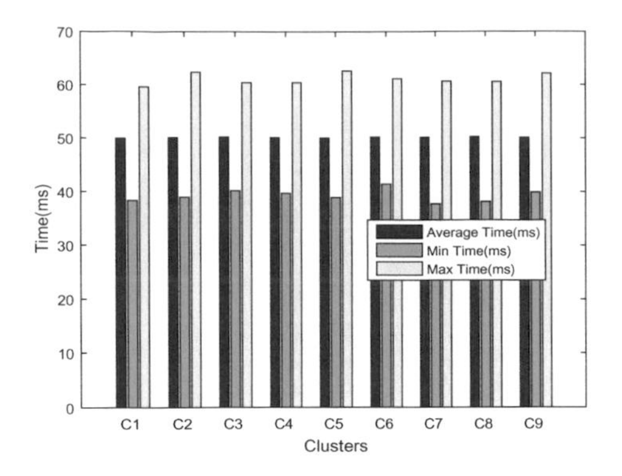

Fig. 8. RT by region using RR

Table 8. RT of region using RR

Cluster	Avg. RT (ms)	Min. RT (ms)	Max. RT (ms)
C1	49.93	38.28	59.63
C2	50.09	38.88	62.39
C3	50.19	40.15	60.4
C4	50.12	39.67	60.4
C5	50.02	38.88	62.63
C6	50.17	41.38	61.14
C7	50.13	37.67	60.66
C8	50.25	38.13	60.57
C9	50.11	39.89	62.14

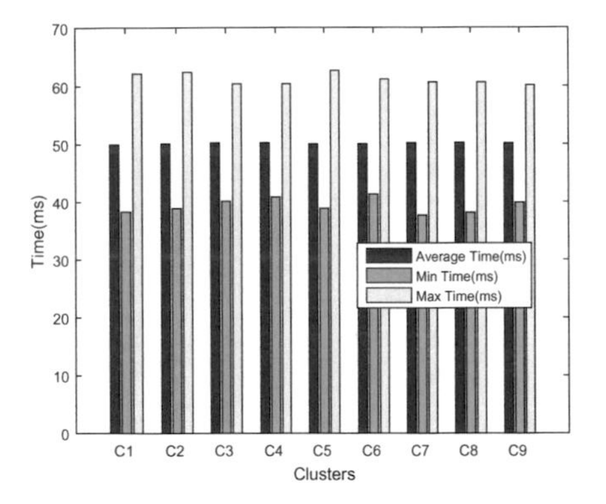

Fig. 9. RT of region using IST

Table 9. RT of region using IST

Cluster	Avg. RT (ms)	Min. RT (ms)	Max. RT (ms)
C1	49.93	38.28	62.14
C2	50.11	38.88	62.39
C3	50.26	40.15	60.4
C4	50.28	40.89	60.4
C5	50.04	38.88	62.63
C6	50.06	41.38	61.14
C7	50.21	37.67	60.66
C8	50.26	38.13	60.63
C9	50.16	39.89	60.13

Fog Cost for both the algorithms is same. Total VM cost is 3.00 ($), total MG cost is 9.88 ($) and total Data Transfer (DT) cost is 0.62 ($) (Table 8).

The graphs in Fig. 8 shows the RT of regions (clusters) using RR-algorithm. Table 9 shows the regions RT for IST.

The above Figures shows that the simulation results using IST and RR-algorithm are nearly equal.

9 Simulation Results

In this section, there is a discussion about simulations and their outcomes. For the simulation results, there are some considerations of price scheme, data and memory storage. Assumed cost per hour for VM at each fog is 0.1$. Storage cost per second considered is 0.1$. Memory cost per GB assumed is 0.1$. Data transfer cost per GB is 0.1$. 1 Physical hardware unit is placed at each fog. 10 users can be treated simultaneously. Instruction length of request is 100 bytes. Number of

VMs at each fog are 5. Image size is 10000 bytes. Bandwidth considered is 1000. Fog memory is 20GB. Number of processors in each fog are 4. Time shared VM policy is used.

The proposed IST technique is compared with RR-algorithm using different SBPs to evaluate the performance of proposed system. The simulations are performed in cloudsim using a system comprising hard drive capacity of 500 gigabytes (GB) and RAM of 4GB memory size. Further details of system are mentioned in Table 3. The outcomes of IST are shown in the plots and tables above which justify that performance of IST is satisfactory and nearly closer to the RR-algorithm which is further able to be improved. RT of DC for RR-algorithm is shown in Fig. 4 and RT of DC using IST is shown in Fig. 5. Overall cost of fog using RR-algorithm is shown in Fig. 6 and overall fog cost using IST is shown in Fig. 7. Figure 8 shows the RT of regions using RR-algorithm and Fig. 9 shows the RT of regions using IST.

Evaluating the results, it is obvious that the cost is almost the same for IST but the RT of IST is greater than that of RT calculated using RR algorithm. The RT of the DCs and Regions can further be minimized by making changes in the proposed scenario like we can introduce a new SBP and compute the performance using IST.

10 Conclusion and Future Work

In this paper, a resource allocation in cloud-fog environment have been discussed. The proposed technique, IST, efficiently allocate the resources to DCs and fulfill the DS requests. In this paper, the focus is on allocation of VMs to DCs. The efficiency of IST is computed by comparing the RT and Cost. It can be seen from the concluded results that the cost of the fog and cluster is almost the same. The RT of IST is also similar to the RT of RR algorithm. Low cost and RT shows the reliability of this system.

The proposed technique has the ability to provide even better results by making small changes in it. In future, we plan to hybrid RR and IST to compute the performance of present scenario and also after making changes in the current one as well. We also plan to compute the performance of IST using some other SBP(s) in future.

References

1. Zhang, Q., Cheng, L., Boutaba, R.: Cloud computing: state-of-the-art and research challenges. J. Internet Serv. Appl. **1**, 718 (2010)
2. Fatima, I., Javaid, N., Iqbal, M.N., Shafi, I., Anjum, A., Memon, U.: Integration of cloud and fog based environment for effective resource distribution in smart buildings. In: 14th IEEE International Wireless Communications and Mobile Computing Conference (IWCMC-2018), pp. 2-6 (2018)
3. Zahoor, S., Javaid, N., Khan, A., Muhammad, F.j., Zahid, M., Guizani, M.: A cloud-fog-based smart grid model for efficient resource utilization. In: 14th IEEE International Wireless Communications and Mobile Computing Conference (IWCMC-2018) (2018)

4. Buyya, R., Yeo, C.S., Venugopal, S., Broberg, J., Brandic, I.: Cloud computing and emerging IT platforms: Vision, hype, and reality for delivering computing as the 5th utility. Future Gener. Comput. Syst. **25**, 599–616 (2009)
5. Subashini, S., Kavitha, V.: A survey on security issues in service delivery models of cloud computing. J. Netw. Comput. Appl. **34**, 111 (2011)
6. Mukherjee, M., Matam, R., Shu, L., Maglaras, L., Ferrag, M.A., Choudhury, N., Kumar, V.: Security and privacy in fog computing: challenges. IEEE Access **5**, 1929319304 (2017)
7. Wickremasinghe, B., Buyya, R.: CloudAnalyst: a CloudSim-based tool for modelling and analysis of large scale cloud computing environments. MEDC Proj. Rep. **22**(6), 433–659 (2009)
8. del Valle, Y., Venayagamoorthy, G.K., Mohagheghi, S., Hernandez, J.-C., Harley, R.G.: Particle swarm optimization: basic concepts, variants and applications in power systems. IEEE Trans. Evol. Comput. **12**, 171–195 (2008)
9. Wang, X., Ning, Z., Wang, L.: Offloading in internet of vehicles: a fog-enabled real-time traffic management system. In: IEEE Transactions on Industrial Informatics, vol. 11 (2018)
10. Dehghanpour, E., Kazemi Karegar, H., Kheirollahi, R., Soleymani, T.: Optimal coordination of directional overcurrent relays in microgrids by using cuckoo-linear optimization algorithm and fault current limiter. IEEE Trans. Smart Grid **9**, 1365–1375 (2018)
11. Thirumalaisamy, B., Jegannathan, K.: A novel energy management scheme using ANFIS for independent microgrid. Int. J. Renew. Energy Res. (IJRER) **6**(3), 735–46 (2016)
12. Zachar, M., Daoutidis, P.: Microgrid/Macrogrid energy exchange: a novel market structure and stochastic scheduling. IEEE Trans. Smart Grid **8**, 178–189 (2017)
13. Huld, T., Moner-Girona, M., Kriston, A.: Geospatial analysis of photovoltaic mini-grid system performance. Energies **10**, 218 (2017)
14. Lin, W.-M., Tu, C.-S., Tsai, M.-T.: Energy management strategy for microgrids by using enhanced bee colony optimization. Energies **9**, 5 (2015)
15. Tumuluru, V.K., Huang, Z., Tsang, D.H.K.: Integrating price responsive demand into the unit commitment problem. IEEE Trans. Smart Grid **5**, 2757–2765 (2014)
16. Tsai, L., Liao, W.: StarCube: an on-demand and cost-effective framework for cloud data center networks with performance guarantee. IEEE Trans. Cloud Comput. **6**, 235249 (2018)
17. Nguyen, Thoai, Nam.: A Genetic Algorithm for Power-aware Virtual Machine Allocation in Private Cloud (2013)
18. Dasgupta, K., Mandal, B., Dutta, P., Mandal, J.K., Dam, S.: A genetic algorithm (GA) based load balancing strategy for cloud computing. Procedia Technol. **10**, 340–347 (2013)
19. Desai, T., Prajapati, J.: A Survey of Various Load Balancing Techniques and Challenges in Cloud Computing (2013)
20. Mirjalili, S.: Dragonfly algorithm: a new meta-heuristic optimization technique for solving single-objective, discrete, and multi-objective problems. Neural Comput. Appl. **27**, 1053–1073 (2015)
21. Cao, Z., Lin, J., Wan, C., Song, Y., Zhang, Y., Wang, X.: Optimal cloud computing resource allocation for demand side management. IEEE Trans. Smart Grid 113 (2016)
22. Adhianto, L., Banerjee, S., Fagan, M., Krentel, M., Marin, G., Mellor-Crummcy, J., Tallent, N.R.: HPCTOOLKIT: tools for performance analysis of optimized parallel programs. Concurrency Comput. Pract. Exp. (2009)

23. Sahni, J., Vidyarthi, P.: A cost-effective deadline-constrained dynamic scheduling algorithm for scientific workflows in a cloud environment. IEEE Trans. Cloud Comput. **6**, 218 (2018)
24. Wang, S., Zhang, J., Huang, T., Pan, T., Liu, J., Liu, Y.: Flow distribution-aware load balancing for the datacenter. Comput. Commun. **106**, 136–146 (2017)
25. Wang, L., Gelenbe, E.: Adaptive dispatching of tasks in the cloud. IEEE Trans. Cloud Comput. **6**, 33–45 (2018)
26. Islam, M.A., Ren, S., Quan, G., Shakir, M.Z., Vasilakos, A.V.: Water-constrained geographic load balancing in data centers. IEEE Trans. Cloud Comput. **5**, 208–220 (2017)
27. Liu, L., Chang, Z., Guo, X., Mao, S., Ristaniemi, T.: Multiobjective optimization for computation offloading in fog computing. IEEE Internet Things J. **5**, 283–294 (2018)

Smart Grid Management Using Cloud and Fog Computing

Muhammad Hassaan Ashraf, Nadeem Javaid(✉), Sadam Hussain Abbasi, Mubariz Rehman, Muhammad Usman Sharif, and Faizan Saeed

COMSATS Univeristy Islamabad, Islamabad, Pakistan
nadeemjavaidqau@gmail.com
http://www.njavaid.com/

Abstract. Cloud computing provides Internet-based services to its consumer. Multiple requests on cloud server simultaneously cause processing latency. Fog computing act as an intermediary layer between Cloud Data Centers (CDC) and end users, to minimize the load and boost the overall performance of CDC. For efficient electricity management in smart cities, Smart Grids (SGs) are used to fulfill the electricity demand. In this paper, a proposed system designed to minimize energy wastage and distribute the surplus energy among energy deficient SGs. A three-layered cloud and fog based architecture described for efficient and fast communication between SG's and electricity consumers. To manage the SG's requests, fog computing introduced to reduce the processing time and response time of CDC. For efficient scheduling of SG's requests, proposed system compare three different load balancing algorithms: Round Robin (RR), Active Monitoring Virtual Machine (AMVM) and Throttled for SGs electricity requests scheduling on fog servers. Dynamic service broker policy is used to decide that which request should be routed on fog server. For evaluation of the proposed system, results performed in cloud analyst, which shows that AMVM and Throttled outperform RR by varying virtual machine placement cost at fog servers.

1 Introduction

Cloud computing provides services to its users through web-based tools and applications. Cloud service provider like Google Cloud Platform, Amazon Web Service, Microsoft Azure and IBM Cloud etc. maintain their own CDC. According to [1] every data center has some physical hardware CPU, memory, storage disc and bandwidth to store and process data called Physical Machine (PM). CDC logically divides one PM into multiple machines, called Virtual Machines (VM). Some resources of PM assigned to every VM that is responsible to handle the user requests. On CDC Resource assignment perform in two ways static or dynamic scheduling. In case of static VM placement, PM divides into equal size of VMs before request generated. While in dynamic placement PM divides into VMs at runtime, when the user generates request system detect the strength of request and assign VM according to the need of that request. Information from

L. Barolli et al. (Eds.): NBiS 2018, LNDECT 22, pp. 624–636, 2019.
https://doi.org/10.1007/978-3-319-98530-5_54

cloud only accessible using internet services. The popularity of cloud computing is increasing gradually because of its services and freedom of access to its associated users. Cloud described in three models shown in figure Fig. 1 on the base of its services:

- Software as a Service (SaaS)
- Platform as a Service (PaaS)
- Infrastructure as a Service (IaaS)

The first model of the cloud SaaS provides software like services (Email, virtual desktop, gaming, communication) to its clients and they can use licensed software via SaaS. In PaaS cloud provides the runtime environment on the internet. IaaS provides a virtual interface that provides a pre-configured interface to the consumers.

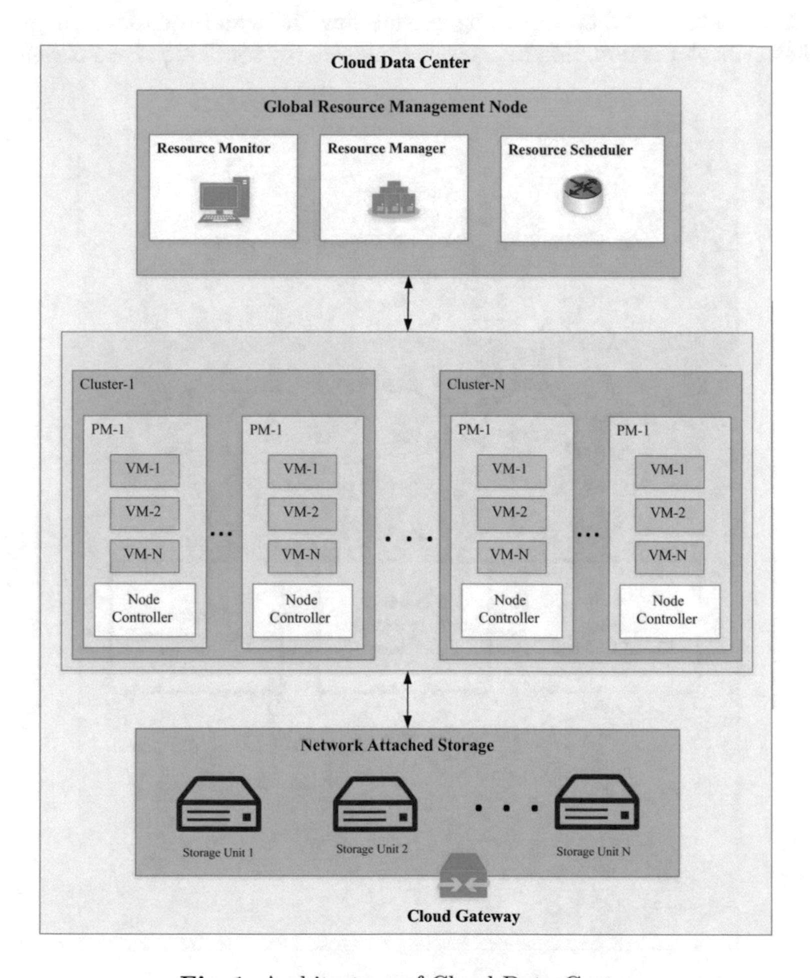

Fig. 1. Architecture of Cloud Data Center

In cloud computing, scalability is a big challenge because connected devices and cloud response time are directly proportional according to [2]. In delay, sensitive applications system need fast communication. The concept of fog computing is recommended to minimize the intensity of such challenges. The fog computing provides same services as cloud computing and acts as an intermediary between cloud servers and consumer to reduce the overall load of cloud servers. The concept of fog computing helps to minimize latency, enhance the flexibility and reliability of the network. In this proposed system, fog computing services such as networking, storage and computation used in SG for efficient energy management of smart cities. Fog computing provides delay sensitive communication between consumer and servers. Information Communication and Technology (ICT) convert Traditional Grid (TG) into SG for two-way communication. SG connected with other SGs and electricity consumers to fulfill the energy requirements. SGs share information with fog servers, fog servers maintain the data of SGs and predict the future load on SGs. Fog servers have information about surplus and deficient SGs of a region. Fog helps to arrange energy for SGs in case of deficiency

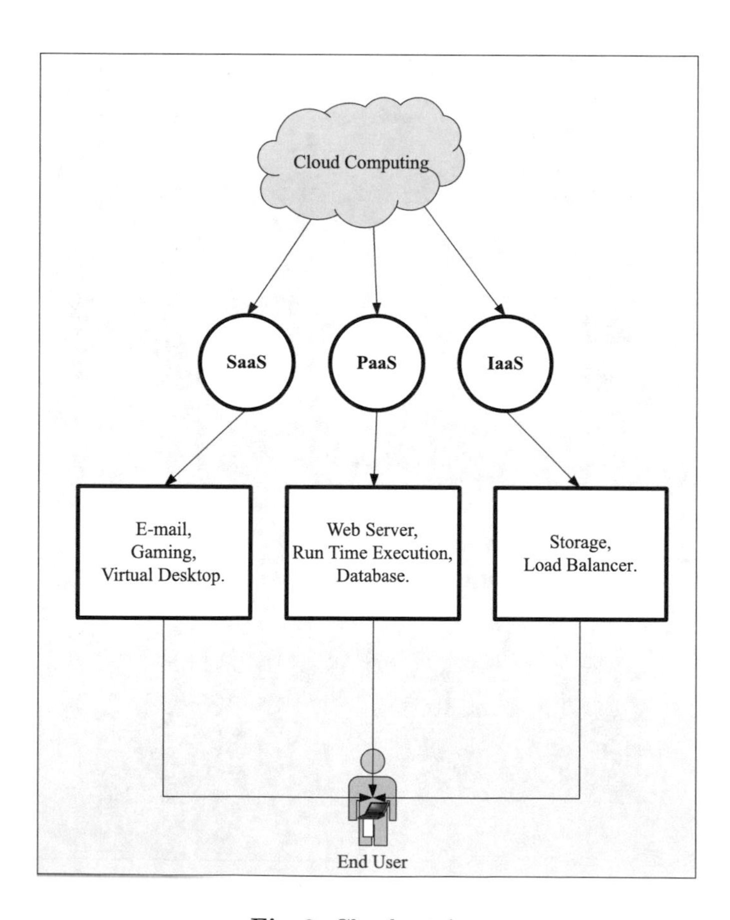

Fig. 2. Cloud services

via cloud and utility. If a whole region becomes energy deficient than fog communicates with cloud and cloud assign utility company to that region to fulfill the energy demand. In this paper, the proposed system covers a large residential and commercial area of six regions. Every region has its own fog servers to manage the requests of SGs. Every building and Power Supply Stations (PSS) has its own SG to fulfill the electricity demand. Every SG communicate with other SGs to make a coordination system for electricity sharing. In case of energy deficiency SG requests to its corresponding fog to fulfill the electricity demand. Fog manages the request and coordinates with other SGs, and if fog found surplus energy in any other SG then assign that surplus energy to deficient energy SG. If there is no surplus energy in all SGs of a region then fog requests to cloud, and cloud server assigns appropriate utility company to deficient SG. To balance the load on fog proposed system use three load balancing algorithms RR, Throttled and AMVM load balancing algorithm (Fig. 2).

1.1 Motivation

Fog computing provides delay sensitive communication between consumer and servers. The demand for electricity is increasing gradually. To minimize the electricity wastage and satisfy the user requirements in the efficient and effective manner we need to schedule its resources and consumption. Random electricity demand by the consumers causes the problem to fulfill the energy requirements. Authors in [3–5] explore the idea of energy scheduling by integrating TG with ICT for two-way communication. For fast, flexible and reliable communication among different consumers, the system requires cloud computing. Authors in [3,5] proposed the idea of cloud computing with the intermediary layer of fog computing to schedule energy demand. Cloud and fog computing help in runtime decision making [6]. The authors in [6], specified that the system needs to predict the demand for energy to minimize the wastage. In [6,7] they proposed load predictor systems using SG. SG use to communicate with other SGs and developed a coordination network between different electricity consumer to guess overall load. Fog computing uses to minimize the load and processing time of cloud. Fog act as an intermediary layer between consumer and cloud servers consist of VMs to perform in-network processing. To make the fog efficient and fast system need load balancing algorithms and service broker policies to distribute the load among different VMs of fog. Authors in [4,5] designed cloud environment for limited users of different regions to check the performance of their proposed load balancing algorithms. Authors emphasis on the optimization of the overall response time of fog servers for residential buildings by overlooking the overall system costs. Proposed system motivated to optimize overall response time and communication cost of fog computing during scheduling of electricity requests received from SGs.

1.2 Contribution

In this paper, the proposed system minimizes the response and processing time of fog computing and reduces the overall system cost.

2 Related Work

For fast and delay sensitive communication, cloud load management is a challenging field for researchers. To minimize the processing time of cloud, authors are working on different mechanisms. They introduce an intermediary layer of fog computing. Fog computing helps to achieve fast response time for the consumer as compared to the cloud. Fog provides the runtime decision-making capabilities with the help of its computing devices, that is used to distribute the overall load of fog. Many algorithms are proposed to balance the load on cloud and fog computing and help to reduce the VMs cost and fog processing time. Most commonly used algorithms are Round Robin (RR), throttled, honeybee foraging and biased random sampling etc. These algorithms are used to schedule the user requests on fog with minimum latency. The authors in [8], described a cloud load balancing algorithm and compare its results with RR, Throttled and honeybee foraging. Their proposed system performs better and balances the load successfully in a multiuser environment.

The researchers in [9], designed a honeybee foraging inspired system that minimizes the overall system processing and waiting time of a request in the queue to assign VM. Their proposed system performs better than adoptive, adaptive-dynamic and heat diffusion based dynamic load balancing algorithms. The authors in [10], proposed Ant Colony optimization (ACO) to overcome the problem of resource allocation in fog. And compare ACO's energy consumption, processing time and standard deviation with RR. They claim that their ACO consume minimum energy with minimum processing time as compare to RR.

Another system that allocates the task efficiently to balance the overall load discussed in [11]. Proposed model allocates low power tasks and improves the system throughput by reducing system delay. In [12], authors compare Cloud-Based Demand Response (CBDR) with Distributed Demand Response (DDR) and show that CBDR performs better than DDR with respect to reliability and scalability of the communication network. DDR is channel dependent and within few iterations returns optimal solution. DDR is unreliable due to information loss. While CBDR is channel independent and cost-efficient. The authors in [5], compare six load balancing algorithms that are RR, throttled, Particle Swarm Optimization (PSO), Artificial Bee Colony (ABC), Ant Colony Optimization (ACO) and sixth is a hybrid of ABC and ACO algorithm called Hydride Artificial Bee and Ant Colony Optimization (HABACO). Their proposed HABACO outperform the other five discussed algorithms.

3 Problem Formulation

The objective of the proposed system is to schedule the SG electricity requests by minimizing the overall processing time and VM placement cost. Suppose cloud data center has 'N' number of PMs,

$$PM = \{pm_1, pm_2, pm_3, ..., pm_N\} \tag{1}$$

and every PM can have multiple VMs,

$$VM = \{vm_1, vm_2, vm_3, ..., vm_j\} \tag{2}$$

Every VM has some resources such as CPU, RAM, network BW, storage At fog computing layer, system has 'N' number of fog nodes to manage consumer requests

$$Fog = \{f_1, f_2, f_3, ..., f_n\} \tag{3}$$

A fog node has maximum electricity requests from SGs

$$Req_{Total} = \sum_{i=1}^{n}(R_i) \tag{4}$$

Load balancer balance VM load by distributing requests equally among all the VMs. When fog servers receive user requests, servers maintain a queue and put that request into a queue and when finding free VM then assign the request to that free VM. Processing time can be calculated as

$$Processing_{Time} = \frac{Number\ of\ Requests}{Capacity\ of\ VM} \tag{5}$$

Total processing time can be estimated as

$$Total_{PT} = \sum_{X=1}^{n}\sum_{Y=1}^{m}((Processing - Time) * A) \tag{6}$$

where A is the arrival time of a specific request, X is the user request assignmnet and K is the VM assignmnet Response time: The RT is a time taken by the fog to receive and response the request.

$$RT = Delay_{Time} + Finishing_{Time} - Arrival_{Time} \tag{7}$$

Objective of proposed system is to minimize total processing by minimizing overall response time of fog servers.

4 Proposed System

The proposed system is designed to fulfill electricity demand of consumers using cloud and fog computing efficiently in term of cost and processing time. This section, describes the architecture of our proposed system, where we use the three-layered architecture that consists of cloud, fog and consumer layers. These three layers are interconnected to share information with each other. The fog computing act as an intermediary layer between cloud servers and consumers. Proposed system deals with two types of electricity demand, electric vehicles energy demands consider as commercial consumption (Power supply stations) and household demand as residential consumption (Buildings, homes and apartments). Every building and power supply station has its own SG to generate and fulfil the energy demand of its connected users. SGs communicate with fog server to share their information about energy generation and demand. For an energy deficient SG, fog server will predict the energy demand and provide energy, from energy surplus (energy generation greater than energy demand) SG. If all the SGs become energy deficient then fog server sends their information to cloud server for utility assignment. Utility company is the electricity provider at large scale.

For simulation, proposed system considers six regions and every region managed by two fog servers to schedule SGs requests. There are hundred SGs in every region and every SG has a maximum of thousand homes. Every home requests to SG for electricity, request sends to fog servers then fog assign electricity to that specific home from SG by maintaining the information of generation and consumption. Every region has an electric vehicle Power Supply Station (PSS) that can fulfill the maximum one thousand vehicles charging or discharging requests in one hour. In this proposed system cloud layer has full information about demand and generation of all the smart grids of six different regions. Electricity supplier (utility) connected to the cloud and SGs. Cloud server has the full information about electricity generation of utility and connected with fog servers of every region. Every fog server has the information about energy consumption in its corresponding region and takes runtime decision to fulfill the energy demand of consumers. Fog servers communicate with the cloud server to assign utility company that can fulfill the request of an electricity deficient SGs. The proposed energy optimization system designed to minimize the cost and latency of fog servers. Latency minimized by reducing the overall response between fog and SGs. And fog server processing cost minimized by efficiently reducing the number of virtual machine and number of fog servers used in the system. Processing speed and cost are inversely proportional. In this research, we designed an optimized system with the minimum trade-off of cost and response time. The proposed system considers two situations of fog communication. In the first scenario proposed scheme uses two fogs in every region to communicate with the cloud, the overall system uses twelve fogs in six regions. Every fog has multiple VMs. Each VM has some hardware resources like processors, temporary storage and software resources like operating system. To minimize fog processing time, we need to assign virtual machine more efficiently. In literature authors [3–5] discuss some load balancing algorithms that are RR, Throttled and Particle Swarm Optimization (PSO) with multiple service broker policies. This proposed system compares RR, Throttled and AMVM load balancing algorithm with new dynamic service broker policy using one and two fogs in each region. Detail of the proposed load balancing algorithms described below.

4.1 RR Load Balancing Algorithm

RR load balancing algorithm maintains a table of the request with respect to time of arrival. It has a scheduler that maintains request with respect to time.

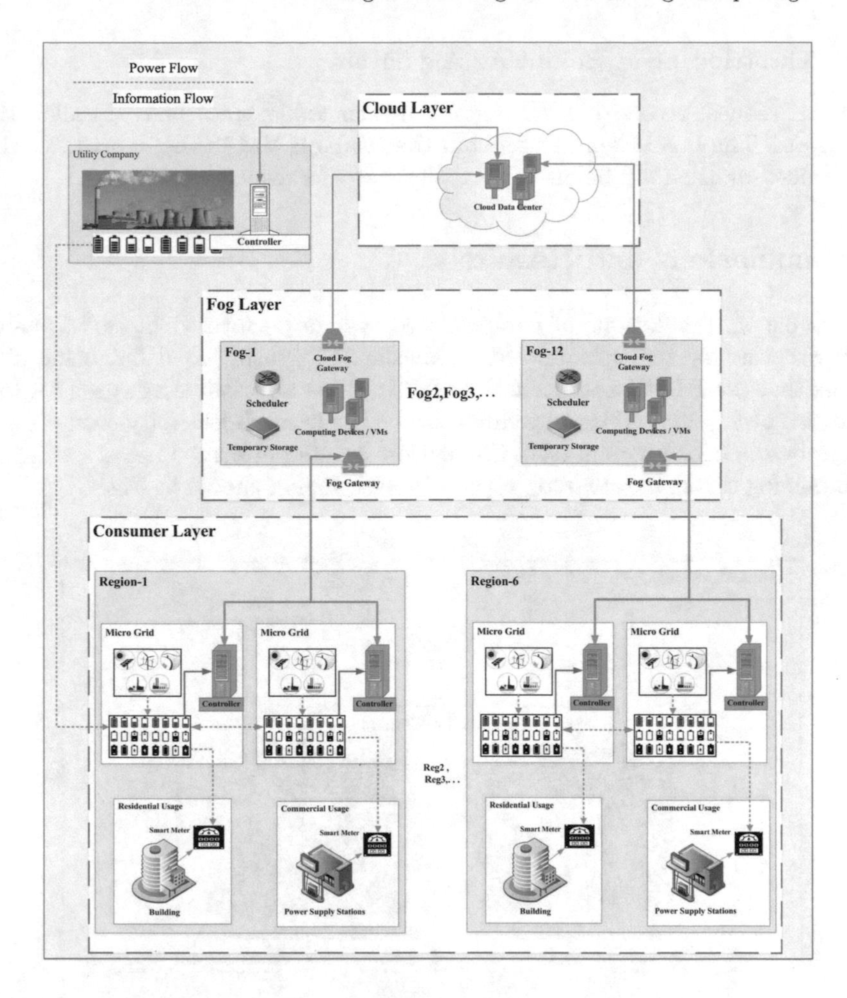

Fig. 3. Proposed system model

RR algorithm allocating multiple requests coming from the users and assign VMs one by one (Fig. 3).

4.2 AMVM Load Balancing Algorithm

AMVM load balancer works just like throttled. Users requests are queued by the data center controller as they are received. First, the availability of the VM is checked and if the VM is available the request is removed from the beginning of the queue and allocated to VM. The status of the VM is changed from available to busy. After the execution of the request, the status of the changed back from busy to available.

4.3 Throttled Load Balancing Algorithm

The user request received by the data center controller and then VM allocation performed. The Load Balancer find out the available VM in the index table that maintained by the load balancer and allocates VM respectively.

5 Simulations and Results

In this paper, simulations of the proposed system performed in cloud analyst tool. Proposed system schedule SGs requests at fog using load balancing algorithms that are RR, Throttled and AMVM. These algorithms are used for load balancing at fog servers. Results simulated on the base of electricity requests generated by the consumers in 24 h. Simulation results performed using six regions with one fog server and two fog servers in each region shown in Fig. 4.

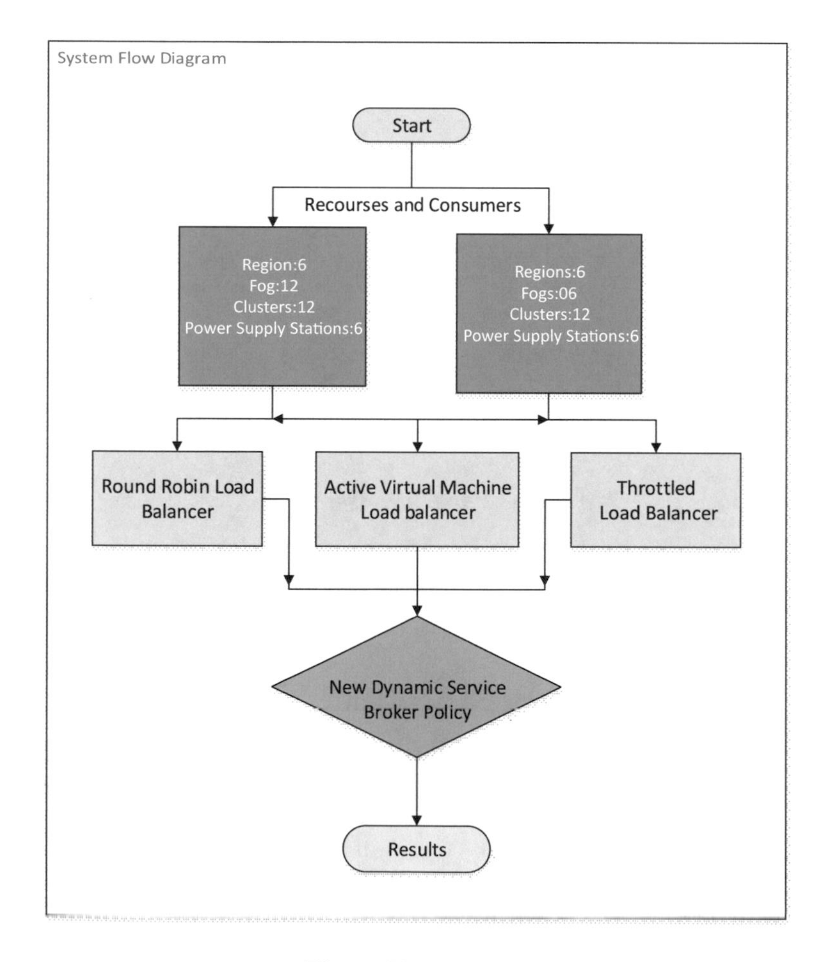

Fig. 4. Flow diagram

5.1 Twelve Fog Servers

In this scenario, two fog servers are considered for each region in the experimental setup. For residential consumption, every region has two hundred buildings and every building has its own SG to fulfill the electricity demand of maximum one thousand apartments at a time. Every region has one SG for PSS, every PSS has maximum 100 electric vehicles capacity for charging and discharging at a time. There is twelve fog serves considered in this scenario. The proposed system optimizes SG request scheduling using fog computing with respect to time and cost. Results of this scenario shown in Fig. 5 show the average response time of six regions using twelve fog servers every region has one PSS and two clusters of SGs. AMVM and Throttled load balancing algorithm outperform RR using two fog servers in each region. Figure 6 shows that average response and processing time of AMVM outperform RR using twelve fog servers.

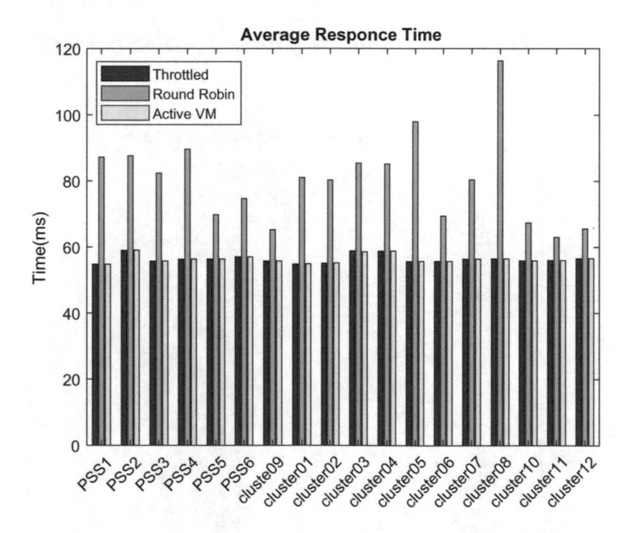

Fig. 5. Average response time by region (12-fogs)

5.2 Six Fog Servers

In this scenario, we try to minimize fog servers cost by minimizing the number of fog servers. Only one fog server used in each region for electricity requests scheduling. The numbers of requests and other system evaluation parameters are same as in the previous scenario. The proposed system optimizes fog and cloud computing with respect to time and cost. Overall system's cost using both scenarios shown in Fig. 7. Comparison of response and processing time using

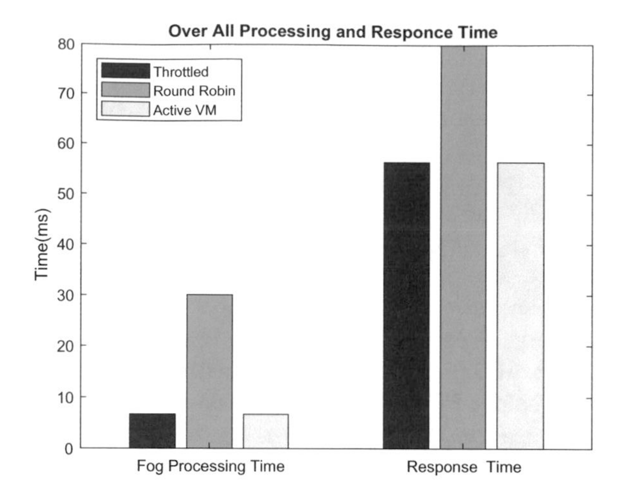

Fig. 6. Overall response and processing time (12-fogs)

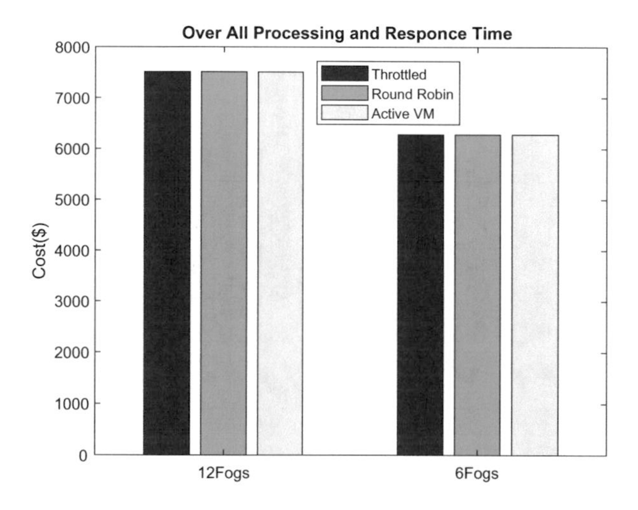

Fig. 7. System total cost

twelve and six fog servers is shown in Fig. 8 that shows that response time and VM placement cost of fog servers are inversely proportional and there is a tradeoff between fog placement cost and processing time.

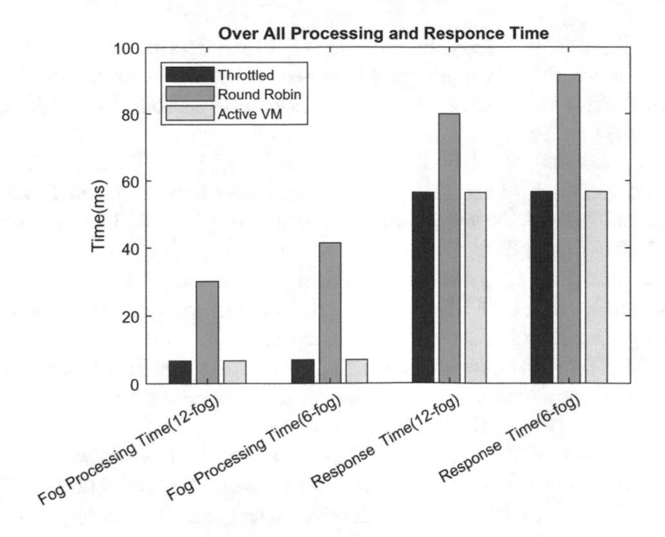

Fig. 8. Response and processing time comparison

6 Conclusion

In this paper proposed system represents a model of SG management using cloud and fog infrastructure. The comparison results of three load balancing algorithms perform. Results show that for efficient processing in fog computing AMVM and throttled algorithm perform batter with limited fog resources. Throttled and AMVM algorithms support system scalability with minimum response and processing time. Throttled and AMVM performs better than RR in both scenarios by varying number of fog servers. Various load balancing algorithms and broker policies are compared to improve and reduce the response time and delay respectively, of the fog servers. The response time and VM placement cost of fog servers are inversely proportional and there is a tradeoff between fog placement cost and processing time. The future work includes such intelligent load balancing algorithms that warn fog ahead of time about the predicted user requests for the resources from different regions of the world.

References

1. Sotiriadis, S., Bessis, N., Buyya, R.: Self managed virtual machine scheduling in Cloud systems. Inf. Sci. **433**, 381–400 (2017)
2. Osanaiye, O., Chen, S., Yan, Z., Lu, R., Choo, K.K.R., Dlodlo, M.: From cloud to fog computing: a review and a conceptual live VM migration framework. IEEE Access **5**, 8284–8300 (2017)
3. Fatima, I., Javaid, N., Iqbal, M.N., Shafi, I., Anjum, A., Memon, U.: Integration of cloud and fog based environment for effective resource distribution in smart buildings. In: 14th IEEE International Wireless Communications and Mobile Computing Conference (IWCMC-2018) (2018)

4. Javaid, S., Javaid, N., Tayyaba, S., Sattar, N.A., Ruqia, B., Zahid, M.: resource allocation using Fog-2-Cloud based environment for smart buildings. In: 14th IEEE International Wireless Communications and Mobile Computing Conference (IWCMC-2018) (2018)
5. Zahoor, S., Javaid, N., Khan, A., Muhammad, F.J., Zahid, M., Guizani, M.: A Cloud-Fog-Based smart grid model for efficient resource utilization. In: 14th IEEE International Wireless Communications and Mobile Computing Conference (IWCMC-2018) (2018)
6. Khalid, A., Javaid, N., Mateen, A., Ilahi, M.: Smart homes coalition based on game theory. In 32-nd IEEE International Conference on Advanced Information Networking and Applications (AINA-2018) (2018)
7. Collotta, M., Pau, G.: An innovative approach for forecasting of energy requirements to improve a smart home management system based on BLE. IEEE Trans. Green Commun. Netw. **1**(1), 112–120 (2017)
8. Chen, S.L., Chen, Y.Y., Kuo, S.H.: CLB: a novel load balancing architecture and algorithm for cloud services. Comput. Electr. Eng. **58**, 154–160 (2017)
9. Gupta, H., Sahu, K.: Honey bee behavior based load balancing of tasks in cloud computing. Int. J. Sci. Res. **3**(6) (2014)
10. Pham, N.M.N., Le, V.S.: Applying Ant Colony System algorithm in multi-objective resource allocation for virtual services. J. Inf. Telecommun. **1**(4), 319–333 (2017)
11. Razzaghzadeh, S., Navin, A.H., Rahmani, A.M., Hosseinzadeh, M.: Probabilistic modeling to achieve load balancing in Expert Clouds. Ad Hoc Netw. **59**, 12–23 (2017)
12. Cao, Z., Lin, J., Wan, C., Song, Y., Zhang, Y., Wang, X.: Optimal cloud computing resource allocation for demand side management in smart grid. IEEE Trans. Smart Grid **8**(4), 1943–1955 (2017)
13. Singh, S.P., Sharma, A., Kumar, R.: Analysis of load balancing algorithms using cloud analyst. Int. J. Grid Distrib. Comput. **9**(9), 11–24 (2016)

Evaluation of Self-actualization Support System by Using Students Independence Rubric

Yoshihiro Kawano(✉)

Department of Informatics,
Tokyo University of Information Sciences, Tokyo, Japan
ykawano@rsch.tuis.ac.jp

Abstract. To actualize own life, proactive action is one of essential skills. However, most students are reactive. Reactive students are characterized in the following no challenge, no thinking, awaiting instructions, and avoidance of trial and error. Our contributions are aiming to provide a proactive action support system for students life, and to evaluate significant of the system. The 7 habits is one of the most powerful schemes for proactive action choice. We are developing proactive action support system by visualization of quadrant II activities called Self-reflector-Plus. Self-reflector-Plus was systematized the first three habits in the 7 habits. Periodic and long-term practice is necessary to gain the significant effect of the 7 habits. In this paper, to examine our system, we have designed rubric to evaluate effect of Self-reflector-Plus. There are 9 components corresponding habit 1 to 3 in the 7 habits.

1 Introduction

Today, graduating students are faced with major decisions involving career choice, diverse values and ways of life, self-analysis, humane growth, and contributing to society. Emphasizing their strengths and demonstrating their potential contributions have become important tasks in their search for employment after finishing their university studies. Recently, job hunting activities are diversified. For example, students usually get jobs using various methods, i.e., through job search to company employee, referral from mentors, internships, using social media, and starting up their own venture. Moreover, promoting these activities, belonging to communities and making human networks are important. To thrive in such social circumstances, people need to actively emphasize their individual strong points for career building.

Though proactive action is necessary for career building, it's difficult to behave proactively for several students. The 7 habits is one of powerful schemes for proactive action choice [1]. In the 7 habits, quadrant II is the most important activity to make our life proactive. However, it's difficult to take proactive action by themselves because quadrant II activity does not come from outside.

In our previous works, we developed proactive action support system by visualization of quadrant II activities called "Self-reflector" [2, 3]. Self-reflector systemized the first three habits in the 7 habits. In recent years, using Internet by mobile devices

L. Barolli et al. (Eds.): NBiS 2018, LNDECT 22, pp. 637–647, 2019.
https://doi.org/10.1007/978-3-319-98530-5_55

was popularized. However, there were few frequencies for which the examinee uses this system, because this system was not applied to mobile use. Periodic and long-term practice is necessary to gain the significant effect of the 7 habits. Our issue is improvement of frequency of the system use for students. In this paper, we apply the system to mobile interface. Our contributions are as follows.

1. Improvement of user experience
2. Promoting of periodic and long-term use
3. Achievement for your mission statement

We propose a mobile application of this system called "Self-reflector-Plus".

2 Self-reflector

2.1 Concept and Functions

Self-reflector has been developed based on the first three habits in the 7 habits. Self-reflector offers us the functions which visualizes plan and action (Fig. 1). Plan function developed as a scheduler like calendar application. Then, in Fig. 2, you can put quadrant II activities into the scheduler first (Habit 3: Put First Things First). Quadrant II activities are extracted based on own mission statement (Habit 2: Begin with the End in Mind). The action function realizes measurement of quadrant II activities (Habit 1: Be Proactive).

Fig. 1. Overview of *Self-reflector*

Before beginning of the action, you can receive a notification e-mail. When a link in the mail is clicked, measurement in the activity time will start. Because of these functions, you can get gap between ideal and real of activities visually by two-line charts (Fig. 3). Figure 4 shows radar chart of quadrant II activities. These activities are divided four parts (that is, physical, mental, social/emotional, and spiritual). Self-reflector supports with visualization of successful experience to approach own ideal.

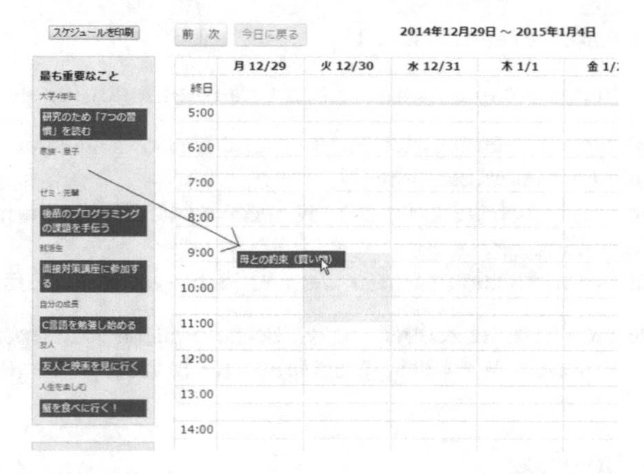

Fig. 2. Put quadrant II activities into *Self-reflector*

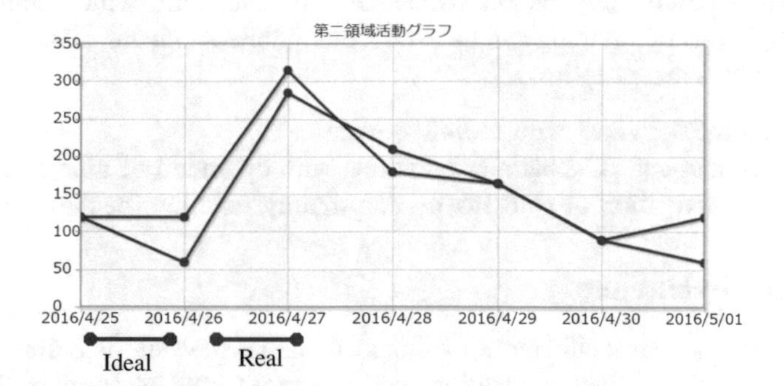

Fig. 3. Line charts of quadrant II activities

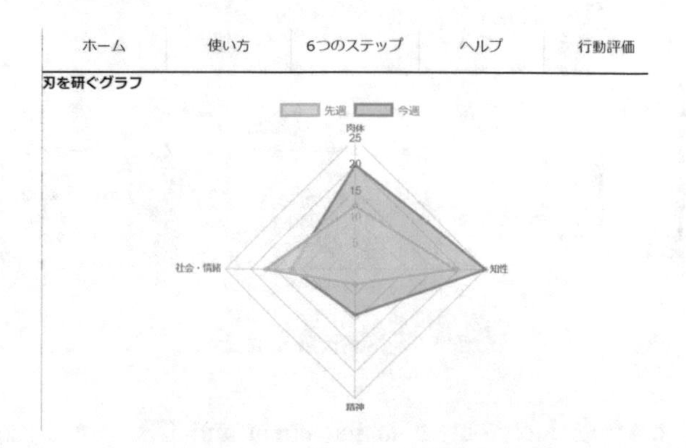

Fig. 4. Radar charts of quadrant II activities

2.2 Problems About Frequency of Us

According to preliminary experiment, we had confirmed following problems.

(1) Many users do not even register mission statement or scheduler activities of the system using computer intentionally.
(2) Some users rarely behave even if they received notification e-mail about own scheduler activities.
(3) Maintenance of motivation for periodic and long-term use is difficult.

These above problems are serious for us, because effect of independence for students cannot be inspected. We redesign Self-reflector to resolve the problems.

3 Self-reflector-Plus

3.1 Concept

We propose a mobile application called Self-reflector-Plus which improved Self-reflector. Functions of anything can be used from mobile application. Basic concepts of Self-reflector-Plus are as follows.

(1) Registers the activities with mobile devices.
(2) Activates start/end of scheduler activities with operation of mobile application.
(3) Improves motivation of students by community function for their activities.

3.2 System Architecture

Figure 5 shows system architecture of this system. This system is composed of Web server, database (DB), client applications and Firebase Cloud Messaging (FCM) which is cloud type messaging platform provided by Google Inc.

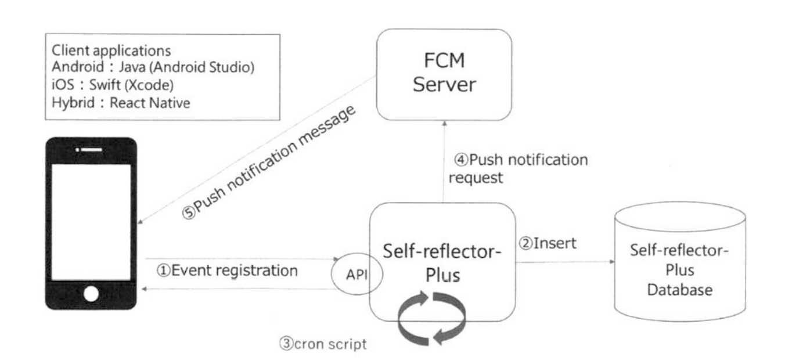

Fig. 5. System architecture

Web server controls input/output to the client and DB, and it requests to FCM server for push notification. Cron script on the server checks nearby events of DB periodically, then the script requests push notification to FCM server. From Table 1,

Table 1. Environment of system development

	Component	Environment	Role
Server	Web system	Django	Core system
	DB	Sqlite3	Data of users and events
	Web API	Ruby	APIs for clients
	Script	Ruby, cron	Push notification request
Client	Android	Java	Android devices
	iOS	Swift	iOS devices
	Hybrid	React Native	Hybrid Web application

we provide three types of client applications, that is, Android, iOS, and Hybrid Web application.

3.3 Implementation

We have developed prototype of Self-reflector-Plus. Figure 6 shows screenshot of the system. In this page, you can register mission statement and your roles. In Fig. 7, you can register predeterminate activities with own mobile devices. Before 15 min when scheduler activity is begun, the system notifies about it to your mobile devices. By using this function, you can measure time of the activity (Fig. 8).

Fig. 6. Screenshot of Self-reflector-Plus

Figure 9 shows top page of community function. There are several threads about activities for students. You can discuss with other members about credentials, skill-up, human network, life style, and career, etc.

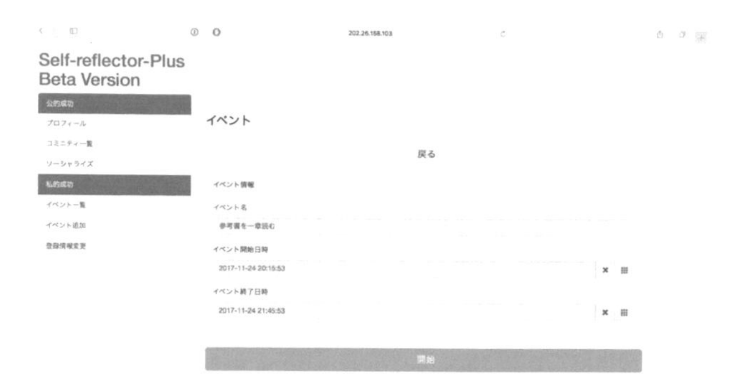

Fig. 7. Events list and registration

Fig. 8. Starting event with mobile devices

Fig. 9. Community functions

4 Subject Experiment

4.1 Purpose

We had subject experiment using this system. Purpose of the experiment is to evaluate improvement of frequency of use and maintenance of continuity. Evaluation items are as follow.

(1) Frequency of use with mobile application
(2) Periodic and long-term use
(3) Degree of improvement of independence state

(1) above is evaluated by system logs such as login or user operation, and users' questionnaires. (2) above is evaluated by usability of UI and use situation of community function. Finally, (3) above is evaluated by rubric for improvement of independence state of students. We have designed rubric to evaluate effect of Self-reflector-Plus (Table 2). In Table 2, there are 9 components corresponding habit 1 to 3. In excellent column shows ideal students. (1)–(3) above is corresponding to purposes (1)–(3) of the study that described in Sect. 1 respectively.

Table 2. Rubric of improvement of independence state

	Independence	Reactive	Dependence
Habit 1	Act to take initiative to solve problems	You can act been instructed to solve the problem	Proactive action choice is difficult
	Concentrate on what you can control	You will try to change the things that you cannot control	You are to worry about the things you cannot control.
	Give a good effect on own surrounding environment	Affected by the surrounding environment	Not interested in the surrounding environment
Habit 2	Create your plan while envision the final goal	Create your plan while envision a near-term target	You cannot act in accordance with the plan
	You yourself want to be is clear	You yourself want to be is not yet clear	You have never thought about yourself want to be
Habit 3	Focus on important and not urgent activities	Work on only urgent activity. Always busy	Enjoy unimportant activity only
	Say "No!" about unimportant activity	Accept if you asked from a trusted person	Accept if you asked from anybody. Yes-man
	Delegation. Delegate work to the others	Not delegate own work to others. All your works are yours	Your work and others it cannot be distinguished
	Execute effective self-management based on want to be yours	Execute self-management, but does not based on want to be yours	You cannot self-management

4.2 Conditions of Experiment

Conditions of this experiment as is follows.

- Examinees: 25 people (3rd or 4th grade students)
- Term: 3 weeks (2017.12.30–2018.1.20)
- Application use:
  Android: Mobile application use (push notification)
  Others: Web application use (without notification)

Rubric shown in Table 2 is considered by three viewpoints of "independence", "reactive", and "dependence". After and before of this experiment, we have evaluated improvement of independence state using the rubric. Questionnaire items are as follow.

- System usability
- Reason that student don't use the system
- Availability of push notification
- Whether student had time to think mission statement
- Whether student have achieved own goal.

4.3 Result and Discussions

Table 3. shows the analysis of the system log. And from Tables 4, 5, 6, 7 and 8 show the questionnaire results for usage. The effective number of respondents of this questionnaire was 22.

Table 3. Analysis of system logs

Items	Numbers
Registered users	25
Active users	14
Registered events of all users	68
Number of users who started event	6
Started events	18
Number of users who finished event at least once	4
Finished events	14
Number of users who used tag search	1
Tag searches	2
Created threads	1
Comments to threads	1
Number of users who registered mission statement	17

From the system logs of Table 3, it has confirmed that out of the 25 registered users, 56% of users keep accessing the system after the registration. Since the total number of registered events are 68 in the experimental period 3 weeks, 14 users had registered about five events on average. Since 6 users had recorded the start time of the event, the actual number of user of the system is not so high. Additionally, the social functions such as tag search, thread creation, and comments were not much used. On the other hand, from the fact that 70% of users had registered a mission statement, it seems that they were conscious to use this system.

As shown in Table 4, 10 users answered that the event registration is bothersome. This answer is most common answer. Therefore, simplification of the event registration procedure is required. Several favorable opinions about the user-friendly interface of the system are also seen. On the other hand, 4 users answered that it is difficult to consider the mission statement. Hence, it is necessary to improve introduction support

Table 4. System usability

Items	Numbers	Rate
Operation easy to understand, satisfactory UI	4	18.20%
Useful, easy to use	3	13.60%
Quickly find function to use	2	9.10%
UI is designed to visually	7	31.80%
Quick feedback response	6	27.30%
Functions are difficult to understand	2	9.10%
Event registration is troublesome	10	45.50%
Mission statement is difficult to understand	4	18.20%

Table 5. Reason that student don't use the system

Items	Numbers	Rate
Used	3	13.60%
No push notification	6	27.30%
Event registration is troublesome	10	45.50%
Event start is troublesome	7	31.80%
No plan of activities	8	36.40%
Purpose of system is unclear	1	4.50%
Complex functions	0	0%
Operation steps are difficult to understand	0	0%
No effect even if using the system	1	4.50%
Forgotten to use	1	4.50%

Table 6. Availability of push notification

Items	Numbers	Rate
No notification because of OS	11	45.80%
D id not notice	0	0%
Noticed	6	25%
Do not use a most	7	29.20%

Table 7. Whether student had time to think mission statement

Items	Numbers	Rate
At least once a week	3	13.60%
About once every two weeks	2	9.10%
First time only	12	54.50%
Not self reflection	5	22.70%

Table 8. Whether student have achieved own goal

Items	Numbers	Rate
Completed activities and achieved own goal	5	22.70%
Did not achieved own goal	5	22.70%
Recognized own goal, but did not acton	7	31.80%
Did not recognized own goal	2	9.10%
No goal	3	13.60%

for users, such as users' manual and description about subject experiment. From Table 5, as reason for not using the system, 10 users answered that the event registration is bothersome. And there was a reply that "I feel that it has no effect either using or without using the system" in free description column of the questionnaire. From this, description about the effect of using this system and the purpose of the experiment was not sufficient for users. As shown in Table 6, 45.8% of iOS users did not receive push notification of the event start time. To resolve this problem, a client application or other mechanism are required for iOS. About Android user, 29.2% of users were not much use of the system whereas 25% of user were much use. Therefore, some approaches are required for promoting the use of the system.

From Table 7, it is confirmed that 54.5% of the users thought about their mission statement only at the first use while the user of about 20% had a chance to look back their mission statement. We consider that it is necessary some methods to motivate for taking the time to self-reflection.

From Table 8, nearly half of the user has answered that they had been able to keep registered schedule. On the other hand, about 40% users has answered that they had not been able to achieve their goal. In order to support the goals of the user, the overall response of the non-system, such as tutorials and group work at the time of introduction is required. In addition, there was a favorable reply in free description column that "I had not been able to use the system practically but had been able to behave toward the first goal. And had a feeling of accomplishment.

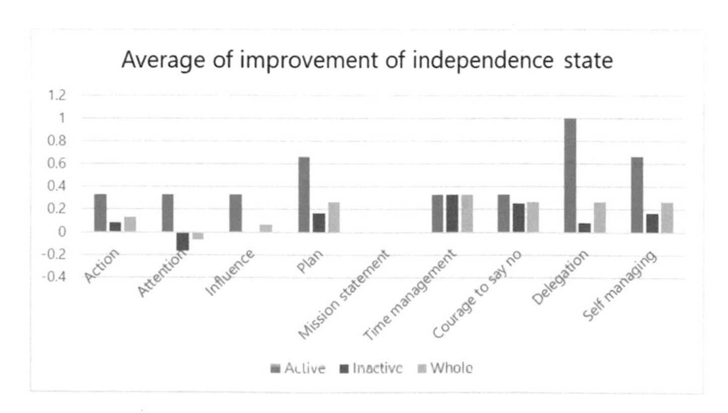

Fig. 10. Average of improvement of independence state

Based on the rubric to assess the independence state of the students shown in Table 2, the average value of the degree of improvement in the pre-post of the subject experiments is shown in Fig. 10. In the results, "Active student" means the user who registered event more than once, "Inactive student" means the user who registered event once or less, and "Whole student" means total of active students and inactive students. As shown in Fig. 10, active student has a higher average value than the inactive student, in the seven items that is action, attention, influence, plan, courage to say no, delegation, and self-managing. In the four items, that is attention, plan, delegation, and self-managing, there was a difference of from 0.5 to 0.9 points by the presence or absence of activity. These four items affect the behavior change for the quadrant II activities. On the other hand, in the two items, that is mission statement and time management, there was no differences depending on the presence or absence of activity. However, in this experiment, only 3 subjects in 15 subjects were able to be observed the changes in the improvement of independence state. 3 subjects are not a sufficient number of samples for performing statistical processing. Therefore, it is necessary to continuous observation.

5 Conclusions

In this paper, we have proposed mobile application called Self-reflector-Plus which improved Self-reflector. This system is applied to mobile interface and has community functions. Functions of anything can be used from mobile application. Basic concepts are registration/activation of activities available with mobile devices, and improvement of motivation by community function for their activities. In addition, we have designed rubric to evaluate effect of Self-reflector-Plus corresponding habit 1 to 3. We had subject experiment to evaluate improvement of frequency of use and maintenance of continuity.

From system logs analysis of the experiment, active users of this system were 24% only. Therefore, to support achievement of the goals of the user, the overall response of the non-system, such as tutorials and group work at the time of introduction is required. According to average of improvement of independence state, in the four items that is attention, plan, delegation, and self-managing, there was a difference of from 0.5 to 0.9 points by the presence or absence of activity. Our future works are improvement of UI/UX, re-experiment with refactoring of the plan.

References

1. Covey, S.R.: The 7 habits of highly effective people. King Bear (1996)
2. Kawano, Y., Obu, Y.: A proposal for personal branding support service in social media time. JCEA J. Contemp. Eastern Asia **12**(2) (2013)
3. Kawano, Y., Yuse, Y.: Development of proactive action support system by visualization of quadrant II activities. In: The 2016 International Conference on Information and Social Science (2016)

Characterizations of Local Recoding Method on k-Anonymity

Waranya Mahanan[1(✉)], Juggapong Natwichai[1], and W. Art Chaovalitwongse[2]

[1] Computer Engineering Department, Chiang Mai University, Chiang Mai, Thailand
waranya_ma@cmu.ac.th, juggapong@eng.cmu.ac.th
[2] Department of Industrial Engineering, University of Arkansas, Arkansas, USA
artchao@uark.edu

Abstract. k-Anonymity is one of the most widely used techniques for protecting the privacy of the publishing datasets by making each individual not distinguished from at least k-1 other individuals. The local recoding method is an approach to achieve k-anonymization through suppression and generalization. The method generalizes the dataset at the cell level. Therefore, the local recoding could achieve the k-anonymization with only a small distortion. As the optimal k-anonymity has been proved as the NP-hard problem, the plenty of optimal algorithm local recoding has been proposed. In this research, we study the characteristics of the local recoding method. In addition, we discover the special characteristic dataset that all generalization hierarchies of each quasi-identifier are identical, called an "Identical Generalization Hierarchy" (IGH) data. We also compare the efficiency of the well-known algorithms of the local recoding method on both $non-IGH$ and IGH data.

1 Introduction

Privacy preserving of the publishing data becomes a growing concern in the past decades. The publishing data must have required scientific guaranteed that it could not re-identify [15] an individual. One of the most mechanism applied to protect the privacy of the publishing dataset is k-anonymity by suppression and generalization [12,15] procedure. The publishing dataset would satisfy k-anonymity constraint if the quasi-identifiers of each tuple is not distinguished from other k-1 tuples.

In this paper, we propose to study the characteristics of the local recoding method on such privacy preservation problem both in theoretical and empirical aspects. In addition, the special characteristic dataset that all generalization hierarchies of each quasi-identifier are identical, called an "Identical Generalization Hierarchy" (IGH) data is analyzed in-depth. The basic concepts are introduced in Sect. 2. Section 3 presents the optimal solution characteristics. The empirical study results are shown in Sect. 4. Finally, the summary and our future work are presented in Sect. 5.

L. Barolli et al. (Eds.): NBiS 2018, LNDECT 22, pp. 648–658, 2019.
https://doi.org/10.1007/978-3-319-98530-5_56

2 Basic Definitions and Concepts

2.1 Basic Definitions

Definition 1 (Quasi-identifier). Is a set of attributes in the given dataset T that can be joined with the external information to re-identify individual records.

Definition 2 (k-anonymity). A dataset satisfies the k-anonymity condition, where $k > 1$, when each combination of quasi-identifiers exist at least k tuples in dataset.

For example, the dataset in Table 1(a), the attribute *Gender*, *Age*, and *Postcode* are considered as the quasi-identifiers. The generalization method is commonly applied to satisfy the k-anonymity condition by replacing more general value to some values in the dataset. The generalization hierarchies [13,14] are provided to feature the generalization structure of each quasi-identifier.

Table 1. An example of the general dataset (a) original data, (b) generalization by global recoding, and (c) generalization by local recoding

Gender	*Age*	*Postcode*
Male	21	53711
Female	25	53712
Male	26	53711
Male	27	53720
Female	27	53712
Male	28	53711
Female	22	53720

(a)

Gender	*Age*	*Postcode*
Person	[20-29]	5371*
Person	[20-29]	5371*
Person	[20-29]	5371*
Person	[20-29]	5372*
Person	[20-29]	5371*
Person	[20-29]	5371*
Person	[20-29]	5372*

(b)

Gender	*Age*	*Postcode*
Person	[20-29]	537**
Female	[25-29]	53712
Male	[25-29]	53711
Person	[20-29]	537**
Female	[25-29]	53712
Male	[25-29]	53711
Person	[20-29]	537**

(c)

Definition 3 (Generalization hierarchy). Let the generalization for an attribute A is a function on A and let $A_0 \xrightarrow{f_0} A_1 \xrightarrow{f_1} \cdots \xrightarrow{f_{n-1}} A_n$ is a function generalization sequence. The generalization hierarchy for A is a set of function $f_h : h = 0, ..., n-1$ such that $A = A_0$ and $|A_n| = 1$.

The generalization hierarchies of the dataset in Table 1(a) illustrate in Fig. 1. To give an idea, *Age* 25 in Table 1(a) can be generalized to [25–29], [20–29], [0–49], or [0–99], *Postcode* 53711 can be generalized to 5371*, 537**, 53***, 5****, or *****.

2.2 Generalization Hierarchy Data

In this paper, we categorize the dataset into 2 types: (1) a non-identical generalization hierarchy ($non-IGH$) data and (2) an identical generalization hierarchy (IGH) data. Generally, the dataset that uses the k-anonymity method to maintain the privacy has the different data type of quasi-identifiers [5]. Therefore, there is some dataset that all quasi-identifiers are the same type and use the same generalization hierarchy. We refer this type of data as an IGH data, while the others called a $non-IGH$ data.

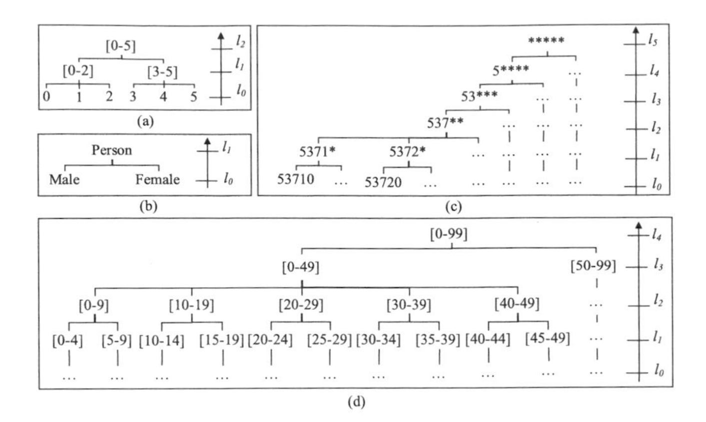

Fig. 1. The generalization hierarchy (a) Movie rating, (b) Gender, (c) Postcode, and (d) Age

Definition 4 (Non-Identical generalization hierarchy data). The general dataset that the set of the generalization hierarchy function of attributes is not identical.

Definition 5 (Identical generalization hierarchy data). Let $H = \{H_1, H_2, ..., H_m\}$ be the set of the generalization hierarchy function of attributes $\{A_1, A_2, ..., A_m\}$ in a dataset T . A dataset T is an identical generalization hierarchy data if and only if $\bigcup_{i=1}^{m} H_i = H_1 = H_1 = ... = H_m$.

Table 2. Customer rating history

ID	Quasi-Identifiers			Sensitive data		
	BM	FZ	FG	Age	Gender	Profession
1	4	0	3	25	M	Police
2	4	0	3	23	F	Professor
3	0	4	3	31	M	Student
4	0	5	1	27	M	Police
5	1	5	0	35	F	Salesman
6	0	4	5	25	M	Student
7	1	4	3	29	M	Salesman

An example of a $non-IGH$ data shown in Table 1(a). Each quasi-identifier of dataset is not the same data type, thus the generalization hierarchy of each quasi-identifier, in Fig. 1, is also not identical. The generalization hierarchy of *Gender*, *Age*, and *Postcode* is in Fig. 1(b), (c) and (d) respectively. An IGH data is defined as a dataset that every quasi-identifier has the same generalization hierarchy. As illustrated in Table 2, is an example of an IGH data. The quasi-identifiers of the dataset are the rating that the individuals give to each movie. Obviously, all quasi-identifiers are the same data type, thus the generalization hierarchy of this dataset, shown in Fig. 1(a), must be identical.

2.3 Global vs Local Recoding Method

Global and local recoding are the method for achieving k-anonymity [9,10]. The global recoding method generalizes data by replacing the more general value on the condition that all value in the quasi-identifier must be generalized to the same generalized level [2]. To illustrate, the dataset that generalized by the global recoding shows in Table 1(b). For achieving 2-anonymization, all value in *Gender* attribute is generalized to the highest level. All information of *Gender* attribute loss, so the global recoding could overgeneralized the dataset. The local recoding is the other way to generalize the dataset. It generalizes only the necessary cells [2,11]. For example, the *Gender* attribute in Table 1(c) generalizes only some cell. The local recoding dataset is not overgeneralized. Thus, the local recoding distorts less information than the global recoding. In the past decades, the various approaches of local recoding have been proposed. The most approaches are the clustering based algorithm [1,3]. The well-known algorithms of the local recoding are Mondrian [10] and k-Member [3]. Mondrian is a top-down greedy algorithm that clustering the data to each group by iteratively choosing an individual tuple. The k-Member clustering the data by selecting $k - 1$ individuals then add to the group.

2.4 Generalization Lattice

For evaluating the utility of k-anonymity using the global recoding, the generalization lattice is adopted. The generalization lattice [6] is a data structure which used to represent the generalization hierarchy of data records. Each node indicates the generalization level of the quasi-identifiers. The nodes in the lowest level of the generalization lattice are the minimal generalization node. The successors of each node is the direct generalization with less specific generalization. The example of the generalization lattice of the dataset in Table 1(a) is illustrated in Fig. 2(a). The lowest level (l_0) node is <000> and the node in the highest level (l_{10}) is <145>. For instance, we show the generalized dataset of node <121> in Table 1(b), *Gender*, *Age*, and *Postcode* are generalized to the level l_1, l_2 and l_1 respectively. In the generalization lattice, the node that satisfied the k-anonymity condition called a $k-anonymous$ node, shown as the shaded nodes in Fig. 2(a), while the node that not satisfied k-anonymity condition called a $non-anonymous$ node. Two properties of the generalization lattice is presented: (1) If a node is a k-anonymous node then all successors in the higher level are also a k-anonymous and (2) If a node is a non-anonymous node, then all predecessors in the lower level are also non-anonymous nodes.

The generalization lattice for the local recoding can be sophisticated. As mentioned earlier, the local recoding generalizes the dataset at the cell level. Therefore, to maintain the two properties of the generalization lattice, each node indicates the generalization level of each tuple. To avoid confusion, from now on, we will call the generalization lattice of global recoding as "global generalization lattice" and the generalization lattice of local recoding as "local generalization lattice". As shown in Fig. 2(c) is the local generalization lattice of *Gender* in

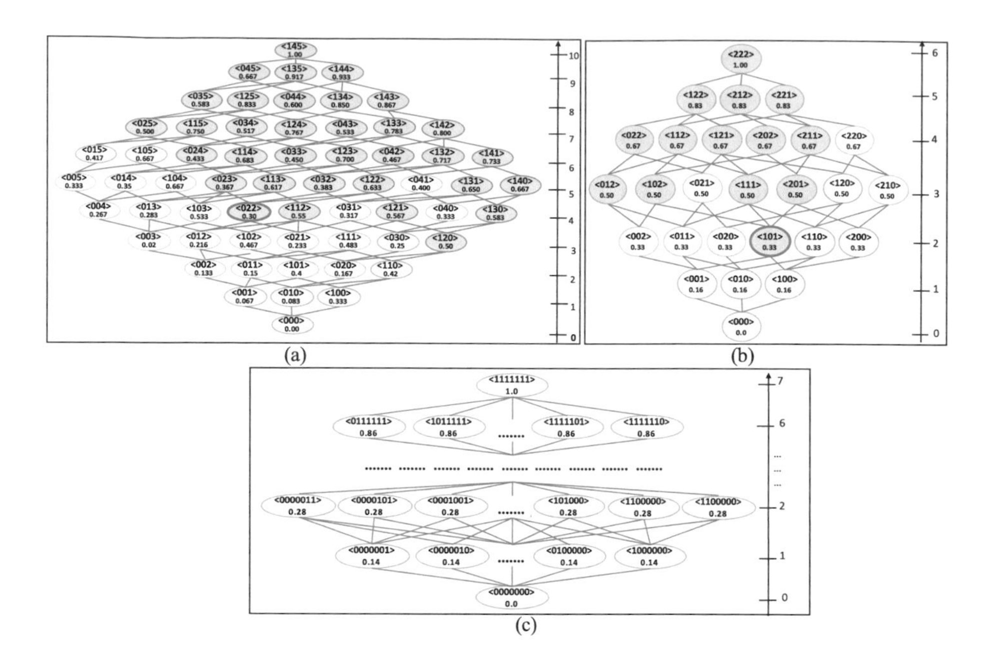

Fig. 2. Lattice of generalization (a) Global generalization lattice of a non-IGH data (b) Global generalization lattice of an IGH data (b) Local generalization lattice of Gender

Table 1(a). The lowest level node is <0000000> which indicates that every tuple of *Gender* is not generalized. Every tuple is generalized to the highest level at highest node <1111111>. In the local recoding, there are multiple generalization lattices. Each lattice represents the generalization hierarchy of each quasi-identifier. For example, Table 1(a) has 3 quasi-identifiers, so there are 3 lattices of local generalization lattice in this dataset. The structure of each local generalization depends on generalization hierarchy of each quasi-identifier. To give an idea, we illustrate the generalized dataset of the node $Gender < 1001001 >$, $Age < 2112112 >$ and $Postcode < 2002002 >$ in Table 1(c).

We found that an IGH data has the special characteristic. In the generalization lattice, the precision [16] is one of the information loss metrics for determining the data utility of each node. The precision of an IGH data in the same level of the generalization lattice is always the same, while a precision of a $non - IGH$ data not. Moreover, for an IGH data, the precision of nodes in the lower level always less than precision in the higher level nodes. For example, the generalization lattice of a $non - IGH$ data shows in Fig. 2(a) and the generalization lattice of an IGH data shows in Figure 2(b). The second line value in the node is the precision. Evidently, the precision of an IGH data of all nodes in level 3 is 0.50 which is identical, while the precision is not identical in a $non - IGH$ data. Furthermore, the precision of an IGH data of the node at level $2 - 0.30$ is always less than the precision of the nodes in the higher level. Unlike the precision of a $non - IGH$ data, at node <022> at level 4 is 0.28 which could more than the precision of the node <120> $= 0.44$ at level 3.

3 Optimal Solution

The optimal k-anonymization has been a concern for ensuring both data privacy and data utility left from the distortion in the k-anonymity model. The data privacy must be protected and the distortion must be minimized. The optimal k-anonymity has been proven to be NP-hard. The problem was originally proved by Meyerson, et al. [11]. Meyerson has shown that the optimal k-anonymity is NP-hard for $k \geqslant 3$, if $|\Sigma| \geqslant |V|$ is allowed and proved the theorem by reducing from the k-Dimensional Perfect Matching problem.

3.1 Optimal k-Anonymity of Non-IGH and IGH Data

The optimal k-anonymity of a $non-IGH$ and an IGH data can be evaluated using the generalization lattice. The k-anonymous node of the generalization lattice with the minimized information loss is the optimal k-anonymity solution. For example, using precision as the information loss metrics, among the k-anonymous nodes of generalization lattice in Fig. 2(a), the optimal solution of a $non-IGH$ data is node <022> with the smallest precision 0.30. As mentioned, for an IGH data, the precision of the nodes in the same level of an IGH data are all the same. And in the higher level nodes, the precision is always higher than the precision in the lower nodes. For this reason, the optimal k-anonymous node of an IGH data will always among the nodes in the lowest level found the k-anonymous nodes. To be illustrated, the optimal solution of an IGH data in Fig. 2(b), the k-anonymous node with the smallest precision, is node <101>.

For the local recoding method, the optimal solution could be evaluated in the similar process. Each node of the local generalization lattice indicates the generalization level of each tuple and each lattice indicates the generalization lattice of each quasi-identifier. Compare with the global generalization lattice, the number of nodes in the local generalization lattice are enormous. For evaluating the optimal solution of the local recoding method, we have to process all node in all lattices in a combination manner. To illustrate, there are 3 quasi-identifiers and 7 tuples in Table 1(a) with the generalization hierarchy in Fig. 1. They must be 3 lattices with 128 nodes, 78125 nodes, and 279936 nodes for $Gender$, Age, and $Postcode$ attribute respectively. Thus, we have to process $128 * 78125 * 279936 = 2.79936E + 12$ nodes for getting the optimal solution. Similar to the global recoding method, the special characteristic of an IGH data still remain. An optimal solution of local recoding method of an IGH data is always among the node in the lowest level found the k-anonymous.

4 Experiment Results

In this section, we conduct experiments to evaluate the efficiency of the well-known local recoding algorithms, Mondrian [10] and k-Member [3].

4.1 Experimental Data and Configuration

We used the real-life datasets to assess the performance of our algorithm. For measuring the efficiency of the algorithm on a $non-IGH$ data, we used the Adult [4] and the Cup [4] dataset. The Adult data is the dataset that extracted from the 1994 Census database. In the experiment, we used the Adult dataset with 32,560 records and the number of quasi-identifier at 8. The Cup dataset is originally used for The Third International Knowledge Discovery and Data Mining Tools Competition which contains a set of wide variety of intrusions simulated in a military network environment. We used 10,000 records and the number of quasi-identifier at 7 of the Cup dataset. For an IGH data, the dataset that used for assessing the performance is MovieLens [8] and Jester [7]. The MovieLens dataset has the rating list that users rated for each movie. The attributes are the list of movies. The rating range is between 0–5. The dataset with 925 records and the number of quasi-identifiers at 9 used in the experiment. The Jester dataset is the anonymous ratings from the Jester online joke recommender system. The rating range is between -10 to $+10$. We use the dataset with 5,000 records and the number of quasi-identifiers at 9.

The algorithm is implemented using Python 2.7. The experiments are proceeded on a 2x Intel X5670 with 24 GB memory running Linux. We average the execution time each configuration three times to obtain stable times of execution.

4.2 Results and Discussion

4.2.1 Utility Comparison

For measuring the utility of k-anonymity of the Mondrian and the k-Member algorithm, the information loss metrics are applied. We study the efficiency of each algorithm on the 3 generally used information loss metrics, Precision ($Prec$), Discernibility Metric (DM) and Average Equivalence Class Size (C_{avg}).

Precision vs k

Precision ($Prec$) was first introduced as an information loss metric in [16]. $Prec$ of a node in the generalization lattice is related to its generalized level in the generalization hierarchy of each quasi-identifier A, which is denoted as h. The precision of GT is given by $Prec(GT) = \frac{1}{m} \cdot \sum_{i=1}^{m} \frac{h_i}{H_i}$, where $GT(A_1, ..., A_m)$ be a generalized table of dataset T and H is the maximum level of the generalization hierarchy.

The lower $Prec$ means the lower information loss. At $Prec(GT) = 0$, all quasi-identifiers are not generalized. In contrast, if each value in the dataset is generalized to the highest level of the generalization hierarchy, then $Prec(GT) = 1$

The experimental result of precision from the Mondrian and the k-Member of each dataset shown in Table 3(a). We vary the k from 2 to 100 in this experiment. It can be seen that, at the same k, the precision of the result from k-Member is less than the precision of the result from the Mondrian in each dataset. For all datasets, the large k causes the higher information loss than the small k. The data trend for every algorithm is in the same way. The result of the MovieLens

Table 3. Utility comparison of information loss metrics (a) Precision, (b) DM, and (c) C_{avg}

k	*Adult*		*Cup*		*MovieLens*		*Jester*	
	Mondrian	*k-Member*	*Mondrian*	*k-Member*	*Mondrian*	*k-Member*	*Mondrian*	*k-Member*
5	10.64	5.73	9.48	4.27	43.37	24.11	13.29	4.40
10	15.85	9.73	15.24	7.31	56.31	37.14	21.27	6.65
40	30.07	18.86	29.77	16.76	72.76	59.26	35.20	12.30
60	38.94	22.50	29.77	20.84	79.76	65.76	35.20	14.18
80	38.94	24.63	39.72	23.88	79.76	70.59	41.76	16.21

(a)

k	*Adult*		*Cup*		*MovieLens*		*Jester*	
	Mondrian	*k-Member*	*Mondrian*	*k-Member*	*Mondrian*	*k-Member*	*Mondrian*	*k-Member*
5	223054	216968	58640	50000	6707	4655	29320	25042
10	444618	348436	141040	100000	13385	9305	70520	50166
40	1776890	1250034	781264	400000	53479	37225	390632	202904
60	3553734	1855498	781264	604000	106955	57577	390632	301600
80	3553734	2457070	1562512	800000	106955	79625	781256	404800

(b)

k	*Adult*		*Cup*		*MovieLens*		*Jester*	
	Mondrian	*k-Member*	*Mondrian*	*k-Member*	*Mondrian*	*k-Member*	*Mondrian*	*k-Member*
5	1.5898	1.2670	1.1062	1.0000	1.4453	1.0065	1.1062	1.0010
10	1.5898	1.1789	1.2755	1.0000	1.4453	1.0076	1.2755	1.0020
40	1.5898	1.1121	1.9531	1.0000	1.4453	1.0109	1.9531	1.0081
60	2.1198	1.1030	1.3021	1.0040	1.9271	1.0356	1.3021	1.0040
80	1.5898	1.0971	1.9531	1.0000	1.4453	1.0614	1.9531	1.0081

(c)

dataset has the most losses compare with the others dataset due to the smallest number of tuples.

Discernibility Metric vs k

Discernibility Metric (DM) is another widely used information loss metric [2]. DM assigns a penalty to each record based on the number of records that are not distinguishable from it. Let $GT(A_1, ..., A_m)$ be a generalized table of dataset T, E be an equivalent class [17] of GT. The DM of GT is given by $DM(GT) = \sum_{\forall E s.t|E| \geqslant k} |E|^2$.

This experiment compares the DM of the result from both algorithms to see how the DM affected by the k. The lower DM means the smaller information loss in the generalized dataset. Since the smaller k obtains the smaller size of an equivalence class, the result in Table 3(b) shows that the DM at the smaller k is less than the DM at the higher k. In the same dataset, the k-Member algorithm produces smaller DM than the Mondrian algorithm. Comparing the DM of each dataset, the Adult dataset has the highest DM due to it has the largest number of tuples followed by Cup, MovieLens, and Jester respectively.

Average Equivalence Class Size vs k

Average Equivalence Class Size (C_{avg})[10] given by $C_{avg}(GT) = \frac{|GT|}{|E| \cdot k}$. It measures the data utility based on the size of equivalence class E of the generalized dataset GT.

We study the effect of C_{avg} on k in this experiment. From the result shown in Table 3(c), the C_{avg} is not affected by k. The C_{avg} of the generalized dataset using in both algorithms is stable when k is increased in every dataset. Since C_{avg} is the average size of the equivalence class, it means that at any k the average size of the equivalence class is similar. The C_{avg} of each dataset is also similar in each algorithm. It means that using C_{avg} as the information loss metric, each algorithm could find the generalized dataset with the similar utility. However, the C_{avg} of the generalized dataset using the k-Member is closer to the optimal solution at 1 than the C_{avg} of the generalized dataset using the Mondrian.

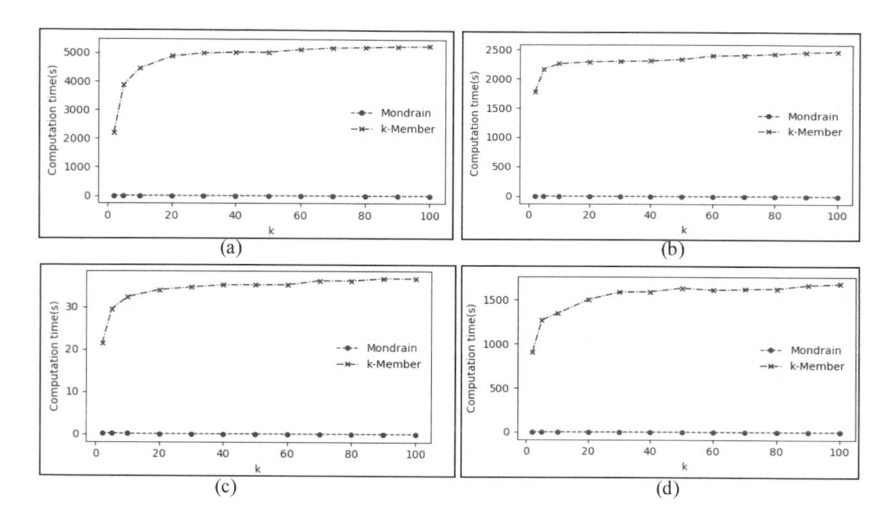

Fig. 3. Computation time comparison of (a) Adult, (b) Cup, (c) MovieLens and (d) Jester dataset

4.2.2 Execution Time Comparison

We compare the performance of both algorithms on Adult, Cup, MovieLens and Jester dataset in Fig. 3(a), (b), (c) and (d) respectively. The k varies from 2 to 100. As can be seen, the Mondrian is a lot faster than the k-Member. The execution time of the Mondrian is decreased when the k is increased. This is because the number of outer iterations of the Mondrian is the number of equivalence classes, so the larger k obtains the smaller number of equivalence classes and faster execution time. In contrast, the execution time of k-Member is increased when the k is increased. The reason is that the k-Member need to iteratively find the nearest k-1 tuples with a selected tuple then add them to each equivalence class, thus the larger k gets the larger number of iteration and slower execution time. Comparing the execution time of each dataset, the Adult take the longest time to find the generalized dataset because it has the largest number of tuples, followed by Cup, Jester, and MovieLens.

5 Conclusion and Future Work

In this paper, we show the characteristic of local recoding method on k-anonymity. We present a special dataset called an "Identical Generalization

Hierarchy" (IGH) data. We also compare the efficiency of the well-known local recoding algorithms on both $non - IGH$ and IGH data. Regarding the experimental results, the k-Member could find the generalized dataset with smaller information loss than the Mondrian algorithm. While the Mondrian algorithm is much faster than the k-Member. Thus, in the future work, the advantageous characteristics of both algorithms are to be combined in order to develop an approximation algorithm based on the local recoding approach.

References

1. Aggarwal, G., Panigrahy, R., Feder, T., Thomas, D., Kenthapadi, K., Khuller, S., Zhu, A.: Achieving anonymity via clustering. ACM Trans. Algorithms **6**(3), 49:1–49:19 (2010)
2. Bayardo, R.J., Agrawal, R.: Data privacy through optimal k-anonymization. In: Proceedings of the 21st International Conference on Data Engineering, ICDE 2005, pp. 217–228. IEEE Computer Society, Washington, DC (2005)
3. Byun, J.W., Kamra, A., Bertino, E., Li, N.: Efficient k-anonymization using clustering techniques. In: Proceedings of the 12th International Conference on Database Systems for Advanced Applications, DASFAA 2007, pp. 188–200. Springer, Heidelberg (2007)
4. Dheeru, D., Karra Taniskidou, E.: UCI machine learning repository (2017). http://archive.ics.uci.edu/ml
5. El Emam, K., Brown, A., AbdelMalik, P.: Evaluating predictors of geographic area population size cut-offs to manage re-identification risk. J. Am. Med. Inform. Assoc. **16**(2), 256–266 (2009)
6. El Emam, K., Dankar, F., Issa, R., Jonker, E., Amyot, D., Cogo, E., Corriveau, J.P., Walker, M., Chowdhury, S., Vaillancourt, R., Roffey, T., Bottomley, J.: A globally optimal k-anonymity method for the de-identification of health data. J. Am. Med. Inform. Assoc. JAMIA **16**, 670–82 (2009)
7. Goldberg, K., Roeder, T., Gupta, D., Perkins, C.: Eigentaste: a constant time collaborative filtering algorithm. Inf. Retr. **4**(2), 133–151 (2001)
8. Harper, F.M., Konstan, J.A.: The movielens datasets: history and context. ACM Trans. Interact. Intell. Syst. **5**(4), 19:1–19:19 (2015)
9. LeFevre, K., DeWitt, D.J., Ramakrishnan, R.: Incognito: efficient full-domain k-anonymity. In: Proceedings of the 2005 ACM SIGMOD International Conference on Management of Data, SIGMOD 2005, pp. 49–60. ACM, New York (2005)
10. LeFevre, K., DeWitt, D.J., Ramakrishnan, R.: Mondrian multidimensional k-anonymity. In: 22nd International Conference on Data Engineering (ICDE 2006), p. 25 (2006)
11. Meyerson, A., Williams, R.: On the complexity of optimal k-anonymity. In: Proceedings of the Twenty-Third ACM SIGMOD-SIGACT-SIGART Symposium on Principles of Database Systems, PODS 2004, pp. 223–228. ACM, New York (2004)
12. Samarati, P.: Protecting respondents identities in microdata release. IEEE Trans. Knowl. Data Eng. **13**(6), 1010–1027 (2001)
13. Samarati, P., Sweeney, L.: Generalizing data to provide anonymity when disclosing information. In: Proceedings of the ACM SIGACT-SIGMOD-SIGART Symposium on Principles of Database Systems 1998 (1998)
14. Sweeney, L.: Achieving k-anonymity privacy protection using generalization and suppression. Int. J. Uncertain. Fuzziness Knowl. Based Syst. **10**(5), 571–588 (2002)

15. Sweeney, L.: k -anonymity: a model for protecting privacy. Int. J. Uncertain. Fuzziness Knowl. Based Syst. **10**(5), 1–14 (2002)
16. Sweeney, L.A.: Computational disclosure control: a primer on data privacy protection. Ph.D. thesis, Massachusetts Institute of Technology, Cambridge, MA, USA (2001). AAI0803469
17. Wong, R.C.W., Li, J., Fu, A.W.C., Wang, K.: (α, k)-anonymity: an enhanced k-anonymity model for privacy preserving data publishing. In: Proceedings of the 12th ACM SIGKDD International Conference on Knowledge Discovery and Data Mining, KDD 2006, pp. 754–759. ACM, New York (2006)

The 9th International Workshop on Intelligent Sensors and Smart Environments (ISSE-2018)

A Cloud-Fog Based Environment Using Beam Search Algorithm in Smart Grid

Komal Tehreem, Nadeem Javaid(✉), Hamida Bano, Kainat Ansar, Moomina Waheed, and Hanan Butt

COMSATS University, Islamabad 44000, Pakistan
nadeemjavaidqau@gmail.com
http://www.njavaid.com

Abstract. Smart Grid (SG) monitor, analyze and communicate to provide electricity to consumers. In this paper, a cloud and fog computing environment is integrated with SG for efficient energy management. In this scenario world is divided into six regions having twelve fogs and eighteen clusters. Each cluster has multiple buildings and each building comprises of eighty to hundred apartments. Multiple Micro Grids (MG's) are available for each region. The request for energy is sent to fog and load balancing algorithm is used for balancing the load on Virtual Machines (VMs). Service broker policies are used for the selection of fog. Round Robin (RR), throttled and Beam Search (BS) algorithms are used with service proximity policy. Results are compared for these three algorithms and from this BS algorithm gives better result.

Keywords: Smart Grid · Micro Grid · Cloud computing
Fog computing · Response time · Processing time · Load balancing

1 Introduction

Humans are making new electrical devices for making their lives more comfortable. The number of electrical devices increase and the requirement of electricity also increases. More power is required to fulfill the demand of electricity. The increase in demand of electricity makes it difficult to manage the load. Traditional grids are not able to fulfill users requirement and carbon emission rate is also increasing that is not good for the environment. Due to these reasons, better energy distribution networks are required [1].

SG is Information and Communication Technology (ICT), which benefits the control and monitor of power systems [2]. The integrated environment of SG with cloud and fog facilitates consumers. Consumers send request to fog for accessing its resources to manage the electricity requirements. Fog fulfills the electricity requirement of the consumer with the help of cloud, MG and macro grid.

Cloud computing is the delivery of on-demand computing resources via web. It provides shared storage, save data permanently and reduces the cost of computation. The large amount of data is stored and executed on the cloud. For this

L. Barolli et al. (Eds.): NBiS 2018, LNDECT 22, pp. 661–672, 2019.
https://doi.org/10.1007/978-3-319-98530-5_57

reason, the load should be balanced among all VMs to utilize resources efficiently and reduce power consumption, execution time and Response Time (RT). Cloud consumer that uses services, storage, computation and appliances of the cloud have to pay a certain amount of cost.

Fog computing is the extension of cloud computing, introduced by Cisco. Due to the rapid increase in number of applications and connected devices, the fog computing is used. The fog saves consumer data temporary. It communicates with cloud to save data permanently and reduces the burden of cloud and.

Three service models for cloud and fog computing are: Software-as-a-Service (SaaS), Platform-as-a-Service (PaaS) and Infrastructure-as-a-Service (IaaS).

The SaaS model delivers the software applications and licensed software. PaaS provide run time environment for the applications deployment, testing, managing and development tools. IaaS is the basic service, provides hardware resources to users like servers, VMs, storage, operating systems etc. It also provide excessive storage.

In this paper, cloud and fog based environment with SG integration is considered. The scenario is divided into three layers. The first layer is the cloud layer, communicate with macro grid. Second is the fog layer which communicates with the cloud layer and MG. The third layer is SG based, where each cluster has multiple smart building. Clusters communicate with fogs and are not allowed to communicate with cloud and MG directly. The fog then communicate with MG to fulfill the user's requirement. If MG do not have sufficient energy, then the fog request cloud for energy.

1.1 Motivation

As in [3], cloud and fog based environment is presented. The fog concept is used to utilize and provide services to the end users to reduce the load on the cloud. In [4,5] different algorithm techniques are used for task management on the demand side. Load balancing techniques are used for evenly distributing the task among different VMs on fog and cloud. In these papers, authors integrated SG with cloud and fog computing environment to overcome the limitations. The main focus of this paper is to reduce the overall RT and for this purpose, BS algorithm is used for load balancing of VMs. The results of BS are compared with RR and throttled, for service proximity policy.

1.2 Contributions

SG is integrated with cloud-fog based environment. The benefits of the proposed scheme are:

- Fogs are placed near to end users to minimize latency rate and response is faster than cloud.
- BS algorithm is proposed in this paper and simulation results are compared with RR and throttled algorithms.
- Minimum RT is achieved with proposed algorithm.

The rest of paper is organized as follows: Sect. 2 discusses related work, Sect. 3 discuss the problem formulation, Sect. 4 explains the proposed modal, Sect. 5 discuss the load balancing algorithms, Sect. 6 presents the simulation results and the conclusion is drawn in Sect. 7.

2 Related Work

Cloud computing is the environment which facilitates its users by providing different resources. Multiple algorithms are applied to balance the load on cloud and fog. Mostly used load balancing algorithms are: RR, throttled, honeybee foraging, active monitoring and biased random sampling. In [3], authors have compared the results of RR and throttled. Authors have proposed new service broker policy and compared the results for already existing policies like Service proximity policy, optimize response time policy and the dynamic reconfigure load. New dynamic service proximity policy gives better results because it selects the fog that is in the same region. Throttled result outstands the RR algorithm result.

The authors in [4], proposed a PSO algorithm for load balancing. This algorithm efficiently balance the load for allocating resources. Results for PSO algorithm are compared with RR and throttled. Simulation shows that the overall performance of PSO is better than other two algorithms. In [5], authors have proposed new service broker policy and compared the results of RR and throttled load balancing algorithms for multiple policies. The results of throttled algorithm are better for new proposed policy. The authors in [6] have compared the load balancing algorithm based on honey bee with different scheduling algorithms like RR, honey bee and modified throttled. The proposed algorithm is efficient in terms of execution and RT. Proposed algorithm is compared with honey bee, modified throttled and RR algorithms.

In [7], authors have presented the demand side load shifting management strategy. A heuristic-based evolutionary algorithm is used in three service areas that are: residential, commercial and industrial. The proposed algorithm reduces the peak load demand. This algorithm also handles large number of devices (Table 1).

In [8], authors proposed genetic algorithm in demand side management for optimizing load redistribution, in industry. The main focus of this paper is to minimize power utilization during peak hours and effectively distribution of energy during off peak hours. Results have shown that power utilization during peak hours is reduced up to 21.91%.

Authors in [9] have proposed constraint measure technique for load balancing. The load from overloaded VMs are reallocated to VMs that are under loaded. RR is used to assign tasks to VMs and dragonfly algorithm is used to reallocate the tasks to under loaded VMs. The results for different threshold values are compared for different number of tasks. The proposed constraint based algorithm is also compared with dynamic load balancing, heuristic dynamic load balancing and honey bee based load balancing algorithms. The simulation results of

Table 1. Related work summary

Reference(s)	Technique(s)	Feature(s)	Achievement(s)
[3]	Throttled, RR, Service proximity policy, optimize response time policy, dynamic reconfigure load and new dynamic service proximity	Load balancing in cloud-fig based SG	Proposed algorithm gives fast response and minimum delay
[4]	Particle Swarm Optimization (PSO), RR and throttled	Load balancing in cloud-fog based SG environment	For effective resource management
[5]	Coalition based on game theory	Distribution of surplus energy between apartments that require more energy	Maintain balance between energy demand and supply
[6]	Load balancing algorithm based on honey bee behavior, RR, honey bee and modifies throttled	Balance load between VMs on cloud and avoid overloading and under loading of resources	Improve execution and RT and minimize data center processing time
[7]	Heuristic-based evolutionary algorithm	Shift load for demand side management in SG	Reduce peak load demand
[8]	Genetic algorithm in demand side management	Moving energy demand to off-peak hours using demand side management in SG in industrial area	Minimize power utilization in on-peak hours by distribution available energy in off-peak hours
[9]	Round Robin, dynamic, heuristic dynamic, Dragonfly and honey bee based load balancing algorithms	Load balancing in cloud by moving tasks from overloaded VMs to underloaded VM	Proposed algorithm has better performance and it migrates minimum no of tasks

dragonfly algorithm outstand in the term of migration of tasks from one VM to another.

3 Problem Formulation

Cloud fog based environment is integrated with SG for efficient energy management. The request for energy is sent to fog and load balancing algorithm is used for balancing the load on VMs. Service broker policies are used for the selection of fog.

$$Fogs = \{F_1, F_2, ..., F_i, ..., F_n\} \tag{1}$$

n is the number of fogs.

$$VM = \{VM_1, VM_2, ..., VM_i, ..., VM_m\} \tag{2}$$

m is the number of VMs which process the k number of requests represented as:

$$R = \{R_1, R_2, ..., R_i, ..., R_k\} \tag{3}$$

The state of response can be defined as:

$$SPT_{ij} = \begin{cases} 1 & If\ VM_i\ is\ assigned\ to\ R_j \\ 0 & otherwise \end{cases} \tag{4}$$

PT is the total time to process a request. Let PT_{ij} be the processing time of the jth request on i^{th} VM represented as:

$$PT_{ij} = \frac{Size\ of\ R_j}{capaicity\ of\ VM_i} \tag{5}$$

Total process time can be calculated as:

$$PT = \sum_{j=1}^{k} \sum_{i=1}^{m} (PT_{ij}\ *\ SPT_{ij}) \tag{6}$$

Let QD be the queuing delay of each request.

$$QD = \sum_{i=1}^{k} QD_i \tag{7}$$

The main objective of this paper is to minimize the RT. RT is the time interval, starts when request is sent to fog till the response is received. It includes the queuing delay and PT. Total RT is defined as:

$$minimize\ RT - PT + QD \tag{8}$$

4 System Model

Cloud and fog based computing environment has been modeled in this paper. Fog computing reduces the load on cloud. Proposed system model is divided into three layers: cloud layer, fog layer and end user layer as shown in Fig. 1.

The first layer is the cloud layer, which has all the services and resources required by the end users. Cloud is connected with macro grid. The middle fog

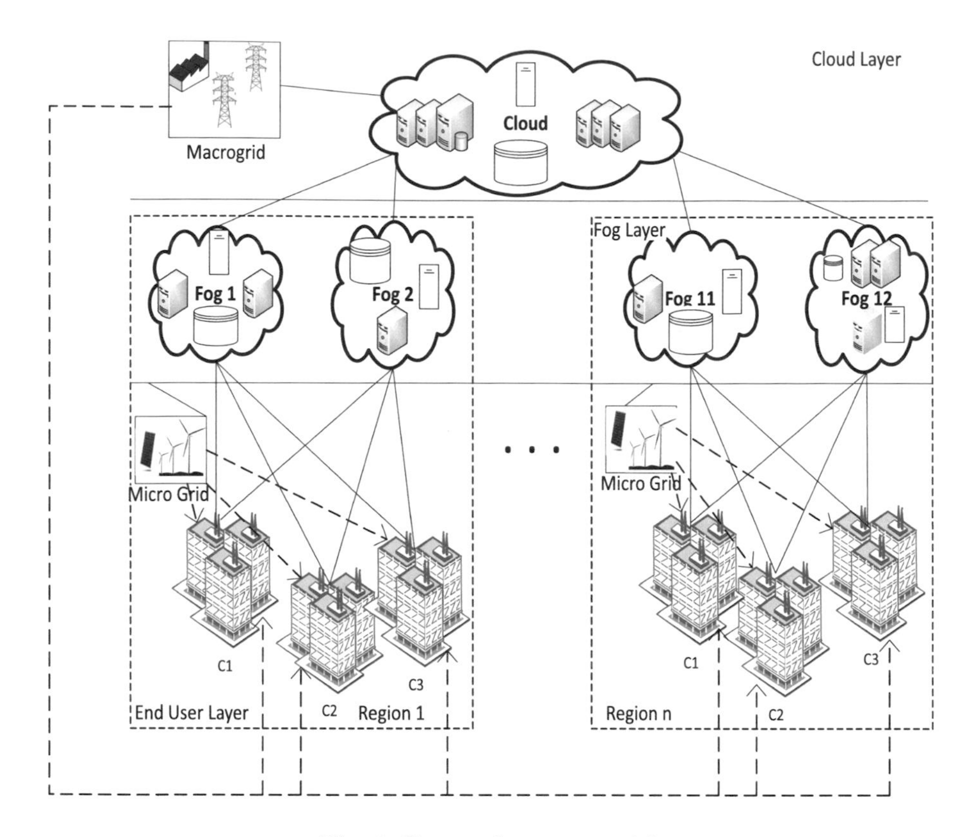

Fig. 1. Proposed system model.

layer concept comes in cloud computing to overcome the delay and latency rate to access the services of the cloud. Fogs are connected with clusters of smart buildings and cloud. The end users do not directly connect or communicate to cloud for accessing the energy resources. It connects to fog for accessing SG facilities. Different number of VMs on fog handle different types of request from the end users.

The bottom layer of cloud computing have smart buildings or clusters. Clusters have different number of smart homes in which there are different number of appliances, which require energy and that energy is provided to the users from MG or macro grid. Each cluster has controller that manages the demand and supply request of each user.

In this paper, world is divided into six regions. These regions are composed of six continents as shown in Fig. 2. Each region has two fogs and three clusters. A single cluster has eighty to hundred smart buildings and each building is composed of eighty to hundred homes. A fog has VMs in the range of ten to fifty shown in Table 2 and fog can forward requests to MG. User request is send to fog and it requests MG for required energy. MG provides energy to fog and it takes decision whether the energy is sufficient for user or not. If supply energy is

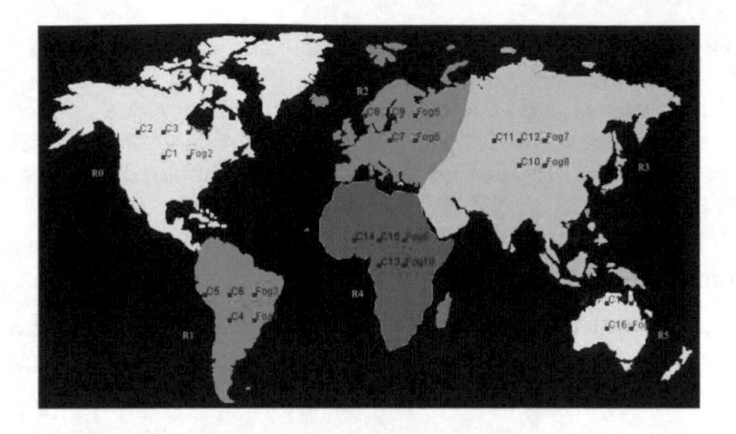

Fig. 2. World division in regions

Table 2. Fogs and clusters distribution

Region name	Region Id	Fogs Id	Numbers of VMs	Clusters Id	Numbers of buildings
North America	0	Fog1	29	C1	80
		Fog2	27	C2	85
				C3	87
South America	1	Fog3	47	C4	90
		Fog4	42	C5	98
				C6	87
Europe	2	Fog5	32	C7	80
		Fog6	23	C8	88
				C9	95
Asia	3	Fog7	30	C10	80
		Fog8	25	C11	87
				C12	99
Africa	4	Fog9	50	C13	90
		Fog10	45	C14	85
				C15	81
Oceania	5	Fog11	35	C16	97
		Fog12	15	C17	98
				C18	94

insufficient then fog request other MG's. In case, demand is too high or supply from MG's is too low then fog send request to cloud to provide macro grid facilities to end users.

5 Load Balancing Algorithm

Load balancing algorithm is used to allocate the VMs, to utilize the resources for better performances. VMs are allocated on the basis of storage, memory and requirements etc. VMs are installed on servers and are allocated to each host on fog for load balancing. Load balancing is used to minimize the Response Time (RT) and cost. Load Balancing algorithms used in cloud Analyst are RR and throttled. Whereas, in this paper, BS algorithm is used as load balancing algorithm. The results for RR, throttled and BS are compared. These algorithms are described below:

5.1 Round Robin

In RR, equal time slots are assigned to each host for assigning equal resources. When a consumer sends request, this algorithm is used to balance the load on VMs, for resource utilization.

5.2 Throttled

Throttled algorithm is the concept of allocating VMs which are able to handle the request of the client. A load balancer selects a VM that can perform operations easily.

5.3 Beam Search

In BS algorithm, load balancer selects the VM which can easily load the request and have minimum pending requests. BS is restricted or modified version of breath first search algorithm. In [10], authors have implemented BS algorithm to solve the shortest common super sequence problem. The experiment results shows that BS algorithm work better for a large and a very large number of instances.

Algorithm 1. BeamSearch

```
1: Input : BeamWidth, minimumAllocation[], VMId[]
2: Output : selectedVm
3: for i=1:Number of VMs do
4:     if i.currentAllocation < minimumAllocation[0] then
5:         minimumAllocation[0] = i.currentAllocation
6:         VMId[0] = i.VMId
7:     else if i.currentAllocation < minimumAllocation[1] then
8:         minimumAllocation[1] = i.currentAllocation
9:         VMId[1] = i.VMId
10:    end if
11: end for
12: if VirtaulMachineState[VMId[0]] == AVAILABLE then
13:     selectedVm = VMId[0]
14: else if VirtaulMachineState[VMId[1]] == AVAILABLE then
15:     selectedVm = VMId[0]
16: else
17:     selectedVm = VMId[0]
18: end if
```

6 Simulation Results

In this paper, simulations are done in 'Cloud Analyst', on JAVA platform. Eclipse environment is installed on 64 bit windows operating system. The processor is core i7.

In this paper, simulation results of three load balancing algorithms are compared. For this purpose, service proximity policy is used as service broker policy, RR, throttled and BS algorithms are used as load balancing algorithms. Results are described for twelve fogs that are connected to centralized cloud.

In the proposed scenario, two fogs and three clusters in a single region are considered. Each fog has the multiple number of VMs. Load balancing algorithms are used to allocate tasks to these VMs. service broker policy decides that which request should be sent to which fog. In service proximity policy, the closest fog is selected which have minimum latency. The request is routed towards the fog, that is selected by service broker policy. We have done simulation for each load balancing algorithm. Input parameters are same for each algorithm simulation and results are than red.

The minimum, average and maximum RT of each cluster of buildings are shown in Table 3.

The minimum, average and maximum request servicing time of each fog is shown in Table 4.

Table 3. Response time

Clusters	Average (ms)	Minimum (ms)	Maximum (ms)
C1	51.79	40.84	68.14
C2	51.55	37.29	66.67
C3	51.75	37.57	65.38
C4	54.05	40.89	68.71
C5	53.86	40.83	66.57
C6	53.88	39.67	69.83
C7	51.42	38.59	63.28
C8	51.39	39.59	63.10
C9	51.38	38.68	63.59
C10	52.79	38.90	67.68
C11	52.67	38.57	67.42
C12	52.79	39.17	70.17
C13	53.27	38.49	70.57
C14	53.94	40.81	68.77
C15	53.17	41.45	64.97
C16	51.89	39.87	67.13
C17	51.86	39.03	67.63
C18	51.51	38.67	62.77

Table 4. Fog request servicing time

Fog	Average (ms)	Minimum (ms)	Maximum (ms)
Fog1	2.79	0.04	4.58
Fog2	1.18	0.02	2.18
Fog3	4.34	0.07	6.97
Fog4	3.96	0.07	6.21
Fog5	2.11	0.03	3.57
Fog6	1.46	0.02	2.56
Fog7	1.64	0.03	2.87
Fog8	4.44	0.07	7.11
Fog9	4.39	0.08	7.25
Fog10	2.91	0.06	4.75
Fog11	3.07	0.05	5.33
Fog12	1.17	0.02	2.08

Each fog has VM cost, data transfer cost, MG cost and total cost. The cost of 24 h is calculated for each fog and is shown in Fig. 3.

Fog 3, 4, 8, 9 and 11 have the more requests servicing time as they process more request. Table 5 shows overall RT and PT comparison between three algorithms. RR and Throttled take maximum RT. However, PT of all algorithms is same. The minimum RT of throttled, average RT of RR and maximum RT of BS gives minimum value.

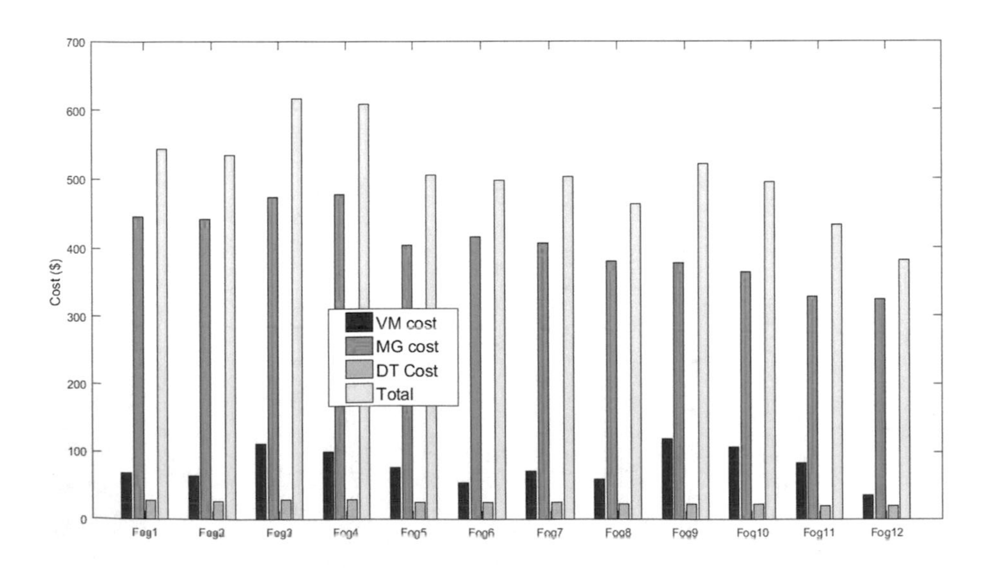

Fig. 3. VM, MG, data transfer cost for each fog

Table 5. Overall response time summary

Algorithm		Average (ms)	Minimum (ms)	Maximum (ms)
BS	RT	52.57	37.29	70.57
	PT	2.83	0.02	7.25
RR	RT	52.56	37.29	74.67
	PT	2.81	0.02	7.25
Throttled	RT	52.57	37.33	74.67
	PT	2.82	0.02	7.25

7 Conclusion

A cloud and fog based environment is integrated with SG. In this paper cloud-fog model is proposed to manage energy requirements for all the six regions of the world. Each region has multiple three clusters and two fogs. Each cluster has residential smart building. For efficient resource management service proximity policy is used as service broker policy and three load balancing algorithms are used. The simulation is performed on a JAVA platform in eclipse. The simulation results are compared and the performance of BS algorithm is better than RR and Throttled for overall RT. Maximum RT is minimum for BS algorithm.

References

1. Okay, F.Y., Ozdemir, S.: A fog computing based smart grid model. In: 2016 International Symposium on Networks, Computers and Communications (ISNCC), Yasmine Hammamet, pp. 1–6 (2016)
2. Yaghmaee Moghaddam, M.H., Leon-Garcia, A., Moghaddassian, M.: On the performance of distributed and cloud-based demand response in smart grid. IEEE Trans. Smart Grid. https://doi.org/10.1109/TSG.2017.2688486
3. Fatima, I., Javaid, N., Iqbal, M.N., Shafi, I., Anjum, A., Memon, U.: Integration of cloud and fog based environment for effective resource distribution in smart buildings (2018)
4. Zahoor, S., Javaid, N., Khan, A., Muhammad, F.j., Zahid, M., Guizani, M.: A cloud-fog-based smart grid model for efficient resource utilization (2018)
5. Javaid, N., Khalid, A., Rahim, M., Mateen, A.: Smart homes coalition based on game theory (2018)
6. Hashem, W., Nashaat, H., Rizk, R.: Honey bee based load balancing in cloud computing. KSII Trans. Internet Inf. Syst. **11**(12), 5694–5711 (2017)
7. Logenthiran, T., Srinivasan, D., Shun, T.Z.: Demand side management in smart grid using heuristic optimization. IEEE Trans. Smart Grid **3**(3), 1244–1252 (2012)
8. Bharathi, C., Rekha, D., Vijayakumar, V.: Genetic algorithm based demand side management for smart grid. Wirel. Pers. Commun. **93** (2017). https://doi.org/10.1007/s11277-017-3959-z
9. Polepally, V., Shahu Chatrapati, K.: Dragonfly optimization and constraint measure-based load balancing in cloud computing. Cluster Comput. (2017). https://doi.org/10.1007/s10586-017-1056-4

10. Gallardo, J.E.: A multilevel probabilistic beam search algorithm for the shortest common supersequence problem. PloS One **7**(12), e52427 (2012)
11. Techopedia.com: What is a smart grid? - Definition from Techopedia (2018). https://www.techopedia.com/definition/692/smart-grid. Accessed 15 Apr 2018
12. Microsoft Azure: What is cloud computing? A beginner's guide — Microsoft Azure (2018). https://azure.microsoft.com/en-in/overview/what-is-cloud-computing/. Accessed 15 Apr 2018

A Microservices-Based Social Data Analytics Platform Over DC/OS

Ming-Chih Hsu and Chi-Yi Lin(✉)

Tamkang University, Taipei, Taiwan, R.O.C.
chiyilin@mail.tku.edu.tw

Abstract. With increasing popularity of cloud services, the microservices architecture has been gaining more attention in the software development industry. The idea of the microservices architecture is to use a collection of loosely coupled services to compose a large-scale software application. In traditional monolithic architecture, by contrast, every piece of code is put together, and the application is developed, tested, and deployed as a single application. Obviously, it is challenging for the traditional architecture to scale properly. In this research, we implemented a social data analytics platform based on the microservices architecture over DC/OS. Specifically, our data analytics service is built by composing many open-source software including Spark, Kafka, and Node.js. On streaming processing, our platform offers a visual interface to show the hottest hashtags of the most popular user posts from an online forum. On batch processing, our platform is able to show the statistics such the top-10 liked or commented posts and the gender counts of the posters. The experimental results show that our data analytics platform can do streaming processing and batch processing successfully and reveal useful analytical results.

Keywords: Cloud computing · Data analytics · Microservices

1 Introduction

In recent years, data analytics has been an extremely hot topic in all kinds of application fields. Some examples include adapting the promotion strategy according to online shoppers' behaviors, targeted advertisement based on social TV analytics [1], and even monitoring the vital signs of the patients with the help of the Internet of Things (IoT) technology in healthcare organizations [2, 3]. In fact, data analytics goes beyond these usage scenarios and can reach every aspect of our daily life. When developing a software system for data analytics, one must choose a software development technique that makes the application easier to understand, develop, and test. An application software could include hundreds of features, and all of these are packaged into a single logical executable in the traditional *monolithic* architecture for software development. To deal with the ever-changing business needs, enterprise applications must be frequently amended in order to add new features or improve the performance. Moreover, enterprises need to scale and manage the applications in an effective way without shutting down the services. With these new challenges, developing software using the traditional monolithic architecture may not be a good choice. For example,

L. Barolli et al. (Eds.): NBiS 2018, LNDECT 22, pp. 673–683, 2019.
https://doi.org/10.1007/978-3-319-98530-5_58

since the application is built as a single unit, the developer must rebuild the whole application even when only a single line of code is modified. It is also risky during the process of upgrading the monolithic application since the whole service may become unusable if the updated application goes wrong and would not start properly. These problems have forced enterprises to seek for new solutions to developing highly manageable and robust software.

In recent years, due to the growth of cloud computing services and the emergence of container technologies, the *microservices architecture* [4, 5] has attracted significant attention in software development communities. With the microservices architecture, a bulky monolithic application is disassembled into a set of independent and lightweight services. These services can then be developed, deployed, and maintained separately. Individual developers may use their own appropriate development framework to implement the services, which is not possible in the monolithic architecture where every piece of code is tightly-coupled with all others. Besides, with a standard protocol such as a RESTful API [6] as the interfaces of all the services, they can be invoked and reused by other services easily. When one of the services is to be updated with a new version, only the specific service needs to be redeployed. Because of its high modularity, availability, and scalability, developing software using the microservices architecture has been quickly accepted by many renowned enterprises such as Amazon, LinkedIn, and Netflix.

Based on the above observations and the potential of open-source software, in this research we follow the microservices architecture to develop a social data analytics platform on top of the Data Center Operating System (DC/OS) [7]. Our platform consists of many frameworks and technologies, including Apache Spark [8], Apache Kafka [9], Apache Mesos [10], Docker [11], MongoDB, and Node.js. To ease the job of scaling the system, we package the services of our data analytics application into a Docker image. Whenever a new node is added into the platform, our application can be deployed to the new node automatically and quickly by executing the Docker image. We choose one of the on-line forums in Taiwan as the target for analysis, and show the discussion trends with some related statistics on the discussion topics.

The rest of this paper is organized as follows. In Sect. 2 we describe the technological background to build a data analytics platform. Section 3 introduces our system architecture, and Sect. 4 describes our preliminary results. Finally, Sect. 5 concludes our work and gives some future directions.

2 Background and Related Work

2.1 Technological Background

Data Center Operating System (DC/OS). DC/OS is an open-source operating system for data centers, released by Mesosphere. It simplifies the job of managing a large and distributed data center for enterprises. At the core of DC/OS is the Apache Mesos, which is a distributed systems kernel that abstracts hardware resources and provides applications with APIs for resource management and scheduling across a data center.

Many modern workloads and frameworks can run on Mesos, including Hadoop [12], Docker, Spark, Kafka, Cassandra [13], etc. In DC/OS, Marathon [14] is used as the container orchestration engine to manage the operations of Docker containers. The high-availability feature of DC/OS is supported by Apache ZooKeeper [15], which offers redundant services to cope with master failures.

Microservices Architecture. Microservices architecture is a software development methodology that tries to decompose a large and complex application into a suite of independently deployable, small, modular services. Each service usually focuses only on a specific goal and runs a unique process. All the services work together to fulfill a business goal by communicating with each other through a well-defined and lightweight mechanism such as RESTful APIs. Since each service can be developed, tested, and deployed independently, we may modify and redeploy the updated services without having to rebuild and redeploy the entire application. However, under the microservices architecture, an application may consists of hundreds of services which must be deployed distributedly. Thanks to the container technologies (such as Docker and LXC [16]) and the orchestration platforms (such as Kubernetes [17] and Mesos), the complexity of deploying of a large number of services is now greatly reduced.

Container Technology. The container technology belongs to the OS-level virtualization, which allows the co-existence of multiple isolated user-space instances called containers. In each container, we can run programs under the allocated resources for that container. Since the container technology does not require a hypervisor to manage the virtual machines, containers are lightweight and outperform hypervisor-based virtual machines in terms of speed and simplicity [18]. *Linux Containers* (LXC) [16] is a toolkit for users to work with containers, which automates the management of containers and enables users to build their own containers. Docker, one of the popular container technologies, was originally based on LXC for resource isolation. Generally speaking, Docker fits well with the microservices architecture because each service can be packaged into a Docker image and then executed in a container.

Apache Kafka. Apache Kafka is a distributed messaging system originally developed by LinkedIn, which has now been widely used by many companies to handle multiple types of data pipes and messaging systems. The operation of Kafka is based on the *publish/subscribe* model. A *producer* publishes messages as a topic to the Kafka cluster consisting of redundant brokers, and multiple *consumers* consume the messages by subscribing to specific topics. Kafka has the advantages of high throughput, low latency, and fault tolerance.

Apache Spark. Apache Spark is a unified analytics engine for large-scale data processing [8]. The core technology of Spark is *Resilient Distributed Dataset* (RDD) [19], which is a distributed memory abstraction for in-memory computations. Spark features a set of libraries for different applications including SQL, machine learning, graph, and Spark Streaming. Among these libraries, Spark Streaming is used for processing real-time streaming data. A variety of data sources can be consumed by Spark Streaming, such as Kafka, HDFS, Twitter, etc.

2.2 Related Work

With regard to the increasing popularity of developing and deploying microservices in the cloud, in [20] Fazio *et al.* discussed some open issues on scheduling and resource management of the microservices. For instance, with heterogeneous performance requirements among the microservices, there exists the problem of heterogeneous configurations and resource allocations. This would require new research to focus on developing techniques to model, represent, and query configurations of microservices and data center resources. In [21] Hill *et al.* proposed a two-stage architecture of a community healthcare management system that utilizes IoT and microservices technologies. Their simulation results showed that with data filtering at the edge of the network, both the data traffic and the time spent on processing the messages can be reduced. They concluded that adopting the microservices architecture has several benefits, including more responsive to emerging requirements, greater software resilience, and enforced data privacy by specific services. In [22] Le *et al.* reconstructed an old monolithic database system with the microservices architecture, which serves the purpose of storing data for an NSF project. The authors concluded that the microservice architecture is suitable to a cloud-based system. In the cloud, each service could run its own instance of a server with the capacity that exactly fits the requirement of the service, which ultimately optimizes hardware usage and efficiency.

In the literature many researchers have implemented big data analytics platforms. Song *et al.* built a storage mining system based on Hadoop and Cassandra [23]. On top of that, they deployed the RHadoop framework to do machine learning. In [24] Shyam *et al.* used Apache Spark to analyze streaming data of a smart grid system. In [25] Nastic *et al.* proposed a unified cloud and edge data analytics platform with a reference architecture and a serverless application execution model, where user-defined functions are seamlessly and transparently hosted and managed by a distributed platform. By discussing use case scenarios from IoT mobile healthcare, the authors demonstrated the need for consolidating cloud- and edge-based data analytics techniques, which is exactly the main objective of their research. [26] is our previous work where we implemented a restaurant recommender system using open-source software including Mesos, Kafka, and Spark Streaming. Here in this work, we move everything onto DC/OS and utilize Docker to package and deploy the services, and then run social data analytics.

3 System Architecture

In this section, we will describe the system architecture of our social data analytics platform in detail. We mainly use DC/OS to manage all the hardware resources in our computer cluster. A number of open-source software including Spark, Node.js, Kafka, and MongoDB are used to build the data analytics platform based on the microservices architecture. The input to our platform is the text-based user posts from the Dcard website. From each user post we can get its associated hashtags, the number of likes, comments, as well as the author's school, department, and gender. With these data, we can do statistical analysis and display the results on a web page.

3.1 Overview

The system architecture is shown in Fig. 1. From this figure we can see that the system mainly consists of two parts, including the *offline processing* and the *streaming processing*. The details of the two parts will be explained in the next two subsections. Here we give a quick overview of the operations in the system. First, a web crawler written in JavaScript is scheduled to fetch the source data from the Dcard website. After the source data is fetched, we do preprocessing on the data to get rid of unnecessary data fields. Then, on the offline processing part, the preprocessed data is sent to MongoDB for persistent storage and creating statistics. On the streaming processing part, we use Kafka and Spark Streaming to analyze the popular hashtags from the user posts in real-time. Finally, we are able to see the statistical results from the database and the visualized popular hashtags at the web client.

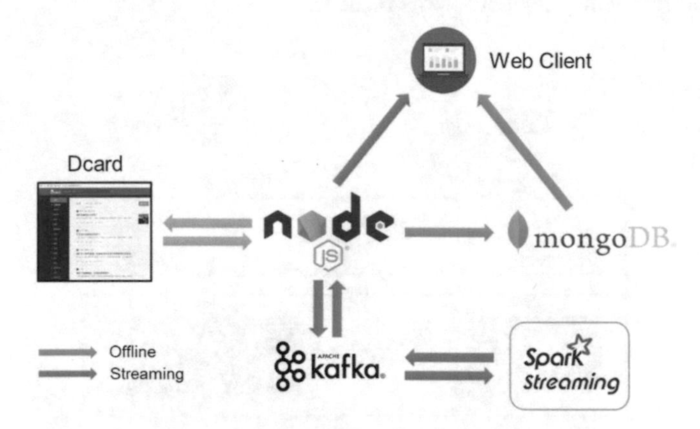

Fig. 1. System architecture.

3.2 Offline Processing

The purpose of the offline processing is to generate statistical results on a daily basis. Therefore, the crawler program is scheduled to fetch the most popular user posts once per day. Figure 2 shows the data flow of the offline processing. On the data flow, our program carries out the following tasks step by step:

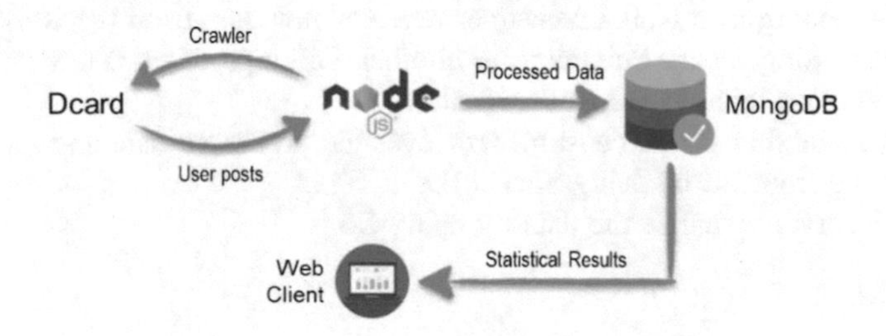

Fig. 2. The data flow of the offline processing.

1. Execute the crawler program to get the list of the article ID of the most popular user posts from the Dcard website.
2. For each article ID, the crawler program further fetches the title and time of the post, gender of the poster, the number of comments, and the number of likes.
3. Store the above information into the MongoDB database.
4. Viewer at the web client specifies specific URLs to query the database and then the statistical results are displayed at the frontend.

3.3 Streaming Processing

The purpose of the streaming processing is to reveal the latest trend on the forum. Therefore, the crawler program is scheduled to fetch the hashtags of the most popular user posts every minute. The data flow of the streaming processing is shown in Fig. 3, in which the following steps are performed:

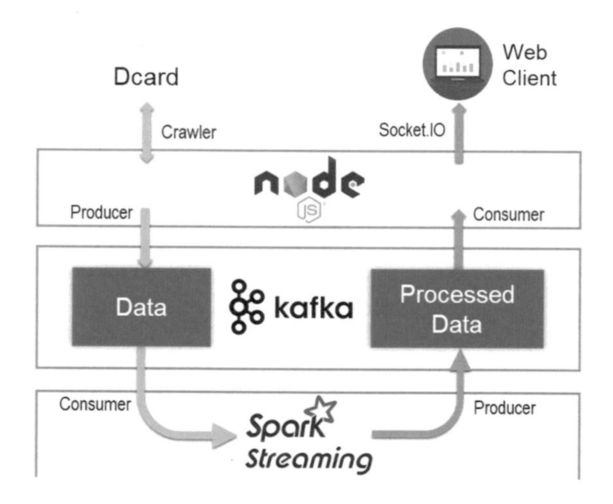

Fig. 3. The data flow of the streaming processing.

1. Schedule the crawler program to fetch the hashtags of the latest most popular user posts from the Dcard website.
2. Node.js as a Kafka producer feeds the raw data as a topic to the Kafka cluster.
3. Spark Streaming as a Kafka consumer receives raw data from the Kafka cluster.
4. After processing, Spark Streaming as another Kafka producer sends the processed data as another topic to the Kafka cluster.
5. Node.js as another Kafka consumer receives the processed data and then forwards them to the frontend by using Socket.IO.
6. The web server visualize the data by using D3.js.

4 Implementation and Preliminary Results

4.1 Our Experimental Cluster

Two desktop PCs and five rack-mount servers constitute our experimental computer cluster. Since servers are more powerful than PCs, we deployed Mesos agents on the servers; one PC serves as the bootstrap node for DC/OS and the Node.js server, while the other PC serves as the Mesos master. Table 1 shows the hardware resources of our computer cluster. The host operating system for the cluster is CentOS 7.3, and on top of that we installed DC/OS. As for the programming languages, we used JavaScript and Python to code our programs for Node.js and Spark, respectively. Besides, in Node.js we created the Kafka producer and the Kafka consumer using the no-kafka library [27]. As for the deployment of Spark and Kafka in the experimental cluster, they can be deployed automatically by DC/OS. Figure 4 shows an overview of the deployed services in our DC/OS cluster.

Table 1. Hardware resources of our computer cluster

Node	CPU (core)	Memory (GB)	Disk (GB)
Bootstrap & Node.js server	4	4	256
Master	4	8	256
Agent 1	8	18.4	256
Agent 2	8	18.4	256
Agent 3	8	14.5	256
Agent 4	8	18.4	256
Agent 5	4	46	1536

4.2 Experimental Results

On the offline processing, the crawler program is scheduled every midnight to fetch all the most popular user posts on that day. Along with the user posts, our crawler program also fetches the number of comments and the number of likes of each user post. Besides, we also use the article ID to get the gender of the poster. During the execution of the crawler program, every piece of the fetched data is reformatted and then inserted into the MongoDB database. From the web client, we can specify specific URLs to show different statistics. This is achieved by using the URL routing [28] technique. In Fig. 5 there are three examples with different URLs, showing the top-30 hottest posts, gender counts of the posters, and the top-10 commented posts.

On the streaming processing, Spark Streaming acting as a Kafka consumer, can get the latest hottest user posts. For each user post, we use the `flatMap` function to generate an array of hashtags. By using the `map` and `reduceByKey` functions, the number of occurrence of each hashtag is calculated. Finally, we use the transform function to output the top-10 popular hashtags, which are then sent to Kafka using the topic of "processed data". Figure 6(a) shows a snapshot of the output from Spark Streaming,

NAME ▲	STATUS		CPU	MEM	DISK
hdfs		Running (13/1)	8.5	34.4 GiB	512.7 GiB
kafka		Running (6/1)	10	18.8 GiB	43.9 GiB
mongodb		Running (1/1)	0.5	1 GiB	0 B
spark		Running (1/1)	1	1 GiB	0 B

Fig. 4. Overview of the deployed services in our DC/OS cluster.

URL: http://127.0.0.1:9000/latest

{"_id":"593e22dc457bac2a6071eb03","id":"226586338","title":"Colourpop如何壓出完美的花紋？！","latest":true,"commentCount":51,"likeCount":1259,"time":"20170611","__v":0},
{"_id":"593e22dc457bac2a6071eb04","id":"226585162","title":"#圖 寶貝妳到底讓我看了什麼...","latest":true,"commentCount":37,"likeCount":1190,"time":"20170610","__v":0},
{"_id":"593e22dc457bac2a6071eb05","id":"226586962","title":"#圖 ㄇㄚˊ幾兔不專業食記分享💕","latest":true,"commentCount":39,"likeCount":1072,"time":"20170611","__v":0},
{"_id":"593e22dc457bac2a6071eb06","id":"226588137","title":"#圖 關於愛情","latest":true,"commentCount":44,"likeCount":1017,"time":"20170611","__v":0},

(a)

URL: http://127.0.0.1:9000/gender

[{"mCount":852,"fCount":1435}]

(b)

URL: http://127.0.0.1:9000/c10

[{"id":"226584775","comments":319},{"id":"226585953","comments":214},{"id":"226584739","comments":122},
{"id":"226586303","comments":118},{"id":"226584801","comments":117},{"id":"226587099","comments":83},
{"id":"226585968","comments":82},{"id":"226585851","comments":77},{"id":"226587155","comments":64},
{"id":"226593341","comments":64}]

(c)

Fig. 5. Statistics displayed with different URLs. (a) top-30 hottest posts (b) poster gender counts (c) top-10 liked posts.

which contains the top-10 hashtags. Note that since the hashtags are Chinese characters, they are garbled in the output. The Node.js server plays the role of a Kafka consumer which subscribes to the topic of "processed data". When the Node.js server receives a new piece of processed data, the data is forwarded to the web client in real-time by Socket.IO. Figure 6(b) shows the partial content of two messages received at the web client through Socket.IO, where name and size refer to a hashtag and its number of occurrence, respectively. At the web client, these top-10 hashtags are then visualized by using D3.JS. Figure 7 shows an example of the visualization result, where only the names of the hashtags are displayed. The number of occurrence of the hashtag will appear if the user moves the mouse cursor over the circle. Also note that whenever a new message is received at the web client, the visualization result will be updated.

```
(u'\u4e00\u500b', 376)
(u'\u4ec0\u9ebc', 342)
(u'\u7684\u4eba', 209)
(u'\u9019\u500b', 161)
(u'\u500b\u4eba', 150)
(u'\u9084\u6709', 133)
(u'\u89ba\u5f97', 108)
(u'\u9019\u7a2e', 100)
(u'\u559c\u6b61', 97)
(u'\u56e0\u70ba', 91)
```

(a)

```
NewMsg: [{"name":"一個","size":272},{"name":"個人","siz
2},{"name":"還有","size":99},{"name":"這個","size":115}
":"因為","size":70},{"name":"那個","size":66}]
NewMsg: [{"name":"什麼","size":303},{"name":"這個","siz
6},{"name":"請問","size":148},{"name":"個人","size":98}
":"覺得","size":92},{"name":"還有","size":7}]
```

(b)

Fig. 6. (a) A snapshot of the output from Spark Streaming (b) Two messages received at the web client through Socket.IO.

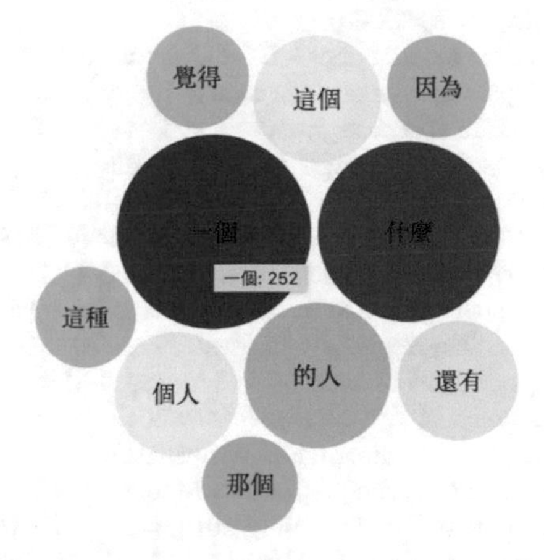

Fig. 7. Two messages received at the web client through Socket.IO.

5 Conclusions and Future Work

In this research, we implemented a social data analytics platform based on the microservices architecture which can do real-time processing of streaming data and offline processing of batch data from an anonymous online forum for college students. On the streaming processing part, the most popular hashtags from the hottest user posts can be visually displayed in real-time. On the offline processing part, statistical results such as the top-30 hottest posts, poster gender counts, and top-10 liked posts can be revealed. Our platform is implemented using open-source software including Spark, Kafka, MongoDB, and Node.js. To scale the capacity of our system, we package the services of our data analytics application into a Docker image, which can be instantiated on demand. In the near future, we would like to improve the user interface that displays the statistical results from batch processing. Moreover, we could try to process other social data such as Facebook or Twitter, to test the system performance when the data size becomes huge.

References

1. Hu, H., Wen, Y., Gao, Y., Chua, T.S., Li, X.: Toward an SDN-enabled big data platform for social TV analytics. IEEE Netw. **29**(5), 43–49 (2015). https://doi.org/10.1109/MNET.2015.7293304
2. Enshaeifar, S., Barnaghi, P., Skillman, S., Markides, A., Elsaleh, T., Acton, S.T., Nilforooshan, R., Rostill, H.: The Internet of Things for dementia care. IEEE Internet Comput. **22**(1), 8–17 (2018)
3. Swedberg, C.: Japanese Hospital tests BLE beacons to track patient–staff interactions. RFID J. (2018). http://www.rfidjournal.com/articles/view?17294. Accessed 25 April 2018
4. Newman, S.: Building Microservices. O'Reilly Media, Inc, Newton (2015)
5. Thönes, J.: Microservices. IEEE Softw. **32**(1), 116 (2015)
6. Fielding, R.T.: Architectural styles and the design of network-based software architectures. University of California, Irvine (2000)
7. The definitive platform for modern apps - DC/OS. https://dcos.io/. Accessed 4 May 2018
8. Apache Spark. https://spark.apache.org/. Accessed 4 May 2018
9. Apache Kafka. https://kafka.apache.org/. Accessed 4 May 2018
10. Apache Mesos. http://mesos.apache.org/. Accessed 4 May 2018
11. Docker. https://www.docker.com/. Accessed 4 May 2018
12. Apache Hadoop. http://hadoop.apache.org/. Accessed 4 May 2018
13. Apache Cassandra. http://cassandra.apache.org/. Accessed 4 May 2018
14. Marathon. https://mesosphere.github.io/marathon/. Accessed 4 May 2018
15. Apache ZooKeeper. https://zookeeper.apache.org/. Accessed 4 May 2018
16. Linux Containers. https://linuxcontainers.org/. Accessed 4 May 2018
17. Kubernetes. https://kubernetes.io/. Accessed 6 May 2018
18. Eder, M.: Hypervisor- vs. container-based virtualization. In: Seminars Future Internet (FI) and Innovative Internet Technologies and Mobile Communications (IITM), Winter Semester 2015/2016, Munich, Germany, July 2016, pp. 1–7 (2016)
19. Zaharia, M., Chowdhury, M., Das, T., Dave, A., Ma, J., McCauley, M., Franklin, M.J., Shenker, S., Stoica, I.: Resilient distributed datasets: a fault-tolerant abstraction for in-memory cluster computing. In: 9th USENIX Conference on Networked Systems Design and Implementation, San Jose, CA, 25–27 April 2012 (2012)
20. Fazio, M., Celesti, A., Ranjan, R., Liu, C., Chen, L., Villari, M.: Open issues in scheduling microservices in the cloud. IEEE Cloud Comput. **3**(5), 81–88 (2016)
21. Hill, R., Shadija, D., Rezai, M.: Enabling community healthcare with microservices. Paper presented at the 16th IEEE international conference on ubiquitous computing and communications, Guangzhou, China, 12–15 December 2017 (2017)
22. Le, V.D., Neff, M.M., Stewart, R.V., Kelley, R., Fritzinger, E., Dascalu, S.M., Harris, F.C.: Microservice-based architecture for the NRDC. In: 2015 IEEE 13th International Conference on Industrial Informatics (INDIN), 22–24 July 2015, pp. 1659–1664
23. Song, Y., Alatorre, G., Mandagere, N., Singh, A.: Storage mining: where IT management meets big data analytics. In: 2013 IEEE International Congress on Big Data, June 27–July 2 2013. pp. 421–422
24. Shyam, R., Ganesh, H.B.B., Kumar, S.S., Poornachandran, P., Soman, K.P.: Apache Spark a big data analytics platform for smart grid. Procedia Technol. **21**, 171–178 (2015)
25. Nastic, S., Rausch, T., Scekic, O., Dustdar, S., Gusev, M., Koteska, B., Kostoska, M., Jakimovski, B., Ristov, S., Prodan, R.: A serverless real-time data analytics platform for edge computing. IEEE Internet Comput. **21**(4), 64–71 (2017)

26. Lee, C.H., Lin, C.Y. Implementation of Lambda architecture: a restaurant recommender system over apache Mesos. In: 2017 IEEE 31st International Conference on Advanced Information Networking and Applications (AINA), 27–29 March 2017, pp. 979–985
27. Apache Kafka 0.9 client for Node. https://github.com/oleksiyk/kafka. Accessed 8 May 2018
28. Routing. https://expressjs.com/en/guide/routing.html. Accessed 25 May 2018

Application of Independent Component Analysis for Infant Cries Separation

Chuan-Yu Chang(✉), Chi-Jui Chen, and Ching-Ju Chen

Department of Computer Science and Information Engineering,
National Yunlin University of Science and Technology, Douliu, Taiwan
{chuanyu,m10517042,cjchen}@yuntech.edu.tw

Abstract. The research on analysing infant crying has received many attentions in recent years. In our prior work, a baby crying translation method called infant crying translator was proposed and showed high recognition accuracy. However, in a real environment, there may be more than one baby crying. These mixed cries will seriously affect the accuracy of recognition. In order to isolate these mixed cries, the independent component analysis was adopted herein. Experimental results show that the proposed method can separate out the mixed cries and greatly improves the recognition rate of infant crying translator. The recognition rate increased from 34% to 68%.

1 Introduction

In recent years, research on sound identification has developed rapidly. Since infants can only express their needs by crying. There are many researchers proposed methods to identify the meaning of crying. However, most of these methods can only identify a baby crying in a quiet environment. In a hospital nursery, more than one baby will cry at the same time. Mixed cries may affect the correctness of the recognition. Blind Sources Separation (BSS) [1] is the way to solve multi-source environment interference. We can use BSS method to restore the original signals even if the characteristics of the original signal and mixed signal are unknown. The most famous method of BSS example is the cocktail party problem [2]. In a noisy environment, it usually contains a variety of sounds, including vocals, music, and ambient noise. People can only pay attention to those interested sound and be able to receive the desired sound from a noisy environment. The independent component analysis [3] with ability to solve the cocktail party problem has been widely applied to a variety of fields, including image processing and audio signal processing. This paper will apply the fast independent component analysis to separate and restore infant cries from the multi-sourced environment. Experimental results show that the proposed method can separate out the mixed cries and greatly improves the recognition rate of infant crying translator. The recognition rate increased from 34% to 68%.

L. Barolli et al. (Eds.): NBiS 2018, LNDECT 22, pp. 684–690, 2019.
https://doi.org/10.1007/978-3-319-98530-5_59

2 Proposed Method

Figure 1 shows the flow chart of the proposed method. There are two major submodules: preprocessing and signal separation. The preprocessing consists of four steps including normalization, endpoint detection, centering, and whitening. The signal separation consists of FastICA, similarity calculation, and crying recognition [4]. Details of these steps are described below.

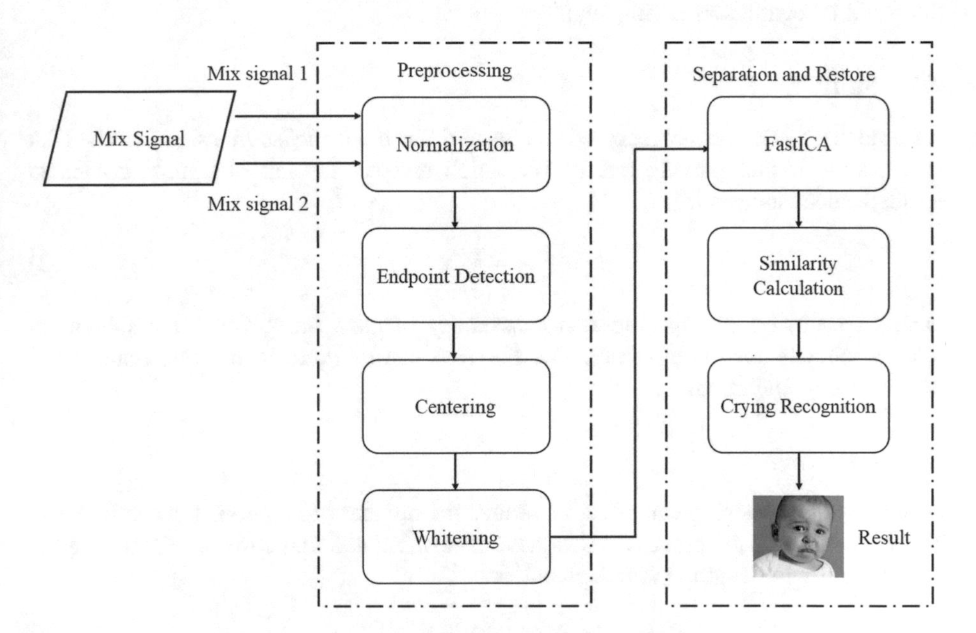

Fig. 1. System flow chart of the proposed method.

2.1 Normalization

The sampling rate and bit resolution of the audio signal are normalized to the same form. In this work, the records crying signals are converted to WAV (Waveform Audio Format), with the bit resolution and sampling rate at 16 bit and 44.1 kHz, respectively.

2.2 Endpoint Detection

Normally, the crying signals usually contain many non-crying segments, such as silence or environmental noises. Hence, end-point detection is adapted to detect the end-points of the cry, thereby eliminating useless signals and establishing crying segments for analysis. This method can reduce the amount of data and computing time.

The intensity of a frame is computed as:

$$Intensity = \sum_{i=0}^{N-1} |S_i| \tag{1}$$

where N is the quantity of sampling point of a frame, and S_i is the intensity of the i-th sampling point in the frame. The frame with *Intensity* higher than a predefined threshold is regarded as crying segment.

2.3 FastICA

Assume that $\mathbf{x}$ is n-channel received signals and $\mathbf{S}$ is n-channel source signals. The ICA is to find a de-multiplexing matrix $\mathbf{W}$, which reverses the mixed signals $\mathbf{x}$ back to n independent sources [5].

$$\mathbf{S} = \mathbf{W}\mathbf{x} \tag{2}$$

FastICA estimates the specific non-gaussianity of each independent component to reduce computational complexity. The FastICA can be divided into two main parts: preprocessing and iterative.

2.4 Centering

Centering and whitening can effective reduce the number of iterations for FastICA [6]. These steps ensure the processed data have zero-mean and unit variance. Centering the i-th channel with length N is calculated as:

$$\bar{x}_i(j) = x_i(j) - E\{x_i\} \tag{3}$$

where j = 1, 2, 3, …, N (sample length). $\bar{x}_i(j)$ is the centered data and $E\{x_i\}$ is the mean of the i-th channel signal.

2.5 Whitening

Whitening is a linear transformation that transforms a set of vectors with a known covariance matrix into a set of new independent vectors. For a $n \times n$ covariance matrix $C_{\mathbf{X}}$ of the centered signals $\bar{\mathbf{x}}$, its eigenvalue decomposition can be defined as

$$C_{\mathbf{X}}=E\{\ \overline{\mathbf{x}\mathbf{x}}^T\} = \mathbf{R}\mathbf{D}\mathbf{R}^T \tag{4}$$

where $\mathbf{R}$ is an orthogonal matrix formed by n eigenvector, and $\mathbf{D}$ = diag(λ_1, λ_2, …, λ_n) is a diagonal matrix formed by n eigenvalues.

Whitened data **Z** is obtained by

$$\mathbf{Z} = \mathbf{D}^{-1/2}\mathbf{R}^{\mathrm{T}}\overline{\mathbf{x}} \tag{5}$$

After the whitening transformation, each channel signal become non-correlated, and the covariance matrix of the whiten data becomes an identity matrix.

2.6 Fast Independent Component Analysis

Whitened data is used in the iterative step to refresh the demixing matrix weight. The weight vector is updated iteratively during a learning process. The negentropy J is approximated by

$$J(\mathrm{w}^{\mathrm{T}}\mathbf{Z}) \approx [E\{G(\mathrm{w}^{\mathrm{T}}\mathbf{Z})\} - E\{G(\mathrm{u})\}]^2 \tag{6}$$

where

$$G(u) = \frac{1}{a}\log\cosh(au) \tag{7}$$

u is a standardized Gaussian variable and G is the non-quadratic function, and a is a constant between 1 and 2. In order to maximize the negentropy, the estimation of non-gaussianity of **w**, differentiation with respect to **w** and further simplifications are performed. The operation of updating a weight vector by Newton iteration and approximate negentropy are represented as Eq. (8).

$$\mathbf{w}^{+} = E\{\mathbf{Z}g(\mathbf{w}^T\mathbf{Z})\} - E\{g'(\mathbf{w}^T\mathbf{Z})\}\mathbf{w} \tag{8}$$

where g is the derivative of G. The initial vector is usually set to a unit one. To prevent vectors of different components from converging to the same maxima, decorrelation is needed when the vector been found. The Gram-Schmidt orthogonalization is represented as Eqs. (9) and (10), respectively.

$$\mathbf{w}_{p+1}^{+} = \mathbf{w}_{p+1} - \sum_{k=1}^{p} (\mathbf{w}_{p+1}^{T}\mathbf{w}_k)\mathbf{w}_k \tag{9}$$

$$\mathbf{w}_{p+1}^{*} - \frac{\mathbf{w}_{p+1}^{+}}{\| \mathbf{w}_{p+1}^{+} \|} \tag{10}$$

where $\mathbf{w}_{p+1}^{*}$ is deduced by eliminating the projections on other weight vectors and then being normalized. Iteration will stops if **w** and **w*** has same direction. According to the whitened data. Those independent components will be separated by the demultiplexing matrix **W**.

3 Experiment Settings

In general, cribs are placed side by side in nurseries of general hospitals or post-natal care centers. Figure 2(a) shows the layout in a real nursery. Since there may be more than one baby crying in the same time, multiple directional microphones and speakers are used to simulate a real environment of a postnatal care center or a nursery as shown in Fig. 2(b). Multiple cries were played at the same time. All the baby cries were collected from the National Taiwan University Hospital Douliou Branch. The recorded cry signals are Waveform Audio format (WAV) with 44.1 kHz sampling rate, 16 bits mono. In this paper, 45 cries were divided into 15 groups. Each group has three cries.

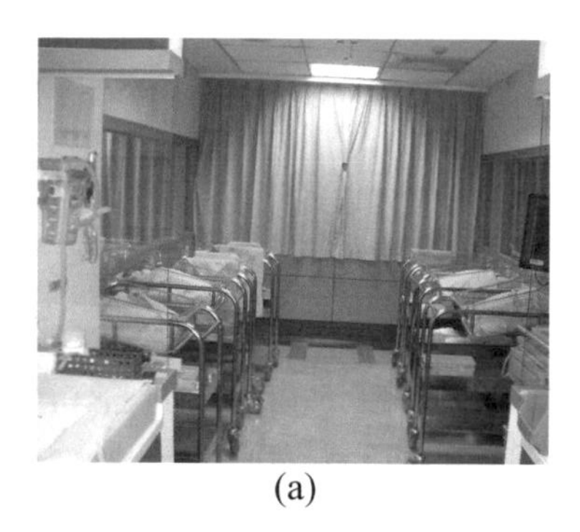

(a)

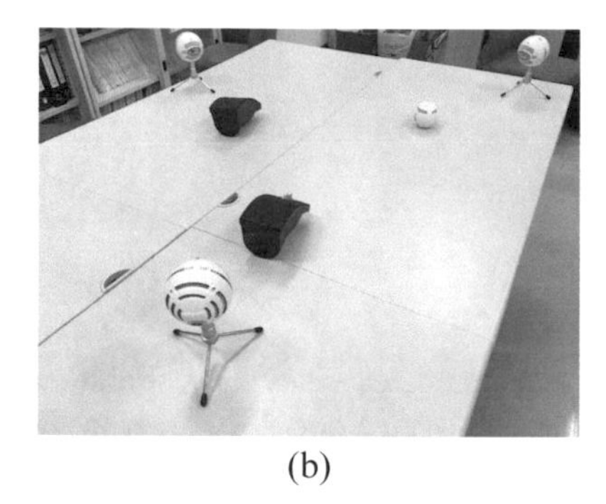

(b)

Fig. 2. (a) Nursery room, (b) simulation environment

3.1 Evaluation Criteria

To evaluate the quality of the restored infant cries, three similarity measurements including Euclidean distance [7], dynamic time warping [8], and structural similarity [9] were adopted. In the time domain, the Euclidean distance between the original signal $\boldsymbol{x}$ and restored signal $\boldsymbol{y}$ can be calculated as Eq. (11).

$$d(x, y) = \sqrt{\sum_{i-0}^{N} (x_i - y_i)^2} \tag{11}$$

where N denotes the number of sampling point. In addition, we can also compare the similarity of the original signal $\boldsymbol{x}$ and restored signal $\boldsymbol{y}$ in frequency domain. The fast Fourier Transform was adopted to convert the signal waveform to the frequency domain.

The dynamic time warping algorithm was widely used to measure the similarity of two signals with different length. The time-normalized distance between the original and restored crying waveform is defined as Eq. (12).

$$D(x,y) = \left[\frac{\sum_{s=1}^{k} d(p_s) \cdot w_s}{\sum_{s=1}^{k} w_s} \right] \tag{12}$$

where $w_S > 0$ is a weighting coefficient, $d(p_s)$ denotes the distance between the s-th point in the original and restored crying waveform. The structural similarity of the original and restored crying spectrums [10] is defined as Eq. (13).

$$SSIM(x,y) = \frac{(2\mu_x\mu_y + C_1)(2\sigma_{xy} + C_2)}{(\mu_x^2 + \mu_y^2 + C_1)(\sigma_x^2 + \sigma_y^2 + C_2)} \tag{13}$$

where μ_x and μ_y denotes the mean value of original (x) and restored (y) crying spectrums, respectively. σ_x and σ_y denotes the standard deviation. σ_{xy} is the covariance of x and y. In this paper, C_1 and C_2 is set as 6.5025 and 58.5225, respectively.

4 Experimental Results

In this paper, the FastICA was adopted to isolate the infant cries from a noisy environment. Figure 3 shows a case of restored signals. Obviously, the waveform of the restored signal is highly overlay to the original signal. This indicates the proposed method can isolate those blind signals well. Table 1 shows the similarities and cry recognition between the restored cry and the original cry of the principal component analysis (PCA) [11] method and the proposed FastICA method. The similarities obtained by the PCA method were lower than 58%. On the other hand, the similarity of the proposed FastICA were all higher than 74% for different measurements. The recognition accuracy of the infant crying translator using the restored crying is 23% and 68% for PCA method and FastICA method, respectively.

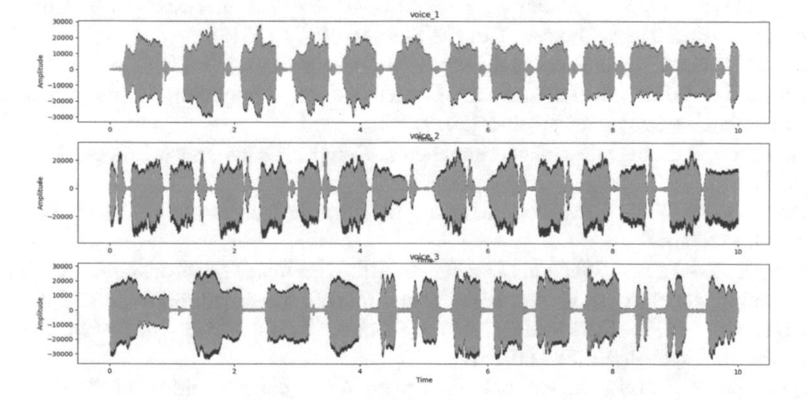

Fig. 3. Overlapping of restored signal and original signal

Table 1. Similarity and cry recognition of PCA and Fast ICA

	Time domain waveform	Frequency domain waveform	DTW	SSIM	Cry recognition
PCA	58.22%	36.88%	58.22%	26.44%	23%
FastICA	87.85%	81.18%	87.85%	74.09%	68%

5 Conclusions

Babies can only express their needs with cries before they speak. In recent years, many methods for analyzing infant crying have been proposed. However, most of these methods can only identify a baby crying in a quiet environment. In a hospital nursery, more than one baby will cry at the same time. Mixed cries may affect the correctness of the recognition. Therefore, this paper proposes a method based on fast ICA to isolate crying. Experimental results show that the proposed independent component analysis method is very effective in separating mixed signals. The separated cry signal was further translated by the infant crying translator, the accuracy rate was greatly improved.

References

1. Cardoso, J.F.: Blind signal separation: statistical principles. Proc. IEEE **86**, 2009–2025 (1998)
2. Bee, M.A., Christophe, M.: The cocktail party problem: what is it? How can it be solved? And why should animal behaviorists study it? J. Comput. Psychol. **122**, 235 (2008)
3. Hyvärinen, A., Hurri, J., Hoyer, P.O.: Independent component analysis. In: Natural Image Statistics, pp. 151–175 (2009)
4. Chang, C.Y., Hsiao, Y.C., Chen, S.T.: Application of incremental SVM learning for infant cries recognition. In: 18th International Conference on Network-Based Information Systems, pp. 607–610 (2015)
5. Yang, C.H., Shih, Y.H., Chiueh, H.: An 81.6 μW FastICA processor for epileptic seizure detection. IEEE Trans. Biomed. Circuits Syst. **9**, 60–71 (2015)
6. Sharma, R.K.: Time efficient architecture implementation of FastICA processor for EEG signals. In: 2017 8th International Conference on Computing, Communication and Networking Technologies, pp. 1–6 (2017)
7. Danielsson, P.E.: Euclidean distance mapping. Comput. Graph. Image Process. **14**, 227–248 (1980)
8. Keogh, E., Chotirat, A.R.: Exact indexing of dynamic time warping. Knowl. Inf. Syst. **7**, 358–386 (2005)
9. Wang, Z., Bovik, A.C., Sheikh, H.R., Simoncelli, E.P.: Image quality assessment: from error visibility to structural similarity. IEEE Trans. Image Process. **13**, 600–612 (2004)
10. Hargreaves, W.A., Starkweather, J.A., Blacker, K.H.: Voice quality in depression. J. Abnorm. Psychol. **70**, 218 (1965)
11. Hyvärinen, A., Jarmo H., Patrik O.H.: Principal components and whitening. In: Natural Image Statistics, pp. 93–130 (2009)

BLE Beacon Based Indoor Position Estimation Method for Navigation

Takahiro Uchiya(✉), Kiyotaka Sato, and Shinsuke Kajioka

Nagoya Institute of Technology,
Gokiso-chou, Showaku, Nagoya 460-8555, Japan
{t-uchiya,kajioka}@nitech.ac.jp,
k-sato@uchiya.nitech.ac.jp

Abstract. Bluetooth Low Energy (BLE) beacons are useful to estimate a user's location in indoor situations. Because our university has numerous BLE beacons, they are easy to use for indoor route navigation. For this study, we propose a position estimation method using BLE beacons. Using this method, we eliminate the existing difficulty of error accumulation of earlier systems and improve indoor location estimation accuracy. We designed an indoor position estimation method and conducted an estimation accuracy evaluation experiment comparing the proposed method and existing methods. The experiment results underscore the effectiveness of the proposed method.

1 Introduction

In recent years, services that use information related to one's current location, such as store search by Google Map or route navigation, are increasingly numerous and diverse. As methods of measuring the current position, satellite navigation such as Global Positioning System (GPS) and Quasi-Zenith Satellite System [1] exist, along with dead reckoning, which estimates relative movement from motion sensor information.

Our developed system [2, 3] supports on-campus tours such as open campus events. A system user can receive information related to the tour site according to the current location through voice dialogue. However, a shortcoming found in earlier research is poor accuracy of indoor position information. Earlier studies used a map function to display the current position and a university destination. Nevertheless, when acquiring location information indoors, an error occurs in the current location information, and the correct location cannot be displayed on the map.

The purpose of this study is to resolve those difficultics of earlier research and to enhance support for route guidance indoors. By introducing an indoor location information acquisition function into the system used earlier, we can resolve the inferior accuracy of the indoor position information, which is a shortcoming of the earlier method. An existing estimation method uses the strength of radio waves received from a wireless LAN. However, if the access points used for estimation are few and if the access point placement is dispersed, then the estimation accuracy might be low. Therefore, we estimate indoor location using Bluetooth Low Energy (BLE) radio waves of the Bluetooth standard. The BLE radio wave transmitters (BLE beacons), of

L. Barolli et al. (Eds.): NBiS 2018, LNDECT 22, pp. 691–699, 2019.
https://doi.org/10.1007/978-3-319-98530-5_60

which many are installed at our university, have lower introduction costs than wireless LAN access points.

2 Earlier Research

2.1 Prototype System

An open-campus event system [3] can support campus visits using an Android application and a server. Two modes exist for this system: one supports a visit during which academic staff members lead visitors; the other is for voice dialogue. This latter mode supports visits by the visitors alone.

After this system obtains the current position outdoors using GPS, it infers the position indoors using dead reckoning. We show the flow of walking path estimation using dead reckoning. For this process, we use the acceleration sensor and angular velocity sensor of the smartphone.

1. Detection of the terminal direction: First, an average acceleration value vector **m** is calculated with respect to the output of the acceleration sensor. Next, the rotation axis $p = (px, py, pz)$ and angle θ for rotating in the vertical direction $g = \{0, -1, 0\}$ are found. Finally, using rotation axis p and the angle θ, the system generates a rotation matrix that transforms the average value vector **m** into a vertically downward vector.
2. Estimation of walking distance: First, the system performs data smoothing. Next, the system performs peak detection. A peak is detected based on whether or not the slope of the average of k points is greater than a threshold value. Finally, the system calculates walking distance wd using the peak number and stride length. For this study, we use 1.39 m as the stride length. Therefore, wd = peak number × 1.39.
3. Estimation of direction of movement: The angular velocity sensor outputs the rotational speed. The system obtains the amount of change from the initial direction by time integration.
4. Estimation of the walking path: The system estimates the walking path using estimation of the walking distance and estimation of the direction of movement.

Through some experimentation related to position estimation, we ascertained that the outdoor position estimation using GPS has good accuracy, but indoor position estimation using dead reckoning has inadequate accuracy. Positional information error increases when the system uses dead reckoning to infer a person's position.

2.2 Revised Version

We proposed a method [4] that corrects positional information using a BLE beacon to resolve difficulties presented by earlier research efforts. We add a mechanism to manage the ID information and related guidance information of the beacon to the server. The smartphone system receives beacon information from a server at the time of starting. Moreover, the system can correct positional information and can obtain voice dialogue information. We implemented this system by extending the previous system. Herein, we developed the following functions.

- Acquisition of BLE beacon information
  We implemented a function for the acquisition of information from a BLE beacon using a smartphone.
- Management of BLE beacon information
  Information related to BLE beacons must be managed to use the proposed method. Therefore, we implemented a mechanism that managed the following information related to a beacon in the server.

Item: Beacon ID, Corrected position, Guidance information, Location information, Tags that a developer can use arbitrarily

- Correction of the current location
  For a location, the prototype system requires coordination of the position and direction. The system corrects the BLE beacon position and corrects the direction using magnetic sensors of a smartphone.

3 Proposed Method

3.1 Overview

Based on the experiment result and the user's opinion of the revised version of system, we did not achieve highly accurate position estimation. Therefore, we surmised that an indoor position estimation system that estimates a user's position without using the motion sensor built in the smartphone is necessary and that presenting the present position on the map showing the indoor composition to the visitor is necessary. We therefore propose a new indoor position estimation system using only information obtained from BLE beacons. The proposed indoor position estimation system estimates the current position from the radio field intensity from the BLE beacon and the indoor BLE beacon arrangement information.

3.2 System Architecture

The system is configured using the Android smartphone application and management server (Fig. 1).

- **Information management server**
  The server manages map images, beacon arrangement information, etc. The information to be managed is presented below.
  - Image data showing the indoor composition
  - Beacon identification number (major value)
  - Information on beacon location in these image data (horizontal value, x; vertical value, y)
- **Smartphone application**
  We explain the function of the class of the implemented application.

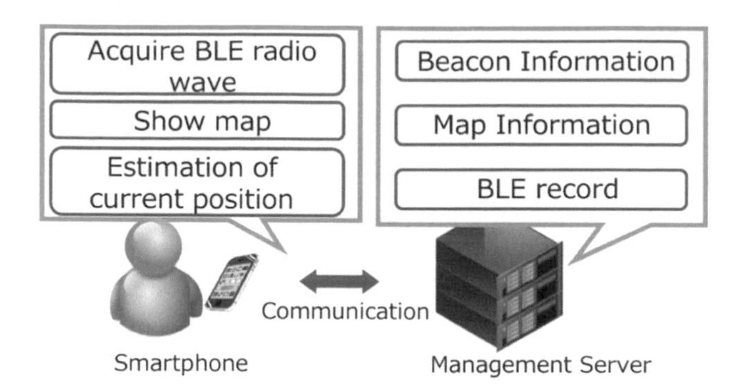

Fig. 1. System architecture.

- Main Activity
  In the core class of the system, the information to be provided to the user is displayed on the user interface.
- BLE
  Class that observes radio waves from BLE beacons and which extracts information from the observed radio waves.
- Estimate Position
  Class for operating the position estimation algorithm
- MapManager
  A class that manages image files, BLE beacon identification numbers, and coordinates.

3.3 Map Construction

In earlier studies, Google map was used to display the map. In addition, a geodetic system similar to the GPS called WGS 84 was used to display the current position. Position estimation by walking route estimation system was performed in this coordinate system. However, when presenting location information indoors using Google Map, it was thought that the indoor structure was not known in Google Map, and that the indoor route guidance support was inadequate.

Therefore, in the proposed method, instead of using Google Map, we prepared image data such that the indoor composition can be known. Indoor position information is presented by presenting the position information related to the image. In an earlier study, the estimated position was expressed by latitudc and longitude using WGS84. However, in the proposed method, pixels of image data are taken as the coordinate system. The beacon arrangement position and estimated position are represented respectively by binary (x, y) with x in the horizontal direction and y in the vertical direction relative to the upper left pixel of the image data of the indoor composition (Fig. 2).

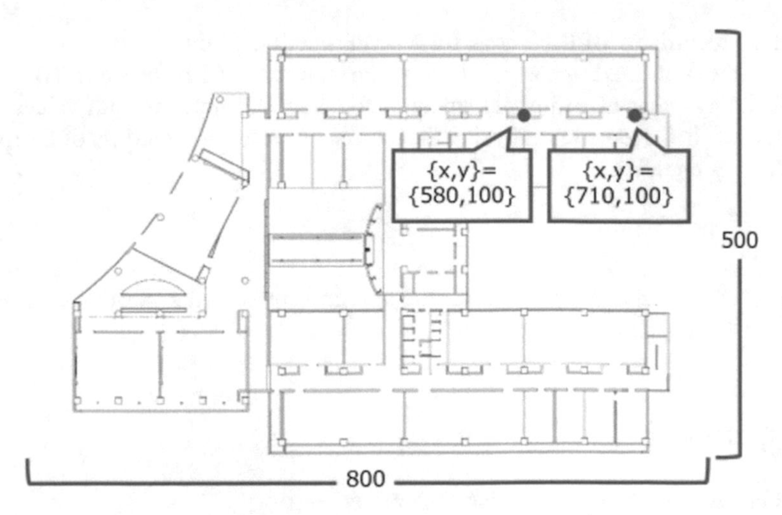

Fig. 2. Format of map and position expression.

3.4 Position Estimation Algorithm

Position estimation in the proposed method uses three values: the large value of radio field strength obtained from a plurality of BLE beacons, beacon arrangement information, and indoor route information. From the value of the radio field strength obtained from the BLE beacon, the system calculates the propagation loss distance and finds the distance from the corresponding BLE beacon. Then, the indoor terminal location is estimated using the beacon arrangement information and the indoor route information.

The flow of position estimation from observation data is shown below (Fig. 3).

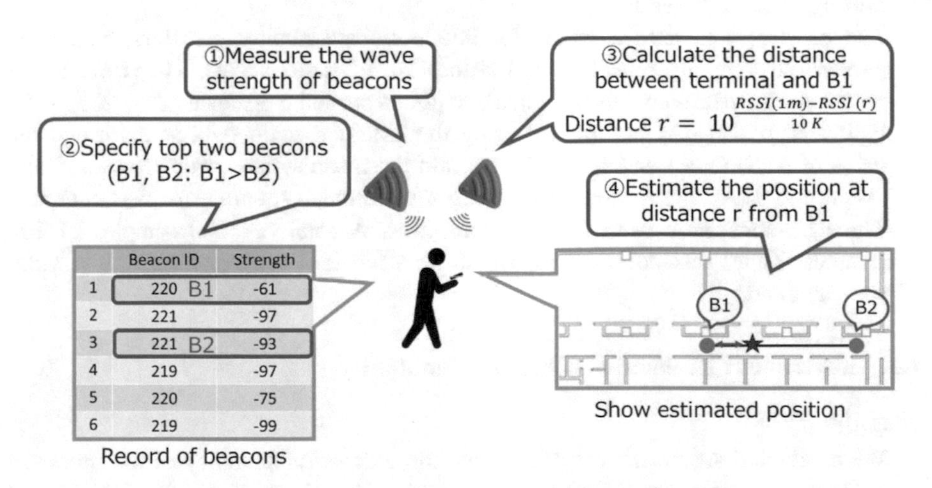

Fig. 3. Overview of the position estimation algorithm.

1. Collect records of BLE beacons for a certain period of time.
2. From the collected records, identify the top two BLE beacons (B1, B2 in descending order of radio field strength) for the radio field strength values.
3. Calculate distance *r* from B1 using the radio wave intensity RSSI (*r*) of B1 from the following equation.

$$r = 10^{\frac{RSSI(1m)-RSSI(r)}{10 \cdot K}}$$

4. On the route between B1 and B2, the system estimates the position at a distance *r* from B1 as the current position.

4 Experiment

4.1 Preliminary Experiment for Parameter Setting

In the position estimation algorithm of the proposed method, the distance from the BLE beacon is calculated from the propagation loss distance. The formula for obtaining distance *r* includes the theoretical value RSSI (1 m) of the radio field intensity observed at the position 1 m distant from the BLE beacon. Furthermore, the formula includes coefficient *K* representing the space through which the radio waves are transmitted. Therefore, a preliminary experiment was conducted to set these two parameters.

- **Setting of RSSI (1 m)**
  To set the RSSI (1 m), the radio field intensity was measured at a position 1 m distant from the BLE beacon. An experiment was conducted to measure 100 samples per beacon using three BLE beacons. The average value −63.35 dBm obtained using this experiment is defined as RSSI (1 m).
- **Setting of parameter *K***
  The measurement sites assumed for this study are mainly corridors, which are affected strongly by reflection and absorption of radio waves. Therefore, to set parameter *K*, measurements were taken at points spaced a distance *r* (*r* = 2, 3, 4, 5). Estimated distance *r*' is obtained using the obtained radio field strength and the value of RSSI (1 m). Subsequently, we find the mean square error between *r*' and distance *r*. Then, the *K* value that minimizes the mean square error was obtained. Three beacons were used for the experiment. A total of 1200 samples of 100 samples at each distance were measured. As a result of the experiment, *K* = 3.200 was obtained.

4.2 Experiment Evaluation of Position Estimation

- **Outline**
  We conducted an experiment to evaluate the estimation accuracy of the proposed method. We compared the proposed method with existing methods such as dead reckoning based position estimation system (DR method) and dead reckoning based position estimation and BLE-based position correction system (DR + BLE).

- **Experiment environment**
  Experiments were conducted by walking the route while holding the terminal operating the estimation system in the hand and fixing it in front of the chest. We define the error distance: the distance from estimated coordinates calculated using the estimation system to the *actual* coordinates of the position.

The experiment site is at Nagoya Institute of Technology Building No. 2 on the second floor. A total of 17 BLE beacons arranged in the corridor on the second floor are used (Fig. 4). Figure 5 shows that 32 measuring points and routes for determining the error were set. The interval between measurement points is 5 m. The BLE record was measured 5 s before the time of arrival at the measurement point. The number of trials for each method was five. The error data obtained in one trial were 32. We calculated the average and standard deviation of 160 total error data obtained from 5 trials. Then we compared the results obtained using the respective methods.

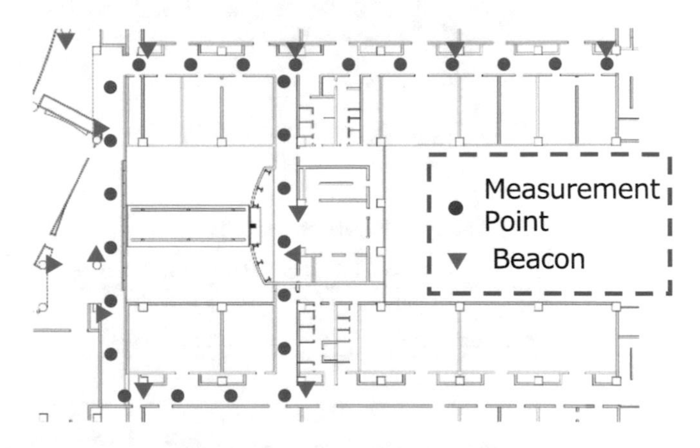

Fig. 4. Arrangement of 17 beacons.

- **Experiment Results and Discussion**
  The experiment results are presented in Table 1. The error average of the proposed method was about 5.6 m. Subsequently, we ascertained that the error mean and the standard deviation are smaller than other methods. The proposed position estimation can be performed stably and accurately.

The average error of each observation point in the proposed method is depicted in Fig. 6. Results demonstrated that the error was more than 15 m at some observation points. The proposed method requires further improvement of position estimation accuracy.

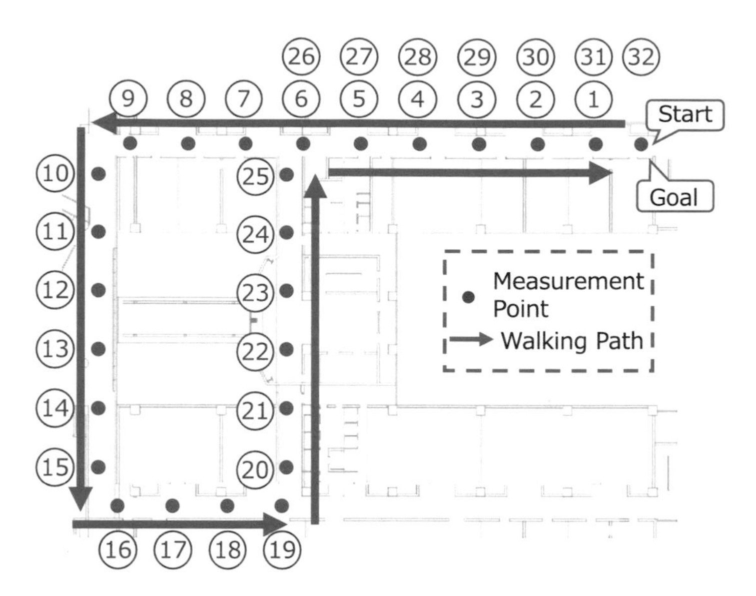

Fig. 5. Measurement points and walking path.

Table 1. Experiment results

	Proposed method	DR	DR + BLE
Average (m)	5.586	10.383	6.937
Standard deviation	5.552	5.971	5.337

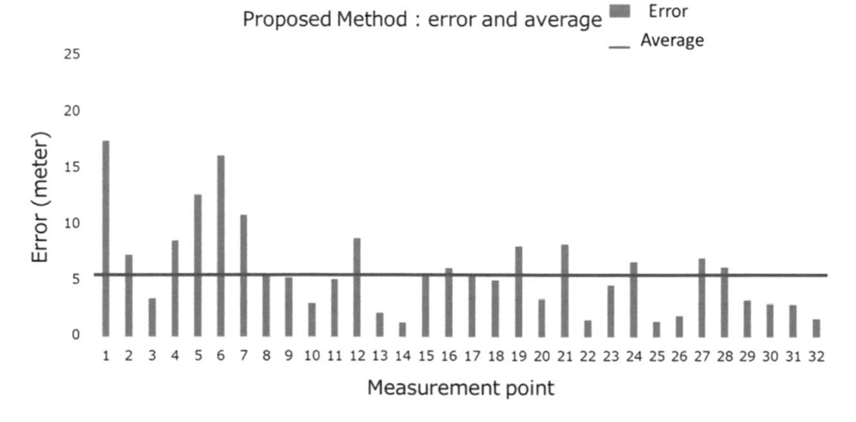

Fig. 6. Average error of each measurement point.

5 Conclusion

For this study, we proposed a position estimation method using BLE beacons. Using this method, we eliminate the difficulty of accumulation error of indoor location in indoor tour support system and improve indoor location estimation accuracy. Based on findings obtained from earlier studies and a radio field strength survey, we designed an indoor position estimation method using the BLE beacon. We conducted estimation accuracy evaluation by comparing the proposed location estimation method with existing methods. From results of the experiment, we confirmed the effectiveness of the indoor position estimation method using the proposed BLE beacon. We expect to use this method as one function of navigation in future work.

Future work is the following.

- Parameter setting by indoor position estimation system using BLE beacons
  In this experiment, two parameters (RSSI (1 m) and K) in the estimation algorithm of the indoor position estimation system were set based on results of preliminary experiments. Future studies must investigate whether setting of these parameters is useful in other environments.
- Evaluation of indoor position estimation system with varied numbers and placement of BLE Beacons
  The accuracy of estimation by indoor position estimation systems using the BLE beacons seems to vary the greatly depending on the beacon number and arrangement. Therefore, it is necessary to arrange the BLE beacons properly. We must investigate the appropriate beacon number and arrangement.

Acknowledgments. This work was supported by JSPS KAKENHI Grant Number JP16H06538.

References

1. Quasi-Zenith Satellite System. http://qzss.go.jp/
2. Yoshida, M., Uchiya, T., Takumi, I.: Development of open-campus application using a spoken dialog system. In: Proceedings of IPSJ-DPSWS 2013 6, pp. 254–261 (2013)
3. Uchiya, T., Yoshida, M., Yamamoto, D., Nishimura, R., Takumi, I.: Design and implementation of open-campus event system with voice interaction agent. Int. J. Mob. Multimed. **11**(3&4), 237–250 (2015)
4. Uchiya, T., Sato, K., Kajioka, S., Yamamoto, D., Takumi, I.: Improvement of indoor position estimation of open-campus event system using BLE Beacon. In: Proceedings of 20th International Conference on Network-Based Information Systems, pp. 845–852 (2017)

Metaheuristic Optimization Technique for Load Balancing in Cloud-Fog Environment Integrated with Smart Grid

Syed Aon Ali Naqvi[1], Nadeem Javaid[1(✉)], Hanan Butt[1], Muhammad Babar Kamal[1], Ali Hamza[2], and Muhammad Kashif[2]

[1] COMSATS University, Islamabad 44000, Pakistan
aonsyed9@gmail.com, nadeemjavaidqau@gmail.com
[2] Government College University, Lahore 54000, Pakistan
http://www.njavaid.com

Abstract. Energy Management System (EMS) is necessary to maintain the balance between electricity consumption and distribution. The huge number of Internet of Things (IoTs) generate the complex amount of data which causes latency in the processing time of Smart Grid (SG). Cloud computing provides its platform for high speed processing. The SG and cloud computing integration helps to improve the EMS for the consumers and utility. In this paper, in order to enhance the speed of cloud computing processing edge computing is introduced, it is also known as fog computing. Fog computing is a complement of cloud computing performing on behalf of cloud. In the proposed scenario numbers of clusters are taken from all over the world based on six regions. Each region contains two clusters and two fogs. Fogs are assigned using the service broker policies to process the request. Each fog contains four to nine Virtual Machines (VMs). For the allocation of VMs Round Robin (RR), throttle and Ant Colony Optimization (ACO) algorithms are used. The paper is based on comparative discussion of these load balancing algorithms.

Keywords: Microgrid · Cloud computing · Fog computing
Renewable energy sources · Demand side management
Ant colony optimization

1 Introduction

An energy management system (EMS) is necessary to control the electricity consumption and generation. Platform for the EMS should be economical and affordable for the ordinary electricity consumer. Microgrid (MG) now made the Power Grids (PGs) more efficient by its unique features like distributed generation and energy storage, MG can connect with PG or can operate independently [1,2]. The generated energy from MG need to be distributed among consumers using Smart Grid (SG) to maintain the balance between energy demand and supply curve in on-peak and off-peak hours [3–5]. Smart meters are used to access

L. Barolli et al. (Eds.): NBiS 2018, LNDECT 22, pp. 700–711, 2019.
https://doi.org/10.1007/978-3-319-98530-5_61

the information of Internet of Things (IoT) devices. IoT includes the number of devices that can sense, act and compute [6]. SG is completely automated infrastructure which controls the energy requirements of these devices in the economical way with the responsibility of rapid response [1,7].

The SG is modern electrical grid, it supports two way communications, which helps to gather the information from different IoT devices using Information and Communication Technology (ICT). The energy generation is managed, devices are monitored and energy is distributed among them by using that gathered information. SG improves reliability, sustainability and efficiency of smart meters. It manages the operations, cost and reduce the peak load for utilities and transmits electricity in the efficient way in economical cost [8].

When billions of Internet of Thing (IoT) devices are in operation at that time the calculation for these devices become burden for the SG. To overcome this issue cloud computing introduced as a centralized computing, integrated with IoT. Cloud computing provides different hardware and software base services over the network. Super computer processing can be achieved by using the cloud computing [9]. It provides its platform for different computational service i.e. user can execute the code or can perform any type of processing. Cloud provides software as a service facility to the users, any one can view his/her profile through browser. User can use Infrastructure of a cloud as a service e.g. server can be used as storage [10].

When the number of devices increased in IoT the processing smart meters may take the extra computing time and response time will be increased due to computing latency [1]. Fog computing introduced to cover this limitation. Fog Computing is laying between IoT devices and cloud data center. Fog provides the storage, computation and networking services between devices and cloud computing. Day by day the number of IoT devices are increasing, the traditional cloud system is not designed for the type of data which IoT devices produce [4]. Other unique features of fog includes; mobility, heterogeneity, location awareness and real time response. The reason for these advantages is nearer to the end user, instead of remote data center.

In this paper, interaction between the SG and the emerging technology fog computing is considered and proposed the Ant Colony Optimization (ACO) algorithm for the load balancing of VMs lying in the fog architecture. At the end results are compared with Round Robin (RR) and throttle algorithms.

1.1 Motivation

For the implementation of EMS, a platform is necessary for the interaction of devices [1]. Author proposed a Fog based EMS, where fog is introduced as a platform for the connectivity of smart devices, which is based on low cost and low power devices. Fog computing provides the facility of separate storage of private and public data because of its distributed nature [11]. The main reason for employing cloud computing is its cheap rates. In [10], considered the problem of charging and discharging of Electrical Vehicles at Supply station, Fogs are working locally under the central cloud. Two different scheduling algorithms are proposed for minimizing the waiting time of vehicles. Author in [12] proposed a cloud-fog based environment and used Particle Swarm Optimization (PSO) and RR for the load balancing of VMs, because burden on a particular VM can lead to high latency, so balancing of load is necessary for effective response.

1.2 Contribution

In this paper, proposed the cloud-fog based infrastructure, integrated with SG to get benefits of cloud computing. Following are the contributions of the paper.

- Six regions are considered from all over the world for managing the electricity consumption using fog layer as a central layer of architecture. Fog layer stores the information of consumer layer at local region level.
- Different numbers of VMs are allocated to the fogs for the simulation purpose ranges from four to nine.
- For the allocation of VMs to manage response time, processing time and cost, Ant Colony Optimization (ACO) technique is used.
- ACO technique is farther compared with RR and throttle algorithm. It has been observed that proposed algorithm performed better than RR in all mentioned aspects and closer to throttle.

The rest of the paper is organized as: Sect. 2 related work, Sect. 3 introduces system model, Sect. 4 based on the simulation and result discussion and at the end conclusion of the paper is drawn.

2 Related Work

To balance the load on cloud and fog multiple algorithms are proposed by the researchers. These algorithms manage the requests of clients and schedule their tasks. The load balancing algorithm assigns task to each VM to utilize all the resources and to avoid the unbalancing between the VM's load. The common algorithms used for the resource allocation are, Round Robin (RR), Shortest Job First (SJF), throttle, Artificial Bee Colony (ABC), Genetic Algorithms (GA) etc.

The authors in [10] formulated the problems of Electrical Vehicles (EVs) charging and discharging. Interaction between the EVs, Electrical Vehicle Power Supply Station (EVPSS) and cloud computing is defined. Scheduling is performed by using two priority assignment algorithms based on calendar and non calendar vehicles. Fog computing platform is used to manage the computation of devices in [1] to get the high response rate with low latency.

In [13] authors considered the efficient management system of mobile phones. Mobile phones always retrieve the location of data from the cloud. However, retrieving always data from the remote cloud, it will increase the response time so the edge computing concept is utilized to reduce latency and get faster response. Approximate approach is utilized to formulate Primal Problem (PP) into three Sub Problems (SPs). Particle Swarm Optimization (PSO) with simulated annealing algorithm is used for the resource allocation in [14] with the new service broker policy to allocate the data center by least estimated response time. Finally proposed technique compared with RR, throttle and Cuckoo search.

Shortest Job First (SJF) is proposed in [15] to tackle the load. The algorithm is implemented in two different scenarios in the paper. In first scenario one data center is allocated to twenty five VMs and in second two data centers are allocated to fifty VMs. Second scenario produces better results than first one. This proposed technique is farther also compared with RR and Equally Spread Current Execution (ESCE) and examined that SJF produced better result than others in the proposed scenario.

In [16] author proposed real-time dynamic pricing model and scheduling the users request on real time environment. Considered problem is of electric vehicles charging and discharging to reduce the peak load and manage requests efficiently. Software define networking technology based decentralized cloud computing architecture is used in the proposed approach. Real load of Toronto electricity is used for the simulations. For reducing load Peak to Average Ratio (PAR) Monte Carlo method is used in [17]. By reducing load from PAR reduces the cost of electricity.

[18] focuses on the novel cost oriented optimization model to allocate the cloud computing resources to the cloud based Information and Communication Technology (ICT) infrastructure. Resources are allocated in a cost efficient and flexible way using the modified priority list algorithms. To demonstrate the effectiveness of the proposed model on the operation cost reduction, comprehensive numerical studies are included.

Authors in [12] used RR, throttle and Particle Swarm Optimization (PSO) for the load balancing of VMs. For the experimental purpose two clusters are considered in two regions. Each cluster linked individually with one fog. Overall performance of PSO is better than the throttle and RR. In [7] three service broker policies; service proximity policy, optimize response time, dynamically reconfigure with load are compared with newly purposed service broker policy. Each cluster linked with two fogs, service broker policy decides the host of requested service from these two fogs. Two load balancing algorithm RR and throttle also used in this paper. Newly proposed service broker policy produces better results with throttle rather than RR.

3 System Model

In the proposed model, 6 regions are considered from all over the world. Each region includes two clusters and each cluster contains four to five buildings. The model is based on three layers, i.e., consumer layer, fog layer and cloud Layer. In a single region, we have two fogs, each fog is linked with two clusters via controller, as shown in Fig. 1. We consider that in each building, we have seven to ten homes. Each home has its own renewable energy resources and battery to save extra energy. When the demand of electricity is higher than the produced energy, then the storage battery will be utilized in order to fulfill the electricity demand. In case, battery does not have enough energy storage as compared to

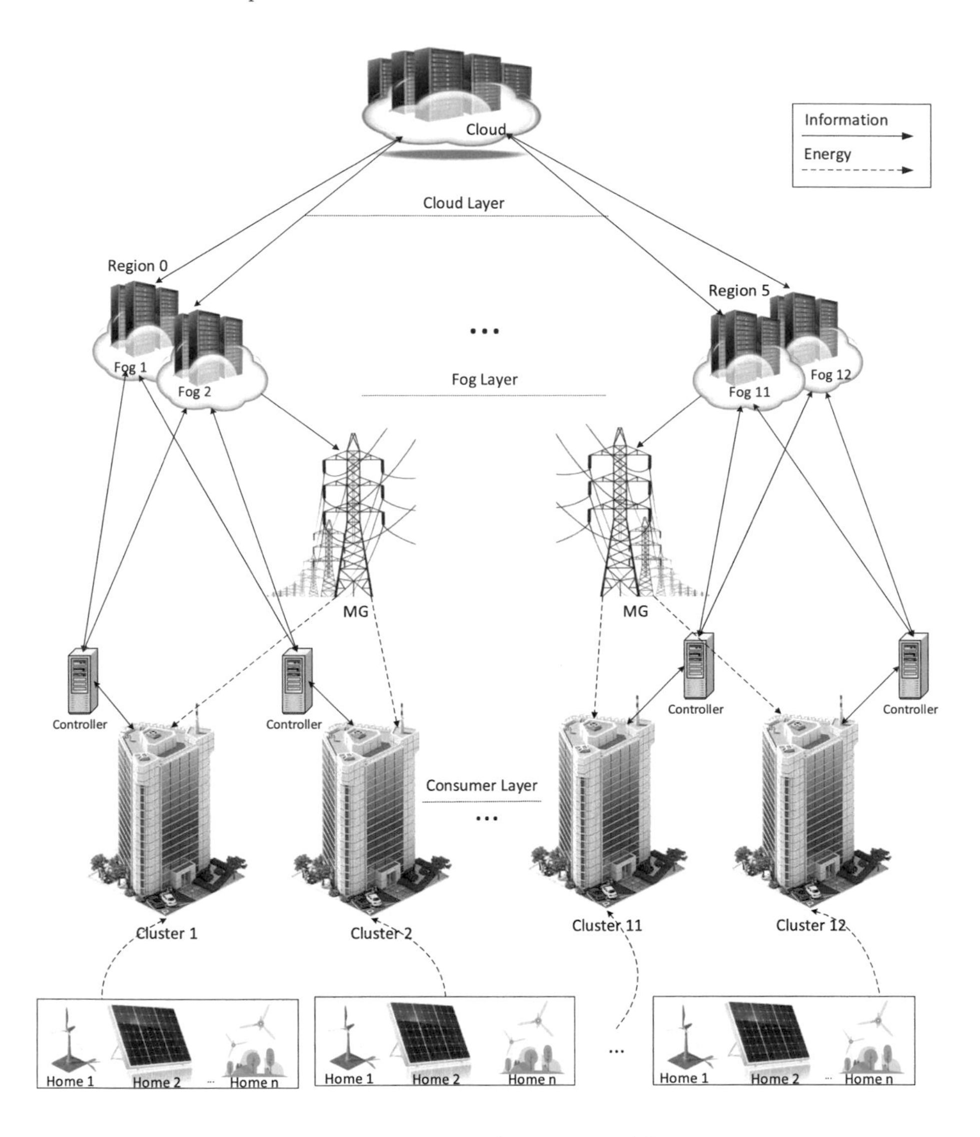

Fig. 1. Proposed system model

user's demand, then the request is send to that fog, by smart meter of a home. Fog transfers the request towards the SG. Numbers of appliances are here to consume electricity.

The fogs are used to allocate the resources in an efficient way. Homes are connected with fog through smart meters. Fog contains different VMs to provide fast response to the consumers. Numbers of different applications are running on the VMs to process the requests of clients.

Cloud layer is on the top of fog layer. Fog provides the facilities of cloud, faster than the cloud because cloud is far away than fog from the user. For the permanent storage of data, fog sends it towards the cloud because it can only save data temporarily. The main reason of using cloud-fog based environment on Smart Grid is for computational purpose. Different load balancing Algorithms are used for the balancing of the load on VMs.

3.1 Problem Formulation

In proposed system model we considered M fogs in each region, each fog contain set of VMs as VM = $\{vm_1, vm_2, vm_3, \ldots, vm_m\}$. Set of VMs is denoted by VM and there are total m numbers of VMs. These VMs deal different tasks which are assigned by the load balancing Algorithm and requested by the users. Set of tasks is represented as T = $\{t_1, t_2, t_3, \ldots, t_n\}$. There are n numbers of total tasks which will be served by different VMs.

The VMs are working in parallel order, each VM have the same capacity to deal with different tasks. These VMs do not share resources with each other. Each VM is assigned by evaluating different performance metrics: response time, processing time, cost of VM, requests per hour. These parameters are formulated using linear programming [19,20].

Processing time of each VM is indicated using PT. To check task is assigned or not, binary representation is given in Eq. 1. where 0 is representing there is no task assigned to VM. 1 is used to indicate that task is assigned to VM.

$$t_{ij} = \begin{cases} 1; & \text{If VM is assigned} \\ 0; & \text{else} \end{cases} \tag{1}$$

Minimizing PT_t is the objective of this proposed work. PT_t represent the total processing time of VM, t_{ij} indicates i requests assigned to j servers in specific time period. PT_{ij} indicates the processing time of each request i on the server j.

$$\text{Minimize } PT_t = \sum_{i=1}^{m} \sum_{j=1}^{n} (PT_{ij} * t_{ij})$$

$$\text{Subject to: } \sum_{i=1}^{m} t_{ij} = 1, j = 1, 2, ..., n \tag{2}$$

$$t_{ij} = 0 \text{ or } 1.$$

The total Response time of VM is calculated by using Eq. 3. CT_x is the completion time of task. The cost of the system is calculated summing up the cost of VMs, MGs and data transfer cost. VM cost is computed by multiple instructions per second and the required task size. Makesspan term is used for describing the total execution time of task. The total cost is computed using the Eq. 4.

$$RT = \frac{\sum VMs(CT_x)}{makesspan * NumofVMs} \tag{3}$$

$$C_t = C_{DT} + C_{VM} + C_{MG} \tag{4}$$

C_t representing the total cost, C_{DT} representing the data transfer cost, C_{VM} representing the cost of VM and C_{MG} representing the cost of microgids.

3.2 Algorithm

The proposed algorithm is ACO, it is used to find the best path from the wide range by selecting optimal path. Ants in real world wander randomly for finding food, after finding food by lying down their pheromone trail return to their colony.

Algorithm 1. ACO

```
1: parameters initialization: pheromone,iteration, routes,
2: for users = 1:15 do
3:     For all appliances ap ∈ n
4:     For all timeslots t ∈ T
5:     Generate population Randomly
6:     Initialization of Pheromone
7:     for it=1:Max-iterations do
8:         for ai=1:N-ants do
9:             For number of tours
10:            For each ant to update pheromone trails
11:             calculate objective function for each ant
12:            for t=1:T do
13:                 load consumption claculation)
14:                Electricity cost calculation
15:                update the pheromone locally for each ant
16:                Choose P_best
17:            end for
18:            update the pheromone globally for each ant
19:        end for
20:        perform maximum iterations
21:    end for
22:    Perceive the demand of users
23:    Get supply from Energy Resources
24:    if Demand<Supply then
25:        Charge the batteries
26:    else if Demand>Supply then
27:        Calculate insufficient power
28:        Calculate Power in batteries
29:        if Insufficient_power! = 0 then
30:            Buy electricity from MG
31:        end if
32:    end if
33: end for
```

The fellow ants, follow the same trail for food by discovering the shortest path.

4 Simulation and Discussion

As mentioned in purposed model, we have considered six regions in the scenario. Regions are representing continents of the world. Total twelve fogs are working under a central cloud. Each Region contains two fogs and two clusters of consumers. These two fogs and two clusters are meshed with each other, in each region. By using them we have simulated our result.

4.1 Simulation Setup

Java platform is used for simulations with Netbeans. By utilizing the 'Cloud Analyst tool' results of ACO are computed and also compared with two other load balancing techniques, i.e., throttle and RR. For the selection of fog three services broker policies are available in 'Cloud analyst'. For plotting Graphs, Matlab is used. Machine for performing the simulation is utilized with processor 'Intel Core i3', RAM 6 GB and storage 500 GB.

4.2 Discussion

The performance measures of proposed scheme are: data center request servicing time, response time and cost.

Figure 2 shows the response time by region, which is provided by fogs to their dependent clusters. RR and Throttle result also shown in Fig with ACO. This graph is plotted using average response time of these three techniques. As shown in Table 1 the overall minimum, maximum and average response time of RR is

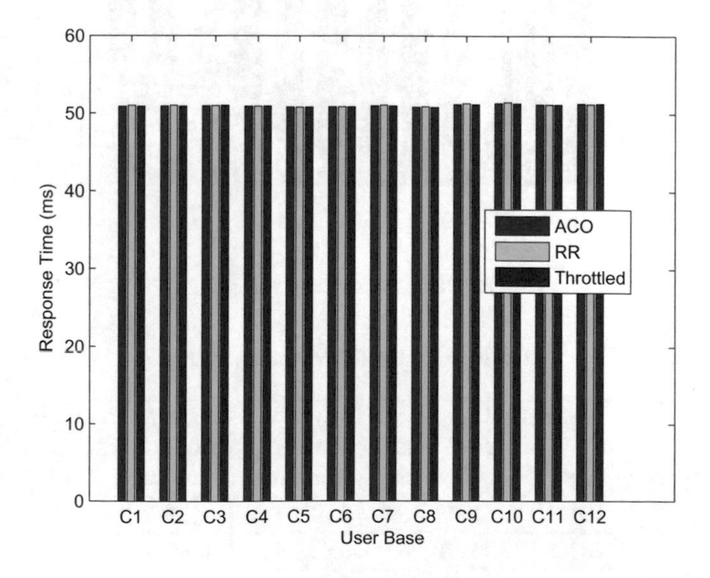

Fig. 2. Response time by region

Table 1. Overall response time

Algorithm	Avg processing time (ms)	Min processing time (ms)	Max processing time (ms)
RR	51.05	35.63	67.62
ACO	51.01	35.72	66.62
Throttle	51.00	35.63	67.62

35.63, 67.72 and 51.05 respectively, with throttled and ACO response time is 35.63, 67.62, 51.00 and 35.72, 66.62 and 51.01 respectively.

Request processing time is depicted in Fig. 3 with RR, throttle and ACO. This graph is also plotted using average processing time of these techniques. In Table 2 we can see that the overall minimum, maximum and average processing

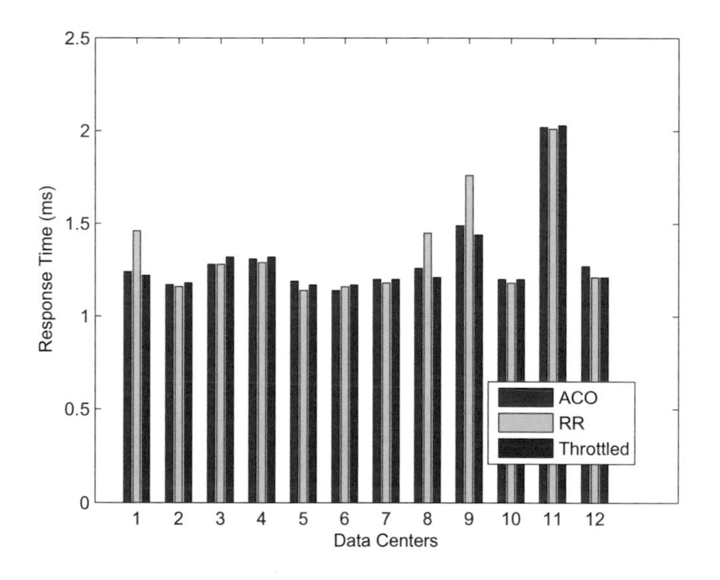

Fig. 3. DC request processing time

Table 2. Overall processing time

Algorithm	Avg processing time (ms)	Min processing time (ms)	Max processing time (ms)
RR	1.35	0.01	7.57
ACO	1.31	0.01	10.62
Throttle	1.30	0.01	5.07

time of RR is 0.01, 7.57 and 1.35 respectively, however, with throttle and ACO processing time is 0.01, 5.07, 1.30 and 0.01, 10.62 and 1.31 respectively. We can observe by considering average processing time that RR is high as compare to throttle and ACO, then ACO has less and throttle has least.

Figure 4 shows the cost comparison, it is plotted using total cost of RR, throttle and ACO. Table 3 showing the total VM cost, data transfer cost and total cost. Firstly of RR is 204.01, 160.49 and 364.50 respectively, secondly throttled 204.01, 160.49, 364.50 and lastly ACO cost is 204.01, 160.49 and 364.50 which same. From the Results we can conclude that the proposed Algorithms ACO is overall performing better than RR and nearer to throttle.

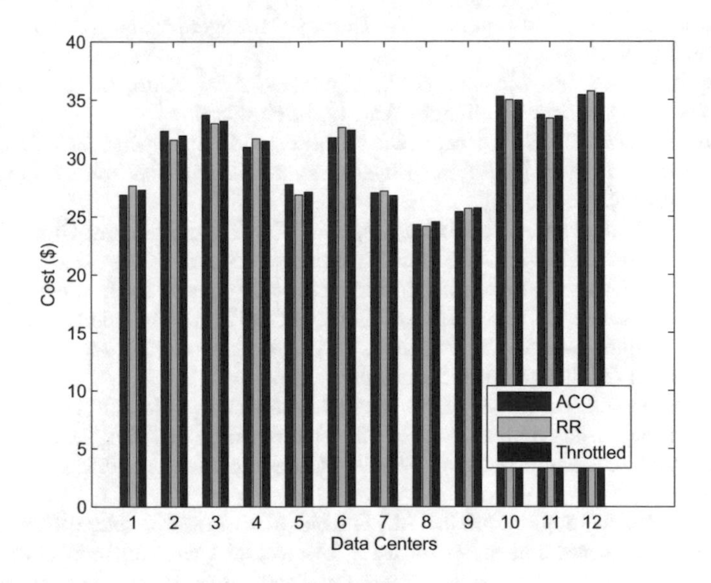

Fig. 4. Total cost

Table 3. Overall cost

Algorithm	VM cost	Data transfer cost	Total cost
RR	204.01	160.49	364.50
ACO	204.01	160.49	364.50
Throttle	204.01	160.49	364.50

5 Conclusion

In this paper, a cloud-fog based architecture is proposed based on three layers; cloud layer, fog layer and consumer's layer. MGs are connected with fogs, user cannot communicate directly with it. Users put their request on fog for the energy requirement. These requests are assigned to fogs by using three different service broker policies. Fog layer compute the energy, consumed by the consumers and process their requests using VMs. Load of VMs is balanced by using RR and throttle and ACO load balancing algorithm. Simulation is performed using Java platform in NetBeans. Simulation results shows that the proposed algorithm ACO performed better than the RR.

References

1. Al Faruque, M.A., Vatanparva, K.: Energy Management-as-a-Service Over Fog Computing Platform
2. Yoldaş, Y., Önen, A., Muyeen, S.M., Vasilakos, A.V., Alan, İ.: Enhancing Smart Grid With Microgrids: Challenges And Opportunities
3. Kumar, N., Zeadally, S., Misra, S.C.: Mobile cloud networking for efficient energy management in smart grid cyber-physical systems. IEEE Wirel. Commun. **23**(5), 100–108 (2016)
4. Kumar, N., Vasilakos, A.V., Rodrigues, J.J.P.C.: A Multi-Tenant Cloud-Based DC Nano Grid for Self-Sustained Smart Buildings in Smart Cities
5. Mallada, E., Zhao, C., Low, S.: Optimal load-side control for frequency regulation in smart grids. IEEE Trans. Autom. Control **62**(12), 6294–6309 (2017)
6. Risteska Stojkoska, B.L., Trivodaliev, K.V.: A Review of Internet of Things for Smart Home: Challenges And Solutions
7. Fatima, I., Javaid, N., Iqbal, M.N., Shafi, I., Anjum, A., Memon, U.: Integration of cloud and fog based environment for effective resource distribution in smart buildings. In: 14th IEEE International Wireless Communications and Mobile Computing Conference
8. Yaghmaee M.H., Moghaddassian M., Garcia A.L.: Power consumption scheduling for future connected smart homes using bi-level cost-wise optimization approach. In: Leon-Garcia, A., et al. (eds.) Smart City 360°, SmartCity 360 2016, SmartCity 360 2015 (2016)
9. Rajeshirke, N., Sawant, R., Sawant, S., Shaikh, H.: Load Balancing In Cloud Computing (2017)
10. Chekired, D.A., Khoukhi, L.: Smart Grid Solution for Charging and Discharging Services Based on Cloud Computing Scheduling
11. Okay, F.Y., Ozdemir, S.: A Fog Computing Based Smart Grid Mode
12. Zahoor, S., Javaid, N., Khan, A., Ruqia, B., Muhammad, F.J., Guizani, M.: A Cloud-Fog-Based Smart Grid Model for Efficient Resource Utilization (2018)
13. Deng, R., Lu, R., Lai, C., Luan, T.H., Liang, H.: Optimal Workload Allocation in Fog-Cloud Computing Toward Balanced Delay and Power Consumption
14. Yasmeen, A., Javaid, N.: Exploiting Load Balancing Algorithms for Resource Allocation in Cloud and Fog Based Infrastructures
15. Javaid, S., Javaid, N., Tayyaba, S.K., Sattar, N.A., Ruqia, B., Zahid, M.: Resource allocation using Fog-2-Cloud based environment for smart buildings. In: 14th IEEE International Wireless Communications and Mobile Computing Conference (IWCMC-2018) (2018)
16. Chekired, D.A., Khoukhi, L., Mouftah, H.T.: Decentralized Cloud-SDN architecture in smart grid: a dynamic pricing model. IEEE Trans. Ind. Inf. (2017)

17. Moghaddam, M.H.Y., Leon-Garcia, A., Moghaddassian, M.L.: On the performance of distributed and cloudbased demand response in smart grid. IEEE Trans. Smart Grid (2017)
18. Yi, G., Kim, H.-W., Park, J.H., Jeong, Y.-S.: Job allocation mechanism for battery consumption minimization of cyber-physical-social big data processing based on mobile cloud computing. IEEE Access **6**, 21769–21777 (2018). ISSN 2169-3536
19. Devi, D.C., Uthariaraj, VR.: Load Balancing In Cloud Computing Environment Using Improved Weighted Round Robin Algorithm For Nonpreemptive Dependent Tasks (2016)
20. Kumrai, T., Ota, K., Dong, M., Kishigami, J., Sung, D.K.: Multiobjective Optimization in Cloud Brokering Systems for Connected Internet of Things

The Optimal Beacon Placement for Indoor Positioning

Ching-Lung Chang(✉) and Chun-yen Wu

Department of CSIE, National Yunlin University of Science and Technology, Yunlin, Taiwan
{chang,m10517001}@yuntech.edu.tw

Abstract. In recent years, based on the development of low-power transmission technologies, the issue of indoor positioning has also received increasing attention. This paper combines grid technology to transform the indoor positioning of Beacon deployment problems into optimization problems. Considering the RSSI signal drift problem, this optimization model converts the RSSI signal strength to Signal Power Ranking (SPR), which is a combination of Simulated Annealing (SA). In order to obtain the location and transmission power of Beacon, to achieve the objective of providing complete identification and the best identification rate with the minimum number of Beacons. At the same time, we also use the IBM ILOG CPLEX optimization tool to verify the SA algorithm. The simulation results show that under different topologies, the SA algorithm can reach the results with the CPLEX tool in a shorter time.

Keywords: Indoor positioning · Beacon · Linear Programming (LP) Simulated Annealing (SA) · COLEX

1 Introduction

Internet of Things has become a basic application. With technology development, some interfaces are applied to communicate with people. Therefore, indoor positioning is a very important [1]. Most of the positioning systems use the Global Positioning System (GPS) of the outdoor positioning system. However, electromagnetic waves do not easily penetrate metal cement structures. Therefore, the indoor positioning system will perform positioning through a number of sensors, such as infrared (IR), radio frequency identification (RFID), Bluetooth, and the like.

The purpose of indoor positioning is usually divided into the following applications. The most common applications are navigational applications, which appear in stores and some public agencies, and guide them to the locations or services of goods to be purchased through indoor positioning. There is also an analysis of human behavior, recording time and path of indoor positioning through known fields. These messages are further analyzed, and this side can usually be used in rehabilitation rooms or general offices for doctors to assist in determining or reminding some office workers of long-term sitting problems.

In [2], the trigonometric concept such as Pythagorean theorem is used to calculate the coordinates. The actual distance is measured according to the RSSI signal sent by

L. Barolli et al. (Eds.): NBiS 2018, LNDECT 22, pp. 712–721, 2019.
https://doi.org/10.1007/978-3-319-98530-5_62

Bluetooth in [3]. In [4], a field based on Filed is set to be grid-like, and the sensed points must be set on the Grid. It uses the optimization model to convert the sensing node deployment problem into Optimization problem. Internal positioning technology, we use RSSI as a basis for positioning. However, in the 2.4 GHz band, the signal is susceptible to RSSI due to interference. Therefore, this paper uses Signal Power Ranking (SPR) instead of RSSI, and uses the concept of grid field to reduce the interference problem [5].

The proposed method in this paper is described below. First, the concept of a grid field is used to construct the testing field. Then, we convert the received signal strength indicator (RSSI) to Signal Power Ranking (SPR) for each kind of Beacon s power. To have the lowest cost and the best deployment, we design a linear programming model satisfying two fundamental demands, complete covered and complete identification. Finally, cplex and the algorithm of simulated annealing are both applied to solve the problem derived from the model.

The rest of this paper is organized as follows. Section 2 presents the proposed system architecture. Section 3 shows experimental environment and results. Section 4 concludes our work.

2 System Architecture

2.1 Signal Power Ranking

This section first constructs the concept of the grid field as shown in Fig. 1 since the final positioning accuracy is determined by the size of the grid field. In Fig. 1, the red circle represents the meaning of built Beacon, and the other 1–15 circles represent the meaning of Beacon. In addition, when constructing the Beacon, we chose to place the Beacon cloth on the ceiling, which was less affected by the disturbance than the cloth was built on the wall. In addition, our Beacon has six kinds of power for deployment.

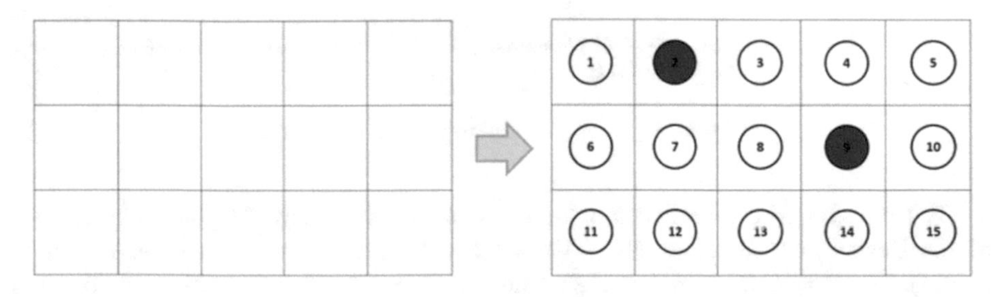

Fig. 1. Grid field.

In RSSI (Received Signal Strength Indicator), we need to do the distance conversion by using Eq. (1).

$$P_d = P_0 - 10 * n * \log_{10}(d) \tag{1}$$

where P_d represents the received RSSI signal strength at d meters; P_0 represents the RSSI signal strength received at 1 m; n represents the attenuation factor in the environment; d is the distance unit in meters.

We can refer to the distance obtained from above equation and convert it into Eq. (2) to convert the signal intensity to the distance curve as shown in Fig. 2.

$$\sum_{r=1}^{r_p} r y_{a_{rp}}\left(d_{m,b}\right) \tag{2}$$

where r_p represents the power ranking under p power; $d_{m,b}$ represents the distance from mesh m to mesh b in meters; a_{r_p} represents the distance from the current RSSI through Eq. (1); $y_{a_{rp}}\left(d_{m,b}\right)$ is the function with $y_{a_{rp}}\left(d_{m,b}\right) = 1$ when $d_{m,b} \in a_{r_p}$ and $y_{a_{rp}}\left(d_{m,b}\right) = 0$, otherwise.

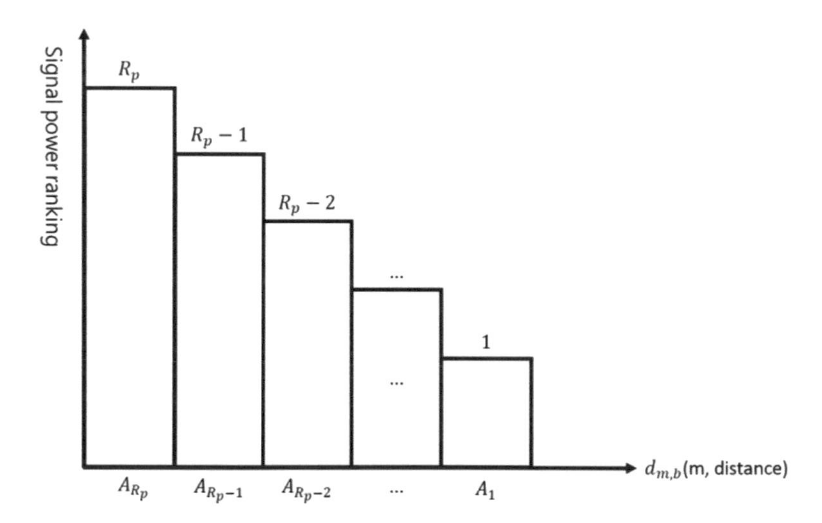

Fig. 2. Signal power ranking conversion curve.

When each grid used for positioning is located on the received signal strength level of the Beacon, a vertical vector (RV) is formed and its table style is as shown in Eq. (3). Where $v_{m,b}$ represents the signal strength level at which the positioning mesh m received the Beacon mesh b. In addition, we use the grid-level vector to distinguish the individual differences, and define the minimum difference as the Hamming distance, as shown in Fig. 3 below. The RV value is shown in Table 1, so the difference in the Hamming distance is calculated. If $RV_1 = (3, 0)$, $RV_4 = (0, 1)$, the calculation method is Hamming distance $= |3 - 0| + |0 - 1| = 4$.

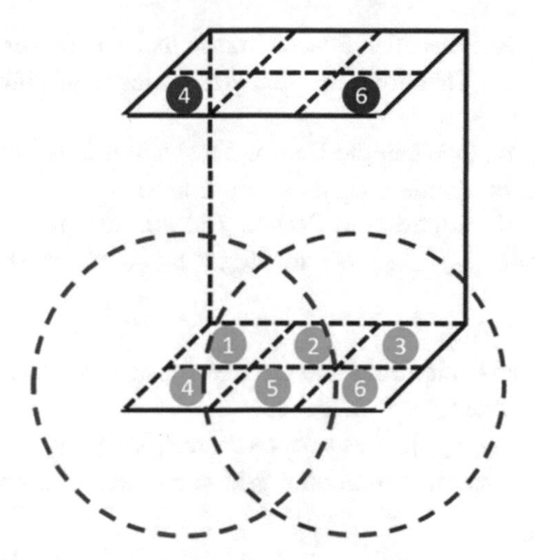

Fig. 3. 3 × 2 grid fields.

Table 1.

	b=4	b=6
V_1	3	0
V_2	1	1
V_3	0	2
V_4	0	1
V_5	4	0
V_6	4	2

$$RV_m = \left(v_{m,1}, v_{m,2}, \ldots, v_{m,b}\right) \tag{3}$$

2.2 Linear Programming

Both Objective Function and Subject to are linear optimization problems and linear programming methods can be used to find the best solution. We use this technique to develop the Beacon-based problem as the mathematically optimized model with the smallest number of Beacon s identifiable locations.

Given Parameters:

$M = \{1, 2, \ldots, m\}$: A set of positionable grids.
$B = \{1, 2, \ldots, b\}$: It may be possible to construct a set of Beacon meshes, where $b \leq$ m.
$P = \{1, 2, \ldots, p\}$: Beacon can choose the power set.

$R_p = \{1, 2, \ldots, r_p\}$: When the Beacon transmits at p power, the corresponding signal strength level. Therefore, the step size varies with different powers, where $p \in \mathrm{P}$.

$A_{r_p} = \{a_1, a_2, \ldots, a_{r_p}\}$: When the Beacon received at different distances transmits at p power, the corresponding signal strength level is.

$d_{m,b}$: The distance from grid m to Beacon grid b in meters.

$y_{a_{rp}}(d_{m,b})$: Function, $y_{a_{rp}}(d_{m,b}) = 1$ for $d_{m,b} \in a_{r_p}$, 0 otherwise.

Decision Variables:

x_b^p: When the value is 1, it means that the signal is sent at p power in Beacon grid b, otherwise it is 0, where $p \in \mathrm{P}, b \in B$.

$RV_m = (v_{m,1}, v_{m,2}, \ldots, v_{m,b})$: The vector of level, where $v_{m,b}$ represents the signal strength level at which the positioning grid m receives Beacon mesh b.

Objective Function:

$$Obj: \max(\alpha * \Phi + (1 - \alpha) * \Psi) \tag{4}$$

Subject to:

$$\sum_{p \in P} x_b^p \leq 1, b \in B \tag{5}$$

$$V_{m,b} = \sum_{p \in P} x_b^p \left(\sum_{r=1}^{r_p} r y_{a_{r_p}}(d_{m,b}) \right) \tag{6}$$

$$, \forall m \in M, b \in B, m \neq b$$

$$\sum_{b \in B} v_{m,b} \geq 1, \forall m \in M \tag{7}$$

$$\sum_{b \in B} (v_{m,b} - v_{n,b}) \geq \Phi, \ \forall m, n \in M, \Phi \geq 1 \tag{8}$$

$$b - \sum_{b \in B} \sum_{p \in P} x_b^p = \Psi \tag{9}$$

$$x_b^p = 0 \ or \ 1, \ \ b \in B, p \in P \tag{10}$$

The Eq. (5) is used to limit a Beacon mesh b to transmit signals with only one p power. The Eq. (6) to calculate whether the received Beacon mesh b is received at the p power in the positioning mesh m. If it is received, it will correspond to $y_{a_{rp}}(d_{m,b}) = 1$ and be converted to Signal strength level. The Eq. (7) ensures that each positioning grid must have received a signal with the above Beacon. The Eq. (8) is used to calculate the recognition degree of the location. When $\Phi = 0$, there are two identical identification grids, so $\Phi \geq 1$. The Eq. (9) calculation has several Beacon meshes that are not built with Beacon. Given an objective function (4), an α weight is used to minimize the number of beacons to be identified.

2.3 Simulated Annealing

Simulated annealing is a probabilistic algorithm that is proven by mathematicians and can be used to find the best solution in a wide range of searches. It is also often used to solve the problem of the best solution (approximate solution) in the region that is most easily encountered in optimization. The concept of the simulated annealing algorithm is that when the substance is at a high temperature, the molecules will be randomly arranged. When the temperature is reduced to the lowest temperature, the convergence can only be achieved by searching for the best position in the week. Figure 4 shows the simulated annealing algorithm.

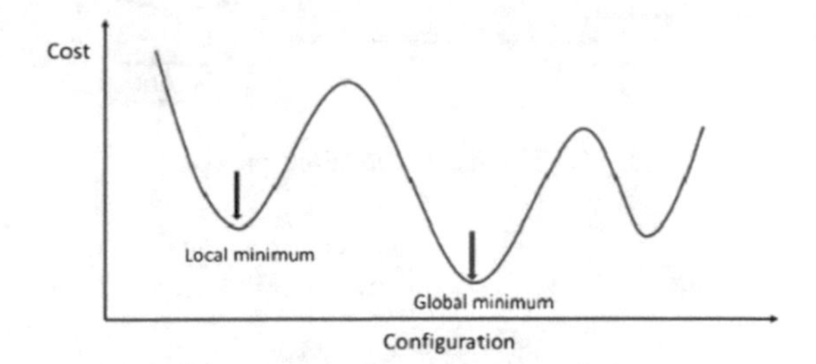

Fig. 4. Simulated annealing schematic.

We use simulated annealing algorithm to solve the linear programming problem derived from the proposed model. The simulation annealing algorithm applied in Beacon deployment is shown in the flow chart of Fig. 5. The procedure is introduced as follows.

Generate the initial energy (E) and initial temperature (T) first. Secondly, set the number of iterations $(K = 0)$ and select the adjacent solution to generate energy (E'). If it is judged $(E' > E)$, the substitution is made; otherwise, the substitution adopts Metropolis criterion probability $\exp\left(-\frac{E'-E}{T}\right)$. After that, increase the number of iterations (*K*++) and determine whether the current temperature iteration $(K \leq L)$ has not been continued. If neccessary, cool down $(T = T * \alpha)$ and finally determine whether the temperature reaches the end temperature $(T < T\_e)$.

In searching neighborhood solution, we propose five rules as follows:

I. Select one of the randomly placed Beacon positions and delete its position.
II. Randomly select a location where Beacon is placed and randomly modify the transmit power.
III. Select a Beacon position at random, add a Beacon position, and set the transmit power at random.

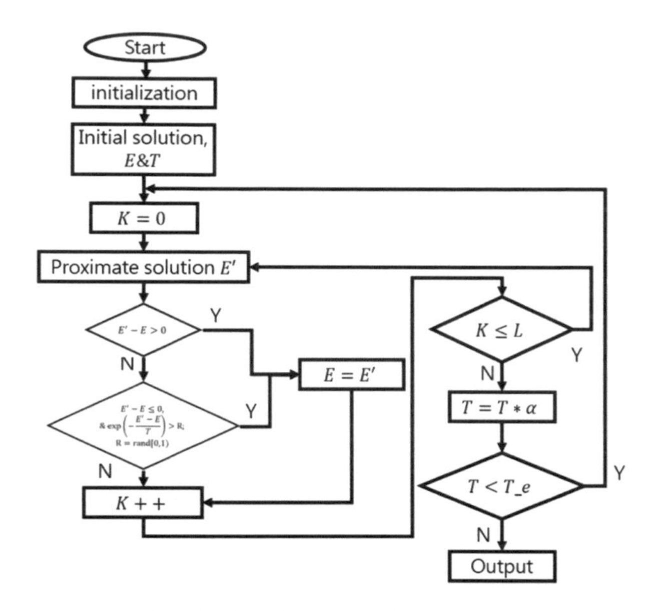

Fig. 5. Simulated annealing process.

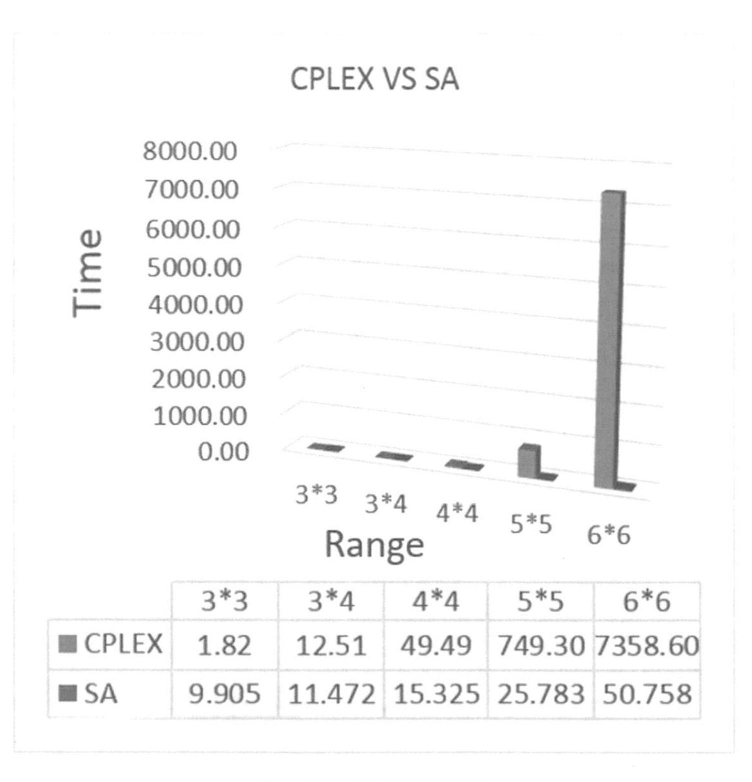

	3*3	3*4	4*4	5*5	6*6
CPLEX	1.82	12.51	49.49	749.30	7358.60
SA	9.905	11.472	15.325	25.783	50.758

Fig. 6. CPLEX VS SA.

IV. Select a location with a Beacon at random, modify its transmit power with the base, and randomly select a location where Beacon is not placed and move it.

Randomly generates a new set of solutions.

3 Experimental Results

By compare to the result obtained from IBM ILOG CPLEX, the result of the simulated annealing algorithm (SA) is efficient. The comparison is shown in Fig. 6. One can see that CPLEX is about 8 s faster than SA when the topology size is small and small. When zoomed in, SA is 1 s faster than CPLEX. However, CPLEX can only run out of 6 * 6 fields on this experimental device, and SA can be up to 140 times faster. It concludes that we can use SA to find the optimal solution efficiently when the field is relatively large.

In Fig. 7, the 25 grids with a size of 5 times 5 are overlaid, and the identified degree and the number of Beacon unplaced curves are calculated through different α weights. From this analysis, we can see that we need to analyze how much a Beacon identification should be, so that we can find the best solution.

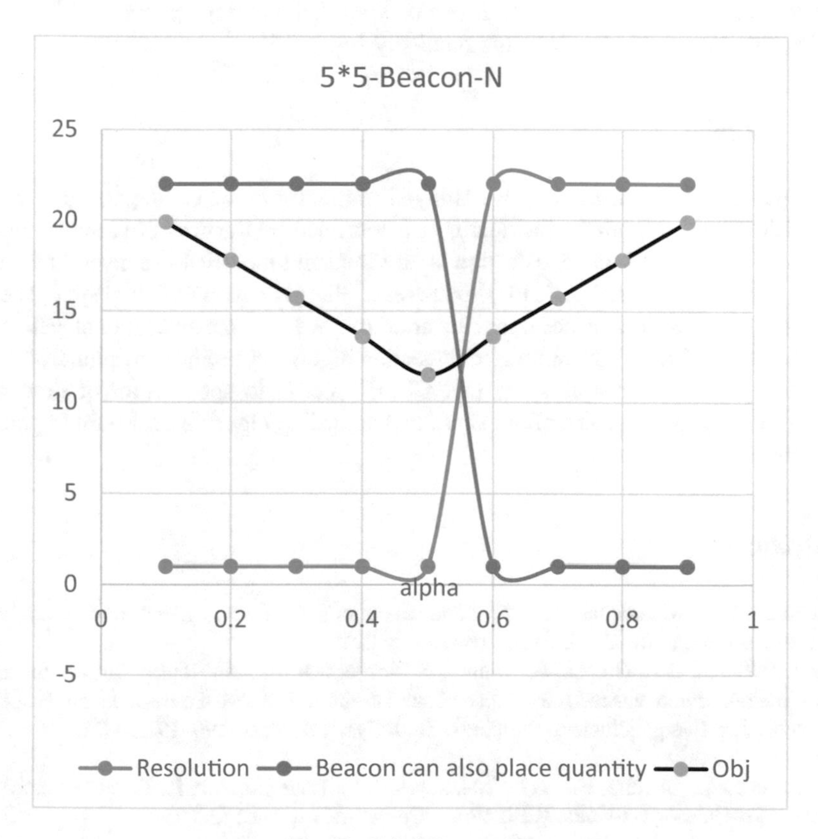

Fig. 7. 5 * 5-Beacon-N.

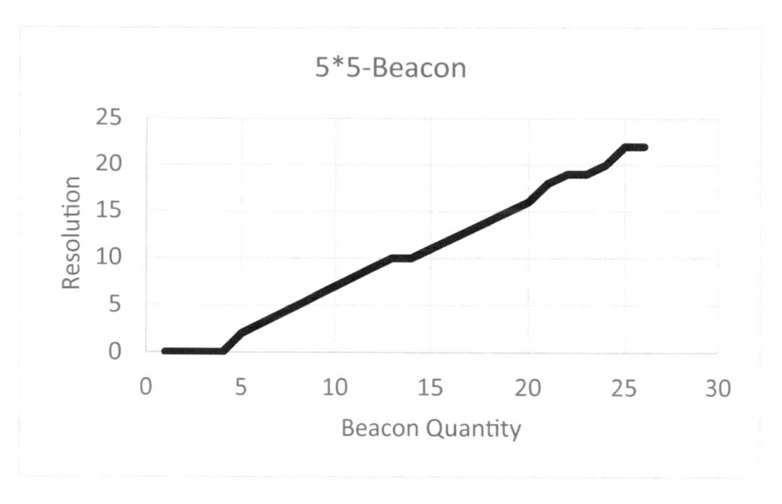

Fig. 8. 5 * 5-Beacon.

Therefore, in Fig. 8 we modify the limit and give a certain amount of Beacon to find the best degree of recognition. It can be seen that when the number of Beacons is smaller, the degree of recognition is relatively low.

4 Conclusion

In this paper, indoor positioning is deployed optimally based on the concept of grid-based field. The fine-grained positioning is determined by the grid size arrangement. In addition, Beacon multi-power selection is provided and the RSSI is converted to signal strength interval conversion (SPR) to increase the grid recognition degree, thereby increasing the positioning accuracy. In addition, we use the educational version of IBM ILOG CPLEX and the simulated annealing algorithm to find the optimal solution. Finally, experimental results show that CPLEX needs to spend a lot of time when solving a large topology. Therefore, simulated annealing algorithm is useful in practical applications.

References

1. Hossain, M.I., Markendahl, J.: IoT-communications as a service: actor roles on indoor wireless coverage. In: Conferences, 18 January 2018
2. Martin, P., Ho, B.J., Grupen, N., Muñoz, S., Srivastava, M.: An iBeacon primer for indoor localization: demo abstract. In: Proceedings of the 1st ACM Conference on Embedded Systems for Energy-Efficient Buildings, BuildSys 2014, pp. 190–191. ACM, New York (2014)
3. Boukerche, A., Oliveira, H.A.B.F., Nakamura, E.F., Loureiro, A.A.F.: Localization systems for wireless sensor networks. IEEE Wirel. Commun. **14**, 6–12 (2007)

4. He, Z., Cui, B., Zhou, W., Yokoi, S.: A proposal of interaction system between visitor and collection in museum hall by iBeacon. In: 2015 10th International Conference on Computer Science and Education (ICCSE), pp. 427–430 (2015)
5. Stojkoska, B.R., Kosović, I.N., Jagut, T.: How much can we trust RSSI for the IoT indoor location-based services? In: Conferences, 23 November 2017

The 9th International Workshop on Trustworthy Computing and Security (TwCSec-2018)

A Secure Framework for User-Key Provisioning to SGX Enclaves

Takanori Machida(✉), Dai Yamamoto, Ikuya Morikawa,
Hirotaka Kokubo, and Hisashi Kojima

Fujitsu Laboratories Ltd., 4-1-1, Kamikodanaka,
Nakahara-ku, Kawasaki, Kanagawa 211-8588, Japan
{m-takanori,yamamoto.dai,morikawa.ikuya,kokubo.hirotaka,
hkojima}@jp.fujitsu.com

Abstract. Intel Software Guard Extensions (SGX) protects user software from malware by maintaining the confidentiality and integrity of the software executed in secure enclaves on random access memory. However, the confidentiality of its stored executable is not guaranteed. Therefore, secret information, e.g. user keys, should be provided to the enclaves via appropriate secure channels. Although one of the solutions is to use *remote attestation* function of SGX, there is a potential risk that user keys are exposed to malicious insiders at the service provider of remote attestation. In this paper, we propose a novel and secure framework for user-key provisioning to SGX enclaves. Our framework utilizes *sealing* function of SGX, and consists of two phases: the provisioning phase and the operation phase. In the provisioning phase, a user key is encrypted by sealing function, and it is stored in storage. Our assumption is that this phase is performed in a secure environment. In the operation phase, the encrypted blob is read from the storage and decrypted. Then, SGX applications can use the user key without exposing it to attackers. We implemented a prototype of our framework using a commercial Intel CPU and evaluated its feasibility.

1 Introduction

Intel Software Guard Extensions (SGX) published in 2013 [1,3,8] is a function included in modern Intel processors and under contemporary focus for several years as one of the hardware-based isolation technologies that protects user software from malware. Intel SGX reserves parts of random access memory (RAM) for "secure enclaves". In secure enclaves, stored data are encrypted with processor internal keys and verified with message authentication codes in the processor. Since the trust boundary of Intel SGX is the processor package, confidentiality and integrity of user software executed in the secure enclaves is ensured, and user software is protected from malware outside these enclaves, even if such malware has OS root privileges.

The preliminary version of this paper was presented in the poster session of the 39th IEEE Symposium on Security and Privacy.

L. Barolli et al. (Eds.): NBiS 2018, LNDECT 22, pp. 725–732, 2019.
https://doi.org/10.1007/978-3-319-98530-5_63

Intel SGX is not the only kid on the block. In 2016, AMD also published a hardware-based isolation technology for cloud environments, called AMD Secure Memory Encryption/Secure Encrypted Virtualization (SME/SEV) [5], which is implemented in modern AMD processors. AMD SME/SEV is based on virtual machine (VM) where parts of RAM reserved for a guest VM are encrypted with a processor internal key unique for each VM. We compare it with Intel SGX, and conclude that one of the advantages of SGX is that it is feasible even in edge and endpoint environments. That is because, only SGX has *sealing* function that realizes a secure storage even in offline environments. In this paper, we focus on the SGX-based protection of applications in edge and endpoint environments.

Beginning in 2017, many researchers published more than a hundred of papers related to Intel SGX and proposed its various applications including not only cloud applications but also edge and endpoint applications, e.g. database management systems [2,9,10], biometric authentication systems [11], smart grids [12], system logs [7], and digital rights management (DRM) [3]. These applications often include secret information such as keys for encryption, decryption, authentication, and signing, which are predefined by an independent software vendor (ISV). Although these keys (user keys) are protected in secure enclaves on RAM, an attacker can get the user keys with binary analysis of SGX executables stored not in RAM but in disk storage before executions of SGX user software; SGX executables have no confidentiality[1]. Consequently, how we set user keys into user software executed in secure enclaves on RAM becomes one of the challenges for Intel SGX. One of the solutions is to use *remote attestation*, a function of SGX. Remote attestation function verifies the trustworthiness of platforms where SGX software is executed, and enables remote users to establish secure communication channels and provide sensitive data with the SGX software through the channels. However, the services of other parties, e.g. Intel, are required for the verification. Therefore, the trust chain of procedure for user-key provisioning using remote attestation inevitably involves other parties. That is, there is a potential risk that user keys are exposed to malicious insiders at other parties. In this paper, we propose a secure framework for user-key provisioning to SGX enclaves, on edge and endpoint environments. Our framework utilizes sealing function of SGX, and consists of two phases.

The rest of the paper is organized as follows. Sections 2 and 3 show the outlines of Intel SGX and AMD SME/SEV, respectively. In Sect. 4, SGX is compared with SME/SEV, and the advantages of SGX are clarified. Section 5 introduces the conventional methods of user-key provisioning to SGX enclaves and their problems, and proposes a novel and secure framework. Finally, we conclude the paper in Sect. 6.

2 Intel SGX

This section explains three functions of Intel SGX.

[1] The integrity of the executable is maintained with a certificate attached to the executable.

Isolated Execution

The prime function of Intel SGX is *isolated execution*, of which outline is introduced in Sect. 1. A computer with SGX is booted, then basic input/output system (BIOS) reserves parts of RAM up to 94 MB for secure enclave that protect user software (a process). The memory controller in the processor prevents malware, even with the privileges comparable to BIOS, device driver, hypervisor, and OS, from accessing the user software executed in the secure enclave. Furthermore, the parts of RAM corresponding to the secure enclave are encrypted with ephemeral keys generated on every boot of the processor, which is based on a symmetric cryptographic scheme. Therefore, SGX is also resistant to physical attacks such as probing attacks to the memory modules.

Sealing/Unsealing

Software, executed in secure enclave, sometimes generates ephemeral data. We might need to reuse the data in the next execution. The sealing function encrypts such data with seal keys, which depend on the processor key, which in turn is based on a symmetric cryptographic scheme. Sealing enables us to transfer data securely from the secure enclave to storage. Unsealing regenerates the seal keys and decrypts the encrypted data. There are two types of seal keys. One is derived from the processor key and hash value of user software executed in the secure enclave on RAM, i.e. it is unique for each software installation. Hence, only when that same software is executed on the same processor as used in the encryption, can data be decrypted correctly. The other seal key is derived from the processor key and hash value of the ISV public key[2] that ISV sets for user software when it is compiled. If ISV sets the same ISV public key into different software, the data can be decrypted correctly since the same seal key can be regenerated. The latter is utilized for sharing sensitive data among software on the same processor.

Remote Attestation

Remote attestation verifies whether software in secure enclave is executed on a genuine Intel processor or not. This function uses a group signature based on the processor key. A verifier requests a certificate, signed with a member secret key derived from the processor key, to the SGX user software (prover) via networks. The Intel attestation verification service verifies the trustworthiness of the certificate anonymously using the group public key. The verifier can obtain the result by asking the service. Remote users can establish secure communication with SGX software by attaching an ephemeral public key to the certificate.

3 AMD SME/SEV

This section reports the outline of AMD SME/SEV, which is based on our survey of papers published by AMD [4–6].

[2] The public key is essentially used for detecting forgeries of executables.

SME encrypts parts of RAM with ephemeral keys generated inside the processor. SEV protects guest VMs from hypervisor and the other VMs by combining AMD Virtualization (AMD-V) with SME. That is, parts of RAM reserved for hypervisor and guest VMs are encrypted with ephemeral keys different for each of VM and hypervisor. Therefore, the trust boundary of AMD SME/SEV is also the processor package, and malware even with hypervisor privileges cannot decrypt the encrypted RAMs of the guest VMs.

The procedure of launching guest VMs on AMD SME/SEV is shown in Fig. 1. This enables guest owners to provide sensitive data with guest VMs securely. For instance, a guest owner launches his VM on a cloud machine according to the following procedure.

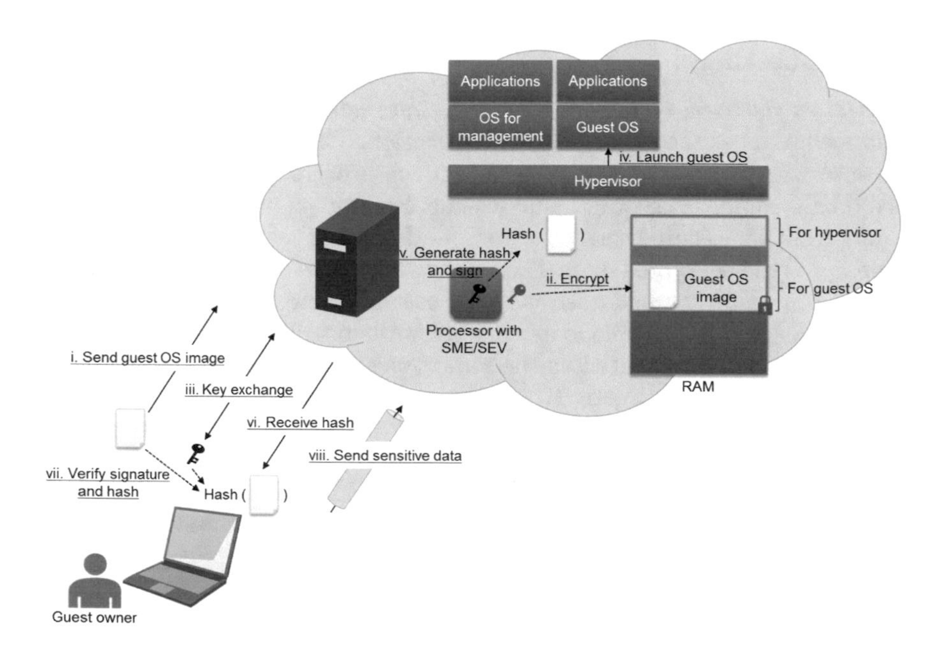

Fig. 1. Procedure of launching a guest VM on AMD SME/SEV

i. A guest owner sends a guest OS image to the hypervisor, then the plaintext of the image is allocated on RAM.
ii. Once the hypervisor requests to launch the image to the processor, an ephemeral key is generated from the random number generator included in the processor, and the image on the RAM is encrypted with the key.
iii. The guest owner exchanges a key, used in both steps v and vii, with the processor, of which trustworthiness is verified by the signature of the owner embedded in advance inside the processor.
iv. The hypervisor launches the image.
v. The processor generates the hash value of the encrypted image, and attaches a signature based on the key obtained in the step iii to the value.

vi. The processor returns the hash value with the signature to the guest owner through the hypervisor.
vii. The guest owner verifies the authenticity of the hash value by using the exchanged key, and checks whether the value coincides with the hash value of the image that the owner sends in the step i.
viii. The owner provides sensitive data such as a user key through a secure channel of Transport Layer Security (TLS) with the guest VM directly. The owner sets encrypted data with the user key to storage, then it can be decrypted and used in the guest VM.

4 Comparison Between Intel SGX and AMD SME/SEV

This section compares Intel SGX with AMD SME/SEV. The differences between them are shown in Table 1. According to the third difference, we conclude that the important advantage of SGX is that it is feasible even in edge and endpoint environments. That is because, the edge and endpoint devices would not operate constantly and be in offline environments. In this paper, we focus on enhancing security of edge and endpoint environments with using SGX.

Table 1. Differences between Intel SGX and AMD SME/SEV

		SGX	SME/SEV
1	What is scope of isolation?	Process	VM
2	Can verify trustworthiness of platforms without using other party services?	No	Yes
3	Can keep secret confidential even in offline environments, i.e. has sealing function?	Yes	No
4	RAM size limitation?	94 MB	-

5 User-Key Provisioning to SGX Enclaves

Although SGX is feasible in edge and endpoint environments, there is a challenge regarding how to provision user keys securely to SGX enclaves. A facile solution is to embed user keys to SGX executables. However, attackers can control the devices physically in these environments, and get the user keys easily with binary analysis of the executables due to no confidentiality.

5.1 Conventional Methods

The first possible solution is the obfuscation of the user keys inside the executables. Although this can be demonstrated at low cost, it cannot keep the user keys completely confidential.

Another solution is to use the remote attestation function. As mentioned in Sect. 2, remote attestation enables users to establish a secure communication channel with an SGX application deployed on a remote platform with using Intel attestation verification service. The users can send their keys securely to the SGX application through the channel. The disadvantage of this method is that the trust chain of this provisioning procedure involves other parties inevitably. That is, there is a potential risk that user keys are exposed to malicious insiders at other parties.

5.2 Our Framework on Edge and Endpoint Environments

Our proposed method can provision user keys to SGX enclave without involving other parties, on edge and endpoint environments. Our framework consists of two phases: the provisioning phase and the operation phase.

Provisioning Phase

This phase provides user keys to secure enclaves. Once the phase is completed, SGX applications using the secure enclaves can utilize the user keys securely. The only requirement of our framework is that devices where an SGX executable dedicated for the provisioning phase is executed are not compromised by malware during this phase. A situation satisfying the requirement is that the devices have not yet been used in real-world environments connected to public networks, e.g. have only been used in environments like a manufacturing factory that makes the devices, or a closed network for administrators. The procedure for the provisioning phase is as follows (also as shown in Fig. 2).

(p-1) The dedicated executable including user keys is set to a device by an ISV.
(p-2) An administrator executes the executable in a secure enclave.
(p-3) The executable performs the sealing function in order to encrypt the user keys with a seal key of the second type mentioned in Sect. 2, and saves the encrypted user keys in storage.
(p-4) The dedicated executable with the user keys must be removed from the device since SGX executables have no confidentiality.

Operation Phase

In this phase, the SGX applications can utilize the user keys sealed in the provisioning phase. The procedure for the phase is as shown below and in Fig. 2.

(o-1) The ISV provides the device with executables of the SGX applications that have the ISV public key identical to the one included in the dedicated executable used in the provisioning phase.
(o-2) An application user or administrator executes the executables.
(o-3) The executables read the encrypted user keys from storage, and perform the unsealing function that decrypts the encrypted user keys with the seal key derived from the processor key and the identical ISV public key.
(o-4) The user key are utilized in the performance of the SGX applications, such as DRM and biometric authentication.

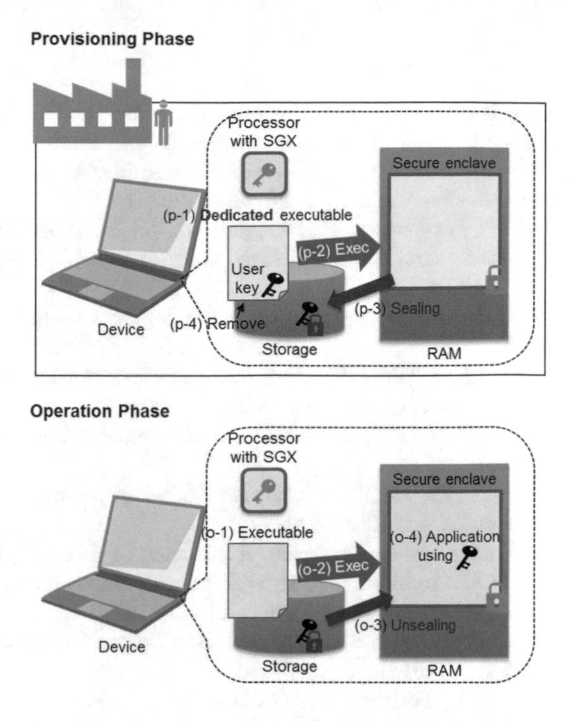

Fig. 2. Our framework of user-key provisioning to SGX enclaves

Our framework is not considered for cloud applications because, in the cloud, it is difficult to demonstrate the provisioning phase in a secure environment for ISV; there is a potential risk that secrets included in the executable dedicated for the provisioning phase fall prey to eavesdropping by insiders at a cloud provider. Consequently, our framework can be used just in edge or endpoint environments because administrators on these environments can manage the dedicated executable without exposing it to attackers. We implemented a prototype of our framework using a commercial Intel CPU (Intel Core i7 6700K) and evaluated its feasibility.

6 Conclusion

In this paper, we compared Intel SGX and AMD SME/SEV, and concluded that Intel SGX was more suitable for edge and endpoint environments than AMD SME/SEV. We proposed a novel framework for user-key provisioning to SGX enclaves on edge and endpoint environments. Our framework utilized sealing function of SGX and achieved high security compared to conventional methods.

References

1. Anati, I., Gueron, S., Johnson, S., Scarlata, V.: Innovative technology for CPU based attestation and sealing. Intel Corporation, White Paper, 14 August 2013
2. Baumann, A., Peinado, M., Hunt, G.C.: Shielding applications from an untrusted cloud with Haven. In: OSDI 2014, pp. 267–283 (2014)
3. Hoekstra, M., Lal, R., Pappachan, P., Phegade, V., del Cuvillo, J.: Using innovative instructions to create trustworthy software solutions. Intel Corporation, White Paper, 14 August 2013
4. Kaplan, D.: AMD x86 memory encryption technologies. LSS 2016, Linux Foundation Events (2016)
5. Kaplan, D., Powell, J., Woller, T.: AMD memory encryption. Advanced Micro Devices (AMD), White Paper, 21 April 2016
6. Kaplan, D., Woller, T., Powell, J.: AMD memory encryption tutorial. In: ISCA 2016, Tutorial (2016)
7. Karande, V., Bauman, E., Lin, Z., Khan, L.: SGX-log: securing system logs with SGX. In: AsiaCCS 2017, pp. 19–30 (2017)
8. McKeen, F., Alexandrovich, I., Berenzon, A., Rozas, C.V., Shafi, H., Shanbhogue, V., Savagaonkar, U.R.: Innovative instructions and software model for isolated execution. Intel Corporation, White Paper, 14 August 2013
9. Priebe, C., Vaswani, K., Costa, M.: EnclaveDB: a secure database using SGX. In: S&P 2018, pp. 264–278 (2018)
10. Schuster, F., Costa, M., Fournet, C., Gkantsidis, C., Peinado, M., Mainar-Ruiz, G., Russinovich, M.: VC3: trustworthy data analytics in the Cloud using SGX. In: S&P 2015, pp. 38–54 (2015)
11. Shepherd, C., Akram, R.N., Markantonakis, K.: Towards trusted execution of multi-modal continuous authentication schemes. In: SAC 2017, pp. 1444–1451 (2017)
12. Silva, L.V., Marinho, R., Vivas, J.L., Brito, A.: Security and privacy preserving data aggregation in cloud computing. In: SAC 2017, pp. 1732–1738 (2017)

Evaluation of User Identification Methods for Realizing an Authentication System Using s-EMG

Hisaaki Yamaba[1(✉)], Kentaro Aburada[1], Tetsuro Katayama[1], Mirang Park[2], and Naonobu Okazaki[1]

[1] University of Miyazaki, Miyazaki, Japan
yamaba@cs.miyazaki-u.ac.jp
[2] Kanagawa Institute of Technology, Atsugi, Japan

Abstract. At the present time, mobile devices such as tablet-type PCs and smart phones have widely penetrated into our daily lives. Therefore, an authentication method that prevents shoulder surfing is needed. We are investigating a new user authentication method for mobile devices that uses surface electromyogram (s-EMG) signals, not screen touching. The s-EMG signals, which are detected over the skin surface, are generated by the electrical activity of muscle fibers during contraction. Muscle movement can be differentiated by analyzing the s-EMG. Taking advantage of the caracteristics, we proposed a method that uses a list of gestures as a password in the previous study. In this paper, we employed support vector machines and attempted to improve the gesture recognition method by introducing correlation coefficient and cross-correlation. A series of experiments was carried out in order to evaluate the performance of the method.

1 Introduction

This paper presents an introduction of support vector machine (SVM) to the user authentication method for mobile devices by using surface electromyogram (s-EMG) signals, not screen touching.

An authentication method that prevents shoulder surfing, which is the direct observation of a users personal information such as passwords, comes to be important. At the present time, mobile devices such as tablet type PCs and smartphones have widely penetrated into our daily lives. So, authentication operations on mobile devices are performed in many public places and we have to ensure that no one can view our passwords. However, it is easy for people who stand near such a mobile device user to see login operations and obtain the users authentication information. And also, it is not easy to hide mobile devices from attackers during login operations because users have to see the touch screens of their mobile devices, which do not have keyboards, to input authentication information. On using a touchscreen, users of a mobile device input their authentication information through simple or multi-touch gestures. These gestures include,

L. Barolli et al. (Eds.): NBiS 2018, LNDECT 22, pp. 733–742, 2019.
https://doi.org/10.1007/978-3-319-98530-5_64

for example, designating his/her passcode from displayed numbers, selecting registered pictures or icons from a set of pictures, or tracing a registered one-stroke sketch on the screen. The user has to see the touch screen during his/her login operation; strangers around them also can see the screen.

To prevent this kind of attack, biometrics authentication methods, which use metrics related to human characteristics, are expected. In this study, we investigated application of surface electromyogram (s-EMG) signals for user authentication. S-EMG signals, which are detected over the skin surface, are generated by the electrical activity of muscle fibers during contraction. These s-EMGs have been used to control various devices, including artificial limbs and electrical wheelchairs. Muscle movement can be differentiated by analyzing the s-EMG [1]. Feature extraction is carried out through the analysis of the s-EMGs. The extracted features are used to differentiate the muscle movement, including hand gestures.

In the previous researches [2–6], we investigate the prospect of realizing an authentication method using s-EMGs through a series of experiments. First, several gestures of the wrist were introduced, and the s-EMG signals generated for each of the motion patterns were measured [2]. We compared the s-EMG signal patterns generated by each subject with the patterns generated by other subjects. As a result, it was found that the patterns of each individual subject are similar but they differ from those of other subjects. Thus, s-EMGs can confirm ones identification for authenticating passwords on touchscreen devices. Next, a method that uses a list of gestures as a password was proposed [3,4]. And also, experiments that were carried out to investigate the performance of the method extracting feature values from s-EMG signals (using the Fourier transform) adopted in [5]. The results showed that the Fourier transform has certain ability to extract feature values from s-EMG signals, but further accuracy was desired

In this paper, support vector machines (SVM) were introduced to identify gestures from s-EMG signals. The false acceptance rate and the accidental acceptance rate of this method were evaluated from the results of the experiments.

2 Characteristics of Authentication Method for Mobile Devices

It is considered that user authentication of mobile devices has two characteristics [2].

One is that an authentication operation often takes place around strangers. An authentication operation has to be performed when a user wants to start using their mobile devices. Therefore, strangers around the user can possibly see the user's unlock actions. Some of these strangers may scheme to steal information for authentication such as passwords.

The other characteristic is that user authentication of mobile devices is almost always performed on a touchscreen. Since many of current mobile devices do not have hardware keyboards, it is not easy to input long character based passwords into such mobile devices. When users want to unlock mobile touchscreen devices, they input passwords or personal identification numbers (PINs) by tapping numbers or characters displayed on the touchscreen. Naturally, users have to look at their touchscreens while unlocking their devices, strangers around them also can easily see the unlock actions. Besides, the user moves only one finger in many cases. So, it becomes very easy for thieves to steal passwords or PINs.

To prevent shoulder-surfing attacks, many studies have been conducted. The secret tap method [7] introduces a shift value to avoid revealing pass-icons. The user may tap other icons in the shift position on the touchscreen, as indicated by a shift value, to unlock the device. By keeping the shift value secret, people around the user cannot know the ture pass-icons, although they can still watch the tapping operation. The rhythm authentication method [8] relieves the user from looking at the touchscreen when unlocking the device. In this method, the user taps the rhythm of his or her favorite music on the touchscreen. The pattern of tapping is used as the password. In this situation, the users can unlock their devices while keeping them in their pockets or bags, and the people around them cannot see the tap operations that contain the authentication information.

3 User Authentication Using s-EMG

The s-EMG signals (Fig. 1) are generated by the electrical activity of muscle fibers during contraction and are detected over the skin surface (Fig. 2) [2]. Muscle movement can be differentiated by analyzing the s-EMG.

In the previous research, the method of user authentication by using s-EMGs that do not require looking at a touchscreen was proposed [3,4]. The s-EMG signals are measured, and the feature values of the measured raw signals are extracted. We estimate gestures made by a user from the extracted features. In this study, combinations of the gestures are converted into a code for authentication. These combinations are inputted into the mobile device and used as a password for user authentication.

1. At first, pass-gesture registration is carried out. A user selects a list of gestures that is used as a pass-gesture. (Fig. 3(a))

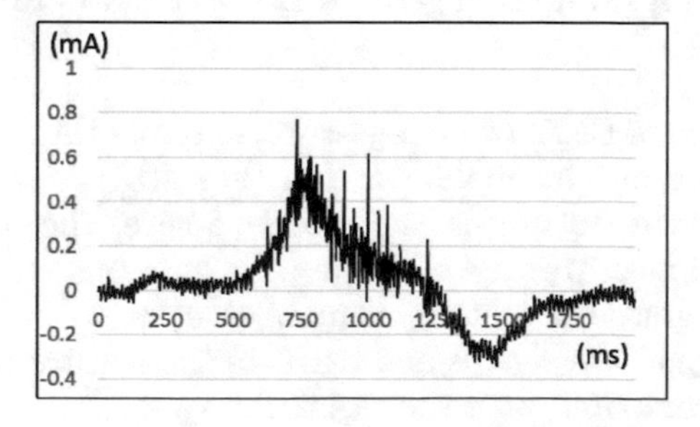

Fig. 1. A sample of an s-EMG signal

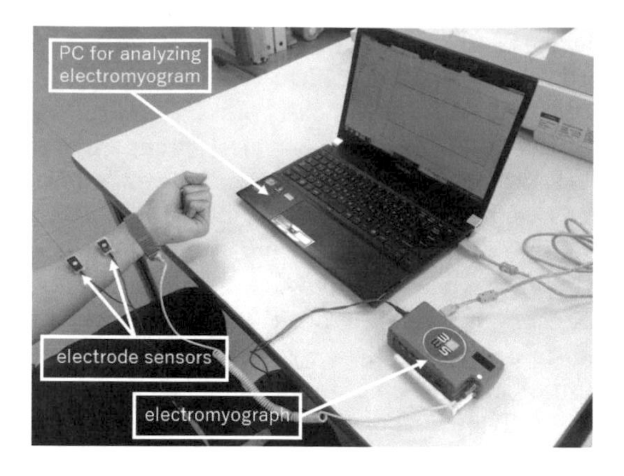

Fig. 2. Measuring an s-EMG signal

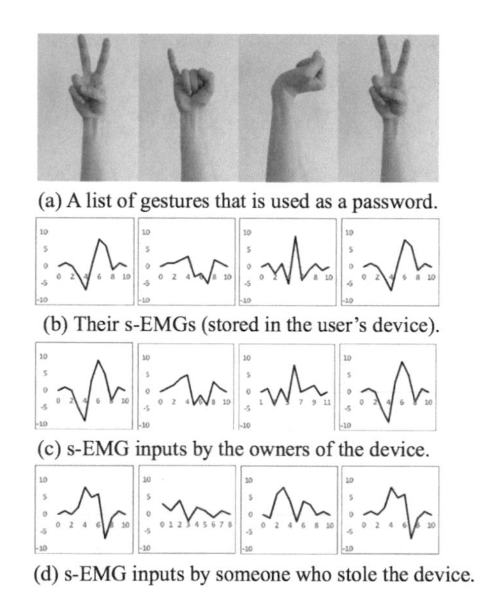

(a) A list of gestures that is used as a password.
(b) Their s-EMGs (stored in the user's device).
(c) s-EMG inputs by the owners of the device.
(d) s-EMG inputs by someone who stole the device.

Fig. 3. A list of gestures used as a password

2. The user measures s-EMG of each gesture, extracts their feature values, and register the values into his mobile device. (Fig. 3(b))
3. When the user tries to unlock the mobile device, the user reproduces his pass-gesture and measures the s-EMG.
4. The measured signals are sent to his mobile device.
5. The device analyzes the signals and extracts the feature values.
6. The values are compared with the registered values.
7. If they match, the user authentication will succeed. (Fig. 3(c))

8. On the other hand, an illegal user authentication will fail because a list of signals given by someone who stole the device (Fig. 3(d)) will not be similar with the registered one.

Adopting s-EMG signals for authentication of mobile devices has three advantages. First, the user does not have to look at his/her device. Since the user can make a gesture that is used as a password on a device inside a pocket or in a bag, it is expected that the authentication information can be concealed. No one can see what gesture is made. Next, it is expected that if another person reproduces a sequence of gestures that a user has made, the authentication will not be successful, because the extracted features from the s-EMG signals are usually not the same between two people. And then, a user can change the list of gestures in our method. This is the advantages of our method against other biometrics based methods such as fingerprints, an iris, and so on. When authentication information, a fingerprint or an iris, come out, the user can't use them because he/she can't change his/her fingerprint or iris. But the user can arrange his/her gesture list again and use the new gesture list.

4 Employment of Support Vector Machines

4.1 Support Vector Machines

Since measured s-EMG signals vary by subject, the extracted features do not show enough performance to correctly differentiate the muscle movement in multiple subjects. Therefore, researchers have explored other methods to improve the performance of feature extraction. Since some methods demonstrate good performance for some subjects but other methods show better performance for other subjects, a feature that can be used to distinguish gestures for everyone is desired. For example, a method that uses the maximum value and the minimum value of raw s-EMG signals was proposed [9].

We introduced support vector machines to identify gestures from s-EMG signal patterns in [4–6]. Support vector machines are one of the pattern recognition models of supervised learning. Linear SVM was proposed in 1963, and extended to nonlinear classification in 1992. A support vector machine builds a classifier for sample data that belong to one of two classes. An SVM trains the separation plane that has the largest margin, and samples on the margin are called support vectors.

4.2 Improvement in the Gesture Recognition Method

In this research, one SVM is prepared for each gesture of one user and trained by data of the corresponding gesture. Theses SVMs decide whether a given s-EMG signal is generated by the corresponding gesture or not. It is expected that SVMs trained under this manner can express the characteristics of the corresponding gegsture. However, they are trained so as not to reject the gesture made by other person. So, we adopted a hard rule to accept a password (a list of gestures) when all the gestures are accepted because false acceptance rates of SVMs may worse than their true rejecting rates. When we compose the passwords of a subject, we select better gestures that have lower a ture rejecting rate in a preliminary experiment.

In order to improve the ability of gesture recognition, we introduced new set of feature values: sum total, standard deviation, mean, sum of squares, skewness, kurtosis, five-number summary. We divide one series of signal data into ten segments and we extract the eleven feature values mentioned above from all the ten segments. We can obtain one hundred and ten values from one series of signal data.

4.3 Selection of Appropreate Data for Training SVMs

In order to improve performance of a SVM, we introduced a selection method of training data from all measured data that can exclude error data and data that deviate from the average. First, we select a specimen waveform from measured data set. Then, a training data set is made from waveforms that are similar with the specimen waveform. To select an appropriate specimen waveform, we

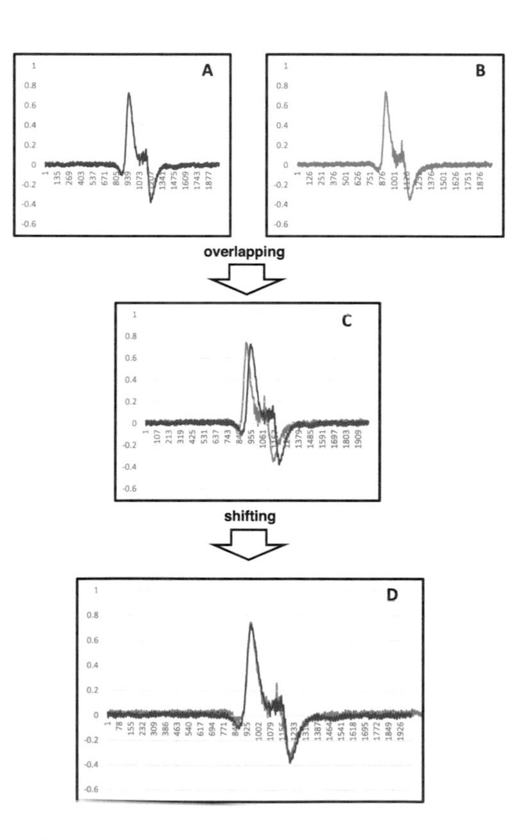

Fig. 4. Alignment of two sets of signal data

employed correlation efficient. A specimen waveform of each gesture of each user is selected from n waveform data (n is the number of measurement). The similarities between each waveform and other $n-1$ waveforms are calculated using correlation coefficient and their averages are obtained. The waveform that has the highest average value is selected as the specimen waveform. Waveforms whose similarity with the specimen waveform are higher are selected as the training data. In the experiments explained in the next section, 20 waveforms from 30 measured waveforms are used as the training data of each gesture.

In order to obtain valid values of correlation coefficient, we have to align the start time of gestures in s-EMG signals (Fig. 4). We used cross-correlation function to achieve this. In signal processing, cross-correlation function is a measure of similarity of two series as a function of the displacement of one relative to the other. For continuous functions f and g, the cross-correlation is defined as:

$$(f * g)(m) = \sum_{n} f(n)g(m-n) \quad (1)$$

We shifted the start time of the measured data of s-EMG signals so as to minimize the value of the cross-correlation function.

5 Experiments

5.1 Objectives

First, we carried out experiments to evaluate the performance of the gesture identification method. Concretely, true rejecting rates (TRR) of each gesture of each subject were investigated. In this experiment, TRR means the rate that the number of waveform inputs of the same gesture of the same subject that are regarded as wrong.

Next, we simulated a scene of user authentication. In this experiment, four gestures were selected from the measured gestures and they were used as a four-character password. We evaluated the FAR (false acceptance rate) of these passwords, and also investigated the resistance against an accidental success.

5.2 Conditions

Eleven students of University of Miyazaki participated as experimental subjects (Subject #1-#11.) And the 12 hand gestures shown in Fig. 5 were introduced. But the eight gestures (B-J) were selected in the experiments because their s-EMG signals were clearer than other gestures.

The set of DL-3100 and DL-141 (S&M Inc.) that was an electromyograph used in the previous researches also used to measure the s-EMG of each movement pattern in this study.

The measured data were stored and analyzed on a PC. The subjects repeated each gesture ten times and their s-EMG signals were recorded. This measurement was carried out 3 times and 30 signals were obtained for each subject and for

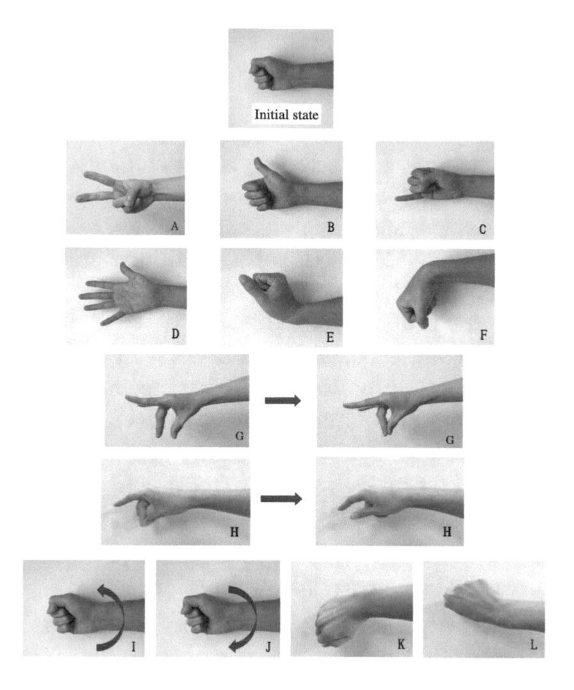

Fig. 5. Gestures used in the experiments

each gesture. 20 signals were selected form the 30 signals and the 20 signals were used to train each SVM. The selected 20 signals were the specimen waveform and the 19 waveforms of s-EMG signals that were more similar with the specimen waveform. The similarity was measured using correlation coefficient as mentioned in Sect. 4.3. In this research, the programming language "R" was adopted to train SVMs.

A password used in the 2nd experiment is made for each of the experimental subjects (six students). Four gestures that composed a gesture were selected according to the results of the 1st experiment. Concretely, the top four gestures whose true rejecting rate were lower were selected. In the 2nd experiment, user authentication was judged to be success when all four gestures in a password match.

It is assumed that an attacker does not know which his gesture generates s-EMG signals that is similar with a specific gesture of a victim user.

5.3 Results

First, a true rejecting rate of each gesture was investigated. Ten signals that were the rest of the measured 30 signals of gestures of a subject were given to the corresponding SVM. Table 1 shows the number of incorrected answers of each gesture. Next, the false acceptance rates of four gestrures passwords were investigated. A password was composed based on the results of the 1st

Table 1. Number of incorrect answers.

Subject	B	C	D	E	F	G	H	I	J
1	$\underline{0}$	2	$\underline{0}$	0	3	2	$\underline{0}$	1	$\underline{0}$
2	$\underline{0}$	10	$\underline{0}$	$\underline{0}$	$\underline{1}$	5	5	4	6
3	0	$\underline{0}$	$\underline{0}$	2	1	1	0	$\underline{0}$	$\underline{0}$
4	1	$\underline{0}$	$\underline{0}$	1	1	$\underline{0}$	2	$\underline{0}$	4
5	$\underline{0}$	0	1	3	$\underline{0}$	$\underline{0}$	6	4	$\underline{0}$
6	$\underline{0}$	$\underline{0}$	2	$\underline{0}$	0	0	0	0	$\underline{0}$

Table 2. FAR of four character passwords.

Subject	FAR	Accidental success rate
1	0%	0.00000000%
2	0%	0.00000102%
3	0%	0.00000000%
4	0%	0.00000174%
5	0%	0.00003196%
6	0%	0.00000007%

experiment. For example, the password of the subject #1 was "BDHJ." The same sequences of the gestures were generated for other subjects and given to the SVMs of the subject #1. Since each subject had 30 measured signals for each gesture, $30 \times 30 \times 30 \times 30 \times 5 = 4,050,000$ inputs (combinations of gestures) were given to each of the subject's password. Table 2 shows the false acceptance rates. All of the passwords could reject all incorrect inputs. Table 2 also shows the results of accidental success. Same with the 2nd experiments, ${}_9P_4 \times 30^4 \times 5$ inputs (four gestures selected from nine, 30 data for each gesture, five other subjects) were given to each of the subject's password.

5.4 Discussion

The results were overall good. The results of true rejection rate of some gestures were not so good for each of the subjects. However, each of the subjects had a gesture that recorded a quite good result from the view point of true rejecting. Since the password of each subject used in the experiment included such gestures, it is considered that the passwords used in the experiments could show the good results.

6 Conclusion

We investigated a new user authentication method that can prevent shoulder-surfing attacks in mobile devices. To improve the performance of the method using SVMs, we adopted correlation coefficient and cross-correlation function and introduced the eleven new candidates of feature values. A series of experiments was carried out to investigate the effectiveness of the improved method. False acceptance rates and resistance against an accidental access were investigated. The results showed the method using SVM is promising. We are planning to introduce another feature value, introduce other gesture candidate, explore appropriate length of pass-gesture, and so on.

Acknowledgements. This work was supported by JSPS KAKENHI Grant Numbers JP17H01736, JP17K00186.

References

1. Tamura, H., Okumura, D., Tanno, K.: A study on motion recognition without FFT from surface-EMG. IEICE Part D **J90–D**(9), 2652–2655 (2007). (in Japanese)
2. Yamaba, H., Nagatomo, S., Aburada, K.: An authentication method for mobile devices that is independent of tap-operation on a touchscreen. J. Robot. Netw. Artif. Life **1**, 60–63 (2015)
3. Yamaba, H., Kurogi, T., Kubota, S., et al.: An attempt to use a gesture control armband for a user authentication system using surface electromyograms. In: Proceedings of 19th International Symposium on Artificial Life and Robotics, pp. 342–245 (2016)
4. Yamaba, H., Kurogi, T., Kubota, S.: Evaluation of feature values of surface electromyograms for user authentication on mobile devices. Artif. Life Robot. **22**, 108–112 (2017)
5. Yamaba, H., Kurogi, T., Aburada, A., et al.: On applying support vector machines to a user authentication method using surface electromyogram signals. Artif. Life Robot. (2017). https://doi.org/10.1007/s10015-017-0404-z
6. Kurogi, T., Yamaba, H., Aburada, A., et al.: A study on a user identification method using dynamic time warping to realize an authentication system by s-EMG Advances in Internet. Data Web Technol. (2018). https://doi.org/10.1007/978-3-319-75928-9_82
7. Kita, Y., Okazaki, N., Nishimura, H., et al.: Implementation and evaluation of shoulder-surfing attack resistant users. IEICE Part D **J97–D**(12), 1770–1784 (2014). (in Japanese)
8. Kita, Y., Kamizato, K., Park, M., et al.: A study of rhythm authentication and its accuracy using the self-organizing maps. Proceedings of DICOMO 2014, pp. 1011–1018 (2014). (in Japanese)
9. Tamura, H., Goto, T., Okumura, D., et al.: A study on the s-EMG pattern recognition using neural network. IJICIC **5**(12), 4877–4884 (2009)

Person Tracking Based on Gait Features from Depth Sensors

Takafumi Mori[1(✉)] and Hiroaki Kikuchi[2]

[1] Graduate School of Advanced Mathematical Sciences, Meiji University, Tokyo 164-8525, Japan
cs172059@meiji.ac.jp
[2] School of Interdisciplinary Mathematical Science, Meiji University, Tokyo 164-8525, Japan
kikn@meiji.ac.jp

Abstract. Gait information is a useful biometric because it is a user-friendly property and gait is hard to mimic exactly, even by skillful attackers. Most conventional gait authentication schemes assume cooperation by the subjects being recognized. Lack of cooperation could be an obstacle for automated tracking of users and many commercial users require new gait identification schemes that do not require the help of target users. In this work, we study a new person-tracking method based on the combination of some gait features observed from depth sensors. The features are classified into three groups: static, dynamic distances, and dynamic angles. We demonstrate with ten subjects that our proposed scheme works well and the accuracy of equal error ratio can be improved to 0.25 when the top five features are combined.

1 Introduction

With the popularization of the Internet, the necessity of safe personal authentication is increasing. Conventional, knowledge-based authentication, i.e., password or PIN, is unsafe because people often forget the confidential information and there is a risk of the leakage of personal information. As a result, biometric authentication, which uses individual biological attributes, is becoming popular. In this study, we focus on a method for gait authentication that uses features of a subject's style of walking.

Gait authentication has three basic classes: machine vision [1], floor sensors, and wearable sensors [2]. A machine vision-based system authenticates people from a camera located at a distant position so that it can be used for wide-range monitoring. This allows people to be observed without being noticed. Because of these features, gait authentication is considered to be an appropriate method for individual tracking.

Authentication tests whether a given user is registered to a system, while tracking distinguishes two users appearing at distinct locations. Authentication is mainly used to prove that a target is a genuine user when the user logs in to

L. Barolli et al. (Eds.): NBiS 2018, LNDECT 22, pp. 743–751, 2019.
https://doi.org/10.1007/978-3-319-98530-5_65

a system or enters a building. However, tracking is used to identify major walking paths and to obtain statistical information about people's flow paths for marketing or crime prevention.

We found the following differences between authentication and tracking.

- People are cooperative with authentication because they are willing to use it. However, tracking is done while the target users are unconsciously.
- Tracking does not require high accuracy because the data are used for statistical information.
- Tracking should address privacy concerns because target users are not always willing to be tracked.
- Data from tracking should be removed when a person refuses to provide his or her information (opt-out).
- A threat to authentication is a malicious adversary who pretends to be the target user. A threat to tracking is to pretend to be someone else.

The differences between authentication and tracking are summarized in Table 1.

Table 1. Differences between tracking and authentication

	Tracking	Authentication
Application	Statistical information	Prove that i am a proper user
Target	Noncooperative	Cooperative
Desired accuracy	Low	High
Matching	$m : n$	$1 : n$
Privacy care	Necessary	Unnecessary
Threat	Recognizing as other person	Pretend to be proper user

In this paper, we study the use of gait information to track people. We propose a new person-tracking method using gait and demonstrate its feasibility with a trial implementation with the Microsoft Kinect V2.

Our main result is that the proposed scheme performs accurately and the equal error rate (EER) is 0.22 in the optimal case when several features are combined.

2 Related Work

A silhouette image is often used in gait authentication. For example, Han et al. proposed the gait energy image (GEI) [3], which is an average silhouette image of one cycle of walking. It requires less processing time and reduces the storage requirement and the robustness against noise.

Shiraga et al. proposed GEINet, an authentication system using GEI [1]. They used a convolutional neural network (CNN), and achieved high accuracy in image recognition when classifying GEI images. They demonstrated that gait can be used to authenticate people with high accuracy.

Muaaz et al. proposed a smartphone-based gait authentication method [2]. They used acceleration vectors observed with a smartphone in a pocket as features. A cycle of walking was used as a template. Multiple templates per subject were stored and registered. In authentication, the dynamic time warping (DTW) distance between a given cycle of walking and each template was calculated and a subject who had more than fifty percent of features for which DTW distances are less than a predetermined threshold was accepted. Moreover, they empirically proved that a mimic attack is impossible in gait authentication.

Some methods have used silhouette images. Andersson and Araujo proposed a gait authentication method using skeleton information from Kinect [4] in 2015. Igual et al. proposed a gender recognition method using depth information [5].

3 Proposed Method for Gait Recognition

In this study, we propose a gait recognition method based on sequences of three-dimensional coordinates of joints in walking. Our proposed method consists of the following five steps:

1. Data capture,
2. Cycle extraction,
3. Sequence extraction,
4. Features calculation, and
5. Identification.

3.1 Data Capture

We use Kinect V2, a motion capture device developed by Microsoft. Kinect was designed for Xbox players who control the Xbox using images of their bodies while playing. Kinect is described as a natural user interface (NUI).

Kinect facilities include an RGB camera, a depth camera, and a microphone. It identifies three-dimensional coordinates of joints of the player to recognize the player's movement. The three-dimensional coordinates captured by Kinect are called skeleton data and can be retrieved via the Kinect Software Development Kit. The specifications of the Kinect V2 are shown in Table 2.

3.2 Cycle Extraction

In this phase, we extract a cycle of walking, which is defined as a series of features in walking at which a foot reaches the same position.

To identify a cycle from continuous skeleton data, we calculate the distance between two ankles and smooth it by taking averages of two neighboring data points. Finally, we identify a cycle that begins at the first peak of distances and ends at the third peak. We show an example of ankle distance and a cycle in Fig. 1. In this figure, a cycle is from point A to point E.

Table 2. Specification of kinect V2

Attribute	Value
RGB resolution	1920 × 1080 pixel
Depth resolution	512 × 424 pixel
Frame rate	30[fps]
Num of observable people	6
Num of observable joints	6
Measurable distance	0.5–4.5 m

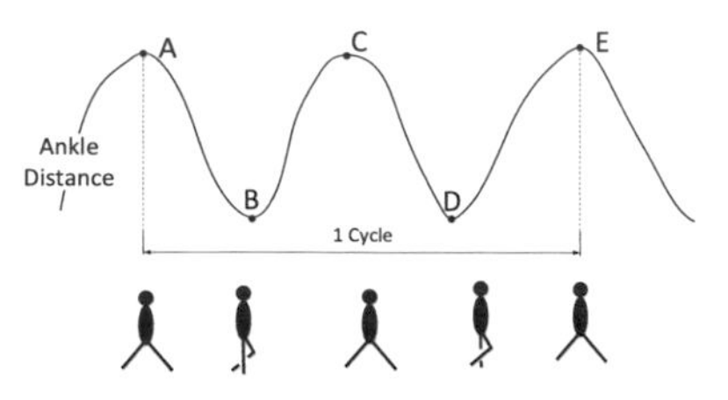

Fig. 1. Sample of one cycle

3.3 Sequence Extraction

We define a total of 36 features that are classified into three groups: static distances, dynamic distances, and joint angles.

Static distances are lengths of between adjacent joints. Because the distance between two adjacent joints is determined by the length of the bone between them, it is a stable quantity. For example, the distance S_1 is the distance between two joints in the left foot and left ankle. The static distances are illustrated with the skeleton model in Fig. 2.

Dynamic distances are distances between two arbitrary joints in a body. The distance varies with movements of the feet and arms while walking. For instance, the distance between the two feet, D_1, fluctuates periodically when a subject is walking. The dynamic distances are illustrated in Fig. 3.

Dynamic Angles are measured between the vertical and a line connecting two joints. They are illustrated in Fig. 3. Dynamic angles are also dynamic quantities.

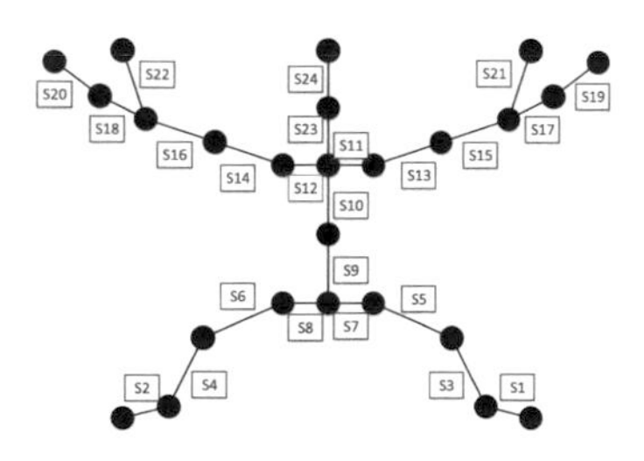

Fig. 2. Diagram of static distances

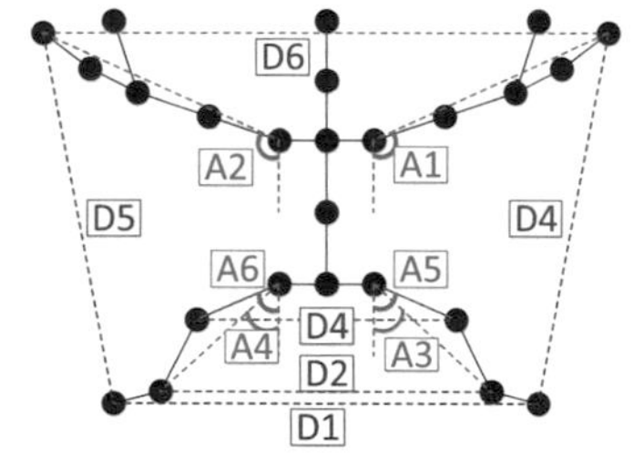

Fig. 3. Diagram of dynamic distances and dynamic angles

3.4 Statistics of Features

In this section, we calculate some useful statistics of features. The features vary in a given cycle; therefore, we use statistics of series of features in a cycle. The statistics include maximum, median, and duration of one cycle in our study.

3.5 Identification

We first consider a simple identification with a single feature and then extend it to the fusion version that combines multiple features.

Let f be a feature of walking, i.e., $f \in \{S_1, \ldots, S_{24}, D_1, \ldots, D_6, A_1, \ldots, A_6\}$. Let $f_{i,k}$ be the kth cycle of the ith subject. Statistics of a series of features $f_{i,k}$ are $\mu(f_{i,k}), median(f_{i,k}), max(f_{i,k})$.

Given two statistics (means) of the kth feature $\mu(f_{i,k})$ and $\mu(f_{j,k'})$, we say that subjects i and j are identical (the same) if:

$$same(i,j) = \begin{cases} T & \text{if } |\mu(f_{i,k}) - \mu(f_{j,k'})| \leq \theta \\ F & \text{otherwise,} \end{cases}$$

where θ is a threshold of matching. The mean can be replaced by other statistics such as the median or maximum.

Next, we extend the simpler identification by combining several features. For example, using Euclidian distance, two features f and g can be tested jointly to identify if i and j are the same:

$$same(i,j) = \begin{cases} T & \text{if } \sqrt{(\mu(f_{i,k}) - \mu(f_{j,k'}))^2 + (\mu(g_{i,k}) - \mu(g_{j,k'}))^2} \leq \theta \\ F & \text{otherwise.} \end{cases}$$

The same steps are applied to the median and maximum values. Note that the Euclidian distance can still be used when combining more than three features.

We may optimize the threshold θ to be the EER. An EER is an error rate at which $\mathrm{FAR}(\theta^*_\ell) = \mathrm{FRR}(\theta^*_\ell)$ at the optimized value θ^*_ℓ, where the false acceptance ratio (FAR) is a fraction of faulty authenticated imposter subjects and the false rejection ratio (FRR) is a fraction of genuine subjects who are wrongly judged as imposters.

4 Experiment

4.1 Experiment Purpose

The purposes of our experiment are as follows:

1. Identify efficient features that can be used to recognize a person accurately,
2. Measure how accurately persons are identified for each set of features, and
3. Find out the best method to combine features to maximize the accuracy of recognizing people.

4.2 Method

We capture ten subjects walking using Kinect V2. Each subjects walks six times. The experiment term was from 5/August/2017 to 17/August/2017. Subjects are uniquely identified with labels U_1–U_{10}.

According to Sect. 3, we set the threshold θ so that FAR is equal to FRR at θ for every feature.

4.3 Results

4.3.1 Data Capture and Calculation of Features

An example of a series of skeleton data is shown in Fig. 4, which shows two-dimensional coordinates of five typical joints, a head, both hands and both ankles, observed from the Kinect sensor over a few cycles.

4.3.2 Calculation of Features and Statistics

The statistics for $\mu(D_5)$ are shown in the bar plot of Fig. 5. Most subjects can be clearly distinguished from each other, except U_9 and U_{10}.

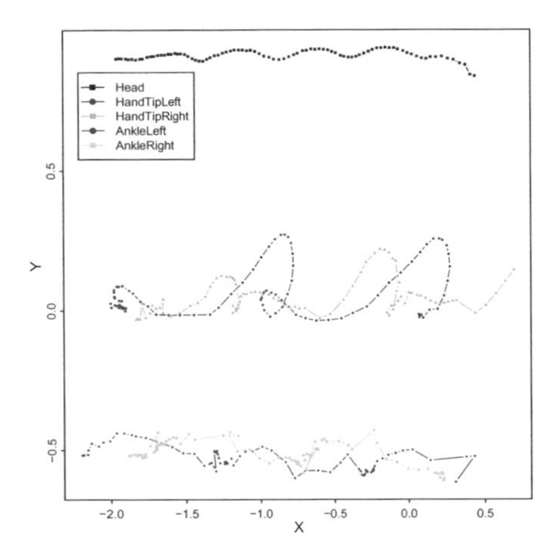

Fig. 4. Example of captured data

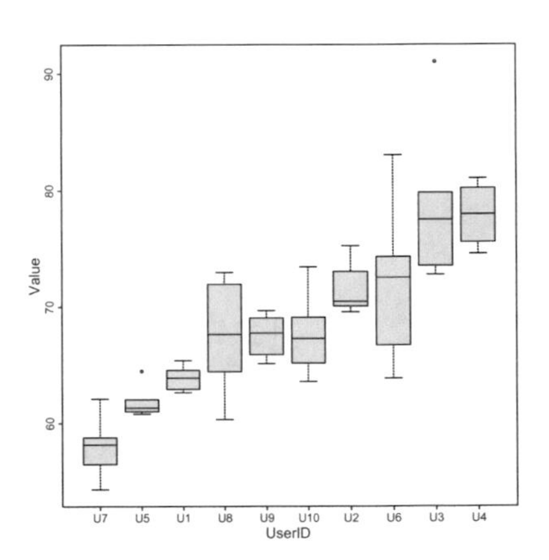

Fig. 5. Distributions of $\mu(D_5)$ for all users

4.3.3 Detection Threshold

A threshold of matching should be carefully determined by looking at the distributions of features. For example, Fig. 6 shows two histograms of $\mu(D_5)$, one for the same subject and the other for between subjects. The figure shows that the variance of distances in the same subject is smaller than that with others.

Receiver operating characteristic (ROC) curves, which show the tradeoff of FAR and FRR for the representative statistics $\mu(S_6)$, $median(D_1)$, and $max(A_2)$ are shown in Fig. 7. We found that $\mu(S_6)$ is the best feature in terms of FAR and FRR for three candidates.

4.4 Evaluation

4.4.1 Comparison of Features

A list of the top 10 EERs in ascending order is shown in Table 3.

Note that there is only one feature using max. Generally, dynamic angles are less useful in identifying subjects and only two features of dynamic angles are in the top 10 list.

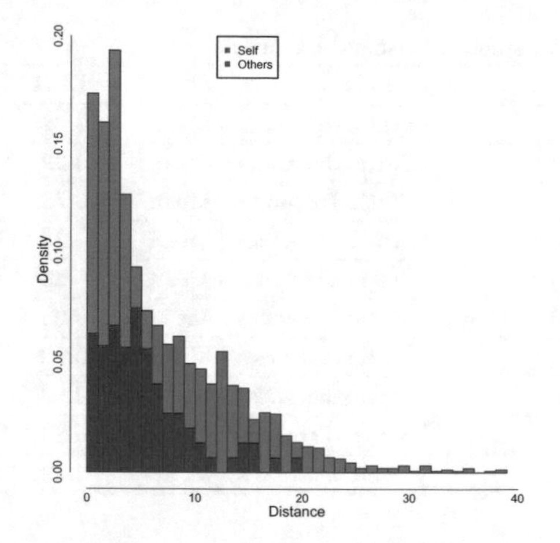

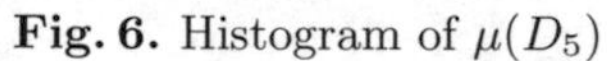
Fig. 6. Histogram of $\mu(D_5)$

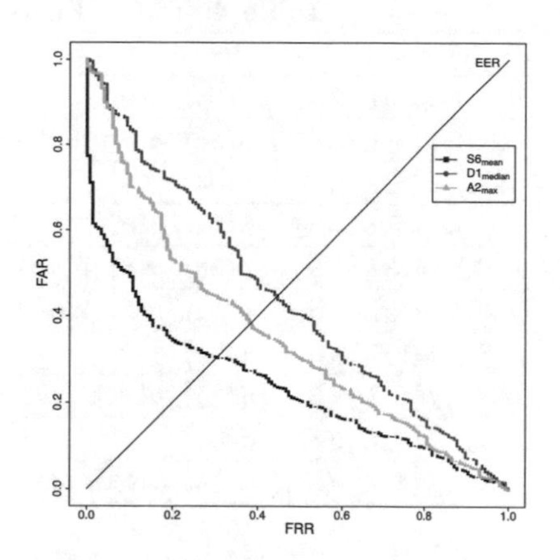

Fig. 7. ROC curves of $\mu(S_6)$, $median(D_1)$ and $max(A_2)$

Table 3. EERs arranged in ascending order

Features	Group	Statistics	EER
$\mu(D_5)$	Dynamic distances	Mean	0.29
$max(D_5)$	Dynamic distances	Max	0.29
$\mu(S_6)$	Static distances	Mean	0.30
$median(D_5)$	Dynamic distances	Median	0.30
$\mu(S_5)$	Static distances	Mean	0.31
$median(A_4)$	Dynamic angles	Median	0.31
$median(S_5)$	Static distances	Median	0.31
$median(S_6)$	Static distances	Median	0.31
$\mu(D_4)$	Dynamic distances	Mean	0.32
$\mu(A_4)$	Dynamic angles	Mean	0.32

4.4.2 Combined Features

We combined multiple features including dynamic distances and dynamic angles. The top 10 EERs listed in ascending order are shown in Table 4.

We show boxplots of EER with respect to the number of features combined using the maxima of dynamic distances in Fig. 8. The EER decreases as the number of combined feature increases, but is saturated at five and no improvement is obtained with six or more features. Combining some features helps to decrease FAR, but increases FRR. Hence, combining too many features could spoil the FAR and result in low EER. Therefore, we conclude that the best number of features to combine is around five in our experiment.

The histogram of maxima of dynamic distances as calculated for self and with others is shown in Fig. 9.

Table 4. Top 10 EERs arranged in ascending order

Features	Group	Statistics	EER
$\mu(S_3)$, $\mu(S_2)$, $\mu(S_{22})$, $\mu(S_9)$, $\mu(S_{16})$	Static distances	Mean	0.22
$\mu(S_3)$, $\mu(S_2)$, $\mu(S_{22})$, $\mu(S_9)$, $\mu(S_{13})$, $\mu(S_{16})$	Static distances	Mean	0.22
$\mu(S_3)$, $\mu(S_2)$, $\mu(S_{22})$, $\mu(S_9)$	Static Distances	Mean	0.23
$\mu(S_3)$, $\mu(S_2)$, $\mu(S_{22})$, $\mu(S_9)$, $\mu(S_{13})$	Static distances	Mean	0.23
$\mu(S_3)$, $\mu(S_2)$, $\mu(S_{22})$, $\mu(S_9)$, $\mu(S_{13})$, $\mu(S_{20})$	Static distances	Mean	0.23
$\mu(S_3)$, $\mu(S_2)$, $\mu(S_{22})$, $\mu(S_9)$, $\mu(S_{13})$, $\mu(S_{16})$, $mu(S_{20})$	Static distances	Mean	0.23
$\mu(S_3)$, $\mu(S_2)$, $\mu(S_9)$, $\mu(S_{13})$	Static distances	Mean	0.23
$\mu(S_5)$, $\mu(S_3)$, $\mu(S_2)$, $\mu(S_{22})$, $\mu(S_9)$, $\mu(S_{13})$	Static distances	Mean	0.23
$\mu(S_3)$, $\mu(S_2)$, $\mu(S_9)$, $\mu(S_{16})$	Static distances	Mean	0.23
$\mu(S_5)$, $\mu(S_3)$, $\mu(S_2)$, $\mu(S_{22})$, $\mu(S_9)$, $\mu(S_{16})$	Static distances	Mean	0.23

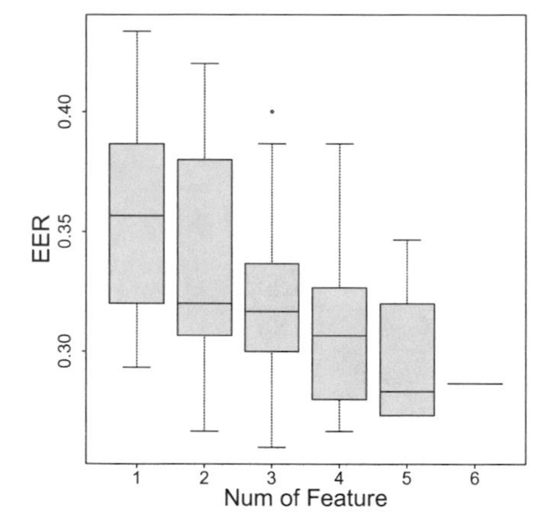

Fig. 8. Correlation of number in combination and EER

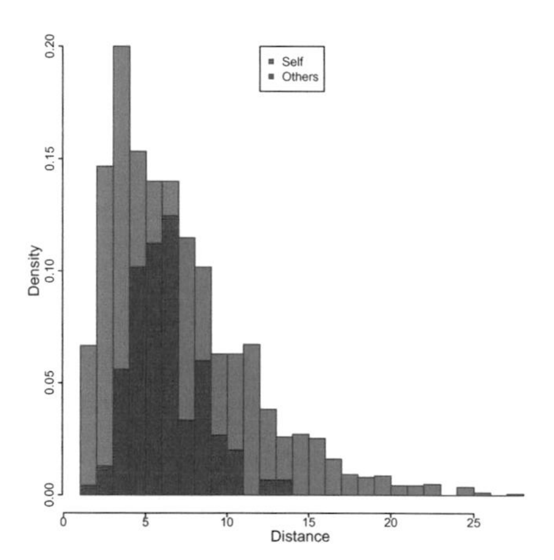

Fig. 9. Histogram of maxima of dynamic distances

4.5 Discussion

Let us consider reasons why some subjects have low accuracy. The cause of outliers in Fig. 5 is considered to be due to measurement errors of the Kinect, which tracks joints based on the image observed from a camera. Hence, if a point is hidden by the body, the Kinect cannot track the joint. Instead, the Kinect tries to estimate the coordinates of hidden joints with some estimation errors.

As for statistics, we found that maximum values are worse than mean and median, as shown in Tables 3 and 4. We think that for maxima, outliers have more significant effects than for the other statistics, which contributes to failure of identification.

We used three kinds of features, static, dynamic and angle. Based on the results in Tables 3 and 4, we found that dynamic distances perform effectively for identification. Although static distances depend on the size of the body and dynamic angles are affected by movements of arms and feet, dynamic distances are affected by both size and movement. We believe that dynamic distances are better features.

5 Conclusions

In this paper, we have proposed a new gait recognition method and demonstrated it using a trial implementation system with Kinect V2. The best EER in our experiment was 0.22 and we conclude that the proposed method can be used for individual tracking in practical situations.

In future work, we plan to study other gait recognition methods that have lower EER or are more robust against walking noise, e.g., shoes and luggage.

References

1. Shiraga, K., Makihara, Y., Muramatsu, D., Echigo, T., Yagi, Y.: GEINet: view-invariant gait recognition using a convolutional neural network. In: Proceedings of the 8th IAPR International Conference on Biometrics (ICB 2016), pp. 1–8, Halmstad, Sweden, June 2016
2. Muaaz, M., Mayrhofer, R.: Smartphone-based gait recognition: from authentication to imitation. IEEE Trans. Mob. Comput. **16**, 3209–3221 (2017)
3. Han, J., Bhanu, B.: Individual recognition using gait energy image. IEEE Trans. Pattern Anal. Mach. Intell. **28**(2), 316–322 (2006)
4. Andersson, V., Araujo, R.: Person identification using anthropometric and gait data from kinect sensor. In: AAAI Conference on Artificial Intelligence (2015)
5. Igual, L., Lapedriza, À., Borràs, R.: Robust gait-based gender classification using depth cameras. EURASIP J. Image Video Process. **2013**(1), 1–11 (2013)

Privacy-Preserving All Convolutional Net Based on Homomorphic Encryption

Wenchao Liu[1,2], Feng Pan[1,2], Xu An Wang[1,2(✉)], Yunfei Cao[3], and Dianhua Tang[3]

[1] School of Cryptographic Engineering, Engineering University of the Chinese People's Armed Police, Xi'an 710086, Shaanxi, China

[2] Key Laboratory for Network and Information Security of Chinese People's Armed Police, Engineering University of the Chinese People's Armed Police, Xi'an 710086, Shaanxi, China
wangxazjd@163.com

[3] Science and Technology on Communication Security Laboratory (CETC 30), Chengdu, China
646589015@qq.com

Abstract. Machine learning servers with mass storage and computing power is an ideal platform to store, manage, and analyze data and support decision-making. However, the main issue is providing security and privacy to the data, as the data is stored in a public way. Recently, homomorphic data encryption has been proposed as a solution due to its capabilities in performing computations over encrypted data. In this paper, we proposed an encrypted all convolutional net that transformed traditional all convolutional net into a net based on homomorphic encryption. This scheme allows different data holders to send their encrypted data to cloud service, complete predictions, and return them in encrypted form as the cloud service provider does not have a secret key. Therefore, the cloud service provider and others cannot get unencrypted raw data. When applied to the MNIST database, privacy-preserving all convolutional based on homomorphic encryption predict efficiently, accurately and with privacy protection.

1 Introduction and Related Work

The neural network aims to solve the classification problem by training the neural network and assigning a correct label to the new sample. The concept of neural network models exist in 1980's and in 2006, deep learning was proposed [1]. Deep neural network constructs a learning model with multiple hidden layers, designs effective learning algorithms and can speed up calculations, so that big data can be processed; higher level features can be obtained through deep neural network, thereby increasing the sample recognition rate or prediction Accuracy. Convolutional neural networks (CNN) are deep neural networks with convolutional layers. CNN's weight-sharing network structure makes it more similar to biological neural networks, which reduces the complexity of the network model and reduces the number of weights. In 2012,

L. Barolli et al. (Eds.): NBiS 2018, LNDECT 22, pp. 752–762, 2019.
https://doi.org/10.1007/978-3-319-98530-5_66

Krizhevsky et al. [2] used CNN called AlexNet to get the best results in the classification task. The AlexNet network has a deeper structure, and ReLU (Rectified linear unit) was designed as a non-linear activation function and avoided overfitting successfully. After AlexNet, the researchers further improved network performance, proposed Region-based CNN [3], Spatial pyramiding net [4], GoogLeNet [5], Visual geometry group [6], etc. In 2014, Springenberg et al. proposed in ACN (All Convolutional Net) [7] that the pooling layer can be easily replaced by a convolutional layer with an increasing stride without loss of accuracy on the image recognition. The network structure is simplified by replacing the pooling layer with a convolutional layer with increasing stride and not affecting the accuracy of the network

Neural networks algorithms are often applied to private data such as medical and financial information. However, the problem of privacy leakage caused by data mining based on neural networks is also becoming more and more serious. In 2006, Barni et al. [8] proposed a neural network based on secure multi-party computation and homomorphic encryption that can protect privacy and Orlandi et al. [9] improve the scheme [8]. Gilad Bachrach et al. [10, 11] proposed a non-interactive privacy preserving neural network, that means no interaction between the client and the server. In 2017, Herve et al. [12] improved the previous algorithm and proposed a privacy preserving neural network which depth is greater than two layers.

Our work is to construct efficient privacy-preserving protocols for an All Convolutional networks. In the proposed protocol, the private information will be encrypted and sent to the cloud service provider in an encrypted form, and the cloud can calculate the predicted result of the encrypted data. The public key of scheme is used for encryption and the private key of scheme is used for decryption. Note that Cloud service provider cannot get raw data and predictions because it does not have a private key. So, this process allows private predictions and because of the convolutional layer is removed, the network structure is also simpler.

2 Neural Networks

2.1 Convolutional Neural Network

The Convolutional Neural Network (CNN) is a feed-forward neural network. Compared with other machine learning algorithms, CNN uses a multi-layer structure to improve generalization and abstraction performance. CNN's main feature is its ability to automatically build complex models, using a series of convolution filters, normalization, nonlinear functions, down-sampling and other operations to extract features from the signal and achieved better performance based on backpropagation training processes and gradient descent optimization. The first level of the network usually represents low-level features. The full-connection layer can combine low-level features to identify complex goals.

There are mainly building block that make up the convolutional neural network. We discuss them separately.

Convolutional Layer

Convolution processing is a method for extracting some primary signal features by local connecting and weight sharing to simulate neural cells with local receptive fields. Local connection means that each neuron on the convolutional layer establishes a connection with the neurons in the fixed area of the previous layer. Weight sharing means that neurons in the same feature are locally connected to the previous layer with a set of same connection strengths, which can reduce network training parameters. The input of each neuron in the convolutional layer comes from the neurons in the fixed area in the previous layer. The size of the area is determined by the size of the convolutional kernel. The convolutional layer is convolved with m learnable convolution kernels adding bias by the input data and m feature maps are obtained through the activation function. The n feature maps of the convolutional layer 2 are respectively convoluted with the m convolution kernels of the convolutional layer 1 by $n \times m$ convolution kernels, combined with the result of the m convolutions, added the bias, through the activation function. The mathematical equation is:

$$y_{mn} = f\left(\sum_{j=0}^{J-1}\sum_{i=0}^{I-1} x_{m+i,n+j} \times w_{ij} + b\right).$$

where: x is the input two-dimensional data; y is the output of $M \times N$; $0 \leq m < M$, $0 \leq n < N$; w is the convolution kernel of size $J \times I$; b is the bias; f is the activation function.

ReLU

ReLU is Rectified Linear Units and an activation function. There are several activation functions in traditional neural networks, for example the tangent $tanh(x)$, $|tanh(x)|$, and the *sigmoid* function $(1+e^{-x})^{-1}$. ReLU is often preferred to other functions as it trains the neural network several times faster [13] and does not reduce accuracy. The ReLU function is a piecewise-linear function that turns all negative values to zero, while the positive value does not change:

$$f(x) = max(0, x)$$

There are three main changes in this model compared to the Sigmoid system: (1) unilateral inhibition; (2) relatively broad excitation boundaries; (3) sparse activation.

ReLU has the following advantages: For linear functions, ReLU is more expressive, especially in deep networks; and for non-linear functions, ReLU has constant gradients due to non-negative intervals, so there is no problem of gradient disappearance (Vanishing Gradient Problem) keeps the convergence speed of the model at a steady state.

Pooling Layer

Another important concept of CNNs is non-linear down-sampling called pooling layer. The feature vector dimension obtained after convolution is large. If these features are used for classification, it will lead to large computational complexity and high complexity. Before using the extracted features to train the classifier, it is necessary to reduce the dimension. Simulation of complex cells screens primary features and combines them into higher-level and abstract features. After the pooling, the number of feature maps does not change, but the computational complexity of the network can be drastically reduced, and the network can have certain invariance in translation and scaling of the object, thus making the network more robust. The pooling operation is

$$y_{mn} = \frac{1}{S_1 S_2} \sum_{j=0}^{S_2-1} \sum_{i=0}^{S_1-1} x_{m \times S_1} + i, n \times S_2 + j$$

where: x is the input after the convolution process, y is the output, and S_1 and S_2 are the down-sampling scales.

Fully Connected Layer

After the convolution and pooling layers are fully connected layers. In order to enhance the nonlinear mapping ability of the network and limit the size of the network at the same time, the network use a fully connected layer after convolutional layer. Different from convolutional layers, the full-connection layer perceives global information and aggregates the local features learned by the convolutional layer to form global features for specific target tasks. Each neuron in this layer is connected with all neurons in the previous layer, with no connection between neurons in the same layer.

The mathematical expression of the fully connected layer is

$$o_j^l = f\left(\sum_{i=1}^{n} x_i^{l-1} \times w_{ji}^l + b^l\right)$$

where: n is the number of neurons in the previous layer, l is the current number of layers, w_{ji}^l is the connection strength between neuron j in this layer and the neuron i in the previous layer, b^l is the bias of neuron j in this layer, f is the activation function.

2.2 All Convolutional Net

Most convolutional neural networks used for object recognition. After assessing the neural network's processing of images, the necessity of the pooling layer was questioned.

ACN [7] was published in 2014 by JT Springenberg and the greatest contribution of this work is to replace the pooling layer in the classical architecture. And, through mathematical formulas and experimental results, it is proved that such replacement has better performance. The author proved that Pooling operation is not necessary. Extracting a wider range of features is important for CNN. The goal of the pooling layer is to extract more higher features by the spatial dimensionality reduction. The network structure is simplified by replacing the pooling layer with a convolutional layer with increasing stride and not affecting the accuracy of the network.

2.3 Modification

Homomorphic encryption only supports addition and multiplication on ciphertext, so only polynomial functions can be directly implemented and it is desirable to limit the calculation to polynomials. Moreover, neural networks usually use floating point numbers. Hence, when developing neural networks, we must adapt protocols accordingly.

Convolutional layer can be evaluated in a straightforward way since it uses only additions and multiplications. In addition, networks use the unencrypted weights and add a bias to the sum of the weight, so the value of weights and bias term can be precomputed directly.

Pooling cannot be computed directly. But ACN removed the pooling layer in the classical architecture CNN and added the convolutional layer with increasing stride and followed by fully connected layers.

Rectified Linear can have a good low-degree polynomial approximation by combining the Taylor series of the ReLU function with the batch normalization.

As we all know, the homomorphic encryption scheme does not support floating-point operations so it's our method that converting floating-point numbers with fixed precision to integers by proper scaling. According to the characteristics of homomorphic encryption operations, all numbers will be modulo some number t. In order to make sure the correctness of the calculation, we must control the scale of the number and increase the modulus t to avoid modulo operation.

3 Homomorphic Encryption

Data encryption is a prominent way to protect data privacy. However, traditional encryption methods have many limitations and cannot support cryptographic operations. This is homomorphism from an algebraic point of view. This problem was raised by Rivest, Adleman, and Dertouzos in 1978 and became an open issue. Fully homomorphic encryption is known as the Holy Grail in cryptography.

Fully homomorphic encryption means that the ciphertext can be arbitrarily calculated without knowing the secret key, that is, for any function f and plaintext m, $f(Enc(m)) = Enc(f(m))$. This special property allows a wide range of theoretical and practical applications for full-homomorphic encryption, such as: cloud computing security, ciphertext retrieval, secure multiparty computing and etc. The first one to propose a fully homomorphic encryption blueprint is Gentry [14] in 2009. Since the first encryption scheme was introduced, numerous FHE schemes have been proposed.

3.1 Partially Homomorphic Cryptosystem

Homomorphic encryption means that the decrypted result of the operation on ciphertext is equivalent to the result of direct operation on plaintext.

Homomorphic encryption is divided into additive homomorphic cryptosystem and multiplicative homomorphic cryptosystem. Partially homomorphic encryption allows only one kind of operation (addition or multiplication) to be evaluated on the encrypted data. Partially homomorphic cryptosystem schemes include: Paillier cryptosystem [15] for additive homomorphic encryption and El Gamal cryptosystem [16] for multiplicative homomorphic encryption.

3.2 Fully Homomorphic Cryptosystem

Partially homomorphic cryptosystem has existed for many years, but Partially homomorphic encryption can only perform one kind of operation, which limits its usage. Fully homomorphic encryption means that the decryption of an arbitrary operation on ciphertext is equivalent to the result of a direct operation of plaintext and it is far more powerful.

Our scheme is based on the BGV [17] homomorphic cryptosystem using key-switch technology and modulus-switch technology, which is one of the most efficient all-homomorphic encryption schemes. HElib [18] is an open source of this scheme.

Due to space constraints, we only give a brief introduction to the BGV program. Detailed information can refer to the original paper. The plaintext space of the BGV scheme operates on a polynomial ring, which is modulo the cyclotomic polynomial $A_t := \mathbb{Z}_t[x]/\phi_m(x)$, where $\phi_m(x)$ is the m-th polynomial. It contains five operations: **KeyGen** represents key generation algorithms, $\mathbf{Enc_{pk}}$ represents encryption, $\mathbf{Dec_{sk}}$ represents decryption, **homAdd** represents homomorphic addition and **homMul** represents homomorphic multiplication. The input of **KeyGen** is three positive integers: m, t, and L where m and t determine the plaintext space, and L denotes the multiplicative depth that the scheme can evaluate and outputs of **KeyGen** is public-private key pair $(pk; sk)$. In this scheme, homomorphic addition and homomorphic multiplication are expressed as

$$\mathbf{Dec}(\mathbf{homAdd}([\![a]\!],[\![b]\!])) = a + b\,mod(\phi_m(x), t)$$
$$\mathbf{Dec}(\mathbf{homMul}([\![a]\!],[\![b]\!])) = a \times b\,mod(\phi_m(x), t)$$

where messages $a, b \in \mathrm{A_t}$.

The current fully-homomorphic encryption schemes encrypt the plaintext by adding noise, but the noise increases with the amount of homomorphic operations. When the noise exceeds a certain threshold, it cannot be decrypted properly. Therefore, when performing homomorphic operations, we must pay attention to the growth of noise so that correctness of decryption can be assured. According to the scheme, the noise growth is more closely related to the homomorphic multiplication, so when evaluating the homomorphic operations of the function, it will give priority to consider the influence of the multiplication operation.

4 Solution and Experimentation

The traditional convolutional neural network has the following structure: $[[CONV \rightarrow ACT]^p \rightarrow POOL]^q \rightarrow [FC]^r$. The part inside brackets can be iteratively calculated.

In our scheme, using the previously mentioned method, ACN has the following structure: $[[CONV \rightarrow BN \rightarrow ACT]^{p} \rightarrow CONV']^{q} \rightarrow [FC]^{r}$, Where ACT represents the polynomial approximation of the ReLU function. BN represents batch normalization; this method guaranteed the operation of the polynomial approximation of the activation function. The pooling layer is replaced by an increased stride convolutional layer, which is similar in function.

4.1 Batch Normalization

We can add a batch normalization layer before each ReLU layer [19]. Due to the limited range of approximate operations, we need to perform an exact polynomial approximation of a part of R around 0. The formula is as follows:

Input: Values of x over a mini-batch: $\mathcal{B} = \{x_{1\ldots m}\}$;

Output: $\{y_i = BN_{\gamma,\beta}(x_i)\}$

Mini-batch mean:

$$\mu_{\mathcal{B}} \leftarrow \frac{1}{m}\sum_{i=1}^{m} x_i$$

Mini-batch variance:

$$\sigma_{\mathcal{B}}^2 \leftarrow \frac{1}{m}\sum_{i=1}^{m} (x_i - \mu_{\mathcal{B}})^2$$

normalize:

$$\hat{x}_i \leftarrow \frac{x_i - \mu_{\mathcal{B}}}{\sqrt{\sigma_{\mathcal{B}}^2 + \epsilon}}$$

Scale and shift:

$$y_i \leftarrow \gamma\hat{x}_i + \beta \equiv BN_{\gamma,\beta} x_i$$

4.2 Approximation of ReLU by Polynomial

The accuracy of full convolutional net depends on the ReLU function. To compute the ReLU function homomorphically, we can use the polynomial approximation of the ReLU on Gaussian Distribution [12].

Input: $X = (X_1, \ldots, X_N)$, a set $Y = (Y_1, \ldots, Y_N)$ and polynomial degree n, X_i are randomly selected from a Gaussian Distribution.

Output: coefficients of $P(X) = (c_0 + c_1 X + c_2 X^2 + \cdots + c_n X^n)$ such that the square error $\epsilon = \sum_{i=1}^{N} (P(X_i) - Y_i)^2$ is minimized.

Taylor series of a smooth approximation of the ReLU function:

$$ln(1+e^x) = ln(2) + \frac{x}{2} - \frac{x^4}{192} - \frac{17x^8}{645120} + \frac{31x^{10}}{14515200} + O(x^{12}).$$

where x is around the point 0. With the help of Batch normalization, 99% of values are in the range of − 3 to + 3.

4.3 Convolutional Layer with Increased Stride

Our network structure is shown in Fig. 1. We construct our scheme on an ACN with the characteristics as described below:

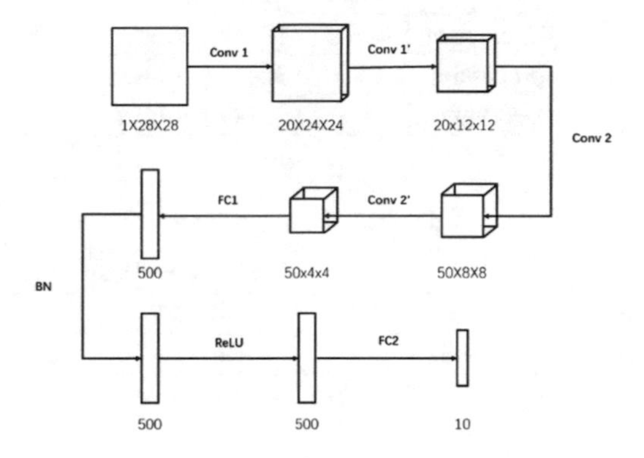

Fig. 1. Our privacy-preserving CAN

(1) The convolutional layers (Conv1 and Conv2).
(2) The network structure is simplified by replacing the pooling layer with a convolutional layer with increasing stride and not affecting the accuracy of the network. For a pooling layer we replace it with a convolutional layer with corresponding stride and kernel size. The number of output channels equal to the input channels.

$$y_{mn} = f\left(\sum_{j=0}^{J-k}\sum_{i=0}^{I-k} x_{m+i,n+j} \times w_{ij} + b\right).$$

(3) The fully connected layers (FC1 and FC2).
(4) There is a batch normalization layer (BN).

Where: k is the stride of normal convolution and other parameters are as described above.

4.4 Empirical Results

We have tested privacy-preserving All Convolutional Net on the MNIST dataset [20]. The data set consists of 60,000 handwritten digital images, all images is written in handwritten digits 0 to 9. Each image has $28 \times 28 = 784$ pixels. The images are all black-and-white images, each pixel is represented by gray level and its range is 0–255. The task of classification is to identify which number is in each image. We use the 50,000 images of this data set for learning and use the remaining 10,000 images for testing. The table shows the results of the test. The accuracy of the training network is 98.97% (it has only 103 mismatches out of 10,000 test cases) (Tables 1, 2 and 3).

Table 1. Performance

Stage	Latency	Throughput
Encryption	80.6 s	18742 per hour
processing	394 s	36455 per hour
Decryption	3 s	20749 per hour

Table 2. Message sizes

	Message size (MB)	Size per instance (KB)
Client → Cloud	357.1	92.164
Decryption	4.51	2.11

Table 3. Accuracy

	Accuracy
State of the art accuracy	99.77%
ACN based FHE	98.97%

5 Conclusion

Today, the convolutional neural network and its variants on cloud services have played a huge role in the medical and financial fields. However, if users want to let cloud service providers provide services, they must send their original information to cloud service providers, resulting in serious personal privacy leakage. The issue of privacy protection of neural networks is an urgent issue to be solved. The accuracy of convolutional neural networks has been widely recognized, so in this article we pay more attention to its privacy protection. We use fully homomorphic encryption to balance the use of cloud service providers' services and personal privacy protection. Because the inefficient of full homomorphic encryption, we need to adjust the convolutional neural network. We use a convolutional neural network with a ReLU activation function and replace the pooling layer with a convolutional layer with an increasing stride. This operation simplifies the structure of the CNN and improves its efficiency.

We have tested privacy-preserving All Convolutional Net on the MNIST dataset. The data set consists of 60,000 handwritten digital images, all images is written in handwritten digits 0 to 9. Each image has $28 \times 28 = 784$ pixels. The images are all black-and-white images, each pixel is represented by gray level and its range is 0–255. The task of classification is to identify which number is in each image. We use the 50,000 images of this data set for learning and use the remaining 10,000 images for testing. The table shows the results of the test. The accuracy of the training network is 98.97% (it has only 103 mismatches out of 10,000 test cases).

Acknowledgements. This work is supported by National Cryptography Development Fund of China under grant number MMJJ20170112, National Natural Science Foundation of China (Grant Nos. U1636114, 61772550, 61572521), National Key Research and Development Program of China (Grant No. 2017YFB0802000), Natural Science Basic Research Plan in Shaanxi Province of China (Grant Nos. 2018JM6028, 2016JQ6037).

References

1. Hinton, G.E., Salakhutdinov, R.R.: Reducing the dimensionality of data with neural networks. Science **313**(5786), 504–507 (2006)
2. Krizhevsky, A., Sutskever, I., Hinton, G.E.: ImageNet classification with deep convolutional neural networks. In: International Conference on Neural Information Processing Systems, pp. 1097–1105. Curran Associates Inc. (2012)
3. Girshick, R., Donahue, J., Darrell, T., et al.: Rich feature hierarchies for accurate object detection and semantic segmentation. In: IEEE Conference on Computer Vision and Pattern Recognition, pp. 580–587. IEEE Computer Society (2014)
4. He, K., Zhang, X., Ren, S., et al.: Spatial pyramid pooling in deep convolutional networks for visual recognition. IEEE Trans. Pattern Anal. Mach. Intell. **37**(9), 1904–1916 (2014)
5. Szegedy, C., Liu, W., Jia, Y., et al.: Going deeper with convolutions. In: Proceedings of the IEEE Conference on Computer Vision and Pattern Recognition, pp. 1–9 (2015)
6. Simonyan, K., Zisserman, A.: Very deep convolutional networks for large-scale image recognition. In: Computer Science (2014)
7. Springenberg, J.T., Dosovitskiy, A., Brox, T., et al.: Striving for simplicity: the all convolutional net. Eprint Arxiv (2014)
8. Barni, M., Orlandi, C., Piva, A.: A privacy-preserving protocol for neural-network-based computation. In: The Workshop on Multimedia and Security, pp. 146–151. DBLP (2006)
9. Orlandi, C., Piva, A., Barni, M.: Oblivious neural network computing via homomorphic encryption. EURASIP J. Inf. Secur. **2007**(1), 1–11 (2007)
10. Xie, P., Bilenko, M., Finley, T., et al.: Crypto-nets: neural networks over encrypted data. In: Computer Science (2014)
11. Dowlin, N., Ran, G.B., Laine, K., et al.: CryptoNets: applying neural networks to encrypted data with high throughput and accuracy. In: Radio and Wireless Symposium, pp. 76–78. IEEE (2016)
12. Chabanne, H., de Wargny, A., Milgram, J., et al.: Privacy-preserving classification on deep neural network. In: IACR Cryptology ePrint Archive, 2017, p. 35 (2017)
13. Krizhevsky, A., Sutskever, I., Hinton, G.E.: ImageNet classification with deep convolutional neural networks. In: Advances in Neural Information Processing Systems, pp. 1097–1105 (2012)

14. Gentry, C.: Fully homomorphic encryption using ideal lattices. In: STOC 2009, vol. 9, no. 4, pp. 169–178 (2009)
15. Paillier, P.: Public-key cryptosystems based on composite degree residuosity classes. In: Proceedings of the Advances in Cryptology—EUROCRYPT 1999, International Conference on the Theory and Application of Cryptographic Techniques, Prague, Czech Republic, 2–6 May 1999. DBLP, pp. 223–238 (1999)
16. Elgamal, T.: A public key cryptosystem and a signature scheme based on discrete logarithms. IEEE Trans. Inf. Theory **31**(4), 469–472 (1984)
17. Gu, C.: New Fully Homomorphic Encryption without Bootstrapping (2011)
18. Halevi, S., Shoup, V.: HElib—An Implementation of Homomorphic Encryption. https://github.com/shaih/HElib/
19. Ioffe, S., Szegedy, C.: Batch Normalization: Accelerating Deep Network Training by Reducing Internal Covariate Shift, pp. 448–456 (2015)
20. LeCun, Y., Cortes, C., Burges, C.J.C.: The mnist Database of Handwritten Digits (1998) http://yann.lecun.com/exdb/mnist/

Verification of Persuasion Effect to Cope with Virus Infection Based on Collective Protection Motivation Theory

Kana Shimbo(✉), Shun-ichi Kurino, and Noriaki Yoshikai

Graduate School of Science and Technology, Nihon University,
1-8-14, Surugadai, Chiyoda-ku, Tokyo, Japan
cskn17012@g.nihon-u.ac.jp

Abstract. It has been reported that many Internet users do not recover personal computers (PCs) that have been infected by computer viruses. To address this problem, we have investigated how to motivate users to remove viruses from their PCs based on the Protection Motivation Theory. Previously, we have reported that the cognitive factors related to response efficacy, responsibility, percentage of performers, and group norm could affect the intention to recover an infected PC. In this study, we created the experimental content to stimulate these cognitive factors and conducted an experiment to verify whether the content would be effective to persuade Internet users to recover infected PCs. Our research confirmed that the content stimulated some cognitive factors and effectively persuaded users to recover their PCs. We also found that some users did not intend to cope with the virus infection because they did not consider the information about the virus infection to be credible.

1 Introduction

It is well known that computer viruses cause serious damages to Internet users, and the spread of viruses is considered a social issue. It is difficult to prevent the virus infection on personal computers (PCs) because tens of billions of new viruses emerge each year. Thus, Internet users must remove computer viruses from their PCs when they become infected.

However, the target of a virus infection is not just PCs. Recently, the number of IoT devices, such as monitoring cameras, that have been infected with viruses have increased, and such devices have been exploited for cyberattacks. In 2016, major services on the Internet, such as Twitter, Amazon, and Netflix were damaged by DDoS attacks by exploiting vulnerable IoT devices [1]. In addition, it has been reported that traffic accidents may occur if automobile systems become infected with viruses because these vehicles could be controlled by viruses [2]. Thus, it is becoming increasingly necessary to recover infected devices with viruses.

However, it has been reported that many users do not recover infected PCs even when they recognize that a virus infection has occurred [3].

Therefore, we investigated an effective way to persuade users to recover their PCs by clarifying a decision-making mechanism to remove computer viruses [4, 5].

L. Barolli et al. (Eds.): NBiS 2018, LNDECT 22, pp. 763–771, 2019.
https://doi.org/10.1007/978-3-319-98530-5_67

Our previous studies have indicated that the risk cognition for a computer virus affects the intention to address the infection. In this paper, we verified our hypothesis using the experimental content with the messages stimulating the risk cognition.

2 Antivirus Projects

To reduce the number of infected PCs on the Internet, some countries have started the antivirus projects that alert users of infected PCs.

Japan has initiated "Cyber Clean Center (CCC) [6]" project (currently "Advanced Cyber Threats response InitiatiVE (ACTIVE)" [7]), which identifies the users of the infected PCs and recommends they remove the viruses from their PCs.

First, CCC detects the computer viruses on the Internet and checks the IP addresses that transmit the virus. Then, CCC reports these the IP addresses to the Internet service providers (ISPs) to identify the users of the infected PCs. The ISPs notify the users that their PCs are infected with the viruses and provide the antivirus software for free. Finally, the users remove the viruses using the antivirus software.

CCC has reported that this has reduced the virus infection rate in Japan [8]. Germany referred to CCC results [9] and began a similar project, i.e., the Botfrei [10]. The Botfrei has reported that the project has been effective in reducing the virus infection rate [11]. Australia has also initiated a similar project [12].

However, the effectiveness of these projects is hampered by two significant problems.

One is that many infected PC users are not motivated to recover their PCs. CCC reports that 70% of users who received the alerts do not recover their infected PCs [3]. If users do not intend to recover their infected PCs, the number of infected PCs might be not reduced even though these users would have been made aware of the virus infection.

The other is that these projects have no proper business model. These projects impose a heavy burden on the antivirus software venders because the venders are required to provide the software to users for free. In Japan, CCC ended the provision of the free antivirus software and suggests that users pay the software costs. As a result, the user motivation to address the virus infection has been reduced further.

Therefore, we focus on the problem of the user motivation relative to recovering PCs. We have applied the psychology to this problem and investigated the effective persuasion to motivate users to recover their PCs.

3 Background

3.1 Collective Protection Motivation Theory

"Threat appeal" [13] is defined as the persuasion to motivate people to cope with a threat by emphasizing the risk. The Protection Motivation Theory explains the factors that influence the intention to address a threat by the threat appeal. In addition, the Collective Protection Motivation Theory (CPMT) [14] proposes the decision-making factors when coping with a threat that requires "collective coping." These factors include eight cognitive factors about the threat, i.e., severity, vulnerability, response efficacy, cost, self-efficacy, responsibility, percentage of performers, and group norm.

3.2 Decision-Making Process Model

Computer virus infections can be considered a threat to the users of infected PCs; thus, CPMT has been adopted in the decision-making process to address computer virus infections, and a decision-making process model for computer viruses as shown in Fig. 1. has been proposed.

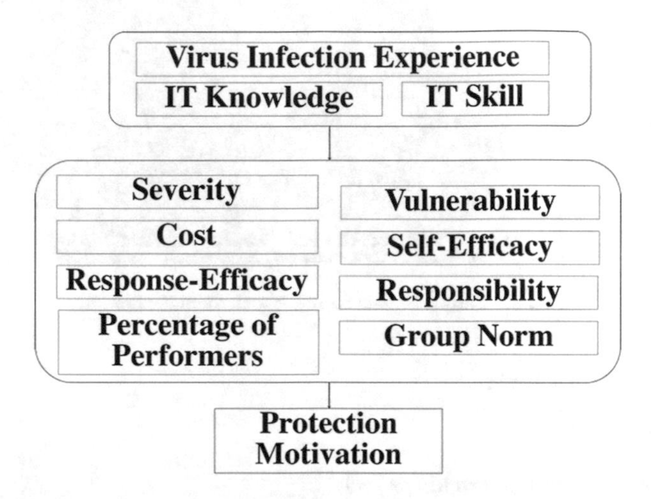

Fig. 1. Decision-making process model for computer viruses

This model comprises eight cognitive factors based on CPMT for the protection motivation and three potential user factors, i.e., virus infection experience, IT knowledge, and IT skill. We surveyed Japanese users by questionnaire to determine each element in the model and analyzed their relations [4].

The results suggested that the intent to resolve a virus infection involved four cognitive factors, i.e., the cognitions for response efficacy, responsibility, percentage of performers, and group norm.

3.3 Purpose

The cognitive factors that affect the intent to resolve a virus infection have been identified. However, it is not confirmed that the persuasion with these factors is effective relative to the protection motivation. Therefore, this paper discusses the effect of persuasion relative to addressing virus infections using messages that stimulate these cognitive factors.

4 Methods

We created the experimental content as shown in Fig. 2. with the messages that stimulate the abovementioned cognitive factors. Then, we investigated the effectiveness of the persuasion using the experimental content.

Fig. 2. Partial scenes in the experimental content

4.1 Experimental Content

4.1.1 Design Conditions

(1) *Messages to stimulate cognitive factors*
Table 1 shows each message used to stimulate each of four cognitive factors; the cognitions for response efficacy, responsibility, percentage of performers, and group norm.

Table 1. Messages to stimulate cognitive factors

Subject of cognition	Message
Response efficacy	Recovering infected PCs can reduce the risk
Responsibility	Using infected PCs without recovering cause problems for other users
Percentage of performers	The percentage of users who intend to cope with infected PCs is considered to increase
Group norm	Recovering an infected PCs is the common courtesy, similar to wearing a mask if you have a cold

(2) *Credibility of experimental content*
To collect the accurate experimental data, it is very important that the content is credible, which depends on the information source. Specifying the well-known information sources, such as the URLs of the government agencies and the

universities, allows people to trust the information [15]. Therefore, our experimental content specifies the source of the information.

(3) *Targets*
University students were targeted in our experiment because the experimental conditions were easy to control in a university lecture.

(4) *Production tool*
We created the experimental content, which comprises a combination of the still images and the voices, for one month using the Windows Movie Maker.

(5) *Duration of experimental content*
The duration of the content was expected to be less than 10 min to ensure that the participants did not become bored or lose interest. The actual duration of the content was seven minutes.

4.1.2 Scenario Outline

The scenario of the experimental content is summarized as follows.

A university student does not recover her PC infected with a computer virus. One day, she watches a video on the Internet that explains the risks of virus infections and measures to take against infections. After watching the video, she recognizes the importance taking measures against computer viruses and is motivated to act.

4.2 Experiment

4.2.1 Experimental Process

(1) The participants were divided into an experimental group (Group 1) and a control group (Group 2).
(2) The participants in Group 1 watched the experimental content and answered a questionnaire (as shown later), and the participants in Group 2 answered the same questionnaire without viewing the content. The two questionnaires were administered simultaneously in different rooms to prevent the interaction between the participants in the two groups.
(3) The questionnaire results were analyzed.

4.2.2 Questionnaire

The questionnaire comprised four parts, i.e., the items about the individual information, the items determining the cognitive factors for the viruses based on CPMT, the items determining the intent to address a virus infection, and the free-writing items (e.g., reason (not) to intend to address the infection). To simulate a realistic situation and determine the intent to address a virus infection, the participants answered the questionnaire under the assumption that they had received the postal mail from an ISP about a virus infection on their PCs.

4.2.3 Analysis

To confirm that the experimental content stimulates the cognitive factors, we statistically evaluated whether there was a significant difference between Groups 1 and 2 relative to the average of each cognitive factor. In addition, to confirm that the experimental content motivates to cope with the virus infection, we also did whether there was a significant difference between Groups 1 and 2 relative to the average of the intention to carry out the measure. Here, the Mann–Whitney U test was adopted (5% significance level) using the R software environment [16].

5 Results

5.1 Conditions

The experiment was carried out at Nihon University during a lecture regarding the information science. The experimental conditions are shown in Tables 2 and 3.

Table 2. Experimental conditions

Condition	Content
Target	Students in College of Science and Technology, Nihon University
Number of participants	157
Age	18–26
Place	Funabashi campus and Surugadai campus in Nihon University
Period	November 7, 2018, 1:20–4:30 p.m.
	November 29, 2018, 1:20–2:50 p.m.

Table 3. Participant data

	Men	Women	Total
Group 2	61	11	72
Group 1	64	21	85
Total	125	32	157

5.2 Results and Analysis

The results obtained for determining the cognitive factors from the questionnaire responses are shown in Fig. 3.

The components in the figure show the results for four cognitive factors for Groups 1 and 2. The p valucs of three factors (i.e., the cognitions for response efficacy, responsibility, and group norm) were less than 0.001, and a significant difference was observed between Groups 1 and 2. The p value relative to the cognition for the percentage of performers was approximately 0.177, and no significant difference was observed between Groups 1 and 2.

The questionnaire results relative to determining the intent to address a virus infection are shown in Fig. 4. Here, the p value was less than 0.001, and a significant difference was observed between Groups 1 and 2.

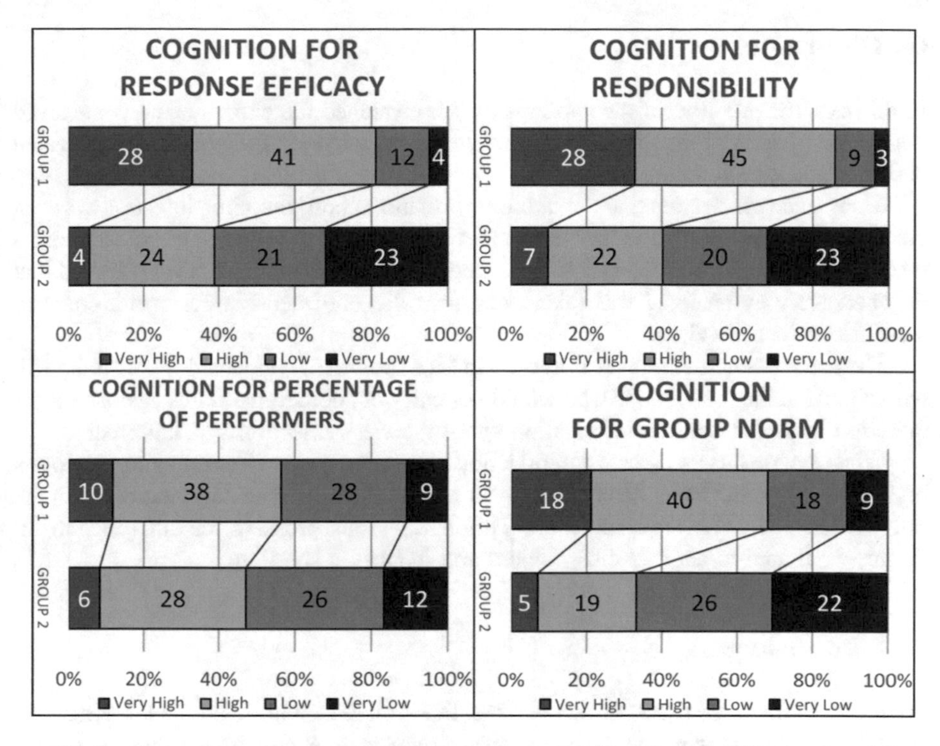

Fig. 3. Comparison of analytical results relative to cognitive factors of Groups 1 and 2

From the analytical results, we can confirm that the experimental content stimulated the cognitions relative to response efficacy, responsibility, and group norm, as shown in Fig. 3. In addition, the experimental content motivated the intent to address a virus infection, as shown in Fig. 4. On the other hand, the experimental content did not significantly affect the cognition relative to the percentage of performers. Thus, it is necessary to clarify why the cognition for percentage of performers was not stimulated and then improve the messages in the experimental content.

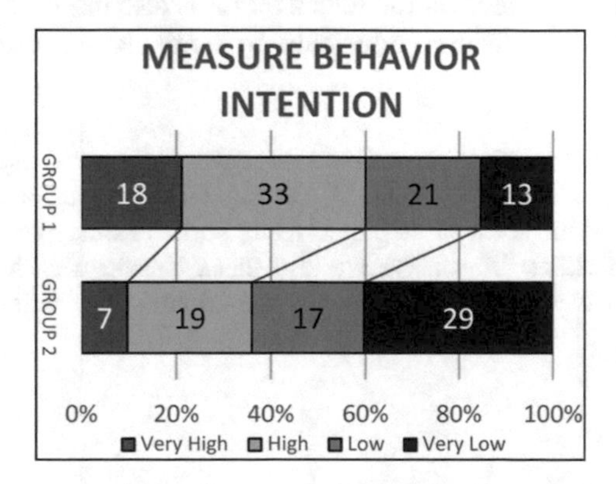

Fig. 4. Comparison of analytical results relative to the intent to address the virus infection of Groups 1 and 2

6 Discussion

In this experiment, 40% of the participants who watched the experimental content did not intend to address the virus infection. We investigated why they were unmotivated based on the questionnaire data.

Many answers demonstrated that the information about the virus infection was not credible. Approximately one-half of the participants who did not intend to carry out a protective measure indicated this lack of credibility. These participants considered that the message they received was fake; thus, they did not intend to perform a countermeasure against the virus.

The same phenomenon was observed in an experiment conducted in Australia [5], where most of the participants who would not carry out countermeasures against a virus infection indicated that the information about the virus infection was not credible.

When Internet users receive a notification and alert about a virus infection, there is a high probability that users will not carry out the measure because they cannot credit the information about the virus infection. This is a serious problem for our research. In future, we intend to address this problem and identify a solution.

7 Conclusions

To motivate Internet users to recover the PCs infected with a computer virus, we created the content with the messages that stimulate four cognitive factors and conducted an experiment to verify whether the content motivates users as desired.

The results confirmed that the content stimulated three cognitive factors and effectively persuaded users to recover their PCs. On the other hand, some users did not intend to recover their PCs because they did not consider the provided information credible. Therefore, it is necessary to consider the credibility of the information when informing users about virus infections.

The educational methods to prevent the virus infection have frequently been discussed. However, it is impossible to prevent the infection completely in advance. Therefore, the education to recover the infected PCs is also required when infected. Our research would help the effective education to motivate the users to recover their infected PCs.

Acknowledgements. We would like to thank Assistant Professor Daisuke Takagi of Tokyo University and Doctor Stavrakakis of Sydney University for their valuable advice and comments. In addition, we thank the members of our project team, Tasuku Komagome, Momomi Takahashi, and Akane Harada of Nihon University. We also thank Yusuke Kanehira and Kota Fujisaki (Nihon University) for their assistance in creating the experimental content.

References

1. 10 Threat of Information security. https://www.ipa.go.jp/files/000058504.pdf
2. Security in IoT era. http://www.jaist.ac.jp/is/labs/tan-lab/projects/160307IoTSecurityWS/201-ito-CCDS.pdf
3. Achievements of Cyber Clean Center in January 2011. https://www.telecom-isac.jp/ccc/report/201101/1101monthly.html
4. Hamatsu, S., Kurino, S., Yoshikai, N.: Measure behavior intention model for information security based on persuasion psychology and its application. Inf. Process. Soc. Jpn. **56**(12), 2200–2209 (2015)
5. Yoshikai, N., Shimbo, K., Stravrakakis, J., Takahashi, T.: Study on persuasion effect of computer virus measures based on collective protection motivation theory. Lecture Notes on Data Engineering and Communications Technologies, vol. 7, pp. 518–528 (2017)
6. About Cyber Clean Center. https://www.telecom-isac.jp/ccc/
7. An anti-malware support project through the public-private partnership. https://www.ict-isac.jp/active/
8. Arimura, K.: Measure of Malware in Japan observed by an anti-bot virus project "Cyber Clean Center". Inf. Process. Soc. Jpn. **51**(3), 275–283 (2010)
9. Countermeasure of information security in MIC. http://www.kagawa-net.org/katsudou/h24/h240525lectureshiryou.pdf
10. Herzlich willkommen bei botfrei. https://www.botfrei.de/de/index.html
11. Statistical evaluation of the impact of the Pilot. https://acdc-project.eu/wp-content/uploads/2015/05/ACDC_D4.2_Statistical_Evaluation_final.pdf
12. Australian Internet Security Initiative. https://www.cert.gov.au/aisi
13. Fukada, H.: Persuasion Psychology Handbook. Kitaooji Shobo Publishing, Kyoto (2004)
14. Tozuka, T., Fukada, H.: A test of collective protection motivation theory in threat appeal persuasion. Jpn. J. Exp. Soc. Psychol. **44**(1), 54–61 (2005)
15. Fogg, B.J.: Persuasive Technology: Using Computers to Change What We Think and Do. Morgan Kaufmann, Los Altos (2002)
16. The R Project for Statistical Computing. http://www.r-project.org/

Zero-Knowledge Proof for Lattice-Based Group Signature Schemes with Verifier-Local Revocation

Maharage Nisansala Sevwandi Perera[1(✉)] and Takeshi Koshiba[2]

[1] Graduate School of Science and Engineering, Saitama University, Saitama, Japan
perera.m.n.s.119@ms.saitama-u.ac.jp

[2] Faculty of Education and Integrated Arts and Sciences, Waseda University, Tokyo, Japan
tkoshiba@waseda.jp

Abstract. In group signature schemes, signers prove verifiers, their validity of signing through an interactive protocol in zero-knowledge. In lattice-based group signatures with Verifier-local revocation (VLR), group members have both secret signing key and revocation token. Thus, the members in VLR schemes should show the verifiers, that he has a valid secret signing key and his token is not in the revoked members list. These conditions are satisfied in the underlying interactive protocol provided in the first lattice-based group signature scheme with VLR suggested by Langlois et al. in PKC 2014. In their scheme, member revocation token is a part of the secret signing key and has an implicit tracing algorithm to trace signers. For a scheme which generates member revocation token separately, the suggested interactive protocol by Langlois et al. is not suitable. Moreover, if the group manager wants to use an explicit tracing algorithm to trace signers instead the implicit tracing algorithm given in VLR schemes, then the signer should encrypt his index at the time of signing, and the interactive protocol should show signer's index is correctly encrypted. This work presents a combined interactive protocol that signer can use to prove his validity of signing, his separately generated revocation token is not in the revocation list, and his index is correctly encrypted required for such kind of schemes.

Keywords: Lattice-based group signatures · Verifier-local revocation
Zero-knowledge proof · Interactive protocol

1 Introduction

Commitment schemes are one of the leading primitives of group signature schemes. Commitment schemes allow a prover (signer) to commit to a value while keeping it in secret and later the prover provides additional information to open the commitment. A commitment scheme has three requirements, namely, *Hiding property*, *Binding property*, and *Viability*. The hiding property requires

L. Barolli et al. (Eds.): NBiS 2018, LNDECT 22, pp. 772–782, 2019.
https://doi.org/10.1007/978-3-319-98530-5_68

the receiver (verifier) cannot learn anything about the committed value. The binding property requires the prover cannot change the committed value after the commit step. The viability ensures if both parties, the signer and the verifier follow the protocol honestly, the verifier will always recover the committed value. Kawachi et al. [2] proposed a simple construction from lattices for string commitment scheme. Let COM be the statistically hiding and computationally binding commitment scheme. The statistically hiding requirement ensures any cheating verifier (adversary) cannot distinguish the commitments of two different strings and the computationally binding requirement ensures any polynomial time cheating signer cannot change the committed string after the commitment phase.

In 2014, Langlois et al. [3] presented the first lattice-based group signature scheme with member revocation while employing most flexible revocation approach, *Verifier-local revocation.* Their scheme operates within the structure of a *Bonsai tree* of hard random lattices, specified by a matrix $\mathbf{A}$ and a vector $\mathbf{u}$. In the proof of knowledge system, the signer with identity d has to prove in zero-knowledge that he knows a vector $\mathbf{z}$ which is a solution to the Inhomogeneous Short Integer Solution instance $(\mathbf{A}_d \cdot \mathbf{z})$ while hiding $\mathbf{z}$, where the vector $\mathbf{z}$ is the Bonsai signature issued on the prover's identity d. In other words, the signer has to prove, $||\mathbf{z}||_\infty \leq \beta$ and $\mathbf{A}_d \cdot \mathbf{z} = \mathbf{u} \mod q$ in zero-knowledge while hiding $\mathbf{z}$, where $\mathbf{A}_d = [\mathbf{A}_0|\mathbf{A}_1^0|\mathbf{A}_1^1|\ldots|\mathbf{A}_\ell^0|\mathbf{A}_\ell^1] \in \mathbb{Z}_q^{n\times(2\ell+1)m}$. As a solution for the above problem, they use a masking method. The masking method extends the given vector by adding zero-blocks. In the scheme in [3], they extended the vector $\mathbf{z}$ by adding ℓ suitable *zero-blocks* of size m to obtain a vector $\mathbf{x} = (\mathbf{x}_0||\mathbf{x}_1^0||\mathbf{x}_1^1||\ldots||\mathbf{x}_\ell^0||\mathbf{x}_\ell^1) \in \mathbb{Z}^{(2\ell+1)m}$, such that $||\mathbf{x}||_\infty \leq \beta$, and $\mathbf{A}\cdot\mathbf{x} = \mathbf{u} \mod q$, where $\mathbf{x}_1^{1-d[1]}, \ldots, \mathbf{x}_\ell^{1-d[\ell]}$ are added zero-blocks. To prove in zero-knowledge the possession of $\mathbf{x}$, Langlois et al. [3] adapted the 'Stren Extension' argument system provided in [4]. Moreover, they have generated the revocation token of each user by using the first block of user's secret signing key and the first block of the corresponding Bonsai tree. Thus the revocation token $\mathbf{grt}$ is $(\mathbf{A}_0 \cdot \mathbf{x}_0) \mod q$. At the time of signing, the signer computes $\mathbf{v} = \mathbf{V} \cdot \mathbf{grt} + \mathbf{e}_1 \mod q$, where $\mathbf{V} \in \mathbb{Z}_q^{m\times n}$ is a uniformly random matrix which is drawn from a random oracle and $\mathbf{e}_1 \in \mathbb{Z}^m$ is a small vector which is sampled from the Learning With Error distribution. At the zero-knowledge argument system, the signer additionally proves that the vector $\mathbf{v}$ is honestly generated.

However, in case of separating the revocation token creation from the secret signing key, the generation of the vector $\mathbf{v}$ cannot prove using the interactive protocol given in [3]. Moreover, the scheme in [3] uses an implicit tracing algorithm which requires to execute Verify with the tracing message-signature pair (M, Σ) for each member until the signer is traced. For a large group, this is not a convenient method. The group manager may require to find the signer quickly using an explicit tracing algorithm. For the explicit tracing algorithm, the signer should encrypt his index d at the time of signing, and he should prove his index d correctly encrypted in a ciphertext $\mathbf{c}$. This yields a new zero-knowledge interactive protocol since the protocol given in [3] cannot satisfy those conditions.

Our Contribution
The underlying interactive protocol in Langlois's scheme allows a signer to prove his validity using two vectors of the witness. One vector is for witnessing his Bonsai signature and the other vector is for witnessing he is not being revoked.

However, for a situation that revocation token is not deriving from the secret signing key, the interactive protocol given in [3] cannot be employed. Moreover, we take into account schemes which require the signers to encrypt their index at the signature generation such that the group manager can open signatures and trace the signers using the explicit tracing algorithm. Thus we need a protocol to show that the given ciphertext is correct encryption of the signer's index. As an answer for those requirements, we present a combined protocol which proves that the signer is a certified group member possessing a signature on his secret index with respect to the Bonsai tree signature, the signer's revocation token is correctly committed via an LWE function, and the signer's index is correctly encrypted based on LWE.

2 Preliminaries

2.1 Notations

For any integer $k \geq 1$, we denote the set of integers $\{1, \ldots, k\}$ by $[k]$. We denote matrices by bold upper-case letters such as $\mathbf{A}$, and vectors by bold lower-case letters, such as $\mathbf{x}$. We assume that all vectors are in column form. The concatenation of matrices $\mathbf{A} \in \mathbb{R}^{n \times m}$ and $\mathbf{B} \in \mathbb{R}^{n \times k}$, is denoted by $[\mathbf{A}|\mathbf{B}] \in \mathbb{R}^{n \times (m+k)}$. The concatenation of vectors $\mathbf{x} \in \mathbb{R}^m$ and $\mathbf{y} \in \mathbb{R}^k$ is denoted by $(\mathbf{x}\|\mathbf{y}) \in \mathbb{R}^{m+k}$. If S is a finite set, $b \xleftarrow{\$} S$ means that b is chosen uniformly at random from S. The Euclidean norm of $\mathbf{x}$ is denoted by $||\mathbf{x}||$ and the infinity norm is denoted by $||\mathbf{x}||_\infty$. Let χ be a b-bounded distribution over $\mathbb{Z}$ (i.e., samples that output by χ is with norm at most b with overwhelming probability where $b \geq \sqrt{n}\omega(\log n)$).

$\mathsf{Secret}_\beta(d)$ and $\mathsf{SecretExt}_\beta(d)$ are specific sets of vectors defined in [3] and obtained by appending ℓ *zero* $-$ *blocks* of size 0^m and 0^{3m} respectively to vectors $\mathbf{x} \in \mathbb{Z}^{(2\ell+1)m}$ and $\mathbf{x} \in \mathbb{Z}^{(2\ell+1)3m}$.

2.2 Lattice-Based Functions

Definition 1 (Learning With Errors (LWE) [5]**).** *For a vector $\boldsymbol{s} \in \mathbb{Z}_q^n$ and χ, the distribution $\mathrm{A}_{\boldsymbol{s},\chi}$ is obtained by sampling $\boldsymbol{a} \in \mathbb{Z}_q^n$ uniformly at random and choosing $\mathrm{e} \leftarrow \chi$, and outputting the pair $(\boldsymbol{a}, \boldsymbol{a}^T \cdot \boldsymbol{s} + \mathrm{e})$, where integers $n, m \geq 1$, and $q \geq 2$.*

Definition 2 (Inhomogeneous Short Integer Solution Problem($\mathbf{ISIS}_{n,m,q,\beta}$) [1]**).** *Given matrix $\mathbf{A} \in \mathbb{Z}_q^{n \times m}$ with m uniformly random vectors $\boldsymbol{a}_i \in \mathbb{Z}_q^n$ and a uniformly random vector $\mathbf{u} \in \mathbb{Z}_q^n$, $\mathbf{ISIS}_{n,m,q,\beta}$ asks to find a vector $\boldsymbol{x} \in \Lambda_{\boldsymbol{u}}^{\perp}(\boldsymbol{A})$ such that $||\boldsymbol{x}|| \leq \beta$.*

3 Underlying Interactive Protocol

Our Stern-like [7] interactive argument system allows a signer (prover) to convince the verifier about his validity in zero-knowledge, that the signer is a valid group member that posses a signature generated using his secret key and both his revocation token and his index are correctly committed via an LWE function.

Let n be the security parameter and ℓ be the message length. Let modulus $q = \omega(n^2 \log n)$ be prime, dimension $m \geq 2n \log q$, and Gaussian parameter $\sigma = \omega(\sqrt{n \log q \log n})$. The infinity norm bound $\beta = \lceil \sigma \cdot \log m \rceil$ s.t $(4\beta + 1)^2 \leq q$ and norm bound for LWE noises is b s.t $q/b = \ell\tilde{O}(n)$. Let $k_1 := m + \ell$ and $k_2 := n + m + \ell$.

- The common inputs: Matrices $\mathbf{A} = [\mathbf{A}_0|\mathbf{A}_1^0|\mathbf{A}_1^1|...|\mathbf{A}_\ell^0|\mathbf{A}_\ell^1] \in \mathbb{Z}_q^{n\times(2\ell+1)m}$, $\mathbf{B} \in \mathbb{Z}^{n\times m}$, $\mathbf{V} \in \mathbb{Z}_q^{m\times n}$, and $\mathbf{P} \in \mathbb{Z}_q^{k_1\times k_2}$ and vectors $\mathbf{u} \xleftarrow{\$} \mathbb{Z}_q^n$, $\mathbf{v} \in \mathbb{Z}_q^m$, and $\mathbf{c} \in \mathbb{Z}_q^{k_1}$.
- The prover's inputs: A vector $\mathbf{x} = (\mathbf{x}_0||\mathbf{x}_1^0||\mathbf{x}_1^1||...||\mathbf{x}_\ell^0||\mathbf{x}_\ell^1) \in \mathsf{Secret}_\beta(d)$ for some secret $d \in \{0,1\}^\ell$, vector $\mathbf{e}_1 \in \mathbb{Z}^m$, vector $\mathbf{r} \in \mathbb{Z}_q^n$, and a vector $\mathbf{e} \in \mathbb{Z}^{k_2}$. We use $\mathbf{f}$ instead of $\mathbf{e}_1$ hereunder to discard the confusing $\mathbf{e}_1$with $\mathbf{e}$.
- The prover's goal is to convince the verifier in zero-knowledge that:
  - $\mathbf{A} \cdot \mathbf{x} = \mathbf{u} \mod q$ and $\mathbf{x} \in \mathsf{Secret}_\beta(d)$.
  - $||\mathbf{f}||_\infty \leq \beta$ and $\mathbf{V} \cdot (\mathbf{B} \cdot \mathbf{r}) + \mathbf{f} = \mathbf{v} \mod q$. (Here the revocation token is created separately with a matrix $\mathbf{B}$ and a vector $\mathbf{r}$ instead of using $\mathbf{A}_0$ and $\mathbf{x}_0$).
  - $||\mathbf{e}||_\infty \leq b$ and $\mathbf{P}\mathbf{e} + (0^{k_1-\ell}||\lfloor q/2 \rfloor d) = \mathbf{c} \mod q$ (b is the norm bound for LWE noises and $\bar{p} = \lfloor \log b \rfloor + 1$).

Before the interaction, both the prover and the verifier form the public matrices: $\mathbf{A}^* \leftarrow \mathsf{MatrixExt}(\mathbf{A})$, $\mathbf{V}^* = \mathbf{V} \cdot \mathbf{B} \in \mathbb{Z}_q^{m\times m}$, $\mathbf{I}^* \in \{0,1\}^{m\times 3m}$ ($\mathbf{I}^*$ is obtained by appending *2m zero-columns* to the identity matrix of order m), $\mathbf{P}^* = [\mathbf{P} \mid 0^{k_1\times 2k_2}] \in \mathbb{Z}_q^{k_1\times 3k_2}$, and

$$Q = \begin{pmatrix} 0^{(k_1-\ell)\times\ell} & 0^{(k_1-\ell)\times\ell} \\ \lfloor q/2 \rfloor \mathbf{I}_\ell & 0^{\ell\times\ell} \end{pmatrix} \in \{0, \lfloor q/2 \rfloor\}^{k_1\times 2\ell}.$$

Then the prover uses the Decomposition-Extension technique provided in [3] with his witness vectors as below.

- Let $\mathbf{z}_1, \ldots, \mathbf{z}_p \leftarrow \mathsf{WitnessDE}(\mathbf{x})$.
- Let $\tilde{\mathbf{f}}_1, \ldots, \tilde{\mathbf{f}}_p \leftarrow \mathsf{EleDec}(\mathbf{f})$, then for each $i \in [p]$, let $\mathbf{f}_i \leftarrow \mathsf{EleExt}(\tilde{\mathbf{f}}_i)$.
- Let $\tilde{\mathbf{r}}_1, \ldots, \tilde{\mathbf{r}}_p \leftarrow \mathsf{EleDec}(\mathbf{r})$, then for each $i \in [p]$, let $\mathbf{r}_i \leftarrow \mathsf{EleExt}(\tilde{\mathbf{r}}_i)$.
- Let $\tilde{\mathbf{e}}_1, \ldots, \tilde{\mathbf{e}}_{\bar{p}} \leftarrow \mathsf{EleDec}(\mathbf{e})$, then for each $i \in [p]$, let $\mathbf{e}_i \leftarrow \mathsf{EleExt}(\tilde{\mathbf{e}}_i)$.

At the interactive protocol, the prover instead convince the verifier that he knows $\mathbf{z}_1, \ldots, \mathbf{z}_p \in \mathsf{Secret}_\beta(d)$, $\tilde{\mathbf{f}}_1, \ldots, \tilde{\mathbf{f}}_p \in \mathsf{B}_{3m}$, $\tilde{\mathbf{r}}_1, \ldots, \tilde{\mathbf{r}}_p \in \mathsf{B}_{3m}$, and $\tilde{\mathbf{e}}_1, \ldots, \tilde{\mathbf{e}}_p \in \mathsf{B}_{3k_2}$, such that:

$$\begin{cases} \mathbf{A}^* \cdot (\sum_{j=1}^{p} \beta_j \cdot \mathbf{z}_j) = \mathbf{u} \mod q; \\ \mathbf{V}^* \cdot (\sum_{j=1}^{p} \beta_j \cdot \mathbf{r}_j) + \mathbf{I}^* \cdot (\sum_{j=1}^{P} \beta_j \cdot \mathbf{f}_j) = \mathbf{v} \mod q. \\ \mathbf{P}^* \cdot (\sum_{j=1}^{\bar{p}} b_j \cdot \mathbf{e}_j) + \mathbf{Q} \cdot d^* = \mathbf{Pe} + (0^{k_1-\ell} || \lfloor q/2 \rfloor d) = \mathbf{c} \mod q. \end{cases}$$

Description of the protocol:

1. **Commitment**: The prover samples randomness ρ_1, ρ_2, ρ_3 for COM and the following uniformly random objects:

$$\begin{cases} c \xleftarrow{\$} \{0,1\}^\ell; \\ \pi_{z,1}, \ldots, \pi_{z,p} \xleftarrow{\$} S; \pi_{f,1}, \ldots, \pi_{f,p} \xleftarrow{\$} S_{3m}; \pi_{r,1}, \ldots, \pi_{r,p} \xleftarrow{\$} S_{3m}; \\ \quad \pi_{e,1}, \ldots, \pi_{e,\bar{p}} \xleftarrow{\$} S_{3k_2}; \tau \xleftarrow{\$} S_{2\ell}; \\ \mathbf{k}_{z,1}, \ldots, \mathbf{k}_{z,p} \xleftarrow{\$} \mathbb{Z}_q^{(2\ell+1)3m}; \mathbf{k}_{f,1}, \ldots, \mathbf{k}_{f,p} \xleftarrow{\$} \mathbb{Z}_q^{3m}; \\ \quad \mathbf{k}_{r,1}, \ldots, \mathbf{k}_{r,p} \xleftarrow{\$} \mathbb{Z}_q^{3m}; \mathbf{k}_{e,1}, \ldots, \mathbf{k}_{e,\bar{p}} \xleftarrow{\$} \mathbb{Z}_q^{3k_2}; \mathbf{k}_d \xleftarrow{\$} \mathbb{Z}_q^{2\ell}. \end{cases} \quad (1)$$

Then the prover sends the following commitment $\mathbf{CMT} = (\mathbf{c}_1, \mathbf{c}_2, \mathbf{c}_3)$ to the verifier.

$$\begin{cases} \mathbf{c}_1 = \mathsf{COM}(c, \{\pi_{z,j}, \pi_{f,j}, \pi_{r,j}\}_{j=1}^{p}), \mathbf{A}^* \cdot (\sum_{j=1}^{p} \beta_j \cdot \mathbf{k}_{z,j}); \\ \quad \mathbf{V}^* \cdot (\sum_{j=1}^{p} \beta_j \cdot \mathbf{k}_{r,j}) + \mathbf{I}^* \cdot (\sum_{j=1}^{p} \beta_j \cdot \mathbf{k}_{f,j}); \\ \quad \{\pi_{e,j},\}_{j=1}^{\bar{p}}; \mathbf{P}^*(\sum_{j=1}^{\bar{p}} b_j \mathbf{k}_{e,j}) + \mathbf{Q}\mathbf{k}_d; \tau; \rho_1), \\ \mathbf{c}_2 = \mathsf{COM}(\{T_c \circ \pi_{z,j}(\mathbf{k}_{z,j}), \pi_{f,j}(\mathbf{k}_{f,j}), \pi_{r,j}(\mathbf{k}_{r,j})\}_{j=1}^{p}; \\ \quad \{\pi_{e,j}(\mathbf{k}_{e,j})\}_{j=1}^{\bar{p}}; \tau(\mathbf{k}_d); \rho_2), \\ \mathbf{c}_3 = \\ \mathsf{COM}(\{T_c \circ \pi_{z,j}(\mathbf{z}_j + \mathbf{k}_{z,j}), \pi_{f,j}(\mathbf{f}_j + \mathbf{k}_{f,j}), \pi_{r,j}(\mathbf{r}_j + \mathbf{k}_{r,j})\}_{j=1}^{p}; \\ \quad \{\pi_{e,j}(\mathbf{e}_j + \mathbf{k}_{e,j})\}_{j=1}^{\bar{p}}; \tau(d^* + \mathbf{k}_d); \rho_3). \end{cases} \quad (2)$$

2. **Challenge:** The verifier sends a challenge $Ch \xleftarrow{\$} \{1,2,3\}$ to the prover.
3. **Response:** Depending on the challenge, the prover sends the response RSP computed as follows.
   - Case $Ch = 1$: Let $d_1 = d \oplus c$. For each $j \in [p]$, let $\mathbf{u}_{z,j} = T_c \circ \pi_{z,j}(\mathbf{z}_j); \mathbf{w}_{z,j} = T_c \circ \pi_{z,j}(\mathbf{k}_{z,j}); \mathbf{u}_{f,j} = \pi_{f,j}(\mathbf{f}_j); \mathbf{w}_{f,j} = \pi_{f,j}(\mathbf{k}_{f,j}); \mathbf{u}_{r,j} = \pi_{r,j}(\mathbf{r}_j); \mathbf{w}_{r,j} = \pi_{r,j}(\mathbf{k}_{r,j})$. For each $j \in [\bar{p}]$, let $\mathbf{u}_{e,j} = \pi_{e,j}(\mathbf{e}_j); \mathbf{w}_{e,j} = \pi_{e,j}(\mathbf{k}_{e,j})$. Let $\mathbf{u}_d = \tau(d^*); \mathbf{w}_d = \tau(\mathbf{k}_d)$. Then send,

$$\begin{aligned} RSP = (d_1, \{\mathbf{u}_{z,j}, \mathbf{w}_{z,j}, \mathbf{u}_{f,j}, \mathbf{w}_{f,j}, \mathbf{u}_{r,j}, \mathbf{w}_{r,j}\}^{p_{j=1}}, \\ \{\mathbf{u}_{e,j}, \mathbf{w}_{e,j}\}_{j=1}^{\bar{p}}, \mathbf{u}_d, \mathbf{w}_d, \rho_2, \rho_3). \end{aligned} \quad (3)$$

   - Case $Ch = 2$: Let $d_2 = c$. For each $j \in [p]$, let $\phi_{z,j} = \pi_{z,j}; \phi_{f,j} = \pi_{f,j}; \phi_{r,j} = \pi_{r,j}; \mathbf{s}_{z,j} = \mathbf{z}_j + \mathbf{k}_{z,j}; \mathbf{s}_{f,j} = \mathbf{f}_j + \mathbf{k}_{f,j}; \mathbf{s}_{r,j} = \mathbf{r}_j + \mathbf{k}_{r,j}$. For each $j \in [\bar{p}]$, let $\phi_{e,j} = \pi_{e,j}; \mathbf{s}_{e,j} = \mathbf{e}_j + \mathbf{k}_{e,j}$. Let $\hat{\tau} = \tau$ and $\mathbf{s}_d = d^* + \mathbf{k}_d$. Then send,

$$\begin{aligned} RSP = (d_2, \{\phi_{z,j}, \phi_{f,j}, \phi_{r,j}, \mathbf{s}_{z,j}, \mathbf{s}_{f,j}, \mathbf{s}_{k,j}\}_{j=1}^{p}, \\ \{\phi_{e,j}, \mathbf{s}_{e,j}\}_{j=1}^{\bar{p}}, \hat{\tau}, \mathbf{s}_d, \rho_1, \rho_3) \end{aligned} \quad (4)$$

– Case $Ch = 3$: Let $d_3 = c$. For each $j \in [p]$, let $\psi_{z,j} = \pi_{z,j}; \psi_{f,j} = \pi_{f,j}; \psi_{r,j} = \pi_{r,j}; \mathbf{h}_{z,j} = \mathbf{k}_{z,j}; \mathbf{h}_{f,j} = \mathbf{k}_{f,j}; \mathbf{h}_{r,j} = \mathbf{k}_{r,j}$. For each $j \in [\bar{p}]$, let $\psi_{e,j} = \pi_{e,j}; \mathbf{h}_{e,j} = \mathbf{k}_{e,j}$. Let $\tilde{\tau} = \tau$ and $\mathbf{h}_d = \mathbf{k}_d$. Then send,

$$RSP = (d_3, \{\psi_{z,j}, \psi_{f,j}, \psi_{r,j}, \mathbf{h}_{z,j}, \mathbf{h}_{f,j}, \mathbf{h}_{k,j}\}_{j=1}^{p}, \\ \{\psi_{e,j}, \mathbf{h}_{e,j}\}_{j=1}^{\bar{p}}, \tilde{\tau}, \mathbf{h}_d, \rho_1, \rho_2) \tag{5}$$

4. Receiving the response RSP, the verifier proceeds as follows:
   – $Ch = 1$: Parse RSP as in (3).
   Check that $\forall \in [p] : \mathbf{u}_{z,j} \in \mathsf{SecretExt}(d_1), \mathbf{u}_{f,j} \in \mathsf{B}_{3m}, \mathbf{u}_{r,j} \in \mathsf{B}_{3m}, \forall j \in [\bar{p}] : \mathbf{u}_d \in \mathsf{B}_{2\ell}, \mathbf{u}_{e,j} \in \mathsf{B}_{3k_2}$, and

$$\begin{cases} \mathbf{c}_2 = \mathsf{COM}(\{\mathbf{w}_{z,j}, \mathbf{w}_{f,j}, \mathbf{w}_{r,j}\}_{j=1}^{p}; \{\mathbf{w}_{e,j}\}_{j=1}^{\bar{p}}; \mathbf{w}_d; \rho_2), \\ \mathbf{c}_3 = \mathsf{COM}(\{\mathbf{u}_{z,j} + \mathbf{w}_{z,j}, \mathbf{u}_{f,j} + \mathbf{w}_{f,j}, \mathbf{u}_{r,j} + \mathbf{w}_{r,j}\}_{j=1}^{p}; \\ \qquad \{\mathbf{u}_{e,j} + \mathbf{w}_{e,j}\}_{j=1}^{\bar{p}}; \{\mathbf{u}_d + \mathbf{w}_d\}; \rho_3). \end{cases} \tag{6}$$

   – $Ch = 2$: Parse RSP as in (4). Check that :

$$\begin{cases} \mathbf{c}_1 = \mathsf{COM}(d_2, \{\phi_{z,j}, \phi_{f,j}, \phi_{r,j}\}_{j=1}^{p}, \mathbf{A}^*(\sum_{j=1}^{p} \beta_j \cdot \mathbf{s}_{z,j}) - \mathbf{u}; \\ \quad \mathbf{V}^*(\sum_{j=1}^{p} \beta_j \cdot \mathbf{s}_{r,j}) + \mathbf{I}^*(\sum_{j=1}^{p} \beta_j \cdot \mathbf{s}_{f,j}) - \mathbf{v}; \\ \quad \{\phi_{e,j}\}_{j=1}^{\bar{p}}; \mathbf{P}^* \cdot (\sum_{j=1}^{\bar{p}} b_j \cdot \mathbf{s}_{e,j}) + \mathbf{Q}\mathbf{s}_d - \mathbf{c}; \hat{\tau}; \rho_1), \\ \mathbf{c}_3 = \mathsf{COM}(\{T_{d_2} \circ \phi_{z,j}(\mathbf{s}_{z,j}), \phi_{f,j}(\mathbf{s}_{f,j}), \phi_{r,j}(\mathbf{s}_{r,j})\}_{j=1}^{p}; \\ \quad \{\phi_{e,j}(\mathbf{s}_{e,j})\}_{j=1}^{\bar{p}}; \hat{\tau}(\mathbf{s}_d); \rho_3). \end{cases} \tag{7}$$

   – $Ch = 3$:Parse RSP as in (5). Check that :

$$\begin{cases} \mathbf{c}_1 = \mathsf{COM}(d_3, \{\psi_{z,j}, \psi_{f,j}, \psi_{r,j}\}_{j=1}^{p}, \mathbf{A}^* \cdot (\sum_{j=1}^{p} \beta_j \cdot \mathbf{h}_{z,j}); \\ \quad \mathbf{V}^* \cdot (\sum_{j=1}^{p} \beta_j \cdot \mathbf{h}_{r,j}) + \mathbf{I}^* \cdot (\sum_{j=1}^{p} \beta_j \cdot \mathbf{h}_{f,j}); \\ \quad \{\phi_{e,j}\}_{j=1}^{\bar{p}}; \mathbf{P}^* \cdot (\sum_{j=1}^{\bar{p}} b_j \cdot \mathbf{h}_{e,j}) + \mathbf{Q}\mathbf{h}_d; \tilde{\tau}; \rho_1), \\ \mathbf{c}_2 = \mathsf{COM}(\{T_{d_3} \circ \psi_{z,j}(\mathbf{h}_{z,j}), \psi_{f,j}(\mathbf{h}_{f,j}), \psi_{r,j}(\mathbf{h}_{r,j})\}_{j=1}^{p}; \\ \quad \{\psi_{e,j}(\mathbf{h}_{e,j}),\}_{j=1}^{\bar{p}}; \tilde{\tau}(\mathbf{h}_d); \rho_2). \end{cases} \tag{8}$$

The verifier outputs Valid if and only if all the conditions hold. Otherwise, he outputs Invalid.

4 Analysis of the Protocol

Theorem 1. *Let* **COM** *be a statistically hiding and computationally binding string commitment scheme. The interactive protocol is a zero-knowledge argument of knowledge with perfect completeness and soundness error 2/3 with* $(\mathcal{O}(\ell m) \log \beta + \mathcal{O}(k_2) \log b) \log q$ *communication cost. Thus it satisfies the followings.*

– *There exists an efficient simulator that, on input* $(\mathbf{A}, \mathbf{u}, \mathbf{B}, \mathbf{V}, \mathbf{v}, \mathbf{P}, \mathbf{c})$*, outputs an accepted transcript which is statistically close to that produced by the real prover.*

- *There exists an efficient knowledge extractor that, on input a commitment CMT and 3 valid responses $(RSP^{(1)}, RSP^{(2)}, RSP^{(3)})$ corresponding to all 3 possible values of the challenging Ch, outputs vectors $(\mathbf{y}, \mathbf{f}', \mathbf{r}', \mathbf{e}')$ such that:*
  1. $\mathbf{y} = (\mathbf{y}_0 \| \mathbf{y}_1^0 \| \mathbf{y}_1^1 \| \cdots \| \mathbf{y}_\ell^0 \| \mathbf{y}_\ell^1) \in \mathsf{Secret}_\beta(d)$ *for some* $d \in \{0,1\}^\ell$, *and* $\mathbf{A} \cdot \mathbf{y} = \mathbf{u} \mod q$.
  2. $\|\mathbf{f}'\|_\infty \leq \beta$ *and* $\mathbf{V} \cdot (\mathbf{B} \cdot \mathbf{r}) + \mathbf{f}' = \mathbf{v} \mod q$.
  3. $\|\mathbf{e}'\|_\infty \leq b$ *and* $\mathbf{P}\mathbf{e}' + (0^{k_1 - \ell} \| \lfloor q/2 \rfloor d) = \mathbf{c} \mod q$.

4.1 Completeness and Soundness

An honest prover, with a valid witness $(\mathbf{x}, \mathbf{f}, \mathbf{r}, \mathbf{e})$ for some $d \in \{0,1\}^\ell$, can always obtain $\mathbf{z}_1, \ldots, \mathbf{z}_p \in \mathsf{Secret}_\beta(d), \mathbf{f}_1, \ldots, \mathbf{f}_p \in \mathsf{B}_{3m}, \mathbf{r}_1, \ldots, \mathbf{r}_p \in \mathsf{B}_{3m}$, and $\mathbf{e}_1, \ldots, \mathbf{e}_{\bar{p}} \in \mathsf{B}_{3k_2}$ via the Decomposition-Extension technique [3]. If he follows the protocol, he should always be accepted by the verifier. In this manner, the protocol has perfect completeness.

The protocol admits a soundness error 2/3, which is natural for typical Stern-like protocols. However, this error can be made negligible by repeating the protocol $t = \omega(\log n)$ times in parallel.

4.2 Communication Cost

The KTX scheme [2] COM outputs an element of $\mathbb{Z}_q^n$. Therefore the commitment CMT has bit-size $3n \log q = \tilde{\mathcal{O}}(n)$. The response RSP is executed by, p permutations in S, p permutations in S_{3m}, $\bar{p}$ permutations in S_{3k_2}, one permutation in 2ℓ, p vectors in $\mathbb{Z}_q^{(2\ell+1)3m}$, p vectors in $\mathbb{Z}_q^{3m}$, $\bar{p}$ vectors in $\mathbb{Z}_q^{3k_2}$, and one vector in $\mathbb{Z}_q^{2\ell}$.

In this manner, the bit size of RSP is bounded by $(\mathcal{O}(\ell m)p + \mathcal{O}(k_2)\bar{p}) \log q$, where $p = \lfloor \log \beta \rfloor + 1$ and $p = \lfloor \log b \rfloor + 1$. Thus the overall communication cost of the protocol is bounded by $(\mathcal{O}(\ell m) \log \beta + \mathcal{O}(k_2) \log b) \log q$.

4.3 Zero-Knowledge Property

If **COM** is statistically hiding, we can prove that, the interactive protocol is statistical zero-knowledge argument.

First, construct a PPT simulator *SIM* interacting with a verifier V such that, by giving only the public inputs, *SIM* outputs with probability close to 2/3 a simulated transcript that is statistically close to the outputs of an honest prover in the real interaction. From the public input ($\mathbf{A}$, $\mathbf{u}$, $\mathbf{B}$, $\mathbf{V}$, $\mathbf{v}$, $\mathbf{P}$, $\mathbf{c}$) given by the protocol, both *SIM* and V acquire matrices, $\mathbf{A}^*, \mathbf{V}^*$, $\mathbf{I}^*$, $\mathbf{P}^*$, and $\mathbf{Q}$. Then *SIM* starts simulation by selecting a random $\overline{Ch} \in \{1,2,3\}$. This is a prediction of the challenge value that V will not choose.

Case $\overline{Ch} = 1$**:** *SIM* computes the vectors $\mathbf{z}_1', \ldots, \mathbf{z}_p' \in \mathbb{Z}_q^{(2\ell+1)3m}$ such that $\mathbf{A}^* \cdot (\sum_{j=1}^p \beta_j \cdot \mathbf{z}_j') = \mathbf{u} \mod q$, $\mathbf{r}_1', \ldots, \mathbf{r}_p' \in \mathbb{Z}_q^{3m}$ and $\mathbf{f}_1', \ldots, \mathbf{f}_p' \in \mathbb{Z}_q^{3m}$ such that $\mathbf{V}^* \cdot (\sum_{j=1}^p \beta_j \cdot \mathbf{r}_j') + \mathbf{I}^* \cdot (\sum_{j=1}^p \beta_j \cdot \mathbf{f}_j') = \mathbf{v} \mod q$, and $\mathbf{e}_1', \ldots, \mathbf{e}_{\bar{p}}' \in \mathbb{Z}_q^{3k}$ and $d' \in \mathbb{Z}_q^{2\ell}$, such that $\mathbf{P}^* \cdot (\sum_{j=1}^{\bar{p}} b_j \cdot \mathbf{e}_j') + \mathbf{Q} \cdot d' = \mathbf{c} \mod q$ by using linear algebra.

Then *SIM* samples objects as in Eq. (1) and sends commitment **CMT** $= (\mathbf{c}'_1, \mathbf{c}'_2, \mathbf{c}'_3)$ to *V*, where

$$\begin{cases} \mathbf{c}'_1 = \mathsf{COM}(c, \{\pi_{z,j}, \pi_{f,j}, \pi_{r,j}\}_{j=1}^{p}, \mathbf{A}^* \cdot (\sum_{j=1}^{p} \beta_j \cdot \mathbf{k}_{z,j}); \\ \quad \mathbf{V}^* \cdot (\sum_{j=1}^{p} \beta_j \cdot \mathbf{k}_{r,j}) + \mathbf{I}^* \cdot (\sum_{j=1}^{p} \beta_j \cdot \mathbf{k}_{f,j}) \\ \quad \{\pi_{e,j}\}_{j=1}^{\bar{p}}, \mathbf{P}^* \cdot (\sum_{j=1}^{\bar{p}} b_j \cdot \mathbf{k}_{e,j}) + \mathbf{Q}\mathbf{k}_d; \tau; \rho_1), \\ \mathbf{c}'_2 = \mathsf{COM}(\{T_c \circ \pi_{z,j}(\mathbf{k}_{z,j}), \pi_{f,j}(\mathbf{k}_{f,j}), \pi_{r,j}(\mathbf{k}_{r,j})\}_{j=1}^{p}; \\ \quad \{\pi_{e,j}(\mathbf{k}_{e,j})\}_{j=1}^{\bar{p}}; \tau(\mathbf{k}_d); \rho_2), \\ \mathbf{c}'_3 = \mathsf{COM}(\{T_c \circ \pi_{z,j}(\mathbf{z}'_j + \mathbf{k}_{z,j}), \pi_{f,j}(\mathbf{f}'_j + \mathbf{k}_{f,j}), \pi_{r,j}(\mathbf{r}'_j + \mathbf{k}_{r,j})\}_{j=1}^{p}; \\ \quad \{\pi_{e,j}(\mathbf{e}'_j + \mathbf{k}_{e,j})\}_{j=1}^{\bar{p}}; \tau(d' + \mathbf{k}_d); \rho_3). \end{cases} \tag{9}$$

For a challenge *Ch* from *V*, *SIM* responds as follows:

- If $Ch = 1$: Output $\perp$ and abort.
- If $Ch = 2$: Send, $\mathrm{RSP} = (c, \{\pi_{z,j}, \pi_{f,j}, \pi_{r,j}, \mathbf{z}'_j + \mathbf{k}_{z,j}, \mathbf{f}'_j + \mathbf{k}_{f,j}, \mathbf{r}'_j + \mathbf{k}_{r,j}\}_{j=1}^{p}, \{\pi_{e,j}, \mathbf{e}'_j + \mathbf{k}_{e,j}\}_{j=1}^{\bar{p}}, d' + \mathbf{k}_d, \tau, \rho_1, \rho_3)$.
- If $Ch = 3$: Send, $\mathrm{RSP} = (c, \{\pi_{z,j}, \pi_{f,j}, \pi_{r,j}, \mathbf{k}_{z,j}, \mathbf{k}_{f,j}, \mathbf{k}_{r,j}\}_{j=1}^{p}, \{\pi_{e,j}, \mathbf{k}_{e,j}\}_{j=1}^{\bar{p}}, \tau, \rho_1, \rho_2)$.

Case $\overline{Ch} = 2$: *SIM* samples randomness ρ_1, ρ_2, ρ_3 for **COM** and

$$\begin{cases} \hat{d} \xleftarrow{\$} \{0,1\}^{\ell}, c \xleftarrow{\$} \{0,1\}^{\ell}; d' \xleftarrow{\$} \mathsf{B}_{2\ell}; \\ \mathbf{z}'_1, \ldots, \mathbf{z}'_p \xleftarrow{\$} \mathsf{SecretExt}(d); \mathbf{f}'_1, \ldots, \mathbf{f}'_p \xleftarrow{\$} \mathsf{B}_{3m}; \mathbf{r}'_1, \ldots, \mathbf{r}'_p \xleftarrow{\$} \mathsf{B}_{3m}; \\ \quad \mathbf{e}'_1, \ldots, \mathbf{e}'_{\bar{p}} \xleftarrow{\$} \mathsf{B}_{3k}; \\ \pi_{z,1}, \ldots, \pi_{z,p} \xleftarrow{\$} S; \pi_{f,1}, \ldots, \pi_{f,p} \xleftarrow{\$} S_{3m}; \pi_{r,1}, \ldots, \pi_{r,p} \xleftarrow{\$} S_{3m}; \\ \quad \pi_{e,1}, \ldots, \pi_{e,\bar{p}} \xleftarrow{\$} S_{3k}; \\ \mathbf{k}_{z,1}, \ldots, \mathbf{k}_{z,p} \xleftarrow{\$} \mathbb{Z}_q^{(2\ell+1)3m}; \mathbf{k}_{f,1}, \ldots, \mathbf{k}_{f,p} \xleftarrow{\$} \mathbb{Z}_q^{3m}; \mathbf{k}_{r,1}, \ldots, \mathbf{k}_{r,p} \xleftarrow{\$} \mathbb{Z}_q^{3m}; \\ \quad \mathbf{k}_{e,1}, \ldots, \mathbf{k}_{e,\bar{p}} \xleftarrow{\$} \mathbb{Z}_q^{3k}; \mathbf{k}_d \xleftarrow{\$} \mathbb{Z}_q^{2\ell}, \tau \xleftarrow{\$} S_{2\ell}. \end{cases}$$

Next *SIM* forms and sends commitment CMT as the same manner as in (9). For a challenge *Ch* from *V*, *SIM* responds as follows:

- If $Ch = 1$: $(\hat{d} \oplus c \; \{T_c \circ \pi_{z,j}(\mathbf{z}'_j), T_c \circ \pi_{z,j}(\mathbf{k}_{z,j}), \pi_{f,j}(\mathbf{f}'_j), \pi_{f,j}(\mathbf{k}_{f,j}), \pi_{r,j}(\mathbf{r}'_j), \pi_{r,j}(\mathbf{k}_{r,j})\}_{j=1}^{p}, \{\pi_{e,j}(\mathbf{e}'_j), \pi_{e,j}(\mathbf{k}_{e,j})\}_{j=1}^{\bar{p}}, \tau(d'), \tau(\mathbf{k}_d))$.
- If $Ch = 2$: Output $\perp$ and abort.
- If $Ch = 3$: Send, RSP computed as in the case $(\overline{Ch} = 1, Ch = 3)$.

Case $\overline{Ch} = 3$: *SIM* samples randomness as in $\overline{Ch} = 2$ and sends the commitment $\mathbf{CMT} = (\mathbf{c}'_1, \mathbf{c}'_2, \mathbf{c}'_3)$ to V, where $\mathbf{c}'_2, \mathbf{c}'_3$ are computed as in (9), and

$$\mathbf{c}'_1 = \mathsf{COM}\ (c, \{\pi_{z,j}, \pi_{e,j}, \pi_{r,j}\}_{j=1}^{p}, \mathbf{A}^* \cdot (\sum_{j=1}^{p} \beta_j \cdot (\mathbf{z}'_j + \mathbf{k}_{z,j})) - \mathbf{u};$$
$$\mathbf{V}^* \cdot (\sum_{j=1}^{p} \beta_j \cdot (\mathbf{r}'_j + \mathbf{k}_{r,j})) + \mathbf{I}^* \cdot (\sum_{j=1}^{p} \beta_j \cdot (\mathbf{f}'_j + \mathbf{k}_{f,j})) - \mathbf{v};$$
$$\{\pi_{e,j}\}_{j=1}^{\bar{p}}; \mathbf{P}^* \sum_{j=1}^{\bar{p}} b_j(\mathbf{e}'_j + \mathbf{k}_{e,j}) + \mathbf{Q}(d' + \mathbf{k}_d) - \mathbf{c}; \tau; \rho_1).$$

For a challenge Ch from V, *SIM* responds as follows:

- If $Ch = 1$: Send, RSP computed as in the case $(\overline{Ch} = 2, Ch = 1)$.
- If $Ch = 2$: Send, RSP computed as in the case $(\overline{Ch} = 1, Ch = 2)$.
- If $Ch = 3$: Output $\perp$ and abort.

Since **COM** is statistically hiding, the distribution of the commitment CMT and the distribution of the challenge Ch from V for every case considered above are statistically close to those in the real interaction. Hence, the probability that the simulator outputs $\perp$ is negligibly close to 1/3. Thus, the simulator *SIM* can successfully imitate the honest prover with probability negligibly close to 2/3.

4.4 Argument of Knowledge

Here we prove that, if COM is computationally binding, then the given protocol is an argument of knowledge. For a given commitment CMT and three valid responses $RSP^{(1)}, RSP^{(2)}, RSP^{(3)}$ to all three possible values of the challenge Ch, a valid witness can be extracted.

$$\begin{cases} \mathbf{c}_1 & = \mathsf{COM}(d_2, \{\phi_{z,j}, \phi_{f,j}, \phi_{r,j}\}_{j=1}^{p}, \mathbf{A}^* \cdot (\sum_{j=1}^{p} \beta_j \cdot \mathbf{s}_{z,j}) - \mathbf{u}; \\ & \quad \mathbf{V}^* \cdot (\sum_{j=1}^{p} \beta_j \cdot \mathbf{s}_{r,j}) + \mathbf{I}^* \cdot (\sum_{j=1}^{p} \beta_j \cdot \mathbf{s}_{f,j}) - \mathbf{v}; \\ & \quad \{\phi_{e,j}\}_{j=1}^{\bar{p}}; \mathbf{P}^* \cdot (\sum_{j=1}^{\bar{p}} b_j \cdot \mathbf{s}_{e,j}) + \mathbf{Q}\mathbf{s}_d - \mathbf{c}; \hat{\tau}; \rho_1) \\ & = \mathsf{COM}(d_3, \{\psi_{z,j}, \psi_{f,j}, \psi_{r,j}\}_{j=1}^{p}, \mathbf{A}^* \cdot (\sum_{j=1}^{p} \beta_j \cdot \mathbf{h}_{z,j}); \\ & \quad \mathbf{V}^* \cdot (\sum_{j=1}^{p} \beta_j \cdot \mathbf{h}_{r,j}) + \mathbf{I}^* \cdot (\sum_{j=1}^{p} \beta_j \cdot \mathbf{h}_{f,j}); \\ & \quad \{\psi_{e,j}\}_{j=1}^{\bar{p}}; \mathbf{P}^* \cdot (\sum_{j=1}^{\bar{p}} b_j \cdot \mathbf{h}_{e,j}) + \mathbf{Q}\mathbf{h}_d; \tilde{\tau}; \rho_1), \\ \mathbf{c}_2 & = \mathsf{COM}(\{\mathbf{w}_{z,j}, \mathbf{w}_{f,j}, \mathbf{w}_{r,j}\}_{j=1}^{p}, \{\mathbf{w}_{e,j}\}_{j=1}^{\bar{p}}, \mathbf{w}_d; \rho_2) \\ & = \mathsf{COM}(\{T_{d_3} \circ \psi_{z,j}(\mathbf{h}_{z,j}), \psi_{f,j}(\mathbf{h}_{f,j}), \psi_{r,j}(\mathbf{h}_{r,j})\}_{j=1}^{p}; \\ & \quad \{\psi_{e,j}(\mathbf{h}_{e,j})\}_{j=1}^{\bar{p}}, \tilde{\tau}(\mathbf{h}_d); \rho_2), \\ \mathbf{c}_3 & = \mathsf{COM}(\{\mathbf{u}_{z,j} + \mathbf{w}_{z,j}, \mathbf{u}_{f,j} + \mathbf{w}_{f,j}, \mathbf{u}_{r,j} + \mathbf{w}_{r,j}\}_{j=1}^{p}; \\ & \quad \{\mathbf{u}_{e,j} + \mathbf{w}_{e,j}\}_{j=1}^{\bar{p}}, \{\mathbf{u}_d \mid \mathbf{w}_d\}; \rho_3) \\ & = \mathsf{COM}(\{T_{d_2} \circ \phi_{z,j}(\mathbf{s}_{z,j}), \phi_{f,j}(\mathbf{s}_{f,j}), \phi_{r,j}(\mathbf{s}_{r,j})\}_{j=1}^{p}; \\ & \quad \{\phi_{e,j}(\mathbf{s}_{e,j})\}_{j=1}^{\bar{p}}, \hat{\tau}(\mathbf{s}_d); \rho_3). \end{cases}$$

The computational binding property of **COM** implies that:

$$\begin{cases} d_2 = d_3; \\ \mathbf{u}_d \in \mathsf{B}_{2\ell}; \hat{\tau} = \tilde{\tau}; \mathbf{w}_d = \tilde{\tau}(\mathbf{h}_d); \mathbf{u}_d + \mathbf{w}_d = \hat{\tau}(\mathbf{s}_d); \\ \forall j \in [p] : \phi_{z,j} = \psi_{z,j}; \mathbf{w}_{z,j} = T_{d_2} \circ \phi_{z,j}(\mathbf{h}_{z,j}) \ and \\ \qquad \mathbf{u}_{z,j} + \mathbf{w}_{z,j} = T_{d_2} \circ \phi_{z,j}(\mathbf{s}_{z,j}); \\ \forall j \in [p] : \phi_{f,j} = \psi_{f,j}; \mathbf{w}_{f,j} = \phi_{f,j}(\mathbf{h}_{f,j}) \ and \ \mathbf{u}_{f,j} + \mathbf{w}_{f,j} = \phi_{f,j}(\mathbf{s}_{f,j}); \\ \forall j \in [p] : \phi_{r,j} = \psi_{r,j}; \mathbf{w}_{r,j} = \phi_{r,j}(\mathbf{h}_{r,j}) \ and \ \mathbf{u}_{r,j} + \mathbf{w}_{r,j} = \phi_{r,j}(\mathbf{s}_{r,j}); \\ \forall j \in [\bar{p}] : \phi_{e,j} = \psi_{e,j}; \mathbf{w}_{e,j} = \phi_{e,j}(\mathbf{h}_{e,j}) \ and \ \mathbf{u}_{e,j} + \mathbf{w}_{e,j} = \phi_{e,j}(\mathbf{s}_{e,j}); \\ \mathbf{A}^* \cdot (\sum_{j=1}^{p} \beta_j \cdot (\mathbf{s}_{z,j} - \mathbf{h}_{z,j})) = \mathbf{u} \mod q; \\ \mathbf{V}^* \cdot (\sum_{j=1}^{p} \beta_j \cdot (\mathbf{s}_{r,j} - \mathbf{h}_{r,j})) + \mathbf{I}^* \cdot (\sum_{j=1}^{p} \beta_j \cdot (\mathbf{s}_{f,j} - \mathbf{h}_{f,j})) = \mathbf{v} \mod q; \\ \mathbf{P}^* \cdot (\sum_{j=1}^{\bar{p}} b_j \mathbf{s}_{e,j}) + \mathbf{Q}\mathbf{s}_d - \mathbf{c} = \mathbf{P}^* \cdot (\sum_{j=1}^{\bar{p}} b_j \mathbf{h}_{e,j}) + \mathbf{Q}\mathbf{h}_d \mod q. \end{cases}$$

For each $j \in [p]$, let $\mathbf{y}'_j = (\mathbf{s}_{z,j} - \mathbf{h}_{z,j})$. Then $T_{d_2} \circ \phi_{z,j}(\mathbf{y}'_j) = T_{d_2} \circ \phi_{z,j}(\mathbf{s}_{z,j}) - T_{d_2} \circ \phi_{z,j}(\mathbf{h}_{z,j}) = \mathbf{u}_{z,j} \in \mathsf{SecretExt}(d_1)$. Thus, $\phi_{z,j}(\mathbf{y}'_j) \in \mathsf{SecretExt}(d_1 \oplus d_2)$. Let $\bar{d} = d_1 \oplus d_2$, then for all $j \in [p]$, $\mathbf{y}'_j \in \mathsf{SecretExt}(\bar{d})$, since the permutation $\phi_{z,j} \in S$ preserves the arrangements of the blocks of $\mathbf{y}'_j$. By removing the last $2m$ coordinates in each $3m$-block of $\mathbf{y}'$ obtain vectors $\mathbf{y}' \sum_{j=1}^{p} \beta_j \cdot \mathbf{y}'_j \in \mathbb{Z}_q^{(2\ell+1)3m}$, and $\mathbf{y} \in \mathbb{Z}^{(2\ell+1)m}$. Now we can declare

$$||\mathbf{y}||_\infty \leq ||\mathbf{y}'||_\infty \leq \sum_{j=1}^{p} \beta_j \cdot ||\mathbf{y}_j||_\infty = \sum_{j=1}^{p} \beta_j \cdot 1 = \beta.$$

Moreover, since $\mathbf{y}'_j \in \mathsf{SecretExt}(\bar{d})$ for all $j \in [p]$, we have that $\mathbf{y} \in \mathsf{Secret}_\beta(\bar{d})$ and, $\mathbf{A} \cdot \mathbf{y} = \mathbf{A}^* \cdot \mathbf{y}' = \mathbf{A}^* \cdot \sum_{j=1}^{p} \beta_j \cdot \mathbf{y}_j = \mathbf{A}^*(\sum_{j=1}^{p} \beta_j \cdot (\mathbf{s}_{z,j} - \mathbf{h}_{z,j})) = \mathbf{u} \mod q$.

For each $j \in [p]$, let $\mathbf{f}'_j = (\mathbf{s}_{f,j} - \mathbf{h}_{f,j})$. Then $\phi_{f,j}(\mathbf{f}'_j) = \phi_{f,j}(\mathbf{s}_{f,j}) - \phi_{e,j}(\mathbf{h}_{f,j}) = \mathbf{u}_{f,j} \in \mathsf{B}_{3m}$, which implies that $\mathbf{f}'_j \in \mathsf{B}_{3m}$. Let $\hat{\mathbf{f}} = \sum_{j=1}^{p} \beta_j \cdot \mathbf{f}'_j \in \mathbb{Z}^{3m}$ and by dropping the last $2m$ coordinates from $\hat{\mathbf{f}}$ obtain $\mathbf{f}' \in \mathbb{Z}^m$. We can declare,

$$||\mathbf{f}'||_\infty \leq ||\hat{\mathbf{f}}||_\infty \leq \sum_{j=1}^{p} \beta_j \cdot ||\mathbf{f}'_j||_\infty = \sum_{j=1}^{p} \beta_j \cdot 1 = \beta.$$

Moreover, for each $j \in [p]$, let $\mathbf{r}'_j = (\mathbf{s}_{r,j} - \mathbf{h}_{r,j})$. Then $\phi_{r,j}(\mathbf{r}'_j) = \phi_{r,j}(\mathbf{s}_{r,j}) - \phi_{r,j}(\mathbf{h}_{r,j}) = \mathbf{u}_{r,j} \in \mathsf{B}_{3m}$, which implies that $\mathbf{r}'_j \in \mathsf{B}_{3m}$. Let $\hat{\mathbf{r}} = \sum_{j=1}^{p} \beta_j \cdot \mathbf{r}'_j \in \mathbb{Z}^{3m}$ and by dropping the last $2m$ coordinates from $\hat{\mathbf{r}}$ obtain $\mathbf{r}' \in \mathbb{Z}^m$. We can declare,

$$||\mathbf{r}'||_\infty \leq ||\hat{\mathbf{r}}||_\infty \leq \sum_{j=1}^{p} \beta_j \cdot ||\mathbf{r}'_j||_\infty = \sum_{j=1}^{p} \beta_j \cdot 1 = \beta.$$

We can obtain the relation:

$$\mathbf{V}^* \cdot \hat{\mathbf{r}} + \mathbf{I}^* \cdot \hat{\mathbf{f}} = \mathbf{v} \mod q \iff \mathbf{V}^* \cdot (\mathbf{B} \cdot \mathbf{r}') + \mathbf{f}' = \mathbf{v} \mod q.$$

Let $d^* = \mathbf{s}_d - \mathbf{h}_d = \hat{\tau}^{-1}(\mathbf{u}_d)$. Then it follows that $d^* \in \mathsf{B}_{2\ell}$. Now let $d^* = (d_1, \ldots, d_\ell, d_{\ell+1}, \ldots, d_{2\ell})$ and let $d = (d_1, \ldots, d_\ell) \in 0, 1^\ell$.

For each $j \in [\bar{p}]$, let $\mathbf{e}'_j = (\mathbf{s}_{e,j} - \mathbf{h}_{e,j})$. Then $\phi_{e,j}(\mathbf{e}'_j) = \phi_{e,j}(\mathbf{s}_{e,j}) - \phi_{e,j}(\mathbf{h}_{e,j}) = \mathbf{u}_{e,j} \in \mathsf{B}_{3k}$, which implies that $\mathbf{e}'_j \in \mathsf{B}_{3k}$. Let $\hat{\mathbf{e}} = \sum_{j=1}^{\bar{p}} b_j \cdot \mathbf{e}'_j$ and by dropping the last $2k$ coordinates from $\hat{\mathbf{e}}$ obtain $\mathbf{e}' \in \mathbb{Z}^k$. We can declare,

$$||\mathbf{e}'||_\infty \leq ||\hat{\mathbf{e}}||_\infty \leq \sum_{j=1}^{\bar{p}} b_j \cdot ||\mathbf{e}'_j||_\infty = \sum_{j=1}^{p} b_j \cdot 1 = b.$$

Now, $||\mathbf{e}'||_\infty \leq b$, and $\mathbf{P}^*\mathbf{e}' + \mathbf{Q}d^* = \mathbf{P}\mathbf{e}' + (0^{k-\ell}||\lfloor q/2 \rfloor d) = \mathbf{c} \mod q$.

In conclusion, the constructed efficient knowledge extractor which satisfies all the conditions stated in Theorem 1 outputs vectors $(\mathbf{y}, \mathbf{f}', \mathbf{r}', \mathbf{e}')$. The vector $\mathbf{f}'(\mathbf{e}_1)$ can extract from $\mathbf{e}$. But, for the ease of understanding and to reduce the complexity we prove the witness of $\mathbf{f}'(\mathbf{e}_1)$ and extraction separately.

5 Conclusion

This paper presents a combined interactive protocol that signer can use to prove his validity of signing, his revocation token which is generated separately without deriving from secret signing key is not in the revocation list, and his index is correctly encrypted. Since the proofs in proposed protocol not relying on each other this can be used in any scheme with slight modifications. For instance, the schemes like [6] can use the proposed protocol by adding member registration.

Acknowledgments. This work is supported in part by JSPS Grant-in-Aids for Scientific Research (A) JP16H01705 and for Scientic Research (B) JP17H01695.

References

1. Gentry, C., Peikert, C., Vaikuntanathan, V.: Trapdoors for hard lattices and new cryptographic constructions. In: ACM 2008, pp. 197–206. ACM (2008)
2. Kawachi, A., Tanaka, K., Xagawa, K.: Concurrently secure identification schemes based on the worst-case hardness of lattice problems. In: ASIACRYPT 2008, LNCS, vol. 5350, pp. 372–389. Springer (2008)
3. Langlois, A., Ling, S., Nguyen, K., Wang, H.: Lattice-based group signature scheme with verifier-local revocation. In: PKC 2014, LNCS, vol. 8383, pp. 345–361. Springer (2014)
4. Ling, S., Nguyen, K., Stehle, D., Wang, H.: Improved zero-knowledge proofs of knowledge for isis problem, and applications. In: PKC 2013, vol. 7778. LNCS, pp. 107–124 (2013)
5. Peikert, C.: A decade of lattice cryptography. Found. Trends Theor. Comput. Sci. **10**(4), 283–424 (2016). https://doi.org/10.1561/0400000074
6. Perera, M.N.S., Koshiba, T.: Achieving almost-full security for lattice-based fully dynamic group signatures with verifier-local revocation. In: ISPEC 2018, LNCS (to appear)
7. Stern, J.: A new paradigm for public key identification. IEEE Trans. Inf. Theory **42**(6), 1757–1768 (1996)

The 8th International Workshop on Information Networking and Wireless Communications (INWC-2018)

A Path Search System Considering the Danger Degree Based on Fuzzy Logic

Shinji Sakamoto[1(✉)], Shusuke Okamoto[1], and Leonard Barolli[2]

[1] Department of Computer and Information Science, Seikei University, 3-3-1 Kichijoji-Kitamachi, Musashino-shi, Tokyo 180-8633, Japan
shinji.sakamoto@ieee.org, okam@st.seikei.ac.jp

[2] Department of Information and Communication Engineering, Fukuoka Institute of Technology, 3-30-1 Wajiro-Higashi, Higashi-Ku, Fukuoka 811-0295, Japan
barolli@fit.ac.jp

Abstract. There are many disasters happening in the world and in general it is difficult to predict them. For this reason, there are many disaster prevention centers where the people learn about information, techniques and the ability to take action in relation to disasters and simulates various disasters in the case of emergencies. It is better that people avoid danger as much as possible in everyday life. The conventional path search systems, such as car navigation systems, mainly consider the length of the path. Thus, the system may recommend a dangerous route such as a place easy to a landslide. In this work, we propose a path search system considering the danger degree by using Fuzzy logic. In our proposed system, we use the data of the hazard map as input parameters to decide the danger degree.

1 Introduction

There are many disasters happening in the world and in general it is difficult to predict them. Conventional path search systems, such as car navigation systems, mainly consider only the length of the path. Thus, the system may recommend a dangerous route such as a place easy to a landslide. This kind of application is useful for people who do not want to take risks. Moreover, the autonomous car, and vehicular network will appear in the future [9]. When the autonomous vehicle are realized, people will use a safer path even if that arrival time is a little bit late.

It is better that people to avoid dangers as much as possible in the ordinary life. Therefore, we propose a path search system which considers the danger degree. However, the danger degree cannot be expressed quantitatively. Thus, we use Fuzzy Logic (FL) to decide the danger degree. We consider using a hazard map as input data for Fuzzy Logic Controller (FLC), and the output is the danger degree.

The rest of the paper is organized as follows. The related works are shown in Sect. 2. The proposed path search system is described in Sect. 3. The conclusions of this paper are given in Sect. 4.

L. Barolli et al. (Eds.): NBiS 2018, LNDECT 22, pp. 785–792, 2019.
https://doi.org/10.1007/978-3-319-98530-5_69

2 Related Works

Many path search algorithms have been proposed, implemented and evaluated. Among them Dijkstra's Algorithm [5,12] can find the shortest path in a graph in real time in the case when the distance on each arc of the graph is non-negative number.

A lot of application of path search systems have been developed. For example, Uchida proposed Tsunami prediction system using Dijkstra's algorithm [15] and he proposed the ray tracing based on Dijkstra's algorithm [14].

Also, this work is deeply related to studies about disasters. Disasters can be classified as Natural disasters and Man-made disasters. Here, we focus on Natural disasters. Below [2] classified the different types of Natural disasters. They are divided into six disaster groups: Biological, Geophysical, Meteorological, Hydrological, Climatological, and Extra-Terrestrial. Each group covers different disaster main types, each having different disaster sub-types.

3 Proposed System

We propose a path search system considering the danger degree. In this section, we explain our proposed system. We describe how to decide the danger degree

Fig. 1. The hazard map portal web-site by Ministry of Land, Infrastructure, Transport and Tourism of Japan. (https://disaportal.gsi.go.jp/)

in Sect. 3.1. After that, Dijkstra's algorithm is explained in Sect. 3.2. Finally, we show our proposed system structure in Sect. 3.3.

3.1 Danger Degree

We show the web interface of the hazard map by Ministry of Land, Infrastructure, Transport, and Tourism of Japan in Fig. 1. The hazard map shows three categories: flood, sediment disasters and tsunami. Because some data of hazard map can be used as open data, our proposed path search system uses the data to find a path to avoid the dangerous area.

However, to decide how an area is dangerous is difficult and the data from the hazard map cannot be use directly as the danger degree because the data cannot be expressed quantitatively. In other words, we need to calculate the danger degree from the provided data of the hazard map. Therefore, we use FL. The danger degree of each area should be decided in order to avoid the dangerous area. We consider the data of the hazard map as input parameters to decide the danger degree.

3.1.1 Hazard Map

The hazard map is a map showing places where disasters are likely to occur. The hazard map (see Fig. 1) is created by overlapping a map as shown in Fig. 2. We can freely use the hazard map which is provided by Ministry of Land, Infrastructure, Transport, and Tourism of Japan. By using the hazard map, we can see risks of flood, landslide and tsunami.

Some data to create the hazard map can be downloaded from the web-server of Ministry of Land, Infrastructure, Transport, and Tourism of Japan. Our proposed system uses the provided data.

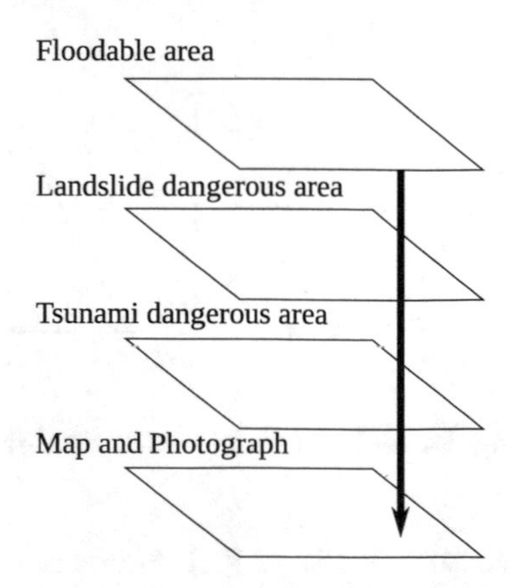

Fig. 2. Overlapping a map in hazard map by Ministry of Land, Infrastructure, Transport and Tourism of Japan.

Almost all provided data[1] format is ESRI (Economic and Social Research Institute) shapefile [1] which is a vector data format. The ESRI shapefile format has three essential files: “shp”, “shx”, and “dbf”. The “shp” has geometry information, the “shx” has the index metadata information and the “dbf” has attribute information which conforms to dBase IV format. The ESRI format is now public [13]. Therefore, we can read the data of hazard map to use in our proposed system.

3.1.2 Fuzzy Logic

FL is a logic based on Fuzzy sets [16], developed by Lotfi A. Zadeh. Fuzzy sets and FL have been developed to manage vagueness and uncertainty in a reasoning process of an intelligent system such as a knowledge-based system, an expert system or a logic control system [4].

The ability of Fuzzy sets and possibility theory to model gradual properties or soft constraints whose satisfaction is a matter of degree, as well as information pervaded with imprecision and uncertainty, makes them useful in a great variety of applications [6,10,11].

In the contents of complex processes, it may turn out to be more practical to get knowledge from an expert operator than to calculate an optimal control, due to modeling costs or because a model is out of reach.

A concept that plays a central role in the application of FL is that of a linguistic variable. The linguistic variables may be viewed as a form of data compression. One linguistic variable may represent many numerical variables. We show triangular and trapezoidal membership functions for FLC in Fig. 3. The x_0 in $f(x)$ is the center of triangular function, $x_0(x_1)$ in $g(x)$ is the left (right) edge of trapezoidal function, and $a_0(a_1)$ is the left (right) width of the triangular or trapezoidal function. We use the linguistic variables for the proposed system because they are suitable for real-time operation [7].

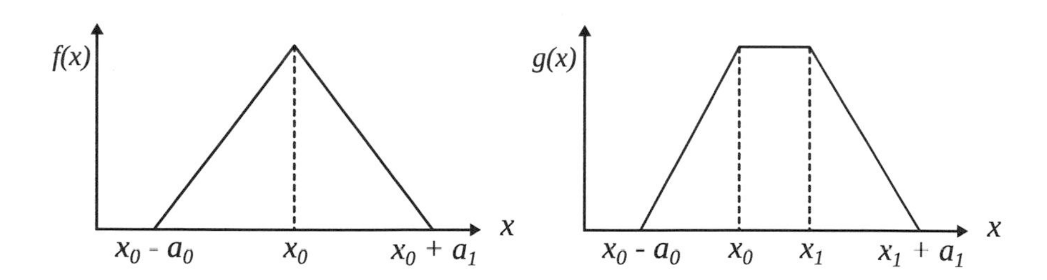

Fig. 3. Triangular and trapezoidal membership functions.

We show the FLC structure in Fig. 4. FLC is one of the most crucial parts of our proposed system to compute the danger degree. FLC consists of the fuzzifier, inference engine (see Fig. 5), Fuzzy Rule Base (FRB) and defuzzifier. The

[1] Some data are in a different format such as text format.

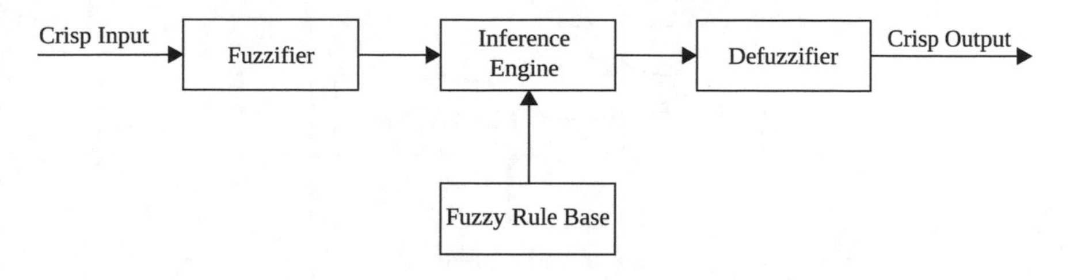

Fig. 4. FLC structure.

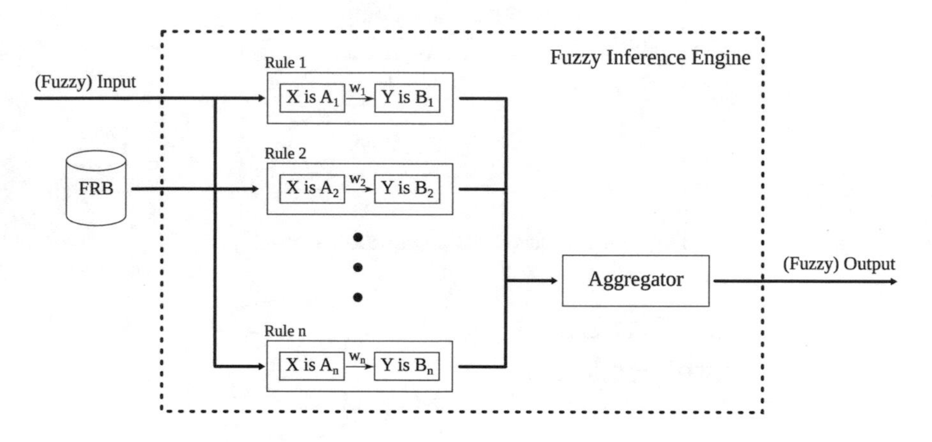

Fig. 5. Structure of Fuzzy inference engine.

fuzzifier fuzzify the crisp input parameters to fuzzy value. Then, the inference engine based on fuzzified parameters and FRB decides an output fuzzy, which is converted to crisp value by defuzzifier.

3.2 Dijkstra's Algorithm

Dijkstra's algorithm [5,12] can find the shortest path in a graph in real time in the case of the distances on each arc of the graph are no-negative numbers. Now, Dijkstra's algorithm is widely used in a car navigation system [3] and OSPF (Open Shortest Path First [8]) which is a routing protocol in the computer network filed.

3.3 System Structure

We show our proposed system structure in Fig. 6. A user makes a request to a server. The Interface Module responds to the request. The Control Module gives a task to the Main Module. We show the Main Module structure in Fig. 7. The Main Module gets the user's metric function from the Control Module. The Main Module can compute the distance from a start point of the user. Also, it

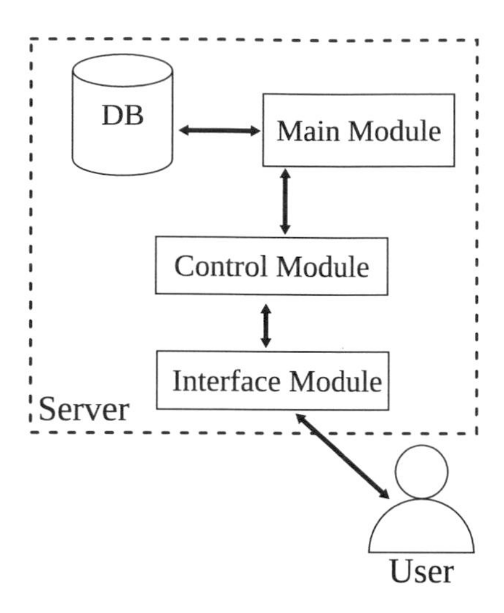

Fig. 6. Structure of the proposed system.

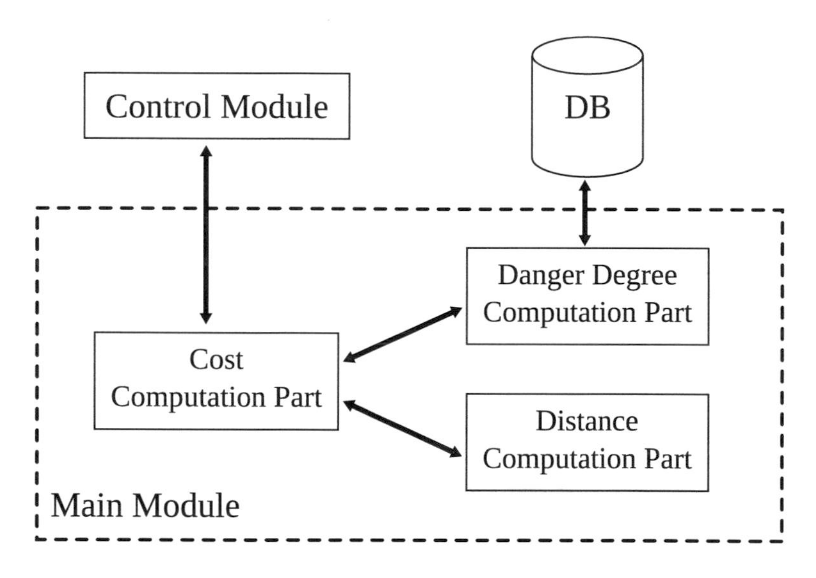

Fig. 7. Main module of the proposed system.

can compute the danger degree of the route. Our proposed path search system is based on Dijkstra's algorithm. By using Dijkstra's algorithm, the system can find a minimum metric path.

The process is as follows. The system computes a danger degree of neighboring cells from the start point. Then, the system calculates a metric function f which is given by:

$$f(l, d) = \alpha l + (1 - \alpha)\, d,$$

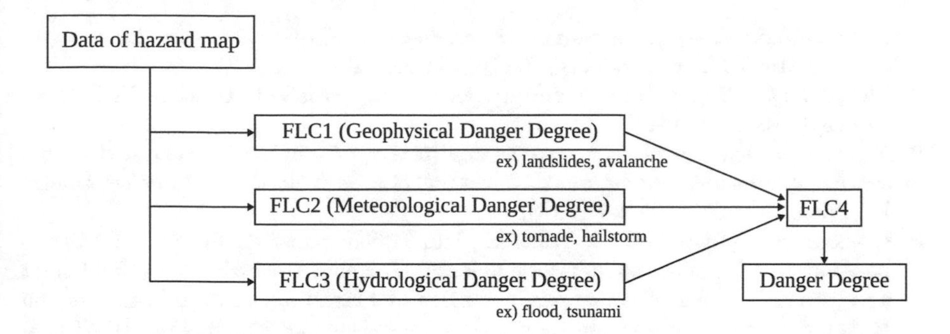

Fig. 8. Structure of the danger degree computation part of the proposed system.

where l is a length (distance) from a neighboring point, d is a danger degree calculated by FLCs as shown in Fig. 8, α is a weight coefficient parameter ($0 < \alpha < 1$). Thus, the system can use Dijkstra's algorithm and can find a path considering the danger degree.

4 Conclusion

In this paper, we proposed a fuzzy-based path search system considering the danger degree. We showed our proposed system structure. Our proposed system uses a hazard map as input data to decide the danger degree. In the future, we would like to implement the proposed path search system and present the evaluation results.

References

1. Barik, R.K., Lenka, R., Sahoo, S., Das, B., Pattnaik, J.: Development of educational geospatial database for cloud SDI using open source GIS. In: Progress in Advanced Computing and Intelligent Engineering, pp 685–695. Springer, Heidelberg (2018)
2. Below, R., Wirtz, A., Guha-Sapir, D.: Disaster category classification and peril terminology for operational purposes. Technical report (2009)
3. Chakraborty, B., Maeda, T., Chakraborty, G.: Multiobjective route selection for car navigation system using genetic algorithm. In: Proceedings of the 2005 IEEE Mid-Summer Workshop on Soft Computing in Industrial Applications (SMCia-2005), pp. 190–195. IEEE (2005)
4. De Silva, C.W.: Intelligent Control: Fuzzy Logic Applications. CRC Press, Boca Raton (2018)
5. Knuth, D.E.: A generalization of Dijkstra's algorithm. Inf. Process. Lett. **6**(1), 1–5 (1977)
6. Liu, Y., Sakamoto, S., Barolli, L., Ikeda, M.: A fuzzy-based system for qualified voting in P2P mobile collaborative team: effects of member activity failure. In: The 31st IEEE International Conference on Advanced Information Networking and Applications (AINA-2017), pp. 639–645. IEEE (2017)

7. Mendel, J.M.: Fuzzy logic systems for engineering: a tutorial. Proc. IEEE **83**(3), 345–377 (1995). https://doi.org/10.1109/5.364485
8. Moy, J.T.: OSPF: Anatomy of an Internet Routing Protocol. Addison-Wesley Professional, Reading (1998)
9. Nascimento, D., Russell, R.: Systems and methods for achieving road action consensus, for example among autonomous vehicles, in a network of moving things. US Patent App. 15/591,937 (2018)
10. Obukata, R., Bylykbashi, K., Ozera, K., Liu, Y., Sakamoto, S., Barolli, L.: A fuzzy-based system for actor node in an ambient intelligence testbed: effects of different parameters on human sleeping conditions. In: The 12th International Conference on Broadband and Wireless Computing, Communication and Applications (BWCCA-2017), pp. 682–690. Springer, Heidelberg (2017)
11. Ozera, K., Inaba, T., Bylykbashi, K., Sakamoto, S., Barolli, L.: Implementation of a new function for preventing short reconnection in a WLAN triage system. In: The 6th International Conference on Emerging Internetworking, Data & Web Technologies (EIDWT-2018), pp. 1–17. Springer, Heidelberg (2018)
12. Skiena, S.: Dijkstra's Algorithm. Implementing Discrete Mathematics: Combinatorics and Graph Theory with Mathematica, pp. 225–227. Addison-Wesley, Reading (1990)
13. Stabler, B.: Shapefiles: Read and Write ESRI shapefiles (2013). https://cran.r-project.org/web/packages/shapefiles/
14. Uchida, K., Barolli, L.: Dijkstra-algorithm based ray-tracing by controlling proximity node mapping. In: The 31st International Conference on Advanced Information Networking and Applications Workshops (WAINA-2017), pp. 189–194. IEEE (2017)
15. Uchida, K., Nogami, S., Takematsu, M., Honda, J.: Tsunami simulation based on Dijkstra algorithm. In: The 17th International Conference on Network-Based Information Systems (NBiS-2014), pp. 114–119. IEEE (2014)
16. Zadeh, L.A.: Fuzzy sets. In: Fuzzy Sets, Fuzzy Logic, and Fuzzy Systems, Selected Papers by Lotfi A Zadeh, pp. 394–432. World Scientific (1996)

A Recovery Method for Reducing Storage Usage Considering Different Thresholds in VANETs

Shogo Nakasaki[1], Yu Yoshino[1], Makoto Ikeda[2(✉)], and Leonard Barolli[2]

[1] Graduate School of Engineering, Fukuoka Institute of Technology, 3-30-1 Wajiro-higashi, Higashi-ku, Fukuoka 811-0295, Japan
tshogonakasakit@gmail.com, mgm17107@bene.fit.ac.jp

[2] Department of Information and Communication Engineering, Fukuoka Institute of Technology, 3-30-1 Wajiro-higashi, Higashi-ku, Fukuoka 811-0295, Japan
makoto.ikd@acm.org, barolli@fit.ac.jp

Abstract. Technologies have been developed for providing higher functionality of on-board unit and providing a communication function with other vehicles and roadside units. Nowadays, vehicles can be called as one communication terminal. However, end-to-end communication is difficult because of the lack of end-to-end connectivity. Delay/Disruption/Disconnection Tolerant Networking (DTN) are used as one of a key alternative network for Vehicular Ad-hoc Networks (VANETs). In this paper, we propose a recovery method for reducing storage usage considering different thresholds in VANETs. From the simulation results, we found that our proposed recovery method has a good performance even for sparse or dense network environment.

Keywords: DTN · VANET · Recovery method

1 Introduction

Recently, with the development of vehicle-infrastructure cooperative system, various services between vehicle and road side unit have been appeared [3,6,7,9,14,23]. Moreover, development of a system that can be installed in on-board-unit by plug-in has attracted attention for legacy vehicle. Vehicle is one of the important network communication node in our lives. Further, research and development on automatic driving has been actively conducted. However, there are many problems in technology, environment, cost, law, and so on. In the future, it will be possible to promote new adaptive communication systems, road traffic support, and safety support by using the big data exchanged between Vehicle-to-Vehicle (V2V) and Vehicle-to-Infrastructure (V2I) communications [11,19].

In Vehicular Ad-hoc Networks (VANETs), end-to-end communication is difficult because of the lack of end-to-end connectivity. Delay/Disruption/Disconnection Tolerant Networking (DTN) are used as one of a key alternative method [21] to provide network services for VANETs. In Vehicular DTN,

L. Barolli et al. (Eds.): NBiS 2018, LNDECT 22, pp. 793–802, 2019.
https://doi.org/10.1007/978-3-319-98530-5_70

consumption of network resources and storage usage of each vehicle becomes a critical problem due to the DTN nodes duplicate messages to other nodes.

In our previous work [10], we have proposed an Enhanced Message Suppression Controller (EMSC) for Vehicular-DTN. The paper reported that the EMSC can reduce the consumption of network resources. But, we used some additional messages to reduce the replicated bundle messages. In addition, we proposed a recovery method for reducing storage usage of each vehicle for VANETs without additional message exchanges [12]. The approach decreased the storage usage in each node, however the performance of delay and delivery ratio were dropped.

In this paper, we propose a recovery method considering different thresholds for reducing delay time and increasing the delivery ratio. We evaluate the proposed method by simulations and consider storage usage, end-to-end delay, and delivery ratio as evaluation metrics.

The structure of the paper is as follows. In Sect. 2, we give the related work. In Sect. 3 is described a proposed recovery method. In Sect. 4, we provide the description of simulation system and the evaluation results. Finally, conclusions and future work are given in Sect. 5.

2 Related Work

Originally, DTN was proposed for space networks [1,8,17,20]. The space networks consider the delay, disconnection and disruption. In DTN, the messages are stored and forwarded by nodes. When the nodes receive messages, they store the messages in their storage. After that, the nodes duplicate the messages to other nodes when it is transmittable. This technique is called message switching. The architecture is specified in RFC 4838 [5].

A number of DTN protocols have been proposed, such as Epidemic [15,22], Spray and Wait (SpW) [18], MaxProp [4] and so on [13].

Epidemic Routing is well-known routing protocol of DTN [15,22]. Epidemic Routing uses two control messages to duplicate messages. Nodes periodically broadcast Summary Vector (SV) message in the network. The SV contains a list of stored messages of each node. When the nodes receive the SV, they compare received SV to their SV. The nodes send REQUEST message if received SV contains unknown messages.

In Epidemic Routing, consumption of network resources and storage usage become a critical problem. Because the nodes duplicate messages to all nodes in its communication range. Moreover, unnecessary messages remain in the storage, because the messages are continuously duplicated even if destination node receives the messages. Therefore, recovery schemes such as timer or anti-packet are needed to limit the duplicate messages. In case of the timer, messages have lifetime. The messages are punctually deleted when the lifetime of the messages are expired. However, setting of a suitable lifetime is difficult. In case of anti-packet, destination node broadcasts the anti-packet, which contains the list of messages that are received by the destination node. Nodes delete the messages according to the anti-packet. Then, the nodes duplicate the anti-packet to other nodes. However, network resources are consumed by the anti-packet.

3 Proposed Recovery Method

3.1 Message Sequence Chart

In our recovery method, we delete the duplicated bundle messages in storage of each node based on the number of neighbors. In Fig. 1, we show the message sequence chart of the proposed recovery method. We consider two functions such as "Receive SV" and "Check" to delete the duplicated bundle messages in node storage. In our method, every node counts the number of neighboring nodes as N from received SVs. In addition, every node periodically check N in intervals of 1 s.

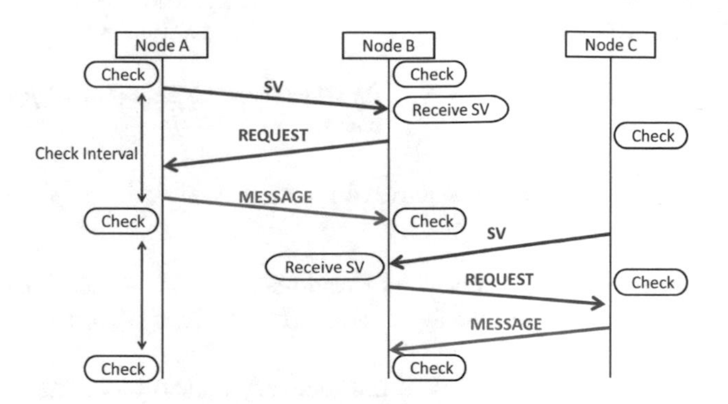

Fig. 1. Message sequence chart.

3.2 Previous Approach

In the checking process by conventional recovery method, the nodes delete bundle messages that are duplicated to other nodes if the Current N is greater than the Previous N.

$$\text{Previous N} < \text{Current N}, \quad \text{Previous N} \neq 0 \tag{1}$$

After that, the nodes store Current N as Previous N, and they set Current N zero. As an exception, the nodes do not delete the message if the Previous N is zero. In this way, the conventional approach decreased the storage usage in each node, however the performance of delay and delivery ratio were dropped.

3.3 Recovery Method Considering Different Thresholds

We present the flowchart of the recovery method in Fig. 2. In the checking process of proposed method, the nodes delete bundle messages that are duplicated to other nodes if the Eq. (2) is satisfied.

$$\frac{\text{Current N}}{\text{Previous N}} > \text{Threshold}, \quad \text{Previous N} \neq 0 \tag{2}$$

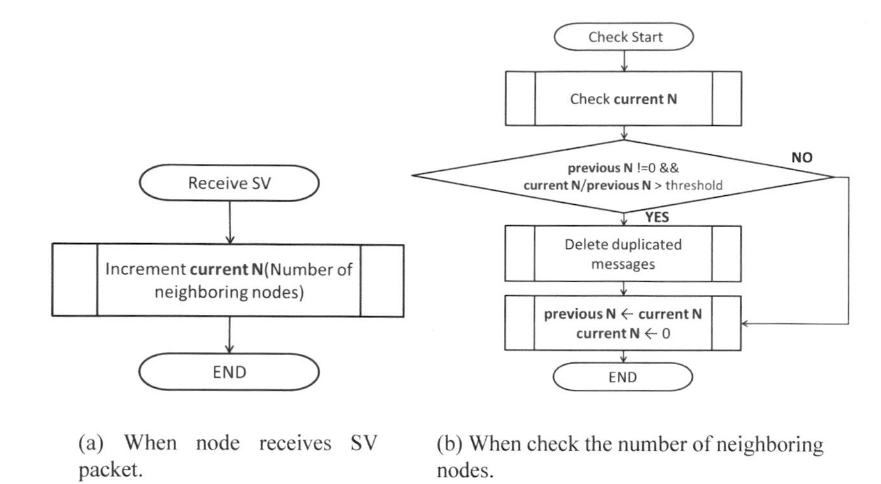

(a) When node receives SV packet.

(b) When check the number of neighboring nodes.

Fig. 2. Flowcharts of recovery method.

Then, the nodes store Current N as Previous N, and they set Current N zero. The nodes do not delete the message if the Previous N is zero. In this paper, we use the threshold from 1 to 3.

Figure 3 illustrates the overview of the recovery method focusing on one node. First phase (see upper figure in Fig. 3) describes the first checking process. The Current N of green vehicle is one. Green vehicle stores the Current N as Previous N. In the second phase, the green vehicle duplicates message A to other vehicles during moving on the road. Third phase describes the checking process for the second time (see bottom figure in Fig. 3). We use the checking intervals of one second. In the checking process, the green vehicle compares the Current N/Previous N with the threshold. Then, the green vehicle deletes message A from its storage because the Current N/Previous N is greater than the threshold. In this case, we set the threshold value as 2. All nodes in the network execute these processes during simulation.

4 Modeling and Simulation Results

In this paper, we evaluate the proposed recovery method compared with conventional Epidemic. For simulations, we consider five different thresholds and three parameters: storage usage, end-to-end delay and delivery ratio for different number of vehicles. We implemented the proposed recovery method on Scenargie [16] network simulator. We consider IEEE 802.11p, which is an approved amendment to the IEEE 802.11 standard to add wireless access in VANETs.

We use two types of routing protocols and send bundle messages from starting-point to end-point considering ITU-R P.1411 propagation model [2]. Therefore, we consider interference of obstacles on 5.9 GHz band.

1. Epidemic (Conventional)
2. Epidemic with proposed recovery method

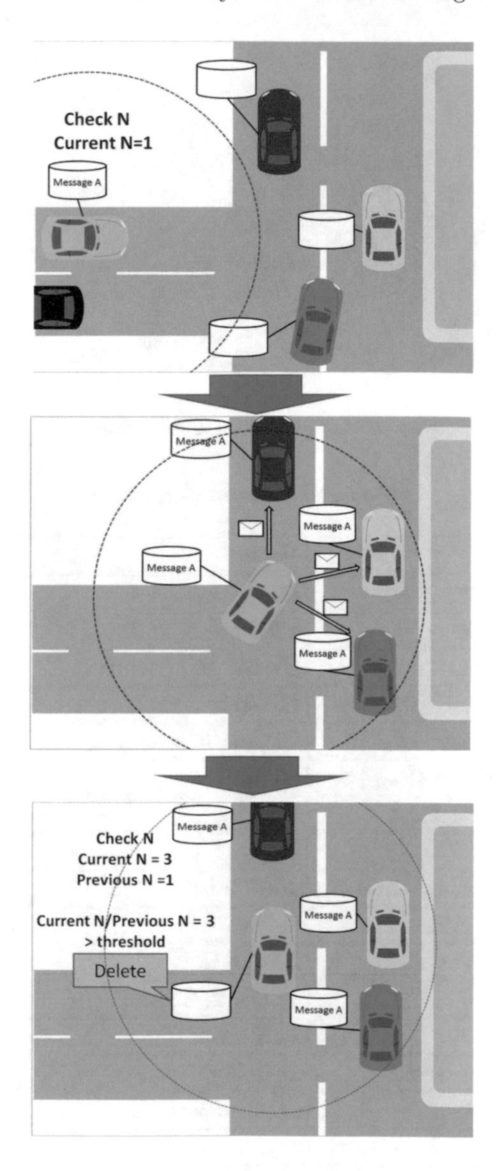

Fig. 3. Overview of recovery method (For example, when we set a threshold value as 2.).

4.1 Scenario Setting

We consider a Manhattan grid scenario with 25, 50 and 100 vehicles. We present the simulation scenario for 50 vehicles in Fig. 4. Table 1 shows the simulation parameters. Both message starting point and the end point are static, and other vehicles move based on GIS-based Random Waypoint (RWP) mobility model.

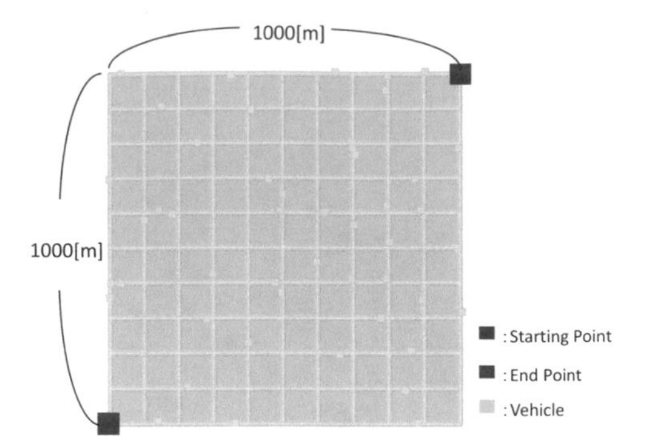

Fig. 4. Road model for 50 vehicles.

Table 1. Simulation parameters.

Parameter	Value
Simulation time ($T_{\max}$)	500 [s]
Area dimensions	1000 [m] $\times$ 1000 [m]
Density of vehicles	25, 50, 100 [vehicles]
Mobility model	GIS-based random waypoint mobility
Minimum speed ($V_{\min}$)	30 [km/h]
Maximum speed ($V_{\max}$)	60 [km/h]
Storage size of nodes	10M [bytes]
SV interval	1 [s]
Message start and end time	10–400 [s]
Message generation interval	10 [s]
Message size	500 [bytes]
Threshold	1, 1.5, 2, 2.5, 3
PHY model	IEEE 802.11p
Propagation model	ITU-R P.1411
Antenna model	Omni-directional
Check N interval	1 [s]

GIS-based RWP mobility model is implemented for being used in VANET and constrains vehicle movement to streets defined by map data for real cities and limits their mobility according to vehicular congestion and simplified traffic control mechanisms.

4.2 Simulation Results

In this section, we evaluate the performance of proposed recovery method with threshold compared with conventional Epidemic.

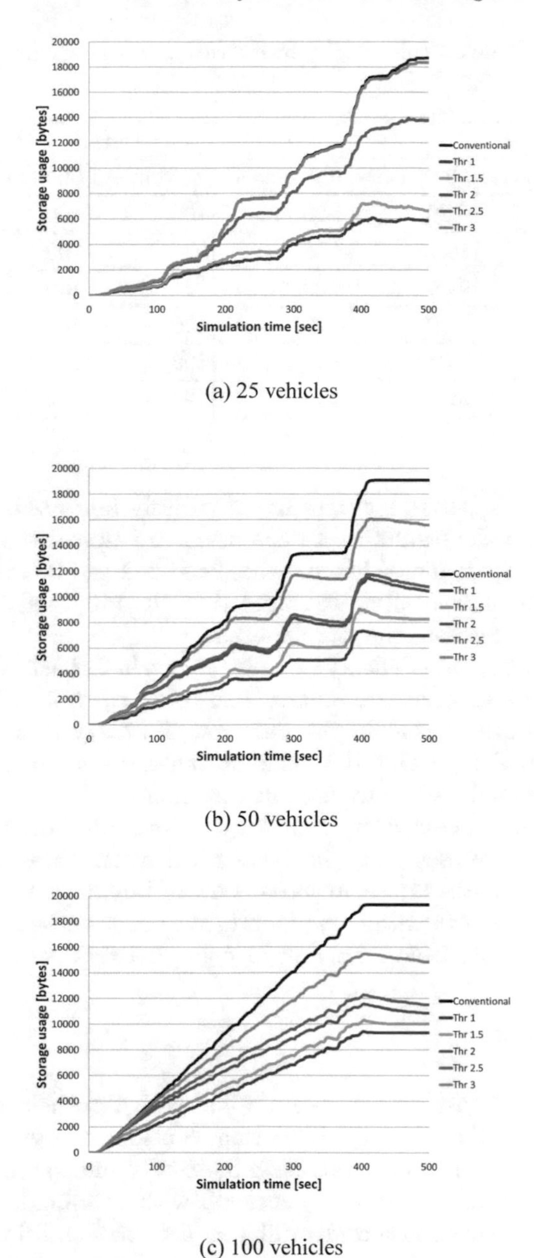

(a) 25 vehicles

(b) 50 vehicles

(c) 100 vehicles

Fig. 5. Results of storage usage for different number of vehicles.

We present the simulation results of storage usage for different number of vehicles in Fig. 5. When 100 vehicles are present in the network, each node received many SVs, thus the difference of performance is remarkable. In this case, we observed that the results of storage usage are linearly increased. In case

Table 2. Results of delay time and delivery ratio. N/A indicates the conventional Epidemic.

Threshold	Delay [s]			Delivery ratio [%]		
	25 vehicles	50 vehicles	100 vehicles	25 vehicles	50 vehicles	100 vehicles
N/A	12.77	9.30	6.20	100	100	100
1	17.32	10.71	6.33	92	100	100
1.5	17.29	9.85	6.22	94	100	100
2	13.60	9.29	6.20	100	100	100
2.5	13.60	9.34	6.19	100	100	100
3	12.77	9.36	6.20	100	100	100

of 25 and 50 vehicles, storage usages are stepwisely increased. For 25 vehicles, when threshold is 3 the results of storage usage are almost the same compared with conventional Epidemic. When the threshold is 2 and 2.5, the storage usage is the same. From these results, we found that the proposed recovery method with threshold can decrease the storage usage.

The results of delay and delivery ratio are shown in Table 2. When the threshold is 1, the results of delay are longer than other cases for different vehicles. When 25 vehicles are present in the network, the delay decreases by increasing the threshold. For 50 and 100 vehicles, our recovery method has a good performance compared with conventional Epidemic.

The results of message delivery ratio for different threshold is 100%, when 50 and 100 vehicles are present. On the other hand, when threshold is 1 and 1.5, delivery ratio decreased. After threshold is 2, all bundle messages reached the end-point. From these evaluation, we found that our proposed recovery method has a good performance even for sparse or dense network environment.

5 Conclusions

In this paper, we proposed a recovery method with threshold in VANETs. We evaluated the proposed method by simulations considering storage usage, end-to-end delay, and delivery ratio as evaluation metrics. From the simulation results, we found that the proposed recovery method with threshold can decrease the storage usage, but we need to consider different network conditions with adaptive threshold.

In the future work, we would like to add new functions and make extensive simulations considering different parameters.

References

1. Delay- and disruption-tolerant networks (DTNs) tutorial. NASA/JPL's Interplanetary Internet (IPN) Project (2012). http://www.warthman.com/images/DTN_Tutorial_v2.0.pdf
2. Rec. ITU-R P.1411-7: Propagation data and prediction methods for the planning of short-range outdoor radiocommunication systems and radio local area networks in the frequency range 300 MHz to 100 GHz. ITU (2013)
3. Araniti, G., Campolo, C., Condoluci, M., Iera, A., Molinaro, A.: LTE for vehicular networking: a survey. IEEE Commun. Mag. **21**(5), 148–157 (2013)
4. Burgess, J., Gallagher, B., Jensen, D., Levine, B.N.: MaxProp: routing for vehicle-based disruption-tolerant networks. In: Proceedings of the 25th IEEE International Conference on Computer Communications (IEEE INFOCOM-2006), pp. 1–11, April 2006
5. Cerf, V., Burleigh, S., Hooke, A., Torgerson, L., Durst, R., Scott, K., Fall, K., Weiss, H.: Delay-tolerant networking architecture. IETF RFC 4838 (Informational), April 2007
6. Cheng, X., Yao, Q., Wen, M., Wang, C.X., Song, L.Y., Jiao, B.L.: Wideband channel modeling and intercarrier interference cancellation for vehicle-to-vehicle communication systems. IEEE J. Sel. Areas Commun. **31**(9), 434–448 (2013)
7. Dias, J.A.F.F., Rodrigues, J.J.P.C., Xia, F., Mavromoustakis, C.X.: A cooperative watchdog system to detect misbehavior nodes in vehicular delay-tolerant networks. IEEE Trans. Ind. Electron. **62**(12), 7929–7937 (2015)
8. Fall, K.: A delay-tolerant network architecture for challenged Internets. In: Proceedings of the International Conference on Applications, Technologies, Architectures, and Protocols for Computer Communications, SIGCOMM 2003, pp. 27–34 (2003)
9. Grassi, G., Pesavento, D., Pau, G., Vuyyuru, R., Wakikawa, R., Zhang, L.: VANET via named data networking. In: Proceedings of the IEEE Conference on Computer Communications Workshops (INFOCOM WKSHPS 2014), pp. 410–415, April 2014
10. Ikeda, M., Ishikawa, S., Barolli, L.: An enhanced message suppression controller for vehicular-delay tolerant networks. In: Proceedings of the 30th IEEE International Conference on Advanced Information Networking and Applications (IEEE AINA-2016), pp. 573–579, March 2016
11. Mahmoud, A., Noureldin, A., Hassanein, H.S.: VANETs positioning in urban environments: a novel cooperative approach. In: Proceedings of the IEEE 82nd Vehicular Technology Conference (VTC-2015 Fall), pp. 1–7, September 2015
12. Nakasaki, S., Yoshino, Y., Ikeda, M., Barolli, L.: A recovery method for reducing storage usage considering number of neighboring nodes in VANETs. In: Proceedings of The 32nd International Conference on Advanced Information Networking and Applications Workshops (WAINA-2018), pp. 130–135, May 2018
13. Nelson, S.C., Bakht, M., Kravets, R.: Encounter-based routing in DTNs. In: Proceedings of the IEEE INFOCOM 2009, pp. 846–854 (2009)
14. Ohn-Bar, E., Trivedi, M.M.: Learning to detect vehicles by clustering appearance patterns. IEEE Trans. Intell. Transp. Syst. **16**(5), 2511–2521 (2015)
15. Ramanathan, R., Hansen, R., Basu, P., Hain, R.R., Krishnan, R.: Prioritized epidemic routing for opportunistic networks. In: Proceedings of the 1st International MobiSys Workshop on Mobile Opportunistic Networking (MobiOpp 2007), pp. 62–66 (2007)
16. Scenargie: Space-time engineering, LLC. http://www.spacetime-eng.com/

17. Schlesinger, A., Willman, B.M., Pitts, L., Davidson, S.R., Pohlchuck, W.A.: Delay/disruption tolerant networking for the international space station (ISS). In: Proceedings of the IEEE Aerospace Conference 2017, pp. 1–14 (2017)
18. Spyropoulos, T., Psounis, K., Raghavendra, C.S.: Spray and wait: an efficient routing scheme for intermittently connected mobile networks. In: Proceedings of the ACM SIGCOMM workshop on Delay-tolerant networking 2005 (WDTN 2005), pp. 252–259 (2005)
19. Theodoropoulos, T., Damousis, Y., Amditis, A.: A load balancing control algorithm for EV static and dynamic wireless charging. In: Proceedings of the IEEE 81st Vehicular Technology Conference (VTC-2015 Spring), pp. 1–5, May 2015
20. Tsuru, M., Uchida, M., Takine, T., Nagata, A., Matsuda, T., Miwa, H., Yamamura, S.: Delay tolerant networking technology - the latest trends and prospects. IEICE Commun. Soc. Mag. **16**, 57–68 (2011)
21. Uchida, N., Ishida, T., Shibata, Y.: Delay tolerant networks-based vehicle-to-vehicle wireless networks for road surveillance systems in local areas. Int. J. Space Based Situated Comput. **6**(1), 12–20 (2016)
22. Vahdat, A., Becker, D.: Epidemic routing for partially-connected ad hoc networks. Technical report, Duke University (2000)
23. Zhang, W., Jiang, S., Zhu, X., Wang, Y.: Cooperative downloading with privacy preservation and access control for value-added services in VANETs. Int. J. Grid Util. Comput. **7**(1), 50–60 (2016)

Butt-Joint Assembly of Photonic Crystal Waveguide Units for Large Scale Integration of Circuits with Variety of Functions

Hiroshi Maeda[1(✉)], Keisuke Haari[2], Xiang Zheng Meng[2], and Naoki Higashinaka[2]

[1] Department of Information and Communication Engineering, Fukuoka Institute of Technology, 3-30-1 Wajiro-Higashi, Higashi Ward, Fukuoka 811-0295, Japan
hiroshi@fit.ac.jp

[2] Graduate School of Engineering, Fukuoka Institute of Technology, Fukuoka, Japan

Abstract. Photonic crystal (PhC) structure is useful for fabrication of highly integrated optical circuit on single substrate. However, it was reported that repetition of chemical etching process to mount various functions damages the structure itself. Then, a new assembly process of functional units is proposed in this paper. For the assembly of butt joint, transmission and reflection on joint surface is critical. Then, measurement results of butt joint in PhC waveguide is evaluated to show the transmission and reflection characteristics. The waveguide shift Δ was situated in the middle of PhC waveguide by composing two separated basement plate of aluminum. It was found that the waveguide shift Δ up to half of waveguide width is not seriously critical for the transmission and reflection. Application of proposed assembly technique by butt joint is hopeful for fabrication of PhC integrated circuit.

1 Introduction

High speed signal routing technology for higher bit-rate data transmission in information network has been intensively studied. Most of modern network system equips optical fiber cable for the backbone, electronic signal processor unit, and opto-electric (O/E) or E/O signal converters. The key point is time consumption in O/E and E/O converters with regeneration of the signal form. Current telecommunication systems hiring hybrid of electric and optical signals must include the E/O and O/E signal conversion circuit in the routing/switching circuits. Those switches are necessary to maintain the original signal waveform for lower bit error ratio (BER), though the signal conversion causes time delay especially for higher bit rate telecommunication. Generally, throughput of the internet and telecommunication systems are regulated by this time delay in the switch.

L. Barolli et al. (Eds.): NBiS 2018, LNDECT 22, pp. 803–809, 2019.
https://doi.org/10.1007/978-3-319-98530-5_71

In optical fiber communication system over several thousand kilometers, Erbium doped fiber amplifier (EDFA) [1] is implemented practically. As EDFAs give gain to the optical signal directly in optical frequency domain, the time delay due to the O/E and E/O conversion was drastically improved with decrease of power consumption in the system. Together with dispersion compensation by fiber grating [1], the pulse distortion can be corrected to obtain higher bit rate.

For further high speed switching, a circuit for routing/switching system without signal frequency conversion to achieve higher throughput is proposed in this paper. For this purpose, we introduce photonic crystal (PhC) structure [2] as the fundamental transmission line and the signal processing circuits.

PhC or electromagnetic band gap (EBG) structure is a key component for optical communication systems, because of the unique propagation characteristics based on photonic band gap theory [2–5]. In the system, cavity structure and its application will provide important role as a filtering device, especially in optical 'wavelength division multiplexing' (WDM) system. For the other application, it is also important, for example, to laser oscillation in active device for all-optical signal processing system [4].

PhC or EBG structure can be realized by a periodic structure of medium with array of pillars or air holes with periodicity of sub-micrometers. It is fabricated by photo lithography technique, like semiconductor IC or LSI fabrication. In fabrication of PhC circuit, air-bridged structure [6] is useful to achieve confinement of the optical field. However, a difficulty of damage to the structure has been pointed out on fragility by repetition of chemical etching processes in the fabrication [6], especially for large scale integration with various functions on it. Then we propose a new fabrication process to fabricate each functional segment under minimum number of etching process. By assembling by butt joint of the segments, it is expected to realize integrated PhC circuits with high functionality as conventional ones. For the assembly, evaluation of transmission and reflection at the joint surface is critical. Therefore, we did some experiments to obtain the characteristics in microwave domain.

As scaling rule generally holds in PhC structure with respect to normalized frequency $\omega P/2\pi c = P/\lambda$, where P is periodicity or lattice period of PhC, c and λ is velocity of light and its wavelength in vacuum, respectively, it is possible to investigate its characteristics in microwave frequency in place of optical frequency. This means that we can make use of magnified models [7,8] in microwave frequency range with lattice period in the order of a few centimeters.

First in this paper, summary of measurement is described, referring to the previous researches [9–11], which is situated in waveguide with some variation, for microwaves in the range from 3.6 to 4.2 GHz. The waveguide is fabricated as a line defect in two-dimensional, square-lattice, pillar-type photonic crystal [7].

The transmission and reflection characteristics of waveguide was measured as S-parameter $|S_{21}|$ and $|S_{11}|$ of microwave circuit measurement. The waveguide shift Δ was situated in the middle of PhC waveguide by composing two separated basement plate of aluminum. It was found that the waveguide shift Δ up to half of waveguide width is not seriously critical for the transmission and reflection. Application of proposed assembly technique by butt joint is hopeful for fabrication of PhC integrated circuit (Fig. 1).

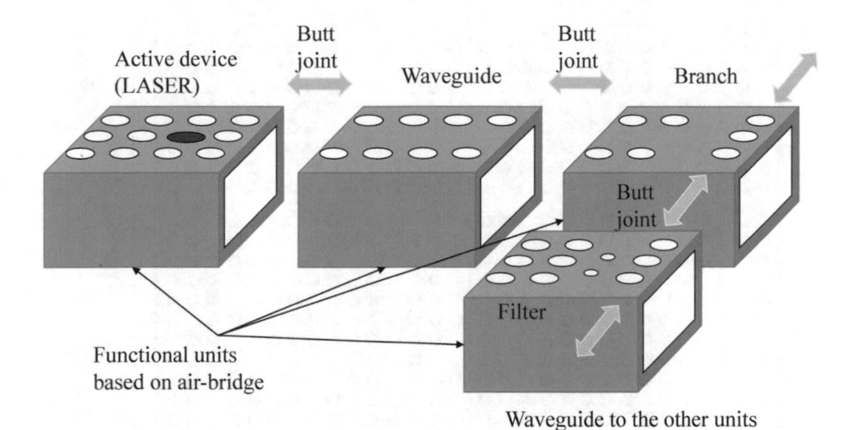

Fig. 1. Assembly of fundamental and functional PhC units of air-bridged type. Each units are fabricated to avoid damage by repetition of chemical etching process.

2 Measurement of Photonic Crystal Waveguide

Experiment is done in accordance with Refs. [7,8,10]. Periodic square array of ceramic dielectric rod with $\varepsilon_r = 36.0$ is put between aluminum plates. The top view is illustrated in Fig. 2(a). The structure is surrounded by aluminum plates and it is grounded to assume that the structure is equivalently two dimensional along with vertical axis. PhC waveguide is depicted in Fig. 2(b) with the parameters. A line of rods were removed to be waveguide [10] in the figure. At the input and output port in Fig. 2(a), S-parameters $|S_{11}|$ as reflection and $|S_{21}|$ as transmission were measured by *Agilent E5071C* vector network analyzer. The transmission and reflection spectra are shown in Fig. 4(a) and (b) as $\Delta = 0$ to show that the structure operates as a waveguide with large transmission and little reflection between 3.9 and 4.2 GHz.

In Fig. 3(a), waveguide discontinuity with butt joint is depicted. In Fig. 3(b), discontinuity with waveguide shift Δ is enlarged. For $\Delta = 0$, the waveguide is continuous. By increasing Δ, we can evaluate the effect of discontinuity as $|S_{21}|$ and $|S_{11}|$.

3 Measurement of Butt-Jointed Waveguide

In Figs. 4(a) and (b), transmission and reflection characteristics are shown, respectively, as function of waveguide shift Δ. The measurement was repeated 3 times, and the averaged data are shown here. In both figures, the spectra do not seem to be affected for $\Delta = 0$ to 10 mm. The transmission curves are tightly

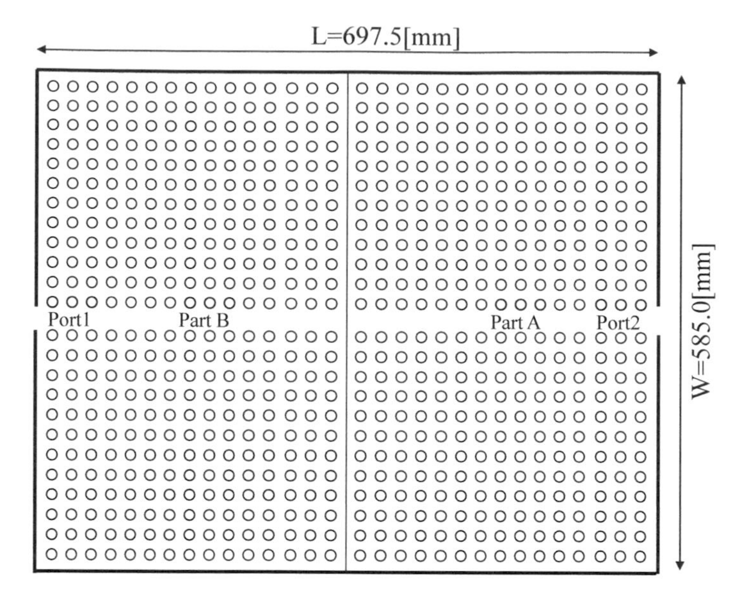

(a) Top view of basement plate to support the photonic crystal waveguide structure. Two separated plates are butt-jointed to realize waveguide shift Δ.

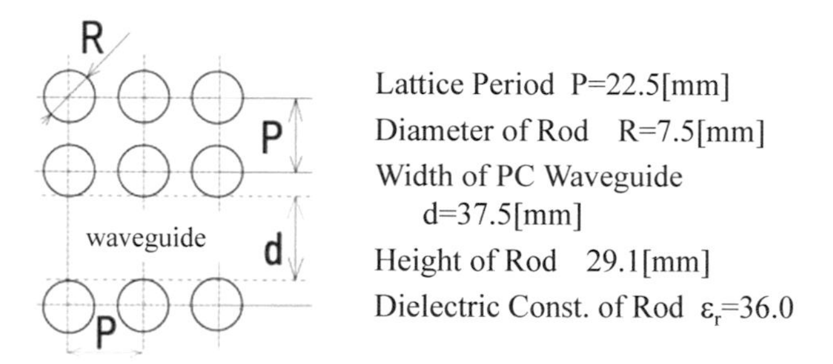

(b) Fundamental periodic structure with the dimension.

Fig. 2. Illustration of top view of photonic crystal waveguide by making use of line defect with the parameters.

overlapping and looks like unique curve for this range of Δ. The reflections also concentrate within difference of a few dB. These results mean that the waveguide shift $\Delta$of 26.7% to the waveguide width d=*37.5* mm is not critical for butt joint.

For larger waveguide shift Δ over 15 mm, the transmission and reflection both affected apparently. In Fig. 4(a), transmission curves for Δ = 15, 20, and 25 mm decrease, while the reflections increase in Fig. 4(b).

Different from slab waveguide or conventional optical fiber based on total internal reflection at core-cladding boundary, wave confinement in PhC waveguide is obtained by PBG. This principle on confinement means that radiation

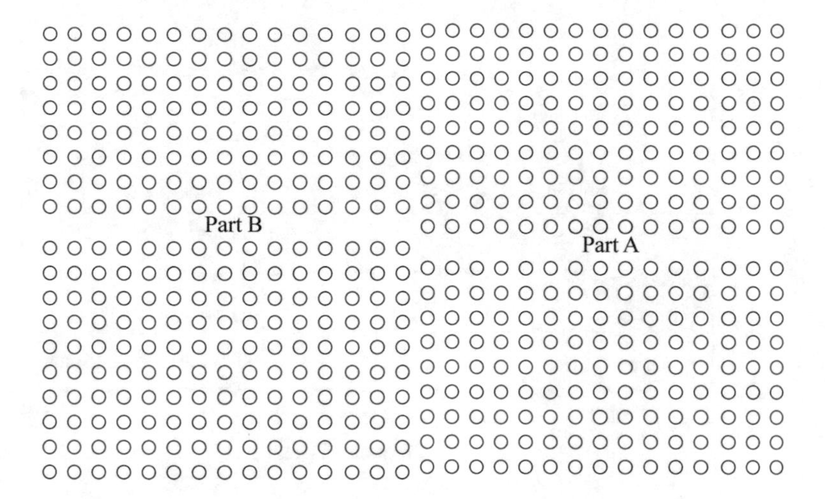

(a) Illustration of butt-jointed waveguides by shifting two basement plates.

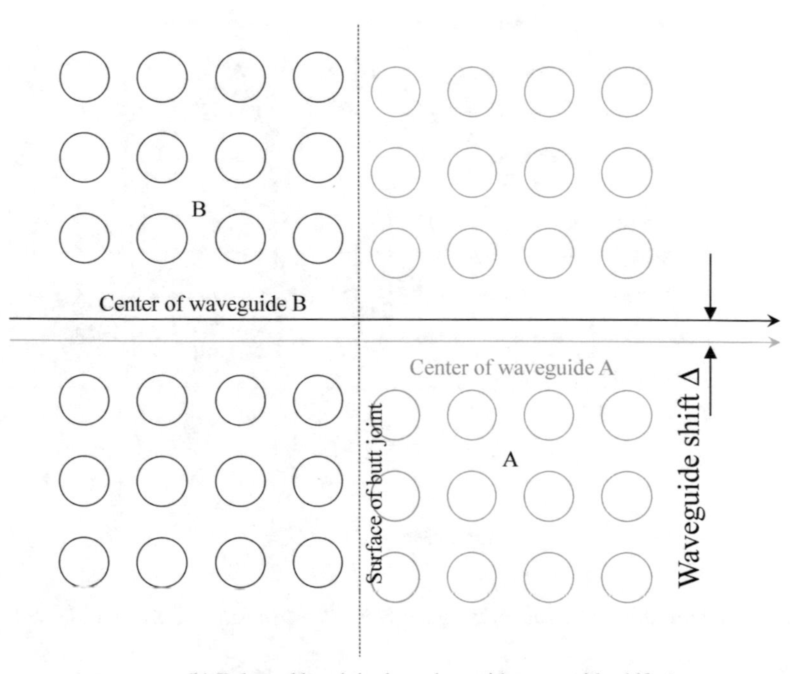

(b) Enlarged butt joint boundary with waveguide shift Δ.

Fig. 3. Butt joint and the waveguide shift Δ.

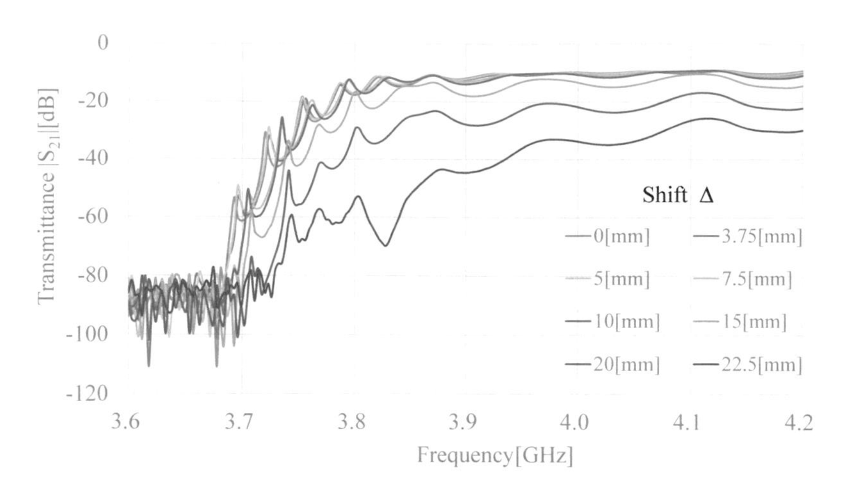

(a) Transmittance $|S_{21}|$.

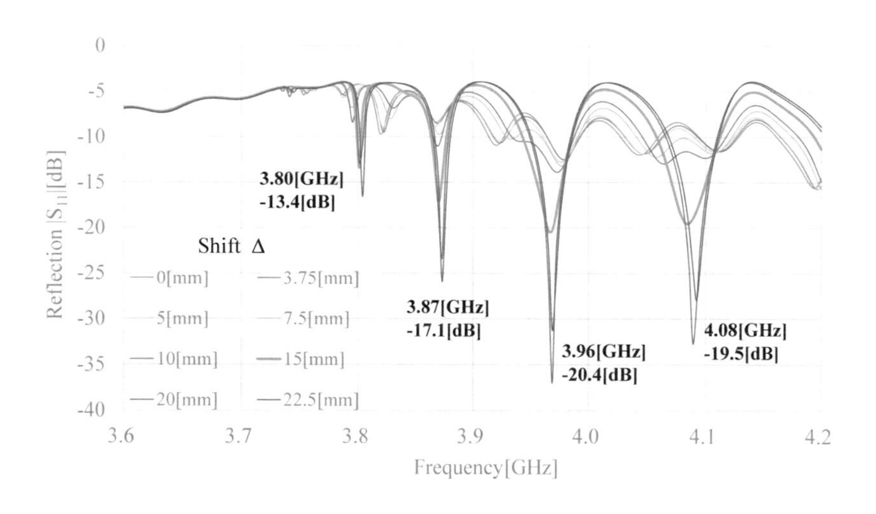

(b) Reflection $|S_{11}|$

Fig. 4. Transmission and reflection spectrum as a function of waveguide shift Δ.

field at waveguide discontinuity can not penetrate into periodic structure due to the PBG as far as the waveguide shift Δ is small with respect to waveguide width d. For larger discontinuity, propagating wave is reflected back due to the discontinuity to reduce the transmission and to increase the reflection.

From above discussion, we can conclude that the proposed butt joint of PhC waveguide components is applicable for small amount of waveguide shift Δ.

4 Conclusion

A new assembly process of functional units by butt joint is proposed in this paper. Then, measurement of the transmission and reflection characteristics of butt joint in PhC waveguide were shown. The waveguide shift Δ was situated in the middle of PhC waveguide by composing two separated basement plate of aluminum. It was found that the waveguide shift Δ up to 26.7% of waveguide width d is not seriously critical for the transmission and reflection. Application of proposed assembly technique by butt joint is hopeful for fabrication of PhC integrated circuit.

Evaluation of transmission and reflection due to discontinuity of gap between two units are our future work.

Acknowledgment. Authors express our appreciation to Mr. H. Ono and Mr. S Furuie of Fukuoka Institute of Technology as part of their undergraduate research under supervision by H. Maeda in 2017–18.

References

1. Okamoto, K.: Fundamentals of Optical Waveguides, 2nd edn. Elsevier, Amsterdam (2006)
2. Yasumoto, K. (ed.): Electromagnetic Theory and Applications for Photonic Crystals. CRC Press, Boca Raton (2006)
3. Inoue, K., Ohtaka, K. (eds.): Photonic Crystals - Physics, Fabrication and Applications. Springer, New York (2004)
4. Noda, S., Baba, T. (eds.): Roadmap on Photonic Crystals. Kluwer Academic Publishers, Boston (2003)
5. Joannopoulos, J.D., Meade, R.D., Winn, J.N.: Photonic Crystals. Princeton University Press, Princeton (1995)
6. Tanabe, T., Suzuki, R., Tetsumoto, T., Kakinuma, Y.: Fabrication and applications of high-Q optical nano- and micro-cavities. Oyou Butsuri **87**(3), 181–186 (2018). A monthly publication of the Japanese Society of Applied Physics
7. Temelkuran, B., Ozbay, E.: Experimental demonstration of photonic crystal based waveguides. Appl. Phys. Lett. **74**(4), 486–488 (1999)
8. Beaky, M.: Two-dimensional photonic crystal Fabry-Perot resonators with lossy dielectrics. IEEE Trans. Microw. Theory Tech. **47**(11), 2085–2091 (1999)
9. Maeda, H., Hironaka, S.: Experimental study on improvement of filtering characteristics of cavities in photonic crystal waveguide. In: Proceedings of ISMOT 2009, pp. 229–232, December 2009
10. Maeda, H., Nemoto, R. Satoh, T.: Experimental study on cavities in 2D photonic crystal waveguide for 4 GHz band. In: Proceedings of ISMOT 2007, pp. 459–462, December 2007
11. Maeda, H., Meng, X., Haari, K., Higashinaka, N.: Signal routing by cavities in photonic crystal waveguide. In: Proceedings of The 6th International Conference on Emerging Internet, Data and Web Technologies, EIDWT 2018. Lecture Notes on Data Engineering and Communications Technologies, vol. 17, pp. 765–772. Springer, March 2018

Clustering in VANETs: A Fuzzy-Based System for Clustering of Vehicles

Kosuke Ozera[1(✉)], Kevin Bylykbashi[2], Yi Liu[1], Makoto Ikeda[3], and Leonard Barolli[3]

[1] Graduate School of Engineering, Fukuoka Institute of Technology (FIT), 3-30-1 Wajiro-Higashi, Higashi-Ku, Fukuoka 811–0295, Japan
kosuke.o.fit@gmail.com, ryuui1010@gmail.com
[2] Faculty of Information Technologies, Polytechnic University of Tirana, Bul. "Dëshmorët e Kombit","Mother Theresa" Square, Nr. 4, Tirana, Albania
kevini_95@hotmail.com
[3] Department of Information and Communication Engineering, Fukuoka Institute of Technology (FIT), 3-30-1 Wajiro-Higashi, Higashi-Ku, Fukuoka 811–0295, Japan
makoto.ikd@acm.org, barolli@fit.ac.jp

Abstract. In recent years, inter-vehicle communication has attracted attention because it can be applicable not only to alternative networks but also to various communication systems. In this paper, we propose a Fuzzy-based system for clustering of vehicles in VANETs. We evaluate the proposed system by simulations. From the simulation results, we found that when DCC parameter is small and SC is high, the possibility that vehicle remains in the cluster is increased.

Keywords: Inter-vehicle communication · VANET · Fuzzy · Cluster

1 Introduction

In recent years, a number of disasters have been occurred around the world. The technologies of disaster management system are improved, however the communication system does not work well in disaster area due to the traffic concentration, device failure, and so on. A key for creating a valuable disaster rescue plan is to prepare alternative communication systems. In disaster area, mobile devices are often disconnected in the network due to network traffic congestion and access point failure. Inter-vehicle communication has attracted attention as an alternative network in disaster situations. In this case, Delay/Disruption/Disconnection Tolerant Networking (DTN) are used as one of a key alternative option to provide the network services [37].

The DTN aims to provide seamless communications with a wide range of network, which have not good performance characteristics [4]. DTN has the potential to interconnect vehicles in regions that current networking protocol

L. Barolli et al. (Eds.): NBiS 2018, LNDECT 22, pp. 810–821, 2019.
https://doi.org/10.1007/978-3-319-98530-5_72

cannot reach the destination. For inter-vehicle communications, there are different types of communication such as Vehicle-to-Vehicle (V2V), Vehicle-to-Infrastructure (V2I), Vehicle-to-Pedestrian (V2P) and Vehicle-to-X (V2X) communications [2,7,10,14,31]. IEEE 802.11p supports these communications in outdoor environments. It defines enhancements to 802.11 required to support Intelligent Transport System (ITS) applications. The technology operates at 5.9 GHz in various propagation environments to high-speed moving vehicles.

There are different works for Vehicular Ad-Hoc Networks (VANETs). In [16], the authors proposed a Message Suppression Controller (MSC) for V2V and V2I communications. They considered some parameters to control the message suppression dynamically. However, a fixed parameter still is used to calculate the duration of message suppression. To solve this problem, the authors proposed an Enhanced Message Suppression Controller (EMSC) [18] for Vehicular-DTN (V-DTN). The EMSC is an expanded version of MSC [16] and can be used for various network conditions. But, many control packets were delivered in the network.

Security and trust in VANETs is essential in order to prevent malicious agents sabotaging road safety systems built upon the VANET framework, potentially causing serious disruption to traffic flows or safety hazards. Several authors have proposed cluster head metrics which can assist in identifying malicious vehicles and mitigating their impact by denying them access to cluster resources [8].

Security of the safety messages can be achieved by authentication [38]. To make the process of authentication faster [17], vehicles in the communication range of an Road Side Unit (RSU) can be grouped to be in one cluster and a cluster head is elected to authenticate all the vehicles available in the cluster. Formation of clusters in a dynamic VANET and selection of cluster head plays a major role is selected.

In [9] is presented a cluster head selection metric based on vehicle direction, degree of connectivity, an entropy value (calculated from the mobility of nodes in the network), and a distrust level (based on the reliability of a node's packet relaying). Vehicles are assigned verifiers, which are neighbors with a lower distrust value. Verifiers monitor the network behavior of a vehicle, and confirm whether it is routing packets and advertising mobility and traffic information that is consistent with the verifier's own view of the neighborhood. The distrust value for nodes which behave abnormally is then automatically increased, while it is decreased for nodes which perform reliably. In this way, the trustworthiness of a node is accounted for the cluster head selection process.

Fuzzy Logic (FL) is the logic underlying modes of reasoning which are approximate rather then exact. The importance of FL derives from the fact that most modes of human reasoning and especially common sense reasoning are approximate in nature [19]. FL uses linguistic variables to describe the control parameters. By using relatively simple linguistic expressions it is possible to describe and grasp very complex problems. A very important property of the linguistic variables is the capability of describing imprecise parameters.

In this paper, we propose a Fuzzy-based system for clustering of vehicles in VANETs. We evaluate the proposed system by simulations. The simulation results show that the possibility of the vehicle to remain in the cluster is increased when SC value is increased and DCC is decreased.

The structure of the paper is as follows. In Sect. 2, we present VANETs and DTNs. We give application of Fuzzy Logic for control in Sect. 3. In Sect. 4, we present our proposed systems. In Sect. 5, we show simulation results. Finally, conclusions and future work are given in Sect. 6.

2 VANETs and DTNs

VANETs are considered to have an enormous potential in enhancing road traffic safety and traffic efficiency. Therefore various governments have launched programs dedicated to the development and consolidation of vehicular communications and networking and both industrial and academic researchers are addressing many related challenges, including socio-economic ones, which are among the most important [15,33].

VANET technology uses moving vehicle as nodes to form a wireless mobile network. It aims to provide fast and cost-efficient data transfer for the advantage of passenger safety and comfort. To improve road safety and travel comfort of voyagers and drivers, Intelligent Transport Systems (ITS) are developed. ITS proposes to manage vehicle traffic, support drivers with safety and other information, and provide some services such as automated toll collection and driver assist systems [22].

In essence, VANETs provide new prospects to improve advanced solutions for making reliable communication between vehicles. VANETs can be defined as a part of ITS which aims to make transportation systems faster and smarter in which vehicles are equipped with some short-range and medium-range wireless communication [3]. In a VANET, wireless vehicles are able to communicate directly with each other (i.e., emergency vehicle warning, stationary vehicle warning) and also served various services (i.e., video streaming, internet) from access points (i.e., 3G or 4G) through roadside units [5,22].

The DTN are occasionally connected networks, characterized by the absence of a continuous path between the source and destination [1,13]. The data can be transmitted by storing them at nodes and forwarding them later when there is a working link. This technique is called message switching. Eventually the data will be relayed to the destination. The inspiration for DTNs came from an unlikely source: efforts to send packets in space. Space networks must deal with intermittent communication and very long delays [36]. In [13], the author observed the possibility to apply these ideas for other applications.

The main assumption in the Internet that DTNs seek to relax is that an End-to-End (E2E) path between a source and a destination exists for the entire duration of a communication session. When this is not the case, the normal Internet protocols fail. The DTN architecture is based on message switching. It is also intended to tolerate links with low reliability and large delays. The architecture is specified in RFC 4838 [6].

Bundle protocol has been designed as an implementation of the DTN architecture. A bundle is a basic data unit of the DTN bundle protocol. Each bundle comprises a sequence of two or more blocks of protocol data, which serve for various purposes. In poor conditions, bundle protocol works on the application layer of some number of constituent Internet, forming a store-and-forward overlay network to provide its services. The bundle protocol is specified in RFC 5050 [34]. It is responsible for accepting messages from the application and sending them as one or more bundles via store-carry-forward operations to the destination DTN node. The bundle protocol provides a transport service for many different applications.

3 Application of Fuzzy Logic for Control

The ability of fuzzy sets and possibility theory to model gradual properties or soft constraints whose satisfaction is matter of degree, as well as information pervaded with imprecision and uncertainty, makes them useful in a great variety of applications.

The most popular area of application is Fuzzy Control (FC), since the appearance, especially in Japan, of industrial applications in domestic appliances, process control, and automotive systems, among many other fields [11,12,20,24–28,35].

3.1 FC

In the FC systems, expert knowledge is encoded in the form of fuzzy rules, which describe recommended actions for different classes of situations represented by fuzzy sets.

In fact, any kind of control law can be modeled by the FC methodology, provided that this law is expressible in terms of "if ... then ..." rules, just like in the case of expert systems. However, FL diverges from the standard expert system approach by providing an interpolation mechanism from several rules. In the contents of complex processes, it may turn out to be more practical to get knowledge from an expert operator than to calculate an optimal control, due to modeling costs or because a model is out of reach.

3.2 Linguistic Variables

A concept that plays a central role in the application of FL is that of a linguistic variable. The linguistic variables may be viewed as a form of data compression. One linguistic variable may represent many numerical variables. It is suggestive to refer to this form of data compression as granulation [21].

The same effect can be achieved by conventional quantization, but in the case of quantization, the values are intervals, whereas in the case of granulation the values are overlapping fuzzy sets. The advantages of granulation over quantization are as follows:

- it is more general;
- it mimics the way in which humans interpret linguistic values;
- the transition from one linguistic value to a contiguous linguistic value is gradual rather than abrupt, resulting in continuity and robustness.

3.3 FC Rules

FC describes the algorithm for process control as a fuzzy relation between information about the conditions of the process to be controlled, x and y, and the output for the process z. The control algorithm is given in "if ... then ..." expression, such as:

If x is small and y is big, then z is medium;

If x is big and y is medium, then z is big.

These rules are called *FC rules*. The "if" clause of the rules is called the antecedent and the "then" clause is called consequent. In general, variables x and y are called the input and z the output. The "small" and"big" are fuzzy values for x and y, and they are expressed by fuzzy sets.

Fuzzy controllers are constructed of groups of these FC rules, and when an actual input is given, the output is calculated by means of fuzzy inference.

3.4 Control Knowledge Base

There are two main tasks in designing the control knowledge base. First, a set of linguistic variables must be selected which describe the values of the main control parameters of the process. Both the input and output parameters must be linguistically defined in this stage using proper term sets. The selection of the level of granularity of a term set for an input variable or an output variable plays an important role in the smoothness of control. Second, a control knowledge base must be developed which uses the above linguistic description of the input and output parameters. Four methods [29,32,39,40] have been suggested for doing this:

- expert's experience and knowledge;
- modelling the operator's control action;
- modelling a process;
- self organization.

Among the above methods, the first one is the most widely used. In the modeling of the human expert operator's knowledge, fuzzy rules of the form "If Error is small and Change-in-error is small then the Force is small" have been used in several studies [23,30]. This method is effective when expert human operators can express the heuristics or the knowledge that they use in controlling a process in terms of rules of the above form.

3.5 Defuzzification Methods

The defuzzification operation produces a non-FC action that best represent the membership function of an inferred FC action. Several defuzzification methods have been suggested in literature. Among them, four methods which have been applied most often are:

- Tsukamoto's Defuzzification Method;
- The Center of Area (COA) Method;
- The Mean of Maximum (MOM) Method;
- Defuzzification when Output of Rules are Function of Their Inputs.

4 Proposed Fuzzy-Based System

The proposed system model is shown in Fig. 1. We consider four parameters: Group Speed (GS), Relative Acceleration (RA), Security (SC) and Distance from Cluster Center (DCC) to decide the Vehicle Remain or Leave the Cluster (VRLC) output parameter. These four parameters are not correlated with each other, for this reason we use fuzzy system. The membership functions are shown in Fig. 2. In Table 1, we show the Fuzzy Rule Base (FRB), which consists of 81 rules.

The term sets of GS, RA, SC and DCC are defined respectively as:

$$\begin{aligned} GS &= \{Slow,\ Middle,\ Fast\} \\ &= \{S,\ M,\ F\}; \\ RA &= \{Decelerate,\ Same,\ Accelerate\} \\ &= \{Dec,\ Sam,\ Acc\}; \\ SC &= \{Weak,\ Middle,\ Strong\} \\ &= \{We,\ Mi,\ St\}; \\ DCC &= \{Near,\ Middle,\ Far\} \\ &= \{Near,\ Mid,\ Far\}. \end{aligned}$$

and the term set for the output $VRLC$ is defined as:

$$VRLC = \begin{pmatrix} Leave \\ Weak\ Leave \\ Very\ Weak\ Leave \\ Not\ Remain\ Not\ Leave \\ Very\ Weak\ Remain \\ Weak\ Remain \\ Remain \end{pmatrix} = \begin{pmatrix} Le \\ WLe \\ VWL \\ NRNL \\ VWR \\ WRe \\ Re \end{pmatrix}.$$

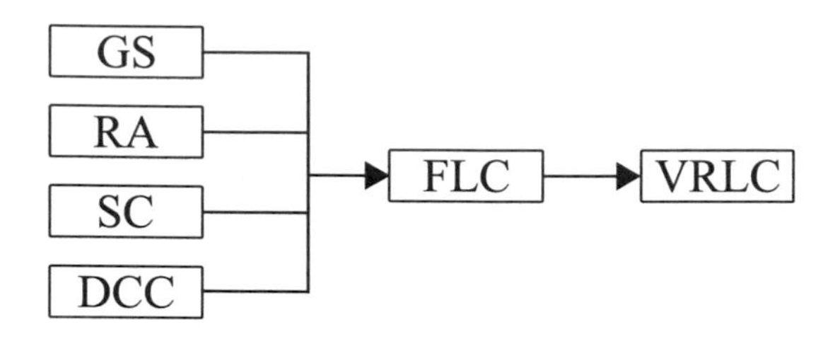

Fig. 1. Proposed system model.

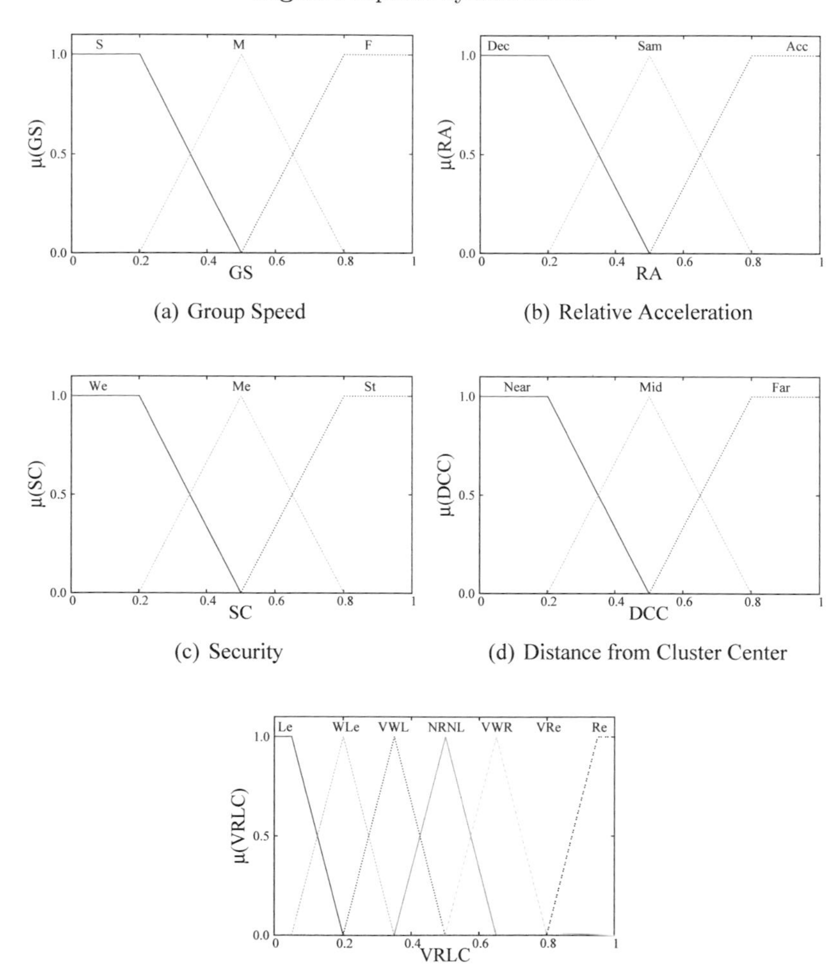

Fig. 2. Membership functions.

Table 1. FRB.

Rule	RA	SC	GS	DCC	VRLC	Rule	RA	SC	GS	DCC	VRLC	Rule	RA	SC	GS	DCC	VRLC
1	Dec	We	S	Near	Le	28	Sam	We	S	Near	VWL	55	Acc	We	S	Near	Le
2	Dec	We	S	Mid	Le	29	Sam	We	S	Mid	WLe	56	Acc	We	S	Mid	Le
3	Dec	We	S	Far	Le	30	Sam	We	S	Far	Le	57	Acc	We	S	Far	Le
4	Dec	We	M	Near	WLe	31	Sam	We	M	Near	VWR	58	Acc	We	M	Near	WLe
5	Dec	We	M	Mid	Le	32	Sam	We	M	Mid	NRNL	59	Acc	We	M	Mid	Le
6	Dec	We	M	Far	Le	33	Sam	We	M	Far	WLe	60	Acc	We	M	Far	Le
7	Dec	We	F	Near	NRNL	34	Sam	We	F	Near	WRe	61	Acc	We	F	Near	NRNL
8	Dec	We	F	Mid	VWL	35	Sam	We	F	Mid	VWR	62	Acc	We	F	Mid	VWL
9	Dec	We	F	Far	Le	36	Sam	We	F	Far	NRNL	63	Acc	We	F	Far	Le
10	Dec	Me	S	Near	WLe	37	Sam	Me	S	Near	NRNL	64	Acc	Me	S	Near	WLe
11	Dec	Me	S	Mid	Le	38	Sam	Me	S	Mid	VWL	65	Acc	Me	S	Mid	Le
12	Dec	Me	S	Far	Le	39	Sam	Me	S	Far	WLe	66	Acc	Me	S	Far	Le
13	Dec	Me	M	Near	VWL	40	Sam	Me	M	Near	WRe	67	Acc	Me	M	Near	VWL
14	Dec	Me	M	Mid	WLe	41	Sam	Me	M	Mid	VWR	68	Acc	Me	M	Mid	WLe
15	Dec	Me	M	Far	Le	42	Sam	Me	M	Far	VWL	69	Acc	Me	M	Far	Le
16	Dec	Me	F	Near	VWR	43	Sam	Me	F	Near	Re	70	Acc	Me	F	Near	VWR
17	Dec	Me	F	Mid	NRNL	44	Sam	Me	F	Mid	WRe	71	Acc	Me	F	Mid	NRNL
18	Dec	Me	F	Far	WLe	45	Sam	Me	F	Far	VWR	72	Acc	Me	F	Far	WLe
19	Dec	St	S	Near	VWL	46	Sam	St	S	Near	VWR	73	Acc	St	S	Near	VWL
20	Dec	St	S	Mid	Le	47	Sam	St	S	Mid	NRNL	74	Acc	St	S	Mid	Le
21	Dec	St	S	Far	Le	48	Sam	St	S	Far	VWL	75	Acc	St	S	Far	Le
22	Dec	St	M	Near	NRNL	49	Sam	St	M	Near	WRe	76	Acc	St	M	Near	NRNL
23	Dec	St	M	Mid	VWL	50	Sam	St	M	Mid	WRe	77	Acc	St	M	Mid	VWL
24	Dec	St	M	Far	WLe	51	Sam	St	M	Far	NRNL	78	Acc	St	M	Far	WLe
25	Dec	St	F	Near	WRe	52	Sam	St	F	Near	Re	79	Acc	St	F	Near	WRe
26	Dec	St	F	Mid	VWR	53	Sam	St	F	Mid	Re	80	Acc	St	F	Mid	VWR
27	Dec	St	F	Far	VWL	54	Sam	St	F	Far	WRe	81	Acc	St	F	Far	VWL

5 Simulation Results

In this section, we present the simulation results for our proposed system. In our system, we decided the membership functions by carrying out many simulations. The simulation results were carried out in MATLAB.

In Figs. 3, 4 and 5, we show the relation between GS, RA, SC and DCC. We consider the GS and DCC as constant parameters.

In Fig. 3, we show the relation between VRLC and RA for different SC values. In Fig. 3(a), GS and DCC are 0.1 units, when the RA is less than 0.2 units. The vehicle acceleration is low so the possibility that vehicle remains in the cluster is low. Also, when RA is higher than 0.8 units, the vehicle acceleration is high, so the vehicle with high possibility leaves the cluster. By increasing SC, the possibility to remain in the cluster is increased. In Fig. 3(b), we increase DCC

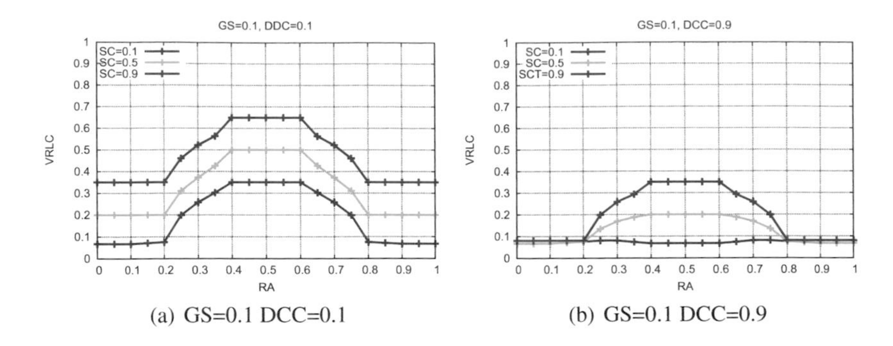

(a) GS=0.1 DCC=0.1 (b) GS=0.1 DCC=0.9

Fig. 3. Simulation results when GS is 0.1.

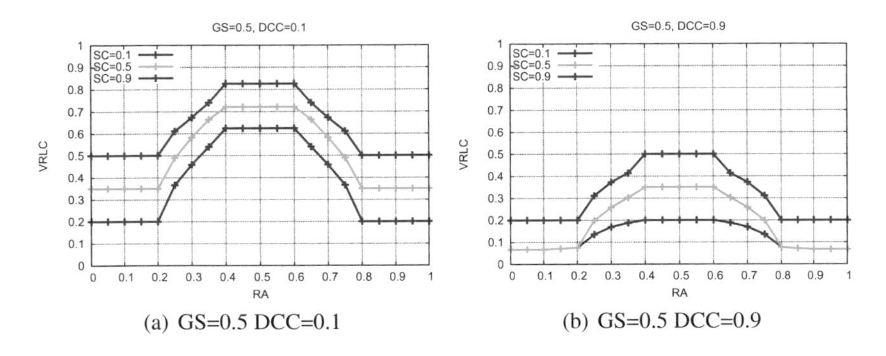

(a) GS=0.5 DCC=0.1 (b) GS=0.5 DCC=0.9

Fig. 4. Simulation results when GS is 0.5.

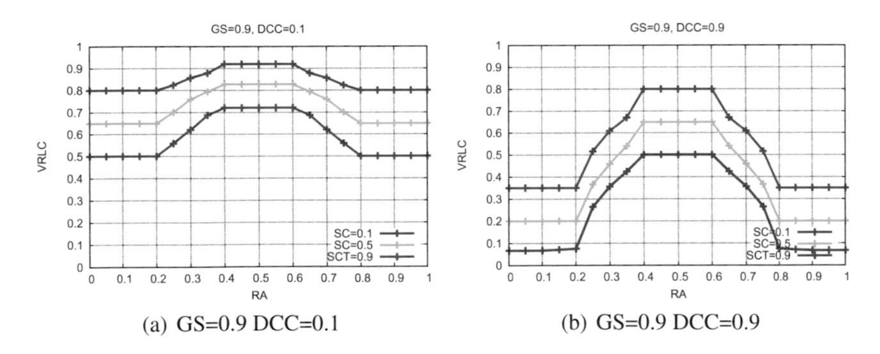

(a) GS=0.9 DCC=0.1 (b) GS=0.9 DCC=0.9

Fig. 5. Simulation results when GS is 0.9.

value to 0.9 units. We see that by increase of DCC, the possibility that the vehicle leaves the cluster is increased.

In Fig. 4, the GS value is 0.5. In Fig. 4(a) and (b), the DCC value is 0.1 and 0.9, respectively. We see that when DCC value is low, the vehicle remains in the cluster. In Fig. 5, we increase the GS value to 0.9. We can see that by increasing

GS value and DCC is low, the possibility that vehicle remains in the cluster is increased.

6 Conclusions

In this paper, we proposed a Fuzzy-based system for clustering of vehicles in VANETs. We considering four parameters: GS, RA, SC and DCC. We evaluated the performance of proposed system by computer simulations. From the simulation results, we found that when DCC parameter is small and SC is high, the possibility that vehicle remains in the cluster is increased.

In the future work, we will consider other parameters for the simulation system and carry out extensive simulations.

References

1. Delay- and disruption-tolerant networks (DTNs) tutorial. NASA/JPL's Interplanetary Internet (IPN) Project (2012). http://www.warthman.com/images/DTN_Tutorial_v2.0.pdf
2. Araniti, G., Campolo, C., Condoluci, M., Iera, A., Molinaro, A.: LTE for vehicular networking: a survey. IEEE Commun. Mag. **21**(5), 148–157 (2013)
3. Booysen, M.J., Zeadally, S., van Rooyen, G.J.: Performance comparison of media access control protocols for vehicular ad hoc networks. IET Netw. **1**(1), 10–19 (2012)
4. Burleigh, S., Hooke, A., Torgerson, L., Fall, K., Cerf, V., Durst, B., Scott, K., Weiss, H.: Delay-tolerant networking: an approach to interplanetary internet. IEEE Commun. Mag. **41**(6), 128–136 (2003)
5. Calhan, A.: A fuzzy logic based clustering strategy for improving vehicular ad-hoc network performance **40**(2), 351–367 (2015)
6. Cerf, V., Burleigh, S., Hooke, A., Torgerson, L., Durst, R., Scott, K., Fall, K., Weiss, H.: Delay-tolerant networking architecture. IETF RFC 4838 (Informational), April 2007
7. Cheng, X., Yao, Q., Wen, M., Wang, C.X., Song, L.Y., Jiao, B.L.: Wideband channel modeling and intercarrier interference cancellation for vehicle-to-vehicle communication systems. IEEE J. Sel. Areas Commun. **31**(9), 434–448 (2013)
8. Cooper, C., Franklin, D., Ros, M., Safaei, F., Abolhasan, M.: A comparative survey of VANET clustering techniques. IEEE Commun. Surv. Tutor. **19**(1), 657–681 (2017)
9. Daeinabi, A., Rahbar, A.G.P., Khademzadeh, A.: VWCA: an efficient clustering algorithm in vehicular ad hoc networks. J. Netw. Comput. Appl. **34**(1), 207–222 (2011)
10. Dias, J.A.F.F., Rodrigues, J.J.P.C., Xia, F., Mavromoustakis, C.X.: A cooperative watchdog system to detect misbehavior nodes in vehicular delay-tolerant networks. IEEE Trans. Ind. Electr. **62**(12), 7929–7937 (2015)
11. Elmazi, D., Kulla, E., Oda, T., Spaho, E., Sakamoto, S., Barolli, L.: A comparison study of two fuzzy-based systems for selection of actor node in wireless sensor actor networks. J. Ambient Intell. Hum. Comput. **6**(5), 635–645 (2015)

12. Elmazi, D., Sakamoto, S., Oda, T., Kulla, E., Spaho, E., Barolli, L.: Two fuzzy-based systems for selection of actor nodes in wireless sensor and actor networks: a comparison study considering security parameter effect. In: Mobile Networks and Applications, pp. 1–12 (2016)
13. Fall, K.: A delay-tolerant network architecture for challenged Internets. In: Proceedings of the International Conference on Applications, Technologies, Architectures, and Protocols for Computer Communications, SIGCOMM 2003, pp. 27–34 (2003)
14. Grassi, G., Pesavento, D., Pau, G., Vuyyuru, R., Wakikawa, R., Zhang, L.: VANET via named data networking. In: Proceedings of the IEEE Conference on Computer Communications Workshops (INFOCOM WKSHPS 2014), pp. 410–415, April 2014
15. Hartenstein, H., Laberteaux, L.: A tutorial survey on vehicular ad hoc networks. IEEE Commun. Mag. **46**(6), 164–171 (2008)
16. Honda, T., Ikeda, M., Ishikawa, S., Barolli, L.: A message suppression controller for vehicular delay tolerant networking. In: Proceedings of the 29th IEEE International Conference on Advanced Information Networking and Applications (IEEE AINA-2015), pp. 754–760, March 2015
17. Huang, J.L., Yeh, L.Y., Chien, H.Y.: ABAKA: an anonymous batch authenticated and key agreement scheme for value-added services in vehicular ad hoc networks. IEEE Trans. Veh. Technol. **60**(1), 248–262 (2011)
18. Ikeda, M., Ishikawa, S., Barolli, L.: An enhanced message suppression controller for vehicular-delay tolerant networks. In: Proceedings of the 30th IEEE International Conference on Advanced Information Networking and Applications (IEEE AINA-2016), pp. 573–579, March 2016
19. Inaba, T., Obukata, R., Sakamoto, S., Oda, T., Ikeda, M., Barolli, L.: Performance evaluation of a QoS-aware fuzzy-based CAC for LAN access. Int. J. Space-Based Situated Comput. **6**(4), 228–238 (2016)
20. Inaba, T., Sakamoto, S., Oda, T., Ikeda, M., Barolli, L.: A secure-aware call admission control scheme for wireless cellular networks using fuzzy logic and its performance evaluation. J. Mobile Multimedia **11**(3&4), 213–222 (2015)
21. Kandel, A.: Fuzzy Expert Systems. CRC Press, Boca Raton (1991)
22. Karagiannis, G., Altintas, O., Ekici, E., Heijenk, G., Jarupan, B., Lin, K., Weil, T.: Vehicular networking: a survey and tutorial on requirements, architectures, challenges, standards and solutions. IEEE Commun. Surv. Tutor. **13**(4), 584–616 (2011)
23. Klir, G.J., Folger, T.A.: Fuzzy Sets, Uncertainty, and Information. Prentice Hall, London (1988)
24. Kolici, V., Inaba, T., Lala, A., Mino, G., Sakamoto, S., Barolli, L.: A fuzzy-based CAC scheme for cellular networks considering security. In: The 17th International Conference on Network-Based Information Systems (NBiS-2014), pp. 368–373 (2014)
25. Liu, Y., Sakamoto, S., Matsuo, K., Ikeda, M., Barolli, L.: Improving reliability of JXTA-overlay platform: evaluation for e-learning and trustworthiness. J. Mobile Multimedia **11**(2), 34–50 (2015)
26. Liu, Y., Sakamoto, S., Matsuo, K., Ikeda, M., Barolli, L., Xhafa, F.: A comparison study for two fuzzy-based systems: improving reliability and security of JXTA-overlay P2P platform. Soft Comput. 1–11 (2015)
27. Liu, Y., Sakamoto, S., Matsuo, K., Ikeda, M., Barolli, L., Xhafa, F.: Improvement of JXTA-overlay P2P platform: evaluation for medical application and reliability. Int. J. Distrib. Syst. Technol. (IJDST) **6**(2), 45–62 (2015)

28. Matsuo, K., Elmazi, D., Liu, Y., Sakamoto, S., Mino, G., Barolli, L.: FACS-MP: a fuzzy admission control system with many priorities for wireless cellular networks and its performance evaluation. J. High Speed Netw. **21**(1), 1–14 (2015)
29. McNeill, F.M., Thro, E.: Fuzzy Logic: A Practical Approach. Academic Press, Boston (1994)
30. Munakata, T., Jani, Y.: Fuzzy systems: an overview. Commun. ACM **37**(3), 68–76 (1994)
31. Ohn-Bar, E., Trivedi, M.M.: Learning to detect vehicles by clustering appearance patterns. IEEE Trans. Intell. Transp. Syst. **16**(5), 2511–2521 (2015)
32. Procyk, T.J., Mamdani, E.H.: A linguistic self-organizing process controller. Automatica **15**(1), 15–30 (1979)
33. Santi, P.: Mobility models for next generation wireless networks: ad hoc, vehicular and mesh networks. Wiley, Chichester (2012)
34. Scott, K., Burleigh, S.: Bundle protocol specification. IETF RFC 5050 (Experimental), November 2007
35. Spaho, E., Sakamoto, S., Barolli, L., Xhafa, F., Ikeda, M.: Trustworthiness in P2P: performance behaviour of two fuzzy-based systems for JXTA-overlay platform. Soft Comput. **18**(9), 1783–1793 (2014)
36. Tanenbaum, A.S., Wetherall, D.J.: Computer Networks, 5th edn. Pearson Education Inc., Prentice Hall (2011)
37. Uchida, N., Ishida, T., Shibata, Y.: Delay tolerant networks-based vehicle-to-vehicle wireless networks for road surveillance systems in local areas. Int. J. Space-Based Situated Comput. **6**(1), 12–20 (2016)
38. Wen, H., Ho, P.H., Gong, G.: A novel framework for message authentication in vehicular communication networks. In: Global Telecommunications Conference, GLOBECOM 2009, pp. 1–6. IEEE (2009)
39. Zadeh, L.A., Kacprzyk, J.: Fuzzy Logic for the Management of Uncertainty. Wiley, New York (1992)
40. Zimmermann, H.J.: Fuzzy Set Theory and Its Applications. Springer, Netherlands (1991)

Movement Detection Methods with Wireless Signals and Multiple Sensors on Mobile Phone for Traffic Accident Prevention Systems

Shoma Takeuchi[1], Noriki Uchida[1(✉)], and Yoshitaka Shibata[2]

[1] Department of Information and Communication Engineering, Fukuoka Institute of Technology, 3-30-1 Wajirohigashi, Higashi-ku, Fukuoka 811-0295, Japan
mgm17102@bene.fit.ac.jp, n-uchida@fit.ac.jp
[2] Faculty of Software and Information Science, Iwate Prefectural University, 152-52 Sugo, Takizawa, Iwate 020-0193, Japan
shibata@iwate-pu.ac.jp

Abstract. In the recent traffic accidents, it has been focused on the pedestrians and the bicycles as well as the automobiles. Especially, it is widely considered that the texting while walking and riding bicycles is extremely dangerous. Thus, this research proposed the mobile traffic accident prevention system with observing the radio signals and various sensors on the smartphone. In the methods, the RSSI levels of IEEE802.11a/b/g/n from others are firstly observed, and the sensors such as the gyro sensor on the smartphone are secondly applied for the modifications of the detection process based on the Markov Chain algorithm. Then, this paper reports the prototype system of the proposed methods, and the experimental results are discussed for the future studies.

1 Introduction

In the recent traffic accidents, it has been focused on the pedestrians and the bicycles as well as the automobiles. According to the National Police Agency in Japan [1] reported that the number of the traffic accidents were about 536,899 in 2015, and it was 36,943 fewer than the last year. On the other hands, the agency also indicated that the number of the death by the accidents have slightly increased in each year because of the environmental changes of the recent traffic system. Then, it is considered that one of the major environmental changes is the wide spread of the smartphones. Actually, it has been widely known the dangerousness of the smartphone's texting while walking and riding bicycles in the current lives.

Therefore, the research proposed the mobile traffic accident prevention system with observing the radio signals from others and the various sensors such as the gyro sensor on the smartphone. The proposed system senses the dangerous movements toward the user by these observations, and the alert is confirmed for the user. In details, the smartphone firstly observes the RSSI levels from other smartphones continuously, and these observed values are calculated by the Markov Chain algorithm for the detection of

L. Barolli et al. (Eds.): NBiS 2018, LNDECT 22, pp. 822–829, 2019.
https://doi.org/10.1007/978-3-319-98530-5_73

the fast approach objects such as the automobiles or bicycles [2]. Then, the sensors on the user's smartphone were also used for the modifications of the detection processes.

In the followings, the Sect. 2 deals with the related studies of the traffic accident prevention system, and the proposed methods with the RSSI levels and the sensors on the smartphone are introduced in the Sect. 3. Then, the implementations of the prototype system are introduced in the Sects. 4 and 5 deals with the experiments by the prototype system. At last, the conclusion and the future studies are discussed in the Sect. 6.

2 Related Works

With the increasing of the developments of ITS (Intelligent Transport System) technologies, many researches of the driving safety systems are confirmed. Although most of previous researches are aiming to the traffic safety systems between the vehicles, there are some previous approaches considered with bicycles or pedestrians other than the vehicles. For instance, the previous paper [3] discussed the electric motor sounds from the eco cars, and it mentioned the effective sound patterns for the pedestrians. Since the wide spread of the eco cars or hybrid cars do not make the engine sounds, it is considered that the rapid approaches to the pedestrians have been focused recently. Therefore, the research suggested the effective artificial sounds in order to alert for the pedestrians.

Secondly, the paper [4] indicated the map-matching approaches for the traffic accident prevention system between the vehicles and the pedestrians. In this research, the V2X (vehicle-to-everything) alert communication system is proposed by the GPS sensors on the smartphones and vehicles, and they also proposed the original map-matching algorithm notifies the status of the upcoming vehicles. The authors mainly focus on the visually handicapped persons from the rapid approaching vehicles, and they evaluated the proposed methods by the computational simulations in the paper.

Moreover, some papers discuss the current wireless network difficulties on the vehicles, and they newly suggest the effective network system for the V2X networks. For example, the research [5] introduced the Bluetooth networks for the near distance V2V (vehicle-to-vehicle) communication systems, and it evaluated the approaching vehicles by the Bluetooth beacons in the proposed alert system. Moreover, the paper [6] evaluated the 5.8 GHz DSRC system compared with the IEEE802.11a/b/g communication system in the V2V networks, and the future results also mentioned the subjects of the future IEEE802.11p networks. At last, the paper [7] proposed the W-CDMA packet control system for the V2V networks. In the research, the vehicle's group communication was focused, and it indicated the effectiveness of the proposed W-CDMA packet scheduling among the vehicles compared with the original W-CDMA packet controls.

However, as mentioned in the previous chapter, it has been focused the traffic accidents between the bicycles and pedestrians, and so it is necessary to considered another approach to the traffic accident prevention system. Especially, the smartphone's texting while walking and riding bicycles are considered as the important subjects in recent with the wide spread with the smartphone.

3 Proposed Methods

The assumed system architectures are shown in Fig. 1. As previously mentioned, this research especially focus on the alert for the pedestrians who are texting while walking and who need to avoid from the texting while riding bicycles. Besides, this research is also assumed to use the proposed system in local areas where the LTE or GSM services are unavailable.

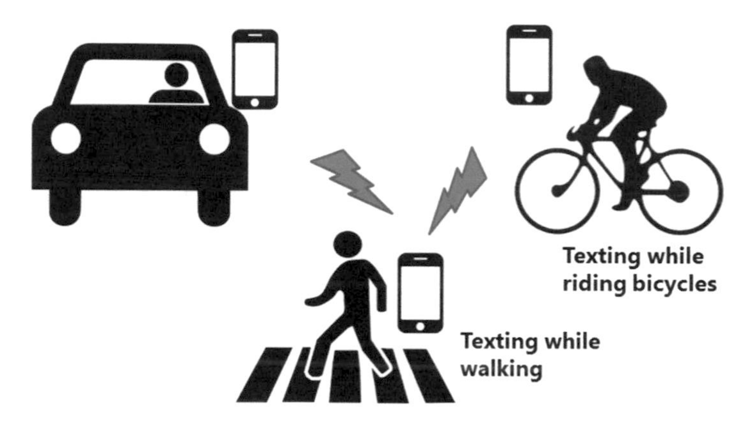

Fig. 1. The proposed system architectures. The smartphone in the pedestrian observes the wireless conditions in order to detect the rapid approaches, and the gyro sensor also used for the modification of the operational status such as texting while walking.

As shown in Fig. 1, the smartphone in the pedestrian continuously observes the wireless signals from other devices in the bicycles or vehicles, and the observed RSSI levels are used for the calculation of the alert detections. In details, the SSID, the radio band, the RSSI are periodically observed by the smartphone. The smartphone also measured the sensors on the smartphone such as the gyro sensor and the velocity meter because of the sensing the pedestrian's conditions. If the pedestrian is decided as the condition of texting while walking, the modified values for the detection process are added for the results of the proposed calculations based on the Markov Chain algorithm [8]. The followings are the proposed alert calculation process for the detection of the rapid approaches.

$$X = \sum_{i=0}^{n} \alpha_i X_i + \delta \tag{1}$$

Here, X_i is the observed value when time sequence is t_i, α_i is the weight value that is satisfied with $\alpha_{1+}\alpha_{2+\ldots+}\alpha_i = 1$, and δ is the adjustment values from the smartphone's sensors such as the gyro sensors and the accelerometer. Then, X_{th} which is the threshold value for the alert is previously set, and the alert is proceeded if the $X > X_{th}$ is satisfied.

Also, in this calculation, the values of the gyro sensors g_i and the accelerometer v_i is calculated for the sum of the pedestrian's movements with *x, y,* and *z* derections.

$$g_i = \sqrt{g_{ix}^2 + g_{iy}^2 + g_{iz}^2} \tag{2}$$

$$v_i = \sqrt{v_{ix}^2 + v_{iy}^2 + v_{iz}^2}$$

Then, if $(g_i, v_i) > (g_{th}, v_{th})$ is satisfied that g_{th} and v_{th} are the threshold values of the gyro and velocity meters previously set, δ in the formula (1) become the larger number in the proposed methods.

4 Implementations

The prototype system is implemented for the evaluations of the proposed methods. The prototype system is made by the Nexus5 that is Android OS ver. 7.0. Then, the system is implemented by the Android Studio 2.1.2, Java, and API 23 of Android OS. Also, the IEEE802.11a/b/g/n from other device is observed by the smartphone. Figure 2 shows the alert message in the prototype system.

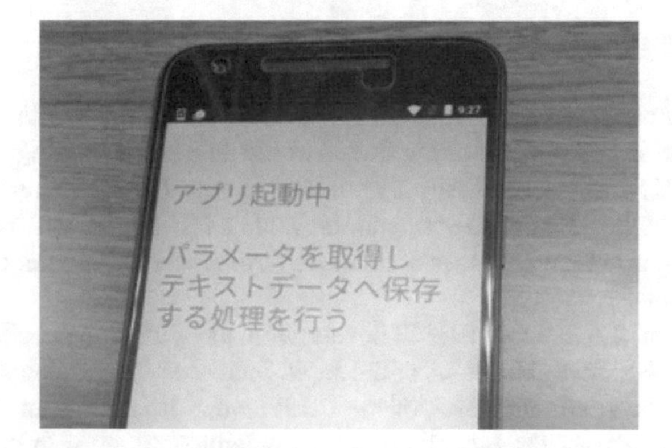

Fig. 2. The screen capture of the implemented system. After starting the application, the RSSI levels are continuously measured, and also the gyro and velocity meters are used for the proposed calculation based on the Markov Chain algorithm. If the calculated value is exceed the previously installed threshold value, this alert message is shown in the smartphone.

Also, Fig. 3 shows the flow chart of the implementations. After starting the application, the smartphone proceed the WiFi scans with the WiFiManager [9] of the android API, and the SSID, the RSSI, and the radio frequency are saved into text files periodically. Next, the observed values from the gyro sensors and the accelerometers are calculated for the modified values, and the alert is proceeded if the calculated value is exceeded than the threshold value.

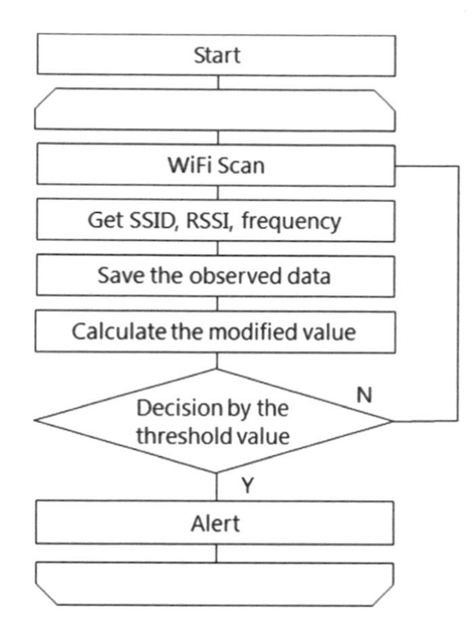

Fig. 3. Flow chart of the implementations by Android APIs. In the implementations, the WiFiManager Class is used for the scanning of the radio signals from others.

5 Experiments

The field experiments are confirmed for the evaluations of the prototype system. The experiments held in the campus of the Fukuoka Institute of Technology, Japan, and the change of the RSSI levels are firstly experimented with the movements of the pedestrian (5 km/h), the bicycle (15 km/h), and the vehicle (40 km/h). One node is statically set in the starting point, and another node moves from the 100 m point to the start point in the experiments.

The results of the observed RSSI levels with the different speeds are shown in Fig. 4. The figure shows that there is the rapid increases in the every speed, and the 40 km/h shows the shape increases of the RSSI levels. It is considered to distinguish the dangerous vehicle or the bicycle, and the implementations of the proposed calculations are now working for the future studies. However, the experiments also indicated that the scanning speed by the Wifi Manager API is not enough for the 40 km/h, and some measured points were missing from the figures. Because the period of the WiFi scanning seems to be depending on the number of WiFi nodes, it might be necessary to consider the alternative implementation methods.

Secondly, the gyro sensors and accelerometers are observed by compared with the pedestrians and the pedestrians who are texting while walking in the experiments. In the experiments, the five persons walk in two ways at the same experimental fields, and one scenario is just walking and another is walking with using the twitter application. Figure 5 shows the results of the accelerometer by walking and texting while walking.

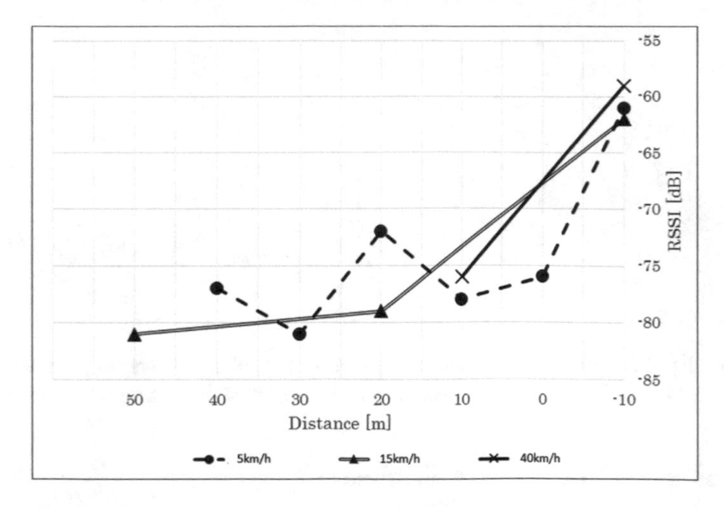

Fig. 4. Results of the RSSI levels in the experiments.

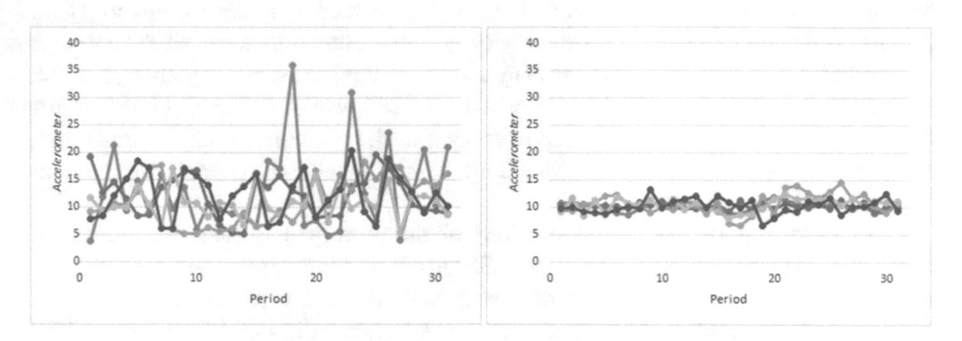

Fig. 5. Results of the accelerometers in the experiments. The left figure shows the observed values of the accelerometer with walking, and the right is the values with texting while walking.

Figure 5 shows the significant differences between just walking and texting while walking. The results indicated that the dangerous condition can be detect by the observations of the accelerometers, and the research of the appropriate threshold values are considered for the future studies.

Next, Fig. 5 shows the results of the gyro meters by walking and texting while walking.

There is the significant differences between just walking and texting while walking according to Fig. 6, and the results indicated that the dangerous condition can be detect by the observations of the accelerometers. There results suggests that the proposed methods is considered to effective for the detection of the pedestrian's conditions, and the future studies include the proper threshold value of the both sensors and the implementations of the proposed calculations.

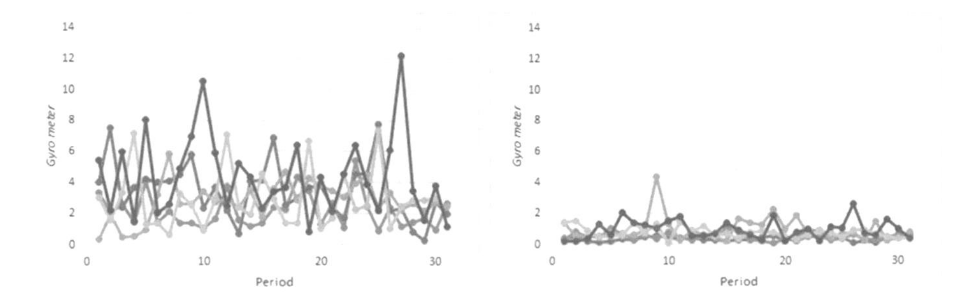

Fig. 6. Results of the gyro meters in the experiments. The left figure shows the observed values of the gyro meters with walking, and the right is the values with texting while walking.

6 Conclusions and Future Studies

In the recent traffic accidents, it has been focused on the pedestrians and the bicycles as well as the automobiles. Especially, it is widely considered that the texting while walking and riding bicycles is extremely dangerous. Thus, this research proposed the mobile traffic accident prevention system with observing the radio signals and various sensors on the smartphone. In the methods, the RSSI levels of IEEE802.11a/b/g/n from others are firstly observed, and the sensors such as the gyro sensor on the smartphone are secondly applied for the modifications of the detection process based on the Markov Chain algorithm.

This paper reports the implementations of the prototype system, and the experiments are held for the evaluations of the RSSI levels and the sensors on the smartphone in order to the effectiveness of the proposed methods. The experimental results indicated the effectiveness of the proposed methods because the rapid increases of RSSI levels are observed from the field experiments of the rapid approaches. Also, the experiments of the gyro sensors and the accelerometers show the significant differences between walking with and without texting on the smartphone.

For the future studies, we are now planning to evaluate the threshold values of the gyro sensors and the accelerometers by the field experiments, and the additional implementations of the alert process in the prototype system.

References

1. The National Police Agency in Japan: The occurrence of a traffic accident in 2015. https://www.npa.go.jp/english/index.html
2. Takeuchi, S., Uchida, N., Shibata, Y.: Proposal of traffic accident prevention system with chronological variations of wireless signals and sensors on mobile nodes, innovative mobile and Internet services in ubiquitous computing. In: Proceedings of the 11th International Conference on Innovative Mobile and Internet Services in Ubiquitous Computing (IMIS-2017), pp. 68–72 (2017)
3. Adachi, Y., Umezu, T., Yamaguchi, H., Higashino, T.: V2X content delivery scheduling for efficient vehicle information sharing. IPSJ SIG Technical Report, pp. 1–8 (2014)

4. Sakamoto, I., Hohuzu. H., Kozjima, R., Matumura, H., Takeuchi, T., Hasegawa, T., Tanaka, Y: Fundamental study for inter-vehicle communication using smartphone. National Traffic Safety and Environment Laboratory Forum 2013, pp. 127–128 (2013)
5. Hamao, K.: Fukushima Prefecture High Tech Plaza, strategic information and communication research promotion business: a research and development of inter-vehicle communications which exchanges road condition sensing and local information with Smartphone: a research and development of inter-vehicle communications by interaction with others using Smartphone's Bluetooth. IEICE Technical Report, ITS 112(72), pp. 19–24 (2012)
6. Kato, S., Minobe, N., Tsugawa, S.: Experimental report of inter-vehicle communications systems: a comparative and consideration of 5.8 GHz DSRC, wireless LAN, cellular phone. IEICE Technical Report, ITS 103(242), pp. 23–28 (2012)
7. Aizawa, T., Shigeno, H., Yashiro, T., Matushita, Y.: A vehicle grouping method using W—CDMA for road to vehicle and vehicle to vehicle communication. IEICE Technical Report, ITS 2000(83(2000-ITS-002)), pp. 67–72 (2000)
8. Serfozo, R.: Basics of Applied Stochastic Processes. Springer Science & Business Media, Berlin (2009)
9. Google Developers: WiFiManager. https://developer.android.com/reference/android/net/wifi/WifiManager

The 7th International Workshop on Advances in Data Engineering and Mobile Computing (DEMoC-2018)

A System to Select Reception Channel by Machine Learning in Hybrid Broadcasting Environments

Tomoki Yoshihisa[1(✉)], Yusuke Gotoh[2], and Akimitsu Kanzaki[3]

[1] Cybermedia Center, Osaka University, Mihogaoka 5-1, Ibaraki, Osaka 567-0047, Japan
yoshihisa@cmc.osaka-u.ac.jp

[2] Graduate School of Natural Science and Technology, Okayama University, Tsushima-naka, Kita-ku, Okayama 700-8530, Japan

[3] Institute of Science and Technology, Academic Assembly, Shimane University, Nishikawatsu-cho 1060, Matsue, Shimane 690-8504, Japan

Abstract. Due to the recent prevalence of the Internet, some TV broadcasting services deliver videos using both electric wave broadcasting systems and the Internet (hybrid broadcasting environments). Video players encounter playback interruptions when they cannot receive a part of video data (video data segment) until the time to play it. The probability to encounter playback interruptions can be reduced by receiving video data segments earlier. However, it is difficult for video players to find from which reception channel (broadcasting system or the Internet) they can receive video data segments earlier since the time required for receiving them depends on various factors such as broadcasting schedules, the number of receiving video players, and so on. To find appropriate reception channels for reducing playback interruptions, we propose a system to select reception channel by machine learning.

1 Introduction

Due to the recent prevalence of the Internet, some TV broadcasting services deliver videos using both electric wave broadcasting systems and the Internet. Video delivery in such hybrid broadcasting environments in that video players (clients) receive video data from hybrid communication channels of broadcasting channels and the Internet channel is one of the hot topics in video delivery research fields and various schemes have been proposed [1–10]. These services often adopt streaming delivery techniques to deliver videos. In video streaming delivery, video data are divided into some video data segments and the server machine for video streaming delivery transmits them to the clients. The clients can play a part of whole video contained in each video data segment. In cases that the clients cannot receive a video data segment until the time to play it, a video playback interruption occurs. To reduce the probability to occur video playback interruptions, existing schemes try to make the clients receive video data segments earlier.

L. Barolli et al. (Eds.): NBiS 2018, LNDECT 22, pp. 833–840, 2019.
https://doi.org/10.1007/978-3-319-98530-5_74

This is a very difficult challenge in hybrid broadcasting environments because the times required for receiving video data segments (reception times) depends on various factors. The reception times for broadcasting systems (broadcasting channel) are shorter as video data segments are broadcast sooner. Smaller numbers of clients further reduce the reception times for the Internet (Internet channel) since the communication speed between the server and each client increases. In cases that clients request and receive video data segments from the Internet, though the reception times for broadcasting systems are earlier, redundant communication traffics arise and the reception times for the Internet lengthen. Therefore, the probability to occur video playback interruptions is reduced by receiving video data segments from the channel from that each client can receive them earlier. Previous researches assume that the clients know the communication speeds of the broadcasting system and of the server via the Internet. In existing schemes, the clients calculate the reception times for the broadcasting system and for the Internet, and select the channel to receive the segment that gives a shorter time.

However, in actual situations, the communication speeds dynamically change and the reception times depends on various factors as we explained above. Therefore, it is difficult to select appropriate reception channels for reducing playback interruptions in actual situations. One of the techniques for decision supports is machine learning. Machine learning has a large possibility to give more appropriate decisions when the appropriateness depends on various factors. Video playback interruptions can be reduced by learning appropriate reception channels under some input items that influences reception times of video data segments.

Hence, in this paper, we propose a system to select reception channel by machine learning in hybrid broadcasting environments. We adopt binary classification technique by logistic regression for machine learning since selecting the broadcasting/Internet channel or not is a binary classification problem. In our proposed system, the clients select the reception channels to receive the next video data segments every they finish receiving a video data segment. We give the time that the target segments are scheduled to be broadcast and the number of the clients that are receiving video data segments from the Internet as the input items. We develop a computer program to learn the appropriate reception channels that gives shorter reception times and investigate the accuracy of the learning results. The paper is organized as follows: we introduce some related work in Sect. 2. We explain our assumed infrastructure and our proposed system in Sects. 3 and 4. Our experimental results for the accuracy of machine learning are presented and discussed in Sect. 5. Finally, we conclude the paper in Sect. 6.

2 Related Work

Some schemes for hybrid broadcasting environments have been proposed.

Similar to our research, some researches propose the systems that use broadcasting systems and the Internet [1–4]. These systems deliver various contents related to the video via the Internet. The video data are broadcast via the broadcasting systems. In our proposed system, different from these systems, the video data are delivered by using both of the broadcasting systems and the Internet.

One of the attractive features of the next generation cellular networks (5G) is multicasting to smartphones. Recently, many researches in the video streaming delivery field focus on the feature and propose schemes to reduce the amount of communication traffic by exploiting multicasting [5–7]. In these schemes, the server uses the same network for multicasting and unicasting. Different from these schemes, our assumed infrastructure has different networks for broadcasting and unicasting such as terrestrial broadcasting systems and LTE networks.

There are some other researches related to hybrid systems. A system to deliver 3-D videos over the environments is proposed in [8]. A hybrid system of CDN (Contents Delivery Networks) and P2P (Peer-to-Peer) is proposed in [9]. Our research is different from these systems in that our assumed infrastructure has different networks for broadcasting and unicasting.

We have proposed a data piece (segments) elimination technique for hybrid broadcasting environments in [10]. The objective of the paper is different from this. We investigate the accuracy of machine learning in the paper.

3 Assumed Infrastructure

In this section, we explain our assumed infrastructure. On the infrastructure, we propose a system to select reception channel by machine learning.

3.1 Hybrid Broadcasting Environment

Figure 1 shows our assumed hybrid broadcasting environment. The clients are video players such as smartphones, laptops, and so on. They play video from the beginning to the end continuously. They connect to the Internet and can receive data from the server via the Internet. Also, they can receive data broadcast from the electric wave broadcasting systems. The broadcasting equipment transmit the electric wave and are managed by the servers. The servers have video data. In the hybrid broadcasting environment, the clients can receive video data both from the broadcasting system and the Internet.

One of the main merits of such hybrid broadcasting environments is that the Internet compensates a demerit of pure broadcasting systems that they do not have uplinks from the clients. The servers can transmit data to all the clients concurrently by using the broadcasting systems.

3.2 Broadcasting Schedule

Video data are divided into N segments, S_1, …, S_N. For example, suppose a news video of that duration is 10 min. When $N = 600$, the duration for each segment is 1 s. For investigating the effectiveness of machine learning, in this paper, we assume that the infrastructure adopts static broadcasting schedule. In the static broadcasting schedule, the server predetermines the broadcasting schedule and cyclically broadcasts the segments in the order of the schedule. We do not fix the scheduling algorithm in this paper since the clients can get the time when each segment is to be broadcast by asking it to the server.

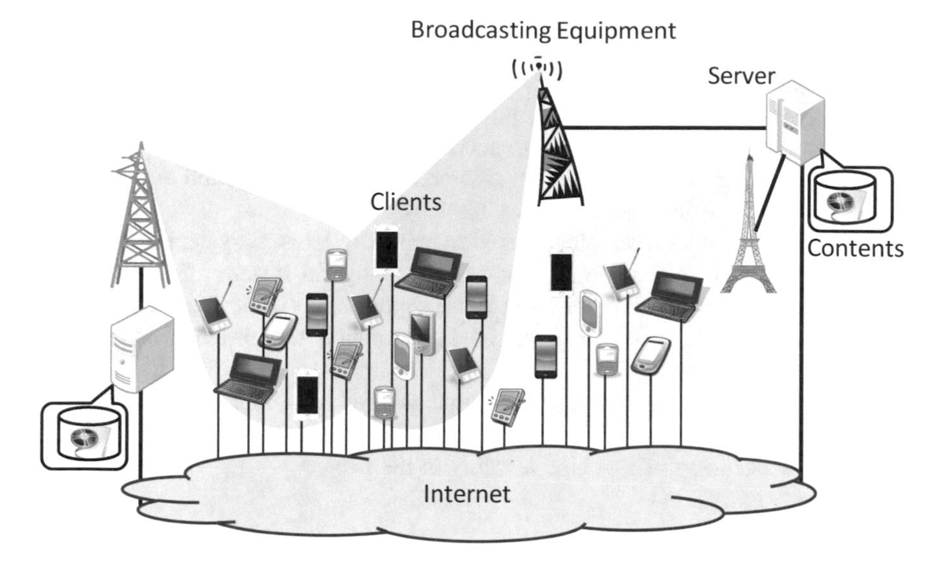

Fig. 1. Our assumed hybrid broadcasting environment

3.3 Communication Schedule

The clients can request the segments to the server and the server sends the requested segments to each client. The time needed to send each segment lengthens if the server sends the segments in parallel since the communication speeds for each segment lowers. Therefore, the server sends the requested segments in accordance with the FIFO (First In First Out) rule. That is, the server manages a request queue and sends the segment that is requested at the oldest time.

4 Proposed System

In this section, we explain our proposed system.

4.1 Channel Selection

The clients receive the segments of a video data sequentially from the first one since they play the video from the beginning. The probability to encounter video playback interruptions for the clients is further reduced by receiving each segment earlier. In cases that clients request and receive video data segments from the Internet, though the reception times for broadcasting systems are earlier, redundant communication traffics arise and the reception times for the Internet lengthen. Therefore, in our proposed system, the clients select the reception channels to receive the next video data segments (target segments) every they finish receiving a video data segment.

To select the appropriate reception channels, our proposed system uses machine learning since the appropriateness depends on various factors. We adopt binary classification technique by logistic regression for machine learning since selecting the broadcasting/Internet channel or not is a binary classification problem.

4.2 Logistic Regression

Logistic regression is a machine learning technique widely used to solve binary classification problems. In the logistic regression, the probability that a situation represented by an input vector X is classified into 1 (the other one is 0), $p(X)$ is given by the following formula. A and B are parameters for the logistic regression which have the same dimension as X.

$$p(X) = \frac{1}{1 + \exp(-(A + B^T X))}$$

By learning more appropriate values of A and B, the formula gives more accurate classifications.

4.3 How to Lean Appropriate Channel

A larger number of input items for machine learning gives a longer convergence time. Therefore, our proposed system gives just two items as input items for the logistic regression. One is for representing the situation of the broadcasting system. The clients can get the time when each segment is to be broadcast by asking it to the server. Therefore, we use the time until the target segment is broadcast from the current time as one of the input items. The other one is for representing the situation of the Internet. The server grasps the number of the clients that are receiving their target segments from the Internet since it sends them. The clients can get the number of the clients. Therefore, we use the number of the clients that are receiving segments from the Internet as one of the input items. Thus, the dimension of the input item X becomes 2 in our proposed system. We set the output item so that 0 means that the client receives the target segment from the broadcasting system and 1 means that from the Internet.

5 Experimental Evaluation

In this section, we present our experimental results for the investigation of the accuracy of machine learning.

5.1 Experimental Setup

We developed a computer program to learn the appropriate reception channels that gives shorter reception times using TensorFlow [11]. Figure 2 shows the data flow network for the program. The input module inputs the training data and the test data to the network. The readout module transforms the data to the tensors to use them in TensorFlow. The optimize module optimizes the parameters A and B.

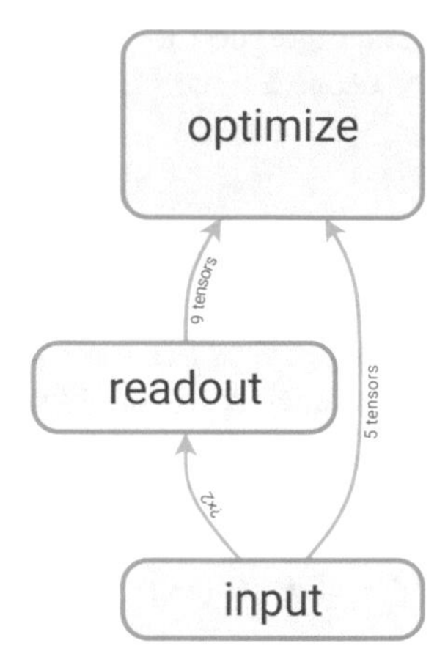

Fig. 2. The data flow network for logistic regression in our proposed system accuracy of machine learning.

To get the training data and the test data, we simulated the behaviors of the clients. In the simulator, the bandwidths of the broadcasting system and the Internet are fixed. The ratio of the bandwidths (bandwidth rate) is the bandwidth of the Internet divided by that of the broadcasting system. For example, the bandwidth rate is 0.1 when the bandwidth of the Internet is 2 Mbps (LTE) and that of the broadcasting system is 20 Mbps. We get the training data and the test data changing the bandwidth rates to investigate the accuracies under some different situations. In the simulator, the clients start playing the video at random timings. Also the number of the clients that are receiving the segments is randomly given. The maximum number of the clients is 10 or 100. We got 10000 training data and 10000 test data by running the simulator.

5.2 Experimental Results

We ran our developed machine learning program and investigated the accuracy of the results. The accuracy is the ratio that the output of the logistic regressions is equal to the correct data given by the test data. The learning step is 10000. The result is shown in Fig. 3.

In the figure, the horizontal axis is the number of the steps and the vertical axis is the accuracy. We show the results when the bandwidth rate is 0.1 or 0.5 though we have run the program under some bandwidth rates to make the graph be easily seen. When the maximum number of the clients is 10, the accuracies increase as the step proceeds under both bandwidth rates. This is because the program can find more appropriate parameters as the step proceeds. On the other hand, when the maximum number of the clients is 100, the accuracies do not increase under both bandwidth rates because the training data is insufficient. Finally, we got the accuracy of 0.99 when the bandwidth rate is 0.5 and the maximum number of the clients is 10. In this case, therefore, the clients can select the appropriate reception channels with high accuracy. On the other hand, when the bandwidth rate is 1.0, the accuracy does not improve.

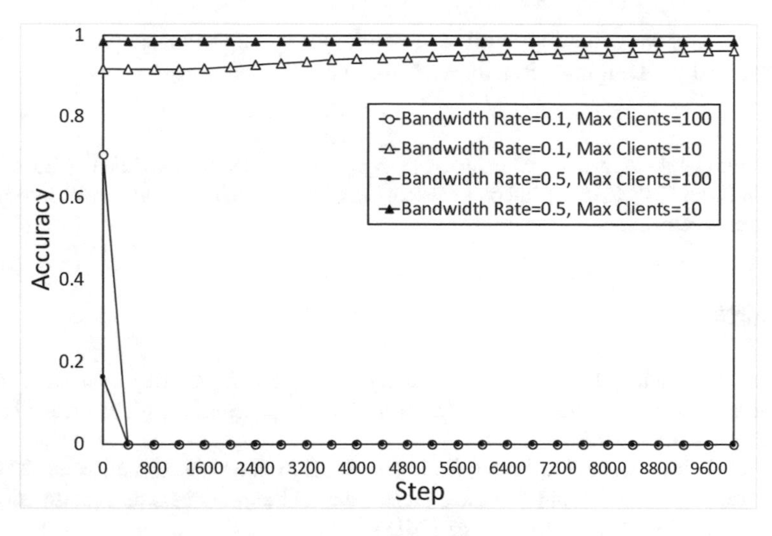

Fig. 3. The accuracy and the step of learning

This is because the broadcasting channel gives shorter reception times in many cases and the program could not learn the more appropriate parameters.

5.3 Discussion

As shown in the result, our proposed system can give the appropriate reception channel when the bandwidth rate is small and the parameters for the machine learning are correctly learned by giving the following two input items:

- the number of the clients that are receiving segments from the Internet
- the time until the target segment is broadcast from the current time

In the experiments, the bandwidths of the broadcasting system and the Internet are fixed. In cases that they change dynamically, more input items will be needed to learn correctly.

6 Conclusion

In this paper, we proposed a system to select reception channel by machine learning in hybrid broadcasting environments. In our proposed system, the clients select the reception channels to receive the next video data segments every they finish receiving a video data segment. We adopted binary classification technique by logistic regression for machine learning. We gave the time that the target segments are scheduled to be broadcast and the number of the clients that are receiving video data segments from the Internet as the input items for machine learning. We investigated the accuracy of the learning results and confirmed that our proposed system gives appropriate reception channels.

In the future, we will adopt on-line learning to select appropriate reception channels for dynamically changing situations. Also, we are planning to develop an actual system.

Acknowledgments. A part of this work was supported by JSPS KAKENHI (Grant Number JP15H02702 and JP18K11316) and by Research Grant of Kayamori Foundation of Informational Science Advancement.

References

1. Agirre, M., Florez, J., Lafuente, A., Tamayo, I., Zorrilla, M.: Deployment of a hybrid broadcast-internet multi-device service for a Live TV programme. IEEE Trans. Broadcast. **64**(1), 153–163 (2018)
2. Boronat, F., Montagud, M., Marfil, D., Luzon, C.: Hybrid broadcast/broadband TV services and media synchronization: demands, preferences and expectations of Spanish consumers. IEEE Trans. Broadcast. **64**(1), 52–69 (2018)
3. Boronat, F., Marfil, D., Montagud, M., Pastor, J.: HbbTV-compliant platform for hybrid media delivery and synchronization on single- and multi-device scenarios. IEEE Trans. Broadcast. Early Access, 26 pp (2017)
4. Christodoulou, L., Abdul-Hameed, O., Kondoz, A.-M.: Toward an LTE hybrid unicast broadcast content delivery framework. IEEE Trans. Broadcast. **63**(4), 656–672 (2017)
5. Araniti, G., Scopelliti, P., Muntean, G.-M., Lera, A.: A hybrid unicast-multicast network selection for video deliveries in dense heterogeneous network environments. IEEE Trans. Broadcast. Early Access. (2018). 11 pages
6. Guo, J., Gong, X., Liang, J., Wang, W., Que, X.: An optimized hybrid unicast/multicast adaptive video streaming scheme over MBMS-enabled wireless networks. IEEE Trans. Broadcast. Early Access. (2018). 12 pages
7. Tian, C., Sun, J., Wu, W., Luo, Y.: Optimal bandwidth allocation for hybrid video-on-demand streaming with a distributed max flow algorithm. ACM J. Comput. Netw. **91**(C), 483–494 (2015)
8. Cruz, L.-A.S., Cordina, M., Debono, C.-J., Assuncao, P.-A.A.: Quality monitor for 3-D video over hybrid broadcast networks. IEEE Trans. Broadcast. **62**(4), 785–799 (2016)
9. Zhang, Y., Gao, C., Guo, Y., Bian, K., Jin, X., Yang, Z., Song, L., Cheng, J., Tuo, H., Li, X.-M.: Proactive video push for optimizing bandwidth consumption in hybrid CDN-P2P VoD systems. In: Proceedings of INFOCOM 2018 (2018). 9 pages
10. Yoshihisa, T.: Data piece elimination technique for interruption time reduction on hybrid broadcasting environments. In: Proceedings of IEEE Pacific Rim Conference Communications, Computers and Signal Processing (PACRIM 2017) (2017). 6 pages
11. TensorFlow: https://www.tensorflow.org/

A Meeting Log Structuring System Using Wearable Sensors

Ayumi Ohnishi[1], Kazuya Murao[2], Tsutomu Terada[1,3](✉), and Masahiko Tsukamoto[1]

[1] Graduate School of Engineering, Kobe University, Kobe, Japan
ayumi@stu.kobe-u.ac.jp, tuka@kobe-u.ac.jp
[2] College of Information Science and Engineering, Ritsumeikan University, Kusatsu, Japan
murao@cs.ritsumei.ac.jp
[3] PREST, JST, Tokyo, Japan
tsutomu@eedept.kobe-u.ac.jp

Abstract. We propose a system that structures a meeting log by detecting and tagging the participants' actions in the meeting using acceleration sensors. The proposed system detects head movement such as nodding of each participant or motion during utterances by using acceleration sensors attached to the heads of all participants in a meeting. In addition, we developed a Meeting Review Tree, which is an application that recognizes a meeting participants' utterances and three kinds of actions using acceleration and angular velocity sensors and tags them to recorded movies. In the proposed system, the structure of the meeting is hierarchized into three layers and tagged contexts as follows: The first layer represents the transition of the reporter during the meeting, the second layer represents changes in information of speakers in the report, and the third layer represents motions such as nodding. As a result of the evaluation experiment, the recognition accuracy of the stratified first layer was 57.0% and that of the second layer was 61.0%.

1 Introduction

For people belonging to organizations or groups, meetings are important for communicating decisions, determining project policies, creating new ideas, and sharing information. At a meeting, the contents of discussions and opinions issued are important information, so taking minutes is common as it enables participants to refer to them at a later date or share information with absentees. However, it takes considerable labor and time to collect the important parts from all utterances and summarize the participants' attitudes toward them. Several systems have been proposed so far that automatically generate minutes, for example, VoiceGraphy [1] and AmiVoice Minutes Preparation Support System [2]. However, these systems need to record all conversations in the meeting and generate them to letters by speech recognition. To look back to important parts

L. Barolli et al. (Eds.): NBiS 2018, LNDECT 22, pp. 841–852, 2019.
https://doi.org/10.1007/978-3-319-98530-5_75

in the meeting such as remarks during which many agreements were obtained or remarks that many argued, it is necessary to review the generated minutes ourselves, thus taking a lot of time and effort. To extract important statements in a conference, it is necessary to recognize the agreement or counterargument proposed by another speaker, to recognize situations in a conversation with multiple people, and to judge situations of the whole meeting. Much research has been dedicated to detecting only utterance and motions such as nodding in conversation [3,4], but recognition of conversations involving multiple people has not yet been addressed. Recognizing a conversational situation requires detecting the followings actions: nodding and its target and degree, tilting a head to mean denial or doubt, forward posture, indicating not hearing a talk, and so on. In this research, we propose a meeting logging system that recognizes situations of the whole meeting by detecting the behavior of multiple people in the conference by using acceleration sensors attached to the heads of all conference participants and tags them. The proposed system is assumed to be used in an educational setting and for sudden meetings, we installed only small wearable sensors attached to participants' heads so that it does not depend on location.

The structure of this paper is as follows: We introduce related research in Sect. 2 and the proposed system in Sect. 3. We conduct an evaluation experiment in Sect. 4, describe the application using the proposed system in Sect. 5, and finally summarize this paper in Sect. 6.

2 Related Research

Many studies on recognition of user behavior in multiple conversations have been conducted. Morency et al. have conducted research to detect nodding by image recognition and to enable robots and others to recognize nonverbal behavior [3]. In this research, it is necessary to record a video with one or more cameras for each person to recognize head movement. Wohler et al. recognized behaviors of nodding, head swinging, head tilting, and attention during a conversation between two people using motion sensors [4]. However, in this system, only the behaviors of individuals were recognized, and conversational situations were not verified. Kawahara proposed the Smart Poster Board, which detects gaze and response from depth information, voice information, image information during poster sessions using Kinect and a microphone array, and predicts utterances [5]. However, this system requires installing a microphone array, Kinect, and six cameras, making it cumbersome when the location of the meeting changes. Sumi et al. proposed the IMADE environment [6] to collect various pieces of information during the conversation such as a subject's movement, gaze, voice, and biometric data. However, it is costly to have the required sensors attached to all meeting participants. Because the conference log acquisition system proposed in this paper is assumed to be used in an educational setting and for sudden meetings, we installed only small wearable sensors attached to participants' heads so that it does not depend on location. An operational judgment of this proposed system is performed based on only the values of the acceleration sensors.

There is also much research on the analysis of the conversational structure such as of combining information of several individuals. Kumano et al. analyzed the relationships among four people, including who smiled and who laughed, by recognizing smiles and smiles from video recorded at the meeting of four people [7]. However, in this research, the motions to be recognized were only smiles, and performing highly accurate recognition necessitates installing a camera in front of each participant, entailing time and cost. Otsuka et al. focused on the line of sight based on the tracking of the head of a person in the image during a dialog involving multiple people, and modeled the conversation structure [8]. However, they used only the movement of the line of sight during the conversation when analyzing the conversation structure and did not pay attention to motions other than eye movement such as nodding. It is difficult to estimate the situation in real time from a single action when creating meeting minutes during which conversations with various degrees of importance are taking place. Therefore, it is necessary to recognize multiple kinds of actions to analyze the relationship between action and conversation in estimating the detailed conversation structure.

In the meeting log structured system proposed in this research, by analyzing the action information obtained by recognizing the utterance state of the individual and the head movement from the acceleration sensors and the angular velocity sensors in combination for a plurality of people, speakers and important parts of the discussion are tagged.

3 Proposed System

This section describes a system that adds tags to conference recording voices and recorded videos. The proposed system uses an acceleration–angular velocity sensor attached to the head of each participant to recognize the movement of the head and tags a recorded video and voice according to recognition results.

3.1 Assumed Environment

Let us assume a meeting of a form in which each participant reports, and we define such a meeting as a report-type meeting. In a report-type meeting, we assume that participants are multiple presenters and facilitators, secretaries, and advisors, but we assume that at least presenters always participate. The most important part of the report-type meeting is that each participant speaks during reports. An approach to recognize speech and a speaker by speech recognition is conceivable, but it is necessary for one utterance to be of sufficient speech length to recognize the speaker with sufficient accuracy; a short comment is difficult to recognize in a confused situation such as a discussion. Therefore, in this study, using an acceleration–angular velocity sensor attached to each participants' head, head movement is detected and the utterance part is estimated. Because the response of other participants to an utterance is also important, we assumed two kinds of circumstances: one is agreement and the other is negation or doubt,

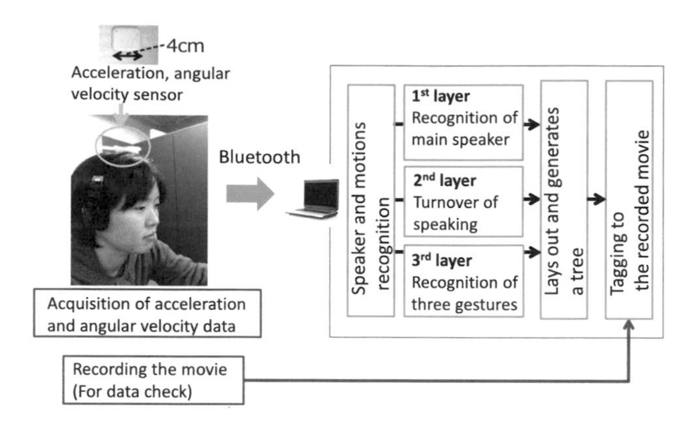

Fig. 1. System configuration

Table 1. Environment of the preliminary experiment

Date	10/2	10/8	10/17	10/25	10/31	11/07	11/14	11/25	12/19	1/10	1/17
Number of participants	5	6	7	7	7	5	5	7	6	3	6
Time [h:min]	2:00	1:50	1:50	1:30	1:30	2:00	1:40	1:50	1:50	2:00	1:20

and, as actions corresponding these situations, two kinds of actions—nodding behavior and head tilting motions—are detected. In addition, we also detect looking-around motion, because this motion is caused when meeting participants would like to check the state of the whole meeting.

3.2 System Configuration

Figure 1 shows the configuration of the proposed system. The proposed system sends data from three-axis wireless acceleration–angular velocity sensors attached to the participants' heads to each participant's PC, collects data, and performs utterance detection and head movement detection. At the same time, to create correct labels, videos from the front of the participant are recorded, and the main recorded video recorded by Meeting Recorder is tagged with detection results. The acceleration–angular velocity sensors used are model WAA-006 [9] manufactured by Wireless Technology Co. Ltd., and have a sampling frequency of 50 Hz. In this paper, we attach a sensor to a headset, but, in the future, the motion sensor will be assumed to be embedded within glasses or attached to the ear. The method of tagging the recorded video with the detected utterance and head movement is described as follows: As shown in the figure, because the whole picture is simply obtained by tagging the detection results and arranging them on a single time series, it is not possible to roughly understand the main flow of the meeting, making it difficult for system users to imagine the meeting situation. In addition, there are too many tags and it takes time to search for

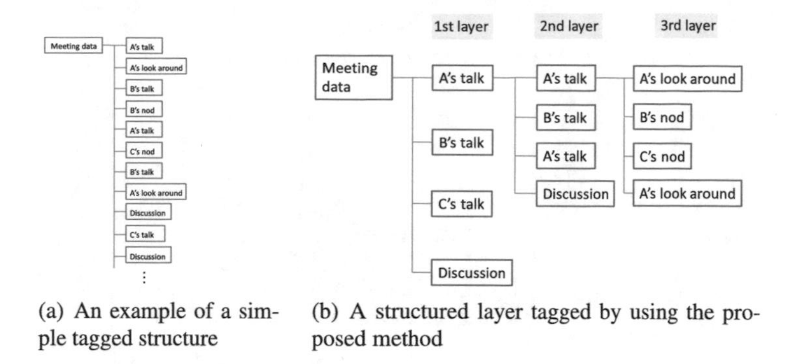

(a) An example of a simple tagged structure

(b) A structured layer tagged by using the proposed method

Fig. 2. Examples of how to tag a meeting structure

the parts users want to see from the recorded video. Therefore, to make it easier to find the scene, we lay out the structure of the meeting as shown in Fig. 2. As a first layer, we tagged the overall flow of the meeting as participant A's reporting time → participant B's reporting time → participant C's reporting time → discussion by everyone → ended. A second layer comprised tagged utterance details caused by other participants' reporting time, for example, A's utterance during the reporting section of A → B's utterance during the reporting section of A → discussion by all. As a third layer, head movement such as nodding of each participant's reactions to A's utterance is tagged. We consider that, by analyzing the third layer in detail, we can estimate the important parts of a discussion, the degree of interest of participants, and the role of participants and contributions to the discussion. In this research, advisors and facilitators are set manually after a meeting.

3.3 Proposed Method

We conducted preliminary experiments to collect data r to construct an algorithm to detect speaking and three types of head movements: nodding, tilting, and looking left and right. The dates of each experiment, the number of participants, and the execution time of the meeting are listed in Table 1.

3.3.1 Utterance Recognition Method

Figure 3 shows a waveform of a variance value of a composite value of an acceleration in a stationary state and an utterance state. The window size used for the calculation of variances was 200 ms. In stationary states, participants sat without doing anything, and no significant change in acceleration was seen. In contrast, in utterance states, subjects talk with gestures such as hand gestures, so the acceleration greatly changed over a long period of time during talking. By considering that the stationary state and the uttered state can be distinguished from the stability and change of the constant time in these waveforms, utterance

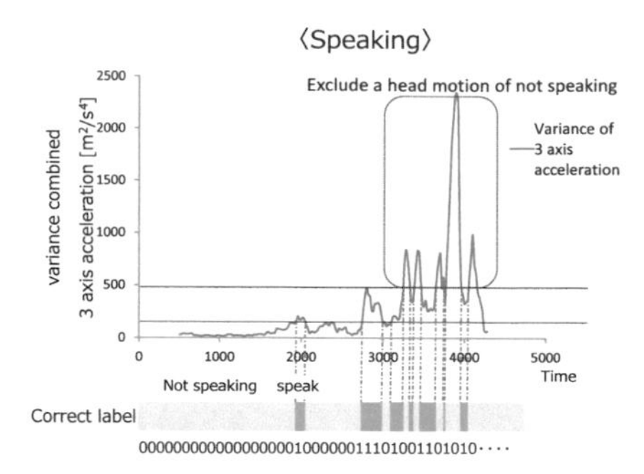

Fig. 3. Speaking recognition method from a variance value waveform of an acceleration

judgment was performed by using the following three kinds of feature amounts, and recognition rates were evaluated.

- Method 1 The utterance state if the variance value V_{acc} in the past 200 ms exceeds the threshold is determined as the utterance state, and other cases if the variance value falls below the threshold are determined as silence states. To exclude head movement other than talking such as nodding, an upper limit is set for the threshold value.
- Method 2 Variance values V_{ang} in milliseconds are used as the threshold value and speech judgment is performed as in Method 1. To exclude head movement other than speech, an upper limit is set for the threshold value.
- Method 3 Values of acceleration and angular velocity are combined and this value is used as a threshold value to make a speech determination. As in Methods 1 and 2, an upper limit is set for the threshold value to exclude head movement other than speaking.

These three methods were compared in evaluation experiments. Also, as a result of utterance judgment, we labeled the silence state as 0 and the utterance state as 1. Motion recognition is performed in parallel as a separate process from the utterance recognition.

Next, the system adds past 10 s values of the results of utterance judgements performed for each participant. The person whose average value of the addition result, L_{ave}, exceeds the threshold is the main speaker (a reporter) in the first layer of the report-type meeting and tags the beginning part of the utterance, and the time until another person exceeds the threshold is set as the speaker section of that person.

Recognition of the end of the speaker section in the case in which the addition result of the utterance labeling of the person judged as the speaker falls below the threshold was not made. This is because the minutes of the meeting were played from the beginning of the talking content and we considered that the system user decides when to finish playing the minutes. In addition, it was judged that the section during which the addition result of the speech labeling of all conference participants exceeds the threshold is a discussion. In this way, after detecting the flow of the first layer's rough meeting and tagging the start of the report of the participant, the section of the first layer is further divided as the second layer for each utterance. Similarly to speaker recognition, when the labeling result is 1 from each participant's utterance determination for each sample, it is counted as an utterance, and a person whose count number of utterance exceeds the threshold within the section is detected. By shortening the window size to be added to the labeling result, it is possible to detect more detailed changes. Therefore, by further hierarchizing the past 2 s, and by adding the values of 0 and 1 of the speech labeling result, the participant whose value exceeds the threshold can be identified as the speaker of the section exceeding the threshold.

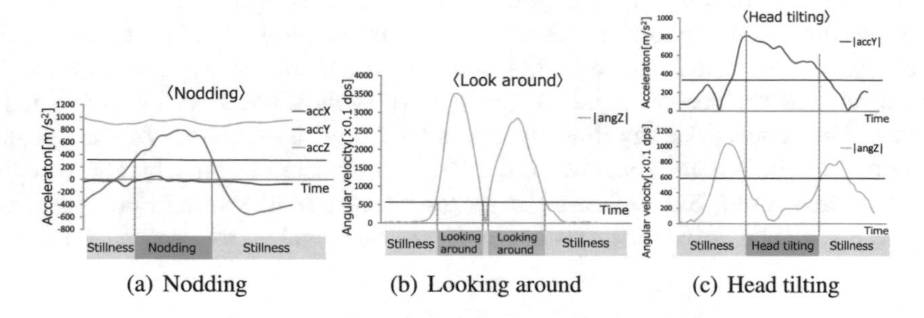

(a) Nodding (b) Looking around (c) Head tilting

Fig. 4. Motion recognition methods

3.3.2 Motion Recognition Method

In motion recognition, the system recognizes three motions—nodding, looking left and right, or tilting the head—along with the still state. In preliminary experiments, waveforms of acceleration–angular velocity as shown in Fig. 4 were seen in the meeting participants when nodding, looking left and right, and tilting the head motions were performed. Based on these waveforms, the conditions for detecting each operation can be determined.

4 Evaluation Experiment

4.1 Evaluation Environment

To evaluate the accuracy of utterance detection and head movement detection by using the proposed method, four people (A, B, C, and D) were seated around a desk, and A, B, and C speak in turn for about 3 min. After that, all members participated in the discussion. Subject D was the facilitator in this experiment. As in the preliminary experiment in Sect. 3.3.1, each subject wore a headset on the head with an acceleration–angular velocity sensor. From the center of the desk, we recorded a video with a 360-degree camera (Meeting Recorder [10]) to obtain the correct motion label. Utterance detection and head activity detection were performed by using the proposed system, and the recall, precision rate, and F value were calculated.

4.2 Evaluation Results and Discussion

4.2.1 Speaking Detection

The detection result of the utterance motion is described next. Table 2 lists the recall rate, precision rate, and F value of the utterance recognition results. These rates were judged from acceleration data only, angular velocity data only, and acceleration data and angular velocity data.

In the hierarchization using utterance recognition, when judging the main speaker of the first layer and the speaker of the second layer, it is preferable that the F value be high because the ratio of the detection result of the utterance within a certain time is used. Therefore, it is considered that it is effective to use the angular velocity data for judging utterance recognition.

Regarding the evaluation results of hierarchization, Table 3 presents the recall, precision, and F value of the speaker judgment for each section in the first layer and the second layer. According to Table 3, the F value of the judgment of the reporter of the first layer was 0.57, and the F value of the speaker judgment of the second layer was 0.61. To confirm the reason why the recognition rate decreased, Fig. 5 shows the judgment results of the first layer and the second layer. The horizontal axis in the figure represents time, and each color of

Table 2. Speaking recognition rate

Subject	Acceleration			Angular velocity			Acceleration + Angular velocity		
	Recall	Precision	F value	Recall	Precision	F value	Recall	Precision	F value
A	0.21	0.60	0.31	0.75	0.51	0.60	0.20	0.63	0.30
B	0.22	0.44	0.29	0.85	0.40	0.54	0.21	0.45	0.29
C	0.23	0.62	0.33	0.44	0.31	0.36	0.10	0.57	0.17
D	0.15	0.31	0.11	0.62	0.17	0.27	0.10	0.34	0.16

Table 3. Evaluation of recognition rate of speaker detection

Layer	Before adjusted delays			After adjusted delays		
	Recall	Precision	F value	Recall	Precision	F value
1	0.63	0.52	0.57	0.70	0.68	0.68
2	0.55	0.70	0.61	0.55	0.70	0.61

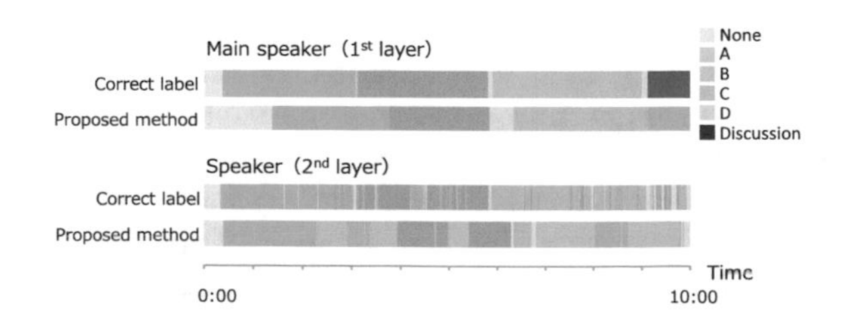

Fig. 5. Recognition results of speaker detection in each layer

the band represents the speaker at that time. According to Fig. 5, the recognition rate declined because of the delay in the detection of the first layer of reporter judgment. Also, at the end of the meeting, the four participants participated in the discussion simultaneously, but in the proposed method it was judged to be A's report. The proposed method judges the section during which the addition result of the utterance labeling of all meeting participants exceeds the threshold as a discussion. However, one person does not exceed the threshold value and is not in the utterance state; even if there are comments from more than one person, it is judged that one of the participants reports. Therefore, in this experiment, during the discussion, it is considered that the frequency of speech was high, and A, who had the most action during utterance, was judged as a reporter.

Regarding the judgment result of the second layer's fine utterance, the utterance of B increased from about 4 min after the correct answer data, whereas in the judgment result there was a judgment delay of about 30 s. Such a delay in judgment was also observed in the judgment of C at the elapsed time of 6 min of the correct data. Because speaker determination uses the labeling result for the past 10 s of utterance determination, accurate detection requires a certain amount of time after an utterance starts, so it is considered that a delay occurs in the determination result. When tagging the first layer, a correction can be made by subtracting this delay. However, in the second layer, regarding the beginning of the 10-min meeting, there is almost no delay in starting A's report, and the start of the utterance had been identical. The reason why this difference occurred is that the second layer set the threshold lower than that of the first layer, and A's report at the beginning of the meeting was not delayed, but the detections of B's and C's reports were delayed because they reported in the middle of the meeting. In other words, at the beginning of the meeting, only A started talking in a situation in which the variance of the combined value of the angular velocity used as the feature amount was small, so there was almost no delay in judgment and it was accurate. However, it is considered that judgment errors of reporter detection and a delay in judgment occurred when the variance of the composite value of the angular velocities of a plurality of people became a value close to the threshold value. For the second layer, the correct answer data have more frequent speaker turnovers than the recognition result. This is because the addition result of the utterance label for 2 s has been used for the judgment of the second layer, making the transition of fine utterance unrecognizable. However, if we had set the addition result of the utterance label to a shorter time, then nodding motion and looking left or right motion would have been erroneously detected.

Next, to correct the delay of judgment between the first layer and the second layer, times of 10 s for the first layer and 2 s for the second layer were subtracted from the judgment result and a reevaluation was performed. The deducted time is the window size used for each judgment. The revaluation results are given in Table 3. From Table 3, we can see that recall, precision, and F value were all higher in the first layer after the correction, but in the second layer, there was no change in recognition rate. From the judgment result, it is considered that the correction of the delay corresponding to the window size did not lead to an improvement in the recognition rate. There are three reasons for this: 1. The correct answer data is finer than the judgment result, 2. the speaker has changed more often, and 3. there are more erroneous judgments in the second layer than in the first layer.

4.2.2 Head Motion Recognition

Table 4 lists the recognition results of head movement. First, the motion of head tilting was not observed in the 10-min evaluation experiment. Because it seems that the contents of the meeting are related, it is necessary to conduct various kinds of meetings and to verify this result in the future. The recall rates of nodding and looking right or left in all subjects are high, but the precision rates are low. This is because the threshold value had been low; in fact, the number of detected results was greater than the number of actual motions. However, there is a tradeoff between the recall ratio and the precision ratio, so it is necessary to adjust the threshold according to the proposed use of a recognition result. In this research, owing to the nature of the minutes we targeted, we considered the case in which it may take time to browse the minutes, but, in cases in which minutes are viewed quickly by users, it is highly likely that important parts of

Table 4. Recognition rate of head motions

Motion	Subject A			Subject B			Subject C			Subject D		
	Recall	Precision	F value	Recall	Precision	F value	Recall	Precision	F value	Recall	Precision	F value
Nodding	0.97	0.15	0.26	1.00	0.30	0.46	0.99	0.26	0.41	0.91	0.28	0.43
Looking around	0.73	0.25	0.39	0.97	0.13	0.23	0.95	0.25	0.40	1.00	0.39	0.56
Head tilting	—	—	—	—	—	—	—	—	—	—	—	—

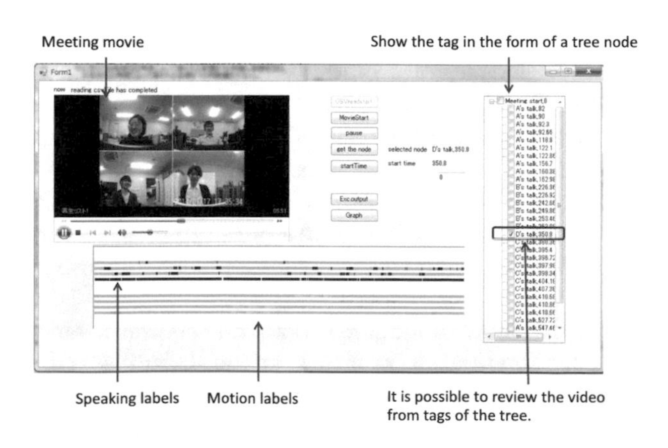

Fig. 6. Screenshot of the application

meetings will not be retrieved. Therefore, although many erroneous tags may be generated, we considered that it is more appropriate to increase the recall rate.

5 Application

We constructed a structured meeting application called the Meeting Review Tree (MRT) to review meeting videos tagged using the proposed method. A screenshot of the MRT is shown in Fig. 6. By reading the acceleration–angular velocity data and recorded video of all participants in the MRT, a recorded video is tagged with three layers by using the proposed method and each layer can be browsed in tree form. The MRT user selects the part of the speaker whom he or she wishes to view from the first layer of the tree and selects a checkbox of a place to be reproduced, such as a speaker's utterances or many nodding points among them; a user can then watch the structured recorded video. In addition, it is possible to confirm the recognition result of the utterance and head movement tags of each participant before and after the selected checkbox.

6 Conclusion

In this research, we proposed a system that recognizes the utterance, nodding, looking around, and head tilting of meeting participants using acceleration and angular velocity sensors and tagging to a recorded video. From the results of the evaluation experiments, the recognition accuracy (F value) of utterance was 0.44, the recognition accuracy (F value) of nodding motion was 0.39, and the recognition accuracy (F value) of looking around was 0.39. The recognition accuracy of the stratified first layer was 57.0% and that of the second layer was 61.0%. In the proposed system, to make it easier to review a meeting video, based on the recognition result, tagging with following three layers; the main speaker, the speaker between them, and head movements of the participant. In addition, we built a minutes meeting application called Meeting Review Tree implementing the proposed method and that can facilitate watching tagged meeting videos. Future tasks include improving recognition accuracy of utterance and head movement. In addition, we plan to propose a method to estimate the degree of understanding and degree of interest of participants in conversation contents from utterance movement and head movement, and we plan to construct a system that automatically extracts and presents parts important to meeting video viewer requirements.

Acknowledgements. This research was supported in part by a grant in aid for Precursory Research for Embryonic Science and Technology (PRESTO) from the Japan Science and by a grant in aid for Scientific Research (18H01059) from the Ministry of Education, Culture, Sports, Science and Technology.

References

1. VoiceGraphy. http://jpn.nec.com/voicegraphy/
2. Advanced Media Inc. http://www.advanced-media.co.jp/products/service/amivoice-speechwriter
3. Morency, L.P., de Kok, I., Gratch, J.: Context-based recognition during human interactions: automatic feature selection and encoding dictionary. In: Proceedings of the 10th International Conference on Multimodal Interfaces (ICMI 2008), pp. 181–188, October 2008
4. Wohler, N., Grosekathofer, U., Dierker, A., Hanheide, M., Kopp, S., Hermann, T.: A calibration-free head gesture recognition system with online capability. In: Proceedings of the 20th International Conference on Pattern Recognition (ICPR 2010), pp. 3814–3817, August 2010
5. Kawahara, T.: Multi-modal sensing and analysis of poster conversations toward smart posterboard. In: Proceedings of the 13th Annual Meeting of the Special Interest Group on Discourse and Dialogue, pp. 1–9 (2012)
6. Sumi, Y., Yano, M., Nishida, T. Analysis environment of conversational structure with nonverbal multimodal data. In: Proceedings of the International Conference on Multimodal Interfaces and the Workshop on Machine Learning for Multimodal Interaction (ICMI-MLMI 2010), p. 44 (2010)
7. Kumano, S., Otsuka, K., Mikami, D., Junji, Y.: Recognizing communicative facial expressions for discovering interpersonal emotions in group meetings. In: Proceedings of the 11th International Conference on Multimodal Interfaces (ICMI-MLMI 2009), pp. 99–106, September 2009
8. Otsuka, K., Yamato, J., Takemae, Y., Murase, H.: quantifying interpersonal influence in face-to-face conversations based on visual attention patterns. In: Proceedings of the 4th ACM Conference on Human Factors in Computing Systems (CHI 2006), pp. 1175–1180, April 2006
9. Wireless Technologies Inc. http://www.wireless-t.jp/
10. MR300. http://www.kingjim.co.jp/sp/mr360/

Hiding File Manipulation of Essential Services by System Call Proxy

Masaya Sato(✉), Hideo Taniguchi, and Toshihiro Yamauchi

Graduate School of Natural Science and Technology, Okayama University, 3-1-1 Tsushima-Naka, Kita-Ku, Okayama 700-8530, Japan
{sato, tani, yamauchi}@cs.okayama-u.ac.jp

Abstract. Security software or logging programs are frequently attacked because they are an obstruction to attackers. Protecting these essential services from attack is crucial to preventing and mitigating damage. Hiding information related to essential services, such as that of the files and processes, can help to deter attacks on these services. This paper proposes a method of hiding file manipulation for essential services. The method makes the files invisible to all services except their corresponding essential services and provides access methods to those files in a virtual machine (VM) environment. In the proposed method, system calls to those files are executed by a proxy process on the other VM. The original system call is not executed in the operating system of the original VM, however, the result of file access is returned to the original process. Thus, the files of essential services are placed on the other VM and other processes on the original VM cannot access to them. Therefore, the proposed method can prevent or deter identification of essential services based on file information monitoring.

Keywords: Security · Virtual machine · File access

1 Introduction

One of the main functions of security software is preventing attacks that use malicious software known as malware. Logging or management tools are also used to detect and analyze attacks and damage caused by malware. This paper defines security software and system management tools as essential services. Protecting these essential services is crucial to preventing attackers and limiting the spread of damage caused by malware [1, 2]. Before attacking, the attacker identifies the target service and prepares the attack. Although some methods to protect essential software have been proposed [3–5], none of these methods considers the identification of essential services. However, this paper asserts that if the identification of essential services were more difficult, then attacking essential services would also become more difficult.

Process information, file information, and communication information may be used to identify an essential service. Previously, we proposed a method to complicate identification of essential services based on process information [6]. This method

L. Barolli et al. (Eds.): NBiS 2018, LNDECT 22, pp. 853–863, 2019.
https://doi.org/10.1007/978-3-319-98530-5_76

replaces process information with dummy information while the process is running, so that other processes cannot identify the process as an essential service. Instead, the process is recognized as providing other services.

However, in our prior method, files and communication information remain visible to attackers. Thus, we propose a method for hiding the file manipulations of essential services. Specifically, the proposed method would only make files visible to their corresponding essential services. In addition, we propose a method to access those files through essential services. The proposed method employs a virtual machine monitor (VMM). The VMM monitors system calls related to file manipulation on a virtual machine (VM) and effectively transfers the information of the system call to the other VM. A proxy process on the other VM then invokes the system calls and returns the results to the essential process on the original VM. The essential process resumes as if it had accessed the required file in local storage. Because file access is interposed by the VMM and execution occurs by the proxy process on the other VM, it becomes much more difficult for attackers to identify essential services by monitoring the file access.

2 Related Work

Protection of security software is proposed. Hsu et al. analyzed the procedure of attacks on security software and proposed an attack-tolerant structure of security software [3]. The ANSS prevents termination of security software by monitoring and controlling system calls that attempt to terminate security software. The ANSS is implemented as a kernel mode driver of Windows operating system (OS). By contrast, the proposed method is implemented using a VMM and another VM. Implementing a security mechanism at a VMM is more secure than an OS kernel.

Malware analysis methods using a VM are proposed. In VMI, security software on one VM monitors the other VM from the outside [4]. Ether is a method that analyzes the behavior of malware on one VM from another VM [7]. Ether monitors system calls and memory access on the monitored VM. Process out-grafting is proposed as a method to analyze malware while keeping the monitoring environment secure [5]. Filesafe is a method to protect files by a VMM [8]. Filesafe controls file access according to a policy held by the VMM. Filesafe is similar to the proposed method in terms of access control. However, the proposed method also hides the existence of files used by essential services.

Deception technique is also applied to malware analysis. A method using deception to analyze and protect attacks is proposed [9]. Specifically, the method uses deception to induce attackers to honeypots, that is, analysis environments that seem like production environment. An advanced analysis method targeting attacks aiming vulnerabilities is proposed [10]. The method responds as if the attack was successful and transfers the request to a honeypot to monitor and analyze advanced attacks. The method aims to analyze advanced attacks. By contrast, the purpose of the proposed method is to hide the security software to protect systems from attack by avoidance.

3 Attacks to Essential Services

3.1 Essential Services

As previously mentioned, we define security software and system management tools as essential services. Correspondingly, a process providing an essential service is defined as an essential process and a file holding sensitive information used by an essential process is an essential file. Security software is an example of a typical essential service. Per the preceding definitions, each process run by security software is an essential process and the configuration files or whitelist files used by the service are considered to be essential files. If an attacker modifies configuration files or whitelist files, as these files are not scanned for malware, the security software cannot detect the malware. This compromises the computer, making the system vulnerable to further attacks. Furthermore, if security software files are visible to attackers, attackers may be able to identify the security software and additional potential exploits. Because attackers often begin by detecting and preparing attacks based on essential services, hiding essential services from attackers reduces the possibility of attacks. Although hiding the existence of process reduces the possibility of the discovery of essential services, the existence of essential files remains a problem. Because attackers can estimate the existence and extent of essential services from the essential files, hiding these files and the processes that manipulate them would provide a valuable enhancement to system security.

3.2 Existing Protection Method of Security Software

Hsu et al. proposed a method to protect security software from termination [3]. The ANtivirus Software Shield (ANSS) protects security software by monitoring and controlling application programming interfaces (API) on Windows. This method adds a kernel-mode driver into Windows and monitors API calls. If an API call attempts to terminate security software, the driver fails the call, which protects the security software from termination by attackers.

Another method for protecting security software uses virtualization technology. Virtual machine introspection (VMI) monitors the inside of a VM from the outside [5]. This method relies on the assumption that the VM and the VMM are isolated from each other. The security software runs on one VM and monitors the other VM. Because each VM is isolated, the security software is safe from attacks on the other VMs. However, integrating the existing security software with the outside VM causes a sematic gap, which is the difference in perspective between the inside and outside of the VM.

Filesafe protects classified files using virtualization technology [8]. Filesafe, which is implemented as a part of a VMM, enforces security policy and controls file access within the VMM. Operating in the VMM, Filesafe is resistant to attacks on the OS, thus, it can prevent leakage of sensitive information.

3.3 Problems

The previously discussed introduce security mechanisms into an OS kernel. However, it is possible to disable the mechanism through kernel-level attacks. Integrating a security mechanism to the outside of the OS kernel is effective, though it is difficult to use existing security mechanisms externally without modification. To monitor the VM from the outside, modification to the existing security mechanisms is required.

We also consider the visibility of files related to essential services. We define visibility as the possibility of attackers discovering the files. Thus, preventing attackers from knowing the names of essential files can impede attacks. Once identified, attackers can estimate the existence of the essential process by monitoring essential files and their manipulation. Thus, hiding the existence of those files may reduce the detection of essential services. Although Filesafe effectively protects files from unauthorized access, the files remain visible to attackers. If these files and their manipulation are visible to attackers, a process accessing those files would be flagged as a potential essential process by attackers. Unlike existing access control approaches, the proposed method considers visibility.

4 File Manipulation Hiding

4.1 Purpose

The proposed method attempts to avoid the identification of an essential service by monitoring files and their manipulation. Process information is also helpful to attackers attempting to identify an essential service. Previously, we proposed a method that hid process information from attackers. However, essential services use files to store configuration information. For example, some daemon programs store process IDs to a file. If an attacker modified a configuration file, this could change or disable the behavior of an essential service. Although antivirus software scans files to detect malware, if the antivirus software uses a whitelist to exclude some files from the scan, attackers can avoid the scan by adding their files to the whitelist.

To address these problems, we propose a method to prevent the identification of an essential service by monitoring essential files. This approach makes files invisible to attackers and only allows essential processes access to the files. The proposed method executes without modifying the essential service. This attribute addresses the problem of existing methods which do require the modification or integration of existing security software.

4.2 Requirements

Requirements. The above problems are resolved as follows:

Requirement 1. Essential files are made invisible to attackers.
Requirement 2. The method that satisfies Requirement 1 is also invisible to normal processes or kernel modules, and thus invisible to attackers as well.

If the method were detected, attackers could estimate the essential service by monitoring the detected method. We assume that attackers insert their programs into normal processes. A normal process is any process except an essential process or kernel modules. Therefore, hiding the method from normal processes and kernel modules is an important aspect of Requirement 2.

4.3 Hiding Method for File Manipulation

Challenges. To fulfill the requirements, the following challenges should be addressed:

Challenge 1. Interposition of access to files related to essential services.
Challenge 2. Control of interposed file access.
Challenge 3. Addressing Challenges 1 and 2 is undetectable by normal processes or kernel modules.

Challenges 1 and 2 are required satisfy Requirement 1. In addition, to satisfy Requirement 2, hiding the method addressing Challenges 1 and 2 from attackers is required.

Interposition of File Access for Essential Files. The following methods are considered by the approach to interposing file access: library call, system call, the processing of a file system, and the processing of device access. In a normal file access procedure, a process calls library functions. The library functions request processing to the kernel by invoking system calls. Finally, the kernel access files.

We monitored file access by interposing system calls. To hide essential files from attackers, interposing library calls is ideal. This is because that we can reduce the chance of monitoring of file manipulations as fast we can interpose. However, it is difficult to interpose all library calls related to file access. Libraries may be dynamically or statically linked to a program. If a library is dynamically linked, we can interpose file access by monitoring function calls to specified memory regions. However, if a library is statically linked, the relevant memory regions differ for each program. Identifying all the library calls to monitor of all programs is impractical. By contrast, it is possible to monitor all access by monitoring file systems or device drivers. However, these procedures occur in the latter part of file access. If an attacker monitors or modifies a file manipulation before it reaches the latter part, essential services may be identified. For this reason, our method interposes system calls for file access.

To handle Challenge 3, we implement the solutions using a VMM. The VMM and VMs are isolated from each other. Thus, non-intrusively interposing file access of a VM from a VMM helps hide the mechanism itself from attackers on the VM.

Control of File Access. As the previous section stated, the proposed method interposes system calls to hide files and file manipulations. This section details the controlling method. File access is made by file handlers. Processes use file handlers to manipulate files. Accordingly, file access is regulated by controlling the acquisition of file handlers. The proposed method interposes a system call to obtain a file handler. If the target file of the system call is an essential file, the proposed method checks the process. If the process is an essential process, the proposed method returns a file handler. If the process is a normal process, the proposed method returns a failure response. To continue the file access related to the returned file handlers, the proposed method controls the file access via the handlers. Thus, only an essential process can access essential files.

File Placement on the Outside the VM. To handle Challenge 3, we employ a VM. While essential processes run on an OS on one VM, essential files are placed on another VM. Because essential files are placed on the other VM, it is difficult to identify or modify the files even if an attacker has root privilege. In addition, an attacker cannot find the essential files by raw device access because the files are not located on the same virtual disk. In this architecture, an essential process cannot access essential files. To enable an essential process to access the essential files, we propose a system call proxy, that is, a method to transfer file access to the other VM. Section 5.5 details the system call proxy mechanism. Because we assume that essential files are files that are only required for essential services, hiding these files from normal processes does not obstruct the normal processes.

5 Implementation

5.1 Environment

We used Xen [11] as a VMM and the VMs were fully virtualized with Intel VT-x. We assumed Linux as a guest OS and system calls were invoked with sysenter instruction in 32bit and syscall instruction in 64bit environment. Figure 1 provides an overview of the proposed architecture. The VMM monitors system calls invoked in the protection target VM. If the system call invoked by the essential process is related to file manipulation, the VMM transfers the information to the proxy process on the proxy VM. The proxy process executes the system call based on the information and returns the result to the VMM. Finally, the VMM returns the processing to the protection target VM.

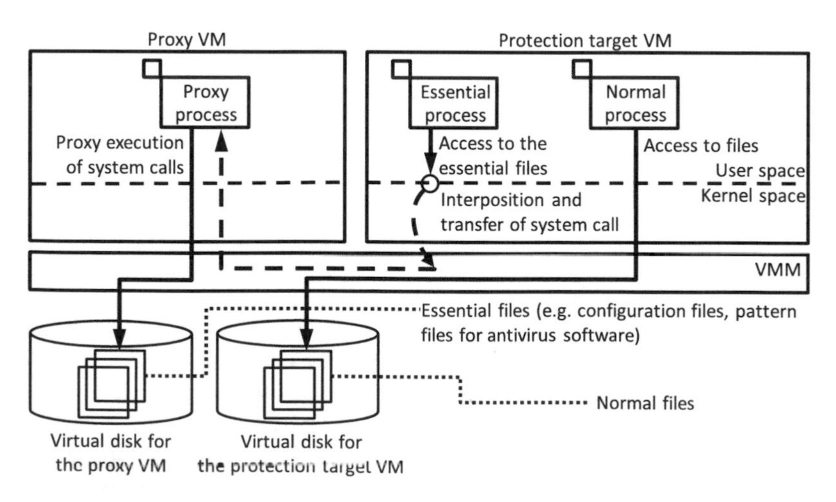

Fig. 1. The overview of the system call proxy.

5.2 System Call Interposition

We used a debug exception to detect invocation of system calls in the VM from the VMM. In the proposed method, the VMM set a hardware breakpoint at the starting address of the system call routine. In addition, the VMM manipulated the VM execution control field to cause VM exit when a debug exception occurred. In this case, when an instruction at the address was executed, a debug exception caused a VM exit. The VMM interposed the system call by handling the VM exit at the designated address.

In addition to the mechanism for system call interposition, classification of system calls is required. The VMM interposes all system calls invoked in a VM, however, what the proposed system requires is the system calls related to file manipulation. To classify the system calls, the VMM chose system calls by a system call number. The VMM acquired the system call number from the RAX register. Accordingly, the VMM interposed system calls of the protection target VM that were specifically related to file manipulation.

5.3 Control of System Calls Related to File Manipulation

Figure 2 shows the flow and mode transition in the proxy execution of system calls. The VMM controls the interposed system call according to the following steps:

1. The VMM classifies the interposed system call. If the system call is invoked by an essential process and is related to file manipulation, the VMM proceeds step 2. If not, the VMM moves to step 5.
2. The VMM acquires information about the system call. To proxy execute the system call, the following information is required: the virtual machine ID (VMID) of the protection target VM, the page directory (PD) of the essential process, the system call number, and the arguments of the system call.

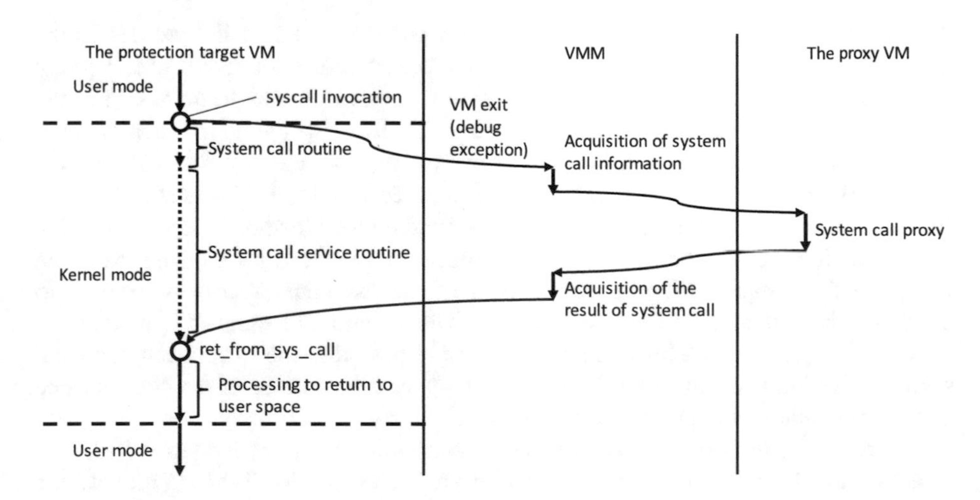

Fig. 2. The flow and mode transition in the proxy execution of system calls.

3. The VMM requests proxy execution of the system call to the proxy process.
4. The VMM receives the result of the system call from the proxy process.
5. Return the processing result to the protection target VM.

After the proxy execution, the proxy process must return the result of the proxy execution. To return the result from the proxy process to the essential process, we added a new hypercall to the Xen hypervisor. Hypercall is an interface to request processing from a VM to a VMM. The proxy process returns the result of the proxy execution by the new hypercall. The VMM receives the hypercall and manipulates the register of the protection target VM to set the return value of the original system call. Finally, the VMM manipulates guest's instruction pointer to skip original system calls.

5.4 Management of Essential Files

In the proposed method, the essential process manipulates normal files and essential files. To use the existing programs as an essential service with no modification, the proposed method must transparently provide normal files and essential files to the essential processes. To address this requirement, we changed the allocation rule of file descriptors. It is ideal to split the table for file descriptors, however, this approach requires modifications to the OS of the protection target VM. Thus, the VMM allocates file descriptors for essential files in descending order. File descriptor numbers for normal files are allocated by the OS in ascending manner (0, 1, 2, …). Conversely, the VMM allocates the file descriptor numbers for essential files in descending order (1023, 1022, 1021, …). If the descriptor number for a normal file is within the range of the numbers allocated to essential files, the VMM aborts the process. Thus, normal files and essential files are distinguished with no modification to the essential service or the OS. Though, we note that the maximum number for the file descriptor is reduced.

5.5 Proxy Execution of System Call

The proxy process executes the system call as a proxy of the essential process. In the proposed approach, a proxy process is allocated to each protection target VM. A proxy execution must address the following two points: the information to be passed to the proxy process and the structure of directory tree. Table 1 lists the information required for proxy execution. The VMID and PD are required to identify the VM and the process. Because the address of PD is unique to process and the value is already saved in VMM when VM exit occurred, we can identify the process without copying its PID. By identifying the process by PD, we can reduce the overhead for copying PID from the protection target VM. The current path of the essential process is required to construct the full path of the essential files. The system call number is required to execute the system call. The arguments are simply passed to the system call. Only the path information is changed by the proxy process, as the structure of the directory tree differs from that of the protection target VM.

First, to handle proxy execution requests from multiple protection target VMs, the directory for the proxy process must be allocated to each VM. Then, to identify the essential service, the directory must be allocated to each essential process. Finally, to

Table 1. Information required to proxy execution of system calls.

	Name	Description
1.	VMID	VMID is required to identify the VM
2.	Address of PD	Address of PD is used to identify the essential process
3.	Current directory	Current directory is used to construct the full path of the target file when a relative path is passed by the system calls
4.	System call number	System call number is used to determine whether the system call is related to file manipulation or not
5.	Arguments	Arguments of the system call are required to request the proxy execution of the system call. If the arguments include the address of the buffer, the VMM copies the buffer between the VM and the VMM

replicate the structure of the directory tree in the protection target VM, the proxy process creates the required directories and files based on the received information.

The VMM identifies whether the file is an essential file or not. Thus, it is required to hold the list of essential files. However, it is difficult to determine which file is an essential file because files are manipulated dynamically. We addressed this problem through the policy. At first, the VMM holds the list of essential files. The list includes files which may be classified as essential files statically. Then, any file newly created by an essential process is handled as an essential file because the file probably includes information identifying the process as an essential process. Finally, a file overwritten by the essential process is also handled as an essential file for same reason applied to newly created files.

6 Evaluation

To evaluate the performance of the proposed method, we measured the time for system calls related to file manipulation. The computer used for evaluation had an Intel Core i7-2600 (3.4 GHz, 4 cores) and 16 GB RAM. The protection target VM had one virtual CPU (VCPU) and 1 GB RAM. The proxy VM comprised three VCPUs and 15 GB RAM. The VCPUs of VMs were pinned to separate physical CPUs.

The evaluation results are provided in Table 2. We compared the performance of open, read, write, and close system calls to a VMM without interposition to those made to a VMM with the system call proxy. In this evaluation, to clarify the overhead of system calls that involved system call proxy, the essential process always accessed the essential files. The size of buffer for both read and write was 10 bytes. The performance degradation in open, read, and write is greater than that of close. We observed substantial degradation in the performance of open. We suggest that the reason for the degradation was memory copy. Read and write have buffers to read or write. These buffers are copied from the protection target VM to the proxy VM. This copy causes a large overhead. The performance of open system call was substantially degraded because of the directory creation. The proxy process creates directories for each protection target VM and its essential processes. Thus, the first access to the essential file

Table 2. Performance of system calls (microseconds).

	Open	Read	Write	Close
W/o interposition	4.68	2.26	6.51	2.32
System call proxy	271	101	110	6.87

by an essential process requires the creation of a directory for the essential process in the proxy VM. Also, to check whether the directory already exists, additional system calls are invoked by the proxy process. These additional system calls cause the large overhead. Thus, preserving the state of directory creation effectively improved the performance of proxy execution. By preserving the states, it is possible to reduce the number of the additional system calls by the proxy process. Additionally, reducing the number of buffer copies between the VMM and VMs also reduces the overhead. These performance improvements constitute our future work.

7 Conclusion

We proposed a method to hide file manipulation for essential services using system call proxy. The method interposes system calls invoked by essential processes and transfers information to the proxy process running on the other VM. The proxy process invokes the system calls and returns the results to the essential processes. This method enables the essential services to access essential files without being monitored by other processes and the OS kernel. Thus, identification of essential services based on file manipulation monitoring becomes difficult. As a result, attacks on essential services become more difficult. In addition, the files related to the essential services are located in the storage of the other VM. Thus, attackers on the protection target VM cannot detect and modify the files. The evaluation results show that the proposed method causes 4.55–267 μs performance overheads for open, read, write, and close system calls. The performance of open system calls was substantially degraded because these calls require the invocation of additional system calls to distinguish files for each essential process. We plan to focus our future work on performance improvement of the system call proxy.

Acknowledgments. This work was partially supported by KAKENHI grant numbers JP18K18051 and JP16H02829.

References

1. Min, B., Varadharajan, V., Tupakula, U., Hitchens, M.: Antivirus security: naked during updates. Softw. Pract. Exp. **44**(10), 1201–1222 (2014)
2. Grégio, A., Afonso, V., Filho, D., Geus, P., Jino, M.: Toward a taxonomy of malware behaviors. Comput. J. **58**(10), 2758–2777 (2015)
3. Hsu, F.H., Wu, M.H., Tso, C.K., Hsu, C.H., Chen, C.W.: Antivirus software shield against antivirus terminators. IEEE Trans. Inf. Forensics Secur. **7**(5), 1439–1447 (2012)

4. Garfinkel, T., Rosenblum, M.: A virtual machine introspection based architecture for intrusion detection. In: Network and Distributed Systems Security Symposium, vol. 3, pp. 191–206 (2003)
5. Srinivasan, D., Wang, Z., Jiang, X., Xu, D.: Process out-grafting: an efficient "out-of-VM" approach for fine-grained process execution monitoring. In: Proceedings of 18th ACM Conference on Computer and Communications Security, pp. 363–374 (2011)
6. Sato, M., Yamauchi, T., Taniguchi, H.: Process hiding by virtual machine monitor for attack avoidance. J. Inf. Process. **23**(5), 673–682 (2015)
7. Dinaburg, A., Royal, P., Sharif, M., Lee, W.: Ether: malware analysis via hardware virtualization extensions. In: Proceedings of 15th ACM Conference on Computer and Communications Security, pp. 51–62 (2008)
8. Wang, J., Yu, M., Li, B., Qi, Z., Guan, H.: Hypervisor-based protection of sensitive files in a compromised system. In: Proceedings of 27th Annual ACM Symposium on Applied Computing, pp. 1765–1770 (2012)
9. Almeshekah, M.H., Spafford, E.H.: Planning and integrating deception into computer security defenses. In: Proceedings of 2014 Workshop on New Security Paradigms Workshop, pp. 127–138 (2014)
10. Araujo, F., Hamlen, K.W., Biedermann, S., Katzenbeisser, S.: From patches to honey-patches: lightweight attacker misdirection, deception, and disinformation. In: Proceedings of 21st ACM Conference on Computer and Communications Security, pp. 942–953 (2014)
11. Barham, P., Dragovic, B., Fraser, K., Hand, S., Harris, T., Ho, A., Neugebauer, R., Pratt, I., Warfield, A.: Xen and the art of virtualization. ACM SIGOPS Oper. Syst. Rev. **37**(5), 164–177 (2003)

Evaluation of Broadcasting System for Selective Contents Considering Interruption Time

Takuro Fujita(✉) and Yusuke Gotoh

Graduate School of Natural Science and Technology,
Okayama University, Okayama, Japan
payc7ift@s.okayama-u.ac.jp, gotoh@cs.okayama-u.ac.jp

Abstract. Due to the recent popularization of digital broadcasting, selective contents broadcasting has attracted much attention. In selective contents broadcasting, although the server delivers contents based on their preferences, users may experience the interruption time while playing their selected contents. To reduce this interruption time, many researchers have proposed scheduling methods. However, since these scheduling methods evaluated the interruption time in simulation environments, we need to evaluate them in network environments. In this paper, we propose a broadcasting system of selective contents and evaluate its effectiveness in network environments.

1 Introduction

Due to the recent popularization of the internet, selective contents, i.e., watching contents selected by users themselves, have attracted great attention [1]. For example, in quiz programs, the server shows several potential answers from which a user selects an answer. If the answer is correct, she watches the video content of the correct answer. Otherwise, she watches the content of the incorrect answer.

In selective contents broadcasting, although the server can deliver several contents to many users, they may experience the interruption time while playing a program. Therefore, many researchers have proposed scheduling methods to reduce the waiting time. These scheduling methods make a broadcast schedule that considers the situation in actual network environments. However, since most scheduling methods in broadcasting schemes evaluate the interruption times in simulation environments, we need to evaluate them using scheduling methods in network environments.

In this paper, we propose a broadcasting system for selective contents and evaluate the interruption time of scheduling methods in network environments.

The remainder of the paper is organized as follows. We explain selective contents broadcasting in Sect. 2 and the delivery system in Sect. 3. Conventional scheduling methods are explained in Sect. 4. Related works are introduced in Sect. 5. We explain our proposed broadcasting system for selective contents in Sect. 6 and evaluate it in Sect. 7. Finally, we conclude in Sect. 8.

L. Barolli et al. (Eds.): NBiS 2018, LNDECT 22, pp. 864–874, 2019.
https://doi.org/10.1007/978-3-319-98530-5_77

2 Selective Contents

Since selective contents have sequences to play them, their structures can be described by state-transition graphs. Therefore, we proposed a play-sequence graph to describe the structure of selective contents [2]. Here we explain play-sequence graphs to make our paper self-contained.

In a play-sequence graph, each node represents a state in which the client plays some content. When the client finishes playing it, the state transits to the next node. For example, a play-sequence graph for a quiz program is shown in Fig. 1. In Fig. 1-A, the user selects an answer from given answers X or Y. Node S_1 is a state where the client plays a video that presents the quiz. S_2 is a state where the client plays a video that explains answers X and Y. The user selects an answer from X or Y while playing the video. If she selects answer X, the state transits to S_3. If she selects answer Y, the state transits to S_4. In this way, the state transits to the next node based on her selections. When the user does not select an answer, the state transits to S_2 again or automatically transits to subsequent nodes S_3 or S_4. S_3 is the state where the user selects answer X, and a video for the correct answer is played. S_4 is the state where the user selects answer Y, and a video for the incorrect answer is played.

Play-sequence graphs can be simplified by applying the following three operations: abbreviate, merge, and split. By applying them, we can simplify the play-sequence graph for a quiz program (Fig. 1-C).

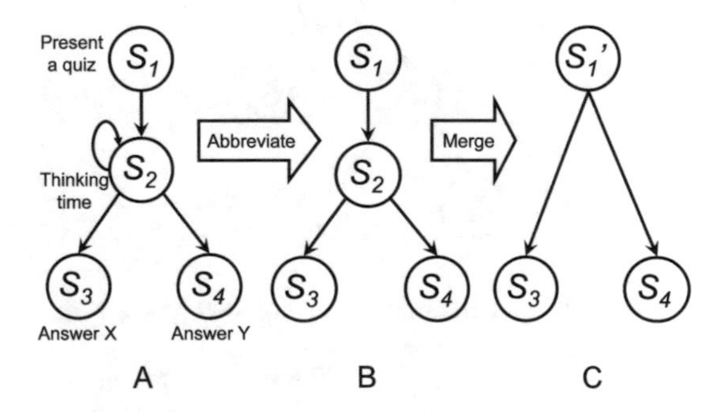

Fig. 1. Simplified play-sequence graph

3 Delivery Systems

In webcasts, there are mainly two types of delivery systems: broadcasting and Video on Demand (VoD). In a broadcasting system for delivering selective contents, the server delivers identical contents data to many clients using a constant bandwidth. The server can reduce the network load. Therefore, systems using broadcast are effective for many clients. However, clients have to wait until their desired data are broadcast.

In the VoD system, the server can set the bandwidth to each client, and users can immediately watch the data. However, the server increases the network load as the number of clients increases. In this paper, we assume a broadcasting system for delivering selective contents.

4 Scheduling Method

4.1 Outline

In selective contents broadcasting that produces a broadcast schedule using play-sequence graphs, we reduce the available bandwidth. However, since the server delivers many contents concurrently, the necessary bandwidth increases. When the available bandwidth is less than the necessary bandwidth, users have to wait to watch the data. Therefore, many researchers have proposed scheduling methods to reduce the waiting time.

4.2 Simple Method

In the simple method, the available bandwidth is equally divided into the same number of channels as the maximum number of nodes at each depth.

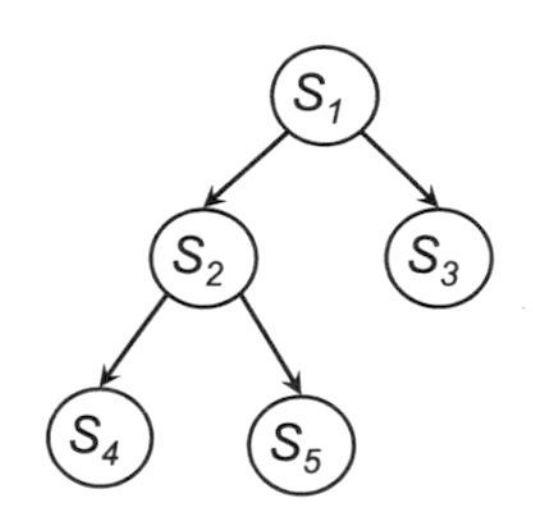

Fig. 2. Play-sequence graph for news program

Time →	60 sec.	60 sec.	60 sec.
C_1 (3.0 Mbps)	S_1	S_2	S_4
C_2 (3.0 Mbps)		S_3	S_5
User plays data (3.0 Mbps)	S_1	S_2	S_5

Fig. 3. Example of broadcast schedule under simple method

When the server broadcasts a program whose play-sequence graph is shown in Fig. 2, the broadcast schedule under the simple method is shown in Fig. 3. The playing time of each content is 60 s. Suppose a case where the consumption rate is 3.0 Mbps and the maximum number of branches is 2: here the necessary bandwidth is $3.0 \times 2 = 6.0$ Mbps. In the simple method, when the number of nodes at a depth is less than the maximum number of nodes at each depth, the server does not broadcast the contents. For example, in Fig. 3, when the server broadcasts S_1 in C_1, it does not broadcast the content in C_2.

When an upper limit exists in the bandwidth, clients have to wait to watch the contents. For example, when the necessary bandwidth is 10 Mbps and the available bandwidth is limited to 4.5 Mbps, since the bandwidth for each channel is as large as $4.5/6.0 = 0.75$ times, the bandwidth for C_1 and C_2 is $3.0 \times 0.75 = 2.25$ Mbps and the data size for the selective contents is $60 \times 3.0/8 = 22.5$ Mbytes. Since the bandwidth for each channel is less than the consumption rate, it takes $22.5 \times 8/2.25 = 80$ s. Since the playing time is 60 s, the user waits $80 - 60 = 20$ s to play S_1. Much the same is true for $S_2, \cdots, S_5$; the user waits for 20 s. For example, when clients continuously watch S_1, S_2, and S_5, the interruption time is 60 s.

4.3 CCB Method

The Contents Cumulated Broadcasting (CCB) method [3] reduces the necessary bandwidth. Its broadcast schedule is shown in Fig. 4. Clients have no waiting time by scheduling contents considering the configuration of play-sequence graphs and the available bandwidth. In Fig. 4, when the bandwidths for C_1 and C_2 are equally 3.0 Mbps, which is the same as the consumption rate, the server schedules S_1 and S_3 to C_1, and S_2 and S_4 to C_2. Since the broadcasting time of S_5 is 120 s, the bandwidth for C_3 is $3.0 \times 60/120 = 1.5$ Mbps. Therefore, the server broadcasts all the contents in 120 s. using the following necessary bandwidth: $3.0 \times 2 + 1.5 = 7.5$ Mbps.

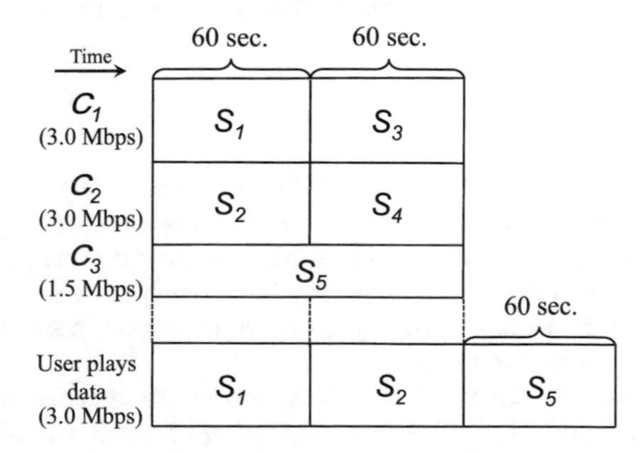

Fig. 4. Example of broadcast schedule under CCB method

When the available bandwidth is limited to 4.5 Mbps, since the necessary bandwidth is 7.5 Mbps, the bandwidth for each channel is as large as $4.5/7.5 = 0.6$ times. The bandwidths for C_1 and C_2 are $3.0 \times 0.6 = 1.8$ Mbps, and for C_3 it is $1.5 \times 0.6 = 0.9$ Mbps. It takes $22.5 \times 8/1.8 = 100$ s to broadcast the content for 60 s. When clients continuously watch S_1, S_2, and S_5, the waiting time is 80 s.

5 Related Works

In selective contents broadcasting, several methods have been proposed to reduce waiting time. The Contents Cumulated Broadcasting-Considering Bandwidth (CCB-CB) method [4] reduces the waiting time by acquiring channel bandwidth that has the identical consumption rate. Several scheduling methods to reduce waiting time for continuous media data have been proposed [5–8]. The Cautious Harmonic Broadcasting (CHB) method [9] divides the data into several segments and frequently broadcasts the first segments.

In on-demand delivery, the Restricted waiting time for Selective Contents (RSC) method [10] reduces the waiting time by setting the upper limit after selecting the content.

Conventional scheduling methods fail to consider the effect for division-based broadcasting systems. To evaluate scheduling methods in actual networks, we proposed a division-based broadcasting system d-Cast [11]. d-Cast introduces several types of conventional broadcasting methods and designs internet broadcasting systems based on a network environment with a server and clients.

6 Broadcasting System for Selective Contents

To evaluate scheduling methods in actual networks, we propose a broadcasting system for selective contents called *Corne* (Contents-based broadcasting system). *Corne* can introduce several types of conventional broadcasting methods as well as design internet broadcasting systems based on a network environment with a server and clients.

6.1 Problems

Next we explain the details of the two problems in *Corne*: the synchronization of timing for delivering contents and sequential playback. First, we must synchronize the starting times of delivering the data in selective contents broadcasting. However, in network environments, when the server uses a scheduling method that needs to synchronize the timing to start delivering segments for several channels, interruption time occurs.

Second, in scheduling methods for selective contents broadcasting, the timing of starting to play the data is identical as starting to receive it. When the server delivers the contents in network environments, we need to implement sequential playback.

6.2 Design Policy

To solve the problems of selective contents, we improved the setting for receiving data.

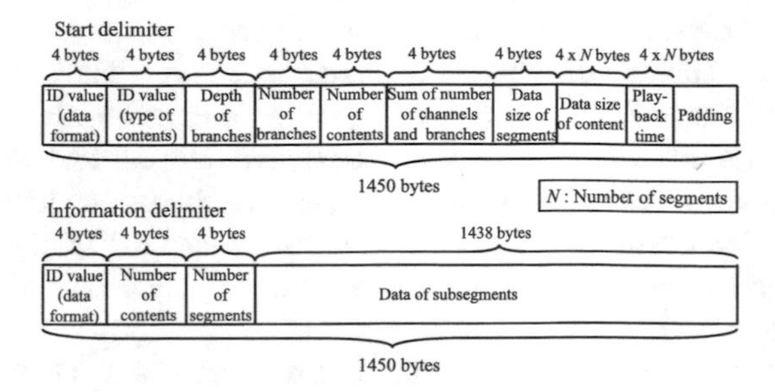

Fig. 5. Data format

The data format of our designed broadcasting system is shown in Fig. 5. There are two types of formats: start-delimiter and information-delimiter headers. The start-delimiter header is added and only delivered in the initial segments. Clients can start receiving data after receiving the start-delimiter header.

6.3 Server Design

Actual networks have many types of communication protocols. In our paper, we design and implement a scheme to synchronize the timing of delivering contents. In it, the delivery processing of each channel and the timing for delivering the

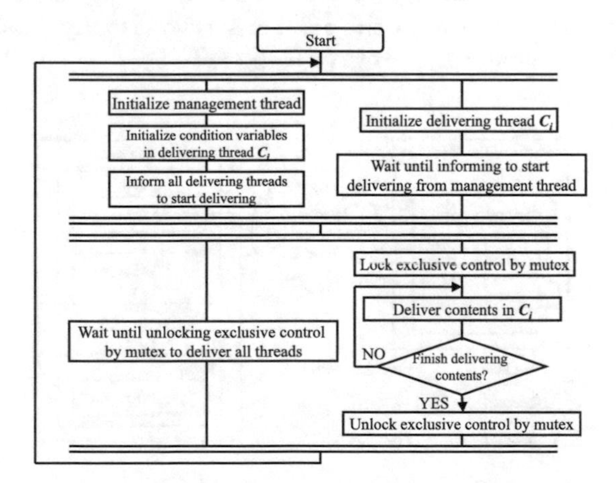

Fig. 6. Flowchart of server

data are performed in parallel. In the mechanism, the server sends start and stop messages in parallel processing synchronized using exclusion control.

The processing flow of the server in our proposed scheme is shown in Fig. 6. The number of channels is n. The server uses broadcast channels C_i ($i = 1, \cdots, n$), sets the processing to manage the timings of the delivered data as managing and delivering threads, and parallelizes them.

The managing thread initializes the variables to send the starting and stopping messages of the parallelized processes called a condition variable. Next the managing thread sends a starting message to deliver the content to all the threads using the condition variable. After sending the message, the exclusion control is released by a mechanism called mutex that manages it. Finally, the managing thread waits for the execution until the process is synchronized among all the delivering threads.

The delivering thread of C_i waits to be executed until the message to start delivering the content is sent from the managing thread. After sending the message from the managing thread, the delivering thread sets the exclusion control for the mutex and delivers contents using C_i. After the segment has been delivered, the delivering thread releases the exclusion control for mutex and waits until the message is sent to restart delivering the segments using a condition variable.

6.4 Sequential Playback

In this paper, we design and implement a sequential playback function using HTTP, which does not need private servers or clients. The configuration of our proposed system is shown in Fig. 7. First, the server divides the data into several segments. Next, it divides each segment into several subsegments and delivers them using several channels. After receiving a HTTP request from the browser, the client starts receiving subsegments using the receiving function, where subsegments are sorted and sent to the HTTP server. Finally, the HTTP server delivers the subsegments to the browser using HTTP.

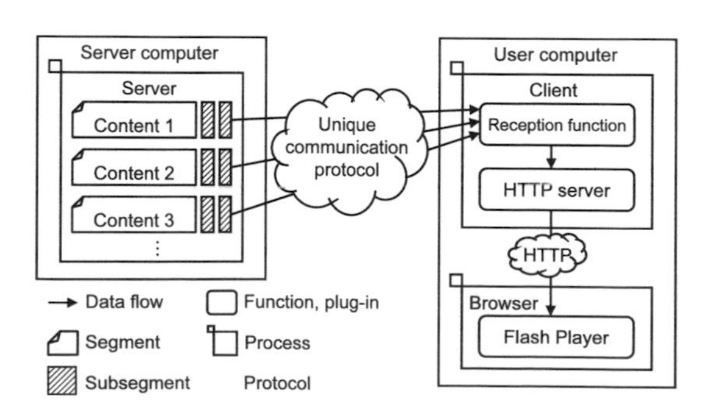

Fig. 7. Configuration of our proposed system

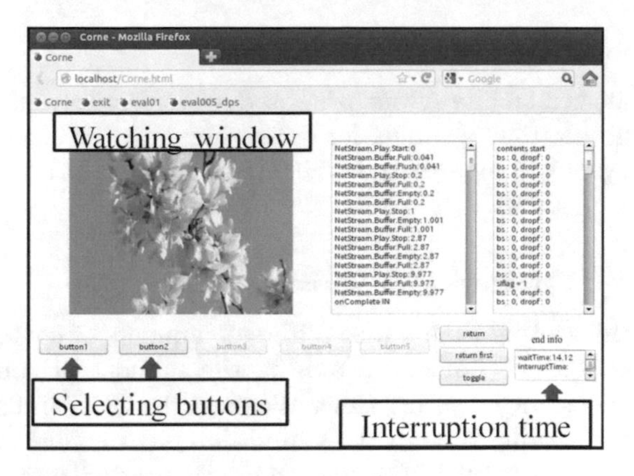

Fig. 8. Screenshot

6.5 Implementation

Based on the solution explained in Subsects. 6.3 and 6.4, we implemented the broadcasting system. A screenshot of our system is shown in Fig. 8.

7 Evaluation

7.1 Evaluation Environment

We constructed an environment that evaluates our proposed system on an actual network. In Fig. 9, the connection between the server and the client is routed through Dummynet that controls the bandwidth. With Dummynet, we can construct an evaluation environment to reproduce various network environments.

In actual broadcasting, we constructed a network environment where the server delivers data to several clients. In our evaluation, to show the effect of a broadcasting schedule, we created an environment in which the server delivers data to one client. In broadcasting, the server can concurrently deliver the same data to many clients in a constant network load. Therefore, the network load is not increased by adding clients.

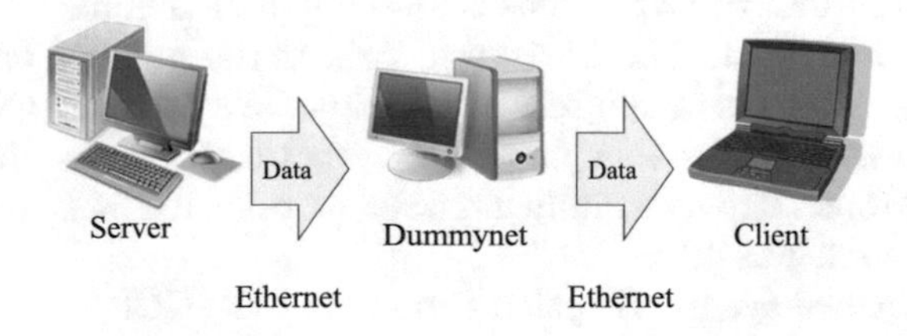

Fig. 9. Assumed network environment

7.2 Evaluation Items

We confirmed the availability of our proposed system to solve the following two problems: synchronization of timing for delivering data and sequential playback. Our proposed system compares the waiting and interruption times using the CCB and simple methods.

7.3 Number of Contents and Waiting Time

We calculated the waiting time under different numbers of contents. The result is shown in Fig. 10. The horizontal axis is the number of contents, and the vertical axis is the average waiting time. We used the CCB and simple methods and evaluated the waiting time in the simulation environment and *Corne*. The available bandwidth was 1.6 Mbps, and the consumption rate was 0.64 Mbps. The maximum number of branches was 2, and the playing time of each content was 10 s.

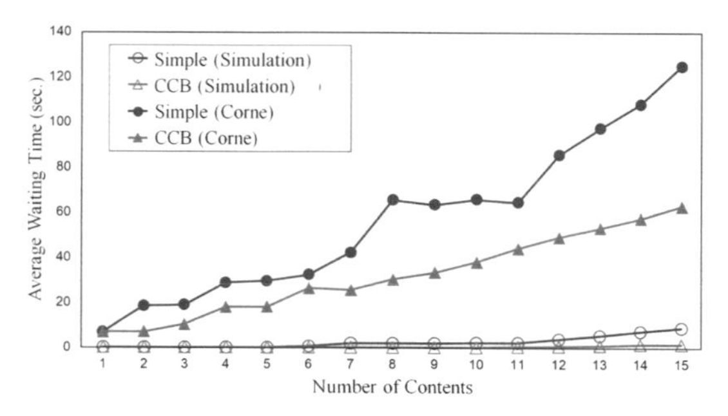

Fig. 10. Average waiting time and number of contents

In Fig. 10, the waiting time on *Corne* is shorter than the simulation environment. In addition, the waiting time under the CCB method are also shorter than the simple method. For example, when the number of contents was eight, the average waiting times of the simple method on *Corne* were 63.5 s and 30.3 s under the CCB method. The average waiting time under the CCB method was reduced to 52.3% more than the simple method.

7.4 Number of Contents and Interruption Time

We calculated the interruption times under different number of contents. The result is shown in Fig. 11. The horizontal axis is the number of contents, and the vertical axis is the average interruption time. We used the CCB and simple methods on *Corne*. The available bandwidth was 1.6 Mbps, and the consumption rate was 0.64 Mbps. The maximum number of branches is 2, and the playing time of each content was 10 s.

In Fig. 11, the average interruption time under the CCB method was shorter than the simple method with an increasing of number of contents. Since *Corne* makes the process by delivering threads, the interruption time under the simple method is lengthened.

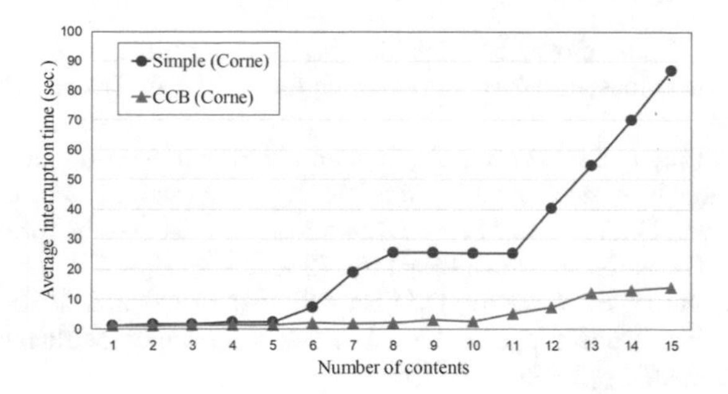

Fig. 11. Average interruption time and number of contents

8 Conclusion

In this paper, to reduce the interruption time while playing them, we designed and implemented a broadcasting system of selective contents called *Corne*. Our evaluation confirmed that our proposed system reduced the waiting and interruption times. For example, when the number of contents was eight, the average waiting times of the simple method on *Corne* were 63.5 s and 30.3 s under the CCB method. The average waiting time under the CCB method was reduced to 52.3% more than the simple method. In the future, we will evaluate the performance of *Corne* under several scheduling methods.

Acknowledgement. This work was supported by JSPS KAKENHI Grant Number 18K11265 and 16K01065. In addition, this work was partially supported by the Telecommunications Advancement Foundation and the Okayama Foundation for Science and Technology.

References

1. WHITE PAPER Information and Communications in Japan (2017). http://www.soumu.go.jp/johotsusintokei/whitepaper/eng/WP2017/2017-index.html
2. Gotoh, Y., Yoshihisa, T., Kanazawa, M.: A scheduling method considering available bandwidth in selective contents broadcasting. In: Proceedings of the IEEE Wireless Communications and Networking Conference (WCNC 2007), pp. 2597–2602 (2007)
3. Yoshihisa, T., Kanazawa, M.: A scheduling method for bandwidth reduction on selective contents broadcasting. In: Proceedings of the IPSJ International Conference on Mobile Computing and Ubiquitous Networking (ICMU 2006), pp. 60–67 (2006)
4. Gotoh, Y., Yoshihisa, T., Kanazawa, M., Takahashi, Y.: A broadcasting scheme for selective contents considering available bandwidth. IEEE Trans. Broadcast. **55**(2), 460–467 (2009)
5. Viswanathen, S., Imilelinski, T.: Pyramid broadcasting for video on demand service. In: Proceedings of the Multimedia Computing and Networking Conference (MMCN '95), vol. 2417, pp. 66–77 (2005)

6. Yoshihisa, T., Tsukamoto, M., Nishio, S.: A scheduling scheme for continuous media data broadcasting with a single channel. IEEE Trans. Broadcast. **52**(1), 1–10 (2006)
7. Mahanti, A., Eager, D., Vernon, M., Stukel, D.: Scalable on-demand media streaming with packet loss recovery. IEEE/ACM Trans. Netw. **11**, 195–209 (2003)
8. Zhao, Y., Eager, D., Vernon, M.: Scalable on-demand streaming of nonlinear media. IEEE/ACM Trans. Netw. **15**, 1149–1162 (2007)
9. Paris, J.-F., Carter, S.W., Long, D.D.E.: Efficient broadcasting protocols for video on demand. In: Proceedings of the ACM International Multimedia Conference (Multimedia 1999), pp. 189–197 (1999)
10. Gotoh, Y., Yoshihisa, T., Kanazawa, M.: A scheduling method for on-demand delivery of selective contents. In: Proceedings of the 4th International Conference on Mobile Computing and Ubiquitous Networking (ICMU 2008), pp. 17–24 (2008)
11. Gotoh, Y., Yoshihisa, T., Kanazawa, M.: d-Cast: a division based broadcasting system for IP networks. In: Proceedings of the IEEE International Conference on Advanced Communication Technology (ICACT 2007), pp. 1902–1907 (2007)

Play Recognition Using Soccer Tracking Data Based on Machine Learning

Tomoki Imai[1(✉)], Akira Uchiyama[1], Takuya Magome[2], and Teruo Higashino[1]

[1] Graduate School of Information Science and Technology, Osaka University, 1-5 Yamadaoka, Suita, Osaka 5650871, Japan
{t-imai,uchiyama,higashino}@ist.osaka-u.ac.jp
[2] Graduate School of Medicine, Osaka University, 2-2 Yamadaoka, Suita, Osaka 5650871, Japan
t-magome@anat2.med.osaka-u.ac.jp

Abstract. In professional football, every play data is recorded such as Pass, Dribble, etc. However, the play data is manually recorded, which requires huge effort. To reduce the human effort, we propose a method to recognize the labels of plays in football games from tracking data. By using features extracted from tracking data, we generate a play classifier model based on machine learning. We have evaluated the proposed method through real tracking data recorded in Japan Professional Football League (J. League). The results have shown that our play recognition is effective for mitigating the heavy workload for play labeling.

1 Introduction

Recently, play data in sports has become more important with the development of information technology. For example, in soccer, the numbers of shoots and corner kicks are recorded for each game as statistical information. In Japan Professional Football League (J. League), for tactical analysis and performance evaluation of players, every play data is recorded such as Pass, Dribble, etc. Each play data contains timestamp, player names, and positions of the ball and players. In addition, the result of each play (success or failure) is recorded. Although the play data is manually recorded by checking video, even professional scorekeepers take around ten hours per game, which is a heavy workload for them. For this reason, a method for recording play data automatically is required.

There are various works to recognize plays in sports. For example, a method for play recognition in rugby games has been proposed [7]. They use acceleration and angular velocity of a wearable sensor and player's position. However, the target is limited to only two types of plays, tackle and scrum, and there are still challenges of play recognition in various sports. On the other hand, sports information analysis based on videos is at the center of attention [5]. In [8,10,11], they propose methods for play recognition in several sports based on deep learning. Ref. [11] proposes a play recognition method for pass, shoot, and dribble in Futsal by convolution neural network (CNN) using videos of multiple cameras. However, the accuracy

L. Barolli et al. (Eds.): NBiS 2018, LNDECT 22, pp. 875–884, 2019.
https://doi.org/10.1007/978-3-319-98530-5_78

is 70% at most, still remaining challenges. In order to improve the performance of recognition it is important to combine multiple information sources such as videos, wearable sensors, and so on.

In this paper, we propose a method to recognize the labels of plays in soccer games based on trajectories of the ball and players. We use tracking data measured by a camera-based system called TRACAB [3]. TRACAB estimates trajectories of the ball and all players in the field by the image recognition. Our method aims to recognize 10 types of plays with ball touch such as Pass, Trap, and Shoot that are actually recorded in J. League games.

For this purpose, there are several challenges for automatically recognizing plays. Firstly, camera-based tracking data includes some position errors due to inaccuracy of the image recognition. In particular, these errors occur frequently in situations where some players gather around the ball, the ball moves out of the pitch, and so on. We design a method that avoids such errors of tracking data by combining carefully designed features. Secondly, there are several plays whose definition are ambiguous among target classes. We tackle with such difficulty by applying machine learning for play recognition.

First, we focus on the fact that the ball's trajectory or speed changes greatly at the moment of the play occurrence. Based on this fact, we detect the occurrence of plays in the tracking data with a rule-based method. Second, we extract the features for each detected play from trajectories of the ball and each player and train a classifier of plays by machine learning. For machine learning, we compare three types of algorithms: Conditional Random Field (CRF), Support Vector Machine (SVM), and Random Forest (RF).

The contributions of this paper are as follows. (1) We propose a method to recognize 10 types of ball touch play recorded in J. League based on trajectories only. (2) In the evaluation, we compare three types of machine learning algorithms (CRF, SVM, and RF) using real data.

Experimental results show that our play detection achieves 86.7% precision and 85.3% recall. For evaluation of play recognition, we compare three types of machine learning algorithms. As a result, we have confirmed that RF shows the best performance on average. Especially, regarding Pass and Trap which account for about 90% of plays in soccer games, our method has achieved over 85% by RF, which is promising to realize mitigating the heavy workload for play labeling.

2 Dataset

Tracking data used in this paper was recorded in J. League games using the camera system called TRACAB by ChyronHego [3]. In TRACAB, multiple cameras are deployed at the top of the stadium. Trajectories of players and the ball are acquired based on image recognition.

We note that the trajectories recorded by TRACAB are not perfect, especially for the ball. This means we sometimes find errors of ball tracking in the following cases: (1) many players gather around the ball, (2) the ball overlaps with a white line, (3) the ball moves out of the pitch, (4) the ball goes up high and falls outside the viewing angle of cameras, and (5) a goalkeeper catches the ball.

Tracking data was recorded at 0.04[Hz]. 2-axis coordinates (x and y) are recorded for players while 3-axis coordinates (x, y, and z) are recorded for the ball. The z axis indicates a height from the ground. Also, the identification numbers of target players are included in tracking data. However, these identification numbers were input manually by operators. To investigate the possibility of workload mitigation, we do not use the identification numbers of players in this paper. Instead, we assume teams of each trajectory are given although their identification numbers are unknown. This is easily recognized by a color feature which is totally different between teams.

3 Proposed Method

3.1 Overview

Figure 1 shows the overview of our method that recognizes the play labels in the soccer game from tracking data. First, we focus on the fact that the ball's trajectory or speed changes greatly at the moment of the play occurrence. Based on this fact, we detect the occurrence of plays in the tracking data. Second, we extract features for each detected play from spatial relationship of the ball and players and train a classifier of plays based on machine learning. For machine learning, we compare three types of algorithms: Conditional Random Field (CRF), Support Vector Machine (SVM), and Random Forest (RF). Table 1 shows our target classes that occur with ball touches.

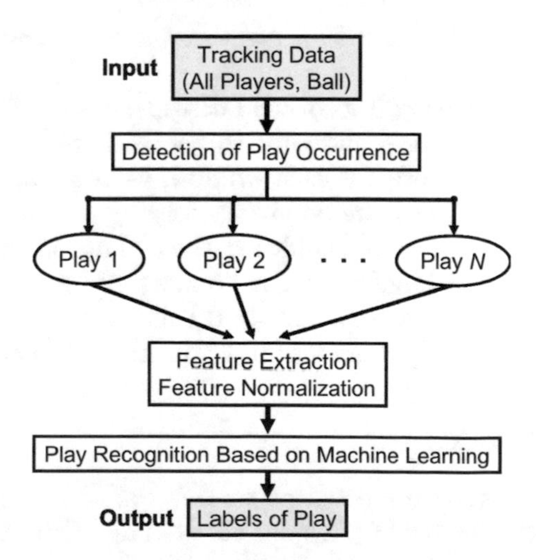

Fig. 1. Overview of proposed method

Table 1. Target classes

Name	Abbr.	Overview
Pass	pa	Kick or head the ball and give it to a teammate
Through ball	tb	A pass aimed at the back of the opponent's backline
Cross	cr	A pass from a wide area towards the center of the field near the opponent's goal
Shoot	sh	The most common play for goals to be scored
Feed	fe	Throw or kick the ball by a goalkeeper
Clearance	cl	Try to keep the ball away from the goal in order to escape from opponent's offense
Trap	tr	Stop the ball and control it
Dribble	dr	Maneuver the ball avoiding defender's attempts to intercept the ball
Catch	ca	A goalkeeper holds the ball
Block	bl	A player obstructs the ball's path with his body to avoid opponent's offense

3.2 Detection of Play Occurrence

In order to detect the play occurrence, we calculate a displacement vector of the ball for each time window W. v_b^t at time t is defined as the vector from p_{t-W}^b to p_t^b where p_t^b is the position of the ball at time t. We slide the window without overlaps and detect the play occurrence in each window if the direction or moving distance of v_b^t changes significantly. In other words, when either one of the following Eqs. (1) and (2) is satisfied, we detect the play occurrence in the window.

$$\frac{v_b^{t-W} \cdot v_b^t}{|v_b^{t-W}||v_b^t|} \leq \cos\theta_{TH} \tag{1}$$

$$|v_b^{t-W}| - |v_b^t| \geq V_{TH} \tag{2}$$

In the above equations, θ_{TH} and V_{TH} are thresholds for angle change and distance, respectively.

In addition, we exclude two exceptional cases as below. First, the ball vector can change even when a player moves with the ball, which is actually not an occurrence of a play. Therefore, for each window, we regard the player closest to the ball as a ball holder. Then, we do not detect any play in the window unless another player becomes a new ball holder at the end of the window. Second, the ball speed tends to change largely when it bounces off the ground largely, which can be wrongly detected as a play. To avoid this, our method does not detect any play when there are no players within 4.0 m from the ball.

3.3 Feature Extraction

At each play occurrence detected at time t_i, we define spatial feature candidates based on coordinates of players and the ball. Table 2 shows all feature candidates. We denote the time of i-th play detected by our method as t_i. Then, the players closest to the ball at t_i and t_{i+1} are represented as p and q, respectively. The positions of p at t_{i-1}, t_i, and t_{i+1} are defined as p_{-1}, p_0, and p_{+1}, respectively. Similarly, we let b and p' be the position of the ball and the position

Table 2. Feature candidates

Type	Value
Position	b_0, b_{-1}, b_1
	Backline of both teams at t_i
	Number of players located within 3 m from b_0
Distance	$D(p_0, p_1)$, $D(p_0, p_{0+s})$, $D(p_0, p_{-1})$
	$D(q_0, q_1)$, $D(q_0, q_{-1})$
	$D(b_0, b_1)$, $D(b_0, b_{0+s})$, $D(b_0, b_{-1})$
	$D(p_1, b_1)$, $D(p_{-1}, b_{-1})$, $D(p_{0+s}, b_{0+s})$
	$D(p_0, p'_0)$
	$D(gk_0, b_0)$, $D(gk'_0, b_0)$
Direction	Direction of the ball movement
	Goal approaching feature
Time	$t_i - t_{i-1}$, $t_{i+1} - t_i$
Situation	A team that possesses the ball at t_i, t_{i-1}, t_{i+1}
	Play type classification results

of the opponent player closest to the ball. Also, gk and gk' are the positions of both of the goalkeepers.

We represent the distance between two objects O_1 and O_2 as $D(O_1, O_2)$. O can be anything such as a player, the ball, a backline, etc. The backline is the x-coordinate position of the player who is the closest to the goalkeeper of his team. We also use moving directions of the ball defined by b_0 and b_{+1}. As a feature specific to Shoot, we define a goal approaching feature as the y-coordinate of the intersection of the nearest goal line and a line defined by b_0 and b_{+1}. The y-coordinate is parallel to the goal line.

To extract features related to moving speed, we use positions p_{0+s} and b_{0+s} of p and b at time $t_i + s$, which is s seconds after t_i. We consider that appropriate values of s are different depending on the types of play. Therefore, we empirically set s to 0.3, 0.6, and 1.0.

Moreover, it may be useful to apply hierarchical classification since our target types of plays are roughly divided into two types: "Ball moving plays" and "Ball holding plays". To imitate such hierarchical classification, we use binary classification results by a decision tree as a feature candidate. "Ball moving plays" are Pass, Through ball, Cross, Shoot, Feed, Clearance, and Block while "Ball holding plays" are Trap, Dribble, and Catch. Two feature values are used for the binary classification, i.e. $D(p_1, b_1)$ and a binary value indicating whether p and q are the same player or not. This is because, for "Ball holding plays", the same player should be the ball holder at the two consecutive plays. Finally, we extract the features shown in Table 2 for each play occurrence and generate a feature vector after normalization.

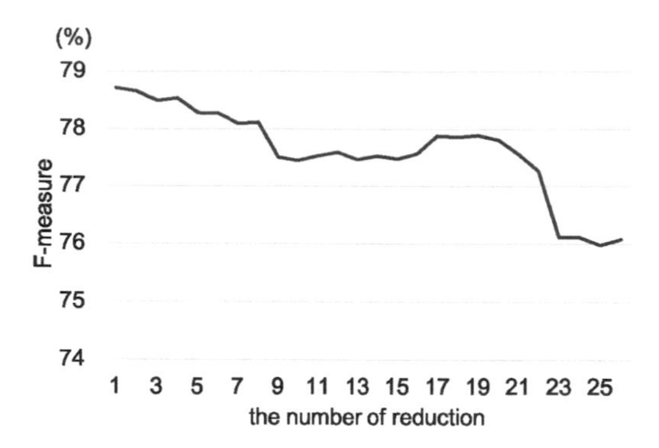

Fig. 2. Result of feature selection

Table 3. Selected features

type	value
Position	b_0, b_{-1}, b_1
	Backline of both teams at t_i
Distance	$D(p_0, p_1)$, $D(p_0, p_{-1})$
	$D(q_0, q_1)$, $D(q_0, q_{-1})$
	$D(b_0, b_1)$, $D(b_0, b_{0+s})$, $D(b_0, b_{-1})$
	$D(p_1, b_1)$, $D(p_{-1}, b_{-1})$, $D(p_{0+s}, b_{0+s})$
	$D(gk_0, b_0)$, $D(gk'_0, b_0)$
Direction	Goal approaching feature
Time	$t_i - t_{i-1}$, $t_{i+1} - t_i$
Play type	Play type classification result

For feature selection, we applied a space dimensionality reduction method [4] based on RF. We repeated removing the feature with the lowest importance until the decrease of the accuracy rate by the removal becomes more than 1%. Figure 2 shows the result of feature selection. The x-axis shows the number of removed features, starting from all the candidates in Table 2. From the result, we have selected 33 features shown in Table 3.

3.4 Play Recognition

We apply three machine learning algorithms: Conditional Random Field (CRF), Support Vector Machine (SVM), and Random Forest (RF). CRF considers sequential order of labels as well as observation. We implemented CRF by PyStruct [1]. SVM is known as one of the best learning models for binary classification [6]. RF is one of ensemble learning algorithms and attracting interest of many researchers since it is more robust to noise than single classifiers [2]. We used scikit-learn [9] for implementation of SVM and RF.

4 Evaluation

4.1 Evaluation Environment

We used tracking data of 20 games recorded by TRACAB in J. League games for evaluation. The ground truth was manually labeled by experts. We removed data in the following situations due to large tracking error: 10 s after kick off and set plays, from the occurrence of a foul to 3 s after restarting by free kick, after the occurrence of ball out, and after scoring a goal until the restart of play.

Since the ground truth was manually labeled, the labeled timings of the play occurrence are not exactly same as the real timings. Therefore, for evaluation, we empirically set the window size W to 0.2 s and allowed the difference within two windows (±0.4 s). We set the parameters of Eqs. (1) and (2) to $\theta_{TH} = 0.62$[radian] and $V_{TH} = 0.6$[m].

For evaluation of the play recognition, we conducted cross validation with 16 games as training data and 4 games as test data. The total number of plays is 28732. As shown in Table 4, the ratios of Pass and Trap account for 51.1% and 38.9% of the total plays, respectively. To remove the bias of unbalanced samples, we set weights of each class to reciprocals of the ratios the numbers of samples.

Table 4. Percentage of each class in evaluation data

Class	Percentage (number)	Class	Percentage (number)
Pass	**51.1% (14688)**	Clearance	1.1% (313)
Through ball	1.6% (454)	Trap	**38.9% (11173)**
Cross	1.7% (486)	Dribble	1.3% (373)
Shoot	0.8% (229)	Catch	0.9% (264)
Feed	0.9% (268)	Block	1.7% (484)

4.2 Result

4.2.1 Detection of Play Occurrence

For the detection of the play occurrence, our method achieved precision of 86.7%, recall of 85.3%, and F-measure of 86.0% on average. For the detailed discussion, recall of the play detection for each class is shown in Fig. 3. In particular, the recall of Block is 45.7%, which is extremely lower than the other classes. Block is the play that prevents the ball kicked by an opponent by using his body. Therefore, Block tends to occur in the same window as the last play, which is the cause of the low recall.

The false negatives are typically observed when the ball's trajectory hardly changes while a play occurs. On the other hand, the false positives are observed when the ball largely bounces off the ground near the player. This is because the speed of the ball greatly changes immediately after the bounce.

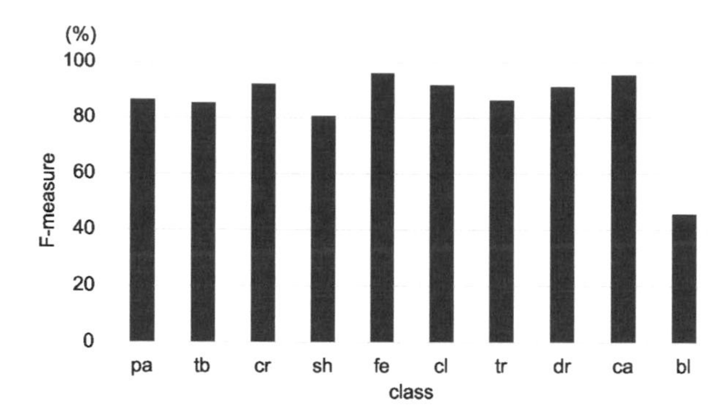

Fig. 3. Result of play detection

The tracking error of the ball is another possible reason lowering both precision and recall. In particular, since we use camera-based tracking data, we confirmed that the possibility of position error tends to increase when many players are close to the ball. In fact, since the recall of Shoot is less than 80%, it can be inferred that position errors in tracking data has a large impact. Therefore, we expect that if the accuracy of tracking data improves, the play detection accuracy will also improve.

4.2.2 Play Recognition

Table 5 shows F-measure of play recognition. We see that RF achieves the best performance. It is notable that the F-measure of Pass and Trap which account for 90% of plays, are over 85%. This highlights the usefulness of our method in terms of play labeling assistance.

For detailed analysis of play recognition, a confusion matrix of RF is shown in Table 6. "not play" in these tables are the cases detected as play due to wrong results of the play detection. From Table 6, there are many cases where Through Ball is wrongly recognized as Pass. This is due to the ambiguity of the definition between Through ball and Pass. The fundamental definition of Through ball is "a pass aimed at the back of the opponent's backline". However, there are many plays in which the label of Pass is given despite satisfying the definition of Through ball. Also, we confirmed that there are many cases where Trap and Dribble are often confused with each other. The both of Trap and Dribble have the same characteristic of moving the ball. Dribble is defined as "play with the

Table 5. F-measure of play recognition

(%)	pa	tb	cr	sh	fe	cl	tr	dr	ca	bl
CRF	85.3	38.1	64.4	38.3	64.0	22.6	83.2	**36.2**	61.4	**4.8**
SVM	85.8	33.7	61.4	27.9	61.9	23.2	83.1	25.6	**70.8**	0.7
RF	**86.0**	**43.2**	**69.2**	**44.4**	**71.9**	**34.3**	**85.9**	30.7	70.5	2.4

Table 6. Confusion matrix of RF

Ground truth \ Estimated	pa	tb	cr	sh	fe	cl	tr	dr	ca	bl
pa	**2162**	4	14	7	14	23	130	0	1	251
tb	**30**	22	1	2	0	0	3	0	0	13
cr	16	0	**63**	3	0	2	3	0	0	10
sh	7	0	2	**16**	0	2	7	0	0	2
fe	7	0	0	0	**46**	1	3	0	2	2
cl	13	1	1	4	0	**18**	6	0	1	6
tr	115	1	0	3	2	3	**1685**	14	3	83
dr	2	0	1	0	0	0	39	**14**	0	2
ca	1	0	1	0	0	1	3	0	**43**	2
bl	11	1	0	0	0	2	**12**	0	0	5
not play	59	2	2	1	5	3	122	5	21	8

intent to dribble past an opponent", which is difficult to distinguish from Trap. Since the definitions of these plays are ambiguous, it is difficult to distinguish them by tracking data only. To solve such problems, we may fuse additional information such as acceleration obtained by wearable sensors.

5 Conclusion

In this paper, we proposed a play recognition method based on tracking data of players and the ball in soccer. Our method focuses on the trajectories of players and the ball and extracts features considering characteristics of plays in soccer. We detect the occurrence of plays based on the trajectory of the ball, followed by play recognition based on machine learning. Through evaluation using real data, we have confirmed that RF achieves the best performance, especially which is highly accurate for recognizing Pass and Trap that account for 90% of plays. This result is promising to mitigate the heavy workload of manual play labeling. Our future work includes improvement of accuracy by information fusion such as acceleration obtained by wearable sensors.

Acknowledgements. This research is a part of the result of Sports Research Innovation Project (SRIP), FY2017 sponsored by Japan Sports Agency. We also thank Data Stadium Inc. for providing the data set.

References

1. Mueller, A.: PyStruct - Structured Learning in Python. https://pystruct.github.io
2. Breiman, L.: Random forests. Mach. Learn. **45**(1), 5–32 (2001)
3. ChyronHego Corporation: ChyronHego: TRACAB Optical Tracking. http://chyronhego.com/sports-data/tracab. Accessed 25 Jan 2018

4. Guyon, I., Elisseeff, A.: An introduction to variable and feature selection. J. Mach. Learn. **3**, 1157–1182 (2003)
5. Honda, M., Ikenaga, T.: Sports information analysis using sensing technology. J. Inf. Process. Soc. Jpn. **57**(8), 738–743 (2016). (in Japanese)
6. Hsu, C.W., Lin, C.J.: A comparison of methods for multiclass support vector machines. IEEE Trans. Neural Networks **13**(2), 415–425 (2002)
7. Kautz, T., Groh, B., Eskofier, B.: Sensor fusion for multi-player activity recognition in game sports. In: 21st ACM SIGKDD Conference on Knowledge Discovery and Data Mining (2015)
8. Mora, S.V., Knottenbelt, W.J.: Deep learning for domain-specific action recognition in tennis. In: Proceedings of IEEE Conference on Computer Vision and Pattern Recognition Workshops (CVPRW), pp. 170–178 (2017)
9. Pedregosa, F.: scikit-learn: machine learning in Python. https://scikit-learn.org/
10. Tora, M.R., Chen, J., Little, J.J.: Classification of puck possession events in ice hockey. In: Proceedings of IEEE Conference on Computer Vision and Pattern Recognition Workshops (CVPRW), pp. 147–154 (2017)
11. Tsunoda, T., Komori, Y., Matsugu, M., Harada, T.: Football action recognition using hierarchical LSTM. In: Proceedings of IEEE Conference on Computer Vision and Pattern Recognition Workshops (CVPRW), pp. 155–163 (2017)

The 7th International Workshop on Web Services and Social Media (WSSM-2018)

A Graphical Front-End Interface for React.js

Shotaro Naiki[1], Masaki Kohana[1], Shusuke Okamoto[2], and Masaru Kamada[1(✉)]

[1] Ibaraki University, Hitachi, Ibaraki 316-8511, Japan
masaru.kamada.snoopy@vc.ibaraki.ac.jp
[2] Seikei University, Tokyo 180-8633, Japan

Abstract. We present a graphical front-end interface for creating dynamical web pages by means of React.js. Its user does not have to write JavaScript codes for specifying the dynamical behavior of the web components but has only to draw state-transition diagrams graphically on the developed graphical editor. Using the graphical editor, the user composes a state transition diagram that specifies the dynamical behavior of each web component in terms of circles representing the states of the component and arrows representing the conditioned transitions among the states. Then the developed translator converts the state transition diagrams into web components of React.js in JavaScript that compose the target web page. This system of the graphical editor and the translator enables general users without knowledge and experiences in programming to create dynamical web pages. Wanna-be programmers may start learning JavaScript and React.js by comparing their diagrams and the translated JavaScript codes.

1 Introduction

It is hard for the general users to start creating dynamical web pages that change elements and contents in response to the user's operation since they are required of knowledges in not only HTML but also PHP, JavaScript and so on. React.js [1], a JavaScript library developed by Facebook, helps the programmer define the components and combine them to form the user interface of the web page. These components may have internal states and event handling programs for shifting the state to change the elements of the web page dynamically.

Islay [2], a tool for authoring interactive animation in terms of state-transition diagrams composed of circles and arrows, has made it possible for general users without programming knowledge to easily create animations and games by intuitive operations. That may be true of creation of dynamical web pages. Namely, the general users may easily create dynamical web pages by intuitive operations with a help from a graphical interface where they can create React.js components by drawing state-transition diagrams.

In this paper, we design such a graphical front-end interface and a translator into JavaScript. Using the graphical editor, the user composes a state transition diagram that specifies the dynamical behavior of each web component in

L. Barolli et al. (Eds.): NBiS 2018, LNDECT 22, pp. 887–896, 2019.
https://doi.org/10.1007/978-3-319-98530-5_79

terms of circles representing the states of the component and arrows representing the conditioned transitions among the states. Then the developed translator converts the state transition diagrams into web components of React.js in JavaScript that compose the target web page. This system of the graphical editor and the translator enables general users without knowledge and experiences in programming to create dynamical web pages. Wanna-be programmers may start learning JavaScript and React.js by comparing their diagrams and the translated JavaScript codes.

2 Preliminaries

2.1 React.js

React.js is a JavaScript library developed by Facebook. The user defines the components and combines them to form a web user interface. These components may have internal states and event handling programs for shifting the state to change the elements such as displayed texts and images.

An example component of React.js is shown in Fig. 1. This component has *count* as its internal state. When "Click" button is pressed, the state is updated by the handleClick function associated with the event handler of the <button> element to increment the number displayed in the <h1> element.

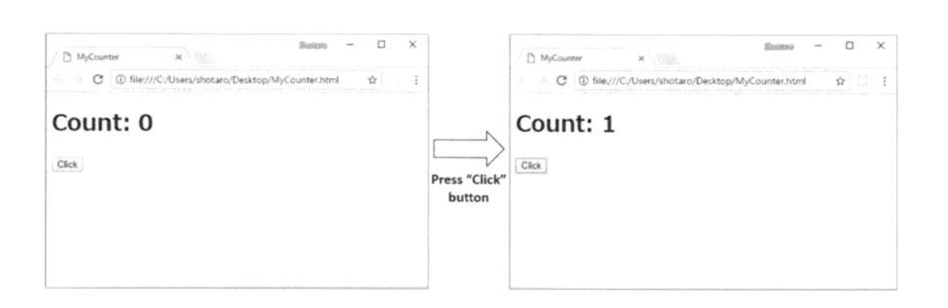

Fig. 1. Execution result of the component

2.2 Islay

Islay is a tool for authoring interactive animations in terms of state transition diagrams. Even general users without knowledge and experience in programming

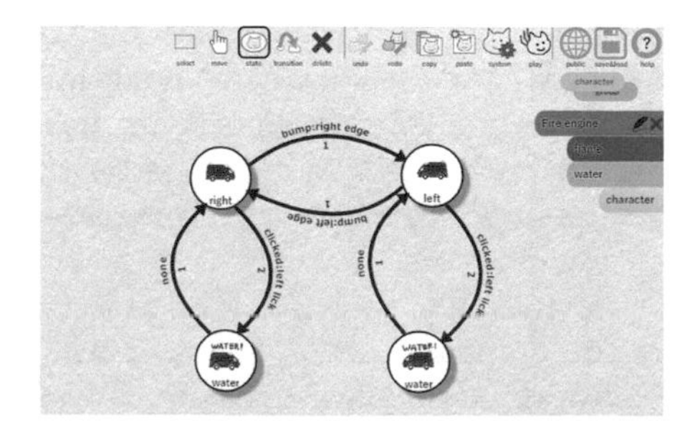

Fig. 2. IslayPub

can create animations and games by drawing state transition diagrams composed of circles and arrows. IslayPub [3] is a web service version of Islay of which a screenshot is shown in Fig. 2.

3 System Design

An overview of the whole system is illustrated in Fig. 3. This system is a web application implemented by means of HTML5, JavaScript and React.js. It is composed of an editor for drawing state-transition diagrams that defines dynamical behavior of the components of web pages, and a translator that converts the state-transition diagrams into JavaScript codes. The internal data structure is saved to a memory in the form of an XML document.

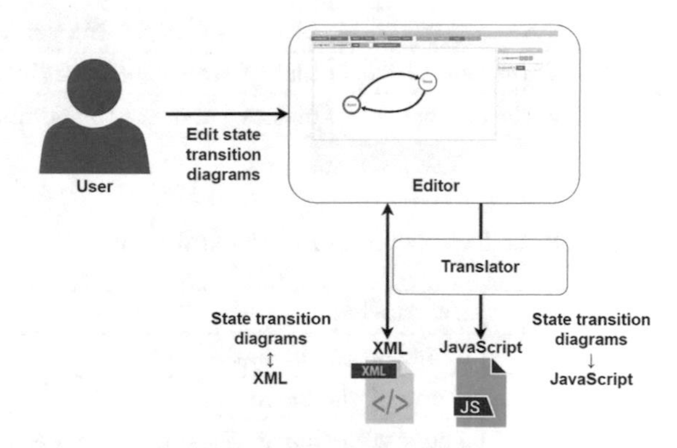

Fig. 3. System overview

3.1 Editor

3.1.1 Data Representation

This system expresses each component constituting a web page and its behavior in terms of a state-transition diagram. A state-transition diagram is made up of *component* objects, *state* objects, *action* objects, *transition* objects and *parts* objects. The component object is the top-level object which relates the id and name of the component to the state objects, transition objects and part objects in the data structure shown in Table 1. A state object represents the states of which the data structure is shown in Table 2. Its data include the id, name and graphical position in the editor canvas and the list of associated actions. An action object represents the action taking place in the state of which the data structure is shown in Table 3. A transition object represents a conditioned transition from a state to another and its data structure is shown in Table 4. A parts object represents a part making up the component of which the data are mainly the HTML element as shown in Table 5.

The whole data of the state-transition diagrams managed by the editor is saved in the XML format. The structure of XML handled in this system is illustrated in Fig. 4. The <component> element represents the component object, the <state> elements represent the states, the <action> elements represent the actions, the <transition> elements represent the transitions, the <parts> elements represent the parts that make up the component, and the <render> element represents the name of the component to render.

Table 1. Data structure of the component object

Attribute	Type	Description
id	string	Identifier of the component
name	string	Name of the component
stateList	array(State object)	List of states for the component
transitionList	array(Transition object)	List of transitions for the component
partsList	array(Parts object)	List of parts for the component

Table 2. Data structure of the state object

Attribute	Type	Description
id	string	Identifier of the state
name	string	Name of the state
x	integer	The x coordinate of the circle representing the state in the canvas
y	integer	The y coordinate of the circle representing the state in the canvas
actionList	array(Action object)	List of actions for the state

Table 3. Data structure of the action object

Attribute	Type	Description
name	string	Name of the action
option	string	Option for the action

Table 4. Data structure of the transition object

Attribute	Type	Description
id	string	Identifier of the transition
path	string	Path data of the arrow representing the transition
from	string	State name of the transition source
to	string	State name of the transition destination
condition	string	Transition condition
option	string	Option for the transition condition

Table 5. Data structure of the parts object

Attribute	Type	Description
name	string	Name of the part
tag	string	HTML element of the part
option	string	Option for the part

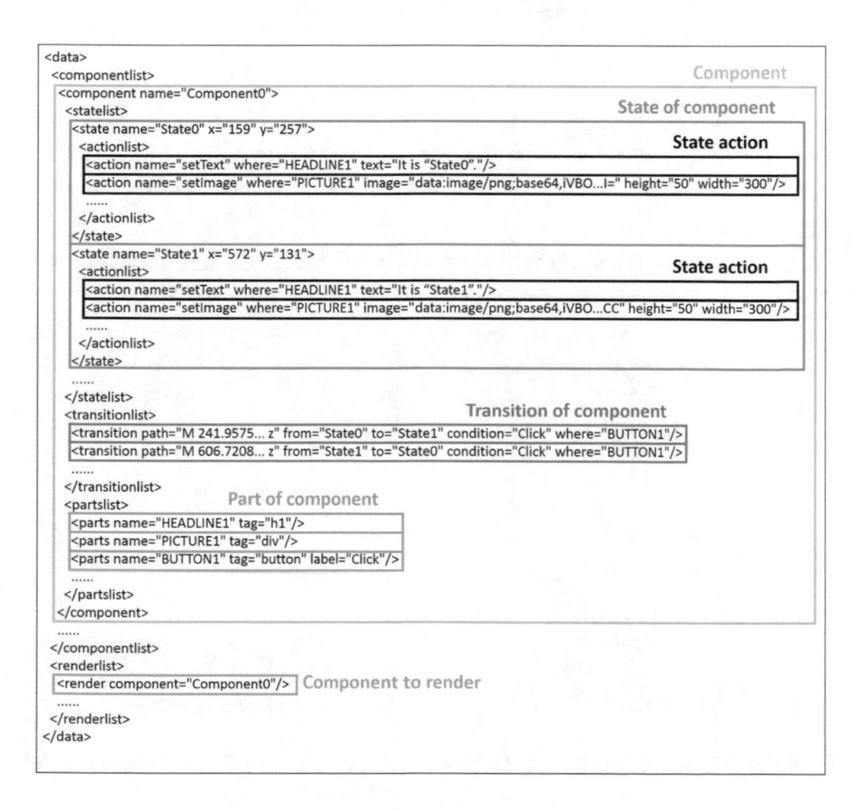

Fig. 4. Structure of XML

3.1.2 Functions

Edit State Transition Diagram

A state-transition diagram of each component is drawn on the canvas of the editor. The user clicks in the empty area on the canvas to create a circle representing a new state. In order to create a transition from a state to another, the user clicks on an existing circle representing the source state and then clicks on another circle representing the destination state. A screenshot of the editor canvas is shown in Fig. 5.

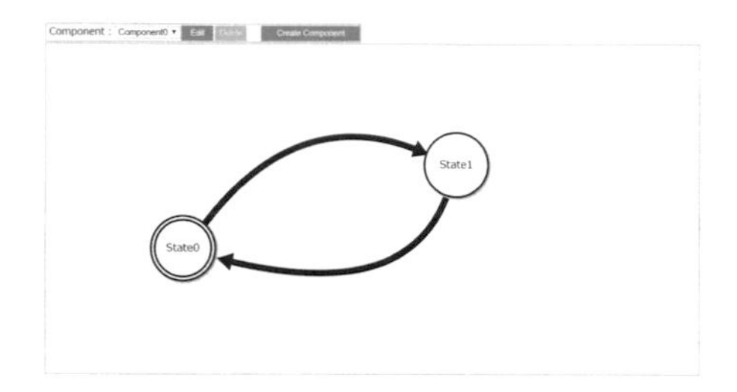

Fig. 5. Canvas of the editor

Edit State

A right click on a circle pops up a dialog for editing properties of the state such as shown in Fig. 6 where the user can set the state name and the actions to be taken at the state such as displaying a text or an image at HTML objects.

Edit Transition

A right click on an arrow pops up a dialog for editing the transition such as shown in Fig. 7 where the user sets the condition for the transition.

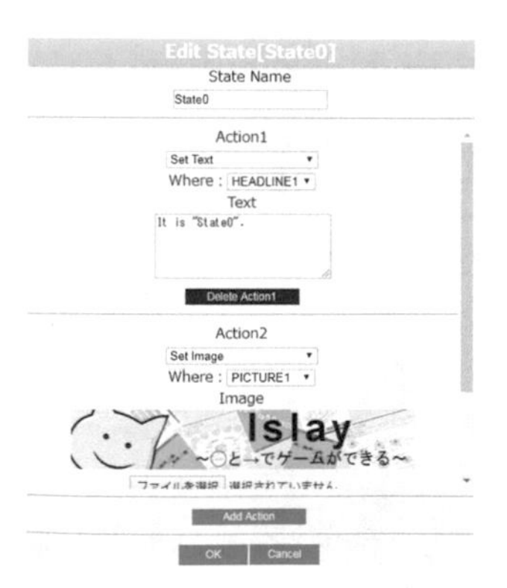

Fig. 6. Dialog for editing actions in the state

Fig. 7. Dialog for editing the condition of transition

3.2 Translator

3.2.1 Parsing XML

The state transition diagrams in the XML format is first parsed to get their internal representation in the data structure illustrated in Fig. 8.

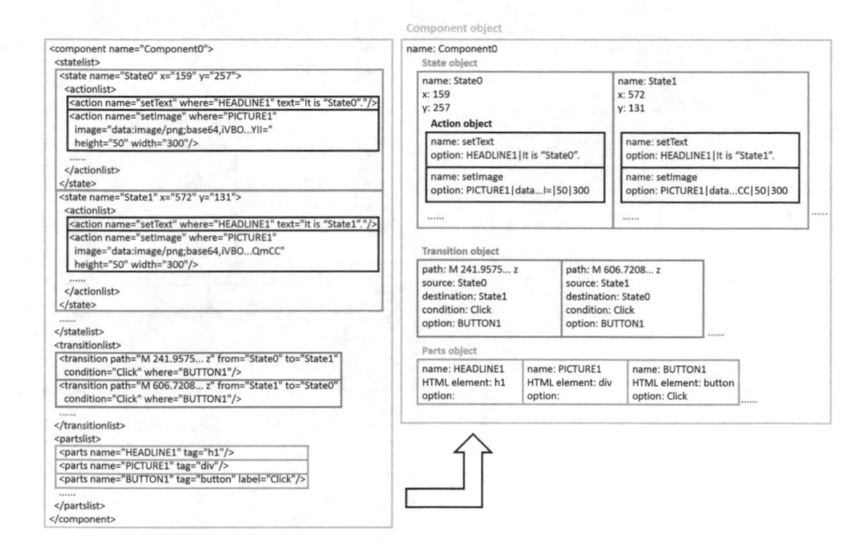

Fig. 8. Correspondence between XML and objects representing state transition diagrams

3.2.2 Code Generation

Then the translator converts the internal data of objects representing state transition diagrams into the JavaScript code in accordance with the notation of React.js. The correspondence between objects representing state transition diagrams and JavaScript codes is illustrated in Fig. 9.

The component object in the internal data and the component class in JavaScript have one-to-one correspondence. On the side of JavaScript, a component class is given its class name the same as the component name and its instance is constructed in the initial state. Each function in the component class is generated from each attribute of state objects, action objects, transition objects, and parts objects. In that way, the JavaScript code for the component class is completed to include the constructor, functions to execute the state actions, functions to transit among the states, and the function to render the HTML component.

Constructor

The constructor sets the name of the initial state. The state name of the first state object in the component object is used for the name of the initial state. The correspondence between the state object and the constructor is illustrated in Fig. 10.

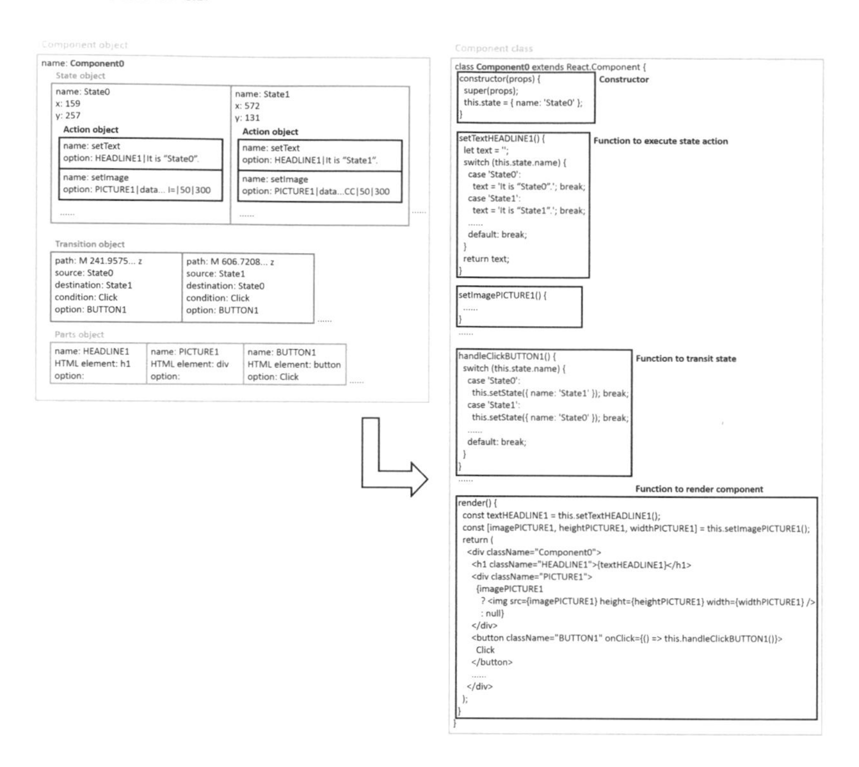

Fig. 9. Correspondence between objects representing state transition diagrams and JavaScript codes

Functions to Execute State Action

The function to execute the state action is generated according to state objects and action objects, and its name is decided by concatenating the type of action such as setText and setImage and the place such as HEADLINE1 or PICTURE1 where the action effects. So each function is associated with a pair of a specific place and an action. A switch statement is used to select the action according to the name of the current state. The correspondence between state objects, action objects and functions to execute the state action is illustrated in Fig. 11.

Function to Transit State

The function to implement transitions to shift a state to another is generated according to the transition objects. This function is associated with each event conditioning the transitions. A switch statement is used to select the source state name according to the name of the current state and to set the destination state as the new current state. The correspondence between transition objects and the function to shift the states is illustrated in Fig. 12.

Function to Render Component

The render function defines the shape of the component consisting of the parts objects as well as the behavior of the component defined by the other functions for state actions and transitions. Each parts object is converted into HTML elements written within the argument of return in Fig. 13. To each HTML element, event listeners such as onClick are attached to detect the events that may condition the transitions.

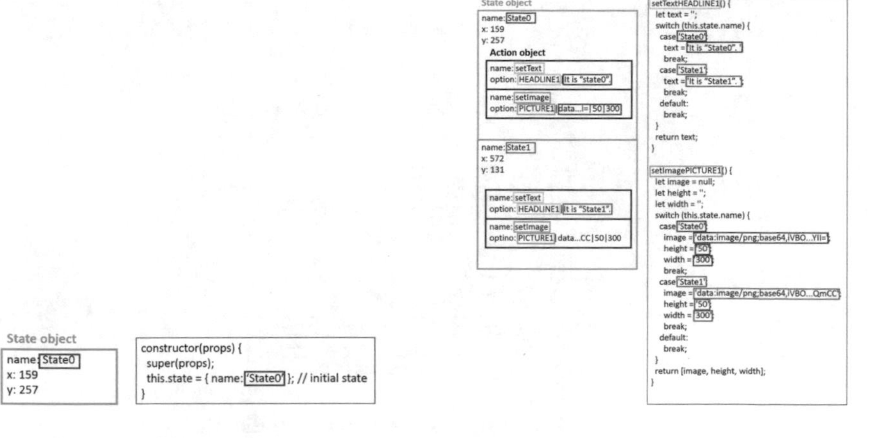

Fig. 10. Correspondence between the state object and the constructor

Fig. 11. Correspondence between state objects, action objects and functions to execute the state action

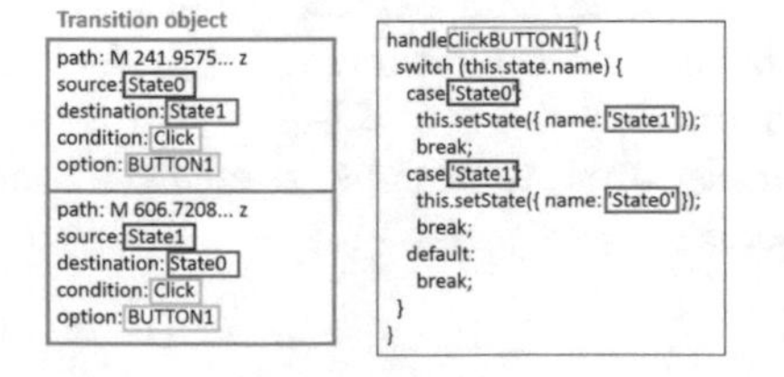

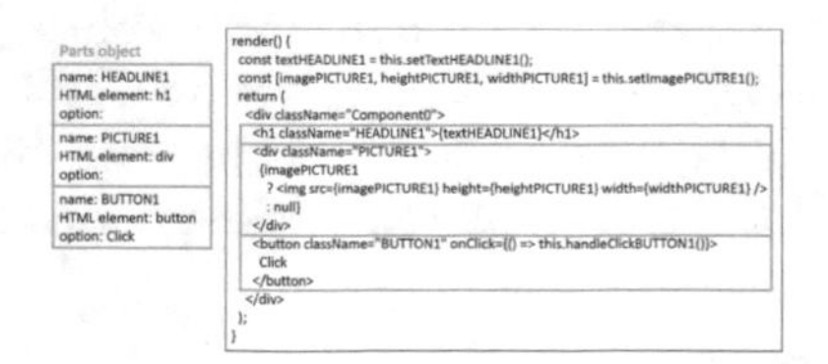

Fig. 12. Correspondence between transition objects and the function to shift the state

Fig. 13. Correspondence between parts objects and the function to render the component

4 Conversion Example

An example of a dynamical web page created by using this system is shown in Fig. 14. Its associated state transition diagram is shown in Fig. 15. This web page has two states. When "Islay" button or "Web System" button is pressed at the upper left corner, the state shifts accordingly to the state named "Islay" or "WebSystem" where text color, text content, image, and headline background color are changed by the specified actions.

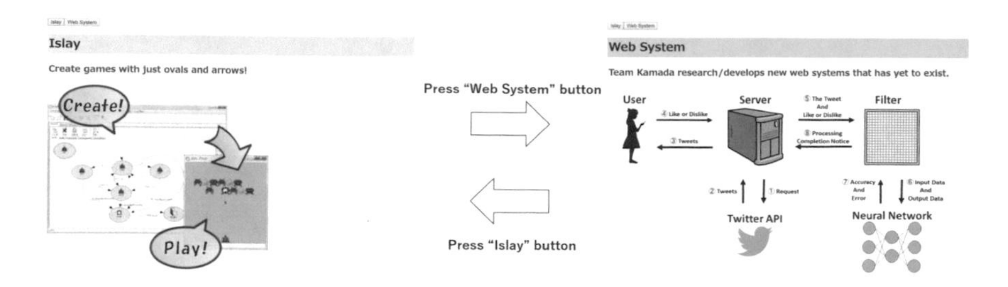

Fig. 14. Example web page

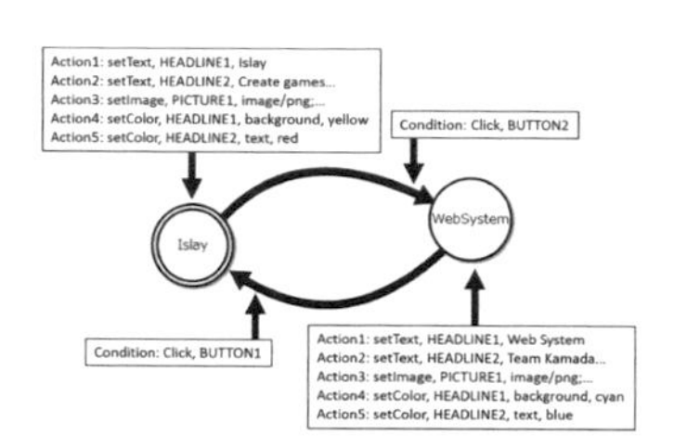

Fig. 15. Example state transition diagram

5 Conclusions

We developed a graphical front-end interface for React.js to create dynamical web pages. Even general users without knowledge and experience in programming can create dynamical web pages by simple operations. It will also help programming beginners start learning JavaScript and React.js.

The future task is to increase the variety of state actions, transition conditions, and parts that make up the component.

References

1. React - A JavaScript library for building user interfaces. https://reactjs.org/. Accessed 23 Jan 2018
2. Okamoto, S., Kamada, M., Nakao, T.: Proposal of an interactive animation tool based on state transition diagram. IPSJ Trans. Program. **46**(SIG1(PRO24)), 19–27 (2005)
3. Suzuki, K., Niibori, M., Rashed, A.S., Okamoto, S., Kamada, M.: Development of IslayPub3.0 – educational programming environment based on state-transition diagrams. In: The 4th International Workshop on Web Service and Social Media (WSSM 2015), Proceedings of the 18th International Conference on Network-Based Information Systems, (NBiS 2015), pp. 702–705, Taipei (2015)

An Attendance Management System Capable of Mapping Participants onto the Seat Map

Shinya Kinoshita, Michitoshi Niibori, and Masaru Kamada(✉)

Ibaraki University, Hitachi, Ibaraki 316-8511, Japan
masaru.kamada.snoopy@vc.ibaraki.ac.jp

Abstract. We present an attendance management system where the student and the seat position are identified by IC cards of the FeliCa standard. Seated in a classroom, each student first touches his/her student ID card of the FeliCa standard with his/her own Android smartphone and then touches another FeliCa card fixed on the desk at his/her seat. Then the developed Android application software sends the student ID and the seat ID to the server. The developed service program produces the seating list with the student names mapped onto the seat map of the classroom. The seating list can be updated instantly as soon as the students touch the cards after getting seated or moving to other seats in the classroom. Looking at the seating list as a web page, the teacher can easily identify and call on each student. The seating list also helps the teacher spot a blank seat or a student missing, or equivalently, cheating in attendance.

1 Introduction

The attendance management systems for schools are required to complete the list of attending students swiftly and easily at every lecture. Many ideas have been presented and implemented to date.

The "IC Card de syussekichosa" [1] is a PC application developed in 2006. Each student touches an external NFC card reader with his/her student ID card so that the application software stores the IDs to a text file. It saved the teacher's labor of calling all the names of students but took much time until all the students finish touching the card reader.

The "i-Compass" [2] is a network-based system developed in 2012. The teacher shows a password of the day for the class to the students. Student send their ID and the password from their own mobile phones to the server for registering their attendance. The list of attending students is instantly completed by employing the mobile phones owned by the students. The server monitors the device identification number of the mobile phone to prevent a second registration attempt from the same phone.

L. Barolli et al. (Eds.): NBiS 2018, LNDECT 22, pp. 897–902, 2019.
https://doi.org/10.1007/978-3-319-98530-5_80

Another system [3] developed in 2015 also employed the students smart phones to shorten the time. Each student touches an Android smart phone capable of NFC communications of his, her or a near-by friend's own to send the student ID to the server. Cheating was prevented by locally broadcasting a secret code via Bluetooth Low Energy Beacon within the classroom.

A different approach [4] has been made in 2016 where a smart phone or tablet is circulated in the classroom on which each student hand-writes signature or takes the portrait of his or her own. The list of students with the signatures and portraits aggregated in the chronological order much helps the teacher spot possible cheating in attendance registration.

A yet another approach [5] has been taken since 2012 where it is supposed that every seat has a unique QR code to identify the seat. Each student scans the QR code pasted on the desk at the seat he or she is sitting by his or her own smart phone to notify the server of the seat position and logs in to identify themself. It is advantageous to have the list of attending students arranged in the shape of the seat map since the teacher can easily identify and call on each student and because the teacher can spot a blank seat or a student missing, or equivalently, cheating in attendance. The only disadvantage of this approach was that the QR codes may be damaged by time.

In this paper, we replace the QR codes of [5] by IC cards of the FeliCa standard [6] and exploit the student ID card of the same standard. Each student touches a smart phone capable of the NFC standard [7] with his or her student ID card and then touches the IC card pasted on the desk at his or her seat to notify the server of the seat position and student ID. No students are required to log in the server so that students out of their smartphone or its battery power can register their attendance via a near-by fried's smartphone.

2 System Design

This system consists of a student's Android application, a web application for management of the courses and the classrooms, and an Android application for management of the classroom seat maps.

The teacher first logs in the web application for management of the courses via a PC or a mobile terminal device, selects "collect attendance" from the function list, and chooses the course name. Then the seat map of the classroom is displayed on the web browser. At the same time, the server gets ready to receive attendance registrations from the student's android application.

Each student selects the course name on the student's Android application, scans his or her student ID card to read the student ID, and then scans the IC card pasted on the desk at the seat to read its FeliCa ID. Then the student's application sends the student ID and the FeliCa ID to the server. Every time the server receives a pair of the IDs, the corresponding seat is marked with a color to indicate that a student is in the seat map displayed on the teacher's web browser.

Figure 1 shows an overview of the system for collecting attendances.

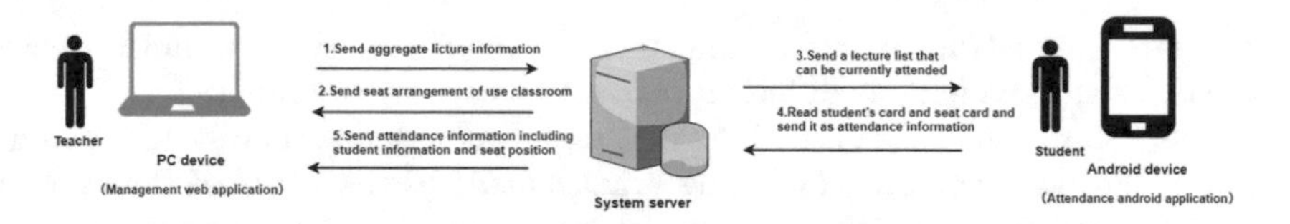

Fig. 1. Overview of the system for collecting attendances

In addition to those two main applications, we have another Android application for management of the classroom seat maps. It is used to scan IC cards newly pasted on the desk and associates their FeliCa IDs to the seats in the seat map stored in the server.

2.1 Student's Android Application

This is an Android application for students to send attendance information to the server. On launch of this application, a message prompts the student to scan a student ID card. The student scans the student ID card by the smart phone that acquires the student information via its NFC function as shown in Fig. 2(a). Then the courses that are available for the students at the time is displayed on the screen.

On selection of a course, a message prompts the student to scan the IC card pasted on the desk at the seat as depicted in Fig. 2(b). Its FeliCa ID that identifies the seat position is acquired in the same manner as in the case of the student card.

The course information, the student information, and the seat position information are simultaneously sent to the server via PHP in the JSON [8] format.

In case an attending student does the same at another seat, the student is supposed to have moved the seat within the classroom.

(a) Scan the student ID card

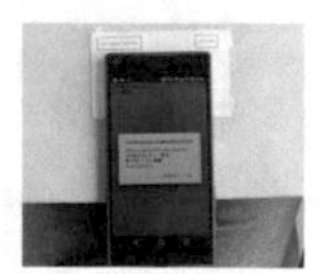
(b) Scan the IC card pasted on the desk

Fig. 2. Steps to register attendance

2.2 Web Application for Management of the Courses and the Classrooms

This is a web application that collects attendance and manages the information processed by the server system. The users are divided into four categories of system administrators, course managers, student supervisors, and ordinary teachers, each of which has different authority to access the functions.

Table 1 shows the list of functions that each user can use. The system administrators can do everything including management of the users of the system. Course managers can enter and edit the seat map of the classrooms and the courses. Student supervisors can check the attendance/absence of all the students across all the courses. Ordinary teachers can collect the attendance status of the course in charge and access the seat map filled with the students, and review its history.

Table 1. Functions that each category of users can use

Users	User management	Register courses	Editing courses	Register classroom	Editing classroom	Attendance collection	Seating list	Attendances /Absences
Administrators	✓	✓	✓	✓	✓	✓	✓	✓
Course managers		✓	✓	✓	✓			✓
Student supervisors						✓	✓	✓
Teachers						✓	✓	

After logging in the system via a web browser, an ordinary teacher selects a course from the list and starts collecting attendances. Then the system gives a web page that shows the seat map of the classroom as Fig. 3. In the initial state, the seat is painted in gray to indicate none is at the seat. Everytime a student scans the student ID card and the IC card on the desk, the seat map is updated by means of the asynchronous communication Ajax [9] to paint the seat in green and put the name of the student sitting in the seat. The teacher can add and edit a memo by double-clicking the seat. The yellow seat has a memo attached to the seat. The screen that shows attendee details is shown in Fig. 4.

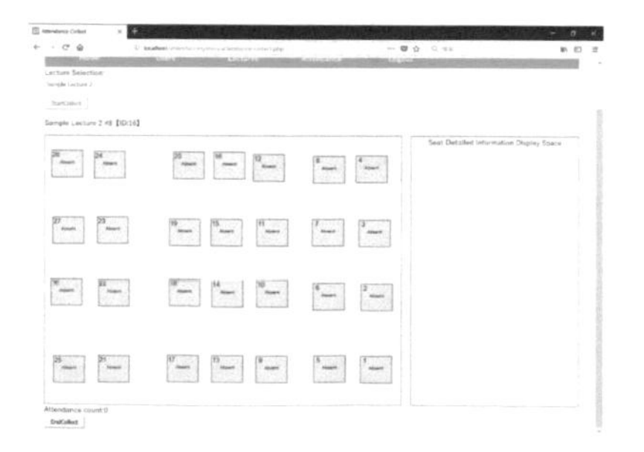

Fig. 3. Seat map screen at the start of attendance collection

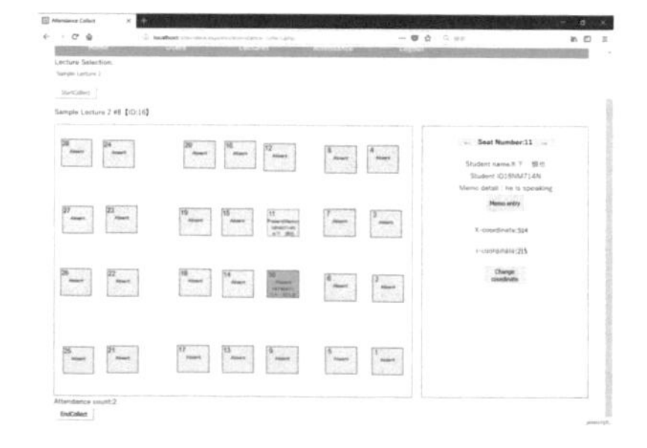

Fig. 4. Screen showing details of attendees

The system administrators enter and edit the list of teachers, course managers, student supervisors who can use this system and the students who are managed in the system.

Course managers can add and edit the layout of seats in the classroom to be displayed as the seat map on the graphical editor shown in Fig. 5. A left click at a blank space creates a new seat. The seats can be dragged around on the canvas to be placed anywhere. A double click on a seat shows its editable properties in detail such as the coordinates in numerics. Each seat will be related to the FeliCa ID of the card pasted on the desk acquired by the following Android application for management of the classroom seat maps.

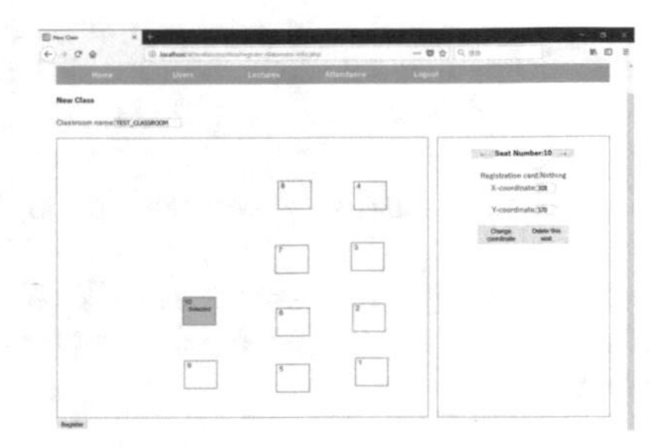

Fig. 5. Screen for placing seats and displaying the detailed information

2.3 Android Application for Management of the Classroom Seat Maps

This is an Android application for relating a new IC card that is being pasted on the desk to the seat in the seat map of the classroom. The system administrators or course managers are supposed to be in the classroom. They use this application on a mobile device to browse a seat map of the classroom. Pressing a seat, the user is asked to register a new card ID to the seat. The user touches a new IC card pasted on the desk to acquire its FeliCa ID (e.g. 12e38f77416998a). In addition to this globally unique ID, we append another ID (e.g. IU001001) for the purpose of systematic management of the card IDs within the school.

3 Conclusion

In this paper, we developed an attendance management system that collects the student IDs from the student ID cards and the seat IDs from the IC cards pasted on the desk via an Android device. No students are required to log in the server so that students out of their smartphone or its battery power can register their attendance via a near-by fried's smartphone. The seating list of the students is presented to the teacher graphically on the web browser.

References

1. The Center of Information Technology, Ibaraki University, IC card de syussekichosa. http://www.ipc.ibaraki.ac.jp/aboutus/development.php. Accessed 7 May 2018
2. i-Compass. http://i-compass.biz/. Accessed 7 May 2018
3. Noguchi, S., Niibori, M., Zhou, E., Kamada, M.: Student attendance management system with Bluetooth Low energy beacon and Android devices. In: Proceedings of the 18th International Conference on Network-Based Information Systems, (NBiS 2015), pp. 710–713, Salerno, Italy (2015)
4. Iio J.: Attendance management system using a mobile device and a web application. In: Proceedings of the 19th International Conference on Network-Based Information Systems, (NBiS 2016), pp. 510–515, Taipei (2016)
5. Honda, N.: Development and practice of attendance management system with QR code and cellular phones. Otemae J. **12**, 253–262 (2011). (in Japanese)
6. Sony Japan — FeliCa. https://www.sony.co.jp/Products/felica/. Accessed 7 May 2018
7. Sony Japan — FeliCa — NFC. https://www.sony.co.jp/Products/felica/NFC/. Accessed 7 May 2018
8. Introducing JSON. http://json.org/. Accessed 7 May 2018
9. Ajax Developer guides. https://developer.mozilla.org/en-US/docs/Web/Guide/AJAX. Accessed 7 May 2018

An Image Source Checker for Educational Presentation Materials

Yasuhiro Ohtaki(✉)

Ibaraki University, Hitachi, Ibaraki, Japan
yasuhiro.ohtaki.1cars@vc.ibaraki.ac.jp

Abstract. In this research, we propose a web-based system to check source URLs of image contained in a presentation material. In higher education, presentation slides are widely used as educational materials in class. These materials are often uploaded to a Learning Management System (LMS) in order to provide them to class students. To make their presentation more impressive, images downloaded from the Internet may be used in the presentation. Under the copyright law of Japan, it is not permitted to upload these materials to a web server, i.e., to LMS without permission of the copyright owner of the image. It would be helpful for a center managing the LMS to have a web based system which can check the presentation materials and show some advice to avoid copyright issues. The proposed system extracts all the images contained in the presentation file uploaded to the system and displays several source URL candidates of the image if they are likely to have been obtained from the Internet. By checking the copyright notice or terms of use indicated on the page of the shown URL, the center can make appropriate advice to avoid copyright issues to the creator of the presentation slides.

1 Introduction

1.1 Background

In higher education, presentation slides are widely used as a teaching materials in class. These materials are usually created by teachers themselves. To make their presentation more impressive, visuals such as pictures, images, icons, graphs are used. However not many teachers are good at taking pictures nor drawing a nice illustration. So they may use pictures downloaded from the internet, which may be copyrighted by others.

In order to provide teaching materials to students, these materials are often uploaded to an e-learning system or LMS (Learning Management System). These systems are usually Web-based systems and students can access uploaded materials with their own devices such as laptop computers and tablets. In general, access to uploaded materials is limited to students taking that class. In some cases, it may be open to the public and may be accessed by an unspecified number of people.

L. Barolli et al. (Eds.): NBiS 2018, LNDECT 22, pp. 903–912, 2019.
https://doi.org/10.1007/978-3-319-98530-5_81

1.2 Copyright Law of Japan

Under the copyright law of Japan as of May 2018, author of a work which use reproductive work copyrighted by others shall inform and pay to the copyright owner of these images a reasonable amount of compensation. The copyright law also limits the right of copyright owners in special situation. Reproduction in schools and other educational institutions are permitted to the extent that they do not violate the legitimate interests of copyright holders. Thus school teachers can reproduce images copyrighted by others in his/her educational materials. This means that he/she is permitted to use these images in his/her slides without obtaining the permission of the copyright holder of the work, and use them in the class.

However, uploading such materials on a web server is not covered by this limitation. Uploading to a web server is considered to be distribution to the general public, and it is believed to be against the legitimate interests of the copyright owner. It does not matter whether access is restricted to a certain number of people. Since e-learning system such as LMS are usually build on a web service, uploading to LMS is handled as transmission to the public even if access control is actually done.

To avoid copyright issues related to such educational materials, the most easiest way is to make sure that the slide creator does not use such problematic images. Faculty members would appreciate it if the LMS Administration Center checks whether the slide contains images with problems when they are going to upload the image to the LMS.

1.3 Purpose

We propose a web-based system to check source URL of images contained in a presentation slide. Figure 1 shows the system overview.

A user uploads a presentation file to the system. The proposed system extracts all the images contained in the presentation file uploaded to the system. Then the system search for the source URL of each image using an external

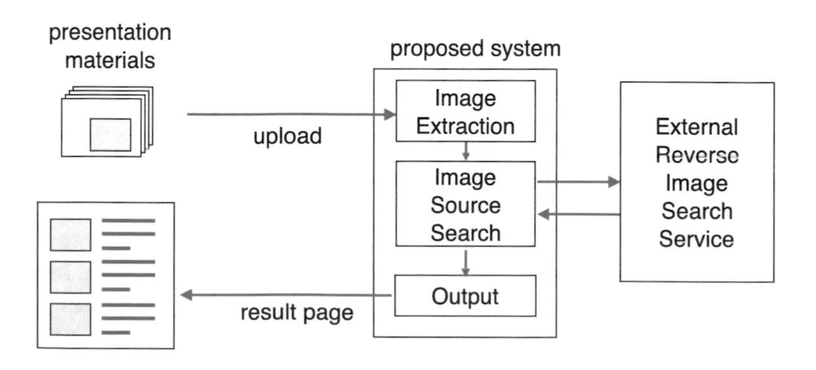

Fig. 1. System overview.

reverse image search service. Finally the system displays several candidate source URLs of the image if they are likely to have been obtained from the Internet.

By checking the copyright notice or terms of use indicated on the page of the displayed URL, the LMS Administration Center can make appropriate advice to avoid copyright issues to the creator of the presentation slides.

We also implemented a prototype system which can handle presentation slides created with Microsoft PowerPoint Application.

2 Preliminaries

2.1 Format of Microsoft `.pptx` file

Microsoft's PowerPoint is the most popular application to create presentations. Presentation files created with PowerPoint have two file formats: `.ppt` file and `.pptx` file. The `.ppt` files are old PowerPoint files which store all presentation data in a single binary file, and are used by Microsoft PowerPoint 97, Microsoft PowerPoint 2000, Microsoft PowerPoint 2002, and Microsoft Office PowerPoint 2003. The `.pptx` files, on the other hand, are used by Microsoft 2010 and later. It is based on the Open XML format and are zip archives containing a separate file for each main component of the presentation along with other files which contain metadata about the overall presentation and relationships between its components. Therefore, by simply changing the file extension from `.pptx` to `.zip`, we can treat a `.pptx` file as a zip file. As shown in Fig. 2, when you unpack the zip file, all the images used in the presentation will be placed in the "media" directory under the "pptx" directory. Note that only one file is saved

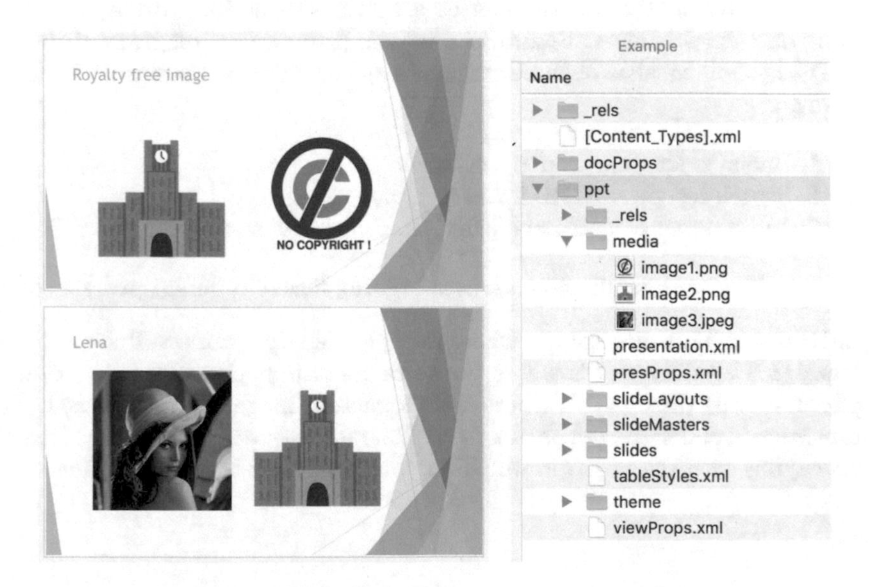

Fig. 2. Slide example and its unpacked structure.

under the "media" directory, even if you copy and use the same image more than once in a presentation file.

2.2 Google Search by Image

To discover the source URL of images, we are using an external reverse image search service, which enables us query with a sample image. One of the famous reverse image search web service is Google Search by Image. This service allows us to discover pages that contain the specific sample image, and also discover manipulated versions and derivative works.

Google once provided the Google Image Search JavaScript API but it has been officially deprecated as of May 26, 2011. Google recommend using a custom search API instead, but only the usual image search API is provided and no reverse image search API was found.

We can still use a regular web interface instead of API. There are two way to pass an input image to the web interface. One is to upload an image directly from the search form, and the other is to pass a URL of the image. If the image is locatable by a URL and accessible via HTTP, the latter is much easier to implement in a server side script. We can easily build a URL in the following form and open it with a HTTP request handler.

http://www.google.co.jp/searchbyimage?image_url=*URL of input image*

3 Prototype Implementation

In this section, we describe our implementation of the prototype. The web application is written in Python running as a server side script with Apache 2.4.6 running on CentOS Linux. Figure 3 shows the process flow diagram of the system. The system consists of three stages: Image extraction, Source URL Search, and Output.

Stage 1: Image extraction from a `.pptx` file
Stage 2: Searching source URL for each image
Stage 3: Building an output page based on the search results

There are three working directories, each holding files of different process stages.

`upload_dir` This directory is where an uploaded `.pptx` file is first placed.
`exp_dir` This directory is where the contents of unpacked file are placed.
`pic_dir` This directory is where the extracted image files are placed. This directory is under the document root of the web system. This means that, by specifying with the appropriate URL, you can access these image files.

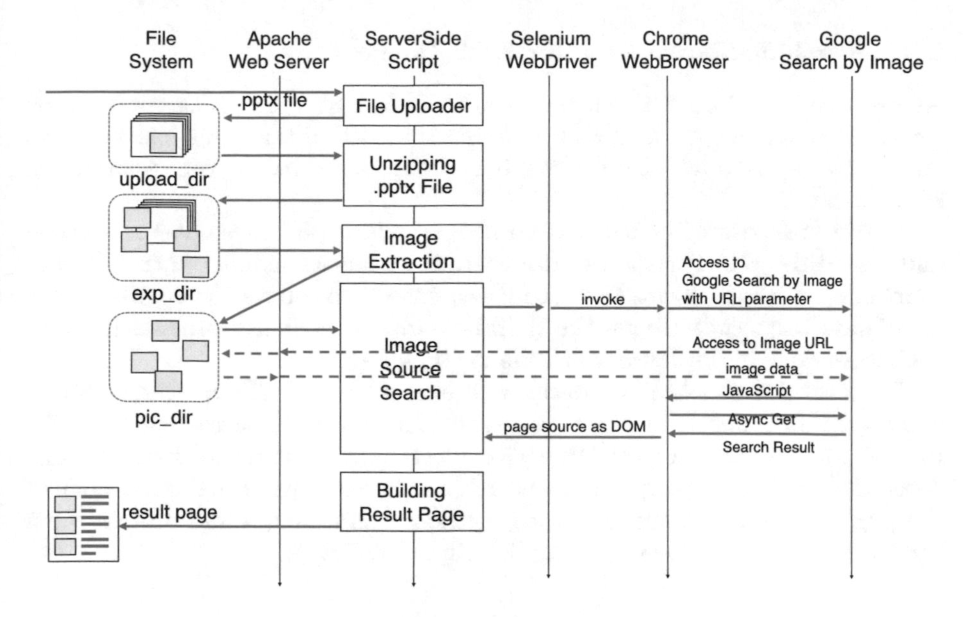

Fig. 3. Process Flow Diagram of the proposed system.

3.1 Image Extraction from Uploaded `.pptx` file

When a user access the top page of our system, a simple form to upload a `.pptx` file is displayed (Fig. 4). The user select a presentation file and click the Submit button. The uploaded file is first placed under the `upload_dir` directory.

As described in Sect. 2.1, it is quite easy to extract images from a `.pptx` file. The system first converts the file extension of the file from `.pptx` to `.zip`, and then unzip the file under the `exp_dir`. Here we have all image files under the "media" directory under the "pptx" directory. Finally, the images are moved from "media" directory to under `pic_dir` directory.

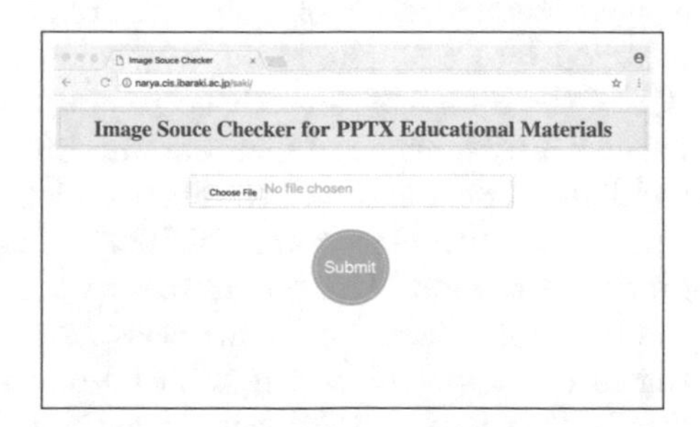

Fig. 4. File upload page.

3.2 Searching Source URL for Each Image

As mentioned in Sect. 2.2, you can use the Google Search by Image service by simply accessing a specific URL containing the URL of the input image. When you input the URL into the address bar of any web browser, you get a search result page.

So we first passed the same URL to a standard HTTP handler in Python and stored the HTTP response. However, the response didn't return the page source of the search result. The HTTP response of the above Get request was a set of JavaScript code. To get the HTML source of the final search result page, we had to execute the JavaScript code in some way.

To solve this problem, we decided to execute the JavaScript code using a standard web browser. In other words, instead of directly accessing the URL from Python, we accessed the URL with a web browser controlled from Python. After the web browser displays the search result, we retrieve the source code of the page as a search result also using Python. Controlling a web browser from Python script can be done by using Selenium WebDriver.

3.2.1 Selenium WebDriver

Selenium WebDriver [3] is a set of library to control a web browser from a remote process. We can control the browser interactively by simulating input to the form, including keystrokes and mouse click operation.

Selenium WebDriver can be used from several languages such as Java, Ruby, Python, JavaScript etc. Libraries are also provided from the Selenium official site. Web browsers which can be controlled from Selenium WebDriver includes Internet Explorer, Chrome, Firefox, Opera and Safari. Selenium also provides several methods to access the source of displayed page directly as a DOM object. These methods include a way to obtain a part of source with a CSS selector (ID name or class name) or tag name to obtain part of the page source.

3.2.2 URL Extraction

The source of the displayed search result of Google Search by Image is a HTML code with layout controlled by CSS. Therefore, we thought that we can easily extract the source URL of the image by directly specifying a proper tag name or class name. Figure 5 shows a screenshot of Chrome web browser with developer tool activated. The highlighted part is the URL source which we want to extract. We can sec that the `cite` tag with class name "iUh30". This class names seem to change randomly for each session. There are other randomly generated class names as shown in the frame in Fig. 5, and same class name is used in multiple places. The closest tag in the upper hierarchy with ID name is far way up and is the one with ID "rso". This makes it difficult to extract the source URL by simply specifying with a fixed ID and trace down the child nodes. Therefore we had to determine adaptively if there was a page which contains the exact same image or not by counting the number of children tags under a specific ID tag.

Fig. 5. Screenshot of web browser with developer tool activated showing the result page of Google Search by Image.

3.3 Building an Output Page

In this stage, the system builds an output page. The page consists of a set of source URL with each unique image. If the result of Google Search by Image contains no URL for the image, then the system shows "No source found" message instead of URLs.

Each source URL is reformatted as an anchor link, so that the user can visit and check the content of the page easily. Figure 6 shows the result page when the `.pptx` file shown in Fig. 2 is uploaded to our system. We are planning to display some more useful message with these URLs, which will discussed later.

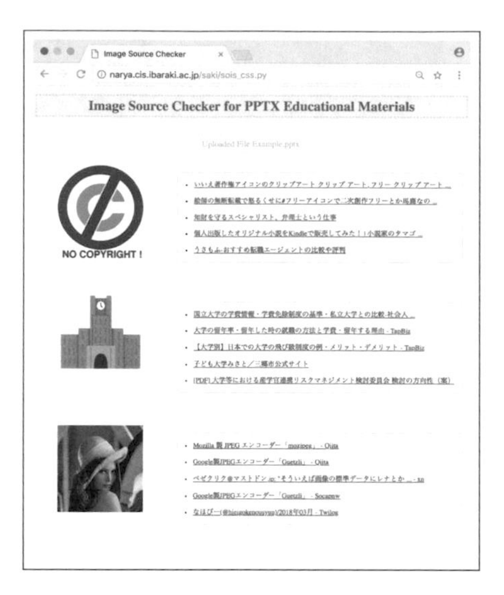

Fig. 6. The result page of the proposed system when the slides shown in Fig. 2 was uploaded.

4 Discussion

4.1 Processing Speed

The prototype is very slow and we need far more work to make this system practical.

The total elapsed time for searching three images was 9.15 s. The most time consuming part is Image Source Search function using external Google Search by Image service. It takes about 4 s for the first reverse image search, including the time to invoke a chrome web browser through Selenium WebDriver. The search for latter images are around 3 seconds per image.

Currently, queries are executed sequentially in order to prevent Google' restrictions. Another way for speeding up is to parallelize inquiries to Google Search by Image.

After our implementation of the prototype, we have found alternative Reverse Image Search Service such as Meta Reverse Image Search API (MRISA) [7] and TinEye Reverse Image Search [6]. These service also provides their API. By directly calling these APIs, it is expected that the execution time can be greatly shortened. Of course, the search result will be affected by changing to another service.

4.2 Reliability of Source URL

In our first prototype, we are using the top 5 URLs as a candidate of page which contains the image. It is widely known that the search results and order of the

URLs differs according to the language setting of browser and OS. As we can see from the result page shown in Fig. 6, the URL listed for the Lena image are all web pages in Japanese web sites. This is because that the search result is affected by the LANG environment of the process that started the Chrome web browser.

In addition, the order of URLs are completely up to Google's scoring. The URL that appears first is not necessarily means that it is the origin of the content. May be we should filter the URLs with another criteria. For example, checking the timestamp of each page, comparing the size of the image (larger image is likely to be close to original), etc.

4.3 More Useful Output

We are now displaying only the URL to the page which considered to be the source of images. But given the purpose of the system, it is necessary to display more helpful messages. It would be more helpful if the system can give us an advice such as "*There are no problem to use this image.*", "*You have to contact the author to get a permission.*", etc. We are now planning to add a module to display these messages in the next prototype.

5 Conclusion

In this research, a web-based system which assist checking the source of image included in educational presentation materials in Microsoft's PowerPoint file has been proposed. Our system uses existing Google Search by Image Service to obtain the source URL of the image extracted from the uploaded `.pptx` file. To access the service and execute the javascript contained in the HTTP response, the system invokes Chrome web browser through Selenium WebDriver. Our system finally displays source URL of each image as a clickable link. By checking the copyright notice or terms of use indicated on the page of the shown URL, the user can make appropriate advice to avoid copyright issues to the creator of the presentation slides.

References

1. Copyright Law of Japan. http://www.cric.or.jp/english/clj/ocl.html. Accessed 25 May 2018
2. Ueda, K., Tominaga, H.: A development and application of similarity detection methods for plagiarism of online reports. In: Proceedings of 9th International Conference on ITHET (2010). https://doi.org/10.1109/ITHET.2010.5480091
3. Selenium WebDriver. https://www.seleniumhq.org/projects/webdriver/. Accessed 25 May 2018
4. Google Image Search API (Deprecated). https://developers.google.com/image-search/. Accessed 29 May 2018

5. Google Custom Search. https://developers.google.com/custom-search/json-api/v1/overview. Accessed 25 May 2018
6. TinEye Reverse Image Search. https://www.tineye.com/. Accessed 25 May 2018
7. MRISA: Meta Reverse Image Search API. http://mrisa.mage.me.uk/. Accessed 25 May 2018

Anonymous Accessible Bulletin Board System with Location Based Access Control Mechanism

Jun Iio(✉), Shogo Asada, and Mitsuhiko Kai

Chuo University, Hachioji-shi, Tokyo 192-0393, Japan
iiojun@tamacc.chuo-u.ac.jp

Abstract. A bulletin board system (BBS) which allows anonymous submission has lower barriers to enter the discussions because everyone can post opinions without any responsibilities. Therefore, there are some risks that "a flaming" occurs, where the discussion is inundated with comments. To suppress emerging such situation, we developed a novel BBS with location-based access control mechanism. The BBS is anonymously accessible by default; however, the access permission to each chat room in the BBS depends on its geospatial information. That is, the chat rooms keep their geographical information when they are created, and users who can access each room must exist near the original location. In this paper, the overview of our system and its applications are described.

Keywords: Bulletin board system · Chat room · Anonymous access
Location-based service · Global positioning system

1 Introduction

In the modern highly networked information society, we communicate with our family, friends, and colleagues using various types of communication tools in our daily life. One of the typical tools is social networking services (SNS), which evolved from bulletin board system (BBS).

We can categorize these communication tools into three classes, systems that do not allow anonymous access, systems that can be accessed anonymously, and systems that can be accessed semi-anonymously. Systems of the former type require registration to start communication, and usually, an account for the system has to be able to identify who the user is. Thus, the users have to register at least their real name. In some cases, gender, age, or some other profile also have to be registered. Facebook, one of the significant SNS categorized into this class.

On the other hand, the second type of systems does not require any personal information. Some BBS systems have internal identification to prevent fake discussion that seems a dialogue among several persons but by only one person in actually. These systems give a private identification label to a user. Figure 1 shows the comments posted by the user with the identification labels.

L. Barolli et al. (Eds.): NBiS 2018, LNDECT 22, pp. 913–921, 2019.
https://doi.org/10.1007/978-3-319-98530-5_82

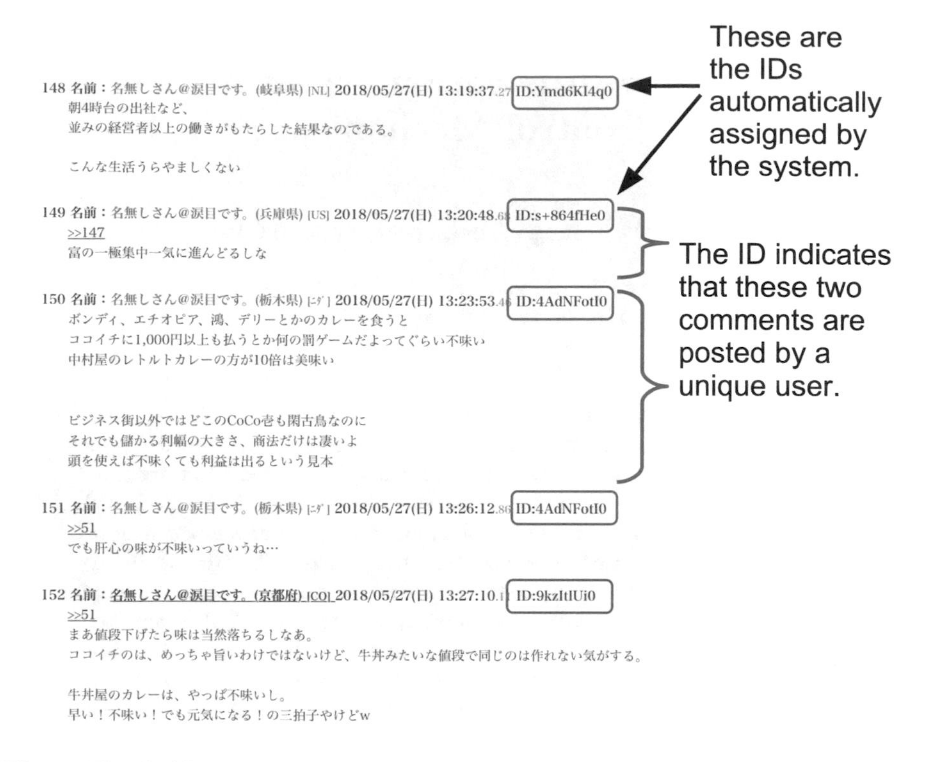

Fig. 1. The "5ch" is a typical anonymous bulletin board system in Japan. It requires no user registration, but each user has an internal identification label. Comments #149 and #150 have an identical ID number automatically assigned to the user, and it shows that the identical user posted these comments.

The last type of systems, which categorized into semi-anonymously accessible BBS, requires registration to start communication, as well. However, any personal profiles do not need to be registered. Users can communicate with each other using their accounts of their choice, which do not require to include any personal information, including their real name. Twitter requires the registration but registering user's personal information is not mandatory; therefore, it is categorized in this class.

In these three classes, a BBS that allows anonymous submission has lower barriers to enter the discussions because everyone can post his or her opinions without any responsibilities. However, these types of systems have some risks that "a flaming" occurs. If the flaming occurs, it could evolve to another risk of defamation and any social problems.

Based on these assumptions, to accelerate communications in the cyberspace, a BBS is expected, which allows anonymous submissions and prevents the emergence of the flaming.

2 Our Proposal

Our proposal for solving the problems previously mentioned is quite simple. The system proposed in this paper has an access control mechanism which is tightly connected with geospatial information. As well as the typical BBS, our system has many chat rooms. The difference between our chat rooms and the ones in the other BBS is our chat rooms have geospatial information on the places where they were created (Fig. 2). We gave the name "Soco" to our proposed system. The name comes from the word "Soco" that means "there" in Japanese.

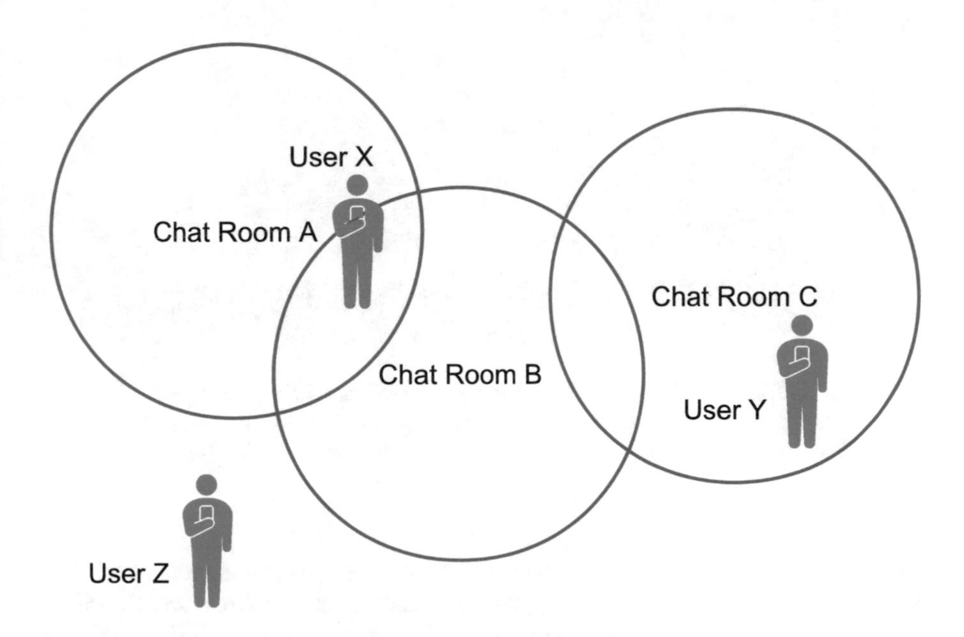

Fig. 2. Each chat room has its available range. The user X can access both the chat room A and B, the user Y can only access the chat room C, and the User Z cannot access any chat rooms in these areas.

Users who want to access the chat room are also associated with their geospatial information. The chat room has its geospatial range in where the users can read and write their message. That is, if users want to access the particular room, they have to move physically to the place near the location where the chat room was created and get into the available range. In another word, if users access the home page of Soco, only the rooms available for them are listed (Fig. 3). Available rooms mean that the rooms which were created at the places within the range of a central focus on the user's current location.

On this screen, a new room can be created by clicking the red button given at the right-bottom corner. The new room created here is associated with geospatial information on the current position of the room creator. Additionally, users can

Fig. 3. The available rooms are listed when a user accesses the home page of "Soco." This screenshot shows that there are three chat rooms. The chat rooms were created at a conference room located at Meguro, at an office at Meguro, and at Tamachi, respectively.

choose the room sizes when they create the new room. In the current implementation, there are three options to choose the sizes of chat rooms; small, middle, and large size are available, and the inner diameters of the available range are 200 m, 2,000 m, and 3,000 m, respectively. As shown in Fig. 3, chat rooms has its name, short description, and one photograph (optional) to represent the character of the room appropriately.

Note that there is a particular chat room named "lounge," which is accessible from any place and it has no access control by the geospatial information. This unique chat room is currently used for information exchange for promoting Soco; therefore, it may be removed shortly.

After creating a new room, or choosing a room that has already been created, the user can enter the room and start communicating by reading comments and submitting new comments. Figure 4 shows a screenshot of the timeline where some comments are listed in the chat room.

A new comment can be posted by clicking the red button at the right-bottom corner, just as creating a new room mentioned previously. In the timeline, not only the message but also photo image can be uploaded. The posted comments and images are listed according to a time sequence, from the bottom to the top (Fig. 4).

Fig. 4. This screenshot illustrates a timeline where some comments are listed according to a time sequence.

3 Applications

In this section, several case examples we considered as a useful application of Soco are explained.

3.1 Communication Among Fans Watching a Sports Game

In this case, we suppose that the fans can exchange their supports and impressions while watching a sports game.

Organizers of the sports event prepare a smartphone application equipped with this service embedded and provide it preliminary to participants. Alternatively, more specifically, the platform of Soco is lent to the organizer of the event. The event can utilize a part of this service provided by Soco.

When the participants of the sports event come to the arena, they access to chat rooms prepared by the organizer. The access to the chat rooms are restricted to the near fields of the arena; therefore, they can only be available from the inner side of the arena, or at least, the surrounding places around the arena.

The several ways of setting chat rooms can be considered. The typical setting is the following: two chat rooms for communicating by fans of team-A and team-B, which are playing the game, are prepared at the location of the stadium. Along with proceeding the game, fans of each team can communicate effectively by posting their opinions, comments, impressions, and support messages. It will be an excellent communication if the message is delivered to or from not only fans but also team members.

3.2 Providing a Discussion Platform for Seminars and Lectures

The second application is to use Soco as the discussion platform for seminars and lectures, where the participants can quickly post their questions and opinions by anonymous submission.

As the sports event previously mentioned, the chat rooms with geospatial information can be utilized when seminars and lectures are held. Usually, the access control to restricted chat rooms for the participants of the seminar are realized by some individual accounts temporary issued. In this service, the feature that only the people who stay in the venue or surrounding places of the venue can access the chat rooms enables access controls that allows the access only by the participants of the seminar. Rigorously, any persons in the surrounding places can access the chat room, even if they are not the participants of the seminar. To eliminate such irregular accesses from non-authorized users, the information on the existence of the chat rooms, including URLs of the services, should be kept a secret among the participants.

In the chat room, the participants of the seminar can exchange their opinions and questions. Not only the participants but also organizers of the seminar, lecturers, assistants can join the discussion via this service. Their cooperative chat enables productive discussion, including the answers to the questions from the participants, and providing additional information if they requested.

The feature that the participants can anonymously submit their opinion promote the active exchange of opinions and questions. Giving their words by raising their hands up is making the participants pause and considered a high bar for them. However, the anonymity of submission in this system relaxes them and lowers the bar for giving their opinion. Therefore, the active communication would be expected by using this system.

3.3 Temporal BBS for Shelters at the Outbreak of a Disaster

If a disaster occurs, an information exchange platform will be needed, in where only the persons concerned can access.

At the outbreak of a disaster, some chat rooms should be prepared for a central focus on the shelters or the places where accidents have occurred. These chat rooms are only available from the near places surrounding the shelters or the venue of accidents.

Restricting areas where can communicate blocks unnecessary information, false rumor, and negative information. Furthermore, the personal information of victims and some sensitive information also need to handle, to help them effectively. Enabling physically restricted access to such information by using this system, it is expected to prevent the diffusion of such information into the internet.

Under the circumstances of confused by the disaster, it is crucial that the systems enable easy access without any account information nor managing passwords. Hence, senior persons and children, who are not accustomed to information services, can also efficiently use this service, so that proper information can be delivered to appropriate victims.

3.4 Utilization for Puzzle-Solving Events, Matching Events, and Escaping Games

The system is used for the events held in the theme-parks or the university campus, supposing to the events where are held in the relatively large area, including puzzle-solving events and escape-games. Alternatively, the events held in the whole town, over the shopping area, can be candidates for utilizing this services.

In the case of these large size area, there are several areas covered by plural ranges of chat rooms, and in accordance of the places to access this service, the list of chat rooms are different from each other.

In each chat room, some hints to solve the problem and some information for the goal of the events are announced. Then, the announcement of information is designed as the users cannot reach the goal if they can gather the information from each chat room in a comprehensive manner.

The participants of the event try to achieve the goal by utilizing this services and walking around within the areas of the event to collect some meaningful information. While a sequence of such behaviors is conducted, the communication among the participants of the event is accelerated by the chat rooms provided by this system.

3.5 Collecting Responses for a Questionnaire from Customers

At the over-the-counter sale and restaurants, sales staffs can efficiently collect the responses to the questionnaire, feedbacks for their services, impressions from customers by using this system. Collecting the responses and opinions via the chat room associated with the geolocation information of the shops and restaurants allows the staffs to acquire the real opinions from the real customers without any noises.

Conversely, having the particular information provided only in the chat rooms announced beforehand on the Internet is also a promotion activity for their customers. It is essential that telling people that there is some information that people do not know unless they come to the store. The business system could be a real promotion for encouraging visitors. Additionally, offering specialized services and other discounts with information provided via the chat rooms that are only accessible around the store enables to promote sales and to attract new customers.

In either case, there is no need to acquire account information or register as a member in advance, so that it is possible to provide a service with a low threshold to ordinary consumers who are candidates for customers.

4 Related Work

Ito and Ogawa [1] have proposed Tweet on Rails, which can connect passengers getting on a train and those waiting at the nearest stations. Geospatial information is acquired every five seconds and sent to the Tweet on Rails server. The locational information is converted to kilometers on the server side, and it enables correct measurement of the user's position. Its messaging function relies on that of Twitter. Attaching a tag to comments posted to Twitter determines whether the posts are submitted to Tweet on Rails or not. These two systems formulate a timeline that is accessible from the passengers on the train and those who are waiting for the train.

Kojima and Tsukada [2] have proposed the information sharing system in shelters. The system constructs some clusters that have several data on the positions of the shelters, information on the facilities and their capacities. It also uses Twitter as its messaging function. The user's comments with geospatial information gotten by the user's device are categorized into several clusters according to the preselected conditions. All comments posted to a particular region can be accessed from all users who are located in the same region.

The aims of these two systems are similar to that of our system Soco. Soco can be used in the train and stations to accelerate the communication among passengers. Moreover, one of the main application of Soco is to prepare some communication platforms in the shelters at the outbreak of a disaster. A clear difference is the communication method. That of two existed system rely on the messaging function provided by Twitter. On the other hand, Soco implements the messaging function by itself.

Yamada *et al.* [3] has proposed "LACom," which is an SNS utilizing augmented reality (AR) and geospatial information. In LACom service, a particular region with a center focus on a user is called "Space," and the user can store text data, images, and web pages into Space. Other users can access Space via AR scope. Space exposes itself in an AR space as a square shaped object. Users are walking around within the AR space. By browsing spaces, it is available that the users send their message to other users who have similar hobbies and curious topics.

5 Conclusions and Future Work

In this paper, a novel bulletin board system was proposed, which has access control mechanism based on the geospatial information. The BBS has many chat rooms that are associated with its locational information, and each room has its real range to check whether a user is permitted to access or not. The chat room is accessible to the user if he or she is located in its range.

It is expected that the feature of the access control using geospatial information can prevent from emerging the flaming situations. Additionally, several applications have been discussed by utilizing such characteristics of our proposed system. The evaluation of this system, especially, measuring the effect on the real use of applications that are proposed in this paper remains as the future work.

References

1. Ito, Y., Ogawa, K.: Tweet on rails: consumer generated media for train riders. In: Proceedings of Interaction 2010. IPSJ (2010). (In Japanese)
2. Kojima, T., Tsukada, K.: Information sharing system for SNS users in the evacuation center based on location information In: Proceedings of the 11th Workshop on Groupware and Network Services. IPSJ (2014). (In Japanese)
3. Yamada, K., Itou, J., Munemori, J.: Proposal of local area communication support system to use location information. In: Proceedings of Multimedia, Distributed, Cooperative, and Mobile Symposium 2011. IPSJ (2011). (In Japanese)

Newly-Added Functions for Video Search System Prototype with a Three-Level Hierarchy Model

Tongjin Lee(✉) and Jun Iio

Chuo University, Hachioji-shi, Tokyo 192-0393, Japan
tongjini@gmail.com, iiojun@tamacc.chuo-u.ac.jp

Abstract. This paper describes newly added functions onto ELVIDS, previously presented the three hierarchical video search system. We have developed this system for fostering scholarly use of videos over the three years. The development in the previous phase was mainly for giving functions for users such as retrieving videos, displaying search results, and playing back the videos. As the next phase, we considered adding functions for metadata registrars. In this paper, the background, sample dataset and a protocol for an authentication are firstly described in the earlier chapters. Then, the functions for metadata registration and the authentication including the methods and the interfaces are introduced. Finally, the conclusions and the further works are mentioned.

Keywords: Video database · OSS

1 Introduction

We have been developing ELVIDS since 2015. ELVIDS is a video search system developed for fostering the scholarly use of videos. The system has the functions enabling users search from three levels of the entire work, scene, shot (see Fig. 1).

The development in the previous phase was to implement the functions to make videos accessible at the three different level through creation of database, interface design, and renewing the design based on the evaluation of the system. The functions are primarily intended for the general users, but, for the actual use of the system, the functions for registering and managing data are necessary.

In this paper, we describe the newly added functions: the metadata registration and the authentication. Before explaining the functions, we first describe the sample dataset and the protocol for the authentication used in ELVIDS; then, we introduce the methods for implementing the functions of the metadata registration and the authentication.

L. Barolli et al. (Eds.): NBiS 2018, LNDECT 22, pp. 922–928, 2019.
https://doi.org/10.1007/978-3-319-98530-5_83

Fig. 1. Search page, search result page, and video playback page of ELVIDS

2 Sample Dataset and Protocol for Authentication

As the datasets, videos entitled Chuo University's "Corridor of Knowledge" and the metadata was used for the development of ELVIDS. The videos, 30 min long for each, are educational program based on joint production of Chuo University and Jupiter Telecommunications Co., Ltd. The videos are distributed both on national cable TV and on YouTube. We selected two the videos and stored the files and the metadata onto the database for testing the registration functions.

To implement the function of authentication, we used OAuth. OAuth is an open protocol for an authorization used for web applications [1]. By adopting OAuth, the registrars can login to the web application with Facebook, Twitter, and GitHub accounts. In other words, they do not need to manage IDs and passwords only for the use of ELVIDS. In addition, there is no need to create the tables on the database for storing and managing personal information for authentication. With the protocol, we added the new functions onto the existing prototype.

3 Added Functions

In this chapter, the method for implementing the newly added functions are explained respectively: the metadata registration and the authentication.

3.1 Metadata Registration

Unlike other systems, ELVIDS has three levels of database structure (Tables 1, 2 and 3), so the registration function must be compatible for the structures. Therefore, we created different input pages for each work, scene, and shot for avoiding discrepancy due to the structure.

In the work registration, a metadata input page, a confirmation page, and a registration completion page are respectively provided (see Fig. 2). Input is manual, but "work id" "level of unit" is set to input values automatically.

In the scenes and shots registrations, only a metadata input page is provided as shown in Fig. 3. Instead, it enables the registrars to confirm, delete, and edit the metadata on a single page.

Table 1. Column name and data type of works.

PK	Logical name	Physical name	Data type
○	id	id	integer
	title	title	text
	creator	wcrt	text
	publisher	wpub	text
	description of a work	wdesc	text
	level of unit	wlev	text
	type of video	vidtyp	text
	duration of work	wdur	integer
	identifier	idt	text
	language	lang	text

Table 2. Column name and data type of scenes.

PK	Logical name	Physical name	Data type
○	work id	work_id	integer
○	id	id	text
	contributor of a scene	sccontr	text
	description of a scene	scdesc	text
	level of unit	sclev	text
	duration of a scene	scdur	text
	start time of a scene (h)	scsth	text
	start time of a scene (min)	scstm	text
	start time of a scene (s)	scsts	integer
	end time of a scene (h)	sceth	text
	end time of a scene (min)	scetm	text
	end time of a scene (s)	scets	text

Table 3. Column name and data type of shots.

PK	Logical name	Physical name	Data type
○	work id	work_id	integer
○	scene id	scene_id	text
○	id	id	text
	description of a shot	shdesc	text
	level of unit	shlev	text
	duration of a shot	shdur	text
	start time of a shot (h)	shsth	text
	start time of a shot (min)	shstm	text
	start time of a shot (s)	shsts	integer
	end time of a shot (h)	sheth	text
	end time of a shot (min)	shetm	text
	end time of a shot (s)	shets	text

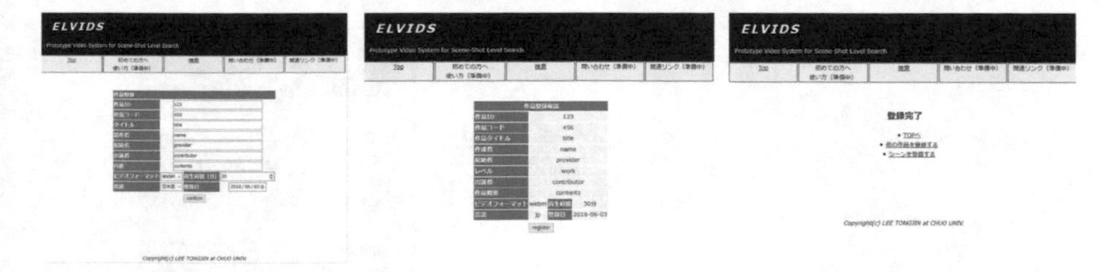

Fig. 2. Work registration pages of ELVIDS

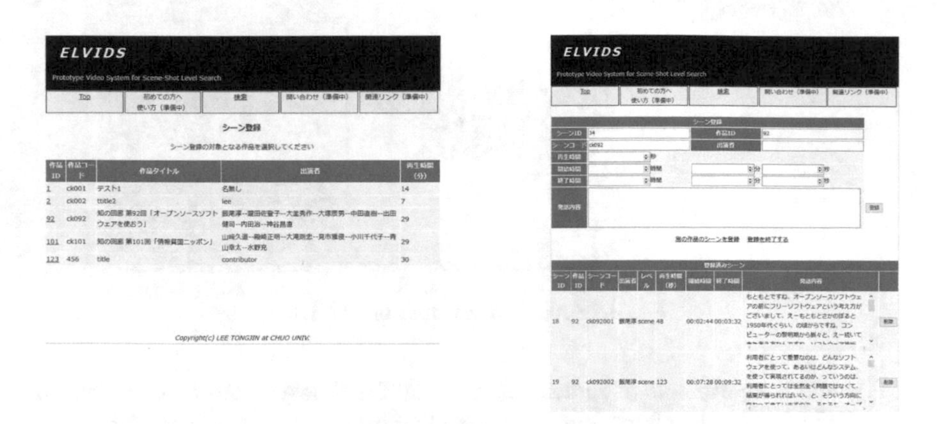

Fig. 3. Scene registration pages of ELVIDS

Unlike the registration of the metadata of work, the registrars need to input a large amount of data because there are multiple scenes and shots in a work. Because of the reason, we did not create the pages for the confirmation and the completion of the registration. Like the work screen, each "id" "level of unit" is set to be automatically entered. By adding the function, the registrars have become able to add videos and the metadata, which are used for searching videos.

3.2 Authentication with OAuth

By adding new pages for the metadata registration, the overall structure of ELVDIS became as shown in Fig. 4.

To avoid access the newly added pages by the general users, we implemented the function of authentication. In concrete, a function that allow the only eligible registrar to access the framed pages in Fig. 4 was added onto the system. To implement the function, we adopted OAuth as an authentication protocol.

When the registrars attempt to access the metadata registration page by typing URL directly onto the address bar, the page asking them to choose one of links to Facebook, Twitter, or GitHub is shown. Clicking one of the links, he/she is required authentication with a set of ID and password. If he/she succeed in authentication, the first page of the metadata registration is shown on the screen.

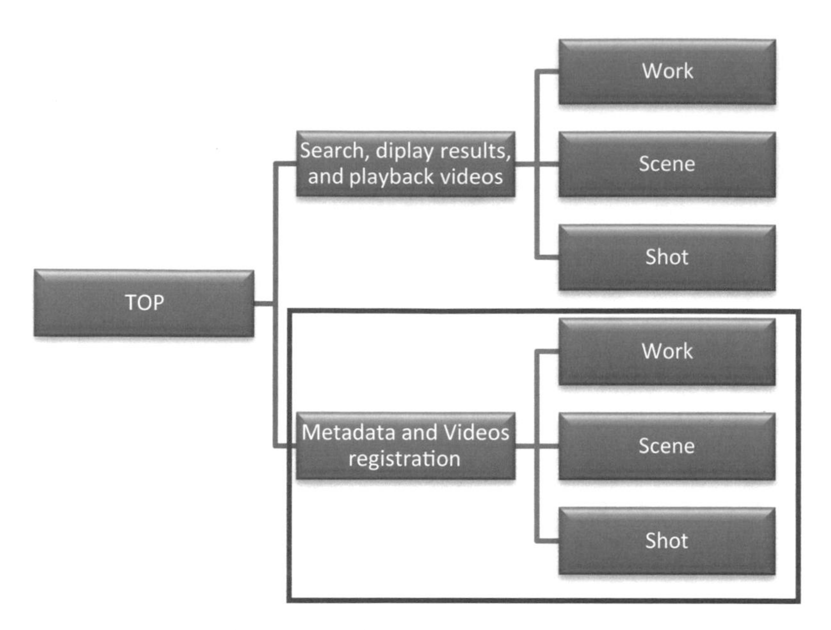

Fig. 4. Simplified structure of ELVIDS.

Figure 5 is an example of authentication using Facebook. In this case, the registrar is required entering the phone number or email address and the password for login on Facebook. Only when the authentication is successful, the screen shifts to the metadata registration screen. By using OAuth, we implemented the functions of authentication for the metadata registration.

Fig. 5. Authentication pages of ELVIDS

4 Related Study

The development of ELVIDS including solving the issue of authentications has been carried out by utilizing Open Source Software (OSS). The decision making for the developing system and the adding functions was based on the following literature.

Tripathi *et al.* [4] conducted survey the systems, which were developed with OSS, such as DSpace, EPrints, Greenstone, and Fedora projects. Their aim of the survey was to discuss which software was suitable for developing Institutional Repository (IR).

They showed that IR development with OSS contributes to elimination of vendor lock-in and reduction of cost with greater flexibility. Also, the software used in the platform mentioned above is capable of capturing, indexing, preserving, and redistributing scholarly research materials.

Zahedi *et al.* [5] identified security issues posted on GitHub repositories. They applied a mixed methods approach, combining topic modeling techniques and qualitative analysis for understanding the presence of the security issues. They found that the majorities of the security issues were topics about identity and cryptography. Also, as one of identified topics, they showed that the issues of the managing user accounts with OAuth were frequently discussed.

Villarrubia and Kim [6] reported building of community system to teach collaborative Software Development with OSS. Their system allows the students to participate in collaborative software development projects and to track their activities. Their paper also explained authentication through Google OAuth 2.0. Note that their system was only compatible with Google OAuth 2.0; conversely, our system enables the registrars to authenticate with one of the accounts of Facebook, GitHub, Google.

5 Conclusions and Further Work

In this paper, we introduced the newly added functions: the metadata registration and the authentication.

The function for metadata registration was added in this phase of the development of ELVIDS. As the has three hierarchical structures enabling the users search from different levels of videos, the function of the registration, which was compatible with the system, was implemented. As a result, two types of the metadata registrations model were created. The work registration has three pages: the input page, the confirmation page, and the registration completion page. Each of scene and shot registration has a single page allowing the registrars to input, confirm, delete, and edit the metadata simultaneously.

The function of the authentication was also implemented with OAuth. By using the protocol, the registrars can avoid managing their IDs and Password. Also, the administrators of the system do not have to create the database tables for storing the registrars' personal information.

By adding the above functions, the basic functions for both the general users and the registrars of the system have been given. As further works, we are going to examine the usability of the system. In addition, partial automation for metadata registration is also going to be carried out in thc next phase of the development.

Acknowledgment. This study was funded by the Chuo University Grant for Special Research. We are grateful to the members of our laboratory for their valuable insights, feedback, and much helpful support.

References

1. OAuth2.0. https://oauth.net/2/. Accessed 7 May 2018
2. omniauth/omniauth-oauth. https://github.com/omniauth/omniauth-oauth2. Accessed 7 May 2018
3. Sinatra. http://sinatrarb.com/. Accessed 7 May 2018
4. Tripathi, D.P., Bhojaraju, G., Pradhan, D.K.: Open source solutions for creation of ETD archives/repository: a case study of central library @ NIT Rourkela. In: 62nd Indian Library Association Conference 2017 (2017). http://hdl.handle.net/2080/2662
5. Zahedi, M., Ali Babar, M.A., Christoph, T.: An empirical study of security issues posted in open source projects. In: Proceedings of the 51st Hawaii International Conference on System Sciences (HICSS), pp. 5504–5513 (2018)
6. Villarrubia, A., Kim, H.: Building a community system to teach collaborative software development. In: 2015 10th International Conference on Computer Science & Education (ICCSE), pp. 829–833 (2015)

A Location-Based Web Browser Network for Virtual Worlds

Masaki Kohana[1(✉)] and Shusuke Okamoto[2]

[1] Ibaraki University, Hitachi, Ibaraki, Japan
masaki.kohana.gopher@vc.ibaraki.ac.jp
[2] Seikei University, Musashino, Tokyo, Japan
okam@st.seikei.ac.jp

Abstract. This paper proposes a way to construct networks among Web browsers. The building a web-based virtual world needs a lot of computing resource, which means that we need a lot of Web servers. However, if we increase the number of Web servers, the financial and the maintenance cost also increases. In this study, we try to use computing resource on Web browsers. In our previous study, we proposed a way to share data among Web browsers. This way has a problem that is the longer data transfer time when the number of users increases. Therefore, we try to construct small Web browser networks based on the location of each players on a virtual world.

1 Introduction

A way to share information about a virtual world is essential for video games and virtual-reality applications. In this type of applications, a user creates and controls an avatar in a virtual world. The user moves around the virtual world and interacts with the other users via the avatar. A client software displays the virtual world and the characters.

As a type of virtual worlds, an open-world virtual world attracts attention. A traditional type of virtual world for online-games divides the virtual world into block. An open-world does not divide the virtual world seemingly. The user can move the world without the reloading map, while the traditional world needs to reload the map when the user reaches the edge of the block. Using an open-world virtual world, the users get the higher degree of freedom of the movement. However, it needs a lot of information and computing resource. Furthermore, load distribution is difficult for online games because the world is not divided.

In our previous work, we used computing resource on Web browsers [5,7,8]. One of the ways to get computing resource is the increase the number of Web servers. However, financial and maintenance cost also increase. On the other hand, the number of Web browsers increases when the number of users increases. We focus on the Web browser to get computing resource.

When we use computing resource on Web browsers, the browsers need to share information about users. To communicate among browsers, we constructed

L. Barolli et al. (Eds.): NBiS 2018, LNDECT 22, pp. 929–936, 2019.
https://doi.org/10.1007/978-3-319-98530-5_84

a network of Web browsers using WebRTC [2,3] and confirmed that we could share information among browsers. However, the network size becomes large with the higher number of users, which leads the longer data transfer time.

This paper proposes a way to construct small networks of Web browsers to resolve the problem of the longer data transfer time. In the massive virtual world, a user can see a part of the world, which means that the user needs only a part of information. Therefore, the user needs to share information with the other users in the neighborhood, and all the users do not need to join the same network. We try to construct Web browser network according to the location of the users in the virtual world.

This paper consists of the following sections. Section 2 introduces some studies related to our study. Section 3 describes an overview of our method to construct networks. Section 4 shows the experimental results and Sect. 5 concludes this paper.

2 Literature Survey

This section introduces some studies that related to our study.

Xhafa et al., investigate several application model for mobile computing [11]. The author focus on the collaborative work with geographically distributed members. This paper introduces peer-to-peer model using WebRTC and central server model and discusses some problems that arise in application migration such as data synchronization and consistency.

Tutschku proposed a method for designing radio networks of cellular mobile communication [10]. It formulates the transmitter locating task as a maximal coverage location problem (MCLP) and they use a greedy heuristic for solving the MCLP problem.

Our study uses WebRTC to communicate among Web browsers. Using WebRTC, we can construct several topologies. Singh et al., evaluate the performance of various topologies using WebRTC [9]. Their result shows that the Receive-side Real-Time Congestion Control (RRTCC) performs by itself. However, it starves when competing with TCP.

Kuhn et al., investigate distributed computation in a dynamic network [6]. In this paper, the network topology changes. Moreover, they consider a worst-case model. The considered model is similar to our assumption that the Web browsers construct networks dynamically.

Ito et al. proposed a webcast system using Web browsers in class [4]. This system shares the screen of a teacher with the students. It constructs a binary-tree topology because the system sends the data to the one-sided.

3 System Overview

This section describes a method to construct networks of Web browsers using the location of the users in the virtual world. First of all, we describe an assumption about a virtual world in this study. We refer to Minecraft that is a sandbox type open-world video game [1].

Figure 1 indicates the relationship between three units for management world. The entire rectangle represents the entire virtual world, and the entire world consists of the small blocks. A chunk is a set of some blocks. In this figure, a chunk consists of 16 (4×4) blocks, and the world consists of 18 (3×6) chunks.

Figure 2 shows an example of the view range. A client loads the information about the other characters included in the view range. In this figure, the elephant and the frog are avatars of the users. An avatar exists in a block. A Web browser loads the information in a unit of a chunk. Moreover, the browser handles the

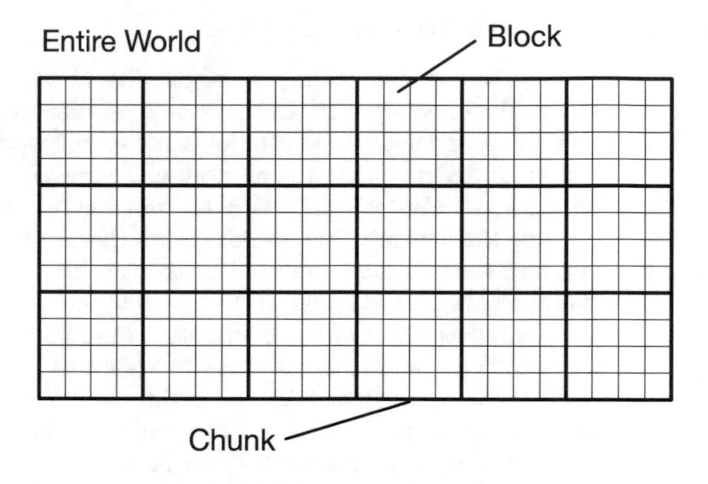

Fig. 1. Relationship between three units

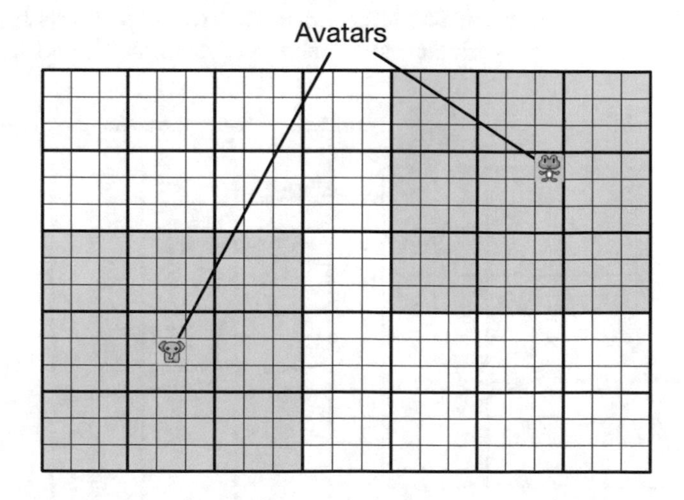

Fig. 2. View range

chunk that the avatar exists in as a central chunk. For example, the elephant is at the center of a chunk. Therefore, the browser loads the information about the chunk. Furthermore, the browser needs more information to display the game world. Therefore, the browser loads the chunks around of the chunk that the avatar exists in. As a result, the Web browser of the elephant loads the gray blocks that around of the elephant. On the other hand, the frog is on a corner of the chunk. However, the browser handles the chunk that the avatar exists in as a central chunk and loads the information in a unit of a chunk. The Web browser about the frog loads the gray blocks that around of the chunk that the frog exists in.

3.1 Construction of Web Browser Networks

As we mentioned before, the user can see a part of the virtual world. All the users do not need to join the same network. In this study, we propose a way to construct Web browser network when a new user joins the virtual world.

The Web server of our system manages the network while the Web browsers connect to the other browsers. First of all, a new browser sends a request to the Web server for joining the network. The server checks the current network topology and picks up a browser that is connected by the new browser.

The user needs to see a part of the virtual world, which means that the Web browser of the user needs only the part of the information to display the world. Therefore, the browser needs to connect to the other browser that exists in the same region of the world. When a new user joins the world, the Web server picks up the users on the same chunk of the new user. Then, the server finds the user that is the nearest for the new user. If no one exists in the same chunk, the server picks up the other users on the chunk around of the chunk that the new user exists.

Figure 3 shows an overview of the network construction. In this figure, there are five users, and the network consists of the four users, B, C, and D. The user

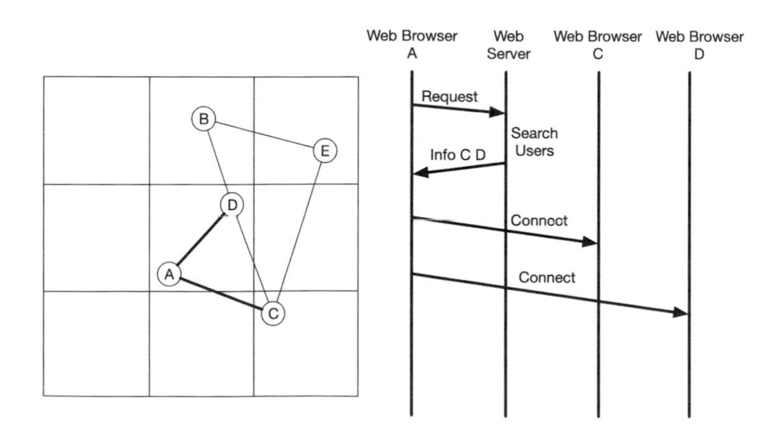

Fig. 3. Network construction

A is a new user. When the user A join the world, the A sends a request to the Web server. The server checks the chunks that the user A requires to display the world. The server picks up the user C and D because the two users are nearest to the user A. The user A receives the information about the users, and the A connects to the user C and D.

3.2 Data Transfer

In our system, we construct a network using WebRTC. Therefore, each user sends user information using data channel of WebRTC, which means that the data transfer does not use the Web server.

Figure 4 shows an example of the data transfer. There are eight users, and the user A is the data source. The gray chunks indicate the view range of the user A. As the step (1), the user A sends information to the user B and C. After the user B and C receive information, the B sends the received information to the user E, and the C also sends information to the user D in step (2). In step (3), the user D sends information to the user E and F. On the other hand, the user E also sends information to the user D and F. The user D and E receive the same information from two users. However, the user does not forward information that already received. Therefore, a user does not send the same information twice. Finally, the user F receives the information. However, the view range of the user F does not include the user A. Therefore, the user F does not forward information to the user G and H.

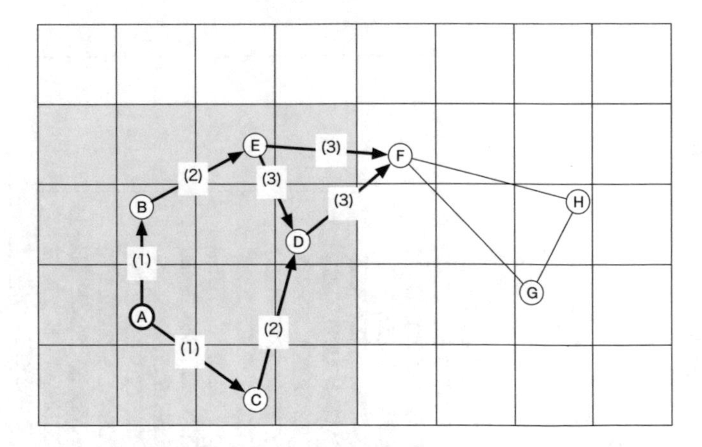

Fig. 4. Data transfer

4 Experimental Result

To evaluate our method, we measure the data cover rate. We perform a simulation-based experiment. The data cover rate shows the number of necessary data that received by a user. If a user needs five user data to display the world and receives three user data, the data cover rate is 3/5 (60%).

The virtual world is 1500×1500, and the chunk size is 15. Therefore, the world consists of 100×100 (10000) chunks. The view range is ten chunks. This range indicates a range for one direction. If the user position is (25, 25), the upper side border is (25, 15) and the lower side is (25, 35). Furthermore, the left side border is (15, 25) and the right side border is (35, 25). Therefore, the total view range includes 441 chunks. For the chunk size and the view range, our configuration refers the default configuration of Minecraft.

Figure 5 shows the data cover rate for each number of users. The number of users is from 1000 to 3000 every 100 users. The x-axis indicates the number of users, and the y-axis indicates the data cover rate. In the most case, the data cover rate is about 0.5 (50%). When the number of users is from 1600 to 2100, the cover rate decreases to about 0.49 (49%).

The 50% cover rate is enough for the virtual world. We consider that the reason for the result is the connection pattern. Figure 6 shows two problem cases. On the left side, there are six users, and all the users join the same network. However, the user C and E is the outside of the view range of the user A. Therefore, the user C does not forward the information whenever the user C receives the information. In this case, despite user D is in the view range of the user A, the user D cannot receive the information.

On the right side, there are six users. We expect the user C and the user D join the same network. However, in our method, the maximum number of connection is two when a new user joins the world. Therefore, the user D connects to the user E and the user F. As a result, the user D does not connect to the user C.

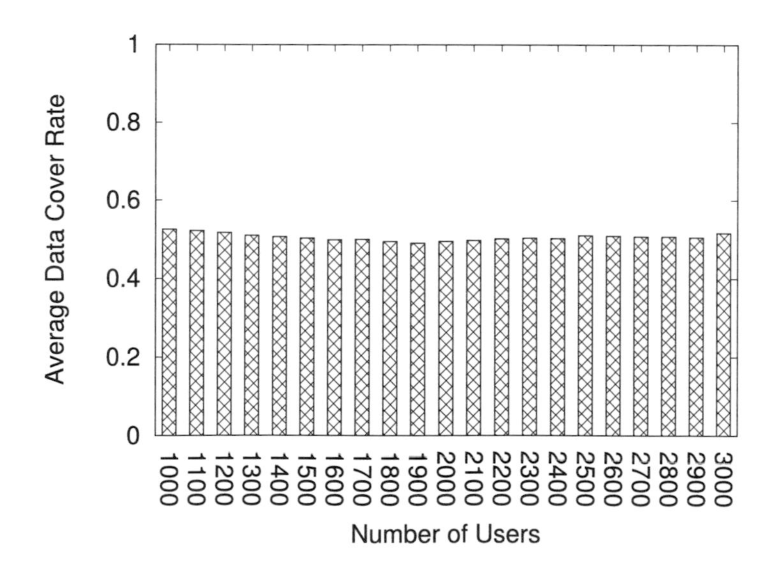

Fig. 5. Data cover rate

One of a way to resolve this problem is the increasing the number of connections. However, the increasing the connection leads to the increasing workload

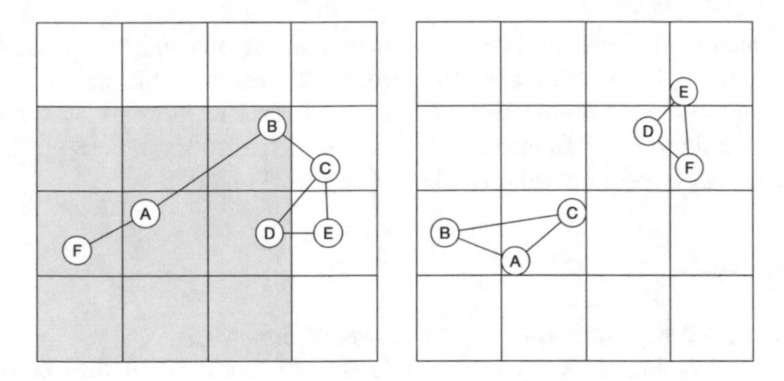

Fig. 6. Problem case

of a browser. Figure 7 shows the average and the maximum number of connections. The x-axis indicates the number of users, while the y-axis indicates the number of connections. In the most case, the average number of connections is two. However, the maximum number of connections increases according to the number of users. To resolve the problem, we can increase the number of connections. However, the workload among browsers might be unbalanced. As a future work, we need to survey a way to increase the number of connections with the balancing workload.

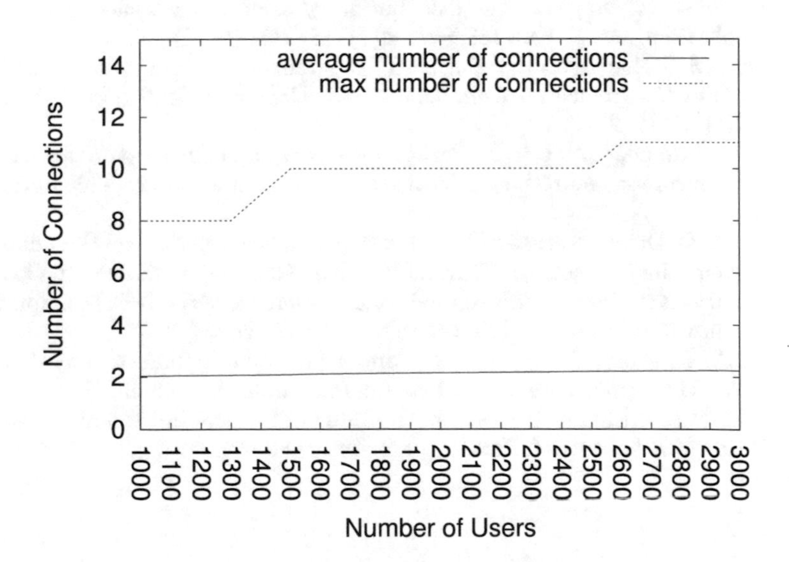

Fig. 7. Number of connections

5 Conclusion

This paper proposed a way to construct networks of the Web browsers based on location in a virtual world. Using this way, we can construct multiple small browser networks. However, the data cover rate of our method is about 50% that is not enough to build a virtual world. To resolve this problem, we can increase the number of connections for a Web browser. However, the increasing connection leads to the increasing workload. Our future work is to survey a way to increase the connection with the balancing workload.

References

1. Minecraft. https://minecraft.net/. Accessed 25 June 2018
2. WebRTC API. https://www.w3.org/TR/webrtc/. Accessed 25 June 2018
3. WebRTC protocol. https://tools.ietf.org/html/draft-ietf-rtcweb-overview-19. Accessed 25 June 2018
4. Ito, D., Niibori, M., Kamada, M.: A real-time web-cast system for classes in the BYOD style. In: The 5th International Workshop on Web Services and Social Media (WSSM-2016) in Conjunction with the 19th International Conference on Network-Based Information Systems (NBiS2016), pp. 520–525 (2016)
5. Kohana, M., Okamoto, S.: A data sharing method using WebRTC for web-based virtual world. In: The 6th International Conference on Emerging Internet, Data & Web Technologies (EIDWT-2018), pp. 880–888 (2018)
6. Kuhn, F., Lynch, N., Oshman, R.: Distributed computation in dynamic networks. In: Proceedings of the Forty-Second ACM Symposium on Theory of Computing (2010)
7. Okamoto, S., Kohana, M.: Load distribution by using web workers for a real-time web application. Int. J. Web Inf. Syst. **4**(7), 381–395 (2011)
8. Okamoto, S., Kohana, M.: Running a MPI program on web browsers. In: 2017 IEEE Pacific Rim Conference on Communciations, Computers and Signal Processing (PACRIM 2017) (2017)
9. Singh, V., Lozano, A.A., Ott, J.: Performance analysis of receive-side real-time congestion control for WebRTC. In: 2013 20th International Packet Video Workshop (2013)
10. Tutschku, K.: Demand-based radio network planning of cellular mobile communication systems. In: Proceedings IEEE INFOCOM 1998, the Conference on Computer Communications. Seventeenth Annual Joint Conference of the IEEE Computer and Communications Societies, Gateway to the 21st Century (1998)
11. Xhafa, F., Zaragoza, D., Caballe, S.: Supporting online/offline collaborative work with WebRTC application migration. In: In: Barolli, L., Xhafa, F., Javaid, N., Enokido, T. (eds.) Innovative Mobile and Internet Services in Ubiquitous Computing, IMIS 2018, Advances in Intelligent Systems and Computing, vol. 773. Springer (2018)

The 6th International Workshop on Cloud and Distributed System Applications (CADSA-2018)

Overview of Digital Forensic Tools for DataBase Analysis

Flora Amato[1(✉)], Giovanni Cozzolino[1], Marco Giacalone[4], Antonino Mazzeo[1], Francesco Moscato[2], and Francesco Romeo[3]

[1] Department of Electrical Engineering and Information Technology, University of Naples "Federico II", Naples, Italy
{flora.amato,giovanni.cozzolino,mazzeo}@unina.it

[2] Department of Scienze Politiche, University of Campania "Luigi Vanvitelli", Caserta, Italy
francesco.moscato@unicampania.it

[3] Department of Jurisprudence, University of Naples "Federico II", Naples, Italy
francesco.romeo@unina.it

[4] Vesalius College, Brussels, Belgium
Marco.giacalone@vub.ac.be

Abstract. The number of digital devices that people use in everyday life has significantly increased. Since they have become an integral part of everyday life, they contain information that are often extremely sensitive. Modern devices use complex data structures to store data (heterogeneous media files, documents, GPS positions and SQLite databases, etc.), therefore, during a forensic investigation, it has been necessary the adoption of specialized acquisition and analysis tools.

1 Introduction

Mobile Forensics is a relative new discipline which originates from the Digital Forensics research topic. Digital Forensics can be defined as the discipline that deals with the identification, preservation, study and analysis of information stored into information systems (strictly respecting the chain of custody principle), in order to highlight the existence of useful evidence to carry out during the detective activities.

The number of digital devices that people use in everyday life has significantly increased, from GPS navigators to modern cameras, e-book readers, notebooks, tablets, smart-phones and so on. Above all, smart-phones and tablets are extremely common devices, as they have become an integral part of everyday life and contain information that are often extremely sensitive. From a technical point of view, these devices are similar to small computers that run their own operating system, more or less specialized and feature-rich.

Modern devices use complex data structures to store data, such as heterogeneous media files, documents, GPS positions and SQLite databases that contain most of the information related to the applications installed on devices. This has led digital forensics investigators to focus more and more their activities on

L. Barolli et al. (Eds.): NBiS 2018, LNDECT 22, pp. 939–943, 2019.
https://doi.org/10.1007/978-3-319-98530-5_85

modern smart devices, like smart-phones, and to Cloud services [1,2] connected to them. Therefore, it has been necessary to translate the investigative activities on these new devices, thus making necessary the adoption of new acquisition and analysis tools. However, mobile devices constantly improve the security and protection of the data they contain [7,12], which makes the work of the Mobile Forensics Expert increasingly difficult and complex.

This work aims to study the different acquisition techniques for a mobile device and the tools used to represent the extracted data. It is structured as the following: in Sect. 2 different acquisition techniques for mobile devices are presented; in Section we present Sect. 3 main hardware and software tools; in Sect. 4 we remark some conclusions.

2 Acquisition Techniques

Depending to the actual considered device and its operating system version, there are different techniques that can be adopted in order to acquire a data stored in it, which vary in complexity, effectiveness and supported devices.

2.1 Physical Acquisition Mode

The physical acquisition mode allows bit-to-bit copy of all partitions of the entire memory, including unallocated space. This is certainly the most advanced, effective and valid technique to adopt in judicial domain. Usually it takes place with the device in "Download" or "Fastboot" mode for Android systems and in DFU mode for iOS systems (prior to the iPhone 4S model), for this reason it is also possible to bypass the device unlock code, if it should be present. The produced output is a memory dump that can be analysed with classic Mobile Forensics analysis tools.

2.2 File System Acquisition Mode

The file system acquisition mode extracts all the files stored in the device memory, including system data, integrated application data and unallocated space within files. However, it allows the recovery of most of the deleted files and often, from the point of view of extracted data, the acquisition is equivalent to an acquisition performed in physical mode.

IOS devices can be acquired in file system mode through the use of the iTunes backup function, which is able to extract all the contents of the device (both present and deleted) as files .ipa (iPhone Application), multimedia content, chat, e-mail and all other information [3,5] from applications [4,6].

On Android devices this mode of acquisition is carried out through the operation of Android backup, which backup all installed applications including installation files apk and shared memory, whether internal or external micro SD card. Starting from Android 5.0 Lollipop to adopt this mode of acquisition root privileges are required, which, of course, is a limitation, as not always it is possible to grant them. As an alternative, to circumvent authorization problems and any password protection or unlock code, this activity could be performed through a custom recovery, a system boot mode that replaces the default recovery mode and which contains many more features, including a full backup of all memory partitions (called NANDroid Backup). The NANDroid backup creates a copy of the individual system partitions and saves them in a shared memory or SD card folder. It will then be possible to analyse this data with one of the programs that will be described in Sect. 3. The use of a custom recovery alters the device and it is not always possible to perform this activity in the judicial field, however during the backup a log file and all the MD5 hashes are generated to guarantee data integrity.

2.3 Logical Acquisition Mode

The logical acquisition mode extracts only the files present and shared by the operating system from the device. It does not perform any carving operation on the memory, but can be useful to demonstrate and certify the presence of some contents received, sent or stored by the analyzed device, during a legal proceeding. For example, this mode can be useful (in the event that the other, more complete, were not supported) for crimes like stalking or child pornography. For this mode the same considerations of the file system mode are valid: it is necessary to have access to the device or to know the unlock code, in order to authorize the PC to perform the data reading.

3 Acquisition and Analysis Tools

Forensic acquisition and analysis techniques are constantly evolving, but the tools adopted for the extraction of data [9–11] from mobile devices are always a step behind the security protection adopted by their operating systems. In fact it has became increasingly complex for engineers who attempt to access and correctly interpret the data stored on memory devices through reverse engineering process [8]. The main difficulty in the creation and development of efficient and reliable forensic analysis tools lies in industrial secrets of manufacturers, the presence of proprietary file systems, new/proprietary operating systems, the lack of knowledge of the investigators and so on. There are several tools, both hardware and software, that allow forensic acquisition of mobile devices, those shown below are the most known and used.

UFED Cellebrite

Cellebrite is the leading company in the Mobile Forensics sector. Its most used tools in the forensic field are UFED Touch, UFED Touch2 and UFED For PC for the acquisition phase and UFED Physical Analyzer and UFED Reader for the analysis phase. These tools support over 20,000 devices and therefore they offer the most complete solution currently on the market. UFED Touch is a small computer, running Windows Compact Edition, on which the Cellebrite acquisition software is executed. On the left side, it hosts USB and Ethernet ports, to which you can connect the device to be acquired, be it a smartphone or tablet, but also an SD card reader. On the right side, instead, there is the destination USB port, to which a mass memory can be connected. The UFED Touch also has a SIM card reader port on the front of the device, a wireless card for software updates and a Bluetooth antenna for acquisitions that require it. There are extensions to the kit, such as UFED Camera used to make photographic acquisitions and brute force unlock of the device, UFED Chinex used to acquire Chinese clone phones and UFED Memory Card reader, for the acquisition of SD cards and memory cards, also equipped with the functionality write block (Write-Block).

Oxygen Forensics

Oxygen Forensics was initially designed for communication between mobile device and PC, then it has been specialized in reading and acquisition of smartphones and tablets. The suite provides a complete kit for acquisition and analysis similar to that proposed by Cellebrite. It contains the Getac F110 tablet PC, the Oxygen Forensics Extractor software, the USB license dongle, the software to be installed on a second PC for the analysis and the cable set for the connection of the devices. As the Cellebrite suite, it allows to acquire a great variety of models, even if it is limited in some aspects, as well as not allowing the acquisition of satellite navigators, drones and other more specific devices.

Magnet Axiom

Thanks to its module Magnet Acquire is able to perform the acquisition of mobile devices running iOS and Android operating system. Magnet does not provide any hardware tools for the acquisition, unlike its rivals, so it supports a limited number of devices. As a suite, it focuses on Computer Forensics, then on the acquisition and analysis of traditional memory devices such as hard disks, pendrive, memory cards, and so on.

Open Source Alternatives

Using open source alternatives is a double-edged sword, because, on the one hand you can save thousands of euros in commercial licenses of software or hardware devices, but on the other hand both the acquisition and the analysis phases become more complicated and cumbersome. Some open source techniques take advantage of features and functionality provided directly by the manufacturer or integrated into the Operating System, such as Android Backup mode, iTunes backup, and also DD copy on external SD card and NANDroid Backup. For Android devices Linux Santoku distribution and the Autopsy software are the most valuable tools, while for iOS it is possible to use the Zdziarski techniques and the iPhone Backup Analyzer software.

4 Conclusions

Mobile Forensics has become a important research and development trend that deals with the identification, preservation, study and analysis of information stored into mobile devices. Modern devices use complex data structures to store data, therefore, it has been necessary to adopt different acquisition and analysis tools compared to those used in Computer Forensics.

In our work, we focused on the study of the acquisition techniques and analysis tools. This study can constitute a starting point for further research, in order to crawl and represent the data extracted form mobile devices.

References

1. Amato, F., Moscato, F.: Pattern-based orchestration and automatic verification of composite cloud services. Comput. Electr. Eng. **56**, 842–853 (2016)
2. Amato, F., Moscato, F.: Exploiting cloud and workflow patterns for the analysis of composite cloud services. Future Gener. Comput. Syst. **67**, 255–265 (2017)
3. Balzano, W., Murano, A., Stranieri, S.: Logic-based clustering approach for management and improvement of vanets. J. High Speed Netw. **23**(3), 225–236 (2017)
4. Balzano, W., Murano, A., Vitale, F.: V2v-en-vehicle-2-vehicle elastic network. Procedia Comput. Sci. **98**, 497–502 (2016)
5. Balzano, W., Murano, A., Vitale, F.: Wifact–wireless fingerprinting automated continuous training. In: 2016 30th International Conference on Advanced Information Networking and Applications Workshops (WAINA), pp. 75–80. IEEE (2016)
6. Balzano, W., Murano, A., Vitale, F.: SNOT-WiFi: sensor network-optimized training for wireless fingerprinting. J. High Speed Netw. **24**(1), 79–87 (2018)
7. Coppolino, L., D'Antonio, S., Mazzeo, G., Romano, L.: Cloud security: emerging threats and current solutions. Comput. Electr. Eng. **59**, 126–140 (2017)
8. Mazzeo, G., Coppolino, L., D'Antonio, S., Mazzariello, C., Romano, L.: Sil2 assessment of an active/standby cots-based safety-related system. Reliabil. Eng. Syst. Saf. **176**, 125–134 (2018)
9. Piccialli, F., Chianese, A.: The internet of things supporting context-aware computing: a cultural heritage case study. Mob. Netw. Appl. **22**(2), 332–343 (2017)
10. Piccialli, F., Chianese, A.: A location-based iot platform supporting the cultural heritage domain. Concurr. Comput. **29**(11) (2017)
11. Piccialli, F., Jung, J.E.: Understanding customer experience diffusion on social networking services by big data analytics. Mob. Netw. Appl. **22**(4), 605–612 (2017)
12. Staffa, M., Sgaglione, L., Mazzeo, G., Coppolino, L., D'Antonio, S., Romano, L., Gelenbe, E., Stan, O., Carpov, S., Grivas, E., Campegiani, P., Castaldo, L., Votis, K., Koutkias, V., Komnios, I.: An openncp-based solution for secure ehealth data exchange. J. Netw. Comput. Appl. **116**, 65–85 (2018)

Remarks of Social Data Mining Applications in the Internet of Data

Salvatore Cuomo[1(✉)], Francesco Maiorano[1], and Francesco Piccialli[2]

[1] Dipartimento di Matematica ed Applicazioni "Renato Caccioppoli", University of Naples FEDERICO II, Naples, Italy
salvatore.cuomo@unina.it, frmaiorano@gmail.com
[2] Dipartimento di Ingegneria Elettrica e Tecnologie dell'Informazione, University of Naples FEDERICO II, Naples, Italy
francesco.piccialli@unina.it

Abstract. Social network analysis attracted interests from both the research and business communities for a strong potential and variety of applications. In addition, this interest has been fuelled by the large success of online social networking sites and the subsequent abundance of social network data produced. A key aspect in this research field is the *influence maximization* in social networks. In this paper we discuss an overview about the models and the approaches widely used to analyse social networks. In this context, we also discuss data preparation and privacy concerns also considering different kind of approaches based on centrality measures.

1 Introduction

For decades, social networks have been studied extensively [1–3] but were limited to extremely small datasets. With the relative recent advent of online social platforms such as Facebook, Twitter, LinkedIn, and Tumblr, social network analysis and mining are observing a remarkable growth due to the availability of large scale social network data. In fact, many other platforms and online social services may describe social interactions in forms that differ from graphs, which would still allow social ties inference to ultimately reconstruct the network or discover interaction patterns.

Many exciting applications of online social network mining have been developed, [4, 5] and this led to the formulation of research problems ranging from the study of social activity, [6] as in social contagion and trust transitivity, [7] to understand information and influence propagation throughout social networks. On the other hand, there exist other approaches to social network analysis based on measures of centrality of individuals in order to understand their degree of participation in social processes.

While in early sociological and anthropological studies, this type of research was carried out on small communities through interviews and questionnaires; recently, social network research has been carried out mainly using data collected from online interactions describing explicit or implicit relationship links among individuals. By social network, we refer to a possibly directed graph, a mathematical entity consisting of a set of nodes, individuals or entities of interest, and a set of edges, which describe connections and relationships among nodes. A graph can be homogeneous, all nodes of

L. Barolli et al. (Eds.): NBiS 2018, LNDECT 22, pp. 944–951, 2019.
https://doi.org/10.1007/978-3-319-98530-5_86

the same type, or heterogeneous, with nodes of more than one type. An example of homogeneous network is described by graphs representing friendships in social networking platforms, such as Facebook.

Social network data mining and the study of information propagation and trust transitivity has found application in several fields [10–12], such as social media analytics, viral marketing, spread of information analysis, [8] measuring the degree of adoption of products or ideas, studying epidemics, behavioural targeting, [9] expert finding, recommendation systems, etc. Network analysis can also be applied to monitoring water distribution networks, [13–15] where accidental or deliberate intrusion of contaminants in the water supply can cause viruses spreading.

In social media, an interesting application is the study of propagation of rumors, information, and influence. When online posts are linked to each other, it is possible to study how stories propagate and identify *experts* or *influencers* for a given topic.

However, many challenges have to be addressed in practice when analyzing social influence. In the case of viral marketing, the key idea is to leverage a small set of influencers to reach a large audience. Sociologists have instead studied the fundamental concept of power in social structures, [16, 17] although there is less agreement on its definition, causes, and consequences. The research community focusing on social network analysis has then developed several concepts aimed at describing and studying power and influence in social networks, with approaches closely related to the idea of centrality. Individuals do not possess some degree of influence or authority in an abstract manner as these are consequences of relations. The idea behind this intuition is that, if a system is loosely coupled (less dense), less influence would be exerted. This idea can be applied locally, considering one or a few individuals, or globally, applying it to the entire population.

We do not present here a full survey but aim at presenting an overview of the techniques used in social network analysis and mining as well as approaches used to tackle the information and influence propagation problem.

2 Data Acquisition and Aggregation in Social Networks

Since the large spread of online social network platforms and blogs, user-generated content is available online with an unprecedented explosion of data. In many cases, user data is automatically produced by wearables, smartphones, or other devices to monitor health, habits, and other kind of activities. Time spent on social network sites is increasing at a faster rate than the overall internet growth [18, 19]; two thirds of the world's Internet population spend more than 10% of their Internet time on social networks.

This explosion and growth of social network data has attracted attention by the research community for several application, ie, marketing purposes, forecasting, expert finding, influence propagation understanding, etc. Formally, a social network is represented as a graph $G = (V, E)$, where V is the set of nodes, representing users, and E is the set of edges that aim at modeling some type of interactions between two given users. The type of relationship determines directed or undirected edges.

In some applications, users explicitly declare their social ties by selecting their *friends* or *following* others activities. In other cases, links between users are implicit but can be inferred by analyzing interactions between users. This can include several actions performed on a social network such as commenting and sharing. Another fundamental reason makes it difficult to retrieve social network datasets in the form of graph (nodes and edges). Many real-world applications do not have access to the network. Graphs built by social networking platforms, such as Facebook or Twitter, have realized it is a valuable asset and will not release it to third parties for the sake of commercial competitive advantage. Trying to reconstruct or infer the network is explicitly forbidden on Facebook. Twitter, instead, actually sells its complete real-time stream of tweets through its service Firehose. Even in this case, trying to infer ties between individuals is forbidden by contract. Raw social network data acquired from online networks may show some degree of noise, such as (i) *duplicate nodes*, a single user created more than one account; (ii) *inactive nodes*, a user no longer uses the platform but did not delete the account; and (iii) *artificial nodes*, in the case of automated agents, such as bots. For these reasons, a pre-processing step is necessary in order to properly clean the data and remove duplicates, verify ranges of values, etc.

In the work of Rahm and Do [20] two main categories define the problem of cleaning and aggregating data, ie, *single-source* and *multiple-source*. The cleaning process itself consists of several phases, ie, analysis, transformation, and mapping and verification.

Another important step in data preparation is related to privacy issues in which the aim is to anonymize the data before further processing and analysis. In this context, Evfimievski et al. [21, 22] proposed a *privacy-preserving data mining* paradigm to perform mining tasks while still protecting individuals information. For instance, even if names or social security numbers are removed from the data, the structure of the graph may help identify individuals [23]. Dalenius [24] formulated the goal of protecting privacy of individuals; since then, many approaches have been proposed such as *query auditing*, [25] *output perturbation*, [26] *secure multiparty computation*, [27] and *data sanitization* [28].

The problem of anonymization in social networks is still under investigation. Three main categories describe methods for graph perturbation. Lappas et al. [29] and Wu et al. [30] proposed a k-anonymity, adding edges via a deterministic procedure. Hanhijarvi et al. [31] and Hay et al. [32] randomly added, deleted, or switched edges in order to prevent identification. Other works [33–35] did not alter the graph but group together nodes into supernodes of size k, ie, a parameter chosen to achieve the desired degree of anonymity.

3 Measure of Centrality

In graph theory and network analysis, the most important vertices are identified by indicators of centrality. Even if these are used to identify influential individuals in a social network, they should not be confused with influence metrics techniques aimed at quantifying the influence of each node in the network. Centrality measures have a sociological origin and are used to assess which individuals have a favoured position with respect to the network topology. Therefore, they try to describe relationships, such as resource exchange or resource sharing. However, centrality-based techniques suffer important limitations in terms of optimality with respect to their field of application as they do not represent universal measures. In fact, they are strongly dependent to the structure of the underlying graph and the topology of the most important nodes. In many cases, the rankings may not apply to other nodes [36, 37]. For these reason, several centrality measures have been developed to try and achieve optimality depending on the case at hand [38]. Moreover, such techniques compute a ranking for each node and they strongly differ from other techniques, as illustrated in the next sections, that instead measure the influence of nodes [39–44].

3.1 A Case of Study on Twitter Data

We provide here a brief analysis of two centrality measures applied to social network data. We choose to analyze the interactions between users during an event (*Comicon*) that took place between April and May 2017. We collect 75 tweets using the Twitter API and identify 50 unique users. We then retrieve the list of followers for each user to ultimately create a directed graph connecting users and their followers. We only keep followers shared among the 50 starting users and discard the rest as we are only interested in analyzing interactions between the starting set. Our goal is to investigate which user shows a certain degree of *power* over propagation of information and influence throughout the sub-network identified over such a small starting set.

We build a directed graph consisting of 3197 nodes and 6240 edges. Each edge starts from one of the followers and ends in one of the users in the starting set. On such network, we compute betweenness and closeness centrality for each node. In Fig. 1, we show the resulting network; here, we highlight the degree of closeness centrality with a warmer color and the degree of betweenness centrality with a bigger node diameter.

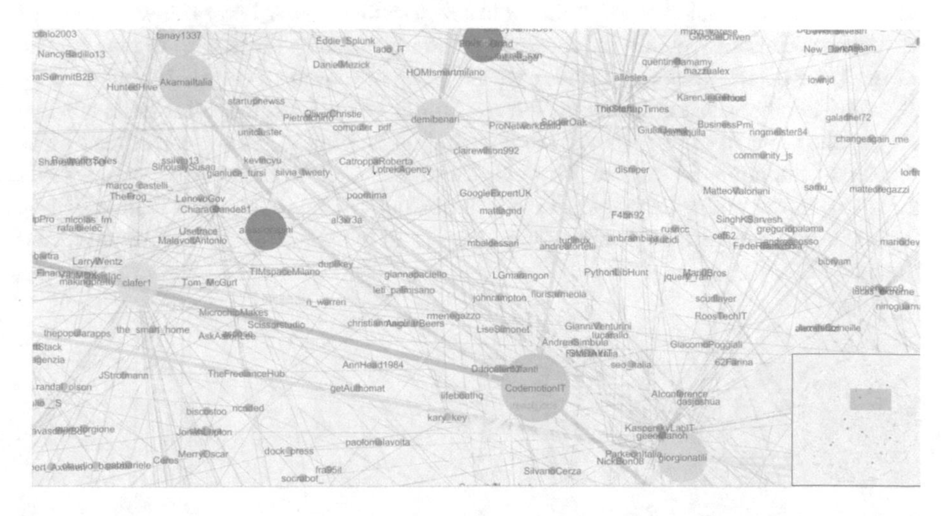

Fig. 1. A detailed view of the main influencers and their interactions from the *Comicon* Tweet data network

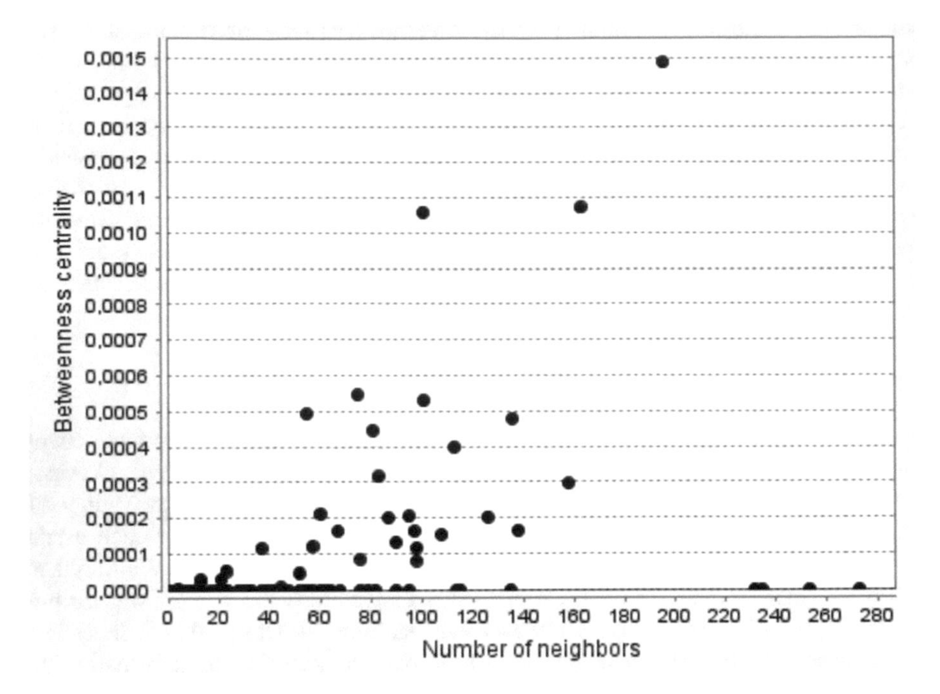

Fig. 2. The betweenness centrality distribution

We also compute these measures on the edges in the network highlighting these characteristics in the same manner.

We show in Figs. 2 and 3 the betweenness and closeness centrality distributions computed on the comicon data network against the number of neighbours. In Fig. 2, we can see how the majority of nodes cluster together toward very low values of betweenness centrality, while influencers can be identified as those with higher values. In Fig. 3, we detect a cluster of nodes with a value of closeness centrality around 0.2–0.4, while another cluster stacks on the maximum value 1. In the latter case, we can interpret the concept of *power* over the information flowing in the network as acting as a center of attention for others. Through these users, information flow might reach others via shorter paths, leveraging favoured positions. This topological feature can be seen as an advantage that enables some degree of power over the information and influence flow.

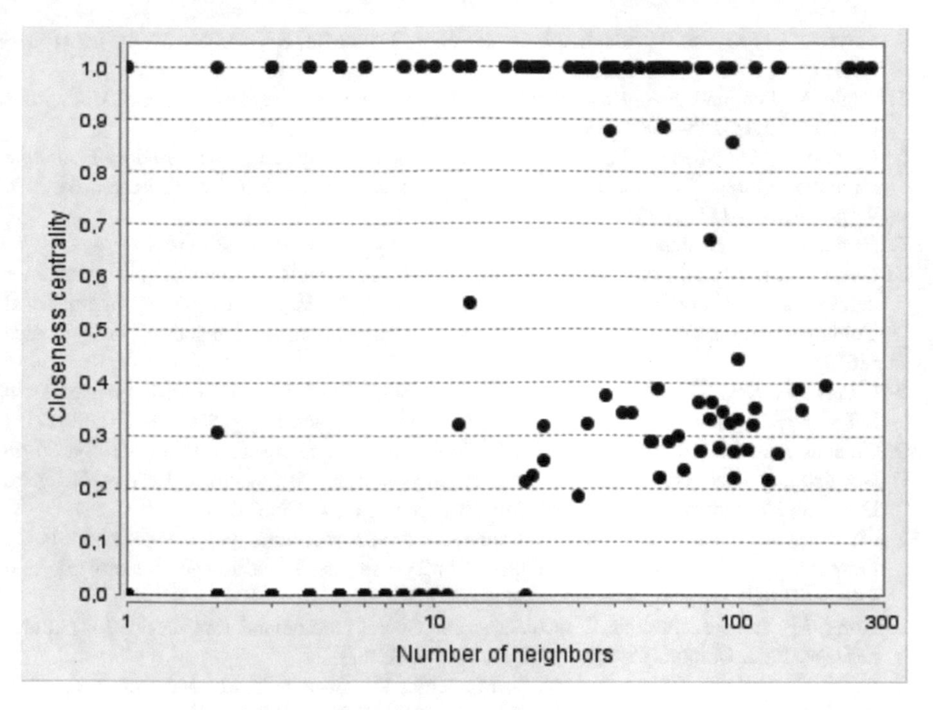

Fig. 3. The closeness centrality distribution

4 Conclusions

The theory and algorithms behind network analysis, information diffusion, and virality has seen great advances. Many research problems are still open and are mainly motivated by real problems.

Much work has to be carried out on the computational intractability of large networks. As we describe in this document, recent approaches are based on forgetting the network structure, trying to infer social contagion from other kind of datasets, namely, propagation logs. The Internet of Things and Internet of Data fields together can benefit from the contributions and developments achieved by the research community. As we now consider networks of individuals, we may as well consider network of artificial intelligent agents in the coming future.

References

1. Degene, A., Forse, M.: Introducing Social Networks. SAGE Publications, Thousand Oaks (1999)
2. Scott, J.: Social Network Analysis: A Handbook. SAGE Publications, Thousand Oaks (2000)
3. Wasserman, S., Faust, K.: Social Network Analysis: Methods and Applications. Cambridge University Press, Cambridge (1994)

4. Fawcett, T., Provost, F.: Adaptive fraud detection. Data Min. Knowl. Discov. **1**(3), 291–316 (1997)
5. Eagle, N., Pentland, A.S.: Reality mining: sensing complex social systems. Pers. Ubiquitous Comput. **10**(4), 255–268 (2006)
6. Cuomo, S., De Michele, P., Piccialli, F., Galletti, A., Jung, J.E.: IoT-based collaborative reputation system for associating visitors and artworks in a cultural scenario. Expert Syst. Appl. **79**, 101–111 (2017)
7. Buskens, V.: The social structure of trust. Soc. Netw. **20**(3), 265–289 (1998)
8. Chianese, A., Marulli, F., Moscato, V., Piccialli, F.: A "smart" multimedia guide for indoor contextual navigation in cultural heritage applications. In: Paper presented at: International Conference on Indoor Positioning and Indoor Navigation, Montbeliard-Belfort, France (2013)
9. Cuomo, S., De Michele, P., Pragliola, M.: A computational scheme to predict dynamics in IoT systems by using particle filter. Concurr. Comput. Pract. Exp. **29**(11), e4101 (2017)
10. Chianese, A., Piccialli, F., Riccio, G.: Designing a smart multisensor framework based on Beaglebone black board. In: Computer Science and Its Applications. Lecture Notes in Electrical Engineering, vol. 330, pp. 391–397. Springer, Berlin (2015)
11. Chianese, A., Piccialli, F.: SmaCH: a framework for smart cultural heritage spaces. In: Paper presented at: 2014 Tenth International Conference on Signal-Image Technology and Internet-Based Systems, Marrakech, Morocco (2015)
12. Hong, M., Jung, J., Piccialli, F., Chianese, A.: Social recommendation service for cultural heritage. Pers. Ubiquit. Comput. **21**(2), 191–201 (2017)
13. Ferlez, J., Faloutsos, C., Leskovec, J., Mladenic, D., Grobelnik, M.: Monitoring network evolution using MDL. In: Paper Presented at: 2008 IEEE 24th International Conference on Data Engineering, Cancun, Mexico (2008)
14. Ostfeld, A., Salomons, E.: Optimal layout of early warning detection stations for water distribution systems security. J. Water Resour. Plan. Manage. **130**(5), 377–385 (2004)
15. Ostfeld, A., Uber, J., Salomons, E., et al.: The battle of the water sensor networks (BWSN): a design challenge for engineers and algorithms. J. Water Resour. Plan. Manage. **134**(6), 556–568 (2008)
16. Freeman, L.: The development of social network analysis. A Study in the Sociology of Science, vol. 1 (2004)
17. Coleman, J., Katz, E., Menzel, H.: Medical innovations: a diffusion study. Soc. Forces **46**(2), 291 (1966)
18. Kuss, D.J., Griffiths, M.D.: Online social networking and addiction: a review of the psychological literature. Int. J. Environ. Res. Public Health **8**(9), 3528–3552 (2011)
19. "Time spent" on these sites growing three times faster than overall internet rate, now accounting for almost 10 percent of all internet time. Nielsen Web site (2009). http://www.nielsen.com/us/en/press-room/2009/social-networks–.html
20. Rahm, E., Do, H.H.: Data cleaning: problems and current approaches. IEEE Data Eng. Bull. **23**(4), 3–13 (2000)
21. Evfimievski, A., Srikant, R., Agrawal, R., Gehrke, J.: Privacy preserving mining of association rules. Inf. Syst. **29**(4), 343–364 (2002)
22. Evfiemski, A., Geherke, J., Srikant, R.: Limiting privacy breaches in privacy preserving data mining. In: Proceedings of the Twenty-Second ACM SIGMOD-SIGACT-SIGART Symposium on Principles of Database Systems, PODS 2003, San Diego, CA (2003)
23. Backstorm, L., Dwork, C., Kleinberg, J.: Anonymized social networks, hidden patterns, and structural steganography. In: Proceedings of the 16th International Conference on World Wide Web, WWW 2007, Banff, Canada (2007)

24. Dalenius, T.: Towards a methodology for statistical disclosure control. Statistik Tidskrift. **15**, 429–444 (1977)
25. Kempe, D., Kleinberg, J., Tardos, É.: Maximizing the spread of influence through a social network. In: Proceedings of the Ninth ACM SIGKDD International Conference on Knowledge Discovery and Data Mining, KDD 2003, Washington, DC (2003)
26. Blum, A., Dwork, C., McSherry, F., Nissim, K.: Practical privacy: the SuLQ framework. In: Proceedings of the Twenty-Fourth ACM SIGMOD-SIGACT-SIGART Symposium on Principles of Database Systems, PODS 2005, Baltimore, MD (2005)
27. Aggarwal, G., Mishra, N., Pinkas, B.: Secure computation of the kth-ranked element. In: Advances in Cryptology - EUROCRYPT 2004. Lecture Notes in Computer Science. Springer, Heidelberg (2004)
28. Agrawal, R., Srikant, R.: Privacy-preserving data mining. In: Proceedings of the 2000 ACM SIGMOD International Conference on Management of Data, SIGMOD 2000, Dallas, TX (2000)
29. Lappas, T., Liu, K., Terzi, E.: Finding a team of experts in social networks. In: Proceedings of the 15th ACM SIGKDD International Conference on Knowledge Discovery and Data Mining, KDD 2009, Paris, France (2008)
30. Wu, W., Xiao, Y., Wang, W., He, Z., Wang, Z.: K-symmetry model for identity anonymization in social networks. In: Proceedings of the 13th International Conference on Extending Database Technology, EDBT 2010, Lausanne, Switzerland (2010)
31. Hanhijarvi, S., Garriga, G.C., Puolamäki, K.: Randomization techniques for graphs. In: Proceedings of the 2009 SIAM International Conference on Data Mining. Sparks, NV (2009)
32. Hay, M., Miklau, G., Jensen, D., Weis, P., Srivastava, S.: Anonymizing social networks. Technical report. University of Massachusetts Amherst, Amherst (2007)
33. Campan, A., Truta, T.M.: A clustering approach for data and structural anonymity in social networks. In: Paper presented at: 2nd ACM SIGKDD International Workshop on Privacy, Security, and Trust in KDD, Las Vegas, NV (2008)
34. Hay, M., Miklau, G., Jensen, D., Towsley, D., Weis, P.: Resisting structural re-identification in anonymized social networks. Proc. VLDB Endow. **1**(1), 102–114 (2008)
35. Zheleva, E., Getoor, L.: Preserving the privacy of sensitive relationships in graph data. In: Privacy, Security, and Trust in KDD. Springer, Berlin (2007)
36. Ž urauskiene, J., Kirk, P.D.W., Stumpf, M.P.H.: A graph theoretical approach to data fusion. bioRxiv preprint (2015)
37. Bauer, F., Lizier, J.T.: Identifying influential spreaders and efficiently estimating infection numbers in epidemic models: a walk counting approach. Europhys. Lett. **99**(6), 68007 (2012)
38. Lawyer, G.: Understanding the spreading power of all nodes in a network: a continuous-time perspective. Proc. Natl. Acad. Sci. **9**, 1–7 (2015). https://arxiv.org/pdf/1405.6707.pdf
39. Centrality. Wikipedia Web site (2017). https://en.wikipedia.org/wiki/Centrality
40. Bonacich, P.: Power and centrality: a family of measures. Am. J. Sociol. **92**(5), 1170–1182 (1987)
41. Borgatti, S.P.: Centrality and network flow. Soc Netw. **27**, 55–71 (2005)
42. Amato, F., Moscato, F.: Exploiting cloud and workflow patterns for the analysis of composite cloud services. Future Gener. Comput. Syst. **67**, 255–265 (2017)
43. Amato, F., Moscato, F.: Pattern-based orchestration and automatic verification of composite cloud services. Comput. Electr. Eng. **56**, 842–853 (2016)
44. Amato, F., Moscato, F.: A model driven approach to data privacy verification in e-health systems. Trans. Data Priv. **8**(3), 273–296 (2015)

An Approach for Securing Cloud-Based Wide Area Monitoring of Smart Grid Systems

Luigi Coppolino, Salvatore D'Antonio, Giovanni Mazzeo(✉), Luigi Romano, and Luigi Sgaglione

University of Naples 'Parthenope', Naples, Italy
{luigi.coppolino,salvatore.dantonio,giovanni.mazzeo, luigi.romano,luigi.sgaglione}@uniparthenope.it

Abstract. Computing power and flexibility provided by cloud technologies represent an opportunity for Smart Grid applications, in general, and for Wide Area Monitoring Systems, in particular. Even though the cloud model is considered efficient for Smart Grids, it has stringent constraints in terms of security and reliability. An attack to the integrity or confidentiality of data may have a devastating impact for the system itself and for the surrounding environment. The main security risk is represented by malicious insiders, i.e., malevolent employees having privileged access to the hosting machines. In this paper, we evaluate a powerful hardening approach that could be leveraged to protect synchrophasor data processed at cloud level. In particular, we propose the use of homomorphic encryption to address risks related to malicious insiders. Our goal is to estimate the feasibility of such a security solution by verifying the compliance with frame rate requirements typical of synchrophasor standards.

1 Introduction

Future generations of Wide Area Monitoring Systems (WAMS) look at commercial cloud architectures as an opportunity to reduce costs, increase data sharing, enhance scalability, and improve availability. At the core of WAMs there are PMUs (Phasor Measurement Units), which are nowadays used in distribution network context to measure and control the status of power grid. PMUs do so through synchrophasors, i.e., time-synchronized numbers that represent both the magnitude and phase angle of the sine waves found in electricity, which are time-synchronized for accuracy. Synchrophasors enable a synchronized evaluation of the phasor through GPS radio clock, and are being extensively deployed together with network-based Phasor Data Concentrator (PDC) applications for providing a precise and comprehensive view of the status of the entire grid. PDC units are in charge of collecting data coming from different PMUs and realize the effective computation.

Prototypal solutions of Cloud-based WAM were proposed [10]. The idea is to capture sensors data on a cloud-computing platform, and leverage its facilities to archive the data into a standard data collection infrastructure, which can include standard databases or grid specific solutions such as OpenPDC [11] to track the system state in real-time by performing the variety of analyses at cloud level.

L. Barolli et al. (Eds.): NBiS 2018, LNDECT 22, pp. 952–959, 2019.
https://doi.org/10.1007/978-3-319-98530-5_87

However, difficulties of sharing data in a secure and low-latency manner limits exploitation of this new powerful technology by bulk electric power grid operators.

In a cloud-based deployment, the PDC application may be exposed to a number of threats that affect both confidentiality and integrity of sensitive data going through the monitoring system. The Insider Threat [12] is a particularly worrying example. It is impersonated by a malicious employee of cloud providers that leverages the privileged position to obtain access to sensitive data [20].

Two solutions are the most accepted against malicious insiders: Trusted Execution (TE) and Homomorphic Encryption (HE). There are works proposing the adoption of TE, in particular Intel SGX, for enabling the cloudification of SCADA systems [15–17]. While the adoption of HE for WAMS was still poorly investigated [18].

In this paper we propose an approach that enables secure synchrophasor data processing in untrusted cloud environments. Our approach is to leverage most recent implementations of HE to create and preserve a chain-of-trust from the field data collection to the cloud processing, and ensures confidentiality, even against malicious insiders. The idea is to encrypt PMU data on field and then transmit such a data to the cloud for subsequent storage or processing Synchrophasor measurements are always kept encrypted and evaluations like phase comparisons will be performed on ciphered data. In this way risks coming from malicious cloud insiders can be addressed and the security is preserved. Unfortunately, HE suffers from a non-negligible performance overhead, which could made the adoption of this technology impossible for synchrophasor data processing, which are characterized from strict frame rate requirements. Most-recent schemes like the one adopted in this paper, i.e., TFHE [13, 14], started to provide very fast results. Our goal is to evaluate the feasibility of HE towards secure synchrophasor data processing in untrusted cloud by estimating supported frame rates.

The remainder of this work is organized as follows. Section 2 provides background on HE. Afterwards, Sect. 3 introduces synchrophasor systems and concepts of Wide Area Monitoring. Then, Sect. 4 defines possible threats in cloud environment. This is followed by Sect. 5 where we provide our solution. Finally, Sect. 6 concludes the document.

2 Background

This section overviews the technology adopted in this paper to enhance the security of cloud-based WAMS, i.e., Homomorphic Encryption (HE). HE is a recent cryptographic method allowing to perform computation directly on encrypted data, without the need of decrypting it. As such, the encryption schemes possessing homomorphic properties can be very useful to construct privacy preserving protocols, in which the confidential data remains secured, not only during exchange and storage, but also during processing. In a context of data outsourcing and of cloud computing, the homomorphic encryption is a mechanism that helps to protect data against intrusions from the cloud provider itself or attacks on the cloud infrastructure. Conventional symmetric and public key cryptosystems encrypt the data such that only the authorized parties can access it. In order to perform operations on this data, one needs to decrypt it first. Contrary to the abovementioned encryptions schemes, homomorphic encryption can be used not only as a method to protect data privacy, but also to execute algorithms directly on encrypted data. The service (cloud) provider processes the received data encrypted with the public key, performs operations over this encrypted data and sends the encrypted result to the end user, owner of the homomorphic secret key. As an example, homomorphic encryption has already been used as a key-tool in the popularization of electronic-based voting scheme. There are various application contexts in which this paradigm can be employed: private searching, keyword search, private storage, anonymous authentication, etc. [19]. During the years, three types of HE algorithms were proposed:

Partially Homomorphic Encryption (PHE) [1, 3], has the ability to carry out just one type of operations (e.g., addition, or multiplication). Clearly, the limitation in the type of executable computations hampered the usage of HE in practical contexts.

Fully Homomorphic Encryption (FHE) [4] in 2009, Gentry, at Stanford, proposes a first credible construction, both in terms of security and theoretical efficiency which proposed the first implementation of a FHE scheme. FHE schemes are capable to perform additions and multiplications over homomorphically encrypted data (ciphertexts) which correspond to addition and, respectively, multiplication operations over the clear text messages (plaintexts). Therefore, since any function can be expressed as a combination of additions and multiplications, FHE cryptosystems could compute in theory any arbitrary function. The first barrier to the adoption of FHE cryptosystems in real-world applications was related to the computational overhead induced by the actual execution on homomorphically encrypted data. In particular, it should be noted that most of the homomorphic encryption schemes provide mainly bit-level operators, thus intrinsically low level. However, making use of recent dedicated compilation and parallelism techniques, it was possible to mitigate the performances overhead for a series of real, yet lightweight, algorithms.

The security of the system is based on the noise introduced into the ciphered text. However such a noise can quickly grow, becoming larger as more homomorphic operations are performed on a given ciphertext. When the noise reaches some maximum amount, the ciphertext becomes undecryptable. Hence, one possible solution is refreshing the ciphertext. That is, performing after each operation a bootstrapping procedure able to remove the noise. This consists in a particular decryption algorithm performed to avoid noise. Unfortunately, this first fully homomorphic cryptosystem, and more generally, any bootstrapping-based cryptosystem known so far (until very recently), are way too costly to have any practical relevance whatsoever.

Somewhat Homomorphic Encryption (SHE), a much more efficient scheme was provided by Van Dijk et al. [5] who proposed a FHE scheme without the heavy bootstrapping procedure, i.e., Somewhat Homomorphic Encryption (SHE) over the integers. The price to pay with SHE is given by the limited number of mathematical operations that can be performed. However, in many real-world applications (e.g., medical, financial) this seems reasonable since – as Naehrig et al. [6] analysis reports – most of the evaluations required, i.e., one-time statistical functions, fits well with SHE constraints. In this work, we pursue the Van Dijk's SHE algorithm. This choice is driven by the need for a library that could be better adapted to HTEE requirements. We had to find a scheme having the following features: limited consumption of memory, fairly good performance, and support for all mathematical operations. SHE fits particularly well for our purposes. The only issue affecting SHE, i.e., the limited number of consecutive executable operations, is not a concern for the Synchrophasor use case.

3 Synchrophasor Systems for Smart Grids

Wide-Area Monitoring Systems (WAMS) of power grids is one important application from Smart Grid infrastructures. WAMS are composed of distributed measurement and control devices, characterized by a hierarchical architecture. A key component of WAMS is the Phasor Measurement Unit (PMU). This is the device in charge of electrical quantities synchronized measurements, e.g., voltage and current phasors, frequency and rate of change of frequency (ROCOF) with an accurate time-tag based on the Universal Coordinated Time generally obtained from a GPS receiver or through IEEE1588 synchronization. PMUs forward the acquired data to a Phasor Data Concentrator (PDC), which collects and aligns the provided measurements before send them next higher PDC levels. Ultimately, data arrives to a control center application where the overall status of the electric grid is evaluated.

Originally, synchronized measurements of WAMS was designed for transmission systems. After the advent of smart grid frameworks, benefits of Synchrophasor technology are moving also to the distribution network. The use of PMUs in the distribution network context represents a new challenge: the stand-alone PMUs and PDCs could be replaced by dedicated functionalities implemented in Intelligent Electronic Devices (IEDs) or by existing measurement devices upgraded in order to build an Internet of Things (IoT) network with synchrophasor functionality. In a Synchrophasor system suitable for distribution grids, several measurement devices will be necessary and, in this new scenario, the classical hierarchical architecture can be inadequate, since it can be unable to manage many PMUs and/or PMU-enabled instruments. A solution can be represented by replacing the hierarchical structure of PDCs with a less expensive and rapidly scalable structure based on cloud computing. The communication systems used by distribution system operators (DSOs) are expected to be shared and/or public. In this case, the bandwidth available for devices involved is strictly dependent on the type of communication channel adopted.

Normally, in transmission system WAMS, PMUs send data at a constant rate of 50–60 frames per second (fps) to guarantee the monitoring of dynamic events. The choice of 50 or 60 frames mean different supported frame rates (Table 1).

Table 1. Synchrophasor frame rate according to IEEE Std C37.118.2™-2011

System frequency	50 Hz			60 Hz					
Reporting rates (F_s–frames per second)	10	25	50	10	12	15	20	30	60

The standard says: "*The actual rate to be used shall be user selectable. Support for other reporting is permissible, and higher rates such as 100 frames/s or 120 frames/s and rates lower than 10 frames/s such as 1 frame/s are encouraged*".

That is, the minimum accepted may be 1 frame per second.

4 Threats in a Cloud Environment

While, on one hand, cloud computing is capable of offering huge benefits to Smart Grid systems in terms of IT cost saving and reliability. On the other hand, it opens to a number of security risks that cannot be underestimated. As evidenced by Coppolino et al. [2], applications running in the untrusted cloud are exposed to well-known attacks, which have been around for years but that gained prominence again because of the large adoption of cloud computing. These include attacks aiming at violating: (i) availability by, e.g., flooding targeted machines (Denial of Service (1)); (ii) data confidentiality/integrity by, e.g., altering communication channels [19] (Traffic Hijacking (2)) or landing on the system to subsequently launch an attack (Account Hijacking (3)). Besides these, the taxonomy considers other attacks, which in turn are typical of the cloud universe. That is, those perpetrated by:

- Internal users who own a Virtual Machine (VM) and exploit flaws in the hypervisor (Shared Technologies Vulnerabilities (4)) to attack another VM instance.
- The Cloud Provider – embodied by disgruntled employees or administrators – that leverages its privileged position to get access to an unprecedented amount of information and on a much greater scale (Malicious Insiders (5)).

The latter is definitely the most worrisome category of attackers who can easily cover their actions and go undetected for years. It is even more worrying in the context of Smart Grid since the impact on the external environment could be destructive. Attacks to the integrity, e.g., could have effects on the capability to provide correct commands to the actuators, and also acquire the right measurements from sensors. This entails that operators may assume that the status of the infrastructure is normal since all measured parameters have the expected values but this is not the case. Equally important is the availability of WAMS applications. Attacks like DoS, in fact, may cause corruptions on the status of the infrastructure, hardware failures, and, more importantly monitoring service outage. In the case of critical infrastructures, e.g., this could mean risks to human lives. Finally, attacks to confidentiality would imply that the adversary either infers the current state of the Smart Grid.

5 Proposed Solution

Figure 1 reports the high-level architecture of our proposed solution. PMUs gather data from sensors deployed on-field and send it to different layers of PDCs up to the PDC gateway, at the top. This unit is in charge of encrypting data – in a homomorphic fashion – and, then, establishing a TLS secure communication with its counterpart at cloud level to create an authenticated channel. Hence, the acquired encrypted measurements are sent via this secure channel. At cloud level, the synchrophasor data is received and sent over a distributed message-based bus (e.g., ZMQ) to different Microservices (MS), i.e., small independent software units that interact each other through messages. Microservices are a new way of conceiving applications architectures, which fit much better for distributed applications like those for cloud platforms. In our work, besides providing and storing measurements, MSs perform also those mathematical functions needed by the WAMS case study.

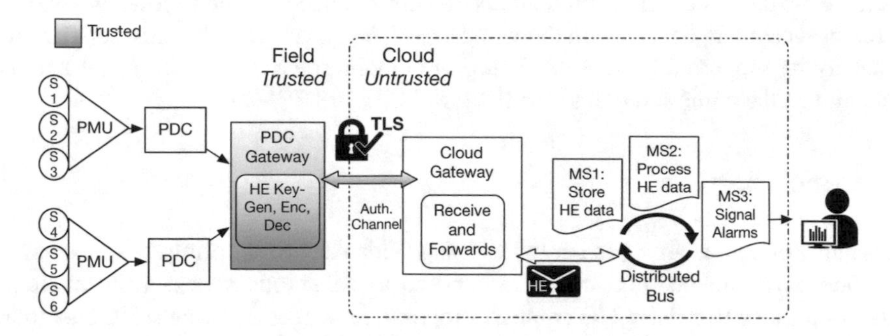

Fig. 1. Proposed Solution

In particular, we identified two mathematical operations to be realized on homomorphic encrypted data for synchrophasor evaluations, i.e.: phase subtraction and phase comparison. From a practical point of view, this meant the definition of a dedicated logical Boolean circuit – composed of gates having homomorphic support – in which ciphered bits will go through to carry out the protected computation.

Figure 2 shows the final scheme of the logical circuit used, which is composed of a full subtractor and a comparator, organized in sequence.

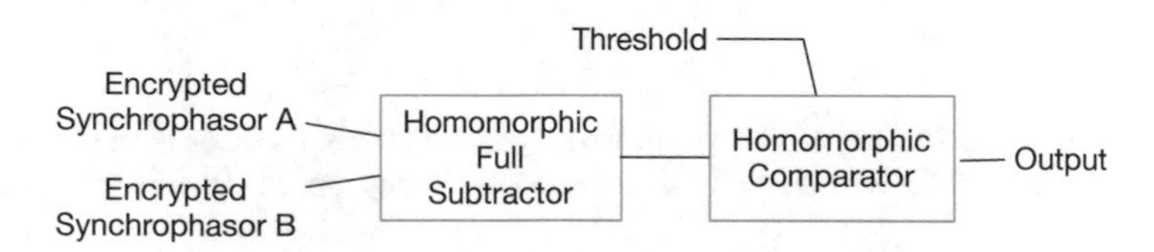

Fig. 2. Logical circuit for HE synchrophasor evaluations

Our implementation used the widely-accepted APIs of TFHE [13], which is based on the Chillotti et al. HE algorithm [14]. It must be noticed that each gate of the logical circuit introduces an overhead. Hence, particular attention must be put on the number of gate levels to be created. The deeper is the circuit, the higher is the overhead of the overall calculus. To have an idea, we evaluated the execution time of most critical logical gates (e.g. AND, OR, MUX). We obtained that binary gates require on average 8.1 ms, while the MUX took 20.4 ms using an Intel Xeon E3-1270 v5 CPU with 4 cores at 3.6 GHz, having 8 hyper-threads (2 per core), 8 MB of cache, and 64 GB of memory.

To improve the performance, we took advantage of the bit-wise nature of homomorphic operations and were able to reduce the amount of computations. The system elaborates group of bits from the most significant to the least significant ones. In this way, only in the worst case, the entire word is evaluated. Moreover, we optimized the system by using a particular implementation of Fast Fourier Transform, i.e., FFTW3, which is $3\times$ times faster than the default Nayuki implementation.

Finally, we ran tests on our developed WAMS. We used the OpenDSS comprehensive electrical power system simulation tool for simulating the WAMS data and properly evaluate the homomorphic computations. Results obtained are the following: 330.4 ms in the best case and 651.3 ms in the worst case. This means that the only standard rate that may be supported by a WAMS having homomorphic encryption is 1 frame per second, i.e., the minimum defined in IEEE Std C37.118.2™-2011.

6 Conclusion

This paper discussed an approach for securing Wide Area Monitoring Systems running in cloud environments, and therefore exposed to dangerous attacks from malicious insiders on sensitive data. The solution proposed leverages homomorphic encryption for executing protected synchrophasor computation. Our goal was to evaluate the feasibility of such an approach and understand the impact on the overall performances. In fact, requirements of data processing rates declared in synchrophasor standards are strict and it is important to verify the compliance of the system that works on homomorphic encrypted data.

Acknowledgements. This project has received funding from the European Union's Horizon 2020 Framework Programme for Research and Innovation under grant agreements No 74071 (COMPACT).

References

1. El Gamal, T.: A public key cryptosystem and a signature scheme based on discrete logarithms. In: Proceedings of CRYPTO 84 on Advances in Cryptology, pp. 10–18. Springer, New York (1985)
2. Coppolino, L., Antonio, S.D., Mazzeo, G., Romano, L.: Cloud security: emerging threats and current solutions. Comput. Electr. Eng. **59**, 126–140 (2017). https://doi.org/10.1016/j.compeleceng.2016.03.004
3. Paillier, P.: Public-key cryptosystems based on composite degree residuosity classes, pp. 223–238. Springer, Heidelberg (1999)
4. Gentry, C.: Fully homomorphic encryption using ideal lattices. In: Proceedings of the Forty first Annual ACM Symposium on Theory of Computing, STOC 2009, pp. 169–178. ACM, New York (2009)
5. van Dijk, M., Gentry, C., Halevi, S., Vaikuntanathan, V.: Fully homomorphic encryption over the integers. Cryptology ePrint Archive, Report 2009/616 (2009). http://eprint.iacr.org/2009/616
6. Naehrig, M., Lauter, K., Vaikuntanathan, V.: Can homomorphic encryption be practical? In: Proceedings of the 3rd ACM Workshop on Cloud Computing Security Workshop, CCSW 2011, pp. 113–124. ACM, New York (2011)
7. Jayaram Masti, R., Marforio, C., Capkun, S.: An architecture for concurrent execution of secure environments in clouds. In: Proceedings of the 2013 ACM Workshop on Cloud Computing Security Workshop, ser. CCSW 2013, pp. 11–22. ACM, New York (2013). http://doi.acm.org/10.1145/2517488.2517489

8. Maene, P., Gotzfried, J., de Clercq, R., Muller, T., Freiling, F., Verbauwhede, I.: Hardware-based trusted computing architectures for isolation and attestation. IEEE Trans. Comput. **99**, 1 (2017)
9. McKeen, F., Alexandrovich, I., Berenzon, A., Rozas, C.V., Shafi, H., Shanbhogue, V., Savagaonkar, U.R.: Innovative instructions and software model for isolated execution. In: Proceedings of the 2nd International Workshop on Hardware and Architectural Support for Security and Privacy, ser. HASP (2013)
10. Anderson, D., et al.: Grid cloud: infrastructure for cloud-based wide area monitoring of bulk electric power grids. In: IEEE Trans. Smart Grid. https://doi.org/10.1109/TSG.2018.2791021
11. Open Source PDC. https://github.com/GridProtectionAlliance/openPDC
12. Claycomb, W.R., Nicoll, A.: Insider threats to cloud computing: directions for new research challenges. In: 2012 IEEE 36th Annual Computer Software and Applications Conference, pp. 387–394 (2012). https://doi.org/10.1109/COMPSAC.2012.113
13. TFHE: Fast Fully Homomorphic Encryption Library over the Torus https://github.com/tfhe/tfhe
14. Chillotti, I., Gama, N., Georgieva, M., Izabachène, M.: Faster fully homomorphic encryption: bootstrapping in less than 0.1 seconds. In: Asiacrypt, pp. 3–33 (2016)
15. Campanile, F., et al.: Cloudifying critical applications: a use case from the power grid domain. In: 2017 25th Euromicro International Conference on Parallel, Distributed and Network-based Processing (PDP), St. Petersburg, pp. 363–370 (2017). https://doi.org/10.1109/PDP.2017.50
16. Brenner, S., Hundt, T., Mazzeo, G., Kapitza, R.: Secure cloud micro services using intel SGX. In: Proceedings of the 17th International IFIP Conference on Distributed Applications and Interoperable Systems (DAIS 2017) (2017)
17. Cerullo, G., Mazzeo, G., Papale, G., Sgaglione, L., Cristaldi, R.: A secure cloud-based SCADA application: the use case of a water supply network. In: Proceedings of the Fifteenth New Trends in Software Methodologies, Tools and Techniques, SoMeT 2016, Larnaca, Cyprus, 12–14 September 2016, pp. 291–301 (2016)
18. Alabdulatif, A., Kumarage, H., Khalil, I., Atiquzzaman, M., Yi, X.: Privacy-preserving cloud-based billing with lightweight homomorphic encryption for sensor-enabled smart grid infrastructure. IET Wireless Sensor Syst. pp. 1–11
19. Amato, F., Moscato, F.: Exploiting cloud and workflow patterns for the analysis of composite cloud services. Future Gener. Comput. Syst. **67**, 255–265 (2017). https://doi.org/10.1016/j.future.2016.06.035. ISSN 0167-739X
20. Amato, F., Moscato, F.: A model driven approach to data privacy verification in E-Health systems. Trans. Data Priv. **8**(3), 273–296 (2015)

Cooperative Localization Logic Schema in Vehicular Ad Hoc Networks

Walter Balzano(✉) and Silvia Stranieri

Dip. Ing. Elettrica e Tecnologie dell'Informazione,
Università di Napoli Federico II, Naples, Italy
wbalzano@unina.it, silviastranieri1047@gmail.com

Abstract. Localization of nodes in wireless sensor networks (WSNs) is typically obtained by exploiting specific systems such as Global Position System (GPS). In this work we consider a GPS-free scenario and we want to provide a way to precisely estimate the nodes position, by exploiting cooperation mechanisms. Many algorithms have been proposed for environment mapping problem considering a single vehicle perspective, but they do not fully exploit the potential of having a network of nodes where the collaboration can seriously speedup the mapping procedure. In this work, we analyze the Vehicular Ad Hoc Networks (VANets) scenario and we propose a strategy for vehicle pose estimation by exploiting a decentralized clustering technique, and some fixed nodes in the environment called anchors, whose position is already known. The power of this approach is in the cooperative nature of vehicles localization, explained by means of Prolog facts and rules that point out the actual inference procedure leading to the pose estimation.

1 Introduction

Wireless Sensor Networks (WSN) are made by a set of nodes exchanging and elaborating data among them wireless. One of the most interesting aspects of such a network is how to distribute the nodes in the environment in order to exploit in the best way possible the power of wireless communication. To this aim, localization techniques are fundamental: Global Positioning System (GPS) is not always the best possible solution to the localization problem, because of potential lack of signal which can lead to a significant decreasing of precision, as explained in [19]. For this reason, many other localization techniques have been proposed lately, some of them exploiting special nodes whose position is already known at the beginning. We are going to use this approach in this work, by using some *anchor* nodes, whose known position is essential in order to infer the localization of any other node in the network.

Clearly, a single anchor node is not enough for the configuration of a network covering a huge environment, as in the case of Vehicular Ad Hoc Networks (VANets), and for this reason a clustering algorithm is considered, so that a single anchor node is responsible for the localization of the nodes in its cluster only. Since we are going to provide a collaborative localization mechanism, we care about the ability of the network in being as autonomous as possible, and this is the reason why we adopt a decentralized clustering algorithm for the cluster creation, exploiting a logic recursive

L. Barolli et al. (Eds.): NBiS 2018, LNDECT 22, pp. 960–969, 2019.
https://doi.org/10.1007/978-3-319-98530-5_88

way to determine what is the cluster the current node belongs to. The base case is provided by the anchor nodes.

The second phase of our approach is based on a pose estimation mechanism, exploiting transformation matrices representing the transition from a reference frame to another one. In particular, a global and a local reference frame are used, where the global one is given by the cluster-head of the cluster, namely the anchor node, while the local one is the one of the nearest neighbor of the current node.

A logic approach is used in order to clearly describe the inference process, with an elegant code in a high-level programming language as Prolog, so that we can abstract from the implementation issues and just focus on the theoretical aspects of the approach.

The rest of the work is organized as follows: in Sect. 2, the state of the art in localization techniques and collaborative approaches is considered; in Sect. 3, our contribution is explained, by differentiating the two phases of our approach, namely clustering algorithm and pose evaluation; eventually in Sect. 4, conclusions of this work are considered.

2 Related Work

One of the main issues in wireless sensor networks is node localization. These are networks made by nodes able to exchange information wireless. Many are the cooperative localization techniques existing nowadays: an overview of them is given in [8]. The application field of auto-organizing mechanisms is huge: it starts from sensor networks [20] and includes topics like ocean-sensing systems [2, 3], and the issues vary from positioning [24] to more specific contexts like the V2V scenario [26] and parking techniques [31, 32]. One of the most used parameter for localization is RSSI (Received Signal Strength Indicator) that is used, for instance, in [17, 18] as localization approach in the traffic control field. The main goal of the authors is to provide a better precision in node position estimation, which is not always guaranteed with GPS. Some techniques used in WSN exploit approaches based on fingerprinting, as shown in [27, 29], or on DDGP3 as in [22, 23], or again DGP algorithm as in [30]. Sometimes, path compression mechanisms are considered, as in [25], and a particular care to resources management and privacy is given as in [33, 34].

In [1], they introduce a collaborative form of localization, since they provide a position estimation according to RNs known positions, by exploiting pairwise RSS measurements. In the same way we will exploit anchor nodes in this work, in [4] they use indoor localization techniques by placing known nodes randomly in the environment, then they estimate the position of unknown nodes by means of received signal strength.

In [5], the authors provide a collaborative localization mechanism based on GMAN (group of mobile anchor nodes), whose position stays constant, that send messages to other nodes, helping in their localization. One of the ideas that mostly inspired this work is [6], where they assume a GPS denied scenario and they propose a distributed algorithm for relative pairwise pose measurement in vehicular context. We are going to extend this approach, by providing a clustering technique and a logic schema to compute vehicles pose, according to a reference frame.

In [7], they propose an algorithm for Concurrent Mapping and Localization (CML) that allows to a set of vehicles to build maps of an unknown environment that they are using, in a collaborative way.

In [12], they highlight how much is important the clusterization process to the aim of improving the wireless sensor network scalability and lifecycle: they propose a method based on cluster-head election, according to the remained energy in each node. Energy is also considered in [28]. One of the main goal of our work and of collaborative localization mechanisms is to make wireless sensor networks (and vehicular ad hoc networks, in particular) as autonomous as possible, and to encourage the self-organizing property of these networks. For this reason, a decentralized clustering algorithm is proposed, as they did in [9], where GDCluster algorithm is provided: nodes gradually build a global data view and they compute a weighted clustering algorithm. Moreover, in [10] a decentralized clustering approach is considered, where each node of the network waits a random amount of time and, according to a local criterion, chooses if build a new cluster or join an existing one. An adaptive and distributed clustering technique is shown in [11], and they demonstrate it to be a robust approach.

3 Two-Phases Cooperative Localization

Cooperative localization systems constitute the basis for self-organizing networks, and they keep improving the process of decentralization of wireless sensor networks in order to make them autonomous systems.

The model we propose in this work exploits the pose computation mechanism based on transformation matrices, and clusterization techniques in order to ensure that the pose detection can be performed starting from a known position (of special nodes called *anchors*).

3.1 Transformation Matrices

The pose (meaning the position and the orientation in 3D space) of a vehicle can be expressed in terms of some transformations, precisely rotations and translations. Figure 1 shows an example of such operations, by indicating in blue the starting frame, and in red the frame after the transformation:

- The rotation of an angle θ on a chosen axis determines a rotation of that angle of the remaining axes, while the rotation axis and the origin of the axes are the same. The rotation is counterclockwise as shown on the left side of Fig. 1.
- Instead, a translation of a reference frame determines a change of the origin of axes, while the relative position of the axes is the same as before, as shown on the right side of the picture.

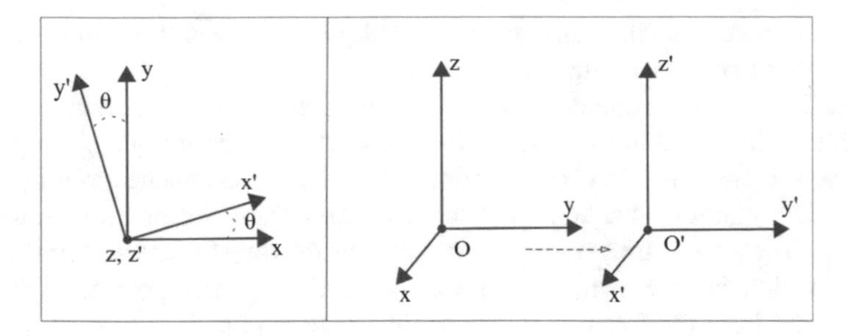

Fig. 1. Example rotation and translation of a reference frame

Any transition from a frame to another can be expressed in terms of these matrices or a composition of them. This is the property that we want to exploit in this work: the position of a vehicle in a vehicular ad hoc network is represented by its pose in the frame attached to it, and this frame can constitute a reference frame for a close vehicle whose position is unknown. In this case, as we will explain later, this frame is said to be local. Instead, when we refer to the frame attached to an anchor node (meaning a node whose position is predetermined) we call it global reference frame. In both cases, the transition between frames is expressed as rotation and translation operations and give us as result the pose of the current vehicle in its frame.

Let's briefly recall what are the transformation matrices needed in order to perform rotation and translation operations.

$$\begin{pmatrix} \cos\theta & -\mathrm{sen}\theta & 0 & 0 \\ -\mathrm{sen}\theta & \cos\theta & 0 & 0 \\ 0 & 0 & 1 & 0 \\ 0 & 0 & 0 & 1 \end{pmatrix} \quad \begin{pmatrix} 1 & 0 & 0 & dx \\ 0 & 1 & 0 & dy \\ 0 & 0 & 1 & dz \\ 0 & 0 & 0 & 1 \end{pmatrix}$$

The first matrix is the rotation of the angle θ around the z axis, while the second one is the translation of a distance (dx, dy, dz). Actually, these two matrices can be merged obtaining a unique matrix where the 3×3 inner matrix always represents the rotation, while the last 4×1 vector represents the translation.

3.2 Single Vehicle Pose Estimation

In this section, we first analyze how the pose computation can be performed, focusing on one single vehicle. Each vehicle has a corresponding cartesian frame, and its pose is given by the position and the orientation of that vehicle in its corresponding frame. We assume that each vehicle is able to estimate the translation and the rotation happening in a time interval, by means of sensors like IMUs (Inertial Measurement Units).

In order to ensure that the pose inference can always be performed, in this model we assume to have some special nodes called *anchors* whose position is already known: they can be infrastructure elements such as traffic lights or any kind of landmark (either artificial or natural). The anchors' frame constitutes the global reference frame for any other vehicles, thus the computed pose for a vehicle will be in terms of the frame of the corresponding anchor node.

Figure 2 shows an example of pose computation for a single vehicle, given the starting frame, indicated as O, which is the frame of the anchor node. At any time, the pose of the vehicle of interest can be estimated as the frame attached to it by means of transformation matrices that represent the transition from the previous frame to the current one. The pose at time 0, can be estimated as the transformation from frame O to frame A (which is essentially a translation); at time 1, the pose is given by the transformation from the frame A to the frame B, and so on.

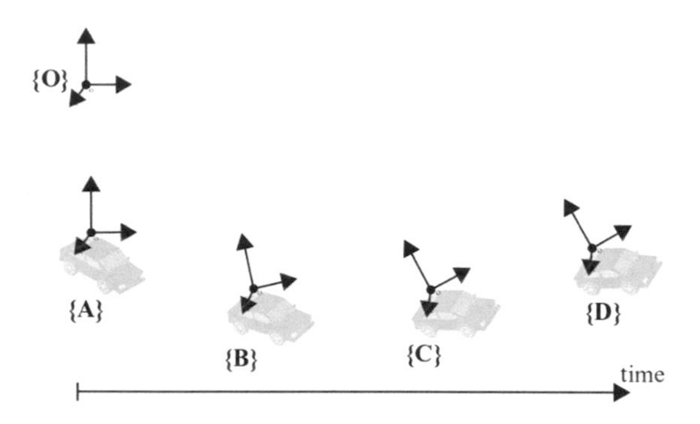

Fig. 2. Single vehicle pose estimation with respect to a reference frame **O**

We indicate with ${}^{i-1}\mathrm{F}_i$ the transition from frame $i-1$ to frame i. A single transformation from a frame to another is typically a combination of a rotation and a translation operation.

3.3 Collaborative Cluster Pose Estimation

In this section, we try to generalize the pose estimation performed in case of a single vehicle for a set of them, by means of collaborative mechanisms that allow the vehicles to help each other in the localization. The model that we are going to describe is based on two phases:

- Clusterization: phase during which some groups of vehicles are built in such a way that each cluster has exactly one anchor node, which gives the *global* reference frame to all the cluster (it operates as a cluster-head);
- Pose estimation: as in the single vehicle case, the pose is computed as transformation from a frame to another, by exploiting the concept of "nearest neighbor", which provides the *local* reference frame.

The idea is the following: inside each cluster, each vehicle computes its pose with respect to the frame of its nearest neighbor, that can be the anchor node or any other vehicle. In the single vehicle case, in order to compute the pose at frame F with respect to the starting frame, it was needed to perform some matrices multiplications in order to computes each single step from the starting frame to the desired one. In this case, the collaborative structure allows us to avoid these multiplication, since we do not need to start from the base frame (which is the global reference frame), but it's enough to have a local reference frame, given by the nearest neighbor (whose position is considered known, since going through the neighbors sooner or later the nearest neighbor will be the very anchor node). For this technique to work, it is needed to ensure that each cluster has an anchor node, since we can state that if a single position in the cluster is known, any other position inside the same cluster can be inferred.

3.3.1 Clusterization

The goal of this paper, and collaborative localization techniques in general, is to make a wireless sensor network as autonomous as possible: we will try to apply this principle to Vehicular Ad Hoc Networks [21], where vehicles can communicate between each other to exchange useful traffic information. The idea of providing a way to estimate vehicle position, by exploiting other vehicles positions, makes the network self-organizing and does not require any centralized entity with a full knowledge of the network configuration and nodes position. For this reason, the clustering algorithm needs to follow this inspiration: any centralized clustering technique would make the network lose the autonomy that we want to obtain, since it would require the presence of a power central server aware of network topology and able to build some clusters according to it. This is the reason why in this work we want to use a decentralized clustering algorithm that allows the nodes to gradually build groups of vehicles, according to the constraint introduced by the anchors node.

We used Prolog programming language in order to represent the inference process of the cluster determination for each node of the network: this is a power tool we have already used in our previous works, since the inference process can be easily explained, leaving aside the implementation details.

The node/cluster association is recursive: the base case is given by the anchor nodes, while the recursive case is applied for any generic node. Since the anchor nodes must be cluster-heads of each cluster, the idea is to provide some Prolog *facts* that associate a fresh cluster identifier to each anchor node (which represent the base case), and some *rules* that infer which is the cluster of the current vehicle *v*, according to the one of its nearest neighbor *near_neg* (which provide the recursive case).

Prolog fracts and rules for identifier cluster estimation

```
IsInCluster(ai,idi).

IsInCluster(v,id):-
    NearestNeighbor(v,near_neg),IsInCluster(nn,id).
```

For any anchor node a_i a corresponding cluster identifier is associated: with *n* anchor nodes we will have *n* Prolog facts expressing the corresponding identifier (this step constitutes our base case). Instead, for any generic vehicle *v* the predicate *IsInCuster* can infer the corresponding cluster identifier, and this recursive case allows us to associate an identifier to any vehicle following the chain of neighbors until an anchor node (whose identifier is known) is reached. This operation is very important in order to understand the global reference frame, which is given by the frame attached to the representative node of the cluster and this information is provided by the cluster *id*. We assume that the *NearestNeighbor* predicate binds each vehicle to its nearest neighbor.

3.3.2 Pose Estimation

Once the clustering algorithm is applied, the vehicle pose estimation can be performed, by means of two parameters:

- The *global* reference frame;
- The *local* reference frame;

The first one is the frame related to the cluster-head of the current cluster, which is the anchor node in that group; while, the second one is given by the nearest neighbor of the current vehicle. In the example shown in Fig. 3, three clusters are built according to the anchors node being in the environment: each cluster has a global reference frame, given by O_1, O_2, and O_3. In the first cluster, for instance, the position of the red car is estimated as its pose in the frame A with respect to the frame C, which is the frame of its nearest neighbor. Similarly, the position of any other vehicle is provided in the cluster, until the nearest neighbor ends up being the very anchor node, and finally the position can be globally defined. The position obtained in the current frame with respect to the local reference frame, is a partial positioning of the vehicle that needs to compute the transformation from the global reference frame to the local one. This is the reason why the cluster identifier is an information that each vehicle needs to keep.

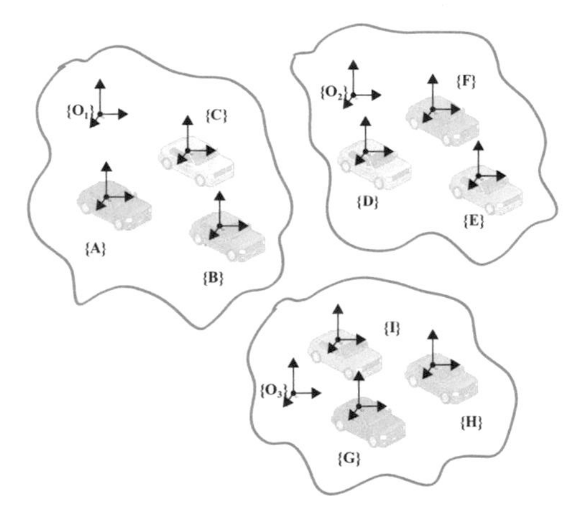

Fig. 3. Example of pose estimation based on clustering: each anchor node has determined a cluster composition, and the pose of any vehicle within it can be provided by exploiting the one of the nearest neighbor vehicle.

We show some Prolog predicates that explain the way the pose estimation is inferred, by first computing global and local reference frame and then using them in order to have the final pose.

Prolog fracts and rules for global and local frame detection, and vehicle pose estimation

```
CurrentFrame(vi,fi).
CurrentFrame(ai,fi).

GlobalReferenceFrame(v,f):-
    IsInCluster(v,id),ClusterHead(id,a),
    CurrentFrame(a,f).

LocalReferenceFrame(v,f):-
    NearestNeighbor(v,near_neig),
    CurrentFrame(near_neig,f).

PartialPose(v,result):-
    LocalReferenceFrame(v,local_frame),
    CurrentFrame(v,current_frame),
    Transformation(current_frame,local_frame,result).

TotalPose(v,result):-
    PartialPose(v,partial_result),
    GlobalReferenceFrame(v,global_frame),
    Transformation(partial_resutl,global_frame,result).
```

We assume to have a fact that binds each vehicle and anchor node to the frame attached to it (for instance, in the first cluster we have the predicate *CurrentFrame (red_car,A)*). The *GlobalReferenceFrame* predicate binds a vehicle to the frame of the cluster-head of the cluster it belongs to (supposing of having a predicate saying which is the cluster-head for a specific cluster identifier). The *LocalReferenceFrame* predicate, instead, binds a vehicle to the frame attached to its nearest neighbor. *PartialPose* predicate associates a vehicle to its pose (*result*) in the current frame with respect to the local reference frame, and *TotalPose* predicate includes an additional comparison between the partial pose and the global reference frame. We assume that *Transformation* predicate includes the transformation matrix that represents the transition from a frame to another.

4 Conclusions

The main contribution of this work is in the logic approach to the collaborative localization technique: we have highlighted the importance of autonomy in a wireless sensor network in general, and in a vehicular ad hoc network more specifically.

Hence, it is needed to go through the problem of decentralizing the possible operations among the nodes of such networks. This aim is observed twice, in this work: in the clustering algorithm for the cluster creation according to the anchor nodes, elected as cluster-heads, and in the pose estimation mechanism, exploiting the transformation matrices expressing the transition from a frame to another with respect to a global reference frame, given by the cluster-heads, and a local reference frame, provided by the nearest neighbor node.

The collaborative impress is due to the fact that any node of the network can find all the information needed to understand the cluster it belongs to and its own position inside the other nodes of the network, and any other centralized entity is unnecessary.

References

1. Li, X.: Collaborative localization with received-signal strength in wireless sensor networks. IEEE Trans. Veh. Technol. **56**(6), 3807–3817 (2007)
2. Mirza, D., Schurgers, C.: Collaborative localization for fleets of underwater drifters. In: OCEANS 2007, pp. 1–6. IEEE, September 2007
3. van Leeuwen, J. (ed.): Computer Science Today. Recent Trends and Developments. Lecture Notes in Computer Science, vol. 1000. Springer, Berlin (1995)
4. Dardari, D., Conti, A.: A sub-optimal hierarchical maximum likelihood algorithm for collaborative localization in ad-hoc network. In: 2004 First Annual IEEE Communications Society Conference on Sensor and Ad Hoc Communications and Networks, 2004, IEEE SECON 2004, pp. 425–429. IEEE, October 2004
5. Zhang, B., Zhang, Z.: Collaborative localization algorithm for wireless sensor networks using mobile anchors. In: Asia-Pacific Conference on Computational Intelligence and Industrial Applications, 2009, PACIIA 2009, vol. 1, pp. 309–312. IEEE (2009)
6. Knuth, J., Barooah, P.: Distributed collaborative localization of multiple vehicles from relative pose measurements. In: 47th Annual Allerton Conference on Communication, Control, and Computing, 2009, Allerton 2009, pp. 314–321. IEEE (2009)
7. Fenwick, J.W., Newman, P.M., Leonard, J.J.: Cooperative concurrent mapping and localization. In: Proceedings of ICRA 2002, IEEE International Conference on Robotics and Automation, vol. 2, pp. 1810–1817. IEEE (2002)
8. Wymeersch, H., Lien, J., Win, M.Z.: Cooperative localization in wireless networks. Proc. IEEE **97**(2), 427–450 (2009)
9. Mashayekhi, H., Habibi, J., Khalafbeigi, T., Voulgaris, S., Van Steen, M.: GDCluster: a general decentralized clustering algorithm. IEEE Trans. Knowl. Data Eng. **27**(7), 1892–1905 (2015)
10. Wen, C.Y., Sethares, W.A.: Automatic decentralized clustering for wireless sensor networks. EURASIP J. Wirel. Commun. Netw. **2005**(5), 686–697 (2005)
11. Khawatmi, S., Zoubir, A.M., Sayed, A.H.: Decentralized clustering over adaptive networks. In: 2015 23rd European Signal Processing Conference (EUSIPCO), pp. 2696–2700. IEEE (2015)
12. Bajaber, F., Awan, I.: Adaptive decentralized re-clustering protocol for wireless sensor networks. J. Comput. Syst. Sci. **77**(2), 282–292 (2011)
13. Balzano, W., Del Sorbo, M.R., Murano, A., Stranieri, S.: A logic-based clustering approach for cooperative traffic control systems. In: International Conference on P2P, Parallel, Grid, Cloud and Internet Computing, pp. 737–746. Springer, Cham (2016)
14. Balzano, W., Del Sorbo, M.R., Stranieri, S.: A logic framework for C2C network management. In: 2016 30th International Conference on Advanced Information Networking and Applications Workshops (WAINA), pp. 52–57. IEEE (2016)
15. Balzano, W., Murano, A., Stranieri, S.: Logic-based clustering approach for management and improvement of VANETs. J. High Speed Netw. **23**(3), 225–236 (2017)

16. Balzano, W., Stranieri, S.: LoDGP: a framework for support traffic information systems based on logic paradigm. In: International Conference on P2P, Parallel, Grid, Cloud and Internet Computing, pp. 700–708. Springer, Cham (2017)
17. Parker, R., Valaee, S.: Vehicular node localization using received-signal-strength indicator. IEEE Trans. Veh. Technol. **56**(6), 3371–3380 (2007)
18. Liu, C., Wu, K., He, T.: Sensor localization with ring overlapping based on comparison of received signal strength indicator. In: 2004 IEEE International Conference on Mobile Ad-hoc and Sensor Systems, pp. 516–518. IEEE (2004)
19. Skone, S., Knudsen, K., De Jong, M.: Limitations in GPS receiver tracking performance under ionospheric scintillation conditions. Phys. Chem. Earth Part A. **26**(6–8), 613–621 (2001)
20. Collier, T.C., Taylor, C.: Self-organization in sensor networks. J. Parallel Distrib. Comput. **64**(7), 866–873 (2004)
21. Yousefi, S., Mousavi, M.S., Fathy, M.: Vehicular ad hoc networks (VANETs): challenges and perspectives. In: 2006 6th International Conference on ITS Telecommunications Proceedings, pp. 761–766. IEEE (2006)
22. Balzano, W., Barbieri, V., Riccardi, G.: Car2Car framework based on DDGP3
23. Balzano, W., Barbieri, V., Riccardi, G.: Smart priority park framework based on DDGP3. In: IEEE AINA 2018 the 32nd IEEE International Conference on Advanced Information Networking and Applications (2018)
24. Balzano, W., Formisano, M., Gaudino, L.: WiFiNS: a smart method to improve positioning systems combining WiFi and INS techniques. In: Cham, S., De Pietro, G., Gallo, L., Howlett, R., Jain, L. (eds.) Intelligent Interactive Multimedia Systems and Services 2017, Smart Innovation, Systems and Technologies, vol. 76 (2017)
25. Balzano, W., Murano, A., Vitale, F.: Hypaco–a new model for hybrid paths compression of geodetic tracks. In: CCPS-2016: The International Conference on Data Compression, Communication, Processing and Security (2016)
26. Balzano, W., Murano, A., Vitale, F.: V2V-En–vehicle-2-vehicle elastic network. Procedia Comput. Sci. **98**, 497–502 (2016)
27. Balzano, W., Murano, A., Vitale, F.: Wifact–wireless fingerprinting automated continuous training. In: 2016 30th International Conference on Advanced Information Networking and Applications Workshops (WAINA), pp. 75–80. IEEE (2016)
28. Balzano, W., Murano, A., Vitale, F.: Eenet: energy efficient detection of network changes using a wireless sensor network. In: Conference on Complex, Intelligent, and Software Intensive Systems, pp. 1009–1018. Springer (2017)
29. Balzano, W., Murano, A., Vitale, F.: SNOT-WiFi: sensor network-optimized training for wireless fingerprinting. J. High Speed Netw. **24**(1), 79–87 (2018)
30. Balzano, W., Vitale, F.: DGP application for support traffic information systems in indoor and outdoor environments. In: International Conference on P2P, Parallel, Grid, Cloud and Internet Computing, pp. 692–699. Springer (2017)
31. Balzano, W., Vitale, F.: DiG-Park: a smart parking availability searching method using V2V/V2I and DGP-class problem. In: 2017 31st International Conference on Advanced Information Networking and Applications Workshops (WAINA), pp. 698–703. IEEE (2017)
32. Balzano, W., Vitale, F.: Pam-sad: ubiquitous car parking availability model based on v2v and smartphone activity detection. In: International Conference on Intelligent Interactive Multimedia Systems and Services, pp. 232–240. Springer (2017)
33. Amato, F., Moscato, F.: Exploiting cloud and workflow patterns for the analysis of composite cloud services. Future Gener. Comput. Syst. **67**, 255–265 (2017)
34. Amato, F., Moscato, F.: A model driven approach to data privacy verification in E-Health systems. Trans. Data Priv. **8**(3), 273–296 (2015)

FiDGP: A Smart Fingerprinting Radiomap Refinement Method Based on Distance-Geometry Problem

Walter Balzano(✉) and Fabio Vitale

Università degli Studi di Napoli Federico II, Naples, Italy
w.balzano@unina.it, fvitale86@gmail.com

Abstract. Localization-based services are having a lot of attention in the latest years, due to the widespread availability of mobile smart devices like smartphones. While in outdoor environments it is possible to use Global Navigation Satellite Systems (GNSS) to obtain accurate user position, in indoor environments where sky visibility is an issue good methodologies are still under research. In this context, WiFi fingerprinting based localization systems are quite interesting as they offer good positional accuracy using available network signals stored in a database named RadioMap, without the need of a secondary localization-only infrastructure. In this paper we present *FiDGP: A smart Fingerprinting radiomap refinement method based on Distance-Geometry Problem*, which exploits wifi-fingerprinting and a DGP-based algorithm in order to provide superior positioning while also keeping the RadioMap updated over time.

Keywords: WiFi fingerprinting · Crowdsourcing
Map generation · DGP

1 Introduction

Indoor localization systems are having a lot of attention in the current literature due to chance of offering location-based services to users, in particular considering the popularity of common and smart devices (like smartphones) which allows pervasive usage of such technologies for many different scopes, like for instance fitness [1] and advertisement.

In outdoor, positioning is commonly available using global satellite-based localization systems (like GPS or GLONASS) as modern personal devices often include such receivers. These solution only work properly in outdoor environments since they require clear sky visibility, and use high amounts of energy, therefore limiting their usage in todays mobile devices.

As said, these systems do not work properly in indoor areas, and therefore it is important to look for different methods. Several methodologies have been proposed in the latest years, some leveraging body mounted sensors (like accelerometers and gyroscopes) to detect user movement, some using hybrid methods based

L. Barolli et al. (Eds.): NBiS 2018, LNDECT 22, pp. 970–978, 2019.
https://doi.org/10.1007/978-3-319-98530-5_89

on sensors and various kind of receivers [7]. Some of these hybrid methods have also been applied to vehicles in Vehicular Ad-hoc Networks [8](VANETs) for localization in large indoor areas [4,10] but also for communication and cooperation purposes [6] and traffic support [14]. However, most widespread solutions nowadays use WiFi signals, as they are commonly available and reliable, using range-finding or fingerprinting.

Fingerprinting-based systems are quite reliable, and able to grant accurate positional accuracy without the need of a separate infrastructure. They do not rely on angles or distances but only on Received Signal Strength (RSS) from available sources. In each place, the set of all the available access points form a unique fingerprint, which can be stored and used for subsequent matching. This information is first gathered in a *training phase* which is often time consuming but mandatory. Moreover, the information stored in the database tend to lose accuracy over time (due to interferences and environmental changes), and is therefore needed to proceed to a new training phase after some time in order to keep the accuracy reasonable.

Distance Geometry Problem is the characterization and study of sets of points based on their relative distance. All the points form a graph, which has a well-determined shape calculated using an algorithm. This graph can then be roto-translated and fixed in space using a set of well-known points. This methodology is used in several distinct fields, like chemistry or biology, to better understand how tiny components (like atoms in a molecule, for instance) interact. However, since DGP calculation is CPU-intensive, it cannot be executed on mobile devices and a proper cloud infrastructure is needed [2,3].

In this paper we present FiDGP, in which we aim to combine wifi-fingerprinting location with DGP in order to provide a constant and automatic improvement in localization accuracy over time, while also reducing the need for subsequent training phases.

2 Related Works

Here we present some work which are closely related to our project.

With regards to wireless-based positioning, in *WiFi Fingerprint Localization in Open Space* [22] authors propose an heatmap-based WiFi fingerprinting method using historical location estimates in order to improve normal fingerprinting-based localization in open spaces. In *WiFi fingerprint indoor positioning system using probability distribution comparison* [20], authors combine normal fingerprinting methodologies using a probabilistic method which calculates user most probable position using a weighted average of the three positions which best match the user measurement. In *Ziloc: Energy efficient wifi fingerprint-based localization with low-power radio* [26] authors analyze a ZigBee based localization: this approach is quite interesting with regards to energy consumption, since ZigBee-based networks are very cheap energy-wise, but also need proper equipment to work (i.e. cannot work on smartphones, but require external hardware). In *Location determination using WiFi-fingerprinting versus WiFi-trilateration* [24] authors provide some insight on difference between a fingerprinting-based approach and trilateration, based on a novel RSSI-to-distance conversion. This is further investigated in *WiFi fingerprinting signal strength error modeling for short distances* [23] which shows that the aforementioned conversion is only valid for short distances.

Sensors-based localization has also been considered in latest years. In *Himloc: Indoor smartphone localization via activity aware pedestrian dead reckoning with selective crowdsourced wifi fingerprinting* [27] authors combine wireless fingerprinting with *Pedestrian Dead Reckoning*, which uses inertial sensors available in mobile devices in order to approximate user location based on previously calculated positions.

DGP has been used in several fields of application [21,25], often related to chemistry [18,19] or biology. Usages of such technologies outside their standard fields of application is currently under research and the results are quite promising.

Finally, in *Hypaco–a new model for hybrid paths compression of geodetic tracks* [9] authors provide a way to compress geodetic hybrid indoor/outdoor positional data with scalable accuracy which can be used to store large amount of data for statistical purposes.

Outline

The rest of the paper is organized as follows: in Sect. 3 we present well-established technologies used by our project. In Sect. 4 we broadly describe our project and its characteristics, and finally in Sect. 5 we present some conclusions and future work opportunities.

3 Core Concepts

Localization systems are having a great deal of attention from literature in the latest years. While for outdoor environment satellite-based localization systems are nowadays quite accurate and widespread, mostly due to the availability of GPS and GLONASS sensors inside smartphones and tablets, a valid, accurate and ubiquitous localization system for indoor and hybrid indoor-outdoor environments is still under research.

For indoor, several methodologies have been proposed in the latest years, based on WiFi and/or sensors embedded in mobile devices (like gyroscopes, accelerometers and compass).

With regards to WiFi localization, one of the most common methodologies is *WiFi fingerprinting*. It is based on two distinct phases: one *training phase* and one *tracking phase*. During the training phase, a device periodically scans nearby networks signal strengths, which form a local unique *fingerprint*, and a user has to manually input its position on a map. This allows building a database, named *RadioMap*, which contains an association between places and their related fingerprint. During the tracking phase, user devices scan the nearby area and determine their position based on the stored RadioMap.

This positioning methodology is quite accurate but also has a downside: RadioMap accuracy tends to decrease over time, due to environmental changes (humidity, pressure and temperature) and local interferences. Therefore is mandatory to repeat the training phase over time. This issue has been already addressed in several papers [11–13], and while the solutions are satisfactory, they also require building a separate infrastructure for training purposes.

Another viable solution for indoor localization is based on *Distance Geometry Problem*, the study of sets of points based only on their relative distance, which allows determination of their shape in space. This shape can be freely rototranslated in space, and is therefore needed to use known fixed points. It is having a lot of attention nowadays in the literature since there are several real life applications like biology, chemistry and more. DGP has also been used in several papers for localization of a set of nodes, using known fixed points of which position is determined beforehand [5,15–17]. Each network element is able to determine its distance from nearby nodes, all the distance information are gathered and analyzed by a centralized system which calculates the shape of the graph, which is then roto-translated in order to match a set of well-known points, and the found positions are sent back to the devices in order to offer location-based services.

4 System Description

The aim of FiDGP is improving current WiFi fingerprinting-based localization systems by reducing the number of training phases needed in order to maintain accuracy over time. In order to achieve such a result, we decided to take in consideration distance between nodes as detected by the devices, building a proper graph on which DGP is applied. Resulting DGP-based positional information is used to correct and improve RadioMap precision over time.

Starting from a base, manually trained RadioMap, we proceed as follows (see Fig. 1):

1. all nodes are positioned using common WiFi-fingerprinting based localization;
2. all nodes calculate their relative distances and send this information over local network for elaboration;
3. a server gathers all the distance information and calculates the corresponding graph, on which is then applied DGP;
4. DGP is roto-translated in order to make the best possible match considering the shape as seen by the fingerprinting-based system;
5. newly-calculated positions are used to update the RadioMap with more accurate values;
6. a new fingerprinting-based localization is calculated and the cycle restarts.

In Fig. 2 we can see an example of the system. Black dots are position as found by the WiFi-fingerprinting localization, with their calculated error. After wireless positioning, the distance between nodes is calculated and DGP is applied. New positions are found inside wifi error areas, and RadioMap is updated accordingly.

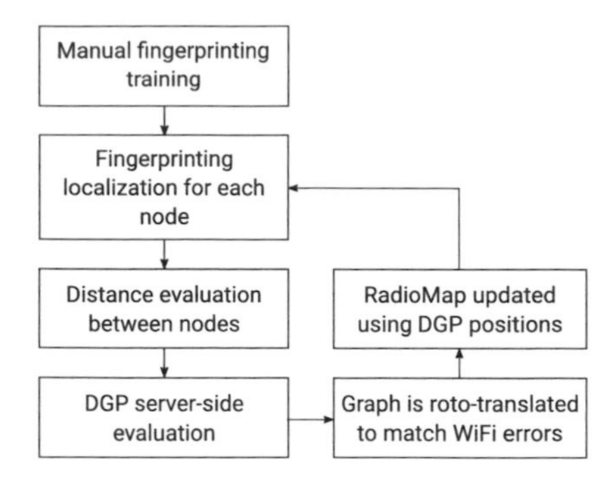

Fig. 1. System behavior: after the first manual initialization step, the system iteratively localizes users and uses DGP in order to improve RadioMap accuracy.

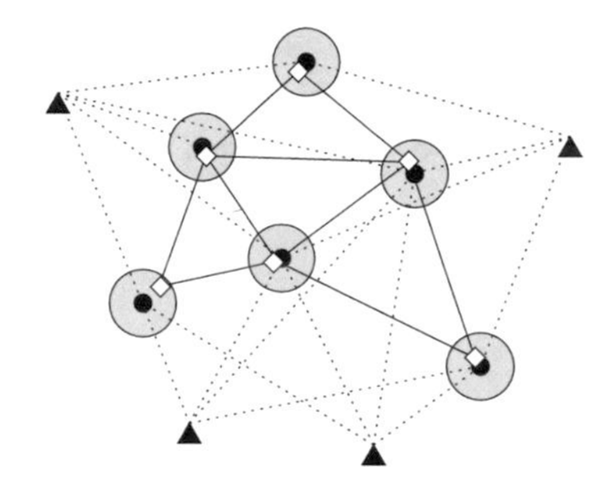

Fig. 2. This graph shows the application of a single iteration of the system. Black triangles show hotspot positions, while circles indicate devices with their respective wifi-fingerprinting error. The graph also shows DGP-corrected positions that are used in order to improve the stored RadioMap.

DGP is able, in other words, to apply a "force" which pushes apart (or pulls) nodes to their most appropriate positions, used to calculate more recent and accurate fingerprints which are then stored in the RadioMap for future position evaluations.

In Algorithm 1 we can see the process executed by client devices.

1. fingerprint is generated from available network hotspots: for each hotspot, its RSS is calculated and stored in a structure (*fingerprint*);
2. distances are calculated from nearby nodes: for each visible node, distance is stored in a structure (*neighbours*);
3. client sends these two piece of information to the server for elaboration (function *sendInformation*);
4. client waits (function *retrievePosition*) positional information from the server;
5. found position is now available to the user: it can be shown or used for any kind of location-based service.

Algorithm 1. Client-side algorithm for position determination

```
Input: visibleHotspots, visibleNeighbours
Output: userPosition
 1: fingerprint := []
 2: neighbours := []
 3: for hotspot ∈ visibleHotspots do
 4:    rss := signalStrength(hotspot)
 5:    fingerprint ← {hotspot, rss}
 6: end for
 7: for neighbour ∈ visibleNeighbours do
 8:    distance := distanceFrom(neighbour)
 9:    neighbours ← {neighbour, distance}
10: end for
11: sendInformation(fingerprint, neighbours)
12: return retrievePosition() // listens for server sendPosition
```

In Algorithm 2 it is possible to see how the data is processed by the server.

1. for each node in the network, server receives its *info* which is a structure containing its *fingerprint* and *neighbours*;
2. fingerprint is stored in an hash (*fingerprints*);
3. fingerprinting-based position is calculated using RadioMap (function *readRadioMap*) and is stored in an hash (*wPos*);

Algorithm 2. Server-side algorithm for position determination and RadioMap correction

```
Input: nodes
 1: graph := {}
 2: wPos := {}
 3: fingerprints := {}
 4: for node ∈ nodes do
 5:    info := receiveInfo(node)
 6:    fingerprint := info.fingerprint
 7:    fingerprints[node] = fingerprint
 8:    wPos[node] = readRadioMap(fingerprint)
 9:    neighbours := info.neighbours
10:    for neighbour ∈ neighbours do
11:       graph[node] ← [neighbour, neighbour.distance]
12:    end for
13: end for
14: dPos := applyDgp(graph, fingerprints) // fingerprints includes error areas
15: for node ∈ nodes do
16:    cPos := correctPosition(dPos[node], wPos[node])
17:    storeRadioMap(cPos, fingerprints[node])
18:    sendPosition(node, cPos) // sends correct position to client
19: end for
```

4. for each node neighbour, a graph edge is built including distance information
5. using the fingerprints (with error estimation as a constraint) and the graph built in the previous steps, DGP is applied and DGP-based positions are calculated (hash *dPos*);
6. for each node, its correct position is calculated (*cPos*);
7. RadioMap is updated (function *storeRadioMap*) using the correct position and its relative node fingerprint;
8. position is sent to client (function *sendPosition*);
9. the process restarts from the beginning.

5 Conclusions and Future Work

In this paper we presented FiDGP, an indoor localization methodology which exploits a smart combination of wireless fingerprinting-based localization with DGP to provide better positioning to users while also reducing the need of manual training phases (after the first one).

Future work may consider some probabilistic method in order to further improve localization and RadioMap accuracy using previously gathered data, or using embedded inertial sensors as an additional source of information.

References

1. Amato, F., Moscato, F.: A model driven approach to data privacy verification in e-health systems. Trans. Data Priv. **8**(3), 273–296 (2015)
2. Amato, F., Moscato, F.: Pattern-based orchestration and automatic verification of composite cloud services. Comput. Electr. Eng. **56**, 842–853 (2016)
3. Amato, F., Moscato, F.: Exploiting cloud and workflow patterns for the analysis of composite cloud services. Future Gener. Comput. Syst. **67**, 255–265 (2017)
4. Balzano, W., Barbieri, V., Riccardi, G.: Car2Car framework based on DDGP3. In: The 23rd International DMS Conference on Visual Languages and Sentient Systems (2017)
5. Balzano, W., Barbieri, V., Riccardi, G.: Smart priority park framework based on DDGP3. In: IEEE AINA 2018: The 32nd IEEE International Conference on Advanced Information Networking and Applications (2018)
6. Balzano, W., Del Sorbo, M.R., Murano, A., Stranieri, S.: A logic-based clustering approach for cooperative traffic control systems. In: Xhafa, F., Barolli, L., Amato, F. (eds.) Advances on P2P, Parallel, Grid, Cloud and Internet Computing. Lecture Notes on Data Engineering and Communications Technologies, vol. 1. Springer, Cham (2017)
7. Balzano, W., Formisano, M., Gaudino, L.: WiFiNS: a smart method to improve positioning systems combining WiFi and ins techniques. In: De Pietro, G., Gallo, L., Howlett, R., Jain, L. (eds.) Intelligent Interactive Multimedia Systems and Services 2017. Smart Innovation, Systems and Technologies, vol. 76. Springer, Cham (2017)
8. Balzano, W., Murano, A., Stranieri, S.: Logic based clustering approach for management and improvement of VANETs. J. High Speed Netw. **23**(3), 225–236 (2017)

9. Balzano, W., Murano, A., Vitale, F.: Hypaco–a new model for hybrid paths compression of geodetic tracks. In: CCPS 2016: The International Conference on Data Compression, Communication, Processing and Security (2016)
10. Balzano, W., Murano, A., Vitale, F.: V2V-EN–Vehicle-2-Vehicle elastic network. Procedia Comput. Sci. **98**, 497–502 (2016)
11. Balzano, W., Murano, A., Vitale, F.: WIFACT–wireless fingerprinting automated continuous training. In: 2016 30th International Conference on Advanced Information Networking and Applications Workshops (WAINA), pp. 75–80. IEEE (2016)
12. Balzano, W., Murano, A., Vitale, F.: EENET: energy efficient detection of network changes using a wireless sensor network. In: Conference on Complex, Intelligent, and Software Intensive Systems, pp. 1009–1018. Springer (2017)
13. Balzano, W., Murano, A., Vitale, F.: SNOT-WiFi: sensor network-optimized training for wireless fingerprinting. J. High Speed Netw. **24**(1), 79–87 (2018)
14. Balzano, W., Stranieri, S.: LoDGP: a framework for support traffic information systems based on logic paradigm. In: Xhafa, F., Caballé, S., Barolli, L. (eds.) Advances on P2P, Parallel, Grid, Cloud and Internet Computing, 3PGCIC 2017. Lecture Notes on Data Engineering and Communications Technologies, vol. 13. Springer, Cham (2018)
15. Balzano, W., Vitale, F.: DGP application for support traffic information systems in indoor and outdoor environments. In: International Conference on P2P, Parallel, Grid, Cloud and Internet Computing, pp. 692–699. Springer (2017)
16. Balzano, W., Vitale, F.: DiG-Park: a smart parking availability searching method using V2V/V2I and DGP-class problem. In: 2017 31st International Conference on Advanced Information Networking and Applications Workshops (WAINA), pp. 698–703. IEEE (2017)
17. Balzano, W., Vitale, F.: Pam-sad: ubiquitous car parking availability model based on v2v and smartphone activity detection. In: International Conference on Intelligent Interactive Multimedia Systems and Services, pp. 232–240. Springer (2017)
18. Lavor, C., Liberti, L., Maculan, N.: Computational experience with the molecular distance geometry problem. In: Pintér, J.D. (ed.) Global Optimization, pp. 213–225. Springer, Boston (2006)
19. Lavor, C., Liberti, L., Mucherino, A.: The interval branch-and-prune algorithm for the discretizable molecular distance geometry problem with inexact distances. J. Glob. Optim. **56**(3), 855–871 (2013)
20. Le Dortz, N., Gain, F., Zetterberg, P.: WiFi fingerprint indoor positioning system using probability distribution comparison. In: 2012 IEEE International Conference on Acoustics, Speech and Signal Processing (ICASSP), pp. 2301–2304. IEEE (2012)
21. Liberti, L., Lavor, C., Maculan, N., Mucherino, A.: Euclidean distance geometry and applications. SIAM Rev. **56**(1), 3–69 (2014)
22. Lu, B., Niu, J., Juny, J., Cheng, L., Guy, Y.: WiFi fingerprint localization in open space. State Key Laboratory of Software Development Environment, Beihang University, Beijing (2013)
23. Moghtadaiee, V., Dempster, A.G.: WiFi fingerprinting signal strength error modeling for short distances. In: IPIN 2012, pp. 13–15 (2012)
24. Mok, E., Retscher, G.: Location determination using wifi-fingerprinting versus wifi-trilateration. J. Locat. Based Serv. **1**(2), 145–159 (2007)
25. Mucherino, A.: On the identification of discretization orders for distance geometry with intervals. In: Nielsen, F., Barbaresco, F. (eds.) Geometric Science of Information, pp. 231–238. Springer, Heidelberg (2013)
26. Niu, J., Lu, B., Cheng, L., Gu, Y., Shu, L.: ZiLoc: energy efficient WiFi fingerprint-based localization with low-power radio. In: WCNC, pp. 4558–4563. IEEE (2013)

27. Radu, V., Marina, M.K.: HiMLoc: indoor smartphone localization via activity aware pedestrian dead reckoning with selective crowdsourced WiFi fingerprinting. In: 2013 International Conference on Indoor Positioning and Indoor Navigation (IPIN), pp. 1–10. IEEE (2013)

Author Index

L. Barolli et al. (Eds.): NBiS 2018, LNDECT 22, pp. 979–982, 2019.
https://doi.org/10.1007/978-3-319-98530-5

Z

Printed in the United States
By Bookmasters